Elements of Elastic Scattering by X-Rays, Neutrons, and Electrons

Hartmut Zabel

Elements of Elastic Scattering by X-Rays, Neutrons, and Electrons

An Introduction

 Springer

Hartmut Zabel (ID)
Fakultät für Physik und Astronomie
Ruhr-Universität Bochum
Bochum, Germany

ISBN 978-3-032-16623-4   ISBN 978-3-032-16624-1   (eBook)
https://doi.org/10.1007/978-3-032-16624-1

Editorial Contact: Ute Heuser
This Springer imprint is published by the registered company Springer Nature Switzerland AG
The registered company address is: Gewerbestrasse 11, 6330 Cham, Switzerland

If disposing of this product, please recycle the paper.

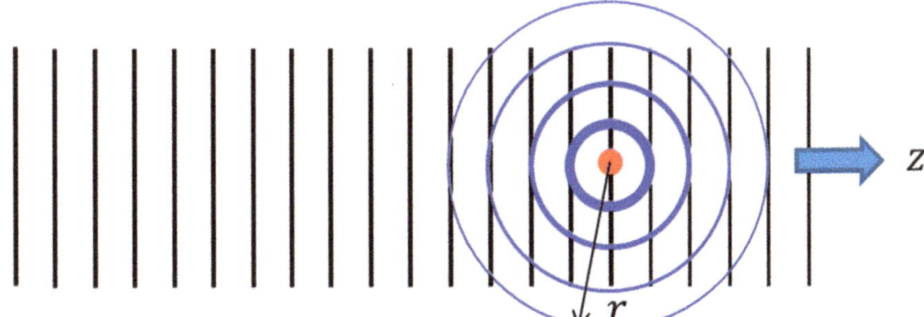

r
z

This book is dedicated to my grandchildren:
Hanna, Henriette, Katharina, Niklas,
Raphael, and Sonja

Preface

This introductory text treats elastic scattering methods for the structural analysis of materials. Information on the atomic structure of materials can be gained by two fundamentally different approaches: microscopy and scattering. Microscopy is performed in real space, and scattering is performed in Fourier space, also known as reciprocal space. Fourier methods filter out the dominant repetitive characteristics and suppress irrelevant clatter. As such, scattering methods are compelling by concentrating on the essential correlations in solids.

This book intends to provide an elementary yet systematic introduction to X-ray, neutron, and electron scattering for the structural analysis of condensed matter materials. These three methods are treated on the same footing by the same mathematical formalism and in a simplified fashion. Neutron scattering is an entirely quantum mechanical process. However, the way neutron scattering is described here does not require any mastery of quantum mechanics. The emphasis is more on the similarity of the three methods than on the canonical treatment of the respective scattering theories. The latter can be found in advanced textbooks listed under "Further Reading." Therefore, I hope that this text will be beneficial also for those scholars who do not have a background in theoretical physics.

The chapters are structured as follows: Chap. 1 introduces the relationship between real and reciprocal space, including translational vector and Miller indices, information that can also be found in condensed matter physics textbooks. Chapter 2 discusses the scattering amplitudes of single and multiple target centers and presents the scattering length for X-ray and neutron scattering. In Chap. 3, the Bragg condition and the scattering intensity of a periodic three-dimensional lattice are calculated, and the structure factor for special lattices is derived. Chapter 4 presents some useful details about X-ray sources, including synchrotron radiation sources and free electron lasers, and continues with discussions of thermal neutron sources and electron sources. Chapter 5 contains information on scattering methods for studying single crystals, which is contrasted in Chap. 6 by the scattering methods focusing on polycrystalline materials. Chapter 7 describes the application of neutron scattering to the study of magnetic materials and also introduces non-resonant and resonant magnetic X-ray scattering. Chapter 8 is dedicated to defects and disordered materials, including order-disorder phase transitions in binary alloys. Chapter 9 focuses on the basics of small-angle scattering with applications to polymers, magnetic nanoparticles, and magnetization reversal processes. Chapter 10 provides an overview of the scattering and reflectivity at thin films and multilayers, including interface roughness. Chapter 11 considers thin magnetic films analyzed by polarized neutron reflectivity and resonant magnetic X-ray scattering. Finally, in Chap. 12, we discuss grazing incidence surface scattering and small-angle scattering methods for the analysis of phase transitions and nanoparticles in surface layers. Of course, there are many more relevant topics that could be discussed. But this is, I hope, a good start for further studies, if needed.

Bochum, Germany   Hartmut Zabel

Acknowledgments

I would like to express my deep and sincere gratitude to all those who taught me scattering physics and stimulated my enthusiasm for this topic. These are, first and foremost, my Ph.D. supervisor, Johann Peisl, and my postdoctoral advisor, Simon C. Moss. Furthermore, I would like to thank my colleagues and coworkers that I have had the pleasure to interact with during many years of practice (in alphabetical order): Robert Feidenhans'l, Christian Gutt, Björgvin Hjörvarsson, Andreas Magerl, Chuck F. Majkrzak, Roger Moret, Ian K. Robinson, John J. Rush, Christian Schüßler-Langeheine, Steve M. Shapiro, Stuart A. Solin, Masatsugu Suzuki, Boris P. Toperverg, Eugen Weschke, as well as many colleagues from whose lectures and discussions I have benefited greatly. In fact, the earliest version of such a text on scattering was planned together with Jeff Lynn (NIST) to include elastic and inelastic scattering. However, the explosive development of the scattering methods in recent years, in particular by using hard and soft X-ray radiation from synchrotron sources, made these early plans obsolete.

Additionally, I would like to thank my excellent Ph.D. students, whom I had the privilege to advise. Some of whom corrected my manuscript at an early stage: John Ankner, Paul Miceli, Dan Neumann, Paul Reimer, David Wiesler, and Xiaomei Zhu at the University of Illinois; Wolfgang Donner, Dmitry Gorkov, Johannes Grabis, Olav Hellwig, Vincent Leiner, Durgamadhab Mishra, Florin Radu, Arndt Remhof, Andreas Schreyer, Ralf Siebrecht, Andreas Stierle, and others at the Ruhr-University Bochum; as well as the Postdocs and research assistances Radu Abrudan, Naoto Metoki, Oleg Petracic, Christoph Sutter, Katharina Theis-Bröhl, and Max Wolff. Furthermore, I am indebted to Frank Schreiber, who critically reviewed the manuscript and provided very helpful suggestions, and Fabian Samad for many helpful corrections. My special thanks go to Mathias Kläui, who not only asked me to provide a text on scattering for materials scientists as a reference for a blended learning course at the Johannes Gutenberg University in Mainz but also worked through many exercises and gave me numerous hints to clarify statements in the text or to extend certain derivations for the benefit of the students. I would also like to thank students attending various courses and lectures for their questions and suggestions. Hildegard Müller did a fantastic job in organizing the feedback from the students. Finally, I would like to thank Ute Heuser and her team of the Springer Nature Verlag for their excellent editorial work.

Last but not least, I would like to thank my wife, Rosemarie, for her support and her unwavering encouragement during the completion of this text.

Bochum, November 2025

Competing Interests The author has no competing interests to declare that are relevant to the content of this manuscript.

Contents

Acronyms Frequently Used

AFM	Atomic force microscopy
CDI	Coherent diffraction imaging
CDW	Charge density wave
CTR	Crystal truncation rod
DAC	Diamond anvil cell
EM	Electromagnetic
FM	Ferromagnetism
GISAS	Grazing incidence small-angle scattering
LEED	Low-energy electron diffraction
MBE	Molecular beam epitaxy
mSLD	Magnetic scattering length density
NP	Nanoparticles
NR	Neutron reflectivity
nSLD	Nuclear scattering length density
PNR	Polarized neutron reflectivity
POL-SANS	Polarized small-angle neutron scattering
r.l.p.	Reciprocal lattice point
r.l.u.	Reciprocal lattice units
RHEED	Reflection high-energy electron diffraction
rms	Root mean square
RSM	Reciprocal space map
SANS	Small-angle neutron scattering
SAS	Small-angle scattering
SASE	Self-amplified spontaneous emission
SAXS	Small-angle X-ray scattering
SDW	Spin density wave
SLD	Scattering length density
TDS	Thermal diffuse scattering
XFEL	X-ray free electron laser
XMCD	X-ray magnetic circular dichroism
XPCS	X-ray photon correlation spectroscopy

XRD	X-ray diffraction
XRMR	X-ray resonant magnetic reflectivity
XRMS	X-ray resonant magnetic scattering
XRR	X-ray reflectivity
xSLD	X-ray scattering length density

Chapter 1
Real and Reciprocal Lattices

1.1 Real Lattice Structures

1.1.1 Unit Cells and Lattice Vectors

Scattering with X-rays, neutrons, or electrons from single crystals is a fundamental method for obtaining information about the structure of matter. In contrast to microscopy, scattering is a reciprocal method that provides us with the Fourier components of the structure. Only through inverse transformation do we obtain information about structures and relationships in real space. Before we delve deeper into these methods, we must first clarify the term "single crystal," which is of central importance for the rest of this and the following Chapters. By the term "single crystal," we understand an extended, idealized periodic arrangement of scattering centers (electrons, atoms, molecules, or atomic nuclei) in space that exhibits perfect translational symmetry and, if necessary, other symmetry elements. This means that all atoms in the single crystal are in their ground state and at their equilibrium position. Deviations from this state lead to excitations and defects, which will be discussed later.

Figure 1.1 shows a sketch of a two-dimensional section of an infinite, periodic arrangement of atoms, symbolized by black dots. First, we define a *unit cell volume* V_{uc} by the unit cell vectors a, b, c of magnitude $V_{uc} = |a \cdot (b \times c)|$. Note that bold italic letters denote vectors. Dots or bullet points between two vectors indicate scalar products and crosses denote vector products. There is an arbitrary choice of how to define the unit cell vectors, as indicated by the four examples in Fig. 1.1. However, once we have selected one particular set, then we ought to stick with it as the basic repeat unit of a single crystalline structure. This also implies that within the extension of a single crystal, the repeated unit cell size and orientation do not change. The unit cell vectors, once chosen, also define the coordinate system for the crystal structure.

H. Zabel, *Elements of Elastic Scattering by X-Rays, Neutrons, and Electrons*,
https://doi.org/10.1007/978-3-032-16624-1_1

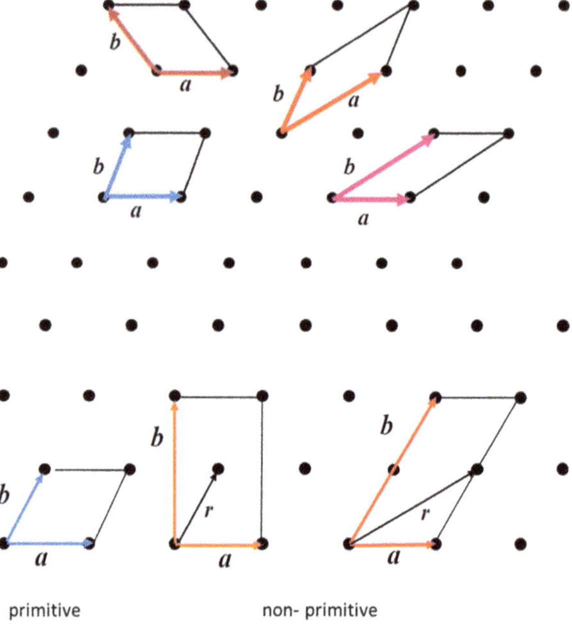

Fig. 1.1 Unit cell vectors of a single crystal in two dimensions.

Fig. 1.2 Primitive (left panel) and nonprimitive unit cells (middle and right panel) with one basis vector r

primitive
unit cell

non- primitive
unit cells

If a unit cell contains exactly one atom per cell, i.e., one atom per corner or exactly one atom inside the unit cell, then it is called a *primitive unit cell*. Sometimes it is convenient to choose a unit cell that contains more than one atom per unit cell. Unit cells with more than one atom per unit cell are called nonprimitive unit cells. In such a case, the crystal lattice is described not only by the three unit cell vectors a, b, c, but also by the *basis vectors* $r = xa + yb + zc$, where $0 \leq x, y, z \leq 1$ are fractions of the unit cell vectors. In the schematics of Fig. 1.2, there is just one basis vector shown. However, depending on the complexity of the crystal structure, unit cells may host many more basis vectors. In protein crystallography, it is not unusual to have several thousand basis vectors per unit cell.

With the help of unit cell vectors and basis vectors, we can now define the lattice vector R_j for the position of any atom "j" in a single crystal lattice. The *lattice vector* R_j is composed of a translation vector T_j and a basis vector r_k within the unit cell:

$$R_j = T_j + r_k, \tag{1.1}$$

where the *translational vector* is defined by

$$T_j = u_j a + v_j b + w_j c \quad \text{with } (u, v, w \in N) \tag{1.2}$$

in the crystal lattice, and the *basis vectors* r_k within any unit cell are

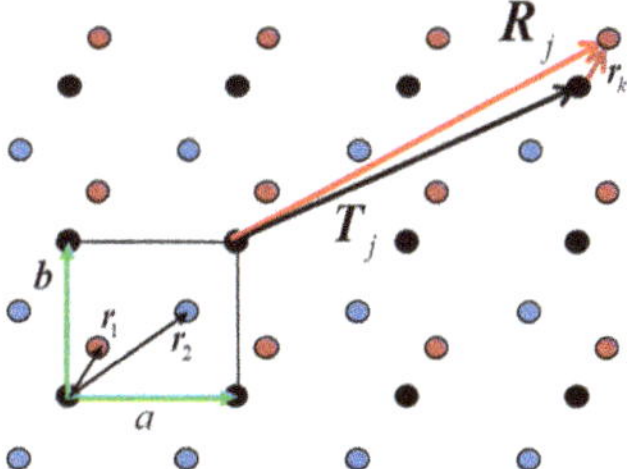

Fig. 1.3 The lattice vector $\boldsymbol{R}_j$ is composed of the translation vector $\boldsymbol{T}_j$, connecting the black points, and the basis vector $\boldsymbol{r}_k$, connecting the blue and red points. These three vectors connect any point in a single crystal lattice

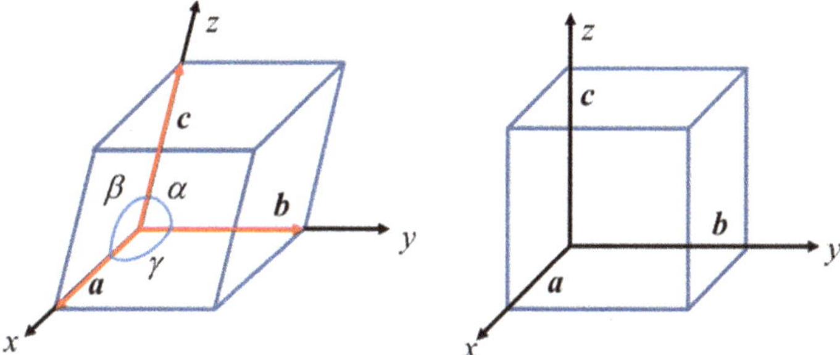

Fig. 1.4 Definition of unit cell vectors and angles between the unit cell vectors for a general lattice in the left panel and a cubic cell in the right-hand panel

$$r_k = x\boldsymbol{a} + y\boldsymbol{b} + z\boldsymbol{c}, \text{ with } (0 \leq x, y, z \leq 1). \tag{1.3}$$

Here j is the label for the unit cells in the lattice, and k is the index for the basis vectors in any unit cell. The translation vector $\boldsymbol{T}_j$ connects all unit cells, whereas the basis vectors $\boldsymbol{r}_k$ connect all atoms belonging to a basis. The lattice vector $\boldsymbol{R}_j$ is the vector sum of the translation vector $\boldsymbol{T}_j$ and the basis vector $\boldsymbol{r}_k$ that connects all atoms uniquely in a single crystal lattice. The spatial origin of the lattice vector is arbitrarily chosen. The geometric relation between the three vectors is shown in Fig. 1.3.

Primitive unit cells contain exactly one atom per cell. Nonprimitive unit cells may be chosen for convenience. Often, primitive unit cells cannot be defined, depending on the complexity of the crystal structure.

In three dimensions, the general definition of unit cells with vectors $\boldsymbol{a}, \boldsymbol{b}, \boldsymbol{c}$ and angles α, β, γ between the unit cell vectors is shown in Fig. 1.4. For cubic cells, the definitions are simplified to $|\boldsymbol{a}| = |\boldsymbol{b}| = |\boldsymbol{c}|$ and $\alpha = \beta = \gamma = 90°$. The moduli of the unit cell vectors $|\boldsymbol{a}|, |\boldsymbol{b}|, |\boldsymbol{c}|$ are referred to as *lattice parameters a, b, and c.*

1.1.2   Bravais Lattices

Single crystals exhibit a wide variety of unit cell sizes, numbers of atoms per unit cell, and crystal symmetries. Crystal structures are categorized into 32 crystal classes and 230 space groups based on the allowed symmetry operations, including inversion centers, rotation axes, mirror planes, glide planes, and combinations thereof. Among them are 14 *Bravais lattices* (named according to the French crystallographer Auguste Bravais (1811–1863)). Bravais lattices exhibit a unique symmetry: each atom in a Bravais lattice is an inversion center. This means that Bravais lattices are invariant under an inversion $R_j \rightarrow -R_j$ at any lattice point. From a practical point of view, inversion symmetry implies that Bravais lattices look identical in the forward direction or backward, viewed from any lattice point chosen and in any arbitrary direction.

The 14 lattices that fulfill the symmetry requirement of Bravais lattices can be subdivided into 7 crystal systems (cubic, tetragonal, orthorhombic, hexagonal, trigonal, monoclinic, triclinic) with 4 types of *unit cells*: primitive (P), body centered (I), face centered (F), and base centered (C). Those 14 lattices are reproduced in Fig. 1.5. The nonprimitive unit cells of Bravais lattices are also known as conventional unit cells, because it is more convenient to discuss and operate with cubic cells that contain a basis of two or four atoms, than to deal with primitive cells of triclinic symmetry. Nevertheless, any Bravais lattice can be represented as a primitive unit cell containing only a single lattice point. In the Exercises 1.2 and 1.3, we consider the primitive unit cells of the cubic I (or bcc) and the cubic F (or fcc) lattices.

In units of the unit cell vectors a, b, c, the coordinates of the basis vectors in primitive and nonprimitive Bravais lattices are as follows:

1. P-lattices: $r_1 = (0, 0, 0)$.
2. I-lattices: $r_1 = (0, 0, 0)$; $r_2 = \left\{\frac{1}{2}, \frac{1}{2}, \frac{1}{2}\right\}$.
3. F-lattices: $r_1 = (0, 0, 0)$; $r_2 = \left(\frac{1}{2}, \frac{1}{2}, 0\right)$; $r_3 = \left(\frac{1}{2}, 0, \frac{1}{2}\right)$; $r_4 = \left(0, \frac{1}{2}, \frac{1}{2}\right)$.
4. C-lattices: $r_1 = (0, 0, 0)$; $r_2 = \left(\frac{1}{2}, \frac{1}{2}, 0\right)$.

> Exactly fourteen known lattice structures meet the symmetry requirements of Bravais lattices, particularly the inversion symmetry.

If the unit cells have cubic symmetry, the P-lattice is also known as simple cubic (sc), the I-lattice is called body-centered cubic (bcc), and the F-lattice is known as face-centered cubic (fcc). Elements that crystallize in the bcc structure are Li, Na, K, Rb, Cs, Ba, V, α-Fe, Nb, Ta, Cr, Mo, and W. Elements adopting the fcc structure are γ-Fe, Cu, Ag, Au, Al, Pt, and Pd.

Tetragonal crystal structures are adopted by β-Sn and β-B. Orthorhombic structures are realized by the black form of phosphorus, by rhombic sulfur, and by black arsenic.

No elements crystallize in the primitive hexagonal structure. All elemental hexagonal structures are close-packed (hcp), which are nonprimitive and also not

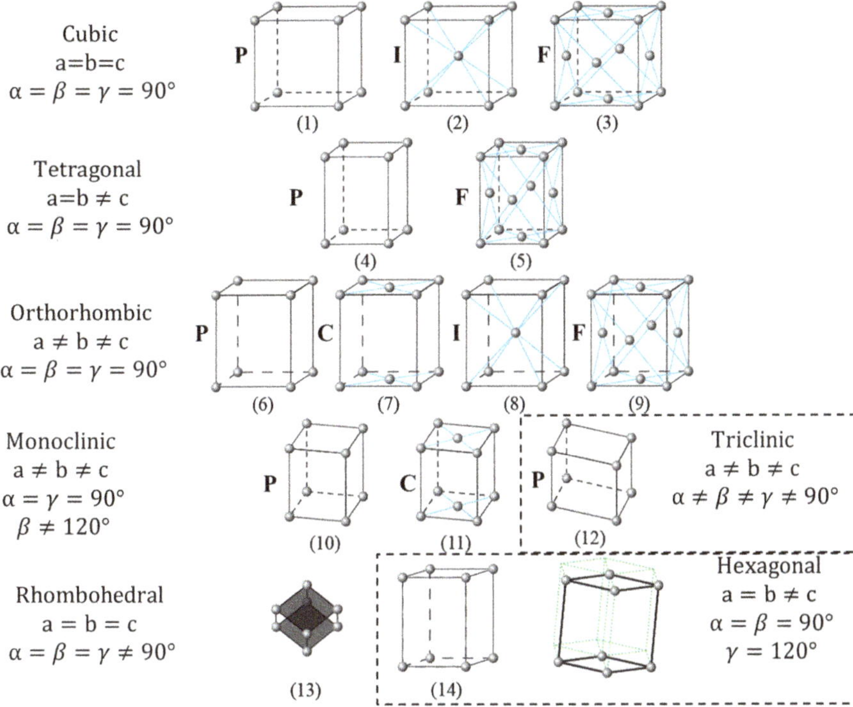

Fig. 1.5 The 7 crystal systems together with their 14 Bravais lattices. The capital letters refer to four types of unit cells: P = primitive, I = body centered, F = face centered, C = base centered. The numbers refer to the crystal systems: (1–3) cubic; (4, 5) tetragonal; (6–9) orthorhombic; (10–12) monoclinic; (13) rhombohedral or trigonal; (14) hexagonal. (Adapted from Wikimedia@commons, Author: Napy1kenobi)

of Bravais lattice type. Examples are Be and Mg of the second period, transition metals such as Sc, Ti, Co, Sn, Y, and Zr, and several rare earth metals. Because of its importance, the *hcp* structure is discussed in more detail further below.

Not all elements crystallize in Bravais lattices. The atoms Si, Ge, and Sn crystallize in the diamond lattice, which is not a Bravais lattice. Carbon crystallizes either in a diamond lattice or in a honeycomb lattice; neither is a Bravais lattice. These special lattices are discussed further below.

All Bravais lattices are occupied by just one elemental atom. By definition, crystal structures with more than one type of atom per unit cell, like in ionic crystals or in alloys, are not Bravais lattices.

1.1.3   Crystal Structures

There are four nomenclatures for characterizing crystal structures and symmetries that are commonly used. Those comprise the *international space group* notation, the

Schoenflies notation, the *Pearson* notation, and the Strukturbericht notation. For our purpose, it is sufficient to use the simplest one, the *Strukturbericht* notation. It consists of a capital letter and a serial number. The letter indicates the number of atoms per chemical formula in a unit cell (A = monatomic; B = diatomic; C = AB_2 compounds; E, F, …K = complex compounds, L = alloys, O = organic compounds, S = silicates), and the serial number does not have a special meaning. For instance, the *fcc* structure is referred to as A1 structure according to the Strukturbericht, the bcc structure is called A2, and hexagonal structures are called A3. The crystal structure of rocksalt (NaCl) has the nomenclature B1 because it has two different atoms in the unit cell.

Table 1.2 lists some common crystal structures together with their symbols according to the Strukturbericht (1. Column) and to the respective space group (4. Column). The space group symbols have the following meaning. The first capital letter stands for the centering in the Bravais lattice (P = primitive, F = fcc, etc.). The following three numbers and lowercase letters refer to symmetry operations (3 = threefold rotational axis, 2_1 = twofold screw axis, m = mirror plane, $\overline{m}$ = mirror plane combined with a glide plane, etc.), which are applied in the three following crystallographic directions:

- The rotation axes and the rotation inversion axes are parallel to the directions listed in Table 1.1.
- The mirror planes are oriented perpendicular to the directions listed in Table 1.1.

Note that different crystal structures according to the Strukturbericht may exhibit the same symmetry elements and therefore cannot be distinguished based on their space group. Some of the *crystal structures* are discussed in more detail in the next section. For instance, the crystals Cu and CaF_2 have the same space group, although their structures are very different. However, they can be distinguished with respect to their Strukturbericht symbol (Table 1.2).

1.1.4 Special Lattice Structures

In addition to Bravais lattices, there are several non-Bravais crystal structures frequently realized in nature and often referred to in the literature by their Strukturbericht notation. Therefore, we discuss a selection of these special crystal structures, which are sketched in Fig. 1.6. First, we distinguish between crystal structures of single elemental atoms and those that contain more than one atom or

Table 1.1 Symmetry operations listed in Table 1.2 are applied according to the three indicated crystallographic directions

Crystal system	1. position	2. position	3. position
Orthorhombic	[100]	[010]	[001]
Tetragonal	[001]	$\langle 100 \rangle$	$\langle 110 \rangle$
Hexagnal	[0001]	$\langle 1000 \rangle$	$\langle 1200 \rangle$
Cubic	$\langle 100 \rangle$	$\langle 111 \rangle$	$\langle 110 \rangle$

Table 1.2 Crystal structures and their designation according to the Strukturbericht (first column) and their respective symmetry according to the space group (fourth column). The capital letters denote the centering in the Bravais lattice according to Fig. 1.5. The remaining three symbols have the following meaning in cubic lattices: the first symbol m denotes a mirror plane perpendicular to the crystal a-axis; $\bar{3}$ denotes a threefold axis parallel to the crystal $\langle 111 \rangle$ directions followed by an inversion; the second symbol m signifies a mirror plane perpendicular to the crystal $\langle 110 \rangle$ directions. The symbol "d" in the diamond structure denotes a special glide plane along $1/4\,(a+b)$

Strukturbericht symbol	Name of crystal structure	Prototype	Space group
A1	fcc	Cu	$Fm\bar{3}m$
A2	bcc	W	$Im\bar{3}m$
A3	hcp	Mg	$Im\bar{3}m$
A4	Diamond	C	$Fd\bar{3}m$
A9	Graphite	C	$Im\bar{3}m$
B1	Sodium chloride	NaCl	$Fm\bar{3}m$
B2	Cesium chloride	CsCl	$Pm\bar{3}m$
B4	Wurtzite	ZnS	$Im\bar{3}m$
C1	Fluorite	CaF_2	$Fm\bar{3}m$
C4	Rutile	TiO_2	$Im\bar{3}m$
C15	Laves phase	Cu_2Mg	$Fd\bar{3}m$
$E2_1$	Perovskite	$CaTiO_3$	$Pm\bar{3}m$
$L1_0$	Intermetallic alloys	CuAu, FePt	$Im\bar{3}m$
$L1_2$	Intermetallic alloys	Cu_3Au, Ni_3Al	$Pm\bar{3}m$
$L2_1$	Heusler alloys	Cu_2MnGe	$Fm\bar{3}m$

ion. The diamond lattice and the hexagonal close-packed lattice are examples of monatomic non-Bravais lattices, displayed in panels (a) and (b), respectively. Diatomic lattices are the so-called zinc sulfite structures, cesium chloride structure, sodium chloride structure, and Cu_3Au structure sketched in panels (c)–(f). Panels (g) and (f) show examples of triatomic structures realized in metal alloys and metal oxides.

(a) **Diamond structure (A4)**: The diamond structure consists of two fcc-unit cells. The first fcc unit cell (fcc1) has the usual basis vectors. The second unit cell (fcc2) is shifted along the space diagonal direction by the amount $\left(\frac{1}{4},\frac{1}{4},\frac{1}{4}\right)$. These two sublattices are occupied by identical atoms and have the following eight basis vectors:

$$\text{fcc1:}\ \boldsymbol{r}_1 = (0,0,0);\ \boldsymbol{r}_2 = \left(\tfrac{1}{2},\tfrac{1}{2},0\right);\ \boldsymbol{r}_3 = \left(\tfrac{1}{2},0,\tfrac{1}{2}\right);\ \boldsymbol{r}_4 = \left(0,\tfrac{1}{2},\tfrac{1}{2}\right).$$
$$\text{fcc2:}\ \boldsymbol{r}_5 = \left(\tfrac{1}{4},\tfrac{1}{4},\tfrac{1}{4}\right);\ \boldsymbol{r}_6 = \left(\tfrac{3}{4},\tfrac{3}{4},\tfrac{1}{4}\right);\ \boldsymbol{r}_7 = \left(\tfrac{3}{4},\tfrac{1}{4},\tfrac{3}{4}\right);\ \boldsymbol{r}_8 = \left(\tfrac{1}{4},\tfrac{3}{4},\tfrac{3}{4}\right).$$

Representatives for this crystal structure are the lattices of C (diamond), Si, Ge, and Sn. The diamond structure does not feature inversion symmetry and is therefore not a Bravais lattice.

(b) **Hexagonal close-packed (*hcp*) structure (A3)**: The *hcp* structure is a monatomic and nonprimitive hexagonal structure with two atoms per basis. The lattice parameters and angles are like in the primitive hexagonal unit cell: $a = b \neq c$; $\alpha = \beta = 90\,°$; $\gamma = 120\,°$. The basis vectors are

$$\boldsymbol{r}_1 = (0,0,0);\ \boldsymbol{r}_2 = \left(\tfrac{1}{3},\tfrac{2}{3},\tfrac{1}{2}\right).$$

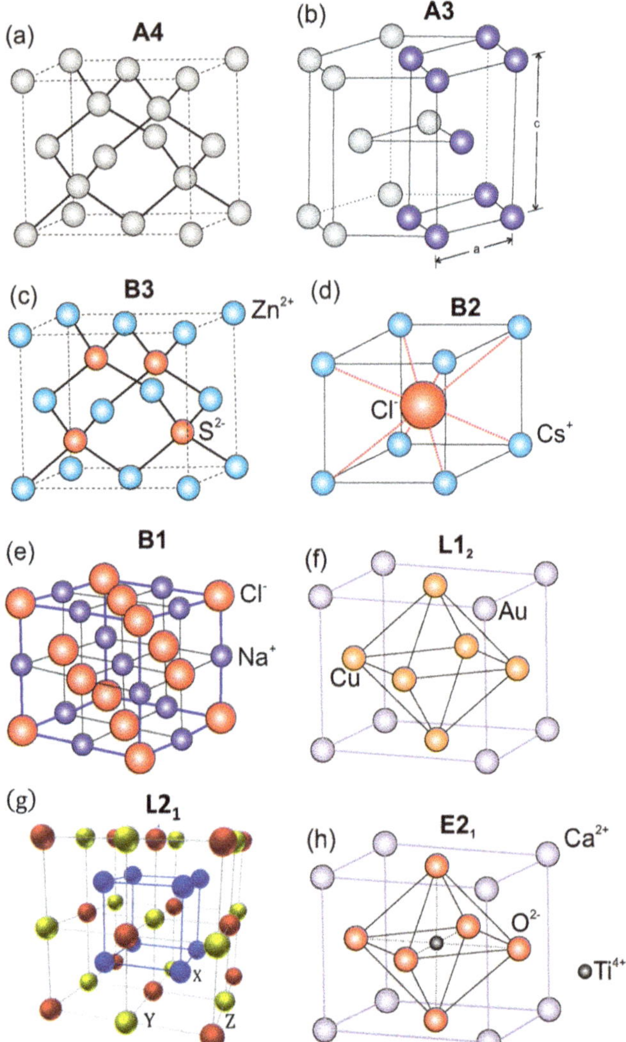

Fig. 1.6 Special lattice structures designated with their nomenclature according to the Strukturbericht. Different colors indicate different atoms or ions. (**a**) diamond structure; (**b**) hcp structure. Here, blue and white spheres represent the same atoms, while blue spheres indicate one hcp unit cell. (**c**) ZnS-structure; (**d**) CsCl-structure; (**e**) NaCl-structure; (**f**) Cu₃Au-structure; (**g**) Cu₃MnSi-structure; (**h**) CaTiO₃-structure. None of the crystal structures schematically shown here fulfilll the symmetry requirement of a Bravais lattice

The atoms belonging to the hcp unit cell are indicated in blue color in Fig. 1.7. The hcp structure arises from the stacking of two hexagonal basal planes in an ABA… sequence. A topographic projection of the A-plane (shown in bluish color) and the B-plane (shown in greenish color) is presented in Fig. 1.7. Plane B

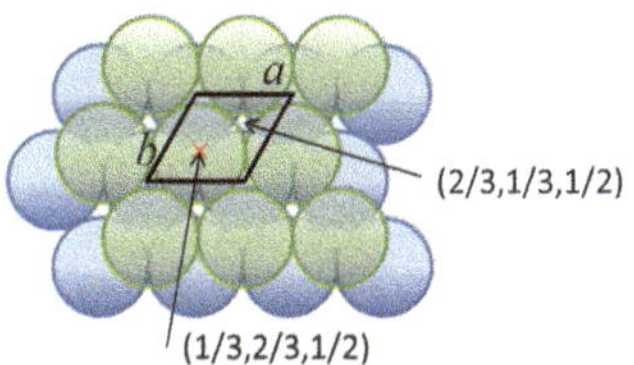

Fig. 1.7 Stacking sequence of hexagonal close-packed basal planes in the hcp structure

is shifted in relation to the A plane over the hole at the position $\left(\frac{1}{3}, \frac{2}{3}, \frac{1}{2}\right)$. There is a second hole at position $\left(\frac{2}{3}, \frac{1}{3}, \frac{1}{2}\right)$, which, however, is not occupied in the hcp structure. In the hcp structure, the *c/a* ratio should ideally be $\sqrt{8/3} = 1.633$. But conventionally, we refer to crystal structures as being hcp even if the actual *c/a* ratio deviates somewhat from the ideal value.

The *hcp* structure is realized by the alkaline earth atoms Be and Mg, by the transition metals Sc, Ti, Zn, Co, Y, Zr, Ru, Cd, Hf, Re, and Os, by the rare earth elements Gd, Tb, Dy, Ho, Er, Tm, and Lu, and by He at high pressures below 2 K.

The *hcp* structure is closely related to the *fcc* structure. The *fcc* structure, viewed along the space diagonal, also consists of a close-packed hexagonal arrangement of atoms. But the stacking sequence of the planes, in contrast to the hcp structure, is ABCA... . The B-plane occupies the holes at $\left(\frac{1}{3}, \frac{2}{3}, \frac{1}{3}\right)$, and the C-plane occupies the holes at $\left(\frac{2}{3}, \frac{1}{3}, \frac{2}{3}\right)$. There are specific Bragg reflections, which are sensitive to the stacking sequence of atomic planes, allowing one to distinguish between hcp and fcc structures. These two stacking sequences may lead to stacking disorder if they become mixed.

(c) **Zinc sulfite (ZnS) structure (B3)**: The ZnS structure, also known as the sphalerite structure, is very similar to the diamond structure. However, the sublattice *fcc*1 with basis vectors r_1 to r_4 is occupied by atoms A, and the sublattice *fcc*2 with basis vectors r_5 to r_8 is occupied by atoms B. The ZnS structure is much more common than the diamond structure and is the crystal lattice of all III–V compounds and also of II–VI compounds. A short list of compounds crystallizing in the ZnS structure is as follows: ZnS, ZnSe, GaAs, GaP, CdS, CdSe, InAs, InSb, CuF, CuCl, AgI, SiC, AlP, and many more.

(d) **Cesium chloride (CsCl) structure (B2)**: This cubic structure is similar to a bcc structure. However, the center atom B is different from the corner atom A. We may consider the CsCl structure as consisting of two simple cubic (sc) lattices, the second one shifted against the first one along the space diagonal to the center position $\left(\frac{1}{2}, \frac{1}{2}, \frac{1}{2}\right)$. The basis vectors of the two sublattices are thus.

sc1: $r_1 = (0, 0, 0)$; sc2: $r_2 = \left(\frac{1}{2}, \frac{1}{2}, \frac{1}{2}\right)$.

The CsCl structure is realized by several ionic crystals and also by several metal alloys: CsI, TlBr, TlI, NH$_4$Cl, AgMg, AlNi, CuZn (β-brass), and BeCu. β-brass is further discussed in Chap. 8 in the context of order–disorder phase transitions.

(e) **Sodium chloride (NaCl) structure (B1)**. The NaCl structure can be considered as two interpenetrating fcc lattices shifted against each other along the x-axis by the amount $\left(\frac{1}{2},0,0\right)$. Thus, the basis vectors of the two sublattices are accordingly:

fcc1: $r_1 = (0,0,0)$; $r_2 = \left(\frac{1}{2},\frac{1}{2},0\right)$; $r_3 = \left(\frac{1}{2},0,\frac{1}{2}\right)$; $r_4 = \left(0,\frac{1}{2},\frac{1}{2}\right)$.

fcc2: $r_5 = \left(\frac{1}{2},0,0\right)$; $r_6 = \left(0,\frac{1}{2},0\right)$; $r_7 = \left(0,0,\frac{1}{2}\right)$; $r_8 = \left(\frac{1}{2},\frac{1}{2},\frac{1}{2}\right)$.

The NaCl structure is mainly realized by ionic crystals where one sublattice is occupied by anions and the other one by cations. Examples are NaCl, LiF, KCl, KBr, AgCl, AgBr, CaO, MgO, and PbS.

(f) **Cu_3Au structure ($L1_2$)**. So far, we have discussed crystal structures with a 1:1 stoichiometry. For other stoichiometries, different structures are expected. An interesting case is an atomically ordered metal alloy of Cu and Au in the ratio 3: 1. Here, the Au atoms occupy the corners of the cubic unit cell, and the Cu atoms occupy the face-centered position of an fcc-like structure. The basis vectors are therefore:

Au: $r_1 = (0,0,0)$; Cu: $r_2 = \left(\frac{1}{2},\frac{1}{2},0\right)$; $r_3 = \left(\frac{1}{2},0,\frac{1}{2}\right)$; $r_4 = \left(0,\frac{1}{2},\frac{1}{2}\right)$.

The intermetallic compound Cu_3Au often serves as a textbook example for order–disorder phase transitions, which are discussed further in Chap. 8 and Sect. 12.1.

(g) **Cu_2MnSi structure ($L2_1$)**. This is an example of a ternary compound. It represents the so-called Heusler compounds with the general composition X_2YZ. The X-positions are occupied by the late 3d transition metals Co, Ni, or Cu; the Y-positions are occupied by the magnetic ions Mn or Fe, and the Z-positions are occupied by elements either from the group 3 or group 4 elements, such as Al, Si, Ge, or Ga. The cubic Heusler structure consists of four interpenetrating fcc sub-lattices. The Z and Y elements occupy the sites of the NaCl lattice. The X elements occupy the two remaining fcc lattices, the first shifted along the space diagonal by $\left(\frac{1}{4},\frac{1}{4},\frac{1}{4}\right)$ and the second one shifted by $\left(\frac{3}{4},\frac{3}{4},\frac{3}{4}\right)$. In the X_2YZ structure, the X-sites appear to have a simple cubic structure in the center. In total, the 16 basis vectors are as follows:

Z: $r_1 = (0,0,0)$; $r_2 = \left(\frac{1}{2},\frac{1}{2},0\right)$; $r_3 = \left(\frac{1}{2},0,\frac{1}{2}\right)$; $r_4 = \left(0,\frac{1}{2},\frac{1}{2}\right)$.

Y: $r_5 = \left(\frac{1}{2},0,0\right)$; $r_6 = \left(0,\frac{1}{2},0\right)$; $r_7 = \left(0,0,\frac{1}{2}\right)$; $r_8 = \left(\frac{1}{2},\frac{1}{2},\frac{1}{2}\right)$.

$X1$: $r_9 = \left(\frac{1}{4},\frac{1}{4},\frac{1}{4}\right)$; $r_{10} = \left(\frac{3}{4},\frac{3}{4},\frac{1}{4}\right)$; $r_{11} = \left(\frac{3}{4},\frac{1}{4},\frac{3}{4}\right)$; $r_{12} = \left(\frac{1}{4},\frac{3}{4},\frac{3}{4}\right)$.

$X2$: $r_{13} = \left(\frac{3}{4},\frac{3}{4},\frac{3}{4}\right)$; $r_{14} = \left(\frac{1}{4},\frac{1}{4},\frac{3}{4}\right)$; $r_{15} = \left(\frac{1}{4},\frac{3}{4},\frac{1}{4}\right)$; $r_{16} = \left(\frac{3}{4},\frac{1}{4},\frac{1}{4}\right)$.

The Heusler alloys play an important role as very promising spintronic materials, showing ferromagnetic properties with expected 100% spin polarization. Typical representatives of this compound are Cu_2MnSi, Cu_2MnGe, Co_2MnAl, Co_2MnSi, and Co_2MnGe. Their strong magnetism could serve as a substitute for rare-earth metals.

(h) **CaTiO3 structure, or perovskite structure ($E2_1$)**. The general chemical formula of the cubic perovskite structure is ABO_3. The A atoms occupy the corner sites $(0,0,0)$, while the B atom sits in the center at $\left(\frac{1}{2},\frac{1}{2},\frac{1}{2}\right)$, surrounded by oxygen atoms on the face-centered sites $\left(\frac{1}{2},\frac{1}{2},0\right)$ and equivalents. The six oxygen ions form an octahedral cage around the central metal ion. The basis vectors are accordingly:

$$A : r_1 = (0, 0, 0).$$

$$B : r_2 = \left(\frac{1}{2}, \frac{1}{2}, \frac{1}{2}\right).$$

$$O : r_3 = \left(\frac{1}{2}, \frac{1}{2}, 0\right); \quad r_4 = \left(\frac{1}{2}, 0, \frac{1}{2}\right); \quad r_5 = \left(0, \frac{1}{2}, \frac{1}{2}\right).$$

The perovskites are a large class of solid-state compounds with a huge variety of physical properties. $SrTiO_3$ shows a paraelectric–ferroelectric phase transition at low temperatures. $Pb(Zr_{1-x}Ti_x)O_3$ is piezoelectric and is used in ultrasound heads for imaging. $YBa_2Cu_3O_7$ (YBCO) is a high-temperature superconductor with a transition temperature of 92 K. $BaTiO_3$ is a ferroelectric, piezoelectric, and pyroelectric material used as a dielectric in capacitors and as a frequency modulator in optics. Further representatives for the perovskite structure are $CaTiO_3$, $BiFeO_3$, $LaMnO_3$, and $LaYbO_3$.

1.1.5   Quasicrystals

In crystallography, the science of ordered crystal structures, a fundamental assumption is made but is not explicitly stated. The entire crystal is believed to consist of the repetition of a single unit cell, either primitive or nonprimitive. With just one unit cell, the entire crystal space can be filled without creating holes or vacancies. Ideally, we neglect the existence of surfaces here. This basic assumption requires that the unit cells should exhibit a rotational symmetry axis $c_n = 2\pi/n$ with the number of positions $n = 1, 2, 3, 4,$ or 6, but not 5 of more than 6. Unit cells that show a *fivefold symmetry axis* cannot fill the space without holes. The proof of this statement is left to Exercise 1.7. Already Johannes Kepler (1571–1630) realized that a flat plane cannot be filled with a regular array of pentagons, without leaving holes, as shown in Fig. 1.8.

However, the space can be filled if we allow two types of unit cells to exist. In the example shown in Fig. 1.8, the white areas may be filled with a different tile, and then we obtain a space-filling periodic pattern of two different cells. This may contradict the rigidity of bond angles in covalent or ionic crystals. But in metal alloys, where bond angles are less important, fivefold symmetry is conceivable and indeed has been observed. The respective metal alloys are known as *quasicrystals*. Quasicrystals consist of an aperiodic array of tiles lacking translational symmetry but following strict tiling rules and displaying side lengths that obey the golden ratio. Quasicrystals and the consequences of a broken translational symmetry are further discussed in Sect. 3.5.4.

Fig. 1.8 It is not possible to fill the plane with a regular array of pentagons. Instead, gaps between the tiles occur frequently. This example is adapted from Johannes Kepler: Harmonices Mundi

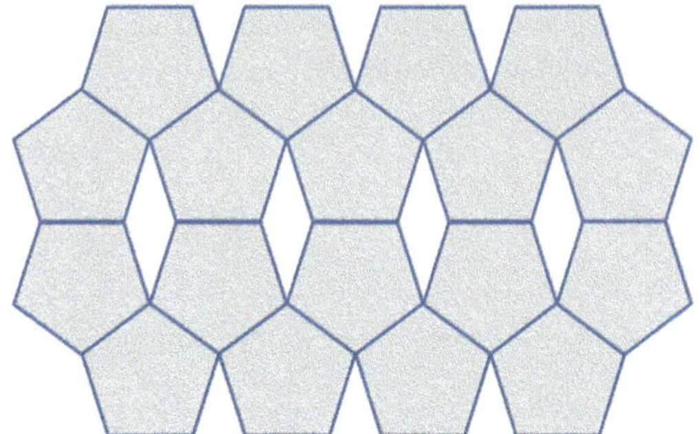

1.2   Reciprocal Lattice

1.2.1   Lattice Planes and Miller Indices

Now we consider "imagined" lattice planes with different orientations in unit cells. They are termed "imagined" because these flat planes are mathematical constructions. To distinguish between different lattice planes in three-dimensional space, they are labeled by a triple of so-called *Miller indices* (*hkl*). A few examples of lattice planes, together with their Miller indices, are shown in Fig. 1.9. The indexing works as follows. Consider, for instance, the (100) plane. This plane cuts the *x*-axis at the distance "*a*" but does not touch the *y* or the *z*-axis. In units of the lattice parameter "*a*," the cuts are therefore 1, 0, 0. The next example is a plane that cuts the *x*-axis at *a*/ 2. The reciprocal value of 1/2 is 2. Accordingly, this plane is labeled (200). If the *y*-axis is cut at *a*/2, the respective label of that plane is (020). Always, the reciprocal fraction of a unit cell parameter is used for labeling the respective lattice plane. Those reciprocal fractions are called Miller indices (*hkl*). These (*hkl*) indices identify lattice planes of the real lattice in the reciprocal space, to be discussed in the next section. In real crystals, these (*hkl*) planes are not necessarily occupied by atoms. For instance, in the *P*-cubic lattice, the (200), (020), and (002) are not occupied by atoms; they are simply mathematical constructions.

In general, the rules for generating (*hkl*) Miller indices are as follows:

- The intercepts of a lattice plane with the crystal axes are inversely proportional to the Miller indices of the plane. If a plane does not cut a crystal axis, the corresponding Miller index is $\lim_{x \to \infty} 1/x = 0$.
- If the reciprocal values of the intercepts do not result in integer numbers, find a common numerator and divide by the reciprocal values of the intercepts. If, for instance, 4, 3, 2 are the intercepts and $\frac{1}{4}, \frac{1}{3}, \frac{1}{2}$ are the reciprocal values, then the common numerator is 12, and the Miller indices for this plane are (3, 4, 6).
- To simplify nomenclature, one writes negative indices as $\bar{n}$, instead of $-n$. The commas between the (*hkl*) values can then be skipped, such as (142) and (111).

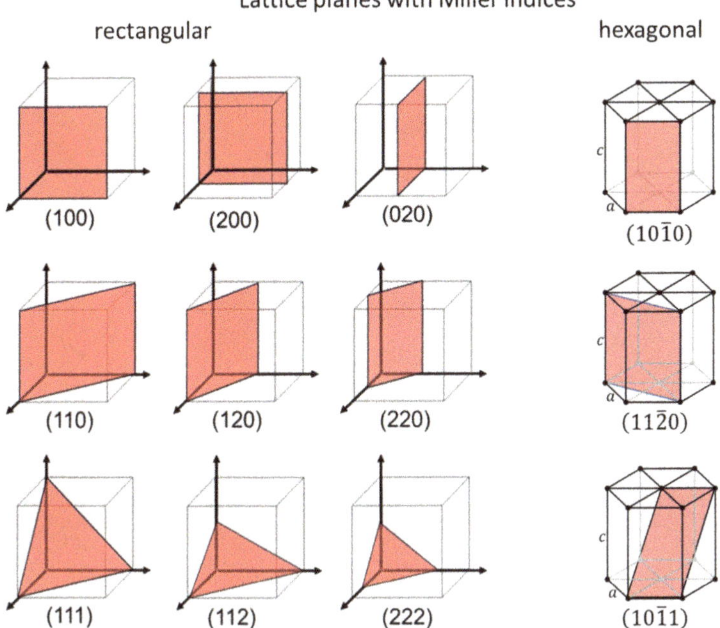

Fig. 1.9 Examples for different lattice planes labeled by their Miller indices in rectangular and hexagonal systems. A bar over a numeral denotes a minus sign

- All symmetry-equivalent families of lattice planes are written with curled brackets { } instead of round brackets ( ). The planes (100), (010), and (001) together form the family of {100} planes.
- A direction $n_1\boldsymbol{a} + n_2\boldsymbol{b} + n_3\boldsymbol{c}$ is similarly written with square brackets [] as $[n_1n_2n_3]$ and a family of directions [100], [010], [100], etc., is denoted as $\langle 100 \rangle$.
- Distances between planes decrease with increasing Miller indices, while the atomic density increases with decreasing Miller indices. Figure 1.10 shows the relation between Miller indices and the spacing between lattice planes d_{hkl}.
- For cubic lattices, the distance between {hkl} planes is (see Eq. 1.14)

$$d_{hkl} = a/\sqrt{h^2 + k^2 + l^2}.$$

Here, "a" is the lattice parameter of the cubic unit cell.
- For crystal structures with *hexagonal symmetry*, it has become common practice to use four Miller indices instead of three: ($hkil$). However, the fourth index i is redundant and results from the negative sum of h and k: $h + k = -i$. A (110) plane, for example, is then denoted $\left(11\bar{2}0\right)$. To save space, the minus sign appears as a bar on top of i. However, the index i is often omitted and replaced by a dot. In the chosen example, the denotation is then: (11.0).

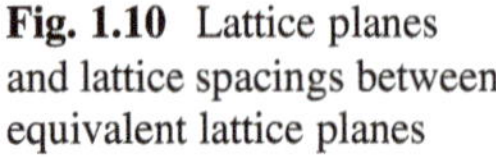

Fig. 1.10 Lattice planes and lattice spacings between equivalent lattice planes

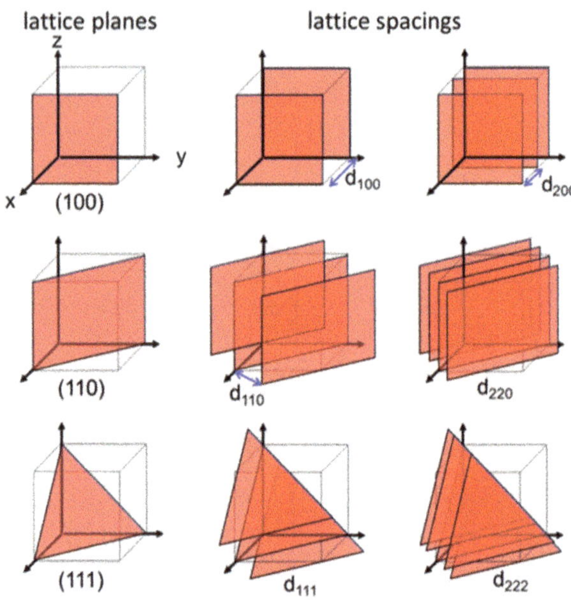

There is an effective method to calculate the spacing of equivalent lattice planes (distances between lattice planes) and angles between nonequivalent lattice planes. This is explained after introducing reciprocal lattice vectors in the next section.

1.2.2 Reciprocal Lattice Vectors

Having familiarized ourselves with lattice planes and lattice spacings in real space, we are now ready to introduce the *reciprocal lattice* with its *reciprocal lattice vectors*. Both lattices, real and reciprocal, are uniquely related like the numerator and denominator of the same quantity. The reciprocal space is also referred to as Fourier space for reasons that become clear later.

The *reciprocal cell vectors* $\boldsymbol{A}$, $\boldsymbol{B}$, $\boldsymbol{C}$ are related to the real space unit cell vectors $\boldsymbol{a}, \boldsymbol{b}, \boldsymbol{c}$ according to

$$A = 2\pi \frac{(\boldsymbol{b} \times \boldsymbol{c})}{\boldsymbol{a} \cdot (\boldsymbol{b} \times \boldsymbol{c})}; \quad B = 2\pi \frac{(\boldsymbol{c} \times \boldsymbol{a})}{\boldsymbol{a} \cdot (\boldsymbol{b} \times \boldsymbol{c})}; \quad C = 2\pi \frac{(\boldsymbol{a} \times \boldsymbol{b})}{\boldsymbol{a} \cdot (\boldsymbol{b} \times \boldsymbol{c})}. \tag{1.4}$$

These expressions can be simplified by recognizing that the unit cell volume is given by $V_{uc} = \boldsymbol{a} \cdot (\boldsymbol{b} \times \boldsymbol{c})$. Then

$$A = \frac{2\pi}{V_{uc}}(\boldsymbol{b} \times \boldsymbol{c}); \quad B = \frac{2\pi}{V_{uc}}(\boldsymbol{c} \times \boldsymbol{a}); \quad C = \frac{2\pi}{V_{uc}}(\boldsymbol{a} \times \boldsymbol{b}). \tag{1.5}$$

We immediately recognize that A is perpendicular to b and c but parallel to a. Therefore, the orthogonality relationships hold:

$$A \cdot a = 2\pi; \quad \text{and} \quad A \cdot b = A \cdot c = 0. \tag{1.6}$$

The modulus of A is

$$|A| = 2\pi / |a|. \tag{1.7}$$

Equivalent relations hold for B and C.

Lattice parameters are usually given in picometers (pm) or nanometers (nm), formerly in Ångströms (Å, $1 \text{ Å} = 10^{-10}$ m). A typical lattice parameter value of 3 Å is 300 pm or 0.3 nm. Correspondingly, the units of |A| are $[A] = \text{m}^{-1}$. For our example, a reciprocal cell value of 2.09 Å^{-1} corresponds to 0.0209 pm^{-1} or 20.9 nm^{-1}.

The volume of a reciprocal cell is related to the real space unit cell volume via

$$V_{rc} = |A \cdot (B \times C)| = \frac{(2\pi)^3}{|a \cdot (b \times c)|} = \frac{(2\pi)^3}{V_{uc}}. \tag{1.8}$$

Using the definitions of the reciprocal cell vectors, we may sketch a reciprocal lattice corresponding to a real lattice and vice versa. Two simple examples are shown in Fig. 1.11 for a two-dimensional projection of a tetragonal and a monoclinic lattice. Note that in both cases, $A \perp b$ and $B \perp a$. The black dots in the real lattice represent lattice points, which may be decorated by atoms. The red dots in the reciprocal lattice represent, however, lattice planes according to their respective Miller indices (*hkl*).

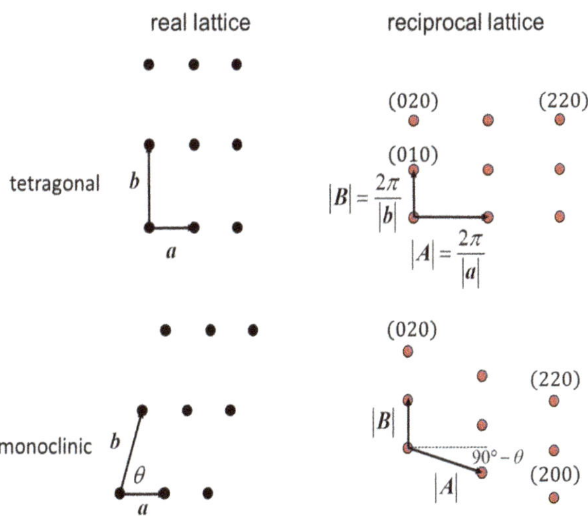

Fig. 1.11 Two examples are shown for the relationship between real space lattice (left panels) and reciprocal space lattice (right panels). Some reciprocal lattice points are labeled with their respective Miller indices

In standard textbooks on Solid State Physics or Condensed Matter Physics, it is shown that the reciprocal lattice of a bcc lattice is an fcc lattice and vice versa. The primitive unit cells of the bcc and fcc lattices are also topics of Exercises 1.2 and 1.3.

Using the reciprocal cell vectors, we can now define a *reciprocal lattice vector*:

$$G_{hkl} = h\mathbf{A} + k\mathbf{B} + l\mathbf{C}. \tag{1.9}$$

Here, h,k,l are the Miller indices designating real space lattice planes. Any G_{hkl} vector refers to a discrete point in the reciprocal lattice. This discrete point represents a periodic array of lattice planes in real space. The reciprocal lattice vector G_{hkl} is of paramount importance for the discussion of scattering experiments with X-rays, neutrons, or electrons on crystalline solids, as we will see in the next chapters. Moreover, G_{hkl} has five very important geometric properties, which we will discuss in the following sections.

1.2.2.1   Product of $G \cdot T$

The product of the reciprocal lattice vector G_{hkl} and the translational vector (Eq. 1.2) T is an integer multiplied by 2π. This can be shown as follows:

$$\begin{aligned} G_{hkl} \cdot T &= (h\mathbf{A} + k\mathbf{B} + l\mathbf{C})(u\mathbf{a} + v\mathbf{b} + w\mathbf{c}) = 2\pi(hu + kv + lw) \\ &= 2\pi N \ (N \in \mathbf{Z}). \end{aligned} \tag{1.10}$$

Here, we used the orthogonality property discussed before. The last equation is a multiple of 2π since u, v, w, h, k, l are all integers, and therefore their products must also be integers, the sum of which is called N.

1.2.2.2   Surface Normal

The reciprocal lattice vector G_{hkl} is oriented orthonormal to the respective (hkl) lattice plane. This can be shown as follows. Each plane is defined by at least two vectors, which lie in the plane. If G_{hkl} is perpendicular to both vectors, then it must be perpendicular to the entire (hkl) lattice plane. A general (hkl) lattice plane is sketched in Fig. 1.12, and two vectors in the plane: $r_1 - r_2 = u\mathbf{a} - v\mathbf{b} = \mathbf{a}/h - \mathbf{b}/k$ and $r_2 - r_3 = v\mathbf{b} - w\mathbf{c} = \mathbf{b}/k - \mathbf{c}/l$. Now, taking the scalar product between G_{hkl} and these two vectors yield:

$$G_{hkl} \cdot \left(\frac{\mathbf{a}}{h} - \frac{\mathbf{b}}{k}\right) = 2\pi - 2\pi = 0; \quad G_{hkl} \cdot \left(\frac{\mathbf{b}}{k} - \frac{\mathbf{c}}{l}\right) = 2\pi - 2\pi = 0. \tag{1.11}$$

Hence, G_{hkl} is perpendicular to both vectors in the (hkl) plane and therefore perpendicular to the entire (hkl) plane. In short: $G_{hkl} \perp (hkl)$. Using this property,

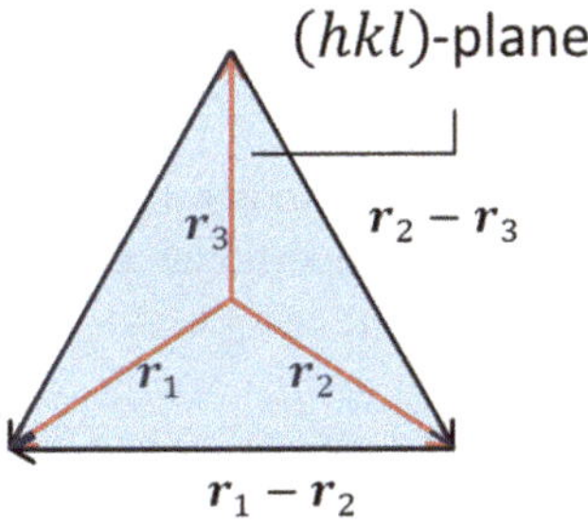

Fig. 1.12 Two vectors $r_1 - r_2$ and $r_2 - r_3$ define the orientation of the *hkl*-plane

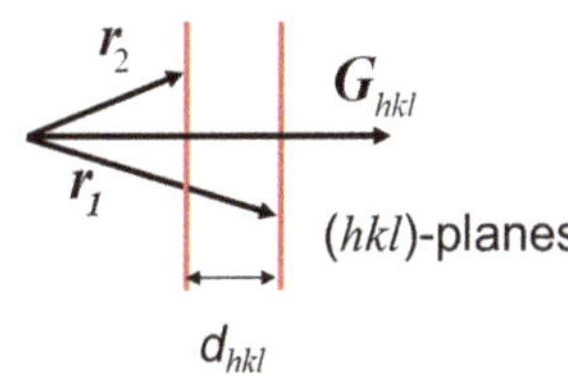

Fig. 1.13 Determination of the lattice plane spacing d_{hkl}

we can now define a normal unit vector to the (*hkl*) plane. Writing $\mathbf{G}_{hkl} = |\mathbf{G}_{hkl}|\widehat{\mathbf{n}}_{hkl}$, the normal unit vector is

$$\widehat{\mathbf{n}}_{hkl} = \frac{\mathbf{G}_{hkl}}{|\mathbf{G}_{hkl}|}. \qquad (1.12)$$

1.2.2.3 Lattice Plane Spacing

The modulus of $\mathbf{G}_{hkl}$ is the reciprocal lattice spacing $2\pi/d_{hkl}$ of a (*hkl*) plane: $|\mathbf{G}_{hkl}| = 2\pi/d_{hkl}$. This can be shown with the help of Fig. 1.13 as follows:

r_1 and r_2 are lattice vectors that connect to the (*hkl*) lattice planes defined by the reciprocal lattice vector $\mathbf{G}_{hkl}$ and the plane normal $\widehat{\mathbf{n}}_{hkl} = \mathbf{G}_{hkl}/|\mathbf{G}_{hkl}|$. If we project the lattice vectors r_1 and r_2 into the direction of $\widehat{\mathbf{n}}_{hkl}$ and take the difference, we will obtain the lattice spacing d_{hkl}:

$$\widehat{\mathbf{n}}_{hkl} \cdot \mathbf{r}_1 = \frac{1}{|\mathbf{G}_{hkl}|}(h\mathbf{A} + k\mathbf{B} + l\mathbf{C}) \cdot (u\mathbf{a} + v\mathbf{b} + w\mathbf{c}) = \frac{2\pi N}{|\mathbf{G}_{hkl}|}.$$

And similarly,

$$\widehat{\mathbf{n}}_{hkl} \cdot \mathbf{r}_2 = \frac{1}{|\mathbf{G}_{hkl}|}(h\mathbf{A} + k\mathbf{B} + l\mathbf{C}) \cdot ((u-1)\mathbf{a} + (v-1)\mathbf{b} + (w-1)\mathbf{c}) = \frac{2\pi(N-1)}{|\mathbf{G}_{hkl}|}.$$

Therefore, the difference is

$$\widehat{\boldsymbol{n}}_{hkl} \cdot (\boldsymbol{r}_1 - \boldsymbol{r}_2) = \frac{2\pi}{|\boldsymbol{G}_{hkl}|} = d_{hkl}.$$

This also shows that the modulus of the reciprocal lattice vector, $|\boldsymbol{G}_{hkl}|$, is inversely proportional to the lattice spacing d_{hkl}:

$$|\boldsymbol{G}_{hkl}| = \frac{2\pi}{d_{hkl}} \tag{1.13}$$

For a cubic lattice, we can now easily determine the lattice spacing d_{hkl} of (hkl) lattice planes:

$$d_{hkl}^{\text{cubic}} = \frac{2\pi}{|\boldsymbol{G}_{hkl}|} = \frac{2\pi}{\sqrt{(h\boldsymbol{A})^2 + (k\boldsymbol{B})^2 + (l\boldsymbol{C})^2}} = \frac{2\pi}{|\boldsymbol{A}|} \frac{1}{\sqrt{h^2 + k^2 + l^2}} = \frac{a}{\sqrt{h^2 + k^2 + l^2}}.$$

$$\tag{1.14}$$

This is the result already quoted above for the cubic lattice. Examples of lattice spacings are $d_{100} = a$; $d_{110} = a/\sqrt{2}$; $d_{111} = a/\sqrt{3}$. For orthorhombic lattices, the d_{hkl} spacings are accordingly:

$$d_{hkl}^{\text{ortho}} = \frac{1}{\sqrt{\left(\frac{h}{a}\right)^2 + \left(\frac{k}{b}\right)^2 + \left(\frac{l}{c}\right)^2}}. \tag{1.15}$$

For hexagonal lattices, the d_{hkl} spacing is

$$d_{hkl}^{\text{hex}} = \frac{1}{\sqrt{\frac{4}{3}\left(\frac{h^2+hk+k^2}{a^2}\right) + \left(\frac{l}{c}\right)^2}} \tag{1.16}$$

1.2.2.4   Angles Between Lattice Planes

It is also easy to calculate the angle between (hkl) lattice planes. Let $\widehat{\boldsymbol{n}}_{hkl}$ be the normal to the (hkl) plane, and $\widehat{\boldsymbol{n}}'_{h'k'l'}$ the normal to the $(h'k'l')$ plane. Then the scalar product of the normal vectors is:

$$\widehat{\boldsymbol{n}}_{hkl} \cdot \widehat{\boldsymbol{n}}'_{h'k'l'} = |\widehat{\boldsymbol{n}}_{hkl}||\widehat{\boldsymbol{n}}'_{h'k'l'}| \cos\theta, \tag{1.17}$$

where θ is the angle between the lattice planes. With the help of the reciprocal lattice vector $\boldsymbol{G}_{hkl}$, the angle θ can easily be calculated. Examples are discussed in the problem section.

1.2.2.5 Translational Invariance

Finally, we discuss a fundamental aspect of reciprocal lattice vectors. Any periodic function of a physical parameter that can be represented as a Fourier transform with Fourier coefficients of the reciprocal lattice is translationally invariant.[1] This can be seen as follows. Let us assume that $n(r)$ is a periodic function in real space, which can be represented by a Fourier series of the type:

$$n(r) = \sum_G n_G \exp i(G \cdot r). \tag{1.18}$$

Here, we replaced $G_{hkl} = G$ for convenience. Now we shift this function from the position r to the position $r + T$, where T is a translational vector of the crystal lattice. Then we find:

$$
\begin{aligned}
n(r + T) \quad &= \sum_G n_G \exp i(G \cdot (r + T)) = \sum_G n_G \exp i(G \cdot r) \exp i(G \cdot T) \\
&= \sum_G n_G \exp i(G \cdot r) = n(r),
\end{aligned}
\tag{1.19}
$$

because $\exp i(G \cdot T) = \exp i(2\pi N) = 1$, where N is an integer. The converse of the above statement is also true: any physical function that is translationally invariant, like the crystal lattice, can be represented as a Fourier series.

Each of the following statements is fundamental to the rest of this text:

1. Each reciprocal lattice point represents an infinite set of equally spaced crystal lattice planes.
2. Each reciprocal lattice point (hkl) is the Fourier transform of a set of crystal lattice planes normal to the reciprocal lattice vector G_{hkl}.
3. Each reciprocal lattice point is representative of the periodicity of the entire crystal lattice in the direction of the reciprocal lattice vector G_{hkl}.

In the following chapters, we will realize more fundamental properties of the reciprocal lattice vector, such as the following:

1. Reciprocal lattice vectors are defined to describe conditions for constructive interference of X-rays, neutrons, and electrons scattered from periodic lattice structures.
2. The reciprocal lattice allows the evaluation of momentum transfers and momentum conservation of waves propagating in periodic lattice structures.

---

[1] A refresher on Fourier transforms is provided in Appendix A1.

Summary

1. Single crystals consist of a periodic arrangement of atoms in space.
2. The smallest repeat unit in single crystals is called a unit cell.
3. Unit cells have lattice parameters a, b, c that form the translation vector:
   $T_j = u_j a + v_j b + w_j c$ (j=index of atom j).
4. Unit cells that contain exactly one lattice point (one atom) are called primitive unit cells.
5. Nonprimitive unit cells contain more than one atom belonging to a base with base vectors r_k.
6. The lattice vector is defined by the sum: $R_j = T_j + r_k$.
7. There exist 14 Bravais lattices featuring inversion symmetry at any lattice site.
8. All Bravais lattices can be represented either as primitive unit cells or by their conventional crystal structure.
9. Non-Bravais lattices have nonprimitive crystal structures and are mostly compounds or alloys.
10. Crystal structures are conveniently labeled by their Strukturbericht letter and number.
11. The reciprocal lattice vector of an (hkl)-plane is $G_{hkl} = hA + kB + lC$.
12. The reciprocal cell vector A is parallel to the unit cell vector a, and perpendicular to the vectors b and c. Similar relations hold for B and C.
13. G_{hkl} is oriented normal to the (hkl) plane.
14. $|G_{hkl}|$ is inversely proportional to the (hkl)-lattice spacing: $|G_{hkl}| = \frac{2\pi}{d_{hkl}}$.
15. The scalar product of G_{hkl} and T is 2π times an integer number: $G \cdot T = \pi N$ ($N \in Z$).
16. Periodic functions, represented as harmonics of the reciprocal lattice, are translationally invariant: $n(r + T) = n(r)$.

Questions

(Note that more than one answer may be correct)

1. **What are the coordinates of the basis vectors for the three unit cells shown in** Fig. 1.2**?**

   (a) $1.(00); 2.(00), \left(\frac{1}{2}, \frac{1}{2}\right); 3.(00), \left(0, \frac{1}{2}\right)$
   (b) $1.(00); 2.(00), (00), 3.\left(0, \frac{1}{2}\right); \left(\frac{1}{2}, \frac{1}{2}\right)$
   (c) $1.(00); 2.(00), \left(\frac{1}{2}, \frac{1}{2}\right); 3.\left(0, \frac{1}{2}\right), \left(1, \frac{1}{2}\right)$

2. **How many atoms (lattice points) are in the** *fcc* **unit cell? The** *fcc* **unit cell is one of the Bravais lattices shown in** Fig. 1.5.

(a) 4

(b) 14

(c) 8

3. **Right or wrong?**

   (a) Bravais lattices contain only one type of atom.
   (b) The primitive unit cell of any Bravais lattice holds only one lattice point.
   (c) A Bravais lattice is a crystal structure that describes the atomic order in alloys.

4. **Miller indices are a convenient way of labeling lattice planes in crystal lattices. What else are Miller indices good for?**

   (a) For calculating the interplanar spacing.
   (b) For calculating angles between planes.
   (c) For calculating the lifetime of a crystal structure.

5. **The reciprocal lattice is a three-dimensional arrangement of reciprocal lattice points. These reciprocal lattice points:**

   (a) Are connected by the translational vector T.
   (b) Show translational invariance.
   (c) Have a symmetry that is unrelated to the real space symmetry.

6. **What are the main geometric properties of the reciprocal lattice vector G_{hkl}?**

   (a) G_{hkl} is normal to the respective (hkl) plane.
   (b) The product of G_{hkl} and the translational vector T_{uvw} is infinite.
   (c) The modulus of G_{hkl} is inversely proportional to the (hkl) lattice spacing.

7. **Any periodic function that can be represented by a Fourier transform shows translational invariance if the Fourier coefficients:**

   (a) Are those of Bravais lattices.
   (b) Are those of the reciprocal lattice.
   (c) Have cubic symmetry.

Exercises

Grades of difficulty: E = easy, M = medium, A = advanced

E 1.1 Inversion symmetry
Show that the diamond lattice lacks inversion symmetry and therefore cannot be a
Bravais lattice.

M 1.2 Primitive unit cell
Consider a body-centered cubic (bcc) lattice with the lattice constant a.

(a) Choose a set of primitive lattice vectors a_1, a_2, a_3, and give the coordinates in units of the bcc lattice.
(b) Determine the volume of the primitive unit cell in terms of the bcc cell.

M 1.3 Miller Indices

Consider the lattice planes with the Miller indices (100) and (001) in the fcc lattice.

(a) Convert the nonprimitive fcc unit cell into a primitive unit cell. Show graphically the primitive unit cell of the fcc lattice.
(b) What are the Miller indices of the (100) and (001) lattice planes now in terms of the primitive unit cell?

M 1.4 Space filling

Consider the hexagonal close-packed structure (*hcp*) sketched in Fig. 1.6b.

(a) Show that for the ideal hcp structure, the ratio of the lattice parameters is $c/a = \sqrt{8/3}$.
(b) Assume that atoms can be treated as hard spheres, which may touch each other when close-packed. Determine the fraction of space that is occupied by the atoms (hard spheres) in a simple cubic lattice (*sc*), in a *bcc* lattice, and a *fcc* lattice.

E 1.5 Angles between *hkl* planes

Consider a cubic unit cell. Determine the angle between the lattice planes with Miller indices (111) and:

(a) (210)
(b) (202)
(c) (301)

E 1.6 Lattice spacings

Consider a cubic unit cell.

(a) What is the lattice spacing of the (111) planes?
(b) What is the lattice spacing of the (221) planes?
(c) What is the lattice spacing of the (300) planes?

A 1.7 Fivefold symmetry axis

In crystallography, it is well known that long-range translational order requires a rotational axis $c_n = 2\pi/n$ with $n = 1, 2, 3, 4, 6$, but not 5 nor higher than 6. Show that this statement is correct.

E 1.8 Reciprocal cell volume

Prove the correctness of the relation in Eq. (1.8): $V_{rc} = (2\pi)^3/V_{uc}$.

Further Reading

W.D. Callister Jr., D.G. Rethwisch, *Materials Science and Engineering: An Introduction*, 10th edn. (John Wiley & Sons, New York, 2018)

H. Ibach, H. Lüth, *Solid-State Physics, An Introduction to Principles of Materials Science* (Springer, Berlin, 2009)

C. Kittel, *Introduction to Solid State Physics*, 8th edn. (John Wiley & Sons, New York, 2004)

Online Dictionary of Crystallography. International Union of Crystallography (2021), https://dictionary.iucr.org/Main_Page. Accessed 17 Apr 2024

Chapter 2
Scattering Amplitudes

2.1   Interaction of Radiation with Matter

Radiation interacts with matter via absorption and scattering. Figure 2.1 illustrates both types of interactions when an incoming stream of particles (photons, electrons, neutrons) hits a target. The particle stream per unit area and unit time is called *intensity I* and has the unit $m^{-2}\,s^{-1}$. The noninteracting fraction of the primary intensity is transmitted in the forward direction.

We assume a constant incident particle intensity (radiation) $I_0(\lambda)$ with a fixed wavelength (energy) λ, as shown in Fig. 2.1, which strikes a target made of high-density material (solid, soft matter, or liquid) of thickness x at normal incidence (90°). A pencil detector behind the target measures the transmitted intensity that is lower than the incoming intensity $I_0(\lambda)$. The attenuation of the transmitted beam intensity $I(x, \lambda)$ in the forward direction is usually described by an exponential decay as a function of thickness x, known as the *Lambert–Beer law*:

$$I(x, \lambda) = I_0(\lambda) \exp(-\alpha(\lambda)x), \tag{2.1}$$

where $\alpha(\lambda)$ is the attenuation coefficient. **Attenuation** has two contributions: **absorption** and **scattering**, and generally depends on the wavelength or energy. Part of the radiation is absorbed in the target, while another part is scattered in different directions. Absorption is usually accompanied by excitation and conversion into various particle types. For example, the photoelectric effect converts a photon into the kinetic energy of an electron.

> Interaction of radiation with matter = Attenuation + Transmission
> Attenuation = Absorption + Scattering

© The Author(s), under exclusive license to Springer Nature Switzerland AG 2026

H. Zabel, *Elements of Elastic Scattering by X-Rays, Neutrons, and Electrons*,

https://doi.org/10.1007/978-3-032-16624-1_2

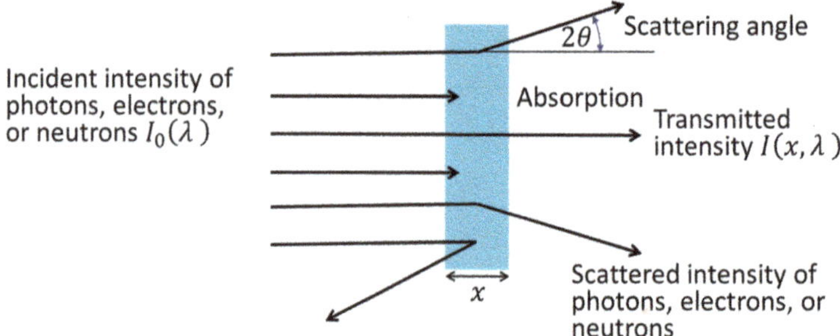

Fig. 2.1 Interaction of a particle beam with a target of thickness x. Some particles are scattered at an angle 2θ. Particles that are not scattered are either absorbed or transmitted

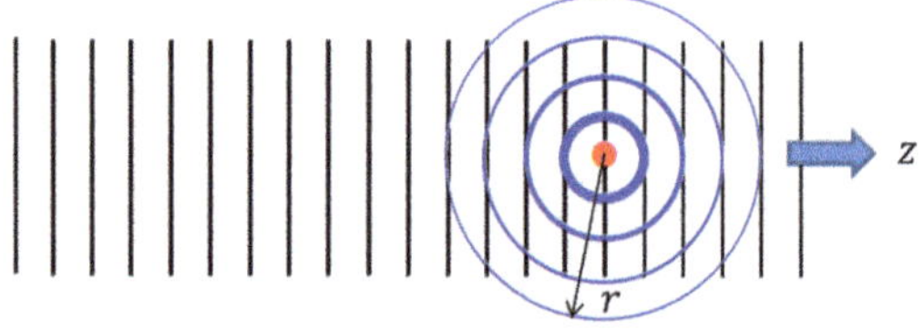

Fig. 2.2 Scattering of a plane wave from a spherical target. z is the direction of the propagating plane wave, r is the distance of the spherical wavefront from the scattering center, and the parallel black lines symbolize the wavelength of the plane wave. The red dot marks a point-like scattering center that generates an outgoing spherical wave whose amplitude decreases, simultaneously attenuating the amplitude of the transmitted plane wave

In the following, we do not consider absorption processes, which have many different causes and require detailed knowledge of all the interrelations. Instead, we focus on scattered radiation in various directions, which is characterized by a scattering angle 2θ between the primary beam and the scattered beam, and a scattering amplitude. The scattering process may be elastic, i.e., energy-conserving, or inelastic due to gain or loss of the initial energy. We restrict our discussions to elastic scattering processes.

2.2 Scattering Amplitude and Scattering Length

To simplify further analysis, we consider an incoming plane wave with a fixed wavelength λ propagating in the z-direction. In Fig. 2.2, the plane wave is represented by a series of straight parallel lines that represent the crests of a sine wave and provide a snapshot of the wave propagating in the z-direction from left to right. Formally, the incoming plane wave can be described by the time-dependent oscillatory amplitude:

$$A_i(z, t) = A_0 \sin(\boldsymbol{k}_i \cdot \boldsymbol{z} - \omega t). \qquad (2.2)$$

Here $k_i = 2\pi/\lambda_i$ is the wavenumber of the incident plane-wave, the index i stands for "incident" or "incoming", ω is the frequency of the wave, t is the time, and A_0 is the time-independent amplitude of the incident wave. All letters in bold are vectors. The scalar product $\boldsymbol{k}_i \cdot \boldsymbol{z}$ is the phase angle of the plane wave. k_i and ω are fixed for a particular wave. While the wave propagates in space along the z-direction, the time t elapses, such that the difference, $(\boldsymbol{k}_i \cdot \boldsymbol{z} - \omega t)$, stays constant. The ratio ω/k_i is the phase velocity. Appendix A1.2 shows that the representation of the scattering amplitude as a trigonometric function is equivalent to the representation as an exponential function with a complex exponent:

$$A_i(z, t) = A_0 e^{i(\boldsymbol{k}_i \cdot \boldsymbol{z} - \omega t)}. \qquad (2.3)$$

The exponential representation of waves is more practical, and ultimately, makes further calculations much easier.

As soon as this wave hits a point-like target fixed in space, the plane wave is scattered by this target, and an isotropic spherical wave propagates outward, as indicated by the two-dimensional projection in Fig. 2.2.

The target may be an electron distribution surrounding a nucleus or a nucleus of an atom, depending on the specifics of the scattering event with X-rays, neutrons, or electrons. But in all cases, we will observe an outgoing spherical wave, whose amplitude can be described in the far-field by

$$A_s(r, t) = A_0 \frac{a}{r} \exp(\boldsymbol{k}_f \cdot \boldsymbol{r} - \omega t). \qquad (2.4)$$

The amplitude of the scattered wave A_s is called the **scattering amplitude**, r is the radial distance of the spherical wave from the target center to the front of the propagating wave, $k_f = 2\pi/\lambda_f$ is the wavenumber of the scattered wave, ω is the frequency, and t is the time that the wavefront of the spherical wave needs to travel the radial distance r. The subscript f stands for "final." The quantity "a" is called the **scattering length** and characterizes the strength of the particle (photon)–target interaction. As A_s is dimensionless, the amplitude a must have the dimension of a length, which explains its name. The $1/r$ dependence of the scattering amplitude is a consequence of energy conservation. The energy flux of the scattered wave is proportional to the square of the amplitude. By integration over the surface of a sphere with radius r, the flux must be constant. Thus, $(1/r)^2 \times 4\pi r^2 = 4\pi = \text{const}$. The scattering amplitude $A_s(r, t)$ is an expression of probability. Not all waves are scattered. But the stronger the interaction, the more photons, electrons, or neutrons are scattered by the target.

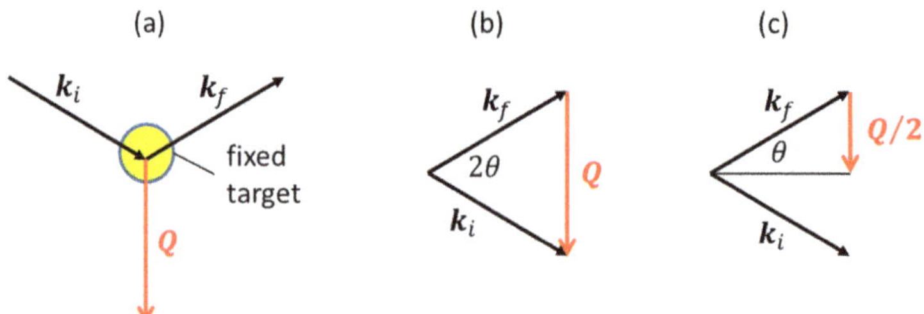

Fig. 2.3 Scattering triangle formed by the incident wave vector k_i, the scattered wavevector k_f, and the scattering vector Q

2.3   Elastic Scattering at a Single Target

In this chapter and throughout the entire book, we focus on **_elastic scattering._**[1] Elastic scattering implies that momentum is transferred to a fixed target, but no energy is exchanged during the scattering process. Therefore, the wavelength of the electrons, neutrons, or photons remains the same before and after the scattering event: $\lambda_i = \lambda_f$, and the modulus of the wavenumbers is identical: $|k_i| = |k_f| = 2\pi/\lambda$. But the direction of the wave propagation changes. Furthermore, momentum conservation requires that the momentum before and after scattering is identical: $p_i = p_f$ or $\hbar k_i = \hbar Q + \hbar k_f$, where $\hbar Q$ is the momentum transfer to the target:

$$\hbar Q = \hbar(k_i - k_f). \tag{2.5}$$

Here $\hbar = h/2\pi$ is the Planck constant ($h = 6.626069 \times 10^{-34}$ Js). Dropping the Planck constant on both sides of the equation, the momentum transfer is given by the vector difference of the incident and scattered wavenumbers:

$$Q = k_i - k_f. \tag{2.6}$$

Scattering vector for elastic scattering: $Q = k_i - k_f$.

The elastic scattering and the momentum transfer to a fixed target are visualized in Fig. 2.3a. Translating the wavevectors, a **_scattering triangle_** is formed, as is shown in panel (b). Panel (c) shows that the scattering vector Q has the magnitude:

---

[1] Elastic scattering is contrasted by inelastic scattering. Inelastic scattering occurs when either energy is transferred from the incident beam to the target or vice versa. Inelastic scattering is important for the study of quasiparticles in solids and liquids, such as phonons, magnons, librons, rotons, etc. Inelastic scattering with X-rays or neutrons is an extensive field that is not covered in this text. Suggestions for further reading on this topic is given at the end of this chapter.

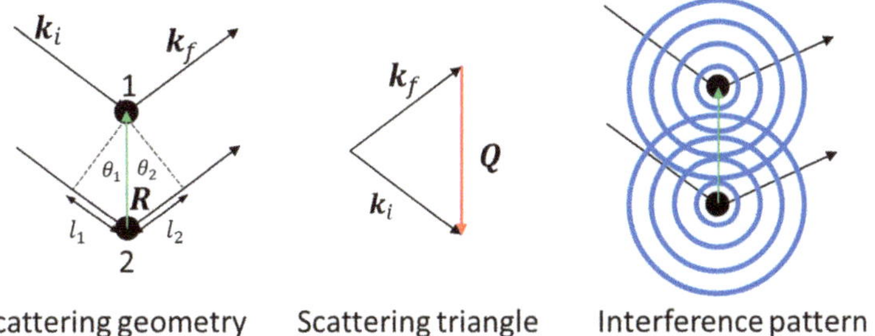

Fig. 2.4 The left panel shows the path difference of spherical waves scattered from two target centers separated by the vector $\boldsymbol{R}$. The middle panel shows the respective scattering triangle formed by the incident and scattered wavevectors $\boldsymbol{k}_i$ and $\boldsymbol{k}_f$. In the right panel, the interference of spherical waves originating from the scattering at different centers is schematically indicated

$$\sin\theta = \frac{Q/2}{k}; \quad Q = 2k\sin\theta = \frac{4\pi}{\lambda}\sin\theta. \tag{2.7}$$

Momentum is conserved only when the triangle is closed. In short, the term "elastic" implies that momentum is transferred to the target (crystal lattice) during the scattering process without any energy transfer. Then the scattering triangle is closed, without the addition of further vectors.[2]

2.4 Scattering at Two or More Targets

Now we consider two targets or even a regular array of target points. When hit by a plane wave, each target point will produce a spherical outgoing wave, and their amplitudes will overlap. The superposition of amplitudes with different phases will result in constructive and destructive interference effects. The condition for **constructive interference** can be related to the path difference or the phase difference. In the situation sketched in Fig. 2.4, the path difference between the plane wave scattered from the first and from the second target separated by the distance $\boldsymbol{R}$ is $l_1 + l_2$.

Constructive interference occurs when the path difference is a multiple of the wavelength λ:

---

[2]In the literature, the scattering vector is often defined as $\boldsymbol{Q} = \boldsymbol{k}_f - \boldsymbol{k}_i$. This definition emphasizes the scattering vector as the difference of the scattered and incident wavenumbers. The definition presented here emphasizes the scattering vector as the momentum transfer to the target. Ultimately, both presentations are equivalent.

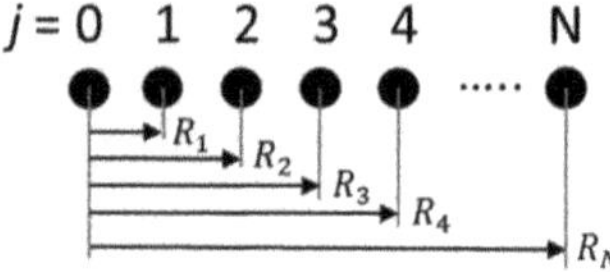

Fig. 2.5 Equidistant scattering centers along a straight line

$$l_1 + l_2 = n\lambda. \tag{2.8}$$

In terms of phase differences, the condition for constructive interference is:

$$\left(\boldsymbol{k}_i - \boldsymbol{k}_f\right) \cdot \boldsymbol{R} = \boldsymbol{Q} \cdot \boldsymbol{R} = n2\pi. \tag{2.9}$$

After scattering, the total scattering amplitude at a distance r from the target is the sum of the amplitudes of rays scattered from target points at the origin $\boldsymbol{R} = 0$ and a distance $\boldsymbol{R}$:

$$A_{s1}(\boldsymbol{Q},t) + A_{s2}(\boldsymbol{Q},t) = A_0 \frac{a}{r}\left(e^{i(k_s \cdot r)}e^{i\boldsymbol{Q}\cdot\boldsymbol{0}} + e^{i(k_s \cdot r)}e^{i\boldsymbol{Q}\cdot\boldsymbol{R}}\right). \tag{2.10}$$

The first term in the bracket of Eq. (2.10) is identical to the term in Eq. (2.4). The second term in the bracket takes into account that the scattering amplitude from the second target is phase-shifted by $\boldsymbol{Q} \cdot \boldsymbol{R}$, regarding the scattering amplitude from the first target. The common factor $e^{i(k_f \cdot r)}$ can be taken out of the bracket:

$$A_{s1}(\boldsymbol{Q},t) + A_{s2}(\boldsymbol{Q},t) = A_0 \frac{a}{r}e^{i(k_f \cdot r)}\left(1 + e^{i\boldsymbol{Q}\cdot\boldsymbol{R}}\right). \tag{2.11}$$

Here, we have neglected the explicit time dependence expressed in the exponent $e^{i\omega t}$, because we consider steady-state conditions. Furthermore, we can safely neglect the common phase factor $e^{i(k_f \cdot r)}$ in front of the bracket. To derive the intensity, the square of the amplitude with its complex conjugate term[3] leads to $e^{i(k_f \cdot r)}e^{-i(k_f \cdot r)} = 1$. Therefore, the simplified amplitude reads now:

$$A_{s1}(\boldsymbol{Q}) + A_{s2}(\boldsymbol{Q}) = \frac{a}{r}\left(1 + e^{i\boldsymbol{Q}\cdot\boldsymbol{R}}\right). \tag{2.12}$$

The scattering amplitude depends on the term in the bracket: $(1 + e^{i\boldsymbol{Q} \cdot \boldsymbol{R}})$. For $\boldsymbol{Q} \cdot \boldsymbol{R} = 2\pi$, the total amplitude is $(1 + 1) = 2$; for $\boldsymbol{Q} \cdot \boldsymbol{R} = \pi$, the total amplitude is $(1 - 1) = 0$. The intensity oscillates accordingly between 0 and 4 in units of $A_0^2(a/r)^2$.

---

[3] Calculations with complex numbers are briefly explained in Appendix A2.

2.4.1 *Extension to N Scattering Centers*

Now we consider an array of scattering centers enumerated by j ($j = 0$ to $N - 1$ number of scattering centers) separated by position vectors $\boldsymbol{R}_j$ (Fig. 2.5). Then the total scattering amplitude becomes:

$$A_{\text{tot}}(\boldsymbol{Q}) = A_0 \frac{a}{r}\left(1 + e^{i\boldsymbol{Q}\cdot\boldsymbol{R}_1} + e^{i\boldsymbol{Q}\cdot\boldsymbol{R}_2} + \ldots\right) = A_0 \frac{a}{r}\sum_{j=0}^{N-1} e^{i\boldsymbol{Q}\cdot\boldsymbol{R}_j}. \tag{2.13}$$

In the case that the scattering centers represent different atoms or isotopes, the scattering length a has to be kept inside the sum:

$$A_{\text{tot}}(\boldsymbol{Q}) = A_0 \frac{1}{r}\sum_{j=0}^{N-1} a_j e^{i\boldsymbol{Q}\cdot\boldsymbol{R}_j}. \tag{2.14}$$

2.4.2 *Orientational Average*

At this point, a split is made regarding the further treatment of Eqs. (2.13) and (2.14). One approach is to consider the sum over the position vectors $\boldsymbol{R}_j$ as part of a regular array with translational symmetry, as in the case of a crystal lattice. In this case, we proceed to Eq. (3.18) and the subsequent equations, which evaluate the scattering amplitude and scattering intensity by taking the lattice sum. Alternatively, the position vectors may be considered random in space, as is the case for disordered systems such as amorphous or liquid materials. In this latter case, an orientational average of the scattering amplitude is taken. Starting with two scattering centers according to Eq. (2.12), we have

$$A_{\text{tot}}(\boldsymbol{Q}) = A_{s1}(\boldsymbol{Q}) + A_{s2}(\boldsymbol{Q}) = \frac{a}{r}\left(1 + e^{i\boldsymbol{Q}\cdot\boldsymbol{R}}\right) \tag{2.15}$$

and the intensity is

$$I_{\text{tot}}(\boldsymbol{Q}) = A_{\text{tot}}(\boldsymbol{Q})A_{\text{tot}}^*(\boldsymbol{Q}) = \left(\frac{a}{r}\right)^2\left(1 + e^{i\boldsymbol{Q}\cdot\boldsymbol{R}}\right)\left(1 + e^{-i\boldsymbol{Q}\cdot\boldsymbol{R}}\right) = \left(\frac{a}{r}\right)^2\left(2 + e^{i\boldsymbol{Q}\cdot\boldsymbol{R}} + e^{-i\boldsymbol{Q}\cdot\boldsymbol{R}}\right). \tag{2.16}$$

We are interested in the orientational average of the last expression, and therefore

$$\langle I_{\text{tot}}(\boldsymbol{Q})\rangle = \left(\frac{a}{r}\right)^2 2\left(1 + \langle e^{i\boldsymbol{Q}\cdot\boldsymbol{R}}\rangle\right), \tag{2.17}$$

where

$$\left\langle e^{i\boldsymbol{Q}\cdot\boldsymbol{R}}\right\rangle = \frac{\displaystyle\int_{-\pi}^{+\pi} d\theta \int_{0}^{\pi} e^{iQR\cos\phi}\sin\phi\, d\phi}{\displaystyle\int_{-\pi}^{+\pi} d\theta \int_{0}^{\pi}\sin\phi\, d\phi} = \frac{\displaystyle\int_{0}^{\pi} e^{iQR\cos\phi}\sin\phi\, d\phi}{4\pi} = \frac{\sin(QR)}{QR}.$$

$$(2.18)$$

In the last equation, we have used the substitution $\cos\phi = x$ and $\sin\phi\, d\phi = -dx$, yielding

$$-2\pi \int_{1}^{-1} e^{iQRx}\, dx = 2\pi\left(\frac{e^{iQR} - e^{-iQR}}{iQR}\right) = 4\pi\frac{\sin(QR)}{QR}. \qquad (2.19)$$

Therefore, the orientationally averaged intensity of two scattering centers is

$$\langle I_{\text{tot}}(\boldsymbol{Q})\rangle = \left(\frac{a}{r}\right)^{2} 2\left(1 + \frac{\sin(QR)}{QR}\right). \qquad (2.20)$$

With more interatomic distances R_{jk} to be considered, we obtain the more general expression:

$$\langle I_{\text{tot}}(\boldsymbol{Q})\rangle = |A(Q)|^{2} = \frac{1}{r^{2}}\sum_{j}\sum_{k} a_{j}a_{k}\frac{\sin(QR_{jk})}{QR_{jk}}. \qquad (2.21)$$

The last equation is known as the **Debye formula**. In Chap. 8 on disordered systems and in Chap. 9 on small-angle scattering, we will encounter these formulas again. Figure 2.6 shows a few examples of the function in Eq. (2.20) for different interatomic distances R. The curves illustrate intensity oscillations as a function of momentum transfer Q and an intensity peak in the forward direction, which becomes more narrow with increasing separation R.

---

[4]The use of the Thomson scattering length to describe X-ray scattering is based on two assumptions: First, the binding energy of electrons in atoms should be so low that they can be considered "free" in the oscillating field of the EM wave. Second, the X-ray photon energy should not be too high to neglect inelastic Compton scattering.

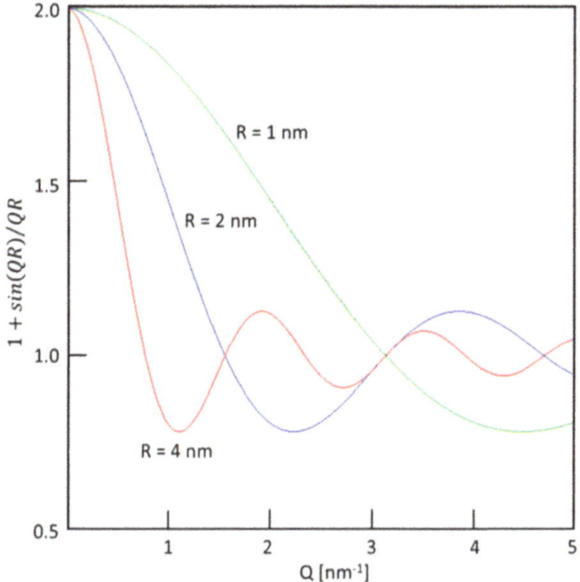

Fig. 2.6 The function $1 + \sin(QR)/QR$ is plotted against the scattering vector Q for three different interatomic distances: $R = 1, 2,$ and 4 nm

2.5   X-Ray Scattering Length, Atomic Form Factor, and Polarization Effects

2.5.1   Thomson Scattering Length

The scattering of X-rays (electromagnetic waves, *EM* waves) at free electrons is known as **Thomson scattering**. The scattering length of Thomson scattering from a single electron is a constant and expressed by the **classical electron radius** (in SI units):[4]

$$r_0 = \frac{1}{4\pi\epsilon_0} \frac{e_0^2}{m_e c^2}, \tag{2.22}$$

and the numerical value of r_0 is 2.817 fm. Here ϵ_0 ($=8.854 \times 10^{-12}$ A^{2}s^4/kg·m^3) is the permittivity of free space, q ($= - |e_0| = 1.6 \times 10^{-19}$ As) is the elementary charge, m_e ($= 9.1 \times 10^{-31}$ kg) is the electron mass and c ($=2.99 \times 10^8$ m/s) is the speed of electromagnetic waves in vacuum. The classical electron radius r_0 is not the true radius of an electron, which is unknown, but an effective radius that describes the interaction of electromagnetic radiation with electrons correctly. Thomson scattering is not isotropic, but shows polarization effects in the directions perpendicular

---

[4]The use of the Thomson scattering length to describe X-ray scattering is based on two assumptions: First, the binding energy of electrons in atoms should be so low that they can be considered "free" in the oscillating field of the EM wave. Second, the X-ray photon energy should not be too high to neglect inelastic Compton scattering.

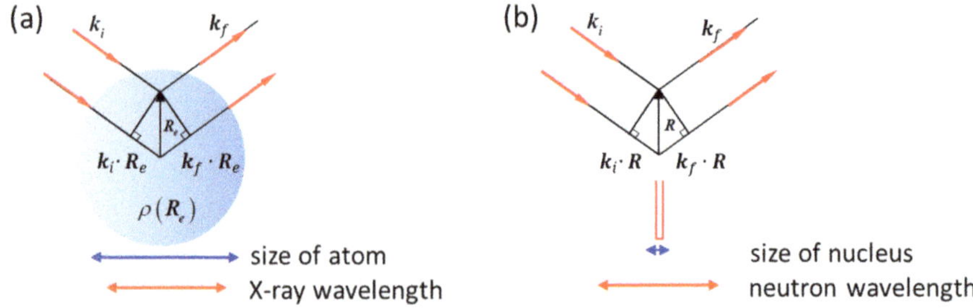

Fig. 2.7 Comparison of X-ray scattering at an electron distribution in an atom (**a**) and neutron scattering at a nucleus (**b**). Note the similarity between atom size and X-ray wavelength and the dissimilarity between nuclear size and neutron wavelength. The nuclear size is, in fact, much smaller than indicated, by a factor of 10^{-5} smaller than the atomic size

to that of the incoming EM radiation, as we will detail further below. Expanding from a single electron to an atom with Z electrons, where Z is the atomic number, the X-ray scattering length is stated by

$$a = -r_0 Z. \tag{2.23}$$

The minus sign in Eq. (2.23) is due to a phase shift of π between the incident EM wave and the re-emitted scattered wave, assuming that the X-ray photon energy $\hbar\omega$ is much larger than any energy close to an absorption edge.

2.5.2   Atomic Form Factor

Now, we must consider that the electron distribution in atoms is not point-like but spread throughout the entire atomic shells. X-rays, scattered at different locations within the atom, therefore interfere more destructively the larger the scattering vector Q (see Fig. 2.7a). The dimensionless atomic form factor $f(Q)$ takes into account the destructive interference with increasing Q. However, at $Q = 0$ (forward scattering), all waves are in phase, and therefore $f(Q = 0) = Z$.

In general, the atomic form factor $f(Q)$ is expressed as an integral over the electron density distribution $\rho_{el}(R_e)$ in an atom. But it is not a simple volume integral. A volume integral $\int \rho_{el}(R_e)dV$ would just recover the total number of electrons in an atom, i.e., the atomic number Z. Instead, the atomic form factor represents the **_Fourier transform_**[5] of the electron density distribution within atoms or ions:

---

[5]Fourier transforms are explained in Appendix A1.

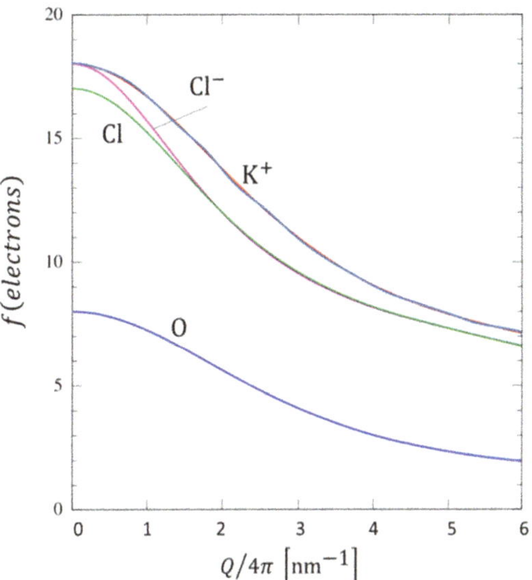

Fig. 2.8 Atomic form factors for oxygen (blue), chlorine (green), Cl⁻ (magenta), and K⁺ (red). Note that K⁺ and Cl⁻ have the same number of electrons but a different electron distribution in the respective atoms. (Adapted from Wikipedia@creative commons, graph by Pieter Kuiper)

$$f(\boldsymbol{Q}) = \int \rho_{\mathrm{el}}(\boldsymbol{R}_e) e^{i\boldsymbol{Q} \cdot \boldsymbol{R}_e} d\boldsymbol{R}_e. \tag{2.24}$$

Here $\boldsymbol{R}_e$ is a radial vector of the electrons within an atom. The Fourier transform of the electron density distribution yields the effective number of electrons that the X-rays are scattered from for increasing scattering vectors $\boldsymbol{Q}$, starting from $\boldsymbol{Q} = 0$ up to higher scattering vectors (or higher scattering angles). The $\boldsymbol{Q}$ dependence of the atomic form factor for all atoms and ions can be found in the X-ray data booklet listed under *General References*. Figure 2.8 shows an example of the atoms oxygen and chlorine, and of the ions Cl⁻ and K⁺. At $\boldsymbol{Q} = 0$, the form factor has the value $Z = 18$. Cl⁻ and K⁺ have the same number of electrons, but a different electron distribution within the atomic shells, which is clearly noticeable by the different shape of their form factors.

In conclusion, the scattering length for unpolarized X-rays and nonresonant conditions is the product of the atomic form factor and the classical electron radius:

$$a_{\mathrm{x-ray}} = -r_0 f(\boldsymbol{Q}). \tag{2.25}$$

Considering that the atomic form factor changes near an absorption edge and that the X-ray absorption increases dramatically once the conditions for characteristic X-ray emission are met, we write the Thomson X-ray form factor in the modified form:

$$f_{\mathrm{Thomson}}(\boldsymbol{Q}, E) = \left(\widehat{\boldsymbol{e}}_i \cdot \widehat{\boldsymbol{e}}_f\right)[f(\boldsymbol{Q}) + f'(E) + if''(E)]. \tag{2.26}$$

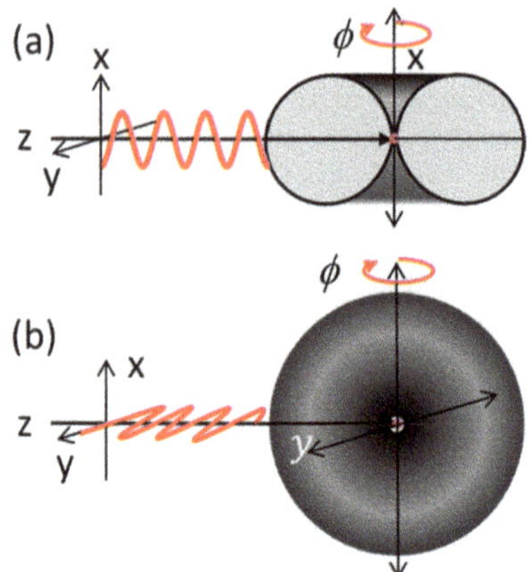

Fig. 2.9 (**a**) Dipole radiation profile of a bound electron oscillating in the x-direction due to an electromagnetic wave polarized in the x, z-plane. (**b**) Same as in (**a**) but for an electromagnetic wave polarized in y, z-plane, causing electrons to oscillate in the y-direction

Here, the real part of the form factor $f(\mathbf{Q})$ depends solely on the scattering vector $\mathbf{Q}$, while the correction $f'(E)$ is exclusively energy dependent. The dispersion correction factor $f'(E)$ changes drastically in the vicinity of absorption edges. This enables a modification of the X-ray scattering contrast - a method utilized in, anomalous X-ray scattering. The imaginary part $if''(E)$ describes the X-ray attenuation process via resonant photoelectric absorption, such as for exciting core level $1s$ electrons to the vacuum level. The prefactor $(\widehat{\mathbf{e}}_i \cdot \widehat{\mathbf{e}}_s)$ consists of unit vectors representing the polarization of the incident and scattered X-ray waves. The scalar product confirms that X-ray charge scattering does not change the original polarization of the incident beam - a point we will discuss in greater detail below.

2.5.3   Polarization Effects of X-Rays

X-rays are transverse electromagnetic *(EM)* waves. Assuming an *EM*-wave propagating along the z-direction, the electric field vector $\mathbf{E}$ and the magnetic field vector $\mathbf{B}$ point along the x, y-directions perpendicular to the z-direction. To make things simple, let us consider first that the *EM*-wave is polarized along the x-direction, meaning that the $\mathbf{E}$ field oscillates in the x, z-plane as indicated in Fig. 2.9a, referred to as σ-polarization. When this *EM*-wave hits an electron, the electron will oscillate parallel to the x, z-plane and emit radiation with an angular distribution typical for dipole radiation. The emitted radiation is cylindrically isotropic about the x-axis. But parallel to the x-direction no radiation is emitted. The toroidal radiation profile is sketched in Fig. 2.9a. In the azimuthal z, y plane, the radiation is homogeneous and independent of the azimuthal angle ϕ.

For an *EM*-wave that is polarized parallel to the y, z-plane, referred to as π-polarization, the radiation profile is cylindrical symmetric with respect to the y-direction; however, no radiation occurs along the azimuth angle $\phi = 90°$, i.e., parallel to the y-axis, as illustrated in Fig. 2.9b. For the π-polarization, the radiation profile can be described by a $\cos^2\phi$-dependence.

An unpolarized incident X-ray beam is represented by the superposition of σ- and π-polarization. The scattered radiation intensity from an electron accordingly has the radiation profile:

$$P(\phi) = \frac{1}{2}\left(1 + \cos^2\phi\right), \tag{2.27}$$

or in terms of the unit polarization vectors,

$$P(\phi) = \frac{1}{2}\left(\left(\widetilde{\boldsymbol{e}}_i^{\sigma} \cdot \widetilde{\boldsymbol{e}}_f^{\sigma}\right)^2 + \left(\widetilde{\boldsymbol{e}}_i^{\pi} \cdot \widetilde{\boldsymbol{e}}_f^{\pi}\right)^2\right). \tag{2.28}$$

$P(\phi)$ is referred to as the **polarization factor** for X-ray scattering. In practice, the polarization factor implies that unpolarized X-rays scattered at $90°$ becomes linearly polarized, its intensity is simultaneously halved. Is linearly polarized X-radiation useful? We will answer this question in Sect. 7.6 ff. regarding magnetic X-ray scattering.

Taking polarization effects into account, the modulus of the **scattering length for X-rays** that enters into the expression for the scattering amplitude, is given by

$$a_{\text{x–ray}} = -r_0 f(Q)\sqrt{P}. \tag{2.29}$$

In the end, we are interested in the X-ray scattering intensity $I(Q)$. The intensity is proportional to the square of the scattering amplitude and is therefore proportional to

$$I(Q) \sim r_0^2 f^2(Q) P(\phi). \tag{2.30}$$

In a scattering experiment, the angle ϕ corresponds to the scattering angle 2θ. The scattered X-ray intensity is the subject of further discussion in Chap. 3.

> The X-ray scattering length is: $a_{\text{x–ray}} = r_0 f(Q)\sqrt{P}.$

2.6   Neutron Scattering Lengths

The discussion of neutron scattering lengths can be confusing because there are five types divided into three groups: coherent, incoherent, and magnetic scattering lengths. An overview is given in Table 2.1. We first introduce the isotope dependence of the coherent scattering length, then we explain the sign of the scattering

Table 2.1 Overview of the different scattering lengths relevant for thermal neutron scattering

Coherent scattering length		Incoherent scattering length		Magnetic scattering length
Positive	Negative	Spin incoherent	Isotope incoherent	Coherent and incoherent

length, and discuss the two contributions to the incoherent scattering length. Finally, we introduce the magnetic form factor and the magnetic scattering length.

2.6.1   Isotope Dependence

For neutron scattering, we consider that the size of a nucleus is extremely small (10^{-6} nm) compared to the wavelength of thermal neutrons (~0.1 to 1 nm). More precisely, the extension of the neutron–nucleus interaction is strong but simultaneously of very short range. So, for neutrons, the scattering occurs at a point-like position fixed in space and described by a δ-function-like interaction potential, referred to as the *Fermi pseudo potential* (see Fig. 2.6b):

$$V_{\mathrm{nuc}}(\boldsymbol{R}) = \frac{2\pi\hbar^2}{m_n} b_{\mathrm{coh}}\delta(\boldsymbol{R}). \tag{2.31}$$

Here $\hbar = h/2\pi$, h is the Planck constant, and m_n is the neutron mass. The δ-function describes the nucleus at position $\boldsymbol{R} = 0$ in space. Roughly speaking, the δ function has infinite height but zero width, and the volume integral yields the value 1. The Fermi pseudo-potential is not the true neutron–nucleus interaction potential. But it is constructed such that it describes correctly thermal neutron scattering in the frame that we discuss here. b_{coh} is a proportionality constant, which will be identified as the neutron scattering length.

In neutron scattering theory,[6] it is shown that the **neutron scattering length** has the form:

$$a_{\mathrm{neutron}} = \frac{m_n}{2\pi\hbar^2} V(Q), \tag{2.32}$$

where $V(Q)$ is the Fourier transform of the Fermi pseudo-potential $V(\boldsymbol{R})$. Therefore, inserting and taking the Fourier transform of the δ function[7] yields

$$a_{\mathrm{neutron}} = \frac{m_n}{2\pi\hbar^2} \frac{2\pi\hbar^2}{m_n} b_{\mathrm{coh}} \int_{-}^{+} \delta(\boldsymbol{R}) \exp(i\boldsymbol{Q}\cdot\boldsymbol{R})d\boldsymbol{R} = b_{\mathrm{coh}} \exp(i\boldsymbol{Q}\cdot 0) = b_{\mathrm{coh}},$$

$$\tag{2.33}$$

which is simply a constant b_{coh}, independent of the scattering vector $\boldsymbol{Q}$. b_{coh} is known as the **coherent scattering length of thermal neutrons** for a fixed nucleus in

---

[6]For example: G.L. Squire, Thermal Neutron Scattering, Cambridge University Press, 3rd edition, 2012, and Dover edition, 1997.

[7]Fourier transforms and δ-functions are explained in Appendix A1.

space, also referred to as the *bound neutron scattering length*. Thermal neutrons are those that have a wavelength between 0.1 and 1 nm. The constancy of the scattering length is also graphically evident. If the extent of the target (nuclei) is negligible in comparison to the neutron wavelength, then destructive interference off a single nucleus cannot occur at any scattering angle; that is, the neutron scattering length must be identical for forward and backward scattering.

Figure 2.10 compares schematically the X-ray atomic form factor and the neutron coherent scattering length as a function of the scattering vector Q. This graph shows us two important properties. First, the X-ray and neutron scattering lengths are on the same order of magnitude, at least for light atoms. Second, the X-ray form factor f drops off at higher Q, whereas the neutron scattering length remains constant independent of Q. This has severe consequences for scattering at high Q values (high-order Bragg reflections), where neutrons have an advantage over X-rays. We come back to this point later.

The coherent neutron scattering length b_{coh} depends on the isotope scattered from and is an irregular function of the mass number $A = N + Z$, where N is the number of neutrons, and Z is the number of protons. Figure 2.11 compares X-ray and neutron scattering lengths for various atoms and isotopes.

X-ray scattering amplitudes scale with the atomic number Z and increase continuously from light to heavy atoms. In contrast, for neutron scattering, we distinguish positive coherent scattering lengths, b_{coh} (light green in Fig. 2.10), negative coherent scattering lengths (dark striped green), and incoherent scattering lengths b_{inc}, marked in yellow. The sign of the coherent neutron scattering length is discussed in Sect. 2.5.2, and the origin of incoherent scattering lengths is explained in Sect. 2.5.3.

Two more properties can immediately be recognized by inspecting Fig. 2.11. First, the isotopes hydrogen and deuterium have opposite **coherent scattering lengths**; hydrogen has a small and negative coherent scattering length (-3.74 fm) while deuterium has a much larger and positive coherent scattering length (6.67 fm). In addition, hydrogen features a huge incoherent scattering length of 25.74 fm. Usually, the **incoherent scattering length** of hydrogen completely dominates over the coherent one. This has important consequences for the investigation of soft and biological matter that contains much hydrogen, as will be discussed in Chap. 9 on small-angle scattering.

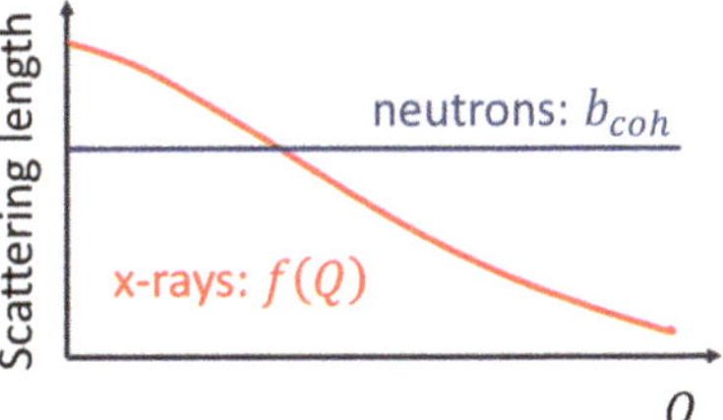

Fig. 2.10 Form factor $f(Q)$ for X-ray scattering and coherent scattering length b_{coh} for neutron scattering as a function of the modulus of the scattering vector $\lceil Q \rceil$

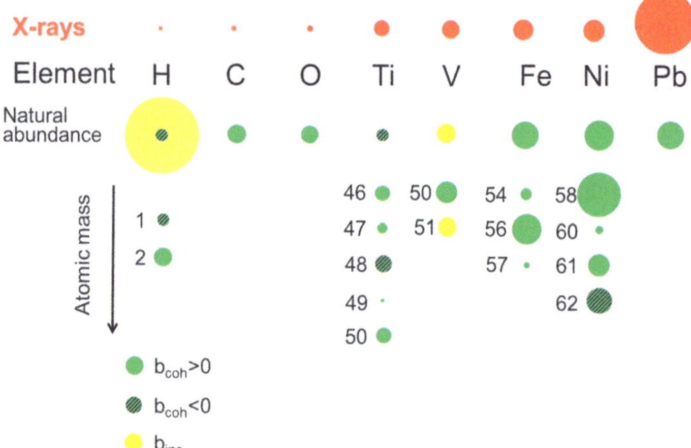

Fig. 2.11 Comparison of X-ray and neutron scattering lengths. The radius of the disks is proportional to the respective scattering length for X-rays in red and for neutrons in green and yellow. Homogeneous green refers to positive coherent scattering lengths, striped green to negative coherent scattering lengths, and yellow to incoherent scattering length. Note that the radius of the disks scales with the atomic number for X-ray scattering, but is irregular for neutron scattering at different isotopes. (Adapted from https://www.ill.eu/fileadmin/user_upload/ILL/1_About_ILL/ Documentation/NeutronDataBooklet.pdf). "Natural abundance" refers to the effective neutron scattering length for the natural isotope mixture of the respective element

Furthermore, the neutron scattering lengths of heavy isotopes do not overwhelm those of light isotopes. Both neutron scattering probabilities are on the same order of magnitude, in contrast to X-ray scattering. This is important for studying light elements like deuterium, carbon, or oxygen in the neighborhood of heavier elements, such as in high-temperature superconducting oxides. When studying, for instance, $YBa_2Cu_3O_7$, the X-ray scattering length of Ba is 158 fm, in contrast to oxygen, which is only 22 fm. However, the neutron scattering lengths are 5.07 and 5.8 fm for Ba and O, respectively. Thus, using thermal neutron scattering, the presence of oxygen becomes clearly "visible". A complete list of neutron scattering lengths for all isotopes is provided on the webpage: https://www.ncnr.nist.gov/resources/n-lengths/, or can be found in the Neutron Data Booklet (https://www.ill.eu/fileadmin/ user_upload/ILL/1_About_ILL/Documentation/NeutronDataBooklet.pdf). A small selection of neutron scattering lengths for a few isotopes is given in Table 2.2 and compared to X-ray scattering lengths.

Table 2.2 Thermal neutron coherent and incoherent scattering lengths and neutron absorption cross sections for a few isotopes. 1 fm $= 10^{-15}$ m, 1 barn $= 10^{-28}$ m^2. "Natural" refers to the natural isotope mixture. The neutron data are taken from the webpage: https://www.ncnr.nist.gov/resources/n-lengths/. The fourth column lists the respective X-ray scattering lengths for comparison

Isotope	b_{coh} (fm)	b_{inc} (fm)	$r_0 Z$ (fm)	Neutron absorption cross section (barn)
Hydrogen (H)	-3.74	25.27	2.82	0.33
Deuterium (D)	6.67	4.04	2.82	0
Oxygen (^{16}O)	5.8	0	22.5	0
Sodium (Na)	3.63	3.59	31	0.53
Chlorine (Cl, natural)	9.57	0	48	33.5
Vanadium (V, natural)	-0.83	6.35	64.8	5.08
Manganese (Mn)	-3.73	1.79	70.5	13.3
Iron (Fe, natural)	9.45	0	73.3	2.56
Nickel (^{58}Ni)	14.4	0	79	4.6

2.6.2   Positive and Negative Coherent Scattering Lengths

In most cases, the scattering length for thermal neutrons is taken as a constant that can be looked up on the internet when needed. But to get some insight into why some scattering lengths are positive, and others are negative, why some are coherent and others are incoherent, we have to go a bit deeper into the analysis of the neutron–nucleus interaction. This requires some basic knowledge of quantum mechanics. However, to keep things simple, we provide here some heuristic arguments.

In general, the kinetic energy of a particle (neutron) in free space is given by (A6):

$$E = \frac{p^2}{2m_n} = \frac{(\hbar k)^2}{2m_n} = \frac{h^2}{2m_n \lambda^2}, \tag{2.34}$$

where p is the momentum and m_n the mass of the neutron. Solving for the wavelength, we obtain:

$$\lambda = \frac{h}{\sqrt{2m_n E_{\mathrm{kin}}}}. \tag{2.35}$$

If a neutron crosses a potential of height V_0, the wavelength becomes affected:

$$\lambda = \frac{h}{\sqrt{2m_n(E_{\mathrm{kin}} \pm V_0)}}. \tag{2.36}$$

Here, the $+$ sign stands for an attractive potential ($-V_0$), and the $-$ sign for a repulsive potential ($+V_0$). This is illustrated for a square well potential in Fig. 2.12.

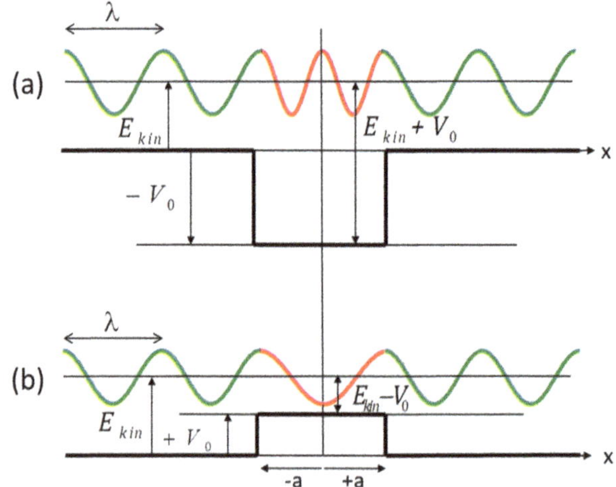

Fig. 2.12 Particle wavelengths crossing a negative attractive potential (**a**) and a positive repulsive potential (**b**)

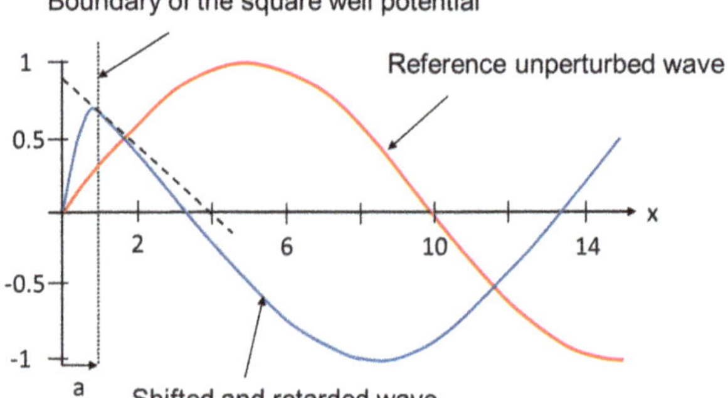

Fig. 2.13 Wave function at the boundary between an attractive square well potential and free space. The boundary condition imposes a negative slope of the wavefunction at the boundary, which translates to a negative neutron scattering length b

A square well potential is constant over a certain width and has vertical walls that extend up to the vacuum level. The particle wavelength decreases over a negative (attractive) potential well and increases over a positive (repulsive) potential well compared to the vacuum level.

At the borders of a square well potential at $x = \pm a$, the particle wavefunction must be continuous and the first derivative should yield the same value for the slope, independent of whether approaching the border from one side or the other. This is referred to the *boundary condition for wave functions*. In Fig. 2.13, we give an example for the case of an attractive potential. The wave inside of the potential well is shorter, and the wave outside is shifted to the left in comparison to the unperturbed wave. This implies that the perturbed wave is retarded compared to the unperturbed wave. At the boundary of this attractive potential, the continuity of the waves inside and outside of the potential well causes the slope to be negative at $x = a$. The negative slope, indicated by the dashed line, determines the negative sign of b_{coh}, here for an attractive potential. The intercept with the x-axis yields the amplitude of the scattering length, here 4 in relative units. In conclusion, the sign of the slope at the boundary determines the sign of the scattering length, and the intercept determines its magnitude. For an attractive potential, b_{coh} is negative; for a repulsive potential, b_{coh} is positive.

There is an even simpler argument for the sign of the scattering length. As $b_{\mathrm{coh}} = (m_n/2\pi\hbar^2)V_{\mathrm{nuc}}$, it is clear that a negative scattering length implies a negative (attractive) neutron–nucleus potential $V_{\mathrm{nuc}} < 0$. Similarly, a positive scattering length implies a positive (repulsive) neutron–nucleus potential $V_{\mathrm{nuc}} > 0$.

Nuclei with a surplus of neutrons act as repulsive potentials, and nuclei with a lack of neutrons display an attractive potential. An illustrative example is given by the hydrogen isotopes. Hydrogen (H) can take up a neutron, forming a stable deuterium isotope, i.e., a (*pn*) nucleus. The *pn* interaction is therefore attractive and the coherent scattering length b_{coh}^H of hydrogen is accordingly negative. However, deuterium cannot take up another neutron to form a stable isotope of tritium (*pnn*). The nucleus of tritium (*pnn*) is radioactive, meaning unstable. Therefore the neutron–deuterium interaction is repulsive. Consequently, the coherent scattering length b_{coh}^D of deuterium is positive. As can be seen from Fig. 2.10, most coherent scattering lengths are positive. Negative scattering lengths can be very useful for studying phase transitions in binary alloys by using the so-called zero matrix method, as explained in Chap. 8.

> The sign of the neutron coherent scattering length is a matter of neutron–nucleus interaction. For attractive interaction, the sign is negative; otherwise positive. Most scattering lengths are positive.

In general, the scattering length b is a complex quantity:

$$b = b_0 + b' + ib'', \tag{2.37}$$

where b_0 is related to the potential scattering discussed above, and b' and b'' are the real and imaginary corrections in resonance scattering, respectively. Resonance scattering takes place in the formation of compound nuclei, leading to strong neutron absorption. In the case of thermal neutron scattering, this occurs only for a few isotopes, such as ^{113}Cd, ^{149}Sm, ^{157}Gd, and a few others. Then the scattering length strongly depends on the initial neutron energy. But in most other cases, the imaginary part of the scattering length can be neglected, and the real part is constant over the range of neutron energies considered here.

2.6.3 *Incoherent Scattering Lengths*

2.6.3.1 Spin Incoherent Scattering Length

Let us assume that the target consists of one isotope with nuclear spin I. During neutron scattering, the neutron with spin $S = 1/2$ and the target nuclei with nuclear spin I form together, for a short period of time, a compound nucleus with a total spin:

$$I + S = I \pm \frac{1}{2}. \tag{2.38}$$

The scattering length depends on the total spin. For an unpolarized neutron beam, both neutron spins $S = \pm \frac{1}{2}$ are present and can combine with the nuclear spin. The respective scattering lengths b_+ and b_- are then associated with the total spin, according to Eq. (2.38).

For an unpolarized neutron beam and an unpolarized target, both combinations contribute to the scattering, but with different probabilities. The probability depends on the number of states associated with the spin $I + \frac{1}{2}$ as compared to the number of states associated with the spin $I - \frac{1}{2}$.

For $I + \frac{1}{2}$, the number of possible quantum states is $2\left(I + \frac{1}{2}\right) + 1 = 2I + 2$. For $I - \frac{1}{2}$, the number of possible quantum states is $2\left(I - \frac{1}{2}\right) + 1 = 2I$. Now we calculate the probability that the scattering will take place with one or the other spin. The probability is the number of states for one or the other spin divided by the total number of possible states. Therefore, we obtain the probability p_+ for the scattering associated with the spin $I + \frac{1}{2}$:

$$p_+ = \frac{2I + 2}{(2I + 2) + 2I} = \frac{I + 1}{2I + 1}, \tag{2.39}$$

and equivalently for the spin $I - \frac{1}{2}$:

$$p_- = \frac{2I}{(2I + 2) + 2I} = \frac{I}{2I + 1}, \tag{2.40}$$

where $p_+ + p_- = 1$. With these definitions, we can now determine the following expectation values:

$$\langle b \rangle^2 = \left(p_+ b_+ + p_- b_-\right)^2 \tag{2.41}$$

and

$$\langle b^2 \rangle = p_+ b_+^2 + p_- b_-^2. \tag{2.42}$$

Now we are prepared to evaluate the ***coherent scattering length***, which is given by the already quoted expression:

$$b_{\text{coh}}^2 = \langle b \rangle^2 = \left(p_+ b_+ + p_- b_- \right)^2, \tag{2.43}$$

and the incoherent scattering length follows from the averaged mean square deviation:

$$b_{\text{inc}}^2 = \left\langle \left(b - \langle b \rangle \right)^2 \right\rangle = \langle b^2 \rangle - \langle b \rangle^2 = p_+ p_- \left(b_+ - b_- \right)^2. \tag{2.44}$$

Note that incoherent scattering occurs only for nuclei with $I > 0$. Therefore, this type of scattering is referred to as ***spin incoherent scattering***. The respective cross sections are:

$$\sigma_{\text{spin,coh}} = 4\pi \, b_{\text{coh}}^2,$$
$$\sigma_{\text{spin,inc}} = 4\pi \left\langle \left(b - \langle b \rangle \right)^2 \right\rangle, \tag{2.45}$$
$$\sigma_{\text{spin,tot}} = \sigma_{\text{spin,coh}} + \sigma_{\text{spin,inc}}.$$

As an example, we consider neutron scattering at a hydrogen (proton) target. Protons have a spin of $I = \frac{1}{2}$, and therefore $p_+ = 3/4$ and $p_- = 1/4$. The scattering lengths b_+ and b_- have to be determined experimentally with polarized beams and turn out to be: $b_+ = 10.8$ fm and $b_- = -47.4$ fm . With these numbers, we obtain for $b_{\text{coh}} = -3.75$ fm and for $b_{\text{inc}} = 25.2$ fm. Often in tables, cross-sections σ are quoted, defined as $\sigma = 4\pi b^2$. Hence, for hydrogen, we obtain the cross-sections: $\sigma_{\text{coh}} = 1.76$ barns, $\sigma_{\text{inc}} = 79.8$ barns. So, in the case of hydrogen, ***incoherent neutron scattering*** completely dominates over coherent scattering. This is in strong contrast to deuterium, which has a sizeable positive coherent scattering length and a lower incoherent scattering length (see Table 2.2). The different scattering lengths for H and D are useful for contrast variation, often applied in soft matter studies and discussed further in Sect. 9.5.

2.6.3.2 Isotope Incoherent Scattering Length

In neutron scattering experiments, often the samples are not isotopically clean, but rather contain a mixture of various isotopes of the same element. Examples are Fe and Ni, which contain isotopes with different scattering lengths and sometimes also a mixture of positive and negative scattering lengths. Considering only two different isotopes with scattering lengths $b_{1,\,2}$ and concentrations $c_{1,\,2}$ with ($c_1 + c_2 = 1$), the average scattering length is:

$$\langle b \rangle_{\text{iso,coh}} = c_1 b_1 + c_2 b_2. \tag{2.46}$$

Now we proceed along the same lines as above for the spin incoherent scattering:

$$b^2_{\text{iso,coh}} = \langle b \rangle^2 = (c_1 b_1 + c_2 b_2)^2, \tag{2.47}$$

and the isotope incoherent scattering length becomes:

$$b^2_{\text{iso,inc}} = \left\langle (b - \langle b \rangle)^2 \right\rangle = \langle b^2 \rangle - \langle b \rangle^2 = c_1 c_2 (b_1 - b_2)^2. \tag{2.48}$$

The argument can easily be extended to more than two different isotopes. Naturally, the isotope incoherent scattering drops to zero for $b_1 = b_2$.

2.6.3.3 Characteristics of Coherent and Incoherent Scattering

"Coherent scattering" is generally understood to be a scattering process in which phase relationships are preserved during scattering, thus resulting in interference effects. Coherent scattering allows the investigation of any pair correlation in space, both short-range and long-range. Typical examples of long-range pair correlations are translation invariance in crystal lattices, which leads to sharp Bragg reflections and is discussed in Chap. 3. Further examples are disordered materials, which still exhibit some short-range order correlation, which can be analyzed using coherent scattering.

In contrast, "*incoherent scattering*" is a scattering process in which the phase relationships are lost and therefore defies any interference effects. Examples of incoherent scattering include the spin-incoherent and isotope-incoherent scattering of neutrons described above, and the incoherent Laue scattering of X-rays and neutrons at an ensemble of scattering centers that express no short- or long-range pair correlations. Incoherent scattering gives rise to a homogeneous and featureless background intensity, proportional to the concentration of scattering centers. Incoherent scattering is particularly useful in neutron scattering to study the self-correlation and self-diffusion of ions in liquids. This type of scattering is called quasielastic neutron scattering and is an important application for the study of diffusion in liquids and excitations in organic and inorganic soft matter. Further information can be found in (Bée 1988).

2.6.4 Magnetic Form Factor and Magnetic Scattering Length

As we have already noted, neutrons carry a spin. Associated with the neutron spin is a ***magnetic moment***:

$$\boldsymbol{\mu}_n = -\gamma_n \mu_N \boldsymbol{\sigma}, \tag{2.49}$$

where the minus sign denotes the antiparallel orientation of both the neutron spin and the magnetic moment. Here μ_N is the **nuclear magneton** ($\mu_N = 5.05783 \times 10^{-27} J/T$), m_n is the neutron mass, $\gamma_n = 1.913$ is the gyromagnetic ratio of neutrons, and $\sigma_n = 2S$ is the Pauli spin-operator with the eigenvalues ± 1. The neutron magnetic moment interacts via dipole–dipole interaction with the unpaired electrons in atomic shells. This interaction gives rise to a magnetic scattering length.

In order to determine the neutron magnetic scattering length, we have to evaluate the neutron magnetic form factor. The **neutron magnetic form factor** $f_m(Q)$ is similar in expression and magnitude to the X-ray atomic form factor:

$$f_{\mathrm{mag}}(\boldsymbol{Q}) = \int \rho_e^{\mathrm{uncomp}}(\boldsymbol{R}) e^{i\boldsymbol{Q}\cdot\boldsymbol{R}} d\boldsymbol{R}. \tag{2.50}$$

However, the Fourier transform of the electron density in Eq. (2.50) does not include all electrons in an atom, as is done for the X-ray atomic form factor. Only those uncompensated electrons in a partially filled electronic shell are taken into account that contribute to the spin and orbital magnetic moments of atoms (or ions). Filled or compensated electronic shells generate no magnetic dipole moment and are discarded in the Fourier integral.

The **neutron magnetic scattering length** b_m is obtained in analogy to the X-ray scattering length by multiplying the magnetic form factor $f_{\mathrm{mag}}(Q)$ with the classical electron radius r_0 and the neutron gyromagnetic ratio γ_n:

$$b_m = \gamma_n r_0 f_{\mathrm{mag}}(Q). \tag{2.51}$$

The magnitude of the prefactor $\gamma_n r_0 = 5.39 \times 10^{-13}$ cm is of the same order of magnitude as the coherent neutron scattering length for most isotopes. This tells us that neutron scattering from nuclei and neutron scattering from magnetic dipole moments yield very similar amplitudes and intensities.

> The neutron magnetic form factor is the Fourier transform of the unpaired electron density in uncompensated atomic orbitals.

Similar to the X-ray atomic form factor, the neutron magnetic form factor $f_{\mathrm{mag}}(Q)$ decreases with increasing modulus of the scattering vector Q. In fact, the reduction is even more drastic than for X-rays, because the Fourier transform does not cover all electrons in an atom but only those in an atomic shell of inner 3d electrons (transition metal atoms) or 4f electrons (rare earth atoms) that carry uncompensated magnetic dipole moments. Figure 2.14 compares schematically both form factors. The fast drop of the magnetic form factor causes problems when studying higher-order magnetic Bragg reflections at high Q values, as we will notice later. Further discussion of the magnetic form factor can be found in Chap. 7 on magnetic scattering.

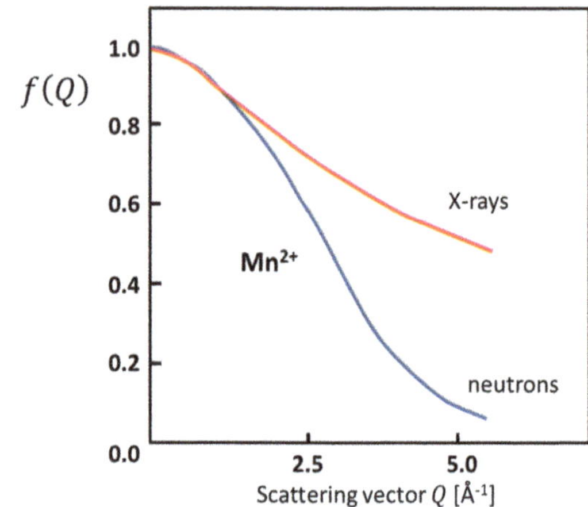

Fig. 2.14 Comparison of X-ray atomic form factor and neutron magnetic form factor for Mn^{2+}. The form factors are normalized at $Q = 0$. (Adapted from Bacon "Neutron Diffraction" Bacon 1975)

2.7 Electron Scattering Form Factor

Low-energy electron scattering (diffraction) is used mainly for studying surface structures of single crystals and the adsorption of gases at surfaces, including their catalytic activity. The elastic scattering of electrons at surfaces follows pretty much the same lines as X-ray scattering, although the scattering process is governed by Coulomb scattering instead of Thomson scattering. The electron scattering form factor $f^e(Q)$ can be related to the X-ray atomic form factor $f^x(Q)$ via the Mott–Bethe formula (Cowley 2006):

$$f^e(Q) = \frac{1}{2\pi^2 a_0} \frac{Z - f^x(Q)}{Q^2}. \tag{2.52}$$

Here a_0 is the Bohr radius and Z is the atomic number. Note that $f^x(Q)$ is dimensionless, whereas $f^e(Q)$ has the dimension of length. It is common to approximate the electron scattering form factor by a finite sum of Gaussians centered at $Q = 0$:

$$f^e(Q) \cong \sum_{i-1}^{5} a_i \exp\left(-b_i \left(\frac{Q}{4\pi}\right)^2\right), \tag{2.53}$$

where the coefficients a_i and b_i are provided in the International Tables for Crystallography for all chemical elements (International Tables for Crystallography n.d.). Figure 2.15 shows a comparison of the X-ray and electron scattering atomic form factors for a few light elements. While $f^x(Q = 0) = Z$, this is not the case for $f^e(Q)$, because of screening effects.

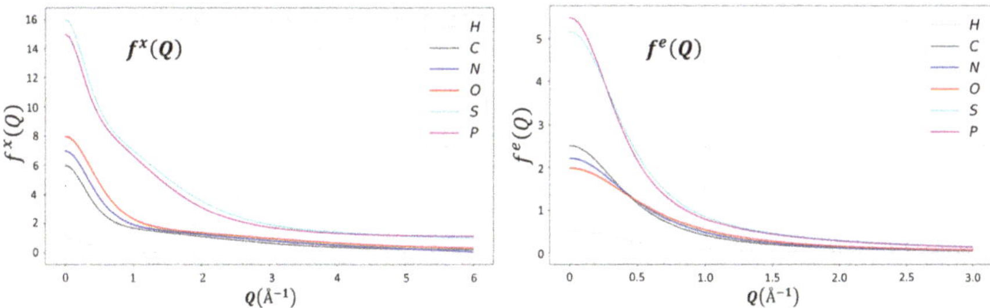

Fig. 2.15 Comparison of the atomic form factors for X-rays and for electrons. Note that the vertical and horizontal scales are different in both plots. The graphs are adapted from the webpage on Software for Macromolecular X-ray Crystallography https://www.ccp4.ac.uk/ccp4-ed/projects/scattering/, which is gratefully acknowledged

Summary

1. Attenuation of a beam of particles that hits a target is composed of absorption and scattering.
2. The scattering amplitude of a spherical wave is the fraction of a plane wave that has been scattered from a point-like target.
3. Elastic scattering implies that the energy and wavelength of a beam of particles or photons are the same before and after scattering.
4. The scattering vector $\boldsymbol{Q}$ is the vectorial difference between the scattering wavenumber $\boldsymbol{k}_f$ and the incident wave number $\boldsymbol{k}_i$.
5. The atomic form factor for X-ray scattering depends on the modulus of the scattering vector $\boldsymbol{Q}$. It is represented by the Fourier transform of the electron density distribution.
6. The atomic form factor decreases with increasing scattering vector Q.
7. The scattering length of X-rays is the product of the classical electron radius, the atomic form factor, and the root of the polarization factor: $r_0 f(Q)\sqrt{P}$.
8. The polarization factor is due to the radiation profile of dipole radiation.
9. The δ-like Fermi pseudo-potential is an effective neutron–nucleus potential that yields a constant scattering length b when calculating the scattering amplitude.
10. The coherent scattering length b_{coh} of neutrons is constant and Q-independent for a given isotope.
11. The coherent scattering length of neutrons b_{coh} depends on neutron–nuclear interaction and varies irregularly with the atomic number.
12. The coherent scattering length b_{coh} of neutrons being a constant is the result of the short range of the neutron–nucleus interaction.
13. The scattering length b can be coherent or incoherent, and it can be positive or negative.
14. Attractive neutron–nuclei potentials produce negative coherent scattering lengths; repulsive neutron–nuclei potentials produce positive coherent scattering lengths.

15. Isotopes with a nuclear spin $I > 0$ give rise to spin incoherent neutron scattering.
16. Neutron magnetic scattering length is proportional to the Fourier transform of the unpaired electron density in uncompensated atomic orbitals, and its shape is similar to the x-ray atomic form factor.
17. The form factor for electron scattering is related to the X-ray atomic form factor via the Mott–Bethe formula.

Questions

(Note that more than one answer may be correct)

1. **What is a necessary condition for elastic scattering?**

   (a) $|\mathbf{k}_i| = |\mathbf{k}_f|$.
   (b) $\mathbf{R} = \mathbf{R}_j - \mathbf{R}_i$.
   (c) The total energy of the X-ray beam is transferred to the crystal lattice.

2. **Why does the coherent scattering length b_{coh} for neutrons not depend on the modulus of the scattering vector $[Q]$?**

   (a) Because neutrons are neutral.
   (b) Because neutrons have a spin.
   (c) Because neutrons scatter from nuclei, which are much smaller than typical neutron wavelengths.

3. **How does the X-ray atomic form factor depend on the modulus of the scattering vector $[Q]$?**

   (a) The form factor is a constant, just like for neutrons and does not depend on Q.
   (b) The form factor increases linearly with increasing Q.
   (c) The form factor is maximal at $Q = 0$ and drops continuously as Q increases.

4. **What are the main effects when radiation hits matter?**

   (a) Part of the radiation is scattered, part is transmitted, and part is absorbed.
   (b) All radiation is transmitted.
   (c) All radiation is being absorbed.

5. **What is the difference between attenuation and absorption?**

   (a) Attenuation is the same as absorption and means a reduction of beam intensity.
   (b) Absorption implies the elimination of a particle and transformation into another type of particle or radiation. Attenuation implies a general diminishing of the primary beam intensity, which can be due to absorption and scattering.
   (c) Attenuation is the opposite of absorption. Absorption reduces the beam intensity; attenuation increases the beam intensity.

6. **What is the condition for constructive/destructive interference?**

   (a) The path difference of two beams must be a multiple of a wavelength for constructive interference; for destructive interference, the path difference should be an odd multiple of half the wavelength.
   (b) For constructive interference, the phase difference of two beams should be a multiple of 2π. For destructive interference, it should be an odd multiple of π.
   (c) By scattering of radiation, one cannot distinguish between constructive and destructive interference.

7. **Why is it necessary to consider a polarization factor for X-ray scattering?**

   (a) Because X-rays are longitudinal electromagnetic waves.
   (b) Because X-ray scattering is based upon dipole radiation of transverse electromagnetic waves.
   (c) Because electrical dipoles do not radiate in the direction perpendicular to the dipole axis.

8. **To perform a neutron diffraction experiment from a crystal lattice, what type of isotopes would you prefer to have in your sample:**

   (a) Isotopes with a large coherent scattering length.
   (b) Isotopes with a large incoherent scattering length.
   (c) Isotopes with a small coherent scattering length.
   (d) Isotopes with a large absorption cross-section.

9. **What is the difference between coherent and incoherent scattering lengths?**

   (a) One is positive, and the other one is negative.
   (b) One is for scattering, and the other one is for absorption.
   (c) Coherent scattering length allows diffraction experiments, and incoherent scattering length is for studying the self-correlation.

Exercises

Grades of difficulty: E = easy, M = medium, A = advanced.

E 2.1 Wave vector and scattering vector

Assuming that the modulus of the incident wavenumber $|k_i| = 62.8 \text{ nm}^{-1}$ and that the angle between k_i and k_f is 90°.

   (a) What is the wavelength of the radiation used?
   (b) What is the modulus of the scattering vector?
   (c) What is the maximum scattering vector that can be reached with the wavenumber still fixed to $|k_i| = 62.8 \text{ nm}^{-1}$?
   (d) If scattering vectors larger than the so far calculated value are required, how can this be achieved?

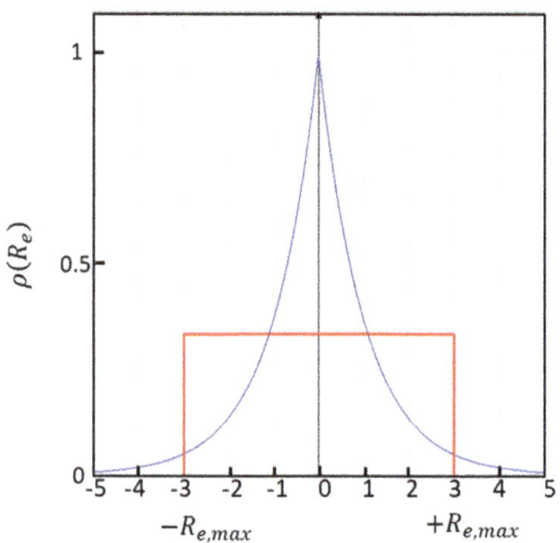

Fig. 2.16 Electron density distributions for a constant electron density (red line) and for an exponential density distribution (blue line)

M 2.2 Atomic form factor

In Appendix A3, the atomic form factor is calculated for an unrealistic constant electron density distribution ρ within a sphere of radius R_{max} (red line in the graph of Fig. 2.16). Here, we want to calculate the X-ray atomic form factor $f(Q)$ for a more realistic electron density distribution $\rho_H(R_e)$ that corresponds to hydrogen atoms (blue line in the graph of Fig. 2.16):

$$\rho_H = \frac{1}{\pi a_B^3}\exp\left(-\frac{2R_e}{a_B}\right).$$

Here a_B is the Bohr radius. Recall that according to Eq. (2.7), the atomic form factor is the Fourier transform of the electron density distribution within an atom.

(a) Calculate the atomic form factor $f(Q)$ for a hydrogen atom. Recall that the spherical integral $\int d\mathbf{R}_e$ can be partitioned into an angular part and a radial part:

$$\int d\mathbf{R}_e = \int_0^{2\pi} d\theta \int_{-\pi}^{\pi} \sin\phi\, d\phi \int_0^{\infty} R_e^2 dR_e.$$

Use the substitution introduced in Appendix A2. To solve the radial part of the integral, use the following equation:

$$\int xe^{ax}\sin(bx)dx = \frac{xe^{ax}}{a^2+b^2}\left(a\sin(bx) - b\cos(bx)\right)$$
$$- \frac{e^{ax}}{\left(a^2+b^2\right)^2}\left[\left(a^2-b^2\right)\sin(bx) - 2ab\,\cos(bx)\right].$$

(b) Which value does the form factor $f(Q)$ approach for $Q \to 0$?

(c) What is the Q − dependence of $f(Q)$ for higher Q values? How does $f(Q)$ depend on Q for $Q > 1$?

E 2.3 Coherent and incoherent scattering length of vanadium

Vanadium consists of two isotopes. The isotope ^{51}V has a nuclear spin of $I = 7/2$ and a natural abundance of 99.75%. The second naturally occurring vanadium isotope is ^{50}V and has a nuclear spin of $I = 6$. The scattering lengths for ^{51}V are as follows: $b_+ = 4.93$ fm and $b_- = -7.6$ fm. Determine the spin coherent and incoherent cross-sections.

References

G.E. Bacon, *Neutron Diffraction* (Clarendon Press, Oxford, 1975)

M. Bée, *Quasielastic Neutron Scattering: Principles and Applications in Solid State Chemistry, Biology, and Materials Science* (Adam Hilger, Bristol, 1988)

J.M. Cowley, Electron diffraction and electron microscopy in structure determination, in *International Tables for Crystallography Volume B*, (Springer, Dordrecht, 2006), pp. 276–345

International Tables for Crystallography (n.d.), http://it.iucr.org/Cb/ch4o3v0001/sec4o3o2/

Further Reading and Recommended Literature

Condensed Matter Physics

H. Ibach, H. Lüth, *Solid-State Physics, An Introduction to Principles of Materials Science* (Springer, Berlin, 2009)

C. Kittel, *Introduction to Solid State Physics*, 8th edn. (Wiley & Sons, New York, 2004)

Scattering Theory (Mostly Elastic)

J. Als-Nielsen, D. McMorrow, *Elements of Modern X-Ray Physics*, 2nd edn. (Wiley, New York, 2011)

G.E. Bacon, *Neutron Diffraction* (Clarendon Press, Oxford, 1975)

D.S. Sivia, *Elementary Scattering Theory: For X-Ray and Neutron Users* (Oxford University Press, Oxford, 2011)

I. Zaluzhnyy, F. Schreiber, *X-ray and Neutron Scattering* (2025), https://www.soft-matter.uni-tuebingen.de/teaching/scattering_lecture/LectureNotes.pdf

Scattering Theory (Mostly Inelastic)

A. Boothroyd, *Principles of Neutron Scattering from Condensed Matter* (Oxford University Press, Oxford, 2020)

F.M.F. de Groot, M.W. Haverkort, H. Elnaggar, A. Juhin, K.-J. Zhou, P. Glatzel, Resonant inelastic X-ray scattering. Nat. Rev. Methods Primers **4**, 45 (2024). https://doi.org/10.1038/s43586-024-00322-6

A. Furrer, J. Mesot, F. Strässle, *Neutron Scattering in Condensed Matter Physics* (World Scientific, New Jersey, 2009)

S.W. Lovesey, *Theory of Neutron Scattering from Condensed Matter* (Clarendon Press, Oxford, 1984) International Series of Monographs on Physics, Vol. 1 (Nuclear Scattering), Vol.2 (Polarization Effects and Magnetic Scattering)

G.L. Squire, *Thermal Neutron Scattering*, 3rd edn. (Cambridge University Press, Cambridge, 2012), and Dover edition, 1997

Inelastic Electron Scattering

F. Hofer, F.P. Schmidt, W. Grogger, G. Kothleitner, Fundamentals of electron energy-loss spectroscopy. IOP Conf. Ser. Mater. Sci. Eng. **109**, 012007 (2016)

Chapter 3
Scattering from Crystal Lattices

3.1 Bragg Reflection

Referring to Eq. (2.8), we recall that the *path difference* between wavefronts 1 and 2 in Fig. 2.4 is

$$\Delta l = l_1 + l_2 = R(\sin\theta_1 + \sin\theta_2). \tag{3.1}$$

The condition for *constructive interference* is

$$\Delta l = s\lambda (s \in N). \tag{3.2}$$

where n is the order of interference. Combining these two equations and assuming a symmetric case, i.e., $\theta_1 = \theta_2$, we find

$$2R\sin\theta = s\lambda. \tag{3.3}$$

Now we replace R with the lattice spacing between atomic planes in a crystal lattice d_{hkl}, where hkl are the Miller indices introduced in Chap. 1. Then we have the **Bragg equation** in the familiar form:

$$2d_{hkl}\sin\theta = \lambda. \tag{3.4}$$

The scattering geometry for this equation is illustrated in the left panel of Fig. 3.1, where $2d_{hkl}\sin\theta$ is the path difference of the wave front scattered at the first and the second point (atom). Note that the order of interference s is incorporated into the Miller indices hkl.

In a scattering experiment, the wavelength of the radiation used is normally known. Then the Bragg equation can be used to determine the *Bragg angle* θ_{hkl} if the lattice spacing d_{hkl} is known:

H. Zabel, *Elements of Elastic Scattering by X-Rays, Neutrons, and Electrons*,
https://doi.org/10.1007/978-3-032-16624-1_3

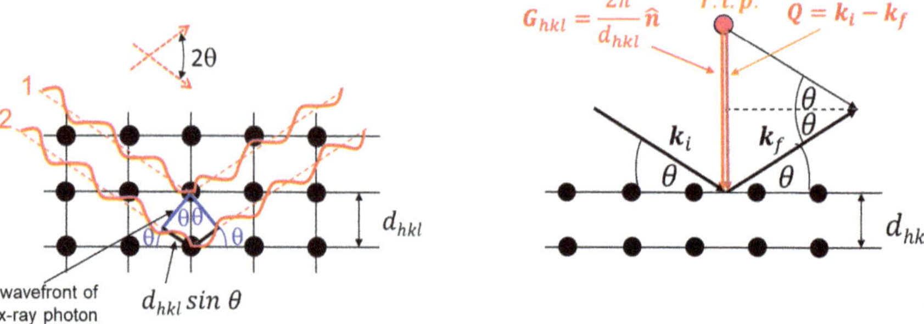

Fig. 3.1 Geometry for constructive X-ray interference in a crystal lattice. Left: Bragg condition in real space. The dots indicate the position of atoms in a crystal lattice, where d_{hkl} is the lattice spacing of (hkl)-planes. Right: Bragg condition in reciprocal space. The upper part shows one reciprocal lattice point according to the lattice spacing shown in the lower part. The orientation of both lattices (real and reciprocal) coincides, and their origins are arbitrary

$$\theta_{hkl} = \sin^{-1}\left(\frac{\lambda}{2d_{hkl}}\right). \tag{3.5}$$

Alternatively, we can determine the *lattice spacing d_{hkl}* if the Bragg angle θ_{hkl} is known:

$$d_{hkl} = \frac{\lambda}{2\sin\theta_{hkl}}. \tag{3.6}$$

For cubic crystals, it can be shown that the interplanar lattice spacing is expressed as

$$d_{hkl} = \frac{a}{\sqrt{h^2 + k^2 + l^2}}, \tag{3.7}$$

where a is the lattice parameter of the cubic unit cell. For a derivation, see Eq. (1.14).

An alternative derivation of the Bragg equation uses the definition of the scattering vector $\boldsymbol{Q}$. According to Eq. (2.6),

$$\boldsymbol{Q} = \boldsymbol{k}_i - \boldsymbol{k}_f,$$

and taking the projections of $\boldsymbol{k}_f$ and $\boldsymbol{k}_i$ on $\boldsymbol{Q}$, we find

$$\boldsymbol{Q} = \boldsymbol{k}_i - \boldsymbol{k}_f = 2|k|\sin\theta\hat{\boldsymbol{n}} = \frac{4\pi}{\lambda}\sin\theta\hat{\boldsymbol{n}}, \tag{3.8}$$

where $\hat{\boldsymbol{n}}$ is a normal vector to the *hkl* plane. The projections are graphically shown in the right-hand panel of Fig. 3.1. The reciprocal lattice vector of the *hkl* plane is

$$G_{hkl} = \frac{2\pi}{d_{hkl}} \widehat{n}. \tag{3.9}$$

Setting $Q = G_{hkl}$, we find

$$\frac{4\pi}{\lambda} \sin \theta \widehat{n} = \frac{2\pi}{d_{hkl}} \widehat{n}, \tag{3.10}$$

or

$$2d_{hkl} \sin \theta = \lambda,$$

which again is the Bragg equation in the familiar form and identical with Eq. (3.4). The condition $Q = G_{hkl}$ is also shown in the right-hand panel of Fig. 3.1 and is called the **Bragg condition**. An equivalent expression for the Bragg condition is $Q \cdot G_{hkl} = (G_{hkl})^2$.

> Bragg condition: $Q = G_{hkl}$.

3.2 Ewald Construction

There is a clever strategy for finding Bragg reflections in the reciprocal space of perfect single crystals. This strategy is known as **Ewald construction** and is demonstrated in Fig. 3.2. It consists of the following five steps:

1. Construct a reciprocal lattice. i.e., a mesh of reciprocal lattice points (*r. l. p.*), based on the known lattice symmetry and the lattice spacing d_{hkl}.
2. Draw an arrow with the length $|k_i| = 2\pi/\lambda$ from an arbitrary point in reciprocal space to one of the *r. l. p.*

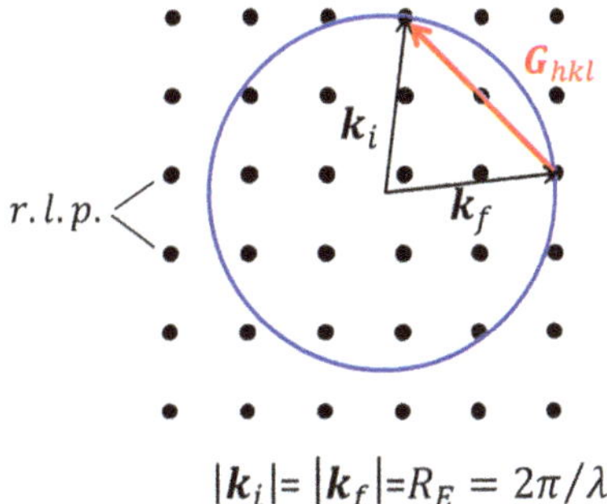

Fig. 3.2 Ewald construction for identifying the Bragg condition in the reciprocal space. The black dots indicate reciprocal lattice points. The blue circle with a radius $R_E = 2\pi/\lambda$ defines the Ewald sphere. If two reciprocal lattice points are located on the Ewald sphere, they fulfill the Bragg condition $Q = G_{hkl}$

3. Span a sphere with a radius $R_E = 2\pi/\lambda$, centered at the origin of the arrow $|\mathbf{k}_i|$. The sphere is called the Ewald sphere. In the projection on a plane, the Ewald sphere is a circle.
4. If the surface of the Ewald sphere (circle) touches another *r. l. p.*, draw a second arrow $|\mathbf{k}_f| = 2\pi/\lambda$ from the center of the sphere (circle) to this *r. l. p.* If the surface of the sphere does not touch another r.l.p., shift the center of the sphere until it touches.
5. Connect both *r. l. p.* on the Ewald sphere (circle). The connecting vector is the reciprocal lattice vector $\mathbf{G}_{hkl}$ and fulfills the Bragg condition $\mathbf{Q} = \mathbf{k}_i - \mathbf{k}_f = \mathbf{G}_{hkl}$.

The Bragg equation and the Ewald construction express the geometric condition in terms of scattering angle θ or scattering vector $\mathbf{Q}$ for identifying potential Bragg reflections according to Bragg's law. But it does not tell us the Bragg reflection *intensities*. The intensity of Bragg reflections can be calculated via the lattice sum, i.e., adding up all amplitudes scattered from atoms or nuclei of a crystal lattice and squaring the sum of the amplitudes. This is done in Sect. 3.3. Before doing so, let us first recall the definitions for the translation vector, the basis vector, and the lattice vector.

In Chap. 1 on crystal structures, we have seen that the **lattice vector** $\mathbf{R}_j$ is composed of a translation vector $\mathbf{T}_j$ and a basis vector $\mathbf{r}_k$:

$$\mathbf{R}_j = \mathbf{T}_j + \mathbf{r}_k, \tag{3.11}$$

where the **translation vector** is

$$\mathbf{T}_j = u_j\mathbf{a} + v_j\mathbf{b} + w_j\mathbf{c}, \quad \text{with } (u, v, w \in N) \tag{3.12}$$

and the **basis vector** is

$$\mathbf{r}_k = x\mathbf{a} + y\mathbf{b} + z\mathbf{c}, \text{ with } (0 \leq x, y, z \leq 1). \tag{3.13}$$

a,b, and **c** are the unit cell lattice parameters. The **reciprocal lattice vector** is defined as

$$\mathbf{G}_{hkl} = \mathbf{G}_h + \mathbf{G}_k + \mathbf{G}_l = h\mathbf{A} + k\mathbf{B} + l\mathbf{C}, \tag{3.14}$$

where $(h, k, l \in N)$ are the Miller indices, and **A,B, C** are the **reciprocal cell vectors** defined by (see also Eq. 1.4)

$$\mathbf{A} = 2\pi\frac{\mathbf{b} \times \mathbf{c}}{\mathbf{a} \cdot (\mathbf{b} \times \mathbf{c})}; \quad \mathbf{B} = 2\pi\frac{\mathbf{c} \times \mathbf{a}}{\mathbf{a} \cdot (\mathbf{b} \times \mathbf{c})}; \quad \mathbf{C} = 2\pi\frac{\mathbf{a} \times \mathbf{b}}{\mathbf{a} \cdot (\mathbf{b} \times \mathbf{c})}. \tag{3.15}$$

Because of the orthogonality properties of $\mathbf{G}$ and $\mathbf{T}$, we find

$$\mathbf{G} \cdot \mathbf{T} = 2\pi(hu + kv + lw) = 2\pi n \tag{3.16}$$

and therefore

$$e^{iG \cdot T} = 1, \tag{3.17}$$

which is another suitable way of expressing the Bragg condition.

3.3 Lattice Sum: One-Dimensional Case

With the stated properties of G and T, we can now determine the total scattering amplitude for a particular (hkl) Bragg reflection and finally calculate the expected Bragg peak intensity. We merely need to sum up all amplitudes, which were scattered from all lattice points in a regular periodic lattice with atoms sitting at the lattice vectors R_j. According to Eq. (2.13), the sum to be evaluated is

$$A_{\text{tot}}(Q) = \frac{a}{r} \sum_{j=1; k=1}^{N_c; N_b} e^{iQ \cdot R_j} = \frac{a}{r} \sum_{j=1; k=1}^{N_c; N_b} e^{iQ \cdot (T_j + r_k)} . \tag{3.18}$$

The double sum goes over all unit cells N_c in a crystal lattice and all atoms N_b that belong to the basis of a unit cell. The double sum can be rewritten as the product of two sums:

$$A_{\text{tot}}(Q) = \frac{a}{r} \left(\sum_{k=1} e^{iQ \cdot r_k} \cdot \sum_{j=1} e^{iQ \cdot T_j} \right). \tag{3.19}$$

The first sum in the bracket determines the **structure factor**:

$$F(Q) = \sum_k e^{iQ \cdot r_k}, \tag{3.20}$$

which we take here as a constant prefactor. The structure factor, not to be mixed up with the form factor, is considered later in more detail in Sect. 3.5. The second sum runs over all unit cells in a lattice. As $T_j = u_j a + v_j b + w_j c$, the second sum can be split up into the product of three sums, one for each of the three lattice directions:

$$A_{\text{tot}}(Q) = \frac{a}{r} \sum_{j=1} e^{iQ \cdot R_j} = \frac{a}{r} F(Q) \left(\sum_u e^{iQ \cdot ua} \sum_v e^{iQ \cdot vb} \sum_w e^{iQ \cdot wc} \right) \tag{3.21}$$

For now, we evaluate only one of the sums for the index u that covers all unit cells in the a-direction of the crystal lattice. The other two sums in the bracket for the indices v and w are treated in the same way.

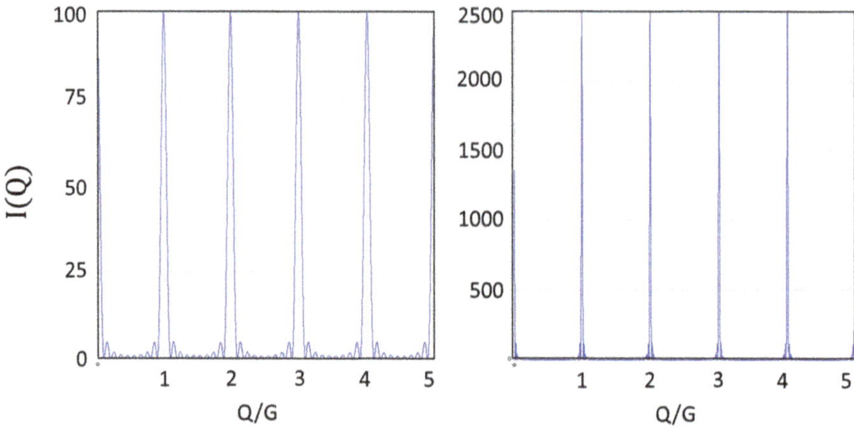

Fig. 3.3 Simulation of main maxima and side maxima for a one-dimensional lattice with a number of unit cells $N_a = 10$ in the left-hand panel and 50 in the right-hand panel

The sum $\sum_u e^{iQ \cdot ua}$ has the form of a geometric progression, explained in the Math Box I. The sum yields the scattering amplitude A_a for the a-direction:[1]

$$A_a(Q) = \sum_{u=0}^{N_a-1} e^{iQ \cdot ua} = \frac{e^{iN_aQ \cdot a} - 1}{e^{iQ \cdot a} - 1}. \tag{3.22}$$

The amplitude A_a, calculated so far for the a-direction, is a complex function. To obtain real intensities, we have to multiply the amplitude by its complex conjugate:[2]

$$A_a \cdot A_a^* = \left(\frac{e^{iN_aQ \cdot a} - 1}{e^{iQ \cdot a} - 1} \right) \left(\frac{e^{-iN_aQ \cdot a} - 1}{e^{-iQ \cdot a} - 1} \right) = \frac{\sin^2 N_aQ \cdot a/2}{\sin^2 Q \cdot a/2}. \tag{3.23}$$

The derivation of the last equation is provided in Math Box II. Equation (3.23) is well known in optics as the **Fraunhofer intensity** of a diffraction grating. The function is plotted in Fig. 3.3 on a linear scale. It has two essential features: **main maxima** and **side maxima** between the main maxima. The conditions for both are explained as follows.

Math Box I Geometric Progression
Geometric progressions (series) can be finite or infinite. A geometric progression up to a finite term has the general form:

(continued)

---

[1] Note that the sum over $u = 0 \dots N_a - 1$ is equivalent with the sum over $u = 1 \dots N_a$, but the former sum corresponds to the definition of the geometric progression.

[2] Calculations with complex functions and its conjugate form are explained in Appendix A2.

$$S = a + ax + ax^2 + \ldots + ax^{N-1} = \sum_{n=0}^{N-1} ax^n .$$

Each term in the sum follows from the previous one by multiplication with a certain term, here by multiplication with x. The finite sum of the geometric progression has a finite well defined value for $x < 1$:

$$\sum_{n=0}^{N-1} ax^n = a\,\frac{x^N - 1}{x - 1}.$$

Math Box II Derivation of Eq. (3.23)
For simplicity we substitute $\boldsymbol{Q} \cdot \boldsymbol{a} = x$ and $N_a \boldsymbol{Q} \cdot \boldsymbol{a} = Nx$. Then

$$A_a \cdot A_a^* = \left(\frac{e^{iNx} - 1}{e^{ix} - 1}\right)\left(\frac{e^{-iNx} - 1}{e^{-ix} - 1}\right),$$

and extracting common factors by using the fact that $e^{ix/2}\,e^{-ix/2} = 1$ and $e^{iNx/2}\,e^{-iNx/2} = 1$, we obtain

$$= \frac{e^{iNx/2}}{e^{ix/2}}\left(\frac{e^{iNx/2} - e^{-iNx/2}}{e^{ix/2} - e^{-ix/2}}\right)\frac{e^{-iNx/2}}{e^{-ix/2}}\left(\frac{e^{-iNx/2} - e^{iNx/2}}{e^{-ix/2} - e^{ix/2}}\right).$$

Finally, for the last equation we get

$$A_a \cdot A_a^* = \left(\frac{e^{iNx/2} - e^{-iNx/2}}{e^{ix/2} - e^{-ix/2}}\right)\left(\frac{e^{-iNx/2} - e^{iNx/2}}{e^{-ix/2} - e^{ix/2}}\right)$$

Next, we realize that $(e^{ix/2} - e^{-ix/2}) = 2i\sin(x/2)$. Therefore,

$$A_a \cdot A_a^* = \frac{\sin(Nx/2)}{\sin(x/2)} \cdot \frac{-\sin(Nx/2)}{-\sin(x/2)} = \frac{\sin^2(Nx/2)}{\sin^2(x/2)},$$

which was to be shown.

3.3.1 Main Maxima

Whenever the arguments of the numerator and the denominator have the values

$$\boldsymbol{Q} \cdot \boldsymbol{a} = \boldsymbol{G}_h \cdot \boldsymbol{a} = \frac{2\pi}{a} ha = 2\pi h, \tag{3.24}$$

and therefore fulfill the condition for a Bragg reflection, then

$$A_a \cdot A_a^* = \frac{\sin^2(N_a \pi h)}{\sin^2(\pi h)} = N_a^2. \tag{3.25}$$

This defines the intensity of the main maxima. It is a nontrivial result because N_a and h are integers and therefore $\sin^2 N_a \pi h = 0$ and $\sin^2 \pi h = 0$ and the ratio $\sin^2 N_a \pi h / \sin^2 \pi h = 0/0$ is undefined. However, applying the rule named after the French mathematician L'Hopital, we derive the result quoted in Eq. (3.25). The proof is given in Math Box III. The final result is remarkable as it tells us that the maximum intensity, i.e., the height of the Bragg reflection, scales with the number of unit cells in the scattering volume to the square.

Math Box III Derivation of Eq. (3.25)

For simplicity we set $\pi h = x$ and $N_a = N$. The equation then becomes

$$\frac{\sin^2(Nx)}{\sin^2 x} = \frac{\sin(Nx)}{\sin x} \frac{\sin(Nx)}{\sin x}.$$

It is sufficient to consider only one of the terms in the product. Next we evaluate the behavior of the ratio $\sin Nx / \sin x$ in the limit of x approaching 0:

$$\lim_{x \to 0} \frac{\sin(Nx)}{\sin x} = \frac{0}{0}.$$

The result is undefined. To overcome this problem, we use the rule of L'Hopital. This rule states that if the limit of two functions $\lim_{x \to 0} f(x) = \lim_{x \to 0} g(x) = 0$, whereas the limits of the first derivatives of these functions are finite: $\lim_{x \to 0} f'(x) \neq 0$; $\lim_{x \to 0} g'(x) \neq 0$, then

$$\lim_{x \to 0} \frac{f(x)}{g(x)} = \lim_{x \to 0} \frac{f'(x)}{g'(x)}.$$

Following this rule, we obtain

$$\lim_{x \to 0} \frac{\sin(Nx)}{\sin x} = \lim_{x \to 0} \frac{N \cos(Nx)}{\cos x} = N.$$

Now we square again the last result and obtain N^2, as quoted in Eq. (3.25).

3.3.2   Side Minima and Side Maxima

First, we realize that the numerator $\sin^2 N_a \mathbf{Q} \cdot \mathbf{a}/2$ in Eq. (3.23) rapidly oscillates as a function of $\mathbf{Q}$ because of the large number N_a of unit cells in a crystal lattice. In contrast, the function $\sin^2 \mathbf{Q} \cdot \mathbf{a}/2$ varies only slowly from one Bragg reflection to the next by incrementing the Miller index h to $h \pm 1$.

In between Bragg reflections, we observe several *side minima* with zero intensity and many *side maxima* with much lower intensity than the main maxima, but still with finite intensity. The number of side minima and maxima depends on the number of unit cells N_a. This can be seen as follows. We first derive the condition for the side minima, followed by the condition for the side maxima.

The first minimum away from the Bragg reflection is reached for the argument in the numerator:

$$N_a(\mathbf{Q}\cdot\mathbf{a})/2 = \pi, \quad \text{or } (\mathbf{Q}\cdot\mathbf{a})/2 = \pi/N_a.$$

However, for this value, the denominator $(\mathbf{Q}\cdot\mathbf{a})/2 = \pi/N_a \neq 0$. Therefore, the first minimum next to the main maximum is characterized by

$$\frac{\sin^2 \pi}{\sin^2 (\pi/N_a)} = \frac{0}{\sin^2 (\pi/N_a)} = 0.$$

This is a true and well-defined intensity minimum. In general, the minima between the main maxima (Bragg reflections) are characterized by the condition:

$$(\mathbf{Q}\cdot\mathbf{a})/2 = k\pi/N_a, \text{with } (k = 1, \ldots, N_a - 1). \tag{3.26}$$

Therefore, there are $N_a - 1$ side minima between any two Bragg reflections.

Between the side minima, there are side maxima, which are defined by the condition:

$$N_a(\mathbf{Q}\cdot\mathbf{a})/2 = \pi/2, \quad \text{or : } (\mathbf{Q}\cdot\mathbf{a})/2 = \pi/2N_a.$$

Therefore, the numerator $\sin^2 N_a \mathbf{Q} \cdot \frac{a}{2} = \sin^2 (\pi/2) = 1$, and the denominator is $\sin^2 \mathbf{Q} \cdot a/2 = \sin^2 (\pi/2N_a) \neq 0$, yielding the ratio:

$$\frac{\sin^2 (\pi/2)}{\sin^2 (\pi/2N_a)} = \frac{1}{\sin^2 (\pi/2N_a)} \neq 0$$

In general, the side maxima follow the condition:

$$(\boldsymbol{Q}\cdot\boldsymbol{a})/2 = (2l+1)\pi/2N_a, \quad \text{with} \quad (l=1,\ldots,N_a-2), \tag{3.27}$$

and therefore, there are $N_a - 2$ side maxima between any two Bragg reflections. This can be seen directly in the left-hand panel of Fig. 3.3, where the intensity is plotted as a function of the scattering vector $\boldsymbol{Q}$ for $N_a = 10$. Usually, the number of unit cells involved in the scattering process is very large. For large N_a, like in single crystals, and since the main maximum grows in intensity proportional to N_a^2, the side maxima play a diminishing role. This can be seen by the simulation in the right-hand panel for $N_a = 50$. The side maxima squeezed between the Bragg reflections become more narrowly spaced, diminish in intensity, and can no longer be resolved experimentally. Note the dramatic increase in the top intensity from 100 to 2500 in arbitrary units, the number of side maxima increasing from 8 to 48 (the latter one cannot be resolved), and the narrowing of the main peak.

For large N_a, the intensity distribution of the main peak can be described by a Gaussian function:

$$\mathop{\mathrm{Lim}}_{\substack{Q\to G \\ N_a \to \text{large}}} \frac{\sin^2(N_a \boldsymbol{Q}\cdot\boldsymbol{a}/2)}{\sin^2(\boldsymbol{Q}\cdot\boldsymbol{a}/2)} = N_a^2 \exp\left(-\frac{N_a^2 a^2 (\Delta Q)^2}{16\pi}\right) \tag{3.28}$$

3.3.3   Full Width at Half Maximum and Integrated Intensity

For the Gaussian function, we can determine the *full width at half maximum* (FWHM) in Q-space. Q_1 is the wavenumber, where the intensity on one side of the curve is half of its top value, and $(\Delta Q_1 = Q_1 - G)$ is the distance from the main maximum:

$$\frac{N_a^2}{2} = N_a^2 \exp\left(-\frac{N_a^2 a^2 (\Delta Q_1)^2}{16\pi}\right).$$

Solving for $\Delta Q_1 = (Q_1 - G)$, we obtain

$$(\Delta Q_1) = \frac{4\pi\sqrt{\ln 2/\pi}}{N_a a} = \frac{4\pi}{\lambda}\cos\theta d\theta.$$

Solving for $d\theta$ and considering that the The FWHM is twice its this value, we obtain:

$$\mathrm{FWHM} = 2d\theta = \frac{2\lambda\sqrt{\ln 2/\pi}}{N_a a \cos\theta} = \frac{0.94\lambda}{L\cos\theta}. \tag{3.29}$$

Fig. 3.4 Bragg reflection intensity calculated for 10 and 100 lattice planes

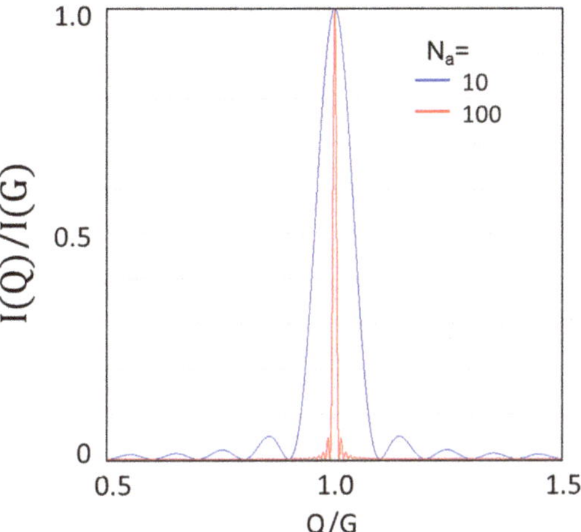

In the last equation, the product $N_a a = L$ is set to the edge length L of a crystallite. Eq. (3.29) is the famous Scherrer equation, allowing to determine the crystal size from the FWHM of a Bragg reflection. The Scherrer equation is also the topic of Exercise 5.1. Note that the FWHM of the main maximum scales with $1/N_a$, and therefore, the integrated intensity of the main maxima is proportional to $N_a^2/N_a = N_a$. The integrated Intensity of a Bragg reflection is therefore proportional to the total number of scattering centers hit by the beam. In Fig. 3.4, the normalized intensity distribution of the main maximum is compared for $N_a=10$ and 100. From this graph, it becomes obvious how fast the FWHM drops with an increasing number of unit cells in the scattering volume. This also implies that a crystallite of about 50 nm edge length is sufficient for generating sharp Bragg reflections. Moreover, an X-ray or neutron coherence length of less than 100 nm is sufficient for generating constructive interference effects.

The side maxima to a Bragg peak are shown again in Fig. 3.5 on a logarithmic scale. They are also referred to as *Laue oscillations*. For normal single-crystal diffraction, they are not important and are invisible because of the large number of atoms participating in the Bragg condition. However, in thin film research, they play a crucial role in determining the film thickness and its structural integrity, which is a topic discussed in detail in Chap. 10.

3.4   Lattice Sum: Three-Dimensional Case

Having treated the one-dimensional case for the lattice sum in the a-direction, we now return to the three-dimensional case. For the other two directions, the condition for constructive interference in the b- and c-directions is equivalent to the one for the

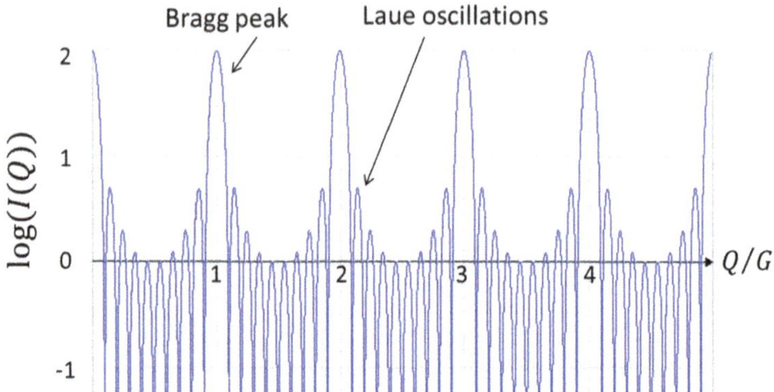

Fig. 3.5 Main maxima (Bragg-peaks) and side maxima (Laue-oscillations) for a thin single crystalline film consisting of ten atomic layers. The atomic form factor is neglected in this and the previous graphs

a-direction, quoted in Eq. (3.24). Therefore, the conditions for constructive interference are

$$\begin{aligned}
Q \cdot a &= G_h \cdot a = 2\pi h, \\
Q \cdot b &= G_k \cdot b = 2\pi k, \\
Q \cdot c &= G_l \cdot c = 2\pi l.
\end{aligned} \tag{3.30}$$

These equations are known as the three **Laue equations** for Bragg reflections in the three-dimensional reciprocal space. h, k, l are the respective Miller indices. The peak intensity of the Bragg reflections is accordingly proportional to

$$I(Q) \sim \frac{\sin^2(N_a Q \cdot a/2)}{\sin^2(Q \cdot a/2)} \frac{\sin^2(N_b Q \cdot b/2)}{\sin^2(Q \cdot b/2)} \frac{\sin^2(N_c Q \cdot c/2)}{\sin^2(Q \cdot c/2)}.$$

Now we seek an expression for the peak intensity under the condition that N_a, N_b, $N_c \to \infty$ as is approximately the case for unit cells in a perfect crystal lattice. First, we treat the a-direction; the other two directions are equivalent. For $N_a \to \infty$, one can show that

$$\lim_{N_a \to \infty} \frac{\sin^2(N_a Q \cdot a/2)}{\sin^2(Q \cdot a/2)} = N_a^2 |A| \delta(Q - G_h). \tag{3.31}$$

Thus, the ratio $\sin^2 N_a Q \cdot a/2 / \sin^2 Q \cdot a/2$ approaches a Dirac δ-function for Q at any reciprocal lattice point $G_h = hA$. The Dirac δ-function has infinite height and zero width at $Q = G_h$. Upon intensity integration, it has the value of one at $Q = G_h$, but is zero otherwise. The peak intensity is proportional to N_a^2, as we have seen before. Here, $|A| = 2\pi/a$ is a normalization factor of the δ-function. We find equivalent expressions for the peak intensity in the other two directions, b and c:

$$\underset{N_b \to \infty}{\mathrm{Lim}} \frac{\sin^2(N_b \mathbf{Q} \cdot \mathbf{b}/2)}{\sin^2(\mathbf{Q} \cdot \mathbf{b}/2)} = N_b^2 |\mathbf{B}| \delta(\mathbf{Q} - \mathbf{G}_k),$$

$$\underset{N_c \to \infty}{\mathrm{Lim}} \frac{\sin^2(N_c \mathbf{Q} \cdot \mathbf{c}/2)}{\sin^2(\mathbf{Q} \cdot \mathbf{c}/2)} = N_c^2 |\mathbf{C}| \delta(\mathbf{Q} - \mathbf{G}_l).$$

Now we collect all terms and get the peak intensity $I_{\max}(\mathbf{Q} = \mathbf{G}_{hkl})$ of a Bragg reflection at any Bragg point in the reciprocal lattice:

$$
\begin{aligned}
I_{\max}(\mathbf{Q}) \quad &= \left(\frac{a}{r}\right)^2 N_a^2 N_b^2 N_c^2 |\mathbf{A}||\mathbf{B}||\mathbf{C}| F^2(\mathbf{G}_{hkl}) \delta(\mathbf{Q}_a - \mathbf{G}_h) \delta(\mathbf{Q}_b - \mathbf{G}_k) \delta(\mathbf{Q}_c - \mathbf{G}_l) \\
&= \left(\frac{a}{r}\right)^2 N^2 V_{rc} \, F^2(\mathbf{G}_{hkl}) \delta(\mathbf{Q} - \mathbf{G}_{hkl}).
\end{aligned}
$$

$$(3.32)$$

In the last line, $N^2 = N_a^2 N_b^2 N_c^2$, N is the total number of unit cells in the lattice, and a is the scattering length, depending on the probe. V_{rc} is the **reciprocal cell volume** defined by

$$V_{rc} = |\mathbf{A}||\mathbf{B}||\mathbf{C}| = \frac{2\pi}{a}\frac{2\pi}{b}\frac{2\pi}{c} = \frac{(2\pi)^3}{abc} = \frac{(2\pi)^3}{V_{uc}}, \qquad (3.33)$$

where $V_{uc} = abc$ is the real space unit cell volume. The structure factor $F(\mathbf{Q} = \mathbf{G}_{hkl})$ is evaluated only at the Bragg points $\mathbf{Q} = \mathbf{G}_{hkl}$, as there is essentially no intensity away from the Bragg reflections (see Sect. 3.5). Furthermore, we use the shorthand notation:

$$\delta(\mathbf{Q} - \mathbf{G}_{hkl}) = \delta(\mathbf{Q}_a - \mathbf{G}_h) \delta(\mathbf{Q}_b - \mathbf{G}_k) \delta(\mathbf{Q}_c - \mathbf{G}_l).$$

In Sect. 3.3, we found that the width of Bragg reflections scales with $1/N$. Therefore, for the integrated total intensity (maximum height × width) of Bragg reflections we find

$$I_{\mathrm{tot}}(\mathbf{Q}) = \left(\frac{a}{r}\right)^2 N V_{rc} \, F^2(\mathbf{G}_{hkl}) \delta(\mathbf{Q} - \mathbf{G}_{hkl}). \qquad (3.34)$$

The total integrated intensity is proportional to the total number of unit cells N and the square of the structure factor $F^2(\mathbf{G}_{hkl})$. The structure factor is discussed in Sect. 3.5. For an sc lattice, $F^2(\mathbf{G}_{hkl}) = 1$ for all reflections. This important result holds in the **kinematic approximation** for an ideal crystal lattice and for X-ray scattering as well as neutron and electron scattering. The limitations of the kinematic approximation are discussed in Sect. 3.7.

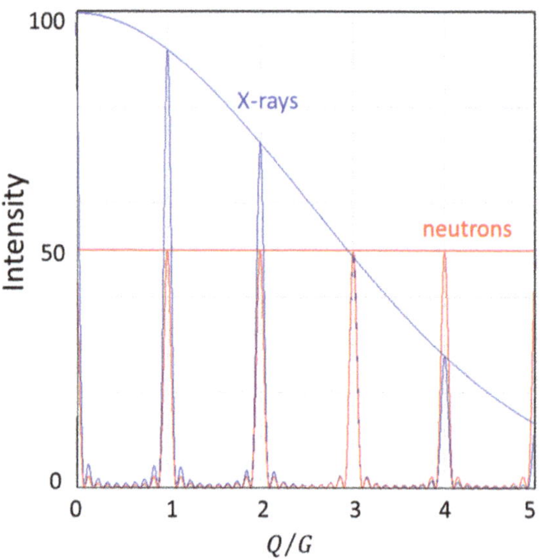

Fig. 3.6 Comparison of Bragg reflections scanned with X-rays (blue) and with neutrons (red). In both cases, the Debye–Waller thermal factor is neglected. The Q scale is normalized by the reciprocal lattice vector G

Specifically, for thermal neutron scattering, the **total Bragg intensity** in the **kinematic approximation** is

$$I_{\text{tot}}^{\text{neutron}}(\boldsymbol{Q}) = \left(\frac{b_{\text{coh}}}{r}\right)^2 NV_{rc}\, e^{-2W} F^2(\boldsymbol{G}_{hkl})\delta(\boldsymbol{Q} - \boldsymbol{G}_{hkl}), \qquad (3.35)$$

and for X-ray scattering, the equivalent expression for the total intensity is

$$I_{\text{tot}}^{\text{x-ray}}(\boldsymbol{Q}) = \left(\frac{r_0}{r}\right)^2 NV_{rc}\, e^{-2W} f^2(\boldsymbol{Q}) P F^2(\boldsymbol{G}_{hkl})\delta(\boldsymbol{Q} - \boldsymbol{G}_{hkl}). \qquad (3.36)$$

Here, we have also included the Debye–Waller factor (e^{-2W}), explained in Sect. 3.6, and the polarization factor P for X-ray scattering. Figure 3.6 compares the intensities of Bragg reflections for X-ray and neutron scattering at a single crystal with the scattering vector $\boldsymbol{Q}$ in the [100] direction of the reciprocal lattice.

In this scan, only the (h00) reflections are recorded. The Q-scale is normalized by the modulus of the reciprocal lattice vector $\boldsymbol{A}$, so that the Bragg peaks occur at multiples of the (h00) Miller indices. Note that the peak at (000) is a true Bragg peak. But in reality, this peak cannot be recorded as it overlaps with the incident beam in the forward direction. The high intensity of the incident beam would destroy the detector.

The location of Bragg peaks in reciprocal space is identical for X-rays and neutrons. But the dependence of the Bragg peak intensity on the scattering vector $\boldsymbol{Q}$ is different due to the different form factors for X-rays and neutrons (compare Fig. 2.7). In the case of neutrons, the intensity is independent of Q; for X-rays, the intensity decreases with increasing Q because of the Q-dependent atomic form factor $f(Q)$. The scale factor between X-ray and neutron peak intensity is chosen arbitrarily.

So far, we have emphasized the equivalence of X-ray and neutron scattering for deriving the lattice sum. For the experts, the quantum mechanical treatment of neutron scattering is briefly outlined in the overview box below. From a more fundamental point of view, X-ray scattering starts from the Maxwell equations, and elastic thermal neutron scattering starts with the Schrödinger equation and calculates the transition rate between an initial and a final state. In the end, we arrive at equations for the intensity or cross-section of X-rays and neutrons, which are distinguished by the particle–target interaction but otherwise have the same expression for the Fourier transform of the pair correlation in the sample. Therefore, X-ray and neutron scattering are indeed comparable and complementary. For each scientific aim, the experimental team has to decide which probe is the most suitable and promising approach.

Quantum Mechanical Description of Neutron Scattering

Neutron scattering implies that particles go from an initial state $|i\rangle$ before interacting with the target, characterized by $|i\rangle = |k_i, E_i, \sigma_i\rangle$, to a final state after interaction with the target, $|f\rangle = |k_f, E_f, \sigma_f\rangle$. Here $|k_{i,\,f}\rangle$ is the initial and final neutron momentum, $E_{i,\,f}$ are the respective energies, and $|\sigma_{i,\,f}\rangle$ are the corresponding spin states.

The transition rate per incident particle from $|i\rangle$ to $|f\rangle$ depends on the interaction potential $V(r)$ and the density of final scattering states per energy interval $\rho(E_f)$. Applying Fermi's golden rule, we find for the transition rate W:

$$W = \frac{2\pi}{\hbar} |\langle f|V(r)|i\rangle|^2 \rho(E_f).$$

After properly normalizing all wave functions and evaluating $\rho(E_f)$, the transition rate yields the elastic cross section in the *Born approximation* (Φ = incident flux):

$$\frac{d\sigma}{d\Omega} = \frac{1}{\Phi}\frac{W}{d\Omega} = \left(\frac{m}{2\pi\hbar^2}\right)^2 |\langle k_f|V(r)|k_i\rangle|^2$$

$$= \left(\frac{m}{2\pi\hbar^2}\right)^2 \left| \int V(r)\exp\left(i(k_f - k_i)\cdot r\right)dr \right|^2$$

$$= \left(\frac{m}{2\pi\hbar^2}\right)^2 \langle V(Q)\rangle^2.$$

$V(Q)$ is the Fourier transform of the neutron–target interaction potential. An alternative expression is

(continued)

$$\left(\frac{d\sigma}{d\Omega}\right)_{\mathrm{coh}} = b_{\mathrm{coh}}^2 N S_{\mathrm{coh}}(\boldsymbol{Q}),$$

where $S_{\mathrm{coh}}(\boldsymbol{Q})$ is known as the coherent elastic scattering function, and N is the number of scattering centers in the target. The scattering function $S_{\mathrm{coh}}(\boldsymbol{Q})$ contains the complete information on the sample in terms of Fourier transforms of pair correlation functions.

In X-ray scattering, it is usual to speak about intensities; in neutron scattering, the term partial or total cross-section is preferred. The relation between the terms is explained in the following gray box.

Intensity or Cross Section?

In general, the intensity I is defined as the number of photons (particles) N per unit area A and time t: $I = N/At$. This definition is proper for the incident intensity of particles hitting the target at the center of a sphere, as indicated in the graph below. However, for the total scattered intensity I_{scat} in 4π we already derived the expression: $I_{\mathrm{scat}} = AA^* = (a/r)^2$. On the other hand, the partial cross-section $d\sigma/d\Omega$ of a scattering event is defined by $\frac{d\sigma}{d\Omega} = \frac{I_{\mathrm{scat}} dS}{d\Omega}$

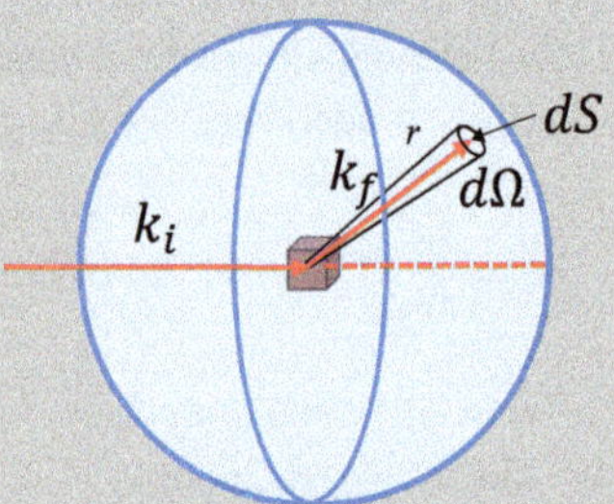

Here $d\sigma$ is the partial flux $I_{\mathrm{scat}} dS$ of particles that is scattered into the solid angle $d\Omega$.

The surface area dS, through which the beam is scattered, is $dS = r^2 d\Omega$. Therefore, we find for the partial scattering cross section:

$$\frac{d\sigma}{d\Omega} = I_{\mathrm{scat}} \frac{dS}{d\Omega} = \left(\frac{a}{r}\right)^2 r^2 = a^2.$$

The partial cross section $d\sigma/d\Omega$ is therefore identical with the scattering length squared, which is $(r_0 f)^2 P$ for X-rays and b^2 for neutrons. In practical terms, the partial cross section $d\sigma/d\Omega$ is the fraction of scattered intensity that

(continued)

passes through the surface area dS of a sphere at distance r from the scattering center, which can be the opening aperture of a detector.

The total cross section integrated over the surface of the sphere is

$$\sigma_{\text{tot}} = \int \frac{d\sigma}{d\Omega} d\Omega = 4\pi a^2$$

in general, and specifically for X-ray scattering by considering the polarization P,

$$\sigma_{\text{tot}}^{\text{X-ray}} = \int \frac{d\sigma}{d\Omega} d\Omega = \frac{8\pi}{3} (r_0 f)^2.$$

3.5   Structure Factors

Now we evaluate the *structure factor* $F(Q) = \sum_k e^{iQ \cdot r_k}$, which is valid for X-ray, neutron, and electron scattering in the kinematic approximation. Here, the sum goes over all phase factors $e^{iQ \cdot r_k}$ with basis vectors r_k. The number of basis vectors in a unit cell N_{uc}, typically considered in solid-state physics, is relatively small: only one in the simple cubic structure, two in the body-centered cubic structure, four in the face-centered cubic structure, and eight in the diamond structure. This is a very small number compared to biological studies, where the number of basis vectors can be up to several thousand.

The structure factor $F(Q)$ is evaluated only at the Bragg points $Q = G_{hkl}$. Evaluation at any other Q value is not meaningful, because for Q-values larger or smaller than G_{hkl} the scattered intensity is essentially zero. Therefore, we evaluate

$$F(Q = G_{hkl}) = \sum_{k=1}^{N_{uc}} e^{iG \cdot r_k}. \tag{3.37}$$

If there is only a single atom at the origin of the unit cell with a basis vector $r_1 = 0$, the structure factor is $F(G) = \sum_{k=1} e^{iG \cdot r_k} = e^{iG \cdot 0} = 1$ for all Bragg reflections. For nonprimitive unit cells with a basis of more than one atom, we need to take the sum of all phase factors $e^{iG \cdot r_k}$:

$$F(Q = G_{hkl}) = e^{iG \cdot 0} + e^{iG \cdot r_2} + e^{iG \cdot r_3} \ldots = 1 + e^{iG \cdot r_2} + e^{iG \cdot r_3} + \ldots$$

3.5.1 Bcc-Structure Factor

As an example, we take the simplest case of a bcc lattice (Fig. 1.5(2)) with two basis vectors: one atom at the origin $r_1 = (0,0,0)$, and one at the center $r_2 = \left(\frac{1}{2}, \frac{1}{2}, \frac{1}{2}\right)$ in units of the lattice parameters a, b, and c.

The structure factor for the **bcc lattice** then is

$$
\begin{aligned}
F_{bcc}(G) \;&=\; \sum_{k=1}^{2} e^{iG \cdot r_k} = \exp(iG \cdot 0) + \exp\left(iG \cdot \left(\frac{1}{2}a + \frac{1}{2}b + \frac{1}{2}c\right)\right) \\
&= 1 + \exp i2\pi\left(\frac{1}{2}hA \cdot a + \frac{1}{2}kB \cdot b + \frac{1}{2}lC \cdot c\right) \\
&= 1 + \exp i\pi(h + k + l).
\end{aligned}
\tag{3.38}
$$

From the last equation, we notice that for an even sum of the Miller indices, the structure factor has the value:

$$
F_{bcc}(G) = 1 + \exp i\pi(2n) = 1 + 1 = 2 \, ,
$$

where n is any integer. However, if the sum is odd, then

$$
F_{bcc}(G) = 1 + \exp i\pi(2n + 1) = 1 - 1 = 0.
$$

Therefore, in the case of a bcc lattice, the Bragg reflections (110), (200), (211) are "allowed" reflections, meaning that they exhibit a finite intensity, whereas Bragg reflections with the indices (111), (100), etc. are "forbidden" and display zero intensity. The simple-to-remember rule for bcc lattices is stated as:

$$
\begin{aligned}
\text{For } (h + k + l) &= \text{even} : F(G_{hkl}) = 2f, \\
\text{For } (h + k + l) &= \text{odd} : F(G_{hkl}) = 0.
\end{aligned}
$$

The physical reason is that for allowed Bragg reflections, all phase factors add up to a constructive interference of waves, whereas for forbidden Bragg reflections, the phase factors cancel due to destructive interference, yielding zero intensity.

3.5.2 fcc-Structure Factor

Next, we consider the **fcc lattice**. There are four basis vectors in the fcc unit cell at positions $r_1 = (0,0,0)$; $r_2 = \left(\frac{1}{2}, \frac{1}{2}, 0\right)$; $r_3 = \left(\frac{1}{2}, 0, \frac{1}{2}\right)$; $r_4 = \left(0, \frac{1}{2}, \frac{1}{2}\right)$ (s. Fig. 1.5). Then the structure factor yields:

Table 3.1 Left panel: (*hkl*)-values and listing of lowest order allowed and forbidden Bragg reflections for bcc, fcc, and crystals with diamond structure. The listed values are those of the structure factor in units of the form factor for X-rays or scattering lengths for neutrons, with zero for forbidden reflections. Right panel: Red vertical lines indicate reflections in the different crystal structures. In the sc lattice, all reflections are allowed; in the diamond lattice, the least number of reflections are allowed

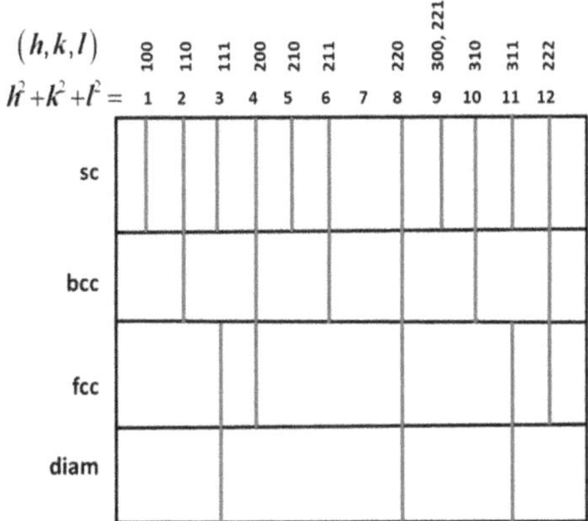

$h^2+k^2+l^2$	(hkl)	F(hkl)		
		bcc	fcc	diam
1	(100)	0	0	0
2	(110)	2	0	0
3	(111)	0	4	$4\sqrt{2}$
4	(200)	2	4	0
5	(210)	0	0	0
6	(211)	2	0	0
8	(220)	2	4	8
9	(300), (221)	0	0	0
10	(310)	2	0	0
11	(311)	0	4	$4\sqrt{2}$
12	(222)	2	4	0

$$F_{fcc}(G) = \sum_{k=1}^{4} e^{iG \cdot r_k} = 1 + \exp i\pi(h + k) + \exp i\pi(h + l) + \exp i\pi(k + l) \quad (3.39)$$

Here, the condition for constructive interference is such that the Miller indices should be all even or all odd, but not mixed. Because if they are mixed, two terms in the sum have the value +1, while the other two terms are -1, and thus add up to zero. However, for unmixed Miller indices, all terms are +1 and add up to 4. Therefore, in the case of *fcc* lattices, Bragg reflections with Miller indices like (111) or (311) are allowed but forbidden for *bcc* lattices.

Allowed and *forbidden* reflections are specific to the crystal structure and its basis. By scrutinizing structure factors, we can determine a crystal structure and distinguish between bcc, fcc, hcp, and other lattices. In Table 3.1, the allowed and forbidden reflections of bcc, fcc, and diamond lattices are compared, and the respective structure factors F(*hkl*) are provided. Note that the F(*hkl*) values for the allowed reflections correspond to the number of atoms in the basis: 2 for *bcc*, 4 for *fcc*, and 8 for the diamond structure. Also note that for the diamond lattice, such as Si, the (111) reflection is allowed, but the second harmonic (222) is forbidden. This has a useful application. Silicon single crystals can be used as monochromators, which do not pass the second harmonic wavelength ($\lambda/2$), and therefore do not contaminate measurements that rely on absolute intensities. The structure factor for the diamond lattice is derived in Exercise 3.2, and for the *hcp* structure in Exercise 3.3.

3.5.3 *CsCl-Structure Factor*

In the discussion so far, we have assumed that the unit cell contains only one type of atom, such as in the bcc structure of Fe or the fcc structure of Cu. In this situation, it is proper to take the scattering length out of the sum and to treat it as a common prefactor. However, if the unit cell contains more than one type of atom, the scattering length has to be kept inside the sum, as is the case for the CsCl- or the NaCl-structure.

For illustration, in the following, we calculate the structure factor for the *CsCl-structure*. The unit cell is shown in Fig. 3.7. It is similar to the bcc structure but with different atoms at the corners and in the center. The basis vectors are Cs^+-ion at $r_1 = (0,0,0)$ and Cl^--ion at $r_2 = \left(\frac{1}{2}, \frac{1}{2}, \frac{1}{2}\right)$.

Assuming X-ray scattering, the structure factor is

$$F(G_{hkl}) = \sum_k f_k e^{iG \cdot r_k} = f_{Cs^+} + f_{Cl^-} \exp i\pi(h + k + l), \qquad (3.40)$$

where f_{Cs^+}, f_{Cl^-} are the respective atomic form factors of the cation and anion. For neutron scattering, the atomic form factors f are replaced by the respective coherent scattering lengths b_{Cs}, b_{Cl}. The conditions for constructive interference are now

$$\text{For } (h + k + l) = \text{even} : F(G_{hkl}) = f_{Cs^+} + f_{Cl^-},$$
$$\text{For } (h + k + l) = \text{odd} : F(G_{hkl}) = f_{Cs^+} - f_{Cl^-}.$$

The formerly forbidden Bragg reflections are now allowed, but they have lower intensity than the main bcc reflections. Bragg reflections can tell a lot about crystal structures, symmetries, and atomic order in crystals. Order versus disorder can immediately be recognized via the intensity of the "forbidden" Bragg reflections. This will be discussed further in Chap. 8 on order–disorder phase transitions. The structure factor for NaCl is determined in Exercise 3.1; the diamond and *hcp* structure factors are treated in Exercises 3.2 and 3.3.

Fig. 3.7 CsCl structure
with different atoms in the
corner and the center

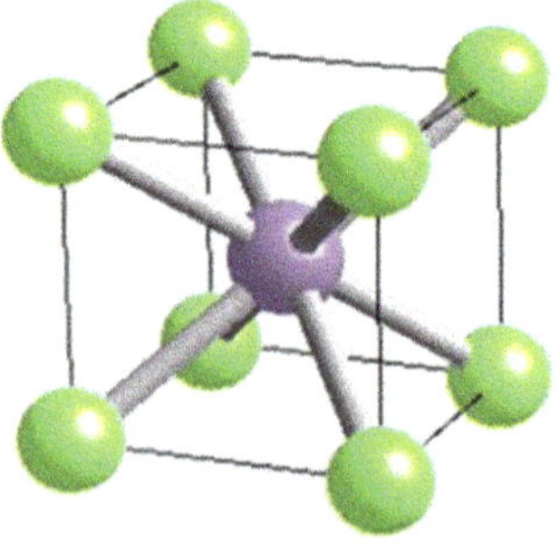

3.5.4   *Quasicrystals*

In older textbooks on crystallography and solid-state physics, we can often read that unit cells with a rotational axis of one-, two-, three-, four-, and sixfold symmetry can fill the space but not those with a fivefold symmetry axis. Proof of this statement is the topic of Exercise E1.7. A *fivefold symmetry* axis, although often encountered in nature, would lack translational symmetry, and therefore, sharp Bragg reflections would not exist. However, in the mid-80s of the last century, alloys were discovered that exhibit a fivefold symmetry axis and simultaneously show sharp Bragg reflections. Crystals with these properties are now known as **quasicrystals**. This obvious contradiction to the standard belief has led to a re-consideration of crystallography and the conditions necessary for sharp and well-defined reciprocal lattice points. The previously accepted symmetry requirement was correct under the premise that the crystal space is filled with only one type of unit cell. If two different cells are tiled with certain rules, space can also be filled. Roger Penrose showed that by tiling a plane with oblate and prolate rhombohedra following strict rules, the two-dimensional space can be packed. Such a pattern shows a fivefold rotational symmetry locally, as sketched in Fig. 3.8, but no translational symmetry (Penrose 1974). The rhombohedra are not arbitrarily chosen but fulfill the rules of the *golden mean* (or golden ratio), implying that the side lengths are proportional to the irrational number $\tau = \left(1 + \sqrt{5}\right)/2 = 1.618$ and that the diagonal inside of the prolate tile is $1 + \tau$. Therefore, short and long distances alternate with fixed rules. Mathematically it turns out that the two-dimensional Penrose pattern can be represented as

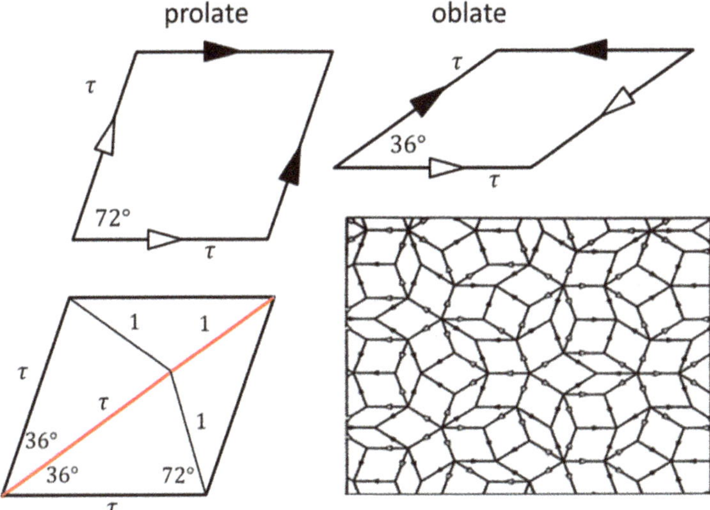

Fig. 3.8 Prolate and oblate rhombohedra are tiled up to match open arrows with open arrows and filled arrows with filled arrows, conserving the arrow directions. The side lengths are all proportional to τ, and the diagonal of the prolate tile is $1 + \tau$

cubes in five dimensions projected into the plane, while cubes in six dimensions result in icosaeders in 3D space, also displaying fivefold symmetry axes.

The **Penrose tiles** satisfy the symmetry conditions of the golden mean and fill all space. The Fourier transform results in a reciprocal lattice that also follows the rules of the golden mean, with r.l.p. representing sharp Bragg reflections. This was confirmed by X-ray and electron scattering. Due to their sixfold dimensionality, six instead of three Miller indices are required to index the Bragg reflections of quasicrystals. Therefore, the view of traditional crystallography must be modified: translational long-range order is not a unique prerequisite for the existence of sharp Bragg reflections. Rather, it is necessary that the tiling, or space filling, follows strictly deterministic rules. Daniel Shechtman received the Nobel Prize in Chemistry in 2011 for the discovery of quasicrystals[3] (Shechtman et al. 1984).

3.5.5   *The Phase Problem*

We have seen in Eq. (1.19) that any periodic function of a physical parameter with the periodicity of the crystal lattice can be represented as a Fourier series with Fourier coefficients of the reciprocal lattice. For instance, the electron density can be represented as

$$\rho(\boldsymbol{r}) = \sum_{G_{hkl}} \rho_G \exp(i\boldsymbol{G} \cdot \boldsymbol{r}). \tag{3.41}$$

Then, the X-ray intensity is proportional to

$$I_G \propto |\rho_G|^2. \tag{3.42}$$

If no absorption occurs and if the spatial distribution of the electron density is not inhomogeneous, then the electron density is a real number with the property:

$$\rho_G = \rho_{\overline{G}}^*. \tag{3.43}$$

If this is correct, it follows for the structure factor of a Bragg peak:

$$F(G) = F^*\left(\overline{G}\right),$$

and for the intensity, it follows:

---

[3]Roger Penrose also received a Nobel Prize in Physics 2020, but for a completely unrelated topic concerning black holes.

$$I(\boldsymbol{G}) = I(\overline{\boldsymbol{G}}).\tag{3.44}$$

The consequence of this important relationship is that structural analysis always assumes the presence of an inversion center. A threefold symmetry axis appears as a sixfold axis, a fivefold axis as a tenfold axis, etc. A polar and a nonpolar axis can only be distinguished if absorption centers are artificially introduced. This fundamental scattering problem is known as the phase problem. The phase loss leads to significant problems in the backtransformation of the recorded intensity distributions to real crystal structures. Examples of this problem can be found in Sect. 5.4 on Laue scattering and in Sect. 5.6 on LEED patterns.

3.6 Thermal Effects

3.6.1 Debye–Waller Factor

According to Eq. (3.18), the scattering amplitude is $A_{\mathrm{tot}} = \frac{a}{r}\sum_{j=0} e^{i\boldsymbol{Q}\cdot\boldsymbol{R}_j}$, with $\boldsymbol{R}_j = \boldsymbol{T}_j + \boldsymbol{r}_k$. $\boldsymbol{T}_j$ and $\boldsymbol{r}_k$ are the translational vector and the basis vectors for a nonprimitive unit cell, respectively. For the following consideration, we neglect the basis vectors and consider, for simplicity, a primitive unit cell. So far, we have assumed that the atoms are fixed and rigid on their lattice sites. However, at any temperature, even at zero Kelvin, they perform lattice vibrations about their equilibrium position (see Fig. 3.9) with a **vibrational amplitude u_j**, the magnitude of which depends on the temperature. On average, $\langle u_j \rangle = 0$, but $\left\langle u_j^2 \right\rangle \neq 0$ and increases with increasing temperature.

This increasing thermal vibrational amplitude leads to an exponential damping of the Bragg peak intensity. In the case of scattering with X-rays or neutrons, we take an *ensemble average* over all atoms in their instantaneous position and within the scattering volume. Each X-ray photon takes a snapshot of all atoms, but a second photon will experience a slightly different arrangement of the atoms as they may have moved between the first and the second probing photon. The detector takes a time average of all positions, which is equivalent to an ensemble average. Thus, we express the **ensemble-averaged** scattering amplitude by angle brackets as

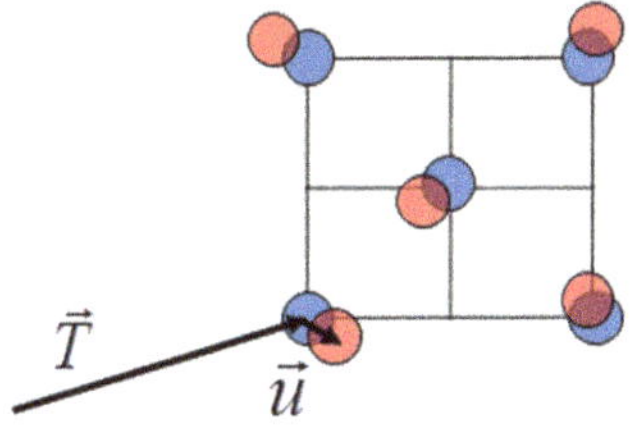

Fig. 3.9 Rigid ideal lattice in blue and thermal deviations in red

$$\langle A(\boldsymbol{Q},T)_{\text{tot}}\rangle = \frac{a}{r}\left\langle \sum_j \exp\big(i\boldsymbol{Q}\cdot(\boldsymbol{T}_j+\boldsymbol{u}_j)\big)\right\rangle = \frac{a}{r}\langle \exp(i\boldsymbol{Q}\cdot\boldsymbol{u})\rangle \sum_j \exp(i\boldsymbol{Q}\cdot\boldsymbol{T}_j)$$

$$(3.45)$$

In the third equation, the term $\langle \exp(i\boldsymbol{Q}\cdot\boldsymbol{u})\rangle$ is removed from the sum, since it is the same on average for all atoms. For the lattice sum itself, $\sum_j \exp(i\boldsymbol{Q}\cdot\boldsymbol{T}_j)$, we do not need to consider an ensemble average, since the translation vectors span the ideal lattice sites. Now we evaluate the thermal prefactor:

$$\langle \exp(i\boldsymbol{Q}\cdot\boldsymbol{u})\rangle = \left\langle 1 + i\boldsymbol{Q}\cdot\boldsymbol{u} - \frac{(\boldsymbol{Q}\cdot\boldsymbol{u})^2}{2!} + \cdots \right\rangle = 1 + i\langle\boldsymbol{Q}\cdot\boldsymbol{u}\rangle - \frac{\langle(\boldsymbol{Q}\cdot\boldsymbol{u})^2\rangle}{2!} + \cdots$$

$$= 1 - \frac{\langle(\boldsymbol{Q}\cdot\boldsymbol{u})^2\rangle}{2} + \cdots = \exp\left[-\frac{1}{2}\langle(\boldsymbol{Q}\cdot\boldsymbol{u})^2\rangle\right] = \exp[-W].$$

$$(3.46)$$

On average, the term $\langle \boldsymbol{Q}\cdot\boldsymbol{u}\rangle = 0$. The term $W = \frac{1}{2}\langle(\boldsymbol{Q}\cdot\boldsymbol{u})^2\rangle$ is known as the **Debye–Waller factor**. The exponent is negative and therefore reduces the Bragg intensity with increasing scattering vector $\boldsymbol{Q}$ and for larger vibrational amplitudes $\boldsymbol{u}$ at higher temperatures.

Knowing the phonon density of states or the specific heat of the material in question, W can be evaluated. In the **high-temperature limit** of the phonon expansion, i.e., for temperatures $T \gg \theta_D$, the Debye–Waller factor can be approximated to

$$W = \frac{3\hbar^2}{2mk_B}\frac{T}{\theta_D^2}Q^2.$$

$$(3.47)$$

Here θ_D is the **Debye temperature**, m is the mass of the lattice atom, and k_B is the Boltzmann constant. The Debye temperature is related to the lattice vibrational frequency via $\theta_D = \hbar\omega_D/k_B$, and ω_D is the so-called cut-off frequency of the acoustic phonon spectrum. θ_D is high for hard materials ($\sim$500 to 1000 K), and low for soft materials ($\sim$100 to 200 K).

Notice that W depends linearly on T and Q^2. Therefore, higher-order Bragg reflections are more affected by the Debye–Waller factor than lower-order Bragg reflections. The temperature dependence of different aluminium Bragg reflections is compared in Fig. 3.10. For a fixed Bragg peak, the intensity drops exponentially with increasing temperature T.

At finite temperature T, we find for the Bragg intensity the following expression:

$$I(Q,T) = \left(\frac{a}{r}\right)^2 NV_{rc}e^{-2W}F^2(\boldsymbol{G}_{hkl})\delta(\boldsymbol{Q}-\boldsymbol{G}_{hkl}).$$

$$(3.48)$$

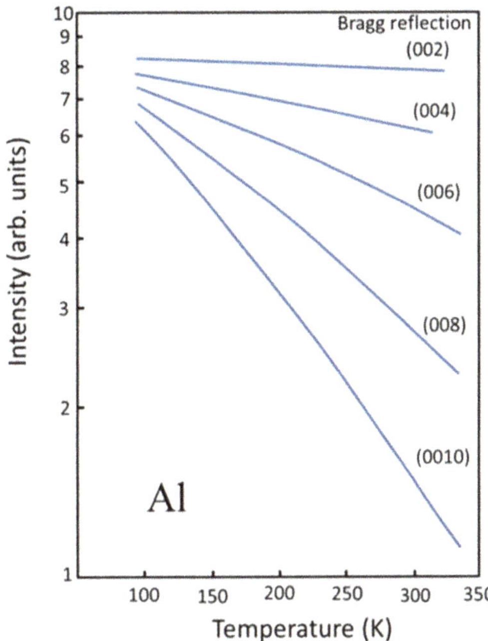

Fig. 3.10 Temperature dependence of the (00l) Bragg reflections. In this schematic, representative of the metal Al, note the exponential decay of the Bragg intensity for a fixed (00l) reflection and the increasing temperature sensitivity of the higher-order Bragg reflections

The **low-temperature limit** of the Debye–Waller factor approaches a constant value, independent of T and Q:

$$W = \frac{3\hbar^2}{8mk_B}\frac{1}{\theta_D}. \tag{3.49}$$

Because of the low temperature approximation, we observe in the logarithmic intensity plot of Fig. 3.10 deviations from the straight lines at low T.

3.6.2 Thermal Expansion

In condensed matter, a temperature change is always accompanied by a volume change. The volume change is the result of anharmonic contributions to the lattice vibration. Not only does the vibrational amplitude increase with rising temperature, but the center of the harmonic potential also shifts. In the quasi-harmonic approximation, the relation between thermal volume expansion $\Delta V/V$ and thermal energy $E_{\text{therm}}(T)$ is as follows:

$$\frac{\Delta V}{V} = \frac{\gamma_G}{B}E_{\text{therm}}(T), \tag{3.50}$$

where γ_G is the Grüneisen parameter and B is the bulk modulus. The Grüneisen parameter is a measure of the anharmonicity of the solid. The thermal expansion $\Delta V/V$ of a cubic lattice is equal in all three directions and therefore can be written as

$$\frac{\Delta V}{V} = 3\frac{\Delta a}{a} = 3\alpha(T)\,T,$$

where $\alpha(T)$ is the thermal expansion coefficient. As $\alpha(T)$ is itself temperature dependent, the linear relationship holds only for small temperature intervals. Thermal expansion shifts the positions of all Bragg reflections according to

$$\frac{\Delta a}{a} = -\frac{\Delta G}{G}. \tag{3.51}$$

Therefore, lattice expansion leads to reciprocal lattice contraction and vice versa. This effect becomes immediately noticeable when performing temperature-dependent scattering experiments. In fact, very high-resolution thermal expansion measurements can be performed using X-ray and neutron scattering methods, which, in turn, allow the determination of the thermal expansion coefficient and the Grüneisen parameter. Assuming the validity of the Dulong–Petit expression for the thermal energy in the high temperature limit ($E_{\text{therm}}(T) = 3N_A k_B T$), we find for the thermal expansion coefficient: $\alpha(T) = N_A k_B \gamma_G/B$. Since the constants N_A and k_B ($N_A =$ Avogadro number) are well known, and B can be determined by other means, the important thermal Grüneisen parameter γ_G can be determined with scattering experiments. Since the ratio $\Delta G/G$ is constant for all Bragg reflections, the shift ΔG increases with the order of the Bragg reflection. Therefore, for high-resolution work, it is advisable to use Bragg reflections with high Miller indices.

3.6.3   *Thermal Vibrations*

There is a third thermal effect to be discussed and considered, which has been neglected so far. Going back to Eq. (3.46), we distinguish between random and isotropic vibrations (u_j) about an equilibrium lattice point (T_j), and nonrandom vibrations. The average of random vibrations is $\langle u_j \rangle = 0$. They yield the Debye–Waller factor. In addition, we find systematic deflections w_j of atoms from their equilibrium position. w_j is due to the presence of lattice vibrations (phonons), which are not isotropic and do not vanish on average. These deflections cause thermal diffuse scattering (TDS) intensity in the neighborhood of Bragg reflections:

$$I(\boldsymbol{Q}, T) \propto \delta(\boldsymbol{Q} - \boldsymbol{G}_{hkl})e^{-2W} + (\boldsymbol{Q} \cdot \boldsymbol{w}(\boldsymbol{q}))^2 e^{-2W}$$
$$= I_{\text{Bragg}}(\boldsymbol{Q}, T) + I_{TDS}(\boldsymbol{Q}, T). \tag{3.52}$$

The total intensity is therefore composed of the Bragg intensity and *TDS*, both affected by the Debye–Waller factor. $w(q)$ is the phonon vibrational amplitude in reciprocal space with wave number q, which can be represented as a Fourier series:

$$w(q) = p_q a_q \sum_j e^{i(Q \pm q) \cdot T_j}. \tag{3.53}$$

Here p_q is the polarization vector of the phonon branch and a_q is the amplitude of the lattice vibration. p_q distinguishes between longitudinal and transverse phonons, and the orientation of the amplitude a_q with respect to the lattice coordinates. Using this representation and taking the lattice sum, the expression for the *TDS* in Eq. (3.52) becomes:

$$(Q \cdot w(q))^2 = (Q \cdot p_q)^2 a_q^2 \sum_G \delta(Q - (G \pm q)). \tag{3.54}$$

Now we further analyze the temperature dependence of the phonon amplitude $a_q^2(T)$ in the high-temperature approximation ($T \gg \theta_D$), yielding

$$a_q^2(T) = \frac{k_B T}{m v_q^2} \frac{Q^2}{q^2}. \tag{3.55}$$

Here m is the mass of the lattice ions, and v_q is the sound velocity. Since the sound velocity of transverse phonons is always lower than that of longitudinal phonons, the transverse phonons contribute the most to the thermal diffuse scattering. The thermal diffuse scattering intensity can now be expressed in the following form:

$$I(Q, T)_{TDS} \propto e^{-2W} \frac{k_B T}{m v_q^2} \frac{Q^2}{q^2} \sum_G \delta(Q - (G \pm q)). \tag{3.56}$$

In the last equation, which is valid for X-rays as well as neutrons with the respective prefactors, we notice that the thermal diffuse scattering is centered at all Bragg points and that it has a Lorentzian $1/q^2$ line shape. At the same time, the *TDS* increases with Q^2, and hence higher order Bragg reflections exhibit more *TDS* than lower order reflections.[4]

In summary, by X-ray or neutron scattering, we experience three thermal effects:

---

[4]Before the invention of triple-axis neutron spectrometers by B.N. Bockhouse, evaluating the TDS was the only method for probing the phonon dispersion in solids. With neutron spectroscopy one notices that the diffuse scattering actually consists of a series of δ-functions in q- and ω-space, where q is the phonon wave number and ω is the lattice vibrational frequency, which can be resonantly excited. Inelastic neutron scattering and triple-axis spectroscopy are not part of this text but can be found elsewhere.

Fig. 3.11 As a function of temperature, three effects are noticeable by X-ray or thermal neutron scattering at crystal reciprocal lattice points: (1) Damping of Bragg intensities by the Debye–Waller factor; (2) Shift of the Bragg peak due to thermal expansion; (3) Increase of thermal diffuse scattering close to the Bragg reflections

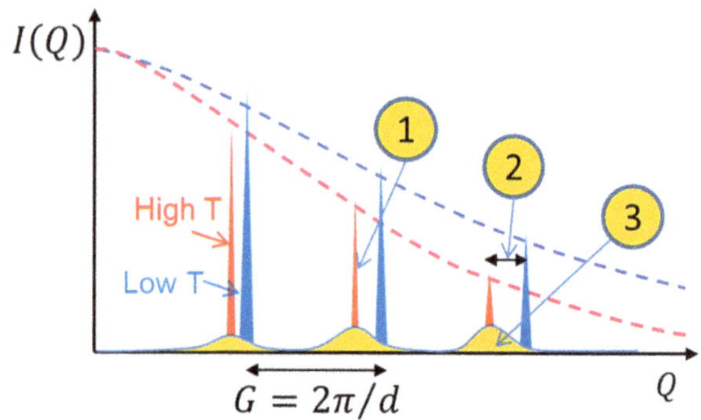

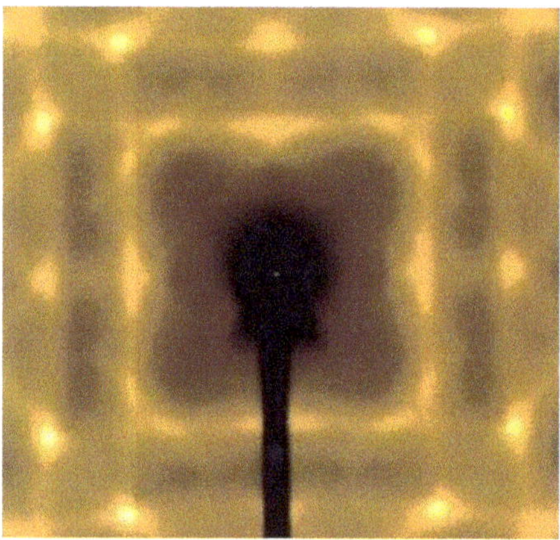

Fig. 3.12 Thermal diffuse scattering recorded at a synchrotron radiation facility in transmission geometry of a Si(100) plate. The bright spots are centered at the allowed Bragg reflections, but the reflections themselves are not recorded in this image as their intensity is much too high. Because of the crystal perfection, rocking the sample by just a hundredth of a degree erases the peak intensity and emphasizes the much broader TDS. The TDS scattering is also seen between the Bragg reflections along the main crystallographic directions. (Reproduced with permission from Holt et al. 1999: M. Holt, et al. Phys. Rev. Lett. **83**, 3317 (1999). DOI: https://doi.org/10.1103/PhysRevLett.83.3317 Copyright (2025) by the American Physical Society)

1. A damping parameter of the Bragg intensity, called the Debye–Waller effect.
2. Thermal expansion that leads to a contraction of the reciprocal space with increasing temperature.
3. Thermal diffuse scattering with a Lorentzian line shape centered at all Bragg reflections and stretching out between Bragg reflections.

Figure 3.11 illustrates all three effects, and Fig. 3.12 shows an experimental confirmation of *TDS* in a Si single crystal using synchrotron radiation. Using thermal neutron scattering, the energy dependence of longitudinal and transverse phonons can be determined as a function of the phonon wavenumber q. However, this analysis requires methods provided by inelastic scattering and is therefore beyond the scope of this text.

3.7 Kinematic Approximation

The way we have treated scattering theory in Sects. 3.1–3.6 is called the ***kinematic approximation***. The kinematic approximation is a set of several approximations whose validity should be tested for each real scattering experiment.

3.7.1 Born Approximation

The first approximation to be considered is the *Born approximation*. The Born approximation implies that each wavetrain (X-rays, neutrons, or electrons) may be scattered only once. If a wavetrain is scattered more than once, the phase information from the first scattering event is lost. But only the first scattering event is controlled via defined incident and exit angles, which allows calculating the scattering vector. For multiple scattering, the scattering vector is mixed up, and structure factor calculations are not possible. The Born approximation is usually valid for weak wave–target interaction and for imperfect crystal lattices. Vice versa, the likelihood of violating the Born approximation increases with the strength of the interaction and with increasing crystal perfection. For instance, electron scattering is likely to violate the Born approximation because of the strong Coulomb interaction between electrons and target ions. Also, X-ray scattering at silicon and diamond crystals is problematic because of the lattice perfection, such that double scattering becomes likely. Neutron scattering at samples containing a high hydrogen concentration is likewise problematic because of the large incoherent cross section. To be on the safe side for a valid application of the Born approximation, the sample should be small or thin and nonperfect.

3.7.2 Primary and Secondary Extinction

Attenuation of the incident beam takes place via internal excitations and by scattering. Attenuation by internal excitations (photoelectric effects) is usually taken into account by a constant absorption factor $1/2\mu$, where μ is the linear absorption coefficient (see Appendix A5). In the kinematic approximation, the ***attenuation*** of the incident beam by scattering is ignored. Attenuation via scattering is known as ***primary extinction*** to make the difference clear from the attenuation by absorption. In the kinematic approximation, it is assumed that the deeper layers experience the same incident intensity as the top layers, as indicated in Fig. 3.13a. Therefore, the

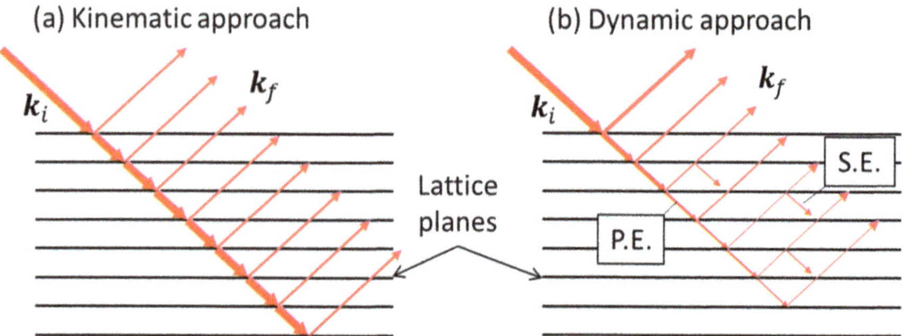

Fig. 3.13 Illustration of the difference between the (**a**) kinematic and (**b**) dynamic approach of scattering. The thick red arrows represent the incident beam intensity, and the thin arrows indicate the scattered beams at different lattice planes. P.E. and S.E. refer to primary and secondary extinction, respectively

scattered intensity should increase with the thickness of the crystal slab from which the beam is scattered. Clearly, this violates the principle of energy conservation. In fact, the intensity as a function of thickness saturates already at a thickness of 1 μm for perfect crystals, such as Si or diamond. Primary extinction refers to the weakening of the incident beam intensity by scattering at successive lattice planes.

Secondary extinction occurs when the scattered wave from one lattice plane is scattered a second time by other lattice planes with the same orientation, see Fig. 3.13b. This may also occur for different crystallites that happen to be oriented properly to fulfill the Bragg condition for the same Bragg angle. The more polycrystalline a material is, the less likely it is to undergo secondary extinction. Reducing secondary extinction increases the intensity of Bragg reflections.

Another, more rigorous scattering theory, called **dynamical scattering theory**, takes care of the shortcomings inherent to the kinematic approximation. For imperfect and small crystallites, like in a powder (polycrystal), the primary and secondary extinction effects can be neglected. If absolute measurements of the form factor or the structure factor are required, the extinction effects have to be seriously considered, since their calculated and expected integrated intensities in the kinematic and dynamic approaches are considerably different. This is discussed further in Appendix A4.

3.7.3   Refractive Index

In the kinematic approximation, the **refractive index** n for X-rays, neutrons, and electrons is set to be one. However, the refractive index is expressed as $n = 1 - \delta + i\beta$, where $\delta \cong 10^{-6}$ for both X-rays and thermal neutrons, and the imaginary part is related to the absorption. The tiny deviation from unity guarantees that in most scattering experiments, refraction effects can be neglected. However, at glancing incident angles, total reflection of X-rays and neutrons occurs, and the reflected or scattered wave becomes distorted. Thus, for surface scattering experiments, the effects due to the refractive index become essential, as considered further in Chaps. 10–12. Furthermore, in ideal crystals, it can be seen that the location of Bragg reflections is slightly shifted from the kinematic value due to the refraction effect.

3.7.4 Coherence

The wavetrains that arrive from the source to the target have no *phase relation*. They may all have the same wavelength (perfect monochromaticity) and they may all travel in the same direction (perfect collimation). Nevertheless, the phase between one wavetrain and the next is arbitrary. This condition implies that constructive interference, as discussed in this section, occurs only for every single wavetrain. The intensity collected by the detector is the sum of all single wavetrains undergoing constructive (or destructive) interference. Partial or full phase overlap of all wavetrains occurs only at synchrotron radiation facilities featuring so-called electron undulators and at X-ray free electron laser facilities that are now available at a few places in the USA, Europe, and Asia, and are briefly described in Chap. 4 (compare Table 4.3).

3.7.5 Interaction Time (Static Approximation)

The interaction time of X-rays with the target atoms is extremely short, i.e., on the femtosecond timescale, because X-rays travel at the speed of light. On the other hand, atoms in crystals are not fixed. They show thermal vibrations on the frequency scale of *THz* and a time scale of picoseconds, which is much slower than the interaction time. Therefore, we conclude that X-ray scattering is equivalent to taking a *snapshot* of the instantaneous configuration of all atoms in the sample. However, each X-ray photon will "see" a slightly different configuration. In the detector, after counting for a couple of seconds at each scattering angle, an average is taken over all these configurations, called the *ensemble average,* which is equivalent to a time average. The ensemble average is expressed in the intensity equation by angle brackets enclosing the structure factor:

$$I_{\text{tot}}(\boldsymbol{Q}=\boldsymbol{G}_{hkl}) = \left(\frac{a}{r}\right)^2 NV_{rc}\left\langle F^2(\boldsymbol{Q}=\boldsymbol{G}_{hkl})\right\rangle \delta(\boldsymbol{Q}-\boldsymbol{G}_{hkl}). \qquad (3.57)$$

The structure factor $\left\langle F^2(\boldsymbol{Q}=\boldsymbol{G}_{hkl})\right\rangle$ is sometimes also referred to as the static structure factor, and the approximation is called accordingly the *static approximation.* The thermal vibrations can be factored out by a *Debye–Waller* (DW) factor, expressed as e^{-2W}. We then obtain the static structure factor in the form:

$$\left\langle F^2(\boldsymbol{Q}=\boldsymbol{G}_{hkl})\right\rangle = F^2(\boldsymbol{Q}=\boldsymbol{G}_{hkl})e^{-2W}. \tag{3.58}$$

In neutron scattering, similar considerations hold. However, the interaction time of neutrons with the target atoms is much longer, and atoms may move while the neutron wavetrain crosses the sample.[5] Typically, thermal neutron velocities are 1000 m/s, i.e., a factor 3×10^5 slower than for X-rays. Therefore, measuring a static structure factor that can be compared to its X-ray counterpart is considerably more difficult. In the case of elastic neutron scattering experiments, it must be ensured that during recording Bragg reflections, an integration over all possible excitation energies (phonon modes) is taking place simultaneously. To guarantee the validity of a static structure factor of neutron scattering, the integration is achieved by recording the Bragg reflection not only concerning the diffraction angle but also by integrating over all possible excitation energies.

In conclusion, the kinematic approximation and, in particular, the Born approximation are appropriate for a weak scattering probability, such that the particles (X-ray photons, neutrons, or electrons) are scattered only once before entering the detector. Weak scattering is achieved either by a small form factor, a low cross-section, or by dilute scattering centers, including those found in a gas. It is good practice to use thin samples or small crystallites, such as powders, to guarantee conditions that likely fulfill the kinematic approximation.

Summary

1. The Bragg equation is $2d_{hkl}\sin\theta = \lambda$, or $\boldsymbol{Q} = \boldsymbol{G}_{hkl}$.
2. The Bragg condition can graphically be constructed via the Ewald sphere.
3. The radius of the Ewald sphere R_E equals the length of the incident or scattered wave number: $R_E = k_i = k_f$.
4. The Ewald sphere is spanned in the reciprocal space of a single crystal.
5. If two reciprocal lattice points touch the surface of the Ewald sphere, the Bragg condition is met: or $\boldsymbol{Q} = \boldsymbol{G}_{hkl}$.
6. The lattice sum adds up all scattering amplitudes from a periodic three-dimensional lattice of unit cells.
7. The Bragg condition for all three orientations of a crystal lattice is expressed by the three Laue equations: $\boldsymbol{G}_h \cdot \boldsymbol{a} = 2\pi h$; $\boldsymbol{G}_k \cdot \boldsymbol{b} = 2\pi k$; $\boldsymbol{G}_l \cdot \boldsymbol{c} = 2\pi l$.

---

[5]This property is used in quasi-elastic neutron scattering to explore the diffusivity of atoms, for instance, in liquids and polymers.

8. The intensity distribution for Bragg reflections has the form: $\frac{\sin^2 N_a \mathbf{Q} \cdot \mathbf{a}/2}{\sin^2 \mathbf{Q} \cdot \mathbf{a}/2}$ (here for the a-direction, similar for the b, c-directions).

9. The intensity distribution quoted under point 7 has a main maximum for $Q = 2\pi h/a$, and $N_a - 2$ side maxima separated by $N_a - 1$ side minima.

10. The maximum intensity scales with the number of unit cells to the square: $(N_a)^2$, the width of the peak scales with $1/N_a$.

11. In the limit $\displaystyle\lim_{N_a \to \infty} \frac{\sin^2 N_a \mathbf{Q} \cdot \mathbf{a}/2}{\sin^2 \mathbf{Q} \cdot \mathbf{a}/2} = N_a^2 |A| \delta(\mathbf{Q} - \mathbf{G}_h)$, the scattering intensity approaches a Dirac δ-function at the position in reciprocal space that fulfills the Bragg condition.

12. The total integrated intensity of Bragg reflections is expressed by
$$I_{tot}(\mathbf{Q}) = \left(\frac{a}{r}\right)^2 N V_{rc}\, F^2(\mathbf{Q} = \mathbf{G}_{hkl}) \delta(\mathbf{Q} - \mathbf{G}_{hkl}).$$

13. The total integrated intensity is proportional to the number of unit cells N in the crystal lattice.

14. The structure factor evaluates all phase factors due to scattering at atoms belonging to the basis of a unit cell. The structure factor is specific to particular crystal structures.

15. The thermal Debye–Waller factor takes into account the thermal vibration of atoms on their lattice sites, which leads to an exponential damping of the Bragg intensity with increasing temperature.

16. As a function of temperature, three effects are noticed by recording Bragg reflections: (1) Thermal expansion of the lattice leads to a contraction of the reciprocal space; (2) Thermal motion of atoms causes an exponential damping of the Bragg peak intensity; (3) Lattice vibrations give rise to thermal diffuse scattering at Bragg reflections and in between.

17. The kinematic approximation to treating scattering from crystal lattices is based on several assumptions. The most important one is that the interaction between the wave and target should be relatively weak, so that multiple scattering of the same particle within the target can be neglected.

Questions

(Note that more than one answer may be correct)

1. **What does the Bragg equation express?**

   (a) The Bragg equation expresses the condition for constructive interference of X-ray waves by considering the difference in path length at different scattering centers.

   (b) The Bragg equation yields the intensity of Bragg peaks.

   (c) The Bragg equations tell us the range of high-energy charged particles in matter.

2. **What are the assumptions for the validity of the Bragg equation?**

   (a) The Bragg equation holds only for charged particles.
   (b) The Bragg equation requires resonant X-ray absorption at the K-edge of the investigated crystal material.
   (c) The Bragg reflection requires an elastic scattering event.

3. **What do you understand by the lattice sum?**

   (a) The lattice sum is the sum of amplitudes from all scattering centers in the crystal lattice.
   (b) The lattice sum is the sum of all phase factors in a crystal lattice.
   (c) The lattice sum is the sum of all scattering amplitudes from scattering centers within a unit cell.

4. **What does the "structure factor" tell us?**

   (a) The structure factor is the Fourier transform of density variations.
   (b) The structure factor tells the difference between lattice symmetries.
   (c) The structure factor is a constant prefactor, sometimes replacing the atomic form factor.
   (d) The structure factor distinguishes between forbidden and allowed reflections.

5. **Under what circumstances can Laue oscillations be observed?**

   (a) Laue oscillations can only be observed for nanosized powder samples.
   (b) Laue oscillations can be observed when studying epitaxial thin films grown on single-crystalline substrates and by scattering in the direction normal to the film.
   (c) Laue oscillations can only be seen by scattering from polymer films.

6. **What does the term "kinematic approximation" imply?**

   (a) Kinematic approximation neglects the attenuation of the incident beam by scattering.
   (b) Kinematic approximation requires single-event scattering of wavetrains.
   (c) Kinematic approximation is fulfilled whenever the interaction of the beam with the target atoms is very strong.

Exercises

Grades of difficulty: E = easy, M = medium, A = advanced.

E 3.1. Structure factor for NaCl

Consider the NaCl crystal structure, which is discussed in Chap. 1. The basis vectors for the NaCl structure can also be found in Chap. 1.

   (a) Calculate the structure factor for the NaCl crystal structure.
   (b) What are the conditions for allowed and forbidden Bragg reflections?

M 3.2. Structure factor for the diamond lattice
Calculate the structure factor for a diamond lattice. Which are the allowed and which
are the forbidden Bragg reflections? The basis vectors for the diamond lattices can
be found in Chap. 1.

M 3.3. Structure factor for the hcp lattice
Determine the general structure factor for the hcp structure and find the conditions
for allowed and forbidden Bragg reflections.

M 3.4. Errors contributing to lattice parameter determination
The precision of measuring a lattice parameter depends mainly on three factors:
(1) scattering angle θ; (2) wavelength dispersion (monochromaticity) $\Delta\lambda$; and
(3) beam divergence $\Delta\theta$. All three contributions should be optimized to arrive at a
minimal relative error for a lattice parameter measurement, i.e., a minimal ratio of
$\Delta a/a = \Delta d/d$.

(a) Determine $\Delta d/d$ as a function of the scattering angle θ and beam divergence
$\Delta\theta$. Hint: take the first derivative of the d-spacing concerning the scattering
angle θ using the Bragg equation $2d \sin\theta = \lambda$.

(b) Now, consider the effect of wavelength dispersion. Hint: Using the same
expression for the d-spacing, take the first derivative with respect to the
wavelength λ.

(c) What is the combined error with respect to the beam divergence $\Delta\theta$ and
wavelength spread $\Delta\lambda$. Hint: Use the equation for error propagation.

(d) Discuss the result and make a recommendation on how to proceed to measure
a lattice parameter with maximum precision.

A 3.5. Debye temperature
The graph below (Fig. 3.14) shows a semilog plot of $(h00)$ Bragg reflection
intensities as a function of temperature in the range of 100 to 300 K for an Al
single crystal. The data were taken by Nicklow and Young using X-ray MoK_α
characteristic radiation (Nicklow and Young 1966). Determine the Debye Temperature θ_D from the slope of the intensity curves. Proceed as follows.

(a) Approximate the slope of the semilog plot by a constant exponential $2W$, and
I_0 is taken as a constant prefactor:

$$I(Q,T) = I_0 e^{-2W}.$$

It is useful to consider the following relationship for the Debye–Waller
factor in the high-temperature approximation for the $(h00)$ reflections:

$$2W = \frac{3\hbar^2}{m_{\mathrm{Al}} k_B} \frac{T}{\theta_D^2} G_{h00}^2.$$

Here, one should not mix the Planck constant and the Miller index.

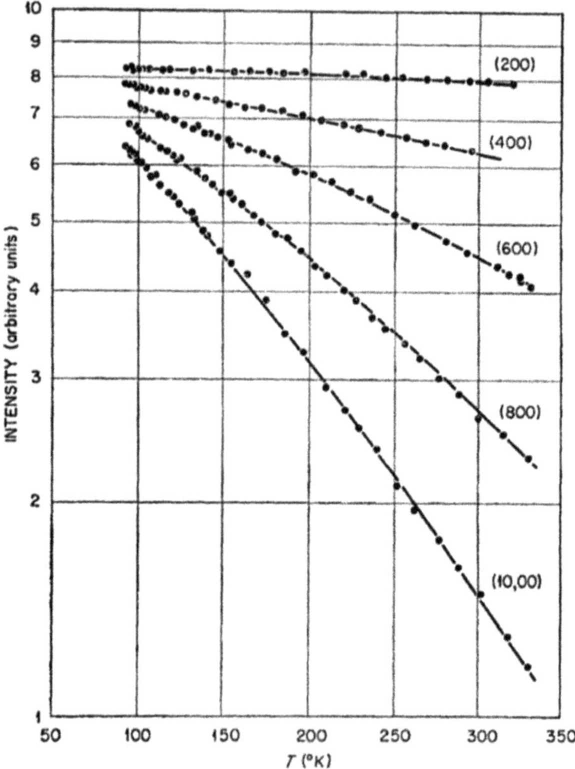

Fig. 3.14 Plot of (h00) Bragg reflection intensities as a function of temperature in the range of 100–330 K for an Al single crystal. The data are reproduced with permission from Nicklow and Young (1966): R.M. Nicklow and R.A. Young, Phys. Rev. 152, 591 (1966). DOI: https://doi.org/10.1103/PhysRev.152.591. Copyright (2025) by the American Physical Society

$a_{\mathrm{Al}} = 0.405$ nm is the lattice parameter of aluminum, and m_{Al} is the mass of aluminum. k_B is the Boltzmann constant, $\hbar$ is the Planck constant, and θ_D is the Debye temperature. Express the Debye–Waller factor explicitly for the (h00) reflections and write it in terms of $2W = Ah^2T$, where A is the slope for a fixed Bragg reflection with the Miller index h.

(b) Next, evaluate the slope A for each Bragg reflection and take the average of all slopes, neglecting the slope for the (200) reflection since it is too difficult to determine from the graph.

(c) From the average slope A, determine the Debye temperature.

(d) Compare your result with the literature value of θ_D for Al and discuss whether the use of the high-temperature approximation is justified after all.

(e) Discuss briefly the significance of the Debye temperature for the thermal properties of solids. Compare Debye temperatures of soft and hard materials.

References

M. Holt, Z. Wu, P. Hawoong Hong, P. Zschack, J.T. Jemian, H. Chen, T.-C. Chiang, Determination of phonon dispersions from X-ray transmission scattering: the example of silicon. Phys. Rev. Lett. **83**, 3317 (1999)

R.M. Nicklow, R.A. Young, Lattice vibrations in aluminum and the temperature dependence of X-ray Bragg intensities. Phys. Rev. **152**, 591 (1966)

R. Penrose, The role of aesthetics in pure and applied mathematical research. Bull. Inst. Math. Appl. **10**, 266ff (1974)

D. Shechtman, I. Blech, D. Gratias, J.W. Cahn, Metallic phase with long-range orientational order and no translational symmetry. Phys. Rev. Lett. **53**, 1951 (1984)

Further Reading

J. Als-Nielsen, D. McMorrow, *Elements of Modern X-Ray Physics*, 2nd edn. (Wiley, New York, 2011)

H. Ibach, H. Lüth, *Solid-State Physics, An Introduction to Principles of Materials Science* (Springer, Berlin, 2009)

C. Janot, *Quasicrystals: A Primer (Monographs on the Physics and Chemistry of Materials)* (Oxford Science Publications, Oxford, 1997)

C. Kittel, *Introduction to Solid State Physics*, 8th edn. (Wiley & Sons, New York, 2004)

W. Langel, Introduction to neutron scattering. ChemTexts **9**, 12 (2023). https://doi.org/10.1007/s40828-023-00184-7

G.L. Squire, *Thermal Neutron Scattering*, 3rd edn. (Cambridge University Press, Cambridge, 2012) and Dover edition, 1997

Chapter 4
Sources of Radiation for Scattering Experiments

Which radiation to be used? For investigating crystal structures, we need either electromagnetic waves or particle waves with a wavelength comparable to typical atomic distances in crystals (0.1–1 nm) and waves that penetrate sufficiently deep into the material without much absorption or damage.

The first condition is met at different energies for X-rays, neutrons, and electrons. For X-ray scattering, we need photons with an energy of about 10 keV; neutron scattering requires thermal neutron energies of about 25 meV; and electron scattering is best performed for electron energies of about 20–200 eV. Table 4.1 provides an overview of the conversions from energies to wavelengths for X-rays, neutrons, and electrons.

Figure 4.1 shows graphically, in a double logarithmic presentation, the energy–wavelength relationship for X-rays, neutrons, and electrons. Note that the energies on the abscissa are different for the three particles. The slopes for the mass particles are the same but different from those of the massless photon. The wavelength–energy relation for neutrons is also explicated in Appendix A6.

The second condition noted above is the penetration depth. The penetration of X-rays in matter depends on the photon energy and the atomic number of the material, but is typically a few micrometers, sufficient for obtaining sharp Bragg reflections. Neutrons are only weakly absorbed by matter. They can penetrate many millimeters to centimeters or even more. However, charged particles such as electrons, protons, or α particles barely penetrate matter at all. α particles and protons, in the energy range needed for scattering experiments, have no penetration of any significance. Electrons with an energy of 100 eV penetrate about 1–2 nm and are therefore very surface sensitive. One may also use neutral hydrogen or He-atoms for scattering experiments. But again, these particles do not penetrate the crystal lattice. They can only be used for studying surface structures and surface phase transitions. Furthermore, these particle beams are by far too elaborate to prepare. So, only X-rays, neutrons, and electrons are suitable for scattering experiments, and those are the particles considered in this text.

© The Author(s), under exclusive license to Springer Nature Switzerland AG 2026

H. Zabel, *Elements of Elastic Scattering by X-Rays, Neutrons, and Electrons*,

https://doi.org/10.1007/978-3-032-16624-1_4

Table 4.1 Conversion of energies into wavelengths for photons, neutrons, and electrons. Here, h is the Planck constant, c is the light velocity, m_n is the neutron rest mass, m_e is the electron rest mass, and λ are the respective wavelengths

Source	Wavelength: general equation	Practical conversions between wavelengths and energies
X-rays	$\lambda = \dfrac{hc}{E_{\text{photon}}}$	$\lambda[\text{nm}] = \dfrac{1240 \text{ eV nm}}{E_{\text{photon}}[\text{eV}]}$
Neutrons	$\lambda = \dfrac{h}{\sqrt{2m_n E_{\text{kin}}}}$	$\lambda[\text{nm}] = \dfrac{0.0286 \text{ nm } [\text{eV}]^{1/2}}{\sqrt{E_{\text{neutron}}[\text{eV}]}}$
Electrons	$\lambda = \dfrac{h}{\sqrt{2m_e E_{\text{kin}}}}$	$\lambda[\text{nm}] = \dfrac{1.2 \text{ nm } [\text{eV}]^{1/2}}{\sqrt{E_{\text{electron}}[\text{eV}]}}$

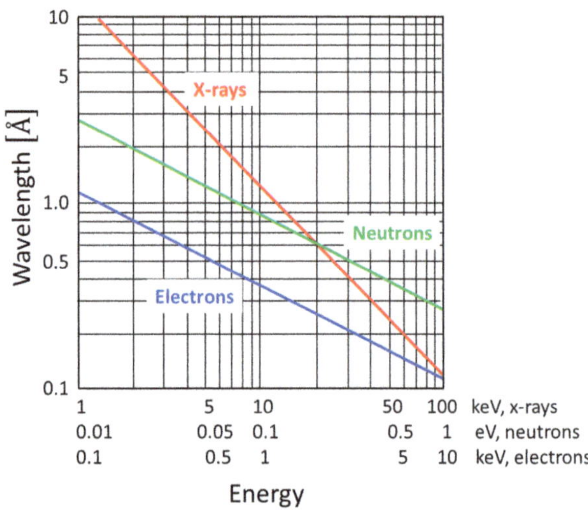

Fig. 4.1 Double logarithmic plot of the energy–wavelength relationship for X-rays, neutrons, and electrons. Note the different energy scales on the abscissa

4.1   X-Ray Sources and Characteristics

For X-ray scattering, photons are generated either by X-ray tubes, by electron synchrotrons, or by free electron lasers. Here, we will give a brief overview of these three radiation sources.

4.1.1   *Laboratory X-Ray Generation*

Standard X-ray tubes, frequently employed in research laboratories, use a high voltage difference between the cathode and anode for accelerating free electrons over a short distance in a vacuum tube. The cathode and anode are concealed in a permanently evacuated glass tube, like traditional light bulbs. Figure 4.2 shows schematically the main features of such a tube. The cathode is connected to a high negative voltage supply between -10 and -100 kV, and the anode is grounded. A

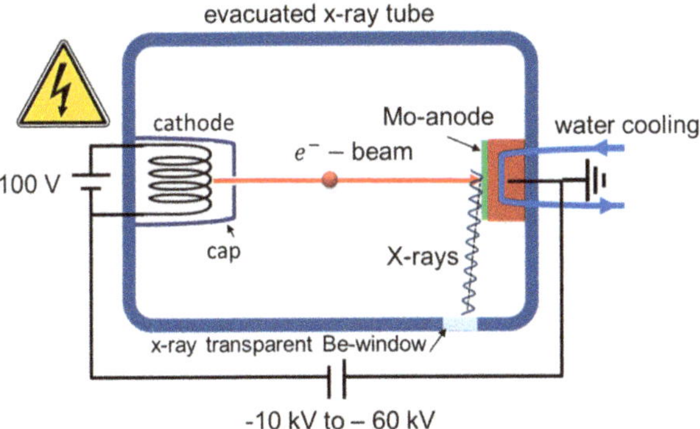

Fig. 4.2 Schematics of an X-ray tube. The cathode is connected to a high-voltage source, and the anode is grounded and water-cooled. An additional circuit supplies the cathode current for heating the filament. Electrons emitted from the cathode are accelerated toward the anode and abruptly stopped in the anode material. X-rays are generated by both the de-acceleration of electrons in the anode material and the photo-excitation of atoms

tungsten filament in the cathode is heated to very high temperatures just below the melting temperature of the wire to generate free electrons via thermionic emission. The current that goes through the filament of the cathode i_{cathode} controls the filament temperature and consequently also the rate of electrons emitted into the vacuum tube. Electrons entering at the high negative potential into the vacuum are immediately accelerated toward the anode, constituting the anode current i_{anode}. A cap with a small aperture surrounding the filament, called a Wehnelt cylinder, acts as an electrostatic lens that keeps the electrons from straying away. There is a close connection between the cathode current heating the filament i_{cathode}, and the anode current i_{anode} hitting the target.

With increasing i_{cathode}, the cathode temperature increases, and electrons start being emitted from the wire into the vacuum of the X-ray tube. This process is referred to as *thermo-ionic emission* and is expressed by *Richardson's law*:

$$i_{\text{cathode}} = AT^2 \exp(- W/k_B T), \qquad (4.1)$$

where A is a constant, T is the absolute temperature, and W is the *work function*, acting as a potential well for electrons escaping from the metal host into the vacuum. By application of the electrostatic potential difference eV_{acc} between the anode and cathode, the work function W becomes effectively lowered and increases the anode current. Electron emission that is controlled by the temperature T and electrostatic potential eV_{acc} is known as **Schottky emission** and expressed by the modified Richardson law:

$$i_{\text{cathode}} = AT^2 \exp(- (W - eV_{\text{acc}})/k_B T). \qquad (4.2)$$

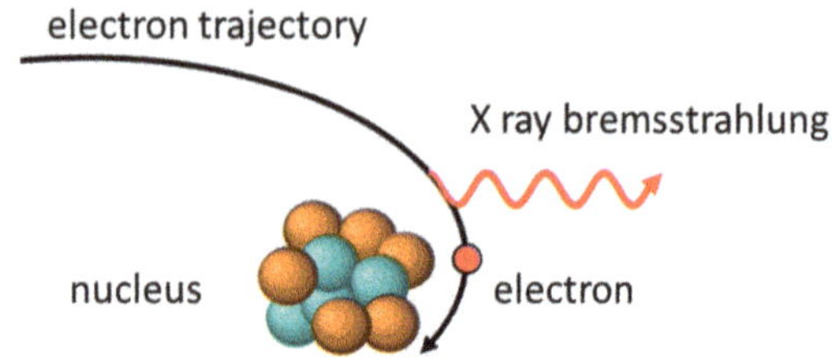

Fig. 4.3 Generation of X-ray bremsstrahlung in the Coulomb potential of target nuclei

The anode (target) consists of a water-cooled Cu block, which may be coated with a film of a different metal. In the target material, the free electrons are rapidly retarded by being trapped in the Coulomb potential of the target nuclei (see Fig. 4.3). While circulating about target nuclei, the electrons emit electromagnetic dipole radiation, called ***bremsstrahlung***. They may emit just one photon, carrying away the total kinetic energy of the electron, or they may split their energy into several photons with a distribution of energies among these photons.

4.1.2 Bremsstrahlung

The de-acceleration of high-energy electrons in the anode causes the emission of X-ray photons with a very broad spectral distribution of energies or wavelengths. This radiation is called ***bremsstrahlung.*** The German expression, common in the literature, can be translated as "de-acceleration radiation." Bremsstrahlung is polychromatic, lacks phase coherence, and is divergent. Such a radiation source is known as an *incoherent broadband photon source*, sometimes also referred to as a "white X-ray source" in analogy to traditional light bulbs. The total integrated intensity over all energies I_{x-ray} is proportional to the product of the anode current i_{anode}, the square of the voltage difference V_{acc} between cathode and anode (***accelerating voltage***), and the atomic number Z of the target material:

$$I_{x-ray} \sim Z i_{anode} V_{acc}^2. \tag{4.3}$$

The intensity of the bremsstrahlung as a function of wavelength and for different accelerating voltages V_{acc} is plotted in Fig. 4.4. This bremsstrahlung spectrum is characterized by an increasing integrated intensity with increasing V_{acc}, a maximum intensity that shifts to shorter wavelengths with increasing V_{acc}, and a sharp drop of the intensity at the ***cut-off wavelength*** λ_{min}:

$$E_{max} = \frac{hc}{\lambda_{min}} = hf_{max} = \hbar\omega_{max} = eV_{acc}. \tag{4.4}$$

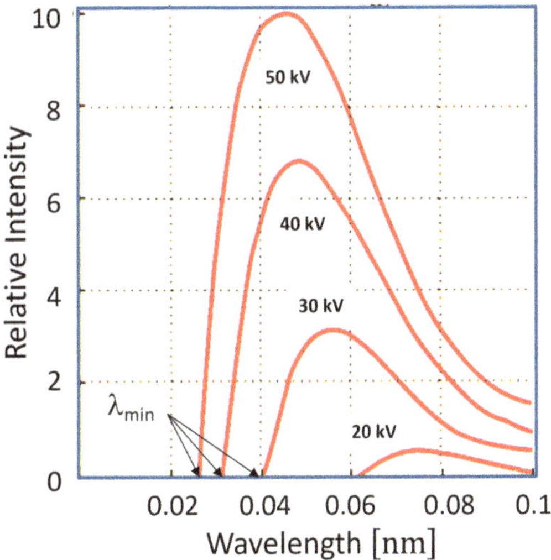

Fig. 4.4 X-ray bremsstrahlung spectrum for different accelerating voltages from 20 to 50 kV. (Reproduced from X-ray Data Booklet n.d.: X-ray Data Booklet @: http://xdb.lbl.gov/)

The cut-off wavelength, respectively, cut-off energy, is defined by the maximum energy that a photon can attain when the complete kinetic energy of the accelerated electron is converted into electromagnetic wave energy of just a single photon.

When operating an X-ray machine, the user has control over the tube voltage, the tube current, and the exposure time. Typical tube voltages for standard sealed-off X-ray tubes are 20–40 kV, anode currents of 10–50 mA, and unlimited exposure times. The standard power of sealed X-ray tubes is 2–3 kW. Standard rotating X-ray tubes (see below) can be operated at higher powers of about 12–18 kW.[1]

4.1.3   Characteristic Radiation

When electrons strike a target, a second process of X-ray generation takes place: Simultaneously with bremsstrahlung, ***characteristic radiation*** is generated, i.e., characteristic for the target material used. The incoming electron may collide with an electron of the inner atomic shell, such as the K-shell of the target material. If the energy of the incoming electron E_{in} is higher than the binding energy of the core–shell electron E_K, it may kick out the core electron leaving a hole in the K atomic shell, as depicted in Fig. 4.5. This *core hole* is sequentially filled with an electron from the next higher shell, and so on until, after a cascade of transitions, the hit atom returns to the ground state. When an electron in an atom makes a transition from a higher to a lower energy shell, it emits the difference in binding energy in terms of

---

[1] Higher power X-ray generartors up to 100 kW have been fabricated in the past. However, they turned out to be not sufficiently reliable and have been replaced by X-rays from electron synchrotrons.

Fig. 4.5 A highly energetic incident electron arriving from the cathode kicks out a core electron in a target (anode) atom, creating a hole in the K shell. An electron from the L shells fills the hole and simultaneously emits an X-ray photon

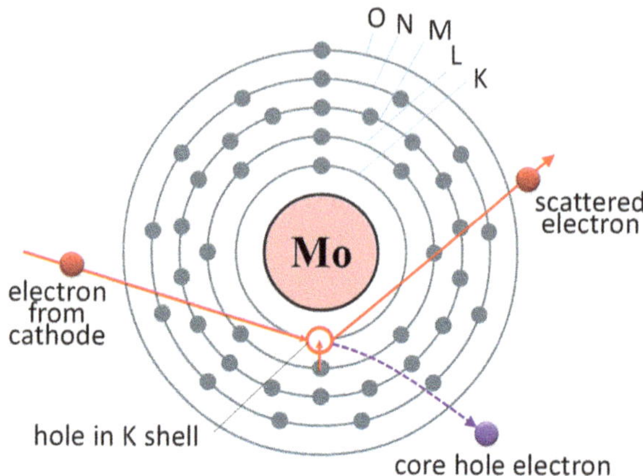

electromagnetic waves, i.e., via dipole radiation. This radiation is called ***characteristic X-ray radiation***. Characteristic X-ray radiation is clearly distinguished from bremsstrahlung radiation by its sharp spectral line.

When the hole in the K-shell is filled with an electron from the L-shell, its extra energy is emitted as a photon with energy:

$$E_K^{\text{photon}} = hf_{K\alpha} = E_K - E_L. \tag{4.5}$$

Here E_K and E_L are the binding energies of electrons in the K and L shells, respectively. The characteristic X-ray radiation for this transition is called K_α - *radiation*.

The described scenario is a simplified representation of the actual process because it does not take into account the dipole ***selection rules*** and the fine structure of the L shell. Due to spin-orbit coupling, the L shell splits up into three sublevels: $L_1 = 2s_{1/2}$, $L_2 = 2p_{1/2}$, and $L_3 = 2p_{3/2}$. The selection rule for dipole transitions requires that the change of the orbital angular momentum Δl must be ± 1 and that the spin angular momentum change Δs must be 0. This excludes a dipole transition from L_1 to K_1. However, the other two transitions $L_2 \rightarrow K_1$ and $L_3 \rightarrow K_1$ are dipole allowed. Their spectroscopic terms are $K_{\alpha1}$ for the $L_3 \rightarrow K_1$ transition and $K_{\alpha2}$ for the $L_2 \rightarrow K_1$ transition. There are also transitions from the M and N-shells to the K-shell, labeled $K_{\beta1}$ and $K_{\beta2}$ etc. A generic ***energy level scheme*** of allowed X-ray emission lines is shown in Fig. 4.6. It is clear that while the general transition process is the same for all atoms, the transition energies are specific for the atom in question. Furthermore, as the transitions are within inner core levels, which do not take part in chemical bonding, the X-ray energies are fingerprints for the atoms independent of their local chemical environment. This is at least true for heavy atoms. In light atoms, energy shifts of core electrons due to the chemical environment can be observed with high-energy resolution X-ray spectroscopy. The respective spectroscopy is termed *X-ray photoelectron spectroscopy* (XPS), or with a focus on light atoms, *electron spectroscopy for chemical analysis* (ESCA).

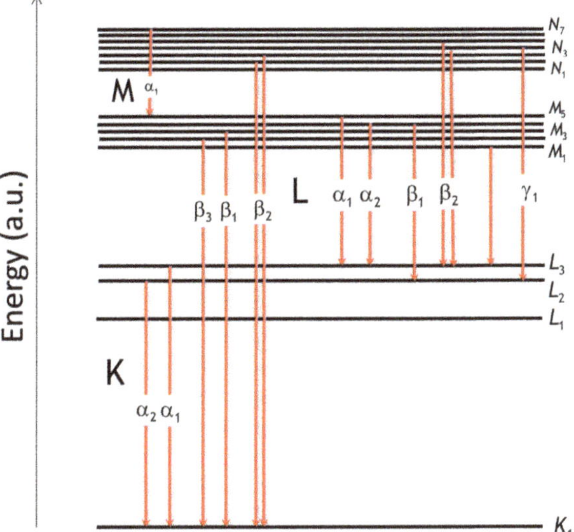

Fig. 4.6 Allowed dipole transitions for characteristic X-ray radiation. (Adapted from the X-ray booklet n.d.: X-ray Data Booklet @: http://xdb.lbl.gov/)

An alternative radiationless process may occur, known as the Auger effect, where the $1s$ hole is filled with a $2s$ electron (L_1 transition), and the transition energy is imparted to the $2p$ electron that escapes to the vacuum. Other electron transitions from higher shells are also possible. The Auger effect is important for light atoms, less so with increasing Z. The Auger effect plays an important role in resonant magnetic scattering for determining the total electron yield at absorption edges, which is further explained in Sect. 7.7.

Figure 4.7 shows the combined X-ray emission spectrum for an Mo target. It consists of the continuous bremsstrahlung spectrum, which is cut off at the minimum wavelength $\lambda_{\min} = 0.35$ nm, corresponding to the maximum energy transfer at 35.4 kV. Superimposed on the bremsstrahlung spectrum are sharp intensity lines due to characteristic $K_{\alpha 1}$ (17.479 keV, $\lambda = 0.0709$ nm $= 70.9$ pm), $K_{\alpha 2}$ (17.374 keV $= 0.0713$ nm $= 71.37$ pm), and K_β (19.61 keV, $\lambda = 0.0632$ nm $= 63.2$ pm) X-ray transitions. The intensity of these lines is much higher than the bremsstrahlung radiation over the same spectral range. Measured and calculated bremsstrahlung spectra can be found in X-ray Data Booklet (n.d.). Some selected data for absorption edges and characteristic emission lines are listed in Table 4.2. Note that the intensity of the K_β spectral line is much weaker than for the K_α lines.

Neglecting the spectroscopic splitting of the K, L, M, etc. spectral lines, the X-ray photon energy can be estimated in analogy to the *Bohr model* for the energy levels in hydrogen atoms. However, one has to take into account that after creating a hole in the K-shell by electron impact, the total charge of the remaining electrons is $Z - S$, where Z is the atomic number, and S is a screening factor. With this modification, the empirical ***Moseley's law*** for the photon energy from the K-shell reads:

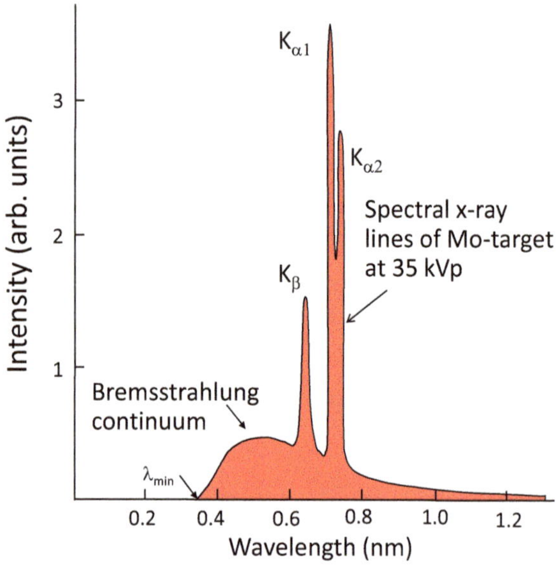

Fig. 4.7 Typical X-ray spectrum for an Mo target. The characteristic X-ray radiation is superposed on the bremsstrahlung spectrum. (Adapted from http://hyperphysics.phy-astr.gsu.edu/hbase/quantum/xrayc.html)

Table 4.2 X-ray absorption edges and characteristic emission lines for some selected elements. Data from http://skuld.bmsc.washington.edu/scatter/AS_periodic.html and from the X-ray data booklet posted at http://xdb.lbl.gov/

Element	K_α absorption edge (keV)	Emission energies (keV)		
		$K_{\alpha 1}$	$K_{\alpha 2}$	K_β
Ni	8.3328	7.478	7.460	8.264
Cu	8.9789	8.047	8.027	8.905
Mo	19.9995	17.479	17.374	19.608
Ag	25.514	22.162	21.990	24.942
W	69.52	59.318	57.981	67.244

$$E_{i \to f} = h f_{i \to f} = R_H (Z - S)^2 \left(\frac{1}{n_f^2} - \frac{1}{n_i^2} \right). \tag{4.6}$$

Here n_i, n_f are the main quantum numbers of the atomic shells for the initial and final state of the electron transition, and R_H ($= 13.6$ eV) is the Rydberg constant of the hydrogen atom. For the K_α transition, the screening factor can be approximated by $S = 1$ and this equation becomes

$$E_{K_\alpha} = h f_{K_\alpha} = R_H (Z - 1)^2 \left(\frac{1}{1^2} - \frac{1}{2^2} \right) = R_H (Z - 1)^2 \frac{3}{4}. \tag{4.7}$$

For the element Mo with $Z = 42$, the estimated K_α X-ray energy is 17.146 keV, which is close to the observed energy of 17.479 keV. Further characteristic energies of X-ray absorption edges and emission lines for targets frequently used in laboratories are listed in Table 4.2.

4.1.4   Sealed Tubes and Rotating Anodes

We already discussed the principal features of sealed-off X-ray tubes; their basic design is shown in Fig. 4.1. The anode consists of a water-cooled Cu block, which can be used for excitation of characteristic Cu-radiation. Otherwise, the Cu anode is coated by another metal sheet to utilize, for instance, Mo or Ag characteristic radiation. If there is only interest in the Bremsstrahlung spectrum but not in the characteristic radiation, then the Cu block can be covered by a thin tungsten (W) film. The characteristic lines of W are in the range of 60 kV, i.e., beyond usual characteristic excitation energies. The bremsstrahlung spectrum can then be used, for example, for energy-dispersive diffraction methods, which are discussed in Sect. 6.4.

Figure 4.8a shows a modern commercial *vacuum-sealed X-ray tube* for use in physics/chemistry/geo-science laboratories. Unlike the basic design shown in Fig. 4.1, it has a flat anode. Only for medical applications, the anode has a wedge shape. In the standard tube, the maximum intensity occurs under an angle of 6° to the surface of the anode. This radiation can penetrate through four beryllium-covered windows, two vertical and two horizontal windows, which are essentially X-ray transparent due to the low atomic number of Be ($Z_{Be} = 4$). The tube is inserted into a tube support that also contains electronically controlled lead shutters for each Be-window. Only those lead shutters are opened that are used for scattering experiments. All others remain closed.

The use of standard vacuum-sealed X-ray tubes is limited by the cooling power of the anode. X-ray generation, as discussed so far, is a very inefficient process.

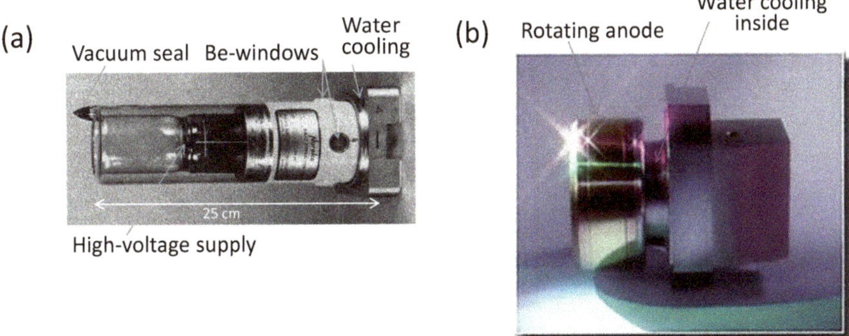

Fig. 4.8 (a) Vacuum-sealed standard laboratory X-ray tube. (b) Disassembled anode of a rotating anode X-ray machine

Approximately 99% of the energy is converted into heat that must be dissipated, while only about 1% is converted into photons. To increase the X-ray intensity, it is not sufficient to just increase the anode current. This would ultimately lead to a meltdown of the anode material. However, the power/intensity can be increased by rotating the anode, thereby spreading the hot spot on the anode over a larger area. The vacuum of rotating anode X-ray machines is produced dynamically and continuously. All other features, such as radiation protection and operation of X-ray windows, are similar. A rotating anode, disassembled from an X-ray machine, is shown in Fig. 4.8b.

4.2   Electron Synchrotron Sources

The second type of machine producing X-rays is an ***electron synchrotron.*** First electrons are brought up to speed with the help of a linear accelerator (linac) and a booster ring. Once the electron energy is ramped up to the GeV range, the electrons are injected in time-separated bunches into an electron storage ring. A lattice of bending magnets keeps the electrons on track in a narrow toroidal tube, which is evacuated to pressures of less than 10^{-9} mbar. Ultrahigh vacuum conditions are required to avoid the collision of electrons with remaining gas atoms, which shortens the storage lifetime. A schematic outline of an electron synchrotron is shown in the inset of Fig. 4.10 and in Fig. 4.11. While electrons in a circular orbit are constantly accelerated toward the center (radial acceleration), they emit electromagnetic radiation like an antenna. At low nonrelativistic energies, the radiation profile has a typical dipole distribution shown in the left panel of Fig. 4.9. However, in storage rings, electrons have a speed close to the speed of light, i.e., they are highly

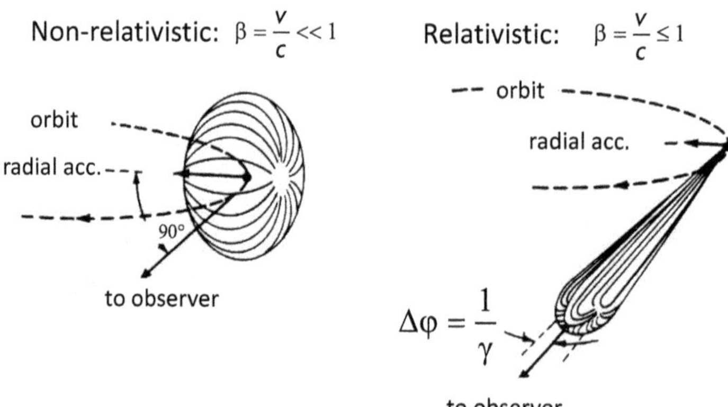

Fig. 4.9 Radiation profile for a classical electron in a circular orbit (left panel) and for a relativistic electron in a synchrotron storage ring (right panel). (Reproduced and adapted with permission from Balerna and Mobilio 2015: A. Balerna, S. Mobilio, Introduction to Synchrotron Radiation, Copyright (2025) Springer Nature Publishing)

relativistic. Then, for an observer in the laboratory frame, the radiation lobes are Doppler shifted, radiating into a narrow radiation cone with an opening angle $\Delta\varphi = 1/\gamma$ in the direction of the traveling electrons, as sketched in the right panel of Fig. 4.9. Since $\gamma = E_{kin}/m_0c^2$, where E_{kin} and m_0c^2 are the kinetic energy and rest energy of the orbiting electron, respectively, and assuming for E_{kin} a typical energy of 6 GeV in a storage ring,[2] we find for the opening angle: $\Delta\varphi = m_0c^2/E_{kin}$=0.5 MeV/1 GeV $= 0.5$ mrad $= 0.03°$. In this narrow radiation cone, the intensity (peak brightness) is extremely high, ten orders of magnitude higher than that of conventional X-ray tubes (see the comparison in Fig. 4.20).

4.2.1   Energy Spectrum

The energy spectrum of synchrotron radiation is continuous and very broad, similar to a bremsstrahlung spectrum, with a cut-off that depends on E_{kin} and the radius R of the storage ring. The radius R of the orbit and the magnetic field B of the bending magnets are related (in practical units) according to Balerna and Mobilio (2015):

$$R[\text{m}] = 3.34 \, \frac{E_{kin}[\text{GeV}]}{B[\text{T}]} . \tag{4.8}$$

In Fig. 4.10, a typical spectrum is shown for $E_{kin} = 1$ GeV and a magnetic field of $B = 1.2$ T. The maximum intensity is reached at the critical photon energy E_c, indicated by a blue arrow. E_c is related to the kinetic electron energy and the magnetic field of the bending magnets via

$$E_c \, [\text{keV}] = 0.665 \, E_{kin}^2[\text{GeV}] \, B[\text{T}]. \tag{4.9}$$

Inserting numbers from above, we find for $E_c = 0.8$ keV. The critical photon energy and the cut-off energy increase with the square of E_{kin}. The total intensity of the synchrotron beam is proportional to the ring current I_e and E_{kin}^2. In practical units, the photon flux $F_{photon}(E_c)$ at the critical energy E_c in a spectral bandwidth $\Delta E/E = 0.1\%$, emitted into a unit solid angle, is for a bending synchrotron ring (Balerna and Mobilio 2015):

$$F_{photon}^{bending} = 1.3 \times 10^{13} \times I_e[\text{A}] \times E_{kin}^2 \left[\text{GeV}^2\right]. \tag{4.10}$$

---

[2]Electron synchrotrons are also called electron storage rings because the kinetic energy of the electrons in the ring remains constant. There is radial, but not translational, acceleration. The energy loss of the electrons due to radiation is compensated by RF resonators placed within the ring.

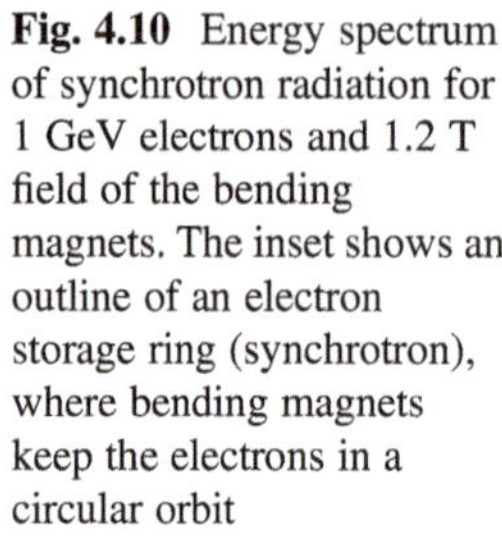
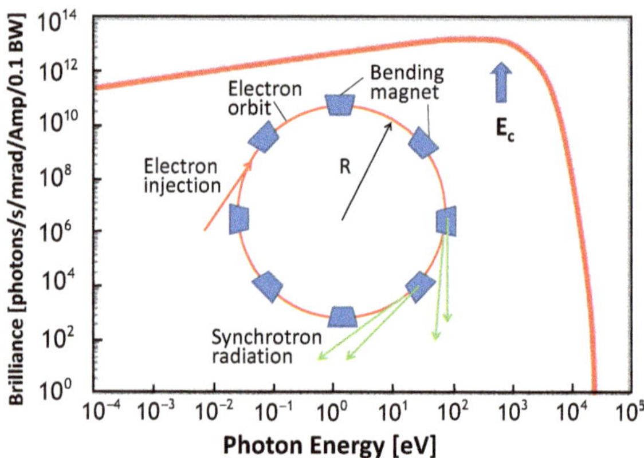

Fig. 4.10 Energy spectrum of synchrotron radiation for 1 GeV electrons and 1.2 T field of the bending magnets. The inset shows an outline of an electron storage ring (synchrotron), where bending magnets keep the electrons in a circular orbit

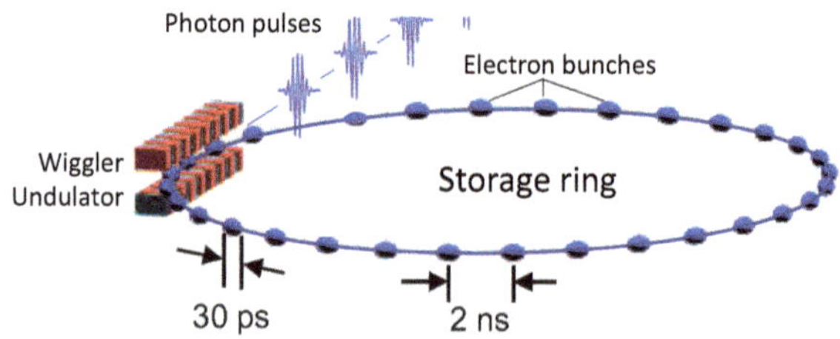

Fig. 4.11 Electron bunches in a synchrotron storage ring, here with a bunch length of 30 ps and a bunch separation of 2 ns. To the left is indicated an insertion device, such as a wiggler or an undulator. Synchrotron facilities usually feature a number of those devices at straight sections of the storage ring, unlike the one shown in the schematics

4.2.2   Electron Bunches

Figure 4.11 shows schematically the electron bunches in a storage ring. A typical arrangement is a bunch length of a few picoseconds and a bunch separation of a few nanoseconds. The number of equally spaced bunches in a ring is arbitrary. In a multi-bunch operation mode, 100–500 pulses per circulation are delivered with a ring current of about 200–400 mA, depending on the circumference. There are various kinds of bunch deliveries: single bunch, multiple bunch, or hybrid operations with a mix of single bunch in one half of the ring and a multibunch in the other half. The actual operation mode is contingent on the user's demand concerning intensity versus time resolution. In the past, the use of synchrotron radiation was limited by the lifetime of the ring current, which was typically a few hours before beam dump and refilling were required. Presently, most synchrotron facilities operate in a so-called *top-up* mode, meaning that the storage ring is frequently filled with electrons in phase with the traveling bunches, allowing a continuous operation without a beam dump. Also indicated in Fig. 4.11 is an insertion device, either a wiggler or an undulator, to be described next.

4.2.3 Insertion Devices

Insertion devices are used to enhance the intensity of the synchrotron radiation, increase the coherence length, and tune the polarization of the emitted light. In a wiggler (Fig. 4.12a), the electron bunches are wiggled in the horizontal plane by passing a double stack of alternating permanent dipole magnets, usually made of NdFeB and magnetized in the perpendicular direction, as seen in Fig. 4.11. The alternating magnets exert an in-plane Lorentz force on the electron's path, causing the electrons to wiggle. The extra acceleration results in a radiation cone that is elliptically shaped and broader than the one from a bending magnet by a factor K. The factor K depends on the gap between the top and bottom magnets and the magnetic induction B. There is no phase relationship among the photons emitted from a wiggler. The photon flux is enhanced in comparison to synchrotrons using only bending magnets by the number of poles in the wiggler: $F_{\text{wiggler}} \propto F_{\text{photon}}^{\text{bending}} \times N_{\text{poles}}$. Furthermore, the radiation from a wiggler is hardened, meaning that the critical energy E_c is shifted to higher values due to the presence of higher magnetic fields in the dipole array of a wiggler.

In undulators (Fig. 4.12b), in contrast, the wiggling amplitude is smaller (smaller K) and the radiation from the first bend is tuned to overlap with the wavefronts of the subsequent bends, resulting in a radiation enhancement by constructive interference. The enhancement factor is proportional to the narrowing of the radiation cone by the factor $1/\sqrt{N_{\text{pol}}}$, where N_{pol} is the number of dipoles in the undulator. The resulting

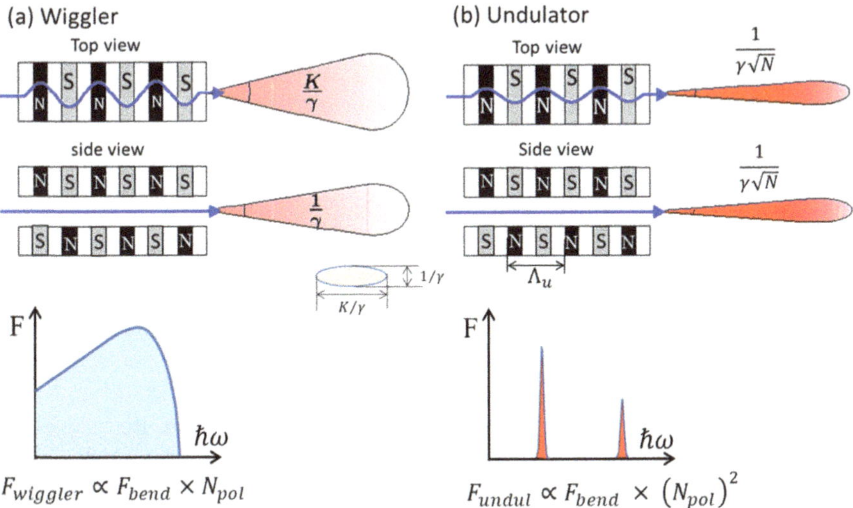

Fig. 4.12 Comparison of (**a**) wiggler and (**b**) undulator insertion devices. Wigglers produce an elliptically shaped radiation cone, which is horizontally broader by a factor K in comparison to the radiation cone of a bending magnet. Undulators exhibit a much narrower radiation cone and higher intensity due to constructive interference of waves generated by magnetic dipoles with a periodicity of Λ_u. The radiation spectra of wigglers and undulators are compared in the bottom panels

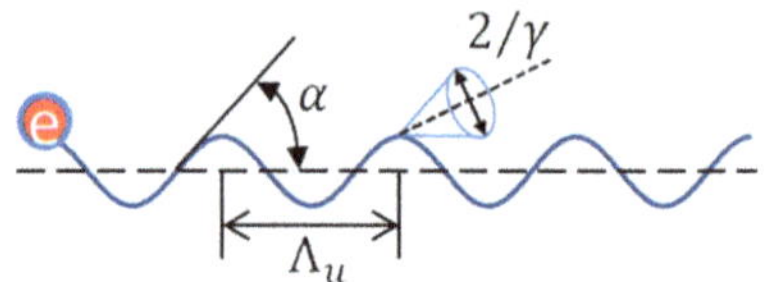

Fig. 4.13 Electron path in an undulator, with undulator periodicity λ_u, deflection angle a, and radiation opening angle $2/\gamma$

flux is proportional to $F_{\text{undulator}} \propto F_{\text{photon}}^{\text{bending}} \times N_{\text{pol}}^2$. Due to the interference condition, the radiation spectrum of an undulator is not broad like for a wiggler, but has sharp spectral maxima, characterized by a main maximum of the fundamental line and higher harmonics. The fundamental depends on the magnet periodicity and other factors that allow for changing and tuning the wavelengths radiated by the undulator. Additional details are discussed further below.

The approximate relationship between the undulator periodicity Λ_u and the emitted X-ray wavelength λ_x is (Balerna and Mobilio 2015; Als-Nielsen and McMorrow 2011):

$$\lambda_x(1, \theta = 0) = \frac{\Lambda_u}{2\gamma^2}\left(1 + \frac{1}{2}K^2\right). \tag{4.11}$$

Here, $\lambda_x(1, \theta = 0)$ is the first fundamental wavelength, $\theta = 0$ implies on-axis radiation, and $K = \alpha\gamma$ is a dimensionless parameter characterizing the deflection of the electrons in the undulator. The smaller the deflection angle α is, the more likely the constructive interference of wavelets from subsequent bends (see Fig. 4.13). Typical K-values for wigglers are 20, and for undulators, less than 1. In contrast to wiggler radiation, the radiation from undulators is nearly monochromatic (with higher harmonics), due to the interference condition expressed in Eq. (4.11). However, some tunability is nevertheless possible by changing the undulator gap, which changes the magnetic induction and hence the K-value.

Wiggler and undulator require a straight section in the synchrotron storage ring. Therefore, modern synchrotron facilities are not circular but feature several straight sections to accommodate insertion devices.

4.2.4 Polarization

There is an additional important property of **synchrotron radiation**. Because of the dipolar character of the orbiting electron bunches, the synchrotron radiation is *linearly polarized* in the plane of the storage ring. This has important consequences for scattering experiments at synchrotron sources. At the scattering angle $2\theta = 90°$, the intensity drops to zero in the horizontal scattering plane. However, in the vertical scattering plane, the intensity is constant. This feature becomes important for magnetic X-ray scattering. Slightly above or below the synchrotron plane, linear in-plane and perpendicular polarization overlap with a phase shift, producing X-ray

Fig. 4.14 The yellow area indicates the radiation cone at a synchrotron. In the center, the polarization of the photons is linear and horizontal. Above and below the horizontal plane, the radiation is partially right and left elliptically polarized

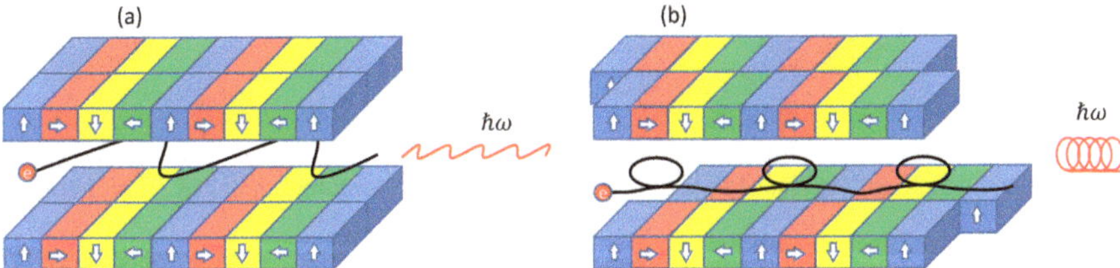

Fig. 4.15 Magnetic dipole arrangement in undulators, which allows for varying the X-ray polarization from linear to circular by shifting the magnetic dipoles against each other. The unshifted arrangement (**a**) produces linear polarization, and the shifted arrangement (**b**) delivers circular polarization

photons that are partially (up to 70%) right (above) and left (below) elliptically *polarized*, as indicated in Fig. 4.14. Therefore, synchrotron radiation sources not only offer much higher X-ray photon intensities compared to conventional X-ray sources but also deliver linearly as well as circularly polarized X-rays. The polarization effects are essential for studying magnetism with X-rays.

Alternatively, undulators can be constructed to generate circularly polarized X-rays, as shown schematically in Fig. 4.15. These undulators consist of four magnet arrays: top and bottom pairs, and side pairs. Linear polarization is achieved when the electron trajectory is wiggled in the orbital plane. However, circular polarization can also be generated by shifting the magnet pairs so that the electron trajectory describes a loop. With the same magnet configuration and the respective shifts, horizontal and vertical linear polarization, or any angle in between, can be selected. Left- and right-hand circular polarization, as well as any ellipticity, are also adjustable. The ability to change the polarization of X-rays is of utmost importance for the study of ferromagnetic, antiferromagnetic, and chiral magnetic systems.

4.2.5   Soft Versus Hard X-Rays

We can distinguish between two distinct scientific communities performing X-ray scattering experiments with synchrotron radiation (aside from spectroscopic studies). One group uses hard X-rays with photon energies of roughly 5–20 keV. The other group uses "soft" X-rays in the range of 0.5–1.5 keV. These two groups are

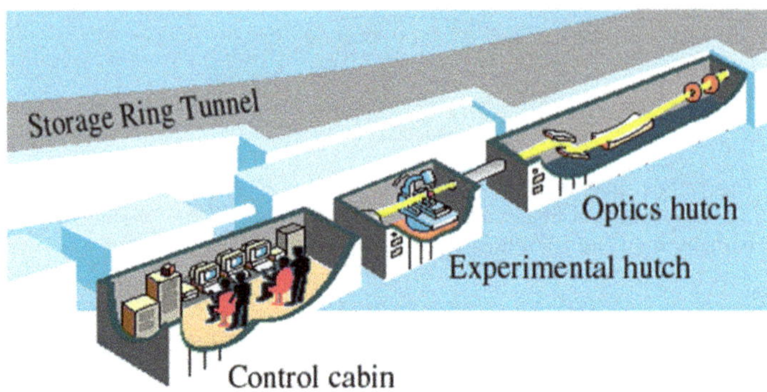

Fig. 4.16 Schematic setup of an operational unit at a synchrotron facility. The optics hutch and the experimental hutch are interlocked for radiation safety and are remotely operated from a control cabin

clearly distinct not only concerning their science focus but also concerning the experimental setup. Hard X-rays are not strongly absorbed in air, in contrast to soft X-rays. Therefore, users of hard X-rays work with an open beam that crosses a diffractometer or goniometer before being recorded in a detector. This open beam arrangement requires the entire experimental setup to be enclosed in a lead-lined hutch for radiation safety. An operational unit is schematically shown in Fig. 4.16, consisting of an optics hutch housing monochromators and focusing devices, an experimental hutch for goniometers and sample environment, and a control cabin for recording and preliminary data analysis. The optics and experimental hutch are interlocked, meaning that the beam shutter can only be opened when all personnel have left the hutch. Any operation inside is remotely controlled from the outside.

In contrast, users of soft X-rays must confine the X-ray beam in vacuum tubes from the storage ring to the experimental chamber. As a rule, the vacuum chamber of the experimental station provides sufficient radiation protection against soft X-rays so that no additional lead-lined hutch is required. To avoid contamination of the vacuum in the storage ring, it is recommended to separate the ring from the experimental station via differential pump units. A schematic of an experimental setup is shown in Fig. 4.17. The soft X-ray beam enters from the left, passing a differential pumping unit and a pinhole before entering the evacuated diffraction chamber. To change samples while maintaining high vacuum conditions, the sample is loaded into the main chamber via a loadlock system. Alternatively, the main chamber is vented after closing the upstream shutters.

4.2.6   Types of Synchrotron Radiation Facilities

Storage rings for hard X-ray science typically have electron energies of 6–8 GeV. Examples are the Advanced Photon Source (APS) at the Argonne National

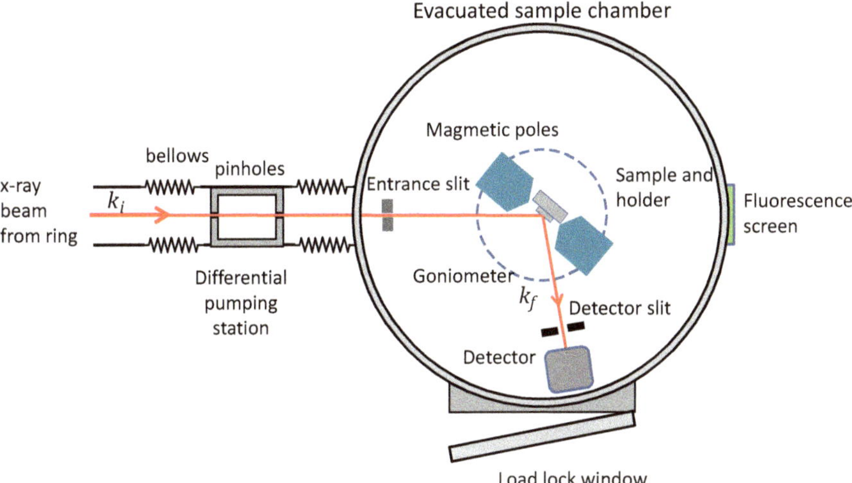

Fig. 4.17 Layout of a vacuum chamber allowing diffraction experiments with soft X-rays. The synchrotron beam enters from the left, passing a differential pumping stage. The sample sits in the center of a goniometer that is controlled from outside the vacuum chamber. Sample change is achieved via a load lock system that conserves the UHV conditions of the soft X-ray beam

Laboratory (USA), the European Synchrotron Radiation Facility (ESRF) in Grenoble, France, the Spring 8 Photon Factory in Japan, and PETRA III at the DESY in Germany. In contrast, for soft X-ray science, it is sufficient to operate a synchrotron source at lower energies of 1.2–1.8 GeV. Examples of soft X-ray facilities are the Advanced Light Source in Berkeley (ALS), USA, the BESSY II facility in Berlin, and the Elletra in Trieste, Italy. The newest synchrotron centers serve both the hard and the soft X-ray communities by making a compromise concerning the electron energy, favoring about 3 GeV machines. Examples are the Swiss Light Source in Villigen, the Diamond synchrotron source at the Rutherford Laboratory, UK, Soleil in Paris, France, and Alba in Barcelona, Spain. There are many more synchrotron facilities in the world. Table 4.2 gives an overview of a few synchrotron facilities. A complete and updated list can be found at https://en.wikipedia.org/wiki/List_of_synchrotron_radiation_facilities.

4.2.7 Brightness, Emittance, Diffraction Limit

Comparing the intensity or flux of synchrotron radiation facilities is usually done in terms of the source brightness. The brightness B, or more precisely the spectral brightness of a light source, is defined by the following expression:

$$B = \frac{\text{Photons/s}}{(\text{source size} \times \text{beam divergence})^2 \times 0.1\%\text{BW}}$$

$$= \frac{\text{Photons/s}}{\left(\sigma_e^r \sigma_e^a \otimes \sigma_{\text{ph}}^r \sigma_{\text{ph}}^a\right)^2 \times 0.1\%\text{BW}} = \frac{\text{Photons/s}}{\left(\varepsilon_e \otimes \varepsilon_{\text{ph}}\right)^2 \times 0.1\%\text{BW}}. \tag{4.12}$$

Here, we distinguish between the radial size of the electron source σ_e^r and the radial size of the photon source σ_{ph}^r. Furthermore, we distinguish between the angular divergence of the electron source and the photon source, σ_e^a and σ_{ph}^a, respectively. The units of $\sigma_{e,\text{ph}}^r$ are mm and the units of $\sigma_{e,\text{ph}}^a$ are mrad. The products $\varepsilon_e = \sigma_e^r \sigma_e^a$ and $\varepsilon_{\text{ph}} = \sigma_{\text{ph}}^r \sigma_{\text{ph}}^a$ are called the emittance of the electron and photon source. Both the electron emittance $\varepsilon_e = \sigma_e^r \sigma_e^a$ and the photon emittance $\varepsilon_{\text{ph}} = \sigma_{\text{ph}}^r \sigma_{\text{ph}}^a$ fill a phase space. The final result is obtained by a convolution between both phase spaces: $\varepsilon_e \otimes \varepsilon_{\text{ph}}$.

BW is the spectral bandwidth $\Delta\lambda/\lambda = \Delta\omega/\omega$. Here, we distinguish between the spectral bandwidth determined at the peak intensity and the spectral bandwidth on average. Often, the term brilliance is used in the literature instead of brightness. Usually, they mean the same. But caution is necessary. The best bet is to compare the units when comparing numbers.

We notice from Eq. (4.12) that the brightness of a photon source increases linearly with the photon flux and quadratically with the inverse emittance. Where is the upper limit of the brightness? The photon flux can be increased with the number of dipoles in the undulator and the overall undulator length L. However, for fixed length $L = N\Lambda_u$, the brightness depends inversely on the square of the emittance.

The electron source size and divergence are given by the accelerator properties and can be considered as fixed within the storage ring. Therefore, we focus on the photon beam divergence $\Delta\theta$, which is approximately given by (Als-Nielsen and McMorrow 2011):

$$\Delta\theta = \sigma_{\text{ph}}^a = \frac{1}{\gamma}\sqrt{\frac{1 + K^2/2}{N}}. \tag{4.13}$$

The term $\sqrt{1 + K^2/2}$ can be related to the X-ray wavelength according to Eq. (4.11):

$$\sqrt{1 + K^2/2} = \sqrt{2\lambda_x \gamma^2 / \Lambda_u}.$$

Then we obtain for the angular divergence:

$$\Delta\theta = \sigma_{\text{ph}}^{a} = \frac{1}{\gamma}\sqrt{\frac{2\lambda_x\gamma^2}{N\Lambda_u}} = \sqrt{\frac{2\lambda_x}{L}} \approx \sqrt{\frac{\lambda_x}{L}}. \tag{4.14}$$

Now, we turn our attention to the photon source size. The photon source size σ_{ph}^{r} cannot become smaller than the value set by the diffraction limit. Beyond the diffraction limit, a reduction of the radial photon source size leads to an increase in the beam divergence, and vice versa. This condition is equivalent to the uncertainty relation in quantum mechanics, which provides a lower bound:

$$\sigma_{\text{ph}}^{r} \times \Delta p_r \geq \frac{\hbar}{2}. \tag{4.15}$$

Here, the transverse momentum uncertainty is $\Delta p_r = \hbar k \Delta\theta$, and $k = 2\pi/\lambda_x$ is the wavenumber. Therefore, for Δp_r follows:

$$\Delta p_r = \hbar \frac{2\pi}{\lambda_x}\sqrt{\frac{\lambda_x}{L}} = \frac{2\pi\hbar}{\sqrt{L\lambda_x}}$$

and

$$\sigma_{\text{ph}}^{r} \geq \frac{\hbar}{2\Delta p_r} = \frac{\hbar}{2}\frac{\sqrt{L\lambda_x}}{2\pi\hbar} = \frac{\sqrt{L\lambda_x}}{4\pi}.$$

The product yields the diffraction-limited "emittance" of the photon beam:

$$\varepsilon_{\text{ph}} = \sigma_{\text{ph}}^{r} \times \sigma_{\text{ph}}^{a} = \frac{\sqrt{L\lambda_x}}{4\pi} \times \sqrt{\frac{\lambda_x}{L}} = \frac{\lambda_x}{4\pi}. \tag{4.16}$$

Thus, the diffraction limit is determined by the X-ray wavelength. A light source is referred to as *"diffraction limited"* when the e-beam emittance is less than that of the radiated photon beam at the aimed X-ray wavelength. Reducing the electron source size further would not improve the photon emittance. In other words, the diffraction limit sets the lower bound for the emittance of a photon beam.

The diffraction limit is not reached with undulators at synchrotron radiation facilities, but it is reached with free-electron lasers, as we will discuss in the next sections after introducing the concept of beam coherence.

4.2.8   Longitudinal and Transverse Coherence

For the following and later discussions of coherent light sources, it is important to define the longitudinal coherence length, transverse coherence length, temporal coherence, phase coherence, and spectral bandwidth. In physics, coherence implies the existence of a fixed relation between two waves over some limited time and/or space. The decay of coherence results in an incoherent relationship that destroys the ability of the considered waves to constructively interfere.

4.2.8.1   Longitudinal Coherence Length

A monochromatic laser light is expected to have an infinitely extended wavetrain in space lasting forever in time. We know that in reality, this is not the case. Even the best laser source has a certain finite bandwidth of wavelengths. We consider two waves with wavelengths λ_1 and λ_2 ($\Delta\lambda = (\lambda_1 - \lambda_2)$), which travel in the same direction, starting at $t = 0$ with same phase. Then, after traveling a distance ξ_c, they will be 180° out of phase and destructively interfere. The distance l_c is called the longitudinal coherence length l_c (Fig. 4.18).

The coherence length is given by the condition

$$N\lambda_1 = \left(N + \frac{1}{2}\right)\lambda_2 = l_c,$$

which yields.

$$N\Delta\lambda = \frac{1}{2}\lambda; \quad \text{or} \quad \frac{l_c}{\lambda}\Delta\lambda = \frac{1}{2}\lambda.$$

Therefore, we obtain the longitudinal coherence length, also known as temporal coherence:

$$l_c = \frac{\lambda^2}{2\Delta\lambda}. \tag{4.17a}$$

The coherence time is accordingly $\tau_c = l_c/c = \lambda^2/2c\Delta\lambda$. The coherence length and the coherence time increase with decreasing bandwidth $\Delta\lambda$, as expected. The more monochromatic the beam, the more extended the longitudinal coherence of the X-ray beam.

Fig. 4.18 Two waves with slightly different wavelengths propagating in the z-direction. The waves are in phase at the beginning, but run 180° out of phase after the distance l_c, which is known as the longitudinal coherence length

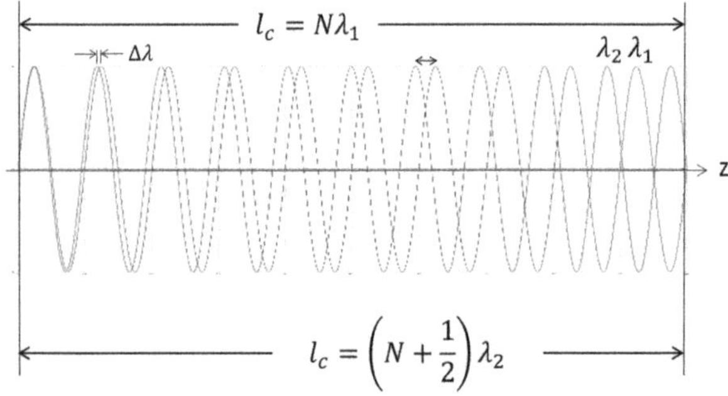

4.2.8.2   Transverse Coherence Length

The coherence length in the transverse direction is important to answer the question of whether two separated points A and B on the wavefront are sufficiently phase related to observing interference effects if passed through two slits, as indicated in Fig. 4.19. Here, z is the distance from the source s with extension d.

According to the sketch in Fig. 4.19, $\theta = t_c/z$. The source s is coherent up to the first diffraction minimum, which is at $\theta = \lambda/d$. Combining both equations, we obtain for the transverse coherence length, also known as the spatial coherence length:

$$t_c \cong \frac{z\lambda}{d}. \tag{4.17b}$$

Hence, the transverse coherence length increases linearly with the distance from the source, with the wavelength of the source, and is inversely proportional to the source size.

There are various methods for determining the longitudinal (temporal) and transverse (spatial) coherence lengths of soft and hard X-rays. They use either pinholes or one-dimensional or two-dimensional gratings. These methods are discussed in Marathe et al. (2016), Gutt et al. (2012) and Takeo et al. (2019).

4.2.9   Free Electron Laser (FEL)

Undulators are approximately 5–10 m long. As their length increases to 100 m and beyond, not only does the X-ray intensity (or brilliance) increase dramatically, but a new phenomenon also occurs, the so-called self-amplifying spontaneous emission (SASE) of radiation (X-ray photons). The SASE principle is a different photon amplification process than that used in a laser. Optical lasers require a resonator sealed at both ends by mirrors. Since back-reflection of X-rays is not possible, cavity-free X-ray lasers were developed. A single electron bunch passing through

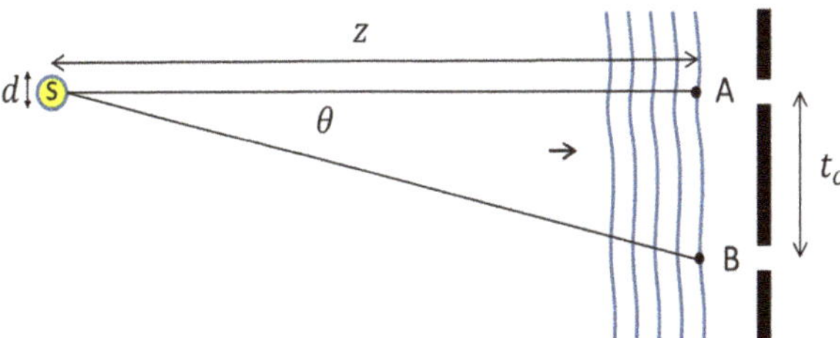

Fig. 4.19 The transverse coherence length determines to what extent two points on the wavefront A and B are in phase and can potentially interfere using a double slit setup. S is the source and d is the source diameter. The further the distance z from the source, the larger the transverse coherence length t_c

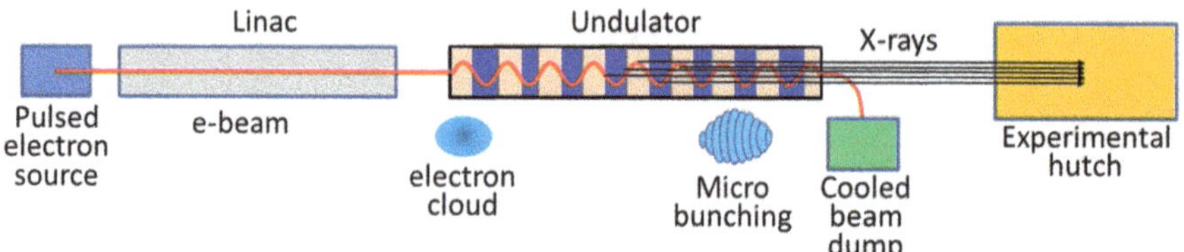

Fig. 4.20 Schematic layout of a free-electron laser. Pulsed electron bunches are produced on the left by a laser and fed into a linac before entering an undulator. At the end of a long undulator, the emitted X-ray intensity is fully coherent, monochromatic, and femtosecond pulsed. The pulsed X-ray beam features tunable photon energy and variable polarization, all depending on the specific design of the undulator

an undulator only once should produce sufficient photon amplification to produce a laser-like X-ray beam, according to the so-called SASE principle explained below. After one pass, the electron bunch is dumped and cannot be recovered, unlike electron bunches in a standard synchrotron storage ring. Figure 4.20 shows a schematic layout of the basic elements of a free-electron laser (FEL).

The SASE principle works as follows: First, electron bunches are generated by a pulsed laser beam striking a target. The emitted electrons are focused, accelerated, and guided through a radio-frequency gun into a linac. In the most advanced version, the linac consists of a series of superconducting acceleration modules that increase the electron energy to several GeV. As the high-energy electrons enter an undulator, they spontaneously emit X-ray photons, which, however, lack phase coherence and monochromaticity. Further down in the undulator, the electromagnetic waves begin to overlap, resulting in a filtering of wavelengths and phases according to the undulator periodicity λ_u. As the X-ray emission progresses, the electric field of the photons becomes so strong that it begins to interact with the advancing electrons. This interaction causes all the electrons to emit radiation in phase with each other. As their path length in the undulator increases, the electrons cluster together into microbunches. The periodicity of the microbunches is determined by the first fundamental X-ray wavelength of the undulator. The formation of the microbunches is because slower electrons are accelerated and faster electrons are decelerated in the strong electromagnetic field to move in phase with the EM wave.

The undulator length L_G required to achieve a SASE radiation gain is given by (Balerna and Mobilio, 2015):

$$L_G = \frac{\lambda_u}{4\pi\rho\sqrt{3}}, \tag{4.17c}$$

where λ_u is the undulator periodicity, and ρ is an FEL efficiency parameter. L_G depends on the actual design of the FEL, but it is typically 2–3 m. Further increase of the undulator length results in an exponential FEL radiation power increase:

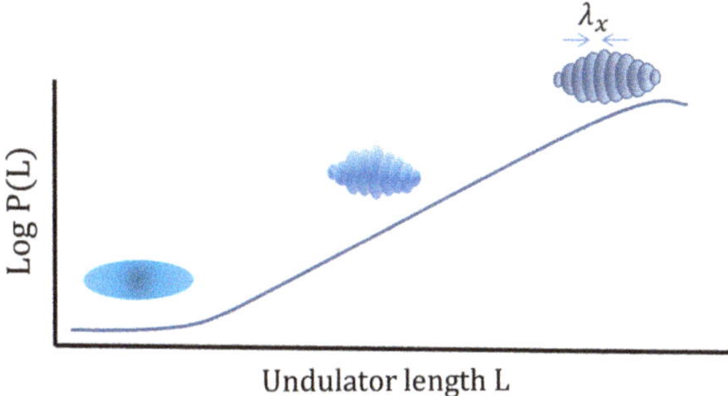

Fig. 4.21 FEL exponential power output as a function of undulator length L. The electrons enter the undulator as an incoherent cloud (left side). Further along the undulator, the electron bunches form micro-bunches with a separation according to the X-ray wavelength. (Adapted from https://photon-science.desy.de/research/students__teaching/.html)

$$P(L) = P_0 \exp(L/L_G). \tag{4.18}$$

As mentioned before, the separation of the microbunches equals the radiation wavelength. This, in turn, results in laser-like properties of the emitted radiation, including monochromaticity, phase coherence, and an exponentially increasing radiation power with increasing length of the undulator before reaching saturation. A cartoon of the SASE process is reproduced in Fig. 4.21. At the maximum radiation power, the peak brightness reaches 8.7×10^{33} in units of photons/(sec $\cdot$ mm^2 $\cdot$ mrad^2 $\cdot$ 0.1%bandwidth). The power increase is restricted by the diffraction limit. While the radiation power reaches extreme values, the longitudinal photon beam coherence is rather poor. This is because the phase of each micropulse is independent of the other. Each pulse evolves from an individual electron. The low coherence is indeed a disadvantage of a SASE FEL. To overcome this problem, seeding methods have been invented, which are discussed in the next section.

4.2.10   Seeding of Free Electron Lasers

The SASE principle of an undulator provides an unfiltered wavelength (energy) spectrum that can be as wide as 30 eV. This wide spectral bandwidth is because SASE starts from shot noise. To obtain a narrower bandwidth and higher longitudinal coherence, an additional filtering process is required. The electron seeding process into micro-bunches can be assisted and improved by passing a strong laser beam through the undulator parallel to the electron trajectory. Alternatively, the undulator may be split into two halves, and the X-ray beam is filtered by a crystal monochromator, while the electron beam is detoured in a chicane. After recombination, a filtered pulse is used for seeding the second undulator. The laser-assisted

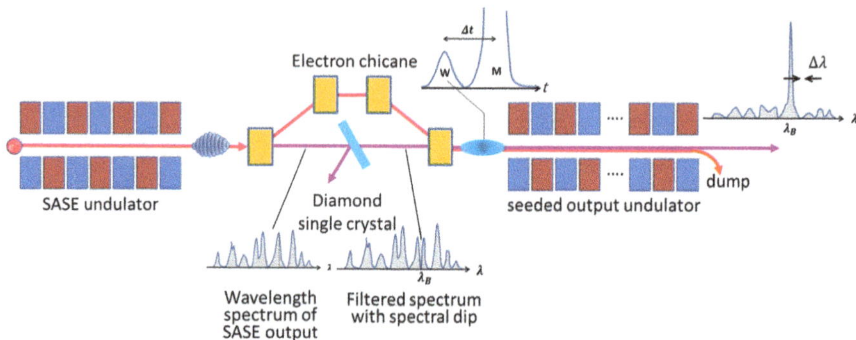

Fig. 4.22 Schematic setup of a self-seeding free electron laser. It consists of two undulators with a spectral filter in between. The first undulator provides a high-intensity SASE pulse that is split at the chicane into an electron pulse and a photon pulse. The photon pulse is filtered by a diamond single crystal set to scatter at the (004) Bragg reflection. The modified spectral distribution is composed of a main peak M and a delayed wake peak W. After the chicane, the electron pulse is recombined with the wake peak, serving as a seed for the SASE operation in the second undulator. The schematic is adapted from Geloni (2016)

option is used for soft X-rays and is known as photon-seeded FEL, whereas the second type is referred to as self-seeding and is used in hard X-ray FELs.

Figure 4.22 is a schematic layout of a self-seeding XFEL, which consists of three parts. In the first part to the left, we have the usual undulator with a SASE pulse at the end. In the middle part, the electrons are deflected by a chicane, while the photons are filtered by a monochromator. Both are recombined before the entrance to the second undulator. The last part consists again of an undulator that serves to generate a seeded and spectrally filtered narrow photon pulse with increased longitudinal coherence at the exit. The chicane has three purposes. First, it is needed to create a beam offset to insert a monochromator for the photon beam. Second, it will smooth out the electron beam microbunches. And third, it serves to tune the delay time between the electron bunch and the photon pulse.

Now we focus on the middle part. As mentioned before, the photon pulse from the first undulator is spectrally filtered by passing through a monochromator. The monochromator is usually a diamond single crystal set to the (004) Bragg reflection. The Bragg reflection punches a narrow spectral hole in the transmitted wavelength distribution at the wavelength λ_B selected by the diamond Bragg peak. At the same time, the monochromator acts as a "band stop," since the photons are reflected back and forth like in a cavity. We may also view this reflection as a secondary extinction effect, as described in Sect. 3.7.2. This causes a delay in the time domain between the main unfiltered transmitted beam "M" and the filtered and monochromatic "wake" peak "W." The delay time between M and W is about 20 to 40 fs (Geloni 2016). The main peak still contains the entire spectral distribution from the first undulator, while the wake peak is spectrally cleaned up.

Behind the chicane, the photon pulse and the electron pulse are recombined so that the electron beam overlaps with the delayed "wake" peak that is fed into the second undulator. This overlap produces a significantly improved monochromaticity and enhanced longitudinal coherence length at the output. The scheme is called self-seeding as it uses the self-produced photon pulse in the first undulator for seeding the second undulator. The schematic layout shown in Fig. 4.22 is only one of several realizations of spectral filtering processes used at FELs for self-seeding. Because of its comparatively simple design and compact construction, it is the most frequently used setup at FELs for hard X-rays between 3 and 12 keV.

4.2.11 Four Generations of Synchrotron Radiation and Free Electron Facilities

The use of synchrotron radiation for X-ray analysis started in the 80s of the last century, mainly in parasitic mode, meaning that a beam line was developed and operated for the condensed matter community at a synchrotron facility dominantly used for high-energy particle physics experiments. Examples are CHESS in Cornell, USA, and Hasylab at DESY in Germany.

In a second generation, dedicated electron synchrotrons were built for materials science and life science experiments, using bending magnets, but no insertion devices were incorporated. This was remedied in the third-generation synchrotron facilities that featured ever more sophisticated insertion devices for enhancing the intensity (brilliance) of the photon beams and improving the coherence of the radiation. Furthermore, various bunch modes could be selected, multi-bunch, single-bunch, or hybrid bunch modes to serve the community performing steady-state and time-resolved pump-probe type experiments. The fourth-generation X-ray light sources are those that feature free electron lasers (FELs) for generating an unprecedentedly brilliant beam with nearly 100% coherence. The development of X-ray and synchrotron radiation sources is schematically illustrated in Fig. 4.23 in terms of brightness versus calendar year. Table 4.3 gives an overview of some existing synchrotron radiation facilities and free-electron lasers. A more complete and updated list can be found at: https://en.wikipedia.org/wiki/List_of_synchrotron_radiation_facilities.

4.2.12 Uses of Synchrotron Radiation

Synchrotron radiation facilities and free electron lasers not only feature orders of magnitude higher intensity than a standard X-ray tube, but they also deliver, in addition, a completely different radiation quality, including:

- High brightness beam focused on a sub-50 × 50 nm^2 spot
- Long transverse and longitudinal coherence lengths of several centimeters.
- Continuously adjustable photon energy from 40 eV to 25 keV.

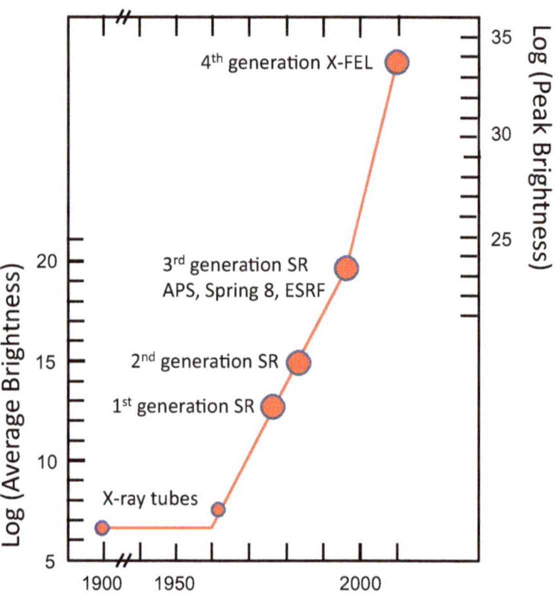

Fig. 4.23 Development of X-ray sources from the early beginning to the present in terms of brightness versus calendar year. (Adapted from Balerna and Mobilio 2015)

- Pulsed time structure with variable pulse length down to femtoseconds and variable repetition rate.
- Tunable linear to circular polarization capability.

These novel X-ray properties enable experiments that are unthinkable with conventional X-ray sources. The main applications of these new capabilities are in spectroscopy, structural analysis, and imaging. We focus on the structural analysis of materials. Some examples are listed below, and further details are explained in the following chapters.

With a highly brilliant focused beam, we can study tiny crystallites in a poly-crystalline matrix or individual tiny crystals in a diamond anvil cell at pressures up to gigabars and high temperatures, which is important for earth and planetary sciences.

With a long coherence length, we can perform X-ray holography and image tiny structural or magnetic domains. Long coherence lengths also enable X-ray photon correlation spectroscopy, which allows us to track electronic and magnetic density fluctuations in real time.

With continuously adjustable photon energies, we can perform resonant X-ray scattering experiments where the energy is tailored to the specific absorption edge of the components in the sample, ranging from the M-edges of 3d metals to the L-edges in 3d and 4d metals, up to the 4f-edges of rare earth metals and compounds.

With a pulsed time structure, we can investigate time-dependent phenomena and, in particular, conduct pump-probe experiments down to the femtosecond time range. Demagnetization and remagnetization processes can be investigated in the time domain, as well as taking movies of fast-moving processes, such as chemical reactions.

Table 4.3 Partial list of operating synchrotron radiation facilities and free electron lasers for the production of far UV light, soft X-rays, and hard X-rays, including some characteristic parameters

	Electron energy [GeV]	Circumference [m]	Beam current in multibunch operation [mA]	First commissioning
Soft X-ray synchrotron facilities				
BESSY II, Berlin, Germany	1.7	240	300	2000
Advanced Light Source, Berkeley, USA	1.9	196.8	500	1993
Elettra, Italy	2.0–2.4	259	300	1994
Intermediate energy synchrotron facilities				
Soleil, France	2.75	354	500	2006
Diamond, UK	3	565.3	500	2007
SLS, Swiss	2.4	288	400	2001
Alba, Spain	3	268.8	200	2011
NSLS-II, USA	3	792	500	2014
ANSTO, Australia	3	216.3	500	2007
SSRF, China	3	432	500	2009
TPS, Taiwan	3	518	500	2014
Max IV, Lund, Sweden	1.5, 3	528	500	2016
Hard X-ray synchrotron facilities				
ESRF, France EU	6	844	100	1994
Advanced Photon Source, Argonne, USA	7	1100	100	1996
Spring 8, Japan	8	1436	100	1998
PETRA III, Germany	6	2304	100	2012
	Electron Energie [GeV]	**Length [m]**	**Wavelength band [nm]**	
Free-electron X-ray laser				
LCLS Stanford, USA	4	1000	0.13–6.2	2009
Fermi, Italy	1.24	110–195	20–100 (FEL I) 4–20 (FEL II)	2011
SwissFEL	6	740	0.1–7	2018
SCALA, Spring 8	8	750	<0.1	
European XFEL, Germany	17.5	2100	0.05–4.7	2017
FLASH, Germany	350–1.25	315	3.4–51	2000

With adjustable linear to circular polarization in combination with adjustable X-ray photon energy, it is possible to study ferromagnetic, antiferromagnetic, and non-collinear spin structures and domains in reflection or scattering geometry.

These are just a few examples of advanced structural and spectroscopic studies of solid-state materials. However, soft and biological materials also benefit from the continuous development of X-ray sources. These new scattering techniques are topics for further investigations, some of which we will present in the following chapters.

4.3   Neutron Sources and Characteristics

4.3.1   Fission-Generated Neutron Sources

4.3.1.1   Basic Layout

Thermal neutrons with wavelengths similar to X-rays are extremely useful for the analysis of structures and excitations of materials. However, producing neutrons is orders of magnitude more difficult than producing X-rays with an X-ray tube or even with a small electron synchrotron ring. Thermal neutron scattering requires an extensive facility, including a *nuclear reactor* that "burns" enriched and fissionable uranium ^{235}U isotopes embedded in a fuel rod, surrounded by a water moderator for slowing down the hot neutrons generated by fission to ambient temperatures, a reflector that reflects the "free" neutrons from the container walls back into the moderator, and beam tubes for guiding thermal neutrons to neutron spectrometers for material analysis. The ^{235}U fission reaction is formally expressed as, where n stands for neutrons (Carlile 2002):

$$n_{\text{thermal}} + {}^{235}\text{U} \rightarrow 2 \text{ fission fragments} + 2.5\, n_{\text{fast}} + 202 \text{ MeV},$$

and is schematically illustrated in Fig. 4.24. This reaction produces 2–3 free neutrons with an initial mean kinetic energy of about 2 MeV each, and additional γ-radiation, β-radiation, and many split-off radioactive isotopes sharing the remaining energy and forming the radioactive waste for safe deposition. The neutrons, after

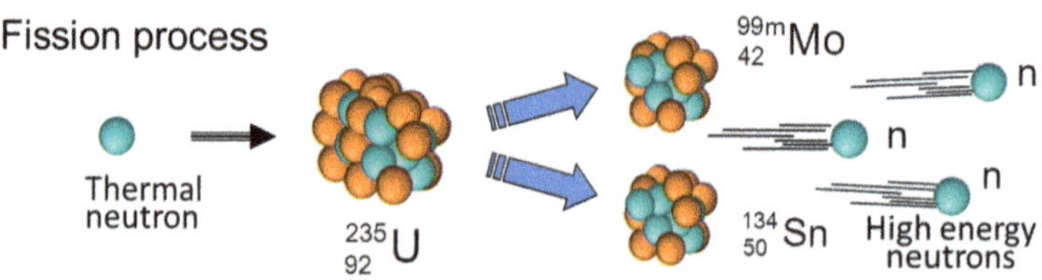

Fig. 4.24 Fission process of ^{235}U after absorption of a thermal neutron. The split-up in ^{99}Mo and ^{134}Sn is one of many other possible fission products

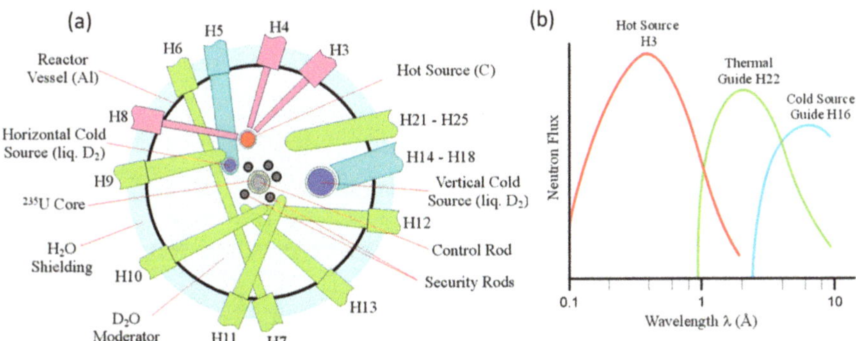

Fig. 4.25 (**a**) Layout of beam tubes at the nuclear research reactor of the Institut Laue-Langevin (Grenoble, France). The beam tubes are color-coded according to their main moderation: red for hot neutrons, green for thermal neutrons, and blue for cold neutrons. The red circle indicates the hot moderator feeding the neutron guides H3, H4, and H8. The two blue circles indicate the cold moderators supplying neurons for the guides H14–H18 and H5. (**b**) Maxwellian distribution of the neutron wavelength according to their moderation. The color coding is identical to the one in (**a**)

moderation to thermal energies of a few meV, are used for sustaining a controlled chain reaction and for various neutron experiments. All this requires a safe off-campus facility that allows control over radiation safety and user operation. There is no such thing as a "home neutron source" in analogy to a laboratory X-ray setup.[3]

The presently largest operating nuclear reactor facility for basic research with the highest thermal neutron flux is situated at the Institut Laue-Langevin in Grenoble (France). The layout of the reactor is shown in Fig. 4.25. The fuel rod containing the fissionable ^{235}U isotopes are located at the center, surrounded by a heavy water (D_2O) moderator tank, reactor vessel, and biological water shielding. The D_2O moderator slows down the fast neutrons to thermal energies by exchange of energy and momentum until thermal equilibrium is reached.[4]

Due to the Maxwell energy (or momentum) distribution for thermal neutrons, the maximum of the neutron wavelengths is approximately 0.2 nm (2 Å, 19.6 meV). The wavelength maximum can be shifted to longer wavelengths (0.6–0.8 nm, $\leq$1.6 meV) by neutron moderation in a cold source (liquid D_2, blue circles in Fig. 4.25a) or to

---

[3]TRIGA (Training, Research, Isotopes, General Atomic) reactors, with a thermal power of approximately 1 MeV, were designed and developed for safe operation and student training. These reactors are widely distributed on university campuses, primarily for educational purposes and isotope production. However, they are less competitive for scattering experiments.

[4]Neutron capture of deuterium results in small amounts of tritium (^{3}H). The slow built-up of radioactive ^{3}H does not pose a problem. Only a few neutron research reactors have a light water (H_2O) moderator, see Table 4.4. Although H_2O is more common and much less expensive, it poses a significant radiation problem due to its high neutron capture cross section, higher by a factor of 10^3 compared to ^{2}H. The corresponding nuclear reaction ^{1}H(n,γ)^{2}H generates 2.2 MeV γ-radiation, which increases the background radiation in the experimental facility significantly.

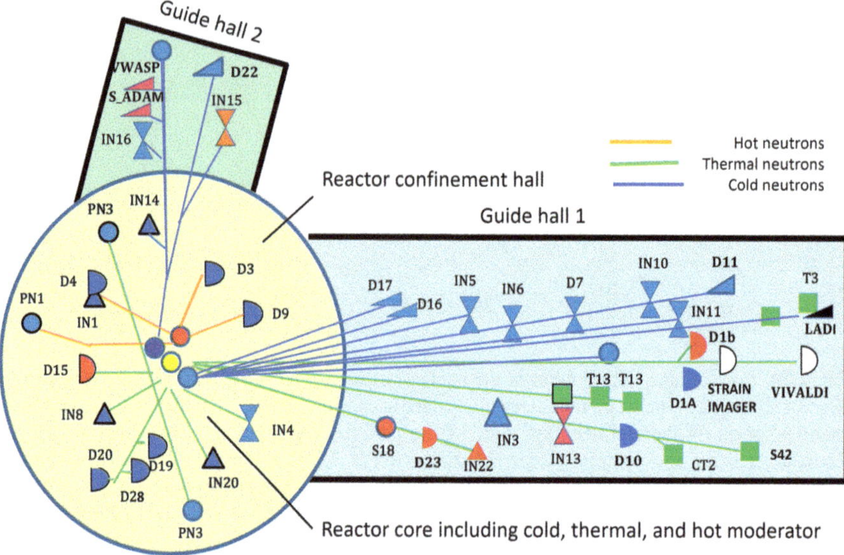

Fig. 4.26 Schematic layout of scattering instruments at the Institut Laue-Langevin (Grenoble, France). A total of 40 instruments are distributed over the central reactor containment hall and in the two neutron guide halls stretching to the right and the top. The various instruments are symbolized by letters and numbers, where "D" stands for diffraction, "IN" for inelastic, "T" for test instruments, and "PF" and "PN" are nuclear spectroscopes. Symbols designate: △ triple-axis instruments; ⋈ time-of-flight high-resolution instruments; ◁ small-angle instruments and reflectometers; □ test instruments. Some instruments have special names, such as LADI and VIVALDI. For didactical reasons, this version dated from 2010 is shown. An updated instrument layout can be found at: https://www.ill.eu/for-ill-users/instruments/instrument-map

shorter wavelengths (0.04 nm, 490 meV) by moderation in a hot graphite block (red circle). The respective Maxwell–Boltzmann distributions of the neutron energies are shown in Fig. 4.25b, red for the hot source, green for the thermal moderator, and blue for the cold source, and are expressed by the following distribution:

$$f(E) = \sqrt{\frac{2E_n/\pi}{(k_B T)^3}} \exp\left(-\frac{E_n}{k_B T}\right). \tag{4.19}$$

Here, E_n is the neutron energy, k_B is the Boltzmann constant, and T is the absolute temperature. The respective neutrons are guided to the instruments by neutron beam tubes: H3, H4, and H8 are used by hot neutrons, the green-coded beam tubes transmit thermal neutrons, and the blue-coded tubes are fed by cold neutrons. A general layout of the neutron scattering instruments distributed over the reactor hall and two additional guide halls is depicted in Fig. 4.26. All instruments using hot neutrons are placed in the reactor hall. All instruments using cold neutrons are guided out to the adjacent guide halls.

Instruments using thermal neutrons are partly in the reactor hall and partly in the guide halls 1 and 2.

4.3.1.2 Guide Tubes and Supermirrors

To accommodate many more instruments than would fit into the reactor containment hall, neutron guide tubes have become indispensable for any modern research reactor layout. Neutron guide tubes act as fiber glass in optics. In evacuated tubes, neutrons are guided over long distances (up to about 100 m) to guide halls adjacent to the central reactor containment building. The neutron guides (Anderson 2002) utilize the reflection of neutrons at ^{58}Ni-coated flat boron glass plates, shaped in rectangular cross sections (Fig. 4.27a). Borated glass is used because of the high absorption cross section of neutrons in boron. The guides are evacuated to avoid absorption and scattering of neutrons by air and moisture (water molecules) inside. The total reflection of neutrons, which works similarly to glass fibers for light, is explained further in Chap. 10. ^{58}Ni is chosen as the coating material because it has the highest critical angle of total reflection of all naturally occurring isotopes. In addition, neutron guides may be bent to avoid direct sight contact with the reactor core to lower the γ-background radiation. Guides may also be tapered for changing their cross section or for focusing to increase the flux on the sample.

Instead of using a plain ^{58}Ni film coating on glass substrates, nowadays *supermirrors* are used that increase the angle of neutron reflection (Mezei 1976). Supermirrors consist of a graded multilayer with alternating ^{58}Ni and Ti layers (Fig. 4.27b). Grading is achieved by varying the multilayer periodicity Λ from top

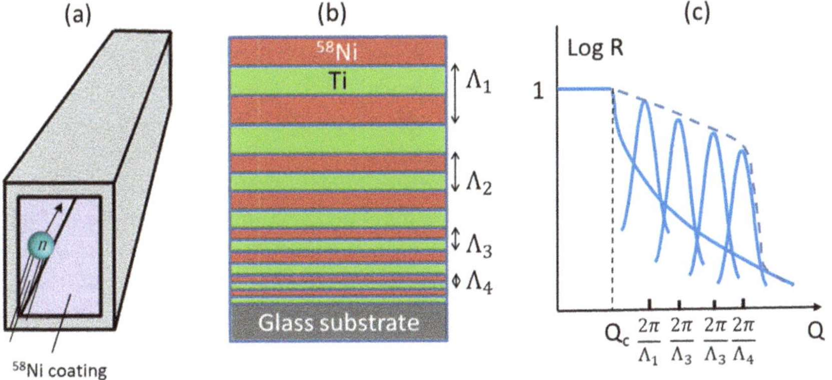

Fig. 4.27 (**a**) Schematics of a guide tube for cold neutrons. The guide is coated inside with the ^{58}Ni isotope for total neutron reflection at glancing angles. (**b**) To increase the reflection angle, instead of a single Ni-coating layer, it is preferred to deposit a graded multilayer with varying multilayer periodicities from top to bottom, here a nickel-titanium multilayer. (**c**) Plotted is the reflectivity R on a logarithmic scale versus the scattering vector Q. The grading of the multilayer has the effect of reflecting the neutrons at varying overlapping Bragg reflections, which effectively increases the reflecting angle to a multiple m of the critical scattering vector Q_c. In the shown example, $m = 4$

to bottom, such that many overlapping Bragg peaks are generated (Fig. 4.27c), shifting the neutron scattering vector for total reflection Q_c to higher effective scattering vectors Q_{eff}. The supermirrors are characterized by their m-factor, where $Q_{eff} = m \times Q_c$, and Q_c is the critical scattering vector for total reflection by the plain ^{58}Ni-coating. m-values of 4–7 are reached with the best supermirrors available.

Neutron guides can also be used to polarize neutrons simultaneously. For ferromagnetic layers, Q_c has a nuclear and a magnetic part (compare Eq. 11.11):

$$Q_c = \sqrt{\left(Q_c^{nuc}\right)^2 \pm \left(Q_c^{mag}\right)^2}. \tag{4.20}$$

Here, the plus sign stands for neutron magnetic moments parallel to the magnetization vector in the multilayer, and the minus sign for antiparallel orientation. As for the supermirrors discussed above, ferromagnetic supermirrors can be produced by alternating ferromagnetic and nonmagnetic spacer layers, such as Fe/W, Fe/Si, or Co/Ti. Then the parallel component is reflected at the higher scattering vector:

$$Q_{eff}^{mag,+} = \sqrt{\left(m \times Q_c^{nuc}\right)^2 + \left(m \times Q_c^{mag}\right)^2}, \tag{4.21}$$

while the antiparallel component is transmitted through the layers and finally absorbed in the borated glass substrate. These ferromagnetic supermirrors can be used as spin filters of one spin orientation, the other being either reflected or transmitted.

4.3.1.3 Nuclear Fuel Elements

The natural isotopic abundance of ^{235}U is only 0.7%. Therefore, the use of neutrons in research reactors requires the enrichment of fissionable ^{235}U isotopes. The standard enrichment method is currently gas centrifugation of uranium hexafluoride (UF_6) in an array of rapidly rotating cylinders.

A partial list of existing and operating steady-state neutron sources can be found in Table 4.4. These research reactors are classified according to their use of highly enriched uranium ^{235}U (HEU), with an enrichment of approximately 80–90%, or low-enriched uranium ^{235}U (LEU), typically enriched to approximately 20%. The long-term political goal is the conversion of all HEU reactors to LEU operation. LEU is preferred because its operation does not pose a threat to nuclear proliferation.

The instrumentation at neutron research reactors that is used for elastic scattering experiments is discussed in the following chapters.

Table 4.4 Operating steady-state thermal neutron sources using fissionable ^{235}U

Name	Thermal power (MW)	Thermal flux n/cm^2 s	Moderator	Guide halls	Number of instruments
ILL, France	56 (HEU)	10^{15}	Cold (L-D$_2$), D$_2$O thermal, hot (graphite)	2	40
NIST NCNR, USA	20 (LEU)	4×10^{14}	Cold (L-H$_2$), D$_2$O thermal	1 + extention	28
High Flux Isotope Reactor, ORNL, USA	85 (HEU)		H$_2$O thermal, cold	1	12
JJRR-3M Tokai, Japan	20	3×10^{14}	H$_2$O	1	26
FRM II, MLZ, Germany	20 (HEU)	8×10^{14}	D$_2$O thermal, cold (L-D$_2$), hot (graphite)	2	31
ANSTO, Australia	20 (LEU)			1	15
DHRUVA, BARC, India	100 (HEU)	1.8×10^{14}	Heavy water	–	

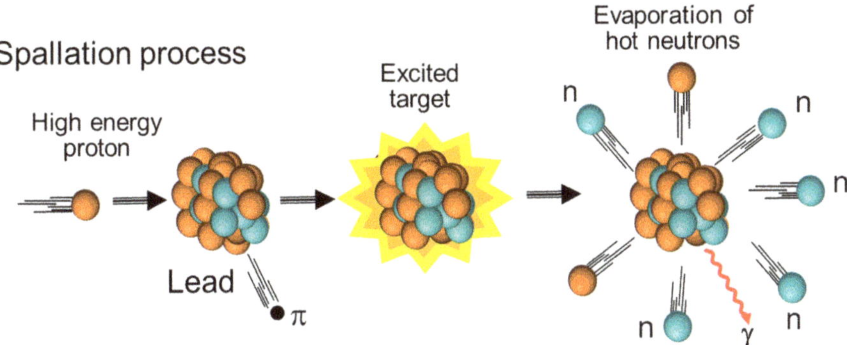

Fig. 4.28 Illustration of the spallation process of heavy nuclei. High-energy charged particles collide with a heavy nucleus, which enters an excited "molten" state. During de-excitation, the heavy nucleus emits many fast neutrons and other radioactive particles. For use in scattering experiments, the neutrons must be moderated to thermal energies

4.3.2 Spallation-Generated Neutron Sources

An alternative method of producing neutrons through fission is the so-called **spallation reaction**. The spallation of nuclei requires the impact of high-energy, heavy charged particles such as protons or α-particles. The nuclei struck enter a highly excited "molten" nuclear state. The fragments then de-excite by emitting neutrons with energies of the order of 1–2 MeV. During de-excitation, the nuclei evaporate 20–30 neutrons and other particles per hit nucleus. Figure 4.28 shows schematically a spallation process of heavy metal nuclei such as Pb or W, adapted from C.J. Carlile (2002).

The production of neutrons by spallation requires a system consisting of a linear accelerator (LINAC) for protons with a kinetic energy of several GeV, a heavy metal target, and a moderator. Currently, two types of targets are used: solid tungsten targets and liquid mercury targets. The future European Spallation Source (ESS) will be equipped with a rotating tungsten target, similar to a rotating X-ray anode machine, but on a much larger scale. The SNS spallation sources at Oak Ridge National Laboratory (USA) and the J-Park facility in Japan use a liquid mercury target circulated by large pumps to cool the thermal energy released during proton impact. The hot neutrons (≤ 2 MeV) emitted after spallation must be moderated to thermal energy to be usable for scattering experiments. The moderation process is similar in fission reactors and spallation sources.

The main difference between neutron production by nuclear fission and spallation is that a nuclear fission reactor requires enriched and fissile ^{235}U isotopes and operates continuously, whereas a spallation source uses non-fissile heavy metals and is pulsed. Depending on the facility configuration, the pulse frequency is between 14 and 60 Hz, the pulse width is in the microsecond to millisecond range, and the pulse height depends, among other factors, on the proton beam current. In both cases, the thermal neutron energy spectrum after moderation is Maxwell-distributed, with a peak maximum depending on the moderator temperature. A detailed description of spallation neutron sources can be found in Bermejo and Sordo (2013) and Conrad (2020).

The layout of a pulsed neutron source is sketched in Fig. 4.29, which is adapted from https://www.isis.stfc.ac.uk/Pages/News21_PracticalGuide.aspx). First, H^-

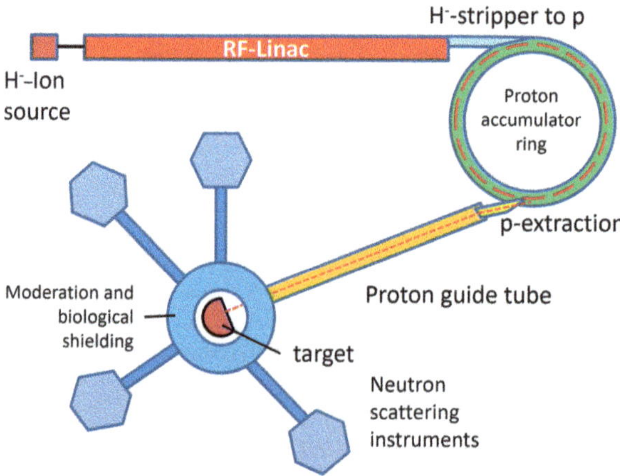

Fig. 4.29 Main components of a neutron spallation source for producing a pulsed neutron beam. After acceleration of H^- ions in a linac, the electrons are stripped off and protons are accumulated in a synchrotron ring, before being kicked into a proton guide that leads to the target station. The target usually consists of a water-cooled heavy solid metal. But liquid mercury is also used as a target material. The spallation not only produces hot neutrons but also many radioactive byproducts. Heavy biological shielding is mandatory for safe operation. The hot neutrons emitted from the target require moderation in a fashion similar to nuclear fission reactors. Instruments for thermal neutron scattering surround the target station at some distance. (Adapted from Conrad 2020)

ions (i.e. two electrons bond to one proton) are accelerated in a linear accelerator to around 0.5 up to 3 GeV, then the hydrogen ions pass through an electron stripper and protons are stirred into a proton accumulator. The accumulated bunches are then ejected by an extractor and the high-energy protons are directed to the target, where they are converted into neutrons. The hot neutrons have to be moderated, like in a nuclear fission reactor.

The intense radiation near the target station must be shielded while the moderated and pulsed neutron beam is directed to instruments surrounding the moderator. There are two options for the proton pulses: either high frequency (high repetition rate (hrr)) combined with short proton pulses or low frequency (low repetition rate (lrr)) combined with long proton pulses (see Fig. 4.30). In hrr-mode the proton pulse length is typically 10–100 μs long; in lrr-mode, the proton pulse length stretches over 1–3 ms. The neutron pulse width is primarily determined by the characteristics of the moderator. Typical moderation times of currently operated spallation sources are 10–100 μs. This corresponds to the proton pulse length in hrr mode and is significantly shorter than the proton pulse width in lrr mode. When developing new spallation sources, the combination of proton pulse length and pulse period must be optimized to maximize the neutron yield for the intended instrument suite. An example is provided for the ESS in Anderson et al. (2020).

The intrinsic time structure of the pulsed neutron beam may not meet all user needs. In such a case, an additional beam chopper can be used for further pulse shaping, as depicted in Fig. 4.27b. Table 4.5 provides an overview of the characteristic parameters of the spallation neutron sources currently in operation. The most recent, the European Spallation Source in Lund, Sweden, is scheduled to start user operation in 2027, featuring a repetition rate of 14 Hz, 2.86-ms-long pulses, and an initial set of 15 instruments.

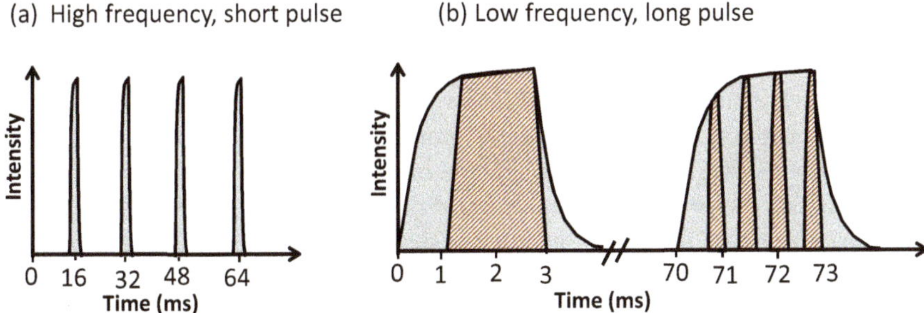

Fig. 4.30 Time structure of the pulsed neutron beams at spallation neutron sources. Two design options are realized: (**a**) high-frequency pulses combined with short pulse lengths, and (**b**) low-frequency pulses combined with long pulse lengths. Further pulse shaping with a chopper is optionally available to change the pulse length and the pulse structure, as indicated by the shaded areas for two different shaping options

Table 4.5 List of spallation neutron sources. TS = target station. A complete list of facilities and instruments can be found at https://www.iucr.org/resources/commissions/neutron-scattering/where-neutrons

	Spallation neutron sources				
	Proton beam energy (GeV)	Thermal power (MeV)	#Targets, target material	Rep. rate, proton pulse length	Instruments
SNS, ORNL, USA	1.3	2	2, liquid Hg	TS1, 60 Hz, 1 μs TS2: 15 Hz, 1 ms	20@T1, 8@T2
ISIS, RAL; UK	0.2	0.128, 0.032	2, Ta-coating W-target	TS1: 50 Hz, 200 μs TS2: 10 Hz, 30 ms	31
J-PARC; Japan	3	1	Liquid Hg	25 Hz, 20– 90 μs	23
ESS, Lund, Sweden	2	5	1, rotating W-target	14 Hz, 2.86 ms	22 planned
	Special designs				
SINQ, PSI Switzerland	0.59		Pb target in a D_2O moderator	cw-operation	14
IBR-2, FLNP, Dubna, Russia		2	PuO_2 with two counter-rotating reflectors	5 and 10 Hz 340 μs	12

4.3.3   Comparison of Continuously Operating and Pulsed Neutron Sources

The thermal neutron flux of steady-state or continuously operating neutron reactors with a flux of about 10^{15} neutrons/cm^2 s has come to a saturation since almost 50 years. The reactors with the highest flux are those in Oak Ridge (HFIR) and Brookhaven (HFBR, decommissioned), and at the Institut Laue-Langevin, Grenoble, France. The situation is different for spallation sources. Here we notice a steady increase in neutron flux since the early prototypes ZING-P, ZING-P′, and IPNS operated at the Argonne National Laboratory,[5] USA. The European Spallation Source (ESS) is expected to have a thermal neutron flux in the peak of the pulses that is by a factor of 10^3 higher than that of the most powerful steady-state sources. The time-average thermal neutron flux is, however, comparable to that of steady state sources. For instruments, designed to exploit the pulsed neutron flux, this is a big indeed a big advantage. Spallation sources are not flux-limited, in contrast to

---

[5]The first neutron spallation source (Intense pulsed neutron source, IPNS) was developed by J.M. Carpenter at the Argonne National Laboratory (USA). The IPNS used a solid W-target and started user operation in 1981. The IPNS was decommisioned in 2008.

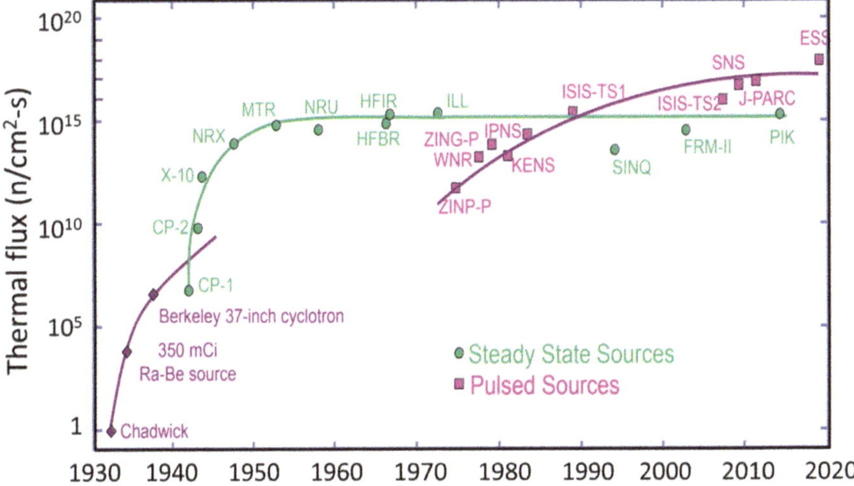

Fig. 4.31 Development of the thermal neutron flux since the discovery of the neutron in 1932 by J. Chadwick. The early sources, which were not used for neutron scattering experiments, are indicated in red diamonds; fission reactor sources for neutron production are shown in green dots; and spallation sources are marked as red squares. All symbols specify the commissioning year of the respective source. The estimated flux of ESS is marked via a red square. The graph is adapted from Andersen and Carlile (2016): K.H. Andersen and C.J. Carlile, J. Phys. Conf. Ser. 746, 012030 (2016) available under @CC

steady-state sources. In the former case, the neutron flux increases with the proton beam power, given by the product of beam current times accelerating voltage. For the ESS, this is 2 mA × 2.5 GV, yielding a power of 5 MW. The neutron flux is eventually limited by the cooling power of the target and operating cost.

Figure 4.31 provides an overview of the development of neutron sources world-wide since the neutron discovery by James Chadwick in 1932. It also shows a flux comparison between steady-state and spallation sources

4.3.4   Uses of Thermal Neutrons

Despite their inherently lower flux, thermal neutrons offer many advantages over X-rays. Neutrons penetrate materials significantly deeper than X-rays, which is useful for studying bulk materials in operation, such as batteries or fuel cells, while discharging. Furthermore, the position and diffusion of light atoms in the matrix of heavy atoms can be visualized because the scattering length is independent of atomic mass. The isotopic exchange of hydrogen and deuterium can be used in soft matter to label either the core or the shell of nanoparticles. The magnetic moment of neutrons is central to studying the spin structure of magnetic materials and magnetic phase transitions. In addition to magnetic moments, neutrons also respond to the magnetic induction. This allows neutrons to visualize magnetic flux lines in superconductors and magnetic density distributions in skyrmion lattices. The extremely high photon flux of synchrotron radiation and free-electron lasers is undoubtedly useful for many novel experiments. However, this high brilliance also leads to sample destruction, which is a major problem for soft matter samples. Neutron beams can activate samples, but they never have the power to destroy them.

Neutron scattering is generally considered a bulk probe. However, neutrons are very sensitive to surfaces and interfaces in thin films and multilayers when examined at small angles close to the angle of total external reflection. Under these conditions, atomic monolayer sensitivity can be achieved.

A broad range of applications for neutron scattering includes inelastic scattering by the excitation of quasiparticles such as phonons, magnons, librons, and others. However, inelastic and quasielastic neutron scattering are not discussed in this text. We refer the reader to relevant literature on this extensive field. Instead, the following chapters demonstrate the usefulness of thermal neutrons for elastic scattering experiments.

4.4 Electron Sources

Low-energy electron diffraction (LEED) requires an electron source that is rather straightforward in design. A schematic and simplified version is sketched in Fig. 4.32 together with a disassembled LEED unit (https://www.specs-group.com/specs/products/detail/erleed-150/). A hot electron gun produces electrons by evaporation from a tungsten filament, similar to the electron production in X-ray tubes. However, the electron energies used are much lower, in the order of a few hundred eV instead of keV in the X-ray case. The electrons are then focused and accelerated to the sample (target) and back-scattered from the target toward a hemispherical fluorescent screen. When the fluorescent screen is hit by electrons, greenish fluorescent light can be seen by the naked eye or may be detected by a CCD camera. In front of the fluorescence screen are mounted two grids. The first grid closest to the screen is a retarding grid. It lets pass only those electrons that have the same kinetic energy before and after scattering. This is because the retarding grid is at a potential opposite to the accelerating potential. Inelastically scattered electrons, which have lost energy to the target, do not have sufficient kinetic energy to pass the retarding grid. Since inelastically scattered electrons increase the background intensity, they are filtered out. A second grid compensates for all electrical fields around the sample to generate spherical scattering conditions (Van Hove 1992).

LEED is very surface-sensitive. The diffraction pattern suffers from contamination at the sample surface. Therefore, all parts are enclosed in an ultrahigh vacuum chamber with a base pressure of about 10^{-9} mbar or better (https://de.wikipedia.org/wiki/Low-Energy_Electron_Diffraction) (Fig. 4.32).

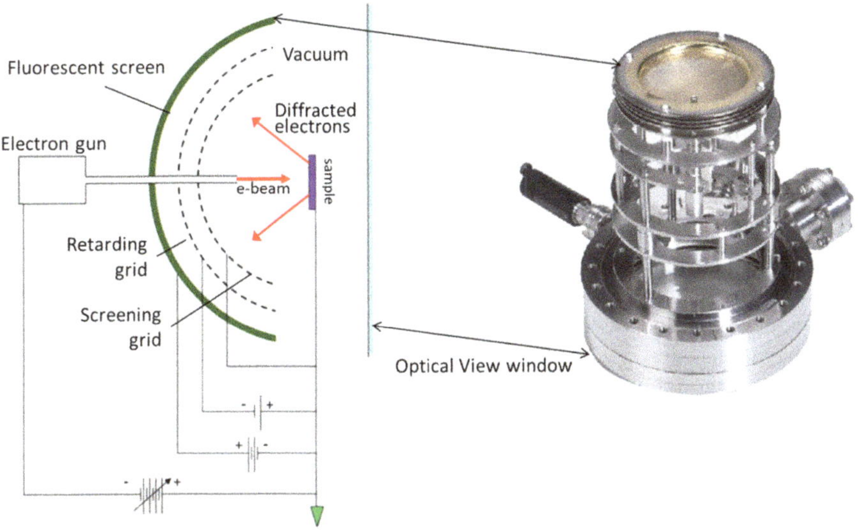

Fig. 4.32 Diagram illustrating a low-energy electron diffraction (LEED) setup. The left section shows a schematic with components like an electron gun, retarding grid, and screening grid directing an electron beam toward a sample. Diffracted electrons are shown moving toward a fluorescent screen. The right section displays a cylindrical UHV-LEED unit with an optical view window, designed for integration into a surface analysis chamber. Arrows indicate the flow of electrons and the positioning of the vacuum and optical components. (Adapted from https://www. specs-group.com/specs/products/detail/erleed-150/)

Summary

1. The conversion between X-ray photons and photon energy is $\lambda\,[\mathrm{nm}] = \frac{1240\,[\mathrm{nm\ eV}]}{E\,[\mathrm{eV}]}$.

2. The X-ray radiation from an X-ray tube consists of Bremsstrahlung and characteristic radiation.

3. The continuous Bremsstrahlung spectrum shows a cut-off energy corresponding to the accelerating energy of the electron.

4. The characteristic X-ray radiation features sharp spectral lines and is due to the excitation of core electrons in atomic shells.

5. An electron synchrotron facility consists of a linear accelerator, a booster ring, and a storage ring.

6. At synchrotron sources, the X-ray beam is confined to a very narrow cone in the forward direction of the orbiting electron bunch.

7. Insertion devices used at synchrotron sources are wigglers and undulators.

8. Wigglers increase the cone of radiation; undulators narrow the cone of radiation and increase the intensity.

9. X-ray free-electron lasers use a very long undulator to induce a self-amplified spontaneous emission (SASE) of X-ray radiation.

10. The SASE principle can be spontaneous or assisted by a seeding laser beam or with the help of a monochromator for self-seeding.
11. Neutrons can be produced by fission of the uranium isotope ^{235}U, or by spallation of heavy atoms, including tungsten and mercury, via the impact of a proton beam.
12. The generation of thermal neutrons suitable for scattering experiments requires the moderation of hot neutrons in a moderator to ambient temperatures, usually in a heavy water tank.
13. Thermal neutron wavelengths can be shifted to longer wavelengths with the help of a low-temperature moderator, or to shorter wavelengths using a high-temperature moderator.
14. Spallation neutron sources are distinguished by their pulse repetition rate as either high-frequency, short-pulse sources or low-frequency, long-pulse sources.
15. The main components of a LEED system are a hot electron gun, retarding grids, and a fluorescent screen.

Questions

(Note that more than one answer may be correct)

1. **Why is it necessary that the wavelength of X-rays, neutrons, or electrons is on the same order of magnitude as the d-spacing of crystal lattices?**

   (a) Because any other wavelength would be strongly absorbed.
   (b) Because the appropriate wavelength has to fulfill the condition for momentum conservation.
   (c) Because for wavelength $\lambda > 2d$ the Bragg condition cannot be fulfilled, and for wavelengths $\lambda \ll 2d$ the scattering amplitude becomes too small.

2. **How is the proportionality expressed between X-ray photon energy and X-ray wavelength λ ?**

   (a) $E_{\mathrm{xray}} \sim \lambda^2$.
   (b) $E_{\mathrm{xray}} \sim 1/\lambda$.
   (c) $E_{\mathrm{xray}} \sim \lambda$.

3. **What is the difference between the bremsstrahlung (BS) and characteristic X-ray (CX) radiation?**

   (a) BS is generated by the de-acceleration of electrons in the target material; CX is due to the photo-effect at core electrons.
   (b) BS occurs during the bending of electrons in the Coulomb potential of nuclei; CX is due to refilling a hole in the core of atomic shells.
   (c) BS occurs by de-acceleration of photoelectrons; CX is produced by Auger electrons.

4. **How can characteristic X-ray radiation be produced? Name at least two types of production.**

   (a) By de-acceleration of electrons in the Coulomb field of nuclei.
   (b) By the photo-electric effect of core electrons.
   (c) By a proton beam hitting a metallic target.
   (d) By inelastic electron scattering at core electrons.
   (e) By accelerating electrons on a circular orbit in a synchrotron.
   (f) By fission of uranium.

5. **What determines the cut-off Energy E_{max} of the Bremsstrahlung spectrum?**

   (a) The accelerating voltage of the X-ray tube.
   (b) The temperature of the cathode filament.
   (c) The current between the cathode and anode.

6. **What are the basic features of a synchrotron radiation source?**

   (a) Linear accelerator, booster, storage ring, magnetic lattice.
   (b) X-ray anode, klystron, vacuum chamber.
   (c) Microwave cavity, quadrupole magnet, ionization chamber.

7. **How can the polarization of synchrotron radiation be changed?**

   (a) Synchrotron radiation is not polarized, and this cannot be changed.
   (b) Synchrotron radiation is linearly polarized in the electron orbiting plane. Circular polarization can be used above or below the synchrotron orbit.
   (c) Undulators can be used to change the polarization of the synchrotron beam to different linear or circular polarizations.

8. **How are thermal neutrons generated for scattering experiments?**

   (a) By fission of U-235 and the slowing down of hot neutrons in a water moderator.
   (b) By heating U-235 to its melting temperature, thereby neutrons are released.
   (c) With a high-energy proton beam hitting a heavy metal target.
   (d) By fission of deuterium with a laser beam and separating protons and neutrons.

9. **Why is it necessary to moderate neutron energies to ambient temperatures after being produced by fission or spallation?**

   (a) Because hot neutrons would escape the reactor walls and are therefore not available for scattering experiments.
   (b) Because thermal neutrons are needed to sustain the fission chain reaction.
   (c) Because thermal neutrons of about 300 K or 25 meV have the proper wavelength of about 2 Å for scattering experiments at crystalline lattices.

10. **What are the essential differences between a continuous (CW) and a spallation neutron source?**

(a) CW operation is for cold neutrons only; spallation sources are dedicated to hot neutrons.
(b) CW neutron sources use fissionable ^{235}U; spallation sources use nonfissionable heavy or liquid metals.
(c) CW operation requires extensive radiation protection; spallation operation requires no protection.
(d) CW uses LEU; spallation sources use HEU.

11. **What is a retarding grid, and what is it used for?**

(a) A retarding grid stops protons from straying.
(b) A retarding grid is a high-energy filter for electrons used in LEED. Only electrons with energy beyond a set energy can pass. Used for reducing the low-energy background.
(c) A retarding grid is a low-energy filter for electrons in LEED optics. Only electrons with energies below a set energy can pass. Used for increasing the intensity of LEED spots.

Exercises

Grades of difficulty: E = easy, M = medium, A = advanced.

E 4.1. Threshold energy
What is the threshold energy required to excite the Cu Kα radiation? How does this compare with the energy of the Cu Kα radiation? Please explain.

E 4.2. Energy resolution
What energy resolution is required for distinguishing between the Cu Kα1 and Cu Kα2 radiation? What is the wavelength resolution required for the same emission lines? Answer in absolute and relative terms.

E 4.3 Cut-off energy
What is the cut-off Bremsstrahlung wavelength if the accelerating anode voltage is 40 kV?

M 4.4 Positive and negative coherent scattering length for neutrons
Discuss what the consequence is of scattering neutrons at nuclei with positive and negative scattering lengths.

M 4.5 Thermal neutron moderator
The process of slowing down fast neutrons by collisions with the moderator material is called moderation. Moderation is essential for the reactor operation and for performing neutron scattering experiments.

(a) How can you explain why water is an excellent moderator for hot neutrons? Construct your arguments by inspecting the coherent cross sections σ_{coh}, responsible for scattering, and the absorption cross sections σ_{abs}, responsible for neutron capture. For your discussion, you may consult Table 2.1, although the listed values are only for thermal neutrons and not for high-energy neutrons. Total cross sections are defined as $\sigma = 4\pi b^2$, where b is the scattering length.

(b) Estimate the number of collisions required to thermalize neutrons. Assume that during moderation, ballistic collisions take place mainly with protons in water. The angle of collision may vary between 0° (no energy transfer) and 180° (complete energy transfer). Assume that, upon each collision, on average 50% of the neutron kinetic energy is transferred to the protons. How many collisions are required before neutrons of 1 MeV energy reach thermal energy of 25 meV?

(c) What will happen if you use a moderator with heavy water (D_2O) instead of light water (H_2O)? If you were responsible for the reactor operation, which moderator would you prefer?

References

J. Als-Nielsen, D. McMorrow, *Elements of Modern X-Ray Physics*, 2nd edn. (Wiley, New York, 2011)

K.H. Andersen, C.J. Carlile, A proposal for a next generation European neutron source. J. Phys. Conf. Ser. **746**, 012030 (2016)

I.S. Anderson, Neutron optics, in *Neutron Data Booklet*, ed. by A.-J. Dianoux, G. Lander, (Old City Publishing, Philadelphia, 2002) https://www.ill.eu/fileadmin/user_upload/ILL/1_About_ILL/Documentation/ILL_brochure/NeutronDataBooklet.pdf

K.H. Anderson et al., The instrument suite of the European spallation source. Nucl. Instrum. Methods Phys. Res., A **957**, 163402 (2020)

A. Balerna, S. Mobilio, in *Synchrotron Radiation, Basics, Methods, and Applications*, ed. by S. Mobilio, F. Boccherini, C. Meeneghini, (Springer Verlag, Berlin, 2015)

F.J. Bermejo, F. Sordo, Neutron sources, in *Neutron Scattering Fundamentals*, Experimental Methods in the Physical Sciences, ed. by F.F. Fernandez-Alonso, D.L. Price, vol. 44, (Academic Press, New York, 2013), pp. 137–243

C.J. Carlile, The production of neutrons, in *Neutron Data Booklet*, ed. by A.-J. Dianoux, G. Lander, (Old City Publishing, Philadelphia, 2002) https://www.ill.eu/fileadmin/user_upload/ILL/1_About_ILL/Documentation/ILL_brochure/NeutronDataBooklet.pdf

H. Conrad, Spallation—neutrons beyond nuclear fission, in *Handbook of Particle Detection and Imaging*, ed. by I. Fleck, M. Titov, C. Grupen, I. Buvat, (Springer, Cham, 2020). https://doi.org/10.1007/978-3-319-47999-6_30-2

G. Geloni, Self-seeded free-electron lasers, in *Synchrotron Light Sources and Free-Electron Lasers*, ed. by E.J. Jaeschke et al., (Springer International Publishing Switzerland, Cham, 2016)

C. Gutt, P. Wochner, B. Fischer, H. Conrad, M. Castro-Colin, S. Lee, F. Lehmkühler, I. Steinke, M. Sprung, et al., Single shot spatial and temporal coherence properties of the SLAC Linac coherent light source in the hard X-ray regime. Phys. Rev. Lett. **108**, 024801 (2012)

S. Marathe, X. Shi, M.J. Wojcik, A.T. Macrander, L. Assoufid, Measurement of X-ray beam coherence along multiple directions using 2-D checkerboard phase grating. J. Vis. Exp. **116**, 53025 (2016). https://doi.org/10.3791/53025

F. Mezei, Novel polarized neutron devices: supermirror and spin component amplifier. Commun. Phys. **1**, 81–85 (1976)

Y. Takeo, H. Motoyama, Y. Senba, H. Kishimoto, H. Ohashi, H. Mimura, Probing the spatial coherence of wide X-ray beams with Fresnel mirrors at BL25SU of SPring-8. J. Synchrotron Radiat. **26**(Pt 3), 756–761 (2019)

M.A. Van Hove, Low-energy electron diffraction and electron holography: experiment and theory, in *Equilibrium Structure and Properties of Surfaces and Interfaces. NATO ASI Science*, ed. by A. Gonis, G.M. Stocks, vol. 300, (Springer, Boston, 1992). https://doi.org/10.1007/978-1-4615-3394-8_9

X-ray Data Booklet (Center for X-ray Optics and Advanced Light Source Lawrence Berkeley National Laboratory) (n.d.), http://xdb.lbl.gov/

Further Reading

X-ray and Synchrotron Sources

A. Balerna, S. Mobilio, Basics of synchrotron radiation, in *Synchrotron Radiation, Basics, Methods, and Applications*, ed. by S. Mobilio, F. Boccherini, C. Meeneghini, (Springer, Berlin, 2015)

E.J. Jaeschke et al. (eds.), *Synchrotron Light Sources and Free-Electron Lasers* (Springer International Publishing Switzerland, Cham, 2016)

S. Mobilio, F. Boscherini, C. Meneghii, *Synchrotron Radiation, Basics, Methods, and Applications* (Springer Verlag, Berlin, 2015)

Neutron Sources

T. Brückel, G. Heger, D. Richter, R. Zorn (eds.), *Laboratory Course Neutron Scattering*, Schriften des *Forschungszentrums Jülich Reihe Materie und Materia/Matter and Materials Band*, vol 38

A.-J. Dianoux, G. Lander (eds.), *Neutron Data booklet* (Old City Publishing, Philadelphia, 2002) https://www.semanticscholar.org/paper/Neutron-Data-Booklet-Dianoux-Lander/fe650823b1 7eb77d7e2aec4403109cadf72dc778

F.F. Fernandez-Alonso, D.L. Price, *Neutron Scattering Fundamentals: Experimental Methods in the Physical Sciences*, vol 44 (Academic Press, New York, 2013)

B.T.M. Willis, C.J. Carlile, *Experimental Neutron Scattering* (Oxford University Press, Oxford, 2013)

Chapter 5
Elastic Scattering Methods: Single Crystal Diffraction

5.1 Scattering Geometries

We consider an ideal three-dimensional crystal lattice (also called a *single crystal* or *mono-crystal*), which we aim to analyze using scattering methods. The analysis is intended to provide information on the lattice parameters, the crystal structure and orientation, and the angular distribution of the microcrystallites within real single crystals, the so-called mosaicity. For performing such an experiment, two options are available: scattering in reflection (Bragg) geometry or transmission (Laue) geometry, both of which are compared in Fig. 5.1. Here, the symmetric case is shown concerning the incident and exit angle, but in experiments, asymmetric geometries are also common. Because of the absorption, X-ray transmission geometry requires a thin enough crystal plate, typically not more than a few micrometers thick. For neutron scattering, this is not an issue in most cases.

Figure 5.2 shows schematically the *reciprocal lattice* of a single crystal. The reciprocal lattice forms a three-dimensional mesh of reciprocal lattice points (red dots), each one labeled by (hkl) Miller indices, representing the (hkl)-planes from which the radiation is scattered. In the schematics of Fig. 5.2, a two-dimensional cut through the three-dimensional reciprocal lattice is taken, showing the ($H0L$) plane. Some of the reciprocal lattice points (r.l.p.s) are labeled; more indices can easily be added. The size of individual r.l.p.s. depends on the crystal size and perfection of the crystal lattice. For an infinitely extended crystal, the r.l.p. shrinks to a tiny delta-like spot; for micro- to nanosize crystals, the area of the r.l.p. increases inversely proportional to the crystal's size. The actual width, determined by an appropriate scan across r.l.p., is folded with the resolution of the testing instrument, discussed later.

To explore the reciprocal lattice of a crystal structure, two types of scans are commonly executed: **radial scans** (or longitudinal scans) and **transverse scans** (or rocking scans), both of which are shown in the left and right panels of Fig. 5.2, respectively.

© The Author(s), under exclusive license to Springer Nature Switzerland AG 2026

H. Zabel, *Elements of Elastic Scattering by X-Rays, Neutrons, and Electrons*,

https://doi.org/10.1007/978-3-032-16624-1_5

(a) Reflection geometry (Bragg)   **(b) Transmission geometry (Laue)**

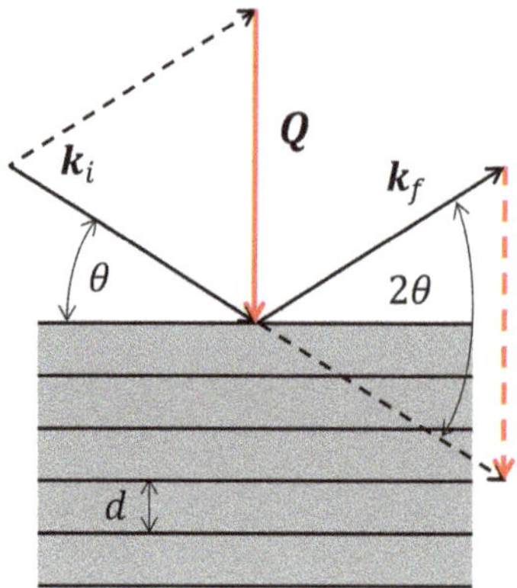
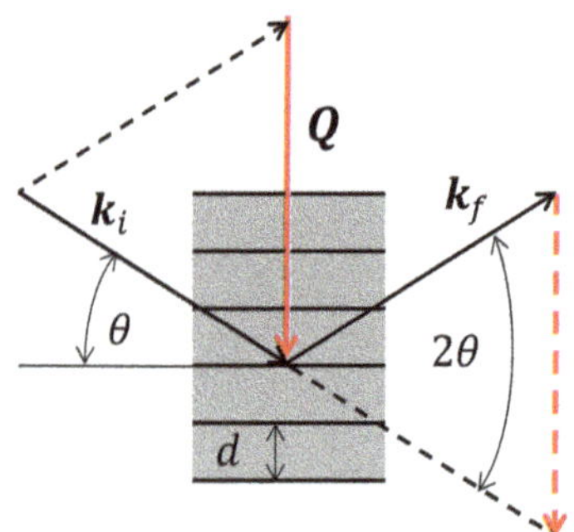

Fig. 5.1 X-ray and neutron scattering at single crystals can be performed either in reflection geometry (**a**) or in transmission geometry (**b**). Aside from the symmetric configuration shown here, asymmetric scattering is common practice. k_i and k_f are the wavevectors of the incident and scattered waves, respectively, and Q is the scattering vector. d is the lattice spacing and θ is the diffraction angle

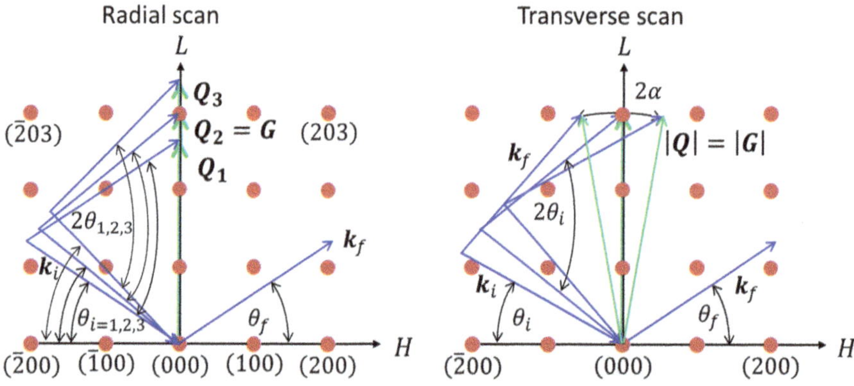

Fig. 5.2 Left panel: radial scan (or longitudinal scan) through the reciprocal lattice. Radial scans proceed by increasing or decreasing the scattering vector Q (green) and crossing one of the reciprocal lattice points by following the green arrows (here across the (003) r.l.p.) while maintaining the orientation of Q. Right panel: transverse scan (or rocking scan) through a reciprocal lattice point. Here, the length of the scattering vector Q is fixed at a reciprocal lattice point $Q = G$, and the incident angle θ_i is scanned across a Bragg peak, while the detector angle $2\theta_i$ is kept constant

In a ***radial scan***, the scattering vector Q is continuously increased along a reciprocal lattice direction (in Fig. 5.2 along the [00L] direction). This is achieved by simultaneously increasing the angle θ_i of the incident wave vector k_i and the angle θ_f of the scattered wave vector k_f with $\theta_i = \theta_f$. If the scattering vector Q passes one of

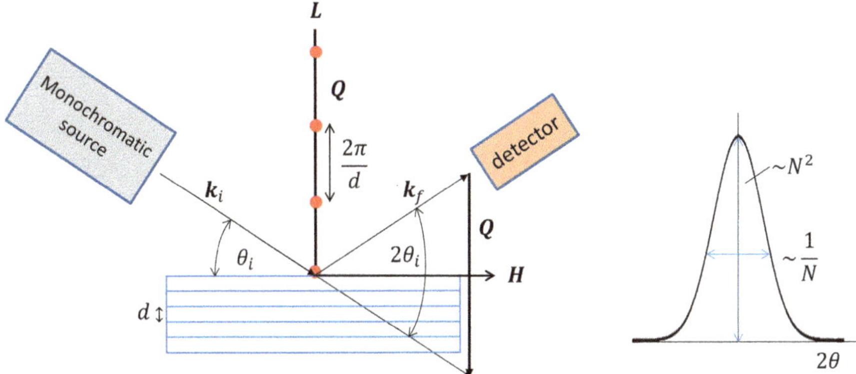

Fig. 5.3 Left panel: Scattering geometry for recording a Bragg reflection from a single crystal lattice. Right panel: Gaussian shape of a Bragg reflection. The maximum intensity is proportional to the number of scattering centers squared (N^2), while the full width at half maximum scales with $1/N$

the r.l.p., a Bragg peak occurs, i.e., an increase in intensity is recorded in the detector up to the center of the r.l.p., followed by a drop in intensity further on. Bragg reflections usually have a Gaussian-shaped intensity distribution with the maximum centered at $|\boldsymbol{Q}| = |\boldsymbol{G}_{hkl}|$.

In a ***transverse scan***, the length of the scattering vector $|\boldsymbol{Q}|$ stays constant and equal to one of the r.l.p.: $|\boldsymbol{Q}| = |\boldsymbol{G}_{hkl}|$. Here, the detector angle $2\theta_i$ is kept fixed at the Bragg position. Then both the incident and exit angles θ_i, θ_f are scanned across the r.l.p. by tilting (rocking) the sample. This scan describes an arc of opening angle 2α in the ($H0L$)-plane with a constant radius $|\boldsymbol{Q}|$. The width of the Bragg peak in the transverse direction yields information on the crystal mosaicity; the details are discussed further below.

The scattering geometry for a radial scan is shown again in Fig. 5.3, visualizing the $\theta : 2\theta$ relation in more detail. In most laboratory X-ray setups, the X-ray source is fixed. The incident angle θ_i is then scanned by rotating the sample and simultaneously rotating the detector angle by $2\theta_i$. As we have seen, this procedure continuously changes the length of the scattering vector $\boldsymbol{Q}$, eventually fulfilling the Bragg condition: $|\boldsymbol{Q}| = |\boldsymbol{G}_{hkl}|$. In some experimental configurations, the sample is fixed horizontally, while the source and the detectors move simultaneously with the angles: $\theta_i = \theta_f$. The horizontal configuration is advantageous for studying soft matter materials.

The intensity versus scattering angle $2\theta_i$ is plotted in the right-hand panel of Fig. 5.3. The ***full width at half maximum*** (FWHM) of the recorded intensity is inversely proportional to the number of scattering centers ($\sim 1/N$) in the direction of the scattering vector $\boldsymbol{Q}$, and the height of the Bragg peak is proportional to the number of scattering centers squared: N^2. Thus, the integrated intensity of a Bragg peak is proportional to the total number of scattering centers N in the beam. The recorded width of the Bragg reflection is folded with the instrumental resolution, and this must be considered when evaluating the intrinsic intensity profile.

When using a neutron beam, the intensity of the Bragg reflection in the kinematic approximation is expected to be in the radial direction:

$$I_{tot}^{neutron}(\boldsymbol{G}) = b_{coh}^2 \frac{N\lambda^3}{V_{uc}} e^{-2W} L |F(\boldsymbol{G})|^2 \tag{5.1}$$

Here V_{uc} is the unit cell volume, N is the number of unit cells in the crystal exposed by the neutron beam, L is the Lorentz factor, and λ is the neutron wavelength. For X-ray scattering, we obtain an equivalent equation for the total integrated Bragg intensity:

$$I_{tot}^{x-ray}(\boldsymbol{G}) = r_0^2 N \frac{\lambda^3}{V_{uc}} e^{-2W} f^2(\boldsymbol{G}) LPA |F(\boldsymbol{G})|^2 \tag{5.2}$$

Here LPA is the Lorentz-polarization-absorption factor. The last two equations are derived in Appendix 4, and are famous as they show that the total intensity depends on the wavelength to the third power and the square of the structure factor:

$$I(\boldsymbol{G}) \propto \lambda^3 |F(\boldsymbol{G})|^2$$

This dependence is typical of the kinematic approximation, which we have adopted here. In the dynamical scattering theory, it is shown that the intensity depends on the square of the wavelength and linearly on the modulus of the structure factor (Als-Nielsen and McMorrow 2011):

$$I(\boldsymbol{G}) \propto \lambda^2 |F(\boldsymbol{G})| \tag{5.3}$$

The conditions for the validity of the kinematic approximation are discussed in Sect. 3.7. In short, the Born approximation is valid in the case of a low scattering probability. Neutron scattering experiments are often flux-limited. Then it helps to double the wavelength and gain a factor of 8 in intensity. Due to primary and secondary extinction effects, the Bragg intensity of perfect crystals turns out to be much lower than for imperfect crystals. This is impressively shown in a bending experiment; the results are displayed in Fig. 5.4, adapted from Warren (1990). A single crystal quartz plate was bent, while the intensity increased from the perfect dynamic limit to the imperfect kinematic limit. An imperfect lattice reflects an order of magnitude higher intensity.

Radial scans can be performed in all directions of the reciprocal lattice. The more Bragg peaks are scanned, the better the crystal structure can be determined. Since each r.l.p. represents a Fourier component of the real lattice, the more Fourier components collected, the more accurately the electron density distribution in the single crystal can be described. This is rather trivial for the standard *bcc*, *fcc*, or *hcp*

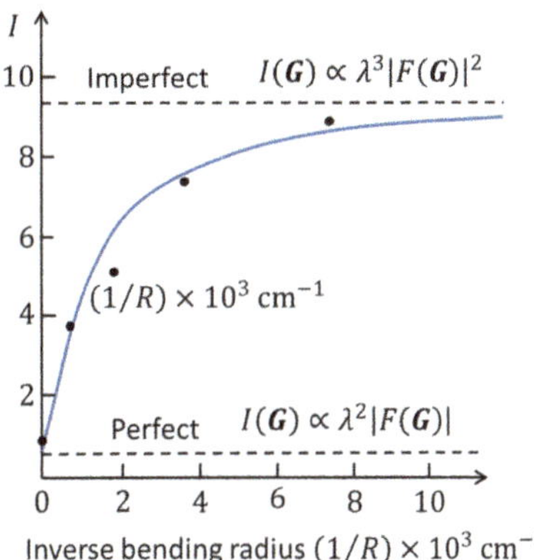

Fig. 5.4 Change of the (202) Bragg peak intensity from the perfect limit to the imperfect limit by continuously bending a quartz crystal plate as a function of the inverse bending radius R. (Adapted from Warren 1990: B.E. Warren, X-ray diffraction, Dover Pub. 1990)

(a) (b) (c)

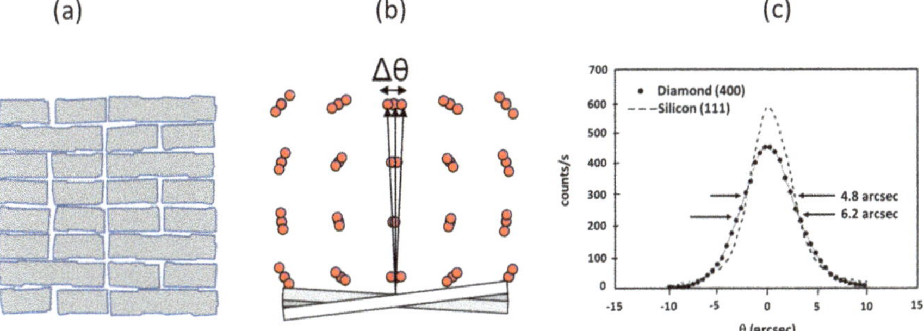

Fig. 5.5 (**a**) Real crystals consist of an ensemble of small crystal blocks with narrow angular distribution and small-angle grain boundaries in between the blocks, like the bricks of a wall. (**b**) Each crystallite has its own reciprocal lattice, which is slightly tilted against the other. (**c**) Experimental determination of the rocking curve width at the (400) reflection of synthetic diamond and the (111) reflection of a silicon single crystal, reproduced from Stoupin et al. (2019): S. Stoupin, et al. Crystals 9, 396 (2019), available via open access

lattices. However, in the case of more complex compounds and protein crystals with thousands of basis vectors, scanning a large number of Bragg reflections is of paramount importance.

Transverse scans, shown in the right-hand panel of Fig. 5.2, are used to determine the crystal orientation. A real single crystal, in contrast to an ideal one that we considered so far, is composed of finite crystal blocks with a narrow angular distribution of crystallites (mosaics), separated by small-angle grain boundaries, as sketched in Fig. 5.5a. Each crystallite has its own reciprocal lattice, which is slightly tilted to one another and rotated about the origin of the reciprocal lattice. The superposition of r.l.p's for different crystallites is schematically indicated in Fig. 5.5b. The mosaic distribution can be determined by an angular or transverse scan, also known as a *rocking curve*, described before. The FWHM of this peak represents the angular distribution of crystallites in the crystal, folded with the instrumental resolution. An example is shown in Fig. 5.3c, comparing the rocking

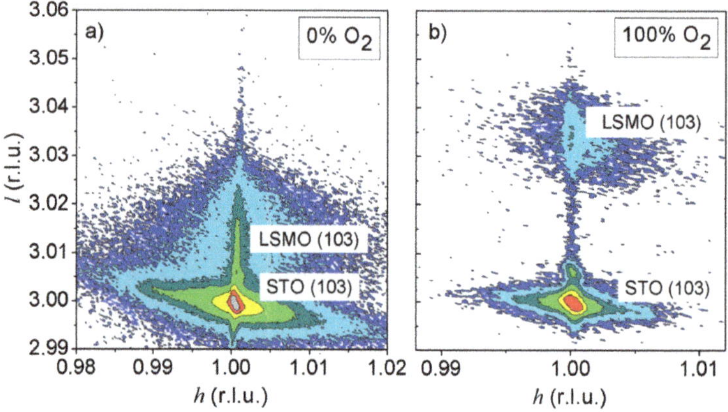

Fig. 5.6 Reciprocal space maps of a La$_{(1-x)}$Sr$_x$MnO$_3$ (LSMO) thin film grown epitaxially on a SrTiO$_3$(001) substrate. Maps were taken before oxygen exposure (**a**) and after exposure (**b**). More information is provided in the text. (Reproduced with permission from Petrisor et al. 2011: T. Petrisor, Jr. et al. J. Appl. Phys. **109**, 123913 (2011), https://doi.org/10.1063/1.3596807. Copyright (2025) by AIP Publishing)

curves of silicon at the (111) reflection and the (400) reflection of synthetic diamond. These almost-perfect crystals display rocking curve widths of only a few arcseconds (1 arcsec = 0.0003°). In contrast, metal single crystals often show a mosaicity of a few hundredths of a degree up to a tenth of a degree. The other extreme is powder samples or polycrystalline materials with a mosaic distribution spanning 4π.

In modern X-ray work, radial scans and transverse scans are often combined by mapping out the area of interest around reciprocal lattice points. This scanning technique is called **reciprocal space mapping (RSM)**. RSM is performed by scanning the scattering vector line by line in a grid-like fashion, while keeping an area detector in a fixed position. An example is shown in Fig. 5.6 from the work of Petrisor et al. (2011). Here, a La$_{(1-x)}$Sr$_x$MnO$_3$ (LSMO) thin film was grown epitaxially on a SrTiO$_3$(001) (STO) substrate. Reciprocal space maps in the ($h0l$)-plane were taken in the neighborhood of the (103) Bragg reflections. Here, ($h00$) is the direction parallel to the surface, and ($10l$) is in the direction perpendicular to the film and substrate surface. The maps were recorded first after growth and annealing in a pure Ar atmosphere (a), followed by annealing in a pure oxygen atmosphere (b). In panel (a), we notice that the (103) Bragg reflections of the film and the substrate are fused almost completely in the l-direction, whereas the peaks match perfectly in the h-direction. This shows that the film and substrate lattices match in perfect epitaxy, while in the l-direction the LSMO lattice is slightly contracted concerning the STO lattice. Furthermore, the STO (103) peak is rather broad in the h-direction and much sharper in the l-direction.

After annealing in an oxygen atmosphere, the LSMO (103) peak moves from $l = 3.01$ up to $l = 3.035$ in reciprocal lattice units, while the in-plane r.l.p. at $h = 1.0$ (r.l.u.) remains unchanged. This indicates that the LSMO lattice shrinks when exposed to an oxygen atmosphere, which is a rather surprising result. The authors explain the lattice contraction by a change of the valence state of Mn^{3+} to Mn^{4+} upon oxygen exposure, which results in a smaller Mn-ionic radius. The example shows that much information can be gained from mapping the reciprocal space of a crystal lattice. This was just one example. Similarly, RSM may reveal strains, strain gradients, epitaxial misfits, thermal expansion, and other effects.

5.2 Single Crystal Instrumentation

Using single-crystal angle-dispersive diffraction methods, a four-circle diffractometer is the most versatile tool for exploring the reciprocal space. It allows scanning any reflection in reciprocal space that satisfies the Bragg or Laue condition, performing line scans between Bragg reflections or reciprocal space mapping of an area of interest around an r.l.p. The main components of a four-circle diffractometer are shown in Fig. 5.7. What we used to call θ-angle is now called ω-circle. Two additional circles are available to align the crystal: the large χ circle and the ϕ circle, which is mounted on the χ circle that holds the sample. The fourth circle is the 2θ detector arc. The goniometer is also known as an **Eulerian cradle**. Four-circle diffractometers for crystal structure determination are mainly used in conjunction with X-ray radiation, and the method is called X-ray crystallography. In some special cases, neutron crystallography is justified and preferable to X-ray crystallography. This is the case when light atoms in a unit cell are close to heavy atoms. For example, when lanthanides or actinides bind to light elements, including H, Li, F, O, N, C, S, or others. Since the neutron cross section is independent of the atomic mass number, light atoms can have similar coherent scattering lengths as heavy atoms and appear just as strong in the diffraction pattern as the heavy elements. Another important application of neutron crystallography is in the field of magnetic materials. Due to the magnetic moment of neutrons, in addition to the crystal structure, ferromagnets,

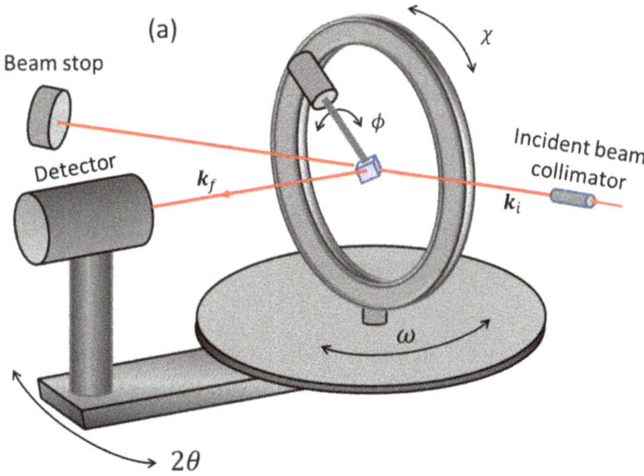

Fig. 5.7 Four-circle X-ray diffractometer featuring 2θ, ω, χ, and ϕ-circles

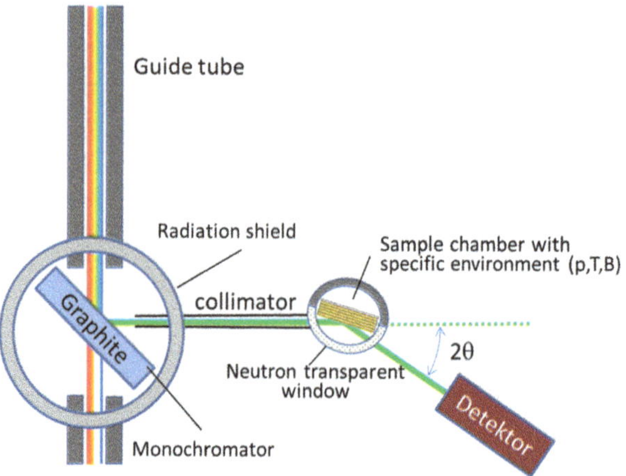

Fig. 5.8 Schematic of an experimental setup for elastic neutron scattering in a guide hall. The graphite monochromator reflects a wavelength band $\Delta\lambda$ to be used for scattering experiments. Other wavelengths can be used by further instruments downstream. The sample inside a sample chamber is exposed to a specific environment, depending on the aim of the experiment

antiferromagnets, and all types of complex magnetic structures can be determined. Magnetic neutron scattering is discussed in Chap. 7.

If only a 2D reciprocal plane is of interest instead of a 3D reciprocal space, a two-circle diffractometer with independent control of the θ (or ω) sample circle and the 2θ detector circle is sufficient. An example of a θ: 2θ setup for neutron diffraction is shown in Fig. 5.8. Here, we assumed that the neutron diffractometer is placed in a guide hall and that a neutron guide provides a wide spectrum of wavelengths. A graphite monochromator reflects out a narrow wavelength band $\Delta\lambda$ for use in an angle-dispersive scattering experiment (here color-coded in green). All other neutron wavelengths pass through the monochromator to be used by subsequent instruments down the hall (downstream). The collimated beam hits the sample inside a chamber providing a particular environment of pressure, temperature, magnetic field, or combinations thereof. The sample chamber is mounted on a goniometer, allowing independent rotation of the sample and the detector while the monochromator Bragg angle remains fixed. The neutrons are detected either by an ion chamber filled with BF_3 gas or ^3He gas. Thermal neutrons react with the isotope ^{10}B, emitting alpha particles, which, in turn, cause ionization of the detector filling gas. Likewise, ^3He strongly absorbs neutrons, transforming ^3He to Triton (^3H) and emitting a proton that again ionizes the filling gas.

5.3   Monochromators for X-Ray and Neutron Diffractometers

To conduct angle-dispersive diffraction experiments, the incident X-ray or neutron beam must be monochromatized. Symmetrical Bragg reflection from Si(111) and Ge (111) single crystals is the most common method for X-ray monochromators. They can consist of a single-crystal plate through which the incident beam passes once

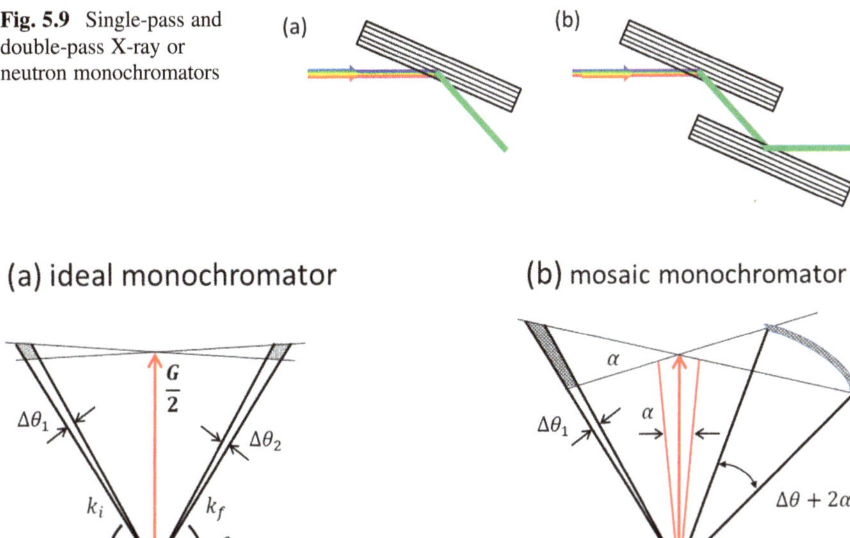

Fig. 5.9 Single-pass and double-pass X-ray or neutron monochromators

Fig. 5.10 Phase space (gray shaded area) of a nearly ideal monochromator (**a**) and a mosaic monochromator with a mosaicity of α (**b**). In both cases, the beam divergence $\Delta\theta$ of the incident and scattered neutron beam is assumed to be the same. (Adapted with permission from Freund 1980: A.K. Freund et al. Lecture Notes in Physics, Springer, Berlin, Heidelberg. Vol 112. (1980). https://doi.org/10.1007/3-540-09727-9_98 (Copyright (2025) by Springer Publishing Company))

(Fig. 5.9a), or of a double-crystal arrangement allowing a double pass (Fig. 5.9b). The first option has the advantage that the monochromator can be bent simultaneously for beam focusing. The second option is preferable if the forward beam direction needs to be maintained except for a small offset. In addition to single- and double-pass monochromators, more complex arrangements are conceivable and are in use. In all cases, the monochromators offer a narrow wavelength bandwidth of $\Delta\lambda/\lambda$ for high-resolution work.

The narrow mosaicity distribution of Si(111) and Ge(111) single crystals is well adapted to the highly collimated beam of synchrotron radiation sources. However, in the case of neutrons, a monochromator with a narrow rocking curve does not make good use of the much larger beam divergence in a neutron guide tube. Here, a monochromator crystal is preferred with a mosaicity matching the beam divergences in the guide tube to make full use of the incident flux. Figure 5.10 compares the usable phase space (angle × momentum) of a nearly perfect monochromator with the increased phase space available for a monochromator with a mosaicity of $\Delta\alpha$ and a beam divergence of $\Delta\theta$ for the incoming and scattered beam. Using the result derived in Exercise 3.4, we obtain the usable wavelength band delivered by the mosaic monochromator:

$$\frac{\Delta\lambda}{\lambda} = \sqrt{(\Delta\alpha + \Delta\theta)^2 \cot^2(\Delta\alpha) + \left(\frac{\Delta d}{d}\right)^2}, \tag{5.4}$$

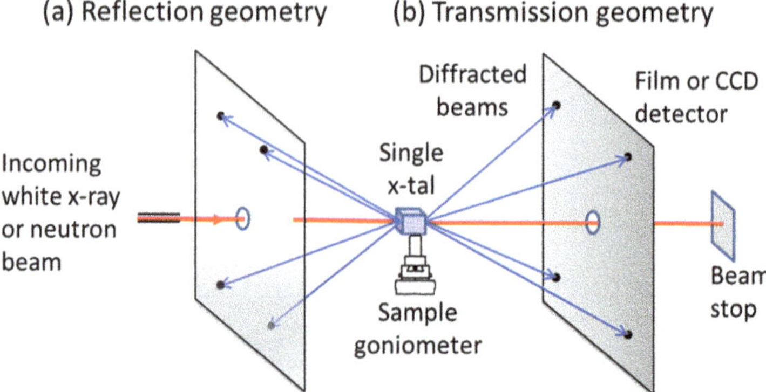

Fig. 5.11 Experimental setup for taking Laue images. The incoming collimated white X-ray or neutron beam enters through a hole in the film and strikes a single crystal. Depending on the orientation, Laue spots are recorded in the forward direction (transmission geometry (**b**)) and backward direction (reflection geometry (**a**)). The recorded Laue spots are characteristic of the crystal symmetry and crystal orientation. In modern versions, the film is replaced by a CCD camera

which is much larger than for the ideal monochromator. Common monochromator crystals for neutron diffraction are highly oriented pyrolytic graphite (HOPG) with a mosaicity of 0.2–0.8°, copper single crystals after some plastic deformation has been performed to increase the mosaicity, and Heusler alloys. More perfect crystals such as Si or Ge can also be used when small crystal plates are arranged in a puzzle-like fashion with small tilt angles in between.[1]

Further considerations for the choice of suitable neutron monochromators are as follows: (1) Large coherent cross section for high reflected intensity; (2) Low incoherent cross section for low background scattering; and (3) Low absorption cross section for low neutron activation. All three conditions are met by the most common neutron monochromators, HOPG and copper.

5.4   Laue Scattering

5.4.1   Basic Setup

Laue scattering is a true single-crystal method. An incident collimated white beam strikes a single crystal of arbitrary symmetry, as indicated in Fig. 5.11. A small goniometer helps align the crystal into a specific orientation. If a particular

---

[1] For high-resolution backscattering neutron spectroscopy, large panels of parabolically arranged Si platelets are used for focusing the scattered neutrons on the detector. This is a topic of inelastic neutron scattering, which is beyond the scope of this text.

wavelength matches the distance and orientation of crystal planes that satisfy the Bragg condition, that wavelength will be diffracted. In the simplest variant, the diffracted beam is recorded by an X-ray film placed perpendicular to the incident beam. A hole in the center of the film prevents overexposure of the film, and the passing primary beam is directed into a beam stop behind the film. Two geometries are used in laboratory equipment: forward scattering or transmission geometry and backward scattering or reflection geometry. Both geometries can be combined as sketched in Fig. 5.11. Backscattering of X-rays is used when the crystal absorption is too high for forward scattering. But the spot intensities are much lower, and longer exposure times are required.

The term "Laue scattering" goes back to the very first X-ray experiment by Max von Laue, Walter Friedrich, and Paul Knipping in 1912, who exposed a ZnS crystal to X-rays generated by a bulky discharge tube. When the transmitted X-rays exposed a film plate, the researchers noticed a blurry image of spots. Despite the low quality of the image, it showed that crystals act like a three-dimensional optical grating with atoms decorating the lattice sites, and second, that the previously unknown radiation consists of waves, electromagnetic waves of very short wavelength. Furthermore, it confirmed the still-fresh idea of the atomic nature of matter and gave a hint at the size of atoms. This discovery earned Max von Laue the Nobel Prize in Physics 2 years later.

The Laue method provides an angle θ of the scattered beam, but the wavelength responsible for a particular spot on the film and the corresponding d-spacing are unknown. Therefore, the Laue method is not practical for crystal structure analysis per se. Instead, the Laue method is primarily used to identify crystal orientations and to align crystals for subsequent experiments. Figure 5.12a shows an example of a Pt (111) surface with a sixfold symmetry axis, recorded in backscattering Laue geometry, reproduced from Arulmozhi and Jerkiewicz (2017). Laue images can be simulated, which is very helpful in identifying the correct crystal orientation. Figure 5.12b shows the Laue simulation of the Cu(111) surface. Both Pt and Cu have the fcc crystal structure and therefore the same basis vectors and the same symmetry. Simulations of Laue patterns are available in several places. The one in Fig. 5.12b is reproduced from Preuss et al. (1974). In fact, the symmetry axis of the fcc(111) plane is threefold. However, since the phase is not preserved in the X-ray and neutron scattering intensity, an inversion center is automatically added to the intensity distribution, causing the fcc(111) surface to appear with a sixfold symmetry axis. In Sect. 5.6, we discuss a similar example of LEED spots from the Ni(111) surface, which shows the true threefold symmetry. Regarding the phase problem, see Sect. 3.5.5.

As already mentioned, the experimental setup of Laue cameras is very simple. It requires only an X-ray or neutron source and collimation of the incident beam. This makes them attractive not only for orienting single crystals in the laboratory, but also for more sophisticated experiments at large-scale facilities. For example, synchrotron radiation can be used to study tiny micrometer-sized crystals with a highly collimated beam, and neutrons can be used to focus on hydrogen- or deuterium-containing crystals in transmission geometry.

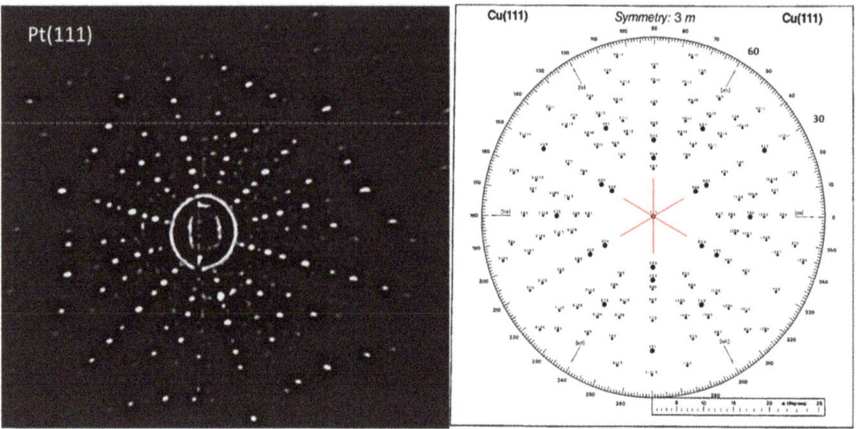

Fig. 5.12 (**a**) Laue image of Pt(111). The circle in the center is the central hole in the film. (Reproduced with permission from the publisher from Arulmozhi and Jerkiewicz 2017 (Copyrighted (2025) by Springer Nature)). (**b**) Simulation of the Laue spots for the Cu(111) surface, which is equivalent to the Pt(111) surface. (Reproduced from Preuss et al. 1974 with permission from Springer Publishing Company). The sixfold symmetry is clearly visible in both cases. The red lines are guides to the eye

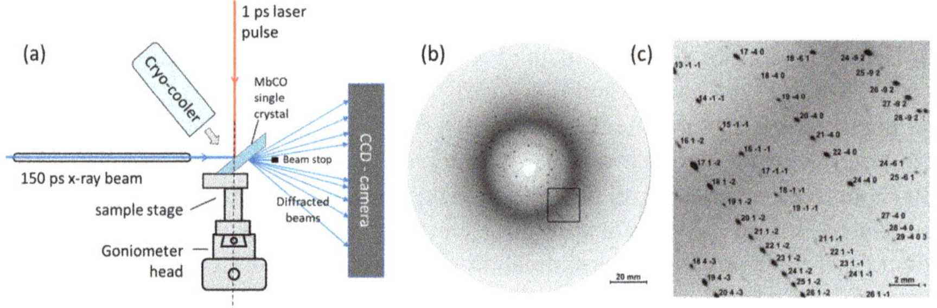

Fig. 5.13 Laue diffraction pattern of crystallized carbon-monoxide bonded to myoglobin (MbCO). Laue patterns are taken with a time delay of 100 ps after photolyzation of CO. (**a**) Experimental setup; (**b**) Laue diffraction pattern; (**c**) enlarged section of the pattern labeled with Miller indices. (Adapted from Schotte et al. 2003 and reproduced with permission from Schotte et al. 2004: F. Schotte et al. Journal of Structural Biology, **147**, 235–246 (2004), with permission from Elsevier)

5.4.2 Protein Functioning with 150-ps Time Resolution

In the following, we discuss briefly a time-resolved Laue investigation of the protein myoglobin (Mb). The experimental setup used by the team at the ESRF facility is outlined in Fig. 5.13a, which consists, in addition to the standard goniometer head and a camera, of a cryo-cooler to keep the sample in a crystalline state and a laser enabling pump-probe experiments (Schotte et al. 2003, 2004). Mb has a structure similar to hemoglobin (Hb). Both proteins contain flat heme molecules that reversibly bind oxygen (O_2) to Fe^{2+} at the center for oxygen storage and transport. Mb is found in muscles, and Hb in the circulatory system. Carbon monoxide (CO) is poisonous for Hb and Mb, because the bonding sites are the same as for O_2, but the bonding strength is much higher. Therefore, the cyclic binding and release of O_2 is hampered. However, CO can be released from its bonding site by an infrared laser pulse, which results in a conformational relaxation of the Mb protein. The entire process can be studied in "life" by laser excitation, followed by time-resolved X-ray Laue diffraction in the single bunch mode operation of the synchrotron source. The *operando* Laue pattern is shown in Fig. 5.13b, c. The Laue pictures were taken 100 ps after the laser excitation, and 32 such recordings were overlaid to collect sufficient intensity. A total of 3000 Laue spots could be recorded for a wavelength band of about 0.072–0.124 nm; some of them are labeled with Miller indices in the enlarged section of panel (c). Taking a Fourier transform of the Laue spots, it can be seen how the conformational relaxation of myoglobin proceeds in time and space for the release of CO.

This example highlights the capabilities of Laue diffraction techniques. Not only can static structures of materials be explored, but in addition, their response to external parameters, such as temperature, pressure, magnetic field, gas exchange, photocatalyzation, and others.

5.5   Coherent Diffraction Imaging (CDI)

Using small single crystals opens up new opportunities for coherent X-ray Bragg diffraction at synchrotron radiation sources. We will discuss briefly two examples, one concerning nanowires and the other focusing on high-pressure studies of nanocrystals.

5.5.1   Nanowire CDI

The first example regards the CDI of an InAs(111) nanowire grown on an InP(111) substrate (Diaz et al. 2009). The wires on the substrate have diameters of about 150 nm and lengths of about a micrometer. They are well separated, such that with a highly collimated X-ray beam, a single wire can be illuminated. The nanowires have hexagonal symmetry with six {10.0} oriented facets.

A coherent X-ray beam with an energy of 8 keV selected from a Si(111) monochromator was used and focused with the help of a Fresnel zone plate on the InAs nanowire. The beam size was $350 \times 400 \ nm^2$, which covers completely the nanowire with a diameter of 150 nm. The experimental setup is shown schematically in Fig. 5.14a.

The incoming and exit beams are color-coded in yellow. The calculated and expected intensity distribution around the (111) Bragg reflection (reciprocal space

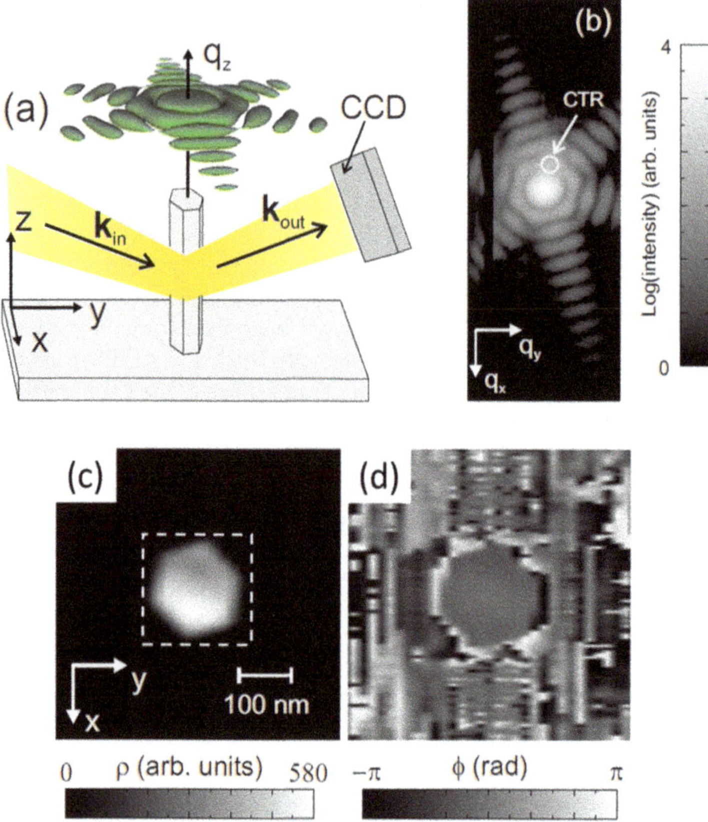

Fig. 5.14 2D reciprocal space map recorded around the (111) Bragg reflection of a single InAs nanowire. (**a**) Schematics of the scattering geometry with a monochromatic incident synchrotron radiation beam. The green-colored areas represent the expected and calculated intensity distribution around the (111) Bragg reflection. (**b**) Measured intensity distribution in the xy-plane, showing sixfold symmetry and intensity fringes up to the tenth order. (**c**) Magnitude of the electron density distribution obtained from back Fourier transformation; (**d**) phase of the scattering amplitude retrieved from oversampling. (Reprinted with permission from Diaz et al. 2009: A. Diaz, et al. Phys. Rev. B **79**, 125324 (2009). DOI: 10.1103/PhysRevB.79.125324. (Copyright (2025) by the American Physical Society))

map) is shown in green. The recorded RSM is reproduced in panel (b), seen as a projection in the xy-plane. The one-to-one correspondence between the calculated and recorded intensity is clearly visible. Finite size Laue fringes up to the tenth order are visible in panel (b), indicative of both high crystal quality and high beam coherence.

The Laue fringes seen in panel (b) are those which we called Laue oscillations in Fig. 3.5 and which we will discuss again in Sect. 10.2. The circle at the center marks a crystal truncation rod from the substrate and can be disregarded.

The back Fourier transform of the intensity yields the magnitude of the electron density, which is shown in panel (c), and the phase is reproduced in panel (d). Inside the wire, the electron density and the phase are constant, which one would expect if there is no strain gradient. Outside, the phase should be zero, but it shows some noise, the origin of which is unclear. These types of experiments have the potential to take high-resolution pictures of physical parameters inside nanoobjects. Therefore, CDI has also been called lensless imaging. In fact, using CDI with a nanosized beam, the boundaries between diffraction and microscopic imaging becomes blurred.

5.5.2   High-Pressure CDI

In another CDI experiment, a very high pressure was applied to a single-crystalline Au(111) nanoparticle, using a diamond anvil cell (DAC) (Yang et al. 2013). The smaller the sample volume is, the higher the maximum pressure that can be applied. The smallest crystals investigated so far had edge lengths on the order of 10 μm. However, with the highly brilliant radiation from an undulator X-ray source, there is sufficient intensity to probe even samples on the sub-micrometer length scale. Yang et al. have used a coherent X-ray diffraction imaging technique to study the internal strain distribution of an Au nanometer-sized single crystal. They used a 10.8 keV (0.114 nm) X-ray beam, selected by a Si(111) double-crystal monochromator and focused on the Au nanoparticle in the DAC. The beam passed either through the Beryllium gasket or through the diamond anvils. Both have low atomic numbers (Z) and therefore absorb only a small fraction of the X-ray beam. The pressure cell is outlined in Fig. 5.15a, scanning electron microscopy pictures of the Au nanocrystals are shown in panel (b), and the scattering setup together with the detector recording at 1 m distance from the sample is reproduced in panel (c). The combination of a single crystal and a coherent X-ray beam in the longitudinal and transverse directions results in a Bragg reflection that displays fringes on either side of the central peak.

Coherent Bragg diffraction allows for the inversion of the three-dimensional diffraction pattern back to real-space images. Prerequisite for the back transformation is the phase retrieval, as the phase is lost during the scattering process (see Sect. 3.5.5 on the phase problem). The algorithms for phase retrieval have been successfully developed in recent years. Hence, coherent diffraction imaging has been called lensless imaging, imaging without using a microscope. In the present experiment, the real-space images were used to follow the structural changes of the Au nanoparticle with increasing pressure and to analyze, in particular, the local strain evolution between pressures of 0.8 and 6.4 GPa (Fig. 5.15).

In contrast to conventional high-pressure experiments with polycrystalline or powder samples, where a broadening of the Bragg reflections is observed with increasing pressure, the width of the Au(111) Bragg reflection surprisingly decreased with increasing pressure. Evidently, isolated grains and grain clusters in polycrystals behave differently under pressure. In the latter case, grain–grain interaction plays an important role, which is absent in the isolated state. For materials science, the combination of coherent X-ray diffraction imaging with individual

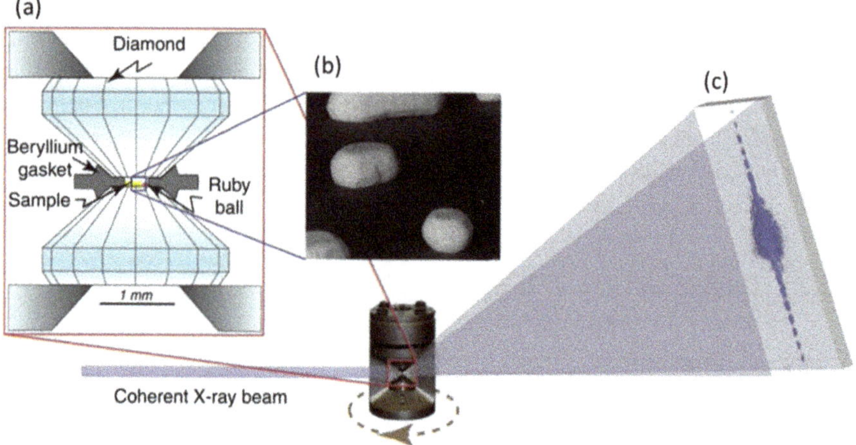

Fig. 5.15 Experimental setup for coherent diffraction at high pressures. (**a**) Coherent diffraction of an Au(111) nanocrystal in a diamond anvil pressure cell, kept in place by a Be gasket. A coherent synchrotron beam, shown in the lower part of panel (**a**), excites the Au(111) Bragg reflection. (**b**) Magnified view of the Au nanoparticle. (**c**) The detector shows the central (111) Bragg peak together with eight fringes on either side. (Reproduced with permission from: Wenge Yang et al. Nat. Commun. **4**, 1680 (2013) (Copyright (2025) by Springer Nature))

nanograins is a novel and promising technique that can only be performed at modern third- or fourth-generation synchrotron radiation sources.

These are just a few examples of modern single-crystal research, which requires the high brilliance, coherence, and narrow beam of undulator radiation. Furthermore, pulse probe experiments enable investigations in the time domain by studying the evolution of structural and electronic parameters that return to equilibrium after an initial ultrashort pulse excitation or trigger a chemical reaction in the femtosecond time range. Further examples are discussed in the next chapters.

5.6   Low-Energy Electron Diffraction (LEED)

Electrons are extremely surface sensitive and cannot penetrate deep under the surface of a crystal. This has two important consequences. First, the three-dimensional reciprocal space has to be adapted to surface conditions. Second, low-energy electron diffraction can be used for studying the surface structures and possible reconstructions of clean single-crystalline surfaces. We discuss the adaptation of the reciprocal lattice to surface conditions first. Referring to Eq. (3.30), the reciprocal lattice points in three dimensions are defined by the three Laue equations:

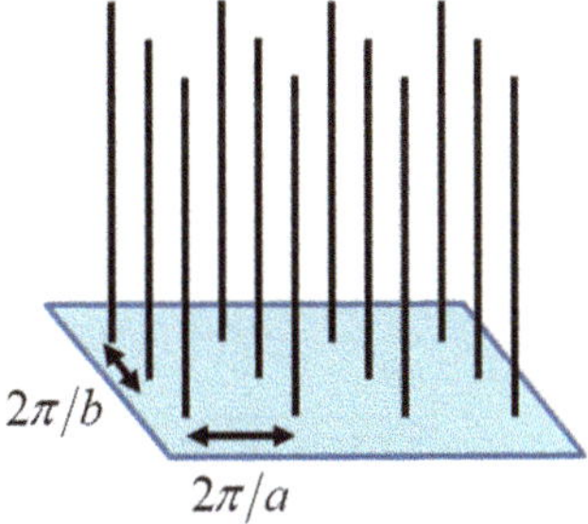

Fig. 5.16 Reciprocal lattice rods describe the structure of a two-dimensional crystal lattice

$$Q \cdot a = G_h \cdot a = 2\pi h,$$
$$Q \cdot b = G_k \cdot b = 2\pi k, \qquad (5.5)$$
$$Q \cdot c = G_l \cdot c = 2\pi l.$$

However, ideal planar surfaces have a dimensionality of two, and therefore one of the Laue equations for the direction normal to the surface (z-direction) is annulled. The remaining two Laue equations with Miller indices h and k are

$$Q \cdot a = G_h \cdot a = 2\pi h,$$
$$Q \cdot b = G_k \cdot b = 2\pi k. \qquad (5.6)$$

These two equations do not define reciprocal lattice points in three dimensions but instead ***reciprocal lattice rods***. The reciprocal lattice rods represent the surface structure of a crystal, as indicated in Fig. 5.16.

In principle, any point along such a rod would fulfill the Bragg condition according to the definition of the two-dimensional reciprocal lattice vector. However, this will eventually violate the momentum conservation for elastic scattering, the condition requiring that the scattering triangle has to be closed, as discussed in Sect. 2.3, Fig. 2.3. The extra condition of momentum conservation can be graphically implemented by defining a so-called **Ewald sphere**, as shown in the right panel of Fig. 5.17.

The Ewald sphere touches the origin of the reciprocal lattice at (00). The incident wave vector k_i is drawn from the origin to the center of the sphere. Thus, the radius of the sphere is determined by the energy or wavelength of the incident electrons. Any outgoing wave vector k_f from the center to the periphery fulfills the condition of momentum conservation for elastic scattering, because in this case $|k_i| = |k_f|$. Both requirements, momentum conservation and Bragg condition, are met whenever the reciprocal lattice rods and the wave vector k_f merge on the Ewald sphere. Those are then the two-dimensional lattice points, which can be observed on the fluorescence screen of a LEED system, as indicated in Fig. 5.17, left panel.

LEED images can be used to assess surface quality, including smoothness, roughness, and contamination. However, it is particularly important to observe surface structures, reconstructions, and phase transitions depending on external

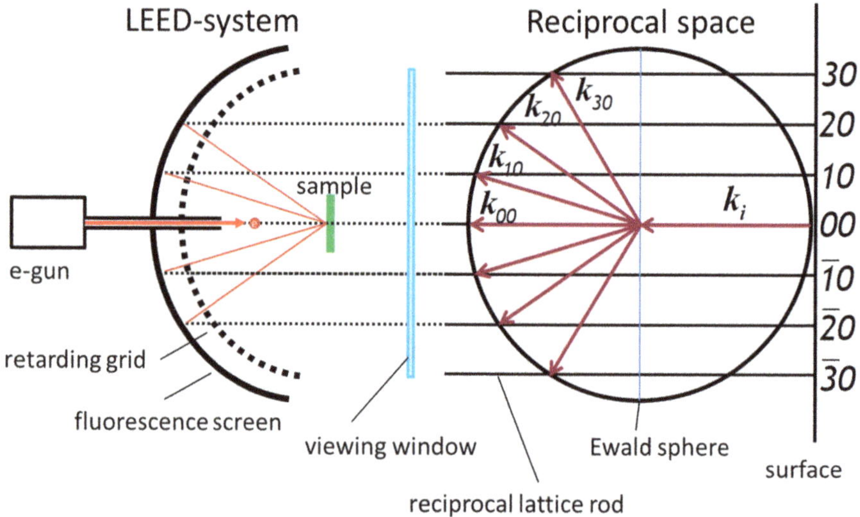

Fig. 5.17 Right panel: reciprocal space for two-dimensional crystal lattices consisting of lattice rods. Superposed on the lattice rods labeled by (*hk*) Miller indices is the Ewald sphere with a radius $|k_i|=|k_f|$. Allowed Bragg reflections occur at the cross points of reciprocal lattice rods and the Ewald sphere. Left panel: Experimental setup for low-energy electron diffraction. Red lines indicate Bragg reflected electrons to be observed on the fluorescence screen

Fig. 5.18 LEED pattern of a Cr(10) surface with fourfold symmetry. (Reproduced with permission from Bödeker et al. 1999: P. Bödeker et al. Phys. Rev. B 59, 9408–9431 (1999) DOI: https://doi.org/10.1103/PhysRevB.59.9408 (Copyright (2025) by the American Physical Society))

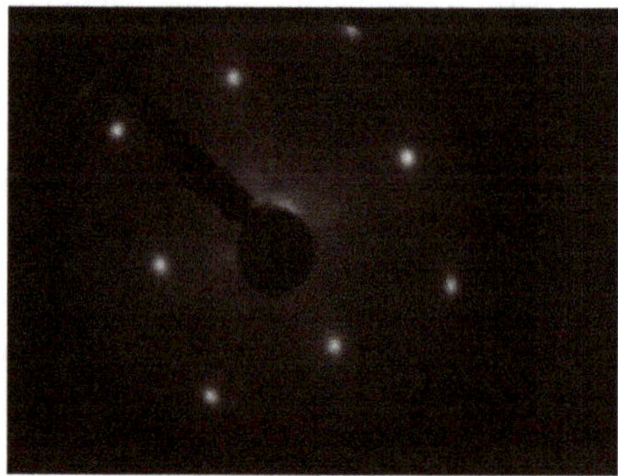

parameters such as temperature or interaction with residual gases. An example of a LEED pattern taken from a clean chromium (100) surface is shown in Fig. 5.18. The fourfold symmetry of the LEED (10) spots is clearly visible. LEED systems are important tools for investigating surfaces not only of cleaned bulk single crystals but also of thin films and heterostructures grown by molecular beam epitaxy (MBE). LEED is usually integrated into an MBE system for post-growth characterization and analysis.

Figure 5.19 shows another example of the (00.1) surface of hexagonal Yttrium (Bödeker et al. 1999). Yttrium has a true sixfold symmetry axis along the *c*-direction,

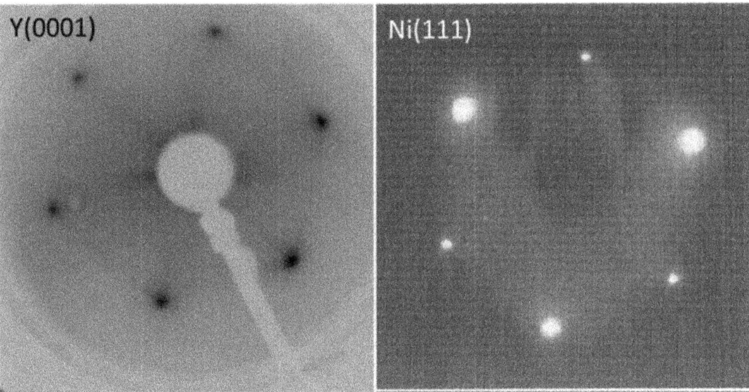

Fig. 5.19 Left panel: LEED image of the Y(0001) surface at an electron energy of 63 eV showing a six-fold symmetry as expected for a hexagonal crystal. (Reproduced from Remhof 1999 (courtesy of A. Remhof)). Right panel: LEED image of the fcc Ni(111) surface, showing the correct threefold symmetry axis, in contrast to the Laue image of the Pt(111) surface in Fig. 5.12. Six Bragg reflections are observed, but with different intensities. (Reproduced from Ibach and Lüth 2009: with permission from Springer Nature)

not a pretended one by adding an inversion center. In contrast, in the LEED image to the right, we observe a threefold symmetry axis of the fcc-Ni(111) surface (Remhof 1999). This is in contrast to the Laue picture of Pt(111) in Fig. 5.12, which revealed a sixfold symmetry axis, although not physically present. The difference is due to the strong absorption of electrons in metals that violates the conditions for the automatic addition of an inversion center.

5.7   Reflection High-Energy Electron Diffraction

For in situ growth analysis, a different but similar electron diffraction method is often used, called reflection high-energy electron diffraction (RHEED). This method differs from LEED in its use of higher-energy electrons in the range of 10 keV up to 50 keV and its scattering geometry. In RHEED, the electrons strike the surface at a glancing angle of approximately 1–3°, in contrast to LEED, where the angle of incidence is 90° (perpendicular to the surface). The higher electron energy and glancing angle scattering geometry are compatible with the in situ growth environment that RHEED is primarily used for to monitor.

The use of natural single crystals is common in geosciences, but not in physical sciences. Natural crystals often lack sufficient purity for fundamental studies. Starting in the 70s of the last century, scientists have developed methods to grow artificially single-crystalline films on substrates. The single-crystalline substrates serve as templates for the epitaxial overgrowth. First, semiconductor films and multilayers were grown. Later, in the 80s, metals were added, and today almost

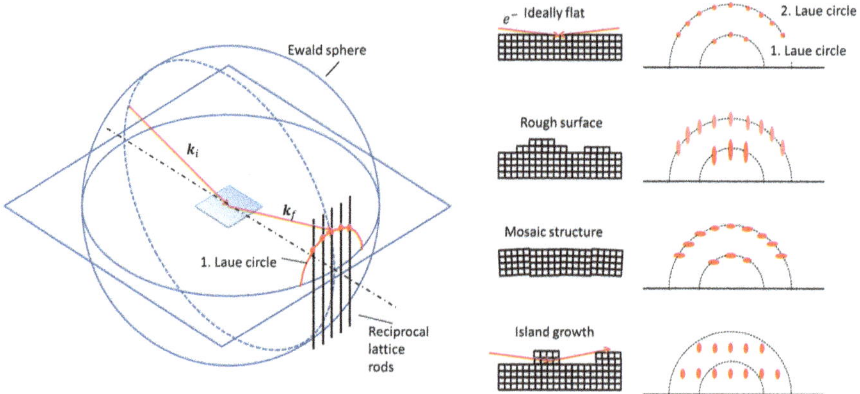

Fig. 5.20 Left panel: Reciprocal lattice rods pierce through the Ewald sphere resulting in a 1. Laue circle. Middle panel: from flat to rough surface morphologies. Right panel: corresponding RHEED patterns

any material in the form of films, heterostructures, or multilayers can be artificially fabricated by various deposition techniques in ultraclean environments. One of the most frequently used growth methods for the preparation of epitaxial single-crystalline films on substrates is ***molecular beam epitaxy*** (MBE) (Herman and Sitter 1996). MBE requires a big ultrahigh vacuum chamber containing so-called Knudsen cells for thermal evaporation of materials at high temperatures and electron beam evaporators for materials with melting temperatures not attainable by Knudsen cells. Included in the MBE chamber is also a substrate holder with a heater on the backside for cleaning and annealing the substrate before deposition and for adjusting an optimized growth temperature during evaporation. Having all this at hand, one would like to observe in situ the quality of the growing film and readjust the growth parameters if needed. X-ray scattering would, in principle, be possible, but it is cumbersome and not fast enough for direct control. Efforts in this direction have largely failed in the past. RHEED, in contrast, meets the growth requirements and allows the electrons to penetrate the evaporating plume while not obstructing the area between evaporation cells and the substrate (Herman and Sitter 1996).

Due to the limited penetration depth of the electrons into the sample, the reciprocal lattice rods are the same for RHEED and LEED. These rods pierce the ***Ewald sphere*** and form a first-order ***Laue circle*** with discrete reflection spots of the surface structure, as shown in the left part of Fig. 5.20. Depending on the electron energies, a second-order Laue circle with Bragg points can also be observed during electron scattering from an ideally smooth surface. However, as the surface roughens, the Bragg reflections on the Laue circles elongate into stripes. Eventually, they form a three-dimensional reciprocal lattice when electron transmission through islands grown on the surface dominates over electron reflection from flat parts. Because RHEED detection is very fast, it is possible to determine immediately and precisely how a film grows on a substrate, whether it has a crystalline structure, how the

Fig. 5.21 RHEED image of the Cr(001) surface along the [100] azimuth. The first and second Laue rings are visible, but show some elongation, indicating the presence of steps on the surface and/or some mosaicity. (Reproduced with permission from Bödeker et al. 1999: P. Bödeker et al. Phys. Rev. B 59, 9408–9431 (1999) DOI: https://doi.org/10.1103/PhysRevB.59.9408 (Copyright (2025) by the American Physical Society))

surface symmetry and orientation relate to the substrate, and whether the film roughens during growth or develops a polycrystalline structure.

By carefully recording the intensity of a Laue spot, one can even monitor in real time the number of atomic layers deposited. A complete layer shows a maximum RHEED spot intensity, whereas a half-filled layer causes destructive interference between reflections from the top layer and the one below, resulting in an intensity drop. The number of intensity oscillations observed equals the number of atomic layers deposited. Figure 5.21 shows an in situ RHEED picture from a Cr(001) surface taken during MBE growth. Further information on RHEED and MBE is available in (Herman and Sitter 1996; Ichimiya and Cohen 2004).

Summary

1. Single crystals feature a perfect lattice periodicity and consist of only one crystal orientation.
2. In single-crystal diffraction with X-rays or neutrons, two main scattering geometries are used: reflection geometry and transmission geometry.
3. When performing scans, we distinguish between radial scans and transverse scans.
4. Radial scans reveal lattice parameters of the crystal lattice; transverse scans probe the angular mosaic distribution of crystallites within a single crystal.
5. In radial scans, the FWHM of Bragg reflections scales with the inverse number of scattering centers. The peak height scales with the square of the scattering

centers. The integrated intensity is proportional to the number of scattering centers.

6. The total intensity of Bragg reflections predicted by the kinematic theory scales with the square of the structure factor and the third power of the wavelength.

7. The perfect crystal lattice has a lower reflected intensity than the imperfect one due to primary and secondary extinction effects.

8. Radial and transverse scans can be combined in reciprocal space maps, where the intensity in the area around a Bragg reflection is scanned point-by-point.

9. Si, Ge, and diamond form nearly perfect crystal lattices with narrow mosaicity. They are useful as X-ray monochromators, but less useful as neutron monochromators.

10. A four-circle diffractometer is the most versatile instrument for single crystal crystallography.

11. We distinguish between single-pass and double-pass monochromators.

12. Neutron monochromators must have a larger mosaicity to make use of a higher beam divergence than is the case for X-rays.

13. Laue scattering methods use a "white" incident beam. Diffracted beams can be observed in transmission geometry and/or reflection geometry.

14. The Laue scattering method is mainly used for single-crystal orientation. As it is a fast recording method, it can also be used for time-dependent recording of sample changes.

15. X-ray beams with high longitudinal and transverse coherence can be used for studying single nanocrystals and for taking coherent diffraction images. The back transformation of these images acts as a lensless microscope image.

16. The reciprocal lattice of crystal surfaces is defined by two Laue equations instead of three. Therefore, the reciprocal lattice of surface structures consists of reciprocal lattice rods instead of reciprocal lattice points.

17. Whenever reciprocal lattice rods touch the Ewald sphere, the Bragg condition is fulfilled.

18. With low-energy electron diffraction (LEED), the surface crystal structure and reconstruction can be probed.

19. The surface sensitivity of electrons can be enhanced by selecting a glancing incident angle of the electrons to the surface. The method is referred to as reflection high-energy electron diffraction (RHEED) and allows in situ and real time studies of crystal growth.

Questions

(Note that more than one answer may be correct)

1. **Which scan is required for determining the lattice spacing d_{hkl} of atomic layers?**

   (a) Radial scan.
   (b) Transverse scan.

    (c) Rocking scan.

    (d) Longitudinal scan.

2. **What is a mosaic crystal?**

    (a) Mosaic crystals are disordered crystals.

    (b) Mosaic crystals lack translational symmetry.

    (c) Mosaic crystals are single crystals composed of many small crystallites, where the crystalline axes exhibit a small angle tilt against each other.

3. **How can the mosaicity of crystals be determined?**

    (a) By transverse scans.

    (b) By radial scans.

    (c) By chemical analysis.

    (d) By rocking a crystal in an X-ray beam.

4. **Why are Si and Ge single crystals used as monochromators for work with X-rays in home laboratories and at synchrotron sources?**

    (a) Because of their small mosaicity

    (b) Because of their high X-ray absorption

    (c) Because of the elimination of the $\lambda/2$ component

    (d) Because of their high reflectivity.

5. **When studying crystal structures with low-energy electron diffraction (LEED), which conservation law needs special attention?**

    (a) Energy conservation.

    (b) Momentum conservation.

    (c) Second law of thermodynamics.

6. **How many Laue equations are relevant to describe LEED diffraction patterns?**

    (a) Three.

    (b) Two.

    (c) Four.

7. **Under what circumstances is neutron scattering preferred over X-ray scattering?**

    (a) When isotope contrast is required.

    (b) Whenever light atoms need to be seen in the neighborhood of heavy elements;

    (c) When the sample is in a liquid state.

    (d) When X-ray absorption is too strong.

Exercises

Grades of difficulty: E = easy, M = medium, A = advanced.

M 5.1. Scherrer equation

The Scherrer equation relates the full width at half maximum (*FWHM*) of a Bragg
peak $B_{FWHM} = \Delta(2\theta)$, determined by a radial scan, to the particle size $L = Na$ of
the medium scattered from:

$$B_{\mathrm{FWHM}} = \Delta(2\theta)_{hkl} = \frac{0.94\lambda}{L\cos(\theta_{hkl})}.$$

Here λ is the X-ray wavelength and θ_{hkl} is the Bragg angle of a (hkl)-reflection. N is
the number of lattice planes contributing to the Bragg reflection, and a is the
lattice parameter.

(a) Show how an approximate expression of the Scherrer equation can be
obtained from Bragg's law. Hint: Start with the definition of the scattering
vector and take the first derivative concerning the scattering angle θ. Also,
make use of the relation $dQ = 2\pi/L$, i.e., the line broadening in Q-space, dQ,
is proportional to the reciprocal crystalline size L in the direction of the
scattering vector.

(b) Small crystallite size is the main contribution to the broadening of Bragg
reflections in the radial direction. What else could lead to line-broadening
effects?

A 5.2. Beryllium filter

In thermal neutron scattering, often the primary neutron beam, after being deflected
from a graphite monochromator, is passed through a so-called *low-pass Be-filter*.
The Be-filter contains polycrystalline beryllium kept at low temperatures. The
scattering geometry is outlined in the schematics of Fig. 5.22. Can you imagine
how such a filter works and how it acts on the neutron beam? The term "filter"
implies that not all neutrons make it through, like in a sieve that stops larger
particles from passing through. Why is the cooling of the filter beneficial?

M 5.3. Silicon monochromator

In X-ray scattering, often the Si(111) reflection is used for monochromatization of
the incident X-ray beam and simultaneously as a *low-pass filter*. The term
'low-pass filter' implies that only lower energy or longer wavelength X-rays
are transmitted, whereas higher energy or shorter wavelength X-rays are stopped.
Can you explain how the Si(111) filter works for higher harmonic X-ray wavelengths? Start your answer by discussing the crystalline structure of Si (see
Chap. 1) and the diamond structure factor (see Exercise 3.2). Which Bragg
reflections are allowed and which ones are forbidden?

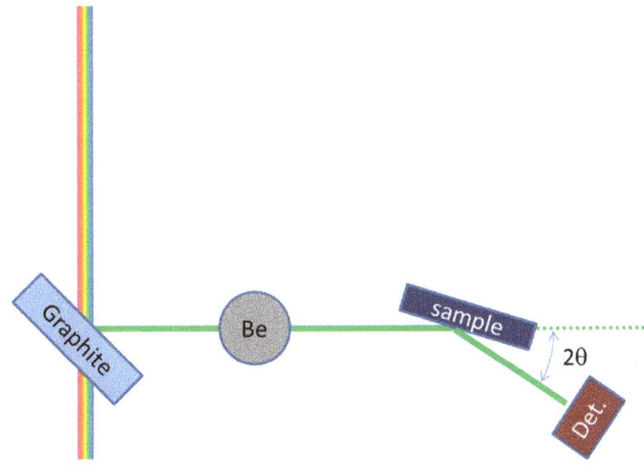

Fig. 5.22 Placement of a Be-filter in a thermal neutron diffraction experiment

References

J. Als-Nielsen, D. McMorrow, *Elements of Modern X-Ray Physics*, 2nd edn. (Wiley, New York, 2011)

N. Arulmozhi, G. Jerkiewicz, Design and development of instrumentations for the preparation of platinum single crystals for electrochemistry and electrocatalysis research. Part 2: orientation, cutting, and annealing. Electrocatalysis **8**, 399–413 (2017). https://doi.org/10.1007/s12678-017-0385-7

P. Bödeker, A. Schreyer, H. Zabel, Spin-density waves and reorientation effects in thin epitaxial Cr films covered with ferromagnetic and paramagnetic layers. Phys. Rev. B **59**, 9408–9431 (1999)

A. Diaz, C. Mocuta, J. Stangl, B. Mandl, D. Combe, et al., Coherent diffraction imaging of a single epitaxial InAs nanowire using a focused X-ray beam. Phys. Rev. B Condens. Matter Mater. Phys. **79**, 125324 (2009)

A.K. Freund, Neutron optics using non-perfect crystals, in *Imaging Processes and Coherence in Physics*, Lecture Notes in Physics, ed. by M. Schlenker, M. Fink, J.P. Goedgebuer, C. Malgrange, J.C. Vieénot, R.H. Wade, vol. 112, (Springer, Berlin, 1980). https://doi.org/10.1007/3-540-09727-9_98

M.A. Herman, H. Sitter, *Molecular Beam Epitaxy—Fundamentals and Current Status* (Springer Verlag, Berlin, 1996)

H. Ibach, H. Lüth, *Solid-State Physics, an Introduction to Principles of Materials Science* (Springer, Berlin, 2009)

A. Ichimiya, P.I. Cohen, *Reflection High Energy Electron Diffraction* (University Press, Cambridge, 2004)

T. Petrisor Jr., M.S. Gabor, A. Boulle, C. Bellouard, C. Tiusan, O. Pana, T. Petrisor, Oxygen incorporation effects in annealed epitaxial $La_{(1-x)}Sr_xMnO_3$ thin films. J. Appl. Phys. **109**, 123913 (2011)

E. Preuss, B. Krahl-Urban, R. Butz, L. Atlas, *Plotted Laue Back-Reflection Patterns of the Elements, the Compounds RX, and RX2* (Springer Verlag, Berlin, 1974)

A. Remhof, Dissertation, Ruhr-Universität Bochum, 1999

F. Schotte, M. Lim, T.A. Jackson, A.V. Smirnov, J. Soman, J.S. Olson, G.N. Phillips Jr., M. Wulff, P.A. Anfinrud, Watching a protein as it functions with 150-ps time-resolved X-ray crystallography. Science **300**(5627), 1944–1947 (2003). https://doi.org/10.1126/science.1078797

F. Schotte, J. Soman, J.S. Olson, M. Wulff, P.A. Anfinrud, Picosecond time-resolved X-ray crystallography: probing protein function in real time. J. Struct. Biol. **147**, 235–246 (2004)

S. Stoupin, T. Krawczyk, Z. Liu, C. Franck, Selection of CVD diamond crystals for X-ray monochromator applications using X-ray diffraction imaging. Crystals **9**, 396 (2019)

B.E. Warren, *X-Ray Diffraction* (Dover Publications, New York, 1990) reprint, originally Addison-Wesley Publ. Comp., 1969

W. Yang, X. Huang, R. Harder, J.N. Clark, I.K. Robinson, H.-k. Mao, Coherent diffraction imaging of nanoscale strain evolution in a single crystal under high pressure. Nat. Commun. **4**, 1680 (2013). https://doi.org/10.1038/ncomms2661

Further Reading

U. Pietsch, V. Holy, T. Baumbach, *High-Resolution X-Ray Scattering: From Thin Films to Lateral Nanostructures*, Advanced Texts in Physics (Springer Verlag, Berlin, 2011)

C.-R. Wie, High resolution X-ray diffraction characterization of semiconductor structures. Mater. Sci. Eng. **R13**, 1–56 (1994)

Chapter 6
Elastic Scattering Methods: Powder Diffraction

6.1 Debye–Scherrer Diffraction

Diffraction from a fine crystalline powder is the other extreme to scattering from an extended single crystal. In crystalline powders, each tiny crystallite (micrometer size) is a single crystal that is randomly oriented in space. For scattering experiments, the powder is confined to a small volume in the scattering center. When exposed to a monochromatic and collimated X-ray or neutron beam, there are always some crystallites that happen to have the correct orientation θ_{hkl} to the incident beam and satisfy the Bragg condition for the (hkl)-Bragg reflection at $2\theta_{hkl}$. These crystallites lie on a cone with an opening angle of $2\theta_{hkl}$, and the respective Bragg angle lies on a cone with an opening angle of $4\theta_{hkl}$, as shown schematically in Fig. 6.1. At a different angle θ, the condition for a different Bragg peak is satisfied, etc. Thus, with this scattering geometry, a set of Bragg reflection cones in the forward and backward directions can be collected simultaneously.

In the past, X-ray diffraction of a powder sample was carried out with a Debye–Scherrer camera and was often practiced in advanced laboratory courses. The scattering process is called *Debye–Scherrer diffraction* (Klug and Alexander 1974). A Debye–Scherrer camera, shown schematically in Fig. 6.2, consists of a closed cylindrical metal box with two opposite holes for the incident beam and the transmitted exit beam, respectively. The transmitted beam goes into a beam stop. On the inside of the metal cylinder, an X-ray film is fixed, and the film is exposed for a couple of minutes to an X-ray beam of fixed wavelength λ to achieve a pronounced blackening that shows a section of the Debye–Scherrer rings from a powder sample. Knowing the wavelength of the incident radiation and measuring the diameter of the rings allows for calculating the respective lattice spacing. The existence of allowed Bragg reflections together with the absence of forbidden reflections provides a clear indication of the crystal structure and symmetry. Powder diffraction patterns are also useful for determining the squared structure factor $|F(G_{hkl})|^2$ of Bragg reflections. Standard Si powder samples with very well-documented structure and lattice

© The Author(s), under exclusive license to Springer Nature Switzerland AG 2026

H. Zabel, *Elements of Elastic Scattering by X-Rays, Neutrons, and Electrons*,

https://doi.org/10.1007/978-3-032-16624-1_6

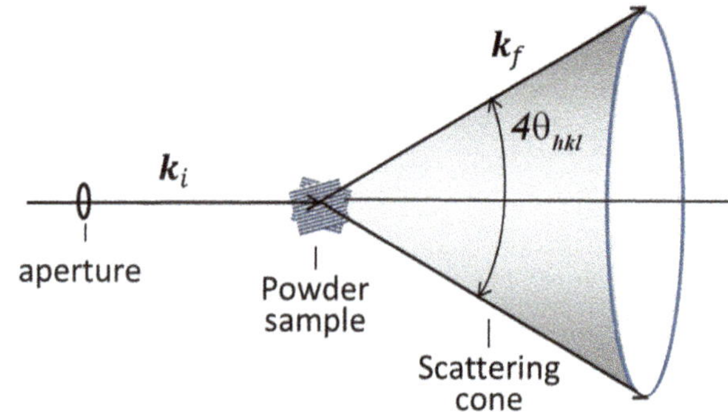

Fig. 6.1 Scattering geometry for X-ray or neutron scattering from a powder sample. The scattered intensity of a Bragg reflection lies on a diffraction cone with the opening angle $4\theta_{hkl}$

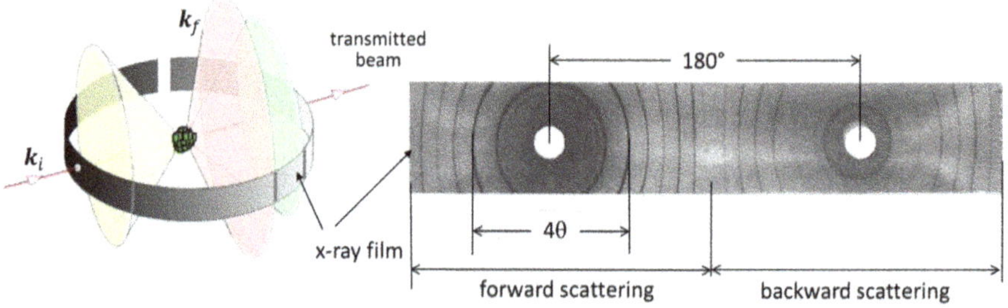

Fig. 6.2 Left panel: Debye–Scherrer X-ray camera for recording Bragg-reflection cones from a powder sample. On the inside of a cylindrical metal box is fixed an X-ray sensitive film. Right panel: flat projection of the exposed X-ray film. Shown is the diffraction pattern of a niobium wire with bcc crystal symmetry, exposed by Cu K_α radiation. The opening of the cone angle 4θ can be determined by scaling with the distance between the holes for the incident and exit beams, corresponding to 180° (left graph adapted from Klug and Alexander 1974)

parameters are useful for reference and calibration (Hubbard et al. 1975). However, the quantitative evaluation of the structure factor is hampered by several difficulties, including absorption effects, sample eccentricity, nonlinear blackening of the film with exposure time, and nonrandom crystallite distributions in the powder sample. Today, Debye–Scherrer cameras are replaced by digital counting techniques, and films are replaced by position-sensitive area detectors, which will be discussed later in this chapter. Nevertheless, for a quick overview, the Debye–Scherrer camera may still be useful.

Frequently, the diffraction pattern of the crystalline powder shows incomplete Debye–Scherrer rings, i.e., rings that have a nonconstant, patchy intensity distribution along the ring, as seen in Fig. 6.3. This may be due to crystallites with a preferred nonrandom orientation. The preferred orientation is called *texture*. Textures are often observed in metal wires. Metal wires and metal sheets are usually polycrystals, i.e., randomly oriented and fused crystallites. Untreated metal wires have a completely random distribution of crystallites. However, after cold rolling or stretching, the crystallites rotate and orient themselves in the stress field, creating a textured structure such as the one seen in Fig. 6.3.

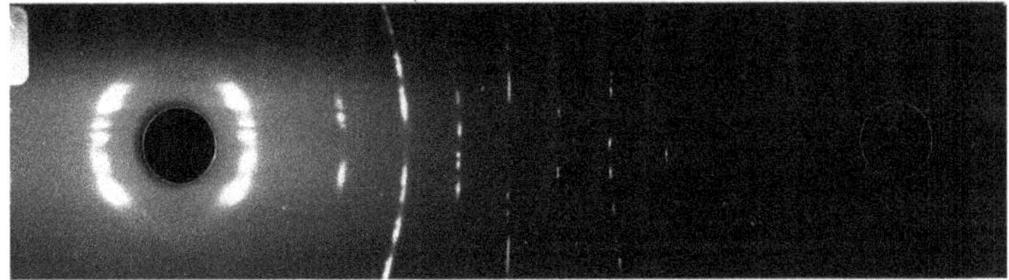

Fig. 6.3 Debye–Scherrer rings of a sample showing a nonrandom and textured distribution of crystallites. The positive print of the X-ray film makes the texture more visible. (Reprinted from Zabel 1978)

In Chap. 5, we have emphasized the importance of single crystals. In powders, each grain is a single crystal. With high brilliance and micrometer-sized synchrotron beams, each individual crystal grain can be investigated and analyzed. This is reasonable for special applications, such as ultra-high-pressure, coherent diffraction imaging, or pump-probe experiments. However, to get a statistical overview of the grain size, grain orientation, chemical composition, kinetics, phase transitions, magnetic structure, reactivity with gases, etc., powder diffraction is indispensable for many science disciplines, such as materials science, metallurgy, ceramics, solid-state chemistry, catalysis, earth and planetary science, archeology, and others.

6.2   Angle-Dispersive X-Ray Powder Diffraction

Powder diffraction is, from an experimental point of view, much simpler than single-crystal diffraction. A one- or two-circle diffractometer is sufficient. In a single scan, all Bragg reflections can be reached without tedious alignment procedures and resetting the sample between Bragg reflections. However, several caveats need to be considered. For one, the powder should be clean and not mixed with crystallites of a different kind. This is a hard-to-fulfill requirement never been met in geological samples. In addition, the crystallites should be neither too small nor too large. Too small crystallites lead to line broadening, and too large crystallites impair the random orientational distribution. Also, powders have a large surface area and are likely to be wetted with moisture. Drying powders and sealing in a vacuum tube before taking X-ray or neutron diffraction patterns is therefore a good practice.

In the following, we discuss in some depth a typical *powder diffraction pattern* taken with X-rays from NaCl (rock salt). The crystal structure of NaCl consists of two fcc lattices, one occupied by Na^+ and the other by Cl^-. Both lattices are shifted against each other along the c-axis by $(0,0,a/2)$, where a is the lattice parameter. Note that Bragg reflections with even (hkl) are much stronger than Bragg reflections with odd (hkl). For even (hkl) the atomic form factors of Na and Cl add up: $(4f_{Na} + 4f_{Cl})^2$, whereas for odd (hkl) the Bragg intensity is proportional to the difference: $(4f_{Na} - 4f_{Cl})^2$. Mixed (hkl) reflections are not allowed—see also Exercise E3.2 for the structure factor calculation. Fig. 6.4 reproduces a powder pattern of NaCl taken

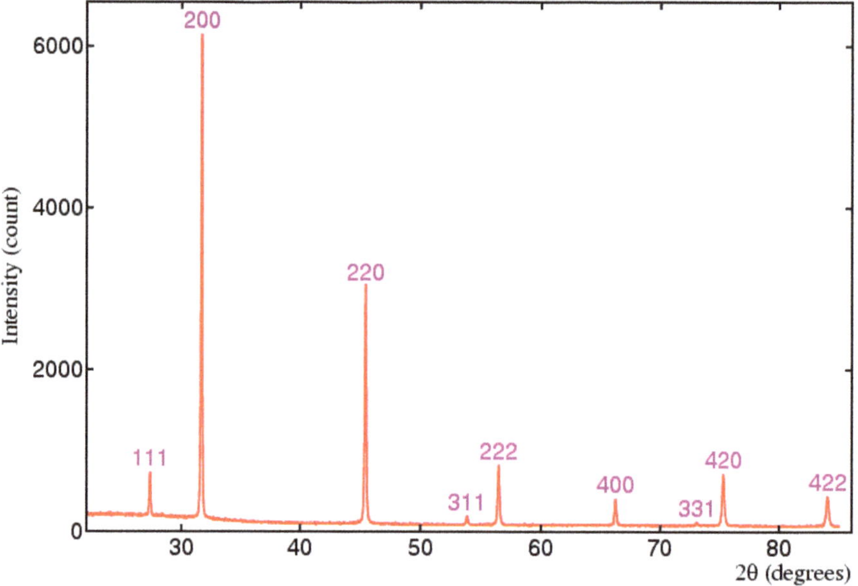

Fig. 6.4 Powder diffraction pattern from NaCl, taken with Cu Kα radiation. Bragg peaks are labeled with their respective Miller indices. Note that peaks with even indices show higher intensity compared to those with odd indices. (Adapted from Klug and Alexander 1974)

with Cu Kα radiation. To calculate the intensity of a powder pattern and to fit the calculated intensities to the measured ones, several additional factors have to be taken into account. The intensity of (hkl) reflections in a powder pattern have the following form (Klug and Alexander 1974):

$$I(G_{hkl}) = \frac{I_0}{r^2} \exp(-2W(Q,T))m_{hkl}|F(G_{hkl})|^2 P(\theta)L(\theta)A. \tag{6.1}$$

Here I_0 is the primary beam intensity of the X-ray source, and r is the distance between the sample and the detector. $F(G_{hkl})$ is the well-known structure factor (see Sect. 3.5), m_{hkl} is the **multiplicity factor**, $P(\theta)$ is the polarization factor, $L(\theta)$ is the Lorentz factor, and A is the absorption factor. These factors are explained below.

The **multiplicity factor** m_{hkl} of reflections gives the number of equivalent (hkl) planes in a crystal structure and therefore reflects the point symmetry of the crystal. For instance, in a cubic crystal, the ($h00$) plane occurs six times, considering that one cannot distinguish between ($h00$), ($0k0$), and ($00l$) planes. Therefore, their multiplicity factor is 6. The ($hh0$) planes and equivalent planes occur 12 times, so $m_{hh0} = 12$, etc. The larger the multiplicity factor is, the higher the probability of a crystallite in the powder sample having just the right orientation for X-ray diffraction.

The **Debye–Waller factor** $\exp(-2W(Q,T))$ takes into account the temperature dependence of Bragg reflections as discussed in Sect. 3.6.1. Since the Debye–Waller

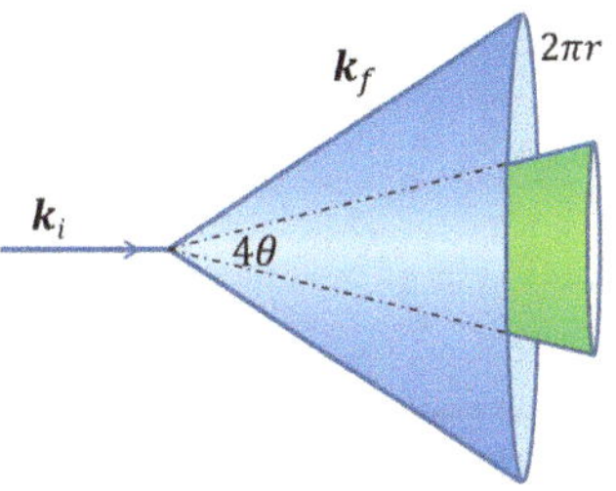

Fig. 6.5 Conical Debye–Scherrer rings. Bragg reflections on larger rings have smaller intensity, just for geometrical reasons

factor is different for Na$^+$ and Cl$^-$ ions due to their different mass, this factor has to be taken into the expression for the structure factor.

The ***polarization factor*** corrects the X-ray intensity concerning polarization effects as a function of the diffraction angle (2θ), which is explained in Sect. 2.5.3. The polarization factor for an unpolarized incident X-ray beam is therefore

$$P(\theta) = \frac{1}{2}\left(1 + \cos^2(2\theta)\right). \tag{6.2}$$

Some polarization may already takes place at the monochromator, which is neglected. Note that the polarization factor is only relevant to X-ray scattering, not to neutron scattering.

The ***Lorentz factor*** is a geometric factor taking into account that Debye–Scherrer rings have different radii, as illustrated in Fig. 6.5. Although the structure factor may be equal if spread out over a larger ring, the intensity in each segment will be lower. As the circle circumference is $2\pi r = 2\pi k_f \sin 2\theta$, to compensate for this effect, a factor $L(\theta) \sim 1/\sin(2\theta)$ is introduced. In addition, we have to consider the density of reciprocal lattice points on this conical ring, which scales with $1/\sin(\theta)$. Taking both factors and the polarization together, the Lorentz-Polarization LP factor becomes

$$LP = \frac{1 + \cos^2 2\theta}{2\sin\theta\sin(2\theta)}. \tag{6.3}$$

The last factor in Eq. (6.1) is the ***absorption factor***. The X-ray absorption depends on the scattering geometry, i.e., whether the powder sample resides in a cylindrical thin-walled glass capillary tube or whether the powder sample consists of a polycrystalline slab. Both ***scattering geometries*** are sketched in Fig. 6.6. In the ***slab geometry***, the absorption factor, derived in Appendix A5, is a constant: $A = 1/2\mu$. The capillary arrangement has the advantage that it can be spun about its axis to ensure a homogeneous orientational distribution of the powder grains.

Taking all factors together, we find for a slab-shaped powder sample the expression for the intensity of (hkl) Bragg reflections:

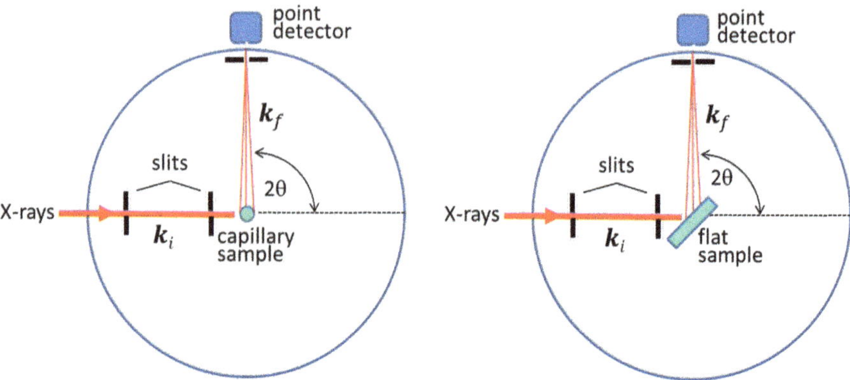

Fig. 6.6 Scattering geometry for recording powder patterns either with a thin-walled capillary tube at the center or some polycrystalline slabθ. In both cases, the diffracted intensity is detected as a function of the 2θ angle using a point detector

$$I(G_{hkl}) = k \frac{I_0}{r^2} m_{hkl} \left| 4f_{Na}(Q)e^{-W_{Na}} \pm 4f_{Cl}(Q)e^{-W_{Cl}} \right|^2 \frac{1}{2\mu} \frac{(1 + \cos^2(2\theta))}{2 \sin(\theta) \sin(2\theta)}. \qquad (6.4)$$

k is an angle-independent constant collecting all remaining factors. The plus sign is for all even Miller indices, the minus sign for all odd indices. As an example, we calculate the intensities for the Bragg reflections shown in the scan of Fig. 6.4 for $T = 300$ K in the high-temperature limit of the Debye–Waller factor and for the case of Cu $K\alpha$ X-ray radiation with a wavelength of $\lambda = 0.154$ nm. The lattice parameter of NaCl is $a = 0.564$ nm. The Bragg angles are calculated via the Bragg equation (Eq. 3.4) using the respective interplanar spacing d_{hkl} (Eq. 3.6). Atomic form factors are taken from X-ray tables. The intensity is normalized by the primary intensity and the sample–detector distance. The numerical values for all factors in Eq. (6.4) are listed in the spreadsheet of Table 6.1.

We find a pretty good agreement between the model calculation and the experimental results. The next task is to run a χ^2-test and to refine the model to get the best results, confirming the crystal structure in question. One of the best-known methods is the so-called Rietveld refinement, which is well documented in the literature: https://en.wikipedia.org/wiki/Rietveld_refinement. For powder patterns taken by neutron scattering, the so-called Fullprof analysis has become the gold standard: https://www.ill.eu/sites/fullprof/.

Back transformation of the powder diffraction pattern to real space is the final step to obtain the radial distribution of the electron density correlation (for X-rays) or the nuclear density distribution (for neutrons). The procedures are introduced and described with examples in Kaduk et al. (2021).

Table 6.1 Spreadsheet for calculating the intensity of an X-ray powder pattern, here for NaCl. 1. Column: *hkl* Miller indices of allowed Bragg-reflections; 2. Column: multiplicity factor; 3. Column: Bragg angle θ calculated for a wavelength of $\lambda = 0.154$ nm; 4. Column: corresponding scattering vector; 5. Column: form factors for Na^+ and Cl^- ions; 6. Column: respective Debye–Waller factors; 7. Column: respective structure factors; 8. Column: respective Lorentz-polarization factors; 9. Column: calculated normalized intensities

1	2	3	4	5	6	7	8	9
hkl	m_{hkl}	θ	Q [nm^{-1}]	f_{Na^+}, f_{Cl^-}	$e^{-W(Na)}, e^{-W(Cl)}$	$\|F\|^2$	LP	$Ir^2/I_0 \times 10^{-4}$
111	8	13.68	19.3	9.0, 13.5	0.92, 0.95	330	8.22	3.78
200	8	15.84	22.27	8.7, 12.7	0.89, 0.94	6197	6.0	29.74
220	12	22.71	31.5	7.65, 10.5	0.8, 0.88	3774	2.71	12.27
311	24	26.92	36.94	7.0, 9.6	0.75, 0.85	135	1.84	0.59
222	8	28.22	38.58	6.75, 9.35	0.73, 0.83	2575	1.657	3.39
400	8	33.09	44.55	6.1, 8.65	0.65, 0.79	1865	1.16	1.13
331	24	36.52	48.56	5.65, 8.3	0.6, 0.75	128	0.95	0.29
420	24	37.63	49.82	5.5, 8.2	0.58, 0.74	1371	0.9	2.96
422	24	41.97	54.57	5.05, 7.85	0.53, 0.70	1068	0.76	1.94

6.3 Angle-Dispersive Powder Diffraction Methods

Modern instrumentation at synchrotron and neutron large-scale facilities allows taking powder patterns very efficiently with high resolution and within short exposure times using large area detectors instead of scanning the 2θ range step by step with a point detector. One of the most powerful *neutron powder instruments* (D20 at the Institut Laue-Langevin, Grenoble) for angle-dispersive detection (ADD) features a detector bank permitting simultaneous detection of intensity over an angular range of about 154°. This speeds up the acquisition time and allows for time-resolved recording, such as during phase transitions, alloy formation, catalytic reactions, etc. The instrument layout is schematically shown in Fig. 6.7, adapted from the ILL Yellow Book (2001).

Assuming circumstances, neutron diffraction by powders or polycrystals follows pretty much the same procedures as X-ray powder diffraction. But in contrast to X-ray scattering, the absorption of neutrons for most of the elements, with a few exceptions, is much less important, and therefore the penetration depth is much larger. Hence, the line shapes are more reliable and less affected by artefacts. Furthermore, the background intensity is, in general, lower and featureless. In contrast, X-ray powder diffraction does not require a synchrotron source but can easily be performed with a sealed or rotating X-ray tube in the home laboratory. The use of synchrotron radiation is only justified if the sample volume is very small, the sample throughput is high, and/or very high data quality is required.

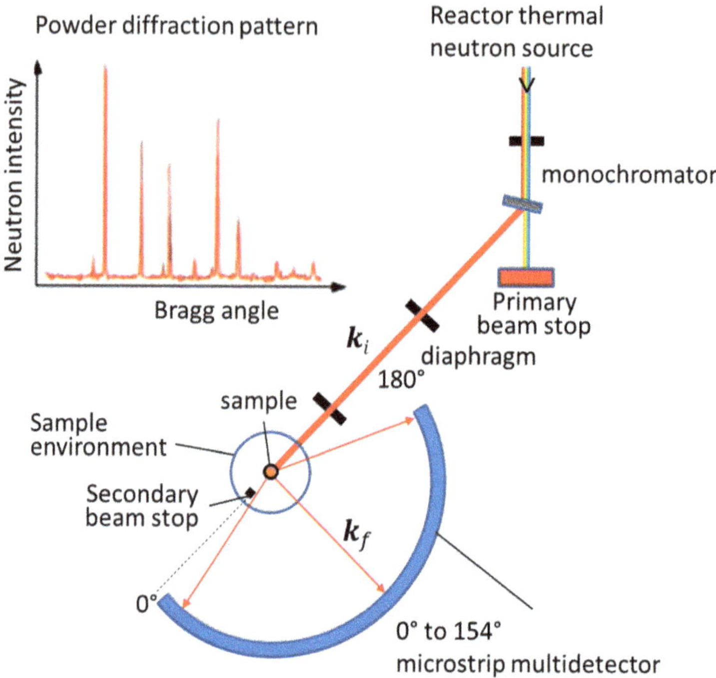

Fig. 6.7 Schematic layout of a neutron powder diffractometer featuring a position-sensitive microstrip multidetector bank spanning almost 180°. The detector bank allows recording scattered intensity simultaneously, greatly reducing the total recording time. The inset shows a typical recorded powder diffraction pattern

6.4   Energy-Dispersive X-Ray Diffraction (EDXRD)

If the radiation source contains a broad wavelength spectrum, angle-dispersive diffraction (ADD) implies that, depending on the resolution of the monochromator, only a narrow wavelength band $\Delta\lambda$ is used for scattering experiments; all other wavelengths outside the wavelength band are discarded.

Energy-dispersive X-ray diffraction (EDXRD) is an alternative method to ADD that is better suited to exploit the characteristics of a "white" X-ray source. EDXRD is performed by using the bremsstrahlung spectrum of a laboratory X-ray anode or the broad X-ray spectrum from a wiggler at a synchrotron facility, and an energy-dispersive detector at a fixed scattering angle 2θ as indicated in Fig. 6.8.

Semiconductor detectors (Si or Ge drifted with Li) exist that have sufficient energy resolution ($\Delta E{\sim}50$ eV) for performing an energy-dispersive X-ray diffraction analysis. To see how this works, we recall that the photon energy is expressed by $E = hc/\lambda$ (see Eq. 4.4 or Table 4.1). Rephrasing the photon energy by using the Bragg equation, we obtain

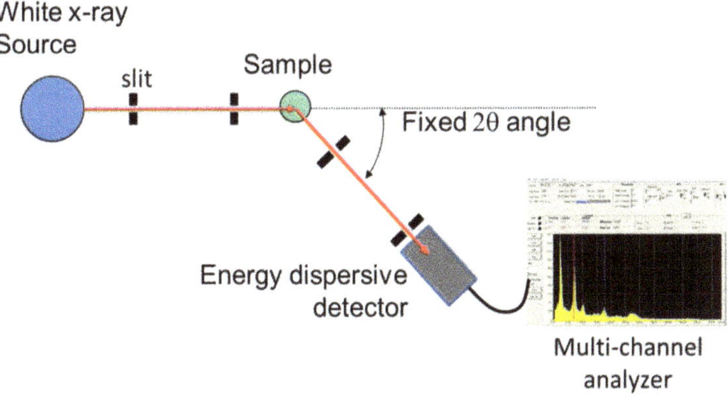

Fig. 6.8 Energy-dispersive X-ray diffraction is performed with a "white" X-ray source and an energy-dispersive detector at a fixed diffraction angle 2θ. Data acquisition is performed with a multichannel analyzer

$$E_{hkl} = \frac{hc}{\lambda_{hkl}} = \frac{hc}{2d_{hkl}\sin\theta} = \frac{\hbar c}{2\sin\theta}|\boldsymbol{G}_{hkl}|. \tag{6.5}$$

Hence, the recorded peak intensity at a particular energy E_{hkl} can be directly related to the reciprocal lattice vector $\boldsymbol{G}_{hkl}$. The resolution of EDXRD is limited by the energy resolution of the detector and is typically lower than for angle-dispersive diffraction. EDXRD has, however, a couple of decisive advantages. By using high photon energies, one can explore much higher (hkl) indices than would be possible in angle-dispersive mode. An example is shown in Fig. 6.9. Here the $(hh0)$ reflections are recorded of an Nb(110) oriented single-crystalline film up to the eighth order at temperatures of 30 K and room temperature. With Mo $K_{\alpha 1}$ characteristic radiation of $\lambda = 0.071$ nm, only the sixth order can be reached at most. The $(hh0)$ reflections up to the eighth order in Fig. 6.9 nicely show the effect of the atomic form factor, the Debye–Waller factor, and the thermal expansion.

The second reason for using EDXRD is the ability to investigate samples in complex environments where access of the beam to the sample and from the sample to the detector is spatially limited. Since no detector movement is required during data acquisition, the experimental setup can be planned and specified in advance.

Frequently, EDXRD is used to study time-dependent properties of polycrystalline samples on a subsecond to minute time scale, such as temperature-dependent sample histories, phase transitions, gas exchange, etc. In addition, energy-dispersive detection can be combined with position-sensitive detection, which records simultaneously the spatial coordinates (Bragg angle) and Bragg energy E_{hkl} of samples in every micrometer-sized pixel of a CCD detector.

As an example, we present an EDXRD experiment performed at the ID09 beamline of the ESRF for studying the crystallization of a supercooled $Ni_{41}V_{59}$ alloy after quenching from the melt (Notthoff et al. 2001). A little pellet was elevated electromagnetically, and the temperature was controlled by convective H_2-He gas

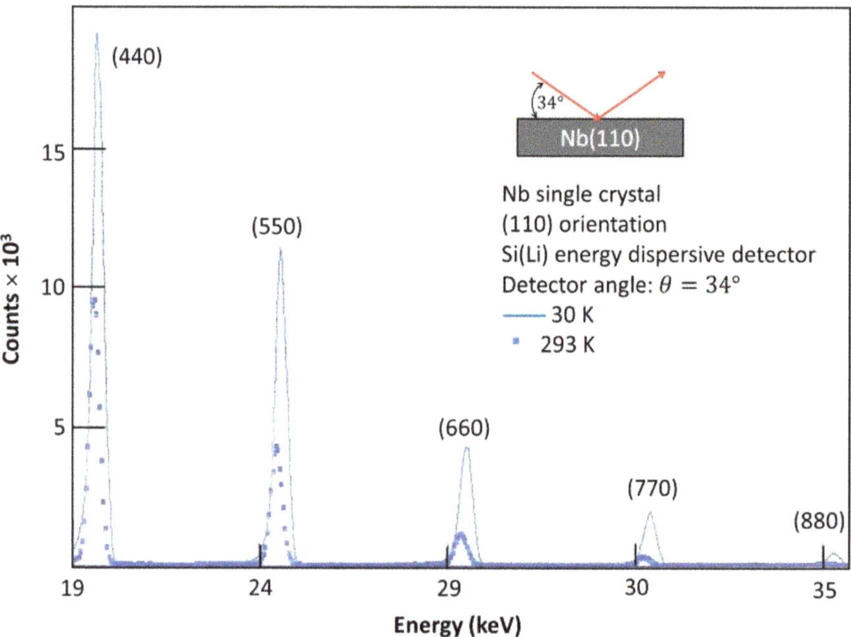

Fig. 6.9 Energy-dispersive detection of the ($hh0$) reflections from an epitaxial Nb film on a sapphire substrate at room temperature and at 30 K. The incident beam angle was set to 34°. (Reproduced from Miceli 1987)

flow. The detector angle was set to $2\theta = 5°$, the integration time was 0.5 s, and the usable energy range was from 7 keV ($\hat{=}0.3\,\text{Å}^{-1}$) to 148 keV ($\hat{=}6.5\,\text{Å}^{-1}$). Figure 6.10 shows the time evolution after quenching for about 60 s from scan 1 to scan 4. Some portion of the undercooled melt displayed in scan 1 crystallizes first from the molten state into a metastable bcc phase, as seen in scans 2 and 3. Immediately after taking scan 3, the alloy transforms to the stable intermetallic σ-phase with tetragonal symmetry, while the bcc phase disappears. The two arrows in the temperature–time plot of the inset mark latent heat development during the phase transformation.

As this example demonstrates, EDXRD is well-suited for the time-resolved recording of structural changes identified by the appearance/disappearance of Bragg reflections. However, the quantitative analysis of the diffraction pattern is significantly more difficult. There are several reasons for this. First, the energy range from 7 to 148 keV covers the K_α absorption edge of Ni at 8.33 keV, which leads to form factor corrections and increased absorption. The K_α absorption edge of V lies below the scanned energy range and therefore does not pose a problem. In general, however, anomalous scattering due to intersecting absorption edges of heavier elements complicates quantitative analysis. Furthermore, if a high-energy X-ray beam strikes the silicon detector, Compton scattering can occur. The X-ray beam

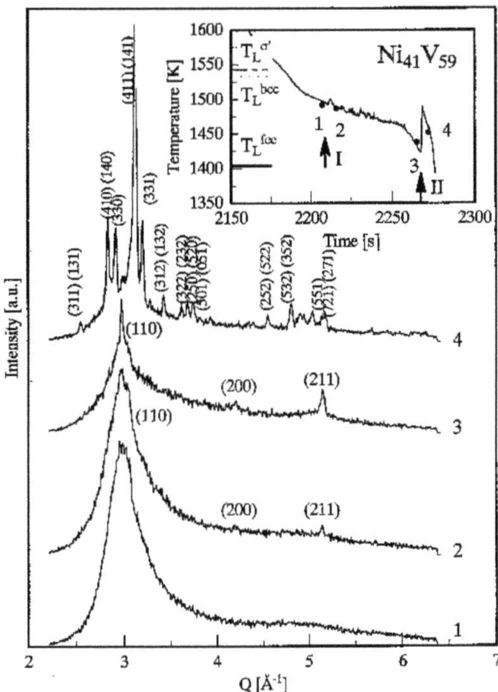

Fig. 6.10 Temperature–time profile during the crystallization of a supercooled Ni$_{41}$V$_{59}$ alloy. The first scan (1) reveals a disordered amorphous alloy. Crystallization of the metastable bcc phase is seen in scans (2) and (3), followed by a transformation to the stable σ-phase in scan (4), about 100 s after quenching. (Reprinted with permission from Notthoff et al. 2001: C. Notthoff et al. Phys. Rev. Lett. **86**, 1038 (2001). DOI: https://doi.org/10.1103/PhysRevLett.86.1038. Copyright (2025) by the American Physical Society)

then loses part of its energy to the Compton electron instead of converting all of the photon energy into electron–hole pairs. When the Compton-scattered X-ray beam leaves the detector, a peak appears in the spectrum whose energy is lower than the original X-ray energy. This peak is called the Compton escape peak. Artifacts of this type can complicate structural analysis, especially when the Compton peak overlaps with other diffraction peaks of a secondary phase.

6.5   Time-of-Flight Neutron Diffraction

Energy-dispersive diffraction of neutrons is not possible due to a lack of suitable energy-dispersive neutron detectors. To overcome this problem, dispersive detection of neutron energy (or wavelength) is performed using *time-of-flight* (ToF) techniques. The continuous white incident neutron beam is chopped into short bursts that travel across the sample and are scattered toward the detector, situated at a fixed 2θ angle. The experimental setup is sketched in Fig. 6.11a. Alternatively, the intrinsic time structure of a spallation source can be used as a source of pulsed neutrons (see Sect. 4.3.2 and adjacent Fig. 4.30). The detector is gated to record the time of arrival, with short wavelengths (high velocity) arriving first and long wavelengths (low velocity) arriving later. The total flight time is $T(\lambda) = L/v_n$, where $L = L_1 + L_2$ is the

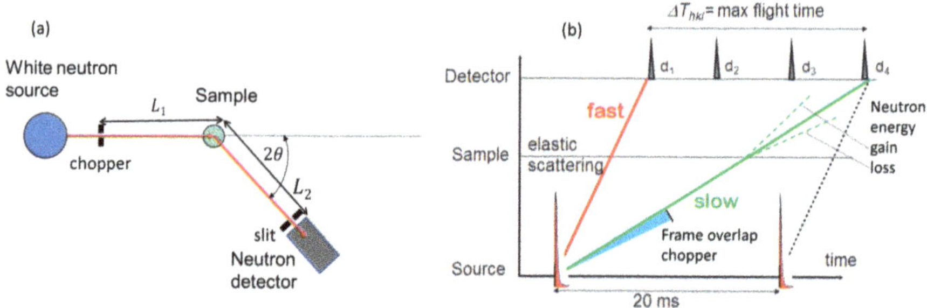

Fig. 6.11 Time-of-flight detection method of neutrons. (**a**) Setup with an initial chopper setting the time clock and flight path L_1 to the sample and L_2 from the sample to the detector. (**b**) Path-time diagram showing the flight path of the extremes, the fastest and slowest neutrons. Indicated are also inelastic processes, where the neutron either gains or loses energy due to the excitation or absorption of quasi-particles in solids or liquids

total length of the neutron flight path from source to sample and from sample to detector, respectively, and $v_n(\lambda)$ is the neutron velocity. Using the definition of the reciprocal lattice vector $\boldsymbol{G}_{hkl}$ and the kinematic relationship for neutrons with mass m_n (see Appendix A6), we find

$$|\boldsymbol{G}_{hkl}| = \frac{2\pi}{d_{hkl}} = \frac{4\pi}{\lambda_{hkl}} \sin\theta_{\mathrm{Det}} = \frac{4\pi m_n v_{hkl}}{h} \sin\theta_{\mathrm{Det}} = \frac{4\pi m_n L}{h T_{hkl}} \sin\theta_{\mathrm{Det}}. \tag{6.6}$$

Now we rearrange the last term in Eq. (6.6) for the flight time T_{hkl} and obtain

$$T_{hkl}(\lambda) = \frac{4\pi m_n}{h} \frac{L \sin\theta_{\mathrm{Det}}}{|\boldsymbol{G}_{hkl}|} = \frac{2m_n}{h} L d_{hkl} \sin\theta_{Det} \propto d_{hkl}. \tag{6.7}$$

Hence, the flight time is linearly related to the lattice spacing d_{hkl}. In practical units, d_{hkl} can be expressed as

$$d_{hkl} = 1.977 \times 10^{-7} \frac{\mathrm{m}^2}{\mathrm{s}} \frac{1}{L(\mathrm{m})\ \sin\ \theta_{Det}} T_{hkl}(\mathrm{s}). \tag{6.8}$$

Small lattice spacings d_1 are probed by the shortest flight time, and larger spacings take longer. The maximum flight time is determined by the pulse separation and flight path. Longer flight times cause a frame overlap, which is suppressed by a special frame overlap chopper that hinders the slowest neutrons from overlapping with the fastest neutrons of the next pulse.

The ΔQ range covered by one detector angle $2\theta_{\mathrm{Det}}$ is rather limited (see Exercise 6.3). For a full scan, it is often necessary to reset the detector angle. This can be seen by the following quick estimate; more details are discussed in Exercise 6.3. We assume a pulse frequency of 50 Hz, providing a wavelength band of approximately $0.5\ \text{Å} < \lambda < 5.0\ \text{Å}$, and a total flight path length of $L_1 + L_2 = 10\ \text{m} + 5\ \text{m} = 15\ \text{m}$. We

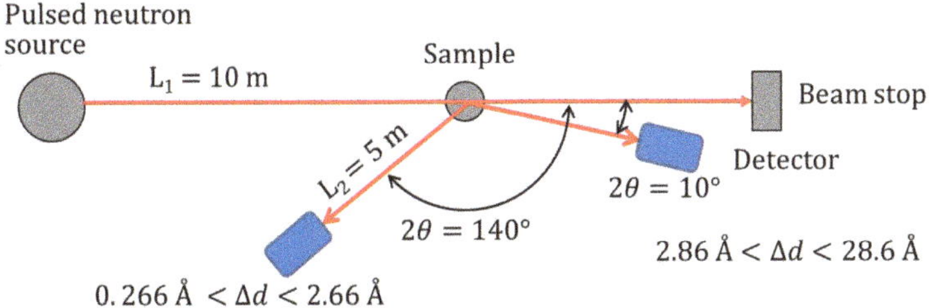

Fig. 6.12 Schematic layout of a ToF instrument with two optional detector positions. The pulsed neutron source on the left side is either the target station of a spallation source or a neutron beam chopper in the case of a continuously operating neutron source. Further details are omitted for clarity

assume that frame overlap choppers are installed. For two detector angles at $2\theta = 10°$ and $140°$, we can expect to detect two ranges of d_{hkl} values, which are very different, as indicated in Fig. 6.12. Therefore, before executing a ToF experiment, the detector angle should be calculated to cover the d-spacings of interest.

Alternatively, a detector bank can be installed around the sample to cover a larger Q or d-range, all counting simultaneously. This is indeed done at all spallation neutron sources. The d_{hkl}-spacing resolution Δd is coupled to the time resolution ΔT. The time resolution is a constant over the scanned Q range, which implies that Bragg peaks do not broaden with increasing Q (decreasing d) unless caused by strain. However, the prefactor in Eq. (6.8) $(L \sin \theta_{\text{Det}})^{-1}$ determines the Δd resolution. The longer the flight path L and the higher the detector angle 2θ is, the higher the Δd resolution.

Neutron ToF methods are well adapted to spallation sources, which intrinsically provide a time structure of the incident white neutron beam. In most cases, the neutron ToF technique is applied for energy-resolved studies, because neutron energy loss and energy gain via interaction with the sample can be detected with very high resolution independent of the crystalline state of the sample. This broad field of applications, known as neutron spectroscopy, is not the subject of the present text. Nevertheless, with the time structure of a spallation neutron source, elastic diffraction experiments are also worthwhile to be performed. Figure 6.13 shows a powder ToF diffraction pattern of the compound $Zn_{1/3}Ge_{1/3}Ga_{1/3}N$, which is a direct gap semiconductor and of interest for optoelectronics (Suehiro et al. 2020). The neutron ToF powder data were taken at the pulsed spallation neutron source of J-PARC using the instrument BL08 Super-HRPD. The detector bank was set to $90°$, and a range of d-spacings 0.4–$5.2\,\text{Å}$ were recorded with a resolution of $\Delta d/d = 0.35 - 0.45\%$. The data were acquired at room temperature and analyzed with the Rietveld refinement method for ToF neutron powder diffraction. We notice that the density of peaks increases with decreasing d-spacings, which is typical for any crystal structure. Because of the high resolution, these peaks can still be resolved.

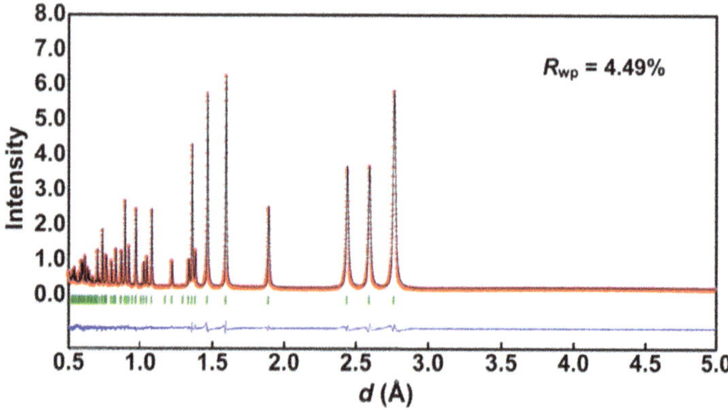

Fig. 6.13 Neutron powder ToF diffraction pattern of $Zn_{1/3}Ge_{1/3}Ga_{1/3}N$, a compound with a wurtzite-like structure. The red line represents the Rietveld refinement analysis, and the blue line is the difference between the experimental and calculated data. (Reproduced with permission from Suehiro et al. 2020: M. Suehiro et al. Applied Physics Express **13**, 115,503 (2020) DOI 10.35848/1882-0786/abc1cb. Copyright (2025) by the IoP Publishing LTD)

The Rietveld refinement method, introduced by Rietveld (1967), uses a least squares approach to refine a theoretical crystal structure calculation, like the one posted in Table 6.1 for NaCl, together with assumed line profiles and other experimental parameters, until it matches the measured profile in an iterative process. The R_{wp} parameter quoted in Fig. 6.12 is the weighted background corrected peak-only profile agreement index. A value of about 4.5% is considered to be very good. This can also be judged by the blue line below the experimental curve that represents the difference between the refinement and the experimental data.

6.6 Si Powder

The Si powder pattern is very well researched, documented in the literature, and available in databases for various X-ray and neutron energies, as well as for different scanning schemes. Figure 6.14 shows a schematic Si diffraction pattern for X-rays with a wavelength of $\lambda = 1.54$ Å, which corresponds to the Cu-K_α characteristic radiation. Due to this excellent documentation, Si powder is used as a "gold standard" for calibrating and adjusting the resolution of the respective instrument. Si powder is also frequently used as a marker, mixed with another powder of unknown crystal structure, and as a reference for line broadening due to particle size and strain. Because of its usefulness, neutron and X-ray powder diffractometers have the highest sample throughput at synchrotron and neutron facilities and produce the most publications in scientific journals.

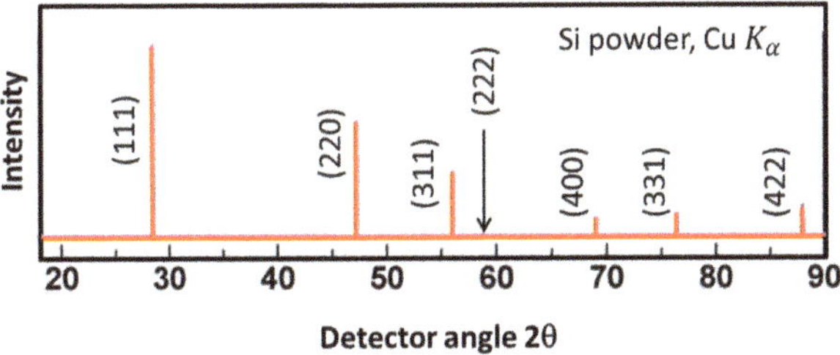

Fig. 6.14 Diffraction pattern of Si powder with a cubic lattice parameter of 5.43 Å. The intensity is plotted versus detector angle 2θ for a fixed wavelength of 1.54 Å, corresponding to the Cu-K$_\alpha$ characteristic radiation. Note the missing intensity at the position of the (222) Bragg reflection due to the diamond structure factor

As a side remark, Si has a diamond crystal structure, and therefore the structure factor features zero intensity at the (222) r.l.p.[1] If any intensity is recorded at the (222) Bragg peak position, we conclude that the incident beam is contaminated with radiation of twice the wavelength. The spectral purity of the radiation is important to know for fitting the powder data.

6.7 Conclusion

In conclusion, powder diffraction is an extremely important, versatile, and valuable analytical tool in many scientific fields. It provides an initial overview of a sample's crystallinity and texture, its chemical composition and stoichiometry, structural and magnetic phase transitions, and chemical reactions and diffusion processes. Neutron powder diffraction is often advantageous because it allows light elements to be "seen" in the vicinity of heavy elements. Furthermore, contrast can be varied by isotope exchange of the same element in the same chemical environment but with different coherent scattering lengths. There is no counterpart to isotope exchange in X-ray diffraction. In X-ray diffraction, contrast can be altered by tuning the X-ray energy to be close to an absorption edge, thereby changing the imaginary part of the atomic form factor. In contrast, isotope exchange changes only the real part of the scattering length.

Summary

1. Angle-dispersive powder diffraction is performed with a collimated X-ray or neutron beam of a constant wavelength.

---

[1] The diamond structure factor for Si is determined in Exercise 3.2.

2. Powder diffraction requires a good statistical distribution of all crystal plane orientations.
3. We distinguish between angle-dispersive diffraction, energy-dispersive diffraction, and time-of-flight detection.
4. Time-of-flight methods are only available for neutrons.
5. Energy-dispersive diffraction is only possible with X-rays.
6. Nonrandom distribution of crystallites can be recognized by spotty Debye–Scherrer rings.
7. Powder patterns can be calculated and compared with experiments for refinement of crystallographic data.
8. Powder samples consist of a large amount of tiny crystallites (<10 μm) with random orientation in space.
9. Powder patterns are traditionally recorded with a Debye–Scherrer camera.
10. In a powder scan, all allowed Bragg reflections are recorded at once.
11. Using neutrons instead of X-rays, the contrast between different atoms in the lattice can be varied by selecting specific isotopes.
12. Energy-dispersive X-ray detection covers a wider Q-range than angle-dispersive scans, but is problematic because of potential artifacts, such as Compton escape peaks.
13. Time-of-flight powder diffraction is best performed at a spallation neutron source that features an intrinsic pulsed time structure of the incident beam.
14. ToF methods use the flight time recorded with a gated detector at a fixed detector angle. For covering an extended Q-range, resetting the detector at several angles is necessary. Alternatively, a detector bank can be installed.
15. Si powder serves as a "gold" standard for calibrating powder diffractometers.

Questions

(Note that more than one answer may be correct)

1. **What precautions need to be considered when taking a powder pattern?**

   (a) The powder grains must be small.
   (b) The powder grains must be randomly oriented.
   (c) The powder grains should be mixed with a standard.
   (d) The powder grains should be protected with a gold film against oxidation.

2. **When calculating a neutron powder pattern, which factors have to be considered?**

   (a) Lorentz factor.
   (b) Debye–Waller factor.
   (c) Half-life of the isotope.
   (d) Polarization factor.

3. **How can the crystal structure of a powder or a polycrystalline sample be analyzed?**

   (a) With the help of a Debye–Scherrer camera.
   (b) With a light microscope.
   (c) With thermal neutron scattering.
   (d) With proton scattering.

4. **How can the texture of polycrystalline samples be recognized?**

   (a) By broken Debye–Scherrer rings.
   (b) By a doubling of all Debye–Scherrer rings.
   (c) By the extinction of certain Debye–Scherrer rings.

5. **Under what circumstances is neutron scattering preferred over X-ray scattering?**

   (a) When isotope contrast is required.
   (b) Whenever light atoms need to be seen in the neighborhood of heavy elements.
   (c) When the sample is in a liquid state.
   (d) When X-ray absorption is too strong.

6. **What does the resolution of a ToF neutron spectrometer depend on?**

   (a) The detector responsiveness.
   (b) The length of the flight path.
   (c) The angle of the detector bank.
   (d) The temperature of the sample.

7. **What is the advantage of energy-dispersive X-ray diffraction?**

   (a) Higher count rate.
   (b) Faster data acquisition.
   (c) Higher Q-range.
   (d) No movement of the detector arm.

Exercises

Grades of difficulty: E = easy, M = medium, A = advanced.

E 6.1 Evaluation of a powder diffraction pattern

Using powder X-ray scattering with Cu-K_α radiation ($\lambda = 0.1541$ nm), the lowest order Bragg reflections of an elemental metal lattice are recorded at the 2θ values: $44.66°$, $65°$, and $82.3°$.

   (a) Assuming that the material scattered from has a cubic crystal structure, determine the interplanar distances d_{hkl} for each 2θ value.

(b) Now try to figure out the cubic crystal structure. Is it sc, bcc, or fcc? It will help if you take ratios of the calculated interplanar spacings and compare these ratios with those expected for certain low-order hkl planes

(c) What is the lattice parameter of the cubic material?

(d) Knowing the lattice parameter, can you identify the metal in question?

E 6.2 Variation of the scattering length

Consider the NaCl powder pattern in Fig. 6.4. How would the intensities of the Bragg peaks change if you set $f_{Na} = |b|$ and $f_{Cl} = -|b|$.

M 6.3 Time-of-flight instrumentation

Confirm all information given in Fig. 6.12 and the adjacent text.

(a) Start with verifying the wavelength band $0.1\,\text{Å} < \lambda < 5.5\,\text{Å}$, provided by a neutron beam chopper or a pulsed neutron source with a frequency of 50 Hz.

(b) Calculate the Δd-band for a detector setting of $2\theta = 10°$.

(c) Repeat the same calculation for a detector setting of $2\theta = 140°$ and compare both settings.

References

C.R. Hubbard, H.E. Swanson, F.A. Mauer, A silicon powder diffraction standard reference material. J. Appl. Crystallogr. **8**, 45 (1975)

J.A. Kaduk, S.J.L. Billinge, R.E. Dinnebier, N. Henderson, I. Madsen, R. van Černý, M. Leoni, L. Lutterotti, S. Thakral, D. Chateigner, Powder diffraction. Nat. Rev. Methods Primers **1**, 77 (2021). https://doi.org/10.1038/s43586-021-00074-7

H.P. Klug, L.E. Alexander, *X-Ray Diffraction Procedures: For Polycrystalline and Amorphous Materials*, 2nd edn. (Wiley, New York, 1974)

P.F. Miceli, Thesis, University of Illinois U-C, 1987

C. Notthoff, B. Feuerbacher, H. Franz, D.M. Herlach, D. Holland-Moritz, Direct determination of metastable phase diagram by synchrotron radiation experiments on undercooled metallic melts. Phys. Rev. Lett. **86**, 1038 (2001)

H. Rietveld, Line profiles of neutron powder-diffraction peaks for structure refinements. Acta Crystallogr. **22**, 151–152 (1967)

T. Suehiro, M. Tansho, M. Hagihala, S. Torii, T. Kamiyama, T. Shimizu, Quaternary nitride system $(1-x)ZnGeN2-2xGaN$ ($x = 1/3$): disordered wurtzite structure revealed by time-of-flight neutron powder diffraction. Appl. Phys. Express **13**, 115503 (2020)

Yellow Book, Institut Laue-Langevin, Grenoble, France, Scientific Coordination Office, 2001

H. Zabel, Thesis, LMU München, 1978

Further Reading

G.E. Bacon, *Neutron Diffraction*, 3rd edn. (Oxford University Press, Oxford, 1975)

J.R.D. Copley, *The Fundamentals of Neutron Powder Diffraction* (National Institute of Standards and Technology, Gaithersburg, 2001)

Chapter 7
Elastic Magnetic Scattering

Magnetic scattering with X-rays and neutrons explores magnetic (spin) structures and spin excitations in bulk samples, thin films, heterostructures, and magnetic nanoparticles. Elastic magnetic scattering is concerned with short-range and long-range spin correlations in magnetic materials. Inelastic magnetic scattering explores spin fluctuations and spin wave excitations in solids, glasses, and liquids. Elastic and inelastic magnetic scattering was solely the domain of neutron scattering up until the early 1970s to 1980s, when in heroic X-ray experiments, the French scientists de Bergevin and Brunel could demonstrate the existence of antiferromagnetic Bragg reflections in NiO single crystals below the Néel temperature, shedding new light on the photon electron spin interaction (De Bergevin and Brunel 1972). In this chapter, we will discuss elastic magnetic neutron scattering of ferromagnetic, antiferromagnetic, and incommensurate magnetic spin structures, and we will present the main ideas of nonresonant and resonant magnetic X-ray scattering.

7.1 Magnetic Neutron Scattering Lengths

Magnetic scattering is one of the hallmarks of neutron scattering. All presently known ferromagnets, antiferromagnets, and complex magnetic structures have been analyzed by magnetic neutron scattering over the past 70–80 years. Whenever a new magnetic material is discovered, magnetic neutron scattering is the method of choice for its analysis. Nowadays, magnetic materials can also be analyzed by resonant and nonresonant magnetic X-ray scattering. However, magnetic neutron scattering is conceptually much more straightforward than magnetic X-ray scattering. We will start with magnetic neutron scattering, and in the last Sect. 7.6 and 7.7, we will touch upon magnetic X-ray scattering.

Neutrons as Fermi-particles carry a spin $S = \pm 1/2\, \hbar$. With the spin, a *magnetic moment* of the neutron is associated:

H. Zabel, *Elements of Elastic Scattering by X-Rays, Neutrons, and Electrons*, https://doi.org/10.1007/978-3-032-16624-1_7

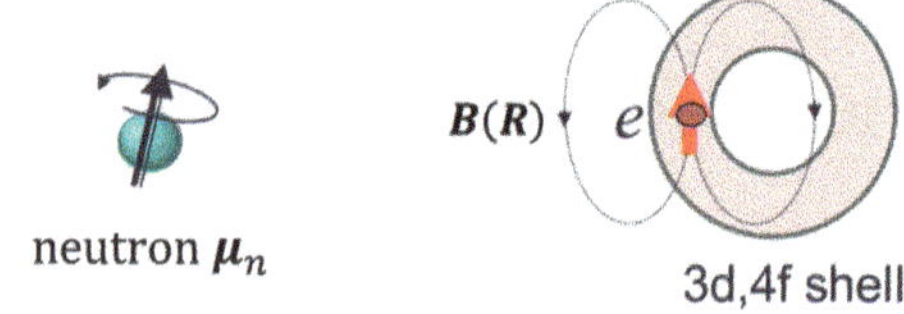

Fig. 7.1 Magnetic dipole interaction between the neutron magnetic moment μ_n, shown to the left, and the magnetic induction $B(R)$, due to stray fields originating from uncompensated magnetic moments in atomic shells at position R, shown to the right

$$\boldsymbol{\mu}_n = -\gamma_n \mu_N \boldsymbol{\sigma}_n, \tag{7.1}$$

where the minus sign denotes the antiparallel orientation of the neutron spin and magnetic moment. Neutrons are magnetic because, although they have no net electric charge, they consist of three quarks: one up quark with a charge of $+2/3$ and two down quarks, each with a charge of $-1/3$. The intrinsic spin and orbital angular momentum of the quarks add up with their internal magnetic fields, thus generating the magnetic moment of the neutron. μ_N in Eq. (7.1) is the **nuclear magneton** defined by

$$\mu_N = \frac{e\hbar}{2m_n} = 5.05783 \times 10^{-27} J/T, \tag{7.2}$$

m_n is the neutron mass, $\gamma_n = 1.913$ is the gyromagnetic ratio of neutrons, and $\boldsymbol{\sigma}_n = 2S$ is the Pauli spin-operator with the eigenvalues ± 1. Thus, the nuclear magneton is smaller than the Bohr magneton $\mu_B = e\hbar/2m_e = 9.274 \times 10^{-24} J/T$ by the ratio of the electron mass to the neutron mass ($m_e/m_n = 1/1833$). In terms of Bohr magnetons, the neutron magnetic moment is $\mu_n = -1.043 \times 10^{-3} \mu_B$, indeed a tiny magnetic moment. This little neutron magnetic moment interacts with the magnetic induction $B(R)$ that emanates from partially filled electronic shells. Figure 7.1 is a cartoon of this interaction.

The interaction between the neutron magnetic moment $\boldsymbol{\mu}_n$ and the local magnetic induction $B(R)$ is of magnetic dipole-dipole character and yields the interaction potential:

$$V_{mag}(\boldsymbol{R}) = -\boldsymbol{\mu}_n \cdot \boldsymbol{B}(\boldsymbol{R}) = -\gamma_n \mu_N \boldsymbol{\sigma}_n \cdot \boldsymbol{B}(\boldsymbol{R}). \tag{7.3}$$

In analogy to the scattering length due to the neutron–nuclear interaction in Eq. (2.19), we express the magnetic scattering length b_m due to the magnetic dipole interaction as the Fourier transform of the magnetic interaction potential $V_{mag}(\boldsymbol{R})$:

$$b_m = \frac{m_n}{2\pi\hbar^2} \int V_{mag}(\boldsymbol{R}) e^{i\boldsymbol{Q}\cdot\boldsymbol{R}} d\boldsymbol{R} = -\frac{m_n}{2\pi\hbar} \int \boldsymbol{\mu}_n \cdot \boldsymbol{B}(\boldsymbol{R}) e^{i\boldsymbol{Q}\cdot\boldsymbol{R}} d\boldsymbol{R}. \tag{7.4}$$

The local magnetic induction is related to the local magnetization $M(R)$ via

$$B(R) = \mu_0 M(R), \tag{7.5}$$

where μ_0 is the permeability of the vacuum. Inserting in Eq. (7.4) yields:

$$b_m = -\frac{m_n \mu_0}{2\pi\hbar^2} \int \mu_n \cdot M(R) e^{iQ \cdot R} dR. \tag{7.6}$$

Now, taking the Fourier transform of the magnetization, the following expression is derived (Squire 1978; Lovesey 1984):

$$B(Q) = \mu_0 \frac{Q \times [M \cdot (Q) \times Q]}{Q^2} = \mu_0 M_\perp(Q) = \mu_0 M(Q) \sin(\angle(Q,M)), \tag{7.7}$$

where

$$M(Q) = \int M(R) e^{iQ \cdot R} dR \tag{7.8}$$

is the Fourier transform of the magnetization that can be split into spin and orbital contributions: $M_{tot}(Q) = M_S(Q) + M_L(Q)$. The expression in Eq. (7.7) shows that the magnetic scattering length is different from zero only for magnetization components, $M_\perp(Q)$, which are oriented perpendicular to the scattering vector Q. This is the geometric condition for probing magnetic structures with neutron scattering. Collecting all terms, we obtain for the magnetic scattering length:

$$b_m = \frac{m_n \mu_n}{2\pi\hbar^2} \sigma_n \cdot B_\perp(Q) = \frac{m_n \gamma_n \mu_N \mu_0}{2\pi\hbar^2} \sigma_n \cdot M_\perp(Q). \tag{7.9}$$

Alternatively, we may determine the **neutron magnetic form factor** $f_m(Q)$, as already introduced in Sect. 2.6.4. The neutron magnetic form factor is similar in expression and magnitude to the X-ray atomic form factor:

$$f_{mag}(Q) = \int \rho_{e,unc}(R) e^{iQ \cdot R} dR. \tag{7.10}$$

The Fourier transform of the electron density in Eq. (7.10) includes, however, only the uncompensated (unc) electrons in atomic shells that give rise to the spin and orbital magnetic moments of atoms. The density of the uncompensated electrons is obviously much lower than the total density, forming a spherical shell with the core electrons at the center.

The **neutron magnetic scattering length** b_m is obtained similarly to the X-ray scattering length by multiplying the magnetic form factor $f_{mag}(Q)$ with the classical electron radius r_0 and the neutron gyromagnetic ratio γ_n:

$$b_m = \gamma_n r_0 \sigma_n f_{mag}(Q). \tag{7.11}$$

Fig. 7.2 Comparison of X-ray atomic form factor and neutron magnetic form factor. The form factors are normalized at $Q = 0$

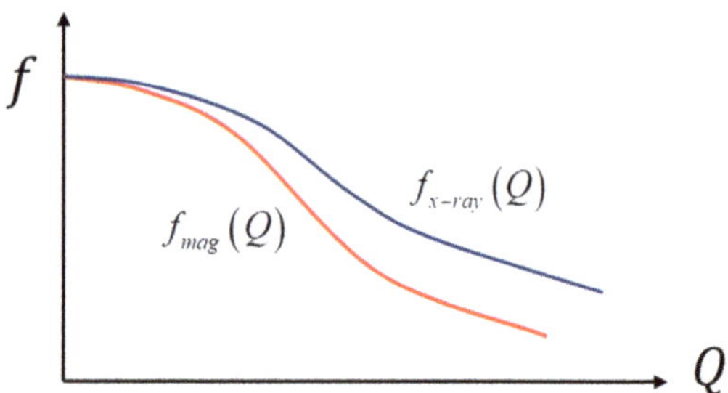

The magnitude of the prefactor $\gamma_n r_0 = 5.39 \times 10^{-13}$ cm is of the same order of magnitude as the coherent neutron scattering length for most isotopes. This tells us that neutron scattering off nuclei and neutron scattering off magnetic dipole moments yield Bragg reflections with very similar amplitude and intensity.

> The neutron magnetic form factor is the Fourier transform of the unpaired electron density in atomic shells.

Similar to the X-ray atomic form factor, the neutron magnetic form factor $f_{mag}(\boldsymbol{Q})$ decreases with increasing scattering vector $\boldsymbol{Q}$. The reduction is even more severe than for X-rays, because the Fourier transform does not cover all electrons in an atom but only those in a spherical shell of inner 3d electrons (transition metal atoms) or 4f electrons (rare earth atoms) that carry uncompensated magnetic dipole moments. Figure 7.2 shows a schematic comparison of both form factors. The fast drop of the magnetic form factor causes problems when studying higher-order magnetic Bragg reflections at high $\boldsymbol{Q}$-values.

Carrying out the integral for electrons in shells with spin angular moment S, orbital angular moment L, and total angular moment $J = L + S$, the general expression for the neutron magnetic form factor becomes (Squire 1978):

$$f_m(\boldsymbol{Q}) \rightarrow \frac{g_J}{2} J f_{mag}(\boldsymbol{Q}), \tag{7.12}$$

where $J(\boldsymbol{Q})$ is the total angular momentum, and g_J is the **_Landé splitting factor_**:

$$g_J = \frac{3}{2} + \frac{S(S+1) - L(L+1)}{2J(J+1)} \tag{7.13}$$

In 3d metals, the orbital angular momentum is nearly quenched, such that $L \cong 0$, $S = J$, and $g_J \leq 2$. Then $J(\boldsymbol{Q}) = S(\boldsymbol{Q})$[1] is again the spin contribution to the magnetic

---

[1] The spin angular moment S should not be mixed with the scattering function $S(\boldsymbol{Q})$ defined at other places, for instance in Eq. (8.9).

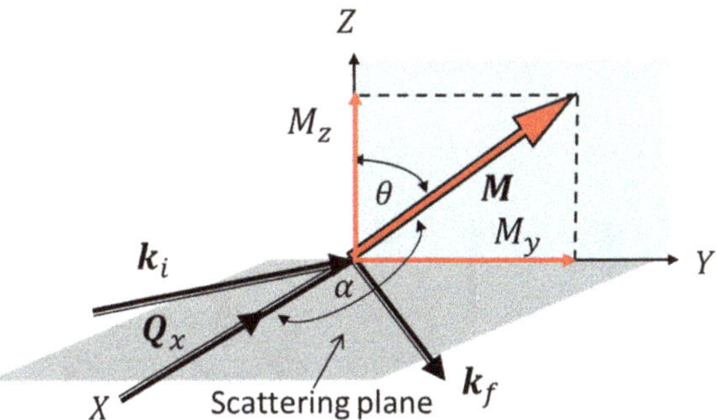

Fig. 7.3 Graphical interpretation of the orientational dependence for magnetic neutron scattering. Only magnetization vectors in the plane perpendicular to the scattering vector Q_x can be probed. This is the bluish-colored $Y - Z$ plane in the graph. Here, the magnetization vector makes an angle of θ against the z-axis and has components M_z and M_y. The $X - Y$ scattering plane is shown in gray. In the graph, the angle $\alpha = 90°$

form factor, i.e., the Fourier transform of all electrons with unpaired spins in the electronic shells. In conclusion, the neutron ***magnetic scattering length*** is now:

$$b_m = \frac{1}{2}\gamma_n r_0 \gamma_n \sigma_n g_J J f_{mag}(\boldsymbol{Q}) = p. \tag{7.14}$$

In the literature on magnetic neutron scattering, often the letter p is preferred for the magnetic scattering length: $b_m = p$.

As already discussed (see Eq. (7.9)), the sensitivity of neutrons to magnetism is only in the plane perpendicular to the scattering vector $\boldsymbol{Q}$, i.e., only for $\boldsymbol{Q} \perp \boldsymbol{M}$. Any magnetic moments parallel to the scattering vector are not recognized by magnetic neutron scattering. This orientational dependence can mathematically be cast into the form:

$$\left(1 - \left(\widehat{\boldsymbol{Q}} \cdot \widehat{\boldsymbol{M}}\right)^2\right) = \left(1 - \cos^2\left(\widehat{\boldsymbol{Q};\boldsymbol{M}}\right)\right) = \sin^2\left(\angle\widehat{\boldsymbol{Q};\boldsymbol{M}}\right) = \sin^2\alpha. \tag{7.15}$$

$\widehat{\boldsymbol{Q}}$ and $\widehat{\boldsymbol{M}}$ are unit vectors in the direction of $\boldsymbol{Q}$ and $\boldsymbol{M}$, respectively, and α is the angle between $\widehat{\boldsymbol{Q}}$ and $\widehat{\boldsymbol{M}}$. This rule is graphically illustrated in Fig. 7.3.

> The geometric condition for magnetic neutron scattering requires that $\boldsymbol{Q} \perp \boldsymbol{M}$.

To observe magnetic moments in all possible directions of a single crystal, the sample has to be realigned so that different crystallographic directions become compatible with the requirement $\boldsymbol{Q} \perp \boldsymbol{M}$. Often ferro- or antiferromagnetic materials show a preferred orientation of their magnetic moments in specific atomic planes, such as the ferromagnetic order in bcc Fe as indicated in Fig. 7.4, the hexagonal basal plane of Gd, or the antiferromagnetic order in the (111) plane of CoO. Using the $\boldsymbol{Q} \perp \boldsymbol{M}$ condition, these planes containing the magnetic moments can be identified.

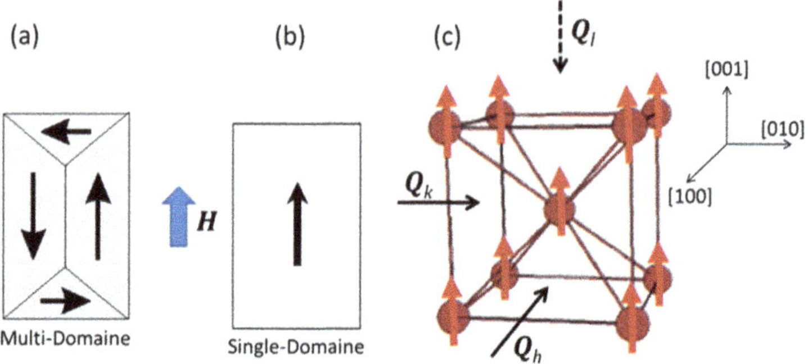

Fig. 7.4 (**a**) Ferromagnetic crystals are usually decomposed in domains with different magnetization orientations; (b) application of a magnetic field generates a single domain state. (**c**) Ferromagnetic spin structure of bcc Fe. The magnetic moments are indicated by arrows on the Fe atoms symbolized by red dots. The ferromagnetic spin orientation can be studied with the scattering vectors Q_h, Q_k but not with Q_l

These considerations do not concern powder or polycrystalline samples because of their random orientation in space.

Figure 7.4 shows an example of how the geometric condition $Q \perp M$ can be applied. In the ferromagnetic state of Fe, the magnetic moments are all parallel to each other and preferentially oriented along one of the cube axes, here [001], as indicated in panel (c), which is called the *easy axis*. However, in the ground state, the magnetic crystal will decompose into a multidomain state to suppress energy-demanding stray fields (panel (a)). Applying a small external magnetic field causes the magnetization vectors in the domains to align with the external field, generating a single-domain state (panel (b)).

The magnetic moments sketched in panel (c) can only be probed by magnetic neutron scattering if the scattering vector Q is oriented perpendicular to the magnetic moments, for instance, with a scattering vector Q_k pointing along [010] or with Q_h oriented along [100]. But for Q_l along [001], there will be no magnetic signal. This missing magnetic intensity can tell us immediately that the magnetic moments must point along the [001] direction.

The magnetization components M_y and M_z in Fig. 7.3 cannot be determined separately via a Q_x-scan, even though the scattering vector Q_x satisfies the perpendicular condition for both components. However, these magnetization components can be determined using polarized neutrons in conjunction with a polarization analysis of the scattered beam. This will be explained further below in Sect. 7.5.

What Are Ferromagnets, and What Are Antiferromagnets?
Ferromagnetism and antiferromagnetism are collective phenomena that concern all magnetic moments in the crystal lattice. The local atomic dipole moments interact to generate either parallel or antiparallel alignment with each other. Which one is preferred is the result of a delicate balance

(continued)

determined by the overlap of wavefunctions from neighboring atoms. But before order among the magnetic dipole moments may occur, entropy-favoring disorder must be lowered. Therefore, ferromagnetism arises below the ordering temperature, known as the ***Curie temperature***; antiferromagnetism appears below the corresponding ordering temperature named ***Néel temperature***. With lowering the temperature, the ferro- and antiferromagnetic order increases. This can be recognized by an increase of the respective magnetic Bragg peak intensity with decreasing temperature. For temperatures above the Curie or Néel temperature, both types of magnets are paramagnets, characterized by zero magnetization but high magnetic susceptibility. The pictures below show a simple ferromagnetic structure in panel (a), and three variants of antiferromagnetic structures in panels b (A-type), c (G-type), and d (C-type). The local magnetic moments are symbolized by arrows (spins). In ferromagnets, the atomic structure and the spin structure have the same periodicity. In antiferromagnets, the up and down spins occupy two different sublattices. Therefore, the spin structure has twice the periodicity of the atomic structure. Well-known ferromagnets are Co, Fe, and Ni, magnetite Fe_3O_4, and hardmagnets, such as $SmCo_5$. Well-known antiferromagnets are the transition metal oxides MnO, CoO, NiO, and Cr_2O_3, and antiferromagnetic metals, such as Cr, Mn, and Mn_2Au.

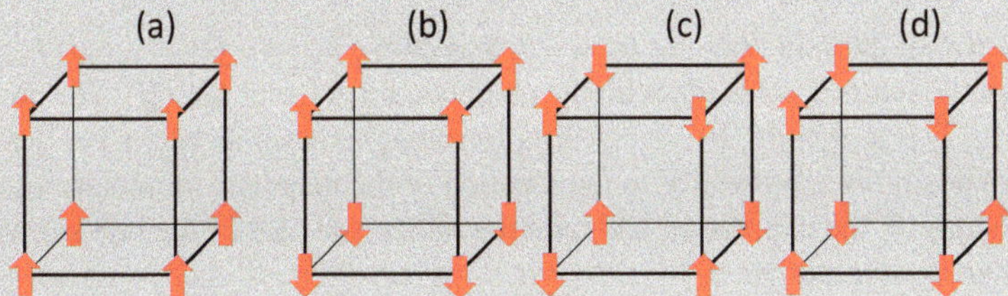

Ferromagnetism is a macroscopic phenomenon. It is easily recognized by the magnetic pull. Because of the spontaneous order below the Curie temperature, ferromagnets do not require an external magnetic field for aligning neighboring magnetic dipole moments. However, in large crystals, the ferromagnetic order may decompose into ferromagnetic domains with completely compensated magnetic flux inside. In such a situation, the ferromagnet is macroscopically "nonmagnetic" and requires an external field to align the domains for driving the magnetization into a single domain state, i.e., into a saturated state. In the case of neutron scattering, it is always a good idea to apply a saturating magnetic field to record the maximum magnetic intensity.

Antiferromagnets, in contrast to ferromagnets, are macroscopically nonmagnetic. Therefore, they have been discovered only late in the twentieth century, first by theoretical prediction and later by neutron scattering. Similar to ferromagnets, antiferromagnets may also decompose into domains. However, aligning antiferromagnetic domains in order to gain a single-domain antiferromagnetic state is much harder than aligning ferromagnetic domains.

The calculations of magnetic form factors and selection rules are lengthy and tedious, often not easy to follow (Squire 1978; Lovesey 1984). However, in the end, we find expressions that look familiar and are easy to interpret using the Born approximation in the framework of the kinematic approximation. The Born approximation is justified under all circumstances as the neutron magnetic interaction is weak, and therefore, the probability for double scattering is negligible. Respective results are discussed in the following.

7.2   Ferromagnetic Neutron Scattering

7.2.1   Long Range Order

We consider neutron scattering from a simple ferromagnetic crystal, such as from a bcc α-Fe single crystal in the ferromagnetic, single domain state. Then, the *scattering amplitude* consists of a nuclear part and a magnetic part with identical locations R_j of the Fe nuclei and Fe spins:

$$A_{tot} = \frac{1}{r}\left(b_n \sum_{j=0} e^{i\mathbf{Q}\cdot\mathbf{R}_j} + \gamma_n r_0 \sigma_n f_m(\mathbf{Q}) \sum_{j=0} e^{i\mathbf{Q}\cdot\mathbf{R}_j} \right) \tag{7.16}$$

Now we set the spin operator $\sigma_n = \pm 1$, and define $b_m = p(\mathbf{Q}) = \gamma_n r_0 \sigma_n f_{mag}(\mathbf{Q})$ as the *magnetic scattering length,* with the magnetic form factor $f_{mag}(\mathbf{Q})$. The numerical value of p is $p = 0.54 f \times 10^{-12}$ cm. In a ferromagnet such as bcc α-Fe, the atomic (nuclear) positions are identical to the location of the magnetic moments. Therefore, we can simplify the expression and take the lattice sum out of the bracket, yielding the scattering amplitude:

$$A_{tot} = \frac{1}{r}(b_n \pm p(\mathbf{Q})) \sum_{j=0} e^{i\mathbf{Q}\cdot\mathbf{R}_j}. \tag{7.17}$$

The total **scattering intensity** (or cross section) contains both nuclear and magnetic contributions. The nuclear and magnetic intensity is accordingly

$$I = A_{tot}A_{tot}^* = \left(\frac{1}{r}\right)^2 NV_{rc}(b_n \pm p(\mathbf{Q}))^2 \sum_{i,j=0} e^{i\mathbf{Q}\cdot(\mathbf{R}_i - \mathbf{R}_j)} \tag{7.18}$$

N is the total number of scattering centers in the scattering volume, and V_{rc} is the reciprocal unit cell volume. The (+) sign stands for neutron polarization parallel to the sample magnetization, and the ($-$) sign for the antiparallel polarization. For an unpolarized neutron beam, we have an equal probability for up- and down neutron polarization, and therefore, we take the average of both polarizations:

$$\frac{1}{2}\left[(b_n + p)^2 + (b_n - p)^2\right] = b_n^2 + p^2. \tag{7.19}$$

This is an important result, telling us that for an unpolarized neutron beam, the intensity (cross section) can be described by the sum of the nuclear and the magnetic intensity without containing a cross term:

$$I(\boldsymbol{Q}) = \left(\frac{1}{r}\right)^2 NV_{rc}\left(b_n^2 + p^2\right) \sum_{i,j=0} e^{i\boldsymbol{Q}\cdot(\boldsymbol{R}_i - \boldsymbol{R}_j)}. \tag{7.20}$$

The sum is taken over all basis vectors of the bcc unit cell and all lattice sites in the crystal. We recall that the sum over the basis vectors provides a structure factor, and the sum over all lattice sites results in a δ-function at Bragg positions, whenever the Bragg condition $\boldsymbol{Q} = \boldsymbol{G}_{hkl}$ is met. Now taking the lattice sum, we find the intensity in the ferromagnetic state:

$$I(\boldsymbol{Q}) = \left(\frac{1}{r}\right)^2 NV_{rc}\left(b_n^2 + p^2\right)\delta(\boldsymbol{Q} - \boldsymbol{G}_{hkl}). \tag{7.21}$$

The last equation implies that both contributions to the intensity, nuclear and magnetic, superimpose and can eventually be separated by using the orientational selection factor. Upon evaluating the lattice sum, Bragg reflections occur only for $\boldsymbol{Q} = \boldsymbol{G}_{hkl}^{struct}$ and $\boldsymbol{Q} = \boldsymbol{G}_{hkl}^{mag}$, and therefore at the location of the Bragg reflections in the reciprocal space, we obtain the total nuclear and magnetic scattering intensity:

$$I(\boldsymbol{Q}=\boldsymbol{G}_{hkl}) = \left(\frac{1}{r}\right)^2 NV_{rc}\left[\lfloor b_n F_n(\boldsymbol{G}_{hkl}^{struct})\rfloor^2 + \left(1 - \left(\widehat{\boldsymbol{Q}}\cdot\widehat{\boldsymbol{M}}\right)^2\right)\lfloor p(\boldsymbol{Q})F_{FM}(\boldsymbol{G}_{hkl}^{mag})\rfloor^2\right],$$

$$\tag{7.22}$$

where $F_n(\boldsymbol{G}_{hkl}^{struct})$ and $F_{FM}(\boldsymbol{G}_{hkl}^{mag})$ are the respective nuclear and ferromagnetic structure factors, and $\left(1 - \left(\widehat{\boldsymbol{Q}}\cdot\widehat{\boldsymbol{M}}\right)^2\right)$ is the orientational dependence for magnetic scattering.

For the *bcc* lattice, we have already evaluated the structure factor in Sect. 3.5. Furthermore, it holds that for the ferromagnetic *bcc* lattice and for allowed Bragg reflections: $F_n(\boldsymbol{G}_{hkl}^{struct}) = F_{mag}(\boldsymbol{G}_{hkl}^{mag}) = 2$. Figure 7.5 shows schematically the Bragg reflections, which are composed of a nuclear and a magnetic part. The part due to nuclear scattering remains constant in $\boldsymbol{Q}$, aside from Debye–Waller factor effects, which are neglected here. In contrast, the magnetic part drops off with increasing $\boldsymbol{Q}$ as expected for the magnetic form factor. The intensity of the magnetic part

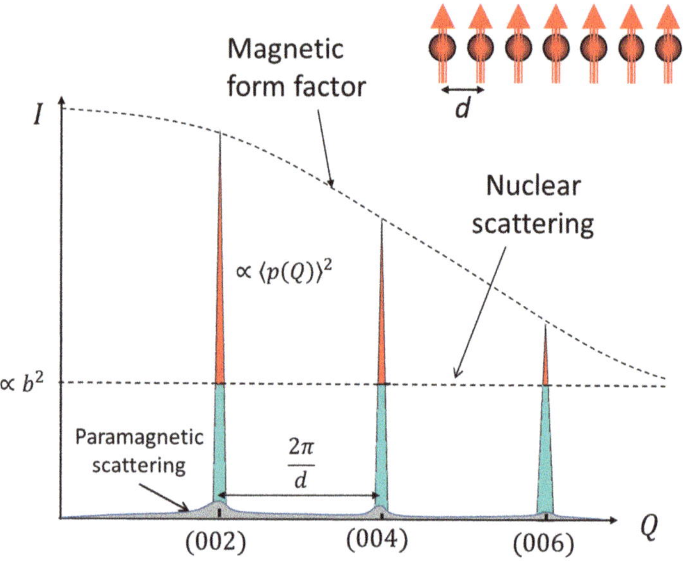

Fig. 7.5 Nuclear (in green) and magnetic intensity (in red) at Bragg reflections of a ferromagnetic material, such as bcc α-Fe. The low intensity between the Bragg reflections, marked in gray color, designates paramagnetic scattering. The inset sketches the ferromagnetic order of magnetic moments with a periodicity matching that of the crystal lattice

$$I_{FM}(\boldsymbol{Q},T) = \left(1 - \left(\widehat{\boldsymbol{Q}}\cdot\widehat{\boldsymbol{M}}\right)^2\right)\left[p(\boldsymbol{Q},T)F_{FM}\left(G_{hkl}^{mag}\right)\right]^2 \propto \langle M_{FM}(T)\rangle^2 \qquad (7.23)$$

is proportional to the ensemble-averaged and squared magnetization $\langle M(T)\rangle^2$. The magnetization is defined as the number of magnetic moments divided by the volume:

$$\langle M_{FM}(T)\rangle = \frac{N}{V}\langle m\rangle, \qquad (7.24)$$

where $\langle m\rangle = g_S\mu_B S$ is the average local magnetic moment, assuming that $J = S$ and $L = 0$. The magnetization itself always depends on the temperature, which will be discussed next.

7.2.2 Temperature Dependence of the Magnetization

The **magnetization** $\langle M\rangle$ is an indication of the magnetic order in the sample and, therefore, is called the **order parameter** of the ferromagnet. In an approximation known as the **molecular field approximation** (MFA), the temperature dependence of the order parameter is described by

$$\langle M(T) \rangle = M_S(-\varepsilon)^{1/2}, \tag{7.25}$$

where M_S is the saturation magnetization at $T = 0$, $\varepsilon = (T - T_C)/T_C$ is the reduced temperature difference, and T_C is the **Curie temperature**. Therefore, we conclude that the root of the magnetic Bragg peak intensity yields direct access to the **ferromagnetic order parameter**:

$$\sqrt{I_{mag}(Q)} \sim \langle M(T) \rangle. \tag{7.26}$$

The magnetization versus temperature is represented schematically for a ferromagnet in Fig. 7.6 by the blue line. In MFA, the exponent of $\langle M(T) \rangle = M_S(-\varepsilon)^{\beta}$ is $\beta = 1/2$. However, more advanced theories predict $\beta = 1/3$ for three-dimensional ferromagnets. With neutron scattering, the critical exponent β can be exactly determined.

The loss of magnetization by approaching T_C from below ($T \rightarrow T_C$) is due to increasing spin fluctuations, counteracting the exchange interaction between the spins and disturbing the long-range order. This is graphically visualized in Fig. 7.7. The higher the temperature, the more entropy wins over order. In the end, the long-range magnetization drops to zero, and paramagnetic (PM) short-range correlation prevails for $T > T_C$.

The model shown in the first three panels of Fig. 7.7 is referred to as transverse spin disorder in the frame of the Heisenberg model. While the local magnetic moments remain constant across the FM–PM phase transition, their orientation in space becomes increasingly random. In contrast, in transition metal ferromagnets, including Fe, Co, and Ni, the exchange splitting of the spin-up and spin-down bands is responsible for the magnetic polarization. Instead of local moments fixed to ions, we find delocalized itinerant moments, whose magnitude depends on the band gap. With increasing temperature, the band gap closes and the modulus of the magnetic moments decreases, until the magnetic moments vanish: $\langle m \rangle = 0$. Then, for $T > T_c$, short-range spin correlations should not exist. The spin fluctuations in 3d transition metals are referred to as longitudinal fluctuations, in contrast to the transverse

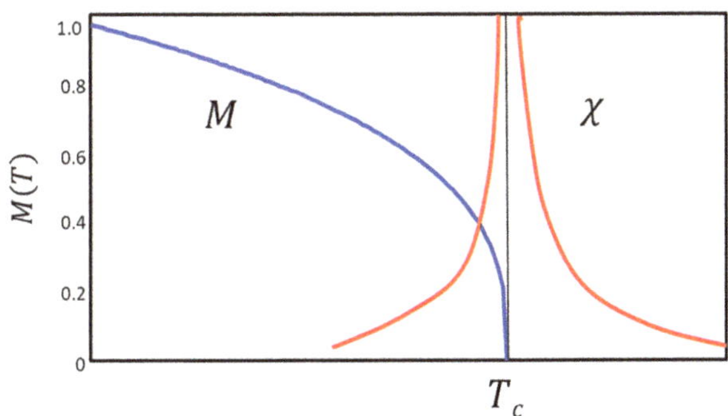

Fig. 7.6 Blue line: magnetization versus temperature for a ferromagnet. Red lines: magnetic susceptibility above and below the Curie temperature T_C

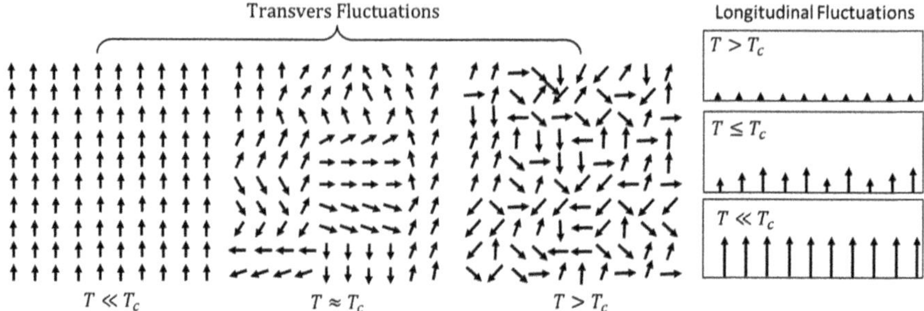

Fig. 7.7 Schematic visualization of increasing spin disorder by approaching the critical temperature, separating the ferromagnetic state from the paramagnetic state. Two different models are contrasted: loss of spin order by transverse fluctuations in the Heisenberg model, and longitudinal fluctuations in the model of band magnetism

Fig. 7.8 Temperature dependence of the magnetization of EuO and EuS. The plot is in reduced magnetization, and in terms of the temperature ratio T/T_c, which renders a comparison between both prototypes of Heisenberg ferromagnets possible. Reproduced with permission from Als-Nielsen et al. (1976). Copyright (2025) by the American Physical Society

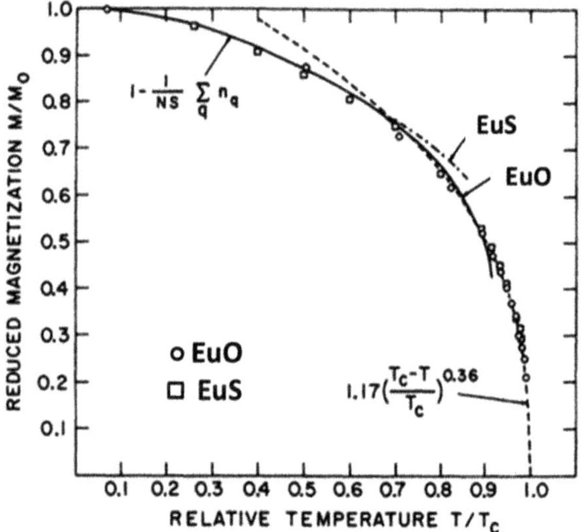

fluctuations of the Heisenberg model. We come back to this point in the section on polarized neutrons in Sect. 7.5.

Figure 7.8 reproduces from the work of Als-Nielsen and coworkers (Als-Nielsen et al. 1976) the neutron scattering results of the magnetization of two ferromagnets EuS and EuO, with Curie temperatures of 16.6 K and 69.1 K, respectively. Both are good representatives of 3D Heisenberg ferromagnets, meaning that the local spin anisotropy is zero and consequently the local spin orientation is isotropic. The reduced magnetization M/M_0 is plotted versus the relative temperature T/T_c for both compounds. The dashed lines represent different theoretical models for the magnetization close to the Curie temperature with a critical exponent of $\beta = 0.36$, in good agreement with theoretical predictions for 3D Heisenberg magnets, while the solid line describes well the low-temperature behavior of the magnetization due to the excitation of spin waves (Stanley 1971).

7.2.3   *Paramagnetic Scattering and Critical Scattering*

Aside from ferromagnetic scattering concentrated in Bragg reflections, the spin–spin correlation gives rise to paramagnetic scattering above and below the Curie temperature T_c. In addition, at T_c, *critical scattering* can be observed due to critical spin fluctuations. Paramagnetic scattering and critical scattering are manifested in the form of diffuse scattering, centered at the respective long-range order Bragg reflections. Disorder scattering and critical scattering are universal features of second-order phase transitions. Therefore, similar observations can also be made at atomic order–disorder phase transitions, which we discuss in Chap. 8.

To describe paramagnetic scattering, we consider the spin–spin pair correlation function $g(\boldsymbol{R})$ of the z-component of spins separated by a distance $\boldsymbol{R}$:

$$g(\boldsymbol{R}) = \left\langle \left(s_0^z - \langle s_0^z \rangle\right)\left(s_R^z - \langle s_R^z \rangle\right)\right\rangle, \tag{7.27}$$

Here $\left(s_0^z - \langle s_0^z \rangle\right)$ describes the difference between the actual value of the spin component s_0^z and the mean value at $\boldsymbol{R} = 0$; for $(s_R^z - \langle s_R^z \rangle)$, the respective difference is evaluated at position R. The brackets $\langle \rangle$ denote space, time, and ensemble averages. Evaluating the product in Eq. (7.27), we obtain:

$$g(\boldsymbol{R}) = \langle s_0^z s_R^z \rangle - 2\langle s_0^z \rangle\langle s_R^z \rangle + \langle s_0^z \rangle\langle s_R^z \rangle = \langle s_0^z s_R^z \rangle - \langle S^z \rangle^2. \tag{7.28}$$

The last term can be rearranged:

$$\langle s_0^z s_R^z \rangle = \langle S^z \rangle^2 + g(\boldsymbol{R}) = \langle M \rangle^2 + g(\boldsymbol{R}). \tag{7.29}$$

In the last equation, we have identified $\langle S^z \rangle^2$ with the long-range order magnetization squared $\langle M \rangle^2$. More precisely, according to Eq. (7.24), $\langle S^z \rangle = (V/N)(1/g_S\mu_B)\langle M \rangle$. Now we plug Eq. (7.29) into our standard expression for magnetic neutron scattering and obtain:

$$I_{mag}(\boldsymbol{Q}) = NV_{rc}(\gamma r_0)^2 \left(\langle M(\boldsymbol{Q}) \rangle^2 \sum_G \delta(\boldsymbol{Q} - \boldsymbol{G}_{FM}) \right.$$

$$\left. + \underbrace{\iint g(\boldsymbol{R},\boldsymbol{R}')e^{i\boldsymbol{Q}\cdot(\boldsymbol{R}-\boldsymbol{R}')}d\boldsymbol{R}d\boldsymbol{R}'}_{g(\boldsymbol{Q})} \right)\left(1 - \left(\hat{\boldsymbol{Q}}\cdot\hat{\boldsymbol{M}}\right)^2\right)$$

$$= NV_{rc}(\gamma r_0)^2 \left(\langle M(\boldsymbol{Q}) \rangle^2 \sum_G \underbrace{\delta(\boldsymbol{Q} - \boldsymbol{G}_{FM})}_{\text{FT of LRO}} + \underbrace{g(\boldsymbol{Q})}_{\text{FT of SRO}} \right)\left(1 - \left(\hat{\boldsymbol{Q}}\cdot\hat{\boldsymbol{M}}\right)^2\right).$$

$$\tag{7.30}$$

In Eq. (7.30), we recognize the superposition of two types of scattering: (1) The Fourier transform of magnetic long-range order, resulting in magnetic Bragg-reflections at the r.l.p.s. of the ferromagnetic spin structure G_{FM}; (2) The Fourier transform of the short-range order pair correlation function $g(Q)$ that gives rise to paramagnetic diffuse scattering above and below the Curie temperature T_c. The last bracket in Eq. (7.30) is the usual orientational dependence for magnetic scattering.

The Fourier transform of the pair correlation function is connected to the Fourier transform of the magnetic susceptibility χ_m via

$$g(Q, T) = k_B T \chi_m(Q, T).$$

Here k_B is the Boltzmann constant and T is the absolute temperature. Expressing the Fourier transforms in terms of deviation from the respective Bragg point: $q = Q - G$, we find for the magnetic susceptibility:

$$g(q, T) = k_B T \chi_m(q, T). \tag{7.31}$$

In statistical mechanics, it is shown that the susceptibility is related to the spin–spin interaction via (Stanley 1971):

$$\chi_m(q, T) = \frac{J_{ex}^{-1}}{\xi^{-2} + q^2}, \tag{7.32}$$

where ξ is the spin–spin correlation length, and J_{ex} is the spin–spin exchange interaction, with

$$J_{eff}^{-1} = \frac{S(S+1)}{3k_B T}. \tag{7.33}$$

The critical scattering has a Lorentzian line shape according to Eq. (7.32) with a strongly temperature-dependent peak centered at the position where the ferromagnetic Bragg reflection will appear below T_c. The height of the critical scattering peak yields the susceptibility $\chi_m(q = 0, T)$, and the width is determined by the correlation length $\xi(T)$. By approaching T_c, the susceptibility and the correlation length diverge. At high temperatures above T_c, the spin–spin correlation diminishes, yielding a flat diffuse background intensity. In the high-temperature limit, the paramagnetic scattering amplitude becomes:

$$A_{PM}(Q) = (\gamma r_0) f_{mag}(Q) NS(S+1) \left\langle 1 - \left(\hat{Q} \cdot \hat{M} \right)^2 \right\rangle = (\gamma r_0) f_{mag}(Q) \frac{2}{3} NS(S+1).$$

$$\tag{7.34}$$

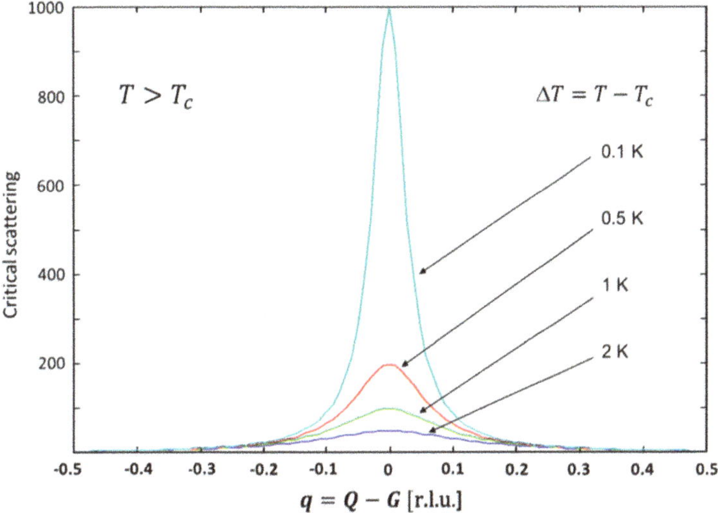

Fig. 7.9 Critical scattering due to short-range order spin correlations close to the critical temperature of a paramagnetic–ferromagnetic phase transition. The critical scattering peak is centered at $Q = G$, i.e., at the position of the ferromagnetic Bragg reflection that occurs below the Curie temperature T_c

Here N is the total number of spins in the scattering volume. The term $\left(\widehat{Q} \cdot \widehat{M}\right)^2$ has to be averaged over all possible spin orientations. If all directions are equally likely, then $\left\langle 1 - \left(\widehat{Q} \cdot \widehat{M}\right)^2 \right\rangle = 2/3$.

Equation 7.34 shows that the paramagnetic scattering amplitude is proportional to the magnetic form factor $f_{mag}(Q)$, and the prefactor $S(S + 1)$ allows us to determine the spin moment independent of any model assumption. Figure 7.9 schematically shows the diffuse scattering for different temperatures just above the Curie temperature. The question now is how the weak paramagnetic scattering above the Curie temperature can be separated from the much stronger nuclear scattering and thermal diffuse scattering. An answer to this question is given below in Sect. 7.5 on polarized neutron scattering.

7.3   Antiferromagnetic Neutron Scattering

In contrast to ferromagnets, in antiferromagnets (AF), the magnetic moments of equal magnitude reside on two different sublattices A and B ($|m_A| = |m_B|$) (see inset in Fig. 7.10), and their orientation alternates from up to down to up, etc. Therefore, macroscopically, AFs are nonmagnetic. Only by neutron scattering, the AF order can

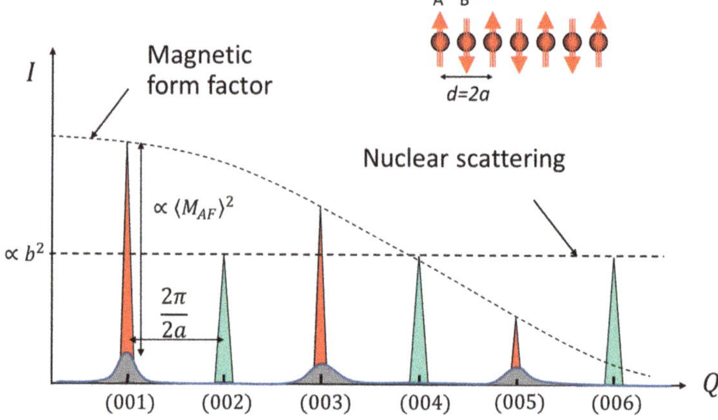

Fig. 7.10 Schematics of neutron diffraction from a single crystalline antiferromagnetic material. The inset shows the alternating spin orientation on the sublattices A and B, which doubles the antiferromagnetic periodicity concerning the crystal lattice periodicity. In the diffraction pattern, the nuclear Bragg peaks (turquoise) and the magnetic Bragg reflections (red) are separated. The diffuse intensity centered at the AF Bragg peaks is due to antiferromagnetic short-range order spin correlations

be revealed, such as in MnO[2] (Shull and Smart 1949). Because of the alternating spin structure in AFs, the magnetic periodicity is doubled compared to the structural (or nuclear) periodicity. This doubling has the effect that AF Bragg reflections and nuclear Bragg reflections no longer overlap but are separated in reciprocal space, as indicated in Fig. 7.10. The ***antiferromagnetic Bragg reflections*** occur at the "forbidden" positions concerning the nuclear structure factor. Sometimes these AF reflections are also referred to as "half-order" peaks or as *"superstructure" reflections*. Therefore, AF Bragg reflections are easy to identify by neutron scattering, and their intensity can be directly related to the ***antiferromagnetic order parameter***.

$$I(\mathbf{Q})=\left(\frac{1}{r}\right)^2 NV_{rc}\left[\left[b_n F_n\left(\mathbf{G}_{hkl}^{struct}\right)\right]^2+\left(p(\mathbf{Q},T)F_{AF}\left(\frac{\mathbf{G}_{hkl}^{struct}}{2}\right)\right)^2\left(1-\left(\widehat{\mathbf{Q}}\cdot\widehat{\mathbf{M}}\right)^2\right)\right],$$

$$(7.35)$$

where the AF contribution is

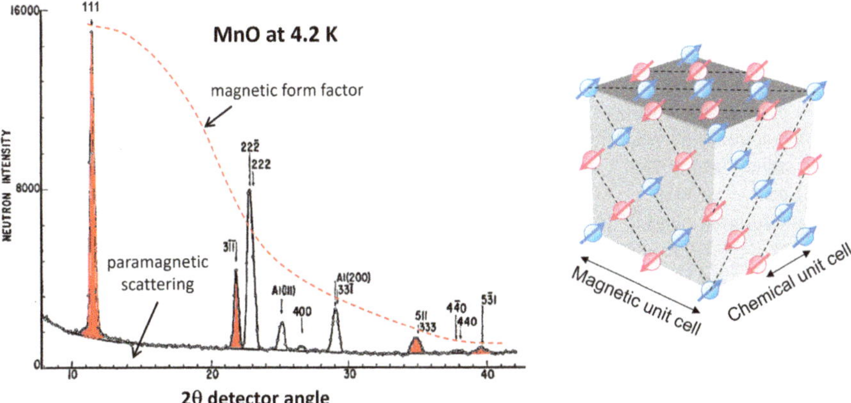

Fig. 7.11 Left panel: Neutron diffraction scan from a powder sample of MnO at 4.2 K in the antiferromagnetic state. The peaks with odd (hkl) indices marked in red are antiferromagnetic reflections; all others with even (hkl) indices are nuclear reflections. Some reflections are from the sample container made of aluminum. The dashed red line indicates the magnetic form factor of Mn^{2+}. Right panel: crystal and spin structure of MnO. Only Mn atoms are shown with spins alternating up and down in the planes perpendicular to the [111] direction. Oxygen atoms are located between the Mn planes. Adapted and reproduced from Shull and Smart (1949). Copyright (2025) by the American Physical Society

$$I_{AFM}(\boldsymbol{Q}) = \left(\frac{1}{r}\right)^2 NV_{rc}\left\langle M_{AF}\left(G_{hkl}^{AF}\right)\right\rangle^2\left(1 - \left(\widehat{\boldsymbol{Q}}\cdot\widehat{\boldsymbol{M}}\right)^2\right), \qquad (7.36)$$

with $G_{hkl}^{AF} = G_{hkl}^{struct}/2$. Usually, the ordering temperature of AFs (known as *Néel temperature*) lies at or below room temperature. The strong temperature dependence of these peaks is another fingerprint of their AF origin. Figure 7.11 reproduces the neutron diffraction pattern from a powder sample of the antiferromagnet MnO (Shull and Smart 1949)[3]. MnO has a NaCl crystal structure and a Néel temperature of 122 K. The diffraction pattern is taken at 4.2 K in the AF state of the sample. The (hkl) indices in Fig. 7.11 are based on the magnetic unit cell ($a_{\mathrm{mag}} = 0.885$ nm), which has a lattice parameter twice that of the chemical unit cell ($a_{\mathrm{chem}} = 0.443$ nm). Most prominent are the AF peaks with Miller indices (111) and (311). These reflections do not exist in the paramagnetic state at temperatures above the Néel temperature, which is a clear sign of the AF nature of the sample and the AF phase transition.

If the Miller indices are based on the chemical unit cell, the AF Bragg reflections occur at half-order positions, such as $\left(\frac{1}{2}\,\frac{1}{2}\,\frac{1}{2}\right)$, $\left(\frac{3}{2}\,\frac{1}{2}\,\frac{1}{2}\right)$, etc. The temperature dependence of the $\left(\frac{1}{2}\,\frac{1}{2}\,\frac{1}{2}\right)$ AF peak is plotted in Fig. 7.12 for MnO from (Sun et al. 2017), and the inset shows the half-order AF Bragg reflection at various temperatures.

---

[3] Neutron scattering on MnO and other antiferromagnetic oxides with higher resolution was later performed by Roth (1958).

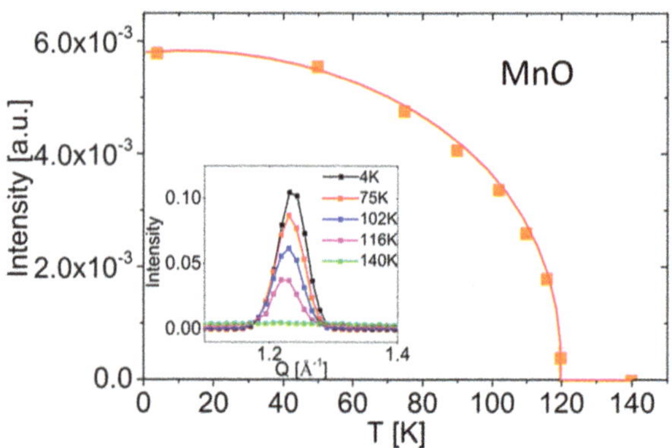

Fig. 7.12 Neutron diffraction of MnO powder sample. The intensity of the magnetic $\left(\frac{1}{2}\,\frac{1}{2}\,\frac{1}{2}\right)$ Bragg peak is plotted as a function of temperature. The inset shows radial scans of the $\left(\frac{1}{2}\,\frac{1}{2}\,\frac{1}{2}\right)$ peak at several temperatures above and below the Néel temperature of 122 K. The red line is a guide to the eye. Reprinted from Sun et al. (2017), open access

The discussion on paramagnetic and critical scattering applies to antiferromagnets in the same way as for ferromagnets. Also in AF, the diffuse scattering due to paramagnetic short-range order correlations and spin fluctuations exists above and below the Néel temperature. Instead of repeating the previous discussion, we show the location of the diffuse scattering at the antiferromagnetic peak positions in Fig. 7.10 schematically, including their Lorentzian line shape.

7.4 Incommensurate Spin Order

The ferromagnetic and antiferromagnetic structures discussed so far are so-called *commensurate magnetic structures,* because the periodicity of the magnetic dipole moments is related to the lattice periodicity by a rational number. For ferromagnetism, the ratio is 1; for antiferromagnets, it is 2. This is not necessarily always the case. The ratio may also be irrational, and then we talk about *incommensurate magnetic structures*. Well-known examples of hosting incommensurate spin structures are the 3d metal chromium, the intermetallic compounds such as Au_2Mn, and the 4f rare earth metals holmium and dysprosium. As a representative of incommensurate spin structures, we will discuss here the spin density wave order in Cr.

Figure 7.13a pictures the commensurate antiferromagnetic spin structure of bcc Cr with a lattice parameter a_{Cr} and a spin periodicity $a_{AF} = 2a_{Cr}$. The spins are oriented parallel to any of the cube edges, which are the magnetically "easy" axes. Cr may adopt a commensurate spin structure due to strain or alloying with Mn. Otherwise, for temperatures below the Néel temperature of 311 K, the Cr spins display an incommensurate spin density wave (ISDW), schematically sketched in panel (b) of Fig. 7.13. In the first approximation, the ISDW can be described by a sinusoidal modulation of the Cr magnetic moments:

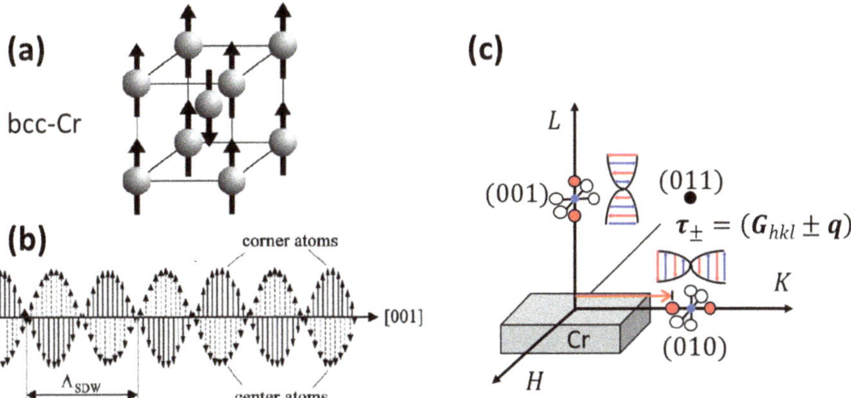

Fig. 7.13 Cr spin structure in real and reciprocal lattice. (**a**) Antiferromagnetic spin orientation in the *bcc* Cr lattice. (**b**) Spin density wave modulation of the magnetization in Cr. The spins are oriented perpendicular to the "propagation" direction in the [001] direction, defining a transverse and incommensurate spin density wave. (**c**) Reciprocal lattice of Cr in the (0*KL*) plane, showing the positions of the commensurate antiferromagnetic peaks (small blue dots) and those of the incommensurate SDW modulation as satellites surrounding the bcc forbidden Bragg reflections. For further explanations, please refer to the text. Half-wave patterns symbolize the propagation direction and the polarization of the incommensurate SDW

$$m(r) = m_0 \sin(\tau_{\pm} \cdot r). \tag{7.37}$$

The amplitude of the ISDW $|m_0|$ is about 0.5 μ_B/atom at 4.2 K (Fawcett 1988) and the in commensurate wave vector $\tau_{\pm}$ is expressed by

$$\tau_{\pm} = \left(\frac{2\pi}{a_{Cr}} \pm \frac{2\pi}{\Lambda_{SDW}} \right) \{100\} = \frac{2\pi}{a_{Cr}} (1 \pm s)\{100\} = G_{\{100\}}(1 \pm s), \tag{7.38}$$

which is in short form:

$$\tau_{\pm} = (G_{hkl} \pm q). \tag{7.39}$$

Here, Λ_{SDW} is the wavelength of the ISDW, $q = 2\pi/\Lambda_{SDW}$ is the respective modulation wavevector, and $s = a_{Cr}/\Lambda_{SDW} \approx 0.05$ is the incommensurability parameter. For $s \rightarrow 0$, the SDW becomes commensurate. Λ_{SDW} increases smoothly from 6 nm at 10 K to about 8 nm at the Néel temperature (Fawcett 1988; Zabel 1999).

The ISDW is a static spin wave that does not propagate in time, only in space. The ISDW may propagate in any of the {100}-directions, giving rise to magnetic satellite reflections at positions $\tau_{\pm}$ near the forbidden (*hkl*) Bragg reflections, i.e., reflections with an odd sum $h + k + l = 2n + 1$. These magnetic satellite reflections are marked as red dots in Fig. 7.13c. In a single-crystalline Cr sample, not all propagation directions may coexist in different domains. Which ISDWs are present and which are not can be determined with neutron scattering. To do so, we have to recall the orientational dependence of magnetic neutron scattering, where the diffracted intensity is proportional to:

$$I(\boldsymbol{Q}) \cong |M(\boldsymbol{Q}=\boldsymbol{\tau})|^2 \sin^2\left(\widehat{\boldsymbol{Q}};\widehat{\boldsymbol{M}}\right) = |M(\boldsymbol{\tau})|^2 \sin^2\alpha, \qquad (7.40)$$

where $M(\boldsymbol{\tau})$ is the Fourier transform of the ISDW, and α is the angle between the scattering vector and the spin direction of the ISDW.

Referring to Fig. 7.13c, we take a scan along the ?-direction. If we find peak intensity at the pair of satellite positions indicated in red, then we conclude that the ISDW propagates (like a standing wave) in the ?-direction with spins oriented perpendicular to K, i.e., in the L- or H-direction. Such a spin density wave is characterized as a transverse SDW (T-ISDW), as outlined by the wavy icon above the (010) r.l.p. The other open circles around the (010) r.l.p.s. either display no intensity or indicate propagation vectors in the H- and L-directions. Similarly, if we find satellite intensity around the (001) reflection by a radial scan in the L-direction, we conclude that a T-ISDW exists in the [001] direction with spins lying in the H, K-plane as symbolized by the half-wave icon.

Figure 7.14 shows neutron scattering results of the SDW in a 300-nm-thick Cr film capped with a 3-nm-thick Fe film, reproduced from (Bödeker et al. 1999). The scan direction is along the K-direction as indicated in the inset. We notice a commensurate peak at $K = 1$ in r.l.u., persisting from low to high temperatures,

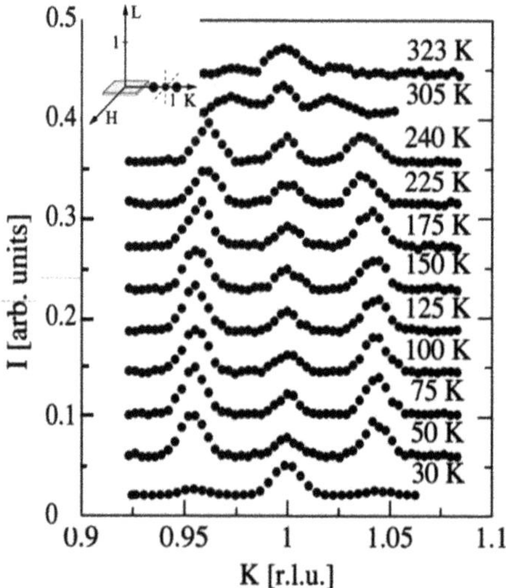

Fig. 7.14 Longitudinal scan with thermal neutrons in the (0K0) direction crossing the (010) forbidden Bragg reflection for different temperatures from 30 K to 323 K. On both sides of the (010) commensurate peak, satellite peaks are observable at $q = 0.95$ and 1.05 in r.l.u. The incommensurate peaks shift with temperature and vanish above 305 K. The commensurate peak appears to be unaffected by the temperature. The inset shows the scan direction. Reproduced with permission from Bödeker et al. (1999). Copyright (2025) by the American Physical Society

and in addition, satellite reflections due to the T-ISDW to the left and the right, as discussed previously. The incommensurate peak positions shift with increasing temperature, and the intensity decreases. At the lowest temperature of 30 K, the satellite peaks are weak, becoming stronger at 50 K. This is an indication of a spin orientational phase transition that takes place between these two temperatures. The central antiferromagnetic peak signals a coexisting commensurate phase persisting up to temperatures even above the Néel temperature, indicative of its strain-induced origin.

So far, we have emphasized neutron scattering for the exploration of the ISDW in Cr. It should, however, be mentioned that the SDW induces a charge density wave (CDW) of the 3d Cr electrons. The CDW has half the period of the SDW and can be detected with X-ray scattering as satellite reflections surrounding the allowed bcc Bragg reflections. The respective X-ray intensity is

$$I_{x-ray}(Q) = |\rho_0(Q)|^2 \delta(Q - G_{hkl}) + \sigma(Q)^2 \delta(Q - (G_{hkl} \pm 2q)). \qquad (7.41)$$

Here G_{hkl} is an allowed bcc reciprocal lattice vector, $\rho_0(Q)$ and $\sigma(Q)$ are the Fourier transforms of the real-space electron density $\rho_0(r)$ and charge density modulation $\sigma(r)$, respectively, and $2q$ is the wave vector of the CDW. Eq. (7.41) yields the fundamental bcc Bragg reflections at $Q = G_{hkl}$ and two satellite reflections at $Q = G_{hkl} \pm 2q$ as well as the higher harmonics at $Q = G_{hkl} \pm 4q$. Using synchrotron radiation, the incommensurate modulation of the Cr lattice by the SDW (CDW) can be made visible, as can be seen in Fig. 7.15, reproduced from the work of Hill and coworkers (Hill et al. 1995).

7.5   Polarized Magnetic Neutron Scattering

Up to now, we have discussed elastic magnetic neutron scattering, disregarding neutron polarization effects. In this section, we briefly present elastic neutron scattering that uses polarized neutrons in the incident beam and a polarization analysis of the scattered beam. The polarized neutron scattering method described here follows the seminal work by Moon et al. (1969). Polarized neutron scattering helps for a better separation of nuclear and magnetic scattering in ferromagnetic and disordered magnetic systems. In particular, it allows us to map out the spin density distribution in materials.

The spin of the neutron has two eigenstates in a magnetic field: parallel or antiparallel to the quantization axis, up or down, or expressed by the spin operator $\sigma = \pm 1$. Let's assume that the incident beam is polarized in the "up" direction. The scattering plane is the X, Y-plane, which is perpendicular to the sample Y, Z-plane (see Figs. 7.3 and 7.16a). The sample magnetization M_Z is oriented parallel to the magnetic field H in the Z-direction, which is the spin quantization axis, as sketched in Fig. 7.16a. Then we expect the scattered neutron spin to be preserved upon scattering in the "up" orientation. The combined nuclear (n) and magnetic (m) scattering amplitude can be expressed in simplified terms as

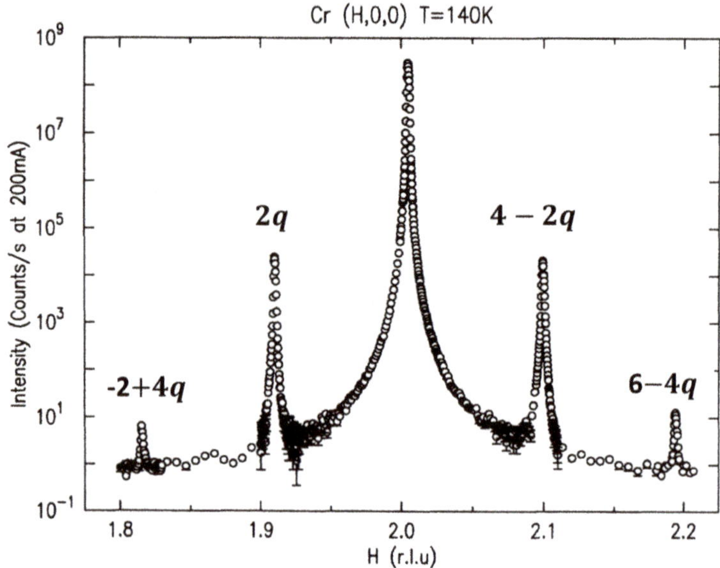

Fig. 7.15 A longitudinal scan along the (*H*00) direction crossing the (200) Bragg reflection of Cr at 140 K. On both sides, $2q$ and $4q$ satellite reflections are observed due to the charge density wave in Cr, accompanying the SDW with half the period. Data were taken at a bending magnet of the National Synchrotron Light Source in Brookhaven, USA, using X-rays at E = 8 keV by employing a Si(111) monochromator. Reproduced with permission from Hill et al. (1995). Copyright (2025) by the American Physical Society

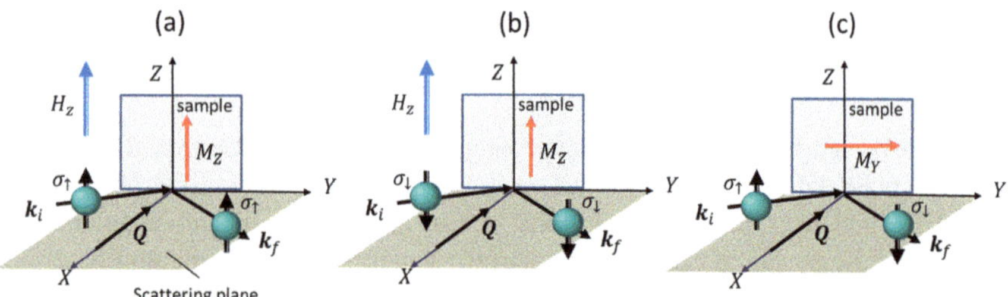

Fig. 7.16 Scattering experiments with a polarized incident neutron beam. In all three cases, the scattering vector is oriented perpendicular to the magnetization vector and parallel to the *X*-direction. (**a**) and (**b**) show non-spin-flip scattering events for up- and down-oriented spins for a magnetization direction parallel to the spin quantization direction Z. In (**c**), the magnetization is turned into the *Y*-direction, causing the spins to flip from up to down or vice versa

$$A_{tot} = A_n(Q) + A_m(Q, \sigma \uparrow\uparrow) = k\left(b_n + p\left\langle \boldsymbol{\sigma}\uparrow \cdot \widehat{\boldsymbol{M}}_z\right\rangle\right) = k\left(b_n + p_z\right). \qquad (7.42)$$

Here p is the magnetic scattering length as defined in Eq. (7.14), $\widehat{\boldsymbol{M}}_z$ is the unit magnetization vector in the Z-direction, and σ is the spin operator. Here we have set $p\left\langle \boldsymbol{\sigma}\uparrow \cdot \widehat{\boldsymbol{M}}_z\right\rangle = p_z$, and the term k collects all prefactors.

Now we turn the incident neutron spin from up to down and study the same magnetic sample, as indicated in Fig. 7.16b. Then the scattering amplitude is

$$A_{tot} = A_n(Q) + A_m(Q, \sigma \downarrow\downarrow) = k\left(b_n - p\left\langle \boldsymbol{\sigma}_\downarrow \cdot \widehat{\boldsymbol{M}}_z \right\rangle\right) = k\left(b_n - p_z\right). \qquad (7.43)$$

It does not matter whether we change the incident neutron spin or the sample magnetization. In both cases, the antiparallel orientation of spin and magnetization results in a minus sign. Both scattering experiments preserve the initial spin state. Only the sign is changed, not the neutron spin orientation during scattering. There-fore, this scattering is called **non-spin-flip scattering**. Note that under special circumstances, the nuclear and magnetic scattering lengths may be equal, and then the bracket $(b_n - p_z)$ vanishes: $(b_n - p_z) = 0$. This will be important later for polarizing the neutron beam.

In the next experiment, we rotate the magnetization vector by $90°$ to be parallel to the Y-direction, as marked in Fig. 7.16c. We start again with a spin-up polarized incident neutron beam and analyze the polarization of the scattered beam. Then we will find that in the scattered beam, the majority of spins are flipped from "up" to "down". Hence, we conclude that the M_Y magnetization acts on the neutron spin, causing the neutron spin to flip by $180°$. This process is known as **spin-flip (SF) scattering**. The opposite spin-flip from "down" to "up" occurs with the same probability. Thus, we can write for the amplitude of the SF-scattering:

$$A_{tot}(SF) = A_m(Q, \sigma \uparrow\downarrow, \sigma \downarrow\uparrow) = p_Y\left\langle \boldsymbol{\sigma}_{\downarrow\uparrow} \cdot \widehat{\boldsymbol{M}}_y \right\rangle = kp_y. \qquad (7.44)$$

There is no nuclear part to the SF-scattering[4]. SF-scattering is entirely magnetic. This conclusion can be generalized. Consider any orientation of the magnetization vector in the Y, Z-plane with M_y and M_z components. Then the M_z projection will always give rise to non-spin-flip magnetic scattering, whereas the M_y component will always cause spin-flip scattering over a certain optical path length. Combining both components, it is possible to determine the orientation of the magnetization vector in the (Y, Z)-plane.

We derive the above results again more formally, using the Pauli spin matrices. Then we express the magnetic interaction potential as

$$V_m = -\,\breve{\boldsymbol{\mu}}_n \cdot \boldsymbol{B} = -\mu_n\,\breve{\boldsymbol{\sigma}} \cdot \boldsymbol{B} = -\mu_n\left(\sigma_x B_x + \sigma_y B_y + \sigma_z B_z\right)$$

$$= -\mu_n\left[\begin{pmatrix} 0 & 1 \\ 1 & 0 \end{pmatrix} B_x + \begin{pmatrix} 0 & -i \\ i & 0 \end{pmatrix} B_y + \begin{pmatrix} 1 & 0 \\ 0 & -1 \end{pmatrix} B_z\right] = -\mu_n\begin{pmatrix} B_z & B_x - iB_y \\ B_x + iB_y & -B_z \end{pmatrix}$$

$$(7.45)$$

---

[4]This statement is correct, aside from nuclear spin flip scattering, occurring for nuclei with an angular momentum $I > 0$, causing spin incoherent scattering (see the discussion in Sect. 2.6.3).

$\breve{\mu}_n$ is the magnetic moment operator, and $\breve{\sigma}$ is the Pauli spin operator. Now we want to combine the magnetic potential with the scalar nuclear potential acting in the X-direction parallel to the scattering vector $Q = Q_x$ and normal to the Y, Z plane:

$$V_n = \frac{2\pi\hbar^2}{m_n} N_A b_n \begin{pmatrix} 1 & 0 \\ 0 & 1 \end{pmatrix} = \frac{2\pi\hbar^2}{m_n} N_A \begin{pmatrix} b_n & 0 \\ 0 & b_n \end{pmatrix}. \tag{7.46}$$

Combining the nuclear and magnetic parts, we obtain

$$V_n + V_m = \frac{2\pi\hbar^2}{m_n} N_A \begin{pmatrix} b_n + p_z & p_x - ip_y \\ p_x + ip_y & b_n - p_z \end{pmatrix}. \tag{7.47}$$

Here $p = \frac{m_n \mu_n}{2\pi\hbar^2} B = \frac{m_n \gamma_n \mu_N \mu_0}{2\pi\hbar^2} M$ is the magnetic scattering length, according to Eq. (7.9). We will not consider magnetization components parallel to the scattering vector. Therefore, the last expression simplifies to

$$V_n + V_m = \frac{2\pi\hbar^2}{m_n} N_A \begin{pmatrix} b_n + p_z & -ip_y \\ +ip_y & b_n - p_z \end{pmatrix}. \tag{7.48}$$

Using this potential, the stationary Schrödinger equation has to be solved for neutron incident plane waves with two spin states, described by

$$\Psi(Q_x) = \begin{pmatrix} \Psi_+(Q_x) \\ \Psi_-(Q_x) \end{pmatrix} = \begin{pmatrix} A_+ e^{iQ_x x} \\ A_- e^{iQ_x x} \end{pmatrix}, \tag{7.49}$$

where the $\pm$ signs indicate spin-up and spin-down neutrons. If the off-diagonal elements are zero, the two spin-states are decoupled, and we obtain the NSF-scattering as described above. With off-diagonal elements, the two spin states are coupled, and the solution of the Schrödinger equation indicates spin-flip scattering.

The scattering intensities (elastic cross-section) for both spin states are

$$I_\pm(Q) = \underbrace{|N(Q) \pm M_{\perp z}(Q)|^2}_{\text{NSF}} + \underbrace{|\pm iM_{\perp y}(Q)|^2}_{\text{SF}}, \tag{7.50}$$

where

$$N(Q) = \sum_j b_j e^{iQ \cdot R_j},$$

$$M_\perp(Q) = \sum_j p_{\perp j} e^{iQ \cdot R_j}$$

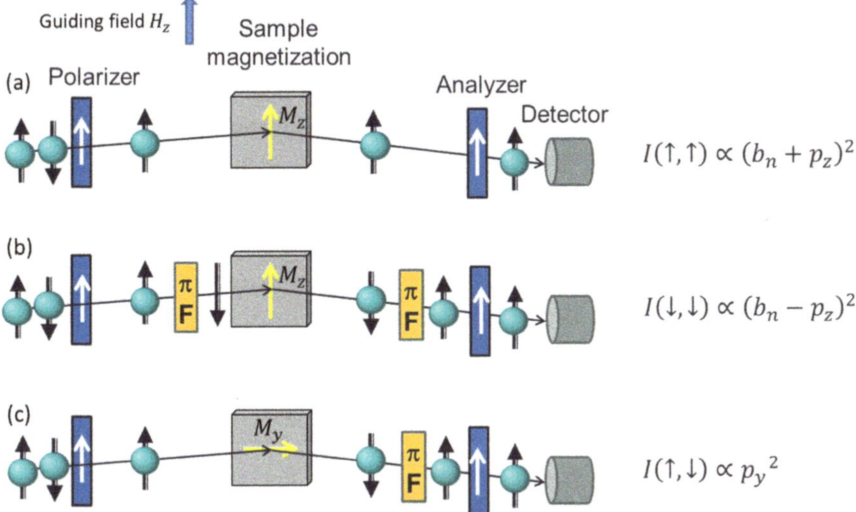

Fig. 7.17 Schematics of neutron scattering with polarization analysis. The polarization analysis requires polarizer and analyzer devices, as well as spin-flippers that can be activated on demand. Activated spin-flippers are marked in ochre. The M_z-component of the sample magnetization is determined with non-spin-flip scattering, and the M_y-component is retrieved through spin-flip scattering

are the Fourier transforms of the nuclear and magnetic real-space densities, and the subscript $\perp$ indicates that $M_{y,\,z} \perp \boldsymbol{Q}$.

A polarization analysis requires a neutron diffractometer with the added capability to polarize the incoming unpolarized beam and to analyze the beam polarization after being scattered at the sample. Traditionally, the polarization of neutron beams is achieved by the (111) Bragg reflection of the Heusler intermetallic compound Cu_2MnAl. For this compound and the structure factor of the (111) Bragg reflection, it turns out that the nuclear coherent scattering length, b_n, equals the magnetic scattering length p: $b_n = p$. Therefore, the intensity of the (111) peak is: $I(\uparrow,\uparrow) \propto (b_n + p)^2 > 0$ and $I(\downarrow,\downarrow) = 0$ for spin-up and spin-down NSF Bragg scattering, respectively. The spin-down neutrons are not scattered but transmitted through the sample, acting like a sieve for the spin-down neutrons. Nowadays, magnetic supermirrors are frequently used for neutron spin polarization instead of single crystals of Heusler compounds. Supermirrors are briefly discussed in Sect. 4.3.1. Figure 7.17 illustrates an experimental setup for polarized neutron scattering. To avoid depolarization of the neutron beam, a guiding magnetic field is provided between the polarizer and the sample, and between the sample and the analyzer.

We first describe the top panel in Fig. 7.17a. The unpolarized incoming neutron beam passes a polarizer. The "up" state is reflected, while the down state is filtered out. The beam is scattered by the sample and subsequently by an analyzer, which is identical to the polarizer. The intensity measured in the detector is then proportional to the nuclear plus magnetic scattering length squared $I(\uparrow,\uparrow) = |N(\boldsymbol{Q}) + M_{\perp z}(\boldsymbol{Q})|^2$.

To test the down polarization, a so-called π-spin flipper must be activated that rotates the neutron spin adiabatically by 180° in a magnetic field perpendicular to the guide field, also known as Mezei-flipper (Mezei 1972). The down polarization interacts with the sample magnetization to yield the intensity for the antiparallel orientation. Before detecting, the neutron spin must be flipped back again to pass the analyzer. Thus, for measuring the "down" polarization component, both 180° spin-flippers must be activated. The measured intensity in the detector is then proportional to: $I(\downarrow, \downarrow) = |N(\boldsymbol{Q}) - M_{\perp z}(\boldsymbol{Q})|^2$.

The third configuration in panel (c) tests the y-component of the magnetization, specifically whether a spin-flip occurred during the scattering process. We start with the "up" polarization, as in the top panel. After the sample, the spin-flipper is activated, and flips back all those spins that have been flipped by the y-component of the magnetization. After being flipped back, those neutron spins can pass the analyzer crystal. As spin-flip is caused by magnetism rather than by coherent nuclear scattering, the spin-flip intensity recorded by the detector is proportional to: $I(\uparrow, \downarrow) = |iM_{\perp y}(\boldsymbol{Q})|^2$. Equivalently, we may first flip the incident neutron spin by activating the π-spin-flipper before the sample and deactivating the flipper after the sample. Then we obtain the down-up spin-flip intensity: $I(\downarrow, \uparrow)$. Both intensities are equal, and we obtain again: $I(\uparrow, \downarrow) = I(\downarrow, \uparrow) = |\pm iM_{\perp y}(\boldsymbol{Q})|^2$.

If the instrument is set up to record spin-flip intensity, what happens to those neutrons that are coherently scattered by nonmagnetic nuclei in the sample? They pass the sample without spin-flip, then become flipped by passing the π-spin-flipper, and are stopped by the analyzer because of the wrong spin state. Hence, spin-flip scattering is purely of magnetic origin and requires an M_y-component. SF scattering filters out nuclear scattering. If one is only interested in magnetic scattering, SF-scattering is the method of choice, requiring a full polarization analysis.

It may be convenient to work with a polarized incident neutron beam, but to spare the polarization analysis. In this case, we add NSF and SF intensities for the "up" polarization:

$$I(\uparrow) = I(\uparrow,\ \uparrow) + I(\uparrow,\ \downarrow) \propto (N + M_z)^2 + M_y^2 = N^2 + M_z^2 + 2NM_z + M_y^2, \quad (7.51)$$

and, equivalently for the down polarization:

$$I(\downarrow) = I(\downarrow,\ \downarrow) + I(\downarrow,\ \uparrow) \propto (N - M_z)^2 + M_y^2 = N^2 + M_z^2 - 2NM_z + M_y^2. \quad (7.52)$$

The difference emphasizes the cross term:

$$I(\uparrow) - I(\downarrow) = 4NM_z, \quad (7.53)$$

where the factor $4N$ acts as an amplification of otherwise weak magnetic scattering.

What we have described here is referred to as XY-polarization analysis, which was first performed by Moon, Riste, and Koehler (Moon et al. 1969). Meanwhile, methods have been developed for 3D XYZ-polarization analysis, which is reviewed by Brown in (Brown 2006). The main aim of using polarized neutron scattering methods is to disentangle nuclear and magnetic scattering whenever both contribute in the same area in the reciprocal space. In ordered magnetic materials, this is mainly the case for ferromagnets. However, in antiferromagnets and modulated magnetic structures, the magnetic reciprocal lattice points do not overlap with those of the crystal lattice, as we have discussed before, and therefore, a polarization analysis is not required. For the investigation of paramagnetic systems, spin glasses, ferrofluids, and amorphous magnetic materials, all displaying widely spread out magnetic density distributions, neutron polarization analysis turns out to be very useful. For a detailed discussion, we refer to the review chapter on Polarized Neutrons and Polarization Analysis by Schweizer (2006).

One example may illustrate the usefulness of SF-scattering. We discussed before in Sect. 7.2.2 transverse and longitudinal spin fluctuations. Transverse fluctuations preserve the modulus of the magnetic moments but change their orientation. Longitudinal fluctuations preserve the orientation but change the modulus. 3d transition metals like Fe, Co, and Ni are so-called "band" magnets. Their magnetism does not originate from local and fixed magnetic moments but is due to an exchange splitting of the 3d spin-up and spin-down bands. As these are nonlocal effects, the respective magnets are also referred to as itinerant magnets. By approaching the Curie temperature, the band splitting closes, and the modulus of the magnetic moments vanishes. Thus, the 3d itinerant magnets in question are expected to show longitudinal spin fluctuations below T_c and no fluctuations or spin correlations above T_c. The question of whether spin fluctuations and even spin wave excitation exist above T_c, was heavily disputed in the community for a long time.

To shed light on the existence of spin fluctuations in itinerant magnets above T_c, the paramagnetic scattering of Fe was measured at 1273 K, which is 250 K above the Curie temperature (Brown et al. 1983). The bcc-ordered α-phase of Fe was stabilized by adding 5% of Si to the alloy. Fig. 7.18 shows a radial scan in the $[hh0]$-direction across the (110) Bragg reflection centered at 3.08 Å^{-1}. The scan was taken with the spin-flipper turned on, yielding only the Y-component of the paramagnetic fluctuations. The nuclear scattering intensity was completely eliminated by the spin-flip scattering.

The scan shows that the paramagnetic scattering is enhanced in the forward direction for $Q \rightarrow 0$ and near the nuclear Bragg peak at 3.08 Å^{-1}. The scattered intensity is much lower toward the zone boundary of the reciprocal lattice at 1.54 Å^{-1}. If Fe were a pure band magnet, paramagnetic moments should not exist above T_c, nor short-range spin correlations that are responsible for the peaked intensity at the zone center, i.e., at the (110) Bragg peak. All this shows that in the itinerant magnetic systems of Fe, short-range order spin correlations exist even far above T_c. From the width of the peak, a correlation length of 1.5 nm can be estimated, sufficient to support spin waves on a local scale. The forward scattering also indicates spin fluctuations or short-range spin waves, hinting at the rather complex magnetism of Fe above T_c. An up-to-date review of magnetic fluctuations and excitations in local and itinerant magnets can be found in the review article by Chatterji (2006).

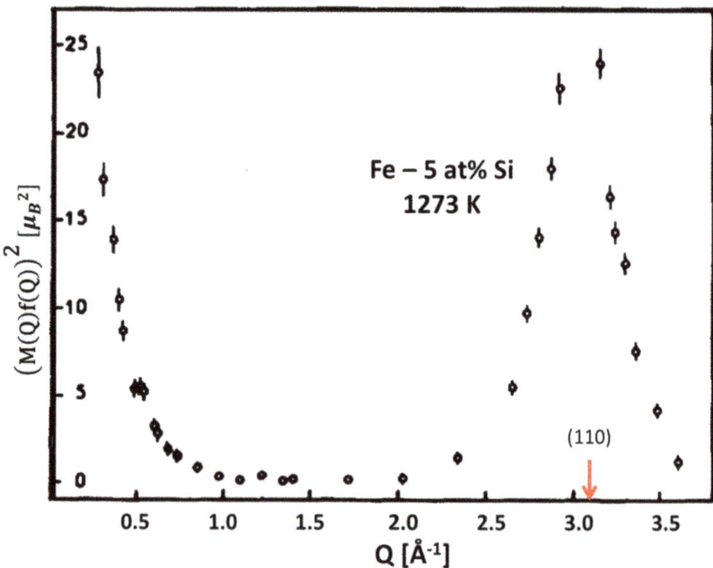

Fig. 7.18 Paramagnetic scattering in the α-phase of Fe at 250 K above $T_c = 1273$ K. Scan in the $(hh0)$-direction with the spin flipper turned on. The position of the (110) Bragg reflection is marked with a red arrow. The nuclear Bragg scattering is completely eliminated by the spin-flip setting. Adapted with permission from Brown et al. (1983). Copyright (2025) by Elsevier, License Number 6053700815364

In the foregoing discussion, we have tacitly assumed that the polarizer/analyzer and the 180° spin flipper work with 100% efficiency. In reality, this is not the case. For quantitative analysis, it is therefore important to determine the maximum achievable polarization efficiencies P of the polarizer and analyzer and the flipping ratio of the spin-flipper F. The polarization efficiency of the instrument is defined as the ratio of up-neutron spins minus down-neutron spins, divided by the sum:

$$P = \frac{N\uparrow - N\downarrow}{N\uparrow + N\downarrow} = \frac{F-1}{F+1}, \tag{7.54}$$

with $0 \leq P \leq 1$, and the flipping ratio defined by $F = N\uparrow / N\downarrow$. Typically, flipping ratios of more than 50 are achieved, and polarization efficiencies of more than 98% are common.

7.6   Nonresonant Magnetic X-Ray Scattering

The fact that X-ray scattering is also sensitive to magnetism was first confirmed by de Bergevin and Brunel (De Bergevin and Brunel 1972) in the late 1970s. The reason for this late verification lies in the small magnetic cross section of X-rays compared to charge scattering. To describe the X-ray photon-spin interaction, we first rephrase the atomic form factor for ***Thomson scattering*** at charges introduced in Chap. 2:

$$f_{\text{Thomson}}(\boldsymbol{Q}, E) = \left(\widehat{\boldsymbol{e}}_i \cdot \widehat{\boldsymbol{e}}_f\right) \left[f(\boldsymbol{Q}) + f'(E) + i f''(E)\right]. \qquad (7.55)$$

Here, $f(\boldsymbol{Q})$ is the atomic form factor, introduced before. $f'(E)$ is a dispersion correction, which becomes important close to absorption edges. And the imaginary part $f''(E)$ describes the photoelectric absorption of X-rays. Most important in the present context is the prefactor $\left(\widehat{\boldsymbol{e}}_i \cdot \widehat{\boldsymbol{e}}_f\right)$, where $\widehat{\boldsymbol{e}}_{i,f}$ are unit vectors in the direction of the X-ray polarization for the incident beam (i) and the scattered beam (f). The product $\left(\widehat{\boldsymbol{e}}_i \cdot \widehat{\boldsymbol{e}}_f\right)$ indicates that Thomson scattering preserves the polarization of the X-ray beam. We may distinguish between σ- and π-polarization. Conventionally, the electric field vector in π-polarization lies in the scattering plane; in σ-polarization, the electric field vector is oriented perpendicular to the scattering plane. Thomson scattering implies that for both polarizations, the scattering is either $\sigma \to \sigma$ or $\pi \to \pi$.

One may ask whether X-rays interact not only with the charge distribution in materials but also with the magnetic distribution of uncompensated spins, similar to neutron scattering. This is a fundamental question that has been treated by several authors (Platzman and Tzoar 1970; Hannon et al. 1988; de Bergevin and Brunel 1981; Blume 1985). While the answer is in general positive, Hannon et al. have derived an ***X-ray magnetic form factor*** for nonresonant scattering, expressed as (Hannon et al. 1988):

$$f_{\text{mag}}(\boldsymbol{Q}) = i r_0 \frac{\hbar \omega}{mc^2} \left[\frac{1}{2} \boldsymbol{L}(\boldsymbol{Q}) \cdot \boldsymbol{A} + \boldsymbol{S}(\boldsymbol{Q}) \cdot \boldsymbol{B}\right]. \qquad (7.56)$$

The terms in the square bracket indicate that magnetic X-ray scattering has two contributions originating from the Fourier transform of orbital magnetism $\boldsymbol{L}(\boldsymbol{Q})$ and the spin magnetism $\boldsymbol{S}(\boldsymbol{Q})$. $\boldsymbol{A}$ and $\boldsymbol{B}$ are polarization vectors determined by $\widehat{\boldsymbol{e}}_{i,f}$ and $\boldsymbol{Q}$. Further calculations show that the orbital contribution to the form factor changes the incident polarization from σ to π, while the incident π polarization is not altered. The $\sigma \to \pi$ flip is a clear sign of magnetic scattering to be distinguished from the polarization-conserving $\sigma \to \sigma$ charge scattering.

The prefactor $\hbar \omega / mc^2 \approx 0.01 r_0$ (for hard X-rays) indicates that the magnetic scattering amplitude is at least a factor of 100 smaller than the Thomson scattering. To observe magnetic X-ray scattering at the position of structural Bragg peaks, as would be the case for ferromagnetic Bragg reflections, is therefore hopeless. If magnetic scattering is to be seen, then at antiferromagnetic reciprocal lattice points, situated between the structural peaks, as was done by de Bergevin and Brunel in their pioneering experiment on AF NiO (De Bergevin and Brunel 1972).

The instructions for probing nonresonant magnetic X-ray scattering are as follows:

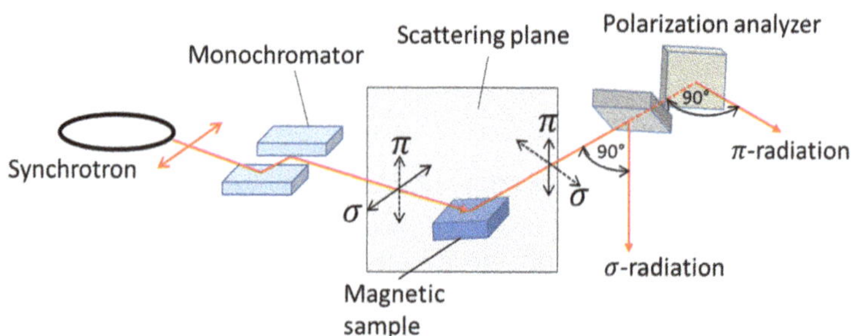

Fig. 7.19 Experimental arrangement for recording nonresonant X-ray magnetic scattering. The synchrotron source delivers linear polarization, which becomes a σ-polarization with respect to the vertical scattering plane. The magnetic sample flips the polarization from σ to π. To confirm the polarization change, an analyzer crystal is used in combination with a detector. Rearrangement of the analyzer allows recording both polarization components

1. Use a polarized incident X-ray beam.
2. Perform a polarization analysis after scattering to ensure $\sigma \rightarrow \pi$ flip;
3. Avoid areas in reciprocal space with high-intensity charge scattering.

Figure 7.19 shows a schematic test arrangement for measuring magnetic X-ray scattering. The synchrotron radiation is linearly polarized in the synchrotron plane, which can be used for magnetic scattering. After passing a monochromator, the beam is reflected upward, which defines a vertical scattering plane. The linear polarization of the synchrotron radiation is now the σ-polarization with respect to the vertical scattering plane. Scattering at the magnetic sample is expected to switch the polarization from $\sigma \rightarrow \pi$, if the magnetization vector is oriented perpendicular to the scattering vector. This can be confirmed by using a polarization analyzer. The analyzer is set to provide a Bragg reflection at the ***Brewster angle*** of $\theta = 45°$ and a Bragg angle of 90°. A common analyzer for hard X-rays is pyrolytic graphite. At an energy of 7.85 keV, the (006) Bragg peak scatters at 90°. The polarizer can be rotated to reflect either σ- or π-polarized X-rays, i.e., always the component that is polarized perpendicular to the scattering plane of the analyzer. Ideally, only π-radiation should be detected in this experiment. Polarization analysis is always required to better analyze and understand the form factors of multipole transitions in resonant and nonresonant magnetic X-ray scattering, as we will discuss in the next section. And to separate charge and magnetic scattering, which provides a much cleaner background for the magnetic signal.

Although the X-ray magnetic form factor in Eq. (7.56) does not distinguish between ferro- and antiferromagnetic materials, from a practical point of view, preferably antiferromagnetic samples are studied, as for ferromagnetic samples, the Thomson scattering entirely dominates the recorded intensity, as already mentioned. Therefore, we will illustrate nonresonant magnetic X-ray scattering by discussing the magnetism of the rare-earth metal holmium (Ho).

Below the ***Néel temperature*** at 130 K, Ho exhibits a spin spiral in the basal plane of the hexagonal close-packed (*hcp*) atomic structure, sketched in Fig. 7.20a

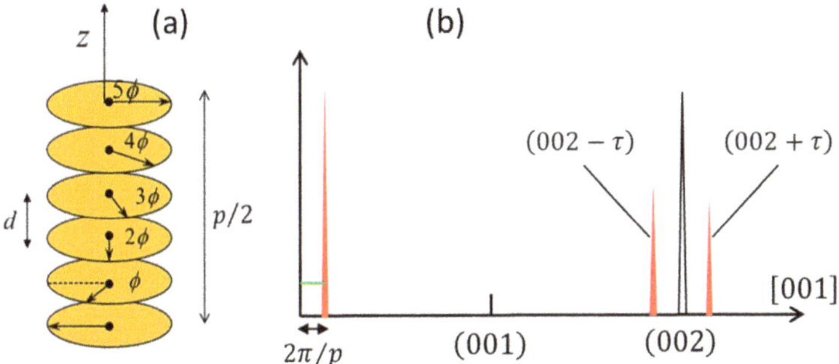

Fig. 7.20 (**a**) Magnetization spiral in Ho below the Néel temperature. (**b**) Magnetic Bragg reflections along the [00L] direction manifested in τ-peaks (in red) left and right to the main structural Bragg peaks (black)

(Koehler et al. 1966). This implies that each Ho (001)-plane is in a ferromagnetic state, but the orientation of the in-plane magnetization vector rotates by an angle φ from one plane to the next, up to a full rotation at $n\varphi = 2\pi$. The relation between the turn angle φ and the period $p = nd$ is $\varphi = 2\pi(d/p)$, where d is the layer spacing (Fig. 7.20a). In commensurate AFs, $p = 2d$. But in Ho the period shrinks from $p = 12d$ and $\varphi = 30°$ below 20 K to $p \approx 10d$ at higher temperatures. The magnetization modulation parallel to the z-axis evokes magnetic satellite peaks, also called τ-peaks, positioned left and right to the main structural reflection at $\boldsymbol{Q} = \boldsymbol{G}_{00l}$. The position of the τ-peaks with the wavenumber $\tau = 2\pi/p$ is

$$\boldsymbol{Q} = \boldsymbol{G}_{00l} \pm \tau \; (l = 0, 2, 4...). \tag{7.57}$$

The position of the τ-peaks is shown schematically in Fig. 7.20b. Because of the spiral type of antiferromagnetism, Ho is often referred to as a **_helimagnetic_** material.

As the **_magnetic τ-peaks_** are separated from the main structural Bragg peaks, the magnetic scattering intensity does not overlap with the Thomson scattering when performing X-ray scattering experiments on Ho. Fig. 7.21 shows the experimental results of scanning the τ-peak at $\boldsymbol{Q} = \boldsymbol{G}_{004} + \tau$ with polarized X-rays at a wavelength of 0.17 nm (7.29 keV) and 0.15 nm (8.26 keV) (Gibbs et al. 1985). Both wavelengths are above and below the Ho-L_{III} absorption edge, which is at 8.068 keV. Hence, these scattering results are classified as nonresonant magnetic scattering. The $+\tau$ peak shifts to higher values with increasing temperature, indicating that the magnetization spiral length shrinks by heating, an observation in complete agreement with results from thermal neutron scattering (Koehler et al. 1966). The polarization analysis of the $(002 + \tau)$, $(004 + \tau)$, and $(006 + \tau)$ peaks showed that the $\sigma \rightarrow \pi$ polarization switch dominates over the $\sigma \rightarrow \sigma$ charge scattering at all three reflections, confirming the magnetic origin of the τ-peaks (Gibbs et al. 1991).

The discovery of nonresonant magnetic X-ray scattering was an important breakthrough. However, the scattering intensity is weak, even at synchrotron radiation sources, and its application is limited to antiferromagnetic or incommensurate

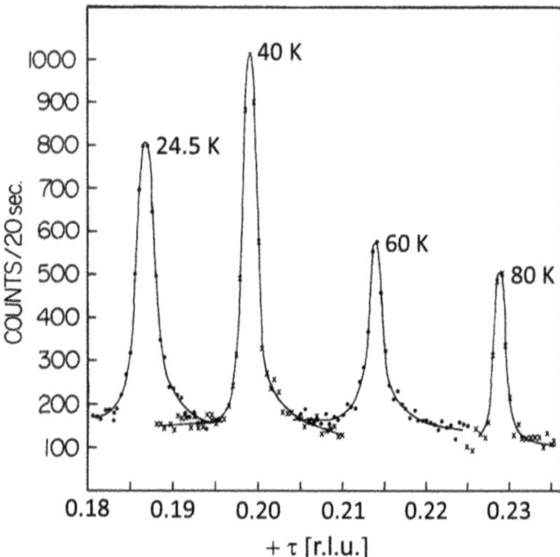

Fig. 7.21 The magnetic $(0, 0, 4 + \tau)$ peak of Ho as a function of temperature measured nonresonantly with hard X-rays of beam energies 7.29 keV and 8.26 keV. The scale of the abscissa is in units of $2\pi/d$, where d is the Ho (001) lattice spacing. The τ peak shifts to higher values with increasing temperature. Adapted with permission from Gibbs et al. (1985). Copyright (2025) by the American Physical Society

magnetic structures. Further developments showed that significantly higher intensities can be achieved by exploiting resonance conditions close to the absorption edges of the magnetic shells, for example, at the L_{II} and L_{III} edges for studying magnetic lanthanides and actinides with hard X-rays, or at the M_{IV} and M_V edges for studying $4f$ electrons in rare-earth metals and compounds with soft X-rays. This is the subject of the following section.

7.7　Resonant Magnetic X-Ray Scattering

It was shown that the charge and magnetic scattering amplitudes under resonant conditions take the following form if only dipole transitions are considered (Hannon et al. 1988; Blume 1985; Gibbs et al. 1985; Thole et al. 1992):

$$f_{res}(\boldsymbol{Q}, E) = \left(\widehat{\boldsymbol{e}}_i \cdot \widehat{\boldsymbol{e}}_f\right)\left(-r_0 Z + F^{(0)}\right) + i\left(\widehat{\boldsymbol{e}}_i \times \widehat{\boldsymbol{e}}_f\right) \cdot \widehat{\boldsymbol{m}} F^{(1)} + \left(\widehat{\boldsymbol{e}}_i \cdot \widehat{\boldsymbol{m}}\right)\left(\widehat{\boldsymbol{e}}_f \cdot \widehat{\boldsymbol{m}}\right) F^{(2)},$$

$$(7.58)$$

where $F^{(0)}$, $F^{(1)}$, and $F^{(2)}$ are defined by:

$$F^{(0)} = \frac{3}{8\pi}\lambda\left[F_1^1 + F_{-1}^1\right]$$

$$F^{(1)} = \frac{3}{8\pi}\lambda\left[F_1^1 - F_{-1}^1\right] \qquad (7.59)$$

$$F^{(2)} = \frac{3}{8\pi}\lambda\left[2F_0^1 - F_{+1}^1 - F_{-1}^1\right]$$

Here, $\hat{e}_i$ and $\hat{e}_f$ are the unit polarization vectors of the incident and scattered X-rays, respectively, and $\hat{m}$ is the unit vector in the direction of the local magnetic moment. Terms proportional to $(\hat{e}_i \cdot \hat{e}_f)$ describe nonresonant and resonant charge scattering, which is equivalent to Eq. 7.56. The term involving $(\hat{e}_i \times \hat{e}_f) \cdot \hat{m}$ is of first order in the magnetization and yields circular dichroism. The term proportional to $(\hat{e}_i \cdot \hat{m})(\hat{e}_f \cdot \hat{m})$ is of second order in the magnetization and causes linear dichroism. $F_{\Delta M}^{L}$ denotes the energy-dependent dipole oscillator strengths for excitations from a core level to an excited unoccupied state above the Fermi energy, where L is the order of transition ($L = 1$, dipole transition) and $\Delta M_J = 0, \pm 1$ is the magnetic selection rule for dipole transitions.

Resonant magnetic scattering promises significantly higher intensities than nonresonant scattering. An increase in the magnetic scattering amplitude by a factor of $100 \times r_0$ compared to nonresonant magnetic scattering, which scales with $0.01 \times r_0$, has been predicted (Hannon et al. 1988; Blume 1985), and even higher values have been observed experimentally (Gibbs et al. 1991; Isaacs et al. 1989). However, these experiments require a significant amount of X-ray absorption spectroscopy, which must be performed in parallel with the scattering experiment. This necessitates a tunable X-ray source, preferably at synchrotron radiation facilities, and the ability to determine the photoelectric absorption coefficient.

We briefly outline the basic idea of X-ray absorption spectroscopy. Figure 2.1 shows a sketch of X-ray attenuation. The absorption part of the X-ray attenuation is mainly due to the photoelectric effect for the energies considered here. The photoelectric absorption cross-section σ_a is defined as the number of excited electrons per unit time, divided by the incident photon flux. The photon flux is defined as the number of photons per unit time and area $I_0 = N/t \times A$. Therefore, σ_a is:

$$\sigma_a = \frac{w_{i \to f}}{I_0}, \tag{7.60}$$

where $w_{i \to f}$ is the probability for the electron transition from the initial core energy level to the final state. Thus, the X-ray absorption cross section is linearly proportional to the number of core holes created during the X-ray absorption. Depending on the photon energy, the core electron will either occupy an empty state in an outer shell or escape into the vacuum. In both cases, electrons from outer shells fill the holes in inner shells created by X-ray absorption. This can proceed in two ways, either by X-ray photon emission (fluorescence decay) or radiationlessly by imparting the transition energy to an electron in an outer shell that, in turn, leaves to the vacuum (Auger process). Only the Auger process is of interest here, since we aim to detect electrons instead of photons during the X-ray absorption process. As these outer Auger electrons have rather low energy, their escape depth into the vacuum is limited to only a few nanometers for 3d and 4f metals. Hence, in contrast to standard transmission experiments, the electron yield measurement is surface sensitive and well adapted to the study of thin films. When collecting all Auger electrons, we

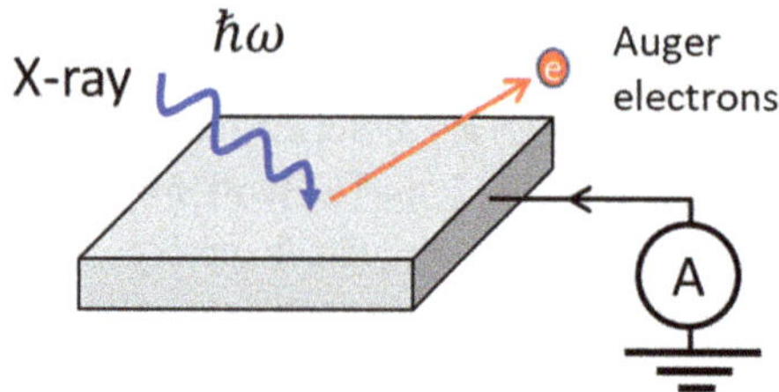

Fig. 7.22 Incident X-ray photons are absorbed and create a core hole, filled subsequently by Auger electrons. The secondary Auger electrons leaving into the vacuum are replaced by electrons from the ground. The process is known as the total electron yield, where an abmmeter measures the currrent between ground and sample

Fig. 7.23 Holmium energy level scheme, indicating two allowed dipole transitions, the L_{II} and L_{III} transitions

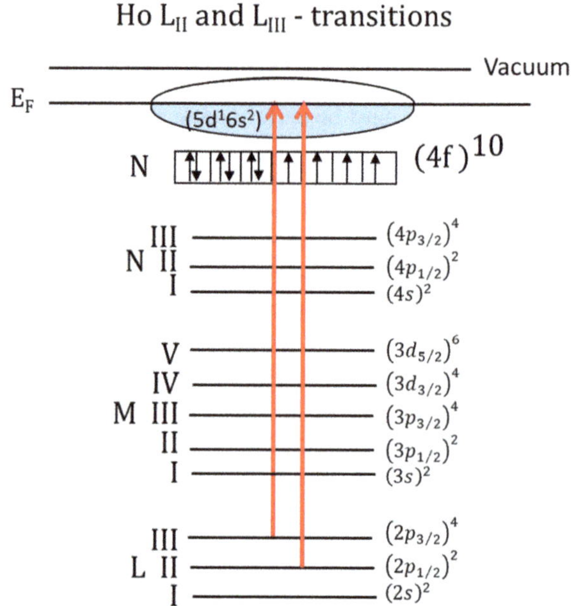

speak about the total electron yield (TEY). Further details on the TEY method are discussed in (Stöhr 1992; Schütz et al. 2007).

The TEY can be measured utilizing the sample drain current, as sketched in Fig. 7.22. It has been shown that the electron yield $Y(E)$ is proportional to the photoelectric absorption cross section σ_a and the absoption coefficient $\mu(E, d)$: $Y(E) \propto \sigma_a \propto \mu(E, d)$ (Stöhr 1992), where d is the sample thickness. Hence, measuring the TEY is an appropriate alternative to a full spectroscopic analysis, employing a costly electron analyzer system. Moreover, it is compatible with the experimental setup that is designed for scattering measurements. The only challenge encountered is the extremely low current level (picoampere) to be recorded, potential noise problems, and charging effects.

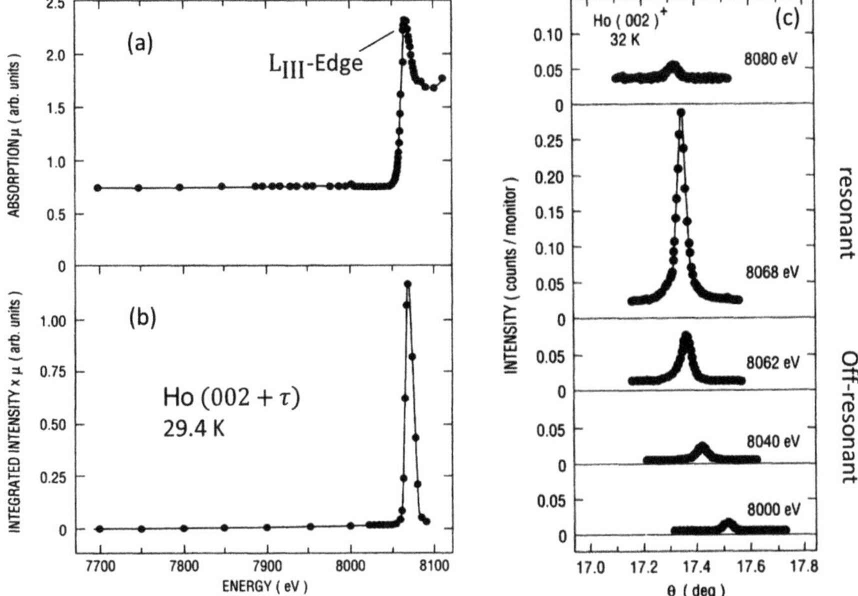

Fig. 7.24 (**a**) X-ray absorption in Ho determined with total electron yield. Resonant absorption occurs at an energy of 8068 eV, corresponding to the L_{III} absorption edge. (**b**) Integrated intensity of the Ho (002 + τ) magnetic satellite peak for the same energy range as in (**a**). Strong intensity enhancement is observed, coinciding with the L_{III} absorption edge. (**c**) Rocking curves of the Ho (002 + τ) magnetic satellite peak for various energies above and below the L_{III} absorption edge. Reproduced with permission from Gibbs et al. (1991). Copyright (2025) by the American Physical Society

Before returning to the magnetic τ peaks of Ho, we first consider the energy level scheme of holmium, as shown in Fig. 7.23. Neglecting the $1s$ state, which is irrelevant for resonant magnetic scattering considered here, we focus on the L shell. Here, dipole transitions from L_{II} and L_{III} to the $5d^{1}6s^{2}$ conduction band is possible. Due to the proximity of the partially filled $4f$ shell, the $5d^{1}6s^{2}$ conduction band is exchange-split and magnetically sensitive. Furthermore, the $5d^{1}$ electrons fulfill the selection rule for dipole transitions from $2p$ to $5d$.

When performing a TEY experiment that covers the energy range from 7.7 keV to 8.1 keV, we find a strong absorption peak at 8068 eV, as shown in Fig. 7.24a. The peak corresponds to the L_{III} absorption edge. The L_{II} absorption edge is located at 8920 eV and therefore out of the energy range displayed in Fig. 7.24a. The magnetic τ peak was recorded at the (002 + τ) position for different X-ray energies as reproduced in Fig. 7.24b. We notice a strong intensity enhancement of the scattered beam intensity for an X-ray energy that corresponds to the resonant L_{III} dipole transition. The enhancement factor is 50-fold increased compared to the off-resonant intensity. Finally, Fig. 7.25c shows rocking curves of the (002 + τ) satellite reflection for different energies above and below the resonance transition at

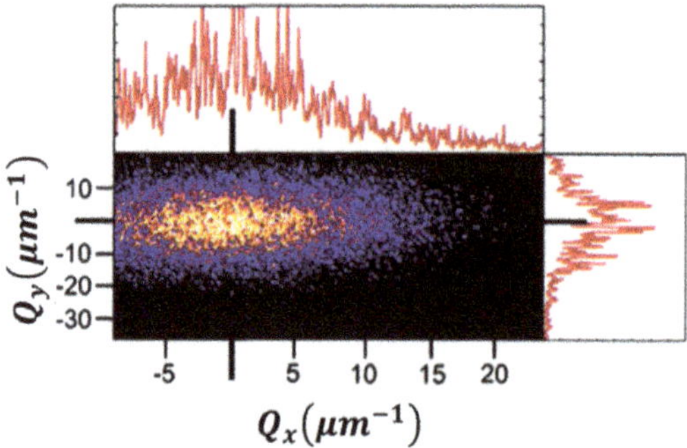

Fig. 7.25 Magnetic τ-peak of the Ho spin spiral recorded at 30 K with a partially coherent and resonant X-ray beam tuned to the M_5 absorption edge. The line scans along the indicated horizontal and vertical black lines show a jagged pattern instead of the usual Gaussian intensity distribution, indicating magnetic fluctuations. Reproduced with permission from Konings et al. (2011). Copyright (2025) by the American Physical Society

8068 eV. These scans confirm the strong enhancement of the peak intensity at the L_{III} absorption edge.

Similar experiments can be performed with softer X-rays using the M_{IV}, M_V absorption edges with a resonance energy of 1355 eV (Ott et al. 2006). Furthermore, it has been shown that ferromagnetic materials can also be studied using circular polarization instead of linear polarization, which will be discussed in Sects. 11.5, 11.6 and 11.7 in the context of thin magnetic films.

7.8   Coherent Resonant Magnetic X-Ray Scattering

So far, we have assumed that the system investigated is spatially homogeneous and temporally constant. This assumption is confirmed by the Gaussian shape of the Bragg peaks that is usually observed using an incoherent X-ray or neutron source, which delivers an ensemble average of the real world. However, modern synchrotron radiation sources and insertion devices deliver (partially) coherent beams. If the coherence is not sufficient, the longitudinal and transverse beam coherence can be further enhanced by spatial filtering with a set of pinholes that the beam has to pass before hitting the sample (compare Sect. 4.2.8). Using such a prepared coherent beam, we will notice that a Gaussian-shaped Bragg peak is in fact composed of many speckles, which originate from tiny spatial or temporal fluctuations, disturbing (or modulating) the constructive interference of the scattered waves. Here we focus again on the τ-peak, representing the spin spiral in Ho. The analysis of the speckle pattern is termed *X-ray photon correlation spectroscopy* (XPCS). The method is well established for studying fluctuations in fluids, polymers, and suspensions with lasers. It is now also available for X-rays with the development of undulators and free-electron lasers. Figure 7.25 shows the speckle pattern of the

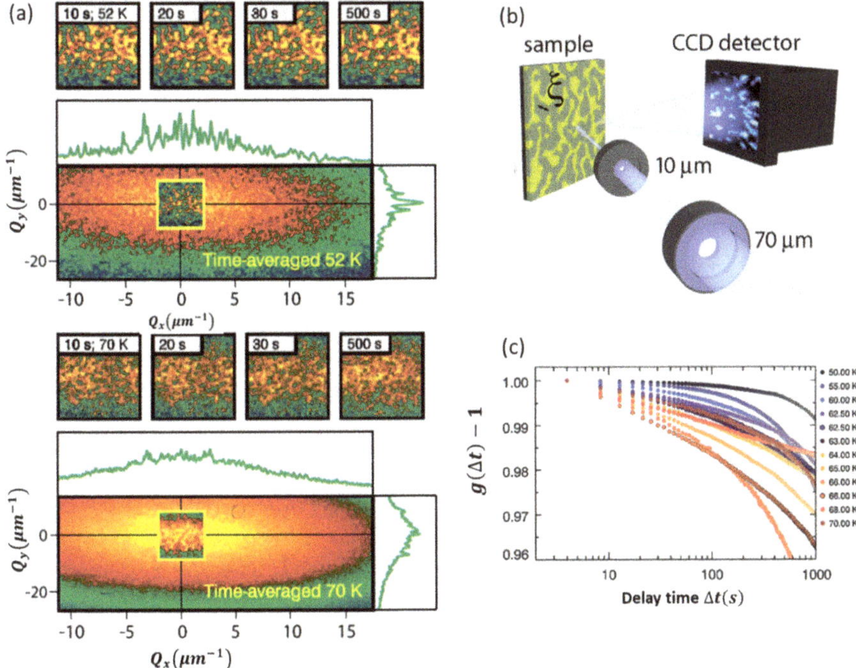

Fig. 7.26 (**a**) Single frames and time-averaged speckle patterns at 52 K and 70 K. The line scans are taken along the black horizontal and vertical lines. (**b**) Experimental setup for the XPCS recording with a partially coherent X-ray beam and a CCD area detector. (**c**) Normalized intensity correlation function plotted as a function of delay time for the indicated temperatures. Note the logarithmic time scale of the abscissa. Reproduced and adapted with permission from Konings et al. (2011). Copyright (2025) by the American Physical Society

$(000+\tau)$-peak at 50 K, recorded resonantly at the M_5 absorption edge with a photon energy of 1344 eV using π-polarization in the horizontal scattering plane (Konings et al. 2011; Konings 2007).

The experimental setup for XPCS is similar to recording a reciprocal space map, such as the one reproduced in Fig. 5.6. But in addition, a partially or fully coherent X-ray beam and a CCD area detector are used. For spatial filtering, and in particular to increase the transverse coherence, the X-ray beam was fed through a 10-µm pinhole, producing a roughly 10 µm spot size on the sample (Fig. 7.26b). Hence, correlations can be probed on the length scale of 10 µm at the center of the peak to about 250 nm at the periphery. Static and dynamic speckles are formed by magnetic domain fluctuations over these length scales that cause constructive and destructive interferences effects in the scattered beam.

A typical procedure is to take a sequence of "snapshots" (single frames) at time intervals Δt_1, Δt_2, etc., and to compare the intensities from the first and the last frame. If the intensities in the pixels are the same, the fluctuations observed are static; otherwise, they are dynamic. Figure 7.26a shows two examples, one at 52 K and the other at 70 K, which is close to the phase transition from the helical state to the paramagnetic state. In the sample considered here, the transition occurs at 80 K due to finite size scaling of the critical temperature in thin films. At 52 K, the time frames with an integration time of 10 s appear to be identical, and even when averaging over all frames, the speckle pattern looks similar, as seen in the yellow box at the center. Hence, the time constant is more than 500 s.

The situation is different for the images taken at 70 K. Here, a change in the speckle pattern is noticeable. The time-averaged pattern in the yellow box appears significantly smoother than the individual images, suggesting that fluctuations smooth the ensemble-averaged diffraction pattern. The time constant is correspondingly shorter. Nevertheless, a certain graininess remains in the pattern, suggesting that some domains are frozen by pinning, most likely at structural defects. Further toward the transition temperature, we expect stronger fluctuations. However, due to the decreasing τ-peak intensity, the fluctuation can no longer be tracked reliably.

For analyzing the speckle pattern, we introduce a normalized intensity correlation function (Dierker et al. 1995):

$$g^{(2)}(Q, \Delta t) = \frac{\langle I(t)I(t + \Delta t)\rangle}{\langle I(t)\rangle^2} = 1 + \alpha(f(Q, \Delta t))^2. \tag{7.61}$$

Here, $I(t)$ is the intensity of a particular pixel at Q and at time t, and $I(t + \Delta t)$ is the intensity of the same pixel recorded after a time increment Δt. The angle brackets indicate a time average over the acquisition time. $g^{(2)}$ is known as the second-order autocorrelation function, where the intermediate scattering function $f(Q, \Delta t)$ is of first order. For an ergodic system, i.e., a system in which the speckle pattern is statistically everywhere the same in time and space at a fixed temperature, the mean scattering function can be represented by a function that decreases with time. Since the correlations usually decay exponentially with a rate Γ, $f(\Delta t)$ is expressed as $f(\Delta t) = \exp(-\Gamma\Delta t)$, with the limiting value $f(\Delta t \to \infty) = 0$ and $g^{(2)}(\Delta t \to \infty) - 1 = 0$ for a complete loss of correlation. In Fig. 7.26c, the autocorrelation function $g^{(2)}(\Delta t, Q) - 1$ is plotted in dependence on the time for different temperatures. A closer analysis shows that the Ho spin spiral is a nonergodic system. Parts of the system decay exponentially with time, while other parts in the speckle pattern remain constant.

This observation raises the question of why domains exist at all in Ho and what distinguishes one domain from another. Two possibilities are conceivable. First, the spin spiral has a handiness. It can be left- or right-handed. If both domains are close to each other, a domain wall must exist between them. If both domains have the same potential energy, we expect—in thermal equilibrium—equal shares of both. Second, the angle of rotation within a spiral depends on the total layer thickness. In this case, the average layer thickness of the Ho layer was 11 monolayers. Thickness variations lead to variations in the turn angle. Spin spirals with different angles of rotation cannot coexist. A domain wall must exist between them, which is tied to the layer thickness variations.

A similar XPCS study was performed by Shpyrko and coworkers on the Cr-SDW domain structure, scanning the charge $2q$ CDW peak (Shpyrko et al. 2007). This work was performed with a partially coherent beam at a hard X-ray undulator in nonresonant mode. With the further advancement of synchrotron sources and free electron lasers, XPCS experiments could be performed with higher coherence and resonantly at the respective absorption edges.

The example on the Ho-spin spiral discussed here required a rather modest frame rate, which was sufficient for the observed slow fluctuations. In other cases, significantly higher frame rates are required, for example, to study diffusion processes, chemical reactions, and pump-probe evolution. The detectors at the European free electron laser XFEL have achieved a maximum frame rate of 4.5 MHz (https://www.xfel.eu/facility/instruments/mid/instrument_specifications/index_eng.html), and the fastest movie with a sequential frame separation of only 50 fs for X-ray imaging was reported by the group of Eisebitt (Günther et al. 2011).

Summary

1. Neutrons are spin ½ Fermi particles.
2. The spin of neutrons is connected to a magnetic moment.
3. The magnetic moment of neutrons is by a factor of 1000 smaller than the magnetic moment of electrons.
4. Magnetic neutron scattering relies on the magnetic dipole–dipole interaction between the neutron magnetic moment and the magnetic field (induction) $\boldsymbol{B}$.
5. The dipole–dipole interaction is comparatively weak, justifying a kinematic treatment of the scattering intensity.
6. The dipole–dipole interaction allows for a rather simple expression of the neutron magnetic form factor.
7. The neutron magnetic form factor is the Fourier transform of all those electrons that feature uncompensated magnetic moments.
8. The orientational dependence for neutron magnetic scattering implies that magnetic information is only obtained for magnetization vectors that are oriented perpendicular to the scattering vector.
9. For an unpolarized neutron beam, the nuclear scattering intensity from the lattice and the magnetic scattering intensity from the magnetic structure contribute additively to the total scattering intensity.
10. For ferromagnetic materials, the conditions for nuclear and magnetic Bragg reflections are identical: $\boldsymbol{Q} = \boldsymbol{G}_{hkl}^{struct} = \boldsymbol{G}_{hkl}^{mag}$.
11. Antiferromagnetic materials consist of two sublattices, one for magnetic dipoles pointing in one particular direction, and the other for magnetic dipole moments pointing in the opposite direction.
12. In antiferromagnets, the magnetic periodicity is doubled with respect to the lattice periodicity.
13. The antiferromagnetic Bragg reflections are separated from the nuclear Bragg reflections and occur at half-order positions.

14. Incommensurate magnetic structures feature wave vectors τ with an irrational ratio to the reciprocal lattice vector G.
15. Nonresonant magnetic X-ray scattering uses polarized X-rays.
16. Magnetic X-ray scattering switches the polarization from σ to π.
17. Nonresonant magnetic X-ray scattering can be used for studying antiferromagnets and incommensurate magnetic structures.
18. Resonant magnetic X-ray scattering intensity is enhanced by a factor of about 10^4 to 10^6 compared to nonresonant magnetic X-ray scattering.
19. Resonant magnetic X-ray scattering uses the dipole transition from a core level to an outer magnetically sensitive valence state. When the energy of incident X-rays is tuned to match an absorption edge of a specified element, resonant enhancement of magnetic scattering occurs.
20. X-ray photon correlation spectroscopy is a time-resolved scattering method with a coherent beam that allows analyzing temporal and spatial correlation fluctuations.

Questions

(Note that more than one answer may be correct)

1. **What is the reason for the magnetic form factor of neutron scattering having a similar shape and Q-dependence as for X-ray scattering?**

   (a) Because the magnetic moments of neutrons interact with the nuclear spin
   (b) Because the magnetic moments of neutrons scatter from spatially extended and partially filled magnetic atomic shells
   (c) Because thermal neutrons have wavelengths similar to X-rays and scatter from nuclei
   (d) Because the magnetic atomic shells have a spatial size similar to the neutron wavelength

2. **Under what conditions can the magnetization of the sample be probed by neutron scattering?**

   (a) Only at very low temperatures
   (b) Only for scattering vectors that are perpendicular to the magnetization
   (c) Only for very small scattering vectors aligned in the forward scattering direction

3. **How can we distinguish between the magnetic and the nuclear part of Bragg reflections in a neutron scattering experiment of a ferromagnetic material?**

   (a) They cannot be distinguished
   (b) Ferromagnetic peaks have a magnetic form factor: nuclear peaks have a constant intensity

   (c) By magnetizing the sample parallel to Q and perpendicular to Q and taking the difference of scans under both conditions

   (d) By comparing a scan in the high-temperature paramagnetic phase and in the low-temperature ferromagnetic phase

4. **The ferromagnetic part of the Bragg peak intensity is proportional to…?**

   (a) The magnetization squared

   (b) The inverse of the orbital moment

   (c) The nuclear spin

5. **Antiferromagnets and ferromagnets can easily be distinguished by neutron scattering, because… .**

   (a) The antiferromagnetic Bragg reflections depend on temperature, whereas the ferromagnetic Bragg reflections show no temperature dependence.

   (b) The ferromagnetic Bragg reflections depend on the orientation of Q and M, whereas the antiferromagnetic Bragg reflections do not.

   (c) In antiferromagnets, the magnetic Bragg reflections are separated from the nuclear Bragg reflections, whereas in ferromagnets they are not.

6. **Nonresonant magnetic X-ray scattering is used to…**

   (a) Study ferromagnets

   (b) Investigate antiferromagnetic and incommensurate magnetic spin structures

   (c) Phase transitions and critical phenomena

7. **Resonant magnetic X-ray scattering exhibits much higher intensity than nonresonant scattering. Why?**

   (a) Resonant scattering uses unpolarized X-rays, which have much higher intensity than polarized X-rays.

   (b) Resonant magnetic X-ray scattering relies on an enhancement effect due to nuclear resonance.

   (c) To observe a strong resonant enhancement, the scattering must involve a dipole transition between a core level and an unfilled outer atomic shell that is magnetically polarized.

Exercises

Grades of difficulty: E = easy, M = medium, A = advanced.

M 7.1. Antiferromagnetic multilayer.

   Let's assume that we can grow an artificial antiferromagnetic multilayer. The multilayer consists of ferromagnetic Fe-layers, followed by paramagnetic Cr-layers, repeated 10 times, as shown in the graph. The arrows indicate the direction of the Fe magnetization, which alternates from layer to layer in an

Fig. 7.27 Artificial
antiferromagnetic multilayer

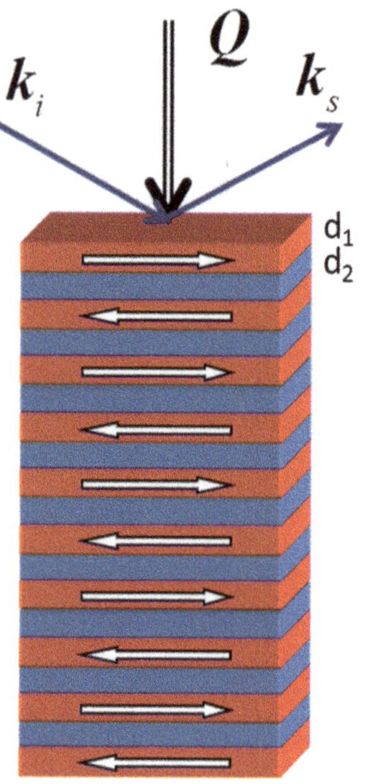

antiferromagnetic fashion. We also assume that the thicknesses of the Fe and Cr
layers are identical: $d_1 = d_2$. We want to study this multilayer by unpolarized
neutron scattering, where the scattering vector is oriented perpendicular to the
multilayer and perpendicular to the magnetization vectors (Fig. 7.27).

(a) Calculate the nuclear structure factor of the multilayer assuming homoge-
    neous thicknesses of the individual layers and sharp interfaces. For simplicity,
    we assume that each layer is ultrathin so you can neglect the finite thickness
    effects of the individual layers. Use a very simplified one-dimensional model
    for the layer sequence in the z-direction. What is the unit cell length in the
    z-direction, or how many layers are in a unit cell?
(b) Calculate the magnetic structure factor, assuming that the magnetic scattering
    length for Cr is zero.
(c) Combine the nuclear and magnetic structure factors and draw schematically
    the Bragg reflections that you expect for this multilayer. Use tabulated nuclear
    ($b_{Fe} = 9.45$ fm; $b_{Cr} = 3.63$ fm) and magnetic ($p_{Fe} = 4.5$ fm at $Q = 0$),
    scattering lengths to estimate the intensity of the Bragg peaks.
(d) Discuss your results: how do the nuclear and magnetic Bragg peaks overlap?
    What is the intensity of the magnetic Bragg reflections in comparison to the
    nuclear Bragg peaks? Are they much weaker, similar in intensity, or much
    stronger? How do the magnetic Bragg peaks depend on the scattering vector?
    What determines the width of the nuclear and magnetic Bragg reflections?

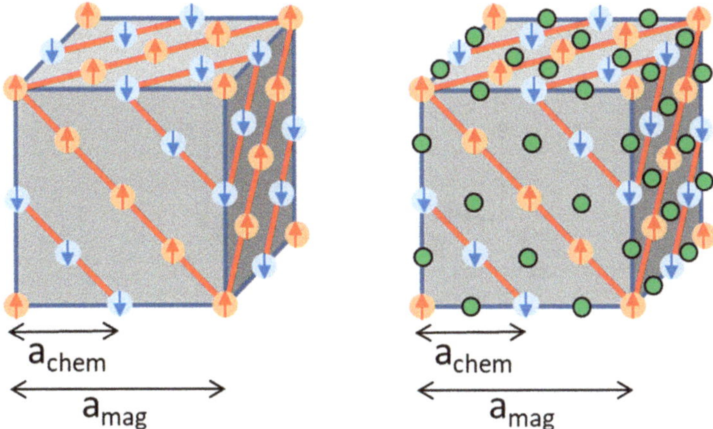

Fig. 7.28 Chemical and spin structure of MnO

A 7.2: **Antiferromagnetic structure of MnO**.

Show that the antiferromagnetic reflections (111) and (311) of MnO would be labeled $\left(\frac{1}{2}\ \frac{1}{2}\ \frac{1}{2}\right)$ and $\left(\frac{3}{2}\ \frac{1}{2}\ \frac{1}{2}\right)$ if the chemical unit cell with a cubic lattice parameter of 0.4425 nm were used instead of the magnetic unit cell with a lattice parameter of 0.885 nm. The magnetic unit cell is shown in the graph to the left. Only the Mn-ions together with their magnetic moments are shown in the left panel. The oxygen ions lie in between, as shown by green circles in the right panel. This is the same structure as the archetypal NaCl structure (see Chap. 1 for more information). Note that all circles are only drawn schematically and do not correspond to the actual ionic sizes. In these figures, a_{chem} designates the chemical unit cell length and a_{mag} the magnetic unit cell length (Fig. 7.28).

Proceed as follows:

(a) Determine the nuclear structure factor of MnO. The chemical unit cell of MnO has a NaCl structure containing 4 Mn-ions and 4 O-ions. For the NaCl, we have already calculated the structure factor in Exercise 3.2.2. However, because of the antiferromagnetism of MnO, we need to consider a unit cell with a lattice parameter: $a_{\text{mag}} = 2a_{\text{chem}}$, such that the magnetic unit cell volume is $V_{\text{mag}} = a_{\text{mag}}^3 = 8a_{\text{chem}}^3$, i.e., 8 times as big as the chemical unit cell volume. This bigger magnetic unit cell contains $4 \times 8 = 32$ Mn-ions and 32 O-ions. To figure out the structure factor, the coordinates of all 64 ions in units of the magnetic lattice parameter a_{mag} must be determined. What is the condition for allowed chemical (nuclear) Bragg reflections?

(b) Now we consider the magnetic structure factor. This is rather difficult. To simplify the calculation we assume that all magnetic moments lie in the Mn-(111) plane and that the orientation of the magnetic moments in these

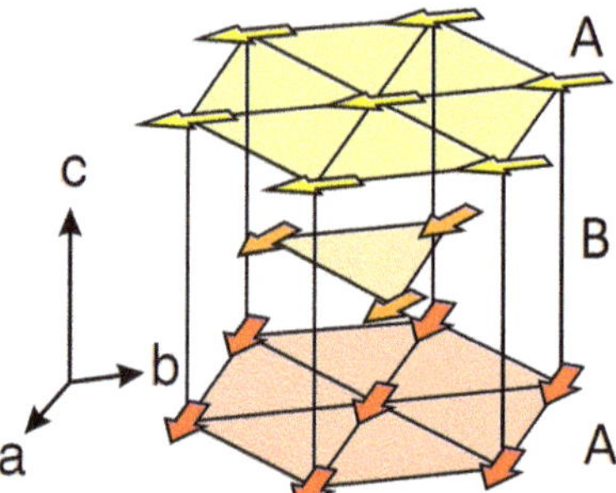

Fig. 7.29 Antiferromagnetic spin structure of hexagonal holmium below 20 K. Each basal plane is indicated by a different color. Small arrows symbolize the local magnetic moments of Ho atoms. Reproduced and adapted with permission from Rettig et al. (2016). Copyright (2025) by the American Physical Society

planes alternates in an antiferromagnetic fashion in the direction normal to the (111) planes. What is the condition for allowed antiferromagnetic Bragg reflections?

(c) Finally, consider reducing the magnetic unit cell to its original size of the chemical unit cell. How do the Miller indices *hkl* be adapted and relabeled to describe the allowed Bragg reflections in terms of the chemical unit cell parameter?

M 7.3 Spin structure of Holmium

Holmium has a hexagonal close-packed (*hcp*) crystal structure with lattice parameters $a = b = 0.357$ nm, $c = 0.561$ nm and a layer sequence ABA.... At temperatures below 20 K, holmium exhibits a commensurate antiferromagnetic structure, which is shown schematically in the figure. It consists of ferromagnetically ordered magnetic moments that lie in the basal plane (in-plane anisotropy). Basal planes A and B have the same ferromagnetic order. However, the magnetic moments in plane B are rotated about the hexagonal c-axis by 30° with respect to plane A. The rotation continues from plane to plane. Therefore, 12 planes or six unit cells are required for a full magnetic turnaround. The magnetic structure and phase transition have been analyzed in detail by neutron scattering. Knowing already the antiferromagnetic spin structure, it is certainly much easier to set up a neutron scattering experiment to confirm this particular spin structure (Fig. 7.29).

Your task is to confirm the antiferromagnetic spin structure of Ho at low temperatures. For this task, you have available a neutron diffractometer using a monochromatic and unpolarized thermal neutron beam of fixed wavelength λ. The diffractometer is equipped with a sample stage that allows turning the sample in any direction in space. Furthermore, you have available a Ho single crystal of reasonable size (1 cm^3) and a cryostat that allows cooling to 5 K. What scans are required to confirm the Ho spin structure in the basal plane and its rotation in the c-direction? An early neutron scattering experiment to explore the spin structure in Ho was carried out at the Oak Ridge National Laboratory by Koehler et al. and is described in (Hannon et al. 1988).

References

J. Als-Nielsen, O.W. Dietrich, L. Passell, Neutron scattering from the Heisenberg ferromagnets EuO and EuS. II. Static critical properties. Phys. Rev. B **14**, 4908–4922 (1976)

M. Blume, Magnetic scattering of X-rays. J. Appl. Phys. **57**, 3615 (1985)

P. Bödeker, A. Schreyer, H. Zabel, Spin-density waves and reorientation effects in thin epitaxial Cr films covered with ferromagnetic and paramagnetic layers. Phys. Rev. B **59**, 9408–9431 (1999)

J.P. Brown, Spherical neutron polarimetry, in *Neutron Scattering from Magnetic Materials*, (Elsevier, Amsterdam, Heidelberg, New York, 2006)

P.J. Brown, H. Capellmann, J. Deportes, D. Givord, K.R.A. Ziebeck, Spatial correlation of magnetisation in the paramagnetic phases of iron and nickel. J. Magn. Magn. Mater. **31**, 295–296 (1983)

T. Chatterji, Magnetic excitations, in T.T. Chatterji, Neutron Scattering from Magnetic Materials, Elsevier, Amsterdam, Heidelberg, New York, 2006. 983) 295–296

F. De Bergevin, M. Brunel, Observation of magnetic superlattice peaks by X-ray diffraction. Phys. Lett. **39A**, 141–142 (1972)

F. de Bergevin, M. Brunel, Diffraction of X-rays by magnetic materials. I. General formulae and measurements on Ferro- and Ferrimagnetic compounds. Acta Crystallogr. A **A37**, 314–324 (1981)

S.B. Dierker, R. Pindak, R.M. Fleming, I.K. Robinson, L. Berman, X-ray photon correlation spectroscopy study of Brownian motion of gold colloids in glycerol. Phys. Rev. Lett. **75**, 449 (1995)

E. Fawcett, Spin density wave antiferromagnetism in chromium. Rev. Mod. Phys. **60**, 209 (1988)

D. Gibbs, D.E. Moncton, K.L. D'Amico, J. Bohr, B.H. Grier, Magnetic X-ray scattering studies of holmium using synchrotron radiation. Phys. Rev. Let. **55**, 234 (1985)

D. Gibbs, G. Grübel, D.R. Harshmann, E.D. Isaacs, D.B. McWhan, D. Mills, C. Vettier, Polarization and resonance studies of X-ray magnetic scattering in holmium. Phys. Rev. B **43**, 5663 (1991)

C.M. Günther, B. Pfau, R. Mitzner, B. Siemer, S. Roling, H. Zacharias, O. Kutz, I. Rudolph, D. Schondelmaier, R. Treusch, S. Eisebitt, Sequential femtosecond X-ray imaging. Nat. Photonics **5**, 99–102 (2011)

J.P. Hannon, G.T. Trammell, M. Blume, D. Gibbs, X-ray resonance exchange scattering. Phys. Rev. Lett. **61**, 1245 (1988)

J.P. Hill, G. Helgesen, D. Gibbs, X-ray-scattering study of charge- and spin-density waves in chromium. Phys. Rev. B **51**, 10336 (1995)

E.D. Isaacs, D.B. McWhan, C. Peters, G.E. Ice, D.P. Siddons, J.B. Hastings, C. Vettier, O. Vogt, X-ray resonance exchange scattering in Uas. Phys. Rev. Lett. **62**, 1671 (1989)

W.C. Koehler, J.W. Cable, M.K. Wilkinson, E.O. Wollan, Magnetic structures of Holmium. I. The Virgin State. Phys. Rev. **151**, 414 (1966)

S. Konings: Pinning of magnetic domains studied with resonant X-rays. Ph.D. thesis, University of Amsterdam (2007)

S. Konings, C. Schüßler-Langeheine, H. Ott, E. Weschke, E. Schierle, H. Zabel, J.B. Goedkoop, Magnetic domain fluctuations in an antiferromagnetic film observed with coherent resonant soft X-ray scattering. Phys. Rev. Lett. **106**, 077402 (2011)

S.W. Lovesey, *Theory of Neutron Scattering from Condensed Matter, Volume 2: Polarization Effects and Magnetic Scattering* (Clarendon Press, Oxford, 1984)

F. Mezei, Neutron spin echo: A new concept in polarized thermal neutron techniques. Z. Phys. **255**, 146 (1972)

R.M. Moon, T. Riste, W.C. Koehler, Polarization analysis of Thernal neutron scattering. Phys. Rev. **181**, 920 (1969)

H. Ott, C. Schüßler-Langeheine, E. Schierle, A.Y. Grigoriev, V. Leiner, H. Zabel, G. Kaindl, E. Weschke, Magnetic X-ray scattering at the M5 absorption edge of Ho. Phys. Rev. B **74**, 094412 (2006)

P.M. Platzman, N. Tzoar, Magnetic scattering of X-rays from electrons in molecules and solids. Phys. Rev. B **2**, 3556 (1970)

L. Rettig, C. Dornes, N. Thielemann-Kühn, N. Pontius, H. Zabel, D.L. Schlagel, T.A. Lograsso, M. Chollet, A. Robert, M. Sikorski, S. Song, J.M. Glownia, C. Schüßler-Langeheine, S.L. Johnson, U. Staub, Itinerant and localized magnetization dynamics in antiferromagnetic Ho. Phys. Rev. Lett. **116**, 257202 (2016)

W.L. Roth, Magnetic structures of MnO, FeO, CoO, and NiO. Phys. Rev. **110**, 1333 (1958)

G. Schütz, E. Goering, H. Stoll, Synchrotron radiation techniques based on X-ray: Magnetic circular dichroism, in *The Handbook of Magnetism and Advanced Magnetic Materials, Vol. 3 - Novel Techniques*, ed. by H. Kronmüller, S.P.S. Parkin, (Wiley, New York, 2007)

J. Schweizer, Polarized neutrons and polarization analysis, in *Neutron Scattering from Magnetic Materials*, (Elsevier, Amsterdam, Heidelberg, New York, 2006)

O.G. Shpyrko, E.D. Isaacs, J.M. Logan, Y. Feng, G. Aeppli, R. Jaramillo, H.C. Kim, T.F. Rosenbaum, P. Zschack, M. Sprung, S. Narayanan, A.R. Sandy, Direct measurement of antiferromagnetic domain fluctuations. Nature (London) **447**, 68 (2007)

C.G. Shull, J.S. Smart, Detection of Antiferromagnetism by neutron scattering. Phys. Rev. **76**, 1256 (1949)

G.L. Squire, *Introduction to the Theory of Thermal Neuton Scattering* (Cambridge University Press, 1978)

H.E. Stanley, *Introduction to Phase Transitions and Critical Phenomena* (Claderon Press, Oxford, 1971)

J. Stöhr, *NEXAFS Spectroscopy*, 2nd edn. (Springer-Verlag, 1992)

X. Sun, E. Feng, Y. Su, K. Nemkovski, O. Petracic, T. Brückel, Magnetic properties and spin structure of MnO single crystal and powder. J. Phys. Conf. Ser. **862**, 012027 (2017)

B.T. Thole, P. Carra, F. Sette, G. van der Laan, X-ray circular dichroism as a probe of orbital magnetization. Phys. Rev. Lett. **68**, 1943 (1992)

H. Zabel, Magnetism of chromium at surfaces, at interfaces, and in thin films. J. Phys. Condens. Matter **11**, 9303 (1999). https://doi.org/10.1088/0953-8984/11/48/301

Further Reading and Advanced Literature

E. Beaurepaire, H. Bulou, F. Scheurer, J.-O. Kappler, *Magnetism: A Synchrotron Radiation Approach*, Lecture Notes in Physics, vol 697 (Springer, Berlin, Heidelberg, New York, 2006)

S. Blundell, *Magnetism in Condensed Matter* (Oxford University Press, Oxford, 2001). https://doi.org/10.1103/PhysRevLett.116.257202

A. Boothroyd, *Principles of Neutron Scattering from Condensed Matter* (Oxford University Press, 2020)

T. Chatterji, *Neutron Scattering from Magnetic Materials* (Elsevier, Amsterdam, Heidelberg, New York, 2006b)

J. Stöhr, H.C. Siegmann, *Magnetism, From Fundamentals to Nanoscale Dynamics*, Springer Series in Solid State Physics (Springer, Berlin, Heidelberg, New York, 2006)

G.L. Squire, *Introduction to the Theory of Thermal Neuton Scattering* (Cambridge University Press, 1978)

Chapter 8
Defects, Disorder, and Order–Disorder Phase Transitions

8.1 Overview

Until now, we have tacitly assumed that single crystals are perfect and free of defects. In reality, all single crystals contain a certain amount of imperfections, such as dislocations, stacking faults, and twin boundaries. Moreover, even the most perfect single crystals we know of, artificially grown silicon and diamond, contain impurity atoms in the ppb range, i.e., atoms with an atomic number different from that of the host lattice. In retrospect, crystal imperfections justify the application of the kinematic approximation for describing X-ray and neutron scattering. Perfect single-crystal structures would require a dynamic theory to properly treat the scattering process.

Defects are not only difficult to avoid but are often necessary to control the functionality of materials. For example, carbon ions in Fe increase mechanical strength, phosphorus ions increase the conductivity of Si, and hydrogen in metals serves to store energy. Work hardening is a necessary process to transform soft metals into tough tools. This requires dislocations, which can be created by hammering a hot, malleable piece of metal—a task formerly performed by blacksmiths. All these examples demonstrate that defects in crystal structures are often crucial to their specific applications. Here, we discuss what we can learn about defects from X-ray and neutron scattering. This is a fairly extensive field of past and current research (Ossi 2006). Therefore, we must limit ourselves to a few aspects: low-concentration and non-interacting point defects in the form of interstitial or substitution defects in crystal lattices; disorder in amorphous materials; and binary alloys exhibiting order–disorder phase transitions.

© The Author(s), under exclusive license to Springer Nature Switzerland AG 2026

H. Zabel, *Elements of Elastic Scattering by X-Rays, Neutrons, and Electrons*,

https://doi.org/10.1007/978-3-032-16624-1_8

8.2 Point Defects

Figure 8.1 provides an overview of different point defects frequently encountered in crystal lattices. They can involve vacancies, substitutional defects, or interstitial defects.

In ionic crystals, we distinguish between the vacancy formation of cations and anions that move to interstitial sites (Fig. 8.2a), known as Frenkel defects. Alternatively, after vacancy formation, the ion may move to the surface, known as a Schottky defect (Fig. 8.2b).

Point defects in metals can be vacancies, interstitial atoms, or substitutional defects. Common interstitial atoms are hydrogen, oxygen, nitrogen, and carbon ions, sometimes referred to as "gase" in metals. Substitutional defects replace regular crystal atoms. They can be larger or smaller than the host metal atoms and cause the corresponding lattice strains. There is a wide variety of such defects that characterize metal alloys.

Point defects cause several elastic responses of the host lattice:

1. The lattice close to the defect becomes distorted with a decaying amplitude with an increasing distance from the defect (Fig. 8.3).
2. The overall lattice expands linearly with increasing defect concentration.
3. The random occupation of interstitial or substitutional sites causes random displacements of host lattice atoms.

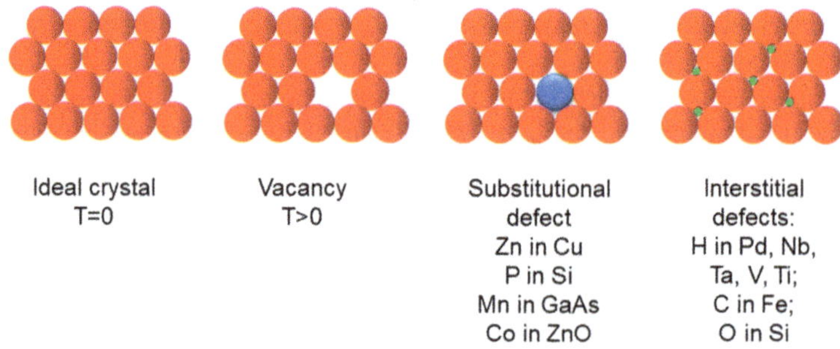

Fig. 8.1 Ideal lattice and examples of point-like defects in crystal lattices in the form of vacancies, substitutional defects, and interstitial defects

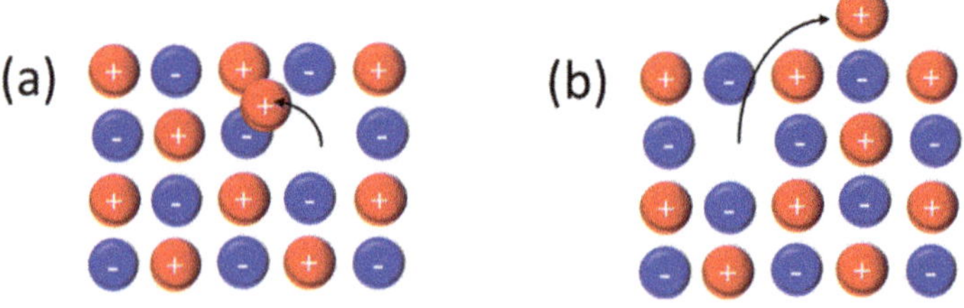

Fig. 8.2 Vacancy formation in ionic crystals. (**a**) Frenkel defect; (**b**) Schottky defect

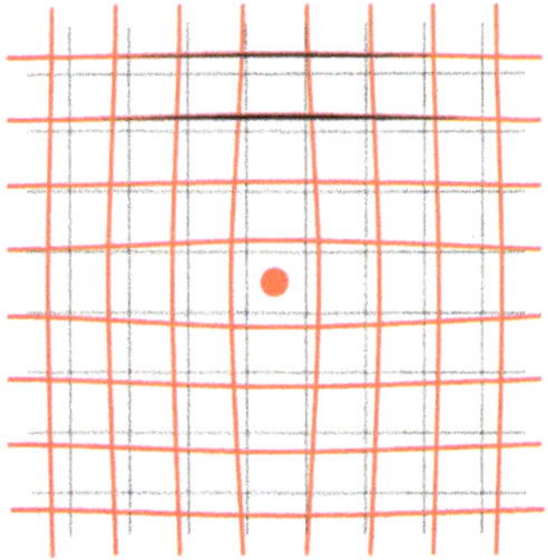

Fig. 8.3 Perfect crystal lattice before lattice distortion (gray lines) and after lattice distortion via uptake of point defects (red lines)

4. At high concentrations, defects may agglomerate and form clusters or precipitates, and vacancies may agglomerate to form voids.

The elastic distortion field close to the defect causes characteristic diffuse scattering near the main Bragg reflections, known as ***Huang diffuse scattering*** (Huang 1947; Dederichs 1973). The diffuse scattering is the Fourier transform of the local distortion field. The local distortion field is proportional to $1/r^2$, where r is the distance to the center of the defect. Taking the Fourier transformation of this ***distortion field***, the diffuse scattering intensity is proportional to: Huang (1947), Dederichs (1973), Krivoglaz (1969)

$$I_{diffuse}(\boldsymbol{q}) \propto x_d \Delta v_d \frac{(\boldsymbol{Q} \cdot \widehat{\boldsymbol{n}}_r)}{q^2}. \tag{8.1}$$

Here, $\boldsymbol{q} = \boldsymbol{Q} - \boldsymbol{G}_{hkl}$ is the deviation from the (hkl) Bragg reflection, x_d is the defect concentration, and Δv_d is the atomic volume of one defect atom in the host matrix (known as defect strength). $\widehat{\boldsymbol{n}}_r$ is a unit vector in the direction of the distortion field. Since the distortion field has radial symmetry, the dot product $\boldsymbol{Q} \cdot \widehat{\boldsymbol{n}}_r$ causes a dumbbell-shaped contour of the diffuse scattering centered at the fundamental Bragg reflections. Experimental verification of the diffuse scattering for hydrogen in metals was provided by Metzger and Peisl (1978). The diffuse scattering is further discussed in Exercise 8.3.

The distortion fields of all point defects overlap, and assuming that the surface is elastically free to expand, this leads to an overall lattice expansion that is linearly proportional to the point defect concentration x_d. The isotropic and linear expansion of a cubic lattice is expressed as (Peisl 1978):

$$\frac{\Delta V}{V} = 3 \frac{\Delta a}{a} = \frac{\Delta v_d}{\Omega} x_d, \tag{8.2}$$

where $\Delta V/V$ is the total volume change due to the defects, $\Delta a/a$ is the lattice parameter change along any of the cubic axes, and Ω is the atomic volume of the host matrix atom. Since $G_{hkl} = 2\pi/a$, it follows that

Fig. 8.4 Blue line: Bragg peak from a defect-free crystal lattice. Red line: Bragg reflection from a crystal lattice containing point defects. Point defects affect the Bragg reflections in a threefold way: 1. the Bragg peak is shifted to lower Q values by $-\Delta \boldsymbol{G}$ from $\boldsymbol{G}$ to $\boldsymbol{G}'$, proportional to the defect concentration; 2. the intensity is lowered due to a static Debye–Waller factor; 3. in the tails of the Bragg peaks, a Lorentzian-shaped diffuse scattering contribution is noticeable

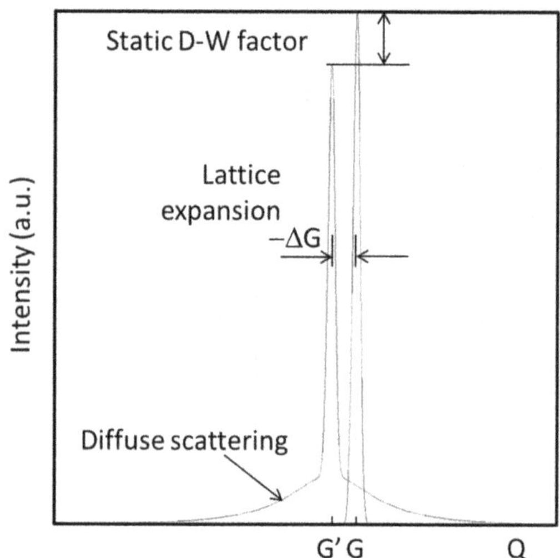

$$\frac{\Delta a}{a} = -\frac{\Delta G_{hkl}}{G_{hkl}} = \frac{1}{3}\frac{\Delta v_d}{\Omega}x_d, \tag{8.3}$$

and therefore the ***lattice expansion*** causes a shift of all Bragg reflections to smaller G_{hkl} values in the reciprocal lattice.

Finally, the random superposition of the defect-induced distortion fields, when probed by scattering methods, also results in a reduction of the Bragg peak intensity, referred to as a static Debye–Waller factor. The ***static Debye–Waller factor*** has the same form as the thermal Debye–Waller factor: e^{-2W} with $W = x_d Q^2 \langle u^2 \rangle$, and u is the average local displacement of the atoms, in complete analogy to the thermal Debye–Waller factor (see Eq. 3.46). The static Debye–Waller factor and the static diffuse scattering are indeed very similar to their thermal counterparts. They can only be distinguished by their concentration dependence on the one hand and their temperature dependence on the other hand.

Vacancies act like interstitial or substitutional point defects. However, their strain fields are negative, leading to an overall contraction of the host lattice. Furthermore, the form factor, or scattering length, of vacancies is the negative value of the atomic form factor of the host lattice, since vacancies are subtracted from the lattice sum.

It is important to note that ***point defects*** distort the crystal lattice locally but do not break its long-range order or symmetry. Therefore, the main Bragg reflections still have a δ-like or Gaussian line shape. The presence of defects is solely expressed by the shift of the Bragg reflection, by a static Debye–Waller factor, and by a diffuse intensity tail. Figure 8.4 schematically illustrates how all three contributions affect the Bragg scattering and the diffuse scattering. Also, notice that the Bragg reflection would shift to higher Q values if the lattice shrinks, which occurs for vacancies and for substitutional defects that have an atomic volume smaller than that of the host lattice. If point defects order, which is often the case via phase transitions at low temperatures, we observe new Bragg peaks reflecting this new order, called super-structure reflections. When randomly distributed, point defects give rise to a constant background intensity known as incoherent Laue scattering.

In addition to zero-dimensional point defects, one-dimensional line defects (edge and screw dislocations) and two-dimensional planar defects (stacking faults and antiphase domains) also occur in crystal structures. All these defects can be investigated qualitatively and quantitatively using scattering methods, as described in the books by Warren and Krivoglaz under "Further Reading." Imaging techniques such as transmission electron microscopy (TEM) are often even better suited for investigating dislocations, their density, and entanglement in crystal structures.

8.3   Disorder

Up to now, we implicitly assumed that our single crystals exhibit long-range order (LRO). LRO requires that two atoms, one at position 1 and another at a distant position 2, are connected by the lattice vector $\boldsymbol{R}_{12} = \boldsymbol{T}_{12} + \boldsymbol{r}_k$ (see Eq. 1.2). This is a rather strict requirement fulfilled only by the class of materials that we denote as single crystals. During a mountain hike, single crystals reveal themselves by their shiny and glittering appearance when exposed to sunlight. Unless you study genuine rocks, all single crystals that are used for physical sciences nowadays are grown artificially in the laboratory, and many new crystals have been synthesized that do not exist in nature. Even proteins need to be crystallized to analyze their molecular structure, a requirement that was valid at least until recently. However, if we scrutinize our environment, we will notice many solids, plastics, liquids, and organic materials that are not crystalline. In fact, the majority of the materials surrounding us are noncrystalline. In Fig. 8.5, two-dimensional models of crystalline and amorphous Si are compared. Both forms are often used in photovoltaics. In amorphous materials, usually bond lengths and bond angles are preserved in comparison to their crystalline counterpart, but the LRO is lost.

What can we learn about noncrystalline materials by scattering experiments? First, we will not be able to find any Bragg reflections. Instead, the scattered intensity (X-rays, neutrons, or electrons) shows with the increasing scattering vector Q first a broad peak, followed at higher Q values by intensity fluctuations with decreasing amplitude. This characteristic shape of the scattered intensity is usually isotropic. In Fig. 8.6, a comparison is shown of the diffraction pattern from a standard crystalline Si powder sample and an amorphous Si sample, both recorded with neutron scattering (https://neutrons.ornl.gov/wand/capabilities; Salmon and Zeidler 2019).

The comparison in Fig. 8.6 shows some similarity in both diffraction patterns. The first peak in the ***amorphous diffraction pattern*** lies close to the first Bragg peak allowed by the diamond structure factor, i.e., the (111) reflection, and even the higher-order Bragg reflections appear to have some bearing on the amorphous structure factor.

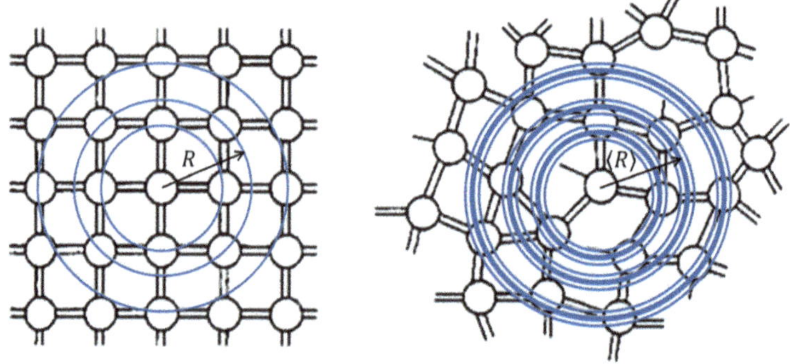

Fig. 8.5 Two-dimensional model of crystalline Si and amorphous silicon containing hydrogen bonds. Double lines represent regular bonds: single lines are unsaturated dangling bonds (adapted from http://www.wikiwand.com/en/Amorphous_silicon). The circles indicate pair correlations at distance R with the nearest and next-nearest neighbors in both structural states. The triple lines on the right indicate radial distance distributions

Fig. 8.6 Diffraction pattern of a Si powder sample (blue solid line) and amorphous Si (red dashed line), both recorded by neutron scattering. Both patterns are adapted from (Ref. (https://neutrons.ornl.gov/wand/capabilities; Salmon and Zeidler 2019)) and scaled to match the same Q-range. The Miller indices refer to the diamond structure factor of Si

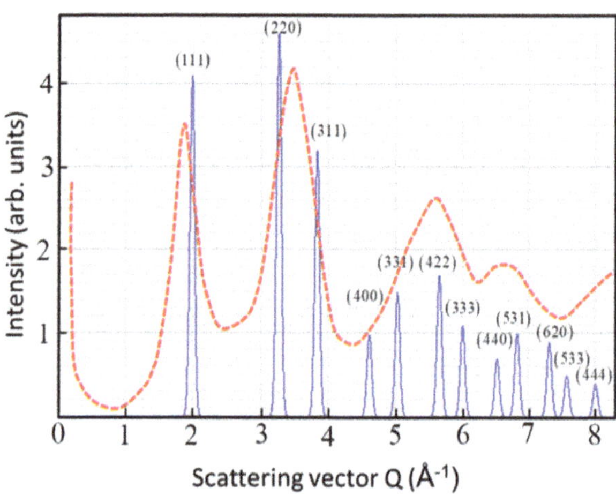

For amorphous materials, no structure factor can be calculated because of the lack of translational symmetry. Instead, the scattering intensity can be expressed by a Fourier transform of pair correlations $g(R)$ between atoms. $g(R)$ can be thought of as the probability of finding a scattering center at a distance R from the origin. In the crystalline case, $g(R)$ is a sharp function with well-defined radial distances of the nearest and next-nearest neighbors from the center, as indicated by the blue circles in Fig. 8.5. For amorphous materials, $g(R)$ exhibits oscillating probabilities but no sharp peaks—a characteristic feature that we will discuss again in connection with Fig. 8.7.

Mathematically, the ***pair correlation function*** is the sum over all pairs ij of atoms at distances $\boldsymbol{R}_{ij}$ within the amorphous scattering volume:

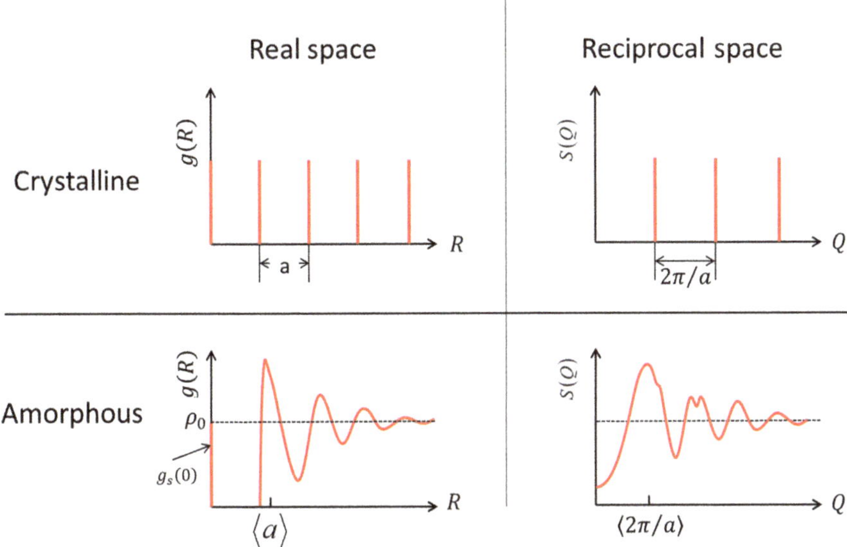

Fig. 8.7 Real-space pair correlation functions of a crystalline solid (**a**) and an amorphous material (**c**). The respective scattering functions (Fourier transforms) are shown in (**b**) for crystalline materials and (**d**) for amorphous or liquid materials

$$g(\boldsymbol{R}) = \frac{1}{N}\sum_{ij}\delta(\boldsymbol{R} - \boldsymbol{R}_{ij}) \tag{8.8}$$

The δ-function has the value 1 if $\boldsymbol{R} = \boldsymbol{R}_{ij}$ and is 0 otherwise. There are N^2 terms but only N atoms to scatter from. Therefore, we normalize the sum by N. Furthermore, there are N terms with $i = j$. Those are the results of the self-correlation $\delta(\boldsymbol{R})$, peaking at $R = 0$; the other terms refer to the distinct correlations $\delta(\boldsymbol{R} - \boldsymbol{R}_{i \neq j})$. In the limit $\lim\limits_{R \to \infty} g(\boldsymbol{R}) = \rho_0$, i.e., the pair correlation function approaches the average density ρ_0. It is common to factor out the average density: $g(\boldsymbol{R}) = \rho_0 \tilde{g}(\boldsymbol{R})$, and to subtract ρ_0, before evaluating the intensity by Fourier transformation:

$$I(Q) = \left(\frac{a}{r}\right)^2 \int (g(\boldsymbol{R}) - \rho_0)e^{i\boldsymbol{Q}\cdot\boldsymbol{R}}d\boldsymbol{R}$$

$$= \left(\frac{a}{r}\right)^2 \int (\delta(\boldsymbol{R}) + \rho_0(\tilde{g}(\boldsymbol{R}) - 1))e^{iQR\cos\phi}R^2 dR d\theta d(\cos\phi)$$

$$= \left(\frac{a}{r}\right)^2 \left(1 + \frac{4\pi\rho_0}{Q}\int_0^\infty (\tilde{g}(\boldsymbol{R}) - 1)R\sin(QR)dR\right) = \left(\frac{a}{r}\right)^2 S(Q) \tag{8.9}$$

The term $\delta(\boldsymbol{R})$ is due to the **_self-correlation_** and becomes "1" by the Fourier transformation. Concerning the spherical integral, we refer to Exercise 2.2. The vector notation is dropped because the correlation function is assumed to be spherically isotropic. The term in the big bracket is referred to as the **scattering function** $S(Q)$, more specifically here the *amorphous scattering function*. Since the oscillatory term in $S(Q)$ scales with $1/Q$, $S(Q)$ approaches 1 at high Q-values.

In the end, we are interested in the real space correlation function, and therefore, we need to back-transform the measured scattering function. Without proof, the back-transformation yields the pair distribution function (Dove and Li 2022):

$$g(R) = \rho_0 + \frac{1}{2\pi^2 R} \int_0^\infty (S(Q) - 1)Q \sin(QR)dQ \tag{8.10}$$

Figure 8.7 shows schematically a comparison between pair correlation functions of crystalline solids with translational symmetry and long-range order (LRO), and of amorphous (or liquid) materials featuring short-range order (SRO). The pair correlation of atoms in crystalline solids can be described by a set of δ-functions, separated by the lattice parameter. The Fourier transform of an infinite series of δ-functions in real space is an infinite series of δ-functions in reciprocal space. In contrast, for disordered amorphous, glassy, or liquid materials, only the **self-correlation function** $G_s(0)$ has a δ-function shape at the origin in real space, which translates into a constant contribution with the value of 1 in the scattering function: $S(Q) = 1$. This is indeed the limiting value of $S(Q)$ at high Q-values. Otherwise, the **pair correlation function** has a lower cut-off for the closest approach of atoms in the material. The first peak in $g(R)$ reflects the average atomic separation and an oscillatory dependence with increasing separation and decreasing amplitude about the average density value ρ_0.

The scattering function $S(Q)$ also oscillates with increasing Q and decreasing amplitude, approaching the limiting value $\lim_{Q \to \infty} S(Q) = 1$, as already noted. The first peak in $S(Q)$ indicates the average reciprocal atomic separation. The limiting value at $Q = 0$ is $S(0) = k_B T/v^2$, where k_B is the Boltzmann constant, T is the absolute temperature, and v is the sound velocity. This limiting value can only be seen when the scattering at small Q values is not dominated by small-angle scattering due to clusters, voids, and inhomogeneities in the material. Small-angle scattering is further discussed in Chap. 9.

For evaluating $g(R)$, it is important to determine the scattering function $S(Q)$ up to high Q-values, up to about 20–25 Å^{-1}. However, the range of Q-values generally accessible is limited by the wavelength of X-rays or neutrons chosen: $Q_{max} = 4\pi/\lambda_{min}$. For instance, for a 1 Å (0.1 nm) wavelength, $Q_{max}=12.5$ Å^{-1} (125 nm^{-1}). Using smaller wavelengths should help but only for neutrons, because for neutrons, the scattered intensity is proportional to b^2, and the scattering length b is independent of the scattering vector Q. For X-rays, in contrast, the intensity $I \propto f^2(Q)$, and the atomic form factor $f(Q)$ strongly limits the accessible Q-range to usually <10 Å^{-1} (100 nm^{-1}).

From an elastic scattering point of view, we cannot distinguish between an amorphous or glassy structure and a liquid. The pair correlations in both are the same. Indeed, glasses are considered frozen liquids. The distinction can, however, be made when extending the scattering techniques to include inelastic scattering. The main difference between liquids and glasses does not lie in local excitations, such as lattice vibrations (phonons), but in the local translational motions, i.e., in the local diffusivity, also denoted as Brownian motion. The local diffusivity can be studied by using quasi-elastic neutron scattering, featuring an energy resolution of about 1 μeV and below. The scattering function is then extended into the energy domain and expressed as $S(Q, \omega)$. Since the present text deals with elastic scattering, we stick with $S(Q)$, and we leave the distinction between liquids and glasses to other texts.

Everything discussed here is based on the assumption of a homogeneous single-component system. Most amorphous systems are more complex and contain different atoms, such as in oxides. Accordingly, the scattering theory must be adapted, as many additional pair correlations between different ions with different scattering lengths must be taken into account.

8.4 Order–Disorder Phase Transition

Now, we return to the discussion of crystalline materials, considering binary alloys. In general, binary alloys of the type $A_{1-x}B_x$, where A and B are metal ions, exhibit two types of phase transitions as a function of temperature: order–disorder or segregation, depending on the pair interaction potential U. If $U_{AA} + U_{BB} - 2U_{AB} < 0$, the alloy will undergo an order–disorder phase transition displaying a superstructure of the type ABAB…; otherwise, the alloy components tend to segregate into A-rich and B-rich domains. In the following, we consider an A_xB_{1-x} alloy that shows an order–disorder phase transition.

The scattering amplitude of a disordered binary alloy A_xB_{1-x}, with random occupation of the lattice sites, can be described by an average atomic form factor:

$$\langle f \rangle = x f_A + (1 - x) f_B \tag{8.11}$$

Now, we want to describe the phase transition from a random to an ordered alloy. Our focus will be on the archetypal CuZn alloy with a 1:1 ratio. As we will see later, this system corresponds to a paramagnetic-to-ferromagnetic transition due to the introduction of a pseudospin operator. The chosen example thus represents an entire class of phase transitions of atomic order or spin order. We attempt to keep the discussion as simple as possible.

The $Cu_{0.5}Zn_{0.5}$ alloy, also known as **beta-brass** or β-CuZn, is a classical example of an order-to-disorder phase transition. The ordered phase shown in the right part of Fig. 8.8 has the B2 (CsCl) structure and exists at temperatures below the ordering temperature $T_c = 740$ K. For $T < T_c$, *long-range atomic order (LRO)* exists on the Cu and Zn sublattices, while for $T > T_c$, only **short-range order** (SRO) prevails. In the perfectly disordered state, the crystal assumes, on average, the bcc structure, with random occupation of all lattice sites, as shown in the left part of Fig. 8.8.

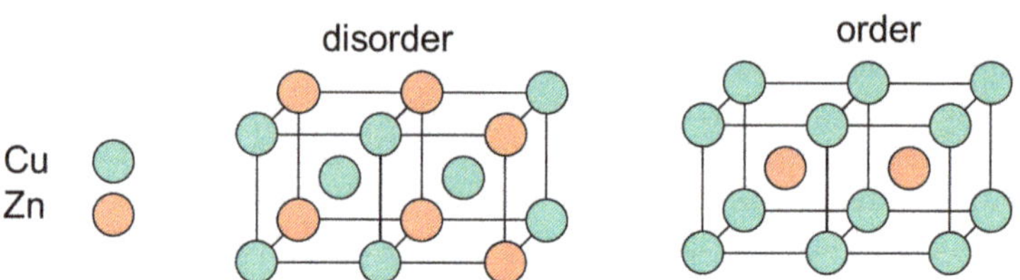

Fig. 8.8 Ordered and disordered structures of the binary alloy β-CuZn with a 50:50 mixture of Cu and Zn metal ions

To describe the order–disorder transition, we introduce a *site occupation number* s_m for the lattice site m, which may also be called a *pseudo-spin number* or pseudo-spin operator. This occupation number has two values:

$$s_m = \begin{array}{l} +1 \text{ if atom A is in position m} \\ -1 \text{ if atom B is in position m} \end{array} \tag{8.12}$$

For an equal number of A and B atoms representing Cu and Zn, the sum taken over all s_m is:

$$\sum_{m=1}^{N} s_m = 0, \tag{8.13}$$

where N is the total number of lattice sites. Now, we introduce a binary form factor (Als-Nielsen and Dietrich 1967a):

$$f_m = \frac{a+b}{2} + \frac{a-b}{2} s_m, \tag{8.14}$$

where a and b are either atomic form factors of the metal ions A and B in the case of X-ray scattering or the respective coherent scattering lengths in the case of neutron scattering. Therefore, $f_m = a$ for $s_m = 1$, and $f_m = b$ for $s_m = -1$. Then, the scattered intensity in units of the prefactor becomes:

$$I(\boldsymbol{Q}) = \sum_{m,n} \left\langle \left(\frac{a+b}{2} + \frac{a-b}{2} s_m \right) \left(\frac{a+b}{2} + \frac{a-b}{2} s_n \right) \right\rangle \exp(i\boldsymbol{Q} \cdot \boldsymbol{R}_{m,n}) \tag{8.15}$$

The scattered intensity takes an ensemble (statistical) average over all lattice sites designated by angle brackets, and the sum is over all lattice sites in the scattering volume. Carrying out the product, we then find:

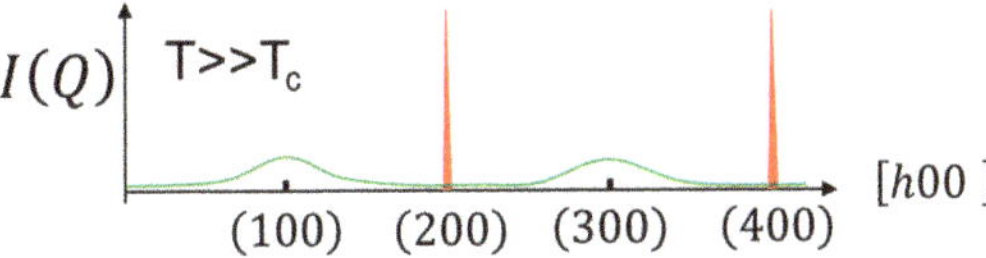

Fig. 8.9 In the disordered state above the ordering temperature T>>T$_c$, sharp Bragg peaks (in red) coexist with diffuse scattering (green line). The diffuse scattering is peaks at positions where, in the ordered phase, superstructure reflections develop. The scattered intensity is shown exemplarily for only one particular direction in the reciprocal lattice, parallel to [$h00$]

$$I(\mathbf{Q}) = \left(\frac{a+b}{2}\right)^2 \sum_{m,n}^{N} \exp(i\mathbf{Q}\cdot\mathbf{R}_{m,n}) + \left(\frac{a-b}{2}\right)^2 \sum_{m,n}^{N} \langle s_m s_n\rangle \exp(i\mathbf{Q}\cdot\mathbf{R}_{m,n})$$

$$(8.16)$$

The cross-term drops out because it involves the product $\langle s_m\rangle\langle s_n\rangle$, which is zero. The first term in Eq. (8.16) is the usual lattice sum that generates Bragg reflections at allowed reciprocal lattice points independent of the sublattice atomic order:

$$I(\mathbf{Q} = \mathbf{G}_{hkl}) = NV_{rc}\left(\frac{a+b}{2}\right)^2 \sum_{hkl}\delta(\mathbf{Q} - \mathbf{G}_{hkl}) \qquad (8.17)$$

These peaks are called the ***fundamental* Bragg reflections**. In the case of β-CuZn, they occur at the positions for allowed bcc reflections, i.e., for reflections with $h + k + l =$ even. The second term in Eq. (8.16) produces diffuse peaks between the fundamental reflections, indicating a new pair correlation of the CsCl type. Those peak positions are at the bcc-forbidden locations in the reciprocal lattice with $h + k + l =$ odd. The allowed Bragg peaks and the location of the diffuse scattering peaks are shown schematically in Fig. 8.9.

To describe the SRO pair correlation in the disordered phase, we introduce probability operators for the positions of atoms A and B:

$$P_m = s_m \exp(i\boldsymbol{\tau}\cdot\mathbf{R}_m). \qquad (8.18)$$

Here, $\boldsymbol{\tau}$ is the wave vector of the ordered structure. The ratio $|\boldsymbol{\tau}|/|\mathbf{G}|$ defines the position of the ***superstructure peak***. For β-CuZn alloys, $|\boldsymbol{\tau}|/|\mathbf{G}| = (n \pm 1)/2$, with n being an integer. If the ratio $|\boldsymbol{\tau}|/|\mathbf{G}|$ is rational, the superstructure is called ***commensurate***; vice versa, if the ratio is irrational, the superstructure is ***incommensurate***. Thus, β-CuZn alloys have a ***commensurate superstructure*** in the ordered state.

Using the operator notation and inserting Eq. (8.18) in the second part of Eq. (8.16), the intensity for the superstructure becomes:

$$I_{superst.}(\boldsymbol{Q}) = \left(\frac{a-b}{2}\right)^2 \sum_{m,n}^{N} \langle P_m P_n \rangle \exp(i(\boldsymbol{Q} - \boldsymbol{\tau}) \cdot \boldsymbol{R}_{m,n}) \qquad (8.19)$$

The ensemble average of P is the long-range order parameter:

$$\langle P_m \rangle = \langle P \rangle, \qquad (8.20)$$

which is independent of the index m. In the high-temperature disordered phase, $\langle P \rangle = 0$; in the low-temperature ordered phase, $0 \leq \langle P(T) \rangle \leq 1$. The *pair correlation function* is defined as:

$$g(\boldsymbol{R}_{m,n}) = \langle (P_m - \langle P \rangle)(P_n - \langle P \rangle) \rangle = \langle P_m P_n \rangle - 2 \langle P_m \rangle \langle P \rangle + \langle P \rangle^2 \qquad (8.21)$$

Since $\langle P_m \rangle = 0$ in the disordered state, the pair correlation function reduces to:

$$g(\boldsymbol{R}_{m,n}) = \langle P_m P_n \rangle + \langle P \rangle^2 \qquad (8.22)$$

Inserting Eq. (8.22) in Eq. (8.19), we find:

$$I_{superst.}(\boldsymbol{Q}) = NV_{rc} \left(\frac{a-b}{2}\right)^2 \langle P \rangle^2 \sum_{hkl} \delta(\boldsymbol{Q} - (\boldsymbol{G}_{hkl} \pm \boldsymbol{\tau}))$$

$$+ \left(\frac{a-b}{2}\right)^2 \sum_{m,n}^{N} \left(g(\boldsymbol{R}_{m,n}) - \langle P \rangle^2 \right) \exp(i(\boldsymbol{Q} - (\boldsymbol{G}_{hkl} \pm \boldsymbol{\tau})) \cdot \boldsymbol{R}_{m,n}) \qquad (8.23)$$

Equation (8.23) contains two terms: the first one refers to long-range superstructure Bragg reflections, characterized by δ-functions at $\boldsymbol{Q} = \boldsymbol{G}_{hkl} \pm \boldsymbol{\tau}$, and the second term refers to short-range superstructure diffuse scattering, which is also centered at $\boldsymbol{Q} = \boldsymbol{G}_{hkl} \pm \boldsymbol{\tau}$.

At high temperatures in the disordered phase, $\langle P \rangle^2 = 0$. Therefore, at $T \gg T_c$, we find only superstructure diffuse scattering due to short-range order pair correlations. At very low temperatures, $T \ll T_c$, the diffuse scattering diminishes. At intermediate temperatures below T_c, the superstructure diffuse scattering and the superstructure Bragg peaks overlap. The scattering intensity is shown for different temperatures schematically in Fig. 8.10. The LRO Bragg reflections are $\delta-$like or Gaussian-shaped, as already mentioned, and the diffuse scattering has a Lorentzian line shape, as explained below. The onset of LRO can be recognized by the appearance of a Gaussian-shaped peak on top of the Lorentzian-shaped diffuse scattering.

In Eq. (8.23), the ***diffuse scattering intensity*** due to SRO is expressed by:

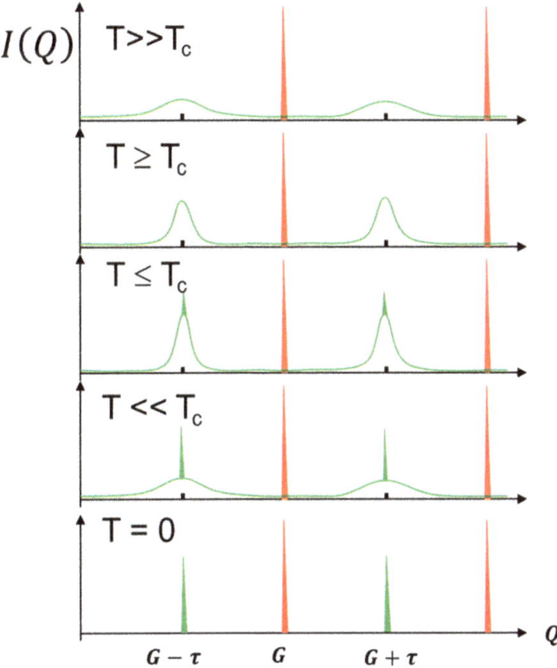

Fig. 8.10 Bragg reflections at G points (in red), long-range order reflections at $G_{hkl} \pm \tau$ points (in green), and short-range order diffuse scattering also centered at $G_{hkl} \pm \tau$ points for an arbitrary direction in reciprocal space. The schematic scattering pattern is shown for high temperatures far above the ordering temperature $T \gg T_c$, above $T > T_c$, just below $T \leq T_c$, far below $T \ll T_c$, and at $T = 0$

$$I_{SRO}(\boldsymbol{Q}) = \left(\frac{a-b}{2}\right)^2 g(\boldsymbol{q}), \tag{8.24}$$

where

$$g(\boldsymbol{q}) = \sum_{m,n}^{N} \left(g(\boldsymbol{R}_{m,n}) - \langle P \rangle^2\right) \exp(i(\boldsymbol{Q} - (\boldsymbol{G}_{hkl} \pm \tau)) \cdot \boldsymbol{R}_{m,n}) \tag{8.25}$$

is the **Fourier transform** of the pair correlation function $g(\boldsymbol{R}_{m,\,n})$, and $\boldsymbol{q} = (\boldsymbol{G}_{hkl} \pm \tau)$.

We summarize the results gained so far. The total coherent and elastic scattering intensity during an order–disorder phase transition consists of three contributions:

1. Fundamental Bragg reflections, independent of local site order:

$$I_f(\boldsymbol{Q} = \boldsymbol{G}_{hkl}) = N V_{rc} \left(\frac{a+b}{2}\right)^2 \sum_{hkl} \delta(\boldsymbol{Q} - \boldsymbol{G}_{hkl}) \tag{8.26}$$

2. Superstructure Bragg reflections due to long-range site order:

$$I_{LRO}(\boldsymbol{Q} - (\boldsymbol{G}_{hkl} \pm \boldsymbol{\tau})) = NV_{rc} \left(\frac{a-b}{2}\right)^2 \langle P \rangle^2 \sum_{hkl} \delta(\boldsymbol{Q} - (\boldsymbol{G}_{hkl} \pm \boldsymbol{\tau})) \qquad (8.27)$$

3. Superstructure diffuse scattering intensity due to short-range pair correlations:

$$I_{SRO}(\boldsymbol{Q} - \boldsymbol{q}) = \left(\frac{a-b}{2}\right)^2 g(\boldsymbol{q}) \qquad (8.28)$$

All contributions are schematically plotted in Fig. 8.10 for different temperatures.

8.5 Short-Range Order and Susceptibility

The short-range order pair correlation function $g(\boldsymbol{R}_{m,\,n})$ can be approximated by an exponentially decaying pair correlation as a function of distance. This form is known as the ***Ornstein–Zernicke correlation function*** in real space (Stanley 1971):

$$g(\boldsymbol{R}_{m,n}) = \frac{1}{\boldsymbol{R}_{m,n}} \exp\left(-\frac{\boldsymbol{R}_{m,n}}{\xi}\right), \qquad (8.29)$$

where $\boldsymbol{R}_{m,\,n}$ is the separation of two scattering points, and ξ is the **correlation length**. The correlations between atoms with distance $\boldsymbol{R}_{m,\,n}$ decay exponentially on the length scale ξ. The correlation length is a function of temperature. Within the molecular field approximation, ξ diverges according to $\xi \sim 1/\sqrt{(T - T_c)}$ close to the critical temperature T_c of the phase transition.

When taking the Fourier transform of $g(\boldsymbol{R}_{m,\,n})$, the pair correlation function has a Lorentzian line shape in the reciprocal space centered at $\boldsymbol{q} = \boldsymbol{G}_{hkl} \pm \boldsymbol{\tau}$:

$$g(\boldsymbol{q}) = \frac{1}{q^2 + \xi^{-2}} \qquad (8.30)$$

Figure 8.11 shows the SRO peak with the LRO peak on top. Hence, this schematically drawn peak captures the situation just below the transition temperature T_c. The integrated intensity of the Gaussian-shaped LRO peak is proportional to $\langle P \rangle^2$, which is known as the **order parameter**. In the case of magnetism, $\langle P \rangle^2$ is identified with the average (sublattice) magnetization squared $\langle M \rangle^2$ (see Eq. 7.24).

The height of the SRO peak is proportional to the Fourier transform of the pair correlation function $g(\boldsymbol{q} = \boldsymbol{\tau})$. In statistical mechanics, it is shown that $g(\boldsymbol{q})$ is related to the Fourier transform of the susceptibility via (Stanley 1971; Schwabl 2006):

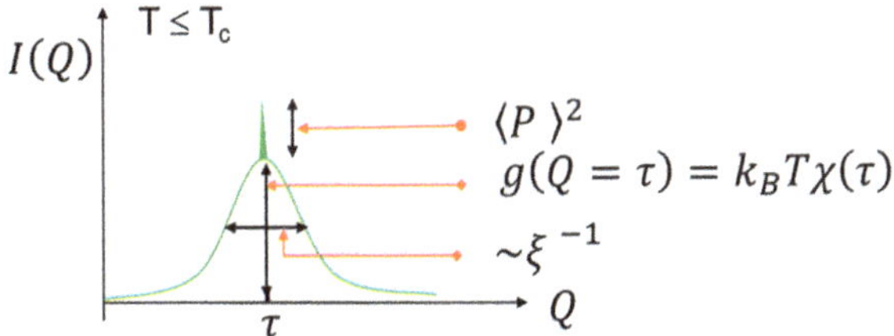

Fig. 8.11 Lorentzian-shaped short-range order peak and Gaussian-shaped long-range order peak. Peak heights and peak widths are related to thermodynamic parameters characterizing the phase transition via susceptibility and correlation length

$$\chi(\boldsymbol{q}) = \beta g(\boldsymbol{q}), \tag{8.31}$$

where $\beta = 1/k_B T$, k_B is the Boltzmann constant, and T is the absolute temperature. The pair correlation expresses thermal fluctuations, and the susceptibility stands for the response of the system to an external field acting on the local order parameter:

$$\chi(\boldsymbol{q}) = \beta g(\boldsymbol{q}) = \beta \frac{1}{q^2 + \xi^{-2}} \tag{8.32}$$

Finally, the width of the Lorentzian-shaped SRO peak is proportional to the inverse correlation length ξ.

Measuring the SRO and LRO peaks at $\boldsymbol{q} = \boldsymbol{G}_{hkl} \pm \boldsymbol{\tau}$ provides access to the essential thermodynamic parameters characterizing a structural phase transition: 1. LRO parameter $\langle P(T) \rangle$; 2. Susceptibility $\chi(\tau, T)$; and 3. Correlation length $\xi(T)$. All three parameters are labeled in Fig. 8.11.

The **susceptibility** of a system characterizes the responsiveness of the order parameter to an external field. The external field is called the ***conjugated field*** (Stanley 1971; Landau and Lifshitz 1980). For instance, in the case of a ferromagnet, the order parameter is the magnetization, and the conjugated field is the applied magnetic field $\boldsymbol{H}$. The conjugated field is the ordering field, in contrast to the temperature or entropy, which causes disorder. The susceptibility tells us how much the magnetization increases with an incremental change in the applied magnetic field. This works well for ferromagnets, but not for antiferromagnets. In the case of antiferromagnets, we need a ***staggered magnetic field*** that acts locally on the up- and down-oriented spins. The susceptibility of antiferromagnets can only be measured with magnetic X-ray or neutron scattering at the position of the antiferromagnetic Bragg reflections, as we have seen in the previous Chap. 7. Thus, scattering methods provide information on the susceptibility of a system at $\boldsymbol{Q}$-values larger than zero, adding much information to macroscopic susceptibility measurements taken at $\boldsymbol{Q} = 0$. For our β-CuAu order–disorder phase transition, the conjugated field is the ***chemical potential*** μ_A and μ_B acting on the lattice sites A and B. The chemical potential is the difference in the interaction between like nearest neighbors AA and BB and unlike neighbors AB. When AB is preferred, we observe a superstructure

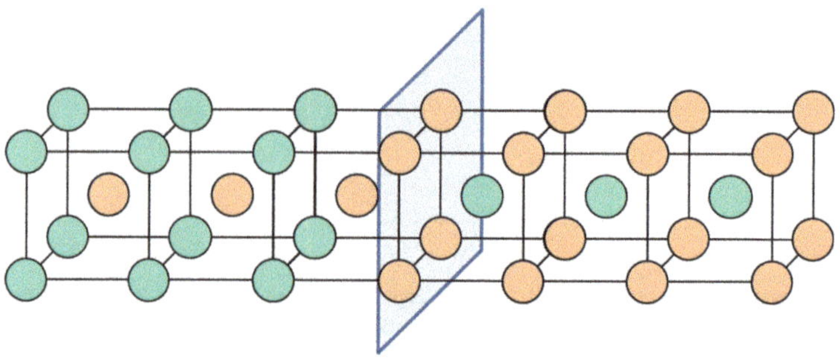

Fig. 8.12 Two domains of the ordered structure, with an antiphase domain wall in between

below the transition temperature; when *AA* and *BB* are preferred, the phase transition is of the precipitation type.

We discussed the width (FWHM) of the SRO peak in relation to the inverse short-range correlation length ξ. However, we did not discuss the FWHM of the LRO superstructure Bragg reflections but only noted that they have a Gaussian shape. The width depends on the LRO domain size. The domain size is limited by the presence of antiphase domains. Antiphase domains arise from the exchange of alloy site occupancy, i.e., when A atoms occupy B sites and vice versa, as sketched in Fig. 8.12. Antiphase domains limit the LRO and broaden the superstructure Bragg reflection. This is even more true the closer the temperature approaches the critical temperature T_c. The antiphase domain fluctuations can be studied in real time using XPCS methods using a coherent X-ray beam, as we discussed in the previous chapter.

In real experiments with X-rays and neutrons, the Debye–Waller factor and extinction effects must be considered for a quantitative analysis of the order parameter, but have been neglected in this presentation.

8.6 Experimental Results for the β-CuZn Order–Disorder Transition

The β-CuZn order–disorder phase transition has been intensively investigated by Dietrich and Als-Nielsen (Als-Nielsen and Dietrich 1967a; Dietrich and Als-Nielsen 1967; Als-Nielsen and Dietrich 1967b; Als-Nielsen 1969; Rathmann and Als-Nielsen 1974); some of their results are shown in the following. The left part of Fig. 8.13 displays the SRO peak at a temperature 9 K above the transition temperature of $T_c = 750\ K$. The line shape of the SRO peak is clearly Lorentzian. The height of this peak (compare Fig. 8.11 with $g(Q - \tau = 0) \propto \chi(\tau)$), which is proportional to the susceptibility $\chi(T, \tau)$, is plotted in the right-hand part of Fig. 8.13 for reduced temperatures $\varepsilon = (T - T_c)/T_c$ above and below the transition temperature T_c and for a constant wavenumber difference $\Delta q = 0.04°A^{-1}$ close to zero. This plot

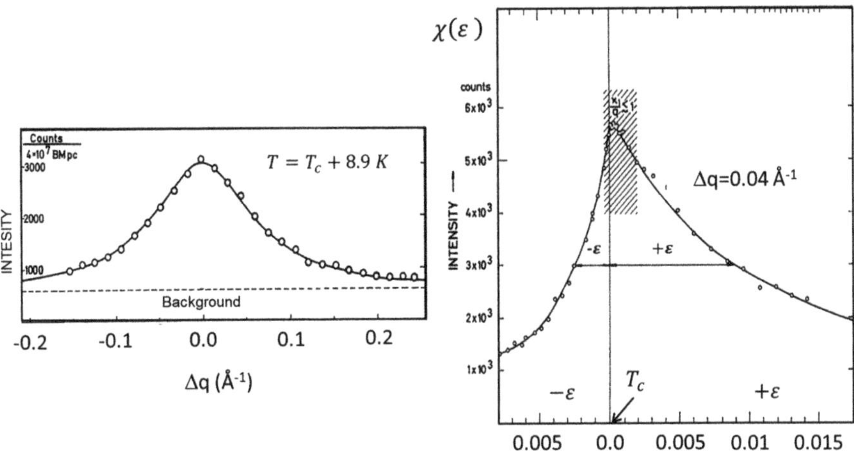

Fig. 8.13 Left panel: Short-range order diffuse scattering centered around the (100) position in the reciprocal lattice of the β-CuZn alloy at a temperature 8.9 K above the ordering temperature (BMpc = beam monitor preset count). Adapted with permission from Als-Nielsen and Dietrich (1967a). Right panel: Temperature dependence of the susceptibility above and below the transition temperature plotted on a reduced temperature scale $\varepsilon = (T - T_c)/T_c$. In the shaded area, the ratio of correlation length to momentum transfer is roughly 1. Reproduced with permission from Als-Nielsen (1969). Copyright (2025) by the American Physical Society

shows that the susceptibility diverges by approaching the transition temperature from above or below. Furthermore, the continuous change of the long-range order parameter with temperature shows that the phase transition is of second order (Fig. 8.14). A first-order phase transition would display a jump of the order parameter at the transition temperature, which is realized by the alloy Cu_3Au (compare Fig. 12.2). Concerning the order of phase transitions, we refer to textbooks on phase transitions, such as Stanley (1971). Note that the temperature dependence of the order parameter in *β-CuZn* is very similar to the paramagnetic–ferromagnetic phase transition presented in Fig. 7.8.

8.7 Remarks on Phase Transitions

In this chapter, we discussed the order-disorder phase transition of βCuZn, exemplary for all other order–disorder transitions and as an example of phase transitions in general. The scattering methods used are not limited to the static and elastic studies discussed here. Many order–disorder phase transitions are the result of phonon softening. In these cases, it is also important to investigate the phonon excitations using inelastic X-ray or neutron spectroscopy. The kinetics of the ordering transition after quenching an alloy from high to low temperatures can be studied in situ using time-resolved X-ray and neutron scattering methods. Synchrotron radiation offers the additional advantage of allowing phase transitions to be

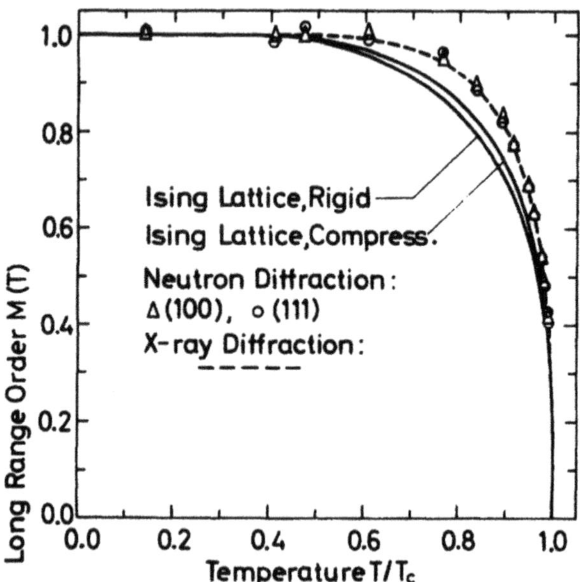

Fig. 8.14 Order parameter of the order-disorder phase transition in β-CuZn derived from the integrated intensity of the superstructure peak at $\tau = (100)$ (triangles) and (111) (circles), plotted as a function of the reduced temperature T/T_c. The temperature dependence of the superstructure peak is compared to different theoretical predictions: rigid Ising lattice and compressible Ising lattice. Reproduced with permission from Rathmann and Als-Nielsen (1974). The dashed curve shows X-ray diffraction data from Chipman and Walker (1972). Copyright (2025) by the American Physical Society

studied in tiny samples in diamond anvil cells as a function of pressure. Furthermore, X-ray photon correlation spectroscopy (XPCS) with coherent light can be used to directly visualize antiphase domain fluctuations in real time. The rapid progress in this field is reflected in the annual reports of the large-scale facilities that exist at numerous synchrotron radiation and thermal neutron source sites worldwide, and whose reports are generally freely accessible.

Summary

1. We distinguish between point defects and extended defects in crystal lattices.
2. Point defects are local defects that do not destroy the long-range order of crystal lattices.
3. Point defects can be either vacancies, substitutional atoms, or interstitial atoms.
4. Vacancies in ionic crystals are characterized as either Frenkel defects or Schottky defects, depending on the location of the ion removed from its lattice site.

5. Local defects cause lattice distortions close to the defect center.
6. Local lattice distortions have three main effects: 1. Lattice expansion; 2. Static Debye–Waller factor; 3. Diffuse scattering.
7. Extended defects lead to the destruction of long-range order.
8. In amorphous materials, short-range correlations can be recognized by diffuse scattering intensity.
9. The scattering function is the Fourier transform of the pair correlation function in disordered materials.
10. At high Q-values, the scattering function of disordered materials approaches a constant value of 1.
11. In general, structural phase transitions can be of the following types: order–disorder, re-orientational, or displacive.
12. The $Cu_{0.5}Zn_{0.5}$ alloy is a well-known system for studying order–disorder phase transitions.
13. The order–disorder transition can be described by introducing a pseudo-spin operator, resembling an antiferromagnetic system.
14. The wavevector τ characterizes the superstructure periodicity. τ is always smaller than G_{hkl}.
15. We distinguish between commensurate and incommensurate superstructures. If the ratio $|\tau|/|G|$ is rational, the system is called commensurate; otherwise incommensurate.
16. The X-ray or neutron scattered intensity consists of three parts: 1. Fundamental Bragg reflections $I_f(Q = G_{hkl})$; 2. Superstructure Bragg reflections due to long-range order $I_{LRO}(Q - (G_{hkl} \pm \tau))$; and 3. Superstructure diffuse scattering due to short-range correlations $I_{SRO}(Q - (G_{hkl} \pm \tau))$.
17. The integrated intensity of LRO-Bragg peaks is proportional to the mean order parameter squared $\langle P \rangle^2$. In the case of magnetism, the order parameter is proportional to magnetization: $\langle P \rangle = \langle M \rangle$.
18. The SRO diffuse scattering is the Fourier transform of the pair correlation function.
19. The SRO peak is characterized by height and a width. The height is proportional to the generalized susceptibility, and the width is proportional to the inverse correlation length.
20. The susceptibility and the correlation length strongly depend on the temperature. Both diverge at the transition temperature.
21. The pair correlation function has a Lorentzian line shape in the reciprocal space centered at $q = G_{hkl} \pm \tau$:
22. The conjugated field to the local concentration fluctuations in the βCuZn alloy is the site-specific chemical potential.

Questions

(Note that more than one answer may be correct)

1. **What is the difference between Schottky and Frenkel defects?**

   (a) Schottky defects occur in liquids, Frenkel defects in solids
   (b) Schottky defects occur in ionic crystals, and Frenkel defects in metals
   (c) Schottky defects are interstitial defects, and Frenkel defects are substitutional defects
   (d) Schottky defects move to interstitial sites, and Frenkel defects move to the surface

2. **How is the X-ray and neutron scattering from crystal lattices affected by point-like interstitial defects?**

   (a) By increased thermal Debye–Waller factor
   (b) By lowering of the overall intensity
   (c) By three effects: lattice expansion, static Debye–Waller effect, and diffuse scattering
   (d) By strong diffuse scattering proportional to the square of the defect concentration

3. **What characterizes amorphous materials?**

   (a) Existence of short-range order and long-range order
   (b) Existence of only short-range order
   (c) Existence of only long-range order

4. **Which order–disorder phase transitions can you think of?**

   (a) Gas–liquid
   (b) Paramagnetic–ferromagnetic
   (c) Martensitic
   (d) Superstructure order in binary alloys

5. **At what temperature does short-range order exist?**

   (a) Above the transition temperature
   (b) Below the transition temperature
   (c) Above and below the transition temperature
   (d) Does not exist at all

6. **What is the long-range order Bragg peak due to?**

   (a) It is a spill-off from the short-range order diffuse scattering.
   (b) It is due to the superstructure ordering of atoms or ions in the crystal lattice.
   (c) Bragg peaks exist only due to the temperature-independent long-range order.

7. **How is a commensurate superstructure defined?**

   (a) It is a superstructure that is in phase with the underlying lattice symmetry.
   (b) It is a superstructure with a wavenumber that is in a rational ratio to the real lattice structure.
   (c) It is a structure with no relation to the crystal structure.

8. **What is an order parameter?**

   (a) A parameter that expresses the probability of the correct atom in a crystal occupying the correct lattice site
   (b) A parameter that defines the long-range order (translational symmetry) of a system
   (c) A parameter that describes the short-range order of a system

Exercises

Grades of difficulty: E = easy, M = medium, A = advanced.

M 8.1 Point Defects
From a scattering point of view, how is it possible to distinguish between Frenkel defects and Schottky defects?

M 8.2 Huang Diffuse Scattering, Fourier Transform of $1/r^2$
Show that the Fourier transform of the local displacement field that drops off with $1/r^2$ is proportional to $1/q$ and show that the *Huang diffuse scattering* is proportional to $\frac{1}{q^2}$.

E 8.3 Form Factor of Vacancies
Figure 8.4 shows that defects have a dramatic effect on the scattering amplitude and the lattice parameter. Discuss what kind of effects on the scattering amplitude you expect for vacancies. Determine the vacancy atomic form factor, namely, the scattering length, and provide arguments for your results.

E 8.4 Thermal Vibration
If you want to add corrections due to the thermal vibration of the atoms, how would you proceed, and what is the corrected expression of the superstructure Bragg reflections?

E 8.5 Susceptibility
What is the expression for the paramagnetic susceptibility, and what is the expression for the generalized susceptibility to describe the order–disorder phase transitions?

E 8.6 Commensurate-Incommensurate Structures
In the graph (Fig. 8.15) is shown a one-dimensional atomic chain on a sinusoidally varying potential in space with a periodicity of d. Decide which of the atomic chains shown in (a) or (b) are ordered commensurately with the potential oscillations and which one is incommensurately ordered.

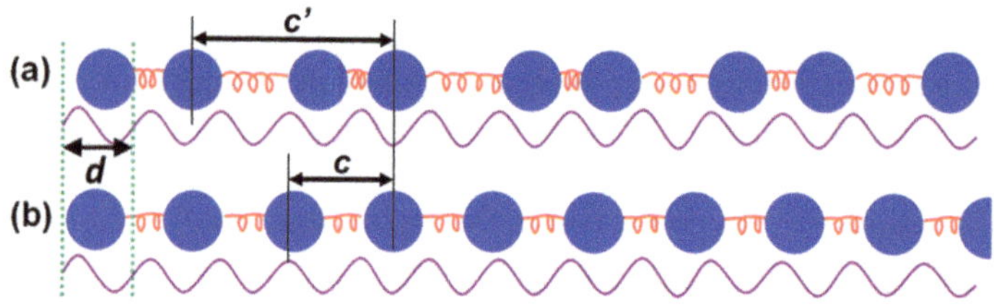

Fig. 8.15 One-dimensional graphical representation of a commensurate structure in panel (**a**) and an incommensurate structure in panel (**b**)

References

J. Als-Nielsen, Investigation of scaling Laws by critical neutron scattering from Beta-Brass. Phys. Rev. **185**, 664–666 (1969)

J. Als-Nielsen, O.W. Dietrich, Pair-correlation function in disordered β-Brass as studied by neutron diffraction. Phys. Rev. **153**, 706 (1967a)

J. Als-Nielsen, O.W. Dietrich, Long-range order and critical scattering of neutrons below the transition temperature in β-Brass. Phys. Rev. **153**, 717 (1967b)

D. Chipman, C. Walker, Long-Range Order in β -Brass. Phys. Rev. B **5**, 3823 (1972)

P.H. Dederichs, The theory of diffuse X-ray scattering and its application to the study of point defects and their clusters. J. Phys. F Metal Phys. **3**, 471 (1973)

O.W. Dietrich, J. Als-Nielsen, Temperature dependence of short-range order in β-Brass. Phys. Rev. **153**, 711 (1967)

M. Dove, G. Li, Review: Pair distribution functions from neutron total scattering for the study of local structure in disordered materials. Nuclear Analysis **1**, 100037 (2022)

K. Huang, X-ray reflexions from dilute solid solutions. Proc. Roy. Soc. A **27**, 134 (1947)

M.A. Krivoglaz, *Theory of X-ray and Thermal Neutron Scattering by Real Crystals*. (Translated from Russian and edited by S.C. Moss) (Plenum Press, New York, 1969)

L.D. Landau, E.M. Lifshitz, *Statistical Mechanics*, 3rd edn. (Butterworth-Heinemann, 1980)

H. Metzger, H. Peisl, Huang diffuse X-ray scattering from lattice strains in high-concentration Ta-H alloys. J. Phys. F Metal Phys. **8**, 39 (1978)

P.M. Ossi, *Disordered Materials, An Introduction* (Springer Verlag, Berlin Heidelberg, 2006)

H. Peisl, Lattice strains due to hydrogen in metals, in *Springer Volume 28, Topics in Applied Physics, Hydrogen in Metals I*, (Springer Verlag, Berlin Heidelberg, New York, 1978)

Rathmann, J. Als-Nielsen, Long-range order in β-Brass studied by neutron diffraction. Phys. Rev. B **9**, 3921–3926 (1974)

P.S. Salmon, A. Zeidler, Ordering on different length scales in liquid and amorphous materials. J. Stat. Mech. **2019**, 114006 (2019)

F. Schwabl, *Statistical Mechanics (Advanced Texts in Physics)*, 2nd edn. (Springer Verlag, Berlin Heidelberg, New York, 2006)

H.E. Stanley, *Introduction to Phase Transitions and Critical Phenomena* (Claderon Press, Oxford, 1971)

Recommended Reading

B.E. Warren, X-ray Diffraction, Dover Books in Physics, 1990.

M.A. Krivoglaz, *Theory of X-ray and Thermal Neutron Scattering by Real Crystals.* (Translated from Russian and edited by S.C. Moss) (Plenum Press, New York, 1969)

A. Furrer, J. Mesot, T. Strässle, *Neutron Scattering in Condensed Matter Physics* (World Scientific, New Jersey, Singapore, Bejing, 2009)

W.H. De Jeu, *Basic X-Ray Scattering for Soft Matter* (Oxford University Press, 2016)

N. Stribeck, *X-Ray Scattering of Soft Matter* (Springer Verlag, 2010)

Chapter 9
Small-Angle Scattering

9.1 General Background

Small-angle scattering (SAS) is an elastic scattering process, similar to X-ray and neutron diffraction. However, the magnitude of the probing scattering vector Q is much smaller by a factor of 10–100 as compared to typical scattering vectors for Bragg reflections. Thus, SAS probes structures on the mesoscopic length scale, ranging from 1 nm to about 1 μm. In this range, the probing objects typically are polymers, micelles, polyelectrolytes, viruses, proteins, fibers, precipitates in alloys, magnetic domains, magnetic nanoparticles, skyrmions, flux lines in superconductors, and others. Small-angle scattering with X-rays (SAXS) or with neutrons (SANS) explores structures and correlations in the direction parallel to the scattering vector Q, as is always the case. In the SAS regime, the scattering vector is normal to the incident wave number k_i, such that $Q \perp k_i$, as sketched in Fig. 9.1a. Furthermore, k_i is normal to the detector plane, and Q is in the detector plane. Figure 9.1b shows the SAS geometry in real space. The scattered beam is usually recorded with a position-sensitive area detector (PSD), sensitive either to X-ray photons or to neutrons. If the sample has an isotropic density distribution, the scattering pattern is circularly symmetric about the incident beam. Any deviation from the circular symmetry may be caused by electric or magnetic fields, by stress and shear fields, or by any other anisotropy-inducing directional correlations.

To set SAS in perspective, we compare in Fig. 9.2 different scattering processes depending on the ratio of the object size d to probing wavelength λ. For $d/\lambda \ll 1$, we observe an isotropic scattering, which is known in optics as ***Rayleigh scattering***. Rayleigh scattering is, for instance, responsible for the blue sky of the horizon. At $d/\lambda \approx 1$, we observe Bragg scattering from periodic structures or high-angle diffuse scattering from disordered and amorphous materials. The region $d/\lambda \gg 1$ is known in optics as ***Mie scattering*** or as ***small-angle scattering*** when probed with X-rays or neutrons. Beyond this range at $d/\lambda \ggg 1$, we find ***shadow formation***, while potential diffraction effects at sharp edges may occur.

© The Author(s), under exclusive license to Springer Nature Switzerland AG 2026

H. Zabel, *Elements of Elastic Scattering by X-Rays, Neutrons, and Electrons*,

https://doi.org/10.1007/978-3-032-16624-1_9

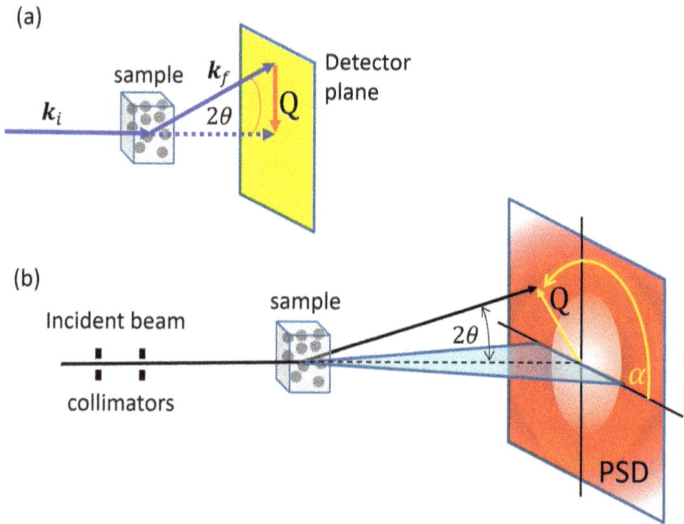

Fig. 9.1 Scattering geometry for SAS. (**a**) SAS geometry in momentum space. The scattering vector $\vec{Q}$ is oriented perpendicular to the incident wavenumber k_i, and the scattering triangle is closed by $k_i - k_f = Q$; (**b**) scattering geometry in real space. The scattered intensity is recorded with an area detector. For isotropic samples, the scattered intensity appears as rings on the detector centered about the incident beam. α is the azimuthal angle of the scattering vector Q

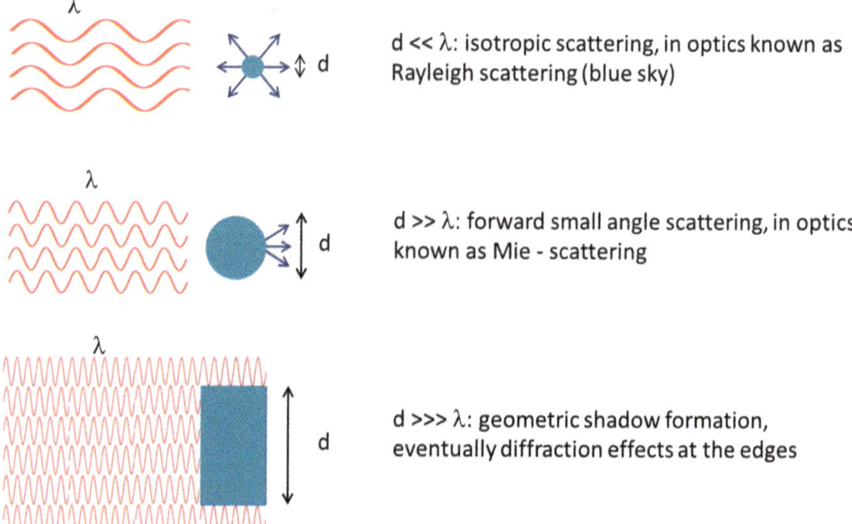

Fig. 9.2 Comparison of Rayleigh scattering, Mie or small-angle scattering, and shadow formation concerning the object size versus the probing wavelength

SAS is an extremely versatile scattering method that can be applied to virtually any system: liquids, soft matter, alloys, magnets, etc. The only limitation to be considered is the Born approximation that demands single-event scattering (see Sect. 3.7.1). This requirement is usually met but must be regarded in the case of SANS on hydrogen-containing samples because of its large scattering length. The risk of multiple scattering can be reduced by decreasing the scattering volume.

In the following, we discuss SAS on polymer samples and spherical particles, and in the second part, we consider SAS on magnetic nanoparticles.

> SAS probes correlations on the mesoscopic length scale in the direction perpendicular to the incident beam direction.

9.2 Contrast and Shape Factor

In small-angle scattering, we consider density–density correlations in space. More specifically, we consider the correlation of some density distribution $\rho(\mathbf{R})$ at position $\mathbf{R}$ and another density distribution $\rho'(\mathbf{R}')$ at position $\mathbf{R}'$ at some distance from position $\mathbf{R}$. The density distributions are illustrated in Fig. 9.3.

As usual in scattering theory, we take the Fourier transform of the **density–density correlation,** and we find for the scattered intensity the general expression[1]:

$$I(Q) = \frac{1}{N}\left(\frac{a}{r}\right)^2 \iint \langle \rho(\mathbf{R})\rho'(\mathbf{R}')\rangle \exp(i\mathbf{Q}\cdot(\mathbf{R}-\mathbf{R}'))dVdV' \qquad (9.1)$$

Here, as before, a stands for the X-ray scattering length $r_0 f(Q,E)$ or for the coherent neutron scattering length b_{coh}, and r is the distance from the scattering center to the detector. The normalization factor N is the total number of scattering

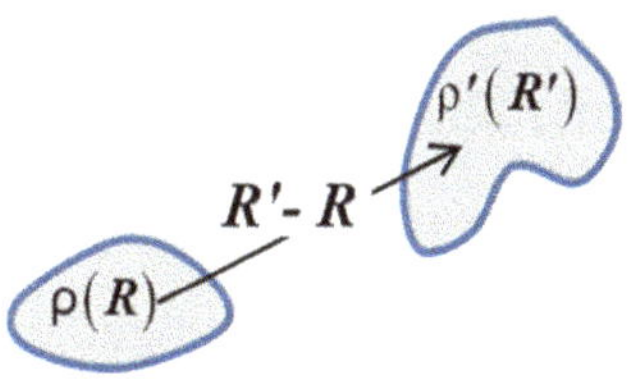

Fig. 9.3 Density distributions at different locations in the sample. Scattering experiments probe whether these density distributions are correlated or not

---

[1] Throughout this chapter we use the notation dV for the volume integration instead of $d\mathbf{R}$ as in the previous chapters, because the notation $d\mathbf{R}$ may lead to confusions.

centers. Eq. (9.1) is a very general expression but not very handy. Therefore, we assume that the density variations in the sample are rather small compared to a background of homogeneous average density. Then, we can decompose the local density into an average part and some deviation from the average density:

$$\rho(\boldsymbol{R}) = \langle \rho \rangle + \delta\rho(\boldsymbol{R}) \tag{9.2}$$

The constant part drops out in the Fourier transform and yields only intensity in the forward direction parallel to the transmitted beam at $Q = 0$. However, for $Q > 0$, only the deviation (contrast) from the average density counts, yielding a scattered intensity:

$$I(Q) = \frac{1}{N} \left(\frac{a}{r}\right)^2 \iint \langle \delta\rho(\boldsymbol{R})\delta\rho'(\boldsymbol{R}') \rangle \exp(i\boldsymbol{Q} \cdot (\boldsymbol{R} - \boldsymbol{R}'))dVdV' \tag{9.3}$$

This expression can be further simplified, assuming that:

$$\langle \delta\rho(\boldsymbol{R}) \rangle = \langle \delta\rho'(\boldsymbol{R}') \rangle = \delta\rho \tag{9.4}$$

Then, we take the density deviation out of the integral and remain with:

$$I(Q) = \frac{1}{N} \left(\frac{a}{r}\right)^2 (\delta\rho)^2 \left[\int \exp(i\boldsymbol{Q} \cdot \boldsymbol{R})dV\right]^2 \tag{9.5}$$

The assumption of a constant density deviation $\delta\rho$ is justified for particles with a density ρ_P embedded in a matrix of constant mean density $\langle \rho \rangle$, as is the case for certain particles (micelles, vesicles, etc.) in a solution, as indicated in Fig. 9.4.

The density deviation $\delta\rho = \langle \rho \rangle - \rho_p$ is also called ***contrast***. With this, the SAS intensity for a total number of N_p particles becomes:

$$I(Q) = \left(\frac{a}{r}\right)^2 \frac{V_p^2 N_p}{N} (\delta\rho)^2 [P(\boldsymbol{Q})]^2. \tag{9.6}$$

Here, N_p is the number of particles, V_p is the particle volume, and $P(\boldsymbol{Q})$ is the ***particle shape factor*** given by:

$$P(\boldsymbol{Q}) = \frac{1}{V_p} \int e^{i\boldsymbol{Q} \cdot \boldsymbol{R}}dV_p. \tag{9.7}$$

The final expression for SAS in Eq. 9.6 is composed of four factors that are easy to identify:

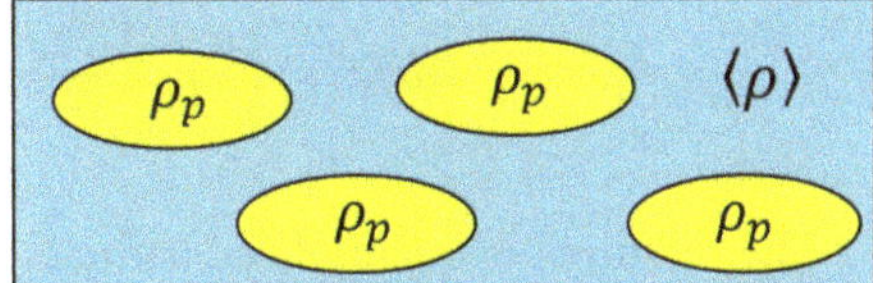

Fig. 9.4 Particles with density ρ_p embedded in a matrix with an average density $\langle\rho\rangle$

$$I(Q) = \underbrace{\left(\frac{a}{r}\right)^2}_{\text{Interaction}} \underbrace{\frac{V_p^2 N_p}{N}}_{\substack{\text{Normalisation} \\ \text{factor}}} \underbrace{(\delta\rho)^2}_{\text{contrast}} \underbrace{[P(\boldsymbol{Q})]^2}_{\substack{\text{Particle} \\ \text{shape factor}}}$$

The most important of these four factors is the density contrast $(\delta\rho)^2$ between particles and the solution. Without contrast, there is no small-angle scattering.

Note that the **particle shape factor** $P(\boldsymbol{Q})$ is the Fourier transform of the particle shape, and the integral is only taken over the particle volume V_P. The big advantage of this expression is the fact that the shape factor can easily be determined for some standard shapes such as spheres, ellipsoids, cylinders, or plates. For instance, for spherical particles of radius R_p, we obtain[2]:

$$P(\boldsymbol{Q}) = \frac{1}{V_p} \int e^{i\boldsymbol{Q}\cdot\boldsymbol{R}} dV_p = 3\frac{\left(\sin(QR_p) - (QR_p)\cos(QR_p)\right)}{(QR_p)^3}$$

$$= \frac{3}{QR} J_1(QR_p), \tag{9.8}$$

where

$$J_1(QR_p) = \frac{\left(\sin(QR_p) - (QR_p)\cos(QR_p)\right)}{(QR_p)^2} \tag{9.9}$$

is the *Bessel-function* of the first kind. This shape factor can be applied to the SAS analysis of surfactant micelles in solution, colloidal particles such as latex beads, ferrofluids, or silica pellets, and spherically shaped proteins in biology.

The expected scattering intensity from spherical particles suspended and uncorrelated in a solution is then:

---

[2]Note that the integral has the same form as the one for the atomic form factor treated in Appendix A3.

$$I(Q) = \left(\frac{a}{r}\right)^2 \frac{V_p^2 N_p}{N} (\delta\rho)^2 \left[\frac{3J_1(QR_p)}{(QR_p)}\right]^2 \tag{9.10}$$

The condition of uncorrelated particles works fine for dilute solutions. However, with increasing particle density, **correlation effects** may occur either due to particle–particle interaction or due to hard-core repulsion. In either case, with increasing particle density, the expression for the scattering intensity has to be modified by an additional scattering function $S(Q)$ that takes into account particle–particle correlation effects:

$$I(\boldsymbol{Q}) = \left(\frac{a}{r}\right)^2 \frac{V_p^2 N_p}{N} (\delta\rho)^2 \left[\frac{3J_1(QR_p)}{(QR_p)}\right]^2 |S(\boldsymbol{Q})|^2, \tag{9.11}$$

where the scattering function is the Fourier transform of the pair correlation function:

$$S(\boldsymbol{Q}) = 1 + N_p \int (g(\boldsymbol{R}) - 1) e^{i\boldsymbol{Q}\cdot\boldsymbol{R}} dV \tag{9.12}$$

In Eq. (9.11) and (9.12), $\boldsymbol{Q}$ is written as a vector, since correlation effects usually lift the isotropic scattering intensity from uncorrelated particles.

The general form of the scattering intensity for dilute solutions of uncorrelated spherical particles, aside from constant prefactors, is expressed in the dependence:

$$I(Q) = const. \left[J_1(QR_p)/(QR_p)\right]^2. \tag{9.13}$$

This function is plotted on a semi-logarithmic scale in Fig. 9.5. Characteristic for the scattering from spherical particles is the intensity oscillation with a period $\Delta Q = \pi/R$ or $\Delta QR = \pi$. The first minimum occurs at $Q_{min}R = 4.493$. Because of this simple relation, small-angle scattering from monodisperse **spherical latex particles** is often used as a "gold standard" that allows us to check the proper functioning of the equipment and to determine the instrumental resolution.

9.3 Guinier Approximation

Instead of deriving an exact expression for a particular particle shape, such as a sphere, we may also derive an approximate expression for an arbitrary particle shape. This is indeed done for the condition $QR < < 1$ and is known as the **Guinier approximation**[3]. For the condition $QR < < 1$, the particle shape factor can be expanded in a power series, yielding:

---

[3] André Guinier (1911–2000).

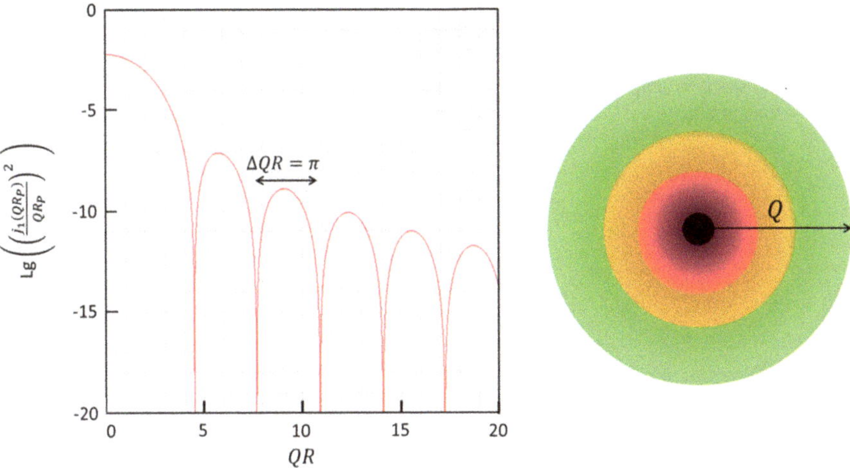

Fig. 9.5 The left panel shows calculated small-angle scattering intensity from spherical particles of a constant radius R. The semi-logarithmic plot shows characteristic oscillations with a period $\Delta QR = \pi$. Right panel: For an isotropic medium, the SAS pattern recorded by an area detector exhibits radial Q dependence and azimuthal symmetry in the detector plane

$$|P(\boldsymbol{Q})|^2 = \left|\frac{1}{V_p}\int e^{i\boldsymbol{Q}\cdot\boldsymbol{R}}dV_p\right|^2 = \exp\left[-\frac{1}{3}(QR_G)^2\right] \tag{9.14}$$

Here, R_G is the **radius of gyration**. The derivation of Eq. (9.13) is provided in the Math Box below and the solutions to Exercise 9.2.

The squared radius of gyration is defined as the squared average radial distance $(r_i)^2$ of any point within a body from its center of mass:

$$R_G^2 = \frac{1}{N}\sum_{i=1}^{N}(r_i)^2. \tag{9.15}$$

The definition of the radius of gyration is reminiscent of the definition of the moment of inertia. This definition holds for a particle with constant density inside its contour. If the density is not equally distributed, we have to take into account the local density and take the volume integral over the particle:

$$R_G^2 = \frac{1}{M}\int \rho(\boldsymbol{r})r^2 dV_p \tag{9.16}$$

For homogeneous spherical particles with radius R_p, the radius of gyration, derived in the Math Box, is:

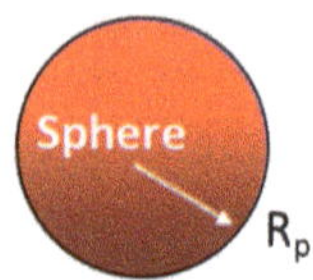
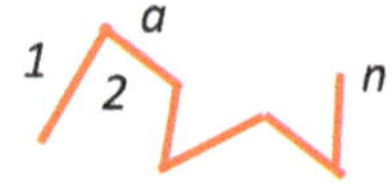

Fig. 9.6 Radius of gyration for spheres and a polymer melt. R_p is the particle radius and a is the segment length of a polymer chain

$$R^2_{G,sphere} = \frac{3}{5} R^2_p \tag{9.17}$$

For a polymer chain in a melt with n segments of length a and a contour length na, the radius of gyration can be calculated as:

$$R^2_{G,polymer} = \frac{na^2}{6} \tag{9.18}$$

Both types of particles are shown schematically in Fig. 9.6. Summarizing, in the Guinier limit, the small-angle intensity is:

$$\lim_{QR \ll 1} I(Q) = \left(\frac{a}{r}\right)^2 \frac{V_p^2 N_p}{N} (\delta\rho)^2 \exp\left[-\frac{1}{3}(QR_G)^2\right] \tag{9.19}$$

and therefore:

$$ln\left(\lim_{QR \ll 1} I(Q)\right) \sim -\frac{1}{3}(QR_G)^2. \tag{9.20}$$

Hence, plotting the logarithm of the intensity versus the squared scattering vector Q^2 will yield from the slope directly the squared radius of gyration R^2_G. An example of a *Guinier plot* is shown in Fig. 9.7. Any higher intensity in this Guinier regime may indicate that the solution is not homogeneous and/or that particles are within the scattering volume larger than those anticipated. To utilize the *Guinier regime*, very good data at small Q-values are mandatory, which requires very good collimation and shielding of the primary beam intensity close to the detector.

Math Box: Expansion in a Power Series In general, an exponential function can be represented as a power series:

$$e^x = 1 + x + \frac{x^2}{2!} + \frac{x^3}{3!} + \frac{x^4}{4!} + \frac{x^5}{5!} \cdots = \frac{x^n}{n!},$$

Setting $x = i\phi$, we obtain:

(continued)

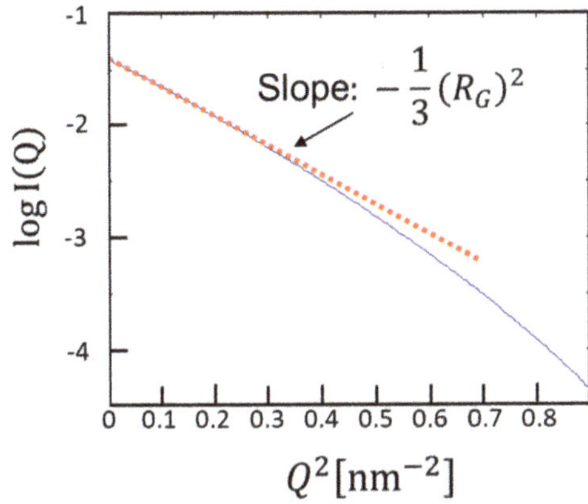

Fig. 9.7 Guinier plot for the determination of the radius of gyration R_G^2. The blue line is the exact expression according to the Eq. (9.13). The red dotted line shows the slope in the Guinier approximation

$$e^{i\phi} = 1 + i\phi - \frac{\phi^2}{2!} - i\frac{\phi^3}{3!} + \frac{\phi^4}{4!} + i\frac{\phi^5}{5!}\ldots$$

Now, we represent the particle shape factor as a power series and take the integral of the individual terms by setting $\phi = \boldsymbol{Q} \cdot \boldsymbol{R}$:

$$P(\boldsymbol{Q}) = \frac{1}{V_p} \int e^{i\boldsymbol{Q}\cdot\boldsymbol{R}} dV_p = \frac{1}{V_p}$$

$$\times \left[V_p + i\boldsymbol{Q} \cdot \int \boldsymbol{R}\, dV_p - \frac{1}{2} \int (\boldsymbol{Q}\cdot\boldsymbol{R})^2 dV_p + \ldots \right].$$

Further calculations are part of Exercise 9.2 and the respective solution.

The Guinier region holds for $QR < < 1$. In this regime, the shape-independent radius of gyration can be determined.

9.4 Porod Approximation

The Porod approximation[4] refers to SAS data taken in the limit $QR \gg 1$. In this regime, $QR \approx \pi$, $\sin(QR) \cong 0$, $\cos(QR) \cong -1$, and the **shape factor** can be approximated by:

---

[4] Günther Porod (1919–1984).

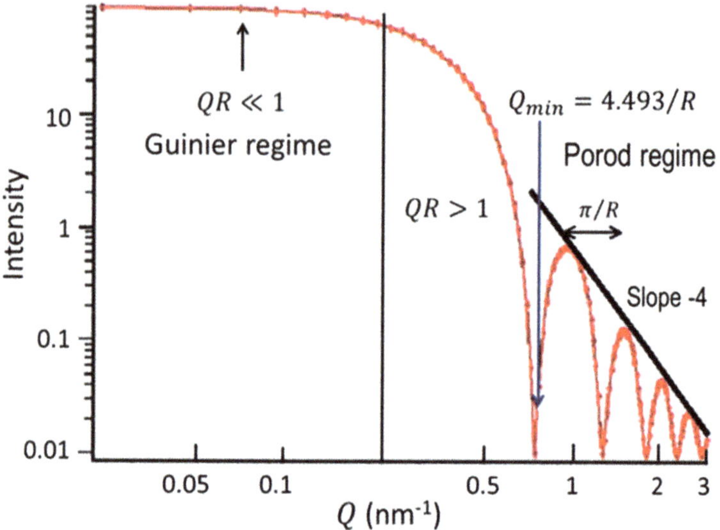

Fig. 9.8 Small-angle scattering from spherical nanoparticles covering the Guinier regime and the Porod regime. Note the double logarithmic plot

$$P(Q) = \frac{1}{(QR)^2} \tag{9.21}$$

Therefore, the scattering intensity becomes:

$$\lim_{QR \gg 1} I(Q) = \left(\frac{a}{r}\right)^2 \frac{V_p^2 N_p}{N} (\delta\rho)^2 \frac{1}{(QR)^4} \sim \frac{1}{Q^4} \tag{9.22}$$

Thus, in the Porod limit, the overall intensity drops off with the inverse scattering vector to the fourth power: $1/Q^4$. This is shown schematically in Fig. 9.8 by a double logarithmic plot of the scattering intensity versus the scattering vector. This graph also indicates the two different scattering regimes: Guinier and Porod. Furthermore, assuming spherical particles, the first oscillation minimum occurs at $Q_{min}R = 4.493$, from which the particle radius can be determined. The oscillation period provides another evaluation of the particle size in addition to the radius of gyration evaluated in the Guinier limit. Note that the $1/Q^4$ dependence is the same as we will encounter for the reflectivity of X-rays or neutrons from flat surfaces, which we will discuss in Sect. 10.4.4.

The Porod regime holds for $QR > > 1$. Here, the radius of spherical particles can be determined on a smaller length scale than is possible in the Guinier regime.

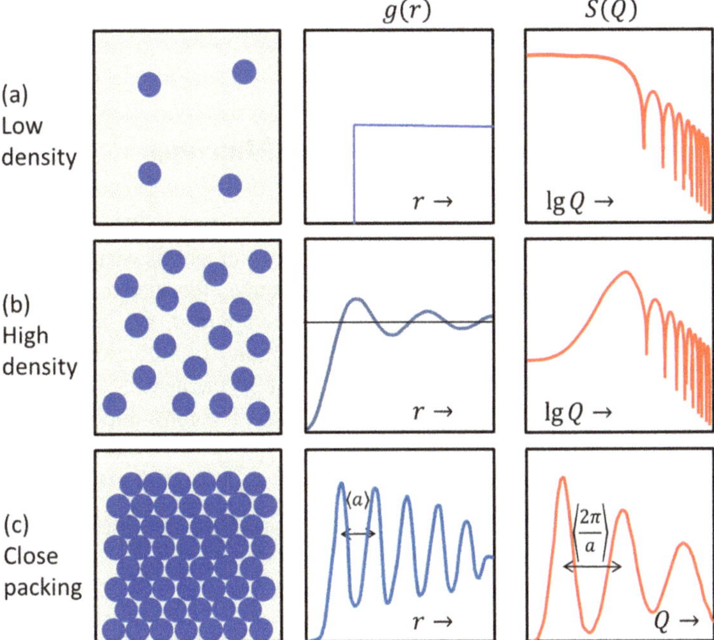

Fig. 9.9 Overview of different particle densities in a solution from low to high, depicted in the left column, causing different pair correlations $g(r)$, shown in the middle column, and their respective scattering functions $S(Q)$ in the right column

Summarizing the previous discussion on SAS concerning small particles and polymer chains, we can distinguish between three different regimes of particle correlation. The first regime (a) in Fig. 9.9 shows the dilute case, where the correlation between particles in a solution is negligible. The only remaining correlation is the hardcore repulsion between otherwise neutral and noninteracting particles. The scattering function reflects the Guinier limit for $QR \ll 1$ as well as the Porod limit for $QR > 1$ with the typical oscillations. For a distribution of particle sizes and shapes, the oscillations can be blurred, but the overall slope remains proportional to Q^{-4}. In the second regime (b), the particle density is significantly higher, and pair correlations are important. The pair correlation function oscillates around the average density ρ_0 and approaches the limit $\lim_{r \to \infty} g(r) = \rho_0$. The scattering function $S(Q)$ is sensitive to the enhanced correlation, especially at low Q values, where the SAS intensity drops with increasing particle–particle correlation. The third region (c) concerns the highest particle density, resulting in a close-packed structure. Typically, the particle packing is such that there is a pronounced correlation with an average lattice parameter $\langle a \rangle$, without establishing true translational long-range order. Therefore, the quasi-Bragg peaks due to this close packing exhibit a strong Debye-Waller factor and peak broadening effects resulting from small domain sizes and short-range order correlations. All this is schematically depicted in panel (c) of Fig. 9.9.

9.5 Contrast Variation and Contrast Matching

SANS has a decisive advantage over SAXS in the analysis of soft matter materials in aqueous solutions. First, X-ray scattering from light elements is weak, and contrast is difficult to obtain for elements that are distinct by only a few electrons. Furthermore, soft X-rays are strongly absorbed by water. In contrast, the neutron coherent scattering lengths for elements most frequently found in biological materials and polymers are big and vary strongly from element to element: ^{12}C (6.64 fm), ^{14}N (9.36), ^{16}O (5.8 fm), ^{31}P (5.13 fm), and ^{32}S (2.85 fm). But most important is the huge difference between the isotopes H (-3.74 fm) and D ($+6.67$ fm), which allows contrast variation over a large range of scattering length densities (Glinka n.d.). The scattering length density (SLD) is defined as:

$$\text{SLD} = \frac{N}{V} \, b_{coh} \tag{9.23}$$

Specifically, for neutrons:

$$SLD_{neutron} = \frac{\sum_i b_i}{V} = n_{nuc} \langle b_{coh} \rangle. \tag{9.24}$$

The sum is taken over all nuclei with scattering length b_i, or alternatively, the average scattering length $<b_{coh}>$ is multiplied by the number density of the nuclei n_{nuc}, which is equivalent to the atomic number density n_A. For X-rays, we have a similar expression when we replace b_i by $r_0 f(Z)$:

$$SLD_{x-ray} = \frac{r_0 \sum_i f_i(Z)}{V} \tag{9.25}$$

Here, r_0 is the classical electron radius, and $f_i(Z) \cong Z$ is the atomic form factor in small-angle approximation, which depends on the atomic number Z. Without taking dispersion effects into account, the X-ray SLD approximates to $SLD_{x-ray} \cong r_0 n_e$, where n_e is the electron number density $n_e = n_A Z$.

Consider a spherical particle consisting of a core and a shell dispersed in an aqueous solution (see Fig. 9.10). Analyzing the SANS results of these particles can be difficult because two different radii overlap, and their scattering pattern is difficult to disentangle. However, in a mixture of H_2O and D_2O, the SLD density of the solvent may be continuously tuned from slightly negative to positive values, as seen in the right panel of Fig. 9.10. The SLD of the solvent depends on the concentration:

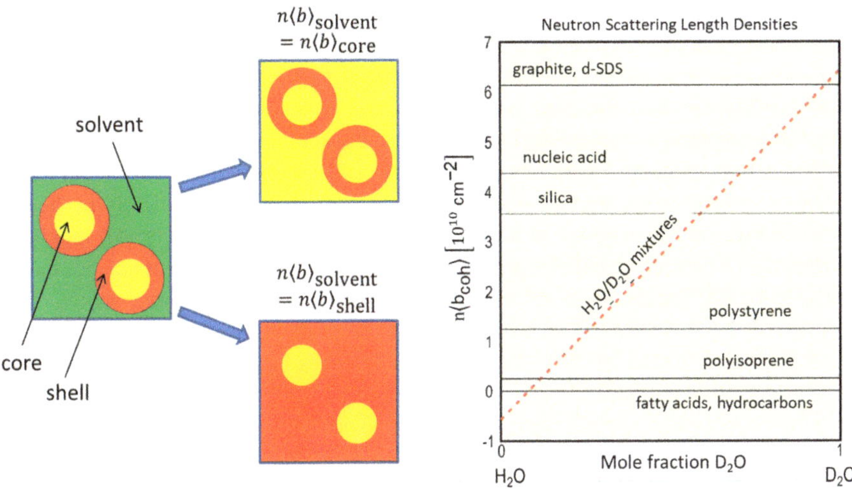

Fig. 9.10 Concept of contrast variation for neutron scattering on particles, consisting of a core and a shell. By contrast variation, the scattering length density of the solvent can be tuned to match either the core or the shell. The neutron scattering length density variation as a function of the mole fraction D_2O is shown in the right panel (Graphics adapted from Glinka (n.d.))

$$SLD^{solvent}_{neutron} = n(xb_{H2O} + (1-x)b_{D2O}) \tag{9.26}$$

where $x = H_2O/(H_2O + D_2O)$. By varying the solvent concentration x, it is possible to match the SLD either to the scattering length density of the shell or the core. If matched to the shell, only the core generates contrast concerning the solvent. Vice versa, if the SLD is matched to the core, only the shell provides contrast. This is a powerful method, but it requires a deuteration laboratory. Furthermore, with increasing H_2O concentration, the incoherent scattering due to hydrogen will also increase simultaneously. Since the incoherent scattering from hydrogen is rather strong and may lead to multiple neutron scattering in the sample that violates the Born approximation, care has to be taken to keep the sample volume sufficiently small when using contrast-matching methods.

9.6 Ferrofluids and Magnetic Nanoparticles

Another important area of SANS activities is magnetism. SANS applications in magnetism are diverse. They concern investigations of magnetic domain structures and sizes, precipitates in magnetic alloys, magnetic nanoparticles in ferrofluids, magnetic colloids, and magnetic glasses. Overviews of these areas are provided in Mühlbauer et al. (2019), Wiedemann (2006) and Vivas et al. (2017). We will discuss here only one aspect of magnetic SANS, representative of other topics, and we will restrict the discussion to SANS with unpolarized neutrons.

Ferrofluids are colloidal liquids that contain nanoscale ferromagnetic particles suspended in organic solvents or water. Each magnetic nanoparticle (NP) is in a single-domain, ferromagnetic state. But the mean magnetization vector varies randomly from one nanoparticle to the next, like in a paramagnet. Therefore, the ferrofluid suspension is classified as superparamagnetic rather than ferromagnetic.

When ferromagnets are reduced to the size of nanoparticles, each nanoparticle contains a single ferromagnetic domain, which is indicated schematically in the left panel of Fig. 9.11 by a series of parallel arrows within a nanoparticle. All magnetic moments in a nanoparticle act together as one giant magnetic moment m_{macro}, called the *macrospin*, indicated in the right panel of Fig. 9.11 by a large arrow. This macrospin is surrounded by a magnetic dipole field. Macrospins can only form in the ferromagnetic state of NPs, that is, at temperatures below the Curie temperature of the respective material.

When superparamagnetic nanoparticles come close to each other, they couple via magnetic dipole–dipole interaction and try to align in a chain-like manner. The agglomeration can be suppressed by sufficiently thick shells surrounding each NP. The shells keep them at a distance where the rapidly decaying magnetic dipole–dipole interaction is effectively reduced.

Next, we consider an ensemble of mNPs, each macrospin having a total magnetic moment m_{macro}. In a zero external magnetic field, the macrospins are randomly oriented, resulting in zero magnetization at remanence; see Fig. 9.12. When turning on a magnetic field, the macrospins respond to the field and partially orient in the field direction, yielding a finite magnetization ($M(T,H) = \chi_m(T)H > 0$). The temperature and field dependence of the magnetization of an ensemble of mNPs can be described by the Langevin function $L(x)$ for noninteracting paramagnetic classical spins:

$$M(T,H) = \frac{N}{V} m_{macro} L(x) \tag{9.27}$$

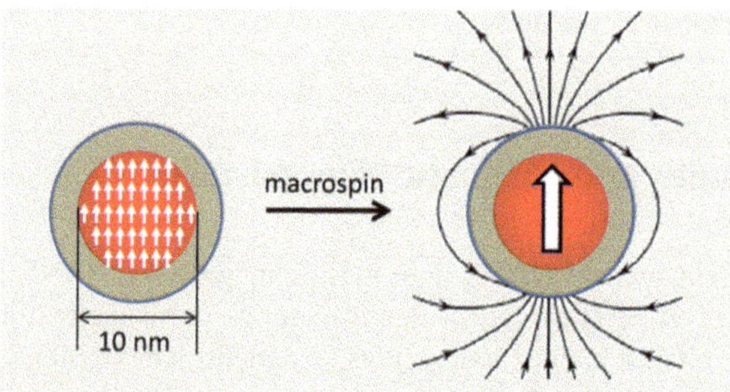

Fig. 9.11 Superparamagnetic nanoparticles consist of a single-domain ferromagnetic core surrounded by an organic shell for oxidation protection and for keeping the particles apart to hinder agglomeration. The atomic magnetic moments are combined in a macrospin, generating magnetic dipole fields

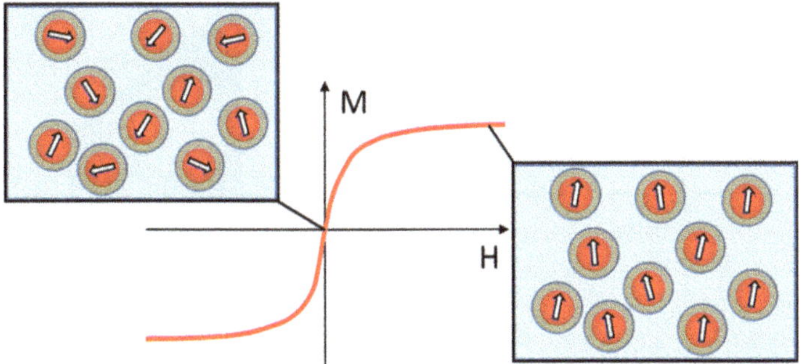

Fig. 9.12 At remanence (H = 0), the magnetization of an ensemble of magnetic nanoparticles is zero (M = 0). Macroscopic magnetization is achieved by applying a magnetic field that aligns the macrospins, similar to spins in a paramagnet. Full alignment (saturation) is never reached due to thermal fluctuations

with

$$L(x) = coth(x) - \frac{1}{x} \tag{9.28}$$

where

$$x = \mu_0 \frac{m_{macro} H}{k_B T} \tag{9.29}$$

is the ratio of the Zeeman energy $\mu_0 m_{macro} H$ to the thermal energy $k_B T$, and μ_0 is the magnetic permeability of the vacuum. $M(T, H)$ is schematicaly plotted in Fig. 9.12. Because of the resemblance to a classical paramagnetic system, the magnetism of nanoparticles is called *superparamagnetism*. However, it is the magnetism of tiny ferromagnetic beads, randomly distributed in the hosting solution that yields a paramagnetic response when exposed to a magnetic field.

Returning to SANS and recalling that for unpolarized neutron scattering, the cross section (or intensity) can be expressed as the sum of nuclear scattering intensity and magnetic scattering intensity (compare Eq. (7.20):

$$I(\mathbf{Q}) = \left(\frac{1}{r}\right)^2 N V_{rc} \left(b_n^2 + (pq)^2\right) \sum_{i,j=0} e^{i\mathbf{Q}\cdot(\mathbf{R}_i - \mathbf{R}_j)}. \tag{9.29}$$

with $q = \sin\alpha$. For the present purpose, this expression can be simplified to:

$$I(\mathbf{Q}) = |A_{\mathrm{nuc}}(\mathbf{Q})|^2 + |B_{\mathrm{mag}}(\mathbf{Q})|^2 \sin^2\alpha \tag{9.30}$$

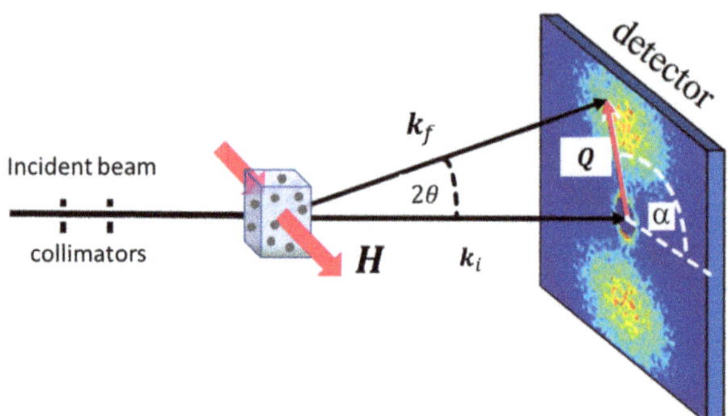

Fig. 9.13 Schematic setup of a SANS instrument for studying magnetic nanoparticles. A magnetic field is applied perpendicular to the incident beam in the horizontal plane. For a scattering vector oriented parallel to the applied field (azimuthal angle $\alpha = 0$, no magnetic intensity is recorded. Adapted from Vivas et al. (2017)

Here, $A_{\mathrm{nuc}}(Q)$ is the nuclear scattering amplitude, and $B_{\mathrm{mag}}(Q)$ is the magnetic scattering amplitude. The term $\sin^2\alpha$ is due to the orientational dependence of magnetic scattering, stating that $I_{\mathrm{mag}}(Q) = 0$ for $Q \parallel M$. In SANS experiments, α is the azimuthal angle, where $\alpha = 0$ is defined by the field direction H, as shown in Fig. 9.13.

If the ferrofluid is at remanence, $M = 0$, and the expected scattering pattern should be isotropic in the azimuthal plane, governed solely by nuclear scattering. Vice versa, if $M = M_{sat}$ by applying a saturating magnetic field, we expect nodes in the scattering pattern at $\alpha = 0$ and $\alpha = 180°$. The magnetic part of the scattering pattern is expected to exhibit a $\sin^2\alpha$ distribution of the intensity, similar to the pattern schematically shown in Fig. 9.13. The difference pattern:

$$I(Q \perp H, \alpha = 90°) - I(Q \parallel H, \alpha = 0°) = \left|B_{\mathrm{mag}}(Q)\right|^2 \tag{9.31}$$

yields the magnetic contribution only, as the nuclear part $I(Q \parallel H, \alpha = 0°)$ cancels. The magnetic contribution contains information on the particle core radius and shell thickness, the particle magnetization, and possible magnetic correlation effects.

In Fig. 9.14 is shown from the work of Wiedemann and coworkers (Wiedenmann et al. 2003) the result of unpolarized SANS on Co nanoparticles suspended in deuterated toluene. The Co particle concentration was 2%, and the particle core radius was determined to be 3.75 nm on average. A saturation magnetic field of 1.1 T was applied in the horizontal plane, normal to k_i. The central black square in the detector map is due to the beam stop of the primary beam. The colored intensity map is coded from blue, corresponding to low intensity, to red, for high intensity. The blue contours show an elliptically distorted pattern. In the horizontal plane, two blue lobes are visible at $Q = \pm\,0.57$ nm^{-1}, which are due to nuclear scattering, as in this region the magnetic intensity vanishes because of the stated condition $Q \parallel H$. The yellow-to-orange and red contours show a hexagonal symmetry with Bragg-

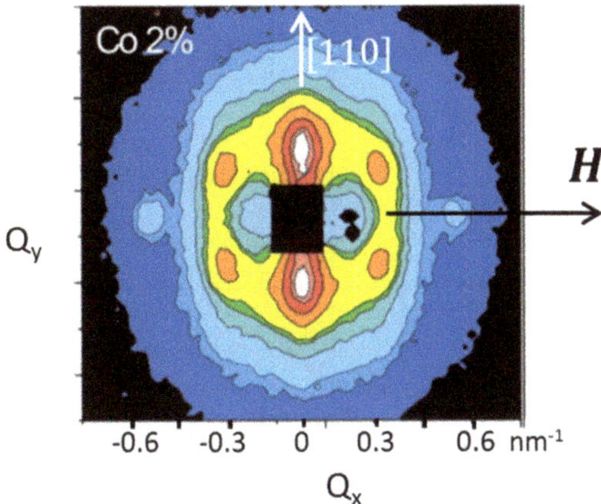

Fig. 9.14 SANS pattern of a 2% Co ferrofluid at room temperature in a horizontal magnetic field of 1.1 T. The pattern shows strong hexagonal correlation effects among the magnetic nanoparticles induced by the applied magnetic field. Adapted with permission from Wiedenmann et al. (2003). Copyright (2025) by the American Physical Society

reflection-like spots in the corners. These spots occur only by application of a high magnetic field. From this, we conclude that a high external magnetic field induces a pseudocrystalline close-packed hexagonal correlation of the cobalt nanoparticles. Using polarized SANS (POLSANS), Wiedemann et al. could show that the magnetization vector of the macrospins in the Co nanoparticle suspension points along the pseudo-hexagonal [110] axis (Wiedenmann et al. 2003).

9.7 Magnetization Reversal

Magnetic thin films play an important role in many areas of science and technology, including data storage and device applications. Although this text is not on the fundamentals of magnetism, we encounter some essential properties of magnetic materials in several chapters. One of these fundamental properties is the magnetic hysteresis. Magnetic hysteresis characterizes magnetic systems and distinguishes ferromagnets from antiferromagnets. A typical ferromagnetic hysteresis is shown schematically in Fig. 9.15a. It is characterized by a saturation magnetization in high applied magnetic fields, remanent magnetization that persists after the applied field is removed, and a coercive field, where the macroscopic magnetization changes sign in either an ascending or a descending field.

Thin magnetic films typically exhibit an in-plane magnetization vector, referred to as the in-plane easy-axis magnetic anisotropy. The shape anisotropy favors in-plane magnetization to lower the stray field energy. However, in some transition-metal alloys with Pd and Pt, the magnetocrystalline anisotropy exceeds the shape anisotropy, resulting in an out-of-plane, or perpendicular, anisotropy. At remanence, the domain structure of films with perpendicular anisotropy differs greatly from that of films with in-plane anisotropy. The two respective domain structures are compared in Fig. 9.15b and d. The perpendicular domains appear as worm-like, meandering stripes separated by narrow domain walls. In contrast,

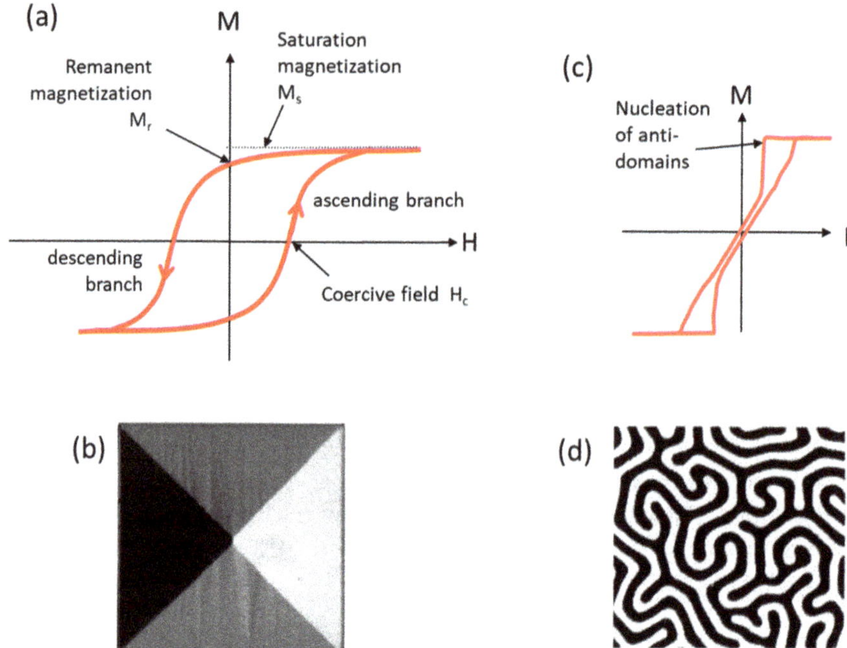

Fig. 9.15 (**a**) Typical ferromagnetic hysteresis, characterized by saturation magnetization, remanent magnetization, and coercive field. (**b**) Typical domain structure, here a Landau pattern, of a thin magnetic film with in-plane magnetization. (**c**) Magnetic hysteresis of a ferromagnetic film with out-of-plane magnetization. Here, the nucleation of opposite domains sets in before field reversal. (**d**) Typical domain pattern of a magnetic film with perpendicular anisotropy at remanence

in-plane magnetized films typically have only a few domains and straight domain walls between them. Magnetization reversal of perpendicularly magnetized films begins with the nucleation of opposite domains even before the field is reversed in a descending magnetic field (see Fig. 9.15c).

The magnetization reversal of a thin CoPt film with perpendicular magnetic anisotropy can be studied using resonant magnetic small-angle X-ray scattering, where the X-ray photon energy is tuned to the Co L_3 absorption edge at 778 eV. At resonance, the photons excite electrons from $2p_{3/2}$ to the 3d band, which provides the magnetic sensitivity. More details on this resonant transition is presented in Sect. 11.6. The experimental setup is shown in Fig. 9.16, featuring a high-intensity incident beam from an undulator, which is further spatially filtered by a 25 μm pinhole to enhance the transverse coherence length, before hitting the sample in transmission geometry. The resonant magnetic SAXS pattern is recorded with a two-dimensional X-ray sensitive CCD detector.

In the magnetic domain state of the sample, the detected intensity map shows a homogeneous ring with a mean radius that is inversely proportional to the mean domain separation $Q_r = 2\pi/L_{domain}$ and some widening of the ring due to domain size fluctuations. There is no Bragg reflection, and therefore, we conclude that there

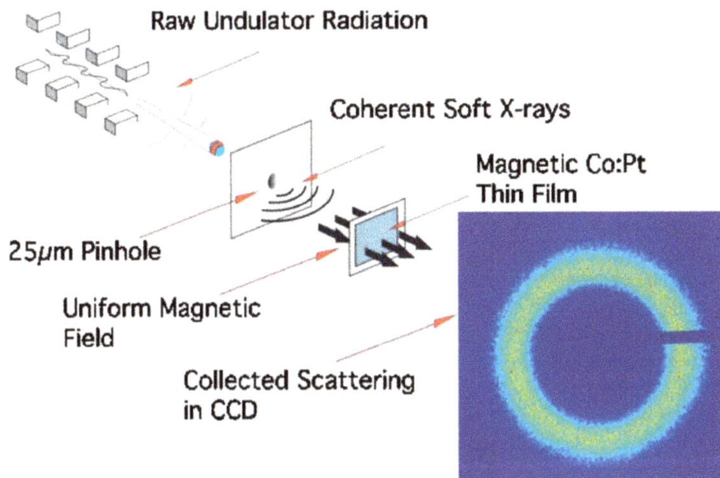

Fig. 9.16 Experimental setup for recording small-angle scattering from perpendicular magnetic domains using undulator radiation tuned to the Co L_3 absorption edge at 778 eV. The raw undulator radiation is spatially filtered by a pinhole to enhance the transverse coherence. The incident beam is parallel to the film normal and to an applied magnetic field. The intensity is recorded with a 2047x2047 pixel CCD detector. The central part is blocked for protection. Reproduced with permission from Pierce et al. (2003) and Pierce et al. (2007). Copyright (2025) by the American Physical Society

is no long-range order, unlike the pattern reproduced in Fig. 9.14. By the application of a magnetic field and without using the pinhole, the magnetization reversal can be followed in an ensemble-averaged manner.

However, when inserting a pinhole to increase the transverse coherence length of the undulator radiation (for a definition of the coherence length, see Sect. 4.2.8.1), it will be seen that the homogeneous ring decomposes into many tiny intensity spots called speckles. The speckles lift the ensemble average, indicating local deviations from the mean. The speckle pattern contains the full unfiltered picture of the sample, and with sufficient intensity and resolution, one could, in general, get a real-space picture of the domain structure by back-transformation. However, for back-transformation to the real space, the phase information is missing, since the modulus of the amplitudes is recorded and not the full complex amplitudes. But by using holographic methods with a reference beam, the real-space picture can indeed be retrieved, as has been shown by Eisebitt and coworkers (Günther et al. 2011).

Returning to the speckled pattern. Speckle patterns are unique for a specific domain structure along the magnetic hysteresis. It is of interest to investigate whether magnetic domains have a memory and whether they always return to the same domain configuration after returning from saturation. The saturation magnetization wipes out all previous domain configurations. As shown schematically in Fig. 9.17, each point on the hysteresis corresponds to a different domain density and domain structure and therefore to a different speckle pattern. In saturation, there is only a single domain, yielding either black or white contrast depending on the

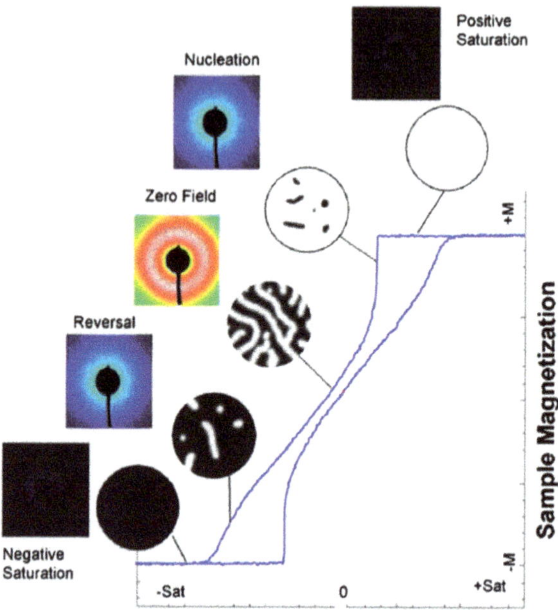

Fig. 9.17 Correspondence between the magnetic hysteresis loop of a sample with perpendicular anisotropy, magnetic domain pattern as observed in magnetically sensitive microscope, and speckle pattern recorded with a partially coherent and resonant X-ray beam tuned to the Co L_3 absorption edge. Characteristic points along the hysteresis loop are highlighted, including positive and negative saturation, domain nucleation, remanence, and reversal field. Reproduced with permission from Pierce et al. (2007). Copyright (2025) by the American Physical Society

magnetization orientation. First domains occur after nucleation, giving rise to weak diffuse scattering. The highest domain density develops at remanence, yielding a clearly defined SAXS ring of maximum intensity. With increasing field reversal, the domain density decreases again until saturation in the opposite magnetization direction is reached.

The speckle pattern that we discuss here is similar to the one we have encountered in Sect. 7.7 for the Ho τ-peak. In the latter case, the speckle pattern is dynamic due to thermal fluctuations. Here, the pattern is static, but it depends on the magnetic field and the magnetization reversal history.

Different domain structures in the CoPt film may result in the same magnetization $M(H)$ on the hysteresis and the same ensemble-averaged magnetization density on the annular SAXS ring, but they will be distinguishable by their different speckle patterns. Therefore, it is important to analyze distinct speckle patterns having different magnetization histories with respect to their autocorrelation and cross-correlation.

The autocorrelation function of pattern A, $F_{AC}(A)$, compares the speckle pattern intensity $A(Q_x, Q_y)$ at positions (Q_x, Q_y) with itself, after executing some specific action in between to be specified below, and another pattern $B(Q_x, Q_y)$ again with itself. The cross-correlation function $F_{CC}(A, B)$ compares both patterns $A(Q_x, Q_y)$ and $B(Q_x, Q_y)$ and normalizes the result with the autocorrelation functions to obtain the generalized correlation function:

$$\rho(A, B) = \frac{\sum (F_{CC}(A, B) - b_{AB})}{[\sum (F_{AC}(A) - b_A) \sum (F_{AC}(B) - b_B)]^{1/2}} \qquad (9.32)$$

Here, b_{AB}, b_A, b_B are the background intensities of the respective patterns, and the sum goes over all pixels in the detector, where the speckles are 2–3 pixels wide. The generalized correlation function takes the value 1 for identical speckle patterns (perfect correlation) and zero for uncorrelated patterns.

Comparing two speckle patterns, recorded at the same point on the hysteresis, one after returning from saturation via a minor loop and the other by taking a full round along the major loop, it turns out that the cross-correlation of the two speckle patterns is zero, i.e., there is no microscopic return point memory (RPM) when taking different paths to return to the same point on the hysteresis. Also, when taking the same path, after the first loop and subsequent loops, the RPM is nearly zero. Although the macroscopic magnetization is the same, the microscopic pattern varies from loop to loop. For the full analysis of this extensive study, we refer to the original publication in (Pierce et al. 2003, 2007).

9.8 Instrumentation

Small-angle scattering is a very powerful method that can be applied to all kinds of matter for analyzing structures, conformations, and phase transitions on the nano- to mesoscopic length scale. For noncrystalline matter, it is as informative as powder diffraction is for crystalline matter. The applications of SAS overlap, in part, with those studied by electron microscopy. However, SAS does not require any special sample preparation, and the scattering results are representative of an ensemble average. SAS can be combined with any kind of sample environment, including high-pressure conditions, extreme temperatures, magnetic or electrical fields, and time-dependent *in-operando* conditions. The prerequisite is a well-constructed small-angle scattering instrument. The principal features of SAXS and SANS instruments are similar. In both cases, angular resolution at very small scattering angles is more important than energy resolution.

Figure 9.18 shows exemplarily the instrument layout of the **SANS instrument** D11 of the Institut Laue-Langevin (ILL) in Grenoble (Günther et al. 2011). A "white" incident beam from the cold neutron source of the ILL arrives from the left side and passes a velocity selector. The velocity selector acts as a "rough" monochromator and guarantees, at the same time, a high neutron transmittance. The wavelength resolution $\Delta\lambda/\lambda$ is only about 10%, but this is sufficient for SANS experiments. Several collimators along the path to the sample reduce the divergence of the incident beam. A selection of different diaphragms assures that the sample is properly exposed to the neutron beam. The sample is positioned on a sample stage that allows fine adjustment in all directions. The sample may also sit inside a chamber providing a special sample environment, such as specific gas pressures,

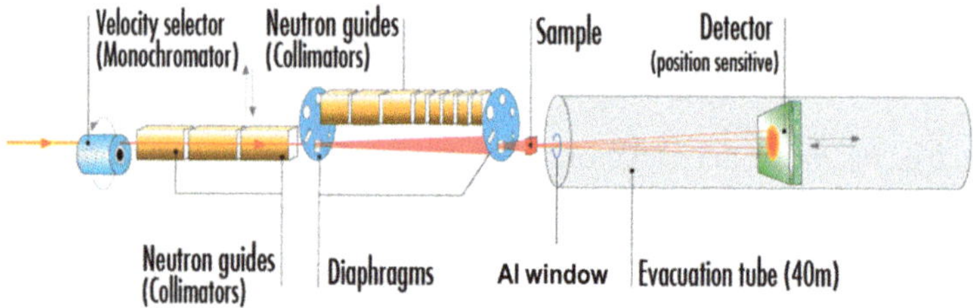

Fig. 9.18 Schematic layout of the SANS instrument D11 at the Institut Laue-Langevin, Grenoble. Reproduced from https://www.ill.eu/fr/users-en/instruments/instruments-list/d11/description/instrument-layout/

temperatures, or magnetic fields. The transmitted and scattered beam enters a long evacuated tube that houses a position-sensitive detector. The beam enters the tube through an aluminum window, which is "transparent" to neutrons and keeps the tank vacuum-tight. Evacuation is necessary to avoid air scattering. The detector inside the long tube (up to 40 m) can be moved back and forth, depending on the resolution requirements. When neutrons fly such a long distance, they actually drop down by gravitational pull, which can be seen on the detector and which needs to be corrected. An effective neutron beam stop in front of the detector is mandatory for protecting the detector and for the detection of weak signals near the primary beam. This is a nontrivial task, as any neutron that is absorbed will generate other types of radiation, which subsequently also need to be eliminated. Instruments for SAXS are designed pretty much along the same principle design features as SANS instruments. Because of the much higher intensity available at synchrotron radiation sources and even at rotating anode X-ray machines in the laboratory, the monochromatization and the collimation of the incident beam can be tightened up, providing higher resolution for studying more extended structures, such as bacteria and larger micelles.

Remarks on SAS

The choice of using SANS or SAXS depends on the sample and the aim of the experiment. For H/D contrast variation, there is no alternative to SANS. Anomalous X-ray scattering provides a bit of contrast variation but is not as powerful as the H/D tunability of neutron cross sections. Thermal neutrons, in contrast to X-ray photons, produce almost no radiation damage, which is an important consideration when studying biological samples. SAXS has an intensity and resolution advantage over SANS and can, in particular, be used for very small samples. Also, resonant conditions can be added, and the transverse coherence of the beam can be adjusted to the needs. In the end, both methods are complementary, and potential users will consider all aspects to make the best use of both.

Summary

1. Small-angle scattering (SAS) is used for probing the structure of materials on the mesoscopic length scale.
2. Typically, in small-angle scattering, the scattering vector Q is very small, and the probing correlation lengths R are quite large, resulting in a product $QR \cong 1$.
3. Materials probed by SAXS or SANS range from grain boundaries, micropores, emulsions, polymers, micelles, polyelectrolytes, viruses, proteins, fibers, precipitates, magnetic domains, skyrmions, to flux lines in superconductors.
4. SAS probes structures in the direction of the scattering vector Q perpendicular to the incident beam: $Q \perp k_i$.
5. SAS is only possible if there is sufficient contrast between particles and the solution or embedding matrix.
6. In general, SAS is proportional to the Fourier transform of the particle shape, which is referred to as the particle shape factor $P(Q)$.
7. The particle shape factor is easily obtained for simple geometric shapes such as spheres, rods, sheets, and for polymer chains of different number of segments.
8. The radius of gyration R_G relates the center of mass of an ensemble of spherical particles to their particle radius R_p.
9. SAS experiments distinguish between two different regimes: Guinier and Porod.
10. The Guinier regime holds for $QR \ll 1$.
11. In the Guinier regime, the logarithms of the SAS intensity plotted against Q^2 drop with $(QR_G)^2/3$, allowing the determination of the radius of gyration R_G.
12. In the Porod regime, $QR > 1$.
13. In the Porod regime, the SAS intensity scales with Q^{-4}.
14. Contrast variation between the solution and solvents is possible, in particular when using SANS. In a aqueous solution, the neutron scattering length density can be tuned from negative to positive values by varying the H_2O/D_2O ratio.
15. SANS is very useful for studying magnetic alloys, domains, and magnetic nanoparticles.
16. Magnetic nanoparticles are characterized by their superparamagnetic behavior in a magnetic field, where each particle contains a macrospin.
17. Due to the angular dependence of magnetic neutron scattering, in the azimuthal plane of SANS, the intensity distribution shows a $\sin^2\theta$ behavior.
18. Magnetic hysteresis with perpendicular anisotropy can be investigated element-specifically using resonant magnetic X-ray scattering.
19. Using coherent synchrotron radiation, the field dependence of speckle patterns can be investigated during magnetization reversal.
20. For SAS instruments, three design features are essential: 1. Very high angular resolution; 2. Relaxed wavelength resolution for increased intensity; 3. Radiation protection of the detector in the forward direction.

Questions

(Note that more than one answer may be correct)

1. **What is the typical Q regime covered by small-angle scattering?**

    (a) 0.001 nm^{-1}–0.05 nm^{-1};
    (b) 0.05 nm^{-1} –5 nm^{-1};
    (c) 5 nm^{-1}–50 nm^{-1}.

2. **When is scattering called small-angle scattering?**

    (a) Whenever the object to be scattered from is much smaller than the probing wavelength.
    (b) Whenever the object to be scattered from is much bigger than the probing wavelength.
    (c) Whenever the object size is comparable to the probing wavelength.

3. **What is the contrast mechanism in small-angle scattering?**

    (a) Contrast is achieved by varying the volume density of particles in solution.
    (b) Small-angle scattering requires local density variations with form factors or scattering length densities higher or lower than the surrounding.
    (c) The highest contrast is achieved when particles and solution have the same scattering length density.

4. **What distinguishes the Guinier region from the Porod region?**

    (a) The Guinier region is at QR<<1; the Porod region is at QR>>1.
    (b) The Guinier region applies to polymers; the Porod region applies to magnetic materials.
    (c) The Guinier region and Porod regions cannot be distinguished.
    (d) The Guinier region scales with $\exp\left[-\frac{1}{3}(QR_G)^2\right]$, and the Porod region scales with Q^{-4}.

5. **What special contrast mechanism can be used in the case of SANS?**

    (a) Isotope mixing
    (b) Inelastic scattering
    (c) H-D contrast
    (d) Neutron absorption

6. **What are the main features of a small-angle instrument?**

    (a) Proton source, linear accelerator, and target.
    (b) Neutron source, velocity selector, neutron guides, diaphragms, target, and 2D area detector.
    (c) X-ray source, monochromator, slits, target, goniometer, and 2D detector

7. **What are magnetic nanoparticles?**

(a) Magnetic nanoparticles are little spheres containing a magnetic fluid.
(b) Magnetic nanoparticles are tiny, single-domain ferromagnets.
(c) Magnetic nanoparticles are incommensurate antiferromagnets on a nanoscale.

Exercises

Grades of difficulty: E = easy, M = medium, A = advanced

M 9.1: Particle Shape Factor.

Show that the particle shape factor for a spherical particle of radius R_p has the form:

$$P(Q) = \frac{1}{V_p} \int e^{i\mathbf{Q} \cdot \mathbf{R}} dV_p = \frac{3}{QR} J_1\left(QR_p\right).$$

Hint: The volume integral in spherical coordinates is given by:

$$\int dV_p = \int_{-\pi}^{+\pi} d\theta \int_{0}^{\pi} \sin\phi\, d\phi \int_{0}^{R_p} R^2 dR.$$

For integration of the angular part, use the substitution:

$$\cos\phi = x \text{ and } \sin\phi\, d\phi = -dx.$$

A 9.2 Particle Shape Factor in the Guinier Approximation

Show that for the condition $QR < <1$, the particle shape factor can be expanded using the approximation for the exponential function, which then yields the expression:

$$|P(Q)|^2 = \left|\frac{1}{V_p} \int e^{i\mathbf{Q} \cdot \mathbf{R}} dV_p\right|^2 = \exp\left[-\frac{1}{3}(QR_G)^2\right]$$

M 9.3 Gaussian Smearing of Dips

Perfectly monodisperse particles do not exist. Each particle has a slightly different radius. Monodispersity is usually expressed in terms of a 1σ-deviation from the mean $\langle R \rangle$, assuming a Gaussian distribution. If the ratio $\sigma/\langle R \rangle = 0.1$, how would this affect the sharp dips observed in Fig. 9.6? Proceed as follows:

(a) Plot the scattering curve $I(Q) = const.\ (J_1(QR_p)/(QR_p))^2$ for different R_p values with a 10% variation. Sum up the scattering curves and note how the behavior of the dips changes with changing R_p. For plotting the respective functions, you may use an online function plotter such as www.mathe-fa.de

(b) Analyze the origin of the dips. What is the origin of the dips in the mathematical description?

(c) Use the results of your answer in b. for estimating the relative spread of dQ/Q as a function of the spread in dR. What conclusion can you draw?

References

Ch. Glinka: Fundamentals of Small Angle Neutron Scattering (Power Point) https://www.ncnr.nist.gov/summerschool/ss04/information/SANS_tutorial.pdf

C.M. Günther, B. Pfau, R. Mitzner, B. Siemer, S. Roling, H. Zacharias, O. Kutz, I. Rudolph, D. Schondelmaier, R. Treusch, S. Eisebitt, Sequential femtosecond X-ray imaging. Nat. Photon. **5**, 99–102 (2011)

S. Mühlbauer, D. Honecker, É.A. Périgo, F. Bergner, S. Disch, A. Heinemann, S. Erokhin, D. Berkov, C. Leighton, M.R. Eskildsen, A. Michels, Magnetic small-angle neutron scattering. Rev. Mod. Phys. **91**, 015004 (2019)

M.S. Pierce, R.G. Moore, L.B. Sorensen, S.D. Kevan, O. Hellwig, E.E. Fullerton, J.B. Kortright, Quasistatic X-ray speckle metrology of microscopic magnetic return-point memory. Phys. Rev. Lett. **90**, 175502 (2003)

M.S. Pierce, C.R. Buechler, L.B. Sorensen, S.D. Kevan, E.A. Jagla, J.M. Deutsch, T. Mai, O. Narayan, J.E. Davies, K. Liu, G.T. Zimanyi, H.G. Katzgraber, O. Hellwig, E.E. Fullerton, P. Fischer, J.B. Kortright, Disorder-induced magnetic memory: Experiments and theories. Phys. Rev. B **75**, 144406 (2007)

L.G. Vivas, R. Yanes, A. Michels, Small-angle neutron scattering modeling of spin disorder in nanoparticles. Sci. Rep. **7**, 13060 (2017)

A. Wiedemann, Chap. 10: Small angle neutron scattering investigations of magnetic nanostructures, in *Neutron Scattering from Magnetic Materials*, ed. by P. Chatterji, (Elsevier Publisher, Amsterdam, Boston, Tokyo, 2006)

A. Wiedenmann, A. Hoell, M. Kammel, P. Boesecke, Field-induced pseudocrystalline ordering in concentrated ferrofluids. Phys. Rev. E **68**, 031203 (2003)

Further Reading and Useful Web Pages

J.S. Higgins, H. Benoit, *Polymers and neutron scattering* (Oxford University Press: Clarendon Press, 1994) Julia S. Higgins and Henri C. Benoit. Polymers and Neutron Scattering. Clarendon Press, 1997

G. Kostorz, Small-angle scattering and its applications to materials science, in *Treatise on Materials Science and Technology*, Neutron Scattering, ed. by G. Kostorz, vol. 15, (Academic Press, 1979)

W.H. De Jeu, *Basic X-Ray Scattering for Soft Matter* (Oxford University Press, 2016)

N. Stribeck, *X-Ray Scattering of Soft Matter* (Springer Verlag, 2010)

G.R. Strobl, *The Physics of Polymers, Concepts for Understanding Their Structures and Behavior, Textbook*, 2nd edn. (Springer, Berlin Heidelberg New-York, 1997)

SANS data can be simulated for various particle shapes using the programs available at: www.ncnr.nist.gov/resources/simulator.html

Animation of SANS D22 at the ILL: https://www.ill.eu/users/instruments/instruments-list/d22/how-it-works/sans-principle-small-angle-neutron-scattering-principles/

Chapter 10
Thin Films and Multilayers

10.1　General Background

Modern research frequently uses artificially grown materials, especially in the form of thin films. First, the structural quality and chemical purity are significantly superior to naturally grown materials, and second, the artificial growth of thin films offers a wide variety of material combinations that would otherwise be impossible to achieve. In this chapter, we will not discuss growth methods for thin films and multilayers, but rather focus on their structural analysis using scattering methods.

To obtain relevant structural information about surfaces, thin films, and multilayers, various scattering methods are available, some of which are presented in the following sections of this chapter. Figure 10.1 provides an overview of the different regions in reciprocal space that are commonly explored to obtain information on film thickness, interface roughness, and correlation effects. These different regions are explained further below. They comprise Laue oscillations, crystal truncation rods, Fresnel reflectivity, Kiessig fringes, off-specular diffuse scattering, in-plane surface scattering, and glancing angle small-angle scattering. All of this takes on additional flavor when adding magnetic sensitivity to the aforementioned X-ray and neutron scattering techniques, as we will consider in Chaps. 11 and 12.

Before proceeding, we note that the nomenclature must be changed in all subsequent equations. Previously, Q_c stood for the scattering vector in the crystallographic c-direction, perpendicular to the a, b plane in a Cartesian system. In the following, Q_c will be used exclusively to denote the critical scattering vector for the total external reflection of X-rays and neutrons. Therefore, we change the directional index from c to z, and the normal to the surface will always be the z-direction.

© The Author(s), under exclusive license to Springer Nature Switzerland AG 2026

H. Zabel, *Elements of Elastic Scattering by X-Rays, Neutrons, and Electrons*,

https://doi.org/10.1007/978-3-032-16624-1_10

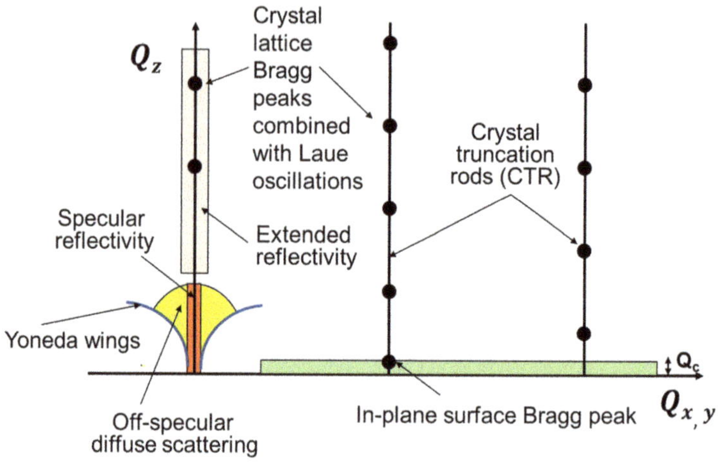

Fig. 10.1 Two-dimensional projection of the reciprocal space covering the regions close to total external reflection of the incident neutron or X-ray beam, the Laue oscillations in the extended reflectivity regime, the crystal truncation rods, and the region of in-plane surface diffraction in the vicinity of the critical scattering vector Q_c

10.2 Thin Crystalline Films (Laue Region)

Laue oscillations were already introduced in Sect. 3.3 in the context of the one-dimensional lattice sum, resulting from a finite number of lattice planes (compare Eq. 3.23 and Fig. 3.5). Assuming that the films are infinitely extended in the xy plane, but contain only a few layers in the perpendicular z-direction, the lattice sum yields the intensity:

$$I(\mathbf{Q}) = \left(\sum_{i=0}^{N-1} e^{i\mathbf{Q}\cdot\mathbf{T}_i} \right)^2 = N_x\delta(\mathbf{Q}_x - \mathbf{G}_h)N_y\delta(\mathbf{Q}_y - \mathbf{G}_k) \times \frac{\sin^2\left(N_z\mathbf{Q}_z\cdot\mathbf{d}_z/2\right)}{\sin^2\left(\mathbf{Q}_z\cdot\mathbf{d}_z/2\right)}, \quad (10.1)$$

where d_z is the interplanar spacing in the z-direction, and N_z is the number of lattice planes. Here, we recognize sharp Bragg reflections in the a, b reciprocal plane and *Laue oscillations* perpendicular to it, as shown in Fig. 3.5. If the number of lattice planes N_z in thin films is not constant but varies locally and statistically, as indicated in Fig. 10.2a, the expression for the intensity is modified accordingly (Stierle et al. 1993):

$$I(\mathbf{Q}_z) \propto \sum_{N_{z,i}=0}^{\infty} \frac{1}{\sqrt{2\pi}\sigma} \exp\left(-\left(\frac{(N_{z,i} - N_0)}{\sqrt{2}\sigma} \right)^2 \right) \frac{\sin^2\left(N_{z,i}\mathbf{Q}_z\cdot\mathbf{d}_z/2\right)}{\sin^2\left(\mathbf{Q}_z\cdot\mathbf{d}_z/2\right)} \quad (10.2)$$

Here, N_0 is the average number of lattice planes, and σ is the width of the Gaussian distribution. Figure 10.2b shows a comparison of Laue oscillations for a

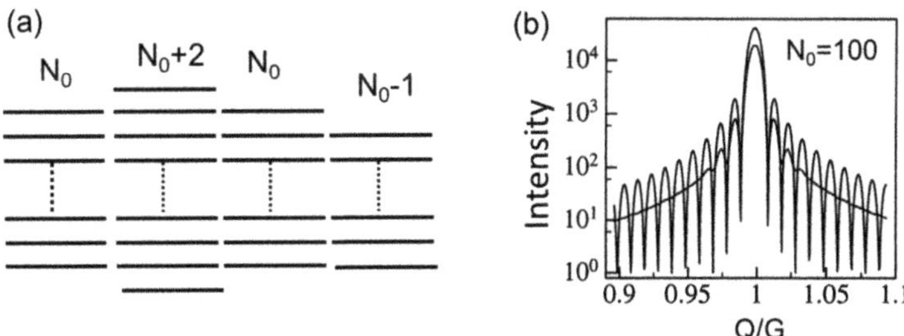

Fig. 10.2 (**a**) Number of lattice planes in a thin film may vary due to growth conditions, (**b**) Calculated intensities for a constant number of $N_0 = 100$ lattice planes, compared to the intensity for a distribution of lattice planes with the mean of $N_0 = 100$ and a variance of 10 planes. Note the logarithmic intensity scale. The oscillations are strongly damped by the distribution. Adapted from Stierle et al. (1993)

Fig. 10.3 Bragg scan of the (0002) peak of cobalt showing typical Laue oscillations for a finite number of coherently scattering lattice planes. The solid line represents a best fit to the data points, yielding an average thickness of 71 planes. Reproduced with permission from Stierle et al. (1993). Copyright (2025) by AIP

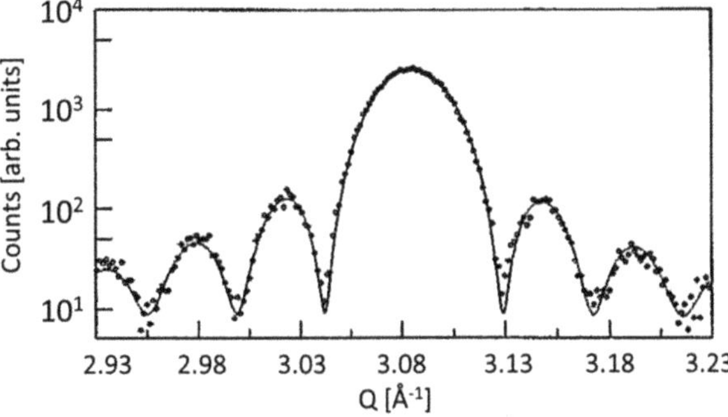

constant number of 100 lattice planes and Laue oscillations of lattice planes with a mean of $N_0 = 100$ planes and a variance of $\sigma = 10$ planes. The distribution leads to a strong damping of the Laue oscillations, which is an important piece of information for characterizing the quality of thin crystalline and epitaxial films.

An example is shown in Fig. 10.3 of an epitaxial Co film grown by molecular beam epitaxy on a sapphire $(Al_2O_3(11\bar{2}0))$ substrate (Stierle et al. 1993; Hellwig and Zabel 2000). The Laue oscillations surround the Co *hcp*(0002) peak and the *fcc*(111) peak, respectively. The solid line is a fit to the data points, yielding a mean number of coherently scattering lattice planes of $N_0 = 71$ with a standard deviation of six lattice planes. According to the main peak position, i.e., the maximum of the Laue oscillations, the interplanar spacing is $d_z=0.2036$ nm. Therefore, the average thickness of coherently grown lattice planes is 71×0.2036 nm $= 14.5$ nm. Reflectivity measurements on the same sample reveal a layer thickness of 14.9 nm. We can therefore conclude that the entire film thickness has a high crystalline perfection of about 97% and, more generally, that the study of Laue oscillations is very well suited to assess the crystalline quality of thin films.

10.3   Crystal Truncation Rods

10.3.1   Basics

The lattice sum introduced in Eq. (3.32) implicitly assumes that crystal lattices are infinitely extended in all three spatial directions. This assumption is generally acceptable but not for the surface from which the beam is reflected. Let us assume that the scattering vector is perpendicular to the surface of a cubic crystal. We recalculate the scattering intensity of a truncated single crystal by taking the sum over a semi-infinite space from the surface to infinity, where w is the atomic layer index in the z-direction. The in-plane lattice sums generate δ-functions at the reciprocal in-plane lattice points G_h and G_k, as considered before in Eq. (10.1). Together, this yields an expression for the scattering intensity of a crystal truncation rod (CTR) similar to Eq. (10.1) but with the number of lattice planes in the z-direction approaching infinity:

$$I_{CTR}\left(Q_{a,b,c}\right) \propto N_x\delta(Q_x - G_h)N_y\delta(Q_y - G_k)\left|\sum_{w=0}^{N_z-1} e^{iwQ_z \cdot d_z}\right|^2$$

$$= N_x\delta(Q_x - G_h)N_y\delta(Q_y - G_k)\lim_{N_z \to \infty}\left(\frac{\sin^2(N_zQ_z \cdot d_z/2)}{\sin^2(Q_z \cdot d_z/2)}\right) \quad (10.3)$$

For a large number of lattice planes N_z, the Laue oscillations—governed by the numerator in Eq. (10.3)—do not play a role. Therefore, we can approximate the numerator by its average value 1/2. Then, the intensity of the crystal truncation rod scattering for $Q_z \cdot d_z \neq 2\pi$ scales with:

$$I_{CTR}\left(Q_z\right) \propto \left|\sum_{w=0}^{\infty} e^{iwQ_z \cdot d_z}\right|^2 = \frac{1}{4\sin^2\left(Q_z \cdot \frac{d_z}{2}\right)}, \quad (10.4)$$

which is, with the help of Math Box II in Chap. 3, equivalent to the expression:

$$I_{CTR}\left(Q_z\right) \propto \left|\sum_{w=0}^{\infty} e^{iwQ_z \cdot d_z}\right|^2 = \frac{1}{2}\left|\frac{1}{1 - e^{iQ_z \cdot d_z}}\right|^2. \quad (10.5)$$

A proof of the last two equations is the subject of Exercise 10.1 and its solutions. We use the latter expression for the intensity when scanning in the normal direction of a flat single-crystal surface instead of a δ-function expression. This emphasizes the importance of the surface through which the beam enters and exits the crystal.

The scattered intensity distribution of a truncated single crystal is shown schematically in Fig. 10.4. As expected, it contains sharp Bragg reflections at the

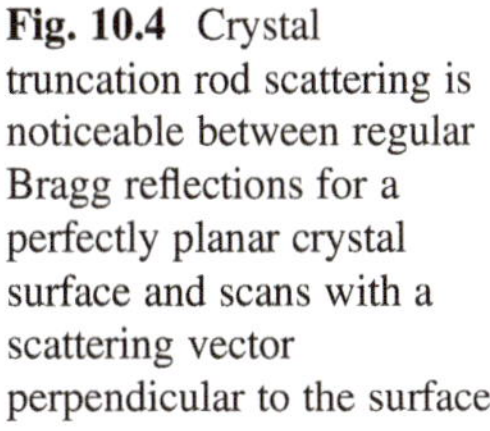

Fig. 10.4 Crystal truncation rod scattering is noticeable between regular Bragg reflections for a perfectly planar crystal surface and scans with a scattering vector perpendicular to the surface

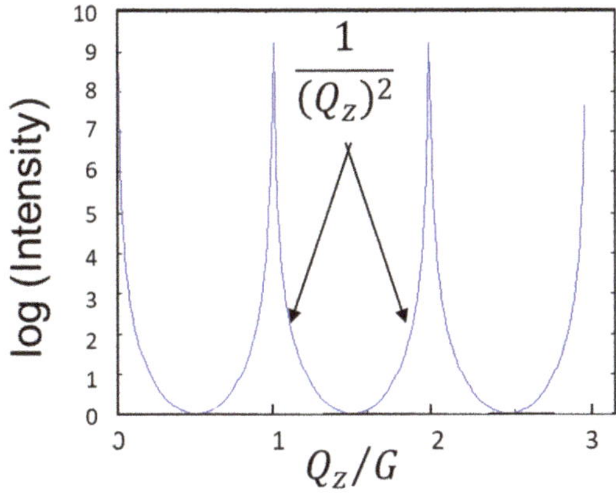

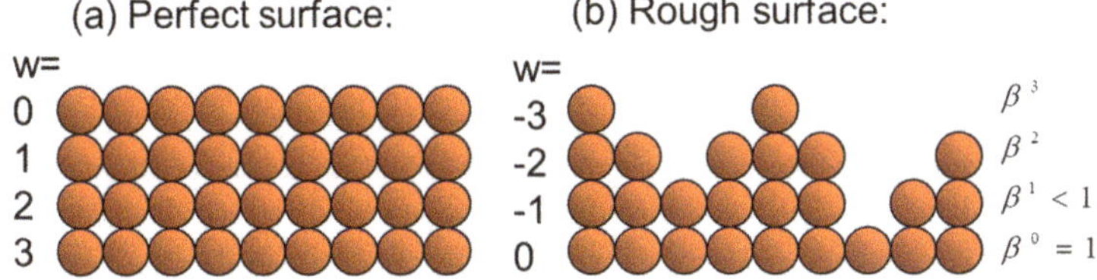

Fig. 10.5 A perfectly smooth surface in panel (**a**) is contrasted to a more realistic scenario of a rough surface in panel (**b**). w is the layer index with a filling parameter $\beta^{|w|}$

reciprocal lattice points. In addition, diffuse scattering can be observed between the Bragg reflections. This diffuse scattering is due to the truncated lattice sum that drops off on either side of a Bragg reflection according to $1/(Q_z)^2$. Therefore, this diffuse intensity is known as **crystal truncation rod** scattering. It has similarities with the reciprocal lattice rods observed by LEED and RHEED for two-dimensional surface layers (see, e.g., Figs. 5.16 and 5.20), and the $1/(Q_z)^2$ dependence is reminiscent of thermal or Huang diffuse scattering. We shall come back to this point later.

10.3.2   Interface Roughness

The truncated lattice sum in Eq. (10.5) is valid for perfect surfaces. However, if we want to take into account surface roughness as indicated in panel (b) of Fig. 10.5, we introduce a filling factor parameter β^w for each layer w, where $w = 0$ is the topmost perfect layer with $\beta^0 = 1$. The filling factor may decreases with each additional atomic plane according to $\beta^1 < 1$, $\beta^2 = (\beta^1)^2$, etc. The limit is $\beta = 0$ for an empty surface layer. With this scheme, we evaluate again the Laue sum for partially filled layers by separating the sum in two parts, one for the perfect bulk, and one for the rough surface. Furthermore, we introduce an attenuation factor $\alpha < 1$ in the second

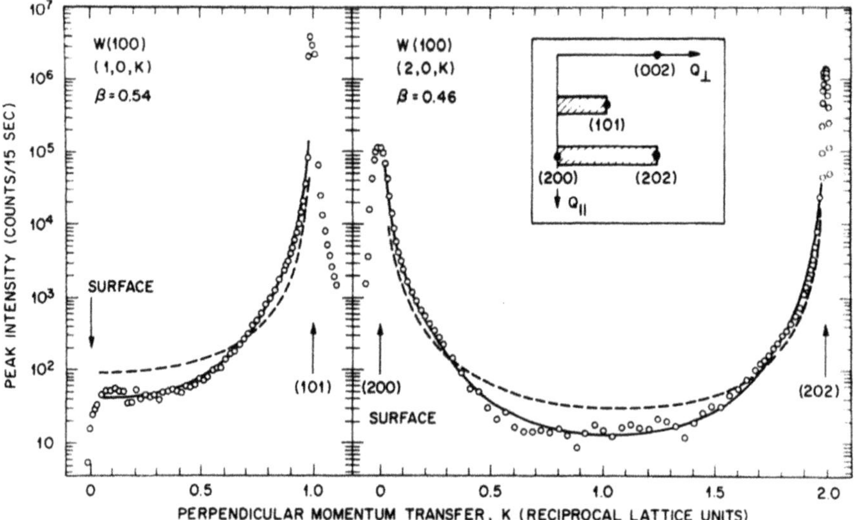

Fig. 10.6 X-ray scans along the truncation rod (10*l*) and (20*l*) directions of tungsten with [001] orientation. The dashed line is the expected intensity for an ideal surface; open circles are data points; and solid lines are fits to the data points with the fit parameter $\beta = 0.46$. Reproduced with permission from Robinson (1986)). Copyright (2025) by the American Physical Society

term in order the sum to converge. With these changes, we obtain: the scattering amplitude in the *z*-direction:

$$A(Q_z) \propto \sum_{w=-\infty}^{0} \beta^{|w|} e^{iwQ_z \cdot d_z} + \sum_{w=1}^{\infty} \alpha^w e^{iwQ_z \cdot d_z} \tag{10.6}$$

The truncation rod scattering intensity, including surface roughness, becomes (Robinson 1986):

$$I_{CTR}^{rough}(Q_z) \propto I_{CTR}^{ideal}(Q_z) \frac{(1-\beta)^2}{1+\beta^2 - 2\beta \cos(Q_z \cdot d_z)} \tag{10.7}$$

The existence of truncation rod scattering has been confirmed by Robinson, probing a clean surface of tungsten with a [001] orientation and scanning the rods along (10*l*) and (20*l*) directions. The results from Robinson (1986) are shown in Fig. 10.6. The dashed line is the calculated intensity according to Eq. (10.5). It does not fit the measured intensity, which is lower between the Bragg points due to surface roughness. The solid line is a better representation of the crystal truncation rod scattering, with β as a fit parameter.

10.3.3   Expanded Surface Layers

The concept of crystal truncation rod scattering is also applicable to surface reconstructions and thin epitaxial films, deposited on crystalline substrates, as schematically shown in Fig. 10.7. For example, we may consider a surface layer that is shifted
by a small amount in the perpendicular direction, introducing a phase shift of the
scattered waves that does not correspond to the crystal lattice sum and therefore must
be accounted for in an extra term, where f and f' are the respective atomic form
factors. The Laue sum with one shifted layer is then:

$$A_{\text{CTR}}^{\text{expand}}\left(Q_z\right) \propto f \sum_{w=0}^{\infty} e^{iwQ_z \cdot d_z} + f' e^{iQ_z \cdot d'_z} = f \frac{1}{1 - e^{iQ_z \cdot d_z}} + f' e^{iQ_z \cdot d'_z} \qquad (10.8)$$

The intensity $I(Q_z) = A_{\text{CTR}}^{\text{expand}} \cdot A_{\text{CTR}}^{*\text{expand}}$ is plotted in Fig. 10.8, assuming that the
atomic form factors for the crystal lattice and the top layer are identical: $f = f'$. The
plots show that the CTR is very sensitive to phase changes at the surface, and it is
easy to distinguish between a compressed and an expanded surface layer.

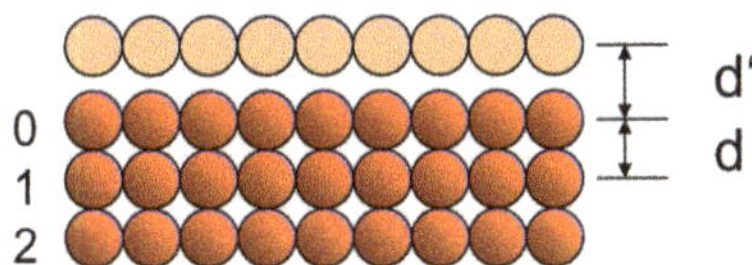

Fig. 10.7 Surface layer on a half-infinite crystal lattice. The top layer may be expanded or
contracted. In both cases, it causes a phase shift that is clearly visible in the CTR intensity, as
shown in Fig. 10.8

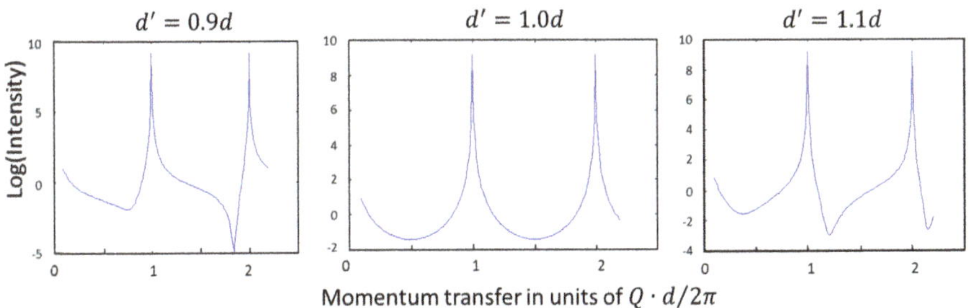

Fig. 10.8 Calculated CTR intensities for three scenarios of the top layer: contracted by 10% in the
left panel, same as the bulk in the middle panel, and expanded by 10% in the right panel

10.3.4   Thin Films

Similar concepts discussed so far can also be applied to the scattering intensity of thin films on a substrate. Then, it is straightforward to show that the expression for the scattering amplitude should be composed of three contributions:

$$
A_{\mathrm{CTR}}^{\mathrm{expand}}(\boldsymbol{Q}_z) \propto \underbrace{\left(f_1 \frac{1}{1 - e^{i\boldsymbol{Q}_z \cdot \boldsymbol{d}_{z1}}}\right)}_{\mathrm{CTR}} + \underbrace{\left(f_2 e^{i\boldsymbol{Q}_z \cdot \boldsymbol{d}_{z2}}\right)}_{\mathrm{Interphase}} \times \underbrace{\left(f_3 \sum_{n=0}^{N-1} e^{i\boldsymbol{Q}_z \cdot \boldsymbol{d}_{z3}}\right)}_{\mathrm{Thin\ film\ Laue}} \tag{10.9}
$$

The substrate is responsible for the CTR, the thin film causes Laue oscillations, and the interface impinges a phase shift between both. f_1, f_2, f_3 are the respective form factors, and d_1, d_2, d_3 are the respective lattice plane spacings. A schematic plot of the intensity according to the expression in Eq. (10.9) is shown in Fig. 10.9. An experimental example is reproduced in Fig. 10.10 taken from Robinson et al. (1988) for a thin NiS_2 film grown epitaxially on a Si(111) substrate. The solid line is a fit according to the expression in Eq. (10.9). Equation (10.9) and the corresponding plot in Fig. 10.9 do not take into account the total reflection of X-rays or neutrons close to $Q \cdot d/2 = 0$. This is discussed in Sect. 10.4.

10.4   Reflectivity

10.4.1   Distinctions and Objectives

Reflectivity, as the name suggests, refers to the reflection of particles (photons, neutrons, or electrons) from surfaces and interfaces. At short wavelengths of

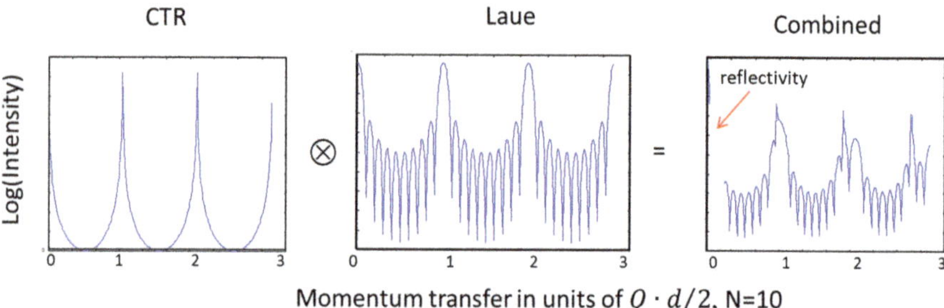

Fig. 10.9 Calculated intensities for a thin film with N = 10 atomic layers on a substrate with an infinite number of layers. The substrate causes CTR intensity, the film gives rise to Laue intensity oscillations, and the interface causes a phase shift. The momentum transfer is assumed to be normal to all lattice planes. The calculation is one-dimensional and corresponds to an extended reflectivity scan (compare Fig. 10.1). The red arrow marks the area of total external reflection, where the intensities are modified, as is discussed in the Sect. 10.4

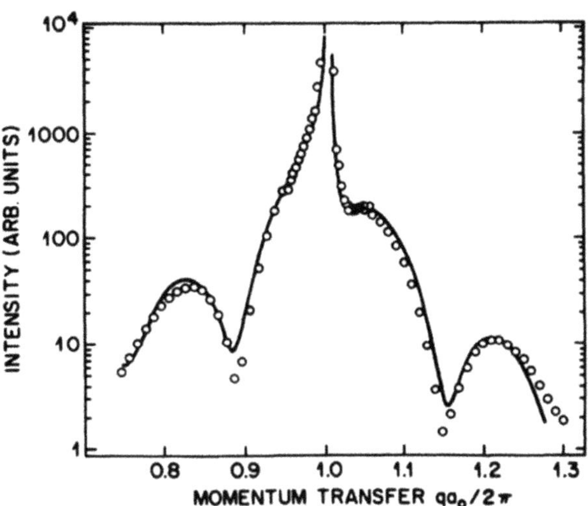

Fig. 10.10 Integrated intensity of the extended reflectivity near the (111) Bragg reflection of the Si substrate, with a NiSi$_2$ thin film deposited on top of the substrate. The solid line represents a fit to the data points, based on Eq. (10.9). Reproduced with permission from Robinson et al. (1988). Copyright (2025) by the American Physical Society

subnanometers considered in this text, specular reflection occurs at small angles. It turns out that the reflected waves contain information about the internal properties of the reflecting material, regardless of its crystallinity, including electron density profiles, layer thicknesses, and interface roughness. Therefore, reflectivity is of interest not only for the study of solid films but also for soft matter materials, especially polymers, liquids, and biomaterials. Furthermore, magnetization density profiles in thin films and multilayers can be investigated using magnetic neutron and X-ray reflectivity methods.

Small-angle scattering (SAS), treated in Chap. 9, and X-ray reflectivity (XRR) or neutron reflectivity (NR) are related, yet the information gained from both scattering techniques is distinctly different. Reflectivity explores density profiles perpendicular to the surface (parallel to $Q_\perp$ in Fig. 10.11a), and averages over structures parallel to the surface. In contrast, SAS explores density profiles in the plane perpendicular to the incoming wavenumber k_i (red shaded area in Fig. 10.11b) and averages over structures parallel to k_i and perpendicular to the scattering vector $Q_\parallel$. Reflectivity measurements require flat samples and interfaces that allow a well-defined glancing angle α of the incident and exit beams. Contrarily, SAS has no restriction on the external sample shape. A comparison of the different scattering geometries is sketched in Fig. 10.11, which emphasizes the fact that structural information is only gained in the direction parallel to the scattering vector Q.

Furthermore, SAS is treated in the framework of the Born approximation, whereas the description of reflectivity requires a distorted wave Born approximation or a dynamical scattering theory. However, reflectivity and SAS can be combined to enhance the surface sensitivity, known as GISAS, which is the topic of Sect. 12.2.

The main objective for using XRR and NR is the exploration of scattering length density (SLD) profiles, which are either electron density profiles probed by X-rays or nuclear density profiles probed by neutrons. Furthermore, periodicities of

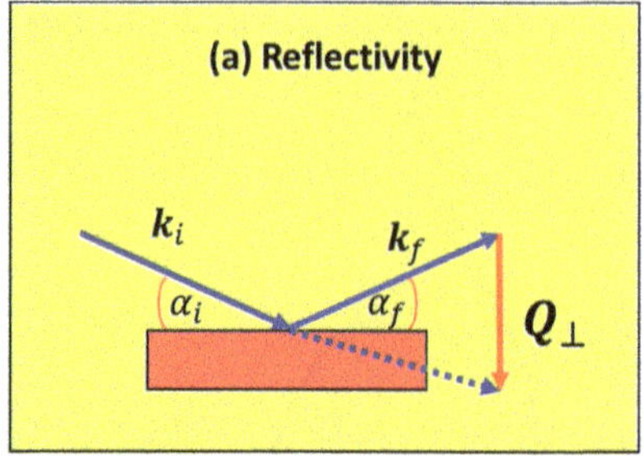

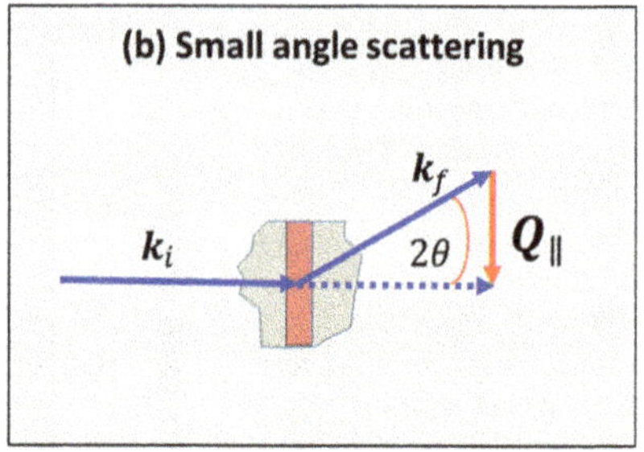

Fig. 10.11 Comparison of the scattering geometry for reflectivity measurements and small-angle scattering. Notice that the scattering vector Q probes correlations perpendicular to the plane in reflectivity scans, but parallel to the slab in small-angle scattering experiments, indicated in red color as part of the sample that otherwise may have an arbitrary shape. Reflectivity explores structures perpendicular to the surface and averages over structures parallel to the surface. SAS explores structures in the plane of the slab, parallell to $Q_\parallel$ and averages over structures perpendicular to the slab. SAS is sensitive to density fluctuations, whereas X-ray and neutron reflectivity are sensitive to density gradients across interfaces

multilayers and roughness correlations in thin films and multilayers are of prime interest. In addition, X-ray resonant magnetic reflectivity (XRMR) and polarized neutron reflectivity (PNR) offer supplementary information on magnetic thin films, interfaces, and magnetization density profiles in ferro- or antiferromagnetic artificial layered structures. The chart in Fig. 10.12 shows some prominent applications explored by XRR, NR, XRMR, and PNR.

Reflectivity is the Fourier transform of the derivative of the scattering length density profile along the scattering vector. Therefore, sudden changes in the density profile have dramatic effects on the reflected intensity. Depending on the sample and material density, total external reflection occurs at angles of incidence smaller than the critical angle. The critical angle depends on the refractive index and is different for X-rays and neutrons. Therefore, we begin this chapter with the derivation of the refractive indices.

10.4.2 *Refractive Index of X-Rays*

The refractive index of X-rays $n(\omega)$ follows from the dielectric constant $\varepsilon(\omega)$ as a function of angular frequency ω of electromagnetic waves in matter that are much higher than any intra-atomic resonance frequency. The refractive index can also be expressed as a function of the atomic form factor $f(Q_z)$:

$$n(Q_z) = 1 - \frac{2\pi r_0}{k_0^2}\rho_A[f(Q_z) + \Delta f] + \frac{i\mu}{2k_0} \tag{10.10}$$

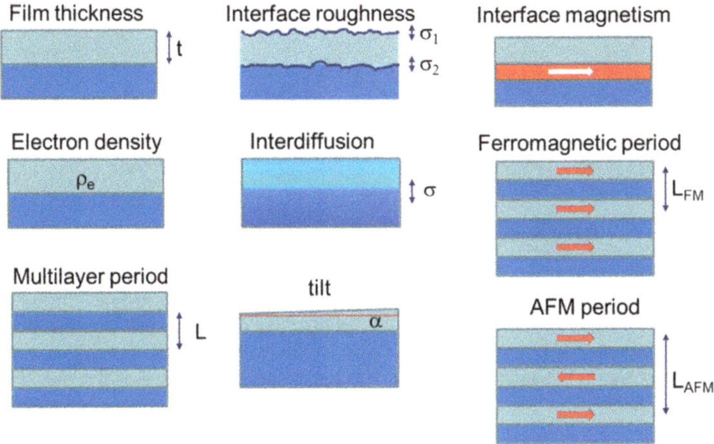

Fig. 10.12 Selection of examples that can be probed with X-ray and neutron reflectivity for the analysis of film thicknesses, interphase roughnesses, multilayer periodicities, and magnetization density profiles

Here, r_0 is the classical electron radius (Eq. 2.13), $k_0 = 2\pi/\lambda$ is the vacuum wavenumber of X-rays, Δf is a dispersion correction, ρ_A is the atomic number density, and μ is the X-ray linear absorption coefficient. As we are mainly concerned with scattering at small angles, $f(Q_z)$ can be approximated by the atomic number Z, and the product $\rho_A Z = \rho_e$ is the electron density in the material investigated. Δf and μ are strongly wavelength-dependent close to X-ray absorption edges. Equation (10.10) is usually written as:

$$n = 1 - \delta + i\beta, \tag{10.11}$$

where:

$$\delta = \frac{2\pi r_0}{k_0^2}\rho_e, \text{ and } \beta = \frac{\mu}{2k_0} = \frac{\mu\lambda}{4\pi}$$

Since the real part $1 - \delta$ is positive and smaller than 1, it turns out that for X-rays, matter is less dense than vacuum by the amount of about $\delta \approx 10^{-5}$. Therefore, at a small critical angle α_c, total external reflection of the X-ray beam occurs. The scattering geometry for reflectivity experiments is sketched in Fig. 10.13. The numerical value of β is on the order of 10^{-8}, unless close to the photoelectric absorption edges.

Snell's law of refraction in terms of the glancing angles $\alpha_i = \alpha_f$ is expressed as:

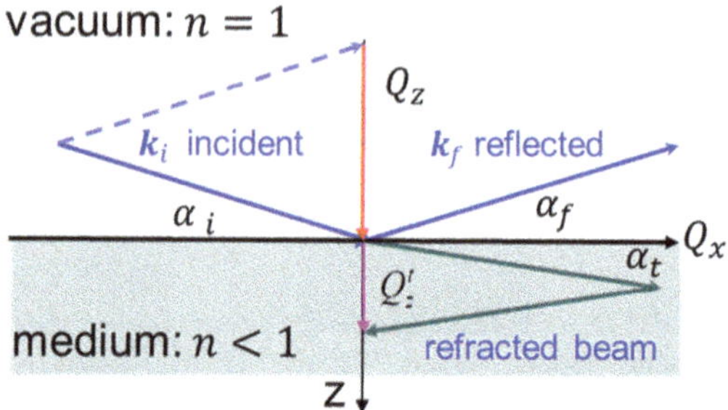

Fig. 10.13 Scattering geometry for reflectivity experiments. We distinguish between the scattering vector Q_z in a vacuum and Q'_z in the medium with a refractive index $n < 1$. α_i, α_f are the glancing angles to the surface for the incident and the exit beam, respectively, and α_t is the glancing angle of the transmitted beam

$$\cos \alpha_{i,f} / \cos \alpha_t = n. \tag{10.12}$$

Total reflection starts to occur when $\alpha_t = 0$ and $\cos\alpha_t = 1$. Then:

$$\cos \alpha_{i,f} = \cos \alpha_c = n. \tag{10.13}$$

The scattering vector in the z-direction normal to the surface is defined as

$$Q_z = (4\pi/\lambda) \sin \alpha_{i,f}.$$

Accordingly, the critical scattering vector for total external reflection is:

$$Q_{z,c}^{x-ray} = (4\pi/\lambda) \sin \alpha_c = 2k_0 \sin \alpha_c \approx 2k_0 \alpha_c. \tag{10.14}$$

As

$$\sin \alpha_c = \sqrt{1 - \cos^2(\alpha_c)} = \sqrt{1 - n^2} = \sqrt{1 - (1 - \delta)^2} \cong \sqrt{2\delta}, \tag{10.15}$$

and neglecting the absorption effect, we find for the critical scattering vector:

$$Q_{z,c}^{x-ray} = 2k_0 \sqrt{1 - n^2} = 2k_0 \sqrt{2\delta} = \sqrt{4k_0^2 2\delta} = \sqrt{16\pi r_0 \rho_e} \tag{10.16}$$

The critical scattering vector is no longer a function of the wavelength. It is entirely determined by the material's property and, in particular, by the electron density ρ_e. The expression in Eq. (10.16) is independent of the state of matter (solid, liquid, amorphous, etc.); only the electron density counts. The product $r_0\rho_e$ is called the X-ray scattering length density (xSLD). It has the units $°A^{-2}$ or nm^{-2}, where $1\ nm^{-2} = 10^{-2}°A^{-2}$.

10.4.3 Refractive Index of Neutrons

For neutrons, we derive the critical scattering vector by considering the potential well between the vacuum and matter. Incident neutrons with mass m_n and a kinetic energy projected in the z-direction of $E_z^i = \hbar^2 k_i^2 / 2m_n$ experience at the vacuum/matter interface a potential step of height $\rho_A V_N$. Here, V_N is the Fermi pseudo-potential with magnitude: $V_N = 2\pi\hbar^2 b_{coh}/m_n$ (compare Eq. (2.31), ρ_A is the atomic number density of the material investigated, and k_i is the incident neutron wavenumber in a vacuum. Inside the material, the transmitted neutron kinetic energy E_z^t is reduced by the potential barrier, as indicated in Fig. 10.14:

$$E_z^t = E_z^i - \rho_A V_N = \frac{\hbar^2 k_i^2}{2m_n}\left(1 - \frac{4\pi b_{coh}\rho_A}{k_i^2}\right) = \frac{\hbar^2 k_t^2}{2m_n} \tag{10.17}$$

The neutron refractive index follows from the ratio of the transmitted to the incident neutron kinetic energy:

$$\frac{E_z^t}{E_z^i} = \frac{E_z^i - \rho_A V_N}{E_z^i} = \frac{k_t^2}{k_i^2} = n_{neut}^2 = 1 - \frac{4\pi b_{coh}\rho_A}{k_i^2} \tag{10.18}$$

The refractive index of neutrons is then approximated to:

$$\frac{k_t}{k_i} = n_{neut} = \sqrt{1 - \frac{4\pi b_{coh}\rho_A}{k_i^2}} \cong 1 - \frac{2\pi b_{coh}\rho_A}{k_i^2} \tag{10.19}$$

The last equation can be compared with the one that we obtained for X-rays:

$$n_{x-ray} = 1 - \frac{2\pi r_0 \rho_e}{k_0^2} \tag{10.20}$$

Here, we immediately notice the similarity to the X-ray refractive index: the classical electron radius, r_0, is replaced by the coherent scattering length b_{coh}, and the

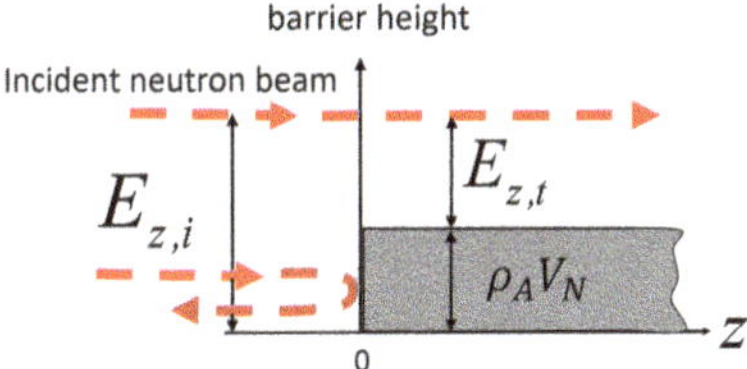

Fig. 10.14 Quantum mechanical potential step for the z-component of the neutron kinetic energy, experienced by neutrons entering from the vacuum into matter. Only neutrons with a kinetic energy higher than the potential step can enter; others are reflected

electron density ρ_e is replaced by the atomic number density ρ_A. While r_0 is always positive, b_{coh} may be positive or negative. Neutron total external reflection occurs only for materials that have a positive and coherent scattering length: $b_{coh} > 1$. Now we find the critical scattering vector of neutrons for total reflection in analogy to Eq. (10.16):

$$Q_{z,c}^{neut} = 2k_i\sqrt{1 - n_{neut}^2} = 2k_i\sqrt{\frac{4\pi b_{coh}\rho_A}{k_i^2}} = \sqrt{16\pi b_{coh}\rho_A}. \qquad (10.21)$$

The product $b_{coh}\rho_A$ is called the *neutron scattering length density* (nSLD). Determining the critical scattering vector Q_c in a reflectivity experiment is particularly useful for X-rays and neutrons. For X-rays, the $Q_{z,\,c}$ value is related to the average electron density ρ_e that allows to determine, for instance, the average composition of an alloy (Miceli et al. 1986). For neutrons, $Q_{z,\,c}$ allows the measuring the coherent scattering length, which is essential for all other neutron scattering experiments.

Critical Scattering Vector for Total Reflection

X-rays: $Q_{z,c}^{x-rays} = \sqrt{16\pi r_0\rho_e}$

Neutrons: $Q_{z,c}^{neut} = \sqrt{16\pi b_{coh}\rho_A}$

While the scattering lengths r_0 and b_{coh} are comparable, the electron number density ρ_e is generally much higher than the atomic number density ρ_A. Therefore, $Q_{z,c}^{neut}$ turns out to be a factor of 5–10 lower than $Q_{z,c}^{x-rays}$. This poses an experimental difficulty in studying the critical edge with neutrons, as finer beam collimation and higher resolution are required for neutron reflectivity (NR) experiments as compared to X-ray reflectivity (XRR), despite the lower intensity of neutron sources. As an example, we use typical values for hard X-ray wavelengths of 0.1 nm and an electron density of bcc-Fe, which results in $Q_c^{x-ray}(Fe) = 0.55$ nm^{-1} and $\alpha_c^{x-ray}(Fe) = 0.25°$. In the case of neutrons, the coherent scattering length of Fe is $b_{coh} = 9.45$ fm, and using the same wavelength, we obtain: $Q_c^{neut}(Fe) = 0.2$ nm^{-1}, and $\alpha_c^{neut}(Fe) = 0.09°$, roughly a factor of 3 lower than for X-rays. However, when using cold neutrons with a wavelength of 0.44 nm, which is usually done, α_c^{neut} increases to $\alpha_c^{neut} = 0.4°$, comparable to X-rays. For neutrons, these critical values are only correct in the case of unpolarized neutrons and nonmagnetic samples. The discussion of polarized neutron reflectivity and accordingly altered refractive indices is postponed to Chap. 11.

10.4.4 Fresnel Reflectivity

Now, we calculate the reflected intensities of X-rays and neutrons from a perfectly flat and infinitely extended surface with unlimited depth, independent of the state of matter (excluding the gaseous state). The reflected intensity is known as *Fresnel reflectivity* according to the French physicist Augustin Jean Fresnel (1788–1827).

First, we determine the Fresnel coefficient for reflection r and transmission t, which follow from the respective electric field amplitudes of the electromagnetic waves. For reflection, the Fresnel coefficient of electromagnetic waves is (Born and Wolf 1980):

$$r = \frac{E_r}{E_i} = \frac{\sin \alpha_i - \sin \alpha_t}{\sin \alpha_i + \sin \alpha_t} = \frac{Q_z - Q_z^t}{Q_z + Q_z^t}, \tag{10.22}$$

and the Fresnel reflectivity for a perfectly flat and half-infinite body is:

$$R_F(Q_Z) = r^2 = \left(\frac{Q_z - Q_z^t}{Q_z + Q_z^t}\right)^2, \tag{10.23}$$

The scattering vector in a vacuum is simply:

$$Q_z = 2k_0 \sin \alpha_i.$$

The scattering vector in the medium, Q_z^t, can be derived by using Snell's law stated in Eq. (10.12) and by carrying out a small-angle approximation in terms of a Taylor expansion:

$$1 - \frac{\alpha_i^2}{2} = n\left(1 - \frac{\alpha_t^2}{2}\right)$$

and:

$$\alpha_i^2 = 2(1 - n) + \alpha_t^2 = 2\delta - 2i\beta + \alpha_t^2$$

where we used $1 - n = \delta - i\beta$ according to Eq. 10.11 and using the approximation $n\alpha_t^2 \cong \alpha_t^2$. Setting $2\delta = \alpha_c^2$, according to Eq. (10.15), we find:

$$\alpha_t^2 = \alpha_i^2 - \alpha_c^2 + 2i\beta \tag{10.24}$$

Then, we obtain for the angle α_t:

$$\alpha_t = \sqrt{\alpha_i^2 - \alpha_c^2 + 2i\beta},$$

and the scattering vector in the medium is:

$$Q_z^t = \sqrt{Q_z^2 - Q_c^2 + 8k_0^2 i\beta}.$$

Inserting the last expression in $R_F(Q_Z)$ yields:

$$R_F(Q_Z) = \left|\frac{\alpha_i - \alpha_t}{\alpha_i + \alpha_t}\right|^2 = \left|\frac{\alpha_i - \sqrt{\alpha_i^2 - \alpha_c^2 + 2i\beta}}{\alpha_i + \sqrt{\alpha_i^2 - \alpha_c^2 + 2i\beta}}\right|^2$$

$$= \left|\frac{Q_z - \sqrt{Q_z^2 - Q_c^2 + 8k_0^2 i\beta}}{Q_z + \sqrt{Q_z^2 - Q_c^2 + 8k_0^2 i\beta}}\right|^2 \tag{10.25}$$

For the transmission, the Fresnel coefficient is (Born and Wolf 1980):

$$t = \frac{E_t}{E_i} = \frac{2\sin\alpha_i}{\sin\alpha_i + \sin\alpha_t} = \frac{2Q_z}{Q_z + Q_z^t} \tag{10.26}$$

This function is plotted in the solution to Exercise 10.6. With $r + 1 = t$, the Fresnel transmissivity is:

$$T_F(Q_Z) = tt^* \frac{Q_z^t}{Q_z} \tag{10.27}$$

and together with the Fresnel reflectivity, we find:

$$R_F(Q_Z) + T_F(Q_Z) = 1. \tag{10.28}$$

Having derived expressions for the Fresnel reflectivity and transmissivity, we need to discuss next several important properties of the Fresnel reflectivity.

1. Fresnel Reflectivity and Approximations

In Fig. 10.15, the ideal Fresnel reflectivity (red line) is plotted, neglecting absorption effects. It is characterized by an intensity plateau of 1 for scattering vectors equal to or smaller than the critical scattering vector:

$$R_F(Q_Z) = 1 \text{ for } \boldsymbol{Q}_z \leq \boldsymbol{Q}_c \tag{10.29}$$

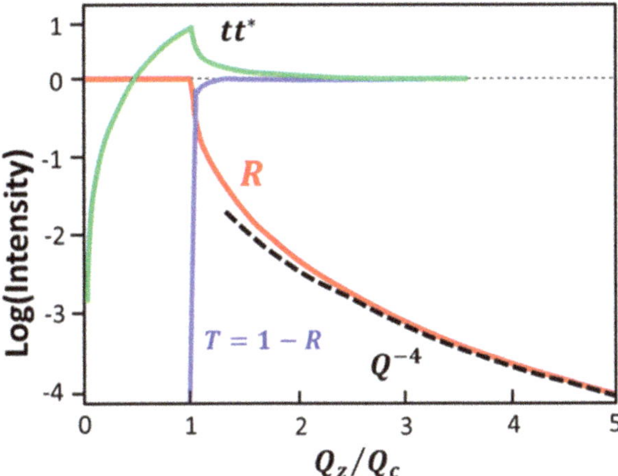

Fig. 10.15 Reflectivity (red), transmission coefficient (green), and transmissivity (blue) are compared in one plot. Adapted from (Stierle 1996)

For $Q_z > Q_c$, the reflected intensity $R_F(Q_Z)$ drops off very fast and over many orders of magnitude. For incident angles $\alpha_i \gg \alpha_c$, the Fresnel reflectivity $R_F(Q_Z)$ can be approximated to:

$$R_F(Q_Z) = r^2 = \left(\frac{Q_z - Q_z^t}{Q_z + Q_z^t}\right)^2 \propto \left(\frac{Q_c}{2Q_z}\right)^4 = \left(\frac{\alpha_c}{2\alpha_i}\right)^4 = \frac{16\pi^2 r_0^2 \rho_e^2}{Q_z^4} = \left(\frac{1 - n^2}{4\alpha_i^2}\right)^2 .$$

$$(10.30)$$

Hence, the intensity decreases with the scattering vector to the fourth power: Q_z^{-4}. Figure 10.15 summarizes the X-ray optical parameters discussed so far: ideal reflectivity without absorption, transmission coefficient tt^*, transmissivity T, and the approximated Fresnel reflectivity Q_z^{-4} as a function of the scattering vector Q_z normalized by the critical scattering vector Q_c and displayed on a logarithmic intensity scale.

2. Absorption

In real experiments, the absorption can never be neglected, in particular for X-rays when the photon energy is tuned close to an absorption edge. Figure 10.16 shows a plot of Eq. (10.25) for different absorption values β on a linear scale. For $\beta = 0$, we obtain a kink at $\alpha = \alpha_c$, as seen before in Fig. 10.15. With increasing β, that is, with increasing absorption, the kink vanishes and becomes increasingly rounded.

3. Experimental Observation

Figure 10.17 shows schematically the Fresnel reflectivity for a perfectly smooth and infinitely extended surface on a logarithmic scale, contrasted by one that is usually observed in real experiments, i.e., finite sample size and roughness at the surface. In the real world, the Fresnel reflectivity is affected by several issues. First, the intensity drops off faster than expected, which is due to surface roughness. The

Fig. 10.16 Linear plot of the reflectivity close to the angle of total reflection according to Eq. (10.29). The critical angle of $\alpha_c = 0.22°$ is typical for hard X-rays (Cu-K_α-radiation) and for reflection from a Si surface. The reflectivity is drawn for zero absorption and for two values of β related to the dispersion δ

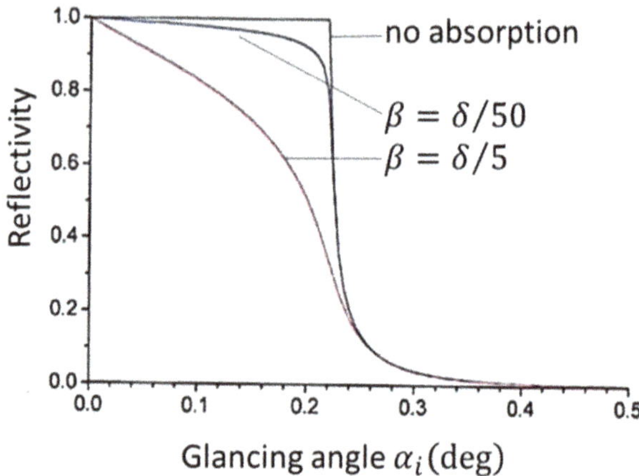

Fig. 10.17 (a) The Fresnel reflectivity curve is plotted versus the momentum transfer. The ideal reflectivity shown in red is representative for X-ray SLDs and spans many orders of magnitude. The experimentally determined reflectivity decreases more rapidly (blue) than for an ideally flat sample due to surface roughness. Total reflection occurs at $Q_z \leq Q_c$. The actual measured total reflection edge at Q_c is affected by the roughness, absorption, and footprint of the beam cross section on the sample

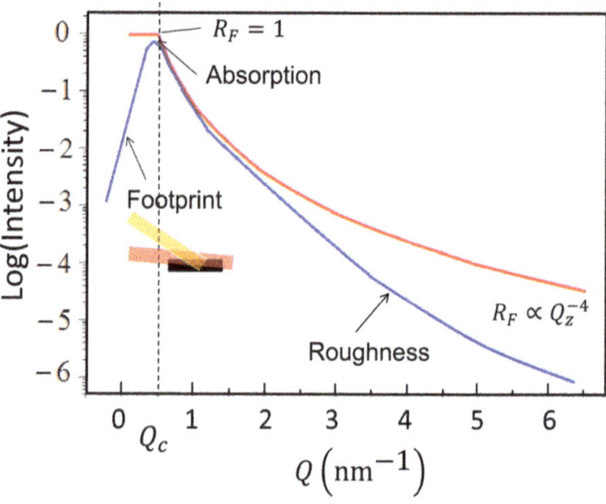

missing intensity is diffusively scattered in the off-specular direction, which we consider later on. Second, the reflectivity of 1 is not reached for $Q_z < Q_c$ because of absorption effects. This can be considerable due to the shallow incident angle. Third, the footprint of the beam cross section on the sample surface exceeds the sample size with a decreasing incident angle. All these factors have to be considered and can be accounted for. While absorption and footprint corrections are easy to perform, surface roughness can be an important material property to be investigated in detail. For instance, in the case of a water surface, the roughness of an otherwise calm water surface is due to capillary waves that have been characterized in detail via XRR (Braslau et al. 1985).

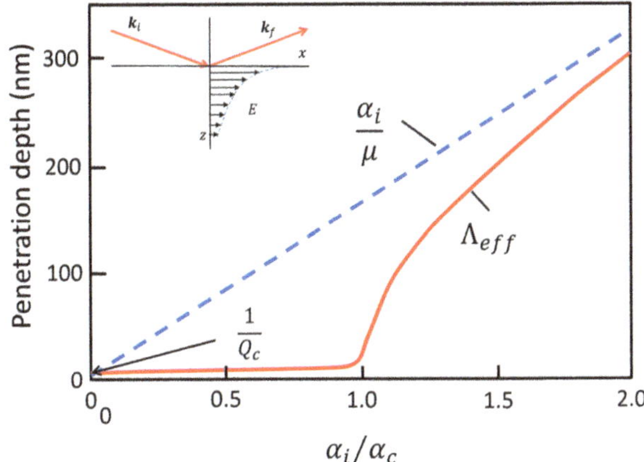

Fig. 10.18 Schematics of the X-ray penetration depth close to the critical angle of total reflection. The inset shows the exponentially damped evanescent electric field vector in the material for $\alpha_{i,f} \leq \alpha_c$. Adapted from Feidenhans'l (1989)

10.4.5 Penetration Depth

Next, we determine the penetration depth of the incident wave depending on the glancing angles $\alpha_{i,f}$. As already noted above, for $Q_z < Q_c$, Q_z^t becomes imaginary, and the scattering amplitude is exponentially damped inside the material. The inside wave is then evanescently propagating parallel to the surface while being exponentially damped below the surface, as indicated in the inset of Fig. 10.18. The Q_z^t component of the scattering vector determines the penetration depth according to $\exp\left(-\,Im\left(Q_z^t\right)\cdot z\right)$. The $1/e$ penetration depth for $z = \Lambda$ therefore, is:

$$\Lambda = \left(Im\left(Q_z^t\right)\right)^{-1} = \left(\sqrt{Q_c^2 - Q_z^2}\right)^{-1} \tag{10.31}$$

This attenuation of the electromagnetic wave occurs in addition to the photoelectric absorption and is much stronger than the latter for $Q_z < Q_c$. The maximal penetration depth for $Q_z \rightarrow 0$ is:

$$\lim_{Q_z \rightarrow 0} \Lambda_{x-ray} = \frac{1}{Q_z^c} = \frac{1}{\sqrt{16\pi r_0 \rho_e}} \tag{10.32}$$

Hence, the maximal penetration depth is determined by the electron density of the material investigated and is less, for instance, for Au (1.3 nm) than for Si (3.2 nm). Taking into account the complete expression including the photoelectric absorption μ, we obtain the effective penetration depth (Vineyard 1982; Dosch 1987; Feidenhans'l 1989):

$$\Lambda_{eff}^{-1}\left(\alpha_{i,f}\right) = \frac{k_0}{\sqrt{2}} \left(\sqrt{\alpha_t^4 + \left(\frac{\mu}{k_0}\right)^2} - \alpha_t^2 \right)^{1/2}, \tag{10.33}$$

where we used the relation: $\alpha_t^2 = \alpha_{i,f}^2 - \alpha_c^2$.

The effective penetration depth $\Lambda_{\textit{eff}}(\alpha_i)$ of the specular beam is plotted schematically in Fig. 10.18 as a function of the ratio $\alpha_{i,f}/\alpha_c$ (red solid line) and is compared to the penetration depth if it were solely determined by the photoelectric absorption (blue dashed line). From this graph, it is obvious that close to the critical angle, the penetration is mainly limited by the imaginary part of the scattering vector. The evanescent waves that travel parallel to the surface within this depth can be used for surface-sensitive X-ray diffraction experiments, as we will discuss in Sect. 12.1.

Similar relations hold for the penetration depth of neutrons. In most cases, neutron-nuclear absorption effects can be neglected, and therefore, only the imaginary part of the scattering vector contributes to the penetration depth below Q_c according to:

$$\Lambda_{neut} = \frac{1}{Q_z^c} = \frac{1}{\sqrt{16\pi b_{coh}\rho_A}} \tag{10.34}$$

10.4.6 Surface Roughness

So far, we have assumed that the surface is perfectly smooth. However, this is rarely the case. On the atomic level, there are always undulations and atomic steps, which distort the wavefront of the X-ray beam, causing a reduction of the reflected amplitude. If the surface roughness gets too large, specular reflection ceases and diffuse scattering takes over. Roughness means that the electron density profile from a vacuum to inside the material, laterally averaged over the surface area, is not a step function but rather a smoothly increasing function, which customarily is described by the density profile:

$$\rho(z) = \langle\rho_1\rangle\mathrm{erf}(z), \tag{10.35}$$

where $\langle\rho_1\rangle$ is the average electron or atomic density of the material, and

$$\mathrm{erf}(z) = \frac{2}{\sqrt{\pi}}\int_0^z e^{-\left(\frac{t}{\sqrt{2}\sigma}\right)^2} dt \tag{10.36}$$

is the error function. The root-mean-square (rms) roughness parameter σ describes the width of the interface for a Gaussian-distributed height fluctuation, depicted in Fig. 10.19.

Fig. 10.19 Sketch of a rough surface with a Gaussian-distributed height fluctuation. The density variation across the surface is described by an error function

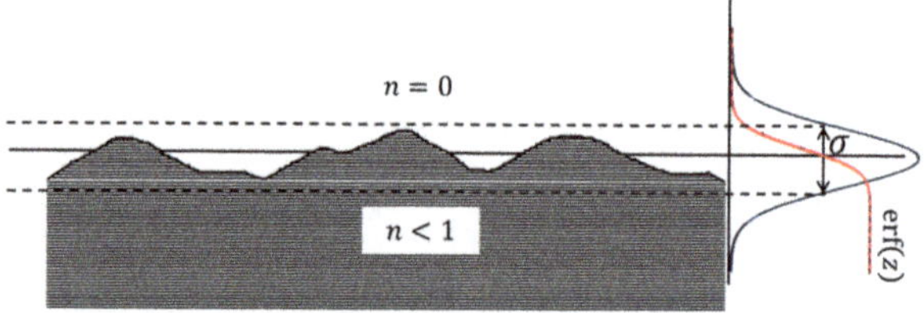

The reflectivity is the Fourier transform of the density profile normal to the surface. For a sharp interface, we already obtained the expression for the Fresnel reflectivity $R_F(Q)_z$. Therefore, we need only to take the Fourier transform of the rough interface that is described by the error function:

$$R(Q_z) = R_F(Q_z)\left|\frac{1}{\langle\rho_0\rangle}\int_{-\infty}^{\infty}\frac{d\rho(z)}{dz}e^{iQ_zz}dz\right|^2 = R_F(Q_z)\exp\left(-Q_z^2\sigma^2\right) \qquad (10.37)$$

Here, $\sigma = \sqrt{(\Delta z)^2}$ is the mean square roughness of the height fluctuations Δz at the interface in the z-direction. More precisely, σ is a roughness parameter for a Gaussian-distributed and uncorrelated rough surface (interphase). In this simple form, the roughness parameter does not yield information on the in-plane correlation of the height fluctuations, nor can it distinguish between a graded interface due to interdiffusion and a rough interface with sharp boundaries. In conclusion, the surface roughness in the Q_z-direction is described by a Debye–Waller-like factor $Q_z^2\sigma^2$, which reduces the Fresnel reflectivity exponentially with increasing Q_z.

10.4.7   Thin Films, Double Layers, and Multilayers

By depositing a thin film onto a semi-infinite substrate, we simultaneously introduce a new interface separating materials with different electron or nuclear densities. Similar to the vacuum/material interface, at the new interface, the electromagnetic wave (or neutron wave) splits again into a transmitted and a reflected wave. The waves reflected from the surface and the interface can overlap and interfere. In a radial scan, we therefore expect an increase in the reflected intensity for certain scattering angles and a decrease for others. In fact, due to the interference effect, we expect intensity oscillations of the reflected waves that are characteristic of the film thickness. Therefore, the reflected intensity from thin films shows the so-called Kiessig fringes[1] (Kiessig 1931) due to the interference of the waves reflected from the surface and the interface, as shown schematically in Fig. 10.20.

---

[1] X-ray reflectivity was applied first by H. Kiessig in 1931 by studying thin Ni coatings on glass.

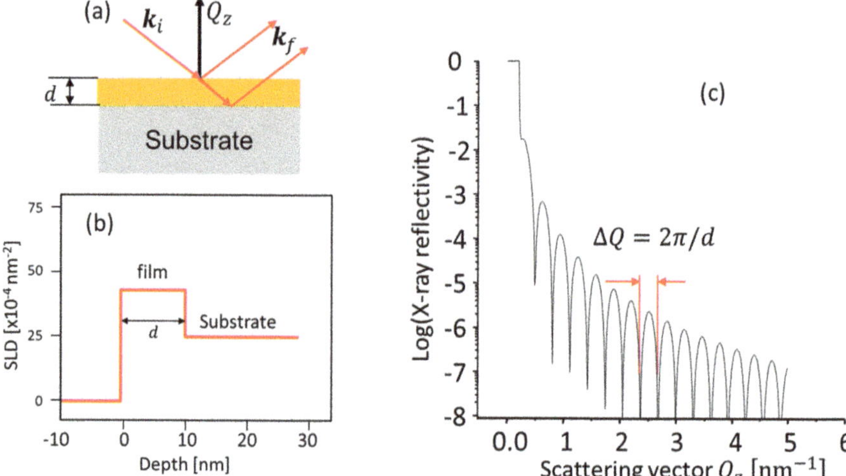

Fig. 10.20 (**a**) Scattering geometry, indicating that the wave is reflected at the surface and at the interface. (**b**) SLD for a thin film of thickness d on a substrate. Here, it is assumed that the film has a higher SLD than the substrate and that all interfaces are smooth. (**c**) X-ray reflectivity of the film showing thickness oscillations, also known as Kiessig fringes, which are due to the interference of waves reflected at the surface and the interface

10.4.7.1   Single Layer

The Fresnel reflectivity from a stratified medium, such as thin films and multilayers, can be calculated exactly with the help of the Parratt recursion formalism[2] (Parratt 1954), outlined below. While this is the exact approach and is to be used for fitting experimental data, the essential physics can be demonstrated in much simpler terms. Here, we present a heuristic expression for estimating the thickness of thin films via reflectivity measurements. The $l-$ th thin-film oscillations of a homogeneous flat film of thickness d occurs at:

$$\left(Q_z^l\right)^2 = l^2\left(\frac{2\pi}{d}\right)^2 + Q_c^2,$$

(10.38)

and a plot of $\left(Q_z^l\right)^2$ versus l^2 yields the film thickness. Alternatively, for $Q_z \gg Q_c$, the difference of the Q_z values from one oscillation minimum (maximum) to the next, as schematically indicated in Fig. 10.20, is proportional to the inverse film thickness d:

---

[2]L.G. Parratt investigated Cu thin films on glass substrates and treated reflectivity, including roughness, in a rigorous manner by using a recursion formalism to account for all transmitted and reflected waves at successive interfaces.

$$\Delta Q_z = \frac{2\pi}{d},$$

where

$$\Delta Q_z = \frac{4\pi}{\lambda}\left(\sin\alpha_2 - \sin\alpha_1\right) = \frac{4\pi}{\lambda}\left(\alpha_2 - \alpha_1\right) = \frac{4\pi}{\lambda}\Delta\alpha \qquad (10.39)$$

Hence, in the small-angle limit, but for $\alpha_i \gg \alpha_c$, the film thickness d follows from:

$$d = \frac{\lambda}{2\Delta\alpha} \qquad (10.40)$$

The thicker the film, the more narrowly the Kiessig fringes are spaced. The maximum layer thickness that can be resolved depends on the angular resolution of the instrument. In general, the resolution is given by $\Delta Q_z = 2k_0\Delta\alpha$. For very thin films or monolayers, the oscillations are widely spaced. They do not require high resolution but high intensity to observe several fringes at high Q values. For thick films, the fringes are squeezed in the forward direction. Here, the demands are low in intensity but high in angular resolution.

As in the case of half-infinite materials discussed in the previous section, the reflected intensity of thin films is affected by the roughness of the surface and the interface. This is illustrated in Fig. 10.21 by reflectivity calculations using the Parratt formalism. We observe clear differences depending on which interface shows more roughness. In the ideal case (a), the Kiessig fringes oscillate about the Fresnel reflectivity curve, indicated by a thin solid line. If the surface is rough and the interface is smooth, as in case (b), less intensity is reflected from the surface, the incident intensity can penetrate further into the film, and is then reflected from the internal interface. Therefore, the Kiessig fringes appear more pronounced, while the general Fresnel reflectivity curve is preserved. Vice versa, if the surface is smooth and the interface is rough, as is assumed in (c), the intensity is less reflected at the interface, and therefore, interference effects are weakened with increasing scattering vector. The last example in panel (d) shows the situation where both the surface and interface are rough. Hence, the Fresnel reflectivity drops off much faster, as already noted in Fig. 10.16. In all cases, the rms roughness is assumed to be $\sigma = 0.5$ nm and is described by a Debye–Waller-like factor for each interface, as quoted in Eq. (10.37).

10.4.7.2 Double Layer

In Fig. 10.22 is plotted another example of a double layer on a substrate with different thicknesses d_1 and d_2. This calculated reflectivity indicates that both thicknesses can be retrieved from the intensity oscillations. The narrow oscillations are due to the thicker layer d_2, and the wider oscillations are due to thinner layer d_1.

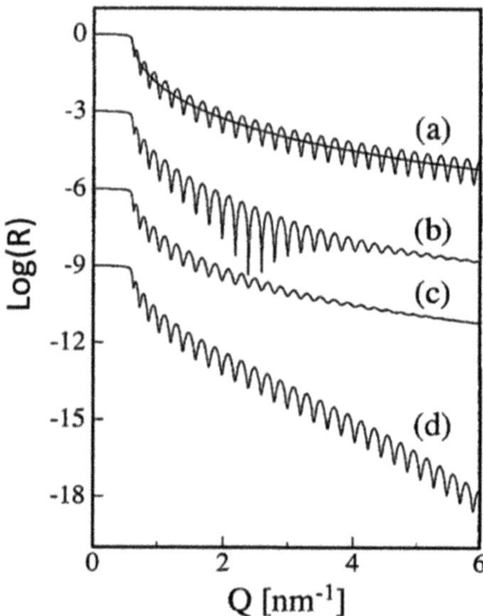

Fig. 10.21 Calculated reflectivities from a 30-nm-thick Co layer on a sapphire substrate and for different roughnesses using the Parratt formalism. (**a**) Reflectivity for an ideal film without surface and interface roughness. The smooth solid line shows the reflectivity if the Co layer were infinitely thick; (**b**) reflectivity for a rough surface and smooth interface; (**c**) reflectivity for a smooth surface and a rough interface; (d) reflectivity for a rough surface and rough interface. In all cases, the rms roughness is $\sigma = 0.5$ nm. The scales of the different reflectivity curves are offset by three orders of magnitude for clarity

10.4.7.3   Multilayer

Figure 10.23 demonstrates the reflectivity of a multilayer. In a multilayer, two layers of different densities alternate with a periodicity that corresponds to the sum of both layer thicknesses $L = d_1 + d_2$. The reflected intensity is rich in information. First, we observe Bragg reflections due to the artificial periodicity. Normally, we would expect the Bragg reflection at the reciprocal lattice points of the multilayer $Q_{z,l} = l\frac{2\pi}{L}$, where l is the order of the Bragg reflection. However, because of the proximity to the critical scattering vector Q_c, the condition for the Bragg reflections is modified, analogous to Eq. (10.38), for thin film oscillations:

$$Q_z^l = \sqrt{\left(l\frac{2\pi}{L}\right)^2 + (Q_c)^2} \tag{10.41}$$

Between the multilayer Bragg peaks, we observe thin-film oscillations due to the total thickness of the multilayer. If the multilayer has N periods, the film thickness is NL, and the Kiessig fringes have (uncorrected) spacings of $2\pi/NL$. These fringes can only be seen if the total multilayer thickness is less than the penetration thickness of the probing radiation. In addition, between any two Bragg peaks, we will find $N - 2$

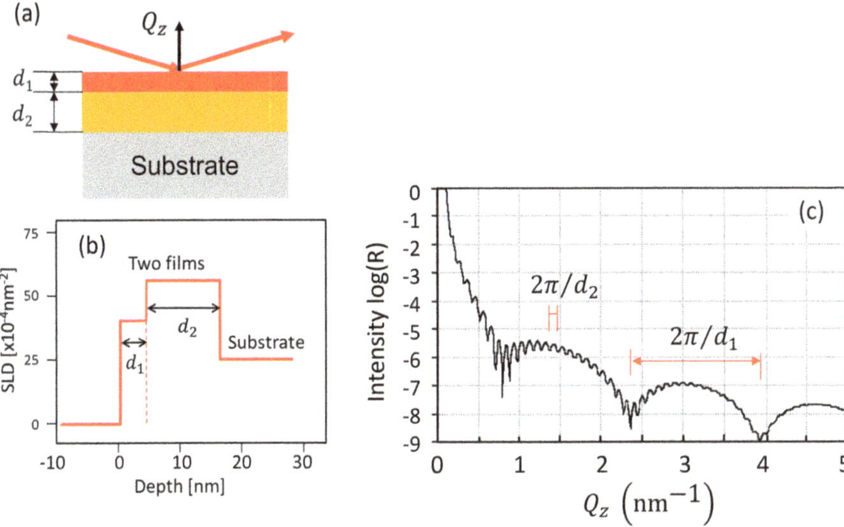

Fig. 10.22   (**a**) Scattering schematics for a double layer on a substrate with different thicknesses d_1 and d_2. (**b**) Respective scattering length density profile, assuming sharp interfaces. (**c**) Calculated reflectivity of two thin films on a substrate. Two oscillations can be recognized with widely different periodicities. The short period oscillation is due to the thicker layer $d_2 = 66$ nm, the wider period oscillation to the thinner layer $d_2 = 4$ nm. A third oscillation due to the sum thickness is also present and can be retrieved via a Fourier analysis of the oscillations, but is not visible to the naked eye

side maxima. Furthermore, a quantitative fit to the experimental data will also reveal the interface roughness and whether the roughness is the same for all interfaces or whether it increases or decreases with the number of layers deposited. An additional issue, which we will discuss later, is the question of whether the roughness is correlated from layer to layer and also whether there exists any in-plane correlation.

Multilayers are fascinating objects to study because of the artificial generation of Bragg reflections, allowing for the analysis of the Fourier components of all charge density and spin density modulations in the periodic structures. Famous examples include the analysis of interdiffusion at the interfaces in (GaAs)/(AlAs) multilayers as a function of temperature (Fleming et al. 1980), the modulation of hydrogen density in Nb/Ta multilayers (Miceli et al. 1985), the propagation of the spin spiral in Ho/Y and other rare earth superlattices (Majkrzak et al. 1991), the exchange coupling in artificial antiferromagnetic superlattices (Lauter-Pasyuk et al. 2002; Schreyer et al. 2002), proximity effects in superconducting/ferromagnetic superlattices (Satapathy et al. 2012), and for the observation of digital magnetization switching in MgO/Fe superlattices (Moubah et al. 2016). Moreover, multilayers do not need to be periodic in order to observe sharp Bragg reflections; they may also be aperiodic, as was shown by the growth of a multilayer with a Fibonacci sequence of thin and thick layers related by the golden mean (Todd et al. 1986).

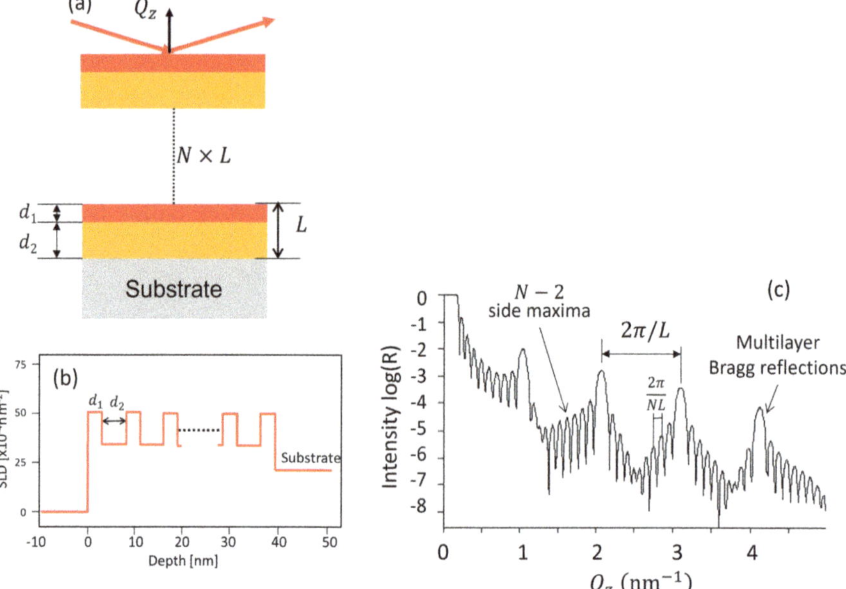

Fig. 10.23 (**a**) Scattering schematics for a multilayer on a substrate with N periods and layer thicknesses d_1 and d_2. (**b**) Respective scattering length density profile, assuming smooth interfaces. (**c**) Calculated scattering intensity for a multilayer with alternating film thicknesses d_1 and d_2 and repeat thickness $L = d_1 + d_2$. Multilayer Bragg reflections occur at multiples of $2\pi/L$ with corrections for the first few Bragg reflections that are shifted by Q_c. Additionally, $N - 2$ side maxima are observed between the Bragg reflections. The Kiessig fringe separation is determined by the total thickness: $\Delta Q = 2\pi/NL$

There are numerous examples published in the literature for all four cases discussed here: plain surfaces, single layers, double layers, and multilayers. We will give a few more examples in Chap. 11 concerning magnetic films and multilayers.

Parratt Recursion Formalism (Parratt 1954)

We consider a stratified medium with m-layers, flat interfaces between the layers, and layer thicknesses d_m. Furthermore, we use the definition of r and t for the reflected and transmitted waves, respectively, according to Eqs. (10.22) and (10.23). Now, let t_m be the amplitude of the electric field transmitted in the medium m and r_m the amplitude of the reflected field in medium m, as indicated in the graph below. Furthermore, we define a phase factor for the field always in the middle between two interfaces:

(continued)

$$a_m = e^{iQ_m d_m/2}$$

where d_m is the thickness of the mth layer, and Q_m is the scattering vector in the medium m. Continuity of the transverse component of the electric field vector at the interface between the m and $m-1$ layers requires:

$$t_{m-1}a_{m-1} + r_{m-1}a_{m-1}^{-1} = t_m a_m^{-1} + r_m a_m,$$

and continuity of the electric held gradient entails:

$$\left(t_{m-1}a_{m-1} - a_{m-1}^{-1}\right)\frac{Q_{m-1}}{2} = \left(t_m a_m^{-1} - r_m a_m\right)\frac{Q_m}{2}$$

The solution of these two equations is achieved by taking their difference and dividing it by the sum:

$$\mathbb{R}_{m-1} = a_{m-1}^4 \left(\frac{R_{m-1,m} + \mathbb{R}_m}{R_{m-1,m}\mathbb{R}_m + 1}\right),$$

where

$$\mathbb{R}_m = \frac{r_m}{t_m}a_m^2$$

is a generalized Fresnel reflectivity for the interface between m and $m+1$, and $R_{m-1,\,m}$ is:

$$R_{m-1,m} = \frac{Q_{m-1} - Q_m}{Q_{m-1} + Q_m}.$$

$R_{0,\,1}$ is the usual Fresnel coefficient for a smooth surface. This can be shown as follows. From $\mathbb{R}_{m-1}$ defined above follows for a single surface

$$\mathbb{R}_0 = a_0^4 R_{0,1}$$

Therefore,

$$|\mathbb{R}_0|^2 = \left(\frac{r_0}{t_0}a_0^2\right)^2 = \left(a_0^4 R_{0,1}\right)^2,$$

such that we obtain the Fresnel reflectivity:

(continued)

$$|\mathbb{R}_0|^2 = R_F \left(\frac{Q_0 - Q_1}{Q_0 + Q_1} \right)^2$$

in agreement with Eq. (10.25). For deriving the reflected intensity of a stratified medium, the recursion method starts at the lowest layer and then works up to the top surface.

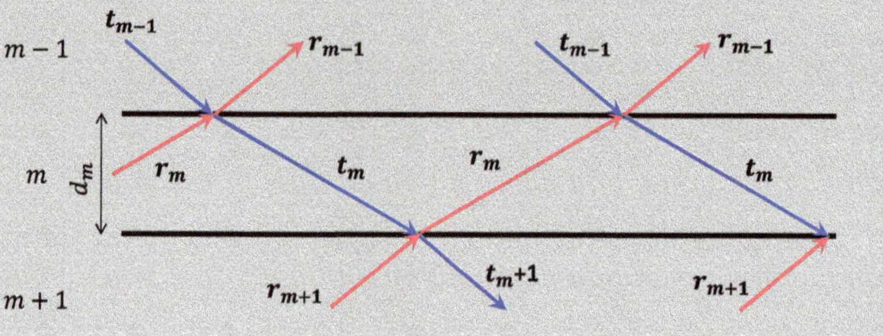

10.4.8   Kiessig Fringes versus Laue Oscillations

The Kiessig fringes or thin-film oscillations discussed above are similar to the Laue oscillations discussed in Chap. 3, Fig. 3.5 and Fig. 10.2. However, these oscillations are distinctly different. The Laue oscillations occur at reciprocal lattice points near Bragg reflections of a single-crystalline solid and are related to a finite number of atomic planes. In contrast, the Kiessig fringes are independent of the crystalline nature of the films and also exist if the films are of soft matter or amorphous material. The only conditions for their occurrence are differences in the electron or nuclear scattering length density and smooth interfaces.

To illustrate the difference between Kiessig fringes at low scattering vectors and Laue oscillations at high scattering vectors, we show in Fig. 10.24 a scan from $Q = 0$ to $Q = 4.165\,\text{Å}^{-1}\,(\triangleq 2\theta = 55°)$ of a 50 nm thick Nb(110) film deposited on a sapphire substrate with $(11\bar{2}0)$ orientation (Hellwig and Zabel 2000). At low scattering angles, we observe the thickness oscillations that are superposed by a longer period modulation. The longer period is due to a thin Nb_2O_5 oxide formation at the niobium surface. Aside from the oxide layer, these measurements yield a thickness for the niobium film, which can be compared with the thickness to be gained from the Laue oscillations near the Nb(110) Bragg reflection. Hence, we have two thicknesses for the same film determined at high and low scattering vectors. What is the difference? At high Q_z, the measured crystalline thickness $d_{crys} = N \times d_{(110)}$ is the number of lattice planes with atomic layer spacing $d_{(110)}$. In contrast, at low Q_z, the measured thickness is the total Nb thickness d_{tot}, regardless of whether it is crystalline or not. But the oxide layer is not included, because it features another electron density and therefore acts as an independent layer. Comparing these two thicknesses, d_{crys} and d_{tot}, bears information on the crystalline

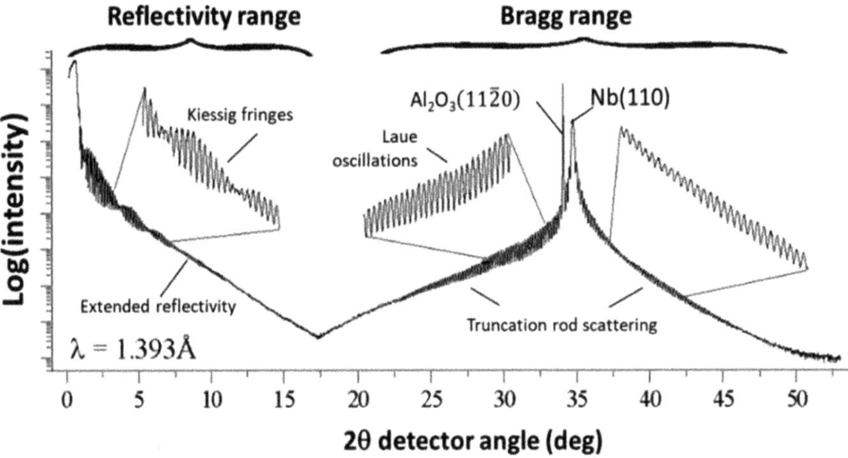

Fig. 10.24 Wide-angle scan covering both the reflectivity region and the Bragg region, yielding different information on the thickness and roughness of thin films and multilayers. Reproduced with permission from Hellwig and Zabel (2000). Copyright © 2025 by Elsevier Science B.V

growth quality and interface roughness, which vary with the growth parameters, such as the growth temperature and annealing procedures. The truncation rod scattering is also indicated in Fig. 10.24, as well as the sharp $(11\bar{2}0)$ peak from the sapphire substrate.

10.4.9 Master Equation

As has been mentioned before, the Parratt recursion formalism is the correct and rigorous form to treat the reflectivity of surfaces, thin films, and multilayers. At the same time, it is tedious to handle and difficult to make qualitative statements. A much more convenient approach taken within the Born approximation is the so-called Master equation, which we have encountered before for describing the surface and interface *rms* roughness. In this approximation, the reflectivity is expressed as the Fourier transform of the density gradient parallel to the surface/ interface normal (Als-Nielsen 1987):

$$R(Q_z) = R_F(Q_z)\left|\frac{1}{\rho_s}\int \frac{\partial \rho(z)}{\partial z}e^{iQ_z z}dz\right|^2 \tag{10.42}$$

Here, $R_F(Q_z)$ is the Fresnel reflectivity of the ideal surface, and ρ_s is the electron density deep in the substrate material. The integral:

$$\frac{1}{\rho_s}\int \frac{\partial \rho(z)}{\partial z}e^{iQ_z z}dz \tag{10.43}$$

is the Fourier transform of the density profile derivative across an interface. To illustrate how this works, we discuss a simple example of a single slab of material with density ρ_0 and thickness d. For simplicity, we furthermore assume that both interfaces are perfectly sharp so that so-called distributional the derivative of the density profile has a δ-function shape, as indicated in Fig. 10.25.

Evaluating the Fourier transform of the interface derivatives, we obtain:

$$\left| \frac{1}{\rho} \int_{-\infty}^{+\infty} \frac{\partial \rho(z)}{\partial z} e^{iQ_z z} dz \right|^2 = \left| \frac{1}{\rho_s} \left(\int \rho_f \delta(0) - \rho_f \delta(d) \right) e^{iQ_z z} dz \right|^2$$

$$= \left(\frac{\rho_f}{\rho_s} \right)^2 \left| e^{iQ_z 0} - e^{iQ_z d} \right|^2 = \left(\frac{\rho_f}{\rho_s} \right)^2 \left| 1 - e^{iQ_z d} \right|^2 = \left(\frac{\rho_f}{\rho_s} \right)^2 \sin^2 \frac{Q_z d}{2} \qquad (10.44)$$

Hence, the reflectivity of the slab is:

$$R(Q_z) = R_F(Q_z) \left(\frac{\rho_f}{\rho_s} \right)^2 \sin^2 \frac{Q_z d}{2} \qquad (10.45)$$

The last term recovers the Kiessig fringes of a film with thickness d, interference minima at $Q_z = l(2\pi/d)$, and separation of the minima $\Delta Q = 2\pi/d$. Instead of sharp interfaces, we can again use the error function $\mathrm{erf}(z)$ to model the roughness of surfaces and interfaces:

$$R(Q_z) = R_F(Q_z) \left(\frac{\rho_f}{\rho_s} \right)^2 e^{-Q_z^2 \sigma^2} \sin^2 \frac{Q_z d}{2} \qquad (10.46)$$

10.4.10   Fitting Procedures

In general, it is straightforward to calculate or simulate the expected reflectivity curve of a layered structure if the layer thicknesses, roughness parameters, and scattering length densities (SLDs) for X-rays or neutrons are provided. The sequence of calculations is illustrated in Fig. 10.26 from steps 1 to 3. However, it is much less

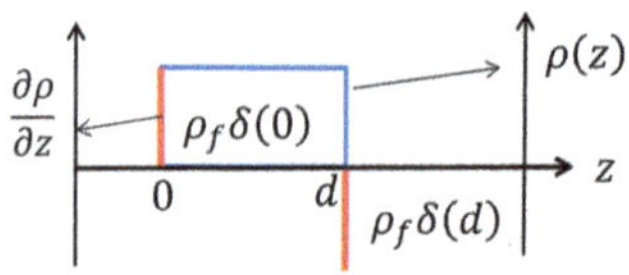

Fig. 10.25 Density profile of a single slab of thickness d. The blue line indicates the electron density $\rho(z)$, and the red bars show the derivative $\partial \rho(z)/\partial z$ at the interfaces

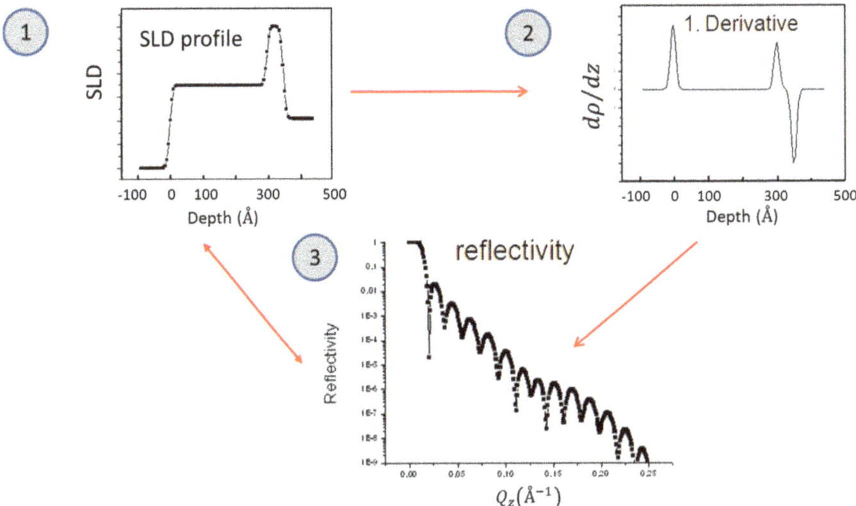

Fig. 10.26 Calculation procedure from scattering length density profiles, derivatives, to reflectivities. The simulation of reflectivities takes the reverse path from measured reflectivities to density profiles

straightforward to retrieve the SLD from a measured reflectivity curve by fitting procedures. Back-transformation, following steps 3 to 1, is the ultimate goal of XRR and NR measurements. However, the back-transformation is hampered by the phase problem, as already stated in Sect. 3.5.5 in the context of crystallography. Due to the loss of the scattering phase, it is per se not possible to distinguish between the top and the bottom layers. This means that a fit to a reflectivity curve may not be unique. Different SLD profiles may yield equally good fits to the data points. To overcome this phase problem, the simplest approach is to use the experimental knowledge about the sample preparation, for instance, by fixing the layer sequence known from the deposition process.

There have been several attempts to solve the phase problem, such as introducing a reference layer of known SLD and location (Majkrzak and Berk 1995), or the use of polarized neutron reflectometry that allows contrast variation by changing the incident neutron polarization (Farhad Masoudi and Pazirandeh 2005). A popular and successful fitting routine was developed by Björk and Anderson (2007), with further refinements by Glavic and Björk (2022; https://aglavic.github.io/genx/). The GenX package, now in the third generation, uses a fast differential evolution algorithm that offers a highly effective and robust optimization procedure. It is an open-source X-ray reflectivity curve fitting software package and can also be used for fitting neutron reflectivity curves. Another approach based on neural network methods was recently presented by Schreiber and coworkers (Munteanu et al. 2024). Here, a novel training procedure was introduced that incorporates dynamic prior boundaries for physical parameters and allows users to flexibly input their own prior knowledge about the sample parameters.

10.5 Off-Specular Diffuse Scattering

10.5.1 Basic Considerations

We realized before that roughness lowers the specularly reflected intensity (see Fig. 10.16 and 10.21). As the total scattered intensity is constant, part of the specularly reflected intensity must be diverted into the off-specular diffuse regime, which is spread out to the left and right of the specular ridge. This is indicated schematically in Fig. 10.27. The shape of the diffuse scattering depends on the number of interfaces, the type of interface roughness, and whether the interface roughness is correlated or not. The diffuse scattering can be analyzed by tilting the film normal in the direction of k_i or k_f, which provides a scattering vector component in the Q_x-direction. The scan direction is equivalent to the transverse scan direction sketched in Fig. 5.2 for single crystals. When the rocking angle is extended further, either the incoming beam on the left-hand side or the reflected beam on the right-hand side will touch the surface. As the glancing angle of the incoming or reflected beams α_i, α_f approach the critical angle α_c, the beam intensity is enhanced and a peak is observed, which is known as the Yoneda enhancement (Yoneda 1963). Since this peak spreads out into a ring, we also call it "Yoneda wing". It is an optical effect and not due to in-plane correlations, but solely due to the squared Fresnel transmission coefficient $|t_{i,f}|^2$, which has a maximum at the critical angle (see Fig. 10.15). The Yoneda effect can be used to enhance certain features, which would otherwise be hard to detect.

Figure 10.28 provides an overview of different scenarios for off-specular intensity as a function of the transverse scattering vector Q_x. The top panel is just a reminder of the already discussed specular (Fresnel) reflectivity from a perfectly flat surface, which is free of diffuse scattering. In the transverse scan, only a single peak occurs by crossing the specular reflectivity ridge at $Q_x = 0$. The middle panel refers to a case of random but limited surface roughness that produces diffuse scattering within a range of angles centered at the specular ridge. The peaks on both sides are due to the Yoneda enhancement. If the surface roughness causes diffuse scattering into the solid angle of 4π, reflectivity could not be observed. What we consider here

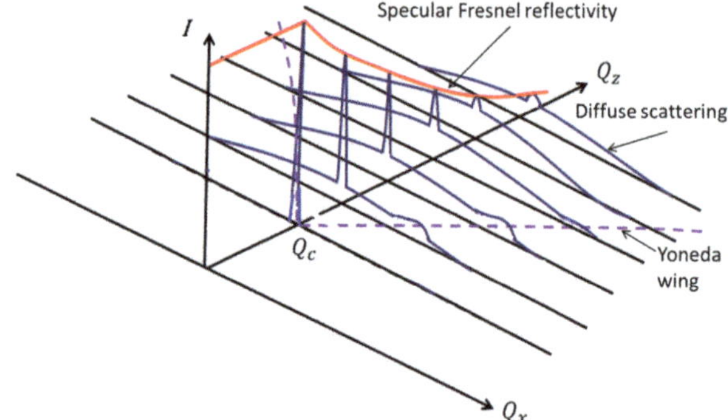

Fig. 10.27 Off-specular diffuse intensity in a perspective view. The specular Fresnel reflectivity ridge is shown in red color, and the off-specular diffuse scattering intensity (blue lines) is enhanced along the circular (dashed) line known as the Yoneda wing. The Yoneda wing also exists on the left side, but is obscured in the schematics

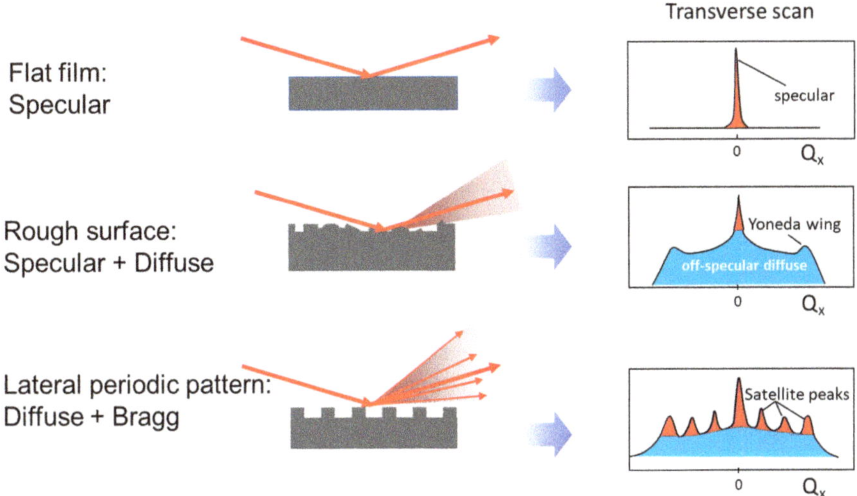

Fig. 10.28 Three surfaces with and without roughness produce different types of diffuse scattering, as discussed in the main text

can be compared with a dusty or milky mirror, which reflects and scatters light simultaneously. But what looks like a perfect mirror for light reflection is rough when probed with much shorter wavelengths of X-rays or neutrons. The bottom panel indicates a periodic array in the film plane, which may occur naturally by a surface reconstruction or artificially by nanostructuring. In any case, the periodicity gives rise to Bragg reflections in the lateral direction, the so-called satellite reflections. The intensity of the third reflection on either side is enhanced by the Yoneda effect.

10.5.2   Height–Height Correlation Function

We describe off-specular diffuse scattering using the Born approximation, which is valid for scattering angles above the critical angle and is justified by the weak diffuse scattering intensity compared to the specular intensity. The following treatment is based on the pioneering work of Sunil Sinha and his group (Sinha et al. 1988). As before in Eq. (8.9), we start with the scattering function being the Fourier transform of the pair correlation function $g(\mathbf{r})$:

$$S_{diff}\left(\mathbf{Q}_{\parallel}\right) = \int g(\mathbf{r}) \exp\left(i\mathbf{Q}_{\parallel} \cdot \mathbf{r}\right) d\mathbf{r} \qquad (10.47)$$

$\mathbf{Q}_{\parallel}$ is a scattering vector parallel to the surface, $\mathbf{r}$ is a vector in the surface plane, and $d\mathbf{r} = \mathbf{dxdy}$. The pair correlation function for a rough interface is mainly a function of the height fluctuations in the z-direction, separated by a lateral distance $r = r_1 - r_2$ in the plane:

$$g(r) = \left\langle (z(r_1) - z(r_2))^2 \right\rangle. \tag{10.48}$$

Here, $z(r_1)$ and $z(r_2)$ are the positions of the interface at locations r_1 and r_2. Evaluating the pair correlation function $g(r)$, we obtain:

$$g(r) = \left\langle 2z^2(r) \right\rangle - 2\langle z(r_1)z(r_2)\rangle = 2\sigma^2 - 2C(r) \tag{10.49}$$

One particular height–height correlation function that has been suggested by Sinha and his group is the following (Sinha et al. 1988):

$$C(r) = \sigma^2 \exp\left[- \left(r/\xi_{\|}\right)\right]^{2h} \tag{10.50}$$

Inserting in Eq. (10.49), we obtain for the pair correlation function:

$$g(r) = 2\sigma^2 \left[1 - \exp\left[- \left(r/\xi_{\|}\right)\right]^{2h}\right] \tag{10.51}$$

Here, σ is the mean-square roughness parameter, already introduced in Eq. (10.25), $\xi_{\|}$ is the lateral correlation length, and h is the so-called *Hurst parameter* (Sinha et al. 1988). The correlation length $\xi_{\|}$ is also known as the *cut-off correlation length,* meaning that in the limit $r \to \infty$, the interace height fluctuations saturate at the value of the rms roughness.

Hence, the upper and lower limits are as follows:

- For $r_1 - r_2 > \xi_{\|}$, the interface appears smooth, with the limit: $\lim_{r \to \infty} C(r) = 0$ and $\lim_{r \to \infty} g(r) = 2\sigma^2$;
- For $r_1 - r_2 < \xi_{\|}$, the interface appears rough and shows fractal behavior, with the limit: $\lim_{r \ll \xi} g(r) = 2\sigma^2 \left(r/\xi_{\|}\right)^{2h}$.

The Hurst parameter h characterizes the *jaggedness* of the interface and is related to the spatial dimensionality D via $h = 3 - D$, where $0 < h < 1$. Therefore, if $h = 1$, $D = 2$, and the interface appears flat and smooth. However, if $h = 0$, $D = 3$, and the interface appears very rough. The difference can be imagined by the appearance of mountain ranges. In one case, we see gentle hills and valleys; in the other case, steep rocks, deep canyons, and sharp cliffs, while the overall elevations are comparable. Figure 10.29 illustrates two cases for $h = 0.9$ and $h = 0.2$ for the same correlation length ξ and mean roughness parameter σ.

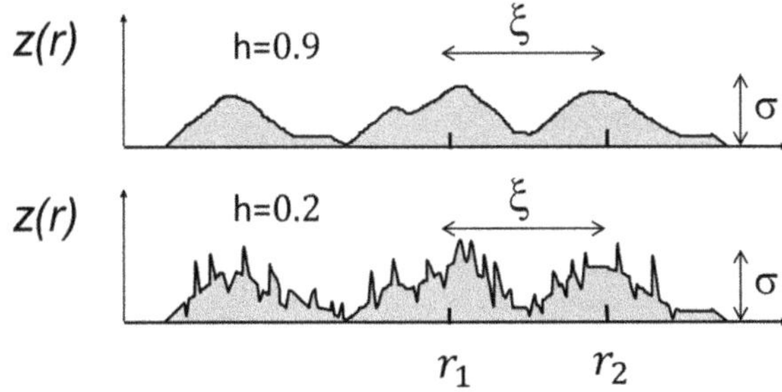

Fig. 10.29 Surface profiles with identical correlation length and mean-square roughness σ but different jaggedness expressed by different Hurst parameters h

10.5.3   Scattering Function for Rough Surfaces and Interfaces

Using the correlation function according to Eq. (10.51), and assuming that $g(r)$ is a Gaussian-distributed random variable, the scattering function, in Born approximation, takes the form (Yoneda 1963):

$$S_{diff}(\boldsymbol{Q}) = \frac{1}{Q_z^2} \int e^{-Q_z^2 g(r)/2} e^{i\boldsymbol{Q}_\parallel \cdot (\boldsymbol{r} - \boldsymbol{r}')} d\boldsymbol{r} d\boldsymbol{r}'. \tag{10.52}$$

Using the expression for $g(r)$ in Eq. 10.49, we find for $S_{diff}(\boldsymbol{Q})$:

$$S_{diff}(\boldsymbol{Q}) = \frac{1}{Q_z^2} e^{-Q_z^2 \sigma^2} \int \left[e^{Q_z^2 C(r)} - 1 \right] e^{i\boldsymbol{Q}_\parallel \cdot (\boldsymbol{r} - \boldsymbol{r}')} d\boldsymbol{r} d\boldsymbol{r}' \tag{10.53}$$

In the last line, -1 has been added in the square bracket to subtract the specular contribution to the scattering function. For small roughness, $e^{Q_z^2 C(r)} \ll 1$, and expanding the exponent, we obtain:

$$S_{diff}(\boldsymbol{Q}) = e^{-Q_z^2 \sigma^2} \int Q_z^2 C(r) e^{i\boldsymbol{Q}_\parallel \cdot (\boldsymbol{r} - \boldsymbol{r}')} d\boldsymbol{r} d\boldsymbol{r}' = \sigma^2 e^{-Q_z^2 \sigma^2} \int \exp[-(r/\xi)]^{2h} e^{i\boldsymbol{Q}_\parallel \cdot (\boldsymbol{r} - \boldsymbol{r}')} d\boldsymbol{r} d\boldsymbol{r}' \tag{10.54}$$

Here, we immediately realize that for $h = 1$, the correlation function has a Gaussian shape, and therefore the Fourier transform of the correlation function is also a Gaussian function:

$$S_{diff}(\boldsymbol{Q}) = \sigma^2 \int \exp\left[-(r/\xi)\right]^2 \exp\left(i\boldsymbol{Q}_\parallel \cdot (\boldsymbol{r} - \boldsymbol{r}')\right) d\boldsymbol{r} d\boldsymbol{r}' = \sigma^2 \sqrt{\pi \xi^2} \exp\left(-Q_x^2 \xi^2/4\right), \tag{10.55}$$

with a FWHM of $4\sqrt{\ln 2}/\xi$. Hence, the shorter the correlation length, the broader the diffuse scattering spreads out.

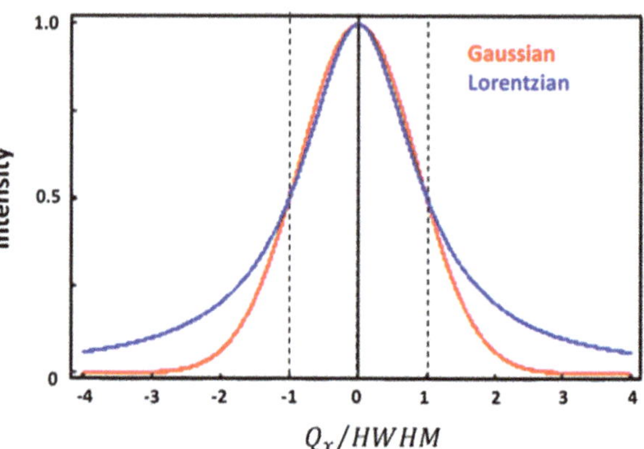

Fig. 10.30 Comparison of a Gaussian- and a Lorentzian-shaped curve. Both curves are normalized to the maximum and the intensity at half-width-at-half-maximum (HWHM). These curves are clearly distinguished, particularly in the shape of the tails

For $h = 1/2$, the correlation function decays exponentially, and the Fourier transform has a Lorentzian shape:

$$S_{diff}(\boldsymbol{Q}) = \sigma^2 \int \exp[-(R/\xi)] \exp(i\boldsymbol{Q} \cdot (\boldsymbol{r} - \boldsymbol{r}')) d\boldsymbol{r} d\boldsymbol{r}' = \sigma^2 \frac{\xi^{-1}}{(\xi^{-1})^2 + Q^2} \quad (10.56)$$

The FWHM of the Lorentzian-shaped curve is $2\xi^{-1}$. In Fig. 10.30, both shapes of the scattering function $S_{diff}(\boldsymbol{Q})$ are compared, normalized to the maximum and the half-width-at-half-maximum (HWHM). The Gaussian curve drops off rather fast and has little intensity beyond the third HWHM. In contrast, the Lorentzian line shape is peaked at the center but then drops off slowly, featuring long tails. Thus, by the visual inspection of the shape of the diffuse scattering, a rough estimate of the Hurst parameter h is possible.

If the diffuse scattering is determined for incident and exit angles close to the critical angle of total reflection, Eq. (10.53) has to be modified by the Fresnel transmission coefficients to take into account the distorted waves. The evaluation of the diffusely scattered intensity from a rough surface within an area $L_x L_y$ yields (Sinha et al. 1988):

$$I(Q)_{diff} = \frac{a^2}{Q_z^2} (L_x L_y) \frac{\left|k_0^2(1-n^2)\right|^2}{16\pi^2} |t(k_i)|^2 |t(k_f)|^2 S_{diff}(\boldsymbol{Q}_t). \quad (10.57)$$

Here, $t(k_i)$, $t(k_f)$ are the Fresnel transmission coefficients defined in Eq. (10.26), $\boldsymbol{Q}_t$ is the wavevector in the medium, $S_{diff}(\boldsymbol{Q}_t)$ is the scattering function for the off-specular diffuse scattering according to Eq. (10.54), and a is the scattering amplitude to be specified for X-rays or neutron scattering.

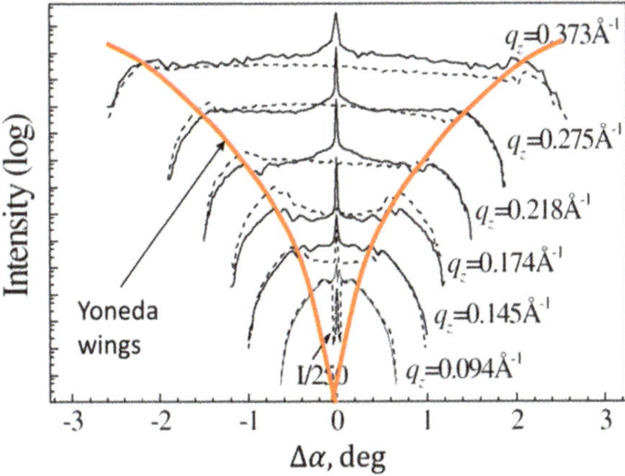

Fig. 10.31 Experimental results of transverse scans for two FePt films deposited on a GaAs substrate. The dashed curves are from the first sample, showing essentially no diffuse scattering, indicating very low roughness. The solid curves represent scans from a second sample at different q_z values (Q_z in the present notation). Curves are shifted vertically by a factor of 10^2 for clarity. The location of the Yoneda wing is clearly visible. For higher Q_z values, Lorentzian-shaped diffuse scattering becomes visible close to the specular peak. Adapted with permission from Nefedov et al. (2002). Copyright (2025) by IoP Publishing

10.5.4   Experimental Examples

An experimental result is shown in Fig. 10.31, reproduced from Nefedov et al. (2002). The transverse scans are taken from an epitaxial FePt(001) alloy layer grown on a GaAs(001) substrate. XRR results reveal a 0.24 nm rms roughness of the GaAs substrate and a 0.29 nm rms roughness of the 10.9 nm thick FePt layer, grown with MBE methods at 350 °C in ultra-high vacuum. As the FePt film roughness is essentially identical to the substrate roughness, the GaAs surface atomic steps are likely imprinted into the film. The specular peak, the Lorenzian-shaped diffuse scattering, and the Yoneda wings are easy to recognize by the solid lines in Fig. 10.31.

Another example concerns diffuse scattering, which can be observed by transverse scans in a multilayer. A distinction is made between the lateral roughness correlation within the layers and the perpendicular roughness correlation between the layers.

The multilayer consists of 50 repeats of 2.3 nm thick Au layers alternating with 2.93 nm thick magnetic Co_2MnGe layers, sketched in the inset of Fig. 10.32. The magnetic layers belong to the group of Heusler alloys with the $L2_1$ structure (compare Fig. 1.6g). In reflectivity mode, we are not concerned about the crystalline structure of the multilayer but focus on the interfaces. Figure 10.32 from Bergmann et al. (2005, 2006) shows an X-ray intensity map using 8 keV X-ray photons. The color-coded intensity map is plotted on a logarithmic scale as a function of glancing incident beam angle α_i versus glancing exit beam angle α_f. The diagonal ridge at

Fig. 10.32 X-ray intensity map of a Co$_2$MnGe/Au multilayer with 50 repeats. $\alpha_{i,\,f}$ are the incident and exit glancing angles. The diagonal $\alpha_i = \alpha_f$ shows the specular reflectivity. Red dots on the specular ridge are due to the multilayer Bragg reflections. The straight lines perpendicular to the specular ridge are due to correlated interface roughness. Reprinted with permission from Bergmann et al. (2006). Copyright (2025) by IoP Publishing

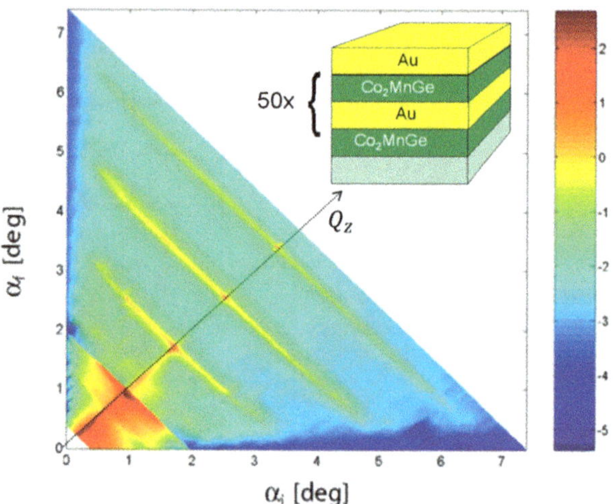

$\alpha_i = \alpha_f$ is the specular reflectivity ridge along Q_z in reciprocal space. At low angles in the lower left corner, the dark red color is due to total reflection below the critical scattering angle $\alpha_c = 0.42°$.

Along the diagonal, we notice red dots with decreasing intensity. Those are the multilayer Bragg peaks. Furthermore, we recognize a much weaker diffuse intensity line emanating from each Bragg peak that spreads out in the direction perpendicular to the specular ridge. These lines of intensity are referred to as Bragg sheets. They occur at all Bragg reflections and signify correlated interface roughness with the same correlation length and amplitude between all layers. Without correlated roughness, these Bragg sheets would vanish, and the diffuse scattering from the still present interface roughness would spread out over the entire reciprocal space. Nevertheless, some diffuse scattering is randomly distributed and contributes to the background scattering. The fitting of the data indicates that the interface *rms* roughness in this multilayer is on the order of $\sigma = 0.4$ nm, and the lateral correlation length $\xi_\parallel$ is on the order of 18 nm with a hurst parameter $h \cong 1$. Numerous further examples exist in the literature on off-specular diffuse scattering due to the surface and interphase roughness.

Summary

1. The reciprocal space of flat single crystals is composed of three main regions: 1. Specular and off-specular reflectivity; 2. Surface scattering using evanescent waves; 3. Crystal truncation rod scattering.
2. Laue intensity oscillations are useful for determining the crystalline thickness and roughness of thin films deposited on substrates.

3. Crystal truncation rod (CTR) scattering is a diffuse scattering ridge between Bragg reflections along reciprocal lattice rods normal to the surface of flat single crystals. The respective scattering intensity scales with $1/Q_\perp^2$, where $Q_\perp$ is the component of the scattering vector normal to the crystal surface.

4. CTR scattering is very sensitive to phase shifts at surfaces and between surface layers and substrates.

5. Reflectivity refers to the reflection of X-ray or neutron waves from surfaces, interfaces, thin films, and multilayers. Reflectivity reveals surface smoothness versus roughness, layer thicknesses independent of the state of matter, and in-plane versus perpendicular correlation effects.

6. We distinguish between Kiessig fringes and Laue oscillations. Laue oscillations require a high crystalline quality of thin films. Kiessig fringes require sharp interfaces but no crystallinity.

7. The refractive index of X-rays and neutrons is smaller than one by a small amount. In the case of neutrons, this statement holds only for isotopes with a positive coherent scattering length.

8. The critical scattering vector Q_z^c for X-rays and neutrons only depends on material properties, not on scattering geometries.

9. The Fresnel reflectivity is 1 below the critical angle for total reflection and drops off with the power of 4 for angles higher than the critical angle.

10. The Fresnel reflectivity is affected by absorption, surface roughness, and footprint problems close to the critical angle.

11. The penetration depth of X-rays below the critical angle for total reflection is determined by the imaginary part of the transmitted scattering vector and equals the inverse of the critical scattering vector Q_z^c.

12. Kiessig fringes reveal the thickness of thin films.

13. The Parratt recursion formalism is the exact method to calculate the reflectivity of thin films and multilayers. The Master formula is an easier approach to calculate approximately the reflectivity of thin films and multilayers.

14. The Master Formula requires taking the Fourier transform of the density profile derivative normal to the surface.

15. Surface and interface roughness cause diffuse off-specular scattering.

16. The off-specular diffuse scattering is probed by transverse Q_x scans.

17. The off-specular diffuse scattering allows the characterization of the in-plane correlation length and the jaggedness of the interface profile.

18. The multilayer Bragg reflections are spread out into parallel Bragg sheets if the interface roughness is correlated in the perpendicular direction.

Questions

(Note that more than one answer may be correct)

1. **What are the three most important parts of the reciprocal lattice for reflectivity and surface work?**

   (a) The reciprocal lattice areas around the (100), (110), and (111) reflections.
   (b) The area close to the total reflection, the off-specular region, and the region for excitation of Bragg reflections with evanescent waves.
   (c) The extended reflectivity area around Bragg reflections, the CTR scattering, and the scattering of the transmitted waves.

2. **What is the difference between Kiessig fringes and Laue oscillations?**

   (a) Kiessig fringes are reflectivity oscillations of thin films close to the angle of total reflection. Laue oscillations are intensity oscillations close to Bragg reflections.
   (b) Kiessig fringes are intensity oscillations due to a finite film thickness, independent of the crystalline state of the film. Laue oscillations require good single crystallinity of thin films.
   (c) Laue oscillations occur for amorphous thin films, and Kiessig fringes are due to liquid films.

3. **What is the penetration depth Λ_{eff} of X-rays depending on above and below the critical angle of total reflection α_c?**

   (a) Below α_c, Λ_{eff} is limited by the imaginary part of the scattering vector. Above α_c, Λ_{eff} is limited by the photoelectric absorption effect.
   (b) Below α_c, Λ_{eff} is limited by the photoelectric absorption effect. Above α_c, Λ_{eff} is limited by the imaginary part of the scattering vector.
   (c) Below α_c, Λ_{eff} is limited by the absorption of neutrons. Above α_c, Λ_{eff} is limited by total reflection.

4. **What do you understand by crystal truncation rod (CTR) scattering?**

   (a) CTR scattering is due to the half-infinite nature of single crystals.
   (b) CTR scattering is diffuse scattering due to surface roughness.
   (c) CTR scattering competes with the Fresnel reflectivity of thin crystalline films.

5. **How can the thickness of thin films be determined by X-ray or neutron reflectivity?**

   (a) By measuring the critical angle of total reflection.
   (b) By measuring the penetration of X-rays or neutrons at the critical angle.
   (c) By measuring the Fresnel reflectivity and determining the periodicity of Kiessig fringes.

6. **How is the roughness of surfaces and interfaces expressed in scattering experiments?**

   (a) Roughness causes an enhanced specular and diffuse scattering.
   (b) Roughness causes a reduced specular scattering and an enhanced diffuse scattering.
   (c) Roughness does not affect XRR and NR.

Exercises

Grades of difficulty: E = easy, M = medium, A = advanced

M10.1 CTR scattering intensity
   Show that the Eqs. (10.4) and (10.5) for CTR scattering are correct.
E10.2 Critical angle
   Determine the critical angle for total reflection for X-rays and neutrons, both with a wavelength of 0.1 nm scattered at a flat silicon surface.
E10.3 Neutron critical scattering vector
   Show that Eq. (10.16) can also be derived from the condition that at Q_c, the kinetic energy of the transmitted wave E_z^t should be zero.
E10.4 Yoneda wings
   Determine the conditions for the Yoneda wings.
E10.5 Fresnel transmission coefficient
   Calculate and plot the Fresnel transmission coefficient.
E10.6 Small-angle approximation
   Show that in small-angle approximation the Fresnel reflectivity scales with Q_z^{-4}.

References

J. Als-Nielsen, in *Structure and Dynamics of Surfaces II*, Topics in Current Physics, ed. by W. Schommers, P. Blanckenhagen, vol. 43, (Springer, Berlin, Heidelberg, 1987), p. 181

A. Bergmann, J. Grabis, B.P. Toperverg, V. Leiner, M. Wolff, H. Zabel, K. Westerholt, Phys., Antiferromagnetic dipolar ordering in Heusler [Co_2MnGe/V]$_N$ multilayers. Phys. Rev. B **72**, 214403 (2005)

A. Bergmann, J. Grabis, A. Nefedov, K. Westerholt, H. Zabel, X-ray resonant magnetic scattering study of [Co_2MnGe/Au]$_n$ and [Co_2MnGe/V]$_n$ multilayers. J. Phys. D. Appl. Phys. **39**, 842–850 (2006)

M. Björck, G. Andersson, GenX: An extensible X-ray reflectivity refinement program utilizing differential evolution. J. Appl. Crystallogr. **40**, 1174–1178 (2007)

M. Born, E. Wolf, *Principles of Optics* (Pergamon, New York, 1980)

A. Braslau, M. Deutsch, P.S. Pershan, A. Wess, J. Als-Nielsen, J. Bohr, Surface roughness of water measured by X-ray reflectivity. Phys. Rev. Lett. **54**, 114 (1985)

H. Dosch, Evanescent absorption in kinematic surface Bragg diffraction. Phys. Rev. B **35**, 2137 (1987)

S. Farhad Masoudi, A. Pazirandeh, Application of polarized neutron in determination of the phase in neutron reflectometry. Phys. B **356**, 21–25 (2005)

R. Feidenhans'l, Surface structure determination by X-ray diffraction. Surf. Sci. Rep. **10**, 105 (1989)

R.M. Fleming, D.B. McWhan, A.C. Gossard, W. Wiegmann, R.A. Logan, X-ray diffraction study of interdiffusion and growth in (GaAs)n(AlAs)m multilayers. J. Appl. Phys. **51**, 357–363 (1980)

A. Glavic, M. Björck, GenX 3: The latest generation of an established tool. J. Appl. Crystallogr. **55**, 1063–1071 (2022)

O. Hellwig, H. Zabel, Oxidation of Nb(110) thin films on a-plane sapphire substrates: An X-ray study. Phys. B **283**, 228 (2000)

H. Kiessig, Untersuchungen zur Totalreflexion von Röntgenstrahlen. Ann. Phys. **10**, 715–769 (1931)

V. Lauter-Pasyuk, H.J. Lauter, B.P. Toperverg, L. Romashev, V. Ustinov, Transverse and lateral structure of the spin-flop phase in Fe/Cr antiferromagnetic Superlattices. Phys. Rev. Lett. **89**, 167203 (2002)

C.F. Majkrzak, N.F. Berk, Exact determination of the phase in neutron reflectometry. Phys. Rev. B. **52**, 10827–10830 (1995)

C.F. Majkrzak, J. Kwo, M. Hong, Y. Yafet, D. Gibbs, C.L. Chien, J. Bohr, Magnetic rare earth superlattices. Adv. Phys. **40**, 99–189 (1991)

P.F. Miceli, H. Zabel, J.E. Cunningham, Hydrogen-induced strain modulation in Nb-ta superlattices. Phys. Rev. Lett. **54**, 917–919 (1985)

P.F. Miceli, D.A. Neumann, H. Zabel, X-ray refractive index: A tool to determine the average composition in multilayer structures. Appl. Phys. Lett. **48**, 24 (1986)

R. Moubah, F. Magnus, T. Warnatz, G.K. Palsson, V. Kapaklis, V. Ukleev, A. Devishvili, J. Palisaitis, P.O.Å. Persson, B. Hjörvarsson, Discrete layer-by-layer magnetic switching in Fe/MgO (001) superlattices. Phys. Rev. Appl. **5**, 044011 (2016)

V. Munteanu, V. Starostin, A. Greco, L. Pithan, A. Gerach, A. Hinderhofer, S. Kowarik, F. Schreiber, Neural network and analysis of neutron and X-ray reflectivity data incorporating prior knowledge. J. Appl. Crystallogr. **57**, 456–469 (2024)

A. Nefedov, T. Schmitte, K. Theis-Bröhl, H. Zabel, M. Doi, E. Schuster, W. Keune, Growth and structure of L10 ordered FePt films on GaAs (001). J. Phys. Condens. Matter **14**, 12273 (2002)

L.G. Parratt, Surface studies of solids by Total reflection of X-rays. Phys. Rev. **95**, 359 (1954)

I.K. Robinson, Crystal truncation rods and surface roughness. Phys. Rev. B **33**, 3830 (1986)

I.K. Robinson, R.T. Tung, R. Feidenhans'l, X-ray interference method for studying interface structures. Phys. Rev. B **38**, 3632 (1988)

D.K. Satapathy, M.A. Uribe-Laverde, I. Marozau, V.K. Malik, S. Das, T. Wagner, et al., Magnetic proximity effect in $YBa_2Cu_3O_7/La_{2/3}Ca_{1/3}MnO_3$ and $YBa_2Cu_3O_7/LaMnO_{3+\delta}$ superlattices. Phys. Rev. Lett. **108**, 197201 (2012)

A. Schreyer, C.F. Majkrzak, T. Zeidler, T. Schmitte, P. Bödeker, K. Theis-Bröhl, A. Abromeit, J.A. Dura, T. Watanabe, Magnetic structure of Cr in exchange coupled FeyCr(001) Superlattices. Phys. Rev. Lett. **89**, 167203 (2002)

S.K. Sinha, E.B. Sirota, S. Garoff, H.B. Stanley, X-ray and neutron scattering from rough surfaces. Phys. Rev. B **38**, 2297 (1988)

A. Stierle, *Dissertation* (Ruhr-Universität Bochum, Germany, 1996)

A. Stierle, A. Abromeit, N. Metoki, H. Zabel, High resolution X-ray characterization of Co films on Al_2O_3. J. Appl. Phys. **73**, 4808–4814 (1993)

J. Todd, R. Merlin, R. Clarke, K.M. Mohanty, J.D. Axe, Synchrotron X-ray study of a fibonacci superlattice. Phys. Rev. Lett. **57**, 1157 (1986)

G.H. Vineyard, Grazing-incidence diffraction and the distorted-wave approximation for the study of surfaces. Phys. Rev. B **50**, 4146 (1982)

Y. Yoneda, Anomalous surface reflection of X rays. Phys. Rev. **131**, 2010 (1963)

Further Reading

J. Als-Nielsen, D. McMorrow, *Elements of Modern X-Ray Physics*, 2nd edn. (Wiley, 2011)

R. Feidenhans'l, Surface structure determination by X-ray diffraction. Surf. Sci. Rep. **10**, 105 (1989)

J. Daillant, A. Gibaud, *X-Ray and Neutron Reflectivity: Principles and Applications* (Springer, Berlin/Heidelberg, 2009)

A.E. Ares (ed.), *X-ray Scattering* (IntechOpen, 2017) open access, peer reviewed volume, publ. https://www.intechopen.com/books/5371

Schreiber, F., Zaluzhnyy, I.: Lectures in: Modern X-ray Scattering: Coherence and its applications, https://www.soft-matter.uni-tuebingen.de/vorlesung_ss22_modern_scattering.html

Chapter 11
Magnetic Thin Films and Multilayers

Magnetic films possess all the physical properties already discussed in Chap. 10 and, in addition, exhibit a magnetization vector whose direction can be reversibly changed by an external magnetic field. Domain structures and magnetization reversal patterns are characteristic of the material selection, interfacial interactions, and film thickness. All of these can be explored using magnetic neutron reflectivity, more specifically, polarized neutron reflectivity that reveals magnetization density profiles normal to the layers. How this works is explained here. The use of resonant magnetic X-ray reflectivity adds element specificity to the magnetization density profiles. The presentation and discussion are introductory and semi-qualitative. For a more detailed presentation, we refer to the reviews on this topic under "Further Reading."

11.1 Magnetic Neutron Reflectivity

We recall from Chap. 7 that the magnetic moment of neutrons, $\boldsymbol{\mu}_n$, interacts with the magnetic induction $\boldsymbol{B}$, providing a magnetic (Zeeman) potential for the neutrons. The corresponding magnetic potential energy is

$$\widehat{V}_m = -(\widehat{\boldsymbol{\mu}}_n \cdot \boldsymbol{B}), \tag{11.1}$$

where the hat indicates the matrix property of the potential and $\widehat{\boldsymbol{\mu}}_n$ is represented by the 2×2 Pauli matrices. In general, the magnetic induction $\boldsymbol{B}$ in a magnetic film consists of an external and applied Oersted field $\boldsymbol{H}$ and a sample magnetization $\boldsymbol{M}$:

$$\boldsymbol{B} = \boldsymbol{H} + \mu_0 \boldsymbol{M} \tag{11.2}$$

We assume that the neutrons interact only with the magnetization vector $\boldsymbol{M}$ inside films and multilayers and that the induction outside the sample can be neglected.

© The Author(s), under exclusive license to Springer Nature Switzerland AG 2026
H. Zabel, *Elements of Elastic Scattering by X-Rays, Neutrons, and Electrons*,
https://doi.org/10.1007/978-3-032-16624-1_11

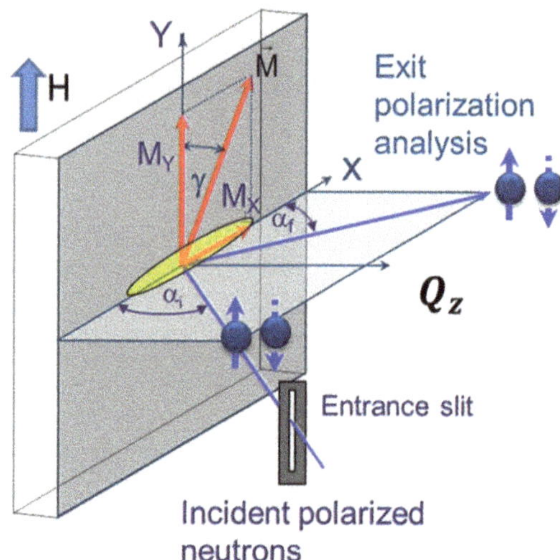

Fig. 11.1 Scattering geometry for polarized neutron reflectivity. The (X, Y) sample plane is shown in grayish color; the (X, Z) scattering plane is indicated in bluish color. The scattering vector Q_z is oriented normal to the sample plane. The polarized incident beam makes a glancing angle α_i with the sample plane, and α_f is the exit glancing angle. The magnetization vector M is tilted in the sample plane and has components in the X- and Y-directions. The Y-axis is the neutron spin quantization axis, maintained by a guiding magnetic field. The incident polarization is either parallel to the guiding field or antiparallel. The exit beam exhibits either the same polarization or a flipped one. The entrance slit, defining the resolution ellipsoid, is marked in yellow and is narrow in the X-direction but wide open in the Y-direction. Adapted from Gorkov et al. (2020). (open access)

Moreover, we further assume that the magnetization vector with components (M_X, M_Y) is coplanar with the film and perpendicular to the scattering vector Q_z, as indicated in Fig. 11.1.

The following discussion is similar to the one in Sect. 7.5, but the coordinate system is adapted to the standard one used in the reflectometry literature. The specularly reflected beam with a scattering vector oriented normal to the surface is customarily designated as the Z-direction. Accordingly, the polarization axis of the incident neutron beam is usually taken as the Y-direction, and the X-direction is perpendicular to it in the sample plane. Typically, the magnetization vector lies in the X, Y plane, and (X, Z) is the scattering plane. We first assume that the sample is in a ferromagnetic and saturated single-domain state, with all magnetic moments represented by the magnetization vector M, and forming an angle γ to the Y- axis in Fig. 11.1.

Considering a magnetization vector $M=(0, M_Y)$, an incident monochromatic neutron beam polarized collinearly with the Y-axis, the potential energy of the neutron–sample interaction can be represented by the sum of nuclear and magnetic contributions:

$$\widehat{V}_\pm = \widehat{V}_{nuc} \pm \widehat{V}_{mag} = \frac{2\pi\hbar^2}{m_n}\rho_A(b_N \pm p_Y) = \frac{2\pi\hbar^2}{m_n}\rho_A\begin{pmatrix} b_N+p_Y & 0 \\ 0 & b_N-p_Y \end{pmatrix} \qquad (11.3)$$

b_N is the nuclear coherent scattering length, and p_Y is the magnetic scattering length. ρ_A is the atomic number density and m_n is the neutron mass. The $\pm$ signs stand for the neutron polarization parallel (antiparallel) to the neutron quantization axis, that is, (anti-) parallel to $\boldsymbol{H}$ in the graph above (neutron guiding field). The magnetic scattering length is defined in Eqs. (7.11) and (7.17) as $p = \gamma r_0 f(Q) = (m_n/2\pi\hbar^2)\mu_n B$.

If, however, the magnetization vector is $\boldsymbol{M}=(M_x, M_Y)$, then the projection along the $X-$axis has to be taken into account, and therefore, the matrix of magnetic interaction V_m is no longer diagonal:

$$\widehat{V}_\pm = \frac{2\pi\hbar^2}{m_n}\rho_A\begin{pmatrix} b_N+p_Y & -p_X \\ p_X & b_N-p_Y \end{pmatrix} = \begin{pmatrix} V_{++} & V_{+-} \\ V_{-+} & V_{--} \end{pmatrix}. \qquad (11.4)$$

Here, $p_X = p\sin\gamma$ and $p_Y = p\cos\gamma$ are the projections of the respective magnetic scattering length components. In this case, the full dynamical theory has to be applied, meaning that the one-dimensional coupled and stationary Schrödinger equation:

$$\left[-\frac{\hbar^2}{2m_n}\frac{\partial^2}{\partial z^2} + V_\pm(z) - E_{kin} \right]\Psi(z) = 0 \qquad (11.5)$$

has to be solved for the neutron plane waves with two spin states:

$$\Psi(z) = \begin{pmatrix} \Psi_+(z) \\ \Psi_-(z) \end{pmatrix} = \begin{pmatrix} A_+ e^{iQ_z z} \\ A_- e^{iQ_z z} \end{pmatrix}, \qquad (11.6)$$

Inserting the spin-dependent wave functions into the Schrödinger equation yields:

$$\frac{\partial^2}{\partial z^2}\Psi_+(z) + \left(\frac{Q^2}{4} - \frac{2m_n}{\hbar^2}V_{++}(z) \right)\Psi_+(z) - \frac{2m_n}{\hbar^2}V_{+-}(z)\Psi_-(z) = 0$$

$$\frac{\partial^2}{\partial z^2}\Psi_-(z) + \left(\frac{Q^2}{4} - \frac{2m_n}{\hbar^2}V_{--}(z) \right)\Psi_-(z) - \frac{2m_n}{\hbar^2}V_{-+}(z)\Psi_+(z) = 0 \qquad (11.7)$$

Written in matrix form, we obtain:

$$\left(\begin{array}{cc} \dfrac{\partial^2}{\partial z^2} + \left(\dfrac{Q^2}{4} - \dfrac{2m_n}{\hbar^2} V_{++}(z) \right) & -\dfrac{2m_n}{\hbar^2} V_{+-}(z) \\[2ex] -\dfrac{2m_n}{\hbar^2} V_{-+}(z) & \dfrac{\partial^2}{\partial z^2} + \left(\dfrac{Q^2}{4} - \dfrac{2m_n}{\hbar^2} V_{--}(z) \right) \end{array}\right)$$
$$\times \left(\begin{array}{c} \Psi_+(z) \\ \Psi_-(z) \end{array} \right) = 0 \tag{11.8}$$

Both equations are coupled for $V_{+-}(z) \neq 0$. Moreover, both spin states split up into transmitted and reflected components when encountering a material with a different refractive index. The solution to these equations has been provided and reviewed by several authors (Felcher et al. 1987; Blundell and Bland 1992; Majkrzak et al. 2006). Here, we will quote some of the most important results. First, we distinguish between spin-flip and non-spin-flip reflectivities.

11.2 Non-Spin-Flip Reflectivity (NSF)

NSF implies that upon scattering and reflection by a magnetic surface or any kind of magnetic slab, the initial polarization does not change. This requires that $\gamma = 0$ in Fig. 11.1, i.e., the neutron spin is collinear with the sample magnetization. Then, the total potential is the one derived already before:

$$V_{\pm} = V_{nuc} \pm V_{mag} = \frac{2\pi\hbar^2}{m_n} \rho_A (\rho_N \pm \rho_Y), \tag{11.9}$$

and the refractive index is similarly derived as in Eqs. (10.17) and (10.18), yielding:

$$n_{\pm}^{neut} = \sqrt{1 - \frac{4\pi}{k_0^2}\left(\rho_A b_N \pm \rho_A' p_Y\right)} \approx 1 - \frac{2\pi}{k_0^2}\rho_A(\rho_N \pm \rho_Y) = 1 - \frac{2\pi}{k_0^2}(b_N \pm p_M) \tag{11.10}$$

Here, ρ_A is the nuclear or atomic number density, and ρ_A' is the magnetic moment density or spin density. The product $\rho_N = \rho_A b_N$ is called the nuclear scattering length density (nSLD), and $\rho_M = \rho_A' p_Y$ is the magnetic SLD (mSLD). ρ_M is also often quoted as $\rho_M = CM_Y$, where the constant C is (see Eq. (7.9)):

$$C = \frac{m_n \gamma_n \mu_N \mu_0}{2\pi\hbar^2} = 2.312 \times 10^{-4} \ \mathrm{nm}^{-2}\mathrm{T}^{-1}$$

In the 3d ferromagnetic transition metals (Co, Fe, Ni), $\rho_A = \rho_A'$, but in compound magnetic materials such as magnetic oxides, they are different, $\rho_A \neq \rho_A'$.

Table 11.1 Table of scattering lengths and critical scattering vectors for three ferromagnetic transition metals

	Fe	Co	Ni
Atomic density (nm^{-3})	84.957	fcc 89.8, hcp 90.2	91.7
Magnetic induction B [T]	2.2	1.72	0.62
Coherent scattering length $[10^{-6}\ \text{nm}]$	9.45	2.49	10.3
nSLD $10^{-4}\ \text{nm}^{-2}$	8.05	2.27	9.40
Magnetic scattering length p $[10^{-6}\ \text{nm}]$	0.6	0.46	0.16
mSLD (ideal) $10^{-4}\ \text{nm}^{-2}$	5.09	3.97	1.43
Nuclear $\langle Q_c \rangle [\text{nm}^{-1}]$	0.20	0.106	0.217
Magnetic $Q_{c+}[\text{nm}^{-1}]$	0.257	0.17	0.23
Magnetic $Q_{c-}[\text{nm}^{-1}]$	0.122	0	0.20

Following the lines after Eq. (10.20), we obtain for the critical scattering vector two values:

$$Q_{c,\pm}^{neutron} = \sqrt{16\pi(\rho_N \pm \rho_M)} = \sqrt{16\pi(\rho_N \pm CM_Y)} = \sqrt{Q_{c,N}^2 \pm Q_{c,M}^2} \qquad (11.11)$$

and two critical angles of total reflection for up- and down-polarized neutrons, which are in good approximation:

$$\alpha_{c,\pm}^{neutron} = \frac{\lambda}{4\pi} Q_{c,\pm}^{neutron} \qquad (11.12)$$

This is an important result. It tells us that there exist two separate critical edges, one for "up-" and the other for "down" polarized neutrons. For instance, while the "up" polarized neutrons are totally reflected at $Q_{c+}^{neutron}$, most of the down-polarized neutrons are still transmitted through the sample until reaching $Q_{c-}^{neutron}$. Hence, "up" and "down" neutrons can be spatially separated, and this is being used for polarizing neutron beams with the help of supermirrors (s. Fig. 4.27). For most isotopes, $\rho_N > 0$, but for $\rho_N < 0$ and down-polarization, the refractive index becomes imaginary; no critical edge for total reflection exists in this case. It may also happen that $\rho_N < \rho_M$. Then, a critical angle exists only for "up" neutrons but not for "down" neutrons, which is the case for Co films (compare Table 11.1).

We can determine the Fresnel coefficients for the up- and down-polarized neutrons in analogy to (10.22), and obtain:

$$r_\pm = \frac{Q_z - \sqrt{Q_z^2 - Q_{c\pm}^2}}{Q_z + \sqrt{Q_z^2 - Q_{c\pm}^2}} \qquad (11.13)$$

With this, and assuming a homogeneously magnetized sample with the magnetization vector tilted by an angle γ with respect to the quantization axis Y, and assuming perfect neutron polarization, we find for the non-spin-flip (NSF) neutron reflectivities the expressions (Toperverg and Zabel 2015):

$$R^{++} = \frac{1}{4}\left|(r_+ + r_-) + (r_+ - r_-)\cos\gamma\right|^2$$
$$R^{--} = \frac{1}{4}\left|(r_+ + r_-) - (r_+ - r_-)\cos\gamma\right|^2$$

(11.14)

NSF reflectivity implies that the spin state of the scattered neutrons is not changed. Spin-up neutrons remain in the "up" orientation before and after reflection at the sample, and the same holds for the "down" spins. Taking the difference of the NSF reflectivities, we obtain:

$$\Delta R^{NSF}(\gamma) = R^{+,+}(\gamma) - R^{-,-}(\gamma) = \left(|r_+|^2 - |r_-|^2\right)\cos\gamma \propto M_Y.$$

(11.15)

The difference is solely determined by the magnetization of the sample and is proportional to the magnetization projection parallel to the Y-direction. By normalizing with the NSF reflectivity for $\gamma = 0°$, we obtain directly $\cos\gamma$:

$$\frac{\Delta R^{NSF}(\gamma)}{\Delta R^{NSF}(0°)} = \frac{\left(|r_+|^2 - |r_-|^2\right)}{\left(|r_+|^2 - |r_-|^2\right)}\frac{\langle\cos\gamma\rangle}{1} = \langle\cos\gamma\rangle$$

(11.16)

The angle bracket indicates an ensemble average by taking the mean magnetization across the entire sample magnetization.

If the magnetization vector were oriented parallel to the X-direction ($\gamma = 90°$), then $\Delta R^{NSF} = 0$. The X-component causes spin-flip scattering, as we will discuss further below. Moreover, if the magnetization vector is oriented perpendicular to the sample plane in the Z-direction, then neither NSF nor SF reflectivity would recognize this magnetization component. For studying perpendicular magnetization components, the neutron reflectivity methods discussed here are not suitable but need to be complemented by grazing incidence methods, as discussed in Sect. 12.2.

Figure 11.2 shows a model reflectivity calculation for a 50 nm thick Fe film on a Si substrate with all interfaces assumed to be ideally flat. The magnetic hysteresis has a square shape as indicated in panel (a), with a coercive field of 10 Oe. In a guide field of slightly above 10 Oe, the sample is magnetically saturated in the Y-direction. In panel (b), we observe for the "up"-polarized neutrons typical thin-film (Kiessig) oscillations and a critical edge at $Q_{c+}^{neutron} = 0.26$ nm^{-1}, higher than the expected $Q_c = 0.20$ nm^{-1} calculated for the nuclear part of Fe in Sect. 10.4.3. The shift to higher momentum transfers is due to the magnetization in the film.

For the down-polarized neutrons, we also observe Kiessig fringes, which, however, are shifted to lower momentum transfers and therefore phase-shifted with respect to the Kiessig fringes of the "up" neutrons. We may take the difference divided by the sum, which is known as the spin asymmetry:

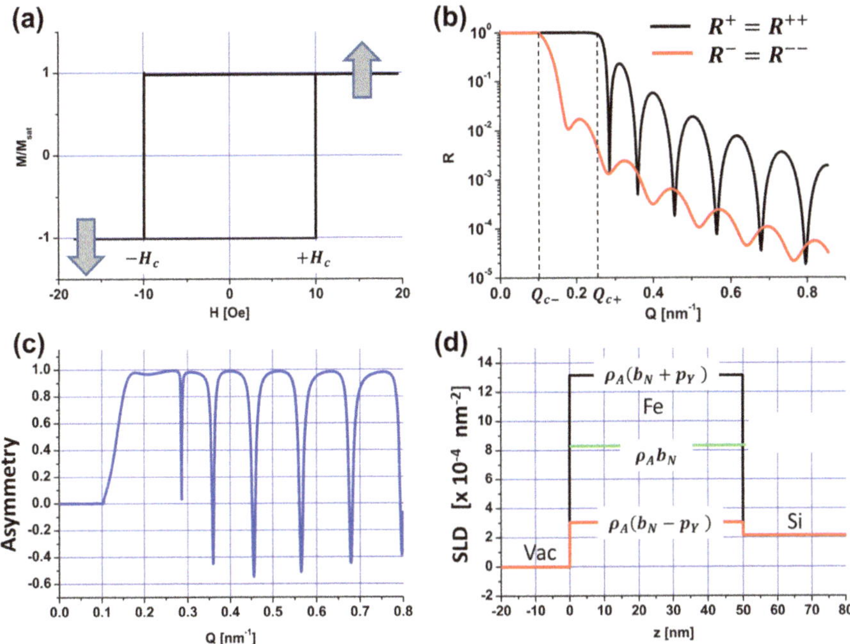

Fig. 11.2 Model calculation for neutron reflectivity from a fully magnetized 50 nm thick Fe film on a Si substrate. (**a**) Magnetic hysteresis with a "square" shape and full saturation at 10 Oe parallel to the Y-direction. (**b**) Non-spin-flip reflectivities R^+ and R^- for neutrons polarized parallel and antiparallel to the magnetization; (**c**) Spin asymmetry as a function of momentum transfer Q_z. (**d**) Scattering length density profiles for the nuclear and magnetic contributions to the reflectivities. Adapted with permission from Toperverg and Zabel (2015), Copyright (2025) by Elsevier

$$A = \frac{R^+ - R^-}{R^+ + R^-} \tag{11.17}$$

The spin asymmetry is plotted for the shown example in panel (c) of Fig. 11.2. In this plot, we can determine the location of the critical scattering vectors at $Q_{c+}^{neutron} = 0.26 \, \text{nm}^{-1}$ and $Q_{c-}^{neutron} = 0.135 \, \text{nm}^{-1}$. The squared difference is according to Eq. (11.11):

$$M_Y = \frac{Q_{c+}^2 - Q_{c-}^2}{32\pi C} \tag{11.18}$$

This yields in our case $M_Y = 2.2$ T, which was to be expected as it corresponds to the model assumption. Here, M_Y is the magnetization averaged over the entire film. Usually, the magnetization is reduced at the surface or interface to the substrate due to roughness, interdiffusion, or interaction with neighboring layers. Then, it is important to evaluate and plot the scattering length density (SLD) profile, obtained from a fit to the experimental reflectivity data. Panel (d) in Fig. 11.2 shows the idealized nuclear and magnetic SLD of the model calculations. The surface is at $z = 0$ and the interface to the Si substrate at $z = 50$ nm. We recognize three SLDs within the film: nuclear (green line), nuclear plus magnetic (+) (black line), and nuclear minus magnetic (−) (red line). We will later compare these idealized calculations with the experimentally obtained results for an Fe film, after discussing spin-flip scattering.

11.3   Spin-Flip Reflectivity (SF)

Any magnetization component perpendicular to the quantization axis causes the neutron spin to turn from "up" to "down", or vice versa. This scattering process is referred to as spin-flip (SF) reflectivity. SF reflectivity is completely magnetic in origin and has no coherent nuclear component.[1] The SF reflectivity is described by (Toperverg and Zabel 2015):

$$R^{+,-}(\gamma) = R^{-,+}(\gamma) = \frac{1}{4}|r_+ - r_-|^2 \sin^2\gamma \propto M_X^2 \qquad (11.19)$$

The SF reflectivities $R^{+,\,-}$, $R^{-,\,+}$ are proportional to the square of the magnetization component parallel to the X-direction. Taking a second scan with known angle $\gamma=90°$, we can normalize the SF reflectivity and obtain the angle γ:

$$\frac{R^{+,-}(\gamma)}{R^{+,-}(90°)} = \langle \sin^2\gamma \rangle \qquad (11.20)$$

Three properties of SF reflectivity are important to note. 1. Because the SF reflectivity depends on the modulus $|M_X|^2$, we do not know whether the magnetization is turned toward the positive or negative X-axis; 2. There is no total reflection nor a critical scattering vector for SF reflectivity. 3. The two reflectivities $R^{+,\,-} = R^{-,\,+}$ are degenerate. Now, we can rephrase the spin asymmetry in Eq. (11.16) by first defining the following quantities:

$$R^+ = R^{+,+} + R^{+,-}; R^- = R^{-,-} + R^{-,+} \qquad (11.21)$$

R^+ and R^- are those reflectivities, which are recorded, if no polarization analysis after the reflection from the sample is performed. It is often referred to as a "simplified" PNR measurement. These simplified reflectivities are composed of a nuclear and a magnetic part, for up-polarization:

---

[1] Nuclear SF scattering occurs only for isotopes that feature a nuclear angular moment $I > 0$. Nuclear SF scattering causes incoherent scattering, such as for the hydrogen isotope ^1H presented in Sect. 2.6.3.

$$R^+ = \underbrace{\frac{1}{2}\left(|r_+|^2 + |r_-|^2\right)}_{\text{nuclear}} + \underbrace{\frac{1}{2}\left(|r_+|^2 - |r_-|^2\right)\cos\gamma}_{\text{magnetic}} \qquad (11.22)$$

and for down-polarization:

$$R^- = \underbrace{\frac{1}{2}\left(|r_+|^2 + |r_-|^2\right)}_{\text{nuclear}} - \underbrace{\frac{1}{2}\left(|r_+|^2 - |r_-|^2\right)\cos\gamma}_{\text{magnetic}} \qquad (11.23)$$

The difference yields the magnetic part only:

$$R^+ - R^- = R^{+,+} - R^{-,-} = \left(|r_+|^2 - |r_-|^2\right)\cos\gamma \qquad (11.24)$$

The sum yields the nuclear part:

$$R^+ + R^- = R^{+,+} + R^{-,-} + 2R^{+,-;\,-,+} = \left(|r_+|^2 + |r_-|^2\right) \qquad (11.25)$$

For the spin asymmetry, we obtain:

$$A = \frac{R^+ - R^-}{R^+ + R^-} = \frac{R^{+,+} - R^{-,-}}{R^{+,+} + R^{-,-} + 2R^{+,-;\,-,+}} = \frac{\left(|r_+|^2 - |r_-|^2\right)}{\left(|r_+|^2 + |r_-|^2\right)}\langle\cos\gamma\rangle \quad (11.26)$$

We may take the SF reflectivity to the extreme by saturating a magnetic film along the Y-direction and then turning the magnetization vector to lie collinear with the X-direction. Then, we expect to observe a purely nonmagnetic and degenerate NSF reflectivity $R^{+,+}=R^{-,-}$ and a purely magnetic SF reflectivity $R^{+,-} = R^{-,\mp}$. The two extremes are compared in Fig. 11.3 by calculating the reflectivities for a 100 nm thick free-standing Fe film. In panel (a), the magnetization angle is $\gamma = 0$; therefore, $R^{+,+} \neq R^{-,-}$, $R^{+,-;\,-,+} = 0$. In panel (b), $\gamma = 90°$ and therefore $R^{+,+} = R^{-,-}$, $R^{+,-} = R^{-,+} > 0$. In real experiments, these extremes may never be realized, as spin canting, spin disorder at interfaces, and edge effects always cause some deviation from the ideal behavior. However, it should be realized that by measuring all four reflectivities versus the momentum transfer Q_z, the modulus of the magnetization ($|M|$) and the orientation of the magnetization vector (γ) can be determined as a function of depth. Hence, NSF and SF reflectivities provide information on the vectorial magnetization depth profile for single layers and multilayers.

The asymmetry A in Eq. 11.26 can, in principle, be used to measure a magnetic hysteresis. However, this is rarely done, as faster and less costly methods exist in laboratories for $M(H)$ measurements, including the magneto-optic Kerr effect (MOKE), vibrating sample magnetometer (VSM), and superconducting quantum interference device (SQUID). Only in special cases, $M(H)$ measurements with PNR

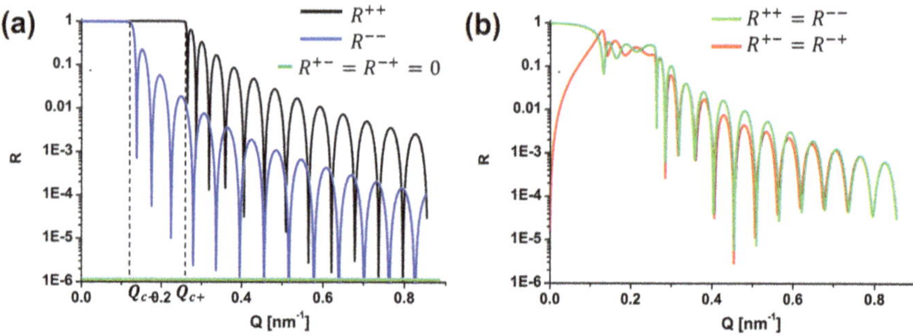

Fig. 11.3 Reflectivities calculated for a free-standing 100 nm thick and magnetically saturated Fe film. (**a**) The angle γ is assumed to be zero, and therefore $R^{+,\,+} \neq R^{-,\,-}$ (black and blue lines), while $R^{+,\,-} = R^{-,\,\mp} = 0$. (**b**) Same as in (**a**) but with $\gamma = 90°$. Therefore, $R^{+,\,+} = R^{-,\,-}$ (green line), $R^{+,\,-} = R^{-,\,\mp} > 0$ (red line). Adapted with permission from Toperverg and Zabel (2015), Copyright (2025) by Elsevier

are justified, for instance, to characterize the magnetization reversal mechanism and to scrutinize the domain formation during reversal via off-specular diffuse scattering. An example is given in Sects. 11.4.2 and 11.4.4.

To complete the model calculations of the NSF and SF reflectivities of magnetic films, we briefly discuss different types of magnetic multilayers. In all three presented cases, the multilayer consists of a periodic sequence of 8 nm thick Fe layers and 1 nm thick Cr layers. The double-layer thickness of $L = 9$ nm is repeated 12 times. All Fe layers are assumed to be in a saturated ferromagnetic state with the magnetization direction as indicated in the insets of Fig. 11.4. All Cr layers are assumed to be nonmagnetic. All interfaces are modeled as perfectly smooth.

The first example in Fig. 11.4a concerns a ferromagnetically ordered multilayer with the magnetization vectors oriented parallel to the NSF Y-axis. Therefore, we observe multilayer Bragg reflections in the NSF reflectivities at $Q_{z,l} = \sqrt{(Q_c)^2 + (2\pi l/L)^2}$, where l is the order of reflections according to Eq. (10.41), a maximum splitting between R^{++} and R^{--}, two critical scattering vectors $Q_c^{\pm}$, and no SF intensity. In the R^{--} reflectivity, the Bragg reflections are suppressed, because the scattering length density of Fe and Cr happens to be almost identical for the spin-down polarization. The Kiessig fringes observed are due to the total layer thickness, i.e., $12 \times L = 108$ nm.

Panel (b) is similar to panel (a) but with every second Fe magnetization rotated by 180°. This causes two main effects. First, the antiferromagnetic arrangement is, on average, nonmagnetic. Therefore, the R^{++} and R^{--} reflectivities are essentially identical. Second, due to the doubling of the magnetic periodicity, half-order Bragg reflections are generated. Panel (c) is similar to panel (b) but with the magnetization vectors rotated by 90° into the SF direction. Therefore, the NSF

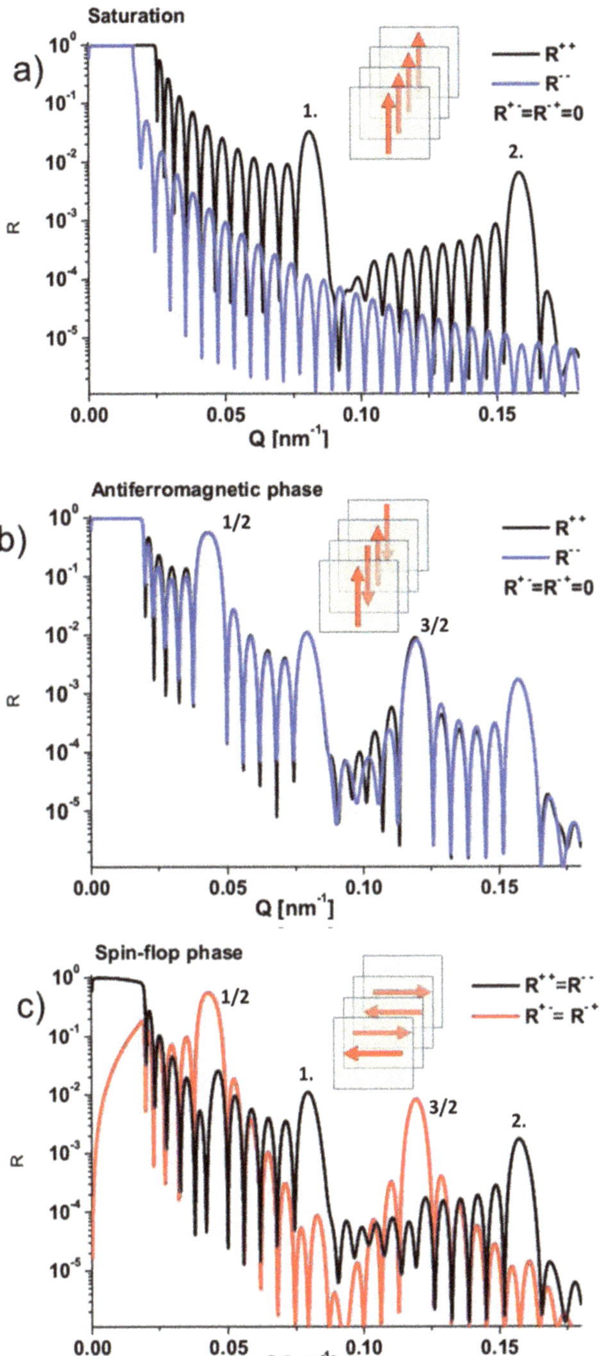

Fig. 11.4 Model calculations of NSF and SF reflectivities for a Fe/Cr multilayer with a double-layer thickness of 9 nm and 12 repeats. The insets show the orientation of the Fe layer magnetization, whereas the Cr layers in between are not marked (**a**) Ferromagnetic alignment of all Fe layer

reflectivities are now nonmagnetic and degenerate, while the SF reflectivities feature the half-order Bragg reflections. In the rotated state, nuclear and magnetic reflectivities are completely separated.

11.4 Experimental Examples

11.4.1 Thin Films

In the following, we present a real experiment that is analogous to the already discussed model calculations. It concerns a 100 nm thick and magnetically saturated Fe film deposited on a MgO substrate. The Fe film is covered with a 12 nm thick Pd protecting layer. Figure 11.5 shows all four measured reflectivities up to 1.45 nm^{-1} and over 5 orders of intensity. The NSF reflectivities are shown in blue and red. The critical edges are separated, and the thickness oscillations are phase-shifted, which are clear signs of the ferromagnetic state of the sample.

For a fully saturated Fe film parallel to the Y-axis, one should not expect any SF intensity. Nevertheless, SF intensity is recorded, but two orders of magnitude lower than the NSF intensities. This is due to the finite efficiency of the polarizer, analyzer, and spin flippers used in this experiment and some spin disorder at the interfaces. Some information on spin flippers can be found in Sect. 7.5 and at the end of this chapter. The Kiessig fringes from the 100 nm thick Fe film are clearly visible and far from the resolution limit of the instrument used.

The inset in Fig. 11.5 shows the nuclear and magnetic SLDs as retrieved from the fit to the experimental data. The magnetic SLD is split for up and down neutrons in the region of the Fe layer. The splitting corresponds to a magnetization of 2.145 T, which is close to the maximum induction of 2.2 T for Fe. Between the vacuum and the Fe film, we notice the unsplit nuclear SLD of the Pd protective layer. At the location of the interfaces, the SLDs vary smoothly as expected for real instead of idealized interfaces. The SF intensity is also an indicator of the background and noise level of the experiment. In the discussed case, the noise level is five orders of magnitude below the total reflected intensity, indicating that measurements beyond 1.5 nm^{-1} are meaningless. Depending on the circumstances, it may be necessary to take data at even higher momentum transfers. This requires, however, a reduction of the noise level. Reducing the background intensity in a neutron environment is not straightforward, as any straying and captured neutron produces new particles (γ, $\beta^{\pm}$) that contribute again to the noise.

Fig. 11.4 (continued) magnetization vectors parallel to the NSF axis, (b) Antiferromagnetic alignment of the Fe magnetization vectors, again parallel to the NSF axis. (c) AF alignment of the Fe magnetization vectors but now parallel to the SF axis. Reproduced and adapted with permission from Toperverg and Zabel (2015), Copyright (2025) by Elsevier

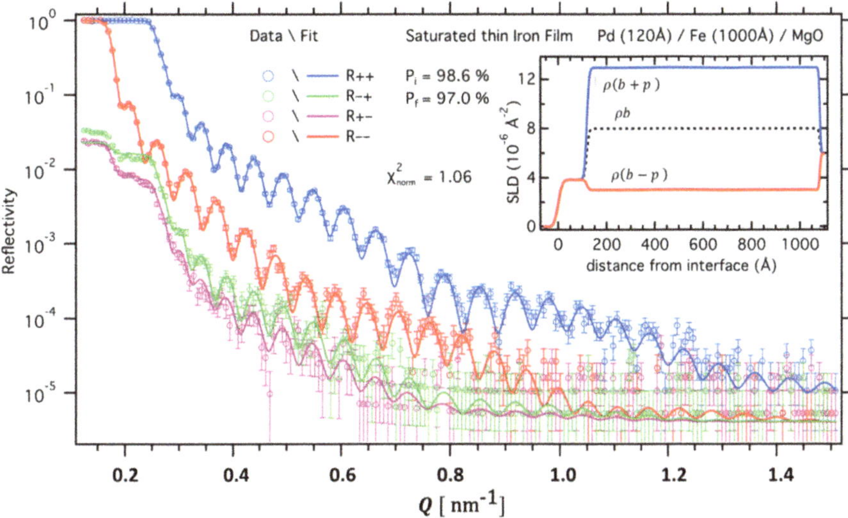

Fig. 11.5 Polarized neutron reflectivity results for a 100 nm thick Fe film on a MgO(001) substrate and covered with a 12 nm thick protecting Pd layer. Experimental reflectivity curves for R++, R−−, R+−, R−+ spin states are recorded for a fully saturated iron film in the Y-direction. Solid lines are the best fits to the experimental data. The inset shows the scattering length profile for up- (blue) and down- (red) polarized neutrons. The dashed line indicates the nuclear part of the scattering length profile of Fe. The nonsplit SLD at the surface is due to the Pd cap layer, and at the interface due to the MgO substrate. Reproduced with permission from Devishvili et al. (2013). Copyright (2025) by AIP

11.4.2 Magnetization Reversal

PNR has become a very versatile tool for the investigation of magnetism in layered and nanostructured materials. This is exemplified by studying the magnetization reversal process in thin films and multilayers. An example is schematically presented in Figs. 11.6 and 11.7 and discussed in the following.

The magnetization reversal from one saturated state to the opposite can proceed in three different ways. These are illustrated in Fig. 11.6 along with the expected PNR results. In panel (a), an opposite domain is nucleated, which propagates through the film and increasingly dominates over the original one. In scenario (b), the magnetization vector is continuously rotated in the external field, and in scenario (c), a domain state is formed at remanence before the opposite domain grows and takes over in the saturated state. These three scenarios can be distinguished using PNR methods, as schematically indicated in the right-hand panels of Fig. 11.6.

In case (a), only up- or down-domains exist, and therefore, the $R^{+,\,+}$ and $R^{-,\,-}$ reflectivities alternate during reversal without generating SF intensities. The domain rotation process in (b) leads to a maximum in SF reflectivity at a rotation angle of $\gamma = 90°$, while the NSF reflectivities go through a minimum. In case (c), the NSF and

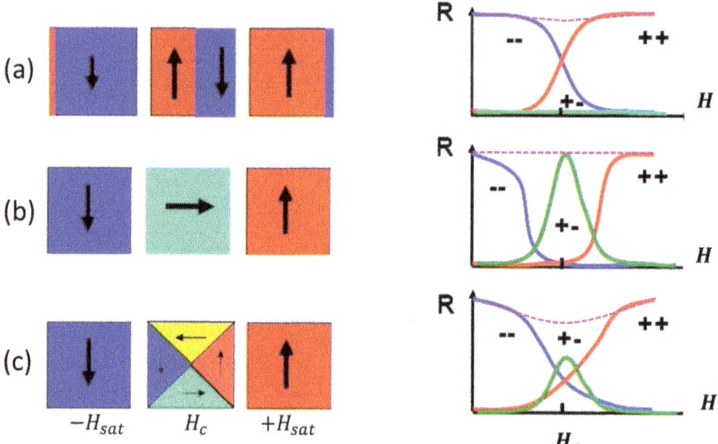

Fig. 11.6 Magnetization reversal processes in a thin magnetic film with in-plane anisotropy. Three scenarios are schematically illustrated: (**a**) domain nucleation and domain wall propagation; (**b**) coherent magnetization rotation; (**c**) domain formation. The right column sketches the respective NSF and SF reflectivities close to the coercive field H_c.

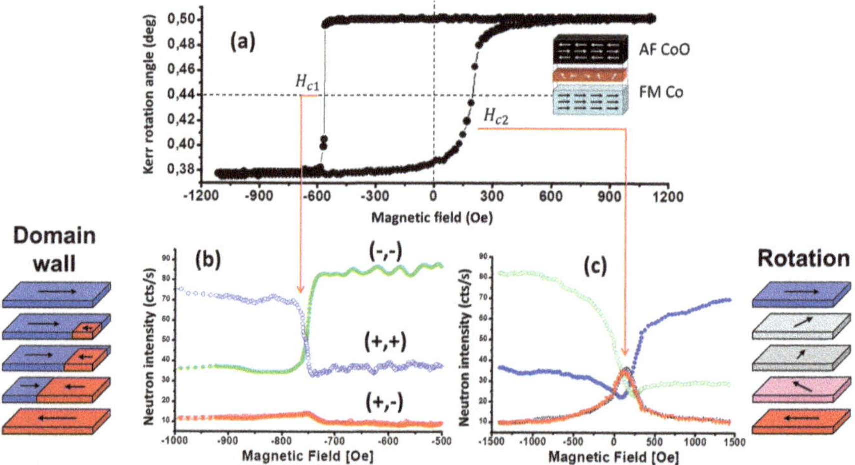

Fig. 11.7 Exchange bias system consisting of an antiferromagnetic CoO layer deposited on a epitaxial Co layer. The F-AF heterostructure is first cooled in a saturating magnetic field to below the Néel temperature of CoO. (**a**) Magnetic hysteresis measured with the magneto-optical Kerr effect. (**b**) All three NSF and SF reflectivities measured during a descending field across the coercive field H_{c1}. (**c**) Same as in (**b**) but in an ascending field across the coercive field H_{c2}. For more explanations, we refer to the main text. Adapted with permission from Radu et al. (2003). Copyright (2025) by APS

SF intensities are approximately equal at the coercive field H_c. The dashed purple line indicates the sum of all four reflected intensities. The missing intensity at H_c for cases (a) and (c) is redistributed into the off-specular diffuse intensity.

An example of the magnetization reversal monitored by neutron SF and NSF reflectivities is shown in Fig. 11.7. The system concerns an exchange bias heterostructure (Radu et al. 2003), consisting of a combination of ferromagnetic (F) and antiferromagnetic (AF) films with a common interface (inset in Fig. 11.7). When the F–AF-bilayer is cooled in a saturation magnetic field of the F-layer below the Néel temperature of the AF-layer, the F–AF interaction at the interface locks the ferromagnet in the field-cooling direction and hinders its magnetization reversal. This causes a shift of the magnetic hysteresis by the exchange bias field, and often also an asymmetric reversal mechanism is observed (Nogue's and Schuller 1999). This can also be seen in panel (a) of Fig. 11.7. Here, the exchange bias system is an antiferromagnetic CoO layer, deposited on a Co(111) epitaxial layer. This F–AF Co-CoO heterostructure was cooled from above the Néel temperature at 290 K in a field of 2000 Oe to low temperatures before recording the magnetic hysteresis curves with the magneto-optic Kerr effect (MOKE).

The descending branch of the hysteresis exhibits a sharp magnetization reversal at the coercive field H_{c1}, while in the ascending branch, the reversal is rounded at the coercive field H_{c2}. With PNR methods, the reversal mechanism can be analyzed in more detail, as is seen in the panels (b) and (c). The NSF R^{++} and R^{--} reflectivities show a large splitting and a sudden switch at the coercive field H_{c1}, while the SF reflectivity intensity stays at a low level, typical for a reversal by domain wall propagation. On the contrary, in the ascending branch at the coercive field H_{c2}, we notice a strong increase in the SF reflectivity, indicative of a reversal via magnetization rotation. This strong disparity in the descending–ascending magnetization reversal is typical for exchange bias systems and of great technological relevance for data storage.

The domain state of magnetic films can be further analyzed by evaluating the dispersion parameter Δ. The asymmetry (Eq. 11.17 and 11.26) provides the average projection of the magnetization component in the Y-direction: $\langle \cos\gamma \rangle$. The SF reflectivities yield the squared average projection of the X-component: $\langle \sin^2\gamma \rangle$. From these two measurements, we find the dispersion of the domain state, which is defined as (Lee et al. 2002):

$$\Delta = \sqrt{\langle \cos^2\gamma \rangle - \langle \cos\gamma \rangle^2} \tag{11.27}$$

The magnetic domain dispersion is a very useful quantity, indicative of magnetic fluctuations and domain formation. For a homogeneous and uniform magnetization, Δ is always zero, independent of the mean rotation angle $\langle \gamma \rangle$. However, if there are some fluctuations in the mean magnetization axis, $\langle \cos\gamma \rangle^2$ is smaller than $\langle \cos^2\gamma \rangle$, and $\Delta > 0$. The dispersion reaches a maximum of 1 for the completely demagnetized state, which is achieved by $\langle \cos\gamma \rangle^2 = 0$ and $\langle \cos^2\gamma \rangle = 1$. Some examples are illustrated in Fig. 11.8.

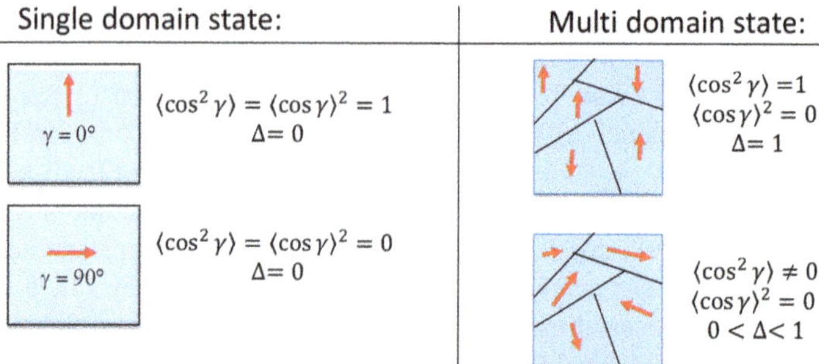

Fig. 11.8 Examples for different domain states in thin ferromagnetic films and the respective values for the dispersion defined as $\Delta = \sqrt{\langle \cos^2 \gamma \rangle - \langle \cos \gamma \rangle^2}$. Adapted with permission from Lee et al. (2002). Copyright (2025) by APS

11.4.3 Magnetic Multilayers

Magnetic multilayers are artificial stacks of alternating magnetic and nonmagnetic layers. The combined thickness of both layers is the repeat unit. The number of repeats may vary from as low as 5 to as many as 1000. A schematics of a magnetic multilayer is shown in Exercise 7.1 and in Fig. 11.11c. Multilayers have the advantage of multiplexing films and interfaces, thereby amplifying the effect to be investigated. Furthermore, the multilayer Bragg reflections are automatically the Fourier components of all periodic structures and modulations in the multilayer. They have served in the past to understand the strength, range, and periodicity of the interlayer exchange coupling of itinerant ferromagnets, the coupling of helical magnetism in rare earth metals across nonmagnetic spacer layers, and proximity effects at the interface of ferromagnets and superconductors. Our next example concerns the proximity effect, which can be made visible by enhancing the interface sensitivity.

The interface sensitivity can effectively be enhanced in a multilayer by choosing equal thicknesses of the bilayers. Then, all even-order Bragg reflections are extinct. This is shown in Exercise 11.2. Then, any tiny deviation from the equal-thickness condition will be noticeable by the Bragg intensity at the forbidden positions.

The example that we will discuss here consists of a high T_c superconducting film ($YBa_2Cu_3O_7$ (short: YBCO)) combined with a ferromagnetic manganite film (Satapathy et al. 2012). Because of the antagonistic nature of superconductivity and ferromagnetism, the combination of both layers in superlattices with common interfaces leads to new and fascinating phenomena. It has been predicted (Buzdin 2005; Bergeret et al. 2005; Efetov et al. 2013) that either the ferromagnetic order parameter close to the interface is reduced via a cryptomagnetic domain-like state by cooling below the superconducting critical temperature T_c or an inverse proximity effect enhances the magnetization in the superconducting layer close to the interface. To investigate these scenarios, two different multilayers were used: one with a

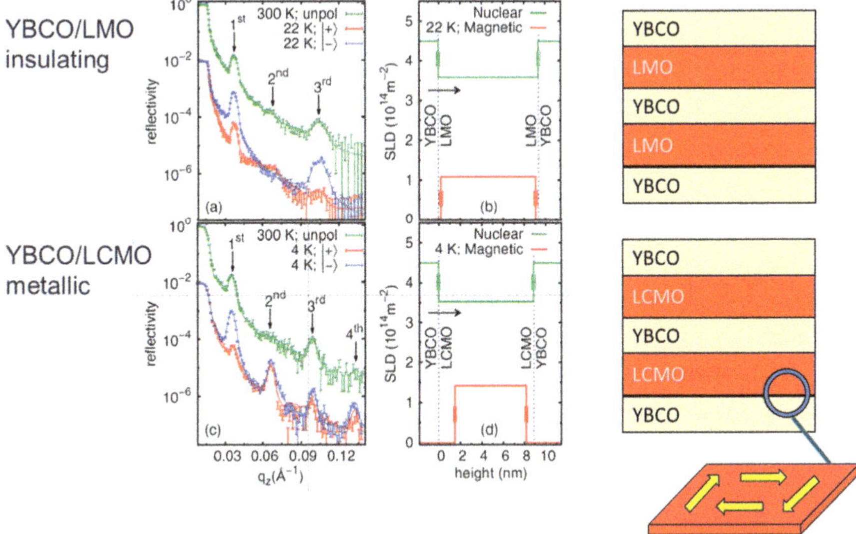

Fig. 11.9 (a) Unpolarized reflectivity scan at 300 K and R^+ and R^- reflectivity scans at 22 K of YBCO/LMO, which is below the superconducting transition temperatures of YBCO and the ferromagnetic transition of LMO. Note that at both temperatures, the even-order Bragg peaks are extinct due to the same thicknesses of the YBCO and LMO layers. (b) Fit results for the nuclear and magnetic SLD profiles for the low-temperature scan. (c) Similar measurements as in (a) but with LCMO replacing LMO. At low temperatures, the even-order Bragg peaks become visible without magnetic splitting, indicating that the effective magnetic thickness of the LCMO layer has changed in the superconducting state of the sample. (d) Fit results of the SLD profiles as in (b) but with a distinctly changed profile for the magnetic layer. The cartoon in the lower right corner visualizes what a cryptomagnetic state may look like. Reprinted and adapted with permission from Satapathy et al. (2012). Copyright (2025) by the APS

metallic ferromagnetic manganite $La_{2/3}Ca_{1/3}MnO_3$ film, labeled YBCO/LCMO-multilayer, and another one with an insulating ferromagnetic manganite film of composition $LaMnO_{3+\delta}$, labeled YBCO/LMO-multilayer (Satapathy et al. 2012). Both multilayers had equal magnetic and superconducting thicknesses of 10 nm each and 10 repeats. The YBCO/LCMO multilayer has a superconducting T_c of 88 K and a ferromagnetic Curie temperature of 200 K. The YBCO/LMO multilayer has respective transition temperatures at 77 K and 140 K.

The PNR results are shown in Fig. 11.9 from the work of (Satapathy et al. 2012). The scans at room temperature were taken with an unpolarized neutron beam. The scans of both multilayers exhibit missing second- and fourth-order Bragg reflections, while first- and third-order Bragg reflections are expressed, confirming the high-precision sample growth with equal YBCO (10 nm), LMO (10 nm), and LCMO (10 nm) layer thicknesses. At low temperatures, the reflectivity scans were taken with R^+ and R^- polarization but without polarization analysis. The first- and third-order reflections are split, which is a clear sign of the ferromagnetic state of the LMO and LCMO layers. For the LMO layer, scanned at 22 K (panel (a)), no second-order peak is present. The magnetic and nuclear SLDs reproduced in panel (b) show sharp interfaces and no layer disproportion. In contrast, the PNR scans of the YBCO/LCMO multilayer taken at 4 K exhibit pronounced second- and fourth-order Bragg peaks. This immediately signifies that the nSLD and mSLD must have different widths in both layers. The nSLD profile, which is identical to the chemical profile, is the same at high and low temperatures. We also notice that the second- and fourth-order Bragg peaks show almost no magnetic splitting, unlike the first- and third-order peaks. From this, we infer that the changes that took place at the conducting YBCO/LCMO interface upon cooling are nonmagnetic in origin. This notion is confirmed by a more detailed fitting procedure, providing the nSLD and mSLD shown in Fig. 11.9d.

Hence, in YBCO/LCMO multilayers, a pronounced proximity effect is observable at low temperatures, which suppresses the ferromagnetic state in the conducting LCMO layer close to the superconducting interface, unlike the insulating LMO layers. The mSLD shows a slight asymmetry, although it is clear that, due to the missing phase information, the exact location of the mSLD depth profiles cannot be determined. However, by repeating the reflectivity scans to even higher Q_z values and by testing many other competing models, the one proposed by the authors of this work has a high confidence level (Satapathy et al. 2012). In any case, these PNR results strongly support the notion of proximity effects at the ferromagnetic/superconducting interface via a quenching of the ferromagnetic spin alignment, most likely via a cryptomagnetic state.

11.4.4   Diffuse Scattering

We already discussed off-specular diffuse scattering in Sect. 10.5 caused by structural roughness at surfaces and interfaces. In magnetic films and multilayers, there is an additional source of diffuse scattering, arising from the finite size of magnetic domains. Whenever the dispersion parameter $\Delta > 0$, a good chance exists of observing off-specular diffuse scattering intensity. Because of its magnetic origin, magnetic diffuse scattering vanishes in saturation at high magnetic fields in a single domain state. Hence, magnetic diffuse scattering can easily be distinguished from structurally induced diffuse scattering by its field dependence.

Before continuing to discuss magnetic diffuse scattering, we need to relate the magnetic domain size to the coherence length of the neutron beam. The coherence length is determined by the setting of the entrance slit, controlling the incident beam divergence. In NR and PNR works, it is usual practice to confine the beam in the X-direction while opening the entrance slit in the Y-direction (see Fig. 11.1). This provides a high-momentum transfer resolution for ΔQ_X and ΔQ_Z but a poor resolution for ΔQ_Y. Accordingly, the neutron coherence length L_n is longer in the X-direction and shorter in the two perpendicular directions, defining a resolution ellipsoid, where the respective axes $L_{n,\,X} \gg L_{n,\,z} \gg L_{n,\,Y}$. If the magnetic domain size D_{mag} is larger than the coherence volume, the reflected intensities from different

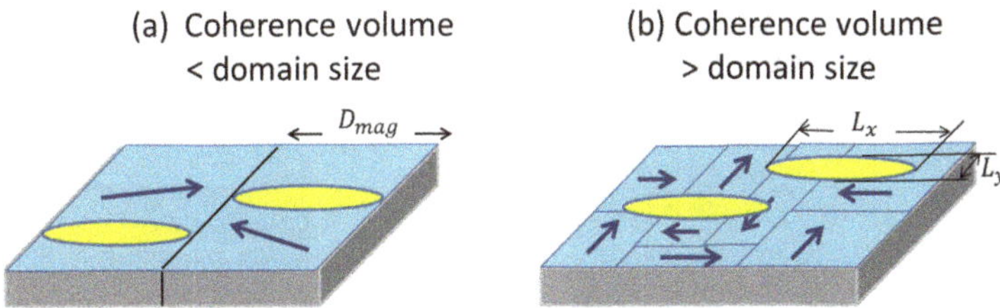

Fig. 11.10 Coherence volume of the neutron beam in relation to magnetic domain sizes. In panel (**a**), the coherence volume is assumed to be smaller than the average domain size, resulting in specular intensity that is sharply peaked at $Q_x = 0$. Panel (**b**) shows the case when the coherence volume covers several domains, causing off-specular diffuse scattering

domains add up incoherently: $I = \sum_i (A_i)^2$, as indicated in Fig. 11.10a, and diffuse intensity does not occur. However, if $D_{mag} < L_n$, the reflected amplitudes add up: $I = \left(\sum_i A_i\right)^2$ (Fig. 11.10b), and their constructive and destructive interferences lead to off-specular diffuse scattering.

The size of magnetic domains can be partially controlled by the applied magnetic field. In saturation, the domain size corresponds to the sample size. In contrast, upon magnetization reversal at the coercive field H_c, the sample is in a domain state, generating plenty of magnetic diffuse scattering in the off-specular regime, limited only by the Yoneda wings.

Figure 11.11 shows an experimental example of the field dependence of magnetic diffuse scattering (Langridge et al. 2000). The system investigated is a Co(2 nm)/Cu (2 nm) multilayer with 50 bilayer repeats. The exhibits multilayer a so-called interlayer exchange coupling (IEC) between the ferromagnetic Co layers, mediated by the Cu spacer layers. The thickness of the Cu spacer layer is chosen to mediate an antiferromagnetic or antiparallel orientation of the in-plane magnetization in adjacent Co layers at remanence, i.e. at an external field $H = 0$. Such an AF-coupled multilayer is schematically sketched in panel (c) of Fig. 11.11. The antiferromagnetic coupling is manifested in a half-order Bragg peak (AF 1/2), centered on the specular ridge at $Q_x = 0$, as can be seen in the intensity map shown in panel (a) of Fig. 11.11. This AF peak has a radial width along the Q_z—direction, which is comparable to the width of the nuclear first-order peak (N 2/2), indicative of similar structural and magnetic correlation lengths normal to the layer planes. This is also shown in the inset of panel (b) by the scan with open circles. Furthermore, the (1/2)-order peak spreads out into the off-specular region at $Q_x \gtrless 0$, indicative of the presence of many small in-plane magnetic domains in the Co layer at remanence.

To confirm the magnetic origin of the diffuse scattering, in panel (b) of Fig. 11.11, the same reciprocal space map is shown, now taken at a saturation magnetic field of 700 Oe. In this field, the 1/2-order Bragg peak together with the off-specular diffuse scattering vanishes, while the nuclear peak is nearly the same as before. The radial scan in the inset of panel (b) shows only the nuclear peak (solid black circles). Panel (d) reproduces Q_x line scans across the 1/2-order peak for various applied magnetic

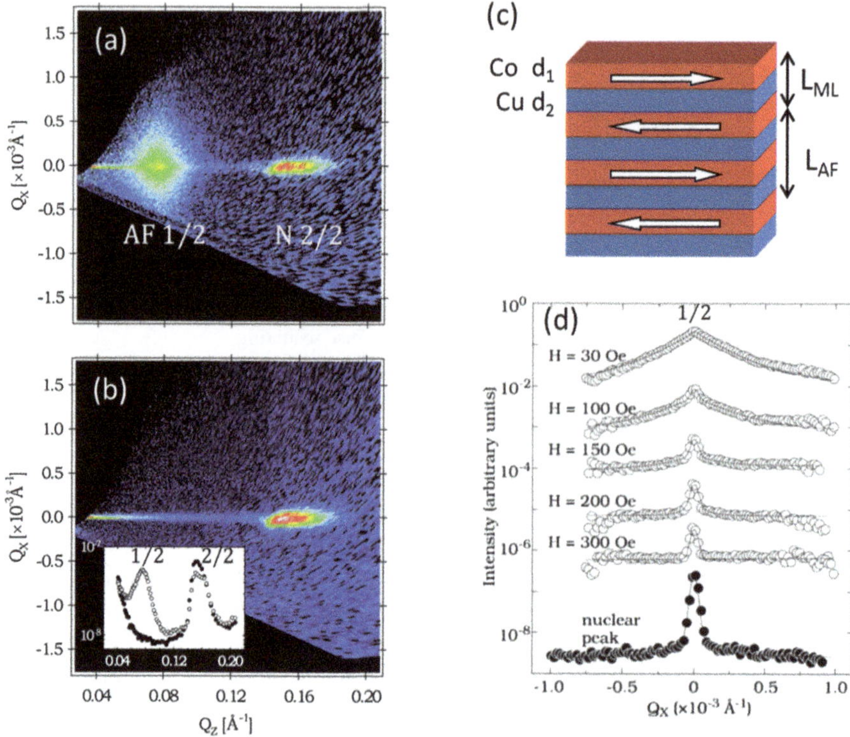

Fig. 11.11 (a) Diffuse scattering map covering the AF peak at $Q_z = 0.075$ Å^{-1} and the nuclear peak at $Q_z = 0.15$ Å^{-1} in a zero field. (b) Same as in (a) but after applying a field of 700 Oe. (c) Schematics of a magnetic multilayer with antiferromagnetic coupling, causing a doubling of the magnetic periodicity and generating accordingly a half-order Bragg peak. The antiferromagnetic repeat unit is twice the multilayer period: $L_{AF} = 2L_{ML}$. (d) Transverse Q_z scans across the AF peak as a function of the applied field. The lowest scan (solid symbols) is an equivalent scan through the nuclear peak at saturation, taken at $Q_z = 0.15$ Å^{-1}, showing the intrinsic and field-independent width of the peak. More information is provided in the main text. Adapted with permission from Langridge et al. (2000). Copyright (2025) by the APS

field values. With the increasing field, the shape of the AF peak changes, and the intensity drops to a low level. At 300 Oe, only a sharp peak remains, which is due to the specular ridge, the shape and width being the same as for the nuclear peak. By the application of 300 Oe, the IEC is overcome. Therefore, all magnetic intensity is now reshuffled to the first-order peak that is composed of nuclear and magnetic contributions.

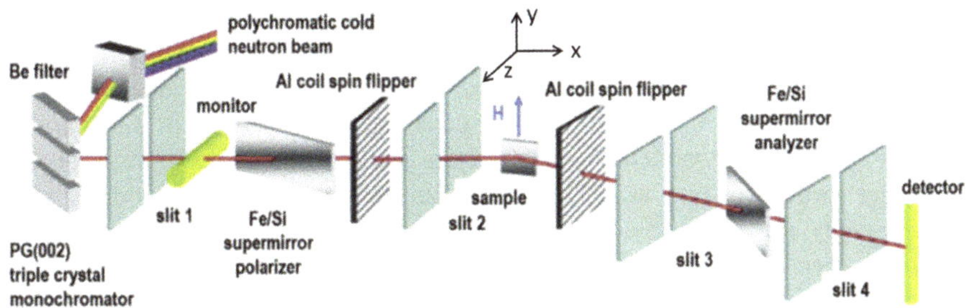

Fig. 11.12 Principal features of a neutron reflectometer for polarized neutron reflectivity experiments. The schematic instrument layout is taken from the NG1 reflectometer at the NIST Center for Neutron Research

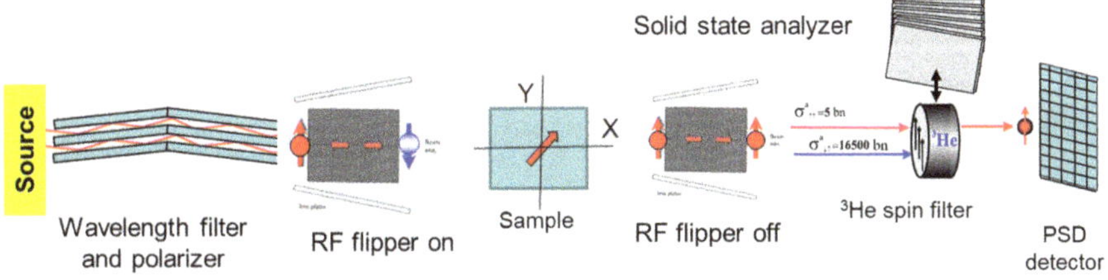

Fig. 11.13 Optical components of a neutron reflectometer capable of detecting specular as well as off-specular diffuse intensity. The Be filter and the polarizer are combined in one wavelength filter and polarizer, the usual Mezei spin flippers are replaced by contactless radio frequency (RF) flippers, and the neutron spin analysis can be either performed by a solid-state fan analyzer or by a polarized ^{3}He spin filter. Detection of specular and diffuse intensity is achieved by a position-sensitive detector (PSD). Reproduced with permission from Toperverg and Zabel (2015), Copyright (2025) by Elsevier

11.4.5   Instrumentation

Instruments for PNR experiments follow the same principles as already discussed for polarized neutron scattering in Sect. 7.5. Therefore, we can keep the discussion short and point out only the differences.

A classical design of a PNR instrument is shown in Fig. 11.12. On the entrance side, it features a pyrolytic graphite (PG) monochromator together with a Be-filter for the removal of higher harmonics. The PG monochromator is split into three slabs and tilted to provide focusing in the vertical direction. Neutron beam polarization is achieved with a Fe/Si supermirror in transmission geometry. Al-coil spin flippers of the Mezei type (Mezei 1972) control the neutron spin state before the sample and before the analyzer and detector. Intensity recording is performed by a zero-dimensional pencil detector.

A more advanced instrumental design is shown in Fig. 11.13 (Toperverg and Zabel 2015; Devishvili et al. 2013). Here, the monochromator/Be filter is replaced by a ferromagnetic neutron wavelength filter, which simultaneously also polarizes the beam. The Al-coil spin flippers are replaced by contactless rf flippers before and after the sample. And the analyzer supermirror, together with the pencil detector, is replaced by a glass flask, containing polarized ^{3}He gas, acting as a wide-angle spin filter. Neutron spins antiparallel to ^{3}He polarization are absorbed, and parallel neutron spins are transmitted. Alternatively, a stack of tilted supermirrors allows recording specular as well as off-specular scattering intensity over a wide angular range. Finally, the pencil detector is replaced by a position-sensitive area detector for recording specular and diffuse intensity over a large solid angle. There are several other designs of neutron reflectometers, in particular for the use of time-of-flight techniques at spallation sources.

Similar to XRR and NR, there are several open-source fitting routines available for PNR. The package developed and maintained at the NIST center for neutron research is the program "gepore.f" (GEneral POlarized REflectivity) (http://www. ncnr.nist.gov/programs/reflect/). It calculates the spin-dependent neutron reflectivities (and transmissions) for model potentials, or scattering length density profiles, assuming specular conditions. To conclude, Table 11.1 is a useful table listing scattering lengths and critical scattering vectors.

11.5   Magnetic X-Ray Reflectivity

With the combination of X-ray magnetic circular dichroism (XMCD) and X-ray resonant magnetic scattering (XRMS), information can be gained on film and interface magnetism similar to PNR. We will first present the basics of this scattering method and then discuss the complementarity between PNR and XRMS. As before with PNR, we will state the main results and refer to the literature for the detailed development of the theoretical background (Macke and Goering 2014; Fink et al. 2013; Schütz et al. 2007). Some experimental examples are discussed in Sect. 11.7.

11.5.1   Refractive Index

We return to the refractive index of X-rays stated in Eq. (10.10):

$$n(Q_z) = 1 - \frac{2\pi r_0}{k_0^2} \rho_A [f_{ch}(Q_z) + \Delta f_{ch}(E)] + \frac{i\mu_{ch}}{2k_0}. \tag{11.28}$$

Here, the subscript ch stands for "charge" to distinguish it from "c" for "critical", and the subscript "m" designates "magnetic". Furthermore, we take into account the resonant magnetic X-ray scattering, stated in Eq. 7.57, with the magnetic scattering lengths (Hannon et al. 1988):

$$f_m(\boldsymbol{Q}, E) = +i(\widehat{\boldsymbol{e}}_i \times \widehat{\boldsymbol{e}}_f) \cdot \widehat{\boldsymbol{m}} F^{(1)} + (\widehat{\boldsymbol{e}}_i \cdot \widehat{\boldsymbol{m}})(\widehat{\boldsymbol{e}}_f \cdot \widehat{\boldsymbol{m}}) F^{(2)}, \tag{11.29}$$

The first term is proportional to $(\widehat{\boldsymbol{e}}_i \times \widehat{\boldsymbol{e}}_f) \cdot \widehat{\boldsymbol{m}}$ and is therefore linear in $\boldsymbol{m}$. The term describes magnetic resonant scattering and is related to the X-ray magnetic circular dichroism. The second term is proportional to $(\widehat{\boldsymbol{e}}_i \cdot \widehat{\boldsymbol{m}})(\widehat{\boldsymbol{e}}_f \cdot \widehat{\boldsymbol{m}})$ and therefore quadratic in $\boldsymbol{m}$. This term is related to the magnetic linear dichroism in X-ray absorption and is of interest to studies of antiferromagnets and noncollinear magnetism. In the following, we will consider only the first term. Combining with the scattering length for charge scattering, we obtain the refractive index for resonant magnetic X-ray reflectivity according to:

$$n(Q_z) = 1 - \frac{2\pi r_0}{k_0^2} \rho_A \left[f_{ch}(Q_z) + \Delta f_{ch}(E) + f_m^{cir}(Q_z, E) \right] + i\frac{(\mu_{ch} + \mu_m)}{2k_0} \tag{11.30}$$

or in short form:

$$n(Q_z) = 1 - (\delta_{ch} + \delta_m) + i(\beta_{ch} + \beta_m) \tag{11.31}$$

Now, the critical scattering vector becomes, in analogy to Eq. (10.16):

$$Q_c^{x-ray} = 2k_0 \sqrt{2(\delta_{ch} + \delta_m)}, \tag{11.32}$$

and the Fresnel reflectivity from a single surface is expressed accordingly Eq. (10.23):

$$R_F(Q_Z) = r^2 = \left| \frac{Q_z - Q_z^t}{Q_z + Q_z^t} \right|^2, \tag{11.33}$$

where:

$$Q_z^t = \sqrt{Q_z^2 - (Q_{c,ch}^2 + Q_{c,m}^2) + 8k_0^2 i(\beta_{ch} + \beta_m)} \tag{11.34}$$

In terms of glancing angles, we can rewrite Eq. (11.34):

$$\alpha_z^t = \sqrt{\alpha_z^2 - 2(\delta_{ch} + \delta_m) + 2i(\beta_{ch} + \beta_m)} \tag{11.35}$$

For a multilayer, the Parratt recursion formalism has to be applied, as discussed before for the charge-only reflectivity.

These equations contain four unknowns: $\delta_{ch,\ m}$ and $\beta_{ch,\ m}$. However, if the absorption coefficients are known, the dispersion coefficients can be evaluated, and vice versa. The optical theorem assures that we can determine the imaginary part of the refractive index by measuring the charge and magnetic absorption cross sections in the forward direction:

$$\sigma_{abs}(Q=0,E) = \frac{4\pi}{k_0} Im\,(\Delta f) = \frac{4\pi}{k_0}(\beta_{ch} + \beta_m) \qquad (11.36)$$

Therefore, X-ray absorption spectroscopy (XAS) is an essential tool of X-ray resonant magnetic scattering. If the imaginary part is determined, the real part can be calculated via the Kramers–Kronig relationship:

$$\delta_{ch,m}(E) = \frac{2}{\pi} \int_0^\infty E' \frac{\beta_{ch,m}}{E'^2 - E^2} dE' \qquad (11.37)$$

or vice versa:

$$\beta_{ch,m}(E) = -\frac{2}{\pi} \int_0^\infty E \frac{\delta_{ch,m}}{E'^2 - E^2} dE' \qquad (11.38)$$

Hence, the Kramers–Kronig integrals relate the real and imaginary parts of a complex function. Here, E and E' are the X-ray energies, and $\mathcal{P}$ denotes the Cauchy principal value. Using this relation, we need to determine only one optical parameter, either β or δ, and obtain the other by the Kramers–Kronig integral. Usually, β is determined conveniently by X-ray absorption spectroscopy, detecting the total electron yield (XAS-TEY), and via X-ray magnetic circular dichroism (XMCD) experiments. Knowing the optical parameters β and δ is mandatory for X-ray resonance magnetic scattering experiments. Therefore, we discuss in the next section the polarization effects and the absorption aspect before returning later to the reflectivity and scattering issues. In any case, using the Fresnel reflectivity and the Parratt formalism, the magnetization profile can be calculated for thin magnetic films and multilayers, and experimental data can be fit with this recursion method.

11.5.2   Polarization Effects and Scattering Lengths

Now, we consider polarization effects for the scattering with linear polarization, expressed in terms of two polarization states π and σ, chosen to be either parallel (π) or perpendicular (σ) to the scattering plane. The four possibilities are then as follows:

$$f_{res}(Q,E) = f_{ch}\left(\hat{e}_i \cdot \hat{e}_f\right) + i f_m^{cir} i\left(\hat{e}_i \times \hat{e}_f\right) \cdot \hat{m} = f_{ch}\begin{pmatrix} e_{\sigma\sigma'} & e_{\pi\sigma'} \\ e_{\sigma\pi'} & e_{\pi\pi'} \end{pmatrix} + i f_m^{cir}\begin{pmatrix} e_{\sigma\sigma'} & e_{\pi\sigma'} \\ e_{\sigma\pi'} & e_{\pi\pi'} \end{pmatrix} \cdot \hat{m}$$

$$(11.39)$$

The polarization unit vectors $e_{\sigma,\pi}$ are indicated in Fig. 11.14 before and after scattering (primed). As already discussed in Sect. 2.5.3, the charge scattering does not change the polarization. Therefore, we can write for the first term:

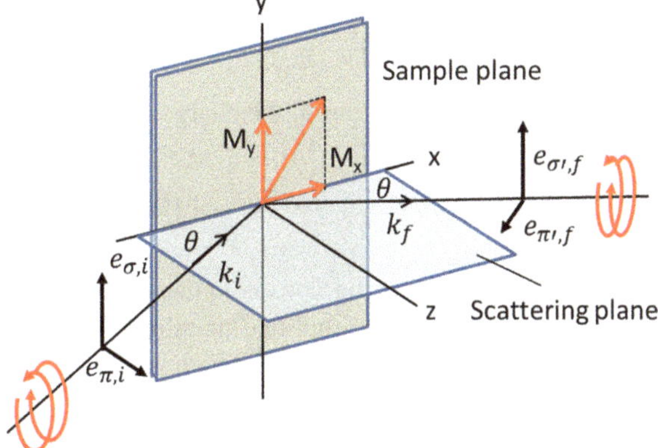

Fig. 11.14 Configuration for soft X-ray specular reflectivity with linear and circular polarizatiion. Note the similarity to PNR in Fig. 11.1

$$f_{ch}\begin{pmatrix} e_{\sigma\sigma'} & e_{\pi\sigma'} \\ e_{\sigma\pi'} & e_{\pi\pi'} \end{pmatrix} = f_{ch}\begin{pmatrix} 1 & 0 \\ 0 & \cos 2\theta \end{pmatrix} \tag{11.40}$$

The second term in Eq. 11.39 has been evaluated by Hill and McMorrow (1996), yielding the sensitivity of the scattering angle θ on the longitudinal component M_x, the transverse component M_y, and the polar component M_z of the magnetization vector:

$$f_m^{cir}\begin{pmatrix} e_{\sigma\sigma'} & e_{\pi\sigma'} \\ e_{\sigma\pi'} & e_{\pi\pi'} \end{pmatrix} \cdot \hat{\boldsymbol{m}} = f_m^{cir}\begin{pmatrix} 0 & M_x\cos\theta + M_z\sin\theta \\ M_z\sin\theta - M_x\cos\theta & -M_y\sin 2\theta \end{pmatrix}$$

$$\tag{11.41}$$

The results show that for the magnetic scattering, there is no $\sigma \to \sigma'$ transition, but $\sigma \to \pi'$ and $\pi \to \sigma'$, which are sensitive to M_x and M_z. However, the dominant contribution comes from the M_x component of the magnetization. Only at high scattering angles 2θ, the sensitivity to the M_y component increases. The M_z component plays only a role at almost normal incidence and is important for small-angle scattering in transmission geometry at samples with perpendicular anisotropy, a geometry that we discussed in Sect. 9.7.

Assuming only in-plane magnetization, i.e., magnetization components in the x, y-film plane, and no z-component, the scattering length for linear polarization is:

$$f_{res}(Q,E) = f_{ch}\begin{pmatrix} 1 & 0 \\ 0 & \cos 2\theta \end{pmatrix} + i f_m^{cir}\begin{pmatrix} 0 & M_x\cos\theta \\ -M_x\cos\theta & -M_y\sin 2\theta \end{pmatrix} \tag{11.42}$$

Both polarization components are combined when using circular polarization with left and right helicities, where e_σ and e_π are perpendicular to each other and phase-shifted by $\pm\pi/2$:

$$e_{l,r} = \frac{1}{\sqrt{2}} \left(e_\sigma \pm i e_\pi \right) \tag{11.43}$$

Using circular polarization, the charge and magnetic scattering length becomes:

$$f_{res}(Q,E) = f_{ch} \frac{1}{2}(1 + \cos 2\theta) + i f_m^{cir}\left(\pm M_x \cos\theta - \frac{1}{2} M_y \sin 2\theta \right) \tag{11.44}$$

The magnetic scattering does not change the circular polarization.

Finally, we present a simplified notation of the scattering length, which helps in the analysis of ordered and disordered systems, as proposed by Korthright et al. (2005). We write the charge and magnetic scattering lengths in short terms:

$$f_{res}^{\pm} = f_{ch} \pm i f_m, \tag{11.45}$$

where the polarization factors are already incorporated. As both scattering lengths have real and imaginary parts:

$$f_{ch} = f_{1ch} + i f_{2ch}; \; f_m = f_{1m} + i f_{2m},$$

it follows for the scattered intensity:

$$I^{\pm}(Q) = [(f_{1ch} + i f_{2ch}) \pm i(f_{1m} + i f_{2m})]^2, \tag{11.46}$$

which yields the expression:

$$I^{\pm}(Q) = f_{ch}^2 + f_m^2 \pm 2(f_{2ch}f_{1m} - f_{1ch}f_{2m}) \tag{11.47}$$

Considering charge correlations g_{ch} and magnetic correlations g_m of ordered or disordered systems, we can derive for the scattering intensity the following expression as a function of scattering vector Q and energy E in the Born approximation (Kortright et al. 2005):

$$I^{\pm}(Q,E) = f_{ch}^2 g_{ch,ch}(Q) + f_m^2 g_{m,m}(Q) \pm 2(f_{2ch}f_{1m} - f_{1ch}f_{2m})g_{ch,m}(Q) \tag{11.48}$$

Here, $g_{c,\,c}$, $g_{m,\,m}$, and $g_{c,\,m}$ are the charge–charge, magnetic–magnetic, and charge–magnetic correlation functions, respectively. The energy dependence is because resonance magnetic X-ray scattering is always performed close to absorption edges. The difference between right and left circular polarization yields the asymmetry:

$$A = I^+ - I^- = 4(f_{2ch}f_{1m} - f_{1ch}f_{2m})g_{ch,m}(Q), \tag{11.49}$$

containing only the cross terms with charge–magnetic cross-correlations, and the sum:

$$S = I^+ + I^- = 2\left(f_{ch}^2 g_{ch,ch}(Q) + f_m^2 g_{m,m}(Q)\right) \tag{11.50}$$

features the squared charge and magnetic amplitudes modified by their conjugate correlation functions. Furthermore, it was shown that the sum S, using circular polarization, is essentially identical to the one obtained for scattering with linear polarization (Kortright et al. 2005). A complete separation of charge and magnetic scattering cross section for resonant magnetic scattering is not possible, in contrast to PNR SF scattering, where nuclear and magnetic scattering can be separated without cross terms.

11.6   Resonant Absorption of Magnetic Transition Metals

11.6.1   Paramagnetic Metals

Figure 11.15 shows schematically the $2s$ and $2p$ core levels of 3d transition elements. The $3s$ and $3p$ levels are omitted for clarity as they play no role in the following discussion. The electron wavefunctions in the $3d$ orbits of paramagnetic 3d transition metals overlap and form a band, the so-called 3d-band that is successively filled from scandium with one electron up to copper with 10 electrons. Together with the $4s$ electrons (also omitted), they contribute to the metal conductivity. The partially filled 3d-band is symbolized by its electron density of states, separately for spin-up and spin-down polarized electrons. The Fermi energy E_F, defining the highest occupied energy level, crosses the 3d-band. Above E_F are the unoccupied valence states.

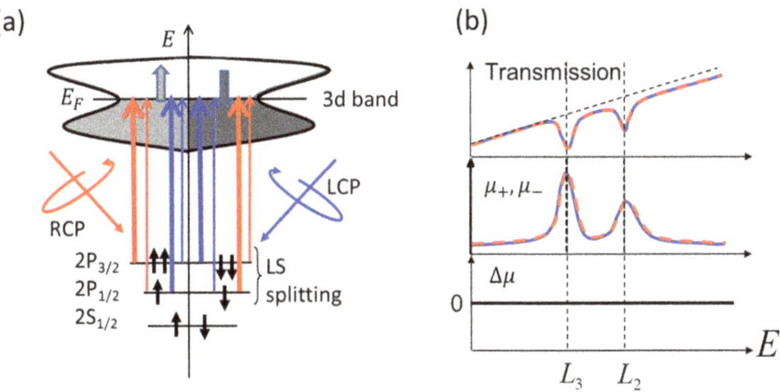

Fig. 11.15 (**a**) Schematic of the atomic levels and electron density of states for paramagnetic transition metals. Core levels and 3d band are indicated; the 4 s band at the Fermi level is omitted for clarity. L_3 and L_2 resonant transitions with right and left circular polarized light are indicated by vertical red and blue arrows. The line thicknesses mark the transition probabilities. (**b**) Top panel: Transmission experiment with RCP and LCP light. At the resonance energies, dips are observed. Middle panel: calculated absorption coefficients. Bottom panel: zero difference of the absorption coefficients as a function of energy

The allowed dipole transitions from $2p$ core levels to $3d$ empty states at the Fermi level are the L_2 ($2p_{1/2} \rightarrow 3d$) and L_3 ($2p_{3/2} \rightarrow 3d$) absorption lines that fulfill the angular momentum change $\Delta l = +1$ and the magnetic quantum number change $\Delta m_l = 0, \pm 1$, both with zero spin quantum number change: $\Delta S = 0$. The $2p_{1/2}$ and $2p_{3/2}$ atomic levels are, in principle, degenerate. However, because of spin–orbit interactions, the degeneracy is lifted. Detailed calculations of the transition probabilities show that right circularly polarized (RCP) photons with positive helicity preferentially excite spin-up core electrons at the L_3 edge and spin-down electrons at the L_2 edge (red arrows in Fig. 11.15a. For left circularly polarized light, the situation is reversed (blue arrows). For a given photon polarization, the sum of contributions at $2p_{3/2}$ and $2p_{1/2}$ shows no spin-polarization. The transition probabilities for right and left circular polarization ($\pm$) to excite up-spins and down-spins ($\uparrow, \downarrow$) are identical at the L_3 edge: $p_+(\uparrow, L_3) + p_+(\downarrow, L_3) = p_-(\downarrow, L_3) + p_-(\uparrow, L_3)$, and equivalently at the L_2 edge: $p_+(\uparrow, 2) + p_+(\downarrow, L_2) = p_-(\downarrow, L_2) + p_-(\uparrow, L_2)$.

When probing paramagnetic $3d$ transition elements, such as Ti, V, or Cr, with LCP and RCP X-rays, no dichroism is observed at the L_2 and L_3 edges. These transition metals are characterized by a broad $3d$ band of conduction electrons, but no splitting between the spin-up and spin-down bands. Therefore, there is no difference in the number of final states. In a transmission experiment, dips are observed at the resonant energies for the L_2 and L_3 transitions, from which the absorption coefficients $\mu_\pm$ for LCP and RCP radiation can be extracted. Since no difference is observed, $\Delta\mu = \mu_+ - \mu_- = 0$, as indicated in panel (b) of Fig. 11.15.

There are a few conclusions that we can draw. 1. If there is no difference between right and left circular polarized light, the metal or material studied is not magnetic. 2. The sum of the L_3 and L_2 absorption cross sections is proportional to the number of holes in the d-band. 3. Noble metals exhibit no L_3 and L_2 absorption signals for two reasons: their d-band is below the Fermi-level, the bands are not exchange-split, and therefore these metals are diamagnetic but not para- or ferromagnetic. 4. The L_3: L_2 absorption cross-section ratio is 2:1, because there are four electrons occupying the $2p_{3/2}$ atomic level compared to two electrons occupying the $2p_{1/2}$ level. The L_3: L_2 ratio is known as the branching ratio.

11.6.2 Ferromagnetic Metals

Now we turn our attention to ferromagnetic transition metals like Fe, Co, and Ni. These metals are characterized by a spin-split $3d$ conduction band, schematically illustrated in Fig. 11.16a. The majority spin-up band and the minority spin-down band are shifted against each other by the exchange interaction in the ferromagnetic state of the metals. The Stoner exchange splitting Δ of the majority and minority bands is indicated in the same panel. The consequence of this splitting is manifested in the so-called band magnetism of these transition metals.

As in the previous case, the resonant absorption of LCP and RCP light probes the density of states in the final state. Due to the shift, the densities for spin-up and spin-

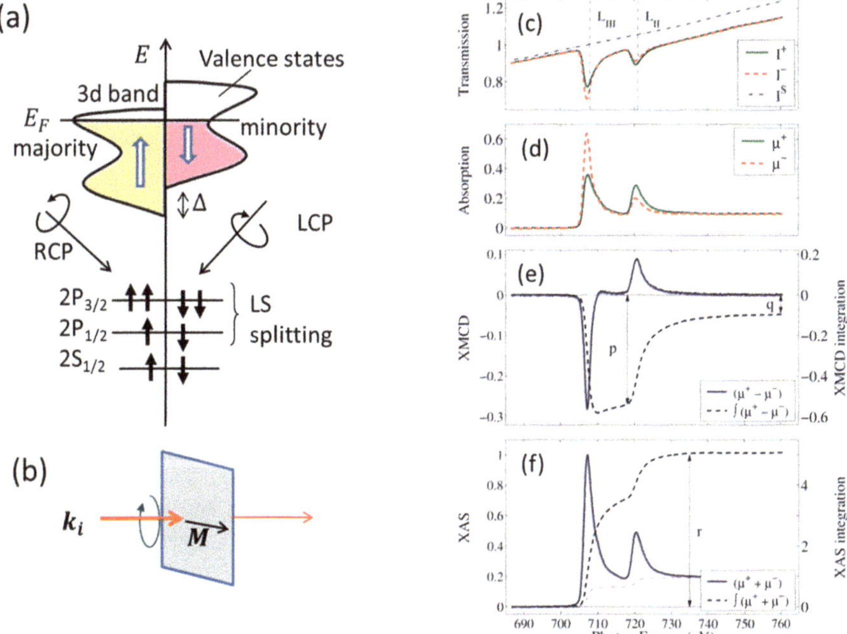

Fig. 11.16 (**a**) Similar to the previous graph, but with a spin-split 3d band for the majority spin-up and minority spin-down bands. The splitting causes different densities of states at the Fermi level, which result in different transmission and absorption coefficients for the L_3 and L_2 transitions, yielding an XMCD effect. (**b**) Experimental configuration for absorption measurements with circular incident photons. The incident wavenumber k_i is normal to a film with an in-plane magnetization M. (**c**) Transmission as a function of energy. (**d**) Absorption cross sections. (**e**) Difference signal. (**f**) Solid line: sum of absorption spectrum, dashed line: intensity integration. The small letters p, q, r indicate different integrations, which are required for determining the spin and orbital contributions to the magnetic moments, here for Fe. Reproduced with permission from Chen et al. (1995). Copyright (2025) by the APS

down electrons in the final states are different. This difference, in turn, gives rise to a dichroic signal in the absorption spectrum, as seen in Fig. 11.16 (c, d). The transmission curves for LCP and RCP are plotted in panel (c), and their corresponding absorption curves μ_+, μ_- are shown in panel (d). The L_3 absorption signal for LCP light is higher than that for RCP, and at the L_2 peak, the signal intensities are reversed. Therefore, their difference, plotted in panel (e), $\mu_+(E) - \mu_-(E) \neq 0$. This is known as the X-ray magnetic circular dichroism (XMCD) signal. The sum $\mu_{XAS} = \mu_+ + \mu_-$, plotted in panel (f), is the total X-ray absorption for LCP and RCP at the L_2 and L_3 resonance transitions. The measurements in panel (c–f) are for Fe metals, reproduced from Chen et al. (1995). Similar results are obtained for Co and Ni metals and their alloys. For Fe, the L_2 and L_3 absorption edges are 12 eV apart.

   The letters p, q, and r in panels (e) and (f) indicate integrals that serve to separate the spin and orbital contributions to the dichroic signal by using sum rules. As this

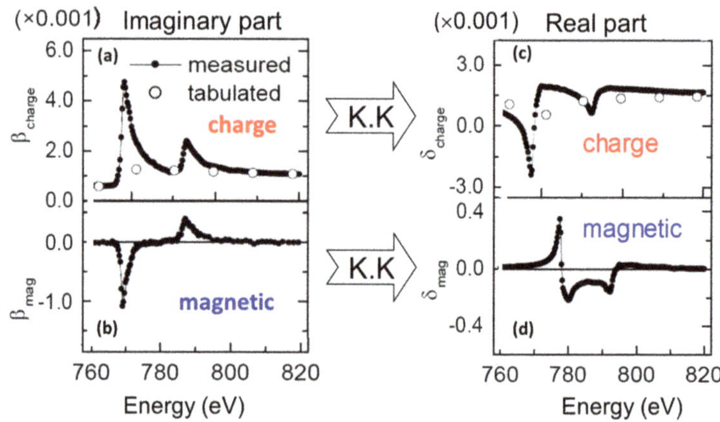

Fig. 11.17 Charge and magnetic absorption corrections (**a**, **b**) and respective dispersion corrections (**c**, **d**) to the refractive index at the $L_{2,3}$ edges of Co. The dispersion corrections are calculated via the Kramers–Kronig relationship. The open circles are from the Henke Table (Henke et al. 1993). Reproduced with permission from Grabis et al. (2005). Copyright (2025) by the APS

does not concern resonant magnetic scattering, we will not discuss it further but refer to the literature and reviews (Schütz et al. 2007; Stöhr and Siegmann 2006; van der Laan and Figueroa 2014).

The maximum dichroic signal is observed if the incident wavenumber k_i is oriented parallel to the sample magnetization in a single domain saturated state: $\Delta\mu = \mu_+ - \mu_- = k_i \cdot M$, as indicated in panel (b) of Fig. 11.16. To determine $\Delta\mu$, it is common practice to reverse the magnetization direction instead of changing the helicity. This is experimentally much easier and delivers equivalent results.[2] Similarly, to measure a magnetic hysteresis, the magnetic field is swept from one saturation field to the opposite, while the helicity is fixed and the X-ray photon energy is locked onto a specific resonant absorption edge. This way, an element-specific hysteresis can be measured for one component in a magnetic heterostructure, instead of averaging over all elements. An example is given Sect. 11.7.2.

By virtue of the optical theorem quoted in Eq. 11.36, the XAS spectra give access to the charge contribution of the absorption coefficient β_{ch}, and the XMCD signal yields the respective magnetic contribution β_m. Using the Kramers–Kronig relationship, the corresponding real parts can be determined: δ_{ch} and δ_m. An example for the Co-edge is shown in Fig. 11.17, reproduced from Grabis et al. (2005).

Having now access to the real and imaginary parts of the refractive index, XRMS allows for the determination of element-specific chemical and magnetic depth profiling of layered structures (Geissler et al. 2002; Jaouen et al. 2002; Tonnerre et al. 1995). These profiles can be obtained by a quantitative analysis of specular reflectivity measurements, usually performed by numerical simulation. The

---

[2]This may change in the future with the availability of fast flipping undulators

calculation of reflectivity needs a dynamical approach, such as the Parratt recursion formalism, as total and multiple reflection effects cannot be neglected in well-defined films and multilayers. A matrix-based formalism for magneto-optics with an arbitrary magnetization direction has been developed by Zak et al. (1990, 1991, 1992) and reviewed in (Macke and Goering 2014). It offers the possibility to calculate the specular reflectivity without any restrictions to the geometry, assuming perfectly smooth interfaces. Roughness is taken into account by assuming a graded interface, subdivided into many slices. This method was used in the analysis of the third example discussed below on Heusler multilayers. The Zak method was extended by Lee et al. (2003a, 2003b) to allow for magnetic scattering from rough interfaces and roughness correlation effects in the sense discussed in Sect. 10.5. The distorted-wave Born approximation was utilized for the analysis of interfaces in magnetic heterostructures, presented in Sect. 11.7.2.

For emphasizing the charge and magnetic contributions to the reflectivity, it is common practice to plot the average sum $S = (I^+ + I^-)/2$ and the asymmetry ratio $A = (I^+ - I^-)/(I^+ + I^-)$, as a function of the scattering vector Q_z or as a function of the energy E, where $I^+(Q_z, E)$, $I^-(Q_z, E)$ are the reflected intensities for right and left circularly polarized X-rays. In the following, the XRMS method is exemplified by the discussion of published work, including exchange-biased magnetic heterostructures, multilayers with Heusler alloys, and the helical magnetism of Ho.

11.7   Case Studies for Resonant Magnetic X-Ray Scattering (XRMS)

11.7.1   Magnetic Reflectivity

We start the discussion with an X-ray resonant magnetic reflectivity study of a FePt/C multilayer featuring a bilayer thickness of 5.5 nm and 20 repeats (Tonnerre et al. 2010). The FePt layers consist of dense nanoparticles embedded in carbon layers. For the analysis of the data, it was assumed that the FePt layers were homogeneous and continuous. The experimental data and fits are reproduced in Fig. 11.18a. The reflectivity was taken with circular polarization at an energy of 706.4 eV, which is close to the L_3 absorption edge of Fe, and recorded as a function of the scattering vector Q_z in positive magnetic field (Ip, black line), and after field reversal in a negative magnetic field (Im, dashed black line) of ± 0.2 T. The calculated reflectivity curves, shown in red and blue, are offset for better visibility. This magnetic reflectivity is completely equivalent to the results expected from PNR measurements. Figure 11.18b shows a specular reflectivity measurement at a constant glancing angle of 6° as a function of energy for plus (blue) and minus (red) magnetic fields. These two curves, measured in reflection geometry, are equivalent to measuring the absorption cross section in transmission geometry. The difference shown in the insets is indeed similar to an XMCD signal. This also underpins the

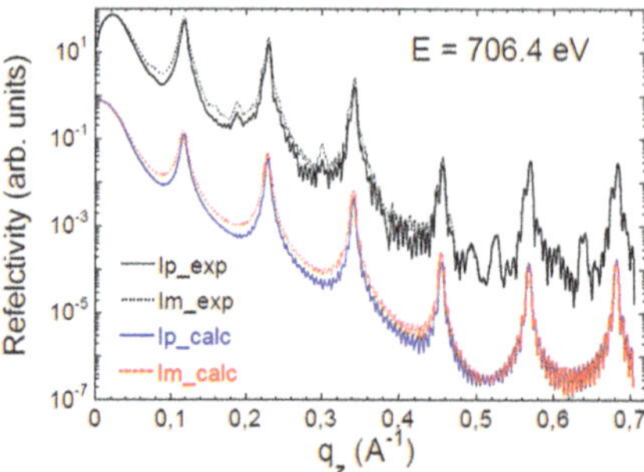

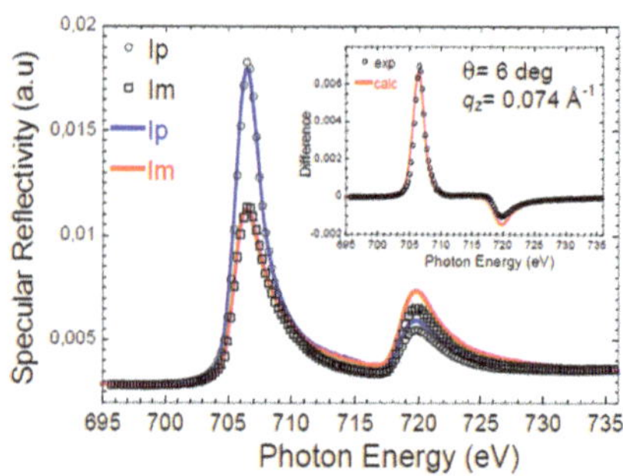

Fig. 11.18 Left panel: Resonant magnetic X-ray reflectivity of an FePt/C multilayer recorded at an X-ray energy of 706.4 eV with two helicities concerning the in-plane magnetization. Black lines are experimental data, and red and blue lines are fits to the recorded reflectivities. Right panel: Energy scans at a glancing angle of 6° covering the range of the $L_{2,3}$ absorption edges. The difference spectrum is equivalent to an XMCD signal taken in transmission geometry. Reproduced from Tonnerre et al. (2010), open access

fact that the magnetic sensitivity is largest for $k_i \parallel M$. From fitting both curves, a Fe magnetic moment of 1.58 μ_B in the FePt clusters could be determined, much lower than that usually observed in FePt alloys (2.8 μ_B). The discrepancy is ascribed to the cluster disorder.

11.7.2 Magnetic Hysteresis and Magnetization Profiles

Next, we present an XRMS study of a magnetic heterostructure with a shared ferromagnetic–antiferromagnetic interface (Roy et al. 2005, 2007). The system is similar to the PNR study presented in Sect. 11.4.2, and here, too, the exchange bias effect provides the physical background for the experiments.

CoO is an antiferromagnet, and permalloy (Py) is a ferromagnetic alloy of the composition $Fe_{0.2}Ni_{0.8}$. The polycrystalline 15 nm thick CoO film was grown on a SiO_2 substrate, followed by a 4 nm thin Py layer, and capped with a SiO_2 protecting layer. The layer sequence is shown in the inset to Fig. 11.19a. The system was cooled through the Néel temperature of CoO to 150 K in a field of 700 Oe to induce an exchange bias effect. The exchange bias effect is usually manifested by an increased and asymmetric coercivity of the ferromagnetic hysteresis. The magnetic hysteresis was measured element-specifically at the Ni and the Co edges as a function of the applied magnetic field. This was achieved by measuring the asymmetry ratio $A = (I^+ - I^-)/(I^+ + I^-)$ at a fixed momentum transfer Q_z. The results are plotted in the inset of Fig. 11.19b.

Without going into the depth of this exchange bias system, a few remarkable observations may immediately be noticed. Ni in Py shows a normal nonshifted hysteresis, which is surprising, as the Py film should be locked magnetically to the CoO antiferromagnet. This usually causes a characteristic shift of the hysteresis in the direction opposite to the field applied during cooling. On the contrary, at the

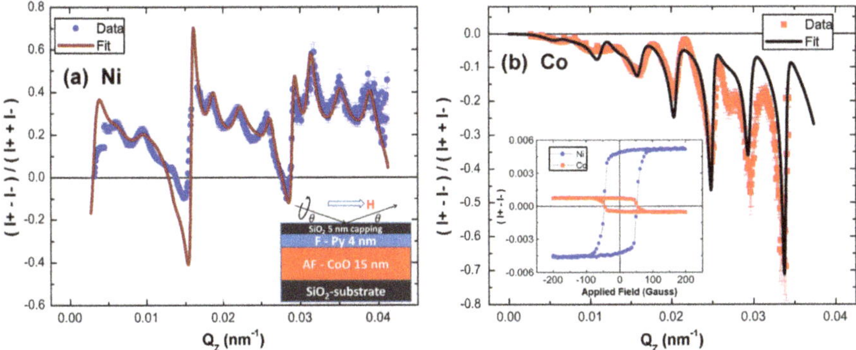

Fig. 11.19 Asymmetry ratios determined at the L_3 edges of Ni (**a**) and Co (**b**). Solid lines are model calculations. Inset in (**a**): layer sequence of the sample; inset in (**b**): magnetic hysteresis of Ni and Co derived from the asymmetry. Reproduced with permission from Roy et al. (2007). Copyright (2025) by the APS

Co-edge, a ferromagnetic hysteresis is also recorded with the same coercive fields but with a smaller and flipped-over asymmetry. Thus, it appears that a thin Co layer at the CoO/Py interface is magnetized, following the magnetization of the Py film instead of being bound to the CoO antiferromagnet. Now, these conjectures can be verified by resonant magnetic reflectivity measurements with circularly polarized light as a function of momentum transfer Q_z and for a fixed energy close to the respective absorption edges of Co and Ni. The respective asymmetries are shown in Fig. 11.19 together with the fit results. From the fits of the reflectivity curves, element-specific magnetization depth profiles can be gained, as shown in Fig. 11.20.

The quantitative analysis of the resonantly recorded reflectivities was achieved by using the kinematic scattering theory in distorted wave Born approximation (DWBA). The complex refractive index $n_{\pm} = 1 - (\delta_{ch} \pm \delta_m) + i(\beta_{ch} \pm \beta_m)$ was evaluated for each layer and for both left and right circular polarization. The fitted values of charge and magnetic scattering factors, f_{ch} and f_m, were then used to calculate the depth-dependent charge dispersion ($\delta_{ch}(z)$) and magnetic dispersion ($\delta_m(z)$), respectively. The results for ($\delta_m(z)$) of Ni and Co are shown in Fig. 11.20. They confirm a homogeneous magnetization within the Py layer and the presence of a thin metallic Co dusting layer at the Py/CoO interface. Most likely, this thin Co layer was not sufficiently oxidized during the deposition process. Because of this, the ferromagnetic interaction between Co and Py was stronger than the antiferromagnetic coupling between Py and CoO, with the consequence that no exchange bias effect was observed. The flipped-over hysteresis is not a sign for an antiparallel magnetization in Ni and in Co but is merely due to the negative asymmetry of the recorded signal (see Fig. 11.19b), which in turn is a question of the optical parameters that are difficult to predict.

This example shows that resonant magnetic scattering with circularly polarized X-rays is capable of probing the magnetization depth profile for different elements separately. Also, the magnetic hysteresis can be measured for each specific element.

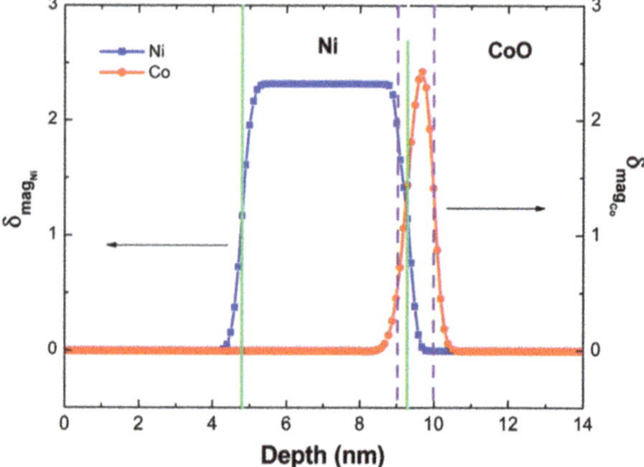

Fig. 11.20 Magnetic dispersion δ_m for Ni (blue) and Co (red) is plotted as a function of depth. Vertical green lines indicate the location of the interfaces between Ni and SiO$_2$ to the left and between Py and CoO to the right. No magnetic signal is seen in the SiO$_2$ cap layer and in the CoO film. Magnetic signals are recorded in the Py layer for Ni and at the interface between Py and CoO of Co. Reproduced with permission from Roy et al. (2007). Copyright (2025) by the APS

However, the absolute magnetization profile in units of μ_B is difficult to obtain, and their relative amplitudes depend on optical parameters that are hard to control. Furthermore, the sign of the hysteresis loops is unreliable. The sign depends on the asymmetry, and the asymmetry varies with the wavevector Q_z along the reflectivity curve. This is in contrast to PNR, which delivers nuclear and magnetic SLDs on an absolute scale as well as magnetic hysteresis loops with the proper sign.

11.7.3 *Magnetization Profiles and Interface Magnetism*

The ferromagnetic order and spin polarization in Heusler alloys are based on the ordered $L2_1$ structure (see Fig. 1.6g). They consist of four interleaved fcc lattices. In the Heusler compound Co$_2$MnGe, the first two fcc lattices are occupied by Co ions, the third by Mn, and the fourth by Ge ions. In the ordered state, this Heusler compound combines a large magnetic moment of 5 μ_B with a 100% spin-polarization at the Fermi level, which is of great interest for spintronic devices, such as giant magneto-resistance (GMR) or tunneling magneto-resistance (TMR) sensors. At the same time, Co$_2$MnGe and similar Heusler compounds show low tolerance to site disorder. Mn has a local moment of 3 μ_B on the regular lattice site but switches to an antiparallel and reduced moment of -1.3 μ_B on a Co antisite. Co has a local moment of 1 μ_B and keeps this moment even on an Mn antisite but lowers the polarization at the Fermi level (Galanakis et al. 2006).

In spintronic devices, very thin Heusler alloy films are used that border nonmagnetic metals or oxides to ensure spin transport through quantum well

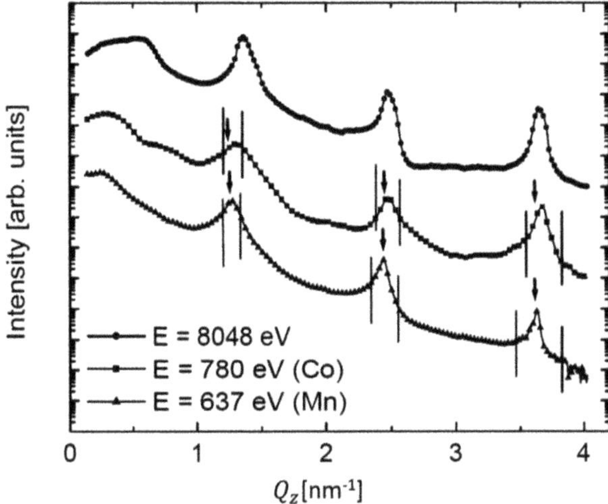

Fig. 11.21 Reflectivity measurements of the Co$_2$MnGe/Au multilayer with 50 repeats versus the momentum transfer Q_z taken with unpolarized hard X-rays and with circular polarized soft X-rays at the Co edge (780 eV) and the Mn edge (637 eV). The scanned Q_z range covers three Bragg reflections, which are further analyzed by XAS and XMCD in the regions denoted by vertical bars. Reproduced with permission from Grabis et al. (2005). Copyright (2025) by the APS

potentials or tunneling barriers. Therefore, an important question is the extent to which spin polarization is preserved in Heusler alloy films up to the interface. To address this question, Grabis et al. grew Co$_2$MnGe/Au multilayers with 50 repeats by sputtering methods in a UHV environment (Grabis et al. 2005). They recorded reflectivity curves with incident circularly polarized X-rays tuned to the respective L_3 absorption edges of Co and Mn. The results are shown in Fig. 11.21 and compared with reflectivity data for nonresonant hard X-rays.

The scattering angles were fixed on one of the multilayer Bragg peaks, and the incident photon energy was scanned over the region of the L_2 and L_3 absorption edges. Since an energy scan also changes the scattering vector, the range of scattering vectors scanned at each Bragg peak is indicated with vertical bars. These energy scans reveal the element-specific magnetic moment density profiles within the Co$_2$MnGe layers. Figure 11.22 shows the intensity sum $(I^+ + I^-)/2$ and the magnetically sensitive asymmetry A for the first three Bragg peaks at the Co absorption edges. Similar data were recorded for the Mn edge. These spectra are structurally very rich due to the convolution of charge and magnetic intensity and are therefore not easy to analyze. However, the sign dependence of the asymmetry allows some simple conclusions to be drawn about the thickness of the nonferromagnetic layer, as model calculations have indicated (Grabis et al. 2005). Based on the asymmetry at the photon energy of 775 eV, which is positive, negative, positive at the first, second, and third-order Bragg peaks, respectively, model calculations suggest that the nonferromagnetic interlayer between Co$_2$MnGe and Au must have a total thickness of approximately 1 nm. It is also characteristic that the energy dependence of the

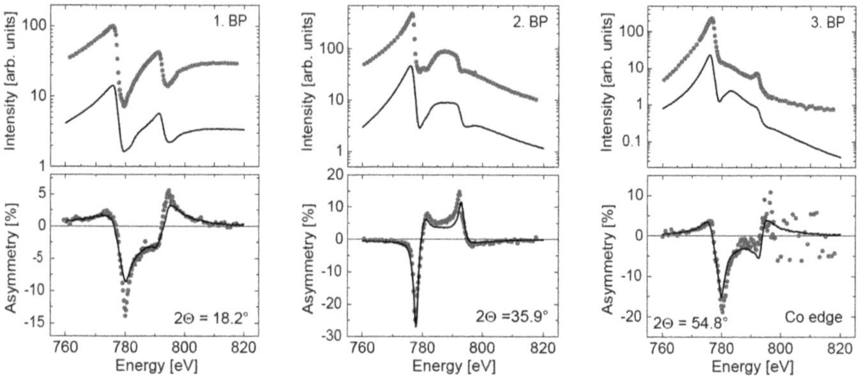

Fig. 11.22 Energy scans at fixed scattering angles $2\theta_1$, $2\theta_2$, $2\theta_3$ of the first three Bragg reflections. The energy scans cover the region of the Co $L_{2,\,3}$ absorption edges. Top row: sum intensities $(I^+ + I^-)/2$ at the three Bragg peaks (BP); Bottom row: asymmetries $(I^+ - I^-)/(I^+ + I^-)$ measured at the respective Bragg peaks. The dots represent measured data; the lines are model calculations. Adapted with permission from Grabis et al. (2005). Copyright (2025) by the APS

asymmetry has two zero crossings for the first and third Bragg reflections, but three zero crossings for the second-order Bragg peak. This is again to be compared with model calculations by varying ferromagnetic and nonferromagnetic layer thicknesses that come to the same conclusions. The final answer is given by a quantitative analysis, as described next.

For a more accurate estimation, of the magnetic and non-magnetic layer thicknesses the energy-dependent intensities and asymmetries were modeled in terms of a magneto-optical matrix formalism performed via the Zak method. The analysis shows that the magnetic moment density profiles determined for Co and Mn are significantly different. Furthermore, the magnetic profiles are narrower than the chemical density profiles, indicating reduced moments at the interfaces, as can be seen in Fig. 11.23. For better comparison, the charge and magnetic dispersion profiles were normalized. For the charge profile, the normalization is:

$$\rho_{ch}(z) = \frac{\left|\delta_{ch}(z) - \delta_{ch,Au}\right|}{\lceil\delta_{ch,Co_2MnGe} - \delta_{ch,Au}\rceil},\qquad(11.51)$$

yielding $\rho_{ch}(z) = 0$ if the dispersion corresponds to the value for bulk Au, and 1 for bulk Co_2AuGe. Similarly, the magnetic density profile is normalized according to:

$$\rho_m(z) = \frac{\delta_m(z)}{\delta_{m,Co_2MnGe}}.\qquad(11.52)$$

The corresponding imaginary part $\beta_{ch,\,m}$ features the same depth profiles as the real part, and if plotted in a normalized fashion, it will carry the same information.

In Fig. 11.23, the magnetic density of Mn appears to be more affected than that of Co by the growth method. In these multilayers, a nonferromagnetic interfacial layer exists at room temperature with a thickness of approximately 0.45 nm at the bottom and 0.3 nm at the top of each Co_2MnGe layer. These results can explain why the

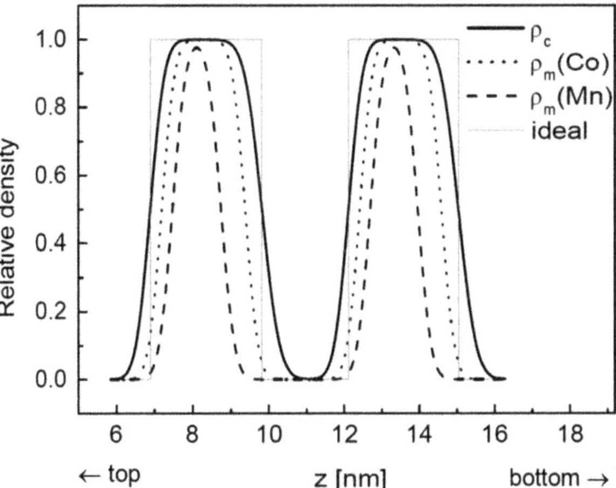

Fig. 11.23 Structural (chemical) and magnetic normalized depth density profiles of Co and Mn in Co$_2$MnGe/Au multilayers from top to bottom as determined from model calculations. Reproduced with permission from Grabis et al. (2005). Copyright (2025) by the APS

tunneling magneto-resistance is found in experiments to be much lower than expected. Further studies are required to improve layer depth profiles to make them more suitable for spintronic devices. This can be done either by optimizing the deposition process or by the application of appropriate post-growth annealing procedures.

The presented example on Co$_2$MnGe/Au multilayers shows that the charge and magnetic asymmetries recorded at the Bragg peaks and close to the Co and Mn edges are very sensitive to the location of the various density profiles. In particular, the thickness of the nonmagnetic "dead" layer at the top and the bottom can be determined uniquely. This appears to contradict the phase problem that one encounters in scattering experiments. However, in the present case, even small phase shifts have a considerable effect on the optical constants, which allows a distinction between top and bottom.

11.7.4   *Magnetic X-Ray Scattering at the M5 Resonance Edge of Ho*

The next example concerns resonant magnetic scattering of the rare earth metal Ho, which we have already discussed in Sect. 7.7. In contrast to the magnetic resonanr signal scattering at the *L*-edges of Ho, we consider here the resonance scattering at the *M*-edge, where a much stronger resonance signal is expected. This example is different from the previous ones in two respects. First, the magnetism of Ho is noncollinear but helical, generating no dichroic signal at the resonance absorptiion edges. Second, as a consequence of the missing circular dichroism, we must use linear polarization in order to gain magnetic sensitivity, as described in Sect. 7.7.

A Ho(001)-film with a thickness of 110 monolayers was epitaxially grown on a W(110) substrate (Ott et al. 2006). Using *linear* polarized synchrotron light with π-polarization in the scattering plane, reflectivity scans were taken at a temperature

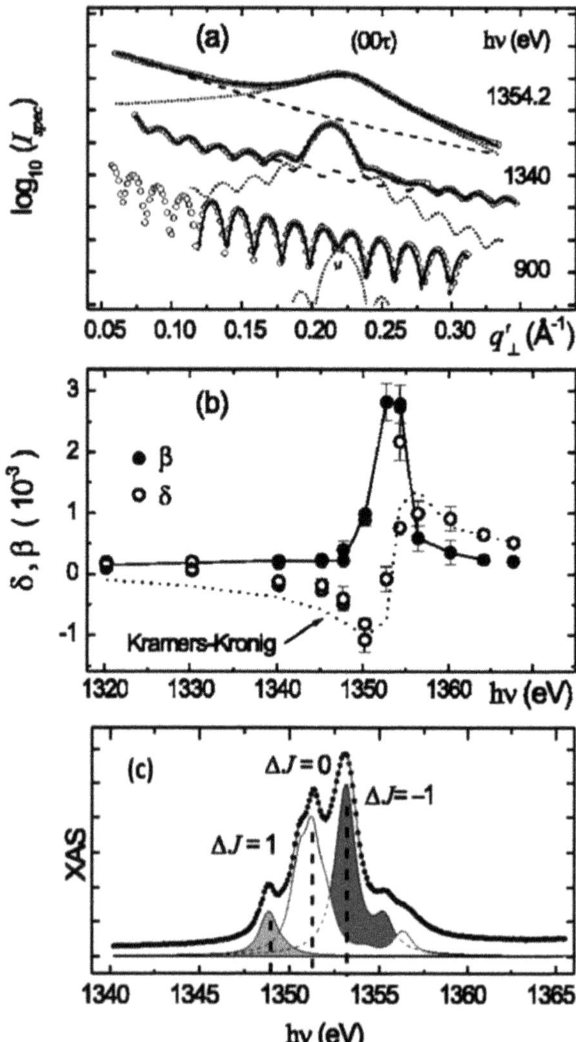

Fig. 11.24 Resonant magnetic X-ray scattering of a Ho(001) film across the τ-peak for different X-ray photon energies. (**a**) Solid lines represent fits to the data points according to the Parratt formalism. The dotted lines denote the contributions from magnetic scattering and the dashed curves from charge scattering, (**b**) Optical parameters β (solid black dots) and δ (open circles) of Ho metal in the M_5 spectral region determined from fitting the resonant reflectivity. The solid line through the data points for β is a guide to the eyes. The dotted line is the calculated δ value via the Kramers–Kronig relationship, (**c**) Absorption spectrum of the Ho M_5 resonance recorded with the sample drain current (TEY). Note the different abscissa scales in (**b**) and (**c**). Reproduced and adapted with permission from Ott et al. (2006). Copyright (2025) by the APS

of 40 K for different X-ray energies near the M_V absorption edge of Ho. This excitation takes place between the $3d^{10}4f^{10}$ initial state to the $3d^9 4f^{11}$ final state with a resonant energy of 1350 eV. For the Ho spectral level scheme, see Fig. 7.23. The first scan shown in Fig. 11.24a was taken at 900 eV, far below the resonant energy. This reflectivity scan shows typical Kiessig fringes of a thin film, but magnetic effects are barely noticeable, because at this energy, the reflectivity is dominated by charge scattering. The same scan taken at X-ray photon energy of 1340 eV exhibits two important changes. First, the Kiessig fringes are less pronounced than that in the previous scan due to an increase in β_{ch} close to the absorption edge, and second, we observe a clear peak at the scattering vector of

$0.2 \, \text{Å}^{-1}$. This peak is the well-known τ-peak, specifically the $(000 + \tau)$ peak, which is due to the spin spiral in Ho, discussed in Sect. 7.6. The peak position agrees very well with the one quoted in Fig. 7.21 for the same temperature. The next scan taken at 1354.2 eV is at the resonance maximum and therefore strongly damped. As the penetration depth is rather shallow under these conditions, the probing number of Ho-atomic layers is small, and therefore, the τ-peak is broadened due to size effects.

The diffraction patterns in Fig. 11.24a are described by a superposition of the Fresnel reflectivity of a Ho slab on a semi-infinite substrate and a magnetic diffraction peak for the spin helix. The Fresnel reflectivity was modeled by using the Parratt recursion formula according to Eq. (11.33), including the optical parameters for absorption β and dispersion δ. In addition to the Kiessig fringes due to the finite Ho-film thickness, the magnetic τ-peak was modeled in Born approximation via the usual lattice sum, that is, by adding up the phase factors of N layers in the film (see Eq. (3.22) and (10.1)) to yield the scattering amplitude:

$$S(Q_z) = Pe^{-Q_z^2 \sigma_m^2} \sum_{u=0}^{N-1} e^{iu(Q_z - \tau) \cdot d} \tag{11.53}$$

Here, in addition, a mean magnetic roughness σ_m and an angle-dependent polarization factor P are included. The result of the intensity calculation $I(Q_z) \propto S^2(Q_z)$ is shown by the solid lines in Fig. 11.24a.

The fit reveals the optical parameters δ and β, which are shown in Fig. 11.24b, as a function of energy across the resonant energy region at 1350 eV. We notice that the absorption, that is, the β-curve, has a sharp maximum at the resonance energy, while the dispersion correction δ goes through a minimum with a step-like behavior at the adsorption edge. As we have seen previously, absorption β and dispersion δ are not independent but connected via the Kramers–Kronig relationship. Therefore, the calculated energy dependence of the dispersion $\delta(E)$ is also shown in Fig. 11.23b by the dashed line, which reproduces well the fitted experimental data for δ.

In Fig. 11.24c, the absorption data are plotted measured with a higher resolution than in panel (b) using the sample drain current, that is, using the TEY method (see Fig. 7.22). The absorption spectrum shows a clearly resolved triplet final state $3d^9 4f^{11}$ with $\Delta J = 0, \pm 1$. Because of the magnetic order, the degeneracy of the J states is lifted and split into M_J sublevels with dipole-allowed $\Delta M_J = 0, \pm 1$ transitions.

This example shows that for noncollinear magnetic structures like the Ho spin helix, we do not observe magnetic circular dichroism. Nevertheless, magnetic sensitivity can be gained by using linear polarization and tuning the X-ray energy to the relevant absorption edge, here the M_V edge. Hence, for studying antiferromagnetic or noncolinear spin structures, the X-ray linear magnetic dichroism (XLMD) has to be used, which is described by the second-order term in Eq. (7.57).

11.8   Comparison of XRMS and PNR

In the previous examples, we have shown how XRMS and PNR methods are capable of analyzing the magnetism of thin films, multilayers, and interfaces. These were but a few examples taken from the extensive literature published on this topic. More examples can be found in the reviews (Macke and Goering 2014; Fink et al. 2013; Schütz et al. 2007).

Concluding, we compare both methods to better understand their strengths and weaknesses. A few characteristics are obvious and are quickly stated.

1. PNR allows layer-resolved vector magnetometry in the plane perpendicular to the scattering vector, that is, usually the sample plane. XRMS is sensitive mainly to only one magnetization component, which is parallel to the incident wavenumber k_i.
2. PNR is independent of resonant conditions; any magnetic induction fluctuation yields neutron magnetic scattering, such as from superconducting flux lines. Magnetic stray fields can affect magnetic neutron reflectivity, which can be both an advantage and also a disadvantage.
3. XRMS allows the determination of the chemical and magnetic depth profiles for different elements separately. The element specificity is one of the strongest hallmarks of XRMS. PNR provides the combined nuclear and magnetic scattering length density profiles. Knowing the chemical composition of individual layers, the nuclear and magnetic profiles can be disentangled.
4. PNR provides all information on an absolute scale, allowing a quantitative analysis of the magnetic induction in layers. XRMS depends on the knowledge of optical parameters, which requires extra spectroscopic measurements.
5. In general, and not only for PNR/XRMS, neutron beams have much lower intensity than X-ray beams. Therefore, the cross section of a neutron beam is bigger, covering a good portion of a 10×10 mm^2 sample, compared to a sub-mm^2 beam size used for X-ray scattering. This has important consequences not only for the sample size to be prepared but also for the design of magnets, cryostats, sample shielding, etc.

Because of the lower flux at cold neutron sources, a typical PNR experiment with different polarizations and a few temperatures requires a week of beam time or more. A typical PNR scan with up- and down-polarization takes about 4–5 h. In comparison, a similar XRMS experiment is finished in less than 10 min. However, when performing an XRMS experiment, many additional XAS scans are to be performed, and much more testing is required before conclusive scans can be started. In the end, the total beam time required for PNR and XRMS experiments for gaining meaningful data is quite comparable.

11.9   Experimental Realization of XRMS

Resonant magnetic X-ray scattering by using "soft" X-rays of the type discussed here is best performed at a synchrotron light source that delivers photons in the energy range from 100 to about 2000 eV. Given a stable electron orbit and a fixed beam size on the sample, we need, in addition, a tunable photon energy and adjustable linear and circular polarization. Both can be achieved with the installation of undulators, as described in Sects. 4.2.3 and 4.2.4. Traditionally, in an energy scan, the polarization was kept constant while scanning the photon energy over the region of interest. In order to obtain an asymmetry curve, two energy scans are required with different polarizations, taken one after the next. If the energy scan requires some time (a few minutes), then it is necessary to keep a constant sample environment between both scans to achieve a reliable comparison. Alternatively, one may switch the polarization at each energy, which is demanding on the undulator, as heavy machinery must be shifted with high precision. A new invention allows flipping the helicity of X-rays from twin elliptical undulators in the MHz frequency range (Holldack et al. 2020). This allows for measuring absorption spectra essentially simultaneously for different polarizations. This new development will likely change the experimental procedures in the future.

An example of a beamline is shown in Fig. 11.25 taken from Sybren Miedema et al. (2016). It is a beamline that has an undulator covering the energy range of 100–1500 eV and allows for variable polarization (circular and linear in any orientation) with a 20–60 μm spot size in the focus. Around the focal spot, any permanent or temporary endstation can be built for specific experiments. Since the beam path from the undulator source to the end station is open, ultra-high vacuum conditions are mandatory everywhere. Usually, the UHV enclosures consist of stainless steel, which simultaneously also provides sufficient radiation protection. An experimental chamber for XRMS is schematically shown in Fig. 4.17.

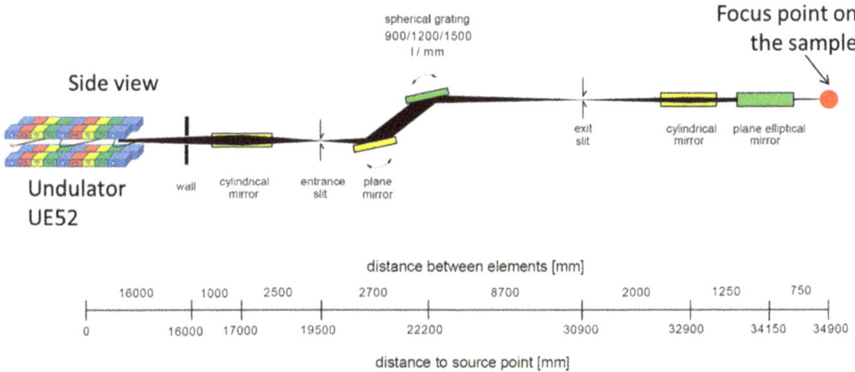

Fig. 11.25 Typical beam line at a soft X-ray synchrotron source, here BESSY II, Germany. The entire 35 m long beam line from the storage ring to the experimental station is enclosed in a vacuum. Reproduced and adapted from Sybren Miedema et al. (2016), licensed under creative common

Summary

1. Polarized neutron reflectivity assumes that the scattering plane is oriented normal to the sample plane.
2. Polarized neutron reflectivity requires at least a polarized incident beam.
3. A full polarization analysis requires a neutron spin polarizer of the incident beam and a spin analyzer of the scattered beam.
4. We distinguish between non-spin-flip (NSF) reflectivity and spin-flip (SF) reflectivity.
5. Non-spin-flip reflectivity is proportional to the magnetization component parallel to the polarization axis.
6. Spin-flip reflectivity is proportional to the square of the magnetization component parallel to the axis in the scattering plane.
7. Spin-flip reflectivities up-down R^{+-} and down-up R^{-+} are identical for all scattering vectors Q.
8. PNR allows for the analysis and distinction of different magnetization reversal processes.
9. Evaluating the dispersion Δ of neutron reflectivity yields information on the magnetic domain state of the sample.
10. Multilayers with equal sublayer thicknesses exhibit no even-order Bragg reflections.
11. Diffuse scattering can only be observed if the neutron coherence ellipsoid is larger than the average domain size.
12. The spectral absorption (β) and dispersion (δ) coefficients are related via the Kramers–Kronig integrals.
13. The optical theorem states that the imaginary part of the refractive index follows from the absorption cross section in the forward direction.
14. X-ray absorption spectroscopy is an essential part of X-ray magnetic scattering.
15. An XMCD signal is only obtained if the final states for up- and down-oriented spins feature different densities.
16. XMCD signals are usually observed for ferromagnetic materials at resonant absorption edges from core levels to spin–split final states.
17. XMCD is observed as a difference in the absorption of right and left circularly polarized X-rays.
18. XRMS combines X-ray scattering with X-ray magnetic circular or linear dichroism to study the magnetic structure of materials.
19. XMCD and XRMS cannot be used for the analysis of antiferromagnets. However, analysis based on XMLD is possible.

Questions

(Note that more than one answer may be correct)

1. **What causes neutron spins to flip?**

   (a) Only Fe magnetization can flip the neutron spin.
   (b) Any magnetization component perpendicular to the neutron polarization axis can flip the neutron spin.
   (c) The neutron spin flip is a combined effort of the magnetization with X- and Y-components.

2. **What causes diffuse off-specular magnetic scattering?**

   (a) Poor monochromatization of the incident beam.
   (b) Magnetic domains that are larger than the neutron coherence length.
   (c) Magnetic domains that are smaller than the neutron coherence length.

3. **What are the conditions for vanishing even-order multilayer Bragg reflections?**

   (a) The form factors of both alternating layers must be identical.
   (b) The Debye–Waller factors of both layers must be identical.
   (c) The layer thicknesses must be identical.

4. **Why is no XMCD signal detected when probing paramagnets and antiferromagnets?**

   (a) Because paramagnets and antiferromagnets are macroscopically not magnetic.
   (b) Because paramagnets and antiferromagnets have ordering temperatures that cannot be experimentally reached.
   (c) Because paramagnets and antiferromagnets have the same density of states for spin-up and spin-down electrons at the Fermi level.

5. **What is the importance of the Kramers-Kronig relationship?**

   (a) The K–K. relationship connects real and imaginary parts of a complex function.
   (b) The K–K. relationship is an expression of causality: the response cannot take place earlier than the stimulus.
   (c) The K–K. relationship connects charge and magnetic dispersions.

Exercise

M 11.1 Multilayer with missing peaks

Show that multilayers with equal layer thicknesses $d_1 = d_2$ and N repeats but different SLDs $\rho_1 b_1 \neq \rho_2 b_2$ exhibit no even-order Bragg reflections.

References

F.S. Bergeret, A.F. Volkov, K.B. Efetov, Odd triplet superconductivity and related phenomena in superconductor-ferromagnet structures. Rev. Mod. Phys. **77**, 1321 (2005)

A.J. Blundell, J.A.C. Bland, Polarized neutron reflection as probe of magnetic films and multilayers. Phys. Rev. B **46**, 3391–3400 (1992)

A. Buzdin, Proximity effects in superconductor-ferromagnet heterostructures. Rev. Mod. Phys. **77**, 935 (2005)

C.T. Chen, Y.U. Idzerda, H.-J. Lin, N.V. Smith, G. Meigs, E. Chaban, G.H. Ho, E. Pellegrin, F. Sette, Experimental confirmation of the X-ray magnetic circular dichroism sum rules for iron and cobalt. Phys. Rev. Lett. **75**, 152–155 (1995)

A. Devishvili, K. Zhernenkov, A.J.C. Dennison, B.P. Toperverg, M. Wolff, B. Hjörvarsson, H. Zabel, SuperADAM: Upgraded polarized neutron reflectometer at the Institut Laue-Langevin. Rev. Sci. Instrum. **84**, 025112 (2013)

K.B. Efetov, I.A. Garifullin, A.F. Volkov, K. Westerholt, Spin-polarized electrons in superconductor/ferromagnet hybrid structures, in *Magnetic Nanostructures*, Springer Tracts in Modern Physics, vol. 246, (Springer, Berlin, Heidelberg, 2013), pp. 85–118

G.P. Felcher, R.O. Hillecke, R.K. Crawford, J. Haumann, R. Kleb, G. Ostrowski, Polarized neutron reflectometer: A new instrument to measure magnetic depth profiles. Rev. Sci. Instrum. **58**, 609 (1987)

J. Fink, E. Schierle, E. Weschke, J. Geck, Resonant elastic soft X-ray scattering. Rep. Prog. Phys. **76**, 056502 (2013)

I. Galanakis, P. Mavropoulos, P.H. Dederichs, Electronic structure and Slater-Pauling behaviour in half-metallic Heusler alloys calculated from first principles. J. Phys. D. Appl. Phys. **39**, 765–775 (2006)

J. Geissler, E. Goering, M. Justen, F. Weigand, G. Schütz, J. Langer, D. Schmitz, H. Maletta, R. Mattheis, Pt magnetization profile in a Pt/Co bilayer studied by resonant magnetic X-ray reflectometry. Phys. Rev. B **65**, 020405.1–020405.4 (2002)

D. Gorkov, B.P. Toperverg, H. Zabel, Artificial magnetic pattern arrays probed by polarized neutron reflectivity. Nanomaterials **10**, 851 (2020)

J. Grabis, A. Bergmann, A. Nefedov, K. Westerholt, H. Zabel, Element-specific characterization of the interface magnetism in $[Co_2MnGe/Au]_n$ multilayers by X-ray resonant magnetic scattering. Phys. Rev. B **72**, 024438-1–11 (2005)

J.P. Hannon, G.T. Trammell, M. Blume, D. Gibbs, X-ray resonance exchange scattering. Phys. Rev. Lett. **61**, 1245–1248 (1988)

B.L. Henke, E.M. Gullikson, J.C. Davis, X-ray interactions: Photoabsorption, scattering, transmission, and reflection at E=50-30000 eV, Z=1-92. At. Data Nucl. Data Tables **54**(2), 181–342 (1993)

J.P. Hill, D.F. McMorrow, Resonant exchange scattering: Polarization dependence and correlation function. Acta Cryst **A52**, 236 (1996)

K. Holldack, C. Schüssler-Langeheine, P. Goslawski, N. Pontius, T. Kachel, F. Armborst, M. Ries, A. Schälicke, M. Scheer, W. Frentrup, J. Bahrdt, Flipping the helicity of X-rays from an undulator at unprecedented speed. Commun. Phys. **3**, 61 (2020)

N. Jaouen, J. Tonnerre, D. Raoux, E. Bontempi, L. Ortega, M. Müenzenberg, W. Felsch, A. Rogalev, H. Dürr, E. Dudzik, et al., Ce 5d magnetic profile in Fe/Ce multilayers for the and -like Ce phases by X-ray resonant magnetic scattering. Phys. Rev. B **66**, 134420.1–134420.14 (2002)

J.B. Kortright, O. Hellwig, K. Chesnel, S. Sun, E.E. Fullerton, Interparticle magnetic correlations in dense Co nanoparticle assemblies. Phys. Rev. B **71**, 012402 (2005)

S. Langridge, J. Schmalian, C.H. Marrows, D.T. Dekadjevi, B.J. Hickey, Quantification of magnetic domain disorder and correlations in antiferromagnetically coupled multilayers by neutron reflectometry. Phys. Rev. Lett. **85**, 4964 (2000)

W.-T. Lee, S.G.E. te Velthuis, G.P. Felcher, F. Klose, T. Gredig, E.D. Dahlberg, Ferromagnetic domain distribution in thin films during magnetization reversal. Phys. Rev. B **65**, 224417 (2002)

D.R. Lee, S.K. Sinha, D. Haskel, Y. Choi, J.C. Lang, S.A. Stepanov, G. Srajer, X-ray resonant magnetic scattering from structurally and magnetically rough interfaces in multilayered systems. I. Specular reflectivity. Phys. Rev. B **68**, 224409.1-19 (2003a)

D.R. Lee, S.K. Sinha, C.S. Nelson, J.C. Lang, C.T. Venkataraman, G. Srajer, R.M. Osgood III, X-ray resonant magnetic scattering from structurally and magnetically rough interfaces in multilayered systems. II. Diffuse scattering. Phys. Rev. B **68**, 224410.1-14 (2003b)

S. Macke, E. Goering, Topical review magnetic reflectometry of heterostructures. J. Phys. Condens. Matter **26**, 363201 (2014)

C.F. Majkrzak, K.V. O'Donovan, N.F. Berk, Polarized neutron reflectometry, in *Neutron Scattering from Magnetic Materials*, ed. by T. Chatterji, (Elsevier, Amsterdam, Heidelberg, New York, 2006)

F. Mezei, Neutron spin Echo: A new concept in polarized thermal neutron Techniques. Z. Phys. **255**, 146 (1972)

J. Nogue's, I.K. Schuller, Exchange Bias. J. Magn. Magn. Mater. **192**, 203 (1999)

H. Ott, C. Schüßler-Langeheine, E. Schierle, A.Y. Grigoriev, V. Leiner, H. Zabel, G. Kaindl, E. Weschke, Magnetic X-ray scattering at the M5 absorption edge of Ho. Phys. Rev. B **74**, 094412 (2006)

F. Radu, M. Etzkorn, R. Siebrecht, T. Schmitte, K. Westerholt, H. Zabel, Interfacial domain formation during magnetization reversal in exchange-biased CoO/Co bilayers. Phys. Rev. B **67**, 134409 (2003)

S. Roy, M.R. Fitzsimmons, S. Park, M. Dorn, O. Petracic, I.V. Roshchin, Z.-P. Li, X. Batlle, R. Morales, A. Misra, X. Zhang, K. Chesnel, J.B. Kortright, S.K. Sinha, I.K. Schuller, Depth profile of uncompensated spins in an exchange bias system. Phys. Rev. Lett. **95**(047201), 1–4 (2005)

S. Roy, C. Sanchez-Hanke, S. Park, M.R. Fitzsimmons, Y.J. Tang, J.I. Hong, D.J. Smith, B.J. Taylor, X. Liu, M.B. Maple, A.E. Berkowitz, C.-C. Kao, S.K. Sinha, Evidence of modified ferromagnetism at a buried Permalloy/CoO interface at room temperature. Phys. Rev. B **75**, 014442 (2007)

D.K. Satapathy, M.A. Uribe-Laverde, I. Marozau, V.K. Malik, S. Das, T. Wagner, et al., Magnetic proximity effect in $YBa_2Cu_3O_7/La_{2/3}Ca_{1/3}MnO_3$ and $YBa_2Cu_3O_7/LaMnO_{3+\delta}$</συβ> superlattices. Phys. Rev. Lett. **108**, 197201 (2012)

G. Schütz, E. Goering, H. Stoll, Synchrotron Radiation Techniques Based on X-ray: Magnetic Circular Dichroism, in *The Handbook of Magnetism and Advanced Magnetic Materials*, Novel Techniques, ed. by H. Kronmüller, S.P.S. Parkin, vol. 3, (Wiley, New York, 2007)

J. Stöhr, H.C. Siegmann, *Magnetism, From Fundamentals to Nanoscale Dynamics*, Springer Series in Solid-State Sciences (Springer Verlag, Berlin, Heidelberg, 2006)

P. Sybren Miedema, W. Quevedo, M. Fondell, The variable polarization undulator beamline UE52 SGM at BESSY II. J. Large Scale Res. Facilit. JLSRF **2**, A70 (2016)

J.M. Tonnerre, L. Seve, D. Raoux, G. Soullie, B. Rodmacq, P. Wolfers, Soft X-ray resonant magnetic scattering from a magnetically coupled Ag/Ni multilayer. Phys. Rev. Lett. **75**, 740–743 (1995)

J.-M. Tonnerre, N. Jaouen, E. Bontempi, D. Carbone, D. Babonneau, M. De Santis, H.C.N. Tolentino, S. Grenier, S. Garaudee, U. Staub, Soft X-ray resonant magnetic reflectivity studies for in and out-of-plane magnetization profile in ultra thin films. J. Phys. Conf. Series **211**, 012015 (2010)

B.P. Toperverg, H. Zabel, Chapter 6: Neutron Scattering in Nanomagnetism, in *Neutron Scattering - Magnetic and Quantum Phenomena, Experimental Methods in the Physical Sciences*, ed. by F. Fernandez-Alonso, D.L. Price, vol. 48, (Academic Press (imprinted of Elsevier), 2015)

G. van der Laan, A.I. Figueroa, X-ray magnetic circular dichroism—A versatile tool to study magnetism. Coord. Chem. Rev. **277**, 95–129 (2014)

J. Zak, E.R. Moog, C. Liu, S.D. Bader, Universal approach to magneto-optics. J. Magn. Magn. Mater. **89**, 107–123 (1990)

J. Zak, E.R. Moog, C. Liu, S.D. Bader, Magneto-optics of multilayers with arbitrary magnetization directions. Phys. Rev. B **43**, 6423–6429 (1991)

J. Zak, E.R. Moog, C. Liu, S.D. Bader, Erratum: Magneto-optics of multilayers with arbitrary magnetization directions. Phys. Rev. B **46**, 5883(E) (1992)

Further Readings

S. Blundell, *Magnetism in Condensed Matter* (Oxford University Press, Oxford, 2001)

J. Stöhr, H.C. Siegmann, *Magnetism, From Fundamentals to Nanoscale Dynamics*, Springer Series in Solid-State Sciences (Springer Verlag, Berlin, Heidelberg, 2006)

H. Kronmüller, S.P.S. Parkin (eds.), *The Handbook of Magnetism and Advanced Magnetic Materials*, Novel Techniques, vol 3 (Wiley, New York, 2007)

T. Chatterji, *Neutron Scattering from Magnetic Materials* (Elsevier, Amsterdam, Heidelberg, New York, 2006)

E. Beaurepaire, H. Bulou, F. Scheurer, J.-O. Kappler, *Magnetism: A Synchrotron Radiation Approach*, Lecture Notes in Physics, vol 697 (Springer, Berlin, Heidelberg, New York, 2006)

Chapter 12
Small- and Wide-Angle Surface Scattering

12.1 Grazing Incidence Surface Diffraction (GID)

12.1.1 Scattering Geometry

The scattering conditions considered for X-ray and neutron reflectivity are also suitable for obtaining crystallographic information in the surface plane. The scattering geometry for reflectivity and surface scattering experiments is depicted and compared in Fig. 12.1.

In reflectivity experiments, the glancing angles $\alpha_{i,\,f}$ are scanned, while the scattering angle 2θ is kept in the zero position: $2\theta = 0$. Specular reflectivity is performed by keeping $\alpha_i = \alpha_f$, resulting in a Q_z-scan; off-specular reflectivity requires $\alpha_i \neq \alpha_f$, which adds a small in-plane component $Q_{x,\,y}$.

In contrast to reflectivity measurements, in surface scattering experiments, the scattering vector lies in the surface plane, indicated by $Q_{\|}$. When changing the length of the scattering vector $Q_{\|}$ with increasing angle 2θ between the incident and exit beams (radial scan) while keeping $\alpha_{i,\,f}$ close to the critical angle α_c, a surface Bragg reflection at high angles will eventually be excited. Those Bragg reflections probe the lattice planes perpendicular to the surface, as schematically indicated in Fig. 12.1. In contrast to bulk scattering, surface Bragg peaks are only sensitive to the crystal structure within the top layers controlled by the penetration depth via tuning the glancing angles $\alpha_{i,\,f}$.

12.1.2 Scattering Intensity

From another perspective, surface scattering is also indicated in Fig. 10.1 and occurs within the green-colored area that signifies the area of total reflection. In this graph, the in-plane surface Bragg peak does not occur at $Q_\perp = 0$, but at $Q_\perp = Q_c$. Surface

© The Author(s), under exclusive license to Springer Nature Switzerland AG 2026
H. Zabel, *Elements of Elastic Scattering by X-Rays, Neutrons, and Electrons*,
https://doi.org/10.1007/978-3-032-16624-1_12

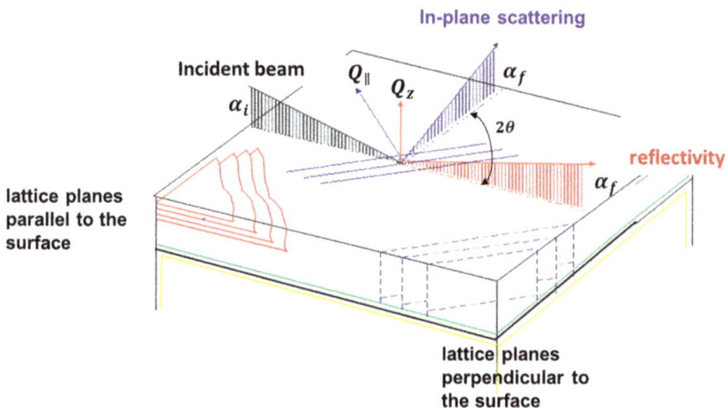

Fig. 12.1 Scattering geometry for reflectivity and surface scattering. In reflectivity mode, the scattering vector is short and oriented perpendicular to the surface (Q_z); in surface scattering mode, the scattering vector is large and oriented parallel to the surface ($Q_\parallel$) with a Q_z component close to the critical scattering vector Q_c. Reproduced and adapted from Dosch (1992), with the permission of the APS

scattering probes the crystal truncation rods at their origin just above the crystal surface. The extended part of the crystal truncation rod scattering was already discussed in Sect. 10.3. We can use the results already obtained for CTR scattering by modifying two aspects. First, we ought to take into account the Fresnel optics due to the surface proximity. This can be done by including the Fresnel transmission coefficients t_i and t_f for the incident and scattered beam, defined in Eq. (10.24). Then, the intensity for the in-plane Bragg reflections becomes:

$$I_{surf}(Q_\parallel) = A_0^2 \left(\frac{r_0}{R}\right)^2 N_x \delta(Q_x - G_h) N_y \delta(Q_y - G_k) |t_i|^2 \left| \sum_m e^{iQ_z \cdot a_{z,m}} \right|^2 |t_f|^2. \quad (12.1)$$

The prefactor $A_0^2 \left(\frac{r_0}{R}\right)^2$ has the usual form. The $\delta(Q_{x,y} - G_{h,k})$—functions express the in-plane Bragg conditions. As the in-plane Bragg peak is defined in three dimensions, the third component is defined by the lattice sum in the z-direction, where $a_{z,m}$ is the lattice spacing in the z-direction.

Second, we need to change the z-component of the scattering vector from Q_z to Q_z^t, because scattering takes place by the evanescent wave inside of the crystal but not in the vacuum. For $Q_z < Q_c$, Q_z^t becomes imaginary and the penetration depth Λ is limited to $\Lambda = (Q_c)^{-1}$. As soon as $\alpha_{i,f} > \alpha_c$, the penetration depth is increasingly governed by the linear attenuation coefficient μ, which was already discussed in Sect. 10.4.5 and shown in Fig. 10.18. With these changes, the surface scattering intensity in the modified kinematic scattering approach becomes:

$$I_{surf}(Q_\parallel) = A_0^2 \left(\frac{r_0}{R}\right)^2 N_x \delta(Q_x - G_h) N_y \delta(Q_y - G_k) |t_i|^2 \left|\sum_m e^{iQ_z^t \cdot a_{z,m}}\right|^2 |t_f|^2, \quad (12.2)$$

with the Fresnel coefficients:

$$t_{i,f} = = \frac{2 \sin \alpha_{i,f}}{\sin \alpha_{i,f} + \sqrt{\sin^2 \alpha_{i,f} - \alpha_c^2}} \qquad (12.3)$$

It can be shown that this yields the following expression for the surface scattering intensity (Vineyard 1982; Dosch 1987; Feidenhans'l 1989):

$$I_{surf}(Q_\parallel) = A_0^2 \left(\frac{r_0}{R}\right)^2 N_a \delta(Q_x - G_h) N_b \delta(Q_y - G_k) \times |t_i|^2 \frac{(2\Lambda/a_z)^2}{1 + \left[(4\Lambda/a_z)\sin\left(Re(Q_z^t)a_z/2\right)\right]^2} |t_f|^2$$

$$(12.4)$$

Here, the ratio Λ/a_z is the number of lattice planes with lattice spacing a_z within the penetration depth Λ. Note that the maximum intensity is no longer located at $Q_z = 0$, as expected from bulk kinematic scattering theory, but at $Q_z^t = 0$, i.e., at $Q_z = Q_c$. Hence, the in-plane bulk Bragg condition is modified due to the X-ray optical properties close to total reflection. The modification is termed "*distorted wave Born approximation*" (DWBA). Aside from the optical coefficient, the expression in Eq. (12.4) is similar to the one derived for the crystal truncation rods in Eq. (10.5).

The essential point of surface scattering is therefore its surface sensitivity and the control over the information depth below the surface. Therefore, with surface scattering methods, wide angle in-plane Bragg reflections can be probed, and the depth profile of structural order parameters near the surface can be determined. An example will be discussed on the order–disorder phase transition of Cu_3Au at the (100) surface in the next section. In short: surface diffraction provides crystallographic information on structures within the penetration depth of the probing X-ray or neutron beam.

12.1.3   Surface Phase Transitions

Cu_3Au exhibits an order–disorder phase transition at a temperature of about 660 K. This phase transition is of first-order, which is to be distinguished from second-order phase transitions. Second-order phase transitions in the bulk are discussed in Chaps. 7 and 8. A characteristic of second-order phase transitions is a continuous change of the order parameter from an ordered state at low temperatures to a disordered state above the critical temperature T_c. The order parameter can be the magnetization of a ferromagnet, the sublattice magnetization of an antiferromagnet,

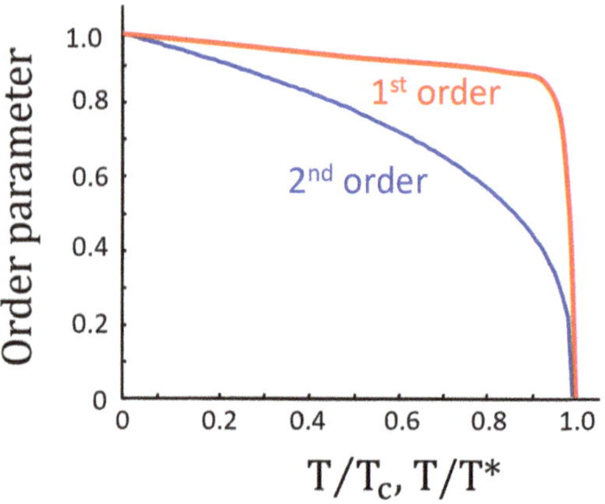

Fig. 12.2 Order parameter as a function of temperature for phase transitions featuring a first-order transition (red line) and a second-order transition (blue line). The temperature, where the order parameter drops to zero is called the critical temperature T_c for second order transitions, or simply transition temperature T^* for first order transitions

or the site occupancy in an order–disorder transition, as discussed for the β-CuZn alloy in Sect. 8.4. In contrast, first-order phase transitions have an order parameter that drops discontinuously from nearly 100% to zero at the transition temperature T^*. Both phase transitions are compared schematically in Fig. 12.2. For a detailed discussion of phase transitions, we refer to the monograph by Stanley (1971).

In general, it is of great interest to explore whether the order parameter that characterizes structural or magnetic phase transitions is the same at the surface and in the bulk or whether it is enhanced or reduced. Theoretically, phase transitions in two and three dimensions are distinctly different, characterized by different critical exponents for the order parameter, the susceptibility, and other thermodynamic properties. However, here, we have the situation that the 2D surface layer is in perfect commensurate registry with the 3D bulk single crystal beneath. Several studies have been conducted in the past on the order–disorder phase transition to analyze these questions. In particular, the alloys $Cu_3Au(001)$ (Dosch et al. 1988), $Fe_3Al(1\bar{1}0)$ (Dosch et al. 1986; Dosch et al. 1992), and $Cu_3Au(111)$ (Zhu et al. 1988), have been investigated by surface-sensitive X-ray scattering methods. In all investigations of this type, it is necessary to ensure that the alloy concentration is constant from the bulk to the surface at all temperatures and that the observed phenomena are not influenced by the depletion of one or another alloy component. This is a serious problem for β-CuZn alloys (β-brass), but less so for CuAu alloys. Furthermore, the investigations must be conducted under ultra-high vacuum conditions to avoid contamination by oxidation.

As an example, we present here briefly the X-ray surface scattering experiment by Dosch and coworkers, who studied the order-parameter profile in $Cu_3Au(001)$ (Dosch et al. 1986) close to the structural phase transition. Cu_3Au has the $L2_1$ structure shown in Fig. 1.6f in the ordered state. The order-parameter S was determined by measuring the intensity of the (001) superstructure reflection and by tuning the incident glancing angle α_i while integrating over the exit angle α_f with the help of a position-sensitive detector. The intensity of the (001) peak is proportional to the square of the order parameter S:

$$I(Q = G_{001}) \propto S^2 \left(f_{Cu} e^{-W_{Cu}} - f_{Au} e^{-W_{Au}} \right)^2, \tag{12.5}$$

where f_{Cu} and f_{Au} are the atomic form factors of Cu and Au, and $W_{Cu,\,Au}$ are the respective Debye–Waller factors. Combining with Eq. (12.4), the superstructure intensity under glancing angle conditions is:

$$I_{surf}(G_{001}) = A_0^2 \left(\frac{r_0}{R} \right) \times |t_i|^2 \left(\frac{S^2 (f_{Cu} e^{-W_{Cu}} - f_{Au} e^{-W_{Au}})^2 (2\Lambda/a_z)^2}{1 + \left[(4\Lambda/a_z) \sin \left(Re(Q_z^t) a_z/2 \right) \right]^2} \right) |t_f|^2 \tag{12.6}$$

In the simplest form, the order parameter S is expressed as the fraction r_{Cu} of copper atoms sitting on the right Cu-sites, minus the fraction w_{Au} of gold atoms residing wrongly on Cu-sites: $S = r_{Cu} - w_{Au}$.

Integration over the exit angle α_f provides sufficient surface sensitivity while enhancing the scattering intensity. With a ratio $\alpha_i/\alpha_c = 0.5$, the penetration depth is small, probing mainly the surface layers, while with a ratio of $\alpha_i/\alpha_c = 1.5$, the intensity is more strongly weighted by the bulk. In panel (a) of Fig. 12.3, the intensity is plotted for both ratios as a function of temperature (Dosch et al. 1988). The surface-weighted intensity (solid dots) appears smoother and more second order-like than the bulk-weighted intensity (open circles). This can be taken as an indication that the surface features another temperature dependence of the order-parameter than the bulk, while the transition temperature appears to be the same. In panel (b) of Fig. 12.3, the scattered intensity is shown for a fixed ratio of $\alpha_i/\alpha_c = 1.0$, which controls the information depth, and variable α_f/α_i for various temperatures close to the transition temperature. Here, we observe the Bragg peak at $\alpha_f/\alpha_i = 1$, as expected from Eq. (12.6). Furthermore, the peak height drops continuously while approaching the transition temperature T^*. Whether the last scan at 660.9 K is due to

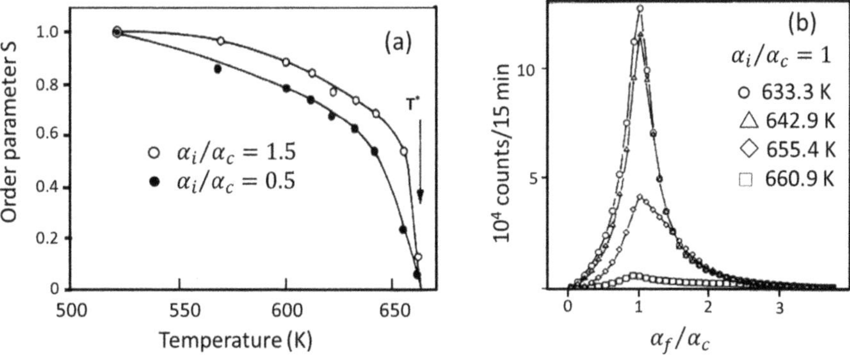

Fig. 12.3 (**a**) a_f-integrated and normalized order parameter S of Cu$_3$Au(100) as a function of the absolute temperature for the incident angles $\alpha_i/\alpha_c = 0.5$ and 1.5. T* is the transition temperature, above which the (001) Bragg intensity vanishes. (**b**) α_f/α_c scans of the (001) intensities for fixed angles $\alpha_i/\alpha_c = 1$ and selected temperatures below T*. Reproduced with permission from Dosch et al. (1988). Copyright (2025) by the APS

a long-range order or short-range order cannot be decided by this plot but requires a transverse scan probing the shape of the peak: Gaussian shape for LRO or Lorentzian shape for SRO. Furthermore, the authors of this work have analyzed the order parameter profile in dependence on the depth below the surface. They concluded that a disordered surface layer exists on top of the ordered bulk for $T < T^*$ and that the disordered layer expands logarithmically further into the bulk by approaching T^* from below.

Similar experiments can, in principle, be performed with neutron scattering, as was demonstrated in Ankner et al. (1989). However, because of the lower flux of neutron sources, equivalent experiments are more tedious to perform and only justified if either special isotopes or magnetic surface structures are to be probed that are not accessible by surface X-ray scattering methods. Nevertheless, any type of surface structure or surface correlation can be investigated with tunable surface sensitivity using the generalized expression:

$$I_{surf}\left(Q_{\parallel}\right) = A_0^2 \left(\frac{r_0}{R}\right)^2 |t_i|^2 S(Q_{\parallel})^2 |t_f|^2, \tag{12.7}$$

where the in-plane structure factor $S(Q_{\parallel})$ is the Fourier transform of the in-plane pair-correlation function:

$$S(Q_{\parallel}) = \int g\left(R_{\parallel}\right) e^{iQ \cdot R_{\parallel}} dR_{\parallel} \tag{12.8}$$

Next, we discuss a similar scattering method, but focusing on small diffraction angles θ.

12.2 Grazing Incidence Small-Angle Scattering (GISAS)

12.2.1 Scattering Geometry

The scattering geometry shown in Fig. 12.1 can also be used and adapted for scattering at small scattering angles θ. At small angles θ, the sensitivity to crystal lattice planes on the atomic scale is lost. Instead, we can investigate density fluctuations in the surface plane on submicrometer length scales, as with normal SAS, but with the added surface sensitivity. Therefore, the surface-sensitive small-angle intensity in DWBA is described as follows:

$$I\left(Q_{\parallel}\right)_{GISAS} = k|t_i|^2 |t_f|^2 \iint \langle \rho(r)\rho'(r')\rangle \exp\left(iQ_{\parallel} \cdot (r - r')\right) dr dr' \tag{12.9}$$

Aside from the transmission coefficients $|t_{i,\,f}|$, Eq. (12.9) has the same form as Eq. (9.1) for small-angle scattering in the bulk. k is an appropriate prefactor specifying X-ray or neutron scattering. Hence, surface-sensitive small-angle scattering can also be performed by taking into account the glancing angle conditions. The respective scattering technique is referred to as grazing incidence small-angle scattering (GISAS) performed with X-rays (GISAXS) or neutrons (GISANS). These methods may be used for studying surface and interface morphologies of sub-micrometer-sized objects, independent of their structural order. Furthermore, GISAS is extensively used for investigating nanoparticles deposited on a substrate by spin-coating or self-organization.

Similar to SAS, when studying nanoparticles with GISAS, the Fourier transform of the particle shape factor $|P(\boldsymbol{Q}_\parallel)|$ must be taken into account:

$$I\left(\boldsymbol{Q}_\parallel\right)_{GISAS} = |t_i|^2|t_f|^2|P(Q_\parallel)|^2 \iint \langle \rho(\boldsymbol{r})\rho'(\boldsymbol{r}')\rangle \exp\left(i\boldsymbol{Q}_\parallel \cdot (\boldsymbol{r}-\boldsymbol{r}')\right)d\boldsymbol{r}d\boldsymbol{r}'. \quad (12.10)$$

Moreover, in GISAS studies of nanoparticles, the film is usually not laterally homogeneous as in reflectivity studies but granular. Then, five scattering scenarios can be distinguished and need to be accounted for in model calculations, as sketched in Fig. 12.4 (Rauscher et al. 1999; Renaud et al. 2009). Those involve 1. pure reflection from the substrate; 2. pure scattering (transmission) from a particle; 3. reflection followed by transmission from a particle; 4. transmission followed by substrate reflection; 5. sequence of reflection, scattering, and reflection. The scattering amplitude has then the following four contributions in addition to the pure reflectivity:

$$F\left(k_i, k_f, Q_x, Q_y\right) = \sum_l \left(F_l^{tt} + F_l^{rt} + F_l^{tr} + F_l^{rr}\right). \quad (12.11)$$

Here, l is the particle index, r stands for reflection, and t for transmission. In DWBA, all distortions of the incoming and scattered waves due to refraction and reflection are accounted for exactly via the transmission coefficients $|t_i|$ and $|t_f|$, while the scattering of those distorted waves from nanoparticles is evaluated in the first Born approximation.

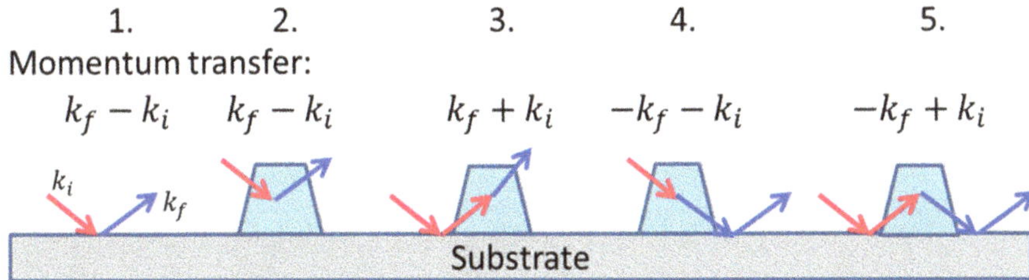

Fig. 12.4 Different scattering and reflection events for particles on a smooth substrate. Reflection occurs from the substrate, and scattering (transmission) takes place in the particles. All five events contribute to the GISAS intensity maps and must be considered in respective simulations. Red color for incident waves, blue color for exit waves

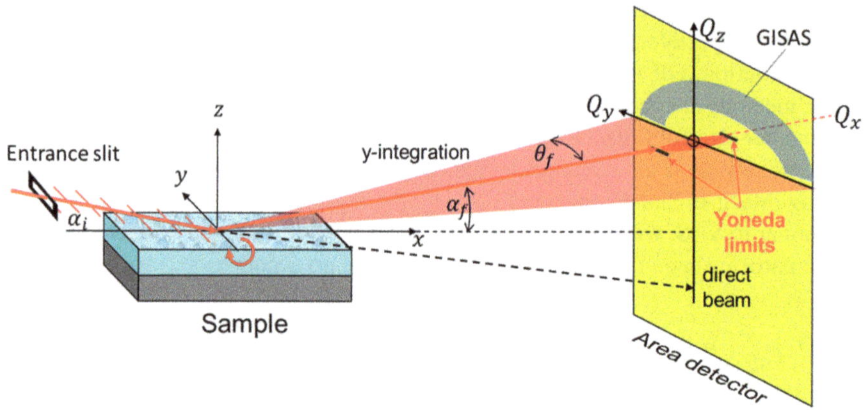

Fig. 12.5 Layout for reflectivity and grazing incidence small-angle scattering experiments detected by an area detector marked in yellow. Further details are provided in the main text

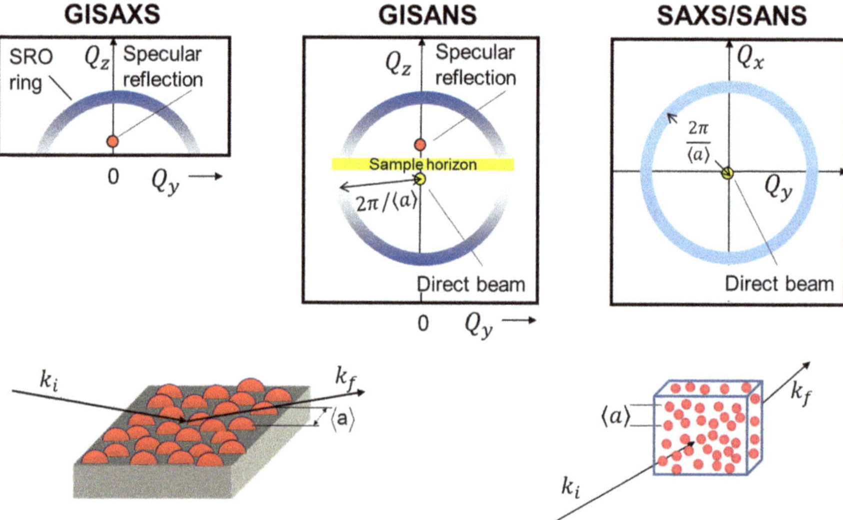

Fig. 12.6 Comparison of GISAXS, GISANS, and SAXS–SANS. The top row shows schematically intensity maps recorded by an area detector; the bottom row indicates the scattering geometry for glancing incident angles (left) and for transmission (right). The sample is assumed to be a solution of nanoparticles with an average nearest-neighbor distance of $\langle a \rangle$, either deposited on a substrate or kept in a cuvette. The half- and full-circle rings are due to the shape factor of the nanoparticles and/or their average separation

Without going into any further details, the connection between GISAS and SAS and with XRR or NR is illustrated in Figs. 12.5 and 12.6.

Figure 12.5 shows the layout for reflectivity and GISAS experiments. The incident beam enters from the left (red arrow) and hits the surface at a glancing angle α_i. Specular reflectivity is achieved by scanning the incident angle α_i and exit angle α_f simultaneously at the condition $\alpha_i = \alpha_f$. This scan is equivalent to a Q_z scan. The specular beam hits the detector at the (Q_x, Q_z) cross-point on the detector, marked by a circle. The incident beam that is transmitted through the sample is absorbed in the substrate for X-rays, but for neutrons, it can usually be detected because of the much lower attenuation. Off-specular diffuse scattering can be probed by rocking the sample about the y-axis, as indicated by the red arrowed circle in the sample plane. The rocking scan adds a Q_x component and is limited by the horizon of the sample that causes Yoneda intensity enhancements, indicated by black bars in front and behind the detector plane. Between the Yoneda limits, off-specular diffuse scattering can be observed due to surface or interface roughness. For reflectivity work, the entrance slit is wide open in the y-direction but narrow in the z-direction. Therefore, reflectivity scans provide high resolution in the Q_z-and Q_x-direction, but not in the Q_y-direction. Any scattering intensity in the y-direction is integrated over. The y-integration is marked as a reddish triangular sheet.

12.2.2   GISAS Maps

To perform GISAS experiments, the wide open entrance slit must be narrowed down for improved resolution in the Q_y-direction. Under these conditions, a SAS ring will be detectable for nanoparticle suspensions, amorphous materials, polymers, or any other material, depending on the correlations probed within the film plane. The scattering ring on the area detector is indicated in gray color in Fig. 12.5.

In GISAS scattering geometry, the scattering vector has the components are (Q_y, Q_z), defined by:

$$Q_y = \frac{2\pi}{\lambda}\left(\sin 2\theta_f\right)\left(\cos \alpha_f\right) \text{ and } Q_z = \frac{2\pi}{\lambda}\left(\sin \alpha_i + \sin \alpha_f\right) \tag{12.12}$$

Figure 12.6 compares expected GISAXS, GISANS, and SAXS–SANS maps for nanoparticle suspensions. In the case of GISAS, the suspension is assumed to be deposited on a substrate and investigated in reflection geometry. In the SAXS–SANS geometry, the same suspension is probed in transmission geometry. The nanoparticles are supposed to be in a noninteracting dilute limit, with a monodisperse spherical shape and an average separation of $\langle a \rangle$. Therefore, in a SAS map, the first correlation ring is expected to appear at a radius of $2\pi/\langle a \rangle$. The scattering ring is superposed on the Porod oscillations, caused by the particle shape factor. This is indeed what is observed with SAXS or SANS. However, in GISAS, the optical properties of the incident and scattered beams close to total reflection are also visible in the maps.

In GISAXS experiments, the specularly reflected beam (red dot in Fig. 12.6, left) is seen above the sample horizon and below the diffuse arc (unless blocked to protect the detector). The incident and transmitted beams are usually absorbed in the sample

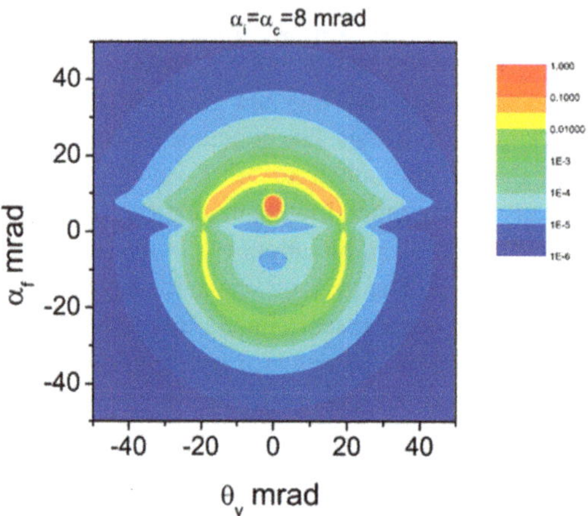

Fig. 12.7 GISANS simulation of spherical Co nanoparticles suspended in a liquid film on a substrate. The intensity map is calculated within the DWBA for the incident angle equal to the critical angle of total reflection. The red spot marks the specular reflected beam. Private communication by B.P. Toperverg (Toperverg, n.d.)

substrate. In GISANS, the almost complete diffuse arc is observable centered about the primary beam. Just above the sample horizon, close to $\alpha_{i,\,f} \cong \alpha_c$, the arc is distorted due to the Yoneda effect. Also, if the specularly reflected beam moves up with increasing Q_z, the diffuse arc becomes distorted in the center at $Q_y = 0$. In most GISANS cases, the primary beam that appears in the intensity map below the sample horizon is weakened but not completely absorbed. In SAXS and SANS, contrary to GISAS, the full diffuse scattering circle is observed without optical distortion, featuring the direct beam at the center, as illustrated in the right panel of Fig. 12.6. Further details can be found in the reviews (Renaud et al. 2009; Müller-Buschbaum 2013).

Figure 12.7 shows a GISANS simulation of 35 nm Co spherical nanoparticles with a concentration of 70% suspended in a liquid layer on a substrate (Toperverg, n. d.). The incident angle is fixed to the critical angle of total reflection: $\alpha_i = \alpha_c$. The totally reflected beam is seen as a red spot in the center of the map at $\alpha_i = \alpha_f$. This beam appears just above the sample horizon. The reflected beam is a mirror image of the incident beam. As this image is not distorted and not broadened, we conclude that the substrate is taken as perfectly smooth on the length scale of the neutron coherence length. The primary beam appears below the sample horizon (light blue circle) and is weakened by attenuation in the substrate.

In the upper and lower parts of the map displayed in Fig. 12.7, we notice a liquid or amorphous scattering arc that is distorted by the critical angle condition and by attenuation. Otherwise, the circle has a top part above the sample horizon (reddish circle) and is continued below the sample horizon (yellow-green). The actual size and shape of the circle is a question of the structure factor $S(Q)$ with a mean correlation length in the suspension, which is equivalent with the mean interparticle separation.

12.2.3   Experimental Examples

12.2.3.1   Metal Clusters

As a visualization of GISAXS, we reproduce in Fig. 12.8 two examples (Babonneau et al. 2000). In the first example, Fe-clusters were embedded in an Al_2O_3 matrix. The second example concerns a granular multilayer with alternating Co-clusters and Al_2O_3 layers. In both cases, sputtering methods were used for preparing the samples. The white vertical stripe in the middle of the maps is due to the beam stop of the synchrotron radiation. Therefore, the position of the reflected beam is not visible in these maps.

The scattering pattern in Fig. 12.8a exhibits a slightly anisotropic arc indicative for Fe-particles that are elongated in the direction parallel to the surface normal. Analysis yielded values of 3.3 nm for the out-of-plane particle size and 2.5 nm for the in-plane diameter, with a 10% size distribution. Also, the mean center-to-center interparticle separation was determined to be 3.5 nm in the plane and 4.5 nm in the

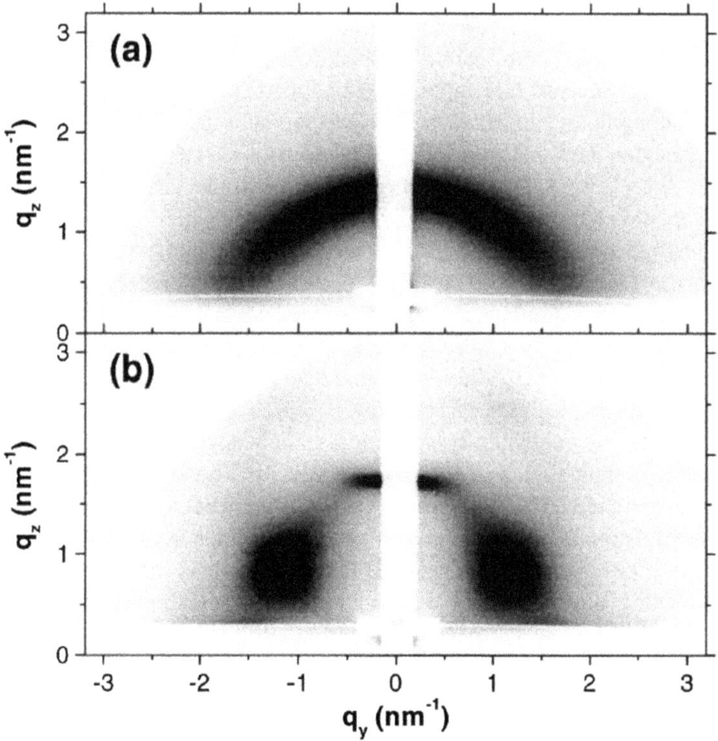

Fig. 12.8   (a) GISAXS map of Fe-clusters with an ellipsoidal shape embedded in an Al_2O_3 matrix. (b) Co clusters stacked in the Co/Al_2O_3 multilayer show signs of in-plane ordering and perpendicular layering. Reproduced from Babonneau et al. (2000). Copyright (2025) by the AIP

direction normal to the surface. From this, the authors conclude that the deposited Fe clusters obtained by co-sputtering are homogeneously and randomly distributed in the alumina matrix.

The GISAXS pattern of the granular Co/Al_2O_3 multilayer (Fig. 12.8b), on the contrary, shows much more characteristic features of particle ordering. The first Bragg peak due to the multilayer periodicity appears at $q_z = 1.745$ nm^{-1} and is partially hidden by the beamstop. The second characteristic feature is the presence of two symmetric lobes arising from the Co cluster network. From their positions, an in-plane and an out-of-plane ordering of the Co clusters is evidenced, induced by the vertical stacking of the layers. In combination with TEM measurements, the authors proposed an fcc-like stacking model for the Co cluster network.

12.2.3.2 Effect of Substrates

When depositing nanoparticles on a substrate, a number of issues require careful attention. Si substrates behave differently from ceramic substrates, like sapphire, or metallic substrates, like aluminum. It is evident that the smoothness/roughness has a great effect on the monolayer formation of nanoparticles versus island growth. If the substrate is first coated with a solvent, it is important to distinguish between wetting and de-wetting solvents. Solvents can either mediate van der Waals interactions between nanoparticles and between NPs and their substrate. Or solvents may hinder direct interaction between NPs. The NPs themselves are usually coated with a protective layer, which also influences the interparticle interaction and the self-assembly of monolayers.

To illustrate these influence factors, the NP organization was compared after deposition on a Si substrate and on a PMMA-coated Si substrate (Mishra et al. 2016). Both substrates were spin-coated with iron oxide NPs in a toluene solution. The NPs had a spherical shape with a mean diameter of 18 nm and a size distribution of about 7%. Figure 12.9 (a) and (b) show scanning electron microscopy images (SEM) for the Si and PMMA substrates, respectively.

The NPs self-assemble into a two-dimensional hexagonal lattice, showing structural defects common in polycrystalline materials. The solvent completely wets the surface, and the NP assembly takes place during the evaporation of the solvent. The hexagonal arrangement in (a) may arise from Van der Waals interactions among the NPs once the solvent concentration is lowered. In panel (b), clustering of NPs is observed with a high degree of disorder within the cluster, as seen in the inset. Although the SEM images provide useful visual information, a more comprehensive and quantitative analysis is possible by taking a Fourier transformation of the NP assembly via GISAXS.

The corresponding GISAXS pattern of NPs deposited on bare Si substrate is shown in Fig. 12.9c. Two remarkable features can be seen in the scattering pattern. First, the intensity is distributed in a circle truncated by the sample horizon. At least three circles can be recognized in this intensity map. Second, superimposed on top of the circles are high-intensity peaks, which modulate the ring intensity. These peaks

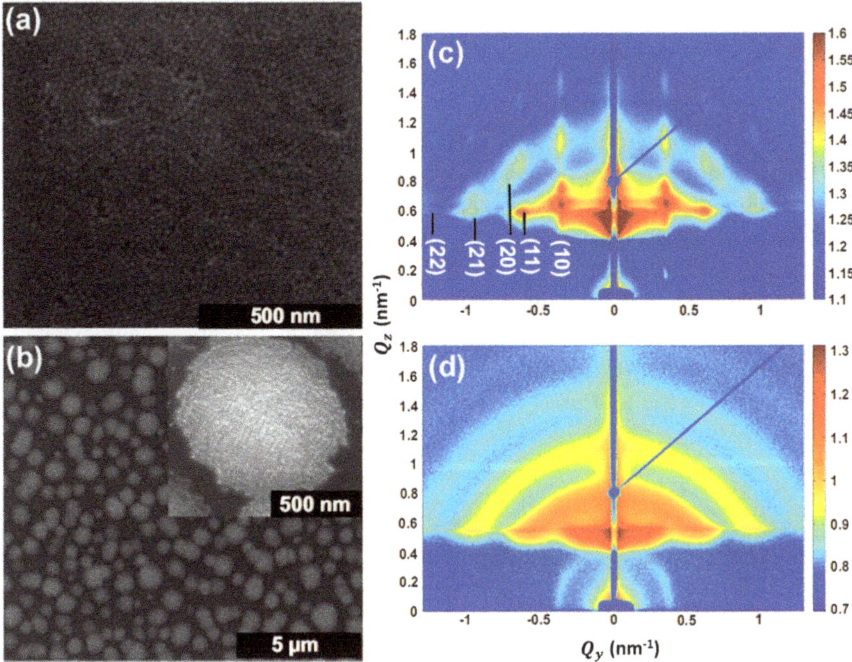

Fig. 12.9 SEM images of NPs spin-coated on (**a**) silicon substrate with native oxide and on (**b**) PMMA-coated silicon substrate, showing different NP arrangements on both substrates. The inset in (**b**) shows one of the NP clusters. The corresponding GISAXS patterns are shown in (**c**) and (**d**), respectively. The intensity is plotted on a logarithmic scale coded in the color scale shown on the right-hand side. The diagonal lines mark the beam stop for the specularly reflected beam. Reproduced with permission from Mishra et al. (2016). Copyright (2025) by IoP Publishing

have a streak-like appearance, being narrow in the Q_y direction, but extended in the Q_z direction. Indeed, the streaks resemble RHEED patterns discussed in Sect. 5.7.

The intensity distribution is the result of the Fourier transform of the in- and out-of-plane electron density variations, where two factors play an essential role: NP shape and NP–NP lateral correlation. The ring-like pattern is due to the Fourier transform of the spherical NPs, while the intense Bragg peaks result from the Fourier transform of the in-plane particle–particle correlation function. In particular, the small width of the peaks in the y-direction indicates the presence of a long-range periodic distribution of the NPs in the substrate surface, while their rod-like line shape in the z-direction is due to the two-dimensional nature of the NP film. Hence, the high-intensity Bragg peaks arise from long-range NP ordering in the GISAXS geometry. In fact, the peak pattern shown in Fig. 12.9c can be assigned to a hexagonal close-packed (hcp) lattice with a lattice parameter of 20.38 nm. Almost the same NP diameter can also be retrieved from the radius of the rings in Fig. 12.9c.

The second system used in this study is illustrated by the image in Fig. 12.9b. The sample was prepared by spin-coating nanoparticles on top of a Si substrate, pre-coated with a few nanometers of polymethyl methacrylate (PMMA). In this case, the NPs present a completely different ordering compared to the previous one. The NPs form islands (mostly disc-like) of approximately 1 µm in size. The inset in panel (b) shows one of the islands, where the NPs are arranged in a close-packed structure. The corresponding GISAXS pattern is shown in Fig. 12.9d. Unlike for NPs spin-coated onto plain Si, the GISAXS pattern of NPs on PMMA does not show any in-plane Bragg peaks or rods, indicating that the NPs within the islands are arranged in a disordered or amorphous fashion. The ring-like structure only arises from the short-range ordering of the NP form factor and does not show any preferred crystallite formation. According to the SEM image, the PMMA-coated substrate forms microscopic droplets with high contact angles. Within the droplets, the NPs are strongly bound to the solvent, and upon evaporation, they try to agglomerate to minimize the surface energy. All this results in a disordered structure (Mishra et al. 2016).

12.2.3.3   POL-GISANS

While GISAXS is widely applied for the analysis of surface structures, GISANS has not often been used and even less so in the polarized version (POL-GISANS). A proof of principle was given by Pannetier et al. (2003), showing a magnetic diffraction pattern in GISANS geometry from a stripe magnetic pattern in $Fe_{0.5}Pd_{0.5}$ thin films with perpendicular magnetic anisotropy. The diffraction pattern yields the average domain size of the up- and down-magnetized domains. Theis-Bröhl and coworkers have employed POL-GISANS to analyze the structural and magnetic parameters of Co NPs with an average diameter of 13.5 nm (Theis-Bröhl et al. 2008, 2011). In Fig. 12.10 are shown the POL-GISANS results of a sample with 18 monolayers of Co NPs deposited by spin coating on a Si substrate. For the measurements, a SANS instrument was used at the Institut Laue-Langevin (see Fig. 9.18) with the incident beam set at 1.1°, well above the critical angle for total reflection. The incident beam was polarized and combined with a spin flipper that allows for measuring I^+ and I^- intensities, but no polarization analysis was

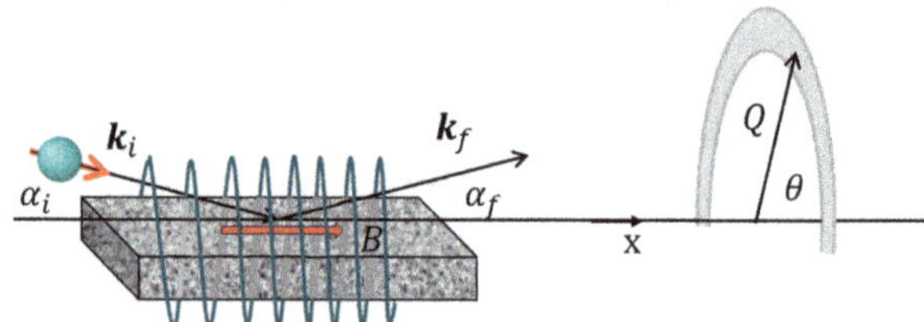

Fig. 12.10 Scattering geometry used for the POL-SANS study. The sample consisting of 18 mono-layers of Co-nanoparticles is exposed to a magnetic field in the direction of the incident beam. A polarized neutron beam with polarization parallel or antiparallel to the incident wave vector probes the sample magnetization and provides I^+ and I^- intensity maps on the area detector. The glancing incident angle is chosen, $\alpha_i > \alpha_c$. Q is the scatteering vector and θ is the azimuthal angle in the detector plane

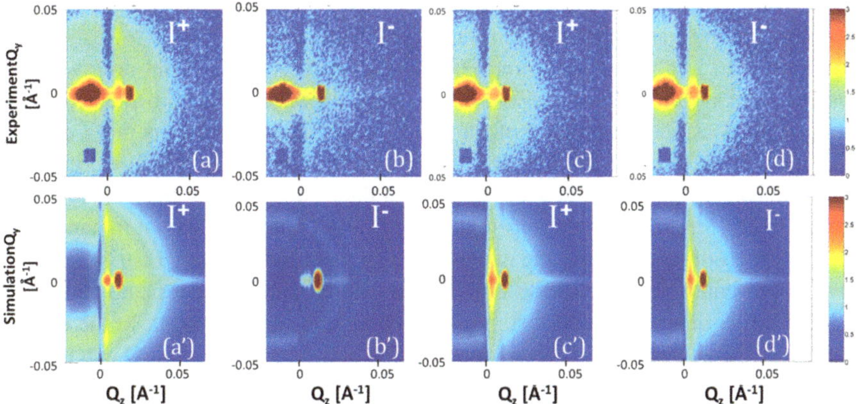

Fig. 12.11 Polarized GISANS of 18 monolayers of Co nanoparticles with an average diameter of 13.5 nm on a Si substrate. A neutron polarizer and a spin flipper were used for selecting I^+ and I^-; a horizontal magnet with a field parallel to the incident beam was applied for the maps in (**a, b**). Top row: Polarized neutron beam probing the sample in zero magnetic field (**c, d**) and in a magnetic field of 110 mT (**a, b**). Bottom row: Respective simulations of the intensity maps using the DWBA approach. Adapted and reprinted from Theis-Bröhl et al. (2008), with the permission of AIP Publishing

performed. Therefore, the intensity maps I^+ and I^- are integrated over the final spin state. An optional magnetic field was applied parallel to the incident beam and in the film plane (along the x-direction). Figure 12.10 illustrates the scattering geometry used for this POL-SANS experiment. Note that for all azimuthal angles θ, the scattering vector Q is oriented perpendicular to the in-plane sample magnetization $M_{\parallel}$. Therefore, the scattering pattern with and without magnetic field is expected to be annularly homogeneous and symmetric, aside from optical effects.

The top row in Fig. 12.11 shows intensity maps I^+ and I^- with the magnetic field on (a, b) and with the magnetic field off (c, d). The bottom row shows the corresponding POL-GISANS simulations within the DWBA. We first discuss the maps with the field off (c,d). Note that all intensity maps are rotated by 90° compared to previously shown GISAS maps.

Without applying a magnetic field, we observe the same intensity distribution in both images, I^+ and I^-, indicating a random orientation of the Co macrospins within the layers. Furthermore, at $Q_z=0.015$ Å^{-1}, we identify the specularly reflected beam, which appears in dark red on the logarithmic color scale. Directly below the specular reflection, the Yoneda peak is visible, which also marks the critical reflection angle α_c. The perpendicular line crossing α_c is the so-called GISANS line. Below the Yoneda peak is the sample horizon at $Q_z=0$. The sample blocks the beam, but below the sample horizon, the transmitted beam reappears as a large red dot.

Furthermore, we notice diffuse scattering filling half a circle and an intensity streak in the Q_z—direction at $Q_y=0$. The streak is the nuclear specular reflectivity of the Co layer, while the diffuse scattering is due to paramagnetic scattering and uncorrelated nuclear scattering from individual nanoparticles.

Figure 12.11 (a, b) are the same POL-SANS images as (c, d) but now with the applied field of 110 mT turned on. The significant difference from the zero-field images is due to the magnetism of the Co NPs and their macrospins preferentially aligned in the magnetic field direction. The majority of the intensity is now located in the I^+ image; the weak ring-shaped intensity in the I^- image arises from the NP form factor. In all cases, the simulations within the DWBA approach shown in panels (a', b',c',d') reproduce very well the experimentally recorded intensity distribution.

The presented example shows how the NPs are arranged in the plane and in the stack. Further analysis would distinguish between the core and shell. In the case of Co, the oxidized shell is antiferromagnetic. The antiferromagnetic CoO can interact with the ferromagnetic core, creating an exchange bias across the common interface. All this and more can be analyzed in detail using the full power of polarized SANS, as Krycka et al. (2010) have shown. The full spin analysis requires a wide-angle spin analyzer behind the sample to detect all diffusely scattered neutrons. This is possible by employing a polarized ^{3}He gas flask that acts as a spin filter for neutron spins, which are antiparallel to the ^{3}He nuclear spin orientation (Petoukhov et al. 2006). More information on POL-SANS and POL-GISANS can be found in the review paper by Mühlbauer et al. (2019).

Summary

1. In surface scattering experiments, the scattering vector lies in the surface plane.
2. In grazing incidence diffraction, the evanescent wave propagating parallel to the surface is used to excite Bragg reflections and to probe reconstructed surfaces and order-parameter profiles close to the surface.
3. In surface scattering experiments, the glancing angles control the penetration of the beam and the information depth of the scattering function.
4. In surface scattering, the distorted wave Born approximation must be applied because of refraction effects close to total reflection.
5. The refraction effects are taken into account via the Fresnel transmission coefficients.
6. In the DWBA, the scattering amplitude is calculated using the distorted incident and outgoing wavefunctions and the full scattering function.
7. Surface scattering methods allow in-plane Bragg reflections to be probed.
8. Surface scattering is of interest to study surface structures, which are different from the bulk (surface reconstruction).
9. With Surface scattering methods, the order parameter profile from the surface to the bulk can be followed during an order-disorder phase transition.
10. The surface scattering geometry can also be used for small-angle scattering with X-rays or neutrons to probe morphologies of sub-micrometers independent of the structural order at surfaces and interfaces.

11. GISAS combines the surface sensitivity of grazing incidence diffraction (surface scattering) with the length scale sensitivity of small-angle scattering.
12. The GISAS technique can be implemented using either neutrons (GISANS) or X-rays (GISAXS).
13. In GISAS, the resulting scattering pattern provides information about the size, shape, and arrangement of nanostructures on or near the surface.
14. In GISAXS, the SAS ring of intensity from nanoparticles is truncated by the surface.
15. In GISANS, the SAS ring from nanoparticles is distorted close to the surface but continues below the sample horizon.
16. Nanoparticles with a two-dimensional close-packed hexagonal order give rise to 2D hcp Bragg reflections.
17. POL-GISANS is performed by taking GISANS images with a polarized incident beam.
18. POL-GISANS yields information on magnetic nanoparticles and the correlation of their macrospin with and without an external magnetic field.

Questions

(Note that more than one answer may be correct)

1. What is the purpose of surface scattering?
a. Surface scattering is useful for studying the density profile of materials.
b. Surface scattering is useful for studying the depth profile of the order parameter.
c. Surface scattering is useful for studying the depth profile of surface reconstructions.

2. What is the similarity and the difference between small-angle scattering (SAS) and reflectivity of X-rays (XRR) or neutrons (NR)?
a. Reflectivity requires flat sample surfaces and interfaces; SAS has no shape restrictions.
b. Reflectivity determines density profiles normal to the surface; SAS determines density-density correlations normal to the incident beam direction.
c. Reflectivity is mainly due to diffuse X-ray and neutron scattering performed at rough surfaces, and SAS is mainly due to specularly reflected intensity from flat surfaces.

3. When is surface scattering of interest?
a. For analyzing alloy compositions.
b. When concentrating on structures near surfaces.
c. For studying the depth profile of surface phase transitions.

4. What distinguishes Grazing incidence diffraction (GID) and grazing incidence small-angle scattering (GISANS)?
a. GID is performed in transmission geometry, GISANS in reflection geometry.
b. GID is surface sensitive; GISANS probes the bulk.

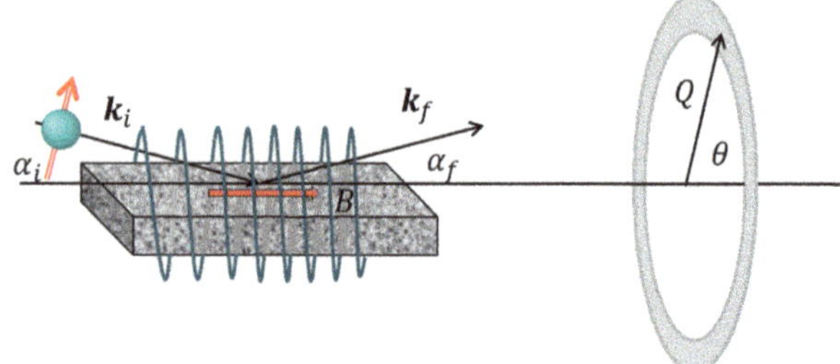

Fig. 12.12 Scattering geometry for for magnetic GISANS measurement with perpendicular polarization

c. GID and GISANS use the same scattering geometry; however, with a wide scattering angle in the first case and a small scattering angle in the second case.

5. What radiation is required to perform GISAS experiments?

a. GISAS can be performed with X-rays, neutrons, or electrons.

b. GISAS can only be performed with electrons and is then called RHEED.

c. GISAS can be performed with He atoms.

6. When is the polarized version of GISANS required?

a. For studying the magnetization of nanoparticles on a substrate, magnetic films, and magnetic domain structures in thin films.

b. For the analysis of structural disorder in surface near regions.

c. For the determination of spin wave dispersion.

Exercises

E12.1. Close packing of nanoparticles.

(a) What is the lattice parameter in the diffraction pattern of Fig. 12.9, assuming an hcp in-plane structure.

(b) Confirm that the NP packing forms an hcp structure, but not a sc structure.

E12.2 POL-SANS.

Assuming GISANS scattering geometry in the polarized version. Furthermore, assuming a Helmholtz coil providing an in-plane magnetic field B that saturates the magnetization of the ferromagnetic nanoparticles in the field direction. What is the expected scattering pattern at a constant scattering vector Q and variable azimuthal angle θ for perpendicular up or down spin polarization? (Fig. 12.12)

References

H. Dosch, *Critical Phenomena at Surfaces and Interfaces*, Springer Tracts in Modern Physics, vol 126 (Springer, Berlin Heidelberg New York, 1992)

G.H. Vineyard, Grazing-incidence diffraction and the distorted-wave approximation for the study of surfaces. Phys. Rev. B **26**, 4146 (1982)

H. Dosch, Evanescent absorption in kinematic surface Bragg diffraction. Phys. Rev. B **35**, 2137 (1987)

R. Feidenhans'l, Surface structure determination by X-ray diffraction. Surf. Sci. Rep. **10**, 105 (1989)

H.E. Stanley, *Introduction to phase transitions and critical phenomena* (Claderon Press, Oxford, 1971)

H. Dosch, L. Mailander, A. Lied, J. Peisl, F. Grey, R.L. Johnson, S. Krummacher, Experimental evidence for an interface delocalization transition in Cu_3Au. Phys. Rev. Lett. **60**, 2382–2385 (1988)

H. Dosch, B.W. Batterman, D.C. Wack, Depth-controlled grazing-incidence diffraction of synchrotron X radiation. Phys. Rev. Lett. **56**, 1144 (1986)

H. Dosch, L. Mailänder, R.L. Johnson, J. Peisl, Critical phenomena at the $Fe_3Al(11\bar{0})$ surface: A glancing angle X-ray scattering study. Surf. Sci. **279**, 367–379 (1992)

X. Zhu, R. Feidenhans'l, H. Zabel, J. Als-Nielsen, R. Du, C.P. Flynn, F. Grey, Grazing incidence X-ray scattering on the Cu_3Au (111) phase transition. Phys. Rev. **37**, 7157 (1988)

J. Ankner, H. Zabel, D.A. Neumann, C.F. Majkrzak, Surface neutron scattering of Si(110). Phys. Rev. B **40**, 792 (1989)

M. Rauscher, R. Paniago, T.H. Metzger, Z. Kovats, J. Domke, H.D. Pfannes, J. Schulze, Eisele, Grazing incidence small angle X-ray scattering from freestanding nanostructures. J. Appl. Phys. **86**, 6763–6769 (1999)

G. Renaud, R. Lazzari, F. Leroy, Probing surface and interface morphology with grazing incidence small angle X-ray scattering. Surf. Sci. Rep. **64**, 255–380 (2009)

P. Müller-Buschbaum, Grazing incidence small-angle neutron scattering: Challenges and possibilities. Polym. J. **45**, 34–42 (2013)

B. P. Toperverg, *GISANS map calculated in DWBA*, private communication.

D. Babonneau, F. Petroff, J.L. Maurice, F. Fettar, A. Vaurès, A. Naudon, Evidence of self-organized growth in granular Co/Al_2O_3 multilayers. Appl. Phys. Lett. **76**, 2892–2894 (2000)

D. Mishra, D. Greving, G.A. Badini Confalonieri, J. Perlich, B.P. Toperverg, H. Zabel, O. Petracic, Growth modes of nanoparticle superlattice thin films. Nanotechnology **25**, 205602 (2016)

M. Pannetiera, F. Ott, C. Fermon, Y. Samson, Surface diffraction on magnetic nanostructures in thin films using grazing incidence SANS. Phys. B **335**, 54 (2003)

K. Theis-Bröhl, M. Wolff, I. Ennen, C.D. Dewhurst, A. Hütten, B.P. Toperverg, Selfordering of nanoparticles in magneto-organic composite films. Phys. Rev. B **78**, 134426 (2008). https://doi.org/10.1103/PhysRevB.78.134426

K. Theis-Bröhl, B.P. Durgamadhab Mishra, H. Toperverg, B. Zabel, A.R. Vogel, A. Hütten, Self organization of magnetic nanoparticles: A polarized grazing incidence small angle neutron scattering and grazing incidence small angle X-ray scattering study. J. Appl. Phys. **110**, 102207 (2011)

K.L. Krycka, R.A. Booth, C.R. Hogg, Y. Ijiri, J.A. Borchers, W.C. Chen, S.M. Watson, M. Laver, T.R. Gentile, L.R. Dedon, S. Harris, J.J. Rhyne, S.A. Majetich, Core-shell magnetic morphology of structurally uniform magnetite nanoparticles. Phys. Rev. Lett. **104**, 207203 (2010)

A.K. Petoukhov, K.H. Andersen, D. Jullien, E. Babcock, J. Chastagnier, R. Chung, H. Humblot, E. Lelièvre-Berna, F. Tasset, F. Radu, M. Wolff, H. Zabel, Recent advances in polarised 3He spin filters at the ILL. Phys. B **385–386**, 1146–1148 (2006)

S. Mühlbauer, D. Honecker, É.A. Périgo, F. Bergner, S. Disch, A. Heinemann, S. Erokhin, D. Berkov, C. Leighton, M. Ring Eskildsen, A. Michels, Magnetic small-angle neutron scattering. Rev. Mod. Phys. **91**, 015004 (2019)

Further Readings

H. Dosch, *Critical Phenomena at Surfaces and Interfaces*, Springer Tracts in Modern Physics, vol 126 (Springer, Berlin Heidelberg New York, 1992)

S. Mühlbauer, D. Honecker, É.A. Périgo, F. Bergner, S. Disch, A. Heinemann, S. Erokhin, D. Berkov, C. Leighton, M.R. Eskildsen, A. Michels, Magnetic small-angle neutron scattering. Rev. Mod. Phys. **91**, 015004 (2019)

A.E. Ares (ed.), *X-ray Scattering* (IntechOpen, 2017) open access, peer reviewed volume, publ. https://www.intechopen.com/books/5371

P. Müller-Buschbaum, in *Applications of Synchrotron Light to Scattering and Diffraction in Materials and Life Sciences*, Lecture Notes in Physics, ed. by M. Gomez, A. Nogales, M. Garcia-Gutierrez, T. Ezquerra, vol. 776, (Springer, Berlin, 2009), p. 61. https://doi.org/10.1007/978-3-540-95968-7_3

G. Renaud, R. Lazzari, F. Leroy, Probing surface and interface morphology with grazing incidence small angle X-ray scattering. Surf. Sci. Rep. **64**, 255–380 (2009)

R. Feidenhans'l, Surface structure determination by X-ray diffraction. Surf. Sci. Rep. **10**, 105 (1989)

Appendices

A1. Delta Function and Fourier Transforms

The Dirac delta function, $\delta(x - x')$, is constructed to have the following properties:

$$\delta(x - x') = 0 \text{ for } x \neq x'$$
$$\delta(x - x') = \infty \text{ finite for } x = x'$$

We may consider the δ-function as a sharp spike at the position $x = x'$ with infinite height and zero width. Integration including the value x' yields:

$$\int_{-\infty}^{\infty} \delta(x - x')dx = 1.$$

From this definition, we find for any arbitruiary function $f(x)$ that is continuous at x':

$$\int_{-\infty}^{\infty} f(x)\delta(x - x')dx = f(x').$$

The defining relationships for the Fourier transform (FT) in integral form, with k as the wavenumber, are:

$$f(x) = \frac{1}{\sqrt{2\pi}} \int_{-\infty}^{\infty} g(k)\exp(ikx)dk$$

H. Zabel, *Elements of Elastic Scattering by X-Rays, Neutrons, and Electrons*, https://doi.org/10.1007/978-3-032-16624-1

$$g(k) = \frac{1}{\sqrt{2\pi}} \int_{-\infty}^{\infty} f(x) \exp(-ikx)dx$$

Applying the Fourier transform to the δ-function in real space $\delta(x)$, with $x' = 0$, we find:

$$FT(\delta(x)) = \frac{1}{\sqrt{2\pi}} \int_{-\infty}^{\infty} \delta(x) \exp(-ikx)dx = \frac{1}{\sqrt{2\pi}} \exp(-ik0) = \frac{1}{\sqrt{2\pi}}.$$

Therefore, the Fourier integral representation for the spike $\delta(x)$ centered at $x = 0$ is a constant for all k-values. Hence, the k-values are spread out as much as possible, as indicated in Fig. A1e. Vice versa, the FT of a δ-spike in k-space is a constant in real space.

$$FT(\delta(k)) = \frac{1}{\sqrt{2\pi}} \int_{-\infty}^{\infty} \delta(k) \exp(ikx)dk = \frac{1}{\sqrt{2\pi}} \exp(-i0x) = \frac{1}{\sqrt{2\pi}}.$$

In other words, the δ-function in k-space is the Fourier-transform of the δ-function in real space and vice versa. With these results, we find for the Fermi-pseudopotential:

$$V(k) = b \int_{-\infty}^{\infty} \delta(x) \exp(ikx)dx = b \ \exp(ik0) = b = const.$$

The Fourier integral is zero except for $x = 0$. For $x = 0$, we have $\exp(ik0) = 1$. This result is interpreted as follows. The Fourier integral of the $\delta(x)$-function is the same for all wavenumbers k. This implies that the Fourier spectrum of a sharp spike in real space has equal amplitude for all wavenumbers in k-space. In other words, a $\delta(x)$-function elicits all wavenumbers from $-\infty$ to $+\infty$ with equal amplitude in k-space. This is quite different from the Fourier transforms of other types of functions. For instance, the Fourier transform of a square centered at $x = 0$ has a $sink/k$-shape (panel a in Fig. A1), just as we derive for the FT of the electron density box model in Appendix A3. The Fourier transform of a Gaussian error function is again a Gaussian function (panel d); the Fourier transform of an exponentially decaying function in real space has a Lorentzian shape in k-space (panel c), etc. Some examples for Fourier transforms are graphically shown in Fig. A1. The last two rows show schematically the Fourier transform of a single δ-function (panel e) and of a periodic series of δ-functions (panel f). The Fourier transform of a periodic sequence of δ-functions in real space is a periodic sequence of δ-functions in k-space (sometimes referred to as δ-combs). Those are the Bragg-reflections discussed in Chap. 3. Thus, X-ray, neutron, or electron scattering by crystal lattices is equivalent to taking a Fourier transform of the real-space structure and periodicity.

Furthermore, the Fourier transform of a periodic sequence of δ-functions in real space is a periodic sequence of δ-functions in k-space (reciprocal space), and those

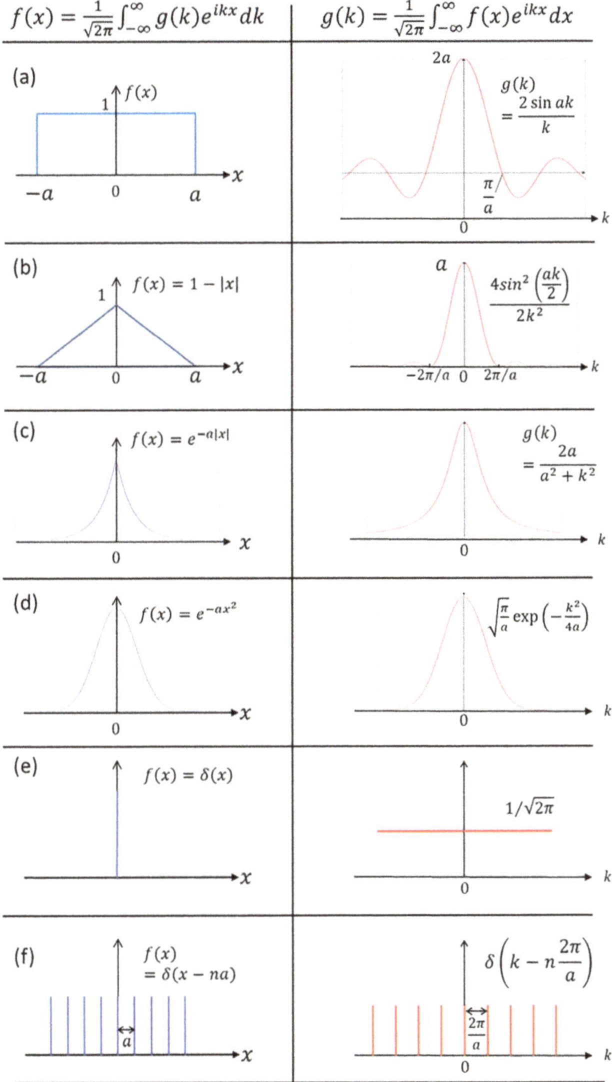

Fig. A1 Graphical representations of some functions in the left-hand column and their respective Fourier transforms in the right-hand column. (**a**) for a rectangular function: (**b**) for a triangular function; (**c**) for an exponentially decaying function; (**d**) for a Gaussian function; (**e**) for a single delta-function; and (f) for a periodic array of delta-functions with period a in real space. Adapted from: https://en.wikipedia.org/wiki/Fourier_transform

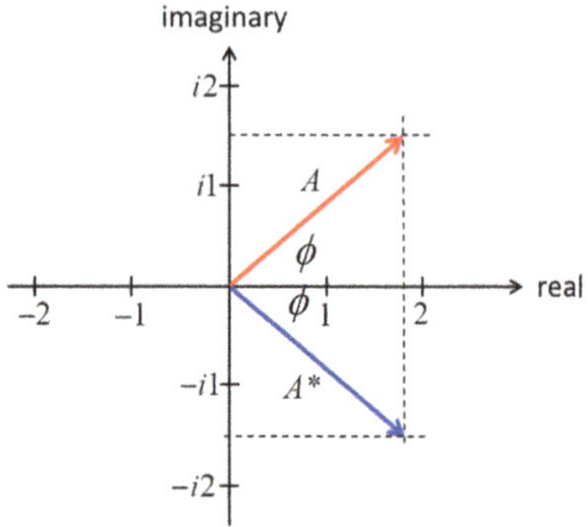

Fig. A2 Graphical representation of complex numbers. A = complex number, A* = complex conjugate number

are the Bragg reflections discussed in Chap. 3. Thus, X-ray, neutron, or electron scattering by a crystal lattice is equivalent to taking a Fourier transform of the real-space structure and periodicity.

A2. Exponential Representation of Amplitudes and Intensities

Structure factor calculations imply taking vector additions of wave amplitudes together with phase angles:

$$A_{tot} = A_1 \sin \phi_1 + A_2 \sin \phi_2 + \dots$$

This sum can mathematically be performed elegantly by representing the amplitudes as complex numbers. A complex number has a real part a and an imaginary part b:

$$A = a + ib$$

Here, i is the imaginary number defined by $\sqrt{-1} = i$, and $i^2 = -1$. Such numbers can conveniently be represented in the complex plane consisting of a real axis (abscissa) and an imaginary axis (ordinate); see Fig. A2. Real numbers are plotted as abscissae, imaginary numbers as ordinates.

For our example, we find with reference to Fig. A2, red arrow:

$$\text{real}(A) = a = |A| \cos \phi$$

$$im(A) = b = |A| \sin \phi,$$

and

$$A = |A|(\cos \phi + i \sin \phi)$$

Now, we express the trigonometric functions in terms of power series:

$$\cos \phi = 1 - \frac{\phi^2}{2!} + \frac{\phi^4}{4!} + \ldots;\ i \sin \phi = i\phi - i\frac{\phi^3}{3!} + i\frac{\phi^5}{5!} + \ldots;$$

Here, we realize that both terms belong to the power series of the exponential function:

$$e^{i\phi} = 1 + i\phi - \frac{\phi^2}{2!} - i\frac{\phi^3}{3!} + \frac{\phi^4}{4!} + i\frac{\phi^5}{5!}\ldots.$$

Thus, we can rephrase:

$$A = |A|(\cos \phi + i \sin \phi) = |A|e^{i\phi}.$$

These relations allow us to represent the amplitudes of the scattered waves by a complex exponential function. In the end, we are interested in the scattered intensity, that is, the square of the absolute amplitude. As the amplitude is expressed in terms of a complex exponential function, this quantity can be found by multiplying the amplitude by its complex conjugate amplitude A* (blue arrow in Fig. A2). The complex conjugate amplitude is obtained by replacing i by $-i$:

$$A^* = |A|(\cos \phi - i \sin \phi) = |A|e^{-i\phi}$$

Now the intensity is:

$$I \sim AA^* = |A|^2(\cos \phi + i \sin \phi)\,(\cos \phi - i \sin \phi) = |A|^2(\cos^2\phi + \sin^2\phi) = |A|^2,$$

or more simply:

$$AA^* = |A|^2 e^{+i\phi}\, e^{-i\phi} = |A|^2$$

Here, we have made use of the fact that:

$$e^{+i\phi}\, e^{-i\phi} = e^{+i(\phi-\phi)} = e^{+i0} = 1$$

Further useful relations are:

$$e^{i2\pi} = \cos(2\pi) + i \sin(2\pi) = 1,\ \text{and similarly}\ e^{i2\pi} = e^{i4\pi} = e^{i6\pi} = 1$$

$$e^{i\pi} = \cos(\pi) + i \sin(\pi) = -1,\ \text{and similarly}\ e^{i\pi} = e^{i3\pi} = e^{i5\pi} = -1$$

In general:

$$e^{in\pi} = (-1)^n \text{ and } e^{in\pi} = e^{-in\pi}$$

The equivalent notation is:

$$\exp(in\pi) = (-1)^n \text{ and } \exp(in\pi) = \exp(-in\pi)$$

Which notation is used is a question of convenience and/or taste.

Using these relations, structure factors can conveniently be calculated. As an example, we consider the structure factor of the bcc unit cell. According to Eq. 3.38, it is:

$$F_{bcc}(G) = f\left(1 + e^{i\pi(h+k+l)}\right)$$

Therefore, if the sum is an even number: $h + k + l = 2n$, $e^{i\pi(h+k+l)} = 1$; if the sum is an odd number: $h + k + l = 2n + 1$, $e^{i\pi(h+k+l)} = -1$.

Summarizing:

$$F_{bcc}(G) = 2f, \text{ for } h + k + l = 2n$$
$$F_{bcc}(G) = 0, \text{ for } h + k + l = 2n + 1$$

The squared structure factor is accordingly:

$$F_{bcc}(G)F^*_{bcc}(G) = f^2\left(1 + e^{i\pi(h+k+l)}\right)\left(1 + e^{-i\pi(h+k+l)}\right) = f^2\left(2 + 2e^{i\pi(h+k+l)}\right),$$

which is either $4f^2$ or 0, depending on the sum $h + k + l$ being even or odd. This is the same results as if we had squared the structure factor directly.

A3. Atomic Form Factors

The atomic form factor is defined by (see Eq. 2.15):

$$f(Q) = \int \rho_{el}(R_e)e^{iQ \cdot R_e} dR_e.$$

In a very simplified form, we assume that the electron density ρ_e is a constant within an atom up to a maximum radius $R_{\max}$ and zero outside. Then, the integral over a sphere of constant density can be written as:

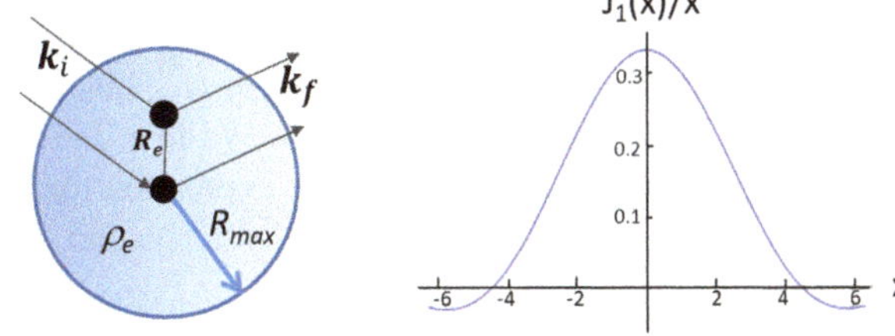

Fig. A3 Left panel: Geometry of X-ray scattering at an atomic cloud of constant density. Right panel: functional dependence of $J_1(x)/x$

$$f(Q) = \rho_e \int\limits_{R_e=0;\,\phi=0}^{R_{\max};\,\pi}\!\!\int \int_{\theta=-\pi}^{+\pi} e^{iQR_e\cos\phi}\,R_e^2\,\sin\phi\,d\phi\,d\theta\,dR_e$$

$$= \rho_e \int\limits_{R_e=0;\,\phi=0}^{R_{\max};\,\pi}\!\!\int e^{iQR_e\cos\phi}\,2\pi R_e^2\,\sin\phi\,d\phi\,dR_e$$

Using the substitution $\cos\phi = x$ and $\sin\phi\,d\phi = -dx$, we find for the angular part:

$$-2\pi\int\limits_{1}^{-1} e^{iQR_e x}\,dx = 2\pi\left(\frac{e^{iQR_e} - e^{-iQR_e}}{iQR_e}\right) = 4\pi\,\frac{\sin QR_e}{QR_e}$$

Now, we return to the radial part:

$$f(Q) = 4\pi\rho_e \int_{R_e}^{R_{\max}} R_e^2\,\frac{\sin(QR_e)}{QR_e}\,dR_e$$

$$= \frac{4\pi\rho_e}{QR_{\max}}R_{\max}^3\left(\frac{\sin(QR_{\max}) - (QR_{\max})\cos(QR_{\max})}{(QR_{\max})^2}\right) = 4\pi\rho_e R_{\max}^3\,\frac{J_1(QR_{\max})}{QR_{\max}}$$

In the last equation, $J_1(QR_{\max})$ is the Bessel function of the first kind. Real electron densities will produce a slightly different atomic form factor, but the general shape is similar (Fig. A3).

A4. Integrated Intensity of Bragg Reflections

In Fig. 5.3, we have plotted the intensity of a Bragg reflection and stated that the total integrated intensity is proportional to the number of scattering centers (atoms or nuclei).

In the following, we want to evaluate the **integrated intensity** of a single-crystal Bragg reflection in the Born approximation. Equations (3.35) and (3.36) are elegant versions to express the intensity at the Bragg points. However, they are not practical for evaluating Bragg intensities and comparing experiments with theory. Therefore, we want to evaluate the total expected integrated intensity of a Bragg reflection when performing a transverse (rocking) scan. We first derive the expression for neutron scattering and then state the equivalent expression for X-ray scattering.

For neutrons, the partial differential cross section, neglecting the Debye–Waller factor and absorption, is:

$$\frac{d\sigma(\boldsymbol{Q})}{d\Omega} = b_{coh}^2 NV_{rc} \, |F(\boldsymbol{G})|^2 \delta(\boldsymbol{Q}-\boldsymbol{G}), \tag{A4.1}$$

The task is now to integrate the partial differential cross section over the solid angle $d\Omega$ via an angular scan:

$$d\sigma(\boldsymbol{Q}) = \sigma \, d\Omega = b_{coh}^2 NV_{rc} \, |F(\boldsymbol{G})|^2 \delta(\boldsymbol{Q}-\boldsymbol{G}) d\Omega.$$

The surface element dS at the detector position is:

$$dS = k_s^2 d\Omega \tag{A4.2}$$

Therefore, the solid angle in reciprocal space is:

$$d\Omega = \frac{dS}{k_s^2} = \frac{dk_y dk_z}{k^2}, \tag{A4.3}$$

where we used Cartesian coordinates with k_s pointing along the x-direction and the surface area perpendicular to it: $dS = dk_y dk_z$ (Fig. A4a. In an angular scan, the crystal is rotated by an angle $d\theta$, which is equivalent to the rotation of the scattering vector Q by an angle $d\theta$.

Therefore, the rotational change of Q can be expressed as:

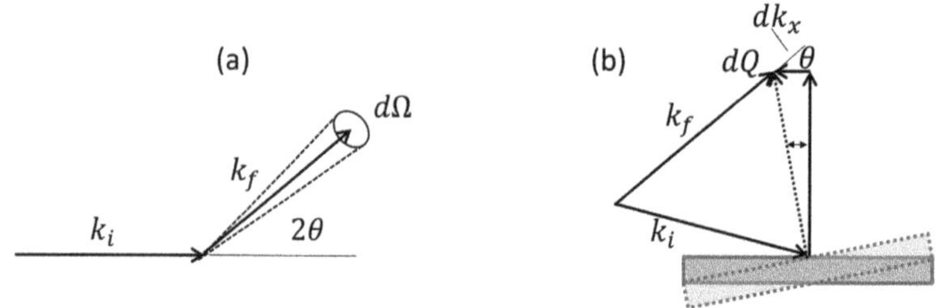

Fig. A4 (**a**) Scattering geometry, indicating the solid angle of the exit beam; (**b**) angles and directions during an angular scan

$$dQ = Qd\theta = Q\omega dt,$$

where ω is the angular frequency. The change of Q in the Cartesian x-direction is:

$$dk_x = dQ \cos\theta$$

In the time interval dt, the sample is rotated by:

$$dt = \frac{dQ}{Q\omega} = \frac{dk_x}{Q\omega \cos\theta} \tag{A4.4}$$

The total number of neutrons N that are diffracted at the Bragg angle θ_B in the time dt, normalized by the incoming flux Φ_0, is:

$$\frac{N(\boldsymbol{G})}{\Phi_0} = \int \sigma\, d\Omega \int dt = \int b_{coh}^2 N V_{rc} |F(\boldsymbol{G})|^2 \delta(\boldsymbol{Q} - \boldsymbol{G})$$
$$\times \frac{dk_y dk_z}{k^2} \int \frac{dk_x}{Q\,\omega\,\cos\theta_B} \tag{A4.5}$$

Reassembling the terms yields:

$$\frac{N(\boldsymbol{G})}{\Phi_0} = \frac{b_{coh}^2 N V_{rc} \left[F(\boldsymbol{G}) \right]^2}{k^2 Q\,\omega\,\cos\theta_B} \int \delta(\boldsymbol{Q} - \boldsymbol{G}) dk_x dk_y dk_z$$
$$= \frac{b_{coh}^2 N V_{rc} |F(\boldsymbol{G})|^2}{2k^3 \omega \sin\theta_B \cos\theta_B} = b_{coh}^2 \frac{N}{\omega} \frac{V_{rc}}{k^3} \frac{|F(\boldsymbol{G})|^2}{\sin 2\theta_B} \tag{A4.5}$$

and therefore:

$$\frac{N(\boldsymbol{G})}{\Phi_0} = b_{coh}^2 \frac{N}{\omega} \frac{\lambda^3}{V_{uc}} \frac{|F(\boldsymbol{G})|^2}{\sin 2\theta_B}$$

Here, V_{uc} is the unit cell volume, N is the number of unit cells in the crystal exposed by the neutron beam, and λ is the neutron wavelength. For X-ray scattering, we obtain an equivalent expression for the total integrated Bragg intensity:

$$I_{tot}^{x-ray}(\boldsymbol{G}) = r_0^2 N \frac{\lambda^3}{V_{uc}} f^2(\boldsymbol{G}) P \frac{|F(\boldsymbol{G})|^2}{\sin 2\theta_B} \tag{A4.6}$$

Here, P is the polarization factor. In both cases, the Debye–Waller factor and attenuations by absorption or scattering have been neglected. The last two equations are famous as they show that the total intensity depends on the wavelength to the third power and the square of the structure factor:

$$I(\boldsymbol{G}) \propto \lambda^3 |F(\boldsymbol{G})|^2$$

This dependence is typical of the Born approximation we have adopted here. In the dynamical scattering theory, it is shown that the intensity depends on the square of the wavelength and linearly on the structure factor (Als-Nielsen and McMorrow 2011):

$$I(\boldsymbol{G}) \propto \lambda^2 |F(\boldsymbol{G})| \tag{A4.7}$$

The validity criteria of the Born approximation are discussed in Sect. 3.7. In short, the Born approximation is valid in the case of a low scattering probability. Neutron scattering experiments are often flux-limited. Then, it helps to double the wavelength and gain a factor of 8 in intensity. Due to primary and secondary extinction effects, the Bragg intensity of perfect crystals turns out to be much lower than for imperfect crystals. This is impressively shown in a bending experiment reproduced in Fig. 5.4, adapted from Warren (1969). A single crystal quartz plate was bent, while the intensity increased from the perfect limit to the imperfect limit.

References

Jens Als-Nielsen and Des McMorrow, Elements of Modern X-ray Physics, 2nd edition, Wiley, 2011

B.E. Warren, X-ray Diffraction, Dover Publications, Inc., New York, reprint, originally Addison-Wesley Publ. Comp., 1969.

A5. Absorption Factor

Absorption effects are more important for X-rays than for neutrons. Considering the symmetric scattering geometry in Fig. A5, the path length of the incident beam in the medium with linear absorption coefficient μ is $z/\sin\theta$ and twice that for the incident and reflected beam: $2z/\sin\theta$. Hence, the attenuation of the intensity along this path will be:

Fig. A5 Absorption of X-rays in a crystalline and reflecting medium. A is the incident beam cross section, θ is the diffraction angle, V is the scattering volume, and z is the depth beneath the surface

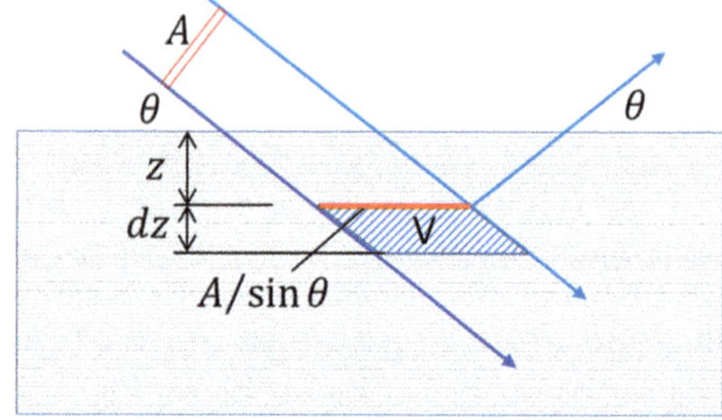

$$dI(z) \propto e^{-2\mu z/\sin\theta}$$

The incident beam with a cross section of Adz is scattered at the depth between z and $z + dz$ in a scattering volume $V \sin\theta$. Therefore, the expected intensity is proportional to the scattering volume while being attenuated through absorption:

$$dI_{sc} \propto \frac{Adz}{V\sin\theta}e^{-2\mu z/\sin\theta}$$

The total intensity is obtained by integrating over the path length z:

$$I_{sc} = \int_{z=0}^{\infty} \frac{Adz}{V\sin\theta}e^{-2\mu z/\sin\theta} = \frac{A}{2\mu V}$$

This result tells us that the absorption effect is proportional to a constant factor independent of the scattering angle. This is because the increasing path length with decreasing scattering angle θ, reducing the intensity, is compensated by an increasing effective beam cross section that increases the intensity.

Including absorption effects, the total scattered X-ray intensity in the kinematic approximation is:

$$I_{tot}^{x-ray}(G) = \frac{1}{2\mu}r_0^2 N \frac{\lambda^3}{V_{uc}}f^2(G)P\frac{|F(G)|^2}{\sin 2\theta_B}$$

A6. Neutron Energy, Momentum, and Wavelength

Neutrons that are produced in a fission reactor may be considered as a "gas" in the moderator with a temperature and energy corresponding to that of the moderator. As is typical for a gas in equilibrium, the energy and velocity of the particles are distributed according to the Maxwell–Boltzmann distribution. The expectation value for the velocity v is then given by:

$$\langle v^2 \rangle = \frac{3k_B T}{m}$$

where m is the neutron mass and k_B Is the Boltzmann constant. Hence, the kinetic energy of neutrons is related to the absolute temperature T via:

$$E = \frac{1}{2}mv^2 = \frac{3}{2}k_B T$$

Table A1 Energy range, temperature range, and wavelength range for neutrons from a cold, thermal, and hot source

Cold source	0.1–10 meV	1–100 K	30–3 Å
Thermal source	5–100 meV	60–1000 K	4–1Å
Hot source	100–500 meV	1000–6000 K	1–0.4 Å

Because of this relation, neutron energies are also quoted in terms of temperature, such as "cold" neutrons, "thermal" neutrons, or "hot" neutrons. The energy ranges of these terms are listed in Table A1.

Now, we express the kinetic energy in terms of the momentum $p = mv$:

$$E = \frac{1}{2}mv^2 = \frac{p^2}{2m}$$

de Broglie (1892–1987) has shown that the momentum of quantum mechanical particles is related to their wavenumber $k = 2\pi/\lambda$ by:

$$p = \hbar k,$$

where $\hbar = h/2\pi$ is the Planck constant. With this, we can rephrase the neutron kinetic energy:

$$E = \frac{p^2}{2m} = \frac{(\hbar k)^2}{2m} = \frac{h^2}{2m\lambda^2}$$

with the de Broglie wavelength:

$$\lambda = \frac{h}{mv}.$$

As we have seen that E, v, λ, T are all interrelated, useful conversions are:

$$\lambda[\text{Å}] = 3.956\,\frac{1}{v[\text{km/s}]} = 9.045\,\frac{1}{\sqrt{E[\text{meV}]}} = 30.81\,\frac{1}{\sqrt{T[\text{K}]}};$$

$$E[\text{meV}] = 0.0862\,T[K] = 5.23v^2\,[\text{km}^2/\text{s}^2] = 81.81\,\frac{1}{\lambda^2[\text{Å}^2]}.$$

Answers to Questions

Chapter 1: 1 c; 2 a; 3 b; 4 a, b; 5 b; 6 a, c; 7 b
Chapter 2: 1 a; 2 c; 3 c; 4 a; 5 b; 6 a b; 7 b; 8 a; 9 c
Chapter 3: 1 a; 2 c; 3 a, b; 4 b, c, d; 5 b; 6 a, b
Chapter 4: 1 c; 2 b; 3 a, b; 4 b, c, d; 5 a; 6 a, c; 7 b, c; 8 a, c; 9 b, c; 10. b; 11. b
Chapter 5: 1 a, d; 2 c; 3 a, d; 4 a, c; 5 b; 6 b; 7 a, b
Chapter 6: 1 a, b, c is optional; 2 a, b; 3 a, c; 4 a; 5 a, b; 6 b, c; 7 b, c, d
Chapter 7: 1 b, d; 2 b; 3 b, c, d; 4 a; 5 c; 6 b, c; 7 c
Chapter 8: 1 d; 2 c; 3 b; 4 b, d; 5 c; 6 b; 7 a; 8 a, b
Chapter 9: 1 b; 2 b; 3 b; 4 a, d; 5 c; 6 b, c; 7 b
Chapter 10: 1 b; 2 a, b; 3 a; 4 a; 5 c; 6 b
Chapter 11: 1 b; 2c; 3c; 4 a, c; 5 a, b
Chapter 12: 1 b, c; 2 a, b; 3 b, c; 4 c; 5 a, c; 6 a

Solutions to Exercises

Chapter 1

E 1.1 Inversion symmetry

We consider the atom with the basis vector $r_5 = (1/4, 1/4, 1/4)$. If we choose this point as our inversion center, then we expect atoms at $r_1 = (0, 0, 0)$ and symmetrically at $r_i = (1/2, 1/2, 1/2)$. However, no atom is located at r_i. Therefore, the diamond structure lacks inversion symmetry, which has many consequences for the Brillouin zones in diamond structures and energy dispersion curves of phonons and electrons. In fact, all molecules and crystals with subunits that display tetrahedral symmetry lack inversion symmetry (Fig. S1).

M 1.2 Primitive unit cell

Consider a body-centered cubic (bcc) lattice with the lattice constant a.

(a) Choose a set of primitive lattice vectors a_1, a_2, a_3 and give the coordinates in units of the bcc lattice constant a.

The primitive unit bcc cell, shown by red arrows, has rhombohedral symmetry and contains only one atom per unit cell (Fig. S2).

The set of primitive lattice vectors a_1, a_2, a_3 in units of the bcc cell are as follows:

$$a_1 = \left(-\frac{1}{2}, \frac{1}{2}, \frac{1}{2}\right)a; \, a_2 = \left(\frac{1}{2}, -\frac{1}{2}, \frac{1}{2}\right)a; \, a_3 = \left(\frac{1}{2}, \frac{1}{2}, -\frac{1}{2}\right)a$$

(a) Determine the volume of the primitive unit cell in terms of the bcc cell.

The volume of the primitive unit cell follows from the product:

© The Editor(s) (if applicable) and The Author(s), under exclusive license to Springer Nature Switzerland AG 2026
H. Zabel, *Elements of Elastic Scattering by X-Rays, Neutrons, and Electrons*,
https://doi.org/10.1007/978-3-032-16624-1

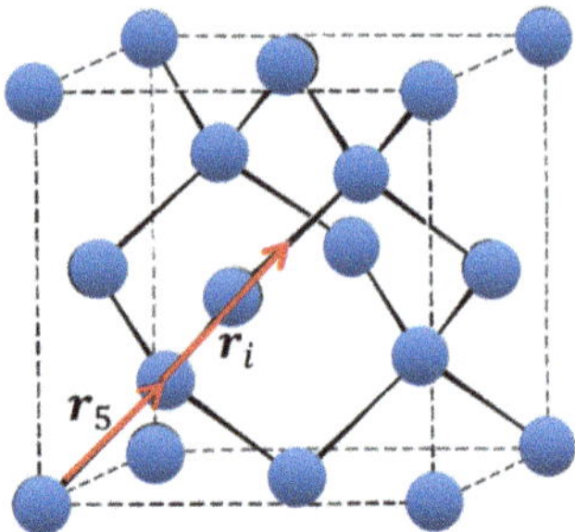

Fig. S1 Diamond crystal lattice in blue and vectors r_5 and r_i

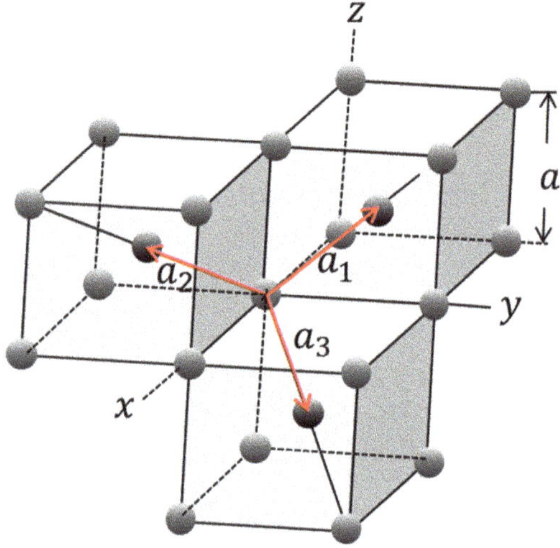

Fig. S2 bcc crystal lattice in gray and basis vectors of the primitive unit bcc cell indicated by red arrows

$$V_{puc} = \boldsymbol{a}_1 \cdot (\boldsymbol{a}_2 \times \boldsymbol{a}_3) = \frac{a^3}{2}$$

The primitive bcc unit cell contains only one atom. Therefore, the volume must be half of the bcc unit cell volume, which contains two atoms.

M 1.3 fcc lattice

Consider the lattice planes with the Miller indices (100) and (001) in the fcc lattice.

(a) Convert the nonprimitive fcc unit cell into a primitive unit cell. Show the primitive unit cell graphically of the fcc lattice, determine the lattice vectors, and the cell volume (Fig. S3).

The unit vectors of the primitive fcc unit cell are:

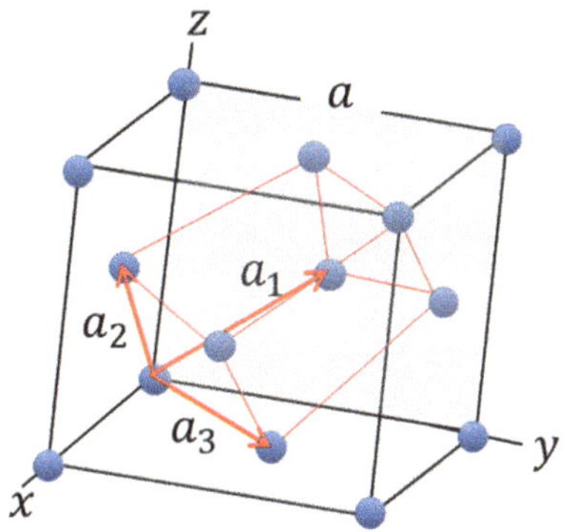

Fig. S3 fcc crystal lattice shown by blue spheres and basis vectors of the primitive unit fcc cell indicated by red arrows

$$\boldsymbol{a}_1 = \left(0, \frac{1}{2}, \frac{1}{2}\right)a; \boldsymbol{a}_2 = \left(\frac{1}{2}, 0, \frac{1}{2}\right)a; \boldsymbol{a}_3 = \left(\frac{1}{2}, \frac{1}{2}, 0\right)a$$

The cell volume is $V_{pc}^{fcc} = a^3/4$.

(b) What are the Miller indices of the (100) and (001) lattice planes now in terms of the primitive unit cell?

The (100) plane becomes a (011) plane in the primitive unit cell volume, and the (001) plane is converted to a (110) plane.

M 1.4 Space filling

Consider the hexagonal close-packed structure (*hcp*) sketched in Fig. 1.6b.

(a) Show that for the ideal hcp structure, the ratio of the lattice parameters is $c/a = \sqrt{8/3}$.

The base of the primitive cell of the *hcp* structure is hexagonal, that is, made up of equilateral triangles of side length a. The spheres of the second layer lie exactly over the centers of these triangles. The spheres of the second layer lie exactly above the centers of these triangles, whereby the atoms of this layer lie at $c/2$ in the c-direction. The center of an equilateral triangle is just $r = a\sqrt{3}/3$ away from the atoms, where r is the radius of the circumcircle. The distance between all nearest neighbors is a. We can now apply the Pythagorean theorem:

$$a^2 = \left(c^2/2 + r^2\right).$$

Solving this equation for the ratio c/a, we obtain $c/a = \sqrt{8/3}$.

(a) Assume that atoms can be treated as hard spheres, which may touch each other when closely packed. Determine the fraction of space that is occupied by the atoms (hard spheres) in a simple cubic lattice (*sc*), in a *bcc* lattice, and in an *fcc* lattice.

In an *sc* lattice, the hard spheres at the corners touch each other. Their radius is half the lattice parameter a. Therefore, their volume is $V_{sphere} = 4\pi(a/2)^3/3$. There are eight spheres in total, but each sphere is shared with eight neighboring cubes. Therefore, the total space that the spheres fill is: $8 \times 4\pi(a/2)^3/3 \times 8 = \pi a^3/6$. Hence, the fraction of space filled is: $\pi/6 = 0.52$ or 52%.

In an *fcc* lattice, there are again eight spheres, each one counting only 1/8. In addition, there are six half-spheres on the faces. All spheres have the same radius of $a\sqrt{2}/4$, where a is the *fcc* lattice parameter and $a\sqrt{2}$ is the face diagonal. Adding up the volumes, we find:

$$V_{spheres\ in\ fcc} = \frac{4\pi}{3}\left(\frac{8}{8} + \frac{6}{2}\right)\left(a\sqrt{2}/4\right)^3 = \frac{\pi}{3}\frac{1}{2^2}2\sqrt{2}a^3 = \frac{\pi}{3\sqrt{2}}a^3 = 0.74a^3$$

Hence, the fraction of space filled is: $\pi/3\sqrt{2} = 0.74$ or 74% for the *fcc* lattice. The same holds for the *hcp* lattice, since both lattices differ only in the stacking sequence but not in the space filling.

E 1.5 Angles between the lattice planes with Miller indices (111) and

(a) (210)

In general, the angles between (*hkl*) planes follow from Eq. 1.17:

$$\widehat{\boldsymbol{n}}_{hkl} \cdot \widehat{\boldsymbol{n}}'_{h'k'l'} = |\widehat{\boldsymbol{n}}_{hkl}||\widehat{\boldsymbol{n}}'_{h'k'l'}|\cos\theta,$$

and is given by:

$$\cos\theta = \frac{\widehat{\boldsymbol{n}}_{hkl} \cdot \widehat{\boldsymbol{n}}'_{h'k'l'}}{|\widehat{\boldsymbol{n}}_{hkl}||\widehat{\boldsymbol{n}}'_{h'k'l'}|}$$

where

$$\widehat{\boldsymbol{n}}_{hkl} = \frac{\boldsymbol{G}_{hkl}}{\lceil \boldsymbol{G}_{hkl} \rceil}; \boldsymbol{G}_{hkl} = h\boldsymbol{A} + k\boldsymbol{B} + l\boldsymbol{C}.$$

Using the orthogonality properties of the reciprocal lattice vectors, we find:

$$\cos\theta = \frac{\boldsymbol{A}\cdot\boldsymbol{A}' + \boldsymbol{B}\cdot\boldsymbol{B}' + \boldsymbol{C}\cdot\boldsymbol{C}'}{\sqrt{A^2 + B^2 + C^2} \times \sqrt{A'^2 + B'^2 + C'^2}}$$

Therefore, the angle between the (111) and the (210) planes is:

$$\cos\theta = \frac{2\cdot1 + 1\cdot1 + 0\cdot1}{\sqrt{4+1+0}\times\sqrt{1+1+1}} = \frac{3}{\sqrt{5}\sqrt{3}} = \sqrt{\frac{3}{5}}; \theta = 39.23°$$

(b) Using the same procedure, the angle between the (202) and the (111) planes is:

$$\cos\theta = \frac{2\cdot 1 + 0\cdot 1 + 2\cdot 1}{\sqrt{4+0+4}\times\sqrt{1+1+1}} = \frac{4}{\sqrt{8}\sqrt{3}} = \sqrt{\frac{2}{3}}; \theta = 35.26\,°$$

(c) Using the same procedure, the angle between the (222) and the (111) planes is:

$$\cos\theta = \frac{2\cdot 1 + 2\cdot 1 + 2\cdot 1}{\sqrt{4+4+4}\times\sqrt{1+1+1}} = \frac{6}{\sqrt{12}\sqrt{3}} = \sqrt{\frac{36}{36}} = 1; \theta = 0\,°,$$

because these planes are parallel.

E 1.6 Lattice spacings

(a) Lattice spacing of the (111) planes in a cubic cell with lattice parameter a is given by:

$$d = \frac{a}{\sqrt{h^2 + k^2 + l^2}}.$$

Therefore, the lattice spacing of the (111)—planes is $d = a/\sqrt{3}$.

(b) The lattice spacing of the (221) planes is accordingly: $d = a/\sqrt{9} = a/3$.

(c) The lattice spacing of the (300) planes is $d = a/\sqrt{9} = a/3$.

   Therefore, in a powder diffraction pattern, the Bragg reflections for the (300) and (221) lattice planes overlap. However, in a single-crystal diffraction experiment, the respective Bragg peaks are located at completely different points in the reciprocal space.

A 1.7 Five-fold symmetry axis

   Consider the two vectors a and b of equal length but opposite direction: $|a| = |b| = d$.

   a is rotated about a vertical axis through point 1 by an angle θ, yielding a'. Similarly, b is rotated about point 2 by the opposite angle $(-\theta)$, yielding b' (Fig. S4).

   For long-range translational symmetry, it is required that the distance $|a' - b'|$ equals:

Fig. S4 Graph illustrating that in classical crystallography, a fivefold symmetry axis is not allowed

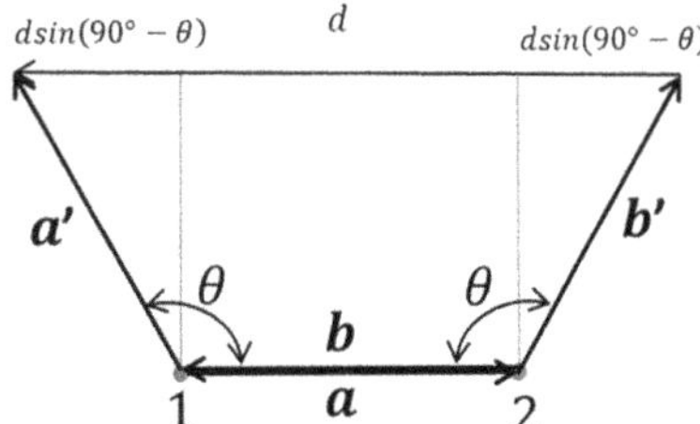

$$|a' - b'| = md = d + 2d \sin(90° - \theta) = d(1 - 2\cos\theta)$$

where $m = 1, 2, \ldots$ and $\theta = 2\pi/n$.
Therefore:

$$md = d\left(1 - 2\cos\frac{2\pi}{n}\right), \text{ and } \left|\cos\frac{2\pi}{n}\right| = \left|\frac{m-1}{2}\right| \leq 1$$

This condition is fulfilled by $m = 3, 2, 1, 0, -1$, requiring n to be: $n = 1, 2,$ 3, 4, or 6, but not 5 or any higher number.

In classical crystallography, a fivefold rotational axis is not allowed. However, in quasi-crystals, the fivefold axis is realized. More about this topic is discussed in Sect. 3.5.4.

E 1.8 Reciprocal cell volume

Show that according to Eq. (1.8), $V_{rc} = (2\pi)^3/V_{uc}$.
The definition of the reciprocal cell volume is:

$$V_{rc} = |\boldsymbol{A} \cdot (\boldsymbol{B} \times \boldsymbol{C})|$$

where

$$\boldsymbol{A} = \frac{2\pi}{V_{uc}}(\boldsymbol{b} \times \boldsymbol{c})$$

Inserting:

$$V_{rc} = \frac{2\pi}{V_{uc}}|(\boldsymbol{b} \times \boldsymbol{c}) \cdot (\boldsymbol{B} \times \boldsymbol{C})|$$

Now, using Lagrange's identity, we find for the combined cross and dot products:

$$(\boldsymbol{b} \times \boldsymbol{c}) \cdot (\boldsymbol{B} \times \boldsymbol{C}) = (\boldsymbol{b} \cdot \boldsymbol{B})(\boldsymbol{c} \cdot \boldsymbol{C}) - (\boldsymbol{b} \cdot \boldsymbol{C})(\boldsymbol{c} \cdot \boldsymbol{B}) = \left(b\frac{2\pi}{b}\right)\left(c\frac{2\pi}{c}\right) - 0 = (2\pi)^2$$

In the last line, we used the orthogonality relations in Eq. (1.6). Therefore:

$$V_{rc} = \frac{2\pi}{V_{uc}}(2\pi)^2 = \frac{(2\pi)^3}{V_{uc}}.$$

Alternatively, valid for rectangular systems, the reciprocal cell volume simply follows from:

$$V_{rc} = |\mathbf{A} \cdot (\mathbf{B} \times \mathbf{C})| = \frac{2\pi}{a}\frac{2\pi}{b}\frac{2\pi}{c}\widehat{\mathbf{A}} \cdot \left(\widehat{\mathbf{B}} \times \widehat{\mathbf{C}}\right) = \frac{(2\pi)^3}{V_{uc}}$$

$\widehat{\mathbf{A}}, \widehat{\mathbf{B}}, \widehat{\mathbf{C}}$ are unit vectors in the respective directions.

Chapter 2

E 2.1 Wavenumbers

Assuming that the modulus of the incident wavenumber is $|\mathbf{k}_i| = 62.8$ nm^{-1} and that the angle between $\mathbf{k}_i$ and $\mathbf{k}_f$ is $90°$.

(a) What is the wavelength of the radiation used?

The wavenumber is defined as $|\mathbf{k}_i| = 2\pi/\lambda$. Thus, the wavelength is $\lambda = 2\pi/|\mathbf{k}_i| = \frac{6.283}{62.8}$ nm $= 0.1$ nm or 1 Å.

(b) The modulus of the scattering vector is $|Q| = 2|\mathbf{k}_i| \sin\theta = 2 \times 62.8$ nm^{-1} $\sin 45° = 88.8$ nm^{-1}.

(c) The maximum scattering vector that can be reached is $|Q| = 2|\mathbf{k}_i| = 125.6$ nm^{-1} at $\theta = 90°$.

(d) If scattering vectors larger than 125.6 nm^{-1} are required, then X-ray radiation or neutron radiation should be used with a wavelength shorter than 0.1 nm.

M 2.2 Atomic form factor

(a) Use Eq. 2.15 for the atomic form factor:

$$f(\mathbf{Q}) = \int \rho_{el}(\mathbf{R}_e) e^{i\mathbf{Q} \cdot \mathbf{R}_e} d\mathbf{R}_e.$$

Thus, for a hydrogen atom:

$$f(\mathbf{Q}) = \frac{1}{\pi a_B^3} \int \exp\left(-\frac{2R_e}{a_B}\right) e^{i\mathbf{Q} \cdot \mathbf{R}_e} d\mathbf{R}_e.$$

Recall that the integral $\int d\mathbf{R}_e$ in spherical coordinates can be partitioned into an angular part and a radial part:

$$\int d\mathbf{R}_e = \int_{-\pi}^{+\pi} d\theta \int_0^{\pi} \sin\phi \, d\phi \int_0^{\infty} R_e^2 dR_e.$$

Therefore, the angular part of the integral is:

$$\int_{-\pi}^{+\pi} d\theta \int_{0}^{\pi} e^{iQR_e \cos\phi} \sin\phi d\phi = 2\pi \int_{0}^{\pi} e^{iQR_e \cos\phi} \sin\phi d\phi$$

Substituting $\cos\phi = x$ and $\sin\phi d\phi = -dx$, we find

$$-2\pi \int_{1}^{-1} e^{iQR_e x} dx = 2\pi \left(\frac{e^{iQR_e} - e^{-iQR_e}}{iQR_e} \right) = 4\pi \frac{\sin(QR_e)}{QR_e}$$

Now, recollecting terms and solving the radial part:

$$f(Q) = \frac{4\pi}{\pi a_B^3} \int R_e^2 \exp\left(-\frac{2R_e}{a_B}\right) \frac{\sin(QR_e)}{QR_e} dR_e$$

$$= \frac{4}{a_B^3 Q} \int_0^{\infty} R_e \exp\left(-\frac{2R_e}{a_B}\right) \sin(QR_e) dR_e$$

According to integration tables (i.e. Bronstein–Semendjajew), the integral has the solution:

$$\int xe^{ax} \sin(bx) dx = \frac{xe^{ax}}{a^2 + b^2} (a\sin(bx) - b\cos(bx)) - \frac{e^{ax}}{(a^2 + b^2)^2}$$

$$\times \left[(a^2 - b^2) \sin(bx) - 2ab \cos(bx)\right]$$

Using this result, we find the X-ray atomic form factor of the hydrogen atom:

$$f(Q) = \frac{4}{a_B^3 Q} \int_0^{\infty} R_e \exp\left(-\frac{2R_e}{a_B}\right) \sin(QR_e) dR_e = \frac{1}{\left(1 + \left(\frac{a_B Q}{2}\right)^2\right)^2}.$$

(b) At $Q = 0$, $f(Q = 0) = Z = 1$ for the hydrogen atom, as expected for an atom with only one electron.

(c) At larger Q-values, the atomic form factor drops off with Q to the power of -4.

E 2.3 Coherent and incoherent scattering length of vanadium

The spin coherent and incoherent cross sections are: $\sigma_{coh} = 4\pi\, b_{coh}^2$ and $\sigma_{inc} = 4\pi b_{inc}^2$, where $b_{coh}^2 = (p_+ b_+ + p_- b_-)^2$ and $b_{inc}^2 = p_+ p_- (b_+ - b_-)^2$. With the scattering lengths quoted: $b_+ = 4.93$ fm and $b_- = -7.6$ fm. Inserting numbers, we obtain: $p_+ = 9/16$, $p_- = 7/16$. $p_+ b_+ + p_- b_- = -0.552$ fm; $b_{coh}^2 = 0.3$ fm^2. $\sigma_{coh} = 3.83$ fm^2 = 0.0038 barn. $b_{inc}^2 = p_+ p_- (b_+ - b_-)^2 = 63/256 \times (12.51)^2$ fm^2 = 38.51 fm^2. Yielding $\sigma_{inc} = 4.84$ barn.

The scattering of neutrons at vanadium is essentially only incoherent. Therefore, it is not possible to record Bragg reflections from vanadium using thermal neutrons. However, with vanadium in the beam, the primary beam intensity of the beamline can be monitored.

Chapter 3

E 3.1. Structure factor for NaCl

(a) Structure factor

The NaCl structure consists of two fcc lattices, one is occupied by Na^+ and the other by Cl^-. Both fcc lattices are shifted against each other by half a lattice spacing in one of the cubic directions, say along the c-direction. Then, we have the basis vectors in units of the lattice parameter:

$$Na^+ : (000); (½ ½ 0); (½ 0 ½); (0 ½ ½);$$

$$Cl^- : (00½); (½ ½ ½); (½ 0 0); (0 ½ 0);$$

Now, we use again the definition of the structure factor. But since the atomic form factors are different for the two sublattices, they have to be taken into the sum:

$$F_{tot}(hkl) = \sum_{k=0} f_{Na^+} e^{iG \cdot r_k} + \sum_{k=0} f_{Cl^-} e^{iG \cdot r_k}.$$

Now, inserting the basis vectors, we get for the Na^+ sublattice the same expression as previously for the generic fcc structure:

$$F_{Na^+}(hkl) = f_{Na^+}(1 + \exp i\pi(h + k) + \exp i\pi(h + l) + \exp i\pi(k + l))$$

Similarly, we find for the structure factor of the Cl^- sublattice:

$$F_{Cl^-}(hkl) = f_{Cl^-}(\exp i\pi(l) + \exp i\pi(h + k + l) + \exp i\pi(h) + \exp i\pi(k))$$

(b) Allowed and forbidden Bragg reflections

For all even (hkl), we find:

$$F_{tot}^{even}(hkl) = 4(f_{Na^+} + f_{Cl^-})$$

For all odd (hkl), we find:

$$F_{tot}^{odd}(hkl) = 4\left(f_{Na^+} - f_{Cl^-}\right)$$

And for mixed (hkl), we have:

$$F_{tot}^{mixed}(hkl) = 0$$

Because of this structure factor, odd-order Bragg reflections turn out to be much weaker than even-order Bragg reflections. As an example, study the powder pattern of NaCl in Fig. 5.8. NaCl-type structures, such as KCl, MgO, MnO, etc., can be easily identified by this characteristic property.

M 3.2. Structure factor for the diamond lattice

(a) Structure factor

Si has a diamond structure, that is, two fcc structures are shifted against each other along the body diagonal by (1/4,1/4,1/4) of the lattice parameter. The basis vectors of the diamond structure are listed in Chap. 1 and are as follows:

1. fcc lattice I: (0, 0, 0) (½, ½, 0) (½, 0,½) (0, ½, ½)
2. fcc lattice II: (¼, ¼, ¼) (¾, ¾, ¼) (¾, ¼, ¾) (¼, ¾, ¾).

Using these basis vectors, the structure factor can be calculated, from which we obtain the information on which of the hkl reflections are allowed and which are forbidden.

We consider the diamond structure as an fcc lattice with two additional basis vectors: $r'_1 = (0,0,0)$ and $r'_2 = \left(\frac{1}{4}, \frac{1}{4}, \frac{1}{4}\right)$. The combined form factor is then

$$F_{Si} = F_{basis} \times F_{fcc}$$

And therefore:

$$F_{Si} = \left(1 + e^{i\frac{\pi}{2}(h+k+l)}\right) \times \left(1 + e^{i\frac{\pi}{2}(h+k)} + e^{i\frac{\pi}{2}(h+l)} + e^{i\frac{\pi}{2}(k+l)}\right)$$

The second bracket is different from zero only for the Miller indices (h, k, l) all even or all odd, yielding $4f$. The first bracket equals 2, when all indices are even and add up to $h + k + l = 4n$ (n = integer). However, the first bracket is zero if $h + k + l = 4n + 2$. Thus, the (220) reflection is allowed, but the (222) reflection is not. All allowed reflections have an amplitude of $F_{Si} = 2 \times 4f_{Si} = 8f_{Si}$ and a corresponding intensity proportional to $F_{Si}^2 = 64f_{Si}^2$.

Now, we turn to the odd Miller indices. If all (h, k, l) are odd, the prefactor is $1 \pm i$. Then, the combined structure factor is $F_{Si} = (1 \pm i) \times 4f_{Si}$, and the intensity is proportional to $F_{Si}^2 = (1 + i)(1 - i)16f_{Si}^2 = 2 \times 16f_{Si}^2 = 32f_{Si}^2$.

Therefore, the (111), (311), (331), (333), etc. reflections are allowed but have a lower intensity compared to the reflections with even Miller Indices.

(b) **Allowed and forbidden Bragg reflections**

Allowed Bragg reflections:

All (hkl) even and the sum: $h + k + l = 4n$, yielding $F_{Si}^2 = 64f_{Si}^2$

All (hkl) odd: $F_{Si}^2 = 32f_{Si}^2$.

Forbidden Bragg reflections:

All (hkl) even and the sum: $h + k + l = 4n + 2$

Mixed (hkl)

M 3.3. Structure factor for the hcp lattice

There are two identical atoms in the hcp structure. Their basis vectors are:

$$r_1 = (0, 0, 0); r_2 = \left(\frac{1}{3}; \frac{2}{3}; \frac{1}{2}\right)$$

With this, we determine the structure factor:

$$F_{hcp} = f\left(1 + \exp i2\pi\left(\frac{1}{3}h + \frac{2}{3}k + \frac{1}{2}l\right)\right)$$

We change to the square of the expression:

$$F_{hcp}F_{hcp}^* = f^2\left(1 + \exp i2\pi\left(\frac{1}{3}h + \frac{2}{3}k + \frac{1}{2}l\right)\right)\left(1 + \exp -i2\pi\left(\frac{1}{3}h + \frac{2}{3}k + \frac{1}{2}l\right)\right)$$

$$= 2f^2\left(1 + \cos 2\pi\left(\frac{(h + 2k)}{3} + \frac{l}{2}\right)\right)$$

Therefore, we obtain for the structure factor of the *hcp* structure:

$$F_{hcp}^2(hkl) = 4f^2 \cos^2 \pi\left(\frac{(h + 2k)}{3} + \frac{l}{2}\right)$$

For the allowed and forbidden reflections, we find four conditions, which are tabulated in Table S1.

Table S1 Conditions for allowed and forbidden Bragg reflections of the *hcp* structure

	$F_{hcp}^2(hkl)$
1. $h + 2k = 3n$; l even	$4f^2$
2. $h + 2k = 3n$; l odd	0
3. $h + 2k = 3n \pm 1$; l even	f^2
3. $h + 2k = 3n \pm 1$; l odd	$3f^2$

M 3.4. Errors contributing to lattice parameter determination

(a) Starting with the equation for the d-spacing: $d = \frac{\lambda}{2\sin\theta}$, we take the first derivative with respect to the angle θ, assuming that the wavelength is well-defined:

$$\frac{\Delta d}{\Delta \theta} = -\frac{\lambda}{2}\frac{\cos\theta}{\sin^2\theta}$$

Dividing by the d-spacing yields:

$$\frac{\Delta d}{d} = -\cot\theta\,\Delta\theta$$

It is sufficient to consider the modulus: $|\Delta d/d| = \cot\theta\,\Delta\theta$. Here, $\Delta\theta$ is the beam divergence.

(b) The other source of error is a lack of monochromaticity. To get an expression for this uncertainty, we take the first derivative of d with respect to the wavelength λ:

$$\frac{\Delta d}{\Delta \lambda} = \frac{1}{2\sin\theta} = \frac{d}{\lambda}.$$

Rearranging:

$$\frac{\Delta d}{d} = \frac{\Delta\lambda}{\lambda}$$

(c) Combining both error contributions by using the formula for error propagation:

$$\frac{\Delta d}{d} = \sqrt{\cot^2\theta(\Delta\theta)^2 + \left(\frac{1}{\lambda}\right)^2(\Delta\lambda)^2}$$

(d) From the last equation, we conclude that $|\Delta d/d|$ decreases with decreasing beam divergence $\Delta\theta$ of the scattered beam and by going to larger θ-values to decrease $\cot\theta$. The second term tells us that we can improve the precision by using larger wavelengths and reducing the wavelength spread $\Delta\lambda$ using a very good monochromator, such as a Si(111) Bragg reflection. Therefore, the correct answer is that lattice parameters can be measured with increasing precision at high-order Bragg reflections. There are several methods to improve the precision of lattice parameter determination. One standard method is to plot Q^2 versus $(h^2 + k^2 + l^2)$ and to determine the factor $(2\pi/a)^2$ from the slope. More methods are described in the book of Klug and Alexander (see book recommendations for Sect. 3.4).

A 3.5 Debye temperature

(a) For the $G_{h00} = 2\pi/d_{h00} = 2\pi h/a_{Al}$ reflections, the Debye–Waller factor can be written in the following form:

$$2W = \frac{3\hbar^2}{m_{Al}k_B}\frac{T}{\theta_D^2}G_{h00}^2 = \frac{3\hbar^2}{m_{Al}k_B}\frac{T}{\theta_D^2}\left(\frac{2\pi h}{a_{Al}}\right)^2 = Ah^2T$$

Therefore, we find for the intensity: $\ln(I/I_0) = -Ah^2T$
For a fixed Miller index h, the temperature dependence can be expressed in terms of a temperature difference, as:

$$\ln\left(\frac{I(T_1)}{I(T_2)}\right) = -Ah^2(T_1 - T_2) = Ah^2\Delta T$$

(b) Hence, the slope A, needed for determining the Debye temperature, is:

$$A = \frac{1}{h^2}\frac{1}{\Delta T}\ln\left(\frac{I(T_1)}{I(T_2)}\right)$$

For ΔT, we take the temperature interval between 100 K and 300 K, that is, $\Delta T = 200\ K$. We neglect the slope for the (200) reflection, as it is too difficult to determine.

Now we graphically evaluate the slope for each $(00l)$ Bragg reflection and divide the slope by l^2, yielding Table S2 for A.

(c) The slope averaged over the four values is $A = 70{,}5 \times 10^{-6}\ K^{-1}$. From A, we can now determine the Debye temperature according to:

Table S2 Evaluating the Debye temperature of aluminum

(h00) reflections	$\frac{1}{\Delta T}\ln\left(\frac{I(T_1)}{I(T_2)}\right) \times 10^{-6}K^{-1}$	h^2	A
400	1200	16	74
600	2400	36	66
800	4750	64	73
10 00	6950	100	69

$$\theta_D^2 = \frac{3h^2}{m_{Al}k_B}\left(\frac{2\pi}{a_{Al}}\right)^2\frac{1}{A}$$

Using: $2\pi\hbar = 6.6 \times 10^{-34}$ Js; $k_B = 1.38 \times 10^{-23}$ J/K; $a_{Al} = 0.405$ nm; $m_{Al} = 27 \times m_p = 27 \times 1.67 \times 10^{-27}$ kg $= 4.5 \times 10^{-26}$ kg; where m_p is the proton mass and 27 is the atomic number of Al; $3\hbar^2/m_{Al}k_B = 0.48 \times 10^{-19}$ Js^{2}K/kg; $(2\pi/a_{Al})^2 = 2.4 \times 10^{20}$m^2, we find for the Debye temperature: $\theta_D^2 = 18 \times 10^4$K^{2}and $\theta_D = 405.5$ K.

(d) The authors Nicklow and Young obtained a room temperature value of $\theta_D = 395$ K, increasing to 405 K at about 100 K. The temperature dependence of the Debye–Waller factor is due to anharmonicities, which are not taken into account with the simple evaluation presented here. The literature value for the Debye temperature at room temperature is about 400 K. So, our value almost matches the literature value.

For evaluating the Debye temperature, we used the high-temperature approximation, which is valid for $T > \theta_D$. However, for determining the Debye temperature, we use a temperature interval located below θ_D. Therefore, using the high-temperature approximation is not really justified. Nevertheless, it is the simpler approach because of the linear relationship between W and T.

(e) The Debye temperature is important in the context of the specific heat of solids. For temperatures above θ_D, the specific heat follows from the classical Dulong–Petit law, which states that the specific heat per mol of atoms is a constant given by $C_V = 3R = 24.94$ J/(mol K), where R is the universal gas constant. For temperatures below θ_D, the specific heat $C_V = BT^3$, where B is a constant. This relationship indicates that atoms in solids at low temperatures behave as quantum mechanical oscillators rather than as classical harmonic oscillators. A second important consequence of this relationship is the fact that $\lim_{T \to 0} C_V = 0$, which is required by statistical mechanics.

In Table S3, are compared Debye temperatures of rather soft metals (Pb, Ag), ionic materials (KBr, NaCl), and hard covalently bonded materials (diamond, graphite parallel to the a-direction). There is a rough relationship between the

Table S3 Table of Debye temperatures and melting temperatures of a few characteristic materials.

	θ_D (K)	Melting temperature (K)
Pb	105 K	600
Ag	225 K	1235
KBr	180 K	
NaCl	280 K	1074
Al	400	933
Diamond	2230 K	Does not melt, it sublimes to vapor directly at high T > 3000 K
Graphite	800 K ∥ c-axis, 2300 K ∥ a-axis	4300 at 100 bar

Debye temperature and the melting temperature. Hard materials tend to have high Debye temperatures and high melting temperatures, and vice versa.

Chapter 4

E 4.1. Threshold energy

The threshold energy corresponds to the Cu Kα absorption energy, which is 8.98 keV (taken from X-ray tables, for instance: X-ray data booklet: http://xdb. lbl.gov/). The Cu Kα characteristic radiation has an energy of only 8.04 keV. The difference is because the absorption brings the electron from the K-shell to the vacuum level, while the characteristic radiation corresponds to the energy difference between the L- and the K-shell, which is smaller. Excitation energy is always larger than the characteristic radiation energy (Fig. S5).

E 4.2 Energy resolution

Cu $K_{\alpha 1}$ radiation has an energy of 8.047 keV and a wavelength $\lambda_{\alpha 1} = 0.154011$ nm. Cu $K_{\alpha 2}$ radiation has an energy of 8.027 keV and a wavelength $\lambda_{\alpha 2} = 0.15439$ nm. Therefore, the absolute energy difference $\Delta E = 0.02$ keV, and the relative energy difference is $\Delta E/E = 0.02/8.03 = 0.0025$ or 0.25%. The wavelength difference is $\Delta\lambda = 3.8 \times 10^{-4}$ nm and $\Delta\lambda/\lambda = 2.5 \times 10^{-3}$. This is the same relative value. As the photon energy is $E = hc/\lambda$ (h = Planck constant, c=speed of light), therefore $\Delta\lambda/\lambda = \Delta E/E$. This resolution requirement is easy to attain using a Si or Ge monochromator. For precision lattice parameter measurements, only the $K_{\alpha 1}$ radiation is used because of the higher intensity, and the $K_{\alpha 2}$ radiation is filtered out by the monochromator in the incident beam.

E 4.3 Cut-off energy

For an accelerating voltage of 40 kV, the maximum photon energy in the bremsspectrum is $E_{max} = 40$ keV. Since $E_{max} = hc/\lambda_{min}$, the cut-off wavelength is $\lambda_{min} = hc/E_{max} = 12.39$ keVÅ/40 keV $= 0.3097$ Å $= 0.03097$ nm.

M 4.4 Positive and negative coherent scattering lengths for neutrons

Positive and negative scattering lengths are only known for neutrons, not for X-rays or electrons. If a substance consists of a single isotope with a negative neutron scattering length, the scattering result is the same as for a positive scattering length. This is because what finally counts is the intensity, which is

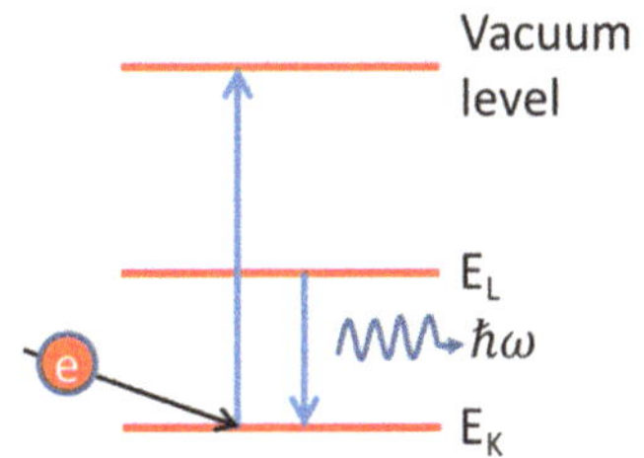

Fig. S5 Threshold energy for X-ray excitation

obtained by taking the square of the form factor, including the scattering length. However, in a binary alloy, the existence of negative scattering lengths provides additional experimental capabilities. Consider the alloy Fe_xNi_{1-x} with a CsCl structure. The fundamental reflections display a structure factor of the type:

$$F_{FeNi}^{fund} = \left(xb_{coh}^{Fe} + (1-x)b_{Ni}\right)$$

All Fe isotopes feature positive coherent scattering lengths of $b_{coh}^{Fe} = 9.45\,fm$ and the Ni isotope ^{62}Ni has a negative scattering length of $b_{coh}^{62Ni} = -8.7\,fm$. Hence, at an alloy concentration of $x = 0.48$ or roughly 50%, the fundamental reflections vanish, whereas the superstructure reflections, with a form factor of the type:

$$F_{FeNi}^{super} = \left(xb_{coh}^{Fe} - (1-x)b_{Ni}\right),$$

become very strong. From this, we notice that negative scattering lengths allow us to de-emphasize fundamental peaks and to emphasize superstructure peaks. This is known as the *null-matrix* method, which is often used to study short-range order correlations in binary alloys.

M 4.5 Thermal neutron moderator:

(a) Light H_2O consists of protons and rather light oxygen atoms. Water has a high density of protons. The collision of neutrons with protons of the same mass should effectively transfer energy from the neutrons to the protons. This should heat up the water, but if kept at a constant temperature by cooling, the moderation continues with the same efficiency. Neutron capture is small but not negligible. The (p, n) nuclear reaction is as follows: $p + n \rightarrow D + \gamma$, and has an absorption cross section of $\sigma_{abs}^{H} = 0.33 \times 10^{-24}$ cm^2 (barn). In contrast, the neutron capture reaction with deuterium (d, n):
$^{2}H + n \rightarrow {}^{3}H + \gamma$ has a much smaller cross section of only $\sigma_{abs}^{D} = 0.0005$ barn. The naturally occurring ^{16}O isotope with an abundance of 99.7% has no neutron capture cross section.

(b) If on each collision the kinetic energy of neutrons is divided in half, we obtain the following series: $E_0,\ E_0/2,\ E_0/2^2,\ E_0/2^3 \ldots\ E_0/2^n = E_t$, where n is the number of collisions and E_t is the thermal energy to be reached after n collisions. Therefore:

$$2^n = \frac{E_0}{E_t} \text{ and } n = \frac{1}{\log 2}\log\left(\frac{E_0}{E_t}\right)$$

Inserting numbers:

$$\frac{E_0}{E_t} = \frac{1 \times 10^6\, eV}{25 \times 10^{-3}\, eV} = 4 \times 10^7$$

and

$$n = \frac{1}{\log 2}\, \log\left(4 \times 10^7\right) = \frac{7 + \log 4}{\log 2} = 25.2$$

This result may be a bit too optimistic: 80–100 collisions before thermalization is more realistic.

(c) If H_2O is replaced by D_2O, the energy transfer per collision will be reduced by a small amount. On the other hand, the absorption cross section of D is almost zero, and therefore fewer neutrons get lost when moderated with heavy water. Furthermore, the γ-background radiation is much less than that for light water moderation. Because of this, moderation with heavy water D_2O is to be preferred over light water moderation. However, heavy water moderation produces a small amount of tritium, which in the long run causes a problem if not removed. Tritium is not a stable isotope with a half-life of 12.3 years. Simply waiting before it decays is not an option.

Chapter 5

M 5.1 Scherrer equation

(a) Simple estimate of the grain or crystallite size, starting with the modulus of the scattering vector:

$$Q = \frac{4\pi}{\lambda} \sin\theta$$

First derivative with respect to the diffraction angle θ:

$$\frac{dQ}{d\theta} = \frac{4\pi}{\lambda}\cos\theta \quad \text{and} \quad dQ = \frac{4\pi}{\lambda}\cos\theta\, d\theta$$

The broadening of a Bragg-reflection dQ in Q-space is due to the finite coherence length L over which constructive interference occurs. Or more simply explained: L is the crystalline size within a mosaic crystal.

$$dQ = \frac{2\pi}{L} = \frac{4\pi}{\lambda}\cos\theta\, d\theta$$

Rearranging:

$$L = \frac{\lambda}{2\cos\theta\, d\theta} = \frac{\lambda}{\Delta(2\theta)\cos\theta_B} = \frac{\lambda}{B_{FWHM}\cos\theta_B}.$$

Here, we have replaced: $2 \times d\theta = d(2\theta) = \Delta(2\theta) = B_{FWHM}$, where d stands for an incremental change and Δ for a difference. The angular difference is identified with the radial width of the Bragg reflection. Therefore:

$$B_{FWHM} = \frac{\lambda}{L\cos\theta_B}$$

This is almost identical to the famous Scherrer equation:

$$B_{FWHM} = \frac{0.94\lambda}{L\cos\theta_B},$$

which was originally derived by calculating the *FWHM* of the Laue intensity (Eq. 3.29), yielding for the prefactor $2[(\ln 2)/\pi]^{1/2} = 0.94$. The Scherrer equation is often quoted as:

$$B_{FWHM} = \frac{K\lambda}{L\cos\theta_B},$$

where K is a dimensionless constant close to 1.

Warning: The Scherrer equation can only be applied after correcting the measured *FWHM* with respect to the beam divergence and wavelength spread, using the equation provided in Exercise 3.3.

(b) Other effects that may lead to a broadening of the Bragg reflection in the radial direction are strain gradients, thermal gradients, composition and doping inhomogeneities, antiphase domain boundaries in ordered alloys such as Cu_3Au, and stacking faults in fcc and hcp lattices. Homogeneous tensile or compressive strain does not lead to the broadening of Bragg reflections but to shifts of the center of Bragg reflections. However, bending load causes a gradient of lattice parameters (strain gradient), ranging from compressed to expanded d-spacings and resulting in reflection broadening. Note that broadening of Bragg reflections in the transverse direction has to be distinguished from broadening in the radial direction. The former is related to the mosaicity of the crystals (tilting angle of crystallites), and the latter is related to the crystalline size and lattice deformation.

A 5.2 Beryllium filter

Be has a hexagonal crystal structure with lattice parameters: a = 0.22858 nm, b = 0.22858 nm, and c = 0.35843 nm. The coherent scattering length b is 7.8 fm, and neutron absorption is negligible. All these data are taken from Internet tables, freely available. Crystal structure and lattice parameters are listed at: http://

Table S4 Scattering lengths and cross sections for Beryllium

Value	Unit	Quantity
Be	–	Isotope
100	%	Natural abundance
7.79	fm	Bound coherent scattering length
0.12	fm	Bound incoherent scattering length
7.63	barn	Bound coherent scattering cross section
0.0018	barn	Bound incoherent scattering cross section
7.63	barn	Total bound scattering cross section
0.0076	barn	Absorption cross section for 2200 m/s neurons

Note: 1 fm $= 1 \times 10^{-15}$ m; 1 barn $= 1 \times 10^{-24}$ cm^2. Data are reproduced from www.ncnr.nist.gov/resources/n-lengths/

periodictable.com/Properties/A/LatticeConstants.html; the coherent scattering length can be found at: https://www.ncnr.nist.gov/resources/n-lengths/. Table S4 lists the scattering lengths and cross sections for Be.

Neutron scattering works like the following. A "white" beam of thermal neutrons is guided from the nuclear reactor source to a graphite monochromator. Graphite has a c-axis lattice parameter of 0.335 nm (neglecting the ABA… stacking). Usually, the (001) reflection is used for monochromatization. Then, at a scattering angle of 45°, the wavelength reflected after the monochromator is 0.473 nm. All other neutron wavelengths will pass through the monochromator without deflection and are either used by other instruments or are removed by an absorber. However, this arrangement cannot impede the second harmonic with a wavelength of 0.2365 nm from being also reflected by the monochromator at the same scattering angle of 45°, corresponding to the (002) reflection of that wavelength. Higher wavelength harmonics mess up a quantitative scattering experiment and must therefore be removed.

This is where the Be filter comes into play. Polycrystalline Be will scatter all neutron wavelengths up to the maximum wavelength of $2d = \lambda_{max}$. With a lattice parameter of $a = 0.228$ nm, this is $\lambda_{max} = 0.456$ nm. Therefore, the Be filter removes the second harmonic $\lambda/2 = 0.2365$ nm, but it will pass unhindered the original wavelength at 0.473 nm. Thus, Be filters are used as low-energy/long-wavelength neutron filters, while high energies/short wavelengths are removed by powder diffraction.

Side remark: the lattice parameter quoted above is the one from the basal plane of the hexagonal crystal structure of Be. The c-axis lattice parameter is $c=0.358$ nm. Should we consider this crystal orientation as well in our discussion? And if so, wouldn't it also remove the original wavelength from the monochromator, because $2c=0.716$ nm > 0.473 nm? The answer is: no. The reason is that in a powder, each Bragg reflection is weighted by the multiplicity factor m_{hkl} of reflections, that is, the number of equivalent (hkl) planes in a crystal structure. For ($00l$)-reflections, the multiplicity factor $m_{00l} = 2$ as compared to

($hk0$)-reflections with $m_{hk0} = 12$. Therefore, the ($00l$)-reflections are statistically not relevant.

In addition, cooling of the Be filter is applied to improve the filter effect by reducing inelastic scattering. Neutrons are very effective in interacting with phonons in crystal lattices. Neutrons can, by inelastic scattering (not discussed in this text), either create or annihilate phonons, that is, quantized lattice vibrations. However, this capability diminishes at low temperatures as the thermal occupation number of phonons decreases exponentially with decreasing temperature, that is, it becomes more difficult to create or annihilate phonons if there are not many around. At high temperatures, the sharp Bragg condition is softened by inelastic neutron scattering, which is to be avoided.

Figure S6 shows the neutron transmission through polycrystalline Be with different thicknesses and for room temperature (RT) and liquid nitrogen (LN) temperature. The Bragg cut-off at about 0.4 nm is clearly visible. Graph and calculations from M. Wahba, Egypt. J. Sol., Vol. (25), No. (2), (2002)

M 5.3 Silicon monochromator

We have seen by the structure factor calculation in Exercise 3.2 that the (111) reflection is allowed, but the (222) is not. At a monochromator angle θ that scatters the wavelength λ via the (111) Bragg reflection, the $\lambda/2$ wavelength would also pass as the (222) reflection. However, this reflection is forbidden by the structure factor, and therefore, this wavelength will not be reflected but transmitted. Thus, Si(111) acts as a monochromator and a filter for the second harmonic at the same time.

Added information, not part of the question: This elegant trick of the Si-monochromator could also, in principle, be used for neutron scattering. However, Si single crystals have a much too narrow crystal mosaicity of about 0.005° and therefore a very small acceptance angle for the incoming neutron beam. In the case of thermal neutrons, the beam is rather wide and has quite a divergence of

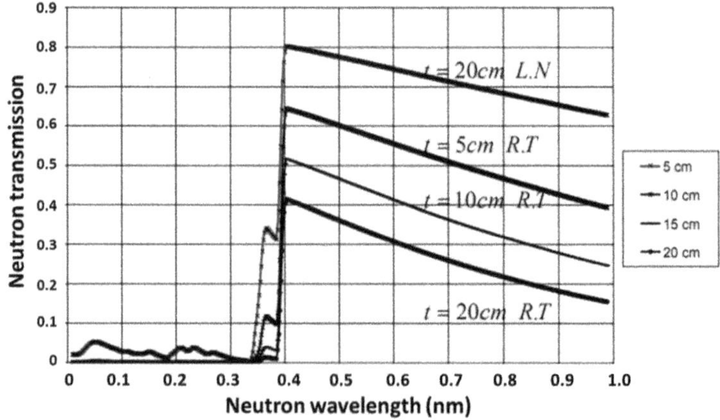

Fig. S6 Neutron transmission through polycrystalline Be with different thicknesses. Reproduced from M. Wahba, Egypt. J. Sol., Vol. (25), No. (2), (2002) with permission

about 0.5°. Using a Si crystal as a neutron monochromator would therefore dramatically reduce the reflected neutron intensity by a factor of $0.005/0.5 = 1/100$. Thus, it is necessary to adapt the monochromator mosaicity to the beam divergence to maximize the intensity on the sample. Typical neutron monochromators are graphite or copper with a mosaicity of about 0.1° (Cu) to 0.5° (highly oriented pyrolytic graphite). For removing higher harmonics, a Be filter is used, as discussed in Exercise 5.2.

Chapter 6

E 6.1 Evaluation of a powder diffraction pattern

(a) First, we calculate the θ values: 22.33°, 32.5°, and 41.15°, and from these, the interplanar spacings $d = \lambda/2\sin\theta$, with the values: $d_1 = 0.202$ nm, $d_2 = 0.143$ nm; $d_3 = 0.117$ nm.

(b) The ratio $d_1/d_3 = \sqrt{3}$, while $d_2/d_3 = \sqrt{3/2}$, and $d_1/d_2 = \sqrt{2}$. This fits the first three reflections of the bcc lattice: (110), (200), and (211).

(c) The lattice parameter follows from $a = d_1 \times \sqrt{2} = 0.2866$ nm, which corresponds, for example, to bulk Fe or Cr.

E 6.2 Variation of the scattering length

Consider the NaCl powder pattern in Fig. 6.4. How would the intensities of the Bragg peaks change if you set $f_{Na} = |b|$ and $f_{Cl} = -|b|$?

The intensity of (hkl) Bragg reflections is then:

$$I(G_{hkl}) = k\frac{I_0}{r^2} m_{hkl} 16 \left| b e^{-W_{Na}} \pm (-b)e^{-W_{Cl}} \right|^2$$

Now, all even (hkl) reflections vanish, and all odd (hkl) reflections have an intensity proportional to:

$$I(G_{hkl}) = k\frac{I_0}{r^2} m_{hkl} 16 b^2 \left| e^{-W_{Na}} + e^{-W_{Cl}} \right|^2$$

The multiplicity of the (111) reflection is 8 and of the other two reflections is 24. The scattering length b does not depend on Q, but the intensity decreases because of the Debye–Waller factors. An estimated intensity plot is shown in Fig. S7.

M 6.3 Time-of-Flight Instrumentation

Confirm all information given in Figure 6.12 and the adjacent text.

(a) We start with verifying the wavelength band $0.5\,\text{Å} < \lambda < 5.0\,\text{Å}$ provided by the chopper with a pulse frequency of 50 Hz.

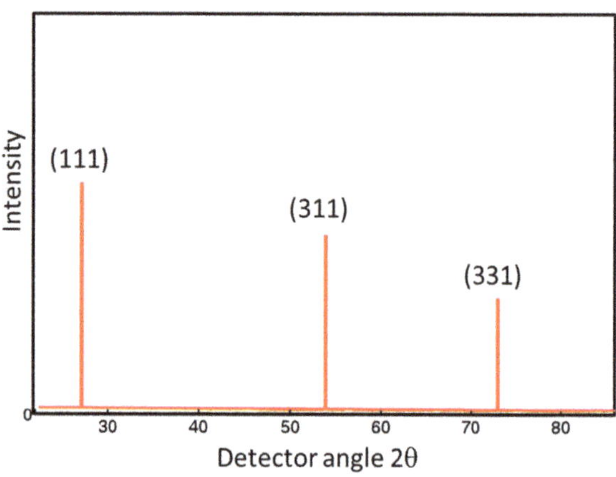

Fig. S7 Expected powder pattern for an NaCl structure with equal modulus but opposite sign of the Na and Cl scattering lengths

The chopper frequency is 50 Hz, meaning that the slowest neutron must travel the distance of 15 m in 20 ms to avoid frame overlap with the fastest neutrons. Therefore, the neutron velocity of the slowest neutron is 750 m/s. From that, we calculate the longest wavelength according to: $\lambda = h/m_{neut}v$, and using the conversion (see Appendix A6):

$$\lambda = \frac{3.958 \; m^2/s}{v_n},$$

the wavelength of the slowest neutrons is 5.27 Å. To be on the safe side, the frame overlap chopper cuts off a bit more to make it 5 Å at the slower end. At the faster end, we can only estimate what neutrons may be available depending on the moderator. A shortest wavelength of 0.5 Å may be reasonable.

(b) Calculate the Δd−band for a detector setting of $2\theta = 10°$.

We only need to use the Bragg equation $d = \lambda/2 \sin \theta$ to obtain for the first setting:

$d = \lambda/2 \sin \theta = 5/2 \sin 5 = 28.7\text{Å};$ the respective Q-value is: $Q = 2\pi/d = 0.22 \, \text{Å}^{-1}$.

for the shorter wavelength: $d = 0.5/2 \sin 5 = 2.87\text{Å}$ and $Q = 2.2 \, \text{Å}^{-1}$. The wavelength band for the low-angle setting stretches from 28.7 to 2.87 Å. Here, $\Delta d = 25.8$ Å and the ΔQ-band is: $\Delta Q = 2 \, \text{Å}^{-1}$.

(c) Repeat the same calculation for a detector setting of $2\theta = 140°$.

$d = 5/2 \sin 70° = 2.66$ Å, $Q = 2.36 \, \text{Å}^{-1}$ and

$d = 0.5/2 \sin 70° = 0.266$ Å, $Q = 23.6 \, \text{Å}^{-1}$. The wavelength band for the high-angle setting stretches from 2.66 to 0.26 Å. Here, $\Delta d = 2.4$ Å and $\Delta Q = 21.2 \, \text{Å}^{-1}$. Note that a large Δd-band is obtained at low angles, and a large ΔQ band is obtained at high angles.

At both detector positions, we use the same wavelength band to probe very different ranges of d-spacings. In the range from 2.66 Å up to 2.87 Å there is some overlap of both detector settings, which is important to confirm the validity of the data.

Chapter 7

M 7.1 Antiferromagnetic multilayer

(a) We consider a very simplified model. Our unit cell is one-dimensional in the z-direction perpendicular to the Fe and Cr layers. We assume that the magnetization is homogeneous within the Fe layers and zero in the Cr-layer. The 1D unit cell comprises two Fe layers and two Cr layers with a lattice parameter of $4d$, and the coordinates are:

Fe layers at coordinates: $z = (0, 2d)$; Cr layers at coordinates: $z = (d, 3d)$.

The nuclear structure factor is then:

$$F_{nuc} = b_{Fe}(\exp(iG \cdot 0) + \exp(iG2d)) + b_{Cr}(\exp(iGd) + \exp(iG3d))$$

With the definition of the reciprocal lattice vector $G = \frac{2\pi}{4d} l = \frac{\pi}{2d} l$ $(l = 1, 2, ..)$ follows for the structure factor:

$$F_{nuc} = b_{Fe}(1 + \exp(i\pi l)) + b_{Cr}\left(\exp\left(i\frac{\pi}{2}l\right) + \exp\left(i\frac{3\pi}{2}l\right)\right)$$

Therefore

$$F_{nuc} = 2(b_{Fe} - b_{Cr}) \text{ for } l = 4n + 2, n \in R$$

$$F_{nuc} = 2(b_{Fe} + b_{Cr}) \text{ for } l = 4n, n \in R$$

$$F_{nuc} = 0 \text{ otherwise}$$

(b) The magnetic structure factor is as follows, where $p = \gamma_n r_0 \frac{g}{2} S(Q)$ is the magnetic scattering length, according to Eq. 7.14:

$$F_{mag} = p_{Fe}(\exp(iG \cdot 0) - \exp(iG2d)) = p_{Fe}(1 - \exp(i\pi l)) = p_{Fe}\left(1 - (-1)^l\right).$$

Therefore:

$$F_{mag} = 2p \text{ for } l \text{ odd}, F_{mag} = 0 \text{ for } l \text{ even}.$$

(c) Using tabulated nuclear scattering lengths ($b_{Fe} = 9.45$ fm; $b_{Cr} = 3.63$ fm) and magnetic scattering length ($p_{Fe} = 4.5$ fm at $Q = 0$), we estimate the Bragg intensities as listed in Table S5.

A schematic plot of the expected Bragg reflections and their intensities from the multilayer is shown in Fig. S8. In red are magnetic reflections. Green and blue peaks refer to the nuclear peaks.

(d) Taking all conditions together, we notice that nuclear Bragg reflections and magnetic Bragg reflections do not overlap. The intensity of the nuclear peaks alternates between low and high. The magnetic Bragg reflections lie in between the nuclear peaks. Their intensity drops off with increasing Q due to the magnetic form factor. The number of repeat units in the multilayer does not enter the equation for the structure factor. However, the number of repeat units determines how narrow Bragg peaks are. With an increasing number of repeat units, the Bragg peaks become narrower, and in a multilayer with an

Table S5 Calculated Bragg intensities for an artificial antiferromagnetic Fe/Cr multilayer

Bragg reflection	Intensity (general)	Intensity in units of fm^2
001	$4p^2$	81
002	$4(b_{Fe} - b_{Cr})^2$	135.5
003	$4p^2$	70
004	$4(b_{Fe} + b_{Cr})^2$	684.3
005	$4p^2$	55
006	$4(b_{Fe} - b_{Cr})^2$	135.5
007	$4p^2$	35

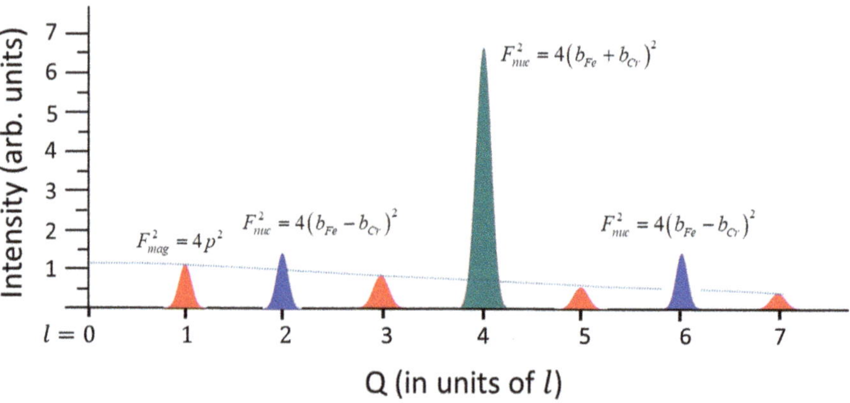

Fig. S8 Schematic plot of the expected (00l) Bragg reflection intensities up to the seventh order

infinite number of repeat units, they would approach a delta function peak shape.

A 7.2 Antiferromagnetic structure of MnO

(a) The chemical unit cell of MnO has a NaCl structure. For NaCl, we have already calculated the structure factor in Exercise 3.2.2. What we need to do is add the magnetic part. If MnO were ferromagnetic with magnetic moments only attached to the Mn^+ ions, the structure factor could be simply written down as:

$$F_{Mn^+}^{chem}(hkl) + F_{Mn^+}^{mag}(hkl) + F_{O^-}^{chem}(hkl) = (f_{Mn} + p_{Mn}) \sum_{i=1}^{4} \exp\left(iG \cdot r_i^{Mn}\right)$$

$$+ f_O \sum_{j=1}^{4} \exp\left(iG \cdot r_j^{O}\right)$$

The selection rules for allowed and forbidden reflections remain the same in the ferromagnetic case. However, MnO is antiferromagnetically ordered with the magnetic moments of the Mn ions alternating up and down. Therefore, the magnetic unit cell doubles compared to the chemical unit cell. If $a_{mag} = 2a_{chem}$, then the basis vectors, taken in units of the doubled unit cell, are listed for the first layer in the sequence indicated by the red circles in Fig. S9:

$$Mn+ : (000); \left(\tfrac{1}{2} 0\, 0\right); \left(\tfrac{1}{4}\, \tfrac{1}{4} 0\right); \left(0\, \tfrac{1}{2}\, 0\right); \left(\tfrac{1}{4}\, \tfrac{3}{4}\, 0\right); \left(\tfrac{1}{2}\, \tfrac{1}{2}\, 0\right); \left(\tfrac{3}{4}\, \tfrac{1}{4}\, 0\right); \left(\tfrac{3}{4}\, \tfrac{3}{4}\, 0\right)$$

Similarly, the 8 O^- ions, belonging to the unit cell in the first layer, are labeled in green and have the coordinates:

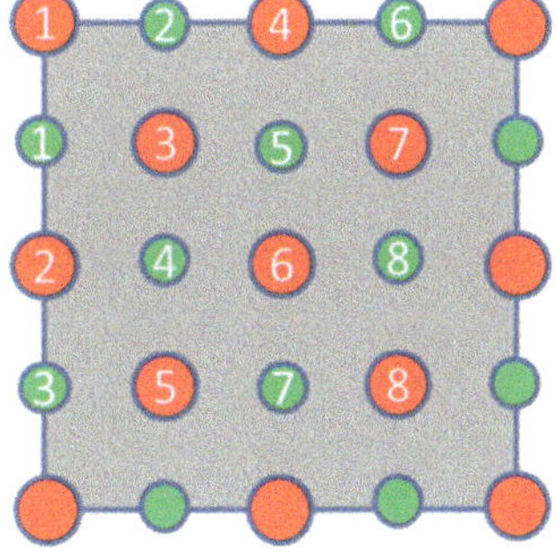

Fig. S9 Enumerating Mn^+-ions (red circles) and O^--ions (green circles) in the first crystal plane

$$O^- : \left(\tfrac{1}{4}00\right); \left(0\tfrac{1}{4}0\right); \left(\tfrac{3}{4}00\right); \left(\tfrac{1}{2}\tfrac{1}{4}0\right); \left(\tfrac{1}{4}\tfrac{1}{2}0\right); \left(0\tfrac{3}{4}0\right); \left(\tfrac{3}{4}\tfrac{1}{2}0\right);$$
$$\left(\tfrac{1}{2}\tfrac{3}{4}0\right)$$

There are 8 Mn^+-ions and 8 O^--ions in the first plane. There are a total of four planes in the magnetic unit cell; see also the schematic drawing of the unit cell within the question section. This makes a total of 32 Mn^+-ions and 32 O^--ions per unit cell.

Calculating the chemical or nuclear structure factor is tedious but straightforward. The result is:

$$F_{nuc}(hkl) = 32(b_{Mn} + b_O) \quad \text{for } h + k + l = 4n \ (n \in R)$$
$$F_{nuc}(hkl) = 32(b_{Mn} - b_O) \quad \text{for } h + k + l = 4n + 2 \ (n \in R)$$
$$0 \qquad\qquad\qquad\qquad\qquad \text{otherwise}$$

Note that for this enlarged unit cell, the nuclear (chemical) (111) reflection is not allowed.

(b) It is much more difficult to calculate the magnetic structure factor because there are myriad spin orientations and spin motifs that could potentially be realized. However, with neutron scattering and using the orientational dependence of magnetic scattering, some preliminary selections can be made. Neutron scattering experiments confirmed that the spins lie in the (111) plane and are stacked antiferromagnetically perpendicular to the [111] direction, see the schematic drawing in Fig. S10.

With this insight, the magnetic structure factor can be calculated as:

Fig. S10 Chemical and magnetic structure of MnO and unit cell lattice parameters

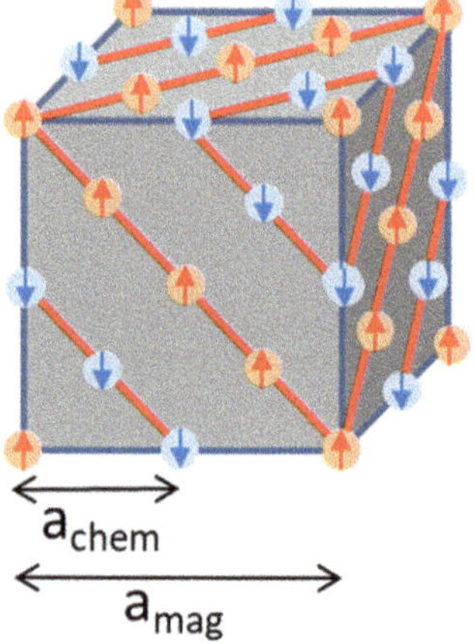

$$F_{mag}(hkl) = 32p_{Mn} \text{ for } (hkl) \text{ all odd and } h+k, \quad h+l, \quad k+l = 4n+2$$

Therefore, the reflection (111) is magnetically allowed, but chemically forbidden. The reflection (311) is chemically and magnetically forbidden. However, the reflection with the indices $(3\overline{1}\overline{1})$ is magnetically allowed.

The diagonal of the magnetic cube has a length $a_{mag}\sqrt{3}$. The (111) reflection has the d-spacing: $d_{111} = a_{mag}/\sqrt{3}$. This is 1/3 the length of the diagonal. Along the diagonal in the [111] direction, there are six of these (111) planes. Therefore, the d-spacing $d_{111} = a_{mag}/\sqrt{3}$ comprises only two of those planes, one for spin-up, the other for spin-down. Constructive interference requires that scattering occurs between these two antiferromagnetically oriented planes.

(c) If we now go back to the original unit cell with half the lattice parameter $a_{chem} = a_{mag}/2$, then the lattice spacing for the antiferromagnetic reflection d_{111} should remain the same. Therefore, we can write:

$$d_{111} = \frac{a_{mag}}{\sqrt{3}} = \frac{a_{mag}}{2\sqrt{3}/2} = \frac{a_{chem}}{\sqrt{\left(\frac{1}{2}\right)^2 + \left(\frac{1}{2}\right)^2 + \left(\frac{1}{2}\right)^2}}.$$

Thus, the (111) reflection of the magnetic unit cell is equivalent to the (½ ½ ½) reflection of the chemical unit cell. The same argument holds for the $(3\overline{1}\overline{1})$ reflection.

M 7.3 Spinstructure of Holmium

First, the single crystal sample of Ho needs to be aligned with respect to the scattering plane of the neutron diffractometer. In particular, the orientation of the c-axis and the basal planes must be identified. Sometimes the crystal orientation is already evident from the crystal shape or crystal habitus, as is the case for quartz crystals. If the crystal shape does not give a clue, it is best to scrutinize the crystal orientation with X-ray methods. This is faster than using neutrons. The standard method is to use a Laue camera in transmission, if the sample is not too thick or in back-reflection for thick and absorbing samples. The respective methods are described in Sect. 5.4. With this preliminary alignment, we can now mount the single crystal on the neutron diffractometer and perform a fine adjustment of the crystal orientation in the neutron beam with the help of a goniometer head.

With the c-axis aligned in the scattering plane and parallel to the scattering vector Q ($Q\|c$), we perform a radial scan from low to high Q-values to cover several (00l) Bragg reflections. This scan is to be performed at room temperature and at 5 K. The simulated result is shown in Fig. S11.

At room temperature (300 K), we observe basically no (001) peak but a strong (002) Bragg reflection. This is to be expected for the hcp structure with an ABA... stacking sequence. The (001) peak is forbidden. If there is some

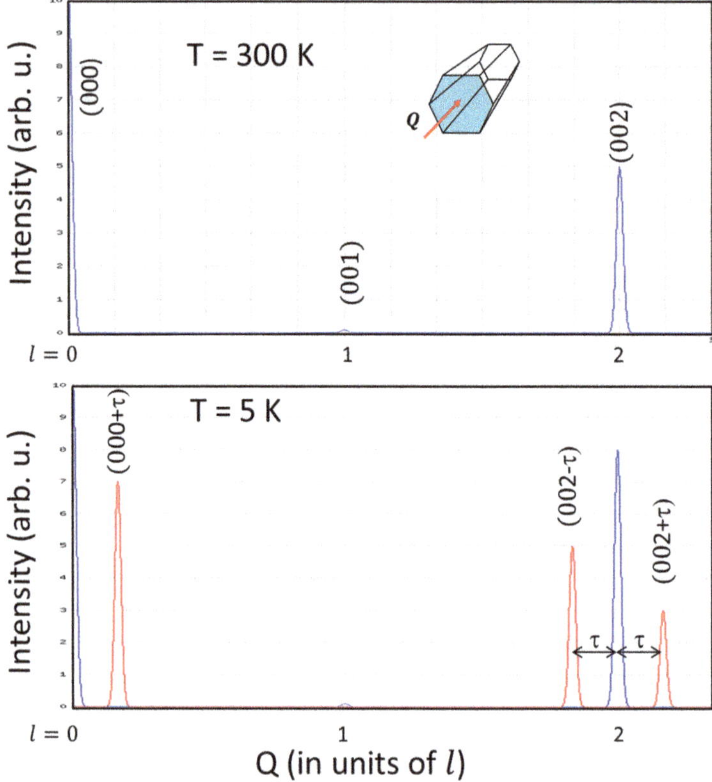

Fig. S11 Diffraction pattern of Ho above and below the Néel temperature.

intensity, it is either due to multiple reflection or to higher harmonics of the neutron wavelength, not properly filtered out by the monochromator and Be filter. The high temperature result thus confirms the hexagonal structure of Ho and the respective c-axis lattice parameter.

At low temperatures (5 K), a rich pattern of Bragg reflections occurs (in red), in addition to the nuclear peaks (in blue). The additional peaks must be of magnetic origin. This can be checked by using the orientational dependence $(1 - (\boldsymbol{Q} \cdot \boldsymbol{M})^2)$ for magnetic neutron scattering. But it can also be argued that any structural phase transition at low temperature would break the hexagonal symmetry of the crystal structure, which in turn would lift the cancellation of the (001) peak intensity. However, the (001) peak remains forbidden even at low temperatures.

Returning to the magnetic peaks. These peaks must originate from magnetic moments that are oriented perpendicular to the scattering vector Q. This means that the Ho magnetic moments must have a component in the hexagonal basal plane. These new reflections appear pairwise to the left and to the right of the fundamental peak position. Furthermore, the distance in reciprocal space between the fundamental peak position and the side peaks is:

$$\tau = \frac{G_{(001)}}{6} = \frac{2\pi}{6c}$$

Therefore, we conclude that the magnetic structure must have a periodicity, which is 6 times as long as the c-axis of the unit cell. This confirms our suspicion of a screw-like periodic structure parallel to the c-axis.

Now, we reorient the sample by 90°, with the c-axis perpendicular to the scattering plane and the basal plane in the scattering plane. We take a radial scan at 77 K along the [h00] direction and stop at the maximum of the (100) in-plane Bragg reflection. Then, with some relaxed detector resolution, we rock the basal plane a few degrees up and down, simulating a (10l)-scan. Then, we notice satellite peaks at the τ-positions on either side of the (100) in-plane peak, labeled (100 $\pm$ τ). The scan performed by Koehler et al. is shown in Fig. S12. From this scan, we can draw several important conclusions:

1. The τ-peaks are of magnetic origin because they exist only at low temperatures.
2. The magnitude of τ in reciprocal space is the same at the (100 $\pm$ τ) and at the (002 $\pm$ τ) positions, describing the same magnetic modulation.
3. Because of the orientational dependence of magnetic neutron scattering expressed by the factor $(1 - (\boldsymbol{Q} \cdot \boldsymbol{M})^2)$, the magnetization vector must be

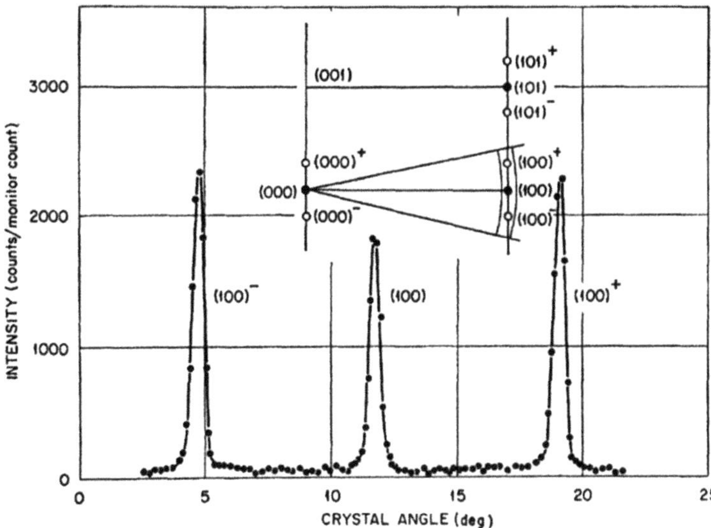

Fig. S12 (10l)-scan (rocking curve) of Ho at 77 K. Two satellite peaks are seen on either side of the hcp (100) peak at (100 $\pm$ τ), which are indicative of the existence of a magnetic spiral propagating in the c-direction. Reproduced from Koehler et al. [7.19] with permission

either in-plane or out-of-plane. The out-of-plane orientation was already excluded by the $(00l \pm \tau)$ scans. Therefore, the magnetic moments must lie in the hexagonal plane.

The magnetic spiral can be described by:

$$m_x(\mathbf{r}_n) = m_0 \cos(\boldsymbol{\tau} \cdot \mathbf{r}_n); \, m_y(\mathbf{r}_n) = m_0 \sin(\boldsymbol{\tau} \cdot \mathbf{r}_n); \, m_z(\mathbf{r}_n) = 0$$

Here, $\mathbf{r}_n$ is the position vector of the magnetic moment in the n-th layer and m_0 is the magnitude of the magnetic moment. The experiments were performed at the high-flux isotope reactor in Oak Ridge, Tennessee.

Chapter 8

M 8.1 Frenkel and Schottky defects

Frenkel defects create vacancies and interstitial ions. Vacancies lead to a shrinkage of the crystal lattice, while interstitial ions lead to lattice expansion. Both effects can compensate for each other. Therefore, it is difficult to reliably identify Frenkel defects through scattering experiments that only measure the lattice parameter.

Schottky defects create vacancies in the lattice and defects that deposit on the surface. The vacancies cause a reduction in the lattice parameter. More importantly, the crystal volume increases due to the surface deposition. Schottky defects are therefore difficult to identify through scattering experiments alone.

Scattering experiments alone cannot clearly distinguish between the two vacancy types. However, in conjunction with additional measurements, this should be possible. A simple method is to measure the density change. Frenkel defects preserve the mass and number density because the number of atoms remains constant in a nearly constant volume. However, Schottky defects reduce the density and simultaneously increase the macroscopic volume.

M 8.2 Huang diffuse scattering

Show that the Fourier transform of $1/r^2$ is proportional to $1/q$ and show that the Huang diffuse scattering is proportional to $1/q^2$.

The Fourier transform is written:

$$FT\left(\frac{1}{r^2}\right) = \int \frac{1}{r^2} e^{i\mathbf{q} \cdot \mathbf{r}} d\mathbf{r}$$

Here, the volume integral is $d\mathbf{r} = d\theta \sin \phi d\phi r^2 dr$. Then, after inserting

$$FT\left(\frac{1}{r^2}\right) = 2\pi \int \frac{1}{r^2} e^{iqr\cos\phi} \sin \phi d\phi \int_0^\infty r^2 dr = 2\pi \iint e^{iqr\cos\phi} \sin \phi d\phi dr$$

Substituting $\cos\phi = x$; $\sin \phi d\phi = -dx$, we find:

$$FT\left(\frac{1}{r^2}\right) = 2\pi \iint\limits_{0,\,-1}^{\infty,\,+1} e^{iqrx}\,dx\,dr = 2\pi \int_0^\infty \left(\frac{e^{iqr} - e^{-iqr}}{iqr}\right)dr,$$

and:

$$2\pi \int_0^\infty \left(\frac{i2\sin qr}{iqr}\right)dr = \frac{4\pi}{q}\int_0^\infty \frac{\sin z}{z}\,dz$$

The last step used the substitution $z = qr, dr = \frac{dz}{q}$. Furthermore, using $\int_0^\infty \frac{\sin z}{z}\,dz = \frac{\pi}{2}$, we finally have:

$$FT\left(\frac{1}{r^2}\right) = \frac{2\pi^2}{q}$$

The diffuse intensity follows from the square of the amplitudes, and therefore, the Huang diffuse scattering is proportional to:

$$I_{diffuse}(q) \propto \frac{1}{q^2}$$

where $q = |\boldsymbol{Q} - \boldsymbol{G}|$. The displacement field around the defect is isotropic for a cubic lattice. However, the diffuse scattering is not isotropic. This is because the intensity is proportional to:

$$I_{diffuse}(\boldsymbol{q}) \propto x_d \Delta v_d \, \frac{\left(\widehat{\boldsymbol{Q}} \cdot \widehat{\boldsymbol{n}}_r\right)^2}{q^2}$$

x_d and Δv_d are the defect concentration and the defect strength, respectively, and $\widehat{\boldsymbol{n}}_r$ is a unit vector in the direction of the displacement field. Since the displacement field is radially isotropic, the dot product causes a dumbbell-like shape of the diffuse scattering at each Bragg point with the highest scattering intensity for $\widehat{\boldsymbol{Q}}_r \parallel \widehat{\boldsymbol{n}}_r$ and zero intensity for $\widehat{\boldsymbol{Q}}_r \perp \widehat{\boldsymbol{n}}_r$, as indicated in Fig. S13. The color code is from red to blue, from high to low diffuse intensity.

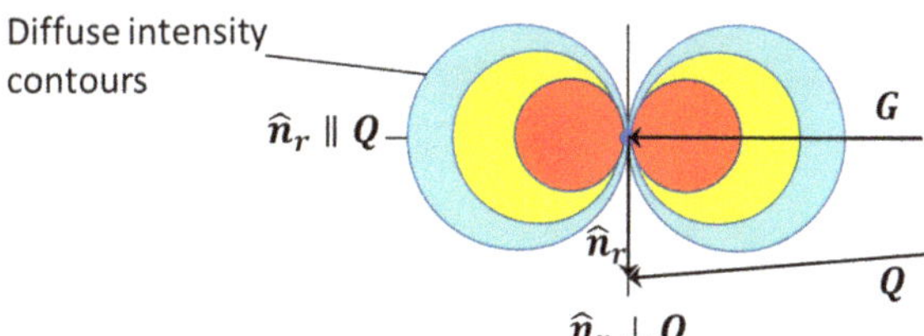

Fig. S13 Contour map of Huang diffuse scattering, evoked by point defects in a crystal lattice

E 8.3 Form factor of vacancies

(a) Discuss what kind of effects on the scattering amplitude you expect for vacancies.

    Aside from diffuse scattering in the tails (Huang diffuse scattering), the shape of the scattering amplitude and the intensity should be unaffected due to the presence of vacancies. For very high vacancy concentrations, the long-range order may be disturbed. The scattering amplitude, though, is expected to decrease because with increasing vacancy concentration, the number of scattering centers decreases.

(b) Determine the vacancy atomic form factor, respectively, the scattering length, and provide arguments for your results.

    The vacancy atomic form factor is the negative form factor of the atom A, as A is removed from the lattice site:

$$f_{vac}(Q) = -f_A(Q)$$

We may treat vacancies as an alloy component in a metal, or more generally, in a crystal lattice. Then, for a random distribution of vacancies with a concentration x_{vac}, we find the average form factor:

$$\langle f(Q)\rangle = (1 - x_{vac})f_A(Q) + x_{vac}f_{vac}(Q) = (1 - x_{vac})f_A(Q) - x_{vac}f_A(Q)$$

Therefore, the average form factor in the presence of random and uncorrelated vacancies is:

$$\langle f(Q)\rangle = f_A(Q) - 2x_{vac}f_A(Q)$$

8.4 Thermal vibration

According to Eq. 7.8, the intensity of the superstructure Bragg reflection is expressed as:

$$I_{superst.}(Q) = \left(\frac{a-b}{2}\right)^2 \sum_{m,n}^{N} \langle P_m P_n\rangle \exp(i(Q - \tau)\cdot R_{m,n})$$

Thermal vibrations and the Debye–Waller factor are connected to the atoms at sites m and n with form factors a and b, respectively. Assuming that the DW factors are identical for all a atoms and for all b atoms, we can simply write:

$$I_{superst.}(\boldsymbol{Q}) = \left(\frac{ae^{-W_a} - be^{-W_b}}{2}\right)^2 \sum_{m,n}^{N} \langle P_m P_n \rangle \exp(i(\boldsymbol{Q} - \boldsymbol{\tau}) \cdot \boldsymbol{R}_{m,n})$$

8.5 Susceptibility

The paramagnetic susceptibility χ_0 is expressed as the partial derivative of the order parameter (here the magnetization) with respect to the externally applied conjugated field (here the Oersted field H):

$$\chi_0 = \frac{\partial M}{\partial H}$$

In magnetism, χ_0 is known as the Curie susceptibility. In the expression for χ_0, the wavenumber dependence is not explicitly expressed, but it is generally assumed that in the laboratory, the macroscopic susceptibility is taken at $Q = 0$. For studying the order–disorder transition via a scattering experiment, the susceptibility can also be determined for $Q \geq 0$:

$$I_{critical\ scat.}(\boldsymbol{q}) = g(\boldsymbol{q}) = k_B T \chi(\boldsymbol{q}),$$

where $\chi(\boldsymbol{q})$ is known as the generalized susceptibility. $\chi(\boldsymbol{q})$ takes the form already discussed in the main text. The macroscopic susceptibility and the generalized susceptibility are related via:

$$\chi(\boldsymbol{q}) = \frac{\chi_0}{1 - \chi_0 J(\boldsymbol{q})}$$

where $J(\boldsymbol{q})$ is the Fourier transform of the pair potential.

8.6 Commensurate-incommensurate structures

In the present context, commensurability is defined as a rational (commensurate) or irrational (incommensurate) relation of one lattice periodicity in relation to the other. In our case, we can express the relationship via:

$$c = pd$$

p is rational for commensurate structures and irrational for incommensurate structures. Therefore, in panel (b), the two periodicities shown are incommensurate. In panel (a) we observe a superstructure with alternating short–long distances of the blue "atoms". Here, the relationship is rational: $c' = 4d$.

In nature, both cases are realized, depending on the strength of the competing interaction potentials: the sinusoidally varying substrate potential and the interaction between the "blue" atoms, represented by springs.

Chapter 9

M 9.1 Particle shape factor

The procedures for solving this problem are, in fact, the same as those for the atomic form factor of the hydrogen atom, assuming a constant density up to a radius R_p and zero beyond. This problem is discussed in Appendix A3, and we can take over the results from there.

First, we write the volume integral in spherical coordinates:

$$\int dV_p = \int_{-\pi}^{+\pi} d\theta \int_0^{\pi} \sin\phi \, d\phi \int_0^{R_p} R^2 \, dR.$$

Then, the particle shape factor becomes:

$$P(Q) = \frac{1}{V_p} \int\limits_{0;\,\phi=0}^{R_p;\,\pi} \int_{\theta=-\pi}^{+\pi} e^{iQR\cos\phi} R^2 \sin\phi \, d\phi \, d\theta \, dR = \frac{2\pi}{V_p} \int\limits_{0;\,0}^{R_p;\,\pi} e^{iQR\cos\phi} R^2 \sin\phi \, d\phi \, dR.$$

Using the substitution $\cos\phi = x$ and $\sin\phi \, d\phi = -\, dx$, we find for the angular part:

$$-\frac{2\pi}{V_p} \int_1^{-1} e^{iQRx} dx = \frac{2\pi}{V_p} \left(\frac{e^{iQR} - e^{-iQR}}{iQR} \right) = 4\pi \frac{\sin QR}{QR}.$$

Now, we return to the radial part:

$$P(Q) = \frac{4\pi}{V_p} \int_0^{R_p} R^2 \frac{Sin(QR)}{QR} dR = \frac{4\pi}{V_p QR_p} R_p^3 \left(\frac{\sin(QR_p) - (QR_p)\cos(QR_p)}{(QR_p)^2} \right)$$

$$= \frac{3}{QR_p} J_1(QR_p)$$

With the Bessel function of the first kind:

$$J_1(QR_p) = \left(\frac{\sin(QR_p) - (QR_p)\cos(QR_p)}{(QR_p)^2} \right)$$

A 9.2. Particle shape factor in the Guinier approximation

Consider the expansion:

$$P(\boldsymbol{Q}) = \frac{1}{V_p} \int_{V_p} e^{i\boldsymbol{Q}\cdot\boldsymbol{R}} dV_p = \frac{1}{V_p}\left[V_p + i\boldsymbol{Q}\cdot\int_{V_p}\boldsymbol{R}dV_p - \frac{1}{2}\int_{V_p}(\boldsymbol{Q}\cdot\boldsymbol{R})^2 dV_p + \ldots\right]$$

The second term drops out because for a random orientation of particles, the scalar product $\boldsymbol{Q}\cdot\int\boldsymbol{R}d\boldsymbol{R}$ is zero. Terms higher than $(QR)^2$ do not need to be considered because of the condition $QR\ll1$. Then, we find:

$$P(\boldsymbol{Q}) = \left[1 - \frac{1}{2V_p}\int_{V_p}(\boldsymbol{Q}\cdot\boldsymbol{R})^2 dV_p + \ldots\right] = \left[1 - \frac{Q^2}{2V_p}\int_{V_p}R^2 dV_p + \ldots\right]$$

Here, the integral

$$\frac{1}{V_p}\int_{V_p}R^2 dV_p = \frac{3}{4\pi}\frac{4\pi}{5}R_p^2 = \frac{3}{5}R_p^2 = R_G^2$$

is the definition for the radius of gyration R_G, and R_p is the radius of the particle in question. Now, we consider that the particle shape factor takes an ensemble average in only one direction parallel to Q. Therefore, a weight factor of 1/3 has to be taken into account. Then, follows:

$$P(\boldsymbol{Q}) = \left[1 - \frac{Q^2}{2V_p}\int_{V_p}R^2 dV_p + \ldots\right] = \left[1 - \frac{1}{2}\left(\frac{1}{3}Q^2 R_G^2\right) + \ldots\right] = \exp\left[-\frac{1}{2}\left(\frac{1}{3}Q^2 R_G^2\right)\right].$$

By squaring, we get:

$$|P(\boldsymbol{Q})|^2 = \exp\left(-\frac{1}{3}(QR_G)^2\right)$$

as required.

M 9.3 Gaussian smearing of dips

(a) The scattering curve for particle size R $=$ 0.95 nm, 1.0 nm, and 1.05 nm is plotted in Fig. S14 in the left panel. This corresponds to the requested 1σ - deviation from the mean radius. The plots are shown separately in the first graph on the left. Each scattering curve for one particular radius R has the dips shifted with respect to the other. So, having different radii present in the sample will smear out the sharp dips. This can be seen clearly by summing up the three scattering curves, shown in the second graph to the right by the black line. The dips become more shallow and rounded, corresponding to the variation in nanoparticle radii. The summation of the three curves is equivalent to an

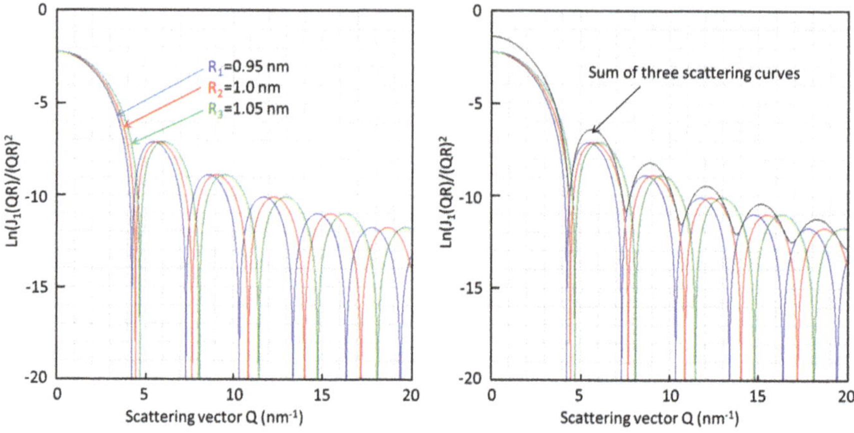

Fig. S14 SAS intensity for different spherical particle sizes in dilute solution

incoherent superposition of scattering intensities, which excludes any correlation effects among the particles. A correlation effect would occur if we had a sequence of alternating small and large-sized particles, as this additionally introduced periodicity, leading to an additional structure in the small-angle scattering curve.

(b) The dips occur at Q-positions, for which $\sin(QR_p) - (QR_p)\cos(QR_p) = 0$ or

$$\sin(QR_p) = (QR_p)\cos(QR_p).$$

This is equivalent to the condition:

$$Q = \frac{1}{R_p}\tan(QR_p),$$

which defines the first minimum at $QR_p = 4.493$. This condition is periodically fulfilled for increasing Q, which is the origin of the periodic occurrence of the dips. At the same time, the overall small-angle scattering amplitude drops off with Q^{-4}, according to Eq. (9.22).

(c) The condition for the dips can also be used to estimate the dip width in dependence on the variation in particle radius. We take the first derivative (dropping the index p for convenience):

$$\frac{dQ}{dR} = \frac{d}{dR}\left(\frac{1}{R}\tan(QR)\right) = -\frac{1}{R^2}\tan(QR) + \frac{1}{R}\frac{Q}{\cos^2(QR)}$$

From this follows, the relative spread dQ/Q:

$$\frac{dQ}{Q} = \frac{R\,dQ}{\tan(QR)} = -\frac{dR}{R} + \frac{Q\,dR}{\cos(QR)\sin(QR)}$$

So, the leading term of the relative error is: $\left|\frac{dQ}{Q}\right| = \left|\frac{dR}{R}\right|$, telling us that the relative spread in Q is proportional to the relative spread in R. The second term indicates that $\left|\frac{dQ}{Q}\right|$ increases linearly with Q. However, any further general statements are difficult to make because of the oscillatory behavior of the denominator $\cos(QR)\sin(QR)$.

Chapter 10

M10.1 Show that the Eqs. (10.4) and (10.5) for CTR scattering are correct

(a) Equation (10.3) was already derived in Chap. 3.3, in the limit of $N \to \infty$, $\lim_{N\to\infty} \left(\sin^2(Nx/2)\right) = 1/2$. This can be shown as follows.

The limits of the oscillatory functions $\lim_{N\to\infty} \sin Nx = \lim_{N\to\infty} \cos Nx = 0$.

Then, using the relation:

$$\sin^2(Nx/2) = \frac{1 - \cos Nx}{2}$$

With the limit:

$$\lim_{N\to\infty} \sin^2(Nx/2) = \lim_{N\to\infty} \frac{1 - \cos Nx}{2} = \frac{1}{2}$$

Therefore, this expression approaches:

$$\lim_{N\to\infty} \frac{\sin^2 Nx/2}{\sin^2 x/2} = \frac{1}{2\sin^2 x/2}$$

which confirms Eq. (10.4). Since $\sin^2 x/2 = |e^{ix} - 1|^2$, we conclude:

$$I_{CTR}(Q) \propto \left|\sum_{w=0}^{\infty} e^{iwx}\right|^2 = \frac{1}{2\sin^2 x/2} = \left|\frac{1}{e^{ix} - 1}\right|^2$$

which is equivalent to Eq. (10.5).

E10.2 Determine the critical angle for total reflection for X-rays and neutrons, both with a wavelength of 0.1 nm scattered at a flat silicon surface

(a) The critical angle for X-ray total reflection is approximately determined by:

$$Q_c = 2k_0 \sin \alpha_c \cong 2k_0\alpha_c = \frac{4\pi}{\lambda}\alpha_c = \sqrt{16\pi r_0 \rho_e}$$

Therefore, the critical angle is given by: $\alpha_c = \frac{\lambda}{4\pi}\sqrt{16\pi r_0 \rho_e}$

Silicon has eight atoms per unit cell, and each atom has 14 electrons in a cube with a lattice constant of 5.41×10^{-10} m. Then, the electron density is $\rho_e = 7 \times 10^{29}$ m^{-3}. $r_0 = 2.82 \times 10^{-15}$ m. Taken all together: $\alpha_c = \frac{1 \times 10^{-10}}{4\pi} \times \sqrt{16\pi \times 2.82 \times 10^{-15}\text{m} \times 7 \times 10^{29}\text{m}^{-3}} = 0.0025 \text{ rad} = 0.144\,°$

(b) In the case of neutrons, we replace the electron density with the atomic number density and the classical electron radius with the coherent scattering length:

$$\rho_a = 5 \times 10^{28}\,\text{m}^{-3}, \, b_{coh}^{Si} = 4.1 \times 10^{-15}\,\text{m}.$$

Then, the critical angle for neutron reflection is:

$$\alpha_c = \frac{1 \times 10^{-10}}{4\pi}\sqrt{16\pi \times 4.1 \times 10^{-15}\text{m} \times 5 \times 10^{28}\text{m}^{-3}} = 0.0008 \text{ rad} = 0.046\,°.$$

Therefore, the critical angle for neutron total reflection is by a factor of 3 smaller than the one for X-ray reflection. This corresponds roughly to the square root ratio of electron density to atomic number density.

E10.3 Show that Eq. (10.21) can also be derived from the condition that at Q_c, the kinetic energy of the transmitted wave E_z^t should be zero.

According to Eq. (10.18), the energy of the transmitted wave is:

$$E_z^t = E_z^i - \rho_A V_N = \frac{\hbar^2 k_0^2}{2m_n}\left(1 - \frac{4\pi b_{coh}\rho_A}{k_0^2}\right)$$

If we set $E_z^t = 0$, then the critical wavenumber for total reflection is $k_{0,c}^2 = 4\pi b_{coh}\rho_A$

The critical scattering vector is $Q_c^2 = (2k_{0,c})^2 = 4 \times 4\pi b_{coh}\rho_A = 16\pi b_{coh}\rho_A$, which was to be shown.

E10.4 Determine the conditions for the Yoneda wings

First, we concentrate on the scattering triangle drawn in black with the scattering vector Q parallel to the z-axis, normal to the surface. α is the glancing incident angle. As usual, the wavenumbers are k_0, and the scattering vector is

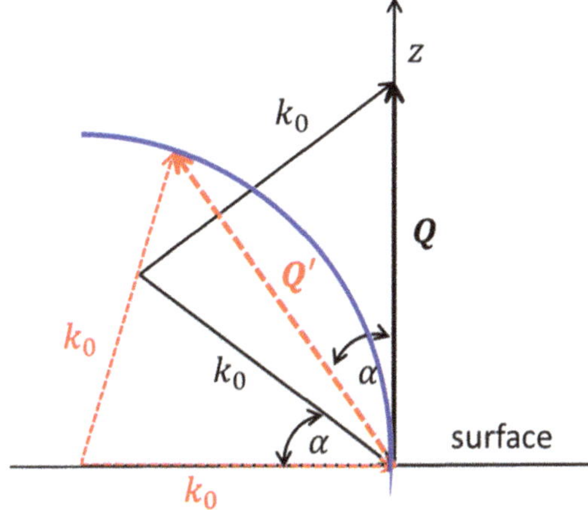

Fig. S15 Construction of the Yoneda wing

$Q = 2k_0 \sin \alpha$. We can tilt the scattering triangle until it touches the horizon of the sample surface. The tilted triangle is shown in red. The maximum tilt angle corresponds to α, or more precisely to $\alpha - \alpha_c$. If we neglect the small deviation due to the critical angle, the location of the points where the tilted scattering triangle touches the surface lies on the circular segment with the radius k_0, shown in blue. This is the Yoneda wing. Now we can repeat the same procedure on the other side and obtain the second Yoneda wing (Fig. S15).

E10.5 Calculate and plot the Fresnel transmission factor

$$t_{i,f} = \frac{E_t}{E_i} = \frac{2 \sin \alpha_{i,f}}{\sin + \sin \alpha_t} = \frac{2 \sin \alpha_{i,f}}{\sin \alpha_{i,f} + \sqrt{\sin^2 \alpha_{i,f} - 2\delta}} = \frac{2 \sin \alpha_{i,f}}{\sin \alpha_{i,f} + \sqrt{\sin^2 \alpha_{i,f} - \alpha_c^2}}$$

For $\alpha_{i,f} < \alpha_c$, the expression becomes imaginary. Therefore, we concentrate on angles above the critical region and make a linear approximation for $\alpha_{i,f} < \alpha_c$. The graph is plotted with the MAFA function plotter (https://www.mathe-fa.de/en#result) (Fig. S16).

E10.5 Show that in the small-angle approximation, the Fresnel reflectivity scales with Q_z^{-4}, according to Eq. (10.25)

$$R_F(Q_z) = r^2 = \left(\frac{Q_z - Q_z^t}{Q_z + Q_z^t}\right)^2 \propto \left(\frac{Q_c}{Q_z}\right)^4$$

We use a short notation: $Q_z = z; Q_z^t = t; Q_c = c$, and the relation: $t^2 = z^2 - c^2$. Then:

Fig. S16 Fresnel transmission factor plotted as a function of the incident angle α_i and normalized by the critical angle α_c

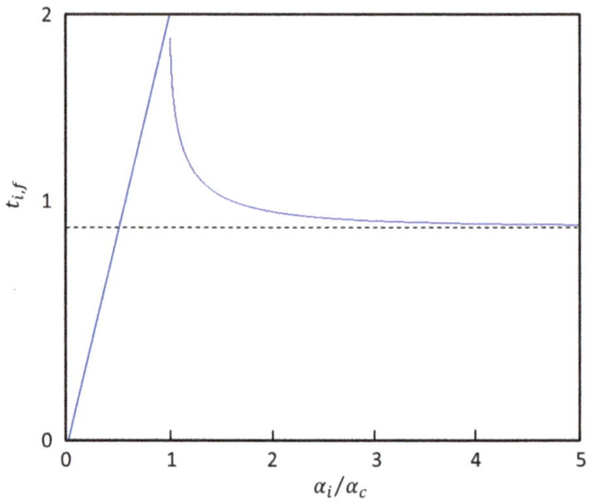

$$R_F(z) = \left(\frac{z-t}{z+t}\right)^2 \cong \left(\frac{z-\sqrt{z^2-c^2}}{2z}\right)^2 = \frac{z}{z}\left(\frac{1-\sqrt{1-\left(\frac{c}{z}\right)^2}}{2}\right)^2$$

$$\cong \frac{1}{4}\left(1-\left(1-\frac{1}{2}\left(\frac{c}{z}\right)^2\right)\right)^2 = \frac{1}{4}\left(\frac{1}{2}\left(\frac{c}{z}\right)^2\right)^2 = \left(\frac{c}{2z}\right)^4 = \left(\frac{Q_c}{2Q_z}\right)^4,$$

which was to be shown. The first approximation $Q_z \cong Q_z^t$ is justified for scattering vectors higher than the critical scattering vector, and the second approximation $Q_c/Q_z \ll 1$ is justified for scattering vectors above the critical edge.

Chapter 11

M11.1 Multilayer with missing peaks

Show that multilayers with equal layer thicknesses $d_1 = d_2$ and N repeats but different SLDs $\rho_1 b_1 \neq \rho_2 b_2$ exhibit no even-order Bragg reflections. Assume that all properties are homogeneous across the layers and that the interfaces are sharp without roughness (Fig. S17).

We calculate the structure factor of a homogeneous double layer by taking the Fourier transform of the combined slab with thickness $= d_1 + d_2$:

$$F(Q) = \int_0^L \rho(z)b(z)e^{iQz}dz = \int_0^{d_1} \rho_1 b_1 e^{iQz}dz + \int_{d_1}^{d_2} \rho_2 b_2 e^{iQz}dz$$

Since the SLDs are constant within the layers, we can simplify:

Fig. S17 Multilayer with defining thicknesses, periodicity, and scattering length density

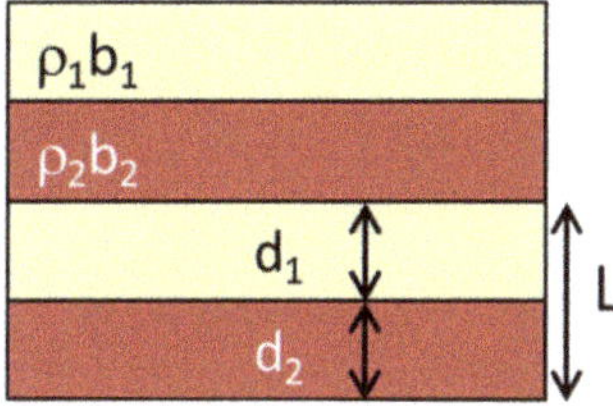

$$F(Q)=\rho_1 b_1 \int_0^{d_1} e^{iQz}dz + \rho_2 b_2 \int_{d_1}^{d_2} e^{iQz}dz = \frac{1}{iQ}\left[\rho_1 b_1\left(e^{iQd_1}-1\right)+\rho_2 b_2 e^{iQd_1}\left(e^{iQd_2}-1\right)\right]$$

$$=\frac{1}{iQ}\left[\rho_1 b_1 e^{iQd_1/2}\left(e^{iQd_1/2}-e^{-iQd_1/2}\right)+\rho_2 b_2 e^{iQd_1}e^{iQd_2/2}\left(e^{iQd_2/2}-e^{-iQd_2/2}\right)\right]$$

$$=\frac{2}{Q}\left[\rho_1 b_1 e^{iQd_1/2}\left(\sin\frac{Qd_1}{2}\right)+\rho_2 b_2 e^{iQd_1}e^{iQd_2/2}\left(\sin\frac{Qd_2}{2}\right)\right]$$

Using $d_1 = d_2 = d$, we find:

$$F(Q)=\frac{2}{Q}\left[\rho_1 b_1 e^{iQd/2}\left(\sin\frac{Qd}{2}\right)+\rho_2 b_2 e^{iQd}e^{iQd/2}\left(\sin\frac{Qd}{2}\right)\right]=\frac{2}{Q}$$

$$\times \left(\sin\frac{Qd}{2}\right)e^{iQd/2}\left[\rho_1 b_1 + \rho_2 b_2 e^{iQd}\right]$$

The factor $e^{iQd/2}$ will be 1 by squaring with the complex conjugate term. For the squared structure factor, we obtain:

$$F(Q)F^*(Q)=\frac{4}{Q^2}\left(\sin\frac{Qd}{2}\right)^2\left[\rho_1 b_1 + \rho_2 b_2 e^{iQd}\right]^2$$

$$=\frac{4}{Q^2}\left(\sin\frac{Qd}{2}\right)^2\left[\rho_1 b_1 + \rho_2 b_2 e^{iQd}\right]\left[\rho_1 b_1 + \rho_2 b_2 e^{-iQd}\right]$$

$$=\frac{4}{Q^2}\left(\sin\frac{Qd}{2}\right)^2\left[(\rho_1 b_1)^2 + (\rho_2 b_2)^2 + 2\rho_1 b_1 \rho_2 b_2 \cos Qd\right]$$

The reciprocal lattice vector for the periodic structure is:

$$G=l\frac{2\pi}{L}=l\frac{2\pi}{2d},$$

where l is the Miller index in the z-direction. Evaluating the structure factor for $l = 1,2,\ldots$, we find:

$$F^2\left(l\frac{2\pi}{2d}\right) = \frac{4d^2}{\pi^2}\left(\sin l\frac{\pi}{2}\right)^2\left[(\rho_1 b_1)^2 + (\rho_2 b_2)^2 + 2\rho_1 b_1\rho_2 b_2\cos l\pi\right]$$

For $l = 1, 3\ldots$ odd:

$$F^2\left(l\frac{2\pi}{2d}\right) = \frac{4d^2}{\pi^2}\left[(\rho_1 b_1)^2 + (\rho_2 b_2)^2 - 2\rho_1 b_1\rho_2 b_2\right]$$

which is:

$$F^2\left(l\frac{2\pi}{2d}\right) = \frac{4d^2}{\pi^2}\left[(\rho_1 b_1) - (\rho_2 b_2)\right]^2$$

For $l = 2, 4\ldots$ even:

$$F^2\left(l\frac{2\pi}{2d}\right) = 0 = \frac{4d^2}{\pi^2}\left(\sin n\pi\right)^2\left[(\rho_1 b_1) + (\rho_2 b_2)\right]^2$$

Hence, for the allowed Bragg reflections, the intensity is proportional to the difference in the scattering length densities. The forbidden Bragg reflections are all those with an even-order Miller index, which was to be shown.

Next, we determine the lattice sum of the multilayer by summing over all N periods. According to (3.22), the lattice sum is:

$$A(\mathbf{Q}) = \sum_{n=0}^{N-1} e^{i\mathbf{Q}\cdot nL} = \frac{e^{iN\mathbf{Q}\cdot L} - 1}{e^{i\mathbf{Q}\cdot L} - 1},$$

where N is the number of double layers. This yields the intensity:

$$I(Q) = A(\mathbf{Q})A^*(\mathbf{Q})F(Q)F^*(Q)$$

$$= \frac{L^2}{\pi^2}\frac{\sin^2 N\mathbf{Q}\cdot L/2}{\sin^2 \mathbf{Q}\cdot L/2}\frac{4}{Q^2}\left(\sin\frac{QL}{4}\right)^2\left[(\rho_1 b_1)^2 + (\rho_2 b_2)^2 + 2\rho_1 b_1\rho_2 b_2\cos\left(\frac{QL}{2}\right)\right]$$

Using the following parameters: $N = 10$; $L=10$ nm, $\rho_1 b_1 = 9$ nm^{-2}; $\rho_2 b_2 = 3$ nm^{-2}, the calculated intensity is plotted in Fig. S18, from which one can recognize that the even-order multilayer Bragg reflections are missing (left side). Finally, we overlay the multilayer reflections with the reflectivity intensity in the high-angle approximation: $1/Q^4$, which is also shown in the figure on the right side.

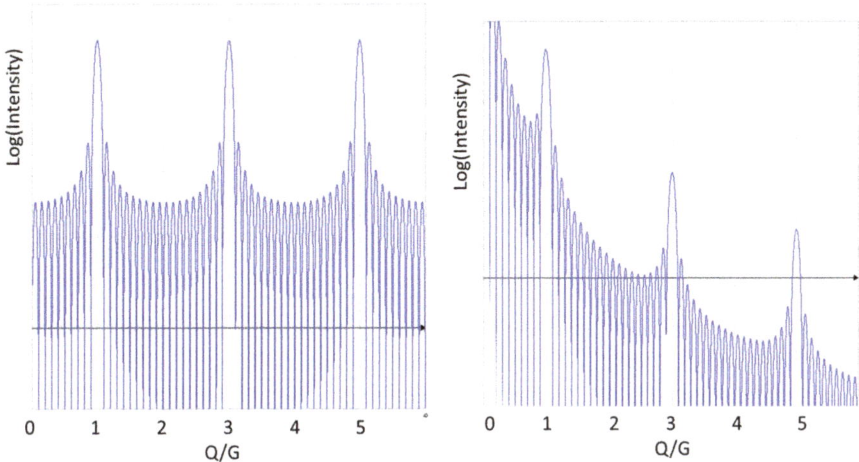

Fig. S18 Calculated diffraction pattern for a multilayer with equal sublayer thickness. Left: low Q reflectivity regime. Right: high Q regime

Chapter 12

E12.1 Close Packing of Nanoparticles

(a) What is the lattice parameter in the diffraction pattern of Fig. 12.9, assuming an hcp in-plane structure?

The lattice spacing for the in-plane hcp structure is:

$$d_{hk} = \frac{a\sqrt{3}/2}{\sqrt{h^2 + hk + k^2}}$$

Since the reciprocal lattice vector for the hcp structure is:

$$G_{hk} = \frac{2\pi}{d_{hk}} = \frac{4\pi}{a\sqrt{3}}\sqrt{h^2 + hk + k^2}$$

Therefore, it follows for the lattice parameter:

$$a_{hcp} = \frac{4\pi}{G_{hk}\sqrt{3}}\sqrt{h^2 + hk + k^2}$$

From the map in Fig. 12.9, we find the following values:

$$G_{10} = 0.36 \ nm^{-1}; \quad \Rightarrow a_{10} = 4\pi/G_{10}\sqrt{3} = 20.15 \ \text{nm}$$
$$G_{11} = 0.611 \ nm^{-1}; \quad \Rightarrow a_{11} = 4\pi/G_{11} = 20.566 \ \text{nm}$$
$$G_{20} = 0.72 \ nm^{-1}; \quad \Rightarrow a_{20} = 8\pi/G_{20} = 20.15 \ \text{nm}$$
$$G_{21} = 0.972 \ nm^{-1}; \quad \Rightarrow a_{21} = 4\pi\sqrt{7}/G_{21}\sqrt{3} = 19.73 \ \text{nm}$$

The average lattice parameter of 20.15 nm is close to the quoted value of 20.38 nm. Deviations are due to the difficulty in reading the scale accurately.

(b) Confirm that the NP packing forms an hcp structure, but not an sc structure.

Assuming an hcp structure, we find a constant lattice parameter for all reflections. Testing whether the close-packed structure could be a simple cubic structure, we perform a similar calculation, now with the cubic d-spacing:

$$d_{sc} = a/\sqrt{h^2 + k^2}$$

and

$$a = (2\pi/G_{sc})\sqrt{h^2 + k^2}$$

The corresponding values are:

$$G_{10} = 0.36 \ nm^{-1}; \quad \Rightarrow a_{10} = 2\pi/G_{10} = 17.45 \ \text{nm}$$
$$G_{11} = 0.611 \ nm^{-1}; \quad \Rightarrow a_{11} = 2\pi\sqrt{2}/G_{11} = 14.54 \ \text{nm}$$
$$G_{20} = 0.72 \ nm^{-1}; \quad \Rightarrow a_{20} = 4\pi/G_{20} = 17.45 \ \text{nm}$$
$$G_{21} = 0.972 \ nm^{-1}; \quad \Rightarrow a_{21} = 2\pi\sqrt{5}/G_{21} = 14.45 \ \text{nm}$$

Assuming an sc lattice, we notice that the lattice parameter is not constant but varies from the (hk) value to the next. This violates the constancy of crystal lattice parameters. Contrarily, with the assumption of an hcp structure, we find a constant lattice parameter. Therefore, sc is the wrong lattice type.

E12.2 POL-SANS scattering geometry

Assume a Helmholtz coil providing an in-plane magnetic field B that saturates the magnetization of the ferromagnetic sample in the field direction. What is the expected scattering pattern at a constant scattering vector Q, for variable azimuthal angle θ, and for perpendicular up or down spin polarization? (Fig. S19)

Fig. S19 Geometry for polarized small-angle neutron scattering. The neutron polarizatioin is perpendicular to the in-plane induction B. The scattering pattern consists of a ring with constant radius Q and homogeneous intensity distribution on the ring, which is due to spin-flip scattering at magnetic nanoparticles.

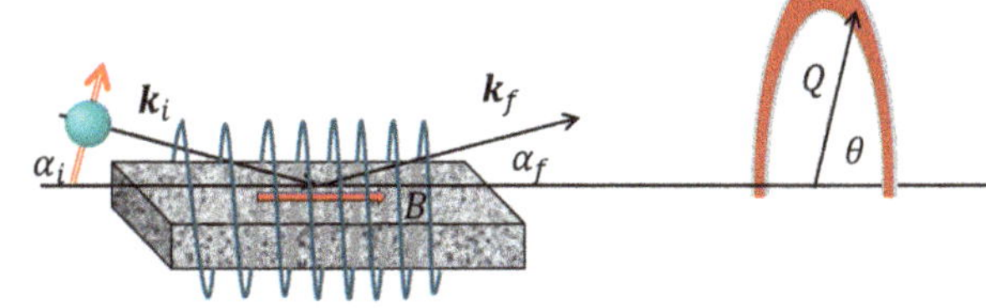

In this geometry, the neutron spin is oriented perpendicular to the magnetic induction B and the sample magnetization M. Therefore, the scattering observed is due to spin-flip scattering. The spin-flip scattering is the same for an up- and down-polarized incident neutron beam and also the same for any orientation around k_i. The scattering pattern, therefore, is radially isotropic but modified for θ close to the critical angle α_c. The lower part of the SANS pattern is weakened or missing because of absorption in the sample and/or the substrate.

Recommendations for Further Reading

X-ray scattering

Jens Als-Nielsen and Des McMorrow, *Elements of Modern X-ray Physics*, 2nd edition, Wiley, 2011
B.E. Warren, *X-ray Diffraction*, Dover Books in Physics, 1990
B.D. Cullity, *Elements of X-ray Diffraction*, Prentice Hall, 2001
Guinier, *X-ray Diffraction*: *In Crystals, Imperfect Crystals, and Amorphous Bodies*, Dover Books in Physics, 1994.
H. P. Klug und L. E. Alexander: *X-ray Diffraction Procedures for Polycrystalline and Amorphous Materials*, 2. Auflage. John Wiley & Sons, New York-Sydney-Toronto 1974
Introduction to X-ray analysis: https://www.iucr.org/education/pamphlets/2/full-text
X-ray Data Booklet (Center for X-ray Optics and Advanced Light Source Lawrence Berkeley National Laboratory). At: http://xdb.lbl.gov/

Neutron scattering

G.L. Squire, *Thermal Neutron Scattering*, Cambridge University Press, third edition, 2012, and Dover edition, 1997.
Stephen W. Lovesey, Theory of Neutron Scattering From Condensed Matter, Volume 1: Nuclear Scattering, Volume 2: Polarization Effects and Magnetic Scattering, Clarendon Press, Oxford, 1984.
Albert Furrer, Joel Mesot, and Thierry Strässle, *Neutron Scattering in Condensed Matter Physics*, World Scientific, 2009.
Fernandez-Alonso and D. L. Price, ed. *Neutron Scattering Fundamentals*, Academic Press—Elsevier, 2013 (Vol 44); Neutron Scattering—Magnetic and Quantum Phenomena, 2015 (Vol 48); *Neutron Scattering Applications*, 2017 (Vol 49).
T. Brückel, G. Heger, D. Richter and R. Zorn (Editors), Laboratory Course Neutron Scattering, Schriften des Forschungszentrums Jülich Reihe Materie und Material / Matter and Materials Band / Volume 38, published 2013 download at: http://

www.iaea.org/inis/collection/NCLCollectionStore/_Public/38/099/38099821.
pdf

X-ray and Neutron Scattering

Ivan Zaluzhnyy and Frank Schreiber, *X-ray and Neutron Scattering*, https://www.
soft-matter.uni-tuebingen.de/teaching/scattering_lecture/LectureNotes.pdf
D.S. Sivia, *Elementary Scattering Theory: For X-ray And Neutron Users,* Oxford
University Press, 2011.

Electron Scattering

Ayahiko Ichimiya and Philip I. Cohen, Reflection High-Energy Electron Diffraction,
Cambridge University Press, 2nd Edition, 2011
Georg Held, Low Energy Electron Diffraction, Crystallography of Surfaces and
Interfces Bunsen-Magazin, page 124, April 2010.

GPSR Compliance
The European Union's (EU) General Product Safety Regulation (GPSR) is a set
of rules that requires consumer products to be safe and our obligations to
ensure this.

If you have any concerns about our products, you can contact us on

ProductSafety@springernature.com

In case Publisher is established outside the EU, the EU authorized
representative is:

Springer Nature Customer Service Center GmbH
Europaplatz 3
69115 Heidelberg, Germany

www.ingramcontent.com/pod-product-compliance
Ingram Content Group UK Ltd.
Pitfield, Milton Keynes, MK11 3LW, UK
UKHW021014080726
473054UK00003B/111

Springer Aerospace Technology

Series Editors

Sergio De Rosa, Department of Industrial Engineering, University of Naples Federico II, Napoli, Italy

Yao Zheng, School of Aeronautics and Astronautics, Zhejiang University, Hangzhou, China

The series explores the technology and the science related to the aircraft and spacecraft including concept, design, assembly, control and maintenance. The topics cover aircraft, missiles, space vehicles, aircraft engines and propulsion units. The volumes of the series present the fundamentals, the applications and the advances in all the fields related to aerospace engineering, including:

- structural analysis,
- aerodynamics,
- aeroelasticity,
- aeroacoustics,
- flight mechanics and dynamics
- orbital maneuvers,
- avionics,
- systems design,
- materials technology,
- launch technology,
- payload and satellite technology,
- space industry, medicine and biology.

The series' scope includes monographs, professional books, advanced textbooks, as well as selected contributions from specialized conferences and workshops.

The volumes of the series are single-blind peer-reviewed.

To submit a proposal or request further information, please contact: Mr. Pierpaolo Riva at pierpaolo.riva@springer.com (Europe and Americas) Mr. Mengchu Huang at mengchu.huang@springer.com (China)

The series is indexed in Scopus and Compendex

Martin Grabe · Georgii Oblapenko ·
Manuel Torrilhon

Editors

Rarefied Gas Dynamics

Proceedings of the 33rd International Symposium

Editors
Martin Grabe
Institute of Aerodynamics and
Flow Technology
Deutsches Zentrum für Luft- und
Raumfahrt (DLR)
Göttingen, Germany

Georgii Oblapenko
Applied and Computational Mathematics
RWTH Aachen University
Aachen, Germany

Manuel Torrilhon
Applied and Computational Mathematics
RWTH Aachen University
Aachen, Germany

ISSN 1869-1730　　　　　　ISSN 1869-1749　(electronic)
Springer Aerospace Technology
ISBN 978-3-032-00093-4　　　ISBN 978-3-032-00094-1　(eBook)
https://doi.org/10.1007/978-3-032-00094-1

This work was supported by Deutsches Zentrum für Luft- und Raumfahrt.

This Springer imprint is published by the registered company Springer Nature Switzerland AG
The registered company address is: Gewerbestrasse 11, 6330 Cham, Switzerland

If disposing of this product, please recycle the paper.

Preface

The 33rd International Symposium on Rarefied Gas Dynamics (RGD33) was held in Göttingen, Germany, from July 15 to 19, 2024. Since the first meeting in Nice in 1958, venues have changed every two years across countries and continents. This marked only the third time the RGD symposium was held in Germany, returning to Göttingen exactly 50 years after it was first hosted there.

Rarefied gas dynamics is a branch of physics mainly concerned with the study of gas flows in which processes at the molecular level play a role. With its diverse applications ranging from astrophysics to micro-flows, it continues to raise challenging questions that require a multidisciplinary effort of mathematicians, physicists, chemists, computer scientists and engineers.

Still under the impression of the fading COVID-19 pandemic, the RGD33 symposium was deliberately planned as a purely face-to-face event in order to promote direct scientific exchange and strengthen the professional community. The response was overwhelmingly positive. With 243 participants from 30 countries, the attendance exceeded expectations. However, for the first time in its history, the symposium was severely affected by sanctions imposed on Russian universities and research institutions as well as numerous other organizations embargoed by the USA and the European Union. Nevertheless, we remain committed to global scientific cooperation, as we are convinced that it also plays an important role in strengthening civil society against authoritarianism worldwide.

A total of 176 papers were presented in three parallel tracks. In addition, 49 posters were presented and put up for discussion in a dedicated session. The organizers took care to include a balance of contributions from mathematics, natural sciences and engineering in order to promote interdisciplinary exchange. Among the topics discussed during RGD33 were:

Boltzmann and Related Equations,
Granular Flows,
Gas-Surface Interactions,
Quasi-Classical Trajectory Simulation,
Moment Methods,

Electric Propulsion Simulation,
Machine Learning,
Air-breathing Electric Propulsion Simulation,
Spectroscopy,
Micro- and Nanoscale Flows,
Plasma Simulation,
Multi-scale Simulations,
CFD for Rarefied Flows,
Modern DSMC,
Fluid Interfaces,
Multi-scale Simulation Methods,
Particle-based Methods,
High-speed/High-enthalpy Flows,
Flows in Vacuum Systems,
Planetary Science and Exploration,
Jets and Plumes,
Thermochemically Non-equilibrium Flows,
Kinetic Methods,
Plasma Experiments.

Four plenary lectures were delivered at the symposium: the Graeme Bird Lecture was given by Alejandro Garcia. Richard Morgan delivered the Lloyd Thomas Lecture, and Vincent Giovangigli spoke in the Harold Grad Lecture. In addition to the three traditional plenary lectures, the Irving Langmuir Lecture was delivered by Anne Bourdon. The lecture was introduced for this symposium to strengthen connections with the related field of plasma research. Incidentally, Irving Langmuir, earned his PhD in Göttingen. In addition to the four outstanding scientists mentioned above, eight other researchers were invited to highlight their recent work: Alina Alexeenko, Christos Tantos, Eunji Jun, Hossein Gorji, Milana Čolić, Zhenning Cai, Tetsuro Tsuji and Philip Varghese.

The *Proceedings of the 33rd International Symposium on Rarefied Gas Dynamics* comprises of 69 full papers that were accepted for inclusion after having been evaluated by at least two independent professional peers.

The organizers thank the Deutsche Forschungsgemeinschaft (DFG) and the DLR Institute for Aerodynamics and Flow Technology for their generous support. We are also grateful to the many members of our community that served as reviewers and session chairs, as well as for the enthusiastic support we received from our DLR colleagues during all the phases of the symposium. In particular, we are indebted to the DLR Event Management Service (Petra Naoum and Rebecca Bartkowski) for smoothly managing the logistics. Last but not least, we thank the members of the International Advisory Committee, who continue to dedicate their time and effort to ensure the venerable *International Symposium on Rarefied Gas Dynamics* remains at the scientific forefront of the field.

<table>
<tr><td>Göttingen, Germany</td><td>Martin Grabe</td></tr>
<tr><td>Aachen, Germany</td><td>Georgii Oblapenko</td></tr>
<tr><td>Aachen, Germany</td><td>Manuel Torrilhon</td></tr>
</table>

Committees

Chair

Martin Grabe, DLR Göttingen

Co-Chair

Manuel Torrilhon, RWTH Aachen University

Local Advisory Committee

Georgii Oblapenko, RWTH Aachen University
Jan Martinez Schramm, DLR Göttingen
Sebastian Karl, DLR Göttingen

Local Organizing Committee

Petra Naoum, DLR Event Management
Rebecca Bartkowski, DLR Event Management
Svetlana Saburova, DLR Event Management
Michaela Welz, DLR Göttingen
Jan Martinez Schramm, DLR Göttingen
Tim Horchler, DLR Göttingen
Tobias Ecker, DLR Göttingen
Sebastian Karl, DLR Göttingen
Divek Surujhlal, DLR Göttingen
Alexander Wagner, DLR Göttingen

International Advisory Committee 2024

Alina A. Alexeenko, USA
Yevgeniy A. Bondar, Russia
Domenico Bruno, Italy
José M. Fernandez, Spain
Michail A. Gallis, USA
Vincent Giovangigli, France
Martin Grabe, Germany
Irina Graur, France
Rakesh Kumar, India
Elena V. Kustova, Russia
Deborah Levin, USA
Thierry Magin, Belgium
Luc Mieussens, France
Rho Shin Myong, South Korea
Sean O'Byrne, Australia
Felix Sharipov, Brazil
Henning Struchtrup, Canada
Quanhua Sun, China
Kojiro Suzuki, Japan
Shigeru Takata, Japan
Vladimir A. Titarev, Russia
Manuel Torrilhon, Germany
Philip Varghese, USA
Kung Xu, China
Yonghao Zhang, China

Prior RGD Symposia

1. Nice, France (1958)
2. Berkeley, USA (1960)
3. Paris, France (1962)
4. Toronto, Canada (1964)
5. Oxford, UK (1966)
6. Cambridge, USA (1968)
7. Pisa, Italy (1970)
8. Stanford, USA (1972)
9. Göttingen, Germany (1974)
10. Aspen, USA (1976)
11. Cannes, France (1978)
12. Charlottesville, USA (1980)
13. Novosibirsk, Russia (1982)
14. Tsukuba, Japan (1984)
15. Grado, Italy (1986)
16. Pasadena, USA (1988)
17. Aachen, Germany (1990)
18. Vancouver, Canada (1992)
19. Oxford, UK (1994)
20. Beijing, China (1996)
21. Marseille, France (1998)
22. Sydney, Australia (2000)
23. Whistler, Canada (2002)
24. Bari, Italy (2004)
25. St. Petersburg, Russia (2006)
26. Kyoto, Japan (2008)
27. Pacific Grove, USA (2010)
28. Zaragoza, Spain (2012)
29. Xi'an, China (2014)
30. Victoria, Canada (2016)
31. Glasgow, UK (2018)
32. Seoul, South Korea (2022)
33. Göttingen, Germany (2024)

Contents

Kinetic Theory

Micro- and Nanoscale Flows

Moment Methods

Boltzmann and Related Equations

Variational Solutions to the Coupled Poiseuille and Thermal Creep Problems on the Basis of the Boltzmann Equation for Hard-Sphere Molecules and Cercignani-Lampis Boundary Conditions

Silvia Lorenzani

Abstract In the present paper, we study some mathematical properties of the closed-form solutions to the Poiseuille and thermal-creep problems, obtained on the basis of the Boltzmann equation for hard-sphere molecules and Cercignani-Lampis boundary conditions.

Keywords Variational method · Poiseuille flow · Thermal-creep problem

1 Introduction

An accurate modeling of the gas-surface interaction is crucial in several applications involving the kinetic theory of rarefied gases. This issue can be addressed by choosing suitable boundary conditions for the Boltzmann equation, which describe the relation between the distribution functions of impinging and re-emitted gas molecules at a solid wall. Although different models of gas-surface interaction have been proposed in the literature, research in this direction has not made much progress over the years and most studies still use Maxwell boundary conditions derived in 1879. The Maxwell scattering kernel is characterized by only one free parameter, which represents the accommodation coefficient of all molecular properties. However, this is an oversimplification, since each physical quantity should have its own accommodation coefficient. Indeed, recent experiments have revealed that this model of gas-surface interaction fails to explain several results related to temperature-driven mass flow rates in microchannels [1, 2]. Because of these limitations, in recent years there has been a revival of interest in another more complex model of boundary conditions, the Cercignani-Lampis scattering kernel, which includes two parameters: α_t, representing the accommodation coefficient of the tangential momentum ($\alpha_t \in [0, 2]$) and α_n, being the accommodation of the kinetic energy due to the velocity normal

S. Lorenzani (✉)
Dipartimento di Matematica, Politecnico di Milano, Milano, Italy
e-mail: silvia.lorenzani@polimi.it

© The Author(s) 2026
M. Grabe et al. (eds.), *Rarefied Gas Dynamics*, Springer Aerospace Technology,
https://doi.org/10.1007/978-3-032-00094-1_1

to the bounding walls ($\alpha_n \in [0, 1]$). Despite the growing application of this model of gas-surface interaction to the study of several problems in rarefied gas dynamics, only a few recent papers investigate the mathematical properties of the solution of the Boltzmann equation with Cercignani-Lampis boundary conditions. Indeed, in [3], Chen has proved the local-in-time well-posedness of the Boltzmann equation in bounded domains with Cercignani-Lampis boundary conditions, where the accommodation coefficients can take any value other than zero (specular reflection). But so far, this author was able to construct the steady and global solution only for accommodation coefficients α_t and α_n close to 1 (diffuse boundary condition). In the present paper, we try to get a deeper insight on this issue, relying on the variational formulation of the thermal-creep problem.

2 Variational Formulation of Coupled Poiseuille and Thermal-Creep Problems

Let us consider the flow of a monatomic gas between two infinite parallel plates fixed at $x = \pm d/2$, due to small gradients of pressure and temperature in the z-direction:

$$k = \frac{1}{p} \frac{\partial p}{\partial z}, \quad \tau = \frac{1}{T} \frac{\partial T}{\partial z} \tag{1}$$

where p and T are the local gas pressure and temperature, respectively. Under these conditions, the gas flow is described by the linearized Boltzmann equation:

$$(D - L)h = S \tag{2}$$

where $Dh = c_x \frac{\partial h}{\partial x}$ is the differential term, $S = -c_z k - c_z \left(c^2 - \frac{5}{2} \right) \tau$ is the source term and Lh is the linearized collision operator. In Eq. (2), $h(x, \mathbf{c})$ is the small perturbation on the basic equilibrium state and $\mathbf{c}$ is the molecular velocity vector expressed in units of $(2RT)^{1/2}$ (with R being the specific gas constant). In the current investigation, we study the Poiseuille and thermal-creep flows on the basis of the exact linearized collision operator for hard-sphere molecules. Then, the Boltzmann equation (2) can be solved by imposing appropriate boundary conditions on the walls of the channel. In the present work, we will focus on the Cercignani-Lampis scattering kernel:

$$h^+(-(d/2)\mathrm{sgn}c_x, \mathbf{c}) = \int_{c_x' < 0} R_{CL}(-\mathbf{c} \to -\mathbf{c}') \, h^-(-(d/2)\mathrm{sgn}c_x, \mathbf{c}') \, d\mathbf{c}', \tag{3}$$

$$R_{CL}(\mathbf{c}' \to \mathbf{c}) = \frac{2c_x}{\pi \alpha_t \alpha_n (2 - \alpha_t)} \exp\left\{ - \frac{[\mathbf{c}_t - (1 - \alpha_t)\mathbf{c}'_t]^2}{\alpha_t(2 - \alpha_t)} \right\}$$

$$\times \exp\left\{ - \frac{[c_x{}^2 + (1 - \alpha_n)c'{}_x{}^2]}{\alpha_n} \right\} I_o\left(\frac{2\sqrt{1 - \alpha_n}}{\alpha_n} c_x c'_x \right) \quad (4)$$

with $\mathbf{c}_t = (c_y, c_z)$ being the two-dimensional vector of the tangential molecular velocity and I_o the modified Bessel function of first kind and zeroth order. Equation (3) can be rewritten in symbolic form as follows: $h^+ = K h^-$, where K is an operator and $h^\pm$ are the restrictions of the function h, defined on the boundary, to positive (negative) values of c_x (that is, h^+ and h^- represent the distribution functions of the reemitted and the impinging molecules on the boundaries, respectively). Since the imposed pressure and temperature gradients give rise to a gas flow and heat transfer in the z-direction, the quantities of physical interest to be computed are the mass $\dot{M}$ and heat flow rate $\dot{Q}$ (per unit time through unit thickness):

$$\dot{M} = \rho \int_{-d/2}^{d/2} v_z(x)dx \quad (5)$$

$$\dot{Q} = \int_{-d/2}^{d/2} q_z(x)dx \quad (6)$$

where ρ is the gas density. In Eqs. (5) and (6), the bulk velocity of the gas $v_z(x)$ and the heat flux $q_z(x)$ are defined as follows:

$$v_z(x) = \pi^{-\frac{3}{2}} \int_{-\infty}^{+\infty} \int_{-\infty}^{+\infty} \int_{-\infty}^{+\infty} e^{-c^2} c_z h(x, \mathbf{c}) \, d\mathbf{c} \quad (7)$$

$$q_z(x) = \pi^{-\frac{3}{2}} \int_{-\infty}^{+\infty} \int_{-\infty}^{+\infty} \int_{-\infty}^{+\infty} e^{-c^2} c_z \left(c^2 - \frac{5}{2} \right) h(x, \mathbf{c}) \, d\mathbf{c}. \quad (8)$$

In the framework of a linearized analysis, the mass and heat flow rates can be written as the sum of the Poiseuille and thermal-creep contributions:

$$\dot{M} = d^2 \, p \, [-G_p k + G_T \tau] \quad (9)$$

$$\dot{Q} = \frac{d^2}{2} \, p \, [Q_p k - Q_T \tau] \quad (10)$$

where G_p, G_T, Q_p and Q_T are dimensionless positive coefficients that refer to the Poiseuille flow, the thermal-creep flow, the mechanocaloric effect and the reduced heat flux, respectively. In this paper, we derive approximate closed-form expressions for all these coefficients, taking into account that the Onsager relation holds: $G_T = Q_p$. To this end, we solve the Boltzmann equation by applying the variational technique presented in detail in [1, 2]. In particular, we introduce the following functional J of the test function $\tilde{h}$:

$$J(\tilde{h}) = ((\tilde{h}, P(D\tilde{h} - L\tilde{h}))) - 2((PS, \tilde{h})) + (\tilde{h}^+ - K\tilde{h}^-, P\tilde{h}^-)_B, \qquad (11)$$

where P is the parity operator in velocity space, defined by $P[h(\mathbf{c})] = h(-\mathbf{c})$, while $((,))$, $(,)_B$ denote two scalar products reported in [1, 2]. According to the variational principle stated in [1], the functional $J(\tilde{h})$ attains its minimum value when $\tilde{h} = h(x, \mathbf{c})$ solves Eq. (2) with appropriate boundary conditions. Inserting $\tilde{h} = h(x, \mathbf{c})$ in Eq. (11), one gets:

$$J(h) = -((PS, h)) = -k \int_{-d/2}^{d/2} v_z(x)dx - \tau \int_{-d/2}^{d/2} q_z(x)dx = -k\frac{\dot{M}}{\rho} - \tau\dot{Q}. \quad (12)$$

Thus, the stationary value of J has a direct connection with the quantities of physical interest for the problem at hand. Since the details of the variational method of solution are given elsewhere [1, 2], below we report only the main steps undertaken to obtain the analytical formula for the coefficients G_p, G_T and Q_T. Relying on the integral form solution of the Boltzmann equation with the simplified Bhatnagar-Gross-Krook collision model, we derive a test function, which depends on three adjustable constants: A, B and C, to be varied in order to obtain the stationary value of $J(\tilde{h})$. Splitting the constants into two parts, due to the linear superposition of the Poiseuille and thermal-creep effects, and normalizing them in a suitable way, the functional $J(\tilde{h})$ reduces to the sum of three functionals, $J^{(1)}(\tilde{h})$, $J^{(2)}(\tilde{h})$, $J^{(3)}(\tilde{h})$, simply grouping together the terms proportional to $(k\theta)^2$, $(\tau\theta)^2$ and $(k\theta)(\tau\theta)$, respectively. For hard spheres of diameter σ, the length parameter θ is given by $\theta = \sqrt{2}/(\pi^{3/2}\sigma^2 n)$, with n being the gas number density. The stationary value of these functionals is obtained by imposing that their derivatives with respect to the adjustable constants A, B and C vanish. Finally, since Eqs. (12) and (9–10) hold, one can derive the following relationships for the Poiseuille flow G_p, the thermal-creep flow G_T, and the reduced heat flux Q_T:

$$G_p = \frac{2}{\delta^2} \min \frac{J^{(1)}(\tilde{h})}{(k\theta)^2} = \frac{2}{\delta^2} \frac{J^{(1)}(h)}{(k\theta)^2} \qquad (13)$$

$$G_T = -\frac{1}{\delta^2} \min \frac{J^{(3)}(\tilde{h})}{(k\theta)(\tau\theta)} = -\frac{1}{\delta^2} \frac{J^{(3)}(h)}{(k\theta)(\tau\theta)} \tag{14}$$

$$Q_T = \frac{2}{\delta^2} \min \frac{J^{(2)}(\tilde{h})}{(\tau\theta)^2} = \frac{2}{\delta^2} \frac{J^{(2)}(h)}{(\tau\theta)^2} \tag{15}$$

where δ is the rarefaction parameter (inverse Knudsen number), $\delta = \frac{d}{\theta}$. By performing an asymptotic expansion of the full variational solution, one can prove that the pressure- and temperature-driven mass flow rates G_p and G_T, respectively, are well approximated, in the range $\delta \geq 3$, by the following closed-form formulas:

$$G_p = \frac{\delta}{\sigma_{0,p}} + \sigma_{1,p} + \frac{\sigma_{2,p}}{\delta} \tag{16}$$

$$G_T = \frac{\sigma_{1,T}}{\delta} + \frac{\sigma_{2,T}}{\delta^2} \tag{17}$$

where the explicit expressions for the coefficients $\sigma_{0,p}$, $\sigma_{1,p}$, $\sigma_{2,p}$ and $\sigma_{1,T}$, $\sigma_{2,T}$ are reported in [2]. On the other hand, the same analysis carried out for the heat flux Q_T did not provide satisfactory results. Therefore, in this case, one should consider the full variational solution, in order to obtain accurate results for $\delta \geq 3$:

$$Q_T = (\sqrt{\pi}\,\delta^2)^{-1} \times [c_{11}c_{22}d_{22} - d_{12}^2 c_{22} - c_{11}d_{23}^2 + 2c_{12}d_{12}d_{23} - c_{12}^2 d_{22}]^{-1}$$
$$\times [-d_1^2 d_{22}c_{22} - c_{11}d_2^2 c_{22} + 2d_1 d_{12}d_2 c_{22} + d_1^2 d_{23}^2 - 2d_1 c_{12}d_2 d_{23} + c_{12}^2 d_2^2] \tag{18}$$

where the coefficients are given by:

$$c_{11} = -\frac{\delta^4}{4}\alpha_t - \frac{4}{3}\sqrt{\pi}\delta^3 + \sqrt{\pi}\delta^3\alpha_t + 4\delta^2 - 4\delta^2\alpha_t - 4\sqrt{\pi}\delta\alpha_n\alpha_t + 12\sqrt{\pi}\delta\alpha_t$$
$$+ 4\sqrt{\pi}\delta\alpha_n + 16\alpha_n\alpha_t - 32\alpha_t - 16\alpha_n - \frac{8}{3\pi}\delta^3 \hat{J}_1 - \frac{32}{\pi}\delta\hat{J}_2 \tag{19}$$
$$+ 16\delta^2 \mathcal{F}_0\alpha_n(1 - \alpha_t) + 16\delta^2 \mathcal{F}_1(1 - \alpha_t)(1 - \alpha_n),$$

$$c_{12} = -\delta^2\alpha_t + 2\sqrt{\pi}\delta\alpha_t - 8\alpha_t, \quad c_{22} = -4\alpha_t, \tag{20}$$

$$c_1 = \frac{\sqrt{\pi}}{6}\delta^3 - \frac{\delta^2}{2}\alpha_t + 2\sqrt{\pi}\delta + \sqrt{\pi}\delta\alpha_t - 4\alpha_t, \quad c_2 = 2\sqrt{\pi}\delta - 2\alpha_t, \tag{21}$$

$$d_{12} = \frac{\delta^2}{4}\alpha_t + \frac{\sqrt{\pi}}{2}\delta\alpha_n\alpha_t - \sqrt{\pi}\delta\alpha_t - \frac{\sqrt{\pi}}{2}\delta\alpha_n - 4\alpha_n\alpha_t + 6\alpha_t + 4\alpha_n + \frac{8}{\pi}\delta\hat{J}_3, \tag{22}$$

$$d_{22} = -2\alpha_t^3 + 6\alpha_t^2 + \alpha_n\alpha_t - \frac{29}{4}\alpha_t - \alpha_n - \frac{2}{\pi}\delta\hat{J}_4, \tag{23}$$

$$d_1 = 2\sqrt{\pi}\delta, \quad d_2 = -\frac{5}{2}\sqrt{\pi}\delta, \quad d_{23} = \alpha_t, \tag{24}$$

with $\mathcal{F}_0 = 0.196079$ and $\mathcal{F}_1 = 0.247679$, $\hat{J}_1 = -1.4180$, $\hat{J}_2 = 1.8909$, $\hat{J}_3 = 0.9449$, $\hat{J}_4 = 4.7252$.

3   Results and Discussion

Since the Poiseuille flow has been widely considered in the literature and, moreover, it is strongly dependent only on the tangential momentum accommodation coefficient, α_t, in the following we will focus on the quantities of interest for the thermal-creep problem, G_T and Q_T. In order to assess the accuracy of the analytical formulas (17) and (18), derived in Sect. 2, we report in Tables 1 and 2 a comparison between our variational outputs and the numerical results obtained in [4, 5]. These tables show a good agreement (the discrepancy lies, on average, within an error of 10%) for all values of the accommodation coefficients α_t and α_n, as the rarefaction parameter varies in the range $\delta \geq 3$. Indeed, the variational estimates for the temperature-driven mass flow rate G_T exhibit a larger deviation from the numerical findings than those for the heat flux Q_T. This is especially true for $\delta = 3$, thus demonstrating that this value of the rarefaction parameter represents the lower limit of validity of our analytical formulas. The closed-form expressions (17) and (18) allow us to investigate deeper the mathematical properties of the solution to the problem at hand. Indeed, a closer inspection of the dependency of $\sigma_{1,T}$ and $\sigma_{2,T}$, in Eq. (17), on the accommodation coefficients has revealed that there is a continuous set of values for (α_t, α_n) leading to the same estimates for the mass flow rate G_T. This remark holds true also for the heat flux Q_T. As an example, Table 3 shows that, for a pair of different values of the accommodation coefficients (relevant for the comparison with recent experimental measurements [1, 2]), the same estimate for G_T, and also for Q_T, is obtained, in terms of the rarefaction parameter δ. In the specific case of the thermal creep problem, we have found that one can get the same values for the mass flow rate G_T (and also for the heat flux Q_T), when both accommodation coefficients range from 0.1 to 0.9. Further investigation is needed to clarify whether these findings are related to the mathematical result obtained by Chen [3], according to which a well-posed steady and global solution of the Boltzmann equation with Cercignani-Lampis boundary conditions can be constructed only when α_t and α_n are close to 1.

Table 1 Thermal-creep flow rate G_T: comparison between our variational outputs [Var(HS)] and the numerical results reported in [4]

δ	α_t	α_n	0.	0.2	0.4	0.5	0.7	0.9	1.
3.	0.1	Var(HS)	0.2512	0.2490	0.2440	0.2405	0.2312	0.2192	0.2121
		[4]	0.247	0.248	0.249	0.250	0.250	0.251	0.251
	0.2	Var(HS)	0.2537	0.2495	0.2431	0.2391	0.2294	0.2174	0.2106
		[4]	0.238	0.239	0.240	0.240	0.241	0.242	0.242
	0.4	Var(HS)	0.2510	0.2451	0.2378	0.2338	0.2246	0.2143	0.2086
		[4]	0.227	0.228	0.229	0.229	0.230	0.230	0.231
	0.5	Var(HS)	0.2468	0.2408	0.2339	0.2301	0.2219	0.2129	0.2080
		[4]	0.224	0.225	0.226	0.226	0.226	0.227	0.227
	0.8	Var(HS)	0.2260	0.2225	0.2189	0.2171	0.2132	0.2093	0.2073
		[4]	0.222	0.222	0.222	0.223	0.223	0.223	0.223
	1.	Var(HS)	0.2072	0.2072	0.2072	0.2072	0.2072	0.2072	0.2072
		[4]	0.223	0.223	0.223	0.223	0.223	0.223	0.223
	1.2	Var(HS)	0.1849	0.1896	0.1942	0.1965	0.2009	0.2051	0.2072
		[4]	0.224	0.223	0.223	0.223	0.223	0.223	0.222
5.	0.1	Var(HS)	0.1524	0.1570	0.1606	0.1620	0.1641	0.1651	0.1653
		[4]	0.152	0.157	0.161	0.164	0.168	0.172	0.174
	0.2	Var(HS)	0.1563	0.1596	0.1621	0.1630	0.1643	0.1648	0.1648
		[4]	0.149	0.153	0.158	0.160	0.163	0.167	0.169
	0.4	Var(HS)	0.1613	0.1628	0.1638	0.1641	0.1644	0.1643	0.1640
		[4]	0.148	0.151	0.154	0.155	0.158	0.161	0.162
	0.5	Var(HS)	0.1628	0.1636	0.1641	0.1643	0.1643	0.1641	0.1638
		[4]	0.148	0.151	0.153	0.154	0.157	0.159	0.160
	0.8	Var(HS)	0.1643	0.1643	0.1642	0.1641	0.1639	0.1637	0.1636
		[4]	0.153	0.154	0.155	0.155	0.156	0.157	0.158
	1.	Var(HS)	0.1635	0.1635	0.1635	0.1635	0.1635	0.1635	0.1635
		[4]	0.158	0.158	0.158	0.158	0.158	0.158	0.158
	1.2	Var(HS)	0.1615	0.1620	0.1625	0.1627	0.1631	0.1634	0.1635
		[4]	0.162	0.161	0.160	0.160	0.159	0.158	0.157
	1.5	Var(HS)	0.1553	0.1575	0.1594	0.1602	0.1617	0.1628	0.1633
		[4]	0.166	0.164	0.161	0.160	0.158	0.156	0.155
	2.	Var(HS)	0.1286	0.1376	0.1454	0.1488	0.1547	0.1593	0.1611
		[4]	0.158	0.155	0.151	0.149	0.145	0.141	0.139
10.	0.1	Var(HS)	0.0768	0.0813	0.0856	0.0876	0.0915	0.0952	0.0969
		[4]	0.0773	0.0817	0.0859	0.0879	0.0918	0.0955	0.0973
	0.2	Var(HS)	0.0796	0.0835	0.0871	0.0888	0.0922	0.0953	0.0968
		[4]	0.0775	0.0813	0.0849	0.0867	0.0901	0.0934	0.0950
	0.4	Var(HS)	0.0847	0.0873	0.0898	0.0910	0.0933	0.0955	0.0966
		[4]	0.0793	0.0819	0.0845	0.0858	0.0883	0.0908	0.0920
	0.5	Var(HS)	0.0869	0.0890	0.0910	0.0920	0.0939	0.0957	0.0965
		[4]	0.0810	0.0830	0.0850	0.0860	0.0880	0.0900	0.0910
	0.8	Var(HS)	0.0929	0.0937	0.0944	0.0947	0.0954	0.0961	0.0965
		[4]	0.0860	0.0868	0.0876	0.0880	0.0888	0.0896	0.0900
	1.	Var(HS)	0.0965	0.0965	0.0965	0.0965	0.0965	0.0965	0.0965
		[4]	0.0900	0.0899	0.0899	0.0899	0.0899	0.0899	0.0899
	1.2	Var(HS)	0.0997	0.0991	0.0985	0.0981	0.0975	0.0968	0.0965
		[4]	0.0938	0.0930	0.0922	0.0918	0.0911	0.0903	0.0899
	1.5	Var(HS)	0.1038	0.1025	0.1011	0.1003	0.0988	0.0972	0.0964
		[4]	0.0984	0.0965	0.0947	0.0937	0.0917	0.0898	0.0888
	2.	Var(HS)	0.1065	0.1050	0.1032	0.1022	0.0999	0.0973	0.0959
		[4]	0.0992	0.0961	0.0927	0.0908	0.0871	0.0833	0.0814

Table 2 Heat flow rate Q_T: comparison between our variational outputs [Var(HS)] and the numerical results reported in [5]

δ	α_t	α_n	0.	0.2	0.4	0.5	0.7	0.9	1.
3.	0.1	Var(HS)	1.1474	1.1258	1.1052	1.0952	1.0759	1.0574	1.0485
	0.2	Var(HS)	1.0899	1.0721	1.0550	1.0466	1.0305	1.0150	1.0074
	0.4	Var(HS)	1.0155	1.0035	0.9918	0.9861	0.9749	0.9641	0.9588
	0.5	Var(HS)	0.9911	0.9815	0.9721	0.9675	0.9584	0.9496	0.9452
		[5]				0.9789			0.9560
	0.6	Var(HS)	0.9724	0.9649	0.9576	0.9540	0.9469	0.9400	0.9365
	0.8	Var(HS)	0.9461	0.9425	0.9391	0.9373	0.9339	0.9304	0.9287
	1.	Var(HS)	0.9277	0.9277	0.9277	0.9277	0.9277	0.9277	0.9277
		[5]				0.9329			0.9329
	1.2	Var(HS)	0.9101	0.9133	0.9166	0.9182	0.9215	0.9249	0.9266
	1.5	Var(HS)	0.8719	0.8794	0.8871	0.8910	0.8989	0.9070	0.9111
		[5]				0.8897			0.9106
	2.	Var(HS)	0.7518	0.7639	0.7765	0.7830	0.7965	0.8105	0.8178
5.	0.1	Var(HS)	0.7068	0.6991	0.6918	0.6882	0.6812	0.6745	0.6712
	0.2	Var(HS)	0.6844	0.6782	0.6723	0.6693	0.6636	0.6581	0.6554
	0.4	Var(HS)	0.6547	0.6506	0.6467	0.6447	0.6409	0.6371	0.6353
	0.5	Var(HS)	0.6450	0.6418	0.6386	0.6371	0.6340	0.6309	0.6294
		[5]				0.6505			0.6410
	0.6	Var(HS)	0.6377	0.6352	0.6328	0.6316	0.6292	0.6268	0.6256
	0.8	Var(HS)	0.6280	0.6268	0.6256	0.6250	0.6239	0.6227	0.6221
	1.	Var(HS)	0.6217	0.6217	0.6217	0.6217	0.6217	0.6217	0.6217
		[5]				0.6320			0.6320
	1.2	Var(HS)	0.6155	0.6167	0.6178	0.6183	0.6195	0.6206	0.6212
	1.5	Var(HS)	0.6008	0.6034	0.6060	0.6074	0.6100	0.6128	0.6141
		[5]				0.6141			0.6232
	2.	Var(HS)	0.5474	0.5512	0.5552	0.5572	0.5613	0.5655	0.5676
10.	0.1	Var(HS)	0.3606	0.3585	0.3566	0.3556	0.3537	0.3519	0.3510
	0.2	Var(HS)	0.3545	0.3528	0.3511	0.3503	0.3487	0.3471	0.3463
	0.4	Var(HS)	0.3459	0.3447	0.3435	0.3429	0.3418	0.3407	0.3402
	0.5	Var(HS)	0.3430	0.3420	0.3411	0.3406	0.3397	0.3388	0.3383
		[5]				0.3516			0.3491
	0.6	Var(HS)	0.3407	0.3400	0.3392	0.3389	0.3381	0.3374	0.3371
	0.8	Var(HS)	0.3377	0.3374	0.3370	0.3368	0.3365	0.3361	0.3359
	1.	Var(HS)	0.3358	0.3358	0.3358	0.3358	0.3358	0.3358	0.3358
		[5]				0.3468			0.3468
	1.2	Var(HS)	0.3339	0.3342	0.3346	0.3348	0.3351	0.3355	0.3356
	1.5	Var(HS)	0.3292	0.3300	0.3308	0.3312	0.3320	0.3329	0.3333
		[5]				0.3422			0.3446
	2.	Var(HS)	0.3100	0.3113	0.3127	0.3134	0.3148	0.3163	0.3170

Table 3 Variational estimates of the mass flow rate G_T and heat flux Q_T for a pair of different values of α_t and α_n

	$\alpha_t = 0.8$	$\alpha_n = 0.15$	$\alpha_t = 0.83$	$\alpha_n = 0.28$	$\alpha_t = 0.88$	$\alpha_n = 0.33$
	$\alpha_t = 0.63$	$\alpha_n = 0.54$	$\alpha_t = 0.63$	$\alpha_n = 0.67$	$\alpha_t = 0.65$	$\alpha_n = 0.77$
δ	G_T	Q_T	G_T	Q_T	G_T	Q_T
3.	0.223	0.94	0.219	0.94	0.215	0.93
5.	0.164	0.63	0.164	0.63	0.164	0.62
10.	0.0935	0.34	0.0943	0.34	0.0951	0.33

Acknowledgements Silvia Lorenzani is supported by GNFM of INDAM, Italy.

References

1. Nguyen NN, Graur I, Perrier P, Lorenzani S (2020) Variational derivation of thermal slip coefficients on the basis of the Boltzmann equation for hard-sphere molecules and Cercignani-Lampis boundary conditions: comparison with experimental results. Phys Fluids 32:102011
2. Missoni T, Yamaguchi H, Graur I, Lorenzani S (2021) Extraction of tangential momentum and normal energy accommodation coefficients by comparing variational solutions of the Boltzmann equation with experiments on thermal creep gas flow in microchannels. Fluids 6:445
3. Chen H (2020) Cercignani-Lampis boundary in the Boltzmann theory. Kinet Relat Models 13:549
4. Basdanis T, Tatsios G, Valougeorgis D (2022) Gas-surface interaction in rarefied gas flows through long capillaries via the linearized Boltzmann equation with various boundary conditions. Vacuum 202:111152
5. Basdanis T (2024) Private communication

Existence and Regularity of Axisymmetric Weak Solutions to the Spatially Homogeneous Landau Equation

Jin Woo Jang and Junha Kim

Abstract We provide a concise summary of our recent findings on measure-valued solutions to the spatially homogeneous Landau equation for hard potentials [6]. Specifically, we prove the existence of axisymmetric measure-valued solutions for any axisymmetric initial data in $\mathcal{P}_p(\mathbb{R}^3)$ ($p \geq 2$). Furthermore, we prove that these solutions become instantaneously analytic for $t > 0$, except when the initial data is a single Dirac mass. For soft potentials and Maxwellian molecules, we show that solutions cannot remain confined to a fixed line, even when starting from line-concentrated initial data.

Keywords Landau kinetic equation · Measure-valued solutions · Axisymmetry

1 Introduction

The spatially homogeneous Landau equation serves as a fundamental model in kinetic theory, capturing the dynamics of particle distributions influenced by grazing collisions. This paper aims to announce the recent establishment of the existence of weak solutions in [6] that preserve axisymmetry, provided the initial data is also axisymmetric.

For $t \in [0, T]$ and $v \in \mathbb{R}^3$ with some $T > 0$, the spatially homogeneous Landau kinetic equation is expressed as:

$$\partial_t f(t, v) = Q(f, f) = \partial_i \int_{\mathbb{R}^3} a_{ij}(v - v_*) \left(\partial_j f(v) f(v_*) - \partial_j f(v_*) f(v)\right) dv_*, \qquad (1)$$

J. W. Jang
Department of Mathematics, Pohang University of Science and Technology (POSTECH), Pohang, South Korea
e-mail: jangjw@postech.ac.kr

J. Kim (✉)
Department of Mathematics, Ajou University, Suwon, Gyeonggi-do, South Korea
e-mail: junha02@ajou.ac.kr

© The Author(s) 2026

M. Grabe et al. (eds.), *Rarefied Gas Dynamics*, Springer Aerospace Technology,
https://doi.org/10.1007/978-3-032-00094-1_2

where $f(t, v)$ is abbreviated as $f(v)$. The kernel a_{ij} is defined by:

$$a_{ij}(z) = \left(\delta_{ij} - \frac{z_i z_j}{|z|^2}\right) |z|^{\gamma+2}, \quad \gamma \in [-3, 1],$$

with δ_{ij} representing the Kronecker delta ($\delta_{ij} = 1$ if $i = j$, and $\delta_{ij} = 0$ otherwise). The case $\gamma \in (0, 1]$ corresponds to hard potentials, $\gamma = 0$ to Maxwellian molecules, and $\gamma \in [-3, 0)$ to soft potentials.

Defining the following auxiliary quantities:

$$b_i(z) := \partial_j a_{ij}(z) = -2z_i |z|^\gamma, \quad c(z) := \partial_i \partial_j a_{ij}(z) = -2(\gamma + 3)|z|^\gamma, \quad \text{for } \gamma > -3,$$

$$c(z) := -8\pi\delta(z), \quad \text{for } \gamma = -3,$$

and their convolutions with f:

$$\overline{a_{ij}(v)} := a_{ij} * f, \quad \overline{b_i(v)} := b_i * f, \quad \overline{c(v)} := c * f,$$

the Landau equation can be reformulated as:

$$\partial_t f = \nabla \cdot (\overline{a}\nabla f - \overline{b}f), \quad \text{or equivalently,} \quad \partial_t f = \overline{a_{ij}}\partial_i\partial_j f - \overline{c}f.$$

It is also noted that the kernel satisfies the bounds:

$$|a_{ij}(z)| \leq |z|^{2+\gamma}, \quad |b_i(z)| \leq |z|^{1+\gamma}.$$

1.1 *Measure-Valued Solutions*

In this paper, we focus on measure-valued solutions to the Landau equation. To this end, we begin by introducing the relevant space of probability measures. Let $\mathcal{P}(\mathbb{R}^3)$ denote the set of probability measures on $\mathbb{R}^3$. For $p > 0$, define

$$\mathcal{P}_p(\mathbb{R}^3) := \left\{ f \in \mathcal{P}(\mathbb{R}^3) \; ; \; \int_{\mathbb{R}^3} |v|^p f(dv) < \infty \right\}.$$

For $p \geq 1$, the space is equipped with the p-Wasserstein distance, given by

$$\mathcal{W}_p(\mu, \nu) := \left(\inf_{\gamma \in \Pi(\mu,\nu)} \int_{\mathbb{R}^3 \times \mathbb{R}^3} \text{dist}(x, y)^p \, d\gamma(x, y) \right)^{\frac{1}{p}},$$

where $\Pi(\mu, \nu)$ denotes the set of all couplings between μ and ν. For $p \in (0, 1)$, $\mathcal{W}_p$ can still be defined by omitting the $1/p$-power in the infimum.

Here, we define the concept of measure-valued weak solutions for the Landau equation in the hard potential case:

Definition 1 (Measure-Valued Weak Solution) Let $\gamma \in (0, 1]$. A function $f(t)$ is called a weak solution to the Landau equation if it satisfies the following conditions:

1. Regularity:

$$f \in L^\infty_{\mathrm{loc}}((0, \infty); \mathcal{P}_2(\mathbb{R}^3)) \cap L^1_{\mathrm{loc}}((0, \infty); \mathcal{P}_{2+\gamma}(\mathbb{R}^3)),$$

where $\mathcal{P}_p(\mathbb{R}^3)$ denotes the set of probability measures with finite p-th moment.
2. Energy Inequality:

$$\frac{1}{2} \int_{\mathbb{R}^3} |v|^2 f_t(dv) \leq \frac{1}{2} \int_{\mathbb{R}^3} |v|^2 f_0(dv), \quad \forall t \geq 0.$$

3. Weak Formulation: For all test functions $\varphi \in C_b^2(\mathbb{R}^3)$ and $T \geq 0$, the following equality holds:

$$\int_{\mathbb{R}^3} \varphi(v) f_T(dv) - \int_{\mathbb{R}^3} \varphi(v) f_0(dv)$$

$$= \int_0^T \int_{\mathbb{R}^3} \int_{\mathbb{R}^3} a_{ij}(v - v_*) \partial_i \partial_j \varphi(v) f_t(dv_*) f_t(dv) \, dt$$

$$+ 2 \int_0^T \int_{\mathbb{R}^3} \int_{\mathbb{R}^3} b_i(v - v_*) \partial_i \varphi(v) f_t(dv_*) f_t(dv) \, dt.$$

Here, $a_{ij}(z)$ and $b_i(z)$ are the coefficients defined in terms of the relative velocity $z = v - v_*$, and f_0 denotes the initial measure.

1.2 Main Theorem

The main establishment is as follows:

Theorem 1 (Theorem 1.6 of [6])
   Let $\gamma \in (0, 1]$ and suppose the initial profile $f_0 \in \mathcal{P}_p(\mathbb{R}^3)$ for $p \geq 2$ is axisymmetric. Then, there exists an axisymmetric measure-valued weak solution to the Landau equation, satisfying

$$f \in C([0, \infty); \mathcal{P}(\mathbb{R}^3)) \cap L^\infty((0, \infty); \mathcal{P}_p(\mathbb{R}^3)) \cap L^1_{loc}((0, \infty); \mathcal{P}_{p+\gamma}(\mathbb{R}^3)).$$

Furthermore, if f_0 is not a single Dirac mass, the solution f becomes analytic for all $t > 0$.

Remark 1 Several observations regarding the main theorem are as follows:

1. **Uniqueness for $p > 2$:** If $f_0 \in \mathcal{P}_p(\mathbb{R}^3)$ is axisymmetric with $p > 2$, the axisymmetric solution f is unique as a weak solution to the Landau equation, starting from f_0. Uniqueness is established in terms of the optimal transportation cost $\mathcal{T}_p(f, \tilde{f})$; see [4, Theorem 8].
2. **Soft potential case:** For $\gamma \in [-3, 0)$, it is shown that there are no axisymmetric solutions concentrated along a line, even if the initial data is line-concentrated (see Lemma 1). Specifically, for $\gamma \in [-3, -2)$, Golse, Imbert, Ji, and Vasseur [5] proved that axisymmetric weak solutions are smooth away from the axis of symmetry.
3. **Moderately soft potential case:** For $\gamma \in (-2, 0)$, Fournier and Guérin [3, Corollary 4] established the existence of weak solutions.

2 Escape of a Line Support

In this section, we introduce a key lemma that demonstrates the solution escapes a line support instantaneously. Namely we have the following lemma.

Lemma 1 (Lemma 2.1 of *[6]*) *Let $\gamma \in [-3, 1]$. Let $f \in L^1_{loc}((0, \infty); \mathcal{P}_{2+\gamma}(\mathbb{R}^3))$ be any weak solution to the Landau equation which is not a single Dirac mass. Let ℓ be any given line in $\mathbb{R}^3$. Then there is no time interval $[t_1, t_2] \in (0, \infty)$ with $t_1 < t_2$ such that $f(t, \cdot)$ is concentrated on a line $\ell \in \mathbb{R}^3$ for a.e. $t \in [t_1, t_2]$.*

Remark 2 Here are some observations regarding the lemma discussed above:

1. Lemma 1 applies not only to hard potentials where $\gamma \in (0, 1]$, but also extends to soft potentials and Maxwellian molecules with $\gamma \in [-3, 0]$.
2. Lemma 1 pertains to a specific, fixed line ℓ.
3. The proof of this lemma is straightforward and significantly simpler compared to [4, Lemma 16]. Additionally, the constraints $\gamma \in (0, 1]$ and $f_0 \in \mathcal{P}_4(\mathbb{R}^3)$ are not required in this case. For reference, Lemma 16 of [4] states:

Lemma 2 (Lemma 16 of *[4]*) *Let $\gamma \in (0, 1]$, $f_0 \in \mathcal{P}_4(\mathbb{R}^3)$, where f_0 is not a single Dirac mass, and let f be a weak solution of the Landau equation starting from f_0. Then, for any given $t_0 > 0$, there exists $t \in (0, \delta)$ such that $f(t)$ is not concentrated on any line $\ell \subset \mathbb{R}^3$.*

4. Our lemma addresses a fixed line $\ell \subset \mathbb{R}^3$. Even if the support of the solution moves away from the line ℓ instantaneously, it may still lie on a different line

(e.g., a rotating line). This does not necessarily imply that the support contains three non-collinear points. In this respect, Lemma 16 of [4] is more general, as it excludes the possibility of the support being confined to any line, including rotating ones. To ensure the support contains a triangle at any given time, we consider axisymmetric solutions when proving existence.

The proof of our main theorem (Theorem 1) fundamentally relies on a pivotal lemma (Lemma 1), which asserts that the support of the solution instantly escapes confinement to a single line.

3 Ellipticity of $\overline{a_{ij}(v)}$

In the paper [2], it is established that the Landau collision operator has the following elliptic property even for a solution with infinite entropy:

Lemma 3 (Lemma 9 of [2]) *Let $T > 0$ and f be a non-zero non-negative function f in $L^\infty([0, T]; L_2^1(\mathbb{R}^3)) \cap C([0, T]; W^{-2,1}(\mathbb{R}^3))$ with $f_{in} \in L_2^1(\mathbb{R}^3)$. Then, there exist constants $\Delta t \in (0, T]$ and $K > 0$ such that*

$$\overline{a_{ij}(v)}\xi_i\xi_j \geq K(1 + |v|^2)^{\frac{2}{2}}|\xi|^2, \quad v, \xi \in \mathbb{R}^3, \ t \in [0, \Delta t].$$

In [2, Remarks of Lemma 9], it is stated that the proof of the lemma allows in fact the initial condition f_{in} to be a measure, provided that it is not concentrated on a single line. In our case for the proof of Theorem 1, by our key lemma (Lemma 1) we observe that the support of a solution f instantaneously escapes a line support for any $t > 0$. For any $t_0 > 0$, on the time interval $[t_0, \infty)$, the measure f_{t_0} can be treated as the initial profile in Lemma 3, allowing us to establish ellipticity.

4 Gain of Moments

We present a theorem on the gain of moments for the hard potential case.

If the initial profile is in $\mathcal{P}_p(\mathbb{R}^3)$ with $p > 2$, we use the following theorem on the gain-of-moments property:

Theorem 2 (Theorems 3 and 6 of [2]) *Let f be any weak solution of the Landau equation (1) with initial datum $f_{in} \in L_2^1(\mathbb{R}^3)$, satisfying the decay of energy*

$$E(f(t, \cdot)) \leq E(f_{in}(\cdot)), \ for \ t \geq 0.$$

Then

1. *For all $s > 0$, if $M_s(f_{in}) < +\infty$, then $\sup_{t \geq 0} M_s(f(t, \cdot)) < +\infty$, and for all $T > 0$,*

$$\int_0^T M_{s+\gamma}(f(t, \cdot))dt < +\infty.$$

2. *For all time $t_0 > 0$, and all number $s > 0$, there exists a constant $C_{t_0} > 0$, depending only on M_{in}, E_{in}, and t_0, such that for all time $t \geq t_0$,*

$$M_s(f(t, \cdot)) \leq C_{t_0}.$$

3. *For any $t \geq 0$, $E(f(t, \cdot)) = E_{in}$: the energy is automatically conserved.*
4. *If $f_{in} \in L^1_{2+\delta}$ for some $\delta > 0$, there exists a weak solution starting from f_{in}.*
5. *If $f_{in} \in L^1_{2+\delta}$ for some $\delta > 0$ and f_{in} is not concentrated on a line, there exists a weak solution starting from f_{in} such that for any $t_0 > 0$, $H(f(t)) < \infty$ for all $t > t_0$ and*

$$\sup_{t \geq t_0} \| f(t) \|_{H^k_s} < \infty, \quad k \in \mathbb{N}, \quad s \geq 0. \tag{2}$$

On the other hand, in the critical case when $p = 2$ and the initial profile is in $\mathcal{P}_2(\mathbb{R}^3)$, we instead argue as follows. If $p = 2$, we apply the de la Vallée Poussin theorem to ensure the existence of a C^2 function $h : [0, \infty) \to [0, \infty)$ satisfying the following properties: $h'' \in [0, 1]$, $h'(0) = 1$, $h'(\infty) = \infty$, and $\int_{\mathbb{R}^3} h(|v|^2) f_0(dv) < \infty$. According to [4, Eq. (29)], there exists a finite constant K_T for any $T > 0$ such that for all $n \geq n_0$:

$$\sup_{t \in [0,T]} \int_{\mathbb{R}^3} h(|v|^2) f_t(dv) + \int_0^T \int_{\mathbb{R}^3} |v|^{2+\gamma} h'(|v|^2) f_t(dv)\, dt \leq K_T.$$

These moments estimates will be used to prove the relative compactness of each measure f_t in $\mathcal{P}(\mathbb{R}^3)$.

5 Proof of the Main Theorem

The proof of the main theorem (Theorem 1) has several steps. The first step is to regularize the measure-valued initial condition f_0 as

$$f_0^n := (\alpha_n^{-1} \mathbf{1}_{|v| \leq n} f_0(dv)) * \sigma_n, \quad n \geq n_0, \tag{3}$$

with the total mass $\alpha_n := \int_{\mathbb{R}^3} \mathbf{1}_{|v| \le n} f_0(dv)$ for some $n_0 \in \mathbb{N}$. In addition, we regularize the equation by adding an additional ellipticity for each $\varepsilon > 0$ as

$$\partial_t f^\varepsilon = Q^\varepsilon(f^\varepsilon, f^\varepsilon) + \varepsilon \Delta_v f^\varepsilon, \tag{4}$$

where Q^ε is a smoothly mollified version of Q such that the coefficients are smooth. The linearized version of (4) has a solution by the standard parabolic theory. Then the solution sequence also has a fixed point f^ε by the Schauder fixed-point theorem with the regularized initial data f_0^n.

Before we pass to the limit $\varepsilon \to 0$ and $n \to \infty$, we can also show that each f^ε is axisymmetric if f_0^n is. This is by showing that $f(Tv)$ satisfies the linear system, where $T = T_\theta$ be a 3×3 rotation matrix such that

$$T := \begin{pmatrix} \cos\theta & -\sin\theta & 0 \\ \sin\theta & \cos\theta & 0 \\ 0 & 0 & 1 \end{pmatrix}, \qquad \theta \in [0, 2\pi).$$

This can be obtained by showing

$$\varepsilon(\Delta f)(Tv) = \varepsilon \Delta(f(Tv)), \;\; \text{for } \theta \in [0, 2\pi),$$

and that

$$Q^\varepsilon(g, f)(Tv) = Q^\varepsilon(g, f(Tv))(v),$$

for a given axisymmetric g.

Then by further obtaining a uniform-in-ε estimate on the energy and the entropy, we have $\{f_t^\varepsilon\}_{\varepsilon > 0}$ is weakly compact in $L^1(\mathbb{R}_v^3)$ for each $t \in [0, T]$ by satisfying the Dunford-Pettis compactness criterion. Moreover, using these moment estimates of the previous section, one can show that by the Arzela-Ascoli theorem, there exists a subsequence $\{f^{\varepsilon'}\}_{\varepsilon' > 0} \subset \{f^\varepsilon\}_{\varepsilon > 0}$ and a limit $f^n \in L^\infty(0, T; L_s^1 \cap L \log L(\mathbb{R}^3))$ that serves as a weak solution to (1), corresponding to the initial data f_0^n specified in (3).

Furthermore, we prove that the solution is indeed axisymmetric if the initial profile is axisymmetric. To demonstrate that the solution f^n retains its axisymmetric property, we confirm that the collision operator $Q(f^n, f^n)$ preserves axisymmetry. For an axisymmetric test function $\phi(v) = \phi(T_\theta v)$, the rotational invariance of $a_{ij}(v - v_*)$ ensures that:

$$\int_{\mathbb{R}^3} Q(f^n, f^n)\phi \, dv = \int_{\mathbb{R}^3} Q(f^n(T_\theta v), f^n(T_\theta v))\phi \, dv,$$

where T_θ represents the rotation operator about the symmetry axis. Given that the initial data is axisymmetric, it follows that $f^n(t, v) = f^n(t, T_\theta v)$ for all $t > 0$.

Lastly, we pass to the limit as $n \to \infty$. For $p \ge 2$, we can proceed to show that for each $t \ge 0$, the family $\{f_t^n\}_{n \ge n_0}$ is relatively compact in $\mathcal{P}(\mathbb{R}^3)$. And together with additional proof on the equicontinuity in time, we use the Arzela-Ascoli theorem to prove that the family $\{f_t^n\}_{n \geq n_0}$ has a limit in $C([0, T]; \mathcal{P}(\mathbb{R}^3))$ in the weak sense. This provides the existence of a solution.

Since the support of our solution instantaneously escapes a line support in any time interval $[t_0, \infty)$ for any $t_0 > 0$, we obtain the ellipticity and the smoothness by the known bootstrap argument $L \log L \to L^2 \to H^1 \to \mathcal{S}$ as in Desvillettes-Villani [2]. Since the axisymmetric solution is now bounded in H^k for any $k \in \mathbb{N}$, we obtain that the solution is smooth and hence is analytic for any $t > 0$ by the result of Chen-Li-Xu in [1].

References

1. Chen H, Li WX, Xu CJ (2010) Analytic smoothness effect of solutions for spatially homogeneous Landau equation. J Different Equat 248(1):77–94. https://doi.org/10.1016/j.jde.2009.08.006
2. Desvillettes L, Villani C (2000) On the spatially homogeneous Landau equation for hard potentials. I. Existence, uniqueness and smoothness. Comm Partial Different Equati 25(1–2):179–259. https://doi.org/10.1080/03605300008821512
3. Fournier N, Guérin H (2009) Well-posedness of the spatially homogeneous Landau equation for soft potentials. J Funct Anal 256(8):2542–2560. https://doi.org/10.1016/j.jfa.2008.11.008
4. Fournier N, Heydecker D (2021) Stability, well-posedness and regularity of the homogeneous Landau equation for hard potentials. Annales de l'Institut Henri Poincaré C, Analyse non linéaire 38(6):1961–1987. https://www.sciencedirect.com/science/article/pii/S0294144921000251
5. Golse F, Imbert C, Ji S, Vasseur AF (2022) Local regularity for the space-homogeneous Landau equation with very soft potentials
6. Jang JW, Kim J (2025) On the axially symmetric solutions to the spatially homogeneous Landau equation

Maxwell-Stefan Diffusion Model Based on Moment Equations for Gas Mixtures

Milana Čolić and Srboljub Simić

Abstract The aim of this paper is to derive an isothermal Maxwell-Stefan model of diffusion from the moment equations of the Boltzmann system describing a mixture of monatomic and polyatomic gases. A comprehensive diffusion asymptotic analysis of the moment equations, corresponding to an isothermal multi-velocity model of Eulerian fluids, is carried out. In particular, the diffusive scaling of the explicit momentum production term–computed from the Boltzmann collision operator with a cutoff hard-potential-type kernel–enables the explicit determination of Maxwell-Stefan diffusion coefficients in terms of mesoscopic parameters. The model is subsequently validated against the Duncan and Toor experiment as a benchmark, confirming its match with experimental data.

Keywords Gas mixtures · Eulerian fluids · Polyatomic gases · Moment equations · Diffusion limit · Maxwell-stefan model

1 Introduction

A mixture is a medium composed of several identifiable constituents whose interactions–chemical, mechanical, thermal, or electromagnetic–pose significant challenges for modeling, even in simplified settings [11]. For example, considering only mechanical interactions still leaves many open problems [8, 16].

Diffusion is a central phenomenon in mixtures, describing mass transport between components. The classical Fick model represents diffusion as a macroscopic flow from regions of high to low concentration. It combines mass balance equations with

M. Čolić
Institute of Mathematics and Scientific Computing, University of Graz, Graz, Austria

M. Čolić (✉) · S. Simić
Department of Mathematics and Informatics, University of Novi Sad, Novi Sad, Serbia
e-mail: milana.colic@uni-graz.at

S. Simić
e-mail: ssimic@uns.ac.rs

© The Author(s) 2026

M. Grabe et al. (eds.), *Rarefied Gas Dynamics*, Springer Aerospace Technology,
https://doi.org/10.1007/978-3-032-00094-1_3

constitutive relations for diffusion fluxes [8], and can also be derived from the Boltzmann equation via Chapman-Enskog asymptotics, under the assumption of local equilibrium [11]. Although widely used in engineering, Fick's law neglects the fact that molecular motion is inherently random and influenced by interactions among different species. This simplification can lead to inaccurate predictions, particularly when cross-diffusion effects are significant, especially in multicomponent systems.

This study focuses on the Maxwell-Stefan diffusion model, which addresses some of Fick's limitations. It is derived from both mass and momentum balance laws under the assumption that macroscopic velocities are negligible compared to molecular velocities (the speed of sound). In the kinetic theory framework, this justifies the neglect of inertia and enables a systematic derivation from moment equations without assuming local equilibrium [9]. Consequently, the Maxwell-Stefan model provides a more accurate description of diffusion in multicomponent mixtures.

Experimental evidence supports this advantage. The Duncan and Toor experiment revealed non-classical diffusion phenomena–such as reverse diffusion, osmotic diffusion, and diffusion barriers–in three-component systems, which the Fick model fails to predict [21]. These effects are accurately captured by the Maxwell-Stefan model, motivating its further study.

The Maxwell-Stefan model has been derived from the Boltzmann system in various settings. For monatomic mixtures, [9] treats Maxwell molecules, while [10] extends the result to general collision kernels, yielding general diffusion coefficients. A higher-order model retaining inertia appears in [17, 18], and kinetic approaches have been extended to non-isothermal [20] and reactive polyatomic mixtures [3]. These derivations rely on computing moments of the Boltzmann collision operator, a task especially challenging in mixtures [19].

This paper builds on the moment system from [12], derived from the Boltzmann equations for monatomic and polyatomic mixtures. We study the isothermal case under a diffusion asymptotics regime–combining small Knudsen and Mach numbers–following [4]. This yields explicit Maxwell-Stefan diffusion coefficients via the momentum production term.

This work includes several novelties: (i) the mixture may include both monatomic and polyatomic components, with cross-collision operators; (ii) diffusion asymptotics is performed on the explicit momentum production term for relevant collision kernels, both mathematically [1, 2] and physically in the single polyatomic case [14, 23]; (iii) the resulting Maxwell-Stefan diffusion coefficients depend on mesoscopic parameters enabling fit to experimental data, here [15].

The paper is structured as follows: Sect. 2 reviews the isothermal multi-velocity model and production terms from [12]. Section 3 presents the scaling, and Sect. 4 carries out the limit, identifying Maxwell-Stefan coefficients.

2 Isothermal Multi-velocity Model for Eulerian Fluids Based on Kinetic Theory

We consider a mixture of $P \geq 1$ Eulerian fluids–i.e., with diagonal pressure tensors and zero heat flux–in the absence of chemical reactions. In the isothermal case, the evolution of each species' mass density ρ_α and velocity $\mathbf{v}_\alpha$, for each $\alpha = 1, \ldots, P$, is governed by the following balance laws [24],

$$\partial_t \rho_\alpha + \nabla_\mathbf{x} \cdot (\rho_\alpha \mathbf{v}_\alpha) = 0, \qquad \partial_t (\rho_\alpha \mathbf{v}_\alpha) + \nabla_\mathbf{x} (\rho_\alpha \mathbf{v}_\alpha \otimes \mathbf{v}_\alpha + p_\alpha \mathbb{I}) = \mathfrak{m}_\alpha, \qquad (1)$$

where p_α is a hydrodynamic pressure of the constituent α, while $\mathbb{I}$ is the identity matrix. Since the source term $\mathfrak{m}_\alpha$ corresponding to momentum exchange satisfies $\sum_{\alpha=1}^{P} \mathfrak{m}_\alpha = 0$, summation of (1) over α recovers conservation laws of mass and momentum density of the whole mixture [24],

$$\partial_t \rho + \nabla_\mathbf{x} \cdot (\rho \mathbf{v}) = 0, \qquad \partial_t (\rho \mathbf{v}) + \nabla_\mathbf{x} (\rho \mathbf{v} \otimes \mathbf{v} + \mathbf{p}) = 0, \qquad (2)$$

after defining the quantities:

$$\rho = \sum_{\alpha=1}^{P} \rho_\alpha; \quad \rho\mathbf{v} = \sum_{\alpha=1}^{P} \rho_\alpha \mathbf{v}_\alpha; \quad \mathbf{p} = \sum_{\alpha=1}^{P} (p_\alpha \mathbb{I} + \rho_\alpha (\mathbf{v}_\alpha - \mathbf{v}) \otimes (\mathbf{v}_\alpha - \mathbf{v})). \quad (3)$$

An important aspect of the modeling concerns the structure of the momentum production term $\mathfrak{m}_\alpha$. It describes the mutual momentum exchange between the components, but its functional form has to be determined. Initially, heuristic arguments were employed–at least in the context of cross-diffusion processes–assuming a linear dependence on the relative velocities $\mathbf{v}_\alpha - \mathbf{v}_\beta$. Another approach was systematically developed within the framework of extended thermodynamics, a macroscopic theory that uses the entropy principle to close the system of balance equations. However, it shares the limitations of all continuum theories–namely, that the phenomenological coefficients cannot be fully determined within this framework. Finally, the production term $\mathfrak{m}_\alpha$ can also be computed using the moment theory for the Boltzmann equation [12, 22, 25], as will be discussed in the following section.

2.1 Explicit Expression for the Momentum Production Term

The system (1) can be derived from Boltzmann-like equations for a gas mixture, under a specific ansatz for the distribution functions: Maxwellians that, for each species, have their own velocity and temperature. Such a distribution function arises as an *exact* solution of the maximum entropy principle for Eulerian fluids [12], and corresponds to an asymptotic regime in which equilibration within each separate component runs faster than equalization of species parameters in the gas as a whole [6, 7]. Moreover, for a specific choice of collision kernel, evaluating the multi-species Boltzmann collision operator yields an explicit expression for the production term $\mathfrak{m}_\alpha$ in terms of mesoscopic parameters.

In this paper, we consider the Boltzmann system for mixtures of monatomic gases (indexed by S_{mono}) and polyatomic gases (indexed by S_{poly}), where the modeling of polyatomic species follows the continuous internal energy approach [2, 5]. For the collision kernel used in recent rigorous analyses of the space-homogeneous problem [1, 2]–corresponding to cutoff hard potentials based on the total collisional energy (kinetic plus internal for polyatomic species)–the production term is computed in [12]. In this paper, we refer to Eq. (B6).

$$\mathfrak{m}_\alpha = -\sum_{\substack{\beta=1 \\ \beta \neq \alpha}}^{P} \frac{\rho_\alpha \rho_\beta}{m_\alpha m_\beta} (\mathbf{v}_\alpha - \mathbf{v}_\beta) \, \mathcal{K}_{\alpha\beta} \, \mathcal{M}_{\alpha\beta}, \tag{4}$$

where $\mathcal{K}_{\alpha\beta} = \mathcal{K}_{\beta\alpha} > 0$ is a dimension constant (24), and $\mathcal{M}_{\alpha\beta} = \mathcal{M}_{\beta\alpha} > 0$ is a dimensionless constant that depends on all mesoscopic parameters, and on whether α and β consists of monatomic or polyatomic gases. The structure of $\mathcal{M}_{\alpha\beta}$ is given below, while all constants are listed in Sect. 6,

$$\begin{aligned}
\mathcal{M}_{\alpha\beta} &= \mathcal{F}_{\alpha\beta}, \quad \alpha, \beta \in S_{\text{mono}}, \\
\mathcal{M}_{\alpha\beta} &= \mathcal{A}_{\alpha\beta} \left(\mathcal{B}_{\alpha\beta} \, \mathcal{F}_{\alpha\beta} + \eta_{\alpha\beta} \, \mathcal{C}_\beta \right), \quad \alpha \in S_{\text{mono}}, \beta \in S_{\text{poly}} \\
\mathcal{M}_{\alpha\beta} &= \mathcal{A}_{\alpha\beta} \left(\mathcal{B}_{\alpha\beta} \, \mathcal{F}_{\alpha\beta} + \eta_{\alpha\beta} \left(\mathcal{C}_\alpha + \mathcal{C}_\beta \right) \right), \quad \alpha, \beta \in S_{\text{poly}}.
\end{aligned} \tag{5}$$

Let us highlight that the term $\mathcal{F}_{\alpha\beta}$ depends non-linearly on $\left| \mathbf{v}_\alpha - \mathbf{v}_\beta \right|^2$, as specified in (25), and originates from the hard potentials in the molecular kinetic energy. Furthermore, the effect of polyatomic species is captured through the constants $\mathcal{A}_{\alpha\beta}$, $\mathcal{B}_{\alpha\beta}$, and $\mathcal{C}_\iota$ ($\iota \in S_{\text{poly}}$), where $\mathcal{C}_\iota$ specifically describes the influence of hard potentials in the molecular internal energy.

In the sequel, various approximations of $\mathcal{F}_{\alpha\beta}$ will be used, motivating the introduction of the following notation in the upcoming remark.

Remark 1 The following shorten notation is used for the coefficients (5),

$$\begin{aligned}
\mathcal{M}_{\alpha\beta}(x) &= x, \quad \alpha, \beta \in S_{\text{mono}}, \\
\mathcal{M}_{\alpha\beta}(x) &= \mathcal{A}_{\alpha\beta} \left(\mathcal{B}_{\alpha\beta} \, x + \eta_{\alpha\beta} \, \mathcal{C}_\beta \right), \quad \alpha \in S_{\text{mono}}, \beta \in S_{\text{poly}} \\
\mathcal{M}_{\alpha\beta}(x) &= \mathcal{A}_{\alpha\beta} \left(\mathcal{B}_{\alpha\beta} \, x + \eta_{\alpha\beta} \left(\mathcal{C}_\alpha + \mathcal{C}_\beta \right) \right), \quad \alpha, \beta \in S_{\text{poly}}.
\end{aligned} \tag{6}$$

Obviously, the choice $x = \mathcal{F}_{\alpha\beta}$ gives (5).

3 Diffusive Scaling

To recover the Maxwell-Stefan model of diffusion, one can rely on heuristic arguments and assume that the order of magnitude of inertia terms is smaller than that of the pressure gradient and source term. A more formal approach, based on scaled (dimensionless) balance laws, transforms the system (1) into a dimensionless set of balance laws with a small parameter. To that end, the following dimensionless variables are introduced for any component $\alpha = 1, \ldots, P$:

$$\hat{t} = \frac{t}{\tau}, \quad \hat{\mathbf{x}} = \frac{\mathbf{x}}{L}, \quad \hat{\mathbf{v}}_\alpha = \frac{\mathbf{v}_\alpha}{V}, \quad \hat{\rho}_\alpha = \frac{\rho_\alpha}{\rho_0}, \quad \hat{p}_\alpha = \frac{p_\alpha}{p_0}, \tag{7}$$

where τ, L, V, ρ_0 and p_0 are reference macroscopic variables. Typically for scaled equations of fluid/gas dynamics, the choice of reference pressure determines the final structure of dimensionless equations. Here, we assume $p_0 = \rho_0 C^2$, where C is the Newtonian speed of sound, which is of the same order of magnitude as the mean molecular velocity. Additionally, the reference macroscopic time is assumed to be $\tau = L/V$. The central assumption of the diffusive scaling is

$$V \ll C \quad \Rightarrow \quad \frac{V}{C} = \varepsilon \ll 1. \tag{8}$$

As a consequence, C may be used as universal reference velocity, which implies the following scaling of time and component velocities

$$\hat{t} = \varepsilon \frac{C}{L} t, \quad \varepsilon \hat{\mathbf{v}}_\alpha = \frac{\mathbf{v}_\alpha}{C}. \tag{9}$$

To scale the production term $\mathfrak{m}_\alpha$ of the form (1), additionally to (7), we assume

$$\hat{m}_\alpha = \frac{m_\alpha}{m_0}, \quad \hat{\mu}_{\alpha\beta} = \frac{\mu_{\alpha\beta}}{m_0}, \quad \hat{\mathcal{K}}_{\alpha\beta} = \frac{\mathcal{K}_{\alpha\beta}}{m_0 B_0}, \tag{10}$$

where the reference value B_0 for the Boltzmann collision kernel is imposed similarly as in [4], implying the scaling for $\mathcal{K}_{\alpha\beta}$ from (24), and finally for $\mathfrak{m}_\alpha$,

$$B_0 = \frac{1}{\varepsilon} \frac{m_0}{\rho_0} \frac{C}{L} \quad \Rightarrow \quad \hat{\mathcal{K}}_{\alpha\beta} = 4\pi(\hat{\mu}_{\alpha\beta})^{1-\frac{\varsigma_{\alpha\beta}}{2}} \hat{k}_{\alpha\beta} \quad \text{and} \quad \mathfrak{m}_\alpha = \frac{\rho_0 C^2}{L} \hat{\mathfrak{m}}_\alpha, \tag{11}$$

where ε is the Knudsen number. In conclusion, these scaling assumptions lead to the following scaled balance laws, for any gas component $\alpha = 1, \ldots, P$,

$$\varepsilon \left(\partial_t \rho_\alpha^\varepsilon + \nabla_{\mathbf{x}} \cdot \left(\rho_\alpha^\varepsilon \mathbf{v}_\alpha^\varepsilon \right) \right) = 0,$$
$$\varepsilon^2 \partial_t \left(\rho_\alpha^\varepsilon \mathbf{v}_\alpha^\varepsilon \right) + \varepsilon^2 \nabla_{\mathbf{x}} \left(\rho_\alpha^\varepsilon \mathbf{v}_\alpha^\varepsilon \otimes \mathbf{v}_\alpha^\varepsilon \right) + \nabla_{\mathbf{x}} p_\alpha^\varepsilon = \mathfrak{m}_\alpha^\varepsilon, \tag{12}$$

where hats are dropped for the simplicity. The production term (4) in this scaling becomes, in the light of Remark 1,

$$\mathrm{m}_\alpha^\varepsilon = -\sum_{\substack{\beta=1 \\ \beta \neq \alpha}}^{P} \frac{\rho_\alpha^\varepsilon \rho_\beta^\varepsilon}{m_\alpha m_\beta} (\mathbf{v}_\alpha^\varepsilon - \mathbf{v}_\beta^\varepsilon)\,\hat{\mathcal{K}}_{\alpha\beta}\,\mathcal{M}_{\alpha\beta}^\varepsilon, \quad \mathcal{M}_{\alpha\beta}^\varepsilon = \mathcal{M}_{\alpha\beta}(\mathcal{F}_{\alpha\beta}^\varepsilon), \tag{13}$$

since the scaling affects only $\mathcal{F}_{\alpha\beta}^\varepsilon$ given in (27). Note that the equations (12) are *exact* in the sense that no terms are neglected. However, the field variables ρ_α^ε and $\mathbf{v}_\alpha^\varepsilon$ are moments of an approximate (local Maxwellian) distribution function that depends on the small parameter ε, as indicated by the superscript.

A key advantage of the kinetic approach is that it allows for the estimation of $\mathrm{m}_\alpha^\varepsilon$ with respect to ε. A polynomial expansion around $\varepsilon = 0$ yields

$$\mathcal{M}_{\alpha\beta}^\varepsilon = \mathcal{M}_{\alpha\beta}^{(0)} + \varepsilon^2 \mathcal{M}_{\alpha\beta}^{(2)} + O(\varepsilon^4), \tag{14}$$

where $\mathcal{M}_{\alpha\beta}^{(0)}$ and $\mathcal{M}_{\alpha\beta}^{(2)}$ depend on the nature of the interacting components, in particular on whether species α and β are monatomic or polyatomic. A brief inspection of the momentum balance law $(12)_2$, in light of (13)–(14), shows that the inertia terms on the left-hand side, of order ε^2, are balanced by corresponding terms in $\mathrm{m}_\alpha^\varepsilon$ of the same order of magnitude. This refined analysis is a particular strength of the kinetic theory framework and cannot be recovered through heuristic macroscopic reasoning.

Moreover, the explicit form of the coefficients $\mathcal{M}_{\alpha\beta}^\varepsilon$ allows for a detailed identification of the terms in the expansion (14). Namely, denoting

$$\bar{\mathcal{F}}_{\alpha\beta} = \lim_{\varepsilon \to 0^+} \mathcal{F}_{\alpha\beta}^\varepsilon = 2^{\frac{\varsigma_{\alpha\beta}}{2}} \frac{4}{3\sqrt{\pi}} \Gamma\left(\tfrac{\varsigma_{\alpha\beta}+5}{2}\right), \tag{15}$$

the following expansion holds

$$\mathcal{F}_{\alpha\beta}^\varepsilon = \bar{\mathcal{F}}_{\alpha\beta}\left(1 + \varepsilon^2 \frac{\varsigma_{\alpha\beta}\,\mu_{\alpha\beta}}{10} \left|\mathbf{v}_\alpha^\varepsilon - \mathbf{v}_\beta^\varepsilon\right|^2 + O(\varepsilon^4)\right). \tag{16}$$

Then, the leading order in ε of (14) is, with the notation of Remark 1,

$$\mathcal{M}_{\alpha\beta}^{(0)} = \mathcal{M}_{\alpha\beta}(\bar{\mathcal{F}}_{\alpha\beta}), \tag{17}$$

while the second order in ε reads

$$\mathcal{M}_{\alpha\beta}^{(2)} = \bar{\mathcal{F}}_{\alpha\beta} \frac{\varsigma_{\alpha\beta}\,\mu_{\alpha\beta}}{10} \left|\mathbf{v}_\alpha^\varepsilon - \mathbf{v}_\beta^\varepsilon\right|^2, \quad \alpha, \beta \in S_{\mathrm{mono}},$$

$$\mathcal{M}_{\alpha\beta}^{(2)} = \mathcal{A}_{\alpha\beta}\,\mathcal{B}_{\alpha\beta}\,\bar{\mathcal{F}}_{\alpha\beta} \frac{\varsigma_{\alpha\beta}\,\mu_{\alpha\beta}}{10} \left|\mathbf{v}_\alpha^\varepsilon - \mathbf{v}_\beta^\varepsilon\right|^2, \quad \alpha \in S_{\mathrm{mono}} \text{ or } S_{\mathrm{poly}}, \beta \in S_{\mathrm{poly}}$$

$$\tag{18}$$

Note that the information on the hard potentials in the molecular internal energy for polyatomic components is suppressed in (18), since the given expressions do not depend on C_ι, $\iota \in S_{\mathrm{poly}}$.

4 Diffusive Limit Recovering the Maxwell-Stefan Model

To recover the Maxwellâ£"Stefan model of diffusion, we consider the asymptotic limit $\varepsilon \to 0^+$. In this limit, system (12) formally reduces to:

$$\partial_t \rho_\alpha + \nabla_\mathbf{x} \cdot (\rho_\alpha \mathbf{v}_\alpha) = 0, \qquad \nabla_\mathbf{x} p_\alpha = \mathfrak{m}_\alpha, \tag{19}$$

where:

$$\rho_\alpha = \lim_{\varepsilon \to 0^+} \rho_\alpha^\varepsilon, \quad \mathbf{v}_\alpha = \lim_{\varepsilon \to 0^+} \mathbf{v}_\alpha^\varepsilon, \quad p_\alpha = \lim_{\varepsilon \to 0^+} p_\alpha^\varepsilon, \quad \text{and}$$

$$\mathfrak{m}_\alpha = \lim_{\varepsilon \to 0^+} \mathfrak{m}_\alpha^\varepsilon = - \sum_{\substack{\beta=1 \\ \beta \neq \alpha}}^{P} \frac{\rho_\alpha \rho_\beta}{m_\alpha m_\beta} (\mathbf{v}_\alpha - \mathbf{v}_\beta) \hat{\mathcal{K}}_{\alpha\beta} \mathcal{M}_{\alpha\beta}^{(0)}, \tag{20}$$

with the coefficients in (17). Let us remark that it is *assumed* that solutions of (12) asymptotically converge, in a suitable sense, to solutions of (19).

This asymptotic limit enables to express dimensionless Maxwell-Stefan diffusion coefficients $D_{\alpha\beta}$ in terms of $\mathcal{M}_{\alpha\beta}^{(0)}$, based on the results from [10],

$$D_{\alpha\beta} = \left(\hat{\mathcal{K}}_{\alpha\beta} \mathcal{M}_{\alpha\beta}^{(0)} \right)^{-1}, \tag{21}$$

which provides an insight into possible practical applications of the analytical results derived in this work. In particular, the coefficients $D_{\alpha\beta}$ (21) are expressed in terms of mesoscopic parameters, which can be adjusted to match experimental data, for instance in the context of the Duncan and Toor experiment [15].

4.1 Benchmark Test: Duncan and Toor Experiment

The Duncan and Toor experiment [15], a benchmark example for the Maxwell-Stefan model of diffusion, involves a mixture of H_2, N_2 and CO_2, with the relevant parameters listed in Table 1. Using these parameters, the coefficients $D_{\alpha\beta}$ defined in (21) become functions of $\zeta_{\alpha\beta}$, $\eta_{\alpha\beta}$ and $\hat{k}_{\alpha\beta}$. The key question is whether a combination of these parameters can be found to match the experimental data for $D_{\alpha\beta}$. Since the problem is undetermined, one possible approach is to express $\eta_{\alpha\beta}$ and $\hat{k}_{\alpha\beta}$ as functions of $\zeta_{\alpha\beta}$, so that $\zeta_{\alpha\beta}$ remains the only free parameter. Using (17), (15), (5)$_3$ and (11), equation (21) is then solved for $\eta_{\alpha\beta}$,

Table 1 Parameters (dimensionless) used in Fig. 1

	Component-1 (H_2)	Component-2 (N_2)	Component-3 (CO_2)
m_α [18]	0.0811	1.1351	1.7838
a_α [23]	-0.0304	0.0036	0.9804
	Interaction 1 and 2	Interaction 2 and 3	Interaction 1 and 3
$D_{\alpha\beta}$ [18]	1.4866	0.2998	1.2136
$\mu_{\alpha\beta}$ [18]	0.0757	0.6937	0.0776

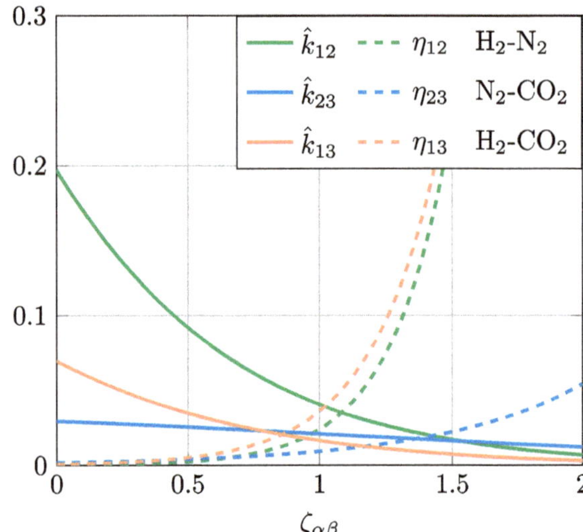

Fig. 1 Parameters $\hat{k}_{\alpha\beta}$ and $\eta_{\alpha\beta}$ as functions of $\zeta_{\alpha\beta}$, adjusted to match the values of $D_{\alpha\beta}$ from the Duncan and Toor experiment through (21)

$$\eta_{\alpha\beta} = \left(\frac{1}{\hat{k}_{\alpha\beta}\, D_{\alpha\beta}\, 4\pi(\hat{\mu}_{\alpha\beta})^{1-\frac{\zeta_{\alpha\beta}}{2}} \mathcal{A}_{\alpha\beta}\mathcal{B}_{\alpha\beta}\, \bar{\mathcal{F}}_{\alpha\beta}} - 1 \right) \frac{\mathcal{B}_{\alpha\beta}\, \bar{\mathcal{F}}_{\alpha\beta}}{(\mathcal{C}_\alpha + \mathcal{C}_\beta)}, \qquad (22)$$

where $\hat{k}_{\alpha\beta}$ is chosen to ensure the positivity of $\eta_{\alpha\beta}$ for all $\zeta_{\alpha\beta} \in [0, 2]$,

$$\hat{k}_{\alpha\beta} = \left(D_{\alpha\beta} 4\pi(\hat{\mu}_{\alpha\beta})^{1-\frac{\zeta_{\alpha\beta}}{2}} \mathcal{A}_{\alpha\beta}\mathcal{B}_{\alpha\beta}\, \bar{\mathcal{F}}_{\alpha\beta} \right)^{-1} - 10^{-4}. \qquad (23)$$

The dependence of such $\hat{k}_{\alpha\beta}$ and $\eta_{\alpha\beta}$ on $\zeta_{\alpha\beta}$ is illustrated in Fig. 1. These results show that the momentum production term computed in [12] is fully compatible, in the diffusion limit, with the Duncan & Toor experiment [15], while still retaining a free parameter ($\zeta_{\alpha\beta}$) that can be tuned to capture additional flow properties.

5 Conclusion and Outlook

Starting from the multi-velocity model of Eulerian fluids (1) derived as moment equations of the Boltzmann system for monatomic and polyatomic gases with an explicit momentum production term (4), and under scaling assumptions corresponding to small Knudsen and Mach numbers, the scaled macroscopic laws (12) are derived. In the limit, these equations formally lead to the Maxwell-Stefan equations (19). The explicit expression for the momentum production term thus provides Maxwell-Stefan diffusion coefficients (21) that depend on mesoscopic parameters of the collision kernel. The model is validated against the Duncan & Toor experiment. An interesting application of this study is to further identify the free parameters that match experimental data, using a strategy similar to [14], and to develop higher-order moment theory [17].

Acknowledgements Authors acknowledge Grants Nos. 451-03-66/2024-03/200125 and 451-03-65/2024–03/200125 from the Ministry of Science, Technological Development and Innovation of the Republic of Serbia.

6 List of Constants

The mesoscopic parameters are:

- free parameter $\zeta_{\alpha\beta} = \zeta_{\beta\alpha} \geq 0$ is the rate of hard potentials,
- $a_\iota > -1$, $\iota \in S_{\mathrm{poly}}$, is related to the polyatomic molecule number of internal degrees of freedom or the specific heat [12, 23],
- free parameter $\eta_{\alpha\beta} = \eta_{\beta\alpha} \geq 0$ controls an influence of the internal energy,
- $k_{\alpha\beta} = k_{\beta\alpha} \geq 0$ is an appropriate dimensional constant whose value is a free parameter $\hat{k}_{\alpha\beta}$.

Let m_α be the molecular mass and let $\mu_{\alpha\beta} = \frac{m_\alpha m_\beta}{m_\alpha + m_\beta}$ be the reduced mass (both scaled according to (10)). Let k be the Boltzmann constant, and Γ denote the usual Gamma function. The constants involved in (5) read

$$\bullet\, \mathcal{K}_{\alpha\beta} = 4\pi\mu_{\alpha\beta} \left(\frac{kT}{\mu_{\alpha\beta}}\right)^{\frac{\zeta_{\alpha\beta}}{2}} k_{\alpha\beta}, \quad \text{with the scaling } \hat{k}_{\alpha\beta} = \frac{k_{\alpha\beta}}{B_0\, C^{-\zeta_{\alpha\beta}}}, \tag{24}$$

$$\bullet\, \mathcal{F}_{\alpha\beta} = 2^{\frac{\zeta_{\alpha\beta}}{2}} \Gamma\left(\frac{\zeta_{\alpha\beta}+5}{2}\right) e_1^{-b_{\alpha\beta}} \tilde{F}_1\left(\frac{\zeta_{\alpha\beta}+5}{2}, \frac{5}{2}, b_{\alpha\beta}\right), \quad b_{\alpha\beta} = \frac{\mu_{\alpha\beta}}{2kT} \left|\mathbf{v}_\alpha - \mathbf{v}_\beta\right|^2, \tag{25}$$

where $_1\tilde{F}_1(a, b, z)$ is the hypergeometric function, defined, for $b > a > 0$, by

$$\Gamma(b-a)\,\Gamma(a)_1\,\tilde{F}_1(a, b, z) = \int_0^1 e^{zt} t^{a-1} (1-t)^{b-a-1}\, \mathrm{d}t, \tag{26}$$

which undergoes the following scaling, with all quantities in dimensionless form,

$$\mathcal{F}_{\alpha\beta}^{\varepsilon} = 2^{\frac{\zeta_{\alpha\beta}}{2}} \Gamma\left(\frac{\zeta_{\alpha\beta}+5}{2}\right) e_1^{-b_{\alpha\beta}^{\varepsilon}} \tilde{F}_1\left(\frac{\zeta_{\alpha\beta}+5}{2}, \frac{5}{2}, b_{\alpha\beta}^{\varepsilon}\right), \quad b_{\alpha\beta}^{\varepsilon} = \varepsilon^2 \frac{\mu_{\alpha\beta}}{2}\left|\mathbf{v}_{\alpha}^{\varepsilon} - \mathbf{v}_{\beta}^{\varepsilon}\right|^2. \tag{27}$$

$$\begin{aligned}
&\bullet\, \mathcal{A}_{\alpha\beta} = \frac{\Gamma\left(a_\beta + \frac{5}{2}\right)}{\Gamma\left(a_\beta + \frac{\zeta_{\alpha\beta}+5}{2}\right)}, \quad \alpha \in S_{\mathrm{mono}}, \beta \in S_{\mathrm{poly}} \\[2mm]
&\bullet\, \mathcal{A}_{\alpha\beta} = \frac{\Gamma\left(a_\alpha + a_\beta + \frac{7}{2}\right)}{\Gamma\left(a_\alpha + a_\beta + \frac{\zeta_{\alpha\beta}+7}{2}\right)}, \quad \alpha, \beta \in S_{\mathrm{poly}}, \\[2mm]
&\bullet\, \mathcal{B}_{\alpha\beta} = \frac{2}{\sqrt{\pi}} \Gamma\left(\frac{\zeta_{\alpha\beta}+3}{2}\right), \\[2mm]
&\bullet\, \mathcal{C}_\iota = \frac{\Gamma\left(a_\iota + 1 + \frac{\zeta_{\alpha\beta}}{2}\right)^2}{\Gamma\left(a_\iota + 1\right)^2}\left(\frac{\mu_{\alpha\beta}}{\sum_{\alpha=1}^{P} m_\alpha}\right)^{\zeta_{\alpha\beta}/2}.
\end{aligned} \tag{28}$$

References

1. Alonso RJ, Čolić M (2024) Integrability propagation for a Boltzmann system describing polyatomic gas mixtures. SIAM J Math Anal 56(1):1459–1494. https://doi.org/10.1137/22M1539897
2. Alonso RJ, Čolić M, Gamba IM (2024) The Cauchy problem for Boltzmann bi-linear systems: the mixing of monatomic and polyatomic gases. J Stat Phys 191:9. https://doi.org/10.1007/s10955-023-03221-4
3. Anwasia B, Bisi M, Salvarani F, Soares AJ (2020) On the Maxwell-Stefan diffusion limit for a reactive mixture of polyatomic gases in non-isothermal setting. Kinet Relat Models 13(1):63–95. https://doi.org/10.3934/krm.2020003
4. Anwasia B, Simić S (2022) Maximum entropy principle approach to a non-isothermal Maxwell-Stefan diffusion model. Appl Math Lett 129(9):107949. https://doi.org/10.1016/j.aml.2022.107949
5. Baranger C, Bisi M, Brull S, Desvillettes L (2018) On the Chapman-Enskog asymptotics for a mixture of monoatomic and polyatomic rarefied gases. Kinet Relat Models 11(4):821–858. https://doi.org/10.3934/krm.2018033
6. Bisi M, Martalò G, Spiga G (2012) Multi-temperature hydrodynamic limit from kinetic theory in a mixture of rarefied gases. Acta Appl Math 122:37–51. https://doi.org/10.1007/s10440-012-9724-0
7. Bisi M, Martalò G, Spiga G (2011) Multi-temperature Euler hydrodynamics for a reacting gas from a kinetic approach to rarefied mixtures with resonant collisions. Europhys Lett 95(5):55002. https://doi.org/10.1209/0295-5075/95/55002
8. Bothe D, Dreyer W (2015) Continuum thermodynamics of chemically reacting fluid mixtures. Acta Mech 226(6):1757–1805. https://doi.org/10.1007/s00707-014-1275-1
9. Boudin L, Grec B, Salvarani F (2015) The Maxwell-Stefan diffusion limit for a kinetic model of mixtures. Acta Appl Math 136(1):79–90. https://doi.org/10.1007/s10440-014-9886-z
10. Boudin L, Grec B, Pavan V (2017) The Maxwell-Stefan diffusion limit for a kinetic model of mixtures with general cross sections. Nonlinear Anal 159:40–61. https://doi.org/10.1016/j.na.2017.01.010
11. Chapman S, Cowling TG (1991) The mathematical theory of non-uniform gases. Cambridge University Press, Cambridge, UK

12. Čolić M (2024) Multi-velocity and multi-temperature model of the mixture of polyatomic gases issuing from kinetic theory. Phys Fluids 36:067134. https://doi.org/10.1063/5.0211158
13. Desvillettes L, Monaco R, Salvarani F (2005) A kinetic model allowing to obtain the energy law of Polytropic gases in the presence of chemical reactions. Eur J Mech B Fluids 24:219–236. https://doi.org/10.1016/j.euromechflu.2004.07.004
14. Djordjić V, Oblapenko G, Pavić-Čolić M, Torrilhon M (2023) Boltzmann collision operator for polyatomic gases in agreement with experimental data and DSMC method. Continuum Mech Thermodyn 35:103–119. https://doi.org/10.1007/s00161-022-01167-8
15. Duncan JB, Toor HL (1962) An experimental study of three component gas diffusion. AIChE J 8(1):38–41
16. Giovangigli V (1999) Multicomponent flow modeling. Birkhäuser Boston Inc., Boston, MA
17. Grec B, Simić S (2023) Higher-order Maxwell-Stefan model of diffusion. La Matematica 2:962–991. https://doi.org/10.1007/s44007-023-00071-0
18. Grec B, Simić S (2024) Numerical study of the higher-order Maxwell-Stefan model of diffusion, trends in mathematics, to appear
19. Gupta VK, Torrilhon M (2015) Comparison of relaxation phenomena in binary gas-mixtures of Maxwell molecules and hard spheres. Comput Math with Appl 70(1):73–88. https://doi.org/10.1016/j.camwa.2015.04.028
20. Hutridurga H, Salvarani F (2017) Maxwell-Stefan diffusion Asymptotics for gas mixtures in non-isothermal setting. Nonlinear Anal 159:285–297. https://doi.org/10.1016/j.na.2017.03.019
21. Krishna R, Wesselingh JA (1997) The Maxwell-Stefan approach to mass transfer. Chem Eng Sci 52(6):861–911. https://doi.org/10.1016/S0009-2509(96)00458-7
22. Pavić-Čolić M (2019) Multi-velocity and multi-temperature model of the mixture of polyatomic gases issuing from kinetic theory. Phys Lett A 383(24):2829–2835. https://doi.org/10.1016/j.physleta.2019.06.009
23. Pavić-Čolić M, Simić S (2022) Kinetic description of polyatomic gases with temperature-dependent specific heats. Phys Rev Fluids 7:083401. https://doi.org/10.1103/PhysRevFluids.7.083401
24. Ruggeri T, Simić S (2007) On the hyperbolic system of a mixture of Eulerian fluids: a comparison between single- and multi-temperature models. Math Methods Appl Sci 30(7):827–849. https://doi.org/10.1002/mma.813
25. Torrilhon M (2016) Modeling nonequilibrium gas flow based on moment equations. Annu Rev Fluid Mech 48(1):429–458. https://doi.org/10.1146/annurev-fluid-122414-034259

Lattice Boltzmann Scheme for Drift Diffusion Equations in Cold Plasma Applications

Nathalie Bonamy Parrilla, Stéphane Brull, and François Rogier

Abstract This paper presents a Lattice Boltzmann Method (LBM) tailored for solving drift-diffusion equations in cold plasma applications. The proposed scheme is aimed to be a first step to address the challenges of simulating cold plasmas, characterized by non-equilibrium conditions and complex interactions among electrons, ions, and electric fields. By employing a parabolic scaling and simplifying assumptions, the method ensures computational efficiency while maintaining accuracy. Validation is performed through numerical test cases. While promising, future work will focus on incorporating energy dynamics and handling variable diffusion coefficients to enhance the method's applicability in diverse plasma scenarios.

Keywords Lattice boltzmann · Drift diffusion · Equivalent equations

1 Introduction

This paper is devoted to the approximation of a drift diffusion equation for cold plasma applications using a Lattice Boltzmann Method (LBM) approach. Cold plasmas are a topic of significant interest due to their numerous industrial applications, ranging from airflow control in aircraft industry to biomedical technologies. Despite their potential, the numerical simulation of cold plasmas poses considerable challenges. Traditional approaches to modeling such systems often rely on finite volumes or similar methods, which, while robust, can be computationally expensive and challenging to implement efficiently. In contrast, the Lattice Boltzmann Method (LBM) offers an attractive alternative due to its inherent simplicity in implementation and natural suitability for parallel computing.

N. B. Parrilla (✉) · S. Brull
Université de Bordeaux, Institut de Mathématiques de Bordeaux UMR 5251, Talence, France
e-mail: nathalie.bonamy-parrilla@math.u-bordeaux.fr

F. Rogier
70 Rue Achille Viadieu, Toulouse, France

© The Author(s) 2026

M. Grabe et al. (eds.), *Rarefied Gas Dynamics*, Springer Aerospace Technology,
https://doi.org/10.1007/978-3-032-00094-1_4

33

Cold plasmas, which are often weakly ionized, consist of neutral particles, ions, and electrons. The dynamics of the charged particles are governed by their interactions with the electric field and the neutral particles which act as a neutral gas background. They are characterized by non-equilibrium conditions where electron temperatures significantly exceed those of heavy particles. Of particular interest are the behaviors of electrons, which share a structural similarity with ions in the chosen modeling equations. In this work, we consider high-pressure cold plasmas, for which a fluid description is appropriate. The equations used to describe these plasmas are drift-diffusion equation for the electron density, coupled with a mass creation equation for positive ions and a Poisson equation for the electric potential. Negative ions are omitted in the current study; however, the proposed scheme can be readily adapted to handle both negative and positive ions. This model is commonly referred as Local Field Approximation (see [4, 6, 7]). The system under study is the following:

$$\begin{cases} \partial_t \rho + \partial_x(\mu\rho\partial_x\phi - D\partial_x\rho) = \nu\rho, \\ \partial_x^2\phi = \frac{e}{\epsilon_0}(\rho - \rho_{ion}), \\ \partial_t \rho_{ion} = \nu\rho. \end{cases} \tag{1}$$

where ρ is the electron density, ρ_i the positive ions density, ϕ the electric potential, ε_0 the dielectric permittivity and e the electronic charge. The transport coefficients, – namely the mobility coefficient μ, the diffusion coefficient D and the ionization frequency ν – are physical quantities that depend on the reduced electric field $|\partial_x\phi|/N$, where N is the neutral particles density. Given the limitations of LBM for diffusion equations, we will neglect the dependency of D in $\partial_x\phi$, which is acceptable for some discharge regimes where the diffusion is secondary compared to the drift and ionization.

In this work, we propose a Lattice Boltzmann scheme tailored to the drift-diffusion equations with source terms, designed for the simulation of cold plasma dynamics.

Our approach will be validated through a series of numerical test cases, demonstrating the accuracy of the proposed scheme. Finally, we will present an application that compares our LBM-based results against those obtained using a finite volume method, highlighting the advantage of using this method in physical applications.

2   Numerical Scheme for Electron Density

This section presents the numerical scheme employed to simulate the electron density (see (1)). We focus only on 1D case, which can be easily generalized to higher dimensions without significant additional complexity.

2.1 Presentation of the Scheme

2.1.1 Space and Time Discretization

We consider a uniform grid $x \in \Delta x \mathbb{Z}$ with a space step Δx. The parabolic scaling is chosen thus the time discretization is given by:

$$\Delta t = \frac{\Delta x^2}{D}$$

which ensures a constant diffusion coefficient D regardless of the mesh size, contrary to the acoustic scaling (see [1]).

Lattice Structure and Distribution Functions

The chosen lattice structure is the so-called D1Q2 model (see [2]). It consists of two discrete velocities given by

$$\lambda_{1,2} = (-1)^{1,2} \frac{\Delta x}{\Delta t}$$

each associated with a distribution function f_i, $i = 1, 2$. The sum of the distribution functions defines the macroscopic electron density ρ:

$$\rho = f_1 + f_2$$

which is the unknown quantity in the drift-diffusion equation. We will denote $f_j(n \Delta t, i \Delta x)$ by $f_{j,i}^n$ from now, and we will use index in a similar way for all the variables (ρ, ϕ...).

2.1.2 Two-Step Scheme

The numerical scheme consists of two main steps:

1. Collision Step: This step involves relaxation toward an equilibrium state and accounts for ionization effects, which introduce mass generation. It is computed as follows

$$f_{1,i}^{n+1/2} = (1 - \omega) f_{1,i}^n + \frac{\omega}{2} \left(\rho_i^n - \frac{\Delta t}{\Delta x} (\mu \rho \partial_x \phi)_i^n \right) + \Delta t \, (\nu \rho)_i^n,$$

$$f_{2,i}^{n+1/2} = (1 - \omega) f_{2,i}^n + \frac{\omega}{2} \left(\rho_i^n + \frac{\Delta t}{\Delta x} (\mu \rho \partial_x \phi)_i^n \right) + \Delta t \, (\nu \rho)_i^n,$$

with ω being the so-called relaxation parameter (see [5]), and where the parameters μ and ν may depend on the modulus of the electric field.

2. Transport Step: In this step, the distribution functions are advected along their corresponding lattice velocities as follows

$$f_{1,i}^{n+1} = f_{1,i+1}^{n+1/2},$$
$$f_{2,i}^{n+1} = f_{2,i-1}^{n+1/2}.$$

The boundary conditions are imposed during the transport step on the boundary cells. Dirichlet boundary conditions on the density are satisfied with the relation:

$$f_{2,0} = \rho_b - f_{1,0},$$
$$f_{1,N} = \rho_b - f_{2,N}.$$

Then the scheme can finally be recast as:

$$f_{1,i}^{n+1} = (1 - \omega)f_{1,i+1}^n + \frac{\omega}{2}(\rho_{i+1}^n - \frac{\Delta t}{\Delta x}(\mu\rho\partial_x\phi)_{i+1}^n) + \Delta t(\nu\rho)_{i+1}^n, \qquad (2)$$

$$f_{2,i}^{n+1} = (1 - \omega)f_{2,i-1}^n + \frac{\omega}{2}(\rho_{i-1}^n + \frac{\Delta t}{\Delta x}(\mu\rho\partial_x\phi)_{i-1}^n) + \Delta t(\nu\rho)_{i-1}^n. \qquad (3)$$

2.1.3 Coupling with Electric Field and Positive Ions

The LBM is coupled with the electric field through Poisson's equation and the equations on ions (see (1)). The transport coefficients ν and μ are then interpolated from a dataset of physical values, depending on the reduced electric field $|\partial_x\phi|/N$.

2.2 Consistency and Equivalent Equations

The study of consistency for a LB scheme, is an important issue. To address this, two methods are commonly used : derivation of the fluid limit of the lattice Boltzmann equation by using Champan-Enskog expansions (see [5]) or the computation of equivalent equations based on Taylor expansions (see [3]). In this paper, we adopt the latter approach. In the following, we compute the equivalent equations.

We start by summing relations (2)–(3) to obtain the update on the density:

$$\rho_i^{n+1} = (1 - \frac{\omega}{2})(f_{1,i+1}^n + f_{2,i-1}^n) + \frac{\omega}{2}(f_{2,i+1}^n + f_{1,i-1}^n)$$
$$+ \frac{\omega\Delta t}{2\Delta x}((\mu\partial_x\phi)_{i-1}^n - (\mu\partial_x\phi)_{i+1}^n) + \frac{\Delta t}{2}((\nu\rho)_{i+1}^n + (\nu\rho)_{i-1}^n),$$

where ρ^n_{i+1} and ρ^n_{i-1} were replaced by their expressions in terms of f. By applying again (2)–(3) to this relation we obtain:

$$\begin{aligned}
\rho^{n+1}_i = {}& (1 - \frac{\omega}{2})[(1 - \frac{\omega}{2})(f^{n-1}_{1,i+2} + f^{n-1}_{2,i-2}) + \frac{\omega}{2}(f^{n-1}_{1,i-2} + f^{n-1}_{2,i+2}) \\
& + \frac{\omega \Delta t}{2\Delta x}((\mu\rho\partial_x\phi)^{n-1}_{i-2} - (\mu\rho\partial_x\phi)^{n-1}_{i+2}) + \frac{\Delta t}{2}((\nu\rho)^{n-1}_{i+2} + (\nu\rho)^{n-1}_{i-2})] \\
& + \frac{\omega}{2}(\rho^{n-1}_i + \Delta t(\nu\rho)^{n-1}_i) \\
& + \frac{\omega \Delta t}{2\Delta x}((\mu\rho\partial_x\phi)^n_{i-1} - (\mu\rho\partial_x\phi)^n_{i+1}) \\
& + \frac{\Delta t}{2}((\nu\rho)^n_{i+1} + (\nu\rho)^n_{i-1}).
\end{aligned} \tag{4}$$

Now, to rewrite this expression only in terms of ρ and not in terms of f we look at the expression of the density in the neighbour cells:

$$\begin{aligned}
\rho^n_{i+1} = {}& (1 - \frac{\omega}{2})(f^{n-1}_{1,i+2} + f^{n-1}_{2,i}) + \frac{\omega}{2}(f^{n-1}_{2,i+2} + f^{n-1}_{1,i}) \\
& + \frac{\omega \Delta t}{2\Delta x}((\mu\rho\partial_x\phi)^{n-1}_i - (\mu\rho\partial_x\phi)^{n-1}_{i+2}) \\
& + \frac{\Delta t}{2}((\nu\rho)^{n-1}_{i+2} + (\nu\rho)^{n-1}_i),
\end{aligned}$$

The sum of this expression with the corresponding one for ρ^n_{i-1} can be identified and replaced in (4):

$$\begin{aligned}
\rho^{n+1}_i = {}& (1 - \frac{\omega}{2})(\rho^n_{i+1} + \rho^n_{i-1} - \rho^{n-1}_i - \Delta t(\nu\rho)^{n-1}_i) + \frac{\omega}{2}(\rho^{n-1}_i + \Delta t(\nu\rho)^{n-1}_i) \\
& + \frac{\omega \Delta t}{2\Delta x}((\mu\rho\partial_x\phi)^n_{i-1} - (\mu\rho\partial_x\phi)^n_{i+1}) + \frac{\Delta t}{2}((\nu\rho)^n_{i+1} + (\nu\rho)^n_{i-1}).
\end{aligned} \tag{5}$$

By performing Taylor expansions on (5), we get:

$$\partial_t \rho - (\frac{1}{\omega} - \frac{1}{2})\frac{\Delta x^2}{\Delta t}\partial^2_x \rho + \partial_x(\mu\rho\partial_x\phi) = \nu\rho + \mathcal{O}(\Delta t),$$

where $(\frac{1}{\omega} - \frac{1}{2})\frac{\Delta x^2}{\Delta t} = D$ thanks to the choice of the parameters. The equivalent equation analysis shows that the scheme is in agreement with the fluid model.

3 Numerical Results

In this section we present numerical results obtained with the LB scheme. Two test cases were considered, the first one being academic and considering the dimensionless stationary problem. In the second test case the LBM scheme is directly compared to a finite volumes scheme.

3.1 Results for the Stationary Convection Diffusion Equation

In this subsection we consider the stationary problem as a way to test the stability of the scheme. The equation on the density is coupled with the Poisson equation through the drift velocity:

$$\rho u - D\partial_x \rho = 0,$$
$$u(x) = \mu \partial_x \phi,$$
$$\partial_x^2 \phi = \lambda(\rho - \rho_{ion}),$$
$$\rho(0) = \rho_{ion}, \ \rho(1) = \rho_{ion} + \delta,$$
$$\phi(0) = 0,$$

on $\Omega = [0, 1]$, where $\rho_{ion} = 1$ and δ to be precised later. The solution for the density is given by:

$$\rho_{sol}(x) = \rho_{ion} \exp\left(\frac{\phi(x)\mu}{D}\right).$$

In order to compute an approximated solution of the Poisson problem, we need an assumption on ϕ. If $|\phi|$ is relatively small, the exponential can be linearized and the Poisson problem then becomes:

$$\partial_x^2 \phi = \frac{\lambda\mu\phi}{D},$$

which then can be solved analytically. This assumption is satisfied if δ is small enough. Finally the quasi analytic solution, taking into account the boundary conditions, is given by:

$$\phi(x) = \ln\left(1 + \frac{\delta}{\rho_{ion}}\right) \frac{D(\exp(-\alpha x) - \exp(\alpha x))}{\mu(\exp(-\alpha x_N) - \exp(\alpha x_N))}, \tag{6}$$
$$\alpha = \sqrt{\frac{\rho_{ion}\lambda\mu}{D}}.$$

The scheme is initialized with a small perturbation of the stationary state

$$\rho(t = 0, x) = \rho_{sol}(x) + \beta \sin(k\pi x), \tag{7}$$

where β is small compared to ρ_{sol}. In Fig. 1 are shown the results for $C = 0.3$, $D = 0.4$, $\mu = 0.3$ and $\delta = 0.01$. In Fig. 1a are displayed the initialization of the scheme (see (7)) and its evolution at time $t = 0.2$ with respect to the stationary state ρ_{sol}. In Fig. 1b is plotted the absolute relative error $\frac{|\rho - \rho_{sol}|}{\|\rho_{sol}\|_{L^\infty}}$. The results are satisfactory with a very small relative error.

3.2 Results for the Instationary Problem

In this subsection we consider the complete system (1). The LB scheme is compared to the results of a finite volume scheme (denoted FV). Both schemes are initialized in the following way :

$$\rho(0, x) = 10^{15}(1 + \exp(-\frac{(x - \frac{L}{2})^2}{L10^{-4}})), \qquad \rho_{ion}(0, x) = \rho(0, x),$$

$$\phi(x) = 0 \,\forall x \in [0, L[, \qquad\qquad \phi(L) = 3.10^4 V,$$

where $x \in [0, L]$ and $L = 0.01m$. In Fig. 2 are shown the results for $C = 0.48$, $D = 0.66$, μ and nu depending on the electric field, at final time $1, 24.10^{-8}s$. In Fig. 2a are plotted the electron densities computed by the two schemes. The LBM and FV solutions show good agreement, especially in the plateau and peak regions. Both methods capture the shape and height of the density peak well, indicating that the numerical accuracy of LBM is comparable to FV for this case. The green curve

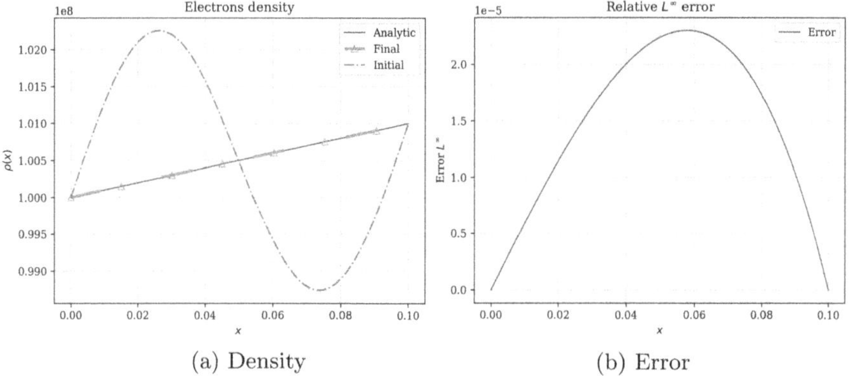

(a) Density   (b) Error

Fig. 1 **a** Initial density in blue, analytical density in orange, and numerical density in magenta at time $t = 0.2$. **b** Absolute relative error at $t = 0.2$

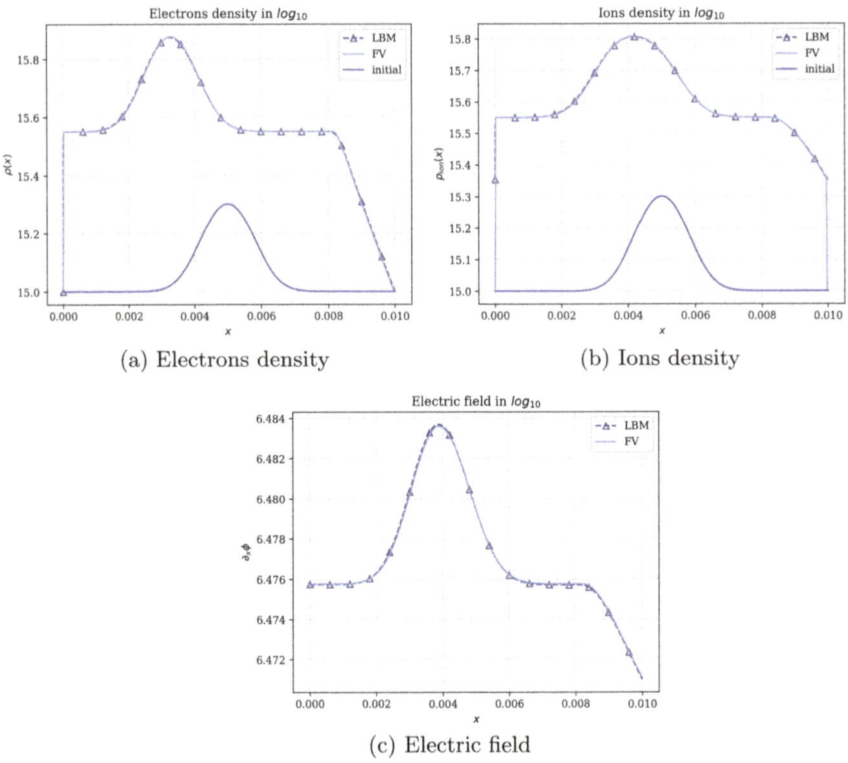

(a) Electrons density

(b) Ions density

(c) Electric field

Fig. 2 **a** Logarithm density computed with LB in blue, with FV in orange, at final time. The initial condition for both schemes is displayed in green. **b** Logarithm of ions density computed with LB in blue, with FV in orange, at final time. **c** Logarithm of electric field coupled to LB in blue, to FV in orange, at final time

(initial condition) is significantly smoother and lower in amplitude than the final density profiles. This reflects the effects of ionisation, electron accumulation and redistribution over time due to electric field-driven transport and diffusion.

4 Conclusion

In this work, we introduced a lattice Boltzmann method scheme with a parabolic scaling tailored for drift-diffusion problems with ionization. The analysis demonstrated that the equivalent equation of the scheme corresponds to a drift-diffusion equation with the correct physical parameters, although with a constant diffusion coefficient determined by the choice of the time step. Numerical results were presented to validate the stability of the scheme in stationary cases. Additionally, promising results were obtained when comparing the scheme with a finite-volume code, highlighting its potential for practical applications. However, there are still several aspects that require improvement. These include the incorporation of a second equation to account for energy dynamics and the development of a strategy to handle variable diffusion coefficients. These enhancements will further extend the applicability and accuracy of the proposed scheme for cold plasma modeling.

References

1. Boghosian B, Dubois F, Graille B, Lallemand P, Tekitek MM (2018) Curious convergence properties of lattice Boltzmann schemes for diffusion with acoustic scaling. Commun Computat Phys 23(4):1263–1278
2. Dellacherie S (2014) Construction and analysis of lattice Boltzmann methods applied to a 1d convection-diffusion equation. Acta Applicandae Mathematicae 131:69–140
3. Dubois F (2008) Equivalent partial differential equations of a lattice Boltzmann scheme. Comput Mathemat Appl 55(7):1441–1449. https://www.sciencedirect.com/science/article/pii/S0898122107006268
4. Hagelaar GJM (2008) Modelling methods for low-temperature plasmas. Ph.D. thesis, Université Toulouse III Paul Sabatier (UT3 Paul Sabatier). https://theses.hal.science/tel-02864696
5. Krüger T, Kusumaatmaja H, Kuzmin A, Shardt O, Silva G, Viggen EM (2016) The Lattice Boltzmann method. Graduate texts in physics, Springer, Cham
6. Markosyan AH, Teunissen J, Dujko S, Ebert U (2015) Comparing plasma fluid models of different order for 1d streamer ionization fronts. Plasma Sources Sci Technol 24(6):065002
7. Matéo-Vélez JC, Degond P, Rogier F, Séraudie A, Thivet F (2008) Modelling wire-to-wire corona discharge action on aerodynamics and comparison with experiment. J Phys D: Appl Phys 41(3):035205. https://doi.org/10.1088/0022-3727/41/3/035205

Experimental Techniques for Non-equilibrium Flows

Development of Neuromorphic Imaging Spectroscopy for Hypersonic Flight Observation

Tamara Sopek, Fabian Zander, Byrenn Birch, and David Buttsworth

Abstract Enhanced capability to collect hypersonic flight data is required to better understand the flow physics around spacecraft entering Earth's atmosphere. Especially daytime recording, while possible, is challenging, requiring compromises due to saturation and environmental noise. Using event-based cameras will allow spectral measurements to be done equally well during the day and night without modifications to instrumentation, overcoming the current major limitation associated with daytime tests. Combining these cameras with high-resolution spectroscopy enables more reliable collection of critical data, such as the flow temperature and species composition. Measurements were performed using a novel neuromorphic spectroscopy system and a range of light sources, obtaining both broadband and discrete line spectra. Dynamic range and long-range spectroscopy results show performance improvement of the event-based system over the camera typically used for the conditions tested, first-time demonstrating that neuromorphic spectroscopy systems are capable of successfully recording emission spectra.

Keywords Neuromorphic · Spectroscopy · Hypersonic

1 Introduction

Visual, photo- and videographic information of hypersonic/space flight events has for a long time played a crucial role in assessing performance, verifying successes and documenting failures. Forensic analysis of photographic records from rocket failures, including those from the Space Shuttle disasters (Challenger and Columbia), has contributed to improved system design.

A strong incentive exists to better characterize the hypersonic flight conditions. However, the only two options for generating realistic aerothermodynamic conditions

T. Sopek (✉) · B. Birch
University of Queensland, St Lucia, Queensland, Australia
e-mail: t.sopek@uq.edu.au

F. Zander · D. Buttsworth
University of Southern Queensland, Toowoomba, QLD, Australia

© The Author(s) 2026

M. Grabe et al. (eds.), *Rarefied Gas Dynamics*, Springer Aerospace Technology,
https://doi.org/10.1007/978-3-032-00094-1_5

are high-speed ground test facilities and flight experiments. Although the ground test facilities [6] are great at replicating individual aspects of the high-speed flows, they cannot fully replicate the real flight environment. The only way possible is through rare and expensive flight testing. Thus, researchers typically conduct a series of much cheaper testing in the ground test facilities as a first stage, followed by more expensive flight tests. Remote observations of high-speed flight events [4, 5] present the best chance to obtain the prime radiation data, also avoiding embedding sensors and extra hardware onto the vehicle.

But, besides the the rarity of hypersonic flight events, additional challenges in data collection exist related to each observation type. In ground-based observations, a large drawback is a considerable amount of atmospheric absorption of radiation, and many of the relevant events occur over the ocean, eliminating the possibility of such observations. While the issue of atmospheric absorption and transoceanic events can be tackled with airborne observations, removing the issue of clouds and accessibility, the issue of sky background brightness remains present in both observation types. Typically used CCD/CMOS cameras struggle in these conditions, and though they can record objects appearing on a clear blue sky, this is difficult, demanding compromises due to saturation and environment noise. Due to these issues, enhanced by the current technology limitations, highly valuable spectroscopic measurement data is scarce. To maximize the collection of data from every hypersonic flight event, the goal of this study was to explore recent advancements in vision technology and its potential improvements over standard cameras for remote observations of high-speed flight events.

Neuromorphic event-based (EB) vision is a new technology with gain in sensor performance compared to conventional cameras. Standard cameras, with the frame-based output dependent on the exposure, and low temporal resolution, bear issues when imaging moving objects and objects on a bright background. In space imaging, they also deal with high transfer rates due to large amounts of low-importance empty space information. EB cameras are better suited for space-oriented imaging with their continuous data stream featuring high temporal resolution, but low amount of highly relevant data and thus low transfer rates. Due to their high dynamic range measurements with EB cameras can be done equally well during day and night without modifications to instrumentation, enabling easier detection and tracking of targets in space environment [1]. Our proposed instrument is new; coupling of the EB camera with a spectrometer for neuromorphic imaging spectroscopy, which has never been trialed before.

2 Event-Based (EB) Cameras

EB cameras are novel dynamic vision devices based on human retina [3] that record only when there is a light intensity change in the field of view. Instead of measuring the "absolute" brightness all the time, each pixel responds to brightness changes in the scene asynchronously and independently. Due to their advanced method of data collection, they have been successfully used for a myriad of applications, especially in robotics [2]. The most relevant application here is their use for star tracking and mapping [1]. As they don't continually record data, the problem of vast data amounts created when imaging empty space is avoided. Also, as they only detect light intensity changes, they can cover a wide range of illumination levels.

While standard camera outputs frames at fixed time intervals, an EB camera outputs visual information about the scene as asynchronous events at microsecond resolution, transmitting data with sub-millisecond latency. An event is generated each time a single pixel detects a light intensity change in the FOV—a given pixel's luminosity change reaches a given threshold, it produces a visual event with an x and y address, a timestamp, and a polarity, either "ON" or "OFF" depending on the change in relative luminosity. The output is a stream of digital "events" of variable speed, shown in Fig. 1, each event corresponding to a pixel's preset level of change in log intensity at a certain time, thus a spatio-temporal pattern rather than conventional frames. However, pseudo images can be created by integrating events stream for a chosen length of time. Mathematically, the output of an EB camera is a stream of N events e_i:

$$e_i = [x_i, y_i, t_i, p_i]_{i=1,\ldots,N} \tag{1}$$

where x and y are the spatial coordinates in pixels, t is the timestamp of the event, and $p \in \{-1, 1\}$ is the polarity (the sign of the log intensity change). This visual information encoding is inspired by the spiking nature of biological visual/optic pathways. After the information is transmitted from the sensor, each pixel memorises its last log intensity while monitoring for a change of set magnitude from this memorised value. Once that change has been observed and recorded, the sensor sends another event. This change is defined as:

$$pC = \Delta L(x, y, t) = L(x, y, t) - L(x, y, t - \Delta t) \tag{2}$$

where C is contrast sensitivity, $L(x, y, t)=\log(I)$ or the logarithmic change of light intensity, t is the timestamp of the event and Δt is the time since the previous event at the same pixel. The logarithmic relationship represents pixels' response to percentage changes in illumination, not the absolute magnitude of the change, resulting in a very wide dynamic range. EB cameras transmit events from the pixel array to periphery and out of the camera via shared digital readout pipeline, with arbiter system determining transfer order. The disadvantages of EB imaging entail the inability to use common image processing algorithms due to asynchronous pixels and no light intensity information. New methods are needed for the spatio-temporal,

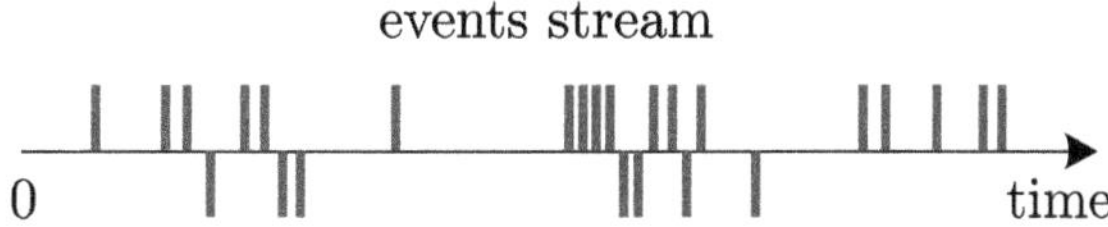

Fig. 1 Events stream output of the EB camera

photometric and stochastic nature of the data, complicating the analysis. With no changes in light intensity (static scene), there is no events creation and the only output is background noise. EB cameras' asynchronous event readout scheme has a limited maximum speed, a potential issue in extremely active scenes. Also, this is new technology, resulting in a few manufacturers, lower sensor resolution, higher cost, and a lean knowledge base.

3 Experimental Setup

An EB camera is paired with a slitless spectrometer to enable collection of spectroscopic data. The slitless spectrometer is a commonly used method to capture spectral emissions of phenomena such as shock layer radiation of a sample return capsule [5]. It typically consists of an order-sorting prism, transmission grating, lens and an optional spectral filter array. The prism refraction angle is mostly chosen to redirect the center wavelength of the observed spectrum back onto the optical axis. Our design was simpler—tilting the recording end of the system at an angle enabled us to eliminate the prism, whilst keeping the image of the spectrum parallel to the focal plane and all colours of the spectrum in focus at the same time. Experiments reported here were done using a Prophesee Gen4 camera (Metavision IMX636 sensor), lens, transmission grating, filter and a light source. The optical arrangement for these experiments was designed to enable the recording of the spectrally and spatially resolved radiation emitted from different light sources. These experiments employed an EB camera and a standard camera system, in the spectral range of approximately 400–900 nm. Three variations of spectrally resolved measurements were performed, using a: (a) Integrating sphere (Labsphere PT-038-PLS), a tunable light source delivers a broadband spectral radiance output, (b) Mercury/argon lamp (Avantes AvaLight-CAL-HgAr), which outputs radiance in the form of discrete spectrum, and (c) streetlight (LED-type and gas-type), readily available light sources for long-range spectroscopy.

Calibration of all spectral data needs to be done to convert the EB camera output into standard light intensity units. This part of the work is ongoing.

4 Results and Analysis

For majority of pseudo images presented in this work integration time was chosen to equal the standard camera exposure time of $\Delta t = 33$ ms except for the dynamic range tests, as described in Sect. 4.2.

4.1   Calibration Light Source Spectra

Broadband Light Source The data reported in Fig. 2 were collected through emission spectroscopy technique using a broadband light source, a tunable LED light source which was used with a 5 mm diameter port, while the output luminance was set to $500\,\text{cd/m}^2$. The recorded broadband spectrum is shown in Fig. 2a, while sections of broadband spectrum are displayed in Fig. 2b. These sections represent spectral regions of the broadband spectrum, obtained using a range of bandpass filters with 10 nm range, which has also helped to determine the bounds of the spectral range achieved. Both figures were created by stitching together pseudo images for different wavelength ranges, to show the full wavelength range achieved with this system. To create light intensity changes with this stationary system, the base plate had to be slightly tapped, also resulting in two spectral lines of different colors and a small spacing between them. The white color in the images corresponds to "ON" events (increase in light intensity), and the blue to "OFF" events (decrease in light intensity). The results shown in this section represent successful demonstration of the basic spectroscopic capability, i.e., ability of the instrument to capture broadband spectrum.

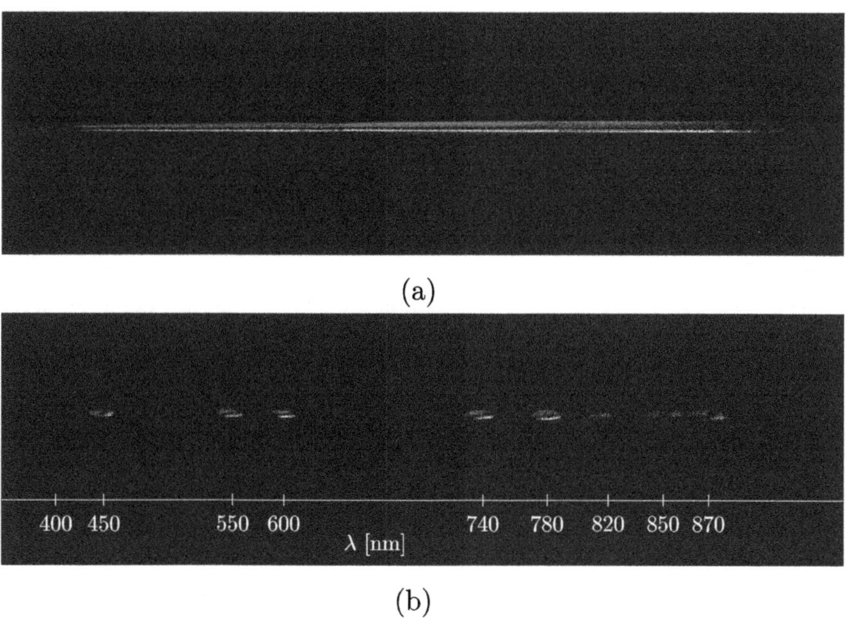

Fig. 2 Example of spectra from a broadband light source: **a** broadband spectrum and **b** sections of broadband spectrum obtained with bandpass filters

Spectral Line Light Source These tests were done using a Mercury-Argon spectral calibration source. The optical power in 600 μm fiber is 1.6 mW. Figure 3a illustrates data obtained with EB camera system; a discrete spectrum, or individual spectral lines, used for wavelength calibration of the EB recordings and corresponding to the light source output, as demonstrated in Sect. 4.1. Figure 3b shows the output of the light source, as provided by the manufacturer. The same calibration light source output was then recorded with a compact spectrometer (Thorlabs CCS175) for a range from 500–1000 nm, shown in Fig. 3c.

4.2 Dynamic Range

EB and standard camera spectrometers were mounted side-by-side and output of the broadband light source (Labsphere PT-038-PLS) recorded at varying luminance levels. These tests were done in a dark room (i.e., lights turned off) with potential sources of light reflection covered. Standard camera spectrometer used settings typically employed in airborne observation missions when recording radiation from a spacecraft re-entering the Earth's atmosphere. Bias values of the EB camera were adjusted, while other were left at default values of the software (Metavision SDK version 4.5.1). The integration time for pseudo images was reduced to $\Delta t = 10$ μs. This time interval was chosen as it corresponds to the average time it takes from when a row of a pixel firing is readout for the first time to the time when the same (or close-by) row of pixel firing is readout again. Since the EB camera records only when there is a change of light intensity in the field of view, an iris was used to block the light source prior to recording and once the recording started, the iris was opened to let the light in. A motorized iris was used for precise and quick control of the aperture. Both camera systems were located at 1 m distance from the light source aperture. The results at specific luminance levels are shown in Fig. 4. The EB data shows a partial spectrum already at the luminance level of 1 cd/m^2, but as not a full spectrum it was not deemed satisfactory. Standard camera images also show spectrum at the lowest luminance setting; but, the SNR is too low for these to be useful. The level of 5 cd/m^2 was determined to be the lower bound for both the EB and the standard camera systems. Images show full spectra at adequate signal-to-noise ratio (SNR) for information extraction. The upper bound of the standard camera system is at 80 cd/m^2, when images become saturated. The upper bound of the EB system is more difficult to find presently as the intensity calibration has not been completed. However, we determined this to be at around 500 cd/m^2, when the shape of the spectrum changes slightly, showing a small bump on the upper side. These results show that while both systems indicate the same lower bound, the standard camera system's upper bound is already at 80 cd/m^2, while the one of the EB system is much higher at 500 cd/m^2. This large difference illustrates the extended dynamic range of the EB camera in the spectroscopy setting. However, the reported values are used

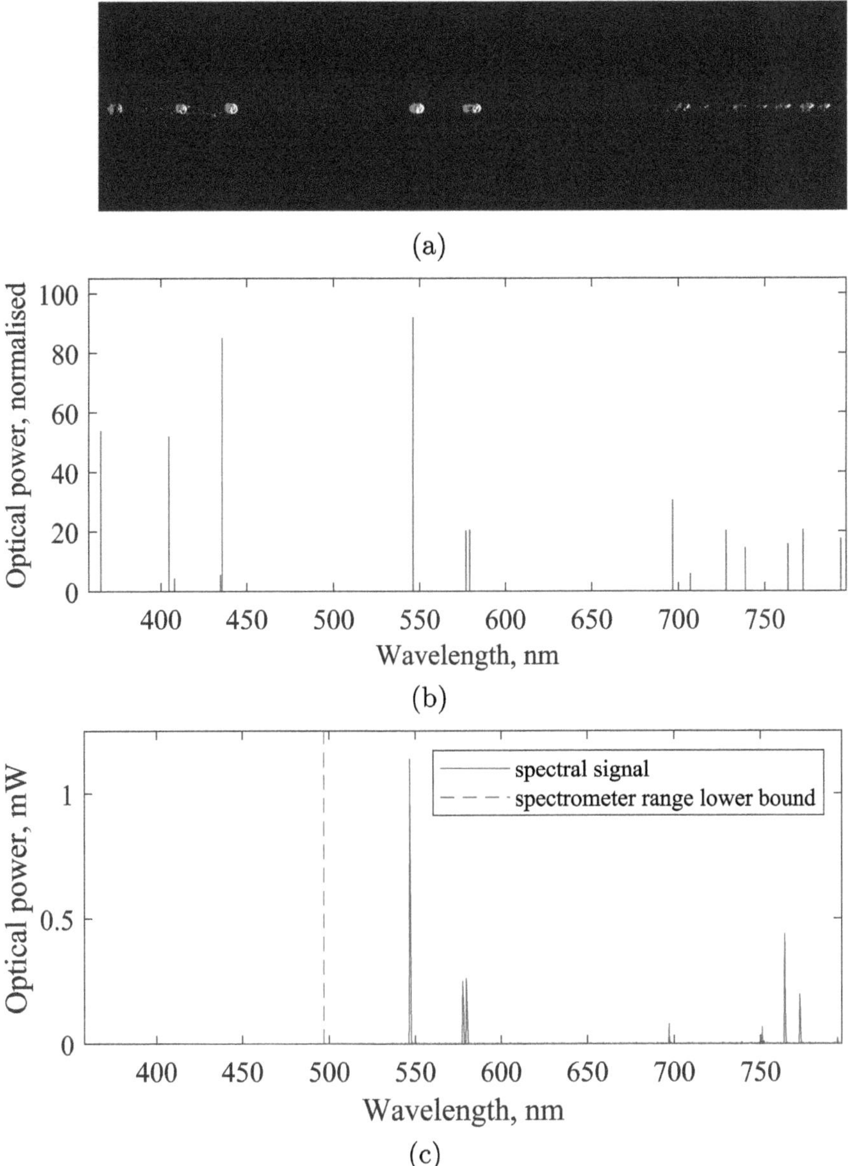

Fig. 3 Comparison of spectra from different sources

here for comparative purposes only, as luminance is a photometric quantity which would need to be converted to its analogue radiometric quantity, radiance, to obtain the amount of light across all wavelengths, and thus a more accurate measurement of the light intensity.

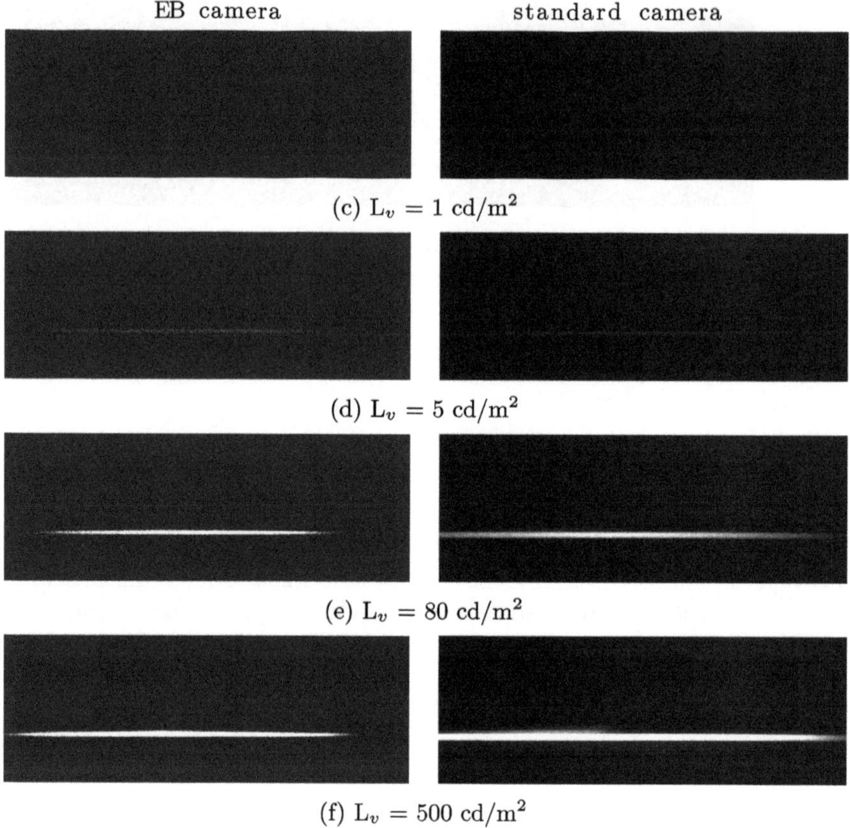

Fig. 4 Demonstration of the dynamic range

4.3 Long-Range Spectroscopy

As objects typically observed during hypersonic flight events are at least many tens of kilometers away from the recording instruments, spectra was imaged from the flagpole beacon light at a distance of approximately 7 km during nighttime. The result is shown in Fig. 5 with another, discrete spectrum consisted of particular spectral lines (observed as bright spots on top of continuous spectrum), thus indicating that it is coming from a gas-type of lamp. The spectral lines locations, originating from the gas-type lamp, were identified and are marked with horizontal lines in the figure. This gas-type lamp with the typical light output color orange-red is most likely a neon lamp, emitting in the range of 600–700 nm. Neon gas discrete spectrum with known spectral line wavelengths has been added for comparison, and the conclusion is that the spectrum of the gas-type lamp is most likely neon spectrum. This also helps us to identify the beacon light spectrum wavelength region as approximately 620–650 nm

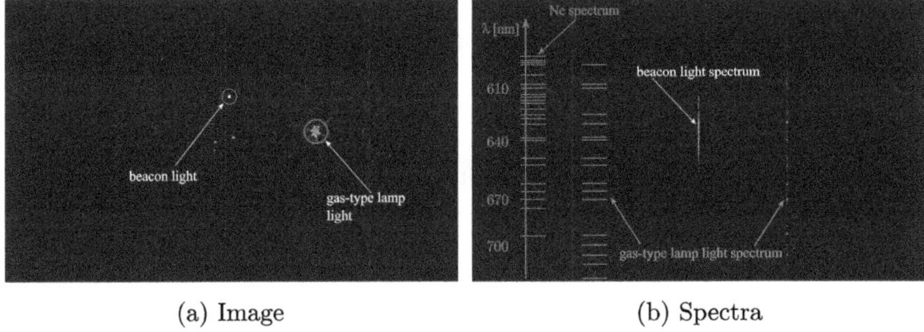

(a) Image   (b) Spectra

Fig. 5 Example of spectra from two different lights at over 5 km distance

5 Conclusion

This work demonstrates the viability of pairing EB cameras and a diffraction grating to record spectra. Measurements with this neuromorphic system were done using a range of light sources and conditions, and the results show the system's ability to capture both broadband and line spectra. By comparison with the spectra produced using a known source and manufacturer's data, specific lines of Hg and Ar were identified. Dynamic range and long-range spectroscopy results show improvement in performance of the EB system over the standard camera. Long-range spectroscopy results show the ability of the EB system to capture spectra from several kilometers distance. The reported results provide confidence in using neuromorphic spectroscopy to obtain hypersonic flight data.

References

1. Afshar S, Nicholson AP, van Schaik A, Cohen G (2020) Event-based object detection and tracking for space situational awareness. IEEE Sensors J 20(24):15117-1-5132. https://doi.org/10.1109/JSEN.2020.3009687
2. Delbrück T, Lang M (2013) Robotic goalie with 3ms reaction time at 4% CPU load using event-based dynamic vision sensor. Front Neurosci 7. https://doi.org/10.3389/fnins.2013.00223
3. Gallego G, Delbrück T, Orchard G, Bartolozzi C, Taba B, Censi A, Leutenegger S, Davison A, Conradt J, Daniilidis K, Scaramuzza D (2020) Event-based vision: a survey. IEEE PAMI 44(1):154–180. https://doi.org/10.1109/TPAMI.2020.3008413
4. Jenniskens M, Gural P, Dynneson L, Grigsby B, Newman K, Borden M, Koop M, Holman D (2011) CAMS: cameras for Allsky meteor surveillance to establish minor meteor showers. Icarus 216:40–61. https://doi.org/10.1016/j.icarus.2011.08.012
5. McIntyre TJ, Khan R, Eichmann TN, Upcroft B, Buttsworth D (2014) Visible and near infrared spectroscopy of Hayabusa reentry using semi-autonomous tracking. J Spacecr Rockets 51(1):31–36. https://doi.org/10.2514/1.A32497

6. Sopek T, Glenn A, Clarke J, Di Mare L, Collen P, McGilvray M (2024) Radiative heat transfer measurements of Titan atmospheric entry in a shock tube. J Thermophys Heat Transf Articles in Adv. https://doi.org/10.2514/1.T6918

Spectroscopic Studies in Non-equilibrium Plasmas

Ruairí J. O'Connor and Philip L. Varghese

Abstract A near-room-temperature capacitively-coupled plasma (CCP) glow discharge cell is used for fundamental studies of low and intermediate pressure Ar plasmas. The cell consists of two cylindrical plate electrodes powered by a 13.56 MHz radio-frequency source, forming a diffuse low-pressure plasma between the electrodes inside a cylindrical glass envelope 10 cm in diameter. The cell is operated at pressures from 0.1 to 10 Torr and electrode voltages up to $\sim$ 2 kV depending on pressure, producing plasmas in which the free electron temperature ranges up to several electron-volts while the gas remains close to ambient temperature ($\sim$ 300K). The emission lines from Ar in the visible and near infra-red (NIR) correspond primarily to transitions from $4p$ states to the metastable and resonant $4s$ Ar states. Calibrated Emission Spectroscopy using Absolute Radiance (CESAR) measurements on these transitions are used to determine population densities of Ar $4p$ states. Tunable laser absorption spectroscopy is used to measure the metastable and resonant Ar $4s$ state populations. These densities are presented in combination with electron density and energy distribution function measurements performed with a Langmuir probe. These measurements enable us to assess the degree of non-equilibrium in the plasma as a function of pressure and electrode voltage. We use Monte Carlo techniques to quantify the uncertainty of all the measurements. The experimental results are used to enhance our understanding of non-equilibrium plasmas and to validate kinetic models of excitation and ionization processes in plasmas.

Keywords Spectroscopy · Non-equilibrium plasma · Emission spectroscopy · Absorption spectroscopy · Langmuir probe

R. J. O'Connor · P. L. Varghese
Department of Aerospace Engineering and Engineering Mechanics, The University of Texas at Austin, Austin, TX, USA

P. L. Varghese (✉)
Oden Institute for Computational Engineering and Sciences, The University of Texas at Austin, Austin, TX, USA
e-mail: varghese@mail.utexas.edu

© The Author(s) 2026
M. Grabe et al. (eds.), *Rarefied Gas Dynamics*, Springer Aerospace Technology,
https://doi.org/10.1007/978-3-032-00094-1_6

1   Introduction

1.1   Background and Motivation

Capacitively coupled plasmas (CCP) are important in many areas of research and industry [2, 4, 8, 17]. Low-temperature glow discharge CCPs present an opportunity for fundamental plasma study with broad applications. The non-equilibrium nature of these plasmas motivates the application of a variety of spectroscopic and probe-based measurements techniques to characterize processes occurring within them.

The basic concept of a CCP is very simple. A plasma is formed in a volume between two capacitively coupled electrodes separated by some gap distance and driven by a radio frequency (RF) oscillating power supply. The bench-top nature of glow discharge devices makes them an attractive target for researchers. Optical and probe access can be readily achieved and a variety of diagnostics applied.

The motivation for this work is to generate data which may be useful to study and validate electron kinetic models and collision cross-sections to be used in advanced plasma simulations. This paper provides an overview of the glow discharge plasma facility and diagnostics currently implemented at UT Austin. The work outlined here, including detailed analysis of the results beyond the scope possible in this paper, will be published in a series of planned, upcoming publications. Optical emission spectroscopy (OES), laser absorption spectroscopy (LAS), and Langmuir probe techniques are applied to a range of low-temperature, low-pressure Ar plasmas. The optical diagnostics were chosen due to their non-intrusive nature and ability to access electronically excited states in the plasma. Results from the OES measurements can be used to correct for stimulated emission in the LAS diagnostic, although this effect is found to be quite small in the observed plasmas. The Langmuir probe allows detailed measurement of the electron energy distribution function (EEDF). Excited state population densities, characteristic temperatures, EEDF and electron density are measured. These results represent a rich set of inputs and validation targets for future non-equilbrium plasma models.

1.2   Glow Discharge Facility

The glow discharge facility at UT Austin consists of the plasma chamber itself as well as the associated control and diagnostic devices. The overall facility is shown schematically in Fig. 1. The plasma volume is formed by two 10 cm diameter parallel plate aluminum electrodes separated by 20 mm and contained within a cylindrical glass chamber. A power supply provides up to 2 kV RF signal at 13.56 MHz via a matching network to the top electrode while the bottom electrode is grounded. Peak-to-peak voltage amplitude is measured using an Impedans Octiv Suite device [10]. Peak-to-peak current amplitude, the phase between voltage and current as well as the RMS power are also recorded. A vacuum pump is used to achieve pressures

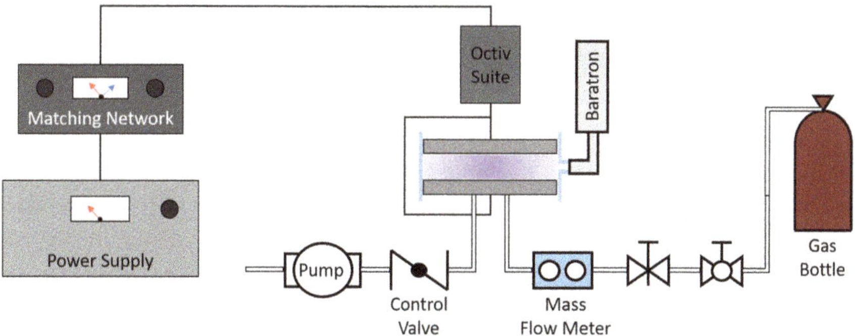

Fig. 1 Schematic diagram of the glow discharge facility

of 0.1 Torr to 10 in the chamber. Ar gas is supplied from lab bottles and regulated using needle valves and a vacuum control valve. Pressure is measured using an MKS Baratron capacitive manometer which is also used in the feedback system for pressure control.

The facility features three diagnostic capabilities, shown in Fig. 2: optical emission spectroscopy (OES), laser absorption spectroscopy (LAS) and a Langmuir probe. These diagnostics are configured to measure time-averaged plasma properties at the center of the plasma volume. Each of these diagnostic techniques are described further in the subsequent sections.

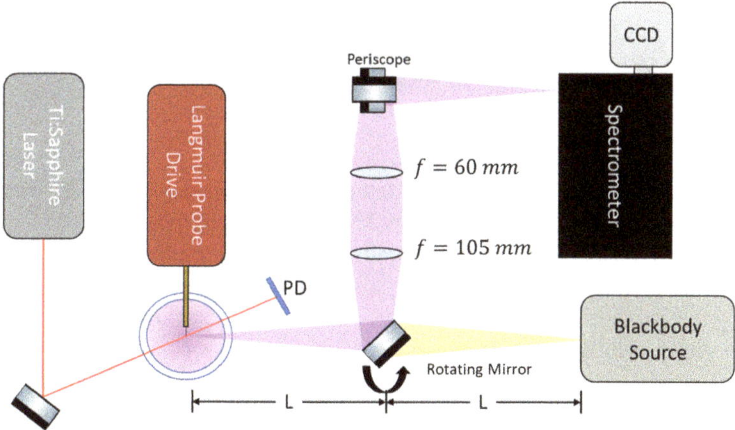

Fig. 2 Schematic diagram of the diagnostics in the glow discharge facility

2 Argon Spectroscopy

2.1 Optical Emission Spectroscopy

OES can be used to measure transitions between electronically excited states of the Ar atom. (Transitions to the ground electronic state are in the vacuum ultraviolet and are not easily measured.) These transitions create a spectrum like the one shown in Fig. 3a. Spectra which are calibrated in both wavelength and absolute spectral radiance can be used to infer the absolute population density of each upper state of the observed transitions. The spectral radiance calibration is performed by measuring a blackbody source at a known temperature of 1473K, as shown in Fig. 2, and comparing the recorded signal to the known spectral radiance of a blackbody given by Planck's equation:

$$B_\lambda(\lambda, T) = \frac{2hc^2}{\lambda^5} \frac{1}{e^{hc/\lambda k_B T} - 1}. \tag{1}$$

Since Ar is a monatomic gas, only electronic energy modes can be excited. The electronic excited states of Ar can be visualized using the Grotrian diagram in Fig. 3 [13]. This diagram emphasizes two excited state groupings of importance to this work. The Ar($4p$) group consists of ten individual states which radiatively decay to the four states in the Ar($4s$) group, emitting light in the visible and near-infra-red and is the basis for our OES measurements. The reverse process, stimulated absorption, is exploited in the LAS measurements discussed in the next section.

In an equilibrium plasma the states should follow a Boltzmann distribution described by Eq. 2:

$$\frac{N_i}{N_j} = \frac{g_i}{g_j} \exp\left(\frac{E_j - E_i}{k_B T}\right), \tag{2}$$

where N_k is the population density, E_k is the energy, and g_k is the degeneracy of state k, where subscript $k = i, j$ refers to the lower and upper states of the transition, respectively. Low-temperature CCPs are known to be in non-equilibrium states, but Eq. 2 provides a useful reference for quantifying the departure from equilibrium in these plasmas. Here B_λ is the spectral radiance of the plasma, λ is the wavelength, T is the temperature, and h, k_B and c are Planck's constant, Boltzmann's constant and the speed of light, respectively.

In order to infer the absolute population density of the upper Ar($4p$) state in each of the observable transitions the following model is used [9]. Assuming uniform conditions along the measurement line-of-sight, a cylindrical measurement volume and an optically thin plasma, the integrated signal intensity of each transition can be related to the upper state population density through

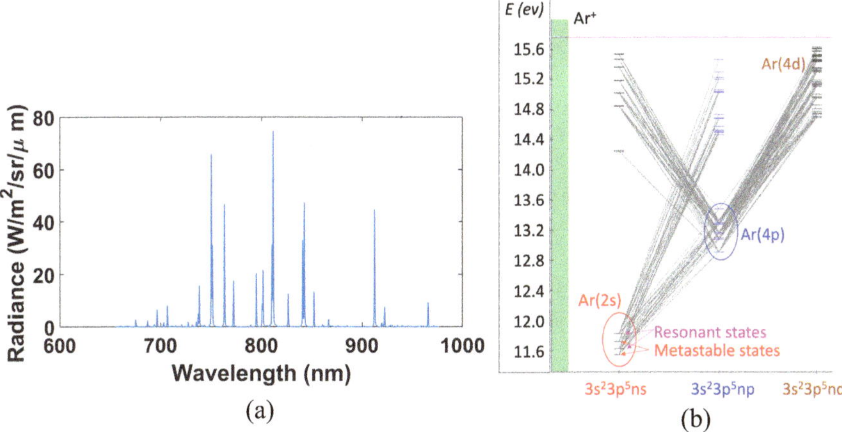

Fig. 3 **a** Spectrum of an Ar plasma at 100 mTorr and 150 V (0.34 W) **b** Grotrian diagram of Ar excited states (reproduced from NIST Atomic Spectra Database [13])

$$I_{ji} = \int_{\lambda_1}^{\lambda_2} \frac{S_{ji}(\lambda)}{C_\lambda} d\lambda = \int_{\lambda_1}^{\lambda_2} n_j \frac{A_{ji}}{4\pi} \frac{hc}{\lambda_{ji}} \Delta L \phi(\lambda, \lambda_{ji}) d\lambda = n_j \frac{A_{ji}}{4\pi} \frac{hc}{\lambda_{ji}} \Delta L. \quad (3)$$

Here A_{ji} is the Einstein spontaneous emission coefficient taken from [13], I_{ji} is the integrated line strength or emission coefficient, ΔL is the depth of field of the imaging system, and ϕ is the line shape function. In this work, ϕ is assumed to be a Gaussian and each measured transition is fit with this function in order to integrate the line profiles analytically.

When this model is applied to each observable transition, the populations of each Ar($4p$) state can be inferred and their distribution visualized in a Boltzmann plot, such as those shown in Fig. 4a and b. The uncertainty bars in these plots are derived from a Bayesian estimate accounting for detector noise as well as the uncertainty in the Einstein A-coefficients [13] and the experimentally determined depth of field. If the states were Boltzmann distributed per Eq. 2, their degeneracy-weighted populations would fall along a straight line in a logarithmic plot against state energy. From Fig. 4 it is clear that this is not the case. The state populations are highly non-equilibrium. The population distributions also change with the plasma conditions indicating that kinetic effects on the excited state populations are important. This type of detailed plasma study using OES can provide a rich dataset for understanding the kinetic processes in the plasma and how those processes change with both pressure and applied voltage.

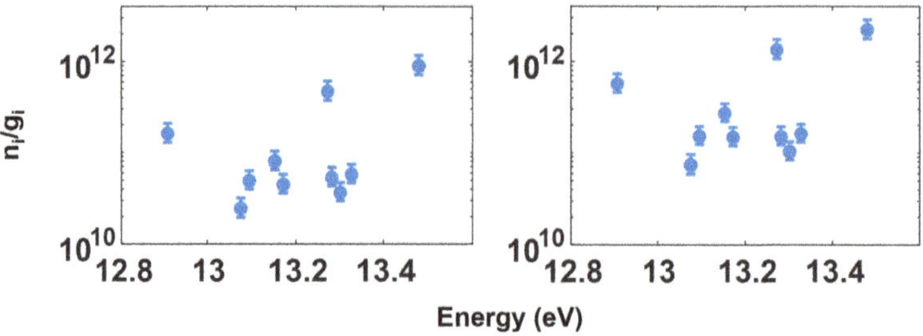

Fig. 4 Boltzmann plots of Ar $4p$ population densities in plasma discharges at 500 mTorr (left) and 1 Torr (right) at 200 V discharge voltage (2.0 and 3.9 W, respectively)

2.2 Laser Absorption Spectroscopy

The same transitions which are studied in emission in Sect. 2.1 can be studied in absorption using tunable laser absorption spectroscopy. As seen in Fig. 3, laser absorption spectroscopy targets transitions from the Ar($4s$) to the Ar($4p$) states. The measurement is achieved by passing a widely tunable Titanium:Sapphire laser through the plasma volume and onto a photodetector as shown in Fig. 1. Some of the laser light is split and passed through an air-spaced etalon of known length and onto a second photodetector to provide a relative wavelength calibration from the fringe pattern. As the laser wavelength is scanned across a transition, the absorption signal is measured and an absolute population density of the lower state in the transition (Ar($4s$) in this case) can be obtained using:

$$\frac{dP_\nu}{P_\nu} = -(n_i B_{ij} - n_j B_{ji})h\nu\phi(\nu)dz, \tag{4}$$

where P_ν is the laser power to the detector, B_{ij} and B_{ji} are the Einstein B-coefficients, ν is the laser frequency and z is in the direction of propagation of the laser beam. The Einstein A- and B-coefficients are related [15] by:

$$g_i B_{ij} = g_j B_{ji} = g_j \frac{c^2}{2h\nu^3} A_{ji}. \tag{5}$$

If conditions are assumed to be uniform along the line of sight the integration is over the plasma slab depth, taken to be the diameter of the discharge D, and gives the transmissivity $\tau(\nu)$:

$$\ln(\tau(\nu)) = -(n_i B_{ij} - n_j B_{ji})h\nu\phi(\nu)D. \tag{6}$$

The population of the upper state in the transition is needed in this equation in order to correct for stimulated emission from the Ar($4p$) states. We cannot use the equilibrium correction for stimulated emission and the population density from the

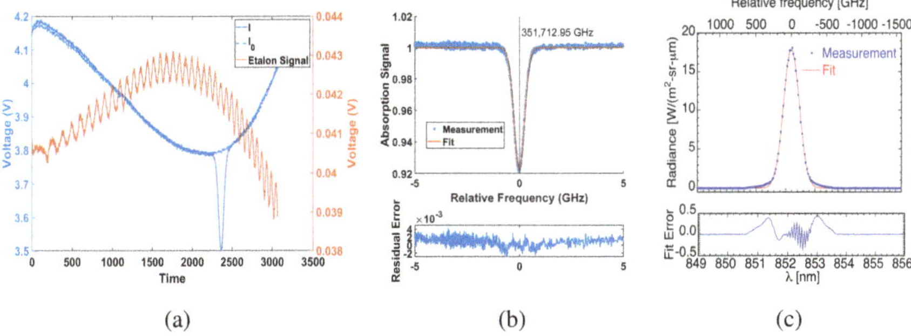

(a)   (b)   (c)

Fig. 5 **a** Raw LAS measurement of Ar $4p \leftarrow 4s$ transition at 852.144 nm recorded in a 1 Torr discharge with 150 V (1.4 W) excitation showing incident I_o, and transmitted I, signals and simultaneously recorded etalon fringe pattern. **b** Calibrated transmissivity plotted as a function of frequency detuning from line center **c** Corresponding emission line under the same conditions. Note the much larger linewidth of the recorded emission line (FWHM $\approx$ 258 GHz) compared to the absorption line (FWHM < 1 GHz.)

OES measurements is used to make this correction. The correction is very small (< 1%) for our experimental conditions. Figure 5 shows a representative measurement of the unabsorbed laser signal (incident intensity recorded with the cell evacuated) and absorbed (transmitted through the cell with discharge operating), as well as the etalon fringe pattern used for the relative frequency calibration. The transmissivity as a function of relative frequency is shown in the second panel. Figure 5 also compares the extremely high spectral resolution of the LAS measurement compared to the OES measurement of the same line at the same plasma condition (852 nm, 1 Torr, 150 V, 1.4 W). The absorption line (full width < 1 GHz) is much narrower than the emission line (full width 258 GHz) showing that the recorded emission line broadening is almost entirely due to instrument broadening effects of the spectrometer.

Some representative results from the LAS measurements are shown in Fig. 6 for one plasma condition. It can be seen that the Ar(4s) states are much closer to Boltzmann distributed than the Ar(4p) states. From the plot we infer that an approximate

Fig. 6 Boltzmann plot showing $4s$ states measured with LAS as well as $4p$ states measured in OES. Condition shown is 1 Torr, 150 V (1.4 W)

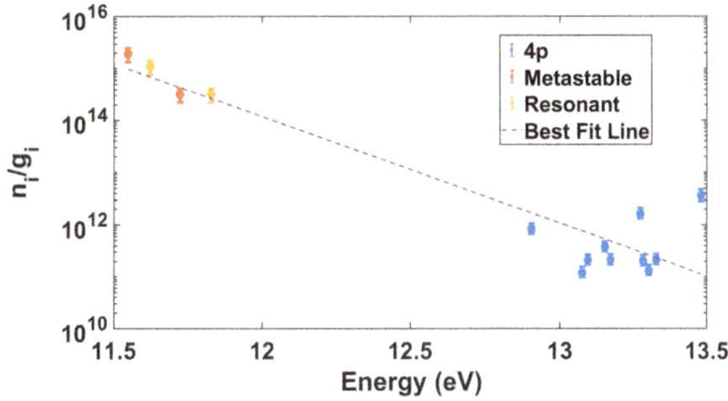

electronic temperature of 0.44 eV describes the relative populations of the Ar($4s$) and Ar($4p$) states, with some of the $4p$ states showing significant non-equilibrium. A detailed comparison of these results with a collisional-radiative model has been submitted for publication [16].

3 Langmuir Probe

A Langmuir probe system from Impedans [12] has been used to measure the EEDF and the electron density and temperature via the moments of the EEDF. The basic concept of the diagnostic is to introduce a thin wire probe tip ($\sim$1mm in diameter) into the plasma volume and bias it relative to the plasma potential [1, 14]. As the bias potential (voltage) is swept the current drawn to the probe tip is measured. At strongly negative biases the current is dominated by ion flux to the probe and at strongly positive biases the current is dominated by electron flux. The total measured current is the sum of the electron and ion current contributions. The variation of electron current drawn by the probe as a function of the bias voltage can be related to the EEDF via the Druyvesteyn model [3] (Eq. 7) assuming (1) an isotropic EEDF. and (2) that the electrons experience no collisions in the sheath around the probe tip. The Impedans probe incorporates RF and DC compensation electronics to minimize well-known perturbative effects of the probe measurement [7, 11]. These consist of self-resonant inductors tuned to the RF driving frequency and its harmonics as well as additional compensation electrodes which capacitively couple to the probe tip to reduce the sheath impedance. A DC reference probe tip is also used to account for local perturbations caused by the probe voltage sweep. The probe geometry is designed to taper down using several steps to contract from the 10mm probe support to the 1mm probe tip to minimize perturbations near the tip.

$$I_e(V - V_p) = \frac{-A_p q_e}{2} \int_{-q_e(V-V_p)}^{\infty} \left(1 + \frac{q_e(V_p - V)}{W}\right) \sqrt{\frac{W}{2m_e}} F(W) dW \qquad (7)$$

Here I_e is the electron current to the probe, V is the bias voltage, V_p is the plasma potential, A_p is the probe tip area, q_e is the electron charge, m_e is the electron mass, W is the electron kinetic energy and $F(W)$ is the EEDF. When the distribution function is found, the electron density, n_e, and mean energy, E_{avg} can be found via Eq. 8.

$$n_e = \int_0^{\infty} F(W) dW \qquad (8a)$$

$$E_{avg} = \frac{1}{n_e} \int_0^\infty W F(W) dW \qquad (8b)$$

In practice the Druyvesteyn model is typically inverted so that the EEDF can be inferred directly from a measured probe I-V trace, shown in Eq. 9. The presence of a second derivative operator on the measured current introduces the complication of noise amplification. Direct derivatives of the measurement can lead to extremely noisy results. To remedy this, we use a Savitzky-Golay filter to smooth and differentiate the probe trace. The resulting smoothed curves and derivatives are shown for a representative measurement in Fig. 7. Also shown is an inferred EEDF. The plasma potential is found as the zero-crossing of the second derivative of the measurement. The results are promising although the use of the Savitzky-Golay filter may introduce artificial features in the distribution function. Measurement uncertainties were not computed for this technique and are assumed to be large (only reliable for order of magnitude estimates of electron density). A more sophisticated model specifically designed for rigorous uncertainty quantification, discussed below, is being developed. Also worth noting, while the Druyvesteyn model is only strictly valid for the component of the measured current coming from electron flux (the measured current is the sum of contributions from both electrons and ions), the second derivative of the ion current is negligible and thus the second derivative of the measured current is considered equal to the second derivative of just the electron component [5–7].

$$\frac{d^2 I_e}{dV^2} = \frac{q_e^2 A_p}{4} \sqrt{\frac{2q_e}{m_e W}} F(W) \qquad (9)$$

Some results are shown in Fig. 8 where it can be seen that the electron density increases with both pressure and voltage. At 2.5 Torr the assumption of collisionless sheaths for the electrons is suspect, and these results are only illustrative of rough trends in the plasma.

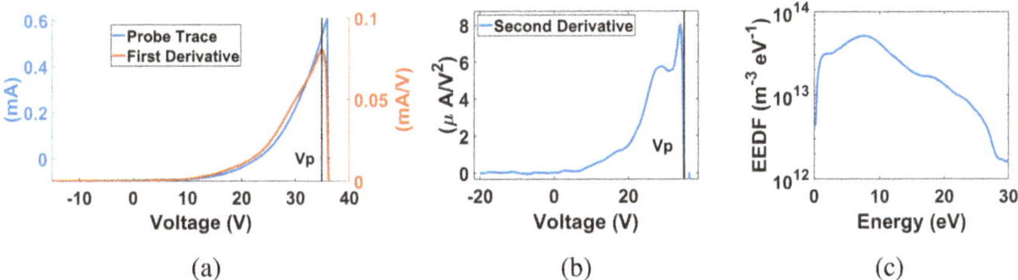

(a)   (b)   (c)

Fig. 7 **a** Electron current measurement for 500 mTorr and 100 V (0.25 W) and Savitzky-Golay filtered first derivative. The ion current has been subtracted using a straight line fit to the ion dominated portion of the probe trace. **b** Savitzky-Golay derived second derivative. **c** The resulting EEDF (n_e: $6.8 \times 10^{14} m^{-3}$, E_{avg}: $7.4 eV$)

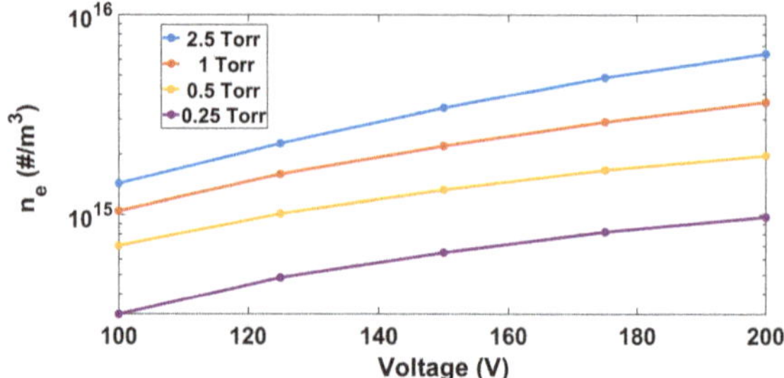

Fig. 8 Trends in electron density for Ar plasmas inferred from Savitzky-Golay filtered Langmuir probe measurements. While the trends are quite accurate, there is an order of magnitude uncertainty in the absolute values due to the derivative analysis technique

The Langmuir probe analysis can be improved by avoiding the differentiation and fitting process entirely. One approach is to fit the Druyvesteyn model directly by representing the EEDF with a set of basis functions and finding the coefficients which best reproduce the experimentally observed current as shown in Eq. 10.

$$I_j = \int_{-qe(V_j-V_p)}^{\infty} G(V_j - V_p, W) \sum_{k=1}^{m} F_k \varphi_k(W) dW + e_j = \mathcal{L}_{jk}(V_p) F_k + e_j \quad (10)$$

Here $\varphi_k(W)$ is the kth basis function, F_k is the kth basis coefficient, G is a function containing the Druyvesteyn model, e is an assumed additive error and j refers to the jth measurement point. A least-squares minimization approach can be used to infer the basis coefficients, with the plasma potential estimated simultaneously. This approach has been tested on analytical test data with a small amount of added noise $O(10^{-5} I_{probe})$. The results are shown in Fig. 9.

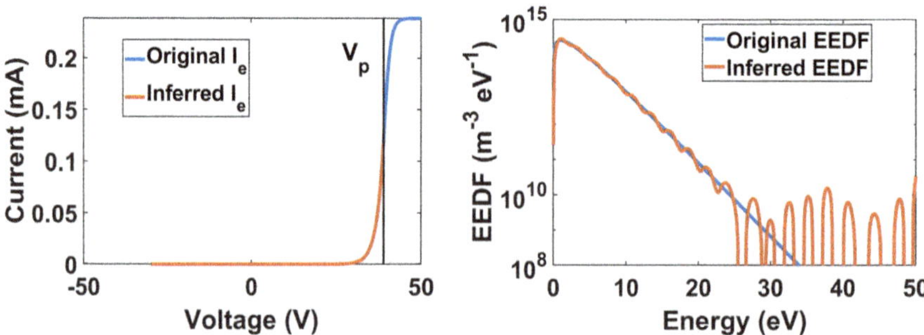

Fig. 9 Initial results of least-squares fitting approach on simulated Langmuir probe measurements. The effect of added noise is clearly seen in the EEDF retrieved

The results are promising, and this approach is being developed further and expanded to a Bayesian uncertainty quantification framework. Page restrictions preclude additional description and full details of this model and the Bayesian inference approach will be discussed in an upcoming publication. Rigorous measurements of EEDFs and electron densities with properly quantified uncertainty will be of immense help in the development and validation of plasma kinetic models under a range of conditions.

4 Conclusion

A pair of non-intrusive optical diagnostics, optical emission spectroscopy and tunable laser absorption spectroscopy, as well as an intrusive probe-based Langmuir probe diagnostic have been applied to Ar plasmas at medium vacuum pressures. Absolute population densities of excited $Ar(4p)$ and $Ar(4s)$ states have been measured using optical techniques. The electron energy distribution function and resulting moments have also been measured with the Langmuir probe. This suite of measurements can provide a rich dataset for validation of plasma models such as collisional-radiative models and reduced-order plasma chemistry models. Their application to non-equilibrium plasmas introduces some diagnostic challenges but provides valuable insight into the kinetics which govern the plasma behavior.

Continued work in this area will lead to improvements in non-equilibrium diagnostic techniques and a greater understanding of this important class of plasmas. Evaluation of non-equilibrium spectra continues to be a challenge. We will extend the optical and probe measurements to low pressure nitrogen plasmas in future work. Langmuir probe models that are valid for a wider range of plasma conditions, as well as techniques to evaluate probe measurements in the context of experimental noise and uncertainty, should be developed. The rarefied gas dynamics community can contribute to the extension of the analysis of probes with collisional sheaths.

Acknowledgements This work is supported by the Department of Energy, National Nuclear Security Administration under Award Number DE-NA0003969. PSAAP III grant: *"Exascale Predictive Simulation of Inductively Coupled Plasma Torches"*

References

1. Chen FF (2012) Lecture notes on Langmuir probe diagnostics. Contribut Plasma Phys 52(4):295–308
2. Crose M, Sang-Il Kwon J, Nayhouse M, Ni D, Christofides PD (2015) Multiscale modeling and operation of pecvd of thin film solar cells. Chem Eng Sci 136:50–61; Control and optimization of smart plant operations https://www.sciencedirect.com/science/article/pii/S0009250915001426
3. Druyvesteyn M (1930) Der Niedervoltbogen. Zeitschrift für Physik 64(11–12):781–798

4. Fang M, Zhang C, Chen Q (2016) Tuning the ito work function by capacitively coupled plasma and its application in inverted organic solar cells. Appl Surf Sci 385:28–33. https://www.sciencedirect.com/science/article/pii/S0169433216310856

5. Godyak VA (1990) Measuring EEDF in gas discharge plasmas

6. Godyak VA, Alexandrovich BM (2015) Comparative analyses of plasma probe diagnostics techniques. J Appl Phys 118(23). https://doi.org/10.1063/1.4937446

7. Godyak V, Alexandrovich B (2002) Electron energy distribution function measurements and plasma parameters in inductively coupled argon plasma. Plasma Sources Sci Technol

8. Horváth B, Derzsi A, Schulze J, Korolov I, Hartmann P, Donkó Z (2020) Experimental and kinetic simulation study of electron power absorption mode transitions in capacitive radiofrequency discharges in neon. Plasma Sources Sci Technol 29(5):055002. https://dx.doi.org/10.1088/1361-6595/ab8176

9. Huddlestone RH, Leonard SL (1965) In: Plasma diagnostic techniques. Academic Press

10. Impedans: Octiv suite [apparatus and software]. https://www.impedans.com (2007)

11. Impedans: Impedans langmuir probe installation and user guide v2.17 (2017)

12. Impedans: Langmuir probe [apparatus and software]. https://www.impedans.com (nd)

13. Kramida A, Ralchenko Y, Reader J, NIST ASD team: NIST atomic spectra database (ver. 5.11), December, National Institute of Standards and Technology, Gaithersburg, MD. [Online]. Available: https://physics.nist.gov/asd

14. Mott-Smith H, Langmuir I (1926) Theory of collectors in. Phys Rev 28:727–763

15. Oxenius J (1986) In: Kinetic theory of particles and photons. Springer

16. Tsagkaridis M, Oliver T, Fries D, O'Connor R, Barberena-Valencia J, Raja L, Varghese P, Moser R (2025) Modeling of low-temperature argon plasma in capacitively-coupled glow discharges with a collisional-radiative model. Manuscript submitted for publication

17. Vass M, Wilczek S, Derzsi A, Horváth B, Hartmann P, Donkó Z (2022) Evolution of the bulk electric field in capacitively coupled argon plasmas at intermediate pressures. Plasma Sources Sci Technol 31(4):045017. https://doi.org/10.1088/1361-6595/ac6361

Initial On-Orbit Cross-Calibration of the Electric Propulsion Electrostatic Analyzer Experiment (ÉPÉÉ)

C. A. Maldonado, T. Eddy, P. A. Fernandes, T. K. Kim, J. Derr, A. Barjatya, S. Debchoudhury, G. R. Wilson, A. J. Rogers, M. Dunn, S. M. Klem, B. Weaver, R. Ulrich, K. Potter, K. Moran, L. Castro, J. McGlown, J. D. Williams, R. L. Balthazor, P. Neal, J. Stauffer, and M. G. McHarg

Abstract The Electric Propulsion Electrostatic Analyzer Experiment is the most recent iteration in a series of laminated electrostatic analyzers that have flown in space. The sensor functions as an ion energy bandpass filter for space plasma, from which the ambient ion density, temperature, and subsequent spacecraft frame charge can be obtained. We present the ÉPÉÉ instrument design and initial on-orbit cross-calibration with the Floating Potential Measurement Unit. The calibrated data are then used to provide observations of the mid-latitude ionosphere during the geomagnetic storms that occurred on November 4th and 5th in 2023.

Keywords Ionosphere · Spacecraft charge · Geomagnetic storm

1 Introduction

The Electric Propulsion Electrostatic Analyzer Experiment (ÉPÉÉ) payload operated on-board the International Space Station (ISS) as a manifested payload on the Space Test Program—Houston 9 (STP-H9) platform from March 15th, 2023, to April 24th,

C. A. Maldonado (✉) · T. Eddy · P. A. Fernandes · T. K. Kim · J. Derr · G. R. Wilson ·
A. J. Rogers · M. Dunn · S. M. Klem · B. Weaver · R. Ulrich · K. Potter · K. Moran · L. Castro ·
J. McGlown
ISR-1, Los Alamos National Lab, Los Alamos, NM, USA
e-mail: cmaldonado@lanl.gov

T. K. Kim · A. Barjatya · S. Debchoudhury
Embry-Riddle Aeronautical University, Prescott, AZ, USA

J. D. Williams · R. L. Balthazor · J. Stauffer
i2 Strategic Services, Colorado Springs, CO, USA

P. Neal · M. G. McHarg
US Air Force Academy Department of Physics, Colorado Springs, CO, USA

© The Author(s) 2026
M. Grabe et al. (eds.), *Rarefied Gas Dynamics*, Springer Aerospace Technology,
https://doi.org/10.1007/978-3-032-00094-1_7

2024. The ÉPÉÉ is a fourth-generation iteration of the family of laminated electrostatic analyzers [1, 2]. The ÉPÉÉ payload is a collaborative effort between the Los Alamos National Laboratory (LANL) and the United States Air Force Academy (USAFA). The sensor provides in-situ energy distribution, density, temperature, and spacecraft charge measurements of the local ionospheric plasma. The operational lifetime of the ÉPÉÉ payload aligns with the ascension towards the solar maximum of Solar Cycle 25, allowing for science observations during a period with frequent and sizeable geomagnetic storms.

2 Instrument Description

The laminated design of ÉPÉÉ is based on previous laminated analyzer designs that have flown in space such as the Flat Plasma Spectrometer (FlaPS), Canary, integrated Miniaturized Electrostatic Analyzer (iMESA) [1, 3], and Automated Plums Sentry (APS) [4] instruments. The ÉPÉÉ laminated analyzer design uses geometry coupled with an electric field to act as an energy band pass filter for charged particle populations. The laminated sensor head is composed of stacked conducting electrode layers with patterns of holes and slots machined in each layer to create any number of analyzer elements [4]. Each element consists of a laminated collimator front-end to set the field-of-view (FOV) and prevent unwanted photons from entering the analyzer. The collimator section is implemented to reduce scatter and prevent off-angle incident ions from entering the detector in addition to preventing photons from entering the analyzer and creating contaminating background signal [5–7]. The collimator is also effective at reducing the effects of spacecraft potential due spacecraft sheath focusing [8]. The front-end collimator is then followed by a laminated ESA cavity where ions are separated based on E/q. The ESA section consists of (1) an entrance aperture, (2) ESA cavity, and (3) an exit aperture. As ions enter the ESA cavity, they are subjected to an electric field created by biasing created by applying a bias to the discriminator plate, V_1, while holding the opposite plate, V_2, at the host vehicle ground. The combination of the applied electric field and analyzer geometry allow ions within a narrow range of a specified bandpass energy to successfully travel through the entrance aperture, become deflected by the transverse electric field in the ESA cavity, and exit the cavity to be counted as signal current on the anode as shown in Fig. 1. A full energy measurement of the plasma is derived by stepping the applied voltage over a range that matches the targeted plasma energies.

3 Initial On-Orbit Cross Calibration with FPMU

The Floating Potential Measurement Unit (FPMU) is a multiprobe payload designed and developed by Utah State University's Space Dynamics Laboratory to measure the floating potential of the ISS in addition to the density and temperature of the local

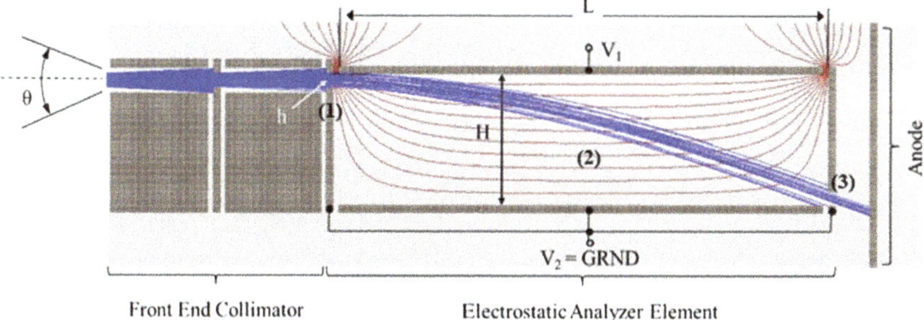

Fig. 1 Magnified cross-sectional view from SIMION for a single laminated electrostatic analyzer element showing the collimator, (1) the entrance aperture plate, (2) electrostatic analyzer cavity, and (3) exit aperture plate

ionospheric plasma environment [9, 10]. The ISS travels at a hyperthermal velocity relative to the ambient ionospheric plasma and the instrument has a very narrow field-of-view. Therefore, the particle flux to the surface of the sensor face can be approximated by a simplified 1-D drifted Maxwellian distribution of the form

$$\phi = nvA \tag{1}$$

where A is the cross-sectional area of the surface. This allows for determination of the signal current measured by the ÉPÉÉ detector

$$I_{\mathrm{meas}} = nu_0 A_{\mathrm{det}} \varepsilon q \tag{2}$$

where A_{det} is the surface area of the ÉPÉÉ detector and ε is a cross-calibration correction factor. Rearranging to solve for the cross-calibration correction factor (ε) between FPMU and ÉPÉÉ yields

$$\varepsilon = \frac{I_{\mathrm{meas}}}{n_{\mathrm{FPMU}} u_0 A_{\mathrm{det}} q} \tag{3}$$

where n_{FPMU} is the number density obtained from FPMU. Assuming plasma quasi-neutrality and FPMU accuracy, ε was calculated to determine the ÉPÉÉ total ion density using data for September 28th and 29th of 2023. The calibration factor was assumed to have a linear response to spacecraft charging; however, off-nominal periods were identified at low and high spacecraft frame potentials. The off-nominal data are removed based on the assumptions that: (1) the ÉPÉÉ data have higher noise at low energies and (2) the higher voltage off-nominal behavior is due to engineering controls used by the ISS to minimize and control higher levels of spacecraft frame charging [11]. The calibration factor as a function of spacecraft charge is generated as shown in Fig. 2.

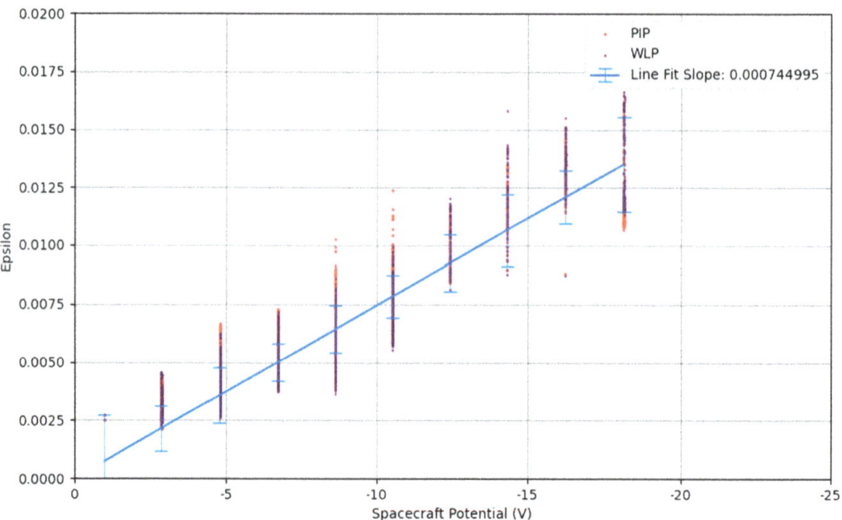

Fig. 2 Linear fit for the cross-calibration correction factor, ε, as a function of ISS frame charge

The ÉPÉÉ analyzer measures the total energy, E_0, of incident ions, which is the sum of the kinetic and potential energy. The potential energy is the electrostatic potential difference between the ambient plasma and the spacecraft frame, ubiquitously referred to as the spacecraft charge. The spacecraft charge is obtained by subtracting the relative kinetic energy of the ISS with respect to the ambient ionospheric plasma from the total ion energy as measured by ÉPÉÉ. The magnitude of the spacecraft charge is due to the net flux of charged particles to and from the spacecraft. The calibrated ÉPÉÉ ion density and spacecraft charge data are shown in Fig. 3. The data are plotted against the FPMU data as a function of orbital time and eclipse conditions. The velocity, altitude, and eclipse conditions are obtained using System Tool Kit.

4   On-Orbit Observations of the 4–5 November 2023 Storms

The ÉPÉÉ data are now used to observe the ionospheric response to two geomagnetic storms which occurred on 4 and 5 November 2023. Figure 4 presents geomagnetic indices and solar wind parameters from 2 to 13 November to provide context for the phases of the geomagnetic storms. An initial minor geomagnetic storm with a magnitude of G1 (on the NOAA scale where $K_p = 5$ [12]) occurred on 4 November 2023. Figure 4a shows meridional component of solar wind induced interplanetary

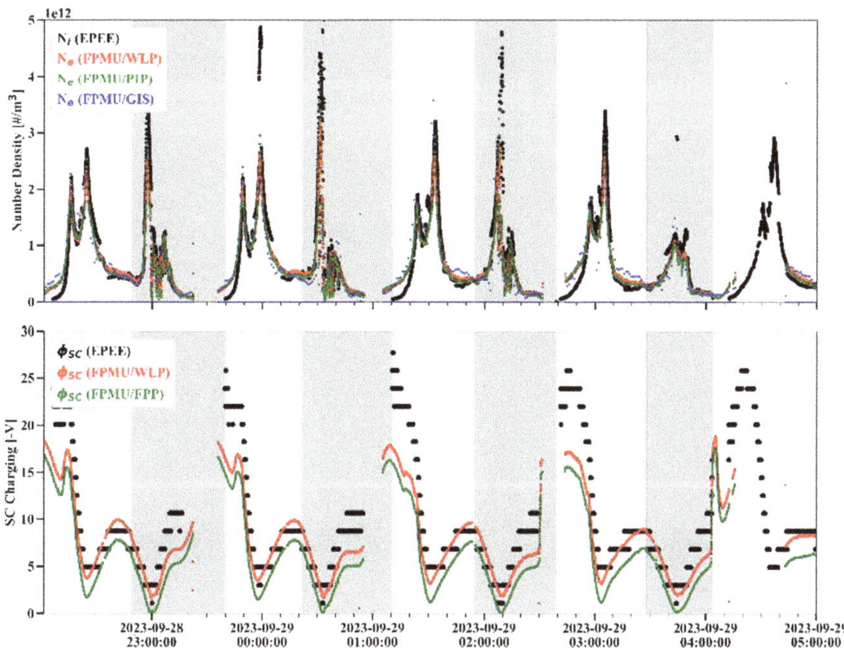

Fig. 3 Comparison of density and spacecraft potential observations as a function of eclipse/sunlight

magnetic field (IMF) in Geocentric Solar Magnetospheric (GSM) coordinate system (B_Z) observed by Advanced Composition Explorer (ACE). As shown in Fig. 4a the main phase of the G1 storm began at ~ 1304 UT on November 4 with a southward turning of the IMF B_Z reaching -11.7 nT. The Kp index shown in Fig. 4d increases to a maximum of 4.7 at ~ 1800 UT and the Dst index shown in Fig. 4e reaches -54 nT at ~ 2258 UT [13].

During the recovery phase of the minor storm a full halo CME originating from an M1.8 solar class flare observed on 3 November 2023 hit the Earth's magnetic field on 5 November 2023 at 0905 UT [13]. The G3 (strong) storm commenced on 5 November with a southward turn of the BZ, reaching a minimum value of $-$ 23.3 nT at 0957 UT and then turned northward reaching a maximum value of 25.5 nT. The BZ component continued to fluctuate between northward and southward through the main phase of the storm [13]. The Kp index reached a maximum value of 7.3 at ~ 1500 UT. After reaching a maximum D_{st} magnitude of $-$ 162.6 nT the storm had a long five-day recovery phase as seen in Fig. 4e.

The ÉPÉÉ data are shown in the panels of Fig. 5 for the period of 3–9 November. This time period was selected to provide observations of the mid-latitude region monitored by ÉPÉÉ (mounted on the ISS) from $-$ 51.2 to 51.2 degrees latitude at approximately 400 km. The Equatorial Ionization Anomaly (EIA) is a set of peaks (crests) in plasma density bounding an equatorial trough in plasma density due to electric fields and neutral winds that push plasma away from the equator towards

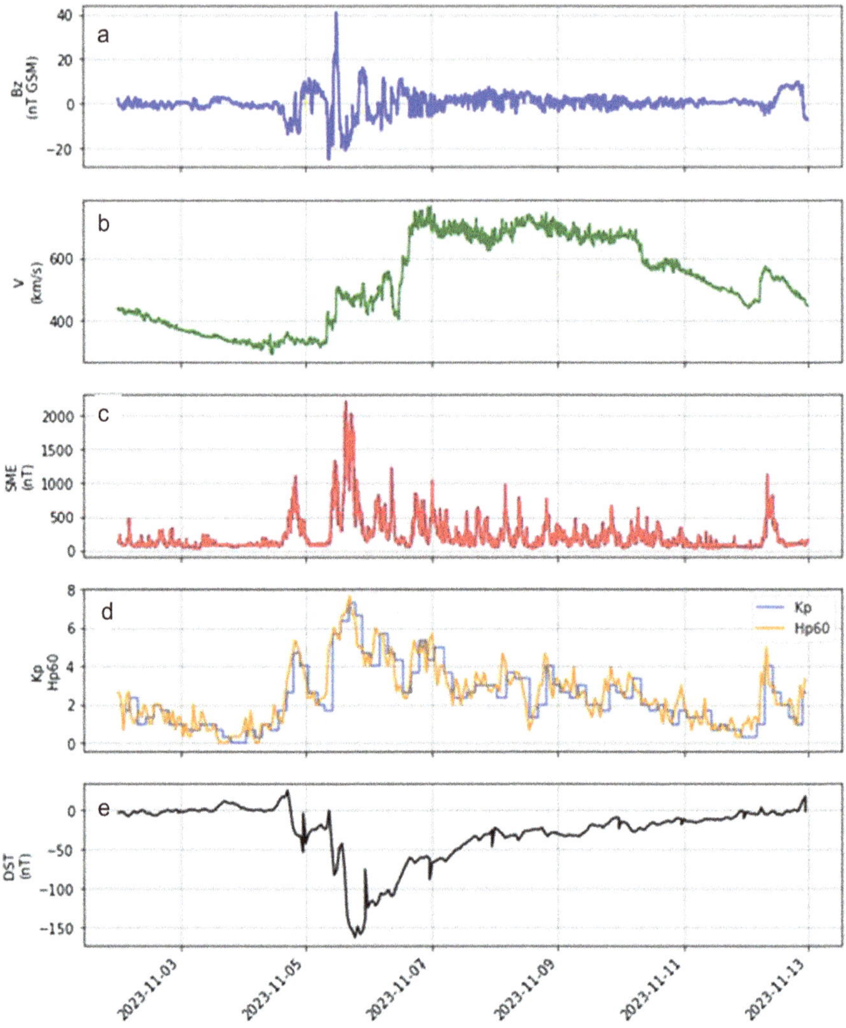

Fig. 4 Solar wind plasma and IMF during the storms of 4–5 November 2023. From top to bottom the panel shows **a** IMF magnitude (B_Z), **b** solar wind proton density, **c** solar wind flow speed, **d** the geomagnetic SME index, **e** the K_p index, and **f** the Dst index

higher and lower geomagnetic latitudes. The panels in Fig. 5 illustrate the (a) average EIA density over this period in addition to the (b) daily density as a function of latitude and longitude. Figure 5c shows plots of the ion density for each day. The data show an enhancement of the EIA crests during the main and recovery phase of the storm on 6 November and 7 November. This increase in ionospheric density is known as a positive ionospheric response to the geomagnetic storm. These positive ionospheric storms are related to an eastward prompt penetration of the electric field (PPEF) that uplifts the F region to higher altitudes, where recombination rates are lower,

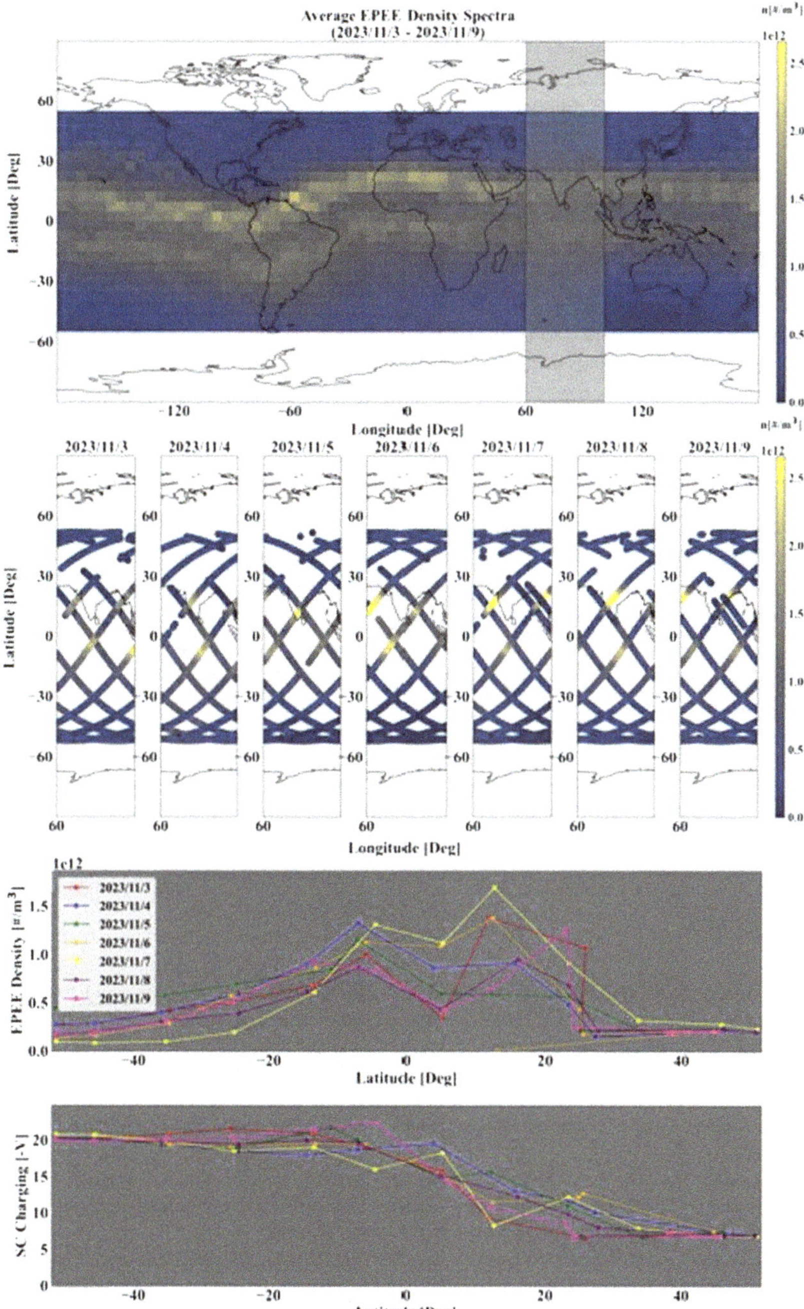

Fig. 5 ÉPÉÉ data over 3–9 November 2023 showing **a** average density and **b** daily density as a function of latitude and longitude. **c** Density and **d** spacecraft charge as a function of day and latitude

thus leading to an enhanced ion density [13]. Figure 5d shows spacecraft charge as a function of day and latitude. The ISS spacecraft charge does not appear to demonstrate an immediate correlation to the November storms however an in-depth investigation of both the density and spacecraft charge will be the subjective of future work.

5 Conclusion

We present initial results from the ÉPÉÉ payload, an ultra-low SWaP plasma spectrometer. The ÉPÉÉ data are cross-calibrated on-orbit using the FPMU WLP and PIP data sets over September 28th and 29th in 2023. The ÉPÉÉ data are then used to observe ISS frame charging and ionospheric density during the peak of Solar Cycle 25, specifically during the G1 and G3 storms in November 4–5, 2023. Future work will include cross-correlation studies using ground-based and Swarm satellite derived total electron content (TEC) data sets. Additionally, the implementation of GUVI data to assess the O/N2 ratio during the November 4–5, 2023, storms will be conducted. This will allow the team to assess the impact of the neutral thermospheric environment on the ionospheric storm morphology.

Acknowledgements Work at Los Alamos National Laboratory is conducted under the auspices of the United States Department of Energy. The authors acknowledge the support of the DoD Space Test Program which provides mission design, spacecraft acquisition, integration, launch and on-orbit operations support for DoD's science and technology (S&T) experiments, and manages all DoD payloads on the International Space Station.

References

1. Maldonado C, McHarg M, Balthazor R, Osiander R (2019) Undergraduate research and science mission opportunities with microtechnology enabled particle detectors. In: Proceedings SPIE 10982, Micro-and nanotechnology sensors, systems, and applications XI, 109820I, Baltimore
2. Balthazor R, Gaefke X, Williams J, Wilcox Z, McHarg M, Neal P, Maldonado CA, Wilson G, Yates W (2024) Anomalous ISS charging observed with Falcon EPEE—early observations. In: Proceedings of the 17th spacecraft charging technology conference, Palais des Papes, Avignon, France
3. Maldonado C, Cress R, Gresham P, Armstrong J, Wilson G, Reisenfeld D, Larsen B, Balthazor R, Harley J, McHarg M (2020) Space radiation dosimetry using the integrated miniaturized electrostatic analyzer—Reflight (iMESA-R). Space Weather
4. Maldonado CA, Ketsdever AD, Balthazor RBNPC, Wilson GR, McHarg MG, Osiander R, Adams RJ (2022) Automated plume sentry observations during international space station thermal control system venting. J Spacecraft Rockets
5. Maldonado C, Wilson G, McGlown J, Morning H, Arnold, D Reisenfeld D, Holloway M (2021) An ultra-low resource ion mass spectrometer for cubesat platforms.In: Proceedings of the 36th annual small satellite conference, Logan, UT

6. Maldonado CA, Wilson G, McGlown J, Morning H, Arnold D, Kim T, Espinoza T, Miller M, Rogers AJ, Reisenfeld D, Holloway M (2022) An ultra-low resource ion mass spectrometer for observation of planetary atmospheres. In: Proceedings of the 37th AIAA/USU small sat conference, Logan, UT
7. Hooks DE, Carpenter B, Hickethier M, Clark C, Brown N, McBride M, Loza-Hernandez I, Stull JA, Maldonado CA (2024) Ultra-black coatings for space instruments: a comparison of traditional Ebonol C processes and a method for future repeatability. J Manuf Process 113(15):230–237
8. Maldonado CA, Reisenfeld D, Fernandes P, Larsen B, Wilson G, Balthazor RL, McHarg MG (2020) The effects of spacecraft potential on ionospheric plasma measurements. J Spacecraft Rockets (submitted)
9. Debchoudhury S, Barjatya A, Minow JI, Coffey VN, Chandler MO (2021) Observations and validation of plasma density, temperature, and abundance from a langmuir probe onboard the international space station. J Geophys Res: Space Phys 126:e2021JA029393
10. Barjatya A, Swenson CM, Thompson DC, Wright KH (2009) Data analysis of the floating potential measurement unit aboard the international space station. Rev Sci Instrum 80:041301
11. Minow JI, Wright KH, Chandler MO, Coffey VN, Craven PD, Schneider TA, Parker LN, Ferguson DC, Koontz SL, Alred JW (2010) Summary of 2006 to 2010 FPMU measurements of international space station frame potential variations. In: 11th Spacecraft charging technology conference, Albuquerque, New Mexico
12. N./.N.S.W.P. Center (2024) NOAA space weather scales. [Online]. Available: https://www.swpc.noaa.gov/sites/default/files/images/NOAAscales.pdf. (Accessed 30 Dec 2024)
13. Agyei-Yeboah E, Fagundes P, Tardelli A, Pillat V, Vieira F, Alves Bolzan M (2025) Global ionospheric response to a G2 and a G3 geomagnetic storms of November 4 and 5 2023. Adv Space Res

Gas-Surface Interactions

Unsteady Stefan Problem with Kinetic Interface Conditions in Rarefied Gas Phase-Transitions

Donat Weniger and Manuel Torrilhon

Abstract We investigate the unsteady phase transition problem with kinetic interface conditions in rarefied gas deposition and sublimation. We couple the R13 equations of macroscopic gas transport with the heat equation to model the interaction between the gas and the solid phase. The kinetic interface conditions are derived from thermodynamic principles and provide an accurate description of the phase transition process. We employ the Finite Element method with a remeshing approach to solve the coupled system and track the moving interface. Our numerical results demonstrate the effectiveness of the proposed method in capturing the complex dynamics of the phase transition process, including the effects of kinetic interface conditions.

Keywords Stefan problem · Kinetic interface conditions · Rarefied gas deposition · Finite element method · Remeshing

1 Introduction

Kinetic effects in phase transition problems are of high importance in research on microelectromechanical devices (MEMS) [5] and in pharmaceutical freeze drying [3]. Those processes are characterized by a rarefied gas phase which undergoes a phase transition. Accurately capturing all relevant kinetic effects is crucial for understanding the system and designing the devices.

The Stefan problem [8], the classical phase transition model, becomes more complex when kinetic interface conditions are considered [11]. In this work, we tackle this problem for a rarefied gas deposition and sublimation process. We couple the R13 equations [9] with the heat equation to model the interaction between the gas and the solid phase. Thermodynamically admissible kinetic interface conditions [1] close the system. We present high-fidelity numerical results based on the classical Finite Element method and a remeshing approach.

D. Weniger (✉) · M. Torrilhon
Applied and Computational Mathematics, RWTH Aachen University, Aachen, Germany
e-mail: weniger@acom.rwth-aachen.de

M. Grabe et al. (eds.), *Rarefied Gas Dynamics*, Springer Aerospace Technology,
https://doi.org/10.1007/978-3-032-00094-1_8

2 Physical Model and Governing Equations

We solve the gas and solid field equations in domains $\Omega_g(t)$ and $\Omega_s(t)$, respectively. The interface $I(t)$ separates the two domains and moves over time with velocity v_n^I. External boundaries $\Gamma_{g/s}$ are fixed. For a general sketch, see Fig. 1.

We assume a monatomic and ideal gas. The R13 equations and the interface conditions for phase transitions are derived for a variety of molecule models, such as Hard-Sphere or Maxwell molecules [1]. They only influence factors of certain coefficients in the final model equations. We therefore do not expect significant influence on the phase transition numerics and will hence restrict ourselves to the simple case of Maxwell molecules.

All equations are non-dimensionalized and linearized around a constant equilibrium state (ρ_0, T_0). Simultaneously, this state is considered to be the saturation condition of the system. Small perturbations from equilibrium by boundary or source terms initiate the phase transition process.

The dimensionless form introduces key parameters into the equations, such as the Knudsen number or the temporal coupling scale between the solid and the gas phase. Analysis of these dimensionless numbers is crucial for gaining insights into the system and choosing appropriate numerical methods.

Fig. 1 General two-phase problem with interfaces separating the two subdomains

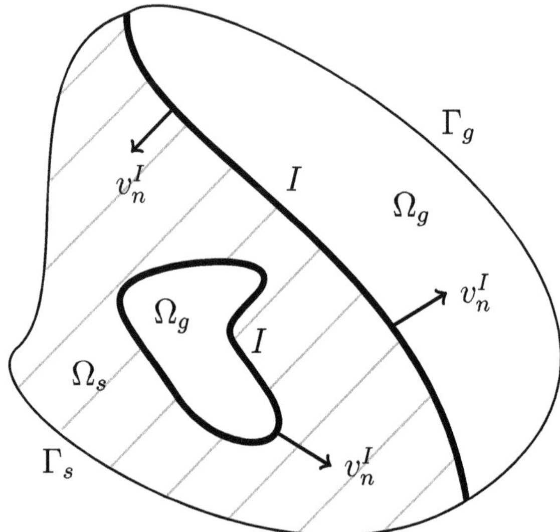

Einstein summation convention is used throughout the paper for the indices i, j, and k. The corresponding unsteady, linearized, and dimensionless R13 equations are given by [9]

$$\partial_t \rho + \partial_{x_j} v_j = 0$$

$$\partial_t v_i + \partial_{x_i} \rho + \partial_{x_i} T + \partial_{x_j} \sigma_{ij} = b_i$$

$$\frac{3}{2}\partial_t T + \partial_{x_j} v_j + \partial_{x_j} q_j = r$$

$$\partial_t \sigma_{ij} + \frac{4}{5}\partial_{x_{\langle j}} q_{i\rangle} + \partial_{x_k} m_{ijk} + 2\partial_{x_{\langle j}} v_{i\rangle} = -\frac{1}{Kn}\sigma_{ij} \tag{1}$$

$$\partial_t q_i + \partial_{x_k} \sigma_{ik} + \frac{1}{2}\partial_{x_k} R_{ik} + \frac{1}{6}\partial_{x_i} \Delta + \frac{5}{2}\partial_{x_i} T = -\frac{2}{3Kn}q_i$$

with the (dimensionless) closure relations

$$m_{ijk} = -2Kn\partial_{x_{\langle k}}\sigma_{ij\rangle}$$

$$R_{ij} = -\frac{24Kn}{5}\partial_{x_{\langle j}} q_{i\rangle} \tag{2}$$

$$\Delta = -12Kn\partial_{x_k} q_k.$$

The heat equation [2] is presented and used in mixed form, i.e., energy balance and Fourier's law of heat conduction. It reads

$$\tau\partial_t T + \partial_{x_j} q_j = r$$

$$q_j + \partial_{x_j} T = 0. \tag{3}$$

The variables are the density ρ, velocity v_i, temperature T, stress deviator σ_{ij}, and heat flux q_i. An external force term b_i or a heat source term r can be added to the system. The higher-order moments m_{ijk}, R_{ij}, and Δ close the system.

Each phase carries its own set of variables and equations. As a consequence, there are two temperatures, T^g and T^s, and two heat fluxes, q_i^g and q_i^s, at the interface. This allows for physical interface conditions that will produce a temperature jump otherwise not possible. Coupling at the interface therefore requires six equations for the R13 equations, one equation for the heat equation, and one more equation for the unknown interface velocity, for a total of eight interface conditions. The phenomenological R13 conditions for phase transitions are given in [1] by

$$v_n - v_n^I = \tilde{\vartheta}\left((T+\rho) - HT^s - \frac{1}{2}(T-T^s) + \frac{1}{2}\sigma_{nn} - \frac{1}{30}\Delta - \frac{1}{10}R_{nn}\right)$$

$$q_n = \xi\left(2(T-T^s) + \frac{1}{2}\sigma_{nn} + \frac{2}{15}\Delta + \frac{2}{5}R_{nn}\right) - \frac{1}{2}(v_n - v_n^I)$$

$$m_{nnn} = \xi\left(\frac{2}{5}(T^s - T) + \frac{7}{5}\sigma_{nn} - \frac{2}{75}\Delta - \frac{2}{25}R_{nn}\right) - \frac{2}{5}(v_n - v_n^I)$$

$$\sigma_{nt} = \xi\left(v_t - v_t^W + \frac{1}{5}q_t + m_{nnt}\right) \tag{4}$$

$$R_{nt} = \xi\left(v_t^W - v_t + \frac{11}{5}q_t - m_{nnt}\right)$$

$$\frac{1}{2}m_{nnn} = \xi\left(\frac{1}{2}\sigma_{nn} + \sigma_{tt}\right) - m_{ntt},$$

assuming that the normal vector n_i points from the gas to the solid phase (i.e., the outward normal considering the gas). Subscripts n and t denote the normal and tangential components, respectively. All quantities are evaluated at the interface on the gas side, except for the solid temperature T^s which introduces coupling between solid and gas. Here, $\tilde{\vartheta} = \sqrt{\frac{2}{\pi}\frac{\vartheta}{2-\vartheta}}$ and $\xi = \sqrt{\frac{2}{\pi}\frac{\vartheta+\chi(1-\vartheta)}{2-\vartheta-\chi(1-\vartheta)}}$ are abbreviations for the classical interface parameters. $\chi \in [0, 1]$ is the accommodation coefficient, and $\vartheta \in [0, 1]$ is the deposition or sublimation coefficient. These are empirical coefficients, potentially dependent on the state of the system, which we do not discuss further in this work. The relative latent heat H enters the system through the Clausius-Clapeyron relation. The interface conditions collapse to the default R13 boundary conditions for $\vartheta = 0$ (i.e. no phase transition), see [6]. The first equation is of Hertz-Knudsen-Schrage type [7], extended to the R13 equations.

The remaining two equations are the Rankine-Hugoniot conditions for the mass balance (mass flux across the interface) and energy balance (Stefan condition). In dimensionless and linearized form, and assuming an external observer, they read

$$-v_n^I = \eta(v_n - v_n^I) \tag{5a}$$

$$v_n^I H = \eta q_n^g - \frac{1}{\tau\gamma}q_n^s. \tag{5b}$$

Here, we assumed that the solid phase is at rest.

The set of dimensionless numbers describing the system is

$$\tau = \frac{t_{0,s}}{t_{0,g}} \qquad \text{(time scale ratio)} \tag{6a}$$

$$Kn = \frac{\mu\sqrt{RT_0}}{\rho_0 RT_0 x_0} \qquad \text{(Knudsen number)} \tag{6b}$$

$$\eta = \frac{\rho_0}{\rho_s} \qquad \text{(density fraction)} \tag{6c}$$

$$H = \frac{H_{\text{latent}}}{RT_0} \qquad \text{(relative latent heat)} \tag{6d}$$

$$\gamma = \frac{R}{c_s} \qquad \text{(heat capacity fraction)}. \tag{6e}$$

The Knudsen number defines the gas regime and is responsible for potential kinetic effects. The time scale ratio τ is a common parameter in multiscale simulations. If $\tau \approx 1$, the dynamics of the solid and the gas are similar. If $\tau \gg 1$, the gas dynamics are much faster than the solid dynamics, and we need a much smaller time step to resolve the gas scale accurately. We will assume $\tau = 1$ for the sake of simplicity, which is usually given only for microscale systems. The density fraction η mainly controls the mass flux across the interface and is small for a solid and a rarefied gas. Effects like porosity can increase η significantly. The relative latent heat H and the heat capacity fraction γ are abbreviations and naturally occur in the Stefan problem. They control the energy flux across the interface but are of less importance to the general dynamics of the system.

The coupled system is solved by the Finite Element method. The discretization is performed as in [10], with an added implicit Euler scheme for the time integration. The moving interface is tracked by the Level-Set method [4], and a remeshing method is used to adapt the mesh to the moving interface. This way, the physical interface always aligns with edges of the computational mesh and all relevant kinetic effects can be captured.

3 Numerical Results

Pharmaceutical freeze-drying aims to remove water from a frozen substrate in a low-pressure environment [3]. One stage of this process involves the slow heating of the substrate to sublimate the water, followed by the deposition of the water vapor on a condenser in a near-vacuum regime with Knudsen numbers $10^{-4} < Kn < 1$. Minimizing the time until the desired drying level is reached is crucial for applications. Simulations are key in understanding and subsequently controlling the process and designing the devices. This scenario motivates our example.

As discussed in Sect. 2, the physical model is only valid for small perturbations around an equilibrium state. This is, for example, given for slow processes, which is the case in freeze-drying. We therefore expect to capture the relevant kinetic effects despite using a linearized model.

We initialize the domain as a square with a solid substrate in the top-right corner, which is to be dried. The condenser is modeled as another block of solid at the bottom. The remaining domain is filled with a gas at $Kn = 0.1$ and is treated as an enclosed container with adiabatic boundary conditions ($\chi \to 0$). The process is initialized with the substrate at $T = 0.01$, the condenser at $T = -0.01$, and the gas at $T = 0$. The gas density and velocity fields are initialized to zero. Dirichlet temperature boundary conditions are applied to the substrate and the condenser to maintain the process.

We let the simulation run for $t \in [0, 1000]$ using 1000 time steps and a triangular mesh with characteristic lengths from 0.01 (close to the interface) to 0.05. For visible growth and decay of the interface, we choose $\eta = 0.01$. The interface parameters are set to $\vartheta = 0.5$ and $\chi = 0.9$ and there are no heat sources or external force terms.

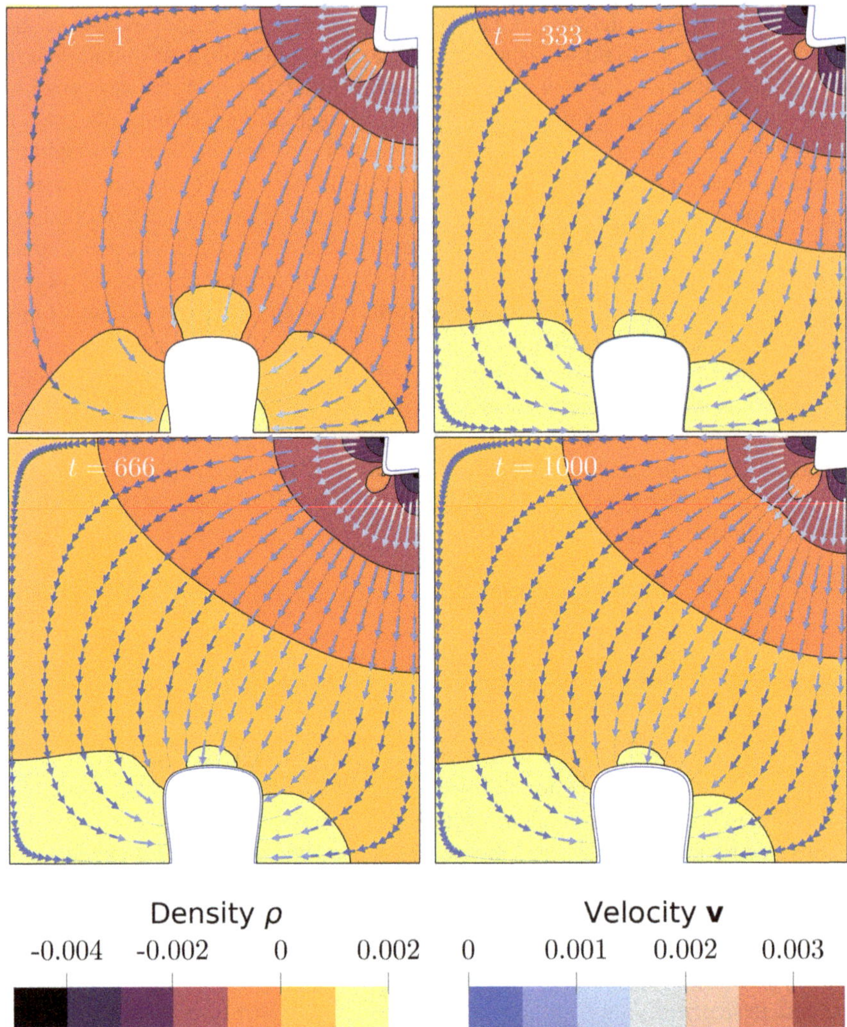

Fig. 2 Combined sublimation and deposition cycle: density contours with velocity streamlines. The four images show the system's state at different times. The top-left after the first time step, the top-right after one-third of all time steps, the bottom-left after two-thirds, and the bottom-right after all time steps. In the top-right corner and on the bottom middle, the initial and final interfaces in blue highlight the total growth and shrinkage

The remaining dimensionless parameters are set to $\tau = H = \gamma = 1$. We present the time evolution after the first time step, after 1/3, after 2/3, and at the final time step.

Figure 2 illustrates the density and velocity fields in the gas phase. The gas is sublimated from the substrate due to the heating. It moves towards the condenser and is deposited there. We observe a denser region around the condenser and a more

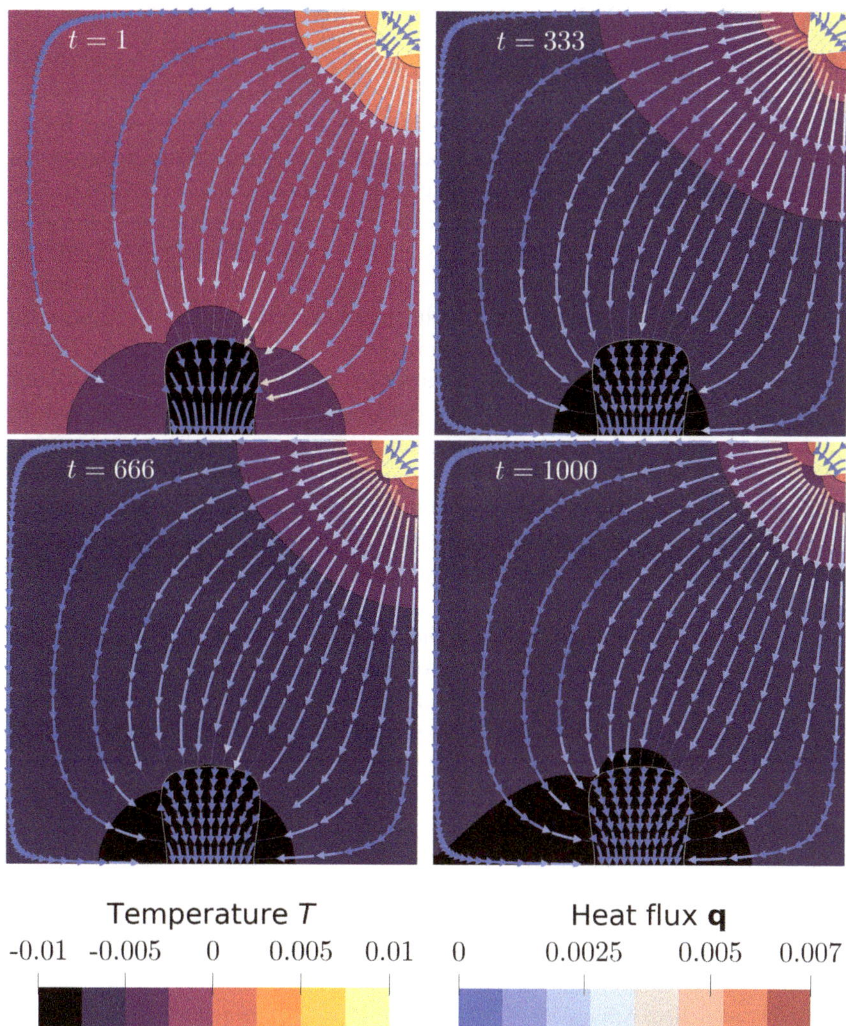

Fig. 3 Combined sublimation and deposition cycle: temperature contours with heat flux streamlines. The four images show the system's state at different times. The top-left after the first time step, the top-right after one-third of all time steps, the bottom-left after two-thirds, and the bottom-right after all time steps

rarefied one near the substrate, resulting in a clear gradient in the density field. Note the shrinkage and growth of the solid regions over time, which is captured by the numerical method. The total degree of movement is highlighted by the blue lines representing the initial and final interfaces.

Figure 3 shows the corresponding temperature and heat flux fields. A clear temperature jump between the substrate and the gas is visible. This jump arises from the kinetic interface conditions and is accurately resolved thanks to the exact remeshing approach.

4   Conclusion and Outlook

We presented a numerical study of the unsteady Stefan problem with kinetic interface conditions in rarefied gas deposition and sublimation. By coupling the transient R13 equations with the heat equation, we modeled the interaction between the gas and solid phases, capturing kinetic effects that are not accounted for in classical models. The kinetic interface conditions allowed for an accurate description of the phase transition process, including phenomena such as temperature jumps at the interface.

Our computational approach employed the finite element method with a remeshing strategy to effectively handle the moving interface. This method ensured that the physical interface aligned with the computational mesh, enabling precise resolution of the kinetic interface conditions. The numerical results demonstrated the capability of the method to capture the complex dynamics of the phase transition process, validating its effectiveness in modeling rarefied gas flows with phase transitions.

The novelty of this work lies in the integration of transient R13 equations with kinetic interface conditions in a unified framework, providing insights into the non-equilibrium effects during phase transitions. By accurately capturing these kinetic effects, our model advances the understanding of processes such as sublimation and deposition in microscale systems.

Future work will focus on exploring the impact of different interface parameters on the phase transition dynamics, such as varying the accommodation coefficient and the deposition or sublimation coefficient, to study their influence on kinetic effects. Additionally, we aim to improve the computational efficiency of the method to enable simulations of larger and more realistic systems. Extending the framework to three-dimensional simulations is a key goal, which will open possibilities for studying complex geometries and more intricate physical phenomena.

References

1. Beckmann A, Rana A, Torrilhon M, Struchtrup H (2018) Evaporation boundary conditions for the linear r13 equations based on the onsager theory. Entropy 20(9). https://www.mdpi.com/1099-4300/20/9/680
2. Donea J, Huerta A (2003) In: Finite element methods for flow problems. Wiley. https://doi.org/10.1002/0470013826
3. Ganguly A, Nail S, Alexeenko A (2012) Rarefied gas dynamics aspects of pharmaceutical freeze-drying. Vacuum Gas Dynam Theory Experim Pract Applications 86(11):1739–1747. https://www.sciencedirect.com/science/article/pii/S0042207X12001716
4. Gibou F, Fedkiw R, Osher S (2018) A review of level-set methods and some recent applications. J Comput Phys 353:82–109. https://www.sciencedirect.com/science/article/pii/S0021999117307441
5. Karniadakis G, Beskok A, Aluru N (2005) In: Microflows and nanoflows—fundamentals and simulation. Springer. https://doi.org/10.1007/0-387-28676-4
6. Rana AS, Struchtrup H (2016) Thermodynamically admissible boundary conditions for the regularized 13 moment equations. Phys Fluids 28(2):027105. https://doi.org/10.1063/1.4941293

7. Schrage RW (1953) In: A theoretical study of interphase mass transfer. Columbia University Press, New York Chichester, West Sussex. https://doi.org/10.7312/schr90162
8. Stefan J (1891) Ueber die Theorie der Eisbildung, insbesondere über die Eisbildung im Polarmeere. Annalen der Physik 278(2):269–286. https://doi.org/10.1002/andp.18912780206
9. Struchtrup H, Torrilhon M (2003) Regularization of Grad's 13 moment equations: derivation and linear analysis. Phys Fluids 15(9):2668–2680. https://doi.org/10.1063/1.1597472
10. Theisen L, Torrilhon M (2021) fenicsr13: a tensorial mixed finite element solver for the linear r13 equations using the Fenics computing platform. ACM Trans Math Softw 47(2). https://doi.org/10.1145/3442378
11. Weniger D, Varghese P, Kowalski J, Torrilhon M (2023) Unsteady Stefan problem with kinetic interface conditions for rarefied gas deposition. Int J Heat Mass Transf 217:124696. https://www.sciencedirect.com/science/article/pii/S0017931023008414

Molecular Dynamics Study of Scattering of Nitrogen from a Cryogenic Nitrogen Surface

Johannes V. Diedrich and Martin Grabe

Abstract The non-equilibrium gas-surface interactions (GSI) between hot nitrogen gas and 4.3 K nitrogen ice were investigated. For this, Molecular Dynamics scattering simulations were carried out, comparing a flat and a rough aggregated surface model. The sticking probability and accommodation coefficients were determined for use in direct simulation Monte Carlo (DSMC) simulations. It was found that the sticking probability is similar for both models, but not unity despite the low wall temperature of 4.3 K. The accommodation coefficients do not indicate diffuse scattering.

Keywords Molecular dynamics · Scattering · Cryopumps

1 Introduction

Little is known about the scattering of reaction control thruster plumes from cryopump surfaces. This unusual system attracted research interest due to its occurrence in the large scale research facility STG-CT (DLR, "High-Vacuum Plume Test Facility for Chemical Thrusters") [1]. In this testing facility for small sattelite thrusters, the dynamics of rarefied gas flows with high velocities can be observed in high vacuum conditions. This is of interest in the field of plume contamination, which is concerned with the assessment of rocket motor plume effects on their environment. To provide the pumping speeds necessary to keep the vacuum during thruster firing, the entire chamber wall is cooled to 4.2–4.8 K with liquid He. The equilibrium vapor pressure of H_2 at wall temperature determines the chamber pressure. It is assumed that all other plume components except He freeze out completely upon first impact.

As the adsorption and scattering processes on the cryopump surface can currently not be studied experimentally in STG-CT, the system is modeled with molecular

J. V. Diedrich (✉)
Georg-August-Universität Göttingen, Institute of Physical Chemistry, Göttingen, Germany
e-mail: johannes.diedrich@chemie.uni-goettingen.de

M. Grabe
German Aerospace Center, Institute of Aerodynamics and Flow Technology, Spacecraft Department, Göttingen, Germany

M. Grabe et al. (eds.), *Rarefied Gas Dynamics*, Springer Aerospace Technology,
https://doi.org/10.1007/978-3-032-00094-1_9

dynamics (MD) scattering simulations. The main objectives are to test the assumption of total freeze-out and—if scattering is a non-negligible process—characterize it parametrically. Hydrazine monopropellant thruster plumes were considered a simple test case with the decomposition products H_2, N_2 and NH_3.

The interaction of rarefied gas flows with surfaces can be modeled using the Direct Simulation Monte Carlo (DSMC) method, using MD to parametrize the scattering kernels [2, 3, 5]. For this purpose, accommodation coefficients are determined as a measure of how much energy the scattering particle loses to or gains from the surface. In this work, the surface is assumed to be fully covered by the plume ice. The system under study is thus the interaction of the plume with its own frozen-out ice. Initially, only one plume species is considered, starting with the heaviest decomposition product, N_2. With the structure of the ice unknown a-priori, the surfaces are built using deposition simulations.

The purpose of this paper is to introduce the problem and document the first approach. The theoretical background and simulation details are discussed in 2, the main results of the deposition and scattering simulations are presented in 3 and finally, conclusions are drawn in 4.

2 Methods and Theory

2.1 Property Sampling

The plume ensemble was discretized by randomly sampling projectile properties from relevant distributions. Representative values for small thrusters in the test environment were chosen for thermodynamic quantities. The speed of the projectile

$$v_{\text{projectile}} = v_{\text{stream}} + v_T. \tag{1}$$

follows the constant stream velocity of the beam $v_{\text{stream}} = 1259\,\text{m s}^{-1}$, modified by a random component v_T drawn from the Maxwell-Boltzmann distribution for one degree of freedom, $f(T)$. This translational temperature was assumed to be $T_{\text{trans}} = 122\,\text{K}$. The rotational temperature was assumed as $T_{\text{trans}} = 159\,\text{K}$, while the direction of rotation was uniformly randomized.

The polar and azimuth angles of the impinging molecule trajectories were drawn from uniform distributions between $0°$–$75°$ and $0°$–$360°$, respectively. The impact positions were likewise uniformly randomized over the whole surface. Both the molecule starting orientation and rotation axis were randomized as well, although the latter was constrained to only allow axes of maximum rotational inertia.

The practice of depositing and scattering particles by sampling random angles and velocities to estimate an ensemble with a defined temperature (c.f. Sect. 2.1) is common practice in the literature (c.f. [3, 5]). The clear advantage is the ease of implementation. Looking at the incident surface-normal and surface-tangential energy components sampled this way, the distributions that result from this procedure show features that may not be expected intuitively from a simple Maxwell-Boltzmann distribution, c.f. Fig. 1. They are likely artifacts introduced by simultaneously drawing from angular and velocity distributions.

2.2 Simulation Details

All MD simulations were performed using LAMMPS [9]. Molecules were treated as rigid rotors within the `fix rigid/nve/small` style in LAMMPS. Time steps were 1 fs in scattering simulations and up to 5 fs in minimization simulations. Simulations were performed with full periodic boundary conditions. The box size in vertical (z) direction was chosen in such a way that, at minimum, the projectile was more than the Lennard-Jones interaction range ($r_{LJ} = 8\,\text{Å}$) away from a surface in any direction. The nitrogen unit cell was derived from the experimental structure by R. W. G. Wyckoff [10]. Following this structure, molecules following the PHAST-N_2 force field ([11]) were constructed with their corresponding pseudo atoms and atomic distances. The unit cells were then scaled to a cell parameter of $a = 5.6\,\text{Å}$. This value lies in-between the experiment ($a = 5.644\,\text{Å}$) and the one found for PHAST-N_2 ($a = 5.485\,\text{Å}$), to allow for relaxation rather than expansion during minimization. The force field used for H_2 [13] was of the same family as PHAST-N_2, the whole family being design to be expandable to other small molecules.

Three types of surfaces were simulated. First, a flat single crystal of N_2 at 4.3 K with dimensions of $10 \times 10 \times 8$ unit cells (approx. $56 \times 56 \times 45\,\text{Å}$) as a simple reference system. The flat surface was constructing by simulating the system in the NPT ensemble to relax the lattice parameter and in NPT at 4.3 K to assume the desired temperature. Secondly, deposition simulations were performed to incrementally build N_2 surfaces using the same projectile property distributions as in the scattering simulations. Starting from the flat model, the surface was truncated, the bottom-most layer was fixed in place and three layers above were thermostatted using a Nosé-Hoover thermostat chain [12]. Molecules were deposited individually by sampling their properties randomly from the distributions described in 2.1, aborting the simulation if they

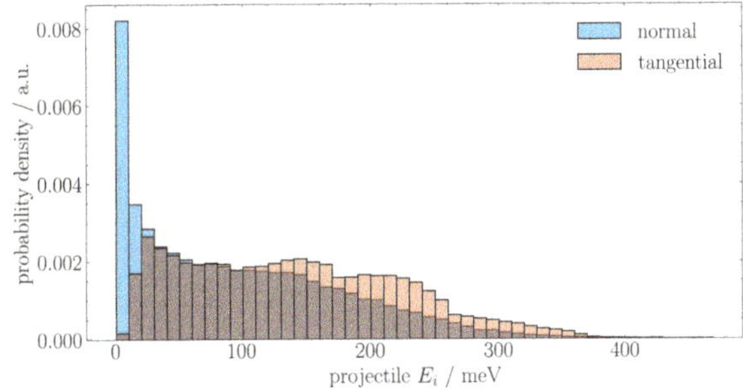

Fig. 1 Distribution of surface-tangential and surface-normal initial projectile energy components, as sampled 5×10^4 times via Eq. 1 and the angular and thermodynamic distributions as described in 2.1

scattered. After any trajectory ending in sticking, the system was cooled using the until the 2 ps average temperature of the accumulated aggregate was reduced below 4.5 K again. This was repeated until the aggregate consisted of the same number of molecules as the flat surface, 3200.

The scattering simulations were performed in the NVE ensemble after a 10 ps thermalization period. 10^4 individual simulations were performed per surface. To randomize the position in phase space of the surface, a random pre-run length of 0–50 ps was added to the thermalization. Trajectories were counted as scattered if the projectile receded from the surface beyond 8 Å. Sticking was defined as the projectile either not leaving the surface within 50 ps. To save computational effort, a stopping criterion for sticking is added, inspired by the study of Bolton et al. [14]. The kinetic plus potential energy of the projectile was averaged over a 1 ps window. Once The simulation was aborted once it dropped below the desorption energy of α-phase nitrogen ice, 69 meV [6]. Such aborted trajectories were counted as sticking, assuming the projectile has not had enough energy to emerge from the potential well for a considerable time. The sticking probability s was defined as the percentage of trajectories classified as sticking. The accommodation coefficients were defined as

$$\alpha = \frac{E_i - E_f}{E_i - E_s}.\tag{2}$$

where E_i and E_f are the initial and final energy of the projectile in any energy component, where α_n, α_t and α_r are defined for surface-normal, surface-tangential and rotational energy, respectively. E_s is a constant equal to the kinetic energy of the most probable velocity in the Maxwell-Boltzmann distribution at surface temperature.

2.3 Uncertainty Estimation

The sticking probability was sampled using Bayesian bootstrapping [15]. For each sample b, given a set of N trajectories, random weights are drawn from the Dirichlet $(0_1, ..., 0_N)$ distribution for each trajectory. New sets of trajectory results (sticking of scattering) are drawn using these weights until B sets of length N exist. In this way, any statistic that can be calculated for the original set of data can be described as a distribution of that statistic over B bootstrapped sets with a mean and standard deviation. The uncertainty given above is the 95% (1.96σ) confidence interval.

The same procedure can be used for the accommodation coefficients, if they are calculated for every individual trajectory. With distributions of initial and final energy in translational and rotational components, the accommodation coefficients were calculated for $B = 10^4$ samples, resulting in AC distributions over which a standard deviation could be calculated.

3 Results and Discussion

The method described in 2.2 was used to build four deposited surfaces to avoid randomization artifacts of a single rough surface. The obtained surfaces are depicted in Fig. 2 and were rendered using Ovito [4]. No defects like vacancies, dislocations or amorphous regions were found in the deposited structures. Only surface roughness from incomplete layers and few vacancies per surface were observed. The extent of the surface roughness was not substantial.

3.1 Scattering Results

The results of performing 10^4 scattering simulations per surface are shown in Table 1. The four deposited surfaces were averaged, reducing statistical errors and uncertainty due to the larger sample size. The sticking probabilities were high ($> 95\%$) for all investigated surfaces. With the approximations taken in the presented simulations (limited species, ice-only surface) we expect the resulting sticking probabilities to be upper limits. This result refutes the assumption of total freeze-out. Compared to previous work on $20\,\mathrm{K}$ methane ice ([5]) and the experimentally determined lower limit of s for N_2 on N_2 at $14\,\mathrm{K}$ ([6]), the values match expectations and approximately follow the surface temperature.

All α_n and α_t are relatively high, indicating that during scattering a significant amount of energy is exchanged with the wall. The α_t results are qualitatively lower than α_n, in accordance with results by [3]. This finding indicates that the scattering is not entirely diffuse, although surface roughness on length scales beyond the $56\,\text{Å}$ simulated in this work may influence this result. The differences in s and ACs between the flat and rough surfaces are minimal and covered by the confidence interval, both for s and the accommodation coefficients. Uncertainties between deposited and flat surface are not quantitatively comparable due to different sample sizes, but

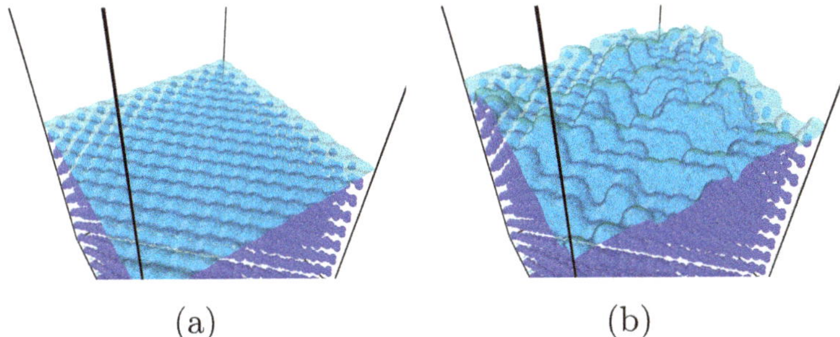

(a)   (b)

Fig. 2 Images of the surfaces. An isosurface mesh is rendered to highlight the three-dimensional structure. From left to right, the flat surface model **a** and one of the deposited surfaces **b**

Table 1 Summary of the sticking probabilities and accommodation coefficients determined in this work

System	T_s / K	s	α_n	α_t	α_r
This work, flat	4.3	0.984(3)	0.92(15)	0.63(29)	0(11)
This work, deposited	4.3	0.982(2)	0.88(26)	0.71(37)	0(47)
N_2 on N_2 [6]	14	≥ 0.85	–	–	–
CH_4 on D_2O [7]	33.5	0.90	–	–	–
CH_4 on CH_4 experiment [5]	20	0.97	–	–	–
CH_4 on CH_4 MD [5]	20	0.91		0.89[b]	11[a]
N_2 on graphene [3][c]	300	0.3	1.06	0.87	–

Results of other, similar systems from literature are listed for comparison

The most applicable system or condition was chosen for comparison when several were available

Uncertainty estimates follow the bootstrapped distributions, c.f. Sect. 2.3

a Inferred from reported E_i and E_f. $\alpha_r = 11$ is due to them setting $E_{i,r} = 0$

b Only combined translational AC reported

c Values for different impact angles reported, the mean is given here. The sticking probability was reported for their $1100\,\mathrm{m\,s^{-1}}$ data set and the ACs for their $750\,\mathrm{m\,s^{-1}}$ data set

qualitatively, uncertainty decreases with sample size. It is therefore notable that the uncertainty of the deposited surface AC results is higher, indicating a broader spread of scattering results. Regarding the rotational AC, the mean zero and very large uncertainty do not allow meaningful comparison between surfaces. The result that rotational energy is gained on the cold surface is in agreement with the result from [5].

3.2 Individual Accommodation Coefficient Distributions

Finding α_r values near or below zero, the underlying distributions were investigated. For this, the ACs were calculated individually for each trajectory. The resulting distributions are shown in Fig. 3. Generally, the distributions range over the entire range 0–1 that the AC is classically defined in. Values outside this range are possible since individual energy components are considered. Such outliers are observed if energy is transferred between modes, like in a molecule gaining more rotational energy in the scattering than it had before. Especially prominent were the high negative values in α_r. The accommodation coefficient has a singularity at $E_i = E_s$, causing extreme outliers when the sampled rotational energy approaches the surface thermal energy (omitted in Fig. 3). Similarly, if the initial energy is zero, the AC loses its meaning except relating the energy gain to the surface energy. This is demonstrated by expressing the rotational energy gain reported by [5] in terms of α_r (c.f. Table 1).

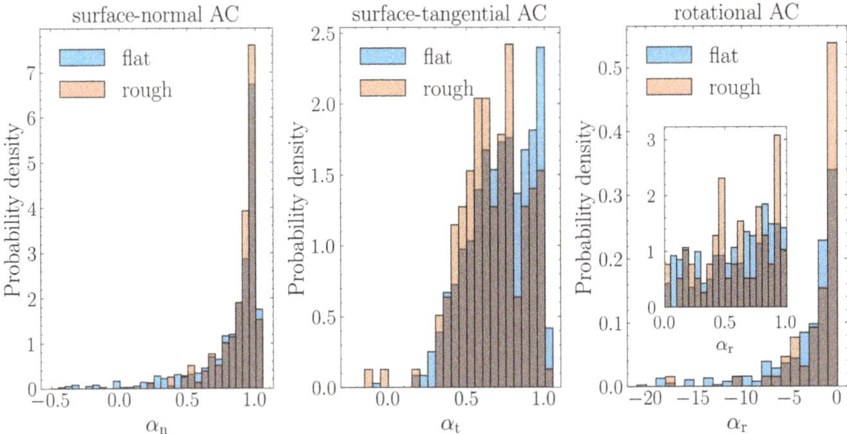

Fig. 3 The distributions of individual accommodation coefficients, in units of probability density, where the bin area corresponds to the fraction of the dataset that falls within the respective bin boundaries. Due to the outliers, only the central 2–98% quantiles of α_r are shown, while the inset zoom shows the 0–1 range for comparison to α_n and α_t. Similarly, in α_t, distant outliers are omitted

The high uncertainty estimates of α_r are in part caused by these mathematical artifacts. However, even in the $0 \leq \alpha_r \leq 1$ region (Fig. 3, inset zoom) it becomes apparent that the distributions of α_r are much broader than of α_n and α_t, ranging far into the negative. As seen in Table 1, the mean α_r values are zero, which in a classical interpretation would indicate no energy change instead of a broad distribution that peaks near $\alpha_r = 1$. The α_t distributions are very broad over the range of 0.4–1, indicating that a single average value in DSMC could be an insufficient representation.

While various other statistical methods for uncertainty estimation are interesting to apply here, many of which result in less extreme values, the bootstrap was chosen here to highlight the broad underlying distributions.

4 Conclusions

The common practice of sampling projectile properties from velocity and angular distributions, it was shown that artifacts occur in the resulting energy component distributions. Further work could investigate how accurately they represent the true energy components of a rocket motor plume impinging on a surface. The proposed deposition scheme might not be necessary for pure species as it produced nearly perfect single crystals. Instead, it is intended and better suited for deposition of mixed ices. The comparison of a flat versus a rough deposited surface showed no significant difference in scattering behaviour. In future work, the deposition method could be used to deposit mixed ices with different concentrations of H_2 and NH_3 and compare their influence on sticking and scattering.

The question of total plume freeze-out, which initially inspired this study, could be addressed with reasonable certainty. At least approx. 1% of the directly impinging molecules scatter at least once. Regarding the scattering parameters, accommodation coefficients were calculated for surface-normal, surface-tangential and rotational energy components, suitable for use in scattering kernels in simulation methods like DSMC. The distributions of especially the rotational AC were found to be very broad. Instead of a singular parameter, incorporating the different distribution shapes might be necessary.

References

1. Grabe M (2016) STG-CT: high-vacuum plume test facility for chemical thrusters. J Large-Scale Res Facil JLSRF 2:A86–A86. https://doi.org/10.17815/jlsrf-2-139
2. Pham TT, To QD, Lauriat G, Léonard C, Hoang VV (2012) Effects of surface morphology and anisotropy on the tangential-momentum accommodation coefficient between Pt(100) and Ar. Phys Rev E 86:051201. https://doi.org/10.1103/PhysRevE.86.051201
3. Mehta NA, Levin DA (2017) Molecular-dynamics-derived gas-surface models for use in direct-simulation Monte Carlo. J Thermophys Heat Transf 31:757–771. https://doi.org/10.2514/1.T4934
4. Stukowski A (2010) Visualization and analysis of atomistic simulation data with OVITO—the open visualization tool. Modell Simul Mater Sci Eng 18:015012. https://doi.org/10.1088/0965-0393/18/1/015012
5. Brann MR, Ma X, Sibener SJ (2023) Isotopic enrichment resulting from differential condensation of methane isotopologues involving non-equilibrium gas–surface collisions modeled with molecular dynamics simulations. J Phys Chem C 127:13286–13294. https://doi.org/10.1021/acs.jpcc.3c02386
6. Bisschop SE, Fraser HJ, Öberg KI, van Dishoeck EF, Schlemmer S (2006) Desorption rates and sticking coefficients for CO and N2 interstellar ices. Astron Astrophys 449:1297–1309. https://doi.org/10.1051/0004-6361:20054051
7. Thompson RS, Brann MR, Sibener SJ (2019) Sticking probability of high-energy methane on crystalline, amorphous, and porous amorphous ice films. J Phys Chem C 123:17855–17863. https://doi.org/10.1021/acs.jpcc.9b03900
8. Price TW, Evans DD (1968) The status of monopropellant hydrazine technology. Contractor Report JPL-TR-32-1227, NASA, Jet Propulsion Lab
9. Thompson AP, Aktulga HM, Berger R, Bolintineanu et al (2022) LAMMPS—a flexible simulation tool for particle-based materials modeling at the atomic, Meso, and continuum scales. Comput Phys Commun 271:108171. https://doi.org/10.1016/j.cpc.2021.108171
10. Wyckoff RWG (1963) Nitrogen, sample at T = 4.2 K. Crystal Struct 1:7–83
11. Cioce CR, McLaughlin K, Belof JL, Space B (2013) A polarizable and transferable PHAST N2 potential for use in materials simulation. J Chem Theory Comput 9:5550–5557. https://doi.org/10.1021/ct400526a
12. Martyna GJ, Klein ML, Tuckerman M (1992) Nosé-Hoover chains: the canonical ensemble via continuous dynamics. J Chem Phys 97:2635–2643. https://doi.org/10.1063/1.463940
13. McLaughlin K, Cioce CR, Belof JL, Space B (2012) A molecular H2 potential for heterogeneous simulations including polarization and many-body van der Waals interactions. J Chem Phys 136:194501. https://doi.org/10.1063/1.4717705

14. Bolton K, Svanberg M, Pettersson JBC (1999) Classical trajectory study of argon-ice collision dynamics. J Chem Phys 110:5380–5391. https://doi.org/10.1063/1.478433
15. Rubin DB (1981) The Bayesian bootstrap. Ann Stat 9:130–134. https://doi.org/10.1214/aos/1176345338

Investigation of Gas-Surface Interactions in Hyperthermal and Hypersonic Gas Flow

Quan Han and Haitao Xia

Abstract This study utilizes molecular dynamics simulations to investigate the interaction between silicon carbide surfaces and argon atoms, focusing on the effects of gas temperature and normal velocity on the energy accommodation coefficient (EAC). The results demonstrate a decrease-increase variation of EAC as gas temperature increases from 600 to 12,000 K, driven by the interplay of different collision types and their respective partial EACs. When the normal velocity of argon atom increases from 400 to 3000 m/s, a decreasing trend in EAC is observed in the low incident energy range, which mirrors the effect of changing gas temperature. However, in the high incident energy range, EAC exhibits a fluctuating increase with rising incident energy, showing clear periodic peaks and valleys. This suggests that while both increasing incident energy, the mechanisms governing gas-surface interactions differ. Specifically, the change in normal velocity results in the lower degree of the energy exchange at the interface. Effectively leveraging this characteristic can provide valuable insights for the aerodynamic and aeroheating design of high-altitude vehicles.

Keywords Gas surface interaction · Hypersonic flow · Energy accommodation coefficient

1 Introduction

Accurately predicting aerodynamic heating in hypersonic vehicles requires a thorough understanding of the energy transfer mechanisms at gas-surface interactions, particularly under high rarefied flow conditions, where the energy accommodation

Q. Han (✉) · H. Xia
School of Mechanical and Electronic Engineering, Nanjing Forestry University, Nanjing, China
e-mail: quanhan@njfu.edu.cn

Q. Han
Jiangsu Key Laboratory for Design and Fabrication of Micro-Nano Biomedical Instruments, School of Mechanical Engineering, Southeast University, Nanjing, China

© The Author(s) 2026

M. Grabe et al. (eds.), *Rarefied Gas Dynamics*, Springer Aerospace Technology,
https://doi.org/10.1007/978-3-032-00094-1_10

coefficient (EAC) governs the efficiency of energy exchange [1, 2]. Despite advancements, significant uncertainties persist due to the complex interplay of factors such as incident energy (including both kinetic energy from hypersonic velocity and internal energy from hyperthermal excitation), surface roughness, and gas–gas interactions [3–6]. These challenges are amplified in hypersonic flows, where extreme velocities and temperatures induce unique gas-surface interaction (GSI) phenomena, including high-energy molecule dynamics and transient adsorption–desorption behaviors.

In 1980, Goodman's theoretical model predicted a qualitative decrease-increase trend in EAC with rising incident energy including the incident velocity and gas temperature [7]. In the later molecular dynamics (MD) studies, various trends in EAC changing with incident energy have been reported [8–16]. For example, Tao and Wang [8] demonstrated that gas–gas intermolecular collisions in "continuous scattering" simulations reduce EAC at lower Knudsen numbers, while high incident velocities limit adsorption time, suppressing energy transfer to surfaces. Han et al. [9] showed that EAC decreases with increasing flight velocity due to reduced gas-surface interaction time but rebounds at hypersonic velocities as interaction depth becomes dominant. Chirita et al. [10]emphasized the role of surface temperature, noting that EAC's sensitivity to thermal fluctuations diminishes at incident energies above 500 meV, where a limiting energy transfer regime emerges.

Regarding the influence of gas temperature, which also increases the incident energy, Sipkens et al. [12] observed through molecular dynamics simulations that EAC increases with rising gas temperature under low-temperature conditions. Similarly, Mane et al. [13] found that EAC between aluminum and inert gases (e.g., helium, argon, xenon) increases with temperature, though the rate of increase is smaller, indicating that solid surface properties may mitigate the temperature effect. However, studies by Peddakotla et al. [14] and Sartori et al. [15] found that the EAC of Pt-Ar and Cu-H_2 systems decreases with increasing gas temperature. This study systematically investigates how variations in incident energy—spanning both kinetic energy (through velocity modulation) and internal energy (via gas temperature)—influence EAC under hypersonic conditions.

2 Methodology

2.1 Setup of the Gas-Surface Interaction Model

The gas surface interaction model between cubic silicon carbide (3C-SiC) and argon molecules was constructed using the molecular dynamics software LAMMPS [17], as shown in Fig. 1. Argon atoms were launched from a fixed position above the SiC surface with velocities sampled from a Maxwell–Boltzmann distribution. The SiC slab measures $30 \times 30 \times 18$ Å3 and exposes its (001) face to the incoming atoms. The surface temperature is 300 K and periodic boundary conditions is applied in the xy-plane. The interactions between SiC atoms are modeled using the Tersoff potential,

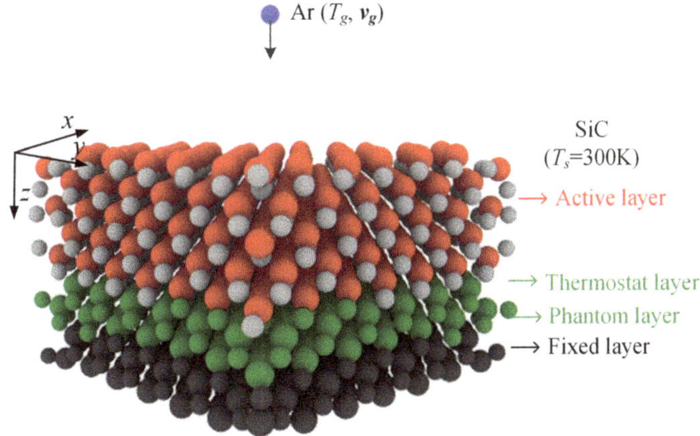

Fig. 1 Schematic diagram of the Ar-SiC interaction model

while the interactions between argon atoms and the C and Si atoms are described using the Lennard–Jones 12–6 potential [18]. All Lennard–Jones interactions have a cutoff radius set to 10 Å (>2.5σ) to balance computational accuracy and efficiency. Ensemble settings were NVE, with a time step of 1 fs and simulation duration of 40 ps, where adsorption is considered if particles remain within the cutoff radius for more than 40 ps.

2.2 Construction of Incident Velocities of Argon Atom

In this study, the construction of incident velocities of argon atom uses sampling methods, performing 10,000 independent simulations per calculation case. The velocity distribution is decomposed into thermal velocities and macroscopic flow velocities (v_g), where the thermal velocity components in the xy-direction follow the Maxwell–Boltzmann equilibrium distribution, with probability density functions as follows:

$$f(v_{x/y}) = \sqrt{\frac{m}{2\pi kT}} e^{-\frac{mv_{x/y}^2}{2kT}} \tag{1}$$

where m is the mass of gas atoms, T is the gas temperature, and k is the Boltzmann constant. The normal velocity (z-direction) adopts a half-normal distribution superimposed with macroscopic flow velocity:

$$f(v_z) = \sqrt{\frac{2m}{\pi kT}} v_z e^{-\frac{m(v_z - v_g)^2}{2kT}} \tag{2}$$

Two sets of conditions were examined: the first set fixed the gas temperature T_g = 300 K and varying the normal flow velocity v_g from 400 m/s to 3000 m/s. In the second set, the gas temperature was increased from 600 K to 12,000 K with v_g = 0 m/s.

2.3 Collision Types During the Gas-Surface Interactions

As shown in Fig. 2, based on trajectory characteristics of gas atoms interacting with the solid surface, collision behaviors can be classified into three types: Immediate Reflection (IR) with single collision with z-direction trajectory reversal, Trapping-Desorption (TD) with multiple collisions, and Permanent Adsorption (PA) with continuous oscillation within cutoff radius for the full simulation duration. In the classical formula for calculating the EAC, the ratio of adsorption ε is introduced to the formula:

$$EAC_{classic} = (1 - \varepsilon)EAC_{collision} + \varepsilon \tag{3}$$

where $EAC_{collision}$ is calculated from the collision events without adsorption. In this study, a modified formula incorporating collision-type contributions is derived[19]:

$$EAC_{modified} = \sum \beta_k \cdot EAC_k \cdot \gamma_k \tag{4}$$

where β_k is the ratio of collision type, k indicates the IR, TD, and PA collisions. EAC_k is the partial EAC of three collision types, and γ_k is the incident energy correction factor:

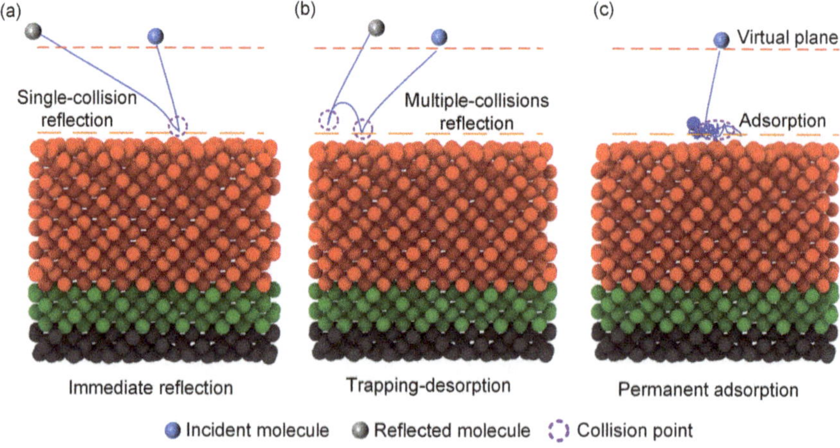

Fig. 2 Collision types during the gas-surface interactions

$$\begin{cases} EAC_k = \dfrac{\langle E_i \rangle_k - \langle E_r \rangle_k}{\langle E_i \rangle_k - \langle E_w \rangle} \\[2ex] \gamma_k = \dfrac{\langle E_i \rangle_k - \langle E_w \rangle}{\langle E_i \rangle - \langle E_w \rangle} \end{cases} \tag{5}$$

where $< E_i >$ is the average energy of the incident molecules, and $< E_r >$ is the average energy carried away by the reflected molecules. $< E_w >$ is the reflected energy if the gas has enough time to reach thermal equilibrium with the surface, which is determined to be $2k_B T$ for the monatomic gases investigated in this study.

3   Results and Discussion

3.1   *Impact of Gas Temperature on EAC*

The EAC shows a clear dependence on incident energy (gas temperature), as illustrated in Fig. 3a. As the gas temperature increases from 600 K to 12,000 K (corresponding to incident energies ranging from 0.10 to 2.08 eV), the EAC initially decreases by 16.7%, from 0.24 to a minimum of 0.2 at 1800 K, and then gradually recovers to 0.32 at 12,000 K. This non-monotonic behavior results from competing collision mechanisms: dominance of IR collisions at high energies versus contributions from TD and PA collisions at lower incident energies. Specifically, the modified formula predicts systematically lower EAC values compared to classical calculations below 0.5 eV, due to the negative correction coefficients ($\gamma_{PA} < 0$) for PA collisions. As shown in Fig. 3b, as the incident energy increases, the ratio of IR collisions rises sharply from 59 to 96%, while the ratio of TD collisions decreases from 33 to 4%, and the ratio of PA collisions drops from 8% to nearly zero. This indicates that, as the gas temperature increases, the thermal velocity of the gas atoms increases, making them more likely to undergo IR collisions, in which they reflect off the surface after a single bounce. Figure 3c displays the trend of partial EAC for the three collision types as a function of incident energy. For PA collisions, the partial EAC is assumed to be in a fully diffusive scattering state, with a value of 1. The partial EAC for IR collisions fluctuates initially and increases with incident energy. In contrast, the EAC for TD collisions decreases rapidly with increasing incident energy before stabilizing. Figure 3d illustrates the variation of the partial incident energy correction factor for the three collision types. In the low incident energy region, the correction factor for IR collisions decreases from 1.4 to 1.1, the correction factor for TD fluctuates around 0.6, and the correction factor for PA collisions is negative, indicating that PA collisions contribute to reducing the overall EAC.

Figure 3e and f show the changes in the average interaction depth and time with incident energy. For the IR collisions, when the incident energy is less than 0.5 eV, the interaction depth increases slightly, while the interaction time decreases significantly. Both effects contribute to fluctuations in the partial EAC for the IR collisions.

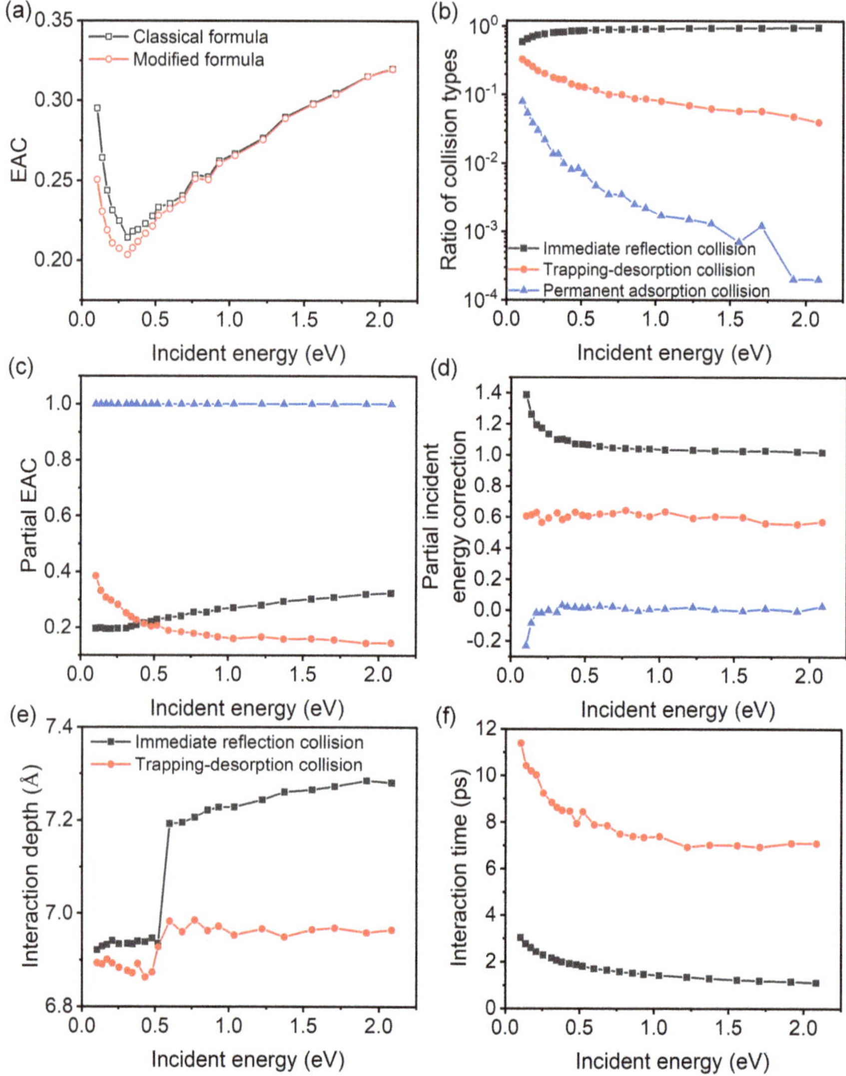

Fig. 3 Effects of temperature-induced changes in incident energy on the EAC and collision details. **a** Overall EAC, **b** Ratio of collision types, **c** Variation of partial EAC for different collision types, **d** Variation of correction factors, **e** Interaction depth, **f** Interaction time

When the incident energy exceeds 0.5 eV, the interaction depth increases sharply and continues to rise with increasing energy, while the interaction time decreases slowly, leading to an increase in the partial EAC for the IR collisions. For the TD collisions, at low incident energies, both the interaction depth and time decrease, causing a rapid decrease in the partial EAC. As the incident energy increases to a higher value, the interaction depth increases sharply and stabilizes, while the interaction time decreases and stabilizes, resulting in a decrease in the partial EAC for the TD collisions to a stable value.

3.2 *Impact of Normal Flow Velocity on EAC*

When a vehicle travels at hypersonic speeds, the gas molecules at the gas–solid interface acquire a normal velocity, which increases the incident energy as the normal velocity rises. As shown in Fig. 4a, the EAC values calculated using both formulas are in good agreement. As the incident energy increases with normal velocity, the EAC also exhibits a non-monotonic increase. In the low incident energy range (0.14–0.39 eV, corresponding to a speed range of 400–1000 m/s), the EAC decreases monotonically, reaching a minimum value of 0.2 at 0.44 eV. Afterward, as the incident energy increases, the EAC gradually rises and reaches a local maximum of 0.39 at 1.04 eV. The EAC then decreases as the incident energy increases and reaches a local minimum of 0.32 at 1.44 eV. Finally, the EAC increases again to a local maximum of 0.45 at 2.04 eV.

Figure 4b displays the variation in collision type proportions with incident energy. As the incident energy increases, the ratio of IR collisions rises significantly from 82 to 100%, the ratio of TD collisions decreases from 16.6% to almost zero, and the ratio of PA collisions decreases to zero. This demonstrates the significant effect of increasing normal velocity on collision type, making the collisions almost entirely IR. Figure 4c shows the trend of partial EAC for the three collision types with incident energy. The partial EAC for IR collisions increases with fluctuation but follows the overall EAC. The EAC for TD collisions decreases rapidly with incident energy. Figure 4d shows the variation of the incident energy correction factor for the three collision types. In the low incident energy region, the correction factor for IR collisions decreases from 1.06 to 1, the correction factor for TD collisions increases from 0.75 to 0.88, and the correction factor for PA increases from 0.3 to 0.57. As incident energy increases, only IR collisions remain, and their correction factor stabilizes at 1.0.

Figure 4e and f show the changes in average interaction depth and time with incident energy. For IR collisions, when the incident energy is less than 0.5 eV, the interaction depth decreases slightly, while the interaction time decreases significantly. Both factors contribute to the reduction in partial EAC for the IR collisions. When the incident energy exceeds 0.5 eV, the interaction depth increases sharply, while the interaction time decreases more slowly, causing the partial EAC for the IR collisions to rise with increasing incident energy.

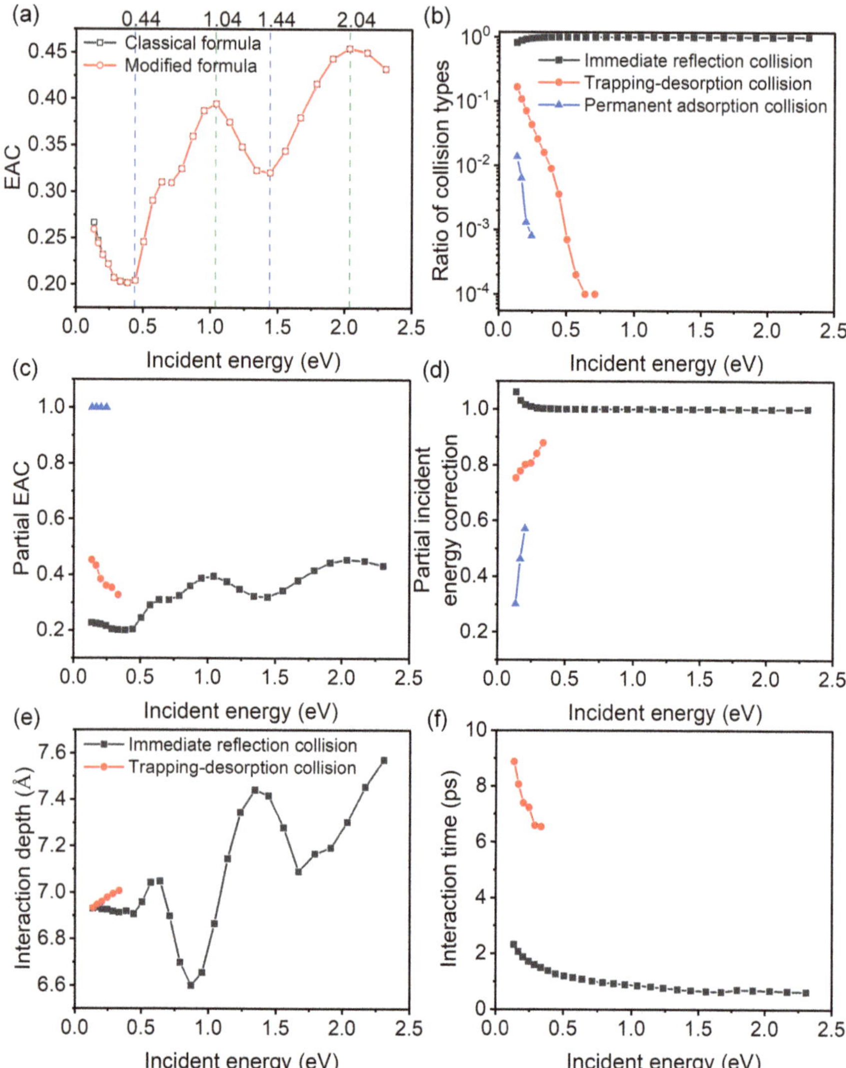

Fig. 4 EAC and collision details varying with normal velocity. **a** Overall EAC, **b** Ratio of collision types, **c** Variation of partial EAC for different collision types, **d** Variation of correction factors, **e** Interaction depth, **f** Interaction time

Acknowledgements This study was supported by the Natural Science Foundation of Jiangsu Province (No. BK20242031), and the China Postdoctoral Science Foundation (2023M740593).

References

1. Liu W, Zhang J, Jiang Y, Chen L, Lee C-H (2021) DSMC study of hypersonic rarefied flow using the Cercignani–Lampis–Lord model and a molecular-dynamics-based scattering database. Phys Fluids 33:072003
2. Mohammad Nejad S, Nedea S, Frijns A, Smeulders D (2022) Development of a scattering model for diatomic gas–solid surface interactions by an unsupervised machine learning approach. Phys Fluids 34:117122
3. Cao B-Y, Sun J, Chen M, Guo Z-Y (2009) Molecular momentum transport at fluid-solid interfaces in MEMS/NEMS: a review. Int J Mol Sci 10:4638–4706
4. Anderson JD (2019) In: Hypersonic and high-temperature gas dynamics. 3rd edn. Aiaa
5. Chen Y, Gibelli L, Li J, Borg MK (2023) Impact of surface physisorption on gas scattering dynamics. J Fluid Mech 968:A4
6. Chen Y, Gibelli L, Borg MK (2024) Impact of random nanoscale roughness on gas-scattering dynamics. Phys Rev E 109:065308
7. Goodman FO (1980) Thermal accommodation coefficients. J Phys Chem 84:1431–1445
8. Ruiling T, Zhihui W (2024) Gas-surface interaction features under effects of gas-gas molecules interaction in high-speed flows. Chin J Aeronaut 37:228–242
9. Han Q, Liu Y, Li Z, Zhang Y, Chen Y (2022) Investigation of energy accommodation coefficient at gas-solid interface of a hypersonic flying vehicle. Aerosp Sci Technol 126:107585
10. Chirita V, Pailthorpe BA, Collins RE (1997) Non-equilibrium energy and momentum accommodation coefficients of Ar atoms scattered from Ni(001) in the thermal regime: a molecular dynamics study. Nucl Instrum Methods Phys Res 129:465–473
11. Prabha SK, Sathian SP (2014) Velocity distribution and velocity correlation of mixture of gases in a nanochannel. Int J Therm Sci 81:52–58
12. Sipkens T, Daun K (2018) Effect of surface interatomic potential on thermal accommodation coefficients derived from molecular dynamics. J Phys Chem C 122:20431–20443
13. Mane T, Bhat P, Yang V, Sundaram DS (2018) Energy accommodation under non-equilibrium conditions for aluminum-inert gas systems. Surf Sci 677:135–148
14. Peddakotla SA, Kammara KK, Kumar R (2019) Molecular dynamics simulation of particle trajectory for the evaluation of surface accommodation coefficients. Microfluid Nanofluid 23:79
15. Sartori E, Brescaccin L, Serianni G (2015) Simulation of diatomic gas-wall interaction and accommodation coefficients for negative ion sources and accelerators. Rev Scient Instrum 87
16. Yamaguchi H, Matsuda Y, Niimi T (2017) Molecular-dynamics study on characteristics of energy and tangential momentum accommodation coefficients. Phys Rev E 96:013116
17. Thompson AP, Aktulga HM, Berger R, Bolintineanu DS, Brown WM, Crozier PS, in 't Veld PJ, Kohlmeyer A, Moore SG, Nguyen TD, Shan R, Stevens MJ, Tranchida J, Trott C, Plimpton SJ (2022) LAMMPS—a flexible simulation tool for particle-based materials modeling at the atomic, meso, and continuum scales. Comput Phys Commun 271:108171
18. Rappe AK, Casewit CJ, Colwell KS, Goddard WA III, Skiff WM (1992) UFF, a full periodic table force field for molecular mechanics and molecular dynamics simulations. J Am Chem Soc 114:10024–10035
19. Han Q, Ma C, Chen W, Wei Z, Zhang Y (2024) Influence of gas-surface and gas-gas interactions on the energy accommodation coefficient in non-equilibrium hypersonic gas flows. Appl Surf Sci 657:159812

Hypersonic and Space Vehicle Aerodynamics

DSMC Computations of Intake Performance for Air Breathing Ion Engine

Takashi Ozawa, Shunsuke Imamura, Masahito Tagawa, and Kumiko Yokota

Abstract At Japan Aerospace Exploration Agency (JAXA), a study of air breathing ion engine (ABIE) has been carried out for future missions in order to improve the duration of satellites in very low earth orbits (VLEO). For the improvement of the ABIE feasibility, intake performance is one of the important factors. In this work, intake configurations with a utilization of a $\mu 10$ ion engine have been investigated for ABIE by conducting direct simulation Monte Carlo (DSMC) simulations, and intake performance at an operational altitude (268 km) has been analyzed. We found that the intake performance can be improved along with the increase of an aspect ratio and a ratio of inflow to outflow cross sections. Several intake configurations were proposed, which can be capable of utilizing the $\mu 10$ ion engine for accelerating satellites in VLEO.

Keywords Rarefied gas dynamics · Hypersonic flows · Aerodynamics · DSMC · Air breathing ion engine

1 Introduction

At Japan Aerospace Exploration Agency (JAXA), a feasibility study of air breathing ion engine (ABIE) [1] has been carried out for future missions in order to improve the duration of satellites in very low earth orbits (VLEO) [2, 3]. In order to improve the ABIE feasibility in VLEO, both ion engine performance with air and intake performance are crucial. Intake performances for air breathing electric propulsion (ABEP) have been studied by several research groups over the last decades [4, 5].

T. Ozawa (✉) · S. Imamura
Japan Aerospace Exploration Agency, Sagamihara Kanagawa, Japan
e-mail: ozawa.takashi@jaxa.jp

M. Tagawa · K. Yokota
Kobe University, Kobe Hyogo, Japan

M. Grabe et al. (eds.), *Rarefied Gas Dynamics*, Springer Aerospace Technology,
https://doi.org/10.1007/978-3-032-00094-1_11

At JAXA, an ABIE system with a utilization of a μ10 ion engine has been proposed for a demonstration of ABIE at VLEO. In this work, we focus on the intake performance analyses by carrying out direct simulation Monte Carlo (DSMC) [6] simulations. From the interface of the μ10 ion engine, we determine prerequisites of intake development, and carry out a parametric study for intake efficiency with respect to configuration (aspect ratio, ratio of inflow to outflow cross sections, etc.) [7] and use of a collimator. The purpose of intake analyses in this work for ABIE is the ignition capability of the μ10 ion engine and operation at 268 km, where the Super Low Altitude Test Satellite (SLATS) [2] had been operated. The corresponding intake performance requirements at 268 km are (1) compression ratio higher than 70 ($P > = 3 \times 10^{-4}$ Pa, based on conservative ignition requirement for the μ10 ion engine), and (2) intake efficiency higher than 0.15.

2 Prerequisites of Intake for ABIE

In order to utilize the μ10 ion engine, our ABIE intake configuration should meet requirements from the interface of the ion engine and the ABIE system. Figure 1 shows an image of intake configuration with the μ10 ion engine. Freestream molecules are inflowed from the outer region of a core body, reflected by a reflector, and collected inside the μ10 waveguide to be accelerated by the ion engine. The prerequisites of the intake configuration in this work are listed in Table 1. A waveguide length, a waveguide radius, and an exit radius are fixed to fulfill the interface with the μ10 ion engine. Molecule transmittances at an exit screen grid (p_{t1}) and at waveguide surfaces (p_{t2}) are set to 0.12 and 0.26 in this work based on the μ10 ion engine information from Nishiyama et al. Also, the system requirement of a maximum diameter of 340 mm, a maximum length of 900 mm, and an inner radius of 120 mm, has been applied. In this work, we do not consider the attitude uncertainty, and thus, both angle of attack and sideslip angle are set to zero in DSMC.

Figure 2 shows a schematic view of the intake configuration with these prerequisites. The inner body cross section, $S_1 = \pi R_1^2 = 0.045$ m^2, and the cross section for ion beam, $A = \pi R_3^2 = 0.01$ m^2, are fixed in this work. In order to investigate intake configuration dependence, we first change a ratio of inflow to outflow cross sections ($a = B/A$) from 2 to 4, where B is a cross section of the intake inlet, defined as $B = \pi(R_2^2 - R_1^2)$ [7]. An intake aspect ratio is defined as $\chi = L_1/(R_2 - R_1)$, and χ is changed from 10 to 15. Also, at $\chi = 10$, we compare cases with and without a collimator. The outer radius, R_2, depends on a with the limitation of 170 mm or less, and the intake length, L_1, depends on χ. Nine types of intake configurations (see Table 2) are proposed in this work and shown in Fig. 3. For Types A2, B2, and C2, a ring collimator is located inside the intake tube with a radius of $(R_2 - R_1)/2$ and a length of L_1.

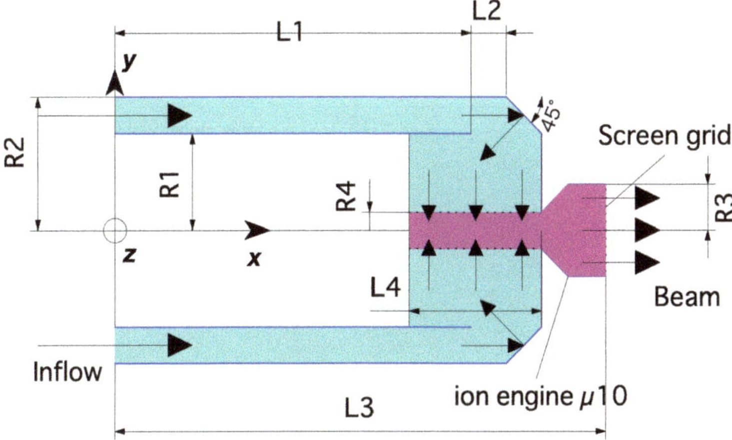

Fig. 1 A proposed intake configuration with a µ10 ion engine

Table 1 Prerequisites of intake parameters in this work

Parameter	Symbol	Value	Unit
Waveguide length	L_4	170	mm
Waveguide radius	R_4	22.5	mm
Exit radius	R_3	57	mm
Transmittance of the exit grid	p_{t1}	0.12	–
Transmittance of the waveguide surface	p_{t2}	0.26	–
Angle of attack	AoA	0	deg
Sideslip angle	$Beta$	0	deg
Maximum diameter	D_{max}	340	mm
Maximum length	L_{max}	900	mm
Inner radius	R_1	120	mm

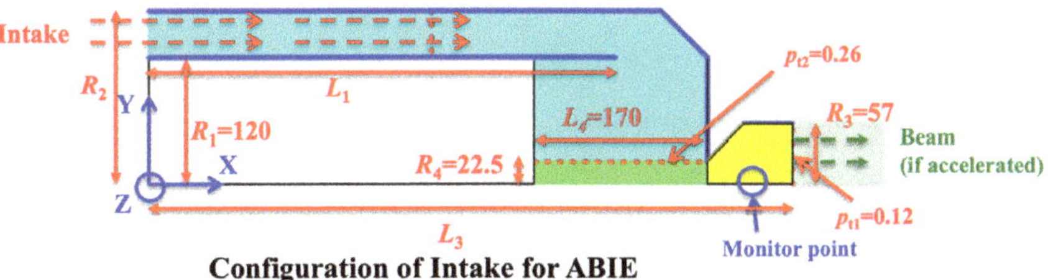

Fig. 2 An image of intake configuration with prerequisites for ABIE

Table 2 Types of intake configurations

	Type 0	Type 1	Type 2
Type A ($\alpha = 2$)	$\chi = 10$	$\chi = 15$	$\chi = 10$ (with a collimator)
Type B ($\alpha = 3$)	$\chi = 10$	$\chi = 15$	$\chi = 10$ (with a collimator)
Type C ($\alpha = 4$)	$\chi = 10$	$\chi = 15$	$\chi = 10$ (with a collimator)

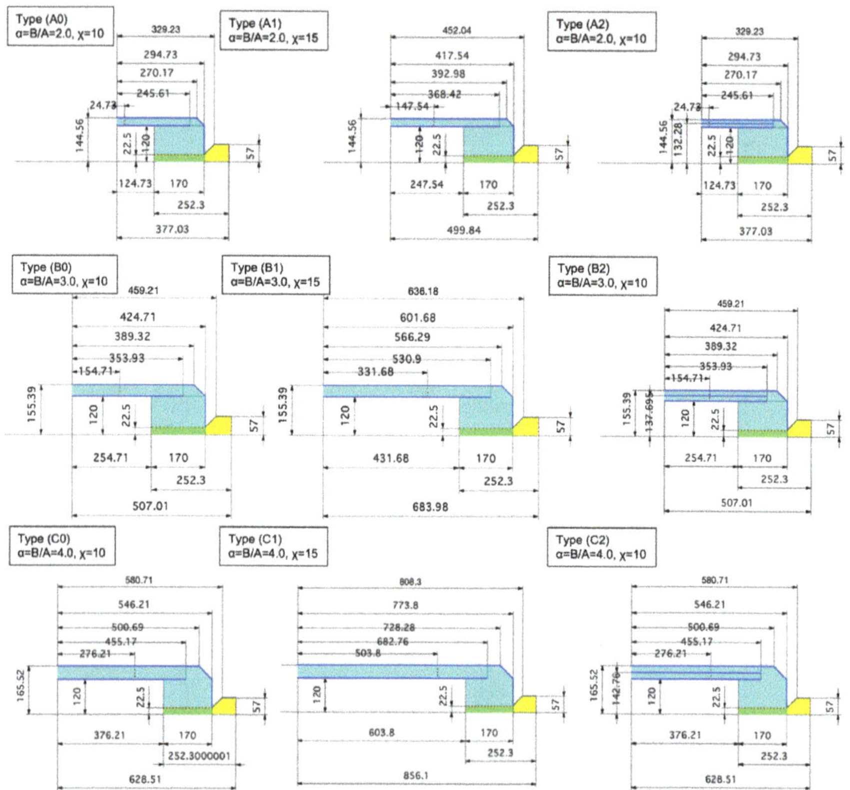

Fig. 3 Comparison of proposed intake configurations for ABIE

3 Computational Modeling

Numerically, we carry out DSMC simulations for hypersonic flows at 268 km around and inside the ABIE intake models. We use a DSMC code, named Modeling Of Transitional-Ionized Flows (MOTIF). General DSMC modeling is employed in the MOTIF code, and details can be found in Refs [8, 9]. In order to apply for ABIE intake analyses, we use the following specific modeling in this work.

(1) For geometry segments at the exit grid, a uniform transmittance (p_{t1}) is employed.

(2) For geometry segments for waveguide surfaces, a uniform transmittance (p_{t2}) is employed.

(3) Neutral particles in the waveguide (green-shaded section in Fig. 2) and discharge chamber (yellow-shaded section in Fig. 2) can be removed based on a ratio of releasing ions to neutral particles (R_{in}).

Note that R_{in} is set to 4 to demonstrate ion acceleration cases with the μ10 ion engine (based on personal communication with Nishiyama et al.). The transmittance parameters determine the ratio of passing the surfaces without collisions to reflection of the surfaces. A dataset of 7 chemical species (N, O, Ar, He, H, N_2, O_2) is used with a freestream condition at 268 km. The freestream speed (V_∞), temperature (T_∞), and number density (n_∞) at 268 km are 7.744 km/s, 921 K, and 1.23×10^{15} m^{-3}, respectively. The mole fractions for N, O, Ar, He, H, N_2, and O_2 are 9.96×10^{-3}, 0.719, 1.21×10^{-4}, 6.05×10^{-3}, 1.59×10^{-4}, 0.256, 9.34×10^{-3}, respectively. No chemical reactions were activated since the impact of chemical reactions is negligible at this altitude. For gas-surface interactions, the Maxwell model is used with a wall temperature (T_w) of 300 K. The surface accommodation coefficient, γ, is basically varied between 0.8 and 1.0 (fully diffuse condition).

We evaluate intake performance using the following 4 output parameters.

(1) Compression ratio: $G = \rho_{ch}$ (density inside discharge chamber) $/ \rho_\infty$ (freestream density)

(2) Intake efficiency 1 [5]: $\eta_{i1} = 1 - \Gamma_1/\Gamma_0$

    Γ_0: particle inflow flux at the inlet,
    Γ_1: particle outflow flux at the inlet.

(3) Intake efficiency 2: $\eta_{i2} = \Gamma_{2i}/\Gamma_0$

    Γ_2: outflow particle flux at the exit (Γ_{2n}: neutral, Γ_{2i}: ion).
    $R_{in} = \Gamma_{2i}/\Gamma_{2n}$.

(4) Ratio of mass flow rate (J [kg/s]) to drag coefficient (CDS [m^2]): J/CDS [kg/s-m^2]

    S: representative cross section ($S = \pi R_2{}^2$).

4 Results and Discussions

4.1 Dependence on α and χ Without Ion Acceleration

First, we conducted DSMC computations with the condition of γ of 1.0 and Rin of 0 for all 9 types of intake configurations. The contour plots of the compression ratio (G) are compared among types A, B, and C in Fig. 4.

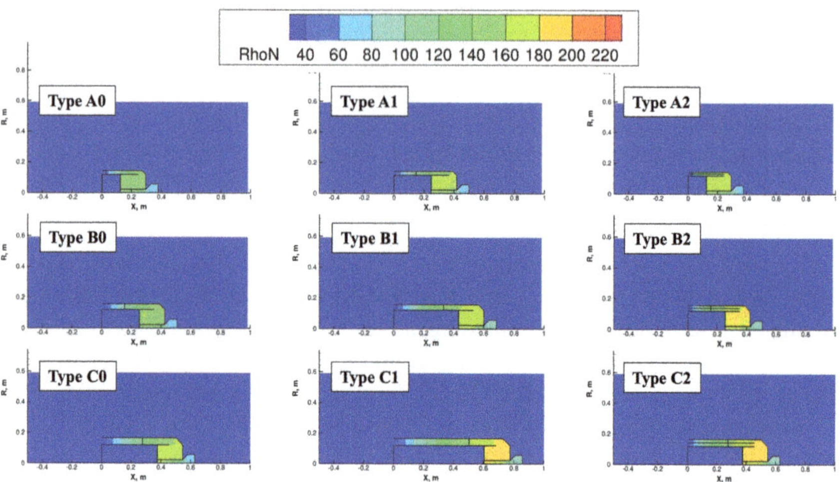

Fig. 4 Comparison of compression ratio contours among 9 intake configurations

Freestream particles entering the intake from the outer region of a core body are reflected by a reflector and collected inside the μ 10 waveguide. The density increases from the inlet to outlet and the density is the highest in the reflected region (outside of the waveguide). The density keeps decreasing from the region outside the waveguide to the region inside the discharge chamber. It can be seen that along with the increase of α and χ, the compression ratio, G, increases inside the waveguide and discharge region. With a collimator, the compression ratio also increases inside the waveguide and discharge region.

The compression ratio inside the discharge chamber is compared in Fig. 5. We found that the compression ratio increases approximately 10–15% along with the increase of α from 2 to 4. Among the types 0, 1, and 2, the type $2(\chi = 10$ with a collimator) shows better compression ratio. Figure 6 shows the dependence of compression ratio and intake efficiency on χ with α of 4. As the intake aspect ratio increases from 10 to 20, both G and $\eta i1$ increases. However, the effect of α is not significant in terms of J/CDS since both the drag and the mass flow rate increases along with the increase of α as you can see in Fig. 7.

4.2 Intake Performance with Ion Acceleration

In this section, we investigate the effect of ion acceleration. Numerically, we demonstrate ion acceleration in the discharge chamber by introducing a ratio of ion outflow flux to neutral outflow flux. According to the study of Nishiyama et al., R_{in} $(=\Gamma_{2i}/\Gamma_{2n})$ is approximately 4 during the μ10 ion engine operation. Therefore, in this study, we compare R_{in} of 0 and 4 cases to figure out the effect of ion acceleration. Figures 8

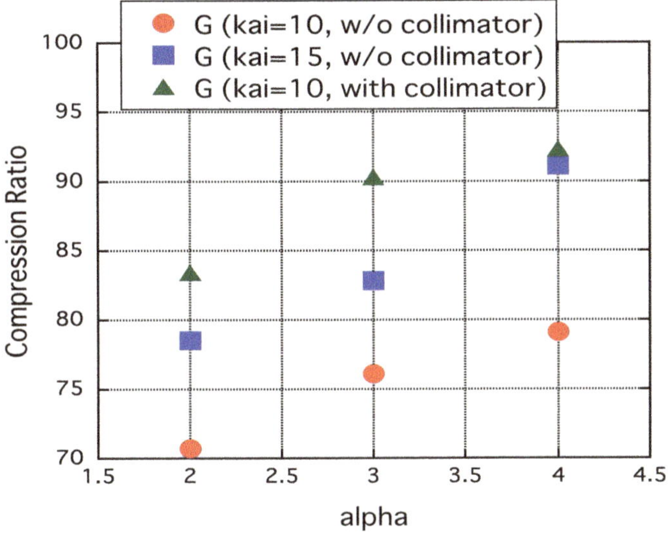

Fig. 5 Dependence of compression ratio (G) on α [$\gamma = 1.0$, $R_{in} = 0$]

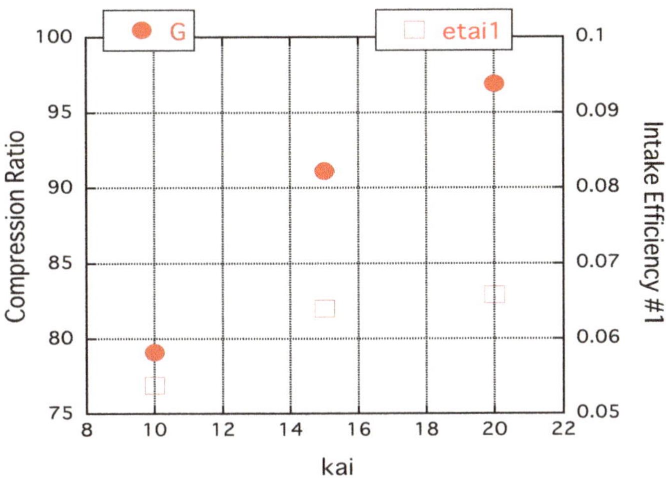

Fig. 6 Dependence of compression ratio (G) on χ [$\gamma = 1.0$, $R_{in} = 0$, $\alpha = 4$]

and 9 show comparisons of the compression ratio and intake efficiency between R_{in} of 0 and 4. We assumed aluminum surfaces for the intakes, whose surface accommodation coefficient is estimated to be 0.8. For both cases, compression ratio increases along with the increase of α. On the other hand, the intake efficiency decreases along with the increase of α. For instance, the intake efficiency is lower than 0.11 for $\alpha = 4$. With the activation of ion acceleration ($R_{in} = 4$ cases), compression ratio

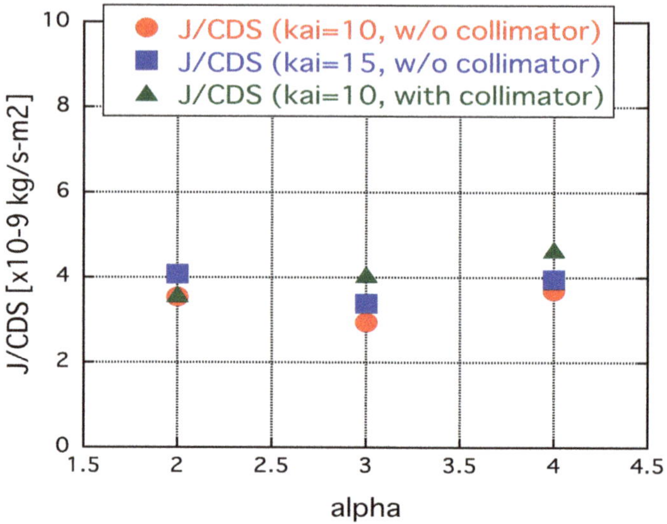

Fig. 7 Dependence of *J/CDS* on α [$\gamma = 1.0$, $R_{\text{in}} = 0$]

decreases. Although the compression ratio for $\alpha = 4$ is higher than 175 without the ion acceleration ($R_{\text{in}} = 0$), the value becomes lower than 80 for the ion acceleration demonstration cases ($R_{\text{in}} = 4$). On the other hand, the intake efficiency increases and is higher than 0.15 for $\alpha = 4$ and $R_{\text{in}} = 4$. In order to fulfill the compression ratio requirement of $G \geq 70$ at 268 km, α should be 4 or higher for the ion acceleration demonstration cases.

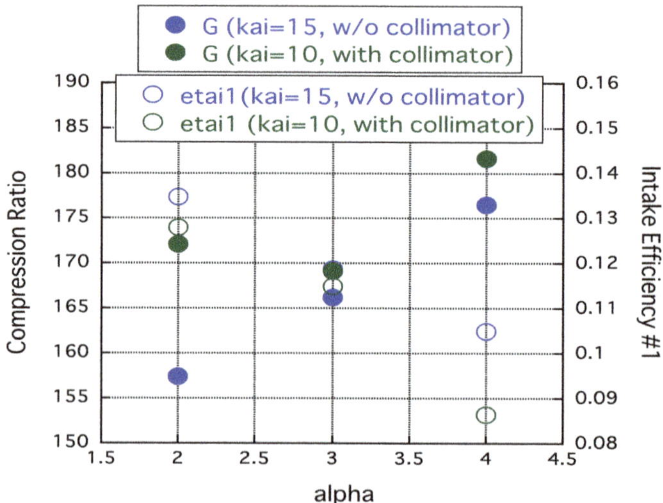

Fig. 8 Comparison of G and η_{i1} with $\gamma = 0.8$, $R_{\text{in}} = 0$ (without ion acceleration)

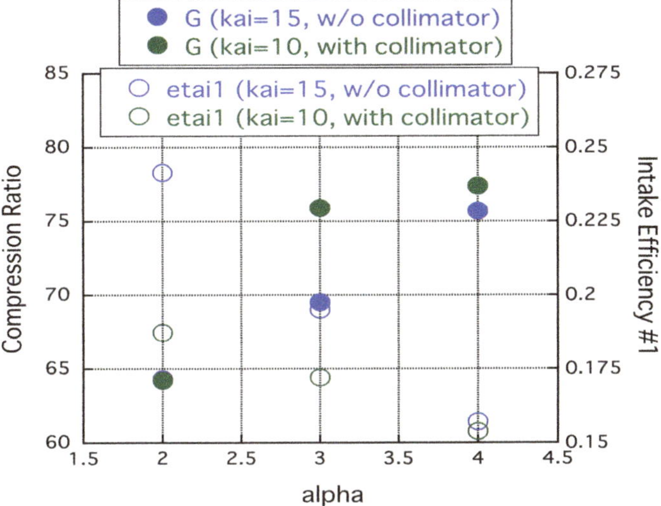

Fig. 9 Comparison of G and η_{i1} with $\gamma = 0.8$, $R_{in} = 4$ (with ion acceleration demonstration)

Figure 10 shows a comparison of η_{i1} and η_{i2} for the ion acceleration demonstration cases. Basically, η_{i2} has similar tendency to η_{i1}. η_{i2} decreases along with the increase of α. Among 6 types, the type A1 ($\alpha = 2$, $\chi = 15$) has the highest intake efficiency of 0.24 while the compression ratio is the lowest. Figure 11 shows a comparison of J/CDS with $\gamma = 0.8$, and $R_{in} = 4$. Similar to the cases without ion acceleration ($R_{in} = 0$), the effect of α is not significant in terms of J/CDS. Note that although J/CDS is not sensitive to α, α should be 4 among the test configurations from the viewpoint of the compression ratio requirement. The types C1 and C2 show similar intake performances, and the selection may depend on length and manufacturing limitations.

Finally, we show the dependence on the aspect ratio in Figs. 12 and 13. α was set to 4, and χ was changed from 10 to 20 (without collimators) with the condition of $\gamma = 0.8$ and $R_{in} = 4$. Both intake efficiencies, η_{i1} and η_{i2}, increase along with the increase of χ, and η_{i1} becomes higher than 0.15 if χ is greater than 14. J/CDS also increases along with the increase of χ.

5 Conclusions

In this work, we have investigated intake performance for ABIE with a utilization of the $\mu 10$ ion engine. Under the prerequisites of interfaces and size limitations, dependence on α (ratio of inflow to outflow cross sections) and χ (intake aspect ratio) have been studied. Along with the increase of α, the compression ratio also increased. On the other hand, intake efficiencies decreased, and J/CDS was not sensitive to α.

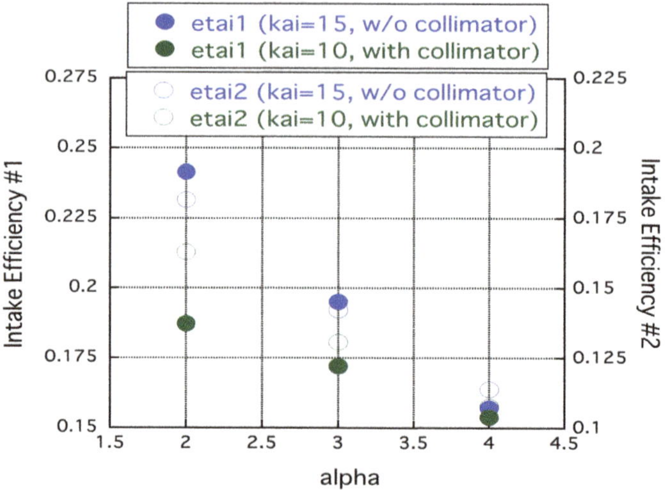

Fig. 10 Comparison of η_{i1} and η_{i2} with $\gamma = 0.8$, $R_{in} = 4$

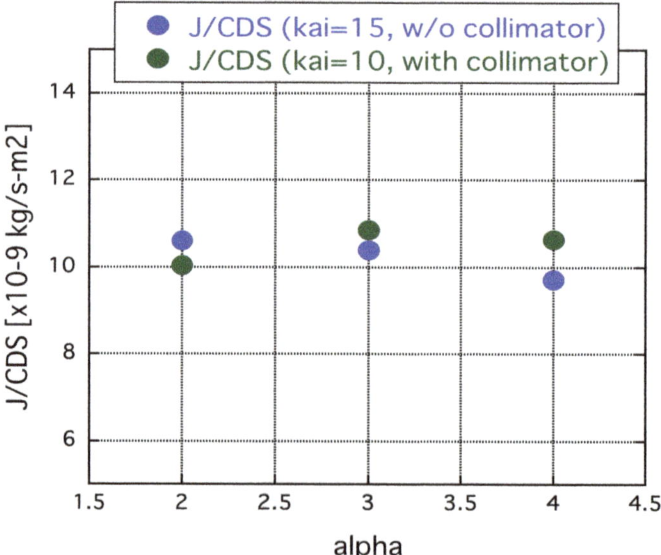

Fig. 11 Comparison of *J/CDS* with $\gamma = 0.8$, $R_{in} = 4$

Along with the increase of χ, both the compression ratio and intake efficiencies increased, which resulted in the *J/CDS* improvement. In addition, with a collimator, both the compression ratio and intake efficiencies became higher. In summary, the following three configurations can be suggested for the ABIE intake under the μ10 ion engine prerequisites in this work.

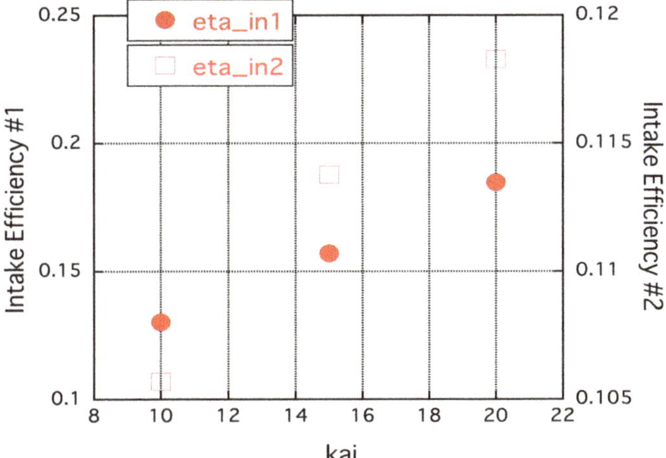

Fig. 12 Comparison of η_{i1} and η_{i2} with $\alpha = 4, \gamma = 0.8, R_{in} = 4$

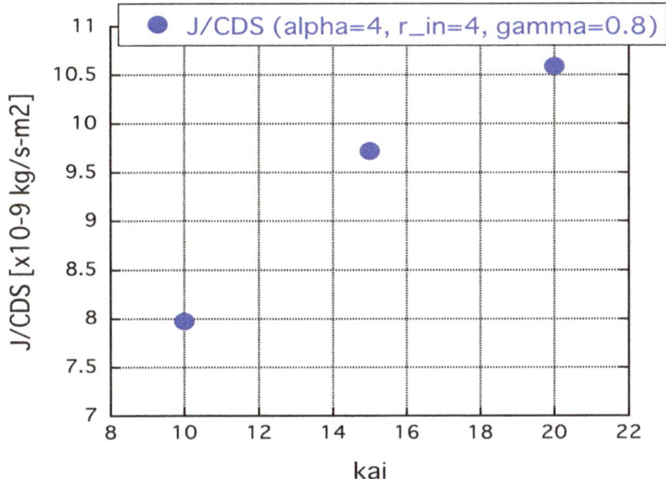

Fig. 13 Comparison of *J/CDS* with $\alpha = 4, \gamma = 0.8, R_{in} = 4$

(1) Length limitation of 500 mm class: Type B2 ($\alpha = 3, \chi = 10$ with a collimator)
(2) Length limitation of 600 mm class: Type C2 ($\alpha = 4, \chi = 10$ with a collimator)
(3) Manufacturing limitation (no collimators): Type C1 ($\alpha = 4, \chi = 15$ without collimators).

Note that these results are dependent on size and interface requirements.

As our future works, we will first study surface accommodation dependence, such as gold-plating, ITO (Indium Tin Oxide), and POSS-Polyimides coatings. Second, we will study the effect of micro-structured surfaces. Third, we will verify intake performance using wind tunnel models in the Hypersonic Rarefied Wind Tunnel (HRWT) at JAXA and in the Laser Detonation Wind Tunnel at Kobe University.

Acknowledgements The research performed at JAXA was partly supported by the Grant-in-Aid for Scientific Research (C), No. 24K07893 from Japan Society for the Promotion of Science (JSPS) whose support is gratefully acknowledged. I would like to acknowledge Prof. Nishiyama of ISAS and the ABIE team at JAXA for useful advice of μ10 ion engine interfaces and specifications.

References

1. Nishiyama K (2003) Air breathing ion engine concept. In; Proceedings of the 54th international astronautical congress, IAC-03-S4-02
2. Sasaki M (2017) Mission for super-low earth orbit, SLATS. Space Res Today 198:10–18
3. Imamura S, Utashima M, Ozawa T, Akiyama K, Sasaki M (2018) Current status of the ongoing orbit transfer of Super Low Altitude Test Satellite (SLATS). In: Proceedings of the 69th international astronautical congress, IAC-18-C1.8.1
4. Binder T, Boldini PC, Romano F, Herdrich G, Fasoulas S (2016) Transmission probabilities of rarefied flows in the application of atmosphere-breathing electric propulsion. AIP Conf Proc 1786(1):190011
5. Romano F et al (2021) Intake design for an atmosphere-breathing electric propulsion system (ABEP). Acta Astronaut 187:225–235
6. Bird GA (1994) Molecular gas dynamics and the direct simulation of gas flows. Clarendon, Oxford, England, U.K.
7. Fujita K (2004) Air intake performance of air breathing ion engines. J Jpn Soc Aeronaut Space Sci 52(610):514–521 (in Japanese)
8. Ozawa T, Suzuki T, Takayanagi H, Fujita K (2011) Analysis of non-continuum hypersonic flows for the Hayabusa reentry. In: AIAA Paper 2011-3311, 42nd AIAA thermophysics conference, Honolulu, Hawaii
9. Ozawa T, Zhong J, Levin DA (2008) Development of kinetic-based energy exchange models for noncontinuum, ionized hypersonic flows. Phys Fluids 20:046102

Toward a Model for Biological Contaminant Dispersal by Spacecraft Operating and Landing in Near-Vacuum

William A. Hoey◍ and John R. Anderson

Abstract When spacecraft operate and land in near-vacuum, as for Earth's Moon or icy moons like Enceladus, they introduce *contaminant* into their environments that may impact mission science at local and global scales. The rarefied gas dynamic plume flows generated by chemical propulsion systems may remove and disperse terrestrial particles borne on such spacecraft, including microbes and spores. These high-speed flows provide a unique vector for transport of *biological contaminant* far from a spacecraft in near-vacuum, i.e. in the absence of atmospheric drag. This paper describes the initial development of gas dynamic and orbital mechanics models for spacecraft-generated plume flows and induced particle transport, and an investigation of necessary boundary conditions to address two key questions: how can plume flows remove and disperse microorganism-laden particles or independent microorganisms borne on a spacecraft? And how can biological contaminants redeposit both locally (to the spacecraft, its instruments, and landing site) and globally (i.e. to regions of scientific interest)?

Keywords Plumes · Planetary protection · Chemical propulsion · Rarefied gas

1 Introduction

Spacecraft that would operate near, or land onto, solar system bodies may carry *biological contaminant* including microbes and spores. The characterization and mitigation of this material is a core responsibility of NASA's Planetary Protection (PP) discipline, which is chartered in part with the protection of solar system bodies from contamination by Earth life [1]. Inadvertent contaminant dispersal during spacecraft operation and landing poses a *forward contamination threat* both locally—to a spacecraft's own instrumentation, its sampling systems, or landing site—and globally to distant regions of high scientific value including pristine sites intended for

W. A. Hoey (✉) · J. R. Anderson
NASA Jet Propulsion Laboratory, California Institute of Technology, Pasadena, CA, USA
e-mail: william.a.hoey@jpl.nasa.gov

© The Author(s) 2026

M. Grabe et al. (eds.), *Rarefied Gas Dynamics*, Springer Aerospace Technology,
https://doi.org/10.1007/978-3-032-00094-1_12

sampling, for instance the open cryovolcanic water vents of Saturn's icy moon Enceladus. Forward contamination could make in situ or returned-sample detection of extraterrestrial biosignatures ambiguous or compromise the integrity of extraterrestrial environments that could harbor life, and is therefore rigorously controlled in spacecraft design, assembly, and operations.

In any powered landing with a chemical propulsion system—whether in atmosphere as for Mars or in near-vacuum as for Earth's Moon or the icy moons—engine plume flows that wash over a spacecraft and its landing site can remove, entrain and disperse biological contaminant of terrestrial origin. The nominal operation of chemical propulsion systems in attitude control, orbital adjustment, descent, and landing exposes portions of a spacecraft to direct and indirect high-speed plumes that can exceed 2000 m/s in core flow. Rarefied gas dynamic plume flows may provide a unique and significant vector for long-distance particle dispersal in near-vacuum environments, where no appreciable atmospheric drag would act to limit transport.

1.1 Toward a Model for Biological Contaminant Dispersal in Plumes

Consider the Europa Lander mission concept, which proposed the use of a Sky Crane landing system derived from Mars Science Laboratory Curiosity (MSL) and Mars 2020 Perseverance (M2020) rover heritage, as shown in Fig. 1, for a landing onto Jupiter's icy moon Europa. This landing would expose a vehicle in bridled descent to a complex plume flowfield produced by four monopropellant hydrazine engines [2, 3]. Comparable Mars operations with the Sky Crane generated dust clouds that partially obscured landing sites and led to particulate deposition onto their rovers [4], showing that even in a continuum atmosphere[1] landing plume flows can directly impact a vehicle payload.

Several important questions of forward contamination are presented by such plume-vehicle interactions at Mars, Europa, or for other solar system bodies, including:

- How can engine plume flows *remove and disperse* biological contaminant (microorganism-laden particles or independent microorganisms) borne by a spacecraft?
- How could these plumes *redeposit* biological contaminant locally to a spacecraft, its instrumentation, and landing site, and globally to surfaces far from its landing site?

---

[1] Plume flow fields in Mars atmosphere will differ substantially from those at Europa given the vast difference in pressure between bodies; at Mars surface (~ 640 Pa) engine plumes remain collimated and continuum, whereas Europa's near-vacuum surface condition (<< 1 μPa) would permit broad expansion through the rarefied and into the free-molecular regime [3].

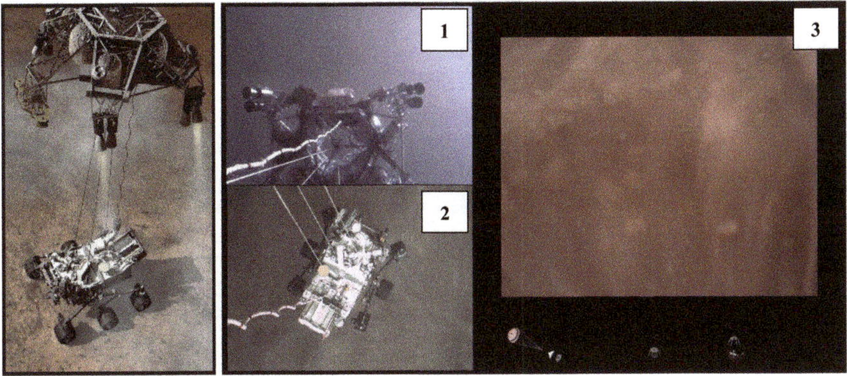

Fig. 1 Sky Crane Mars landings. *Left:* an artist's concept of the Mars Science Laboratory Curiosity rover's landing. *Right:* images from Mars 2020 Perseverance's Sky Crane landing video including *(1)* views upward from the rover's top deck; *(2)* downwards from the Sky Crane; and *(3)* downward from the rover's bottom deck [4]. Surface obscuration is due to the erosion and transport of Mars dusts in the engine plume flow field. *Image credits NASA/JPL-Caltech*

This paper describes the development of an integrated model to answer such questions by delivering information about the distances and angles to which biological contaminant particles will transport as functions of spacecraft configuration (e.g. geometry and engine operation); of initial loadings of contaminant (e.g. shapes, sizes, strains, and associations between microbiology and larger inert particles); and of landing environment (e.g. atmospheric or airless condition, local gravity and rotational reference frame, etc.) Fig. 2 documents the model's intended workflow including key inputs and deliverable results, which derive from similar engineering models developed for the dispersal of lunar regolith in landing plume-surface interactions [5]. The paper proceeds through the columns of Fig. 2; Sect. 2 will address inputs to the framework and the preliminary status of their collection for representative spacecraft in near-vacuum; Sect. 3 will address plume gas dynamic and particle orbital trajectory models under development, and will exhibit results of rarefied gas dynamic plume simulations relevant to the dispersal model. More detailed methodological descriptions and examples of the authors' CFD-DSMC[2] approach can be found in Hoey et al. [3, 5].

---

[2] CFD: computational fluid dynamics; DSMC: direct simulation Monte Carlo.

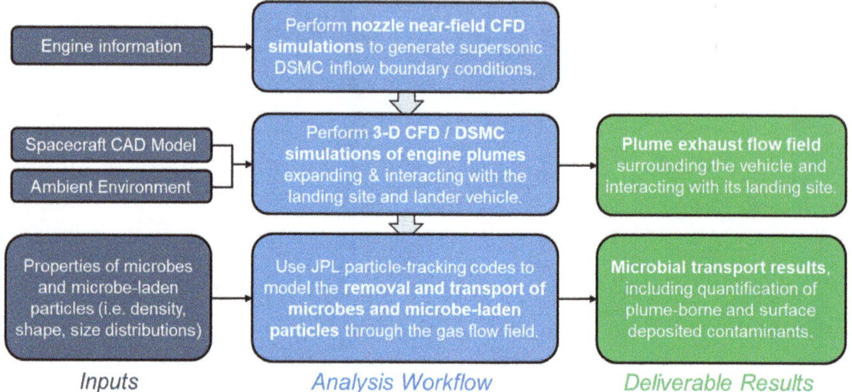

Fig. 2 Model workflow for transport analyses of spacecraft-borne microbes and microbe-laden particles mobilized by chemical propulsion plume flows

2 Model Inputs

2.1 Chemical Propulsion Systems for Mars, Lunar and Icy Moon Cases

Spacecraft chemical propulsion systems intended for attitude control, orbital adjustments, and even primary thrust for probes or landing vehicles often employ monopropellant (N_2H_4) or bipropellant (e.g. MMH/NTO)[3] hydrazine for their stability and performance as detailed by Grabe and Soares [6]. For example, MSL and M2020 Sky Crane architectures used eight monopropellant hydrazine engines to hover and four during bridled descent; the Europa Lander concept proposed a similar design [2, 3]. Other NASA Mars landers to use monopropellant hydrazine systems included Viking, Phoenix, and Insight. Recent mission concepts have studied monopropellant hydrazine in applications including orbiter and lander concepts to Saturn's icy moon Enceladus [7]. Bipropellant hydrazine systems are used both for primary thrust and attitude control on NASA's Europa Clipper mission; see common use on ISS visiting vehicles and similar planned vehicles for NASA's Lunar Gateway space station; and have been proposed for future spacecraft including CLPS (Commercial Lunar Payload Services) uncrewed lunar landers. Other cryogenic propellant mixtures planned for future crewed lunar landers are liquid oxygen / liquid hydrogen and liquid oxygen / liquid methane.

Any propulsion system planned for use in near-surface operations or landings onto solar system bodies could expose these bodies to biological contaminant dispersed

---

[3] MMH: monomethylhydrazine, fuel, CH_6N_2; NTO: dinitrogen tetroxide, oxidizer, N_2O_4.

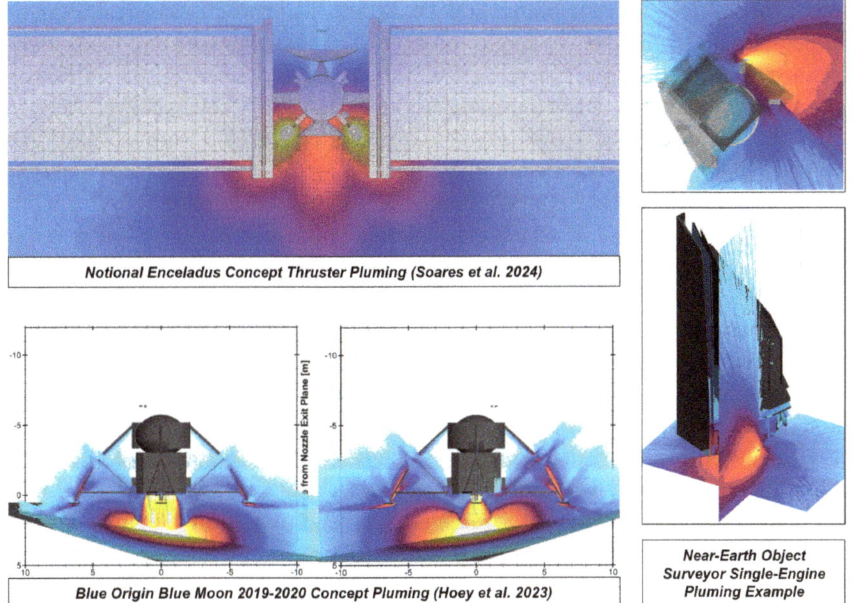

Fig. 3 JPL investigations of self-pluming for space and landing applications with CFD-DSMC methods; models for inert particle transport in such environments are being extended to study microorganism transport in near-vacuum. Representative contours show pressure on log-scales

in plumes, and is therefore of potential interest to NASA's planetary protection discipline. Rarefied gas dynamic plume simulations in near-vacuum conditions will typically require inputs including nozzle contours from throat to exit plane and sonic throat thermodynamic conditions like pressure, temperature, and species mole fraction (alternately, equivalent conditions at the nozzle exit); as well as engine positions, orientations, and firing histories for single- and multiple-engine firings [5]. The authors have performed CFD-DSMC simulations for a range of fuels and thrust levels that already inform the particle dispersal framework; these are introduced in Sects. 3.1 and Fig. 3.

2.2 Spacecraft Geometries and Environments

Any study of particle dispersal by spacecraft-generated plume flows must predict the forces applied to various vehicle surfaces by single- and multiple-engine firings, which will be dependent on the specific configuration both of the spacecraft and its engines. Therefore, spacecraft geometric models and derived meshes of sufficient detail to represent *significant features* are necessary—i.e. both (1) features of interest to a dispersal simulation and onto which loads must be queried and (2) all other features that influence plume flow fields onto, and particle dispersal, from the former—as are rarefied gas dynamic solvers that can treat meshes and multi-plume firings in 3-D. Fortunately, several DSMC solvers are available for 3-D simulations onto spacecraft meshes; the authors apply both NASA JSC's DAC (DSMC Analysis Code) and Sandia's open-source SPARTA (Stochastic PArallel Rarefied-gas Time-accurate Analyzer) [8, 9]. The authors have developed an internal library of relevant spacecraft meshes for NASA and commercial missions including for prior plume simulations; the process of developing a suitable mesh is described in detail by publications including Hoey et al. [5].

In near-vacuum, spacecraft often experience inadvertent plume loading by gas-phase expansion into the backflow region behind a nozzle during single-engine firings, even when engines are oriented to prevent 'direct' plume impingement. In some cases, spacecraft experience 'direct' plume impingement in their nominal oper-ation as plume core flows impinge onto exposed vehicle surfaces within ~ 30° of nozzle centerline (as for Europa Clipper's magnetometer boom and as would occur in bridled descent for a Sky Crane Europa Lander concept); this is also a primary and well-studied mode of plume loading for space stations during the approach and departure of visiting vehicles. Multiple simultaneous firings can produce complex flow fields that impinge and impose appreciable loads onto surfaces not in the direct line-of-sight of nozzle centerlines. Solar array plume impingement can be diffi-cult to avoid, particularly when multi-plume interactions occur—and for icy moon probe and orbiter missions, solar arrays will have exceptionally large surface areas given their distance from the Sun (~ 5.2 AU for Jupiter and Europa; ~ 9.5 AU for Saturn and Enceladus) [7]. During spacecraft landings, plume-surface interactions add additional complexity to the plume-plume and plume-vehicle interactions that occur in orbit, including potential recirculation between the vehicle and landing site and surface erosion / deformation that can influence plume flows [3, 5]. In any case, the system of plume-plume, -vehicle, and -surface interactions for spacecraft oper-ating near, or landing onto, solar system bodies creates many opportunities to impose loads that could remove laden biological contaminant.

Spacecraft atmospheric and exospheric environments may also be significant inputs to models for plume flows and particle transport; e.g. plume simulations at Mars must represent that body's continuum atmosphere. This paper focuses primarily on 'near-vacuum' applications to Earth's Moon and the icy moons of Europa and Enceladus, for which ambient surface-level exospheres are many orders-of-magnitude less dense than any spacecraft-generated plume flows and may therefore be neglected.

2.3　Properties of Relevant Microorganisms and Inert Particles

The shapes, sizes, quantities, and attachment properties of microorganisms to spacecraft surfaces are another key input to a dispersal model. The properties of certain individual microorganisms may be known and their quantities measured on spacecraft surfaces and in assembly cleanrooms. However, relatively few such microorganisms are *standalone*, i.e. not attached to larger, inert particulate. This adds complexity to dispersal analyses: particle removal is strongly influenced by shape and size and so larger inert particles may be relatively easier to remove from surfaces under spacecraft gas dynamic, gravitational, and vibrational loads than standalone microbes. Mikellides et al. (2018) reported on models developed for the removal and transport of biological contaminant from the Mars 2020 Perseverance rover under gravitational acceleration and aerodynamic loads, although not necessarily plume-induced loads, and studied the association between *viable organisms* (VO) and inert particles sampled in cleanroom environments. They found VOs frequently attached to smaller inert particles, at between 0.1–1 per particle of size 1–50 μm diameter, but with no VOs found attached to particles larger than 50 μm [10]. Related work by Malli Mohan et al. [11] expands the study of biological-to-inert-particle attachment to identify chemical composition of inert particles, visualizations of standalone biological contaminant, and results from cultivations of identified VOs. During the Europa Clipper Assembly, Test and Launch Operations (ATLO) campaign, which concluded with a successful launch in October of 2024, tape-lift particle samples were collected by spacecraft contamination control (CC) engineers and analyzed for size distributions and quantities with respect to mission CC requirements. Spacecraft planetary protection (PP) engineers isolated and grew cultures from particle samples taken at similar locations and times, providing information about types and quantities of sampled microorganisms. Future work in support of the particle dispersal model framework described in this paper, already underway, will compare the archived samples and analysis results collected between Europa Clipper's CC and PP teams to augment the Mars-2020-era models of [10, 11].

3　Modeling Approaches

3.1　CFD-DSMC Simulations of Plume Flows

The authors have performed rarefied gas dynamic pluming simulations of relevant thrusters and engines with CFD-DSMC techniques. Monopropellant hydrazine applications feature in support to Europa Lander [2, 3] and Enceladus orbiter [7] mission concepts; a 2022 joint JPL-DLR vacuum chamber thruster test campaign; and in NASA flight project applications including to the upcoming Near-Earth Object Surveyor mission. Bipropellant hydrazine engines feature in support to a NASA

CLPS lander and in pending work related to a 2024 JPL-JSC-DLR test campaign (which provided pressure and heat flux data critical to validating DSMC plume models), while other propellant mixtures have been studied e.g. in validation cases for the Apollo Lunar Module Descent Engine [2] and in collaboration with Blue Origin for a Blue Moon lander concept [5]. More information on the methodology of these simulations can be found in publications like [3, 5]. Example results are demonstrated in Fig. 3.

Future work will integrate these and comparable pluming simulations with biological contaminant properties (Sect. 2.3) and an orbital trajectory model under development for arbitrary solar system bodies (Sect. 3.2) to study long-distance particle dispersal.

3.2 Development of an Orbital Trajectory Model for Biological Contaminant

A particle dispersal model is in development for the biological contaminant dispersal task to generate number density, column density, and body-surface deposition results for 3-D particle fields dispersed by plume firings. The authors apply a model system with a spherical gravitational body (i.e. no surface topography) and, initially, an inertial three-dimensional reference system (neglecting non-inertial effects at first-order). The origin of a right-handed reference system is placed at the center of the spherical body as shown in Fig. 4. Particle trajectories in a gravitational force field are determined using Newton's laws of gravity and motion. The authors have implemented a detailed mathematical formulation for long-distance biological particle transport along orbital trajectories in Mathematica. Future work will document this framework, move it to a compiled language, and link it to source terms for pluming and biological contaminant.

4 Conclusions and Discussion

Spacecraft alter their environments and may compromise their engineering or science objectives as they generate and disperse contaminant (inert or biological). NASA's planetary protection discipline is concerned with limiting the *forward contamination* of solar system bodies with uncontrolled, spacecraft-borne biological contaminant. The action of chemical propulsion system plumes is a mechanism for the removal and long-distance transport of biological contaminant (as particulate, e.g. microbes), particularly for operations and landing in near-vacuum environments. The authors have proposed, and are developing, a framework to model the dispersal of biological contaminant by plumes intended to inform planetary protection requirements and

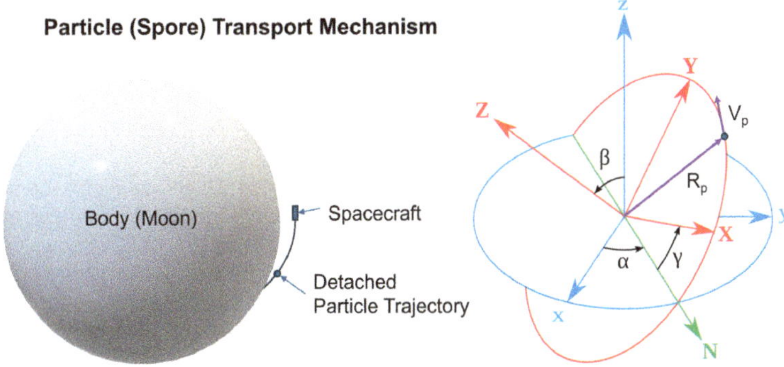

Fig. 4 Particle transport schematic and body-centered (in blue) vs. orbit-plane (in red) coordinate frame references. R_p is the vector from origin to the particle orbital location; V_p, particle velocity. In the body inertial reference frame $R_p = (x_p, y_p, z_p)$ and $V_p = (V_{xp}, V_{yp}, V_{zp})$. The normal vector to the orbit plane is $R_p \times V_p$, the X vector is toward periapsis, and the Y vector orthogonal to $Z \times X$

policy for future missions. This framework builds on established engineering capability for the rarefied gas dynamic simulation of gaseous plumes [2, 3, 7] and plumes carrying particulate [5], and on lessons-learned from the design and assembly of prior spacecraft [4, 6, 10, 11].

Acknowledgements © 2025. California Institute of Technology. Government sponsorship acknowledged. This work was performed at the Jet Propulsion Laboratory, California Institute of Technology, under a contract with NASA. Research was primarily funded by ROSES grant 22-PPR22-0007, PI Dr. W.A. Hoey. High performance computing resources were provided by funding from the JPL Information and Technology Solutions Directorate.

References

1. NASA Office of Safety & Mission Assurance, Planetary Protection Discipline Homepage, https://sma.nasa.gov/sma-disciplines/planetary-protection. Accessed 20 Dec 2024
2. Lam R, Maghsoudi E, Hoey W (2019) Numerical study of lander engine plume impingement on the surface of Europa. In: Proc. 66th JANNAF Propulsion Mtg., 37th exhaust plume and signatures Mtg., Dayton, OH, USA. 2014/46431
3. Hoey W, Lam R, Wong A, Soares C (2020) Europa lander engine plume interactions with the surface and vehicle. In: Proceedings of 2020 IEEE aerospace conference, big sky, MT, USA. https://doi.org/10.1109/AERO47225.2020.9172679
4. NASA Scientific Visualization Studio, Perseverance Rover's Descent and Touchdown on Mars, https://svs.gsfc.nasa.gov/31250. Accessed 20 Dec 2024
5. Hoey W, Martin M, Alred J, Soares C, Ababneh M (2023) Analyses of Blue origin blue moon lunar landing descent engine plume effects. In: Proceedings of 52nd international conference on environmental systems, no. 42, Calgary, Alberta, Canada. 2346/94493
6. Grabe M, Soares C (2019) Status and future of research on plume induced contamination. In: Proceedings of 70th international astronautical congress, IAC-19-D5.3.11, Washington DC, USA

7. Soares C et al (2024) Experimental characterization of spacecraft thruster plume contaminant composition for Enceladus orbiter and lander mission concepts. In: 55th Lunar and planetary science conference, no. 2215, The Woodlands, TX, USA

8. LeBeau G, Lumpkin F (2001) Application highlights of the DSMC Analysis Code (DAC) software for simulating rarefied flows. Comput Methods Appl Mech Eng 191:595–609. https://doi.org/10.1016/S0045-7825(01)00304-8

9. Gallis M, Torczynski J, Plimpton S, Rader D, Koehler T (2014) Direct simulation monte Carlo: the quest for speed. In: Proceedings 29th international symposium rarefied gas dynamics, AIP Conf. Proc. 1628, pp 27–35, Xi'an, China. https://doi.org/10.1063/1.4902571

10. Mikellides I et al (2018) Modelling and laboratory testing of particle resuspension and transport for the assessment of terrestrial-borne biological contamination of the samples on the Mars 2020 mission. In: Proceedings of 14th International Symposium Materials in the Space Environment, Biarritz, France. 2014/48799

11. Malli Mohan G, Cooper M, Venkateswaran K (2019) Microscopic characterization of biological and inert particle associated with spacecraft assembly cleanroom. Nat Sci Rep 9:14251. https://doi.org/10.1038/s41598-019-50782-0

Comparison Between the Kinetic Fokker-Planck and TAU Navier-Stokes Simulations of Hypersonic Air Flow Around the the RFZ-ST2 Upper Stage

Moritz Ertl and Leo Basov

Abstract We are developing a kinetic Fokker-Planck (FP) method to complement our existing computational fluid dynamics (CFD) code for aerodynamic and aerothermal simulations of space applications. The FP method is capable of simulating problems at low atmospheric densities, where the continuum assumptions of the CFD lose their validity. The FP method solves the Boltzmann equation and has the additional benefit of being computationally effective for low Knudsen numbers when compared to the Direct Simulation Monte Carlo method (DSMC). The FP method has been extended to model internal degrees of freedom using the Master Equation Ansatz for diatomic molecules as well as mixtures. In order to validate the new FP implementations and to better understand the limits of CFD and FP, we are simulating an upper stage, as a large test case. In this work we continue previous investigation with CFD and FP of upper stages, by expanding the modelling from single species diatomic N_2 gas to representing air as a gas mixture. We introduce the underlying simulation methods and explain the relevant differences. We compare the resulting flow fields.

Keywords CFD · Particle method · Fokker-Planck · Aerodynamics · Aerothermodynamics · Hypersonic flow · Upper stage

1 Introduction

Predictions of the flow around a vehicle play an important role in the development of space applications. Simulations can support the design process in many ways, from performance analysis, over load predictions to control system design. Computational fluid dynamics (CFD) are well established and well suited for applications in denser atmospheres, such as airplanes or first stages or reusable vertical takeoff

M. Ertl (✉) · L. Basov
Deutsches Zentrum für Luft- und Raumfahrt, Göttingen, Germany
e-mail: moritz.ertl@dlr.de

L. Basov
Applied and Computational Mathematics, RWTH Aachen, Aachen, Germany

© The Author(s) 2026

M. Grabe et al. (eds.), *Rarefied Gas Dynamics*, Springer Aerospace Technology,
https://doi.org/10.1007/978-3-032-00094-1_13

landing (VTVL) stages [8]. However, the underlying continuum assumptions lose their validity for higher Knudsen numbers, which are the typical regime for upper stages at higher altitudes and lower atmospheric densities. In these higher Knudsen (Kn) number regimes the flow can be described by the Boltzmann equation. Our approach is to numerically solve the Boltzmann equation using the kinetic Fokker-Planck (FP) method [9]. It fulfills our requirement of being computationally efficient both for high Knudsen numbers and when approaching the continuum limit.

In order to validate the new FP implementations, to better understand the limits of CFD and FP and to investigate the differences, our team is currently looking into several test cases. One of the more applied and large cases we selected is the simulation of an upper stage. The aim is to validate the methodologies, to gain insights into their limitations and to establish best practices on where and how to apply each method for large scale engineering applications. For the CFD simulations the DLR TAU code is used. It is well established for the simulation of space applications [10]. For the FP simulations a DLR in-house development of the method has been implemented using the DSMC code SPARTA [16] developed at the Sandia National laboratories. The FP method has been extended to model internal degrees of freedom using the Master Equation Ansatz for diatomic [13] and polyatomic molecules [2] as well as mixtures [12, 13]. For the upper stage simulation, the second stage of the RFZ model was selected. The RFZ model is an initiative of the Spacecraft department of the Institute of Aerodynamics and Flow Technology at the German Aerospace Center in Göttingen, and aims to to provide an open source, common research model for reusable launch vehicles [6]. The specifics of the upper stage are presented in ref [7]. In this work we continue our previous investigation with CFD and FP of the upper stage[1], by expanding the modelling from considering only single species diatomic N_2 gas to representing air as a gas mixture. First we shortly introduce the underlying simulation methods and explain the relevant differences. We then compare the resulting flow fields and pressure distribution and discuss relevant differences.

2 Numerical Methods

The governing equation used to describe flows in a wide range range of regimes is the Boltzmann equation:

$$\frac{\mathrm{D}f}{\mathrm{D}t} = S_{\mathrm{Boltz}}, \tag{1}$$

where f is the scalar velocity distribution function, t is time, and S_{Boltz} is the Boltzmann collision integral. A common approach to numerically solve Eq. (1) is the DSMC method proposed by Bird [3] which is a stochastic model that solves the spatial and temporal evolution of simulation particles, each of which represents a large number of real molecules. The interactions between particles are assumed to be binary collisions with instantaneous changes of velocity. These assumptions require grouping of particles in cells with a size in the order of the mean free path and time steps in the order of the collision frequency. The DSMC method has become very mature over the last decades and was shown to be applicable for complex problems with thermal and chemical non-equilibrium. However, applying the model in a flow regime close to continuum makes it prohibitively expensive due the aforementioned resolution criteria (see [4, 14] and references therein).

One approach to make Eq. (1) more manageable is to approximate the collision operator S_{Boltz} by a Fokker-Planck collision operator S_{FP}:

$$S_{\text{Boltz}} \approx S_{\text{FP}} = -\frac{\partial}{\partial V_i}(A_i f) + \frac{\partial^2}{\partial V_j \partial V_j}\left(\frac{D^2}{2}f\right) \tag{2}$$

where V is the molecule velocity with indices given in the Einstein notation. Equation (2) can be reformulated as stochastic differential equation which in turn can be solved using a particle method similar to the DSMC approach. The collision step is then replaced by a velocity update of the particles which does not rely on the building of collision pairs making the computational cost of the FP method become independent of the Kn number. The drift coefficient A_i and the diffusion coefficient D of Eq. (2) are model parameters chosen in such a way that production terms calculated using the Boltzmann collision operator are reproduced by the production terms using the FP collision operator [15] with the extension for diatomic gas mixtures based on the master equation ansatz [11]. For the simulations presented in this paper we use the FP cubic model [9].

The CFD simulations are done using the well established DLR TAU code [17] with the spacecraft extensions [10]. TAU solves the compressible Navier-Stokes equations on finite volumes. We use a AUSMDV upwinding with a with a carbuncle fix combined with least squares gradient reconstruction and a 3-stage explicit Runge-Kutta scheme to achieve second order accuracy in space and time.

3 Numerical Setup and Geometry

3.1 Definition of the Rocket Geometry and the Simulation Grids

The RFZ-ST2 model is a generic model of an upper stage for a VTVL reusable launch vehicle which is shown in Fig. 1. It is part of the RFZ model suite of open source rocket models created to provide a common research geometry for simulating VTVL reusable launch vehicles. The geometry sources and trajectory data are freely available at [5].

Both simulations are exploiting the rotational symmetry of the configuration by reducing the computational domain to a 2D axi-symmetric mesh. The CFD mesh is a body fitted triangular mesh with rectangular sublayer. The FP uses a cartesian hirachichaly refine mesh with the geometry represented as per cell surface elements.

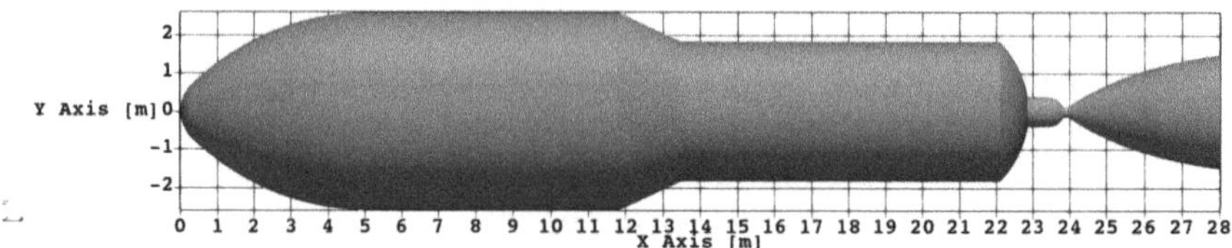

Fig. 1 Visualisation of the RFZ-ST2 upper stage geometry

The mesh refinement functionality of TAU and SPARTA are used to obtain well resolved grids. The final mesh used in this work consists of 1.15×10^6 grid points. The FP simulation domain is a rectangular box with dimensions $L_x = 41$ m, $L_y = 12$ m, and $L_z = 1$ m in x, y, and z direction respectively. The volume grid was generated containing $820 \times 240 \times 1$ cells and was locally adapted during a pre-run using SPARTA's build in grid adaptation method until each cell contained at least 10 numerical particles for a total of 21×10^6 volume cells. The surface mesh contains 3192 faces.

3.2 Settings and Boundary Conditions

A trajectory point within the applicable region of the CFD method is selected from [5] for an altitude of $h = 70$ km. The trajectory point is used to define the farfield boundary condition for both simulations with a velocity in x-direction $v_\infty = 2249.72$ m s^{-1}, a density $\rho_\infty = 1.2 \times 10^{-4}$ kgm^{-3} and a temperature $T_\infty = 226.9$ K. The gas is a mixture of diatomic nitrogen N_2 and oxygen O_2 with a mass fraction ratio of $x_{N_2} : x_{O_2} = 0.767 : 0.233$. A Knudsen number of $Kn = 2.0 \times 10^{-4}$ was calculated for this flight regime using the largest diameter of the rocket $L_{ref} = D_{fairing} = 2.6$ m as the reference length with the mean free path being calculated using the VHS model with parameters taken from [3].

For the CFD the boundary condition for surfaces of the upper stage are set as isothermal viscous walls. The FP simulation uses a diffuse boundary condition for the surface with an accommodation coefficient of 1.0. Both simulations use a wall temperature $T_w = 300$ K. The FP simulation was performed using a time step of $\Delta t = 1 \times 10^{-6}$ s with a particle weight of $w = 2 \times 10^{15}$. The particle weight was chosen so that a minimum of 8 particles are present in each FP cell.

4 Results

The comparison of the flow fields is visualised as a x-y-slice with the FP results displayed in the upper half and the CFD in the lower half. The density distribution is given in Fig. 2 with a logarithmic scale and the Mach number distribution show in Fig. 3.

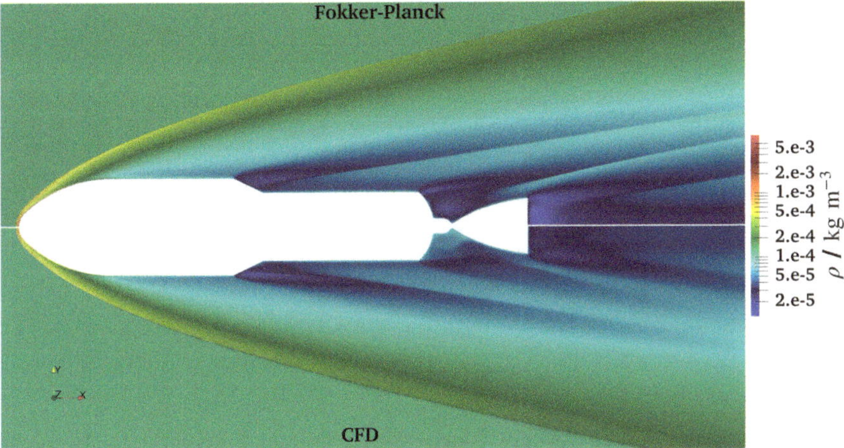

Fig. 2 Comparative visualisation of the RFZ-ST2 upper stage density ρ distribution on a logarithmic scale for the Fokker-Planck simulation (upper half) and the CFD simulation (lower half)

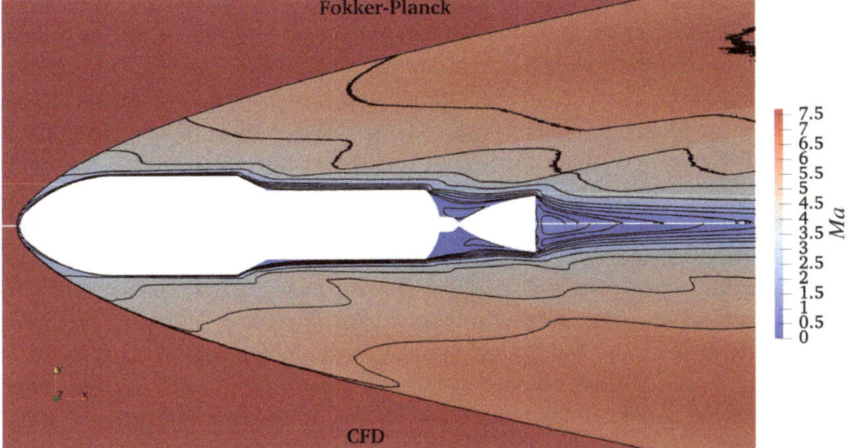

Fig. 3 Comparative visualisation of the RFZ-ST2 upper stage velocity Ma distribution for the Fokker-Planck simulation (upper half) and the CFD simulation (lower half)

The flow field visualisations exhibit a good agreement between the two methods. The structures and angles of shocks in the density visualisation show a good match, as do the the boundary layers and wake regions in the Mach number visualisation. The contour lines in the Mach plot reveal some differences especially in the wake region behind the vehicle. Here, the CFD struggles to resolve the weak shocks in the rarefied regions as sharply as the FP simulation. A difference can be observed in the shock stand-off distance and in the density behind the base plate. A plot of the translational temperature for the mixture and of the vibrational temperatures for both

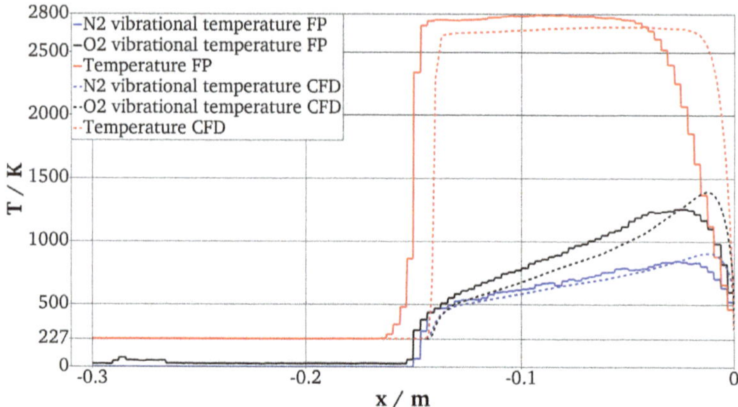

Fig. 4 Comparative Plot of the RFZ-ST2 upper stage temperature T distribution along a center line to the stagnation point for the Fokker-Planck simulation and the CFD simulation

N_2 and O_2 over a center line along the x-axis into the stagnation point is provided in Fig. 4. The steps in the FP results are due to the grid.

The vibrational temperatures of the CFD converge by methodical construction towards an equilibrium with the translational temperature, leading to a value of $T_{trans} = T_{vib} = T_\infty = 227\,\mathrm{K}$ in the free stream. In the FP model the vibrational energies are modeled as discrete level. The probability of a particle changing its vibrational energy level is a function of the translational temperature. Hence the vibrational temperatures tends towards zero in the free stream, due to the low probability of a vibrational mode being excited. The difference in shock stand-off distance is visible in the translational temperature jump with FP predicting the shock at $x_{S,\mathrm{FP}} \approx 0.15\,\mathrm{m}$ and CFD predicting it at $x_{S,\mathrm{CFD}} \approx 0.14\,\mathrm{m}$. This difference is assumed to be due to the modelling of the species mixtures in the FP method. The shock prediction of the DLR TAU is well established and validated and the authors have shown in [1] that the shock stand-off distance matches very well for FP and CFD for a single species free stream. The maximum temperatures of both translational and vibrational temperatures compare well. The relaxation also starts slightly sooner in the FP simulations.

5   Conclusions

We compared simulations of the open source generic upper stage RFZ-ST2 with a N_2 and O_2 species mixture free stream for a trajectory point at $70\,\mathrm{km}$ altitude. We simulated using the DLR CFD code TAU and our implementation of the kinetic Fokker-Planck particle method in the open source code SPARTA. The goal is to validate our FP method for use with industrial size cases of space vehicles at high altitudes. Our results showed a general good agreement between the flow fields, with shock positions and angles comparing very well. Small differences were observed in the shock stand-off distance, which we attribute to the difference in Prandtl number from the current FP implementation. Other possible sources for the differences are the resolution in time and space as given by the cell size and the time step size, as well as the number of simulation particles given by the particle weight. Over all the study shows, that we have a first working implementation of a usable tool for simulating real scale upper stages in high atmospheres, while also pointing out some areas of possible improvements. One of our next steps is the development of a polyatomic model for FP which reproduces a correct Prandtl number for single species gases as well as mixtures. Additionally we are looking into increasing the complexity of our simulations by investigating upper stages with engine plumes.

Acknowledgements The authors gratefully acknowledge the scientific support and HPC resources provided by the German Aerospace Center (DLR). The HPC system CARO is partially funded by "Ministry of Science and Culture of Lower Saxony" and "Federal Ministry for Economic Affairs and Climate Action".

References

1. Basov L, Ertl M, Bykerk T (2024) Comparison of Focker-Planck and CFD simulations of the rfz-st2 upper stage. In: HiSST: 3rd international conference on high-speed vehicle science and technology, 14–19 April 2024, Busan, Korea
2. Basov L, Grabe M (2024) Modeling of polyatomic gases in the kinetic Fokker-Planck method by extension of the master equation approach. In: AIP Conference proceedings. vol 2996. AIP Publishing, p 060004. https://doi.org/10.1063/5.0187658. https://pubs.aip.org/aip/acp/article-lookup/doi/10.1063/5.0187658
3. Bird GA (1994) Molecular gas dynamics and the direct simulation of gas flows. Oxford University Press, New York
4. Bird GA (1998) Recent advances and current challenges for DSMC. Comput Math Appl 35(1):1–14. https://doi.org/10.1016/S0898-1221(97)00254-X. https://linkinghub.elsevier.com/retrieve/pii/S089812219700254X
5. Bykerk T (2023) The rfz model - a standard model for the investigation of aerodynamic and aerothermal loads on a re-usable launch vehicle. https://zenodo.org/communities/rfz-model. Accessed 16 Sept 2023
6. Bykerk T (2023) A standard model for the investigation of aerodynamic and aerothermal loads on a re-usable launch vehicle. In: Aerospace Europe conference 2023—10th EUCASS—9th CEAS, Lausanne, Switzerland
7. Ertl M, Bykerk T (2024) A standard model for the investigation of aerodynamic and aerothermal loads on a re-usable launch vehicle—second stage geometry. In: HiSST: 3rd international conference on high-speed vehicle science and technology, 14–19 April 2024, Busan, Korea. Busan, Korea
8. Ertl M, Ecker T, Klevanski J, Krummen S, Dumont E (2022) Aerothermal analysis of plume interaction with deployed landing legs of the Callisto vehicle. In: 9th European conference for aeronautics and space sciences
9. Gorji MH, Torrilhon M, Jenny P (2011) Fokker–Planck model for computational studies of monatomic rarefied gas flows. J Fluid Mech 680:574–601. https://doi.org/10.1017/

jfm.2011.188. https://www.cambridge.org/core/product/identifier/S0022112011001881/type/journal_article

10. Hannemann K, Martinez Schramm J, Wagner A, Karl S, Hannemann V (2010) A closely coupled experimental and numerical approach for hypersonic and high enthalpy flow investigations utilising the HEG shock tunnel and the DIR tau code. Tech. rep, DLR

11. Hepp C (2022) A kinetic Fokker-Planck algorithm for simulating multiscale gas flows

12. Hepp C, Grabe M, Hannemann K (2020) A kinetic Fokker–Planck approach for modeling variable hard-sphere gas mixtures. AIP Adv 10(8):085219. http://aip.scitation.org/doi/10.1063/5.0017289

13. Hepp C, Grabe M, Hannemann K (2020) Master equation approach for modeling diatomic gas flows with a kinetic Fokker-Planck algorithm. J Comput Phys 418:109638. https://doi.org/10.1063/5.0017289. https://linkinghub.elsevier.com/retrieve/pii/S0021999120304125

14. Ivanov MS (2003) Current status and prospects of the DSMC modeling of near-continuum flows of non-reacting and reacting gases. In: AIP Conference proceedings. vol 663. AIP, pp 339–348. https://doi.org/10.1063/1.1581568. https://pubs.aip.org/aip/acp/article/663/1/339-348/1010897. ISSN: 0094243X

15. Jenny, P., Torrilhon, M., Heinz, S.: A solution algorithm for the fluid dynamic equations based on a stochastic model for molecular motion. J Comput Phys 229(4):1077–1098. https://doi.org/10.1016/j.jcp.2020.109638. https://linkinghub.elsevier.com/retrieve/pii/S0021999109005531

16. Plimpton SJ, Moore SG, Borner A, Stagg AK, Koehler TP, Torczynski JR, Gallis MA (2019) Direct simulation Monte Carlo on petaflop supercomputers and beyond. Phys Fluids 31(8):086101. https://doi.org/10.1063/1.5108534. https://pubs.aip.org/aip/pof/article/1059035

17. Schwamborn D, Gerhold T, Heinrich R (2006) The dlr tau-code: recent applications in research and industry. In: ECCOMAS CFD 2006 CONFERENCE

Numerical Analysis of a High Field Magnetic Heat Shield for Re-Entry Applications

Vishnu Asokakumar Sreekala, Jakub Glowacki, Tulasi Parashar, Ben Parkinson, Nicholas Strickland, Steve Smart, Joseph Bailey, Xiyong Huang, and Nicholas Long

Abstract Hypersonic flight is important for both military and civilian purposes. A major challenge at these high speeds is the extreme heat threatening the safety and performance of the vehicle. Recent advances in high temperature super conducting magnets (HTS) and high performance computing have renewed interest in this subject. We simulate an upcoming experiment to place an HTS magnet in the Gottingen shock tunnel. We show that the 2 T HTS magnet can increase the shock standoff distance by 1.5 cm (a 60% increase) and reduce the heat flux on the front surface by up to 12%.

Keywords Hypersonics flow · Lorentz force · Magnetic heat shield

1 Introduction

Hypersonic flight technology is gaining significance for its potential to revolutionize defense and commercial sectors with unprecedented speeds and capabilities. However, intense aerodynamic heating at hypersonic speeds is a major challenge. During flight, weakly ionized plasma forms behind the bow shock. The conducting nature of the plasma enables flow control via an external magnetic field. In the 1950s, Resler and Sears introduced magnetohydrodynamic (MHD) flow control, applying a magnetic field to generate a Lorentz force within the plasma [1]. This force slows the flow and pushes the shock away from the surface, potentially reducing heat load and providing thermal protection [2].

With advancements in hypersonic vehicle design and superconducting materials in the 1990s, interest in MHD as a thermal protection technology renewed. Early studies

home page: https://www.wgtn.ac.nz/.

V. A. Sreekala (✉) · T. Parashar
SCPS, Victoria University of Wellington, Wellington, New Zealand
e-mail: vishnu.asokakumarsreekala@vuw.ac.nz

J. Glowacki · B. Parkinson · N. Strickland · S. Smart · J. Bailey · X. Huang · N. Long
Robinson Research Institute, Lower Hutt, New Zealand

© The Author(s) 2026

M. Grabe et al. (eds.), *Rarefied Gas Dynamics*, Springer Aerospace Technology,
https://doi.org/10.1007/978-3-032-00094-1_14

by Bush [3] used simplified models, suggesting that an external magnetic field could reduce aerodynamic heating. Experiments by Gülhan et al. [4] in high-enthalpy wind tunnels confirmed that an external magnetic field could lower surface heat flux at stagnation point. Shock tubes have long been utilized in MHD experiments, starting with Zeimer [5] pioneering demonstration of enhanced shock stand-off in an arc-driven shock tube. Martinez Schramm and Hannemann [6, 7] achieved a fivefold increase in shock stand-off using a 4.5 T pulsed magnetic field in a 22 MJ/kg air test flow. Recent advancements in steady state HTS magnets have enabled the exploration for such applications [8].

This paper aims to numerically investigate the effect of high magnetic fields in hypersonic flows, focusing on a blunt body in an upcoming experimental campaign at DLR-HEG [9]. This preliminary study explores flow dynamics responsible for variations in shock stand off distance as well as aerodynamic heating and provides a systematic understanding of heat flux mitigation in the presence of an external high-field HTS magnet. In Sect. 2 we describe the experimental design as well as the computational method, Sect. 3 described our results, and in Sect. 4 we discuss our conclusions and future directions.

2   Methodology

2.1   *Magnet Design and Assembly*

The HTS magnet has been custom designed and constructed specifically for the partially-ionized flow experiments at the DLR-HEG(Deutsches Zentrum für Luft-und Raumfahrt High Enthalpy Shock Tunnel Göttingen.). The main magnet housing forms the pre-specified rounded-cylinder blunt-body profile, aiming to maximize the magnetic field at the front surface with constraints on a total length of 600 m of HTS tape and thermal isolation of the cryogenic coils. The magnet is placed at the radial center of the HE chamber and conductively cooled by a cryocooler at the periphery, requiring an extended thermal bus and current leads. A high-vacuum cryostat is used to house the assembly, as the HEG chamber pressure exceeds the experimental requirements. The Fig. 1 shows the experimental test section for the current campaign.The flow field is facing the HTS coil casing in x direction. The final magnet design consists of two symmetrical double pancake coils, each with a conductor inner diameter of 30 mm and outer diameter of 122 mm, made from 4 mm wide HTS tape. A short section of tape was tested for the critical current as a function of temperature, field magnitude, and orientation. Each coil has approximately 600 turns and is cooled by a copper thermal bus connected to a Sumitomo CH208L cryocooler, capable of extracting 10 W of heat at 20 K. The current leads are HTS tapes, with the warm end anchored to the cryocooler's first stage and copper leads extending to room temperature, carrying up to 350 A. In testing, the coils could be cooled to 15 K. The field distribution model is a 2D axisymmetric model with

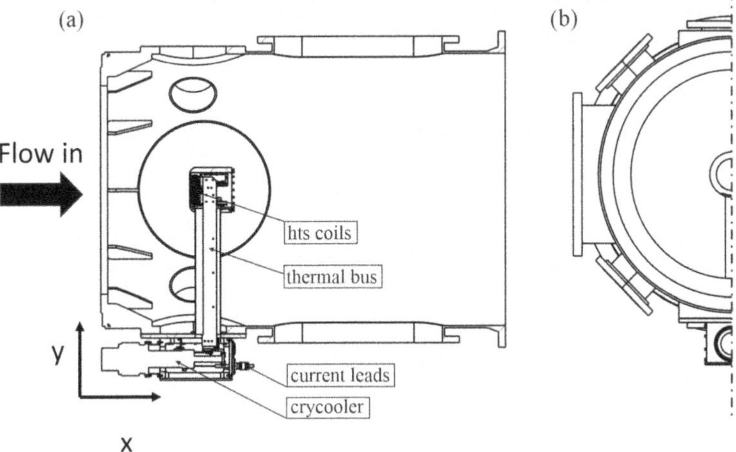

Fig. 1 Experimental test section with the proposed HTS magnet: **a** Side view of the chamber and test cross-section. **b** Front view of the chamber and magnet probe. The flow is in x direction

conductors only, where the field is proportional to the current and does not change with field magnitude. The field map, calculated using finite-element software or the Biot-Savart law, shows a field of 2.0 T at the magnet housing surface for 100 A. Based on the critical current measurements, the magnet could maintain over 300 A at 15 K, but magnetic forces might induce coil degradation at higher currents. A current of 100–200 A is expected to be supported, with higher currents requiring more sophisticated mechanical support.

2.2 Numerical Methods

The influence of magnetic field on bow shock formation was analyzed using the Finite Volume Method in the OpenFOAM framework. This simulation involves two stages: first, calculating the magnetic flux density vector, and second, assessing how the magnetic field affects the flow field and bow shock stand-off distance. The initial magnetic field profile is computed in a custom tool written in OpenFoam that solves the Helmholtz equation for the vector potential.

$$\nabla^2 \mathbf{A} = -\mathbf{J}_\mu \tag{1}$$

The initial magnetic field profile is then computed as the curl of the vector potential $\mathbf{B} = \nabla \times \mathbf{A}$. where $\mathbf{J}$ is the current density (for a 100 A operational current) and μ is the magnetic permeability of free space.

The next step utilizes a two-temperature, density-based solver "hy2Foam" to perform the flow simulation over the magnet assembly [10], specifically developed for modeling hypersonic reactive flows typical of reentry scenarios. The solver is built upon the extended Navier-Stokes-Fourier framework, enabling it to incorporate magnetic field effects. In the current numerical simulations, the inviscid flux vectors are discretized using the Kurganov scheme to achieve second-order accuracy. For the viscous flux vectors, a second-order central-difference scheme is utilized. Time integration is performed using the implicit Euler method. These effects are accounted for by including the Lorentz force as a source term in the momentum equation. The electric current density J within the plasma is derived using Ohm's law, under the assumption that the contribution of the electric field is negligible, and expressed as follows.

$$F_{\text{em}} = J \times B \tag{2}$$

$$J = \sigma\,(U \times B) \tag{3}$$

where U is the velocity field and σ the electrical conductivity expressed using Bush's formulation . In the Bush model [3], the electrical conductivity field, denoted by σ, is given by:

$$\sigma = \sigma_0 \cdot \left(\frac{T_t}{T_0}\right)^n \tag{4}$$

where: σ_0 is the reference electrical conductivity in S/m taken as 5600 S/m, T_0 is the reference temperature in K, taken as 10000 and n is the temperature exponent, taken as 2 in this study, T_t is the local trans-rotational temperature in K.

The high-temperature air's chemical kinetic model features a two-temperature chemical non-equilibrium approach involving seven species (N_2, O_2, NO, N, O, NO^+, e^-) and 11 elementary reactions. Detailed thermodynamic parameters for each species and the rate constants for the reactions are provided by Gupta et al. [11]. Furthermore, the equation for Y_i is discretized with a second-order implicit Euler scheme and solved using the successive over-relaxation method. Additionally, the species reactions are governed by an irreversible Arrhenius reaction mechanism [12] . This approach models the rate of chemical reactions as a function of temperature, enabling a realistic simulation of high-temperature reaction kinetics within the flow field.

2.3 Computational Domain

The flow domain and magnetic field distribution are shown in Fig. 2a. The blunt body features a 100mm cylindrical radius (r_a) with 30mm rounded edges (r_b). A 3D mesh was generated by rotating a 2D gMsh-generated mesh 5° about the stagnation line (L_A), creating 4500 structured cells with wall/shock refinements. As shown in Fig. 2b, the geometry follows Bush et al. [3] with axial symmetry, featuring a dipole field (2T at body surface, positioned 2 cm from stagnation) aligned with flow. Field mapping used linear interpolation across three cross-sections: axial line L_A, L_1 (3 cm

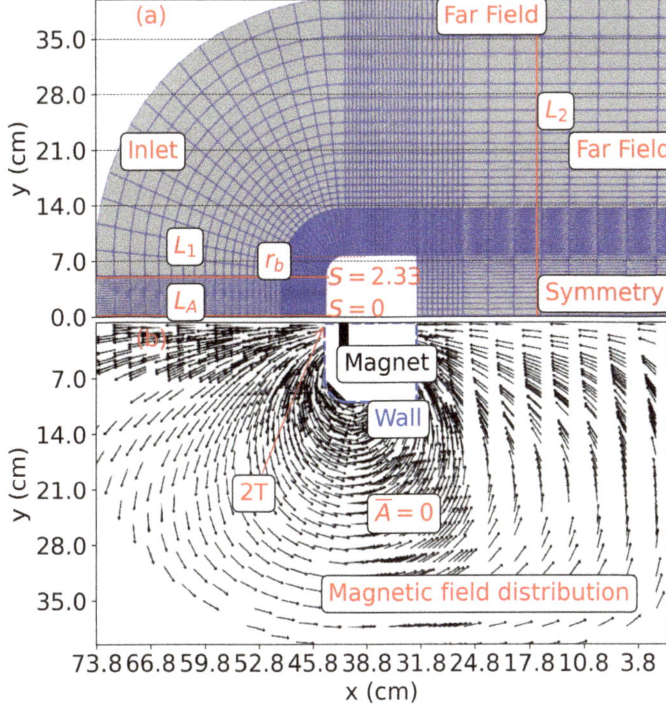

Fig. 2 **a** Computational domain showing the mesh, boundaries, and annotated dimensions. **b** Magnetic field distribution, where the magnetic field strength is calculated to achieve a 2T stagnation point

above L_A), and shoulder line L_{sh} (7 cm above L_A). Surface distance S is measured from stagnation.

2.4 Boundary Conditions and Initialization

The boundary conditions applied for the computational domain is given below in Table 1. The inlet boundary condition is set to be Dirichlet conditions. The farfield, wall and symmetry follow a Neumann conditions.

The initial conditions are set to be the same as the expected experimental values in the DLR-HEG shock tunnel experiment. The inlet velocity is $U = 5994$ m/s, the inlet density of $\rho = 2$ g/m^3 and the chemical composition defined by mass fractions: $Y_N = 10^{-10}$, $Y_{N_2} = 0.756$, $Y_{NO} = 0.0221$, $Y_O = 0.1973$, and $Y_{O_2} = 0.0238$ [13]. The inlet translational and vibrational temperatures are set to (following Dirichlet boundary conditions), $T_t = 1187$ K and $T_v = 100$ K, respectively. The post shock region reaches an ionization close to 10^{-4}. The surface temperature of the magnet

Table 1 Boundary conditions used in the simulation

Boundary	Type	Condition
Inlet	Dirichlet	$\phi = $ Specified
Farfield	Neumann	$\frac{\partial \phi}{\partial n} = 0$
Symmetry	Neumann	$\frac{\partial \phi}{\partial n} = 0$
Fixed wall	Neumann	$\frac{\partial \phi}{\partial n} = 0$

$\phi = $ is the flux crossing the cells

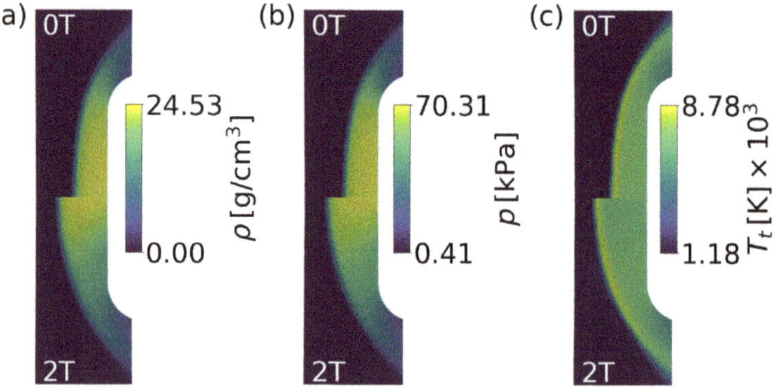

Fig. 3 2D plot for **a** density **b** pressure, **c** translational temperature for $0T$ (top) and $2T$ (bottom)

chamber was computed with a Neumann boundary condition, assuming non-catalytic behavior. The boundary conditions are modified with a slip velocity proportional to the gradient of velocity at the wall, adjusting for the Knudsen layer effects. The slip velocity

$$u_s = \frac{2 - \alpha}{\alpha} \cdot \frac{\lambda}{u} \cdot \left(\frac{\partial u}{\partial y}\right)_{\text{wall}} \tag{5}$$

where α is the accommodation coefficient, which represents the fraction of momentum accommodated by the wall. λ is the mean free path of the gas molecules and $\frac{\partial u}{\partial y}$ is velocity gradient of boundary layer. Hence, Velocity along the walls was modeled with a slip boundary condition, while species mass fractions followed a Neumann boundary condition. The chemical reaction mechanism employed in the model uses to the scheme outlined in [12].

3 Result

Figure 3a shows a zoomed in view of density distribution near the body. The top half shows the no-magnetic field case while the bottom half shows the case with 2T magnetic field at the surface of the body. The presence of the magnetic field pushes

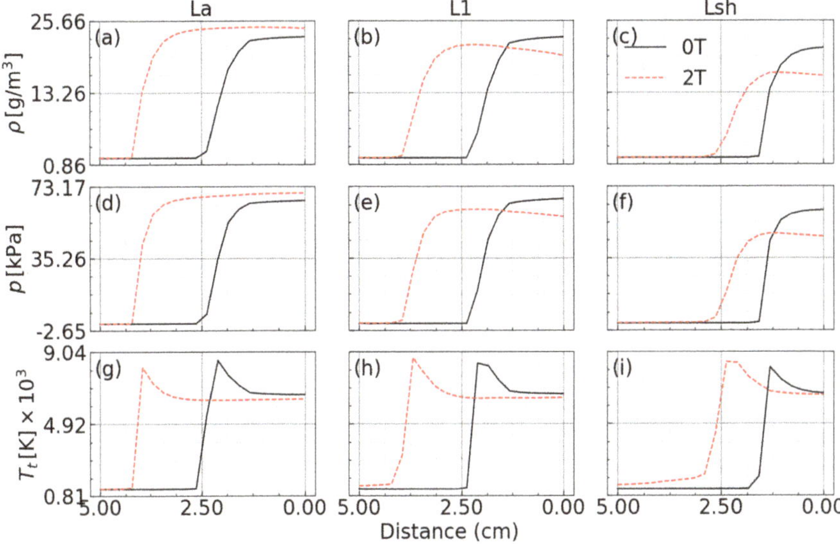

Fig. 4 1D plot of density, pressure and translational temperature along L_A, L_1, L_{sh} for 0T and 2T

the shock significantly further, showing a measurable impact. Figure 3b, c show total pressure and the translational temperature. Both show similar, extended pattern to the density profile. The temperature shows a peak at the shock before the deceleration layer. This feature moves away from the body with the shock in the presence of the magnetic field.

Figure 4 shows the impact of a magnetic field on hypersonic flow properties cross-sectional cuts across the simulation domain. The black lines represent the case without a magnetic field (0T), and the red dashed lines indicate the (2T) case. The plots compare density ρ, pressure p, and translational temperature T_t at different spatial locations (L_A, L_1, L_{sh}). The magnetic field significantly alters the flow dynamics, compressing the flow and affecting shock structure. In the 2T case Fig. 4a, along L_A post-shock density increases from 23.2 g/m^3 (0T) to 24.5 g/m^3(2T), while pressure rises from 66 KPa (0T) to 70 kPa(2T) as shown in Fig. 4d. Translational temperature peaks at 8107 K (0T) and 8000 K (2T) as shown in Fig. 4g. Along L_1, the 2T case shows reduced density (19.1 g/m^3) near the body , with T_t behaving similarly to L_A as shown in Fig. 4b, e, h respectively. At L_{sh}, density and pressure further decrease, with a 0.5 cm reduction in shock stand-off distance compared to L_A. The lowering of the density and pressure away from the symmetry axis likely happens because of the "funneling" effect introduced by the magnetic field. The curving of the magnetic field lines near the magnetic poles create a pathway for the plasma created at the shock to easily flow into the polar region. This funnel can be appreciated in the density and pressure plots shown in Fig. 3.

Figure 5 shows a comparison of the heat flux along the wall. In the 0T case, heat flux reaches close to its peak at the stagnation point along the symmetry, then remain

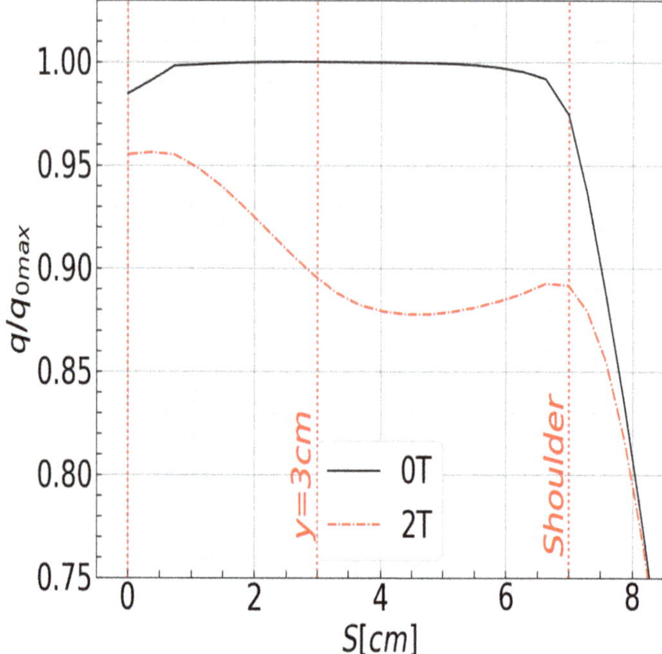

Fig. 5 Wall heat flux comparison for for 0T and 2T. The distance S is measured along the surface of the body, with 0 at the stagnation point

constant as the distance from this point increases. A drop in heat flux is observed before L_{sh}. At stagnation (S = 0), near-isentropic flow thickens the boundary layer, reducing heat flux. A $2T$ field's Lorentz force pushes the shock, diverting plasma (reducing convective heating) and funneling ions poleward. Combined with non-catalytic wall effects suppressing recombination, this creates the observed heat flux dip. As the magnetic field is introduced, the heat flux at the stagnation point initially reduces by up to 5%. Further from the stagnation point the heat flux reduced by roughly 12%.

4 Conclusion

This preliminary study supports an upcoming experimental campaign by Robinson Research Institute and HEG-DLR, examining a 2T HTS magnet in a blunt body configuration. Numerical results show significant flow-field changes under a strong axial magnetic field, with increased shock stand-off distance at higher fields. Although the flow remains isenthalpic from L_A to L_{sh}, post-shock thermal pressure reduction lowers density in the 2T case. A 15 % heat load reduction occurs till L_{sh}, likely due to decreased pressure and density balancing the stronger field. The altered sheath structure modifies the flow field, potentially affecting heat-load redistribution and vorticity. Such ionized flows are also relevant for interplanetary travel, where the magnet could shield against energetic particles.

References

1. Sears WR, Resler JR (1964) Magneto-aerodynamic flow past bodies. Adv Appl Mech 8:1–68
2. Meyer RX (1959) Magnetohydrodynamics and aerodynamic heating. ARS Journal 29(3):187–192
3. Bush WB (1958) Magnetohydrodynamic-hypersonic flow past a blunt body. J Aerospace Sci 25(11):685–690
4. Gulhan A, Esser B, Koch U, Siebe F, Riehmer J, Giordano D, Königorski D (2009) Experiments on heat-flux mitigation by electromagnetic fields in ionized flows. In: 40th AIAA plasmadynamics and lasers conference, p 3725
5. Ziemer RW, Bush WB (1958) Magnetic field effects on bow shock stand-off distance. Phys Rev Lett 1(2):58
6. Martinez Schramm J, Hannemann K (2019) Study of MHD effects in the high-enthalpy shock tunnel Göttingen (HEG) using a 30 T-pulsed magnet system. In: 31st international symposium on shock waves 2: applications, vol 31. Springer International Publishing, pp 617–623
7. La Rosa Betancourt MA, Collier-Wright MR, Herdrich G (2021) A superconductor-based magnetohydrodynamic shielding system for hypersonic re-entry: MEESST. In: AIAA Scitech 2021 Forum, AIAA 2021–4138, published online
8. Huang X, Parkinson B, Strickland N, Smart S, Bailey J, Sreekala VA, Parashar T, Glowacki J, Long N, Weijers H (2024) Investigation of magnetic heat shielding during spacecraft re-entry using hts magnet–preliminary experimental design. IEEE Trans Appl Superconductivity
9. Hannemann K, Martinez Schramm J, Wagner A, Ponchio Camillo G (2018) The high enthalpy shock tunnel Göttingen of the German aerospace center (DLR). J Large-Scale Res Facilities JLSRF 4(A133):1–14
10. Casseau V (2022) Github repository of the hyStrath platform. *Release Fleming, Commit 984e300*. Retrieved from (https://github.com/hystrath/hyStrath/)
11. Gupta RN, Yos JM, Thompson RA, Lee KP (1990) A review of reaction rates and thermodynamic and transport properties for an 11-species air model. In: NASA Technical Memorandum 101528. NASA, pp 1–50
12. Park C (1993) Review of Chemical-Kinetic Problems of Future NASA Missions, I: Earth Entries. J Thermophys Heat Transfer 7(3):1–50
13. Share RX (2018) The high enthalpy shock tunnel Göttingen of the German Aerospace Center (DLR), J Large-Scale Res Facilities (JLSRF) 4:133. https://doi.org/10.17815/jlsrf-4-168. License: CC BY 4.0
14. Lani C (2023) Magnetohydrodynamic flow control: recent developments and applications. Elsevier
15. Poggie J (2000) Magnetic field effects on fluid flow: theoretical and experimental studies. Springer
16. Geuzaine C, Remacle JF (2009) Gmsh: A 3-D finite element mesh generator with built-in pre- and post-processing facilities. Int J Num Methods Eng 79(11):1309–1331
17. Casseau V, Espinoza DE, Scanlon TJ, Brown RE (2016) A two-temperature open-source cfd model for hypersonic reacting flows. part two: multi-dimensional analysis. Aerospace 3(4):45
18. Bityurin A, Bocharov A, Lineberry J (2004) Results of experiments on mhd hypersonic flow control

Effect of Structured Surface on Pressure Measurement in Hypersonic Rarefied Wind Tunnel

Ryunosuke Endo, Takashi Ozawa, Masahito Tagawa, and Kumiko Yokota

Abstract Research at the Japan Aerospace Exploration Agency (JAXA) on the Air-Breathing Ion Engine (ABIE) aims to extend satellite operational duration in very-low Earth orbit (VLEO). This study evaluates the impact of micro-structured surfaces on the intake performance using Hypersonic Rarefied Wind Tunnel (HRWT) experiments and computational analyses. We have developed numerical schemes for micro-structured surfaces in DSMC, and our numerical results were consistent with experimental results in HRWT.

Keywords Rarefied gas dynamics · Hypersonic flows · Aerodynamics · DSMC · Air breathing ion engine · Surface treatment

1 Introduction

At the Japan Aerospace Exploration Agency (JAXA), the Air-Breathing Ion Engine (ABIE) [1] has been proposed as a propulsion system for operations in very-low Earth orbit (VLEO) [2]. The ABIE system intakes atmospheric molecules, ionizes them in a discharge chamber, and generates thrust. The performance of the intake section is critical for maintaining pressure in the discharge chamber and preventing molecular backflow, which significantly impacts the overall system efficiency. To improve intake performance, Shoda et al. [3] proposed micro-structured surfaces on the intake internal surface.

The configuration and flow conditions where micro-structured surfaces are effective in rarefied flow regime is not well-known. In order to validate the improvement

R. Endo (✉)
Waseda University, Shinjuku-Ku, Tokyo, Japan
e-mail: ryukitokito@akane.waseda.jp

T. Ozawa
Japan Aerospace Exploration Agency, Sagamihara, Kanagawa, Japan

M. Tagawa · K. Yokota
Graduate School of Engineering, Kobe University, Kobe, Hyogo, Japan

© The Author(s) 2026

M. Grabe et al. (eds.), *Rarefied Gas Dynamics*, Springer Aerospace Technology,
https://doi.org/10.1007/978-3-032-00094-1_15

of intake performance with micro-structured surfaces, it is necessary to compare pressure increase with micro-structured surfaces between numerical results with measured results.

The objectives of this study are as follows. First, we develop numerical schemes for micro-structured surfaces in direct simulation Monte Carlo (DSMC) [4]. The increase in number of micro-structured surface segments results in increase of computational cost. Thus, we develop a simplified analytical surface reflection model for micro-structured surfaces to improve computational efficiency. Second, we evaluate the effect of micro-structured surfaces on pressure in wind tunnel models under hypervelocity rarefied wind tunnel (HRWT) conditions by carrying out DSMC computations. Note that flow conditions are different between HRWT and ABIE flight conditions in terms of speed ratio, rarefaction level, and so forth. Thus, we need to investigate suitable model configuration to evaluate the micro-structured surface effect under HRWT flow conditions.

Intake performance is determined by downstream pressure. However, due to the experimental environment, which differs from the real environment, the difference in results with and without micro-structured surfaces will be checked, rather than specific target values. In previous studies, the analysis suggested that the pitch dependence of micro structured surfaces may be small and the surface coefficient dependence may be large. In the latter part of this study, a simplified smooth geometry model is used, which also made the analysis cost efficient.

2 Experimental Setup

2.1 Model with Micro-Structured Surfaces

The micro-structured surface proposed by Shoda et al. [3] is shown in Fig. 1. It consisted of micro-structured walls of 20 and 74°. This geometry is designed to control the direction of molecular reflection and promote the movement of molecules downstream in the intake section. As shown in Fig. 2, the average reflected angle from the overall wall surface with an assumption of inflow uniform angular distribution is approximately $- 77°$. This indicates that without molecular collisions, micro-structured surfaces may improve pressure in downstream regions even in thermally equilibrium conditions. Laser etching technology was used to manufacture experimental models of micro-structured surfaces. The experiments involved preliminary test models with micro-structured surfaces of varying pitch widths (50, 100, 500 μm) and a smooth surface model (see Fig. 3). The model is made of aluminum and α is about 0.9–1.0.

Figure 4 shows the intake model used in the experiment and the analysis. Molecules flow in from the left side. The pressure at 45 mm from the inlet, the same point as the measurement section of the wind tunnel test, is summarized as a

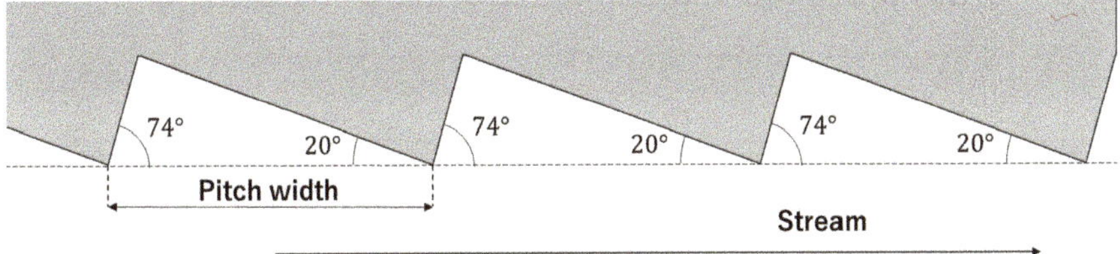

Fig. 1 Schematic of the micro-structured surfaces

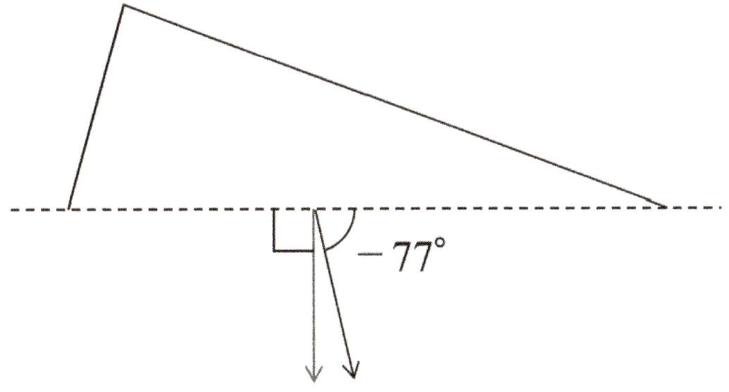

Fig. 2 The image of the average overall reflected angle

Fig. 3 Photograph of experiment model

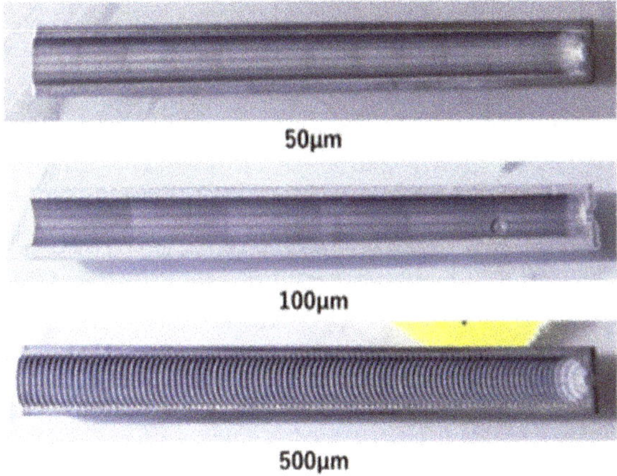

result. For analyses that vary the Knudsen number, the pressure ratio at 45 mm to the inlet pressure is used as the result.

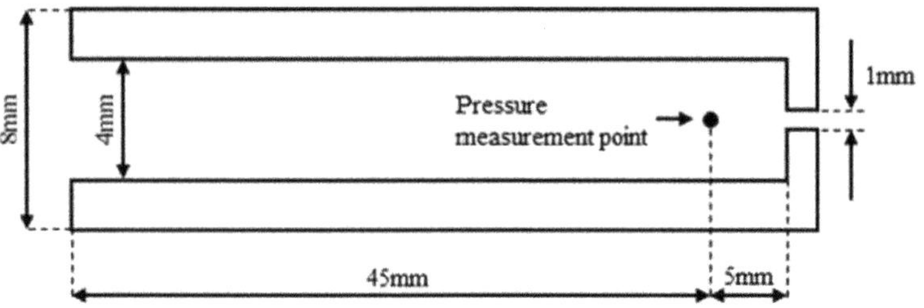

Fig. 4 Image of analytical model for DSMC

Table 1 The DSMC flow field conditions

Parameter	Value	Unit
Speed	750	m/s
Temperature	11	K
Number density	1.1×10^{21}	m^{-3}
Wall temperature	290	K

2.2 Hypersonic Rarefied Wind Tunnel

The HRWT (Hypersonic Rarefied Wind Tunnel) has been developed at JAXA Chofu, designed to generate hypersonic rarefied flows [5, 6]. The HRWT consists of a vacuum chamber, air supply system, heating system, nozzle section, model support section, and measurement devices. In this work, pure nitrogen (N_2) is used as the test gas, and rarefied flow is generated through a 45° conical nozzle. The main test conditions for HRWT (17 mm from the nozzle exit on centerline with stagnation pressure of 16.2 kPa and stagnation temperature of 280 K) are listed in Table 1. Details of HRWT can be found in Ref. [4]. Figure 5 shows a photograph of the model (described in Sec. 2.1) mounting. Note that the Knudsen number range of the HRWT is lower than that of expected ABIE flight conditions. Taking the inner diameter of the model shown in Fig. 4 as the representative length, the Knudsen number in HRWT is approximately 0.08 with a mass flow rate of 0.08 g/s.

2.3 DSMC Analyses

In our previous work, HRWT nozzle flows were investigated [5] and the flow conditions in the HRWT test section were obtained as shown in Table 1. Using these inflow conditions, we performed DSMC calculations for flows inside and outside the models. The test gas was 100% nitrogen (N_2). The chemical reaction and electronic excitation models were not activated in the MOTIF DSMC code [7] as the temperature was below 1000 K. The Maxwell reflection model was used for gas-surface interactions, with the surface accommodation coefficient (α) used to represent the ratio of

Fig. 5 Photograph of model
mounting in HRWT

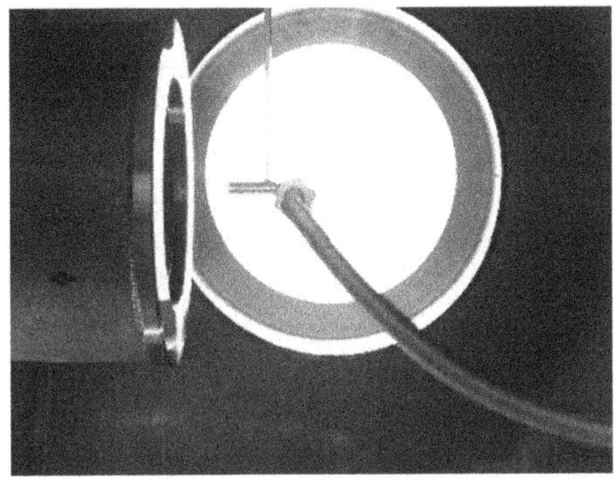

diffuse to specular reflection. The nitrogen molecule diameter was set to 5.1 Å, and
the wall temperature in the simulations was assumed to be 290 K, consistent with the
experimental conditions. The mean free path (mfp) of the freestream is 0.3 mm and
the number density is 1.1×10^{21} m^{-3} at 0.08 g/s. The Knudsen number was varied
by changing the value of the number density. Typical computational parameters are
as follows. The system was assumed to reach thermal equilibrium within 3 ms. The
computational domain was discretized into 50,000 cells, with a resolution of 2 cells
per mfp to ensure accurate results. The distance from the pressure measurement
point to the pressure sensor is approximately 10 cm, and this study does not consider
rarefaction effects in this area.

3 Results and Discussion

3.1 Evaluation of the Impact of Micro-Structured Surfaces in the HRWT Environment

In this study, the effects of micro-structured surfaces on the pressure distribution in
the intake models were evaluated using the HRWT. The test conditions were set with
a gas flow rate of 0.08 g/s, using pure nitrogen (N_2) gas. Table 2 shows the pressure
measurement results under nominal mass flow rate conditions.

The wind tunnel test results revealed that the micro-structured surfaces did not
show a significant impact on pressure distribution in the preliminary test models.
No significant differences in pressure values were observed across any of the flow

Table 2 Experimental results at HRWT

mdot(g/s)	Pressure [Pa]			
	Pitch = 50 μm	Pitch = 100 μm	Pitch = 500 μm	NP
0.08	22.2 ± 1	22.6 ± 1	22.1 ± 1	22.7 ± 1

rates, and no trends related to pitch width were found. Additionally, while pressure increased with higher flow rates, this is consistent with the general relationship between flow rate and pressure in rarefied flow environments, independent of the presence of micro-structured surfaces. This result suggests that, in the HRWT environments, the control of reflection direction due to the micro-structure surfaces is not sufficiently effective due to intermolecular collisions. However, these results are specific to the conditions of the HRWT, and different reflection behaviors may occur under actual very-low Earth orbit (VLEO) conditions or in environments with higher Knudsen numbers. Therefore, to complement the wind tunnel results and more thoroughly investigate the effects of micro-structured surfaces, we further carry out DSMC computational analyses.

3.2 Analyses of Actual Geometry Model

The dependence of pressure variations on surface coefficient and Knudsen number was verified by analyses using an actually simulated model of the micro-structured surfaces of the intake section. The inflow conditions were obtained from the HRWT wind tunnel test conditions. For the analytical conditions for the validation of the surface accommodation coefficient dependence, the surface coefficient varied from 0 to 1 in steps of 0.2. Figure 6 shows a comparison of the pressure analysis results for a pitch width of 100 μm at 45 mm when the surface accommodation coefficient was varied.

The analysis results showed that in the range of surface coefficients from 0.8 to 1.0, the pressure values from the actual geometry model were close to the wind tunnel test results from the HRWT (see Table 2). This indicates that the actual geometry model is an effective tool for accurately reproducing wind tunnel test results. Overall,

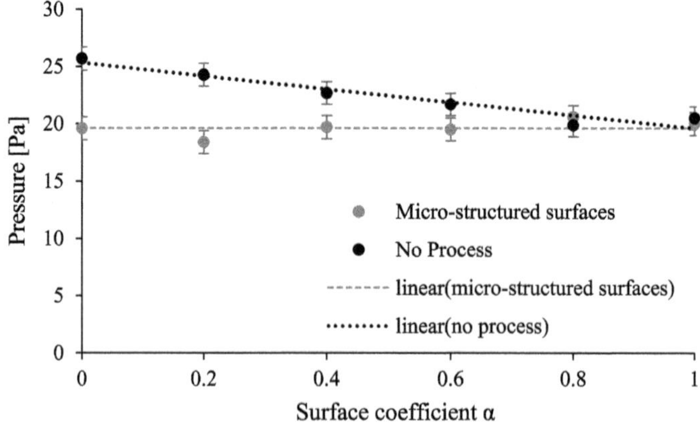

Fig. 6 The pressure dependence on the surface coefficient by analyzing

without micro-structured surfaces, the pressure increases as the α decreases. On the other hand, with micro-structured surfaces, no significant change in pressure was observed when the α was changed. The results without surface processes indicate that specular reflection increases the pressure in the downstream region. On the other hand, with the micro-structured surfaces, results are basically similar to the diffuse case ($\alpha = 1.0$) for all surface accommodation cases. This is because in the near-continuum flow condition, micro-structured surfaces increase intermolecular collisions inside the intake model, and this results in higher pressure gradient from inlet to outlet similar to the full diffuse reflection condition. Considering that realistic surface coefficients are greater than 0.5, it is suggested that a material with $\alpha = 0.6$ could be used to evaluate the influence of micro-structured surfaces in wind tunnel experiments.

The analytical conditions for the Knudsen number dependence study were carried out by varying the Knudsen number from 0.08 to 80 with a surface accommodation coefficient of unity (full diffuse reflection condition).

We analyzed the change in the pressure ratio as the Knudsen number is varied. The pressure ratio is the ratio of the pressure at the measurement point to the freestream pressure. Figure 7 shows the results and it was found that the higher the Knudsen number, the higher the pressure ratio. Comparing the results with and without micro-structured surfaces from, it is observed that the difference in pressure ratio increases as the Knudsen number increases. The reason for this is that the higher Knudsen number allows the molecules to flow without intermolecular collisions. Molecular motion at high Knudsen number is dominated by wall collisions, and the effect of micro-structured surfaces is expected to be more significant. Note that the development of a simplified smooth model and dependence of computational models are presented in Sect. 4.

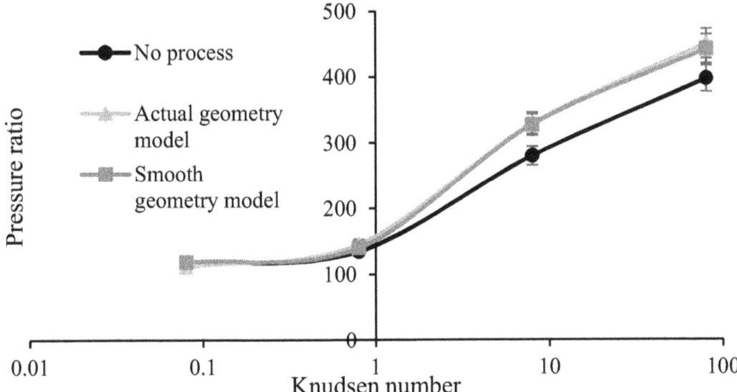

Fig. 7 Increased influence of micro-structured surfaces when Knudsen number is increased by analyzing

4 Comparative Verification of Actual and Smooth Geometry Models

4.1 Simplified Smooth Geometry Model

The actual geometry model faces the challenge that the computational cost becomes higher along with the increase of the number of micro-structured surface segments. Thus, in this section, we develop a simplified smooth model for micro-structured surfaces to improve computational efficiency. A simplified smooth model reproduces the effect of micro-structured surfaces on a flat surface. For this purpose, it provides information on the angle of the reflecting wall according to the angle of incidence of the molecule. Considering a single unit of the micro-structured surfaces, the behavior of molecules can be classified into four types as shown in Fig. 8. Assuming type 1 as "20° wall-reflecting molecules," type 4 as "74° wall-reflecting molecules," and the sum of type 2 and 3 as "iterated molecules," these distributions vary with the angle of incidence of the molecules. The simplified smooth model uses a flux ratio database to determine which wall surface is reflected by the angle of incidence. To generate a database of flux ratios, the type 1–type 4 ratios in Fig. 8 were calculated geometrically from the average path of particles assumed to reflect equally and completely diffusely at each point on the wall surface. For iteration particles, an average angle of $-70°$ was applied, which is the average angle of particles reflecting out of the region from the two wall surfaces (see Fig. 9.).

Figure 10 shows the database of the wall flux ratio to the angle of incidence for perfect diffuse reflection($\alpha = 1$). These distributions are calculated geometrically.

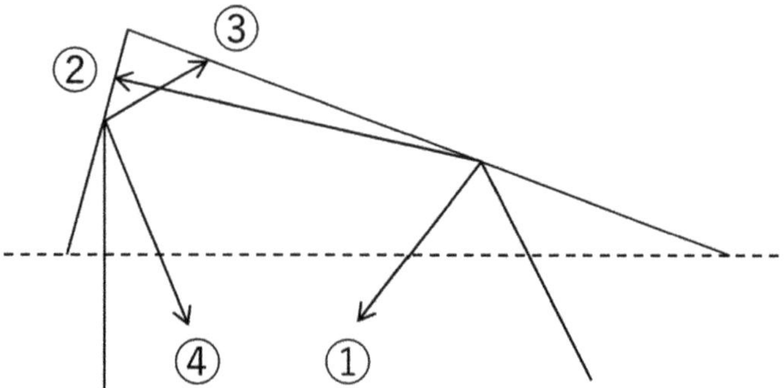

Fig. 8 Classification of molecular reflection

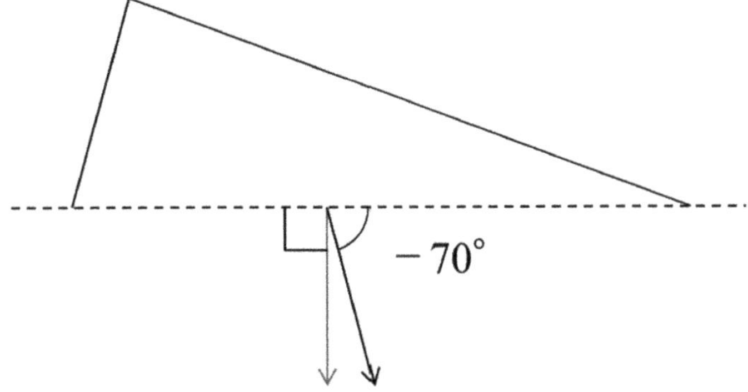

Fig. 9 Wall normal direction for iteration molecules

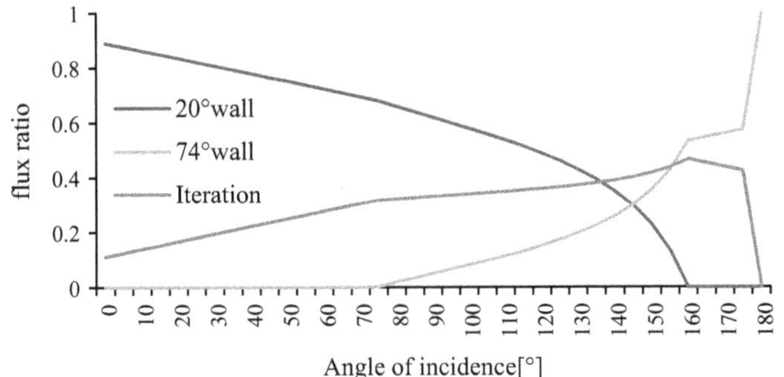

Fig. 10 Flux ratio database at $\alpha = 1.0$

4.2 *Comparison of Actual and Smooth Geometry Models*

A comparison of the Actual geometry model with a simplified smooth model using the database in Fig. 10 was performed. The Knudsen number varied from 0.08 to 80 and the pitch width of the micro-structured surfaces was set to 100 μm. Figure 7 shows the comparison of pressure ratio at 45 mm among models. The results show that the simplified smooth model agrees well with the actual geometry model at each Knudsen number. In addition, the simplified smooth model reduced the computation time by approximately 90% compared to the actual geometry model. Therefore, the simplified smooth model is considered to be useful for improving the efficiency of future research to investigate the effect of micro-structured surfaces for various conditions.

5   Conclusion

In this study, the effect of the micro-structured surfaces applied to the preliminary intake models was evaluated through wind tunnel tests in HRWT and computational analyses. The measured results showed that no significant increase in pressure due to the micro-structured surfaces was observed, and no clear effect of pitch width change was detected. This indicates that the Knudsen number was not sufficiently high to confirm the surface effect in the wind tunnel environments. To complement this, actual geometry model analyses were performed. The numerical results with the surface accommodation coefficient (α) in the range of 0.8 to 1.0 are basically consistent with the measured pressure. Namely, the actual geometry model has potential to simulate molecular reflection behaviors. Investigation of the Knudsen number dependence also showed that the effect of the micro-structured surfaces increases as the Knudsen number increases. This indicates the potential advantage of micro-structured surfaces in rarefied environments. In the latter part of this paper, we developed a simplified smooth model to improve the efficiency of the analysis. Consequently, the simplified smooth model predicts similar pressure results to the actual geometry model in approximately 90% less time. In our future work, we will investigate the applicability and accuracy of the simplified smooth model for other surface accommodation coefficient cases and flow conditions. We will also optimize wind tunnel model configurations to verify the effect of micro-structured surfaces in HRWT.

References

1. Nishiyama K (2003) Proceedings of the 54th international astronautical congress, IAC-03-S4-02
2. Sasaki M (2017) Space Res Today 198:10–18
3. Shoda K et al (2023) CEAS Space J 15:403–411
4. Bird GA (1994) Molecular gas dynamics and the direct simulation of gas flows. Clarendon, Oxford, England, U.K.
5. Ozawa T, Fujita K, Suzuki T (2015) Development of an aerodynamic measurement system for hypersonic rarefied flows. Rev Sci Instrum 86(1):015105
6. Ozawa T, Suzuki T, Fujita K (2015) Aerodynamic measurements and computational analyses in hypersonic rarefied flows. AIAA J 53(11):3327–3337. https://doi.org/10.2514/1.J053889
7. Ozawa T et al (2011) AIAA Paper 2011-3311. In: 42nd AIAA thermo-physics conference, Honolulu, Hawaii, June 27–30

Identifying the Continuum Breakdown of a Plume Surface Interaction in Near Vacuum Environment for a CFD-DSMC Approach

Jannis Petersen, Jonas Göbel, Martin Propst, Bradley Craig, Theodor Heutling, Konstantinos Kontis, Craig White, Martin Tajmar, Jeroen Van den Eynde, and Christian Bach

Abstract Within the ESA-funded activity, LUNAR In-Situ LANding Structures (LUNAR ISLANDS), the feasibility of in-situ manufactured structures is explored. These aim to decrease the exposure of surrounding infrastructure and personnel to high-velocity particles of lunar regolith, which result from the interaction of the exhaust plume of the lander and the surface, which is covered by loose regolith. For the numerical investigation of Plume Surface Interaction (PSI), a hybrid modelling strategy is employed, combining a Navier-Stokes (NS)-based Computational Fluid Dynamics (CFD) solver for the continuum regions of the nozzle and plume with a Direct Simulation Monte Carlo (DSMC) solver for the rarefied regions. This contribution describes the automated identification of the breakdown interface and necessary division of the mesh based on that interface to allow for the use of a single mesh.

Keywords Plume surface interaction · DSMC · CFD · Continuum breakdown

1 Introduction

During the Apollo missions, regolith particles released due to PSI caused various issues, including clogging mechanical parts, instrument interference, radiator overheating, and visibility problems [13, 23]. With future lunar missions planned, strategies to mitigate these effects are being developed to ensure safe landings and operations on the lunar surface [1, 6].

J. Petersen (✉) · J. Göbel · M. Propst · T. Heutling · M. Tajmar · C. Bach
Technische Universität Dresden, Chair of Space Systems, Dresden, Germany
e-mail: jannis.petersen@tu-dresden.de

B. Craig · K. Kontis · C. White
University of Glasgow, School of Engineering, Glasgow, UK

J. Van den Eynde
ESTEC European Space Agency, AZ Noordwijk, The Netherlands

© The Author(s) 2026

M. Grabe et al. (eds.), *Rarefied Gas Dynamics*, Springer Aerospace Technology,
https://doi.org/10.1007/978-3-032-00094-1_16

One proposed solution is the construction of lunar landing pads [5, 19]. Solid, low-erodibility surfaces can significantly reduce the particle release during descent [15, 25]. Given the high costs and logistical challenges of transporting materials to the Moon, In Situ Resource Utilisation (ISRU) methods using regolith with additive manufacturing are preferred [11, 12].

Within the LUNAR ISLANDS project, potential manufacturing techniques for such structures are being tested, supported by numerical simulations to complement experimental data [10].

Roberts [20] proposed an erosion theory based on the key assumption that the erosion rate is proportional to the shear stress, making viscous erosion a central mechanism. More recent studies have shown that other regimes–such as deep cratering, diffuse fluidisation, and non-viscous momentum transfer–also play a significant role [14]. These mechanisms are expected to differ in relevance for landing pads, where particles may be bonded through sintering or binders, or where molten surfaces eliminate permeability. Nonetheless, plume flow properties such as velocity, pressure, and temperature are expected to remain critical for predicting regolith response. For a landing pad, this requires resolving the flow features of a supersonic jet impacting a flat surface. A schematic of the PSI is shown in Fig. 1.

In a lunar PSI scenario, the jet plume expands into a near-vacuum, transitioning from a continuum regime to a rarefied gas flow as the mean free path between particles increases. Once this rarefaction exceeds a threshold, the continuum hypothesis underlying the Navier-Stokes equations becomes invalid, reducing accuracy and potentially causing solver divergence or breakdown [9].

Under lunar vacuum conditions, conventional NS-based CFD solvers struggle to resolve regions beyond continuum flow, necessitating the use of stochastic methods like DSMC for plume analysis.

The DSMC method, a stochastic particle-based approach, is widely used for simulating rarefied gas flows in non-equilibrium Knudsen number regimes [18, 26]. It

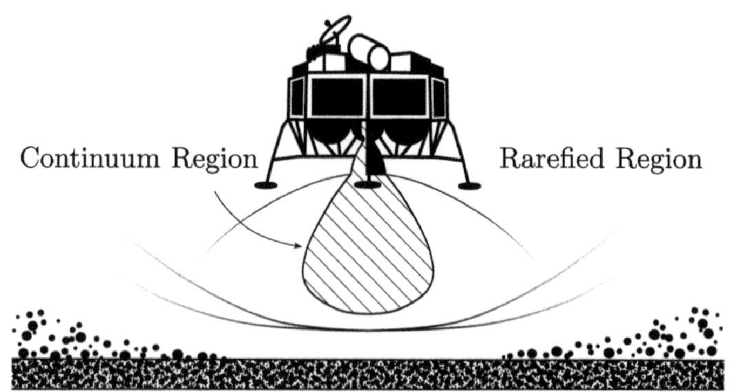

Fig. 1 Schematic illustration of the Plume-Surface Interaction in rarefied conditions

models particles representing many real molecules, significantly reducing computational costs compared to fully deterministic methods like Molecular Dynamics, while effectively capturing gas behaviour.

However, due to the high computational demand of DSMC for dense flows [9], hybrid approaches combining CFD and DSMC have proven valuable for scenarios involving both continuum and rarefied conditions. These approaches leverage the strengths of both methods, enabling detailed simulations across multiple pressure scales or in supersonic and hypersonic flows. Hybrid methods have also been applied successfully to PSI flows, demonstrating their efficiency in these contexts [4, 7, 16, 27].

For hybrid simulations, an interface must be defined to transfer data between NS-CFD and DSMC domains. One option is to position the interface within the continuum region, ensuring all rarefied regions are handled by DSMC. Alternatively, a breakdown criterion for the continuum assumption can define the interface, optimising the division of computational effort by maximising the region handled by the CFD solver.

2 CFD-Setup

The NS-CFD case utilised as a basis for the development of a hybrid approach consists of a conical nozzle, whose jet is impinging a flat wall. The geometric domain and boundary conditions are chosen in a way, that they are in line with a test setup at the University of Glasgow [28]. There, a number of tests for the PSI are run, to gain insight into the interaction of an impinging jet, with loose regolith simulant and sintered regolith samples. For a flat plate, ground pressure measurements are performed, which will be used to compare the numeric results.

The domain is depicted in Fig. 2. The used conical nozzle has a throat diameter of 1mm and an exit diameter d_e of 8.94mm. The Stand-off height (SOH) is set to 8.95 d_e, which is in line with one of the experimental cases.

The total pressure p_T and temperature T_T are set to 1MPa and 780K respectively. The pressure at the outlets is set to 1Pa, as this is the lowest pressure the vacuum chamber can reach. The working gas is pure nitrogen. As the flow is highly compressible, non-reflective boundary conditions were chosen for the inlet and outlet, to prevent compression waves to be reflected and propagate through the domain.

The solver selected for this calculation is OpenFOAM's rhoPimpleFoam. This choice is motivated by the project's parallel development of a coupling procedure between the NS-CFD solver and a Discrete Element Method, aimed at capturing the interaction between the gas flow and regolith particles for the continuum region and therefore for low SOHs. An existing interface for this coupling is already available for rhoPimpleFoam, making it the practical choice. rhoPimpleFoam is a pressure-based, density-variable solver suitable for compressible flows at low to moderate Mach numbers, employing a segregated solution algorithm based on the PIMPLE (merged PISO-SIMPLE) algorithm. In contrast, rhoCentralFoam is a density-based solver

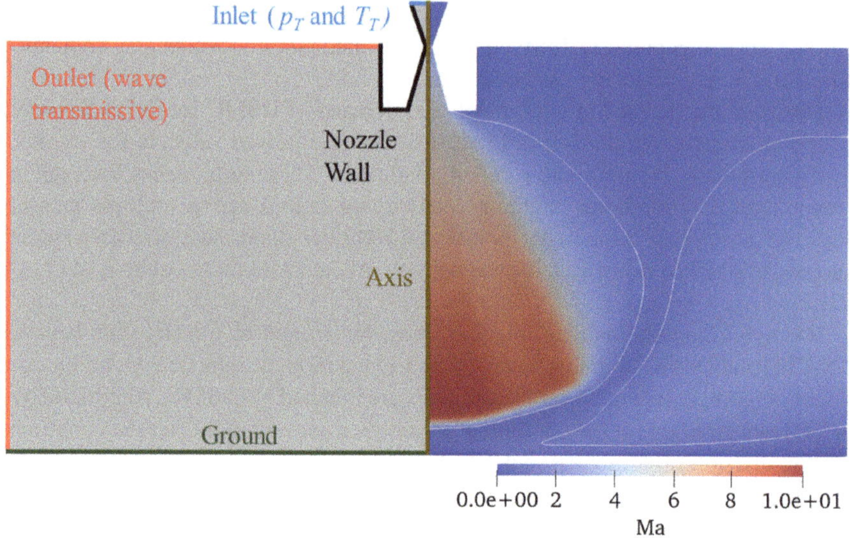

Fig. 2 Domain and boundary conditions used for the NS-CFD approach (left) and Mach contour of the steady-state NS-CFD solution, with $Ma = 1$ isoline highlighted in white (right)

that uses central-upwind schemes designed for high-speed compressible flows, offering better shock-capturing capabilities due to its conservative formulation and low numerical dissipation. Although rhoCentralFoam may yield more accurate results in terms of shock resolution, the chosen methodology, as outlined in Sect. 3, remains compatible with either solver.

Fig. 2 shows the Mach number distribution of the case once it has reached a steady-state. The general flow structure is in line with what would be expected. After the flow expansion through the nozzle, an underexpanded plume forms. The flow is stagnated on the ground at the centre, leading to a redirection in radial direction. Under the formed bow shock the radial flow expands to supersonic speeds, due to the divergent shape of the shock.

3   Breakdown Identification

In fluid dynamics, various semi-empirical parameters have been developed to identify the transition from continuous flow to free molecular flow. When higher-order terms in the distribution function become significant—indicating that translational temperatures and internal energy modes cannot equilibrate quickly due to infrequent particle collisions—the Navier-Stokes equations lose their validity. This validity can be evaluated by examining the magnitude of the coefficients in the first-order solution. The continuum equations fail physically when the velocity distribution function deviates substantially from equilibrium. However, there is no theory that precisely defines the permissible size of higher-order terms in the power series or the extent of deviation from equilibrium before the Navier-Stokes equations become invalid [17, 21].

The Bird's parameter P is one of the most extensively investigated breakdown parameters in the literature. It was initially defined using the density and its gradient along the streamline s as flow property [2, 24]. Nevertheless, it is believed that for more complex flows, flow properties Q, such as the temperature and velocity, should also be taken into account [24]. Equation 1 shows the generalized version, where any of those flow properties can be set for Q. Additionally, the Mach number Ma, the ratio of heats γ and the mean free path λ are taken into account by the equation.

$$P_Q = Ma \sqrt{\frac{\pi \gamma}{8}} \frac{\lambda}{Q} \left| \frac{dQ}{ds} \right| \tag{1}$$

The value for the Bird's continuum breakdown parameter P of about 0.02 can be employed for the breakdown identification of steadily expanding flows [2]. In general, the Bird's parameter has been frequently utilised for the identification of the flow transition in cases of supersonic expansion into vacuum [17].

Another frequently employed breakdown parameter is the gradient-length local Knudsen number, whose definition can be seen in Eq. 2. This formulation also utilises Q for any flow parameter that can be used, the mean free path λ of the molecules and l, the distance between two points in the flow domain.

$$Kn_{GLL} = \frac{\lambda}{Q} \left| \frac{dQ}{dl} \right| \tag{2}$$

Boyd et al. [3] found that the Gradient-Length Local (GLL)-Knudsen number provides more accurate breakdown interface identification than Bird's parameter P for compressible, hypersonic flows. They demonstrated that placing l along the steepest property gradients yields the best results. In addition to commonly used breakdown parameters, Sun mentions the Tiwari's criterion $\|\phi\|$ [22] and Garcia's parameter B [8], both derived from coefficients in the first-order Chapman-Enskog solution. Another approach, a non-equilibrium quantifier N_δ, has been applied by Rahimi et al. [19].

Defining a cut-off value for breakdown parameters remains challenging. Some researchers use a resolution parameter where discrepancies between continuum and kinetic solutions exceed 5%, while others simply assume small values. The cut-off is typically determined through numerical tests and depends on the tolerance for hybrid-to-kinetic solution differences, which may vary by flow configuration. A conservative limit is often advisable [21].

Wang et al. [24] proposed modified parameters Kn_{max} and P_{max}, based on Boyd's Kn_{GLL} and Bird's P_Q, shown in Eqs. 3 and 4. Although still experimental, these parameters have shown that the best prediction of breakdown occurs when Kn_{max} exceeds 0.05.

$$Kn_{\max} \equiv \max(Kn_\rho, Kn_T, Kn_U) \tag{3}$$

$$P_{\max} \equiv \max(P_\rho, P_T, P_U) \tag{4}$$

The footnotes of $Kn_{\max}$ and $P_{\max}$ indicate the various flow properties that are included. As the breakdown is also dependent on viscosity and heat transfer, these are taken into account by using the density ρ, the velocity magnitude U and the translational temperature T [24].

For the above-mentioned reasons, the $Kn_{\max}$ with a value of 0.05 is chosen for the following approach, as it is widely used in the literature. However, the selection of the breakdown criterion only affects the location of the interface, not the coupling procedure itself. This ensures that the approach can be applied to a wide variety of flows or with different parameters and values, once comparisons are made and an optimal choice is determined.

The approach is aimed to enable a one-way coupling of the solvers. Meaning, that after reaching a steady-state solution of the CFD, the DSMC calculation gets initialised using those results. After the DSMC calculation reached a steady-state, the solution can be assembled by the two domains. The DSMC solver used in this case is dsmcFoam+ [26], yet the breakdown identification is not tied to the solvers used. One mesh shall be used for the calculation, to make the coupling automatable. By means of mesh-refinement, the requirements to the mesh can be complied with.

Based on the location of this initial breakdown, the mesh is split into three different regions: the CFD-domain, the DSMC-domain and the interface-domain. The CFD-domain consists of all the cells whose $Kn_{\max}$-value is below the threshold, the DSMC-domain of those whose value is above. The interface domain consists of those cells, which are cut by the $Kn_{max} = 0.05$ contour, as shown in Fig. 3. The interface-domain is used to initialise the DSMC calculation based on the results of the CFD calculation.

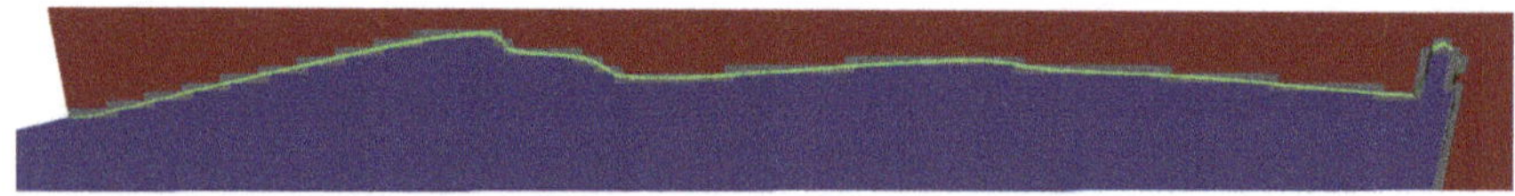

Fig. 3 Divided mesh with CFD-domain (blue), DSMC-domain (red), interface-domain (white) and the $Kn_{\max} = 0.05$ contour (green)

4 Conclusion

A NS-based CFD case was set up for a PSI scenario in near-vacuum conditions using the OpenFOAM environment. Several semi-empirical parameters for identifying the breakdown of the continuum region were discussed, and one was selected and applied to the aforementioned simulation.

Based on a $Kn_{max} = 0.05$ threshold, the mesh was divided into regions, enabling the calculation of rarefied areas with a DSMC solver for more accurate predictions of the flow field where the continuum assumption fails. This approach, which divides a single mesh into regions, allows for automation, as no new mesh is required after the initial CFD simulation. The interface location is recalculated for each case, ensuring computational effort remains low and the DSMC method is applied on a smaller region.

It needs to be noted that the approach is limited to capture the initial continuum breakdown. Due to a recompression of the flow on ground, some of the radial flow is in the continuum region as well. This however will be analysed via the DSMC approach as well.

Acknowledgements We gratefully acknowledge funding for this project from ESA (Modelling and Testing of Plume/Regolith Interaction on Lunar Landing Pads and Berms, ESA Contract No. 4000139663/22/NL/MG) and also want to thank all the students involved.

References

1. ASI, CSA, ESA, JAXA, NASA (206) Dust mitigation gap assessment report
2. Bird GA (1970) Breakdown of translational and rotational equilibrium in gaseous expansions 8(11):1998–2003
3. Boyd ID, Chen G, Candler GV (1995) Predicting failure of the continuum fluid equations in transitional hypersonic flows 7(1):210–219
4. Cai G, Liu L, He B, Ling G, Weng H, Wang W (2022) A review of research on the vacuum plume 9(11):706, pII: aerospace9110706
5. Campbell AI, Carson HC, de Soto M, English K, Fiske MR, Martin L, Murai VR, Ramirez F, Romo E, Schang KE, Smith K (2021) Lunar pad - on the development of a unique isru-based planetary landing pad for cratering and dust mitigation. In: AIAA Scitech 2021 Forum. American Institute of Aeronautics and Astronautics
6. Cannon KM, Dreyer CB, Sowers GF, Schmit J, Nguyen T, Sanny K, Schertz J (2022) Working with lunar surface materials: Review and analysis of dust mitigation and regolith conveyance technologies 196:259–274, pII: S0094576522001965
7. Espinoza D (2018) An open-source hybrid CFD-DSMC solver for high-speed flows. Dissertation
8. Garcia AL, Bell JB, Crutchfield WY, Alder BJ (1999) Adaptive mesh and algorithm refinement using direct simulation monte Carlo 154(1):134–155, pII: S0021999199963052
9. Ghazanfari V, Akbar Salehi A, Reza Keshtkar A, Mahdi Shadman M, Hossein Askari M (2021) Investigation of the continuum-rarefied flow and isotope separation using a hybrid cfd-dsmc simulation for uf6 in a gas centrifuge 152:107985, pII: S0306454920306812

10. Heutling T, Propst M, Petersen J, Tajmar M, Patzwald JJ, Stoll E, Schilde C, Lamping T, Kostas K, White C, Craig B, Hijlkema J, van den Eynde J, Bach C (2023) In-situ manufactured landing pads and berms to enable sustainable operations on the lunar surface
11. Imhof B, Urbina DA, Weiss P, Sperl M, Hoheneder W, Waclavicek R, Madakashira HK, Salini J, Govindaraj S, Gancet J, Mohamed MP, Gobert T, Fateri M, Meurisse A, Lopez O, Preisinger C (2017) Advancing solar sintering for building a base on the moon
12. Isachenkov M, Chugunov S, Akhatov I, Shishkovsky I (2021) Regolith-based additive manufacturing for sustainable development of lunar infrastructure—an overview 180:650–678, pII: S0094576521000060
13. Jaffe LD (1971) Blowing of lunar soil by apollo 12: surveyor 3 evidence. SCIENCE 171:4
14. Metzger PT (2024) Erosion rate of lunar soil under a landing rocket, part 1: identifying the rate-limiting physics 417:116136. https://www.sciencedirect.com/science/article/pii/S0019103524001969, pII: S0019103524001969
15. Metzger PT, Autry GW (2022) The cost of lunar landing pads with a trade study of construction methods. http://arxiv.org/pdf/2205.00378.pdf, 50 pages, 15 figures
16. Morris AB, Goldstein DB, Varghese PL, Trafton LM (2016) Lunar dust transport resulting from single- and four-engine plume impingement 54(4):1339–1349
17. Morris AB (2012) Simulation of rocket plume impingement and dust dispersal on the lunar surface. Dissertation
18. Pellicani F (2016) Atmosphere re-entry simulation using direct simulation Monte Carleo (DSMC) method. Master thesis
19. Rahimi A, Ejtehadi O, Lee KH, Myong RS (2020) Near-field plume-surface interaction and regolith erosion and dispersal during the lunar landing 175:308–326, pII: S0094576520303258
20. Roberts L (1963) The action of a hypersonic jet on a dust layer
21. Sun Q, Boyd ID, Candler GV (2004) A hybrid continuum/particle approach for modeling subsonic, rarefied gas flows 194(1):256–277, pII: S0021999103004698
22. Tiwari S (1998) Coupling of the Boltzmann and Euler equations with automatic domain decomposition 144(2):710–726, pII: S0021999198960119
23. Wagner SA (2006) The apollo experience lessons learned for constellation lunar dust management
24. Wang WL, Boyd ID (2003) Predicting continuum breakdown in hypersonic viscous flows 15(1):91–100
25. Watkins RN, Metzger PT, Mehta M, Han D, Prem P, Sibille L, Dove A, Jolliff B, III DPM, Barker DC, Patrick E, Kuhns M, Laine M, Radley CF (2021) Understanding and mitigating plume effects during powered descents on the moon and mars 53(4)
26. White C, Borg MK, Scanlon TJ, Longshaw SM, John B, Emerson DR, Reese JM (2018) dsmcfoam+: An openfoam based direct simulation Monte Carlo solver 224:22–43
27. White C, Scanlon TJ, Merrifield JA, Kontis K, Langener T, Alves J (2016) Numerical and experimental capabilities for studying rocket plume-regolith interactions. In: AIP conference proceedings
28. White C, Zare-Behtash H, Kontis K, Ukai T, Merrifield J, Evans D, Coxhill I, Langener T, van den Eynde J (2019) Test facility to investigate plume-regolith interactions. In: International conference on flight vehicles, aerothermodynamics and re-entry missions and engineering

Modelling Heat Transfer at Low-Density Hypersonic Spacecraft Flight Regimes

Vladimir V. Riabov

Abstract The effects of rarefaction and nonequilibrium processes on hypersonic rarefied-gas flows over blunt bodies are studied by the Direct Simulation Monte-Carlo technique (DSMC) and by solving the full Navier–Stokes equations and the equations of a thin viscous shock layer (TVSL) under the conditions of wind-tunnel experiments and hypersonic-vehicle flights in the Earth atmosphere at altitudes from 60 to 110 km. The nonequilibrium, equilibrium and "frozen" flow regimes are examined for various physical and chemical processes in air and hydrogen. The influence of similarity parameters (Reynolds number, temperature factor, catalysis parameters, and gas injection rates) on the flow structure near the blunt bodies (a sphere and a cylinder) and on their heat transfer characteristics in hypersonic streams of ionizing air is studied.

Keywords Hypersonic flow · Chemical nonequilibrium · Heat · Gas injection

1 Introduction

The areas of aerodynamic heating and protecting techniques have brought renewed interests in the design of modern hypersonic vehicles [1]. Conservation equations and various physical models are used in the analyses of the structure of the flows of a chemical nonequilibrium multicomponent gas (air, carbon dioxide, hydrogen) near a blunt body [2–8]. In this paper, the flows near spherical and cylindrical bodies under the hypersonic re-entry conditions at altitudes from 110 to 60 km are studied by using the Direct Simulation Monte Carlo (DSMC) technique [6] and numerical solutions of the full Navier–Stokes equations [8] and their approximation, a model of a thin viscous shock layer (TVSL) with the generalized Rankine-Hugoniot conditions [9–11].

The flow characteristics are calculated for different degrees of catalytic activity on the probe surface. It is found that the size of the catalytically influenced zone

V. V. Riabov (✉)
Rivier University, Nashua, NH, USA
e-mail: vriabov@rivier.edu

© The Author(s) 2026

M. Grabe et al. (eds.), *Rarefied Gas Dynamics*, Springer Aerospace Technology,
https://doi.org/10.1007/978-3-032-00094-1_17

of the flow near the body surface depends on the values of rarefaction parameters (the Knudsen and Reynolds numbers), mechanisms of chemical reactions, relaxation rates and diffusion characteristics [11–14]. Calculations of the Stanton number in the stagnation point of the spherical one-meter-radius body, with dependency on the Reynolds number along the Space Shuttle trajectory, agree with flight data [15]. In addition to catalysis, other methods of the heat flux reduction are briefly examined in this paper: the injections of inert gases (helium and nitrogen) [6, 16] and hydrogen from the cylindrical surface to the hypersonic airflow [4].

2 Validation of Computational Results

The fast numerical procedure developed in [11, 12, 14, 16] was used for solving the TVSL equations using Keller's two-point second-order scheme [11, 14]. Calculations were made in the whole range of chemical reaction rates up to the equilibrium values. The iteration process demonstrates a rapid convergence of the second order towards the solution. This property of the numerical algorithm is important when the influence of recombination processes on the flow structure is essential.

The Stanton numbers St at the stagnation point of a sphere were calculated under wind-tunnel conditions at various Re_{of}, Mach number $M_\infty = 15$, and temperature factor $t_w = 0.15$. The comparison between TVSL results (triangles) and solutions of the Navier–Stokes equations [8] with slip (filled squares) and non-slip (empty squares) boundary conditions is shown in Fig. 1 (*left*). The results correlate well with experimental data [13, 17, 18] (circles) and DSMC data [6] (gradients). The TVSL results (triangles) for a cylinder also correlate well with experimental data [13] (circles) (see Fig. 1, *right*). A comparison between the TVSL results, experimental data [18] and DSMC data [6] along the spherical surface is shown in Fig. 2 (*left*).

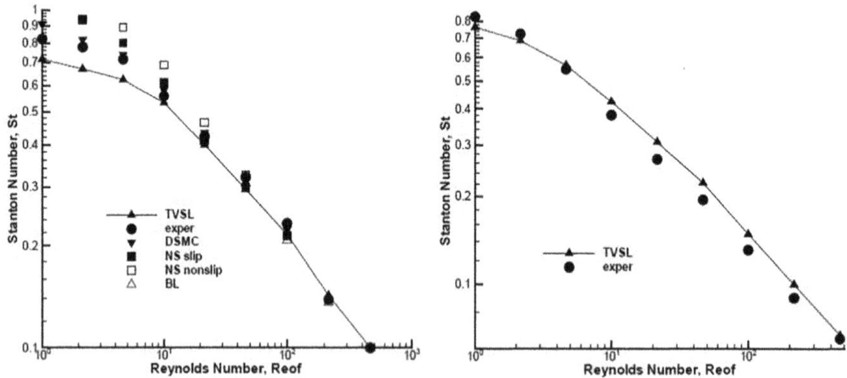

Fig. 1 Stanton numbers *St* versus Reynolds numbers Re_{of} for different medium models and various wind-tunnel experimental conditions [13, 17, 18] for a sphere (*left*) and a cylinder (*right*)

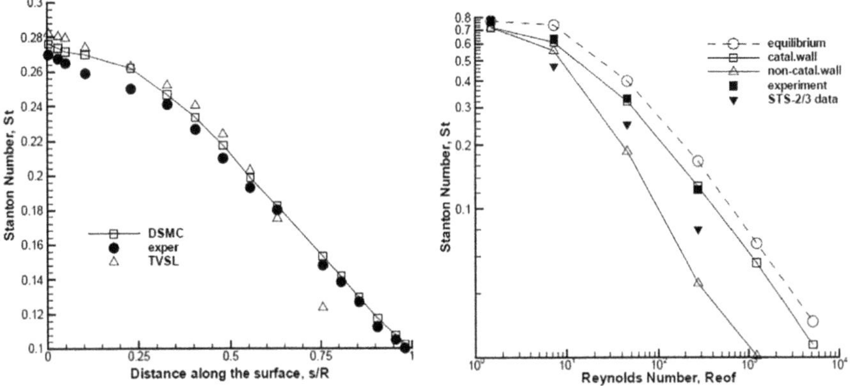

Fig. 2 Left: Stanton numbers St along the spherical surface coordinate s/R at $Re_{\mathrm{of}} = 46.38$, M_∞ $= 6.5$, $t_{\mathrm{w}} = 0.31$, $\gamma = 1.4$. *Right*: Stanton numbers St versus Reynolds numbers Re_{of} for a sphere of radius $R = 1$ m along the Space Shuttle trajectory and different medium models

3 Computational Results for the Flight Conditions

Calculations were carried out for descent flight conditions of a blunt body in the Earth atmosphere at altitudes $h = 110$, 100, 90, 80, 70, and 60 km and Reynolds numbers (at stagnation conditions) $Re_{\mathrm{of}} = 1.49$, 6.97, 47.3, 230, 1220, and 5130 per 1 m, correspondingly. The values of the Stanton number $St = q/(\rho_\infty U_\infty (H_0 - H_{\mathrm{w}}))$ in the critical point of a sphere along the Space Shuttle trajectory [15] are shown in Fig. 2 (*right*).

The surface catalysis significantly influences heat flux q. The values of q under the flight conditions at 80 km ($Re_{\mathrm{of}} = 230$, $U_\infty = 7.9$ km/s, $R = 1$ m) differ by factor of three for various catalytic surfaces due to the nonequilibrium character of physical and chemical processes in the VSL. This fact is confirmed by the STS-2/ 3 flight data [15] (filled triangles, Fig. 2, *right*). At the altitude $h = 67.5$ km this difference reaches 240 percent. The effect is less pronounced for catalytic surfaces [11, 14]. The nonequilibrium TVSL results near catalytic surfaces correlate with the data (circles) [11] obtained for equilibrium VSL (see Fig. 2, *right*).

The catalysis does not noticeably influence pressure, VSL thickness, and coefficient of friction at $1.49 \le Re_{\mathrm{of}} \le 5130$. The calculated values of heat flux and species distributions for flight conditions at $h = 60$–110 km are displayed in Figs. 3 and 4.

The degree of catalysis significantly influences distributions of mass concentrations $\alpha_i = \rho_i / \rho$ in the VSL (see Fig. 3, *left*). The measure of this influence is the width of the catalytically influenced zone d, which is characterized by the difference in distributions α_i for two extreme cases: ideally catalytic and absolutely noncatalytic surfaces. The flight conditions fully define the degree of dissociation of O_2, NO, N_2, concentrations of O and N atoms, and the size of the zone d.

The calculations of α_i on the external boundary of the TVSL at $1.49 \le Re_{\mathrm{of}} \le$ 5130 indicate that the catalysis influences the full width of the VSL at altitudes above

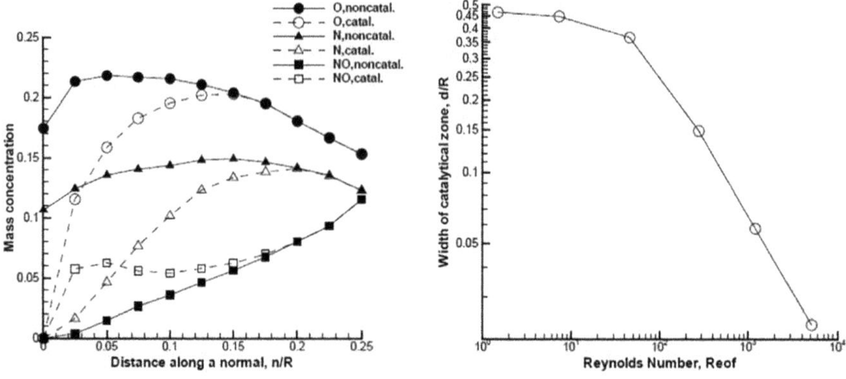

Fig. 3 Left: Mass concentrations αi of air components in the TVSL at $Re_{of} = 230$. *Right*: The width of the catalytically influenced zone d versus Reynolds number Re_{of}

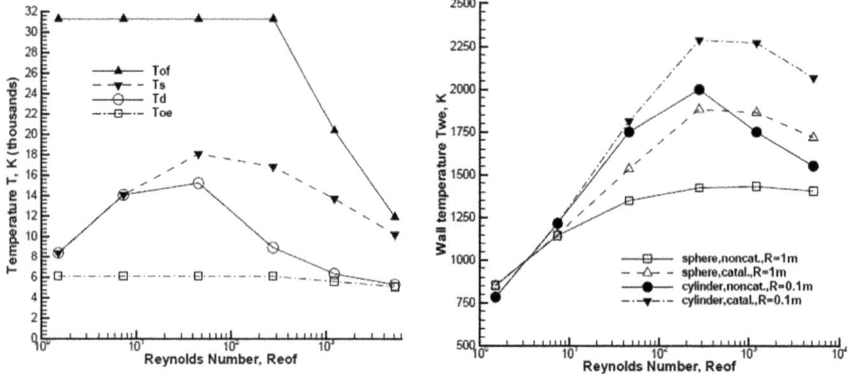

Fig. 4 Left: The values of temperatures T_s, T_d, T_{of}, and T_{oe} versus Reynolds number Re_{of}. *Right*: Equilibrium temperature T_{we} of spherical ($R = 1$ m) and cylindrical ($R = 0.1$ m) surfaces

85 km. At the lower altitudes, species concentrations on the external boundary of the layer reach their values in upstream equilibrium flow [10, 11]. It is found that under the flight conditions at $h = 60$ km, instead of the generalized Rankine-Hugoniot relations, it is possible to use standard boundary conditions [9, 10] at the shock wave. The estimated values d are shown in Fig. 3 (*right*).

Temperature values T_s at the external boundary of TVSL (gradients), stagnation temperature T_{of} (triangles), and values T_d (circles) on the catalytically influenced zone boundary d are shown in Fig. 4 (*left*). At $h \leq 90$ km, a decrease of T_d is observed and its values merge with the temperature T_{oe} (squares) estimated at the equilibrium dissociated-air state behind the shock wave. As Re_{of} increases, T_s rises at h from 110 to 90 km. As the altitude falls below 80 km, due to the vehicle deceleration [15, 17], a monotonous decrease of T_s and T_{of} is observed.

4 Equilibrium Temperature of the Vehicle Surface

Large values of the heat flux towards the body surface led to high level of equilibrium surface temperature $T_{we} = (q/\varepsilon\sigma)^{1/4}$, where ε is emissivity, and σ is the Stephan-Boltzmann's constant. The values T_{we} with $\varepsilon = 0.85$ are shown in Fig. 4 (*right*), for the critical point of a sphere ($R = 1$ m) and on the critical line of a cylinder ($R = 0.1$ m) for two extreme cases of surface catalysis. Using noncatalytic materials (solid lines) leads to a significant decrease in T_{we}. The calculations also indicate that values T_{we} on the cylindrical surface monotonously decrease as the swept angle increases and only slightly depends on the mechanism of catalysis [11, 14].

5 The Binary-Scaling Similitude Law

Numerical results [12] show that parameters in the TVSL are frozen at $Re_{of} < 20$; recombination processes are negligible, and the binary-scaling similitude law [12], $\rho_\infty R = const$, can be applied at $U_\infty = const$. Calculations performed for the critical streamline assuming $U_\infty = 7.8$ km/s, $\rho_\infty R = 5.35 \times 10^{-7}$ kg/m^2 ($Re_{of} = 7.33$), and nose radii $R = 1$ m and 0.005 m show that distributions of flow parameters for these cases correlate satisfactorily. The temperature and electron concentration N_e profiles near a noncatalytic surface are shown in Figs. 5 (*left*) and (*right*), correspondingly.

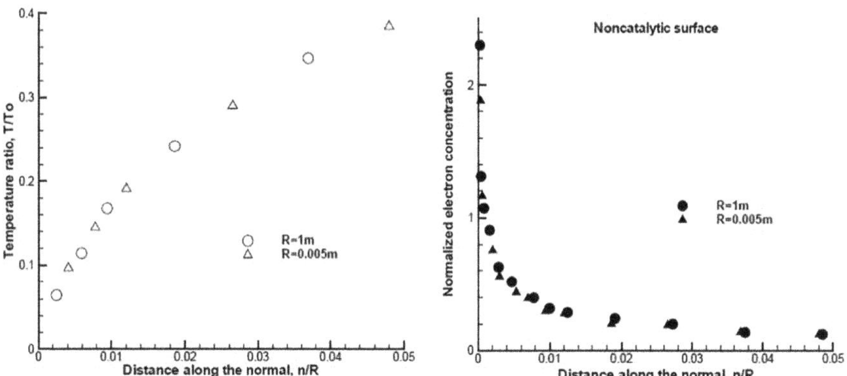

Fig. 5 Temperature T/T_0 (*left*) and normalized electron concentration $N_e \times R \times 10^{-14}$ m^{-2} (*right*) at the stagnation streamline of the sphere at $Re_{of} = 7.33$, $U_\infty = 7.8$ km/s, $\rho_\infty R = 5.35 \times 10^{-7}$ kg/m^2

6 Gas Injection Effects

6.1 Influence of the Air Blowing Factor G_w

The flow pattern over sphere is significantly sensitive to the blowing parameter G_w, which is the ratio of counterflow outgas mass flux to upstream mass flux. The influence of G_w on the flow structure is studied for hypersonic flow of air at $M_\infty = 6.5$ and $Kn_{\infty,R} = 0.0163$ ($Re_{o,R} = 92.8$), and the temperature factor is equal to 0.31. The flow conditions are the same as suggested by Botin [18] for experiments with air blowing in a vacuum chamber at stagnation temperature $T_s = 1000$ K. The sphere radius is $R = 0.015$ m. Air is blowing diffusively from the orifice with the diameter of 0.002 m, which is in the front critical point of the sphere. The factor G_w varies from 0 to 1.5.

In the case of the strong blowing ($G_w > 0.3$), the temperature field is significantly disturbed in the vicinity of the orifice in the subsonic area [6]. The distributions of the Stanton number St along the spherical surface at various blowing factors are shown in Fig. 6. For the considered transitional flow regime conditions, when the mass injection rate equals 0.7 of the freestream mass flux, the viscous layer is blown completely off the surface, and the heat transfer is zero. The displacement effect spreads both in the upstream direction and along the surface. The width of the displacement zone can be characterized by the normalized surface coordinate $(s/R)_{max}$, where the local heat transfer is maximum St_{max} (see Fig. 6). At the same value of the factor G_w, this effect is more pronounced for injections of light inert gases (e.g., He), as shown in Fig. 6.

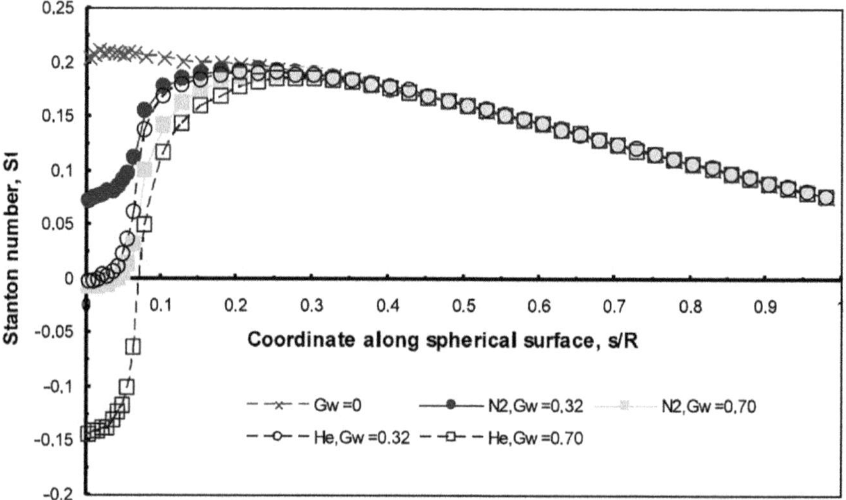

Fig. 6 Stanton number along the spherical surface at $Kn_{\infty,R} = 0.0163$ ($Re_{o,R} = 92.8$) and various air-to-air and helium-to-air mass blowing factors G_w [6]

6.2 Combustion in Hypersonic Air Flows with H_2 Injection

The recent analysis of nonequilibrium processes of hydrogen combustion in hypersonic airflows inside the scramjet engine is based on the numerical solutions of TVSL equations and boundary conditions [4], modeling the flow in the viscous shock layer near a parabolic cylinder at slot and uniform injections. Mathematical description of the nonequilibrium process of hydrogen combustion considered 11 gas components (O_2, N_2, NO, H_2O_2, HO_2, H_2O, OH, H_2, H, O, N) and 35 chemical reactions [19].

The slot injection from the body surface was simulated by using the injection parameter $G = G_0 \exp(-\alpha_w/X)$, where G_0 is the value of the parameter at the critical cylinder line, $\alpha_w = 30$, and X is marching coordinate along the surface. The computational results indicate the presence of significant amounts of nonreacting molecular hydrogen near the body. For $Re_0 > 100$, the diffusion of the "hot" reaction products (water and OH) occurs from the reaction zone to the surface. This phenomenon leads to the increase of heat flux values at the surface, $Q = 2q/(\rho_\infty U_\infty^2)$ towards the body surface along the coordinate X (see Fig. 7, left). The heat from combustion reactions does not prevail over the cooling effect of injection at the considered Reynolds number, $Re_0 = 628$. The heat flux is significantly lower than Q at $G_0 = 0$ near the injection zone at moderate injections ($0.02 \le G_0 \le 0.05$). A large quantity of injected hydrogen leads to effective cooling of the surface at a large distance from the injection zone at a high injection level (see diamonds in Fig. 7, left). The combustion zone moves from the surface towards the TVSL external boundary as the injection intensity increases.

Cooling the porous surface of the parabolic cylinder was modelled at $Re_{of} = 628$ and the constant intensity of hydrogen injection, $G = G_0 = const$ along the surface. More significant decrease of the heat flux is found at the entire body surface (see Fig. 7, right), compared to the case of slot injection. For the weak injections (squares in Fig. 7, right), the heat flux at the entire computational region is less than Q in the absence of the injection ($G_0 = 0$). The distribution of the heat flux becomes noticeably distinct at moderate (triangles, $G_0 = 0.02$) and strong (circles, $G_0 = 0.03$)

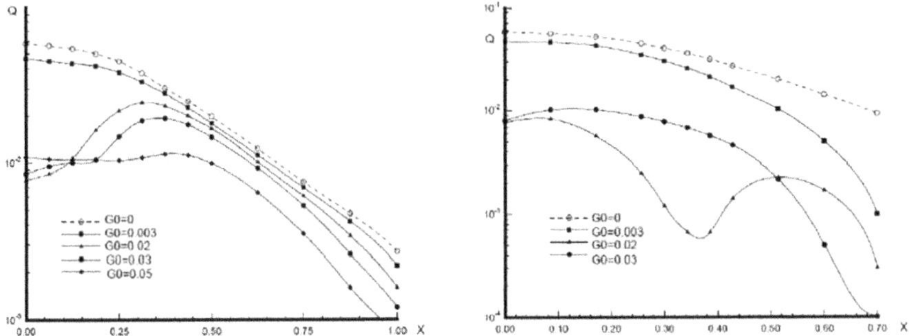

Fig. 7 Heat flux on parabolic cylinder (radius $R = 0.015$ m) at slot (*left*) and uniform (*right*) injections of H_2 to air at $Re_{of} = 628$, velocity $U_\infty = 2933$ m/s, and $T_\infty = 230$ K

injections, as the result of mutual influence of thermophysical properties of hydrogen, high enthalpy values of upstream flow, and the presence of combustion zone in the shock layer.

7 Conclusions

The computational tests were conducted as a model problem for preliminary design of heat protection systems of hypersonic vehicles. The results validate the TVSL model for calculating complex nonequilibrium multicomponent gas flow near blunt bodies under low-density flight and wind-tunnel test conditions. The study of characteristics of the catalytically influenced zone can be used in developing approximation methods of predicting heat fluxes at a catalytic surface of a vehicle. The binary similitude law $(\rho_\infty R = const)$ [12] can be applied in the studies of the nonequilibrium flows at small values of the Reynolds number, $Re_{of} < 20$. Gas injections of N_2, air, He, or H_2 can significantly reduce the heat fluxes on the hypersonic-probe surfaces. Strong influences of the blowing factor G_w (the ratio of outgas mass flux to upstream mass flux) and the rarefaction criterion (the Knudsen number or the Reynolds number) on the flow structure around a sphere and a cylinder (temperature fields, the configuration of mixing flow zones) and on heat-flux distributions along the surface have been found.

References

1. Cheng H (1993) Perspectives on hypersonic viscous flow research. Annu Rev Fluid Mech 25:455–484
2. Gnoffo PA, Gupta RN, Shinn JL (1989) Conservation equations and physical models for hypersonic air flows in thermal and chemical nonequilibrium. NASA TP-2867
3. Gupta R, Yos J, Thompson R, Lee K (1990) A review of reaction rates and thermodynamic and transport properties for an 11-species air model for chemical and thermal nonequilibrium calculations to 30,000K. NASA Reference Publication No. 1232
4. Riabov VV, Botin AV (1995) Hypersonic hydrogen combustion in the thin viscous shock layer. J Thermophys Heat Transfer 9(2):233–239
5. Moss JN, Bird GA (1984) Direct simulation of transitional flow for hypersonic reentry conditions. AIAA Paper 84–0223
6. Riabov VV (2004) Heat transfer on a hypersonic sphere with diffuse rarefied-gas injection. J Spacecr Rocket 41(4):698–703
7. Gupta R, Simmonds A (1986) Hypersonic low-density solutions of the Navier-Stokes equations with chemical nonequilibrium and multicomponent surface slip. AIAA Paper 86–1349
8. Ryabov VV (1980) Numerical investigation of the flow of nitrogen past a sphere with allowance for rotational relaxation. Fluid Dyn 15(2):320–324
9. Cheng HK (1963) The blunt-body problem in hypersonic flow at low Reynolds number. IAS Paper 63–92
10. Moss JN (1974) Reacting viscous-shock-layer solutions with multicomponent diffusion and mass injection. NASA TR 411

11. Provotorov VP, Riabov VV (1983) Effect of chemical reactions on the flow of air in a viscous shock layer. Fluid Mech Soviet Res 12(6):17–25
12. Gusev VN, Provotorov VP, Riabov VV (1981) Effect of physical and chemical nonequilibrium on simulation of hypersonic rarefied gas flows. Fluid Mech Soviet Res 10(5):123–135
13. Botin AV, Gusev VN, Provotorov VN, Riabov VV, Chernikova LG (1990) Study of the influence of physical processes on heat transfer toward the blunt bodies in flows with small Reynolds numbers. Trudy TsAGI 2436:134–144
14. Riabov VV, Provotorov VP (1994) The structure of multicomponent nonequilibrium viscous shock layers. AIAA Paper 94–2054
15. Throckmorton D (1982) Benchmark aeroheating data from the first flights of the Space Shuttle Orbiter. AIAA Paper 82–0003
16. Riabov VV, Provotorov VP (1996) Exponential box schemes for boundary-layer flows with blowing. J Thermophys Heat Transfer 10(1):126–130
17. Nomura S (1983) Correlation of hypersonic stagnation point heat transfer at low Reynolds number. AIAA J 21(11):1598–1600
18. Botin AV (1987) A study of local heat transfer to the spherical surface with gas injection at small Reynolds numbers. Uchenyye Zapiski TsAGI 18(5):41–47
19. Dimitrov V (1982) Simple kinetics. Nauka, Moscow

DSMC Simulation of a Non-axisymmetric Hypersonic Vehicle at 90 km Altitude

Angelos Klothakis, Ioannis K. Nikolos, and Vassilis Theofilis

Abstract This study presents the numerical investigation of a non-axisymmetric hypersonic vehicle, modeled after the X-51 concept, flying at an altitude of 90 km. Due to the low atmospheric density at this altitude and the transitional flow regime, the Direct Simulation Monte Carlo (DSMC) method is employed to accurately capture rarefied effects and molecular collisions. The vehicle's geometry, specifically its inlet design, is examined for its ability to ensure sufficient mass flow and pressure required for scramjet operation under at that altitude. Key outputs include the shock structures and positions and local flow field properties, which elucidate the influence of non-axisymmetric design on the inlet's performance. Results indicate that inlet efficiency, characterized by uniform flow distribution and total pressure increase.

Keywords Hypersonic flows · Hypersonic vehicle · High-altitude flight

1 Introduction

Hypersonic flight, defined as speeds exceeding Mach 5, has long represented a frontier in aerospace engineering. Landmark programs such as the X-15, X-43A, and X-51A have successively advanced our understanding of hypersonic aerodynamics, thermal protection, and propulsion. Among these, scramjet engines (supersonic combustion ramjets) have emerged as a promising solution for sustained high-speed flight at high altitudes. Unlike traditional engines, scramjets maintain supersonic airflow throughout the engine, using the vehicle's forward motion for air compression and enabling efficient combustion at extreme velocities [1–4]. The X-43A's record-setting Mach 9.6 flight and subsequent tests of the X-51A Waverider have validated key elements of scramjet propulsion. These efforts, largely driven by NASA and

A. Klothakis (✉) · I. K. Nikolos
Technical University of Crete, Chania, Crete, Greece
e-mail: aklothakis@gmail.com

V. Theofilis
Technion Israel Institute of Technology, Haifa, Israel

© The Author(s) 2026

M. Grabe et al. (eds.), *Rarefied Gas Dynamics*, Springer Aerospace Technology,
https://doi.org/10.1007/978-3-032-00094-1_18

the U.S. Air Force, have shaped modern hypersonic research by bridging theoretical models with flight-demonstrated performance [5–7].

Recent computational studies have advanced simulation methods for hypersonic vehicles, employing techniques such as direct numerical simulation (DNS) and Direct Simulation Monte Carlo (DSMC) to resolve non-equilibrium and rarefied effects. Qi et al. [8] examined boundary layer transition via DNS, Li et al. [9] evaluated multi-regime aerodynamics of reentry vehicles, and Reji and Lal [10] used DSMC to model rarefied hypersonic flow over capsule geometries.

Building upon this foundation, the present study applies the DSMC method to simulate a three-dimensional, non-axisymmetric vehicle inspired by the X-51A Waverider at 90 km altitude. A geometry was reconstructed from open-source technical drawings [11], allowing detailed investigation of oblique shocks, density gradients, and vortex interactions in a high-altitude regime where continuum approaches become inadequate. This work contributes to the growing body of knowledge on rarefied hypersonic aerothermodynamics.

2 Geometry Definition

Two-dimensional schematics of an X-51A-like configuration were sourced from publicly available materials [11], serving as a reference for constructing the three-dimensional model. In developing this geometry, particular care was taken to account for features that critically affect performance in the hypersonic regime. Figure 1 provides a perspective view of the resulting vehicle design, showcasing how the overall shape of the original X-51A concept recognized for its effective hypersonic aerodynamic characteristics has been preserved. In addition, a simplified representation of the scramjet inlet, located beneath the vehicle, was incorporated to capture the complex flow phenomena typical of hypersonic inlets.

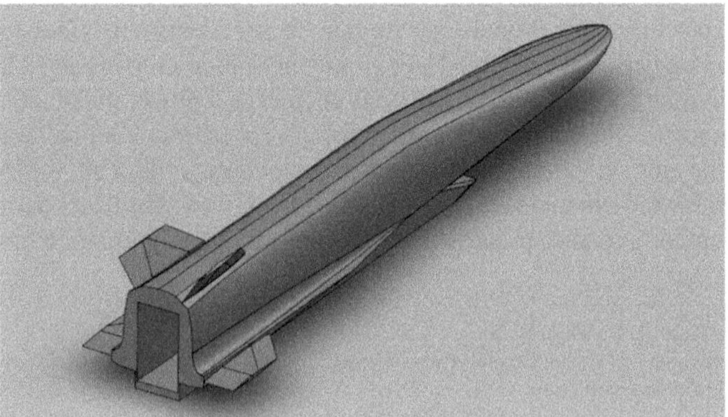

Fig. 1 Overview of the three-dimensional model

Taking all previous considerations into account, the final design is a hypersonic vehicle measuring slightly over 4 m in length and up to 0.86 m in width when including the horizontal fins. Situated beneath the fuselage, the engine extends approximately 2.7 m and features an intake opening of 0.23 m. As illustrated in the plans [11], this engine incorporates a diverging exhaust nozzle, an essential component for achieving optimal performance under the extreme temperatures and pressures characteristic of hypersonic flight. The engine is integrated into the vehicle's structure, with the forward underside acting as a compression ramp to support scramjet functionality. Furthermore, the operating conditions used in the simulations are derived from the U.S. Standard Atmosphere for an altitude of 90 km [12], ensuring realistic environmental parameters for high-altitude hypersonic analysis.

3 Results

The simulation was conducted using the Direct Simulation Monte Carlo (DSMC) method as implemented in the open-source SPARTA solver [13]. The vehicle was assumed to fly at Mach 9 and an altitude of 90 km, where rarefied gas effects are non-negligible. At these conditions, the Knudsen number computed using the relation $Kn = \lambda/L$ where λ is the mean free path and L is the vehicle's length, was found to be 0.00359, placing the flow in the slip regime. This justifies the use of DSMC for accurately resolving velocity slip, temperature jump, and shock structure. The freestream parameters corresponding to this flight point are summarized in Table 2.

The simulation parameters are summarized in Table 1 while the free-stream conditions can be seen in Table 2. Because of the intricate nature of the hypersonic flow surrounding this vehicle, a carefully designed grid refinement scheme was essential. Areas featuring steep gradients particularly near shock waves and within the engine's interior demanded a high-density mesh to capture swift variations in flow properties. As indicated in Table 1, the simulation employed over five billion particles, illustrating the necessity of fine resolution in regions where temperature and pressure can vary substantially. Shocks, expansion waves, and other complex phenomena inside the engine required an even more refined approach than that used for the external flow.

The grid refinement method was guided by local mean free path, ensuring that each cell did not exceed two times the mean free path. To accurately model particle collisions, the near-neighbor scheme [14] was applied, where collision partners were

Table 1 DSMC computational parameters

Number density, (#m^3	Timestep (s)	Transient period	Sampling period	Number of particles	Number of cells	CPU hours
2.00×10^2	3.00×10^7	50,000	50,000	5.8×10^9	~ 114,500,000	165,888

Table 2 Free-stream conditions

Velocity (m/s)	Temperature (K)	Number density (#/m^3)	Wall temperature (K)	Mach
2530	196	2.00×10^{20}	300	9.0

selected within a sphere whose radius equaled the distance a particle travel in a single timestep. Although cell sizes can exceed one mean free path, prior work has demonstrated that this approach remains accurate up to four mean free paths [15] given that a large number of particles per cell is used. Collisions were simulated using the Variable Soft Sphere (VSS) model [16] and molecular nitrogen (N_2) was selected as the simulation gas. The vehicle wall was modeled as isothermal at 300 K with full thermal accommodation. The computational domain used in this work, measured 18 m in the streamwise direction, 12 m normal to the flow, and 8 m in the spanwise direction. The timestep was set to 0.3 μs which is significantly smaller than the mean collision time. On average after refinement a particle needs about 10 timesteps to cross one cell ensuring smooth properties transport. The flow initially was given a transient period of 50,000 timesteps to evolve; an additional 50,000 samples were then recorded to reduce statistical noise.

Figure 2 shows the final adapted grid, focusing on shock regions, the engine inlet, and the combustion chamber. The pressure in the combustion chamber increases by approximately a factor of 40 relative to the freestream, indicating a compression level consistent with typical scramjet inlet design expectations [3], even though combustion is not modeled in the present study. The Mach number contours in Fig. 3 reveal the external flow structures and confirm that the flow remains supersonic around the engine inlet, aligning with scramjet requirements. Along the fuselage, Mach number progressively decreases as successive oblique shocks interact with the boundary layer. At the scramjet inlet, these contours demonstrate the vehicle's effectiveness in capturing and compressing high-speed flow for combustion even at high altitude highlighting the design's efficiency.

Furthermore, Fig. 4 depicts the pressure contours around the vehicle, illustrating the significant flow compression that occurs behind the nose as a result of oblique

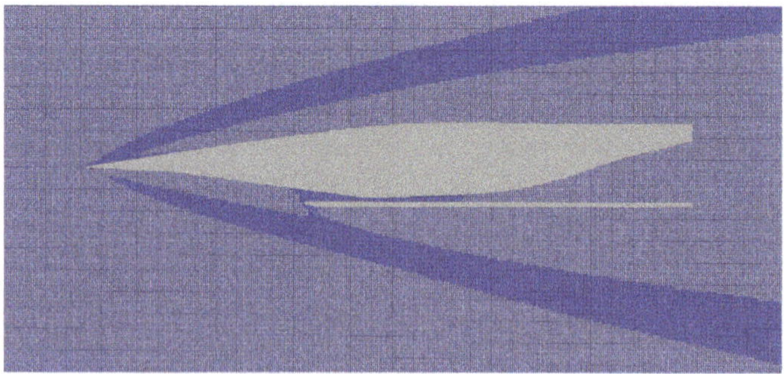

Fig. 2 Flow domain adapted grid

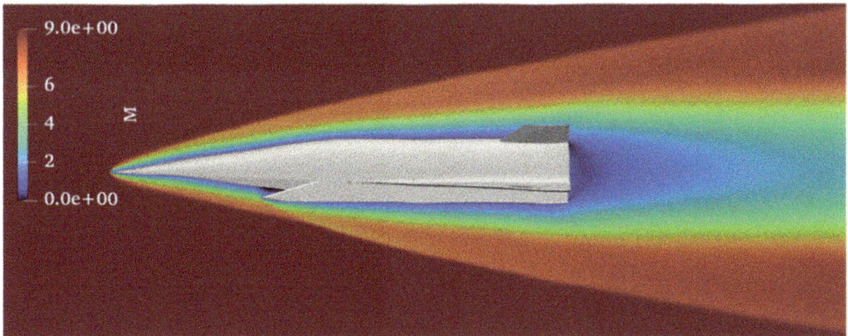

Fig. 3 Mach number contours on the symmetry plane

shock formation. Additional oblique shocks develop at the sharply defined engine inlet edge, elevating local pressures to nearly two orders of magnitude above freestream conditions, an essential characteristic for scramjet performance. Such pressure increases not only facilitate efficient air compression necessary for high-speed combustion but also introduce complex thermal and mechanical loads on the vehicle. This intense pressure rise is accompanied by corresponding density gradients, particularly in the plateau region upstream of the central cavity, where compression stabilizes before further increasing inside the combustor zone. Although no fuel injection is modeled and the precise internal geometry remains idealized, the simulation indicates peak pressures within this cavity, underscoring the importance of understanding compressive loading and structural integrity in high-speed propulsion systems. Downstream, the nozzle expansion converts thermal and pressure energy into kinetic energy, leading to re-acceleration of the flow and a sharp pressure drop. These high-pressure regions correlate with significant thermal and mechanical stresses, which must be accounted for in future design and material selection efforts. It should be noted that the present simulation excludes combustion and employs an approximate geometry, serving as a first-order assessment of flow structure and pressure loading under rarefied, high-Mach conditions (Figs. 4, 5).

In Figs. 6 and 7, Q-criterion is used to visualize both oblique shock structures and vortex-dominated regions, offering vital insight into the flow physics that governs scramjet performance. In Fig. 6 the (curved) leading edge shock that engulfs the vehicle is seen to be followed by a shock system which originates on the vehicle upper and lower surface at the location of laminar-turbulent transition. Past that location a turbulent boundary layer develops and extends to the end of the vehicle, as indicated by the large-scale vortices contained underneath the secondary shock system. Analysis is underway to define the transition location on the basis of linear stability theory in which velocity slip and temperature jump are taken into account [17, 18]. Planar cuts highlight circular blue lines where oblique shocks form, while red areas represent coherent vortex cores. Three-dimensional iso-surfaces further reveal a primary oblique shock extending from the vehicle's leading edge and a secondary shock from the forward boundary of the engine duct both generating

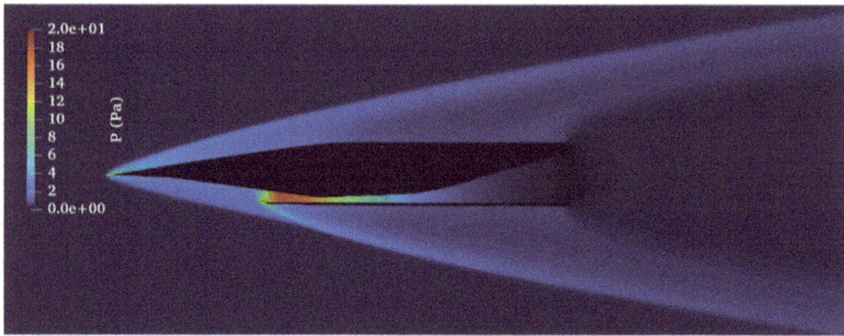

Fig. 4 Pressure contours around the vehicle and inside the engine

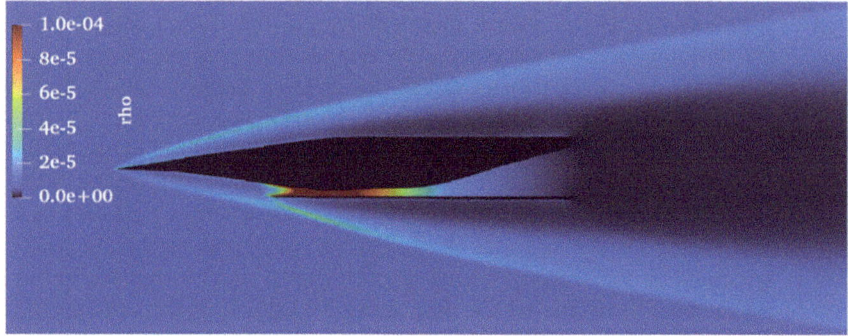

Fig. 5 Density contours around the vehicle and inside the engine

pressure gradients that influence heat transfer and structural loading. Multiple vortical structures also emerge from the vehicle's leading edge, underscoring the interplay between compressible flow features and the importance of precise geometric design to regulate shock-vortex interactions.

Further examination focuses on a major vortex at the ramp leading into the engine inlet. Although this vortex has the potential to disrupt inlet flow uniformity, the vehicle's lower-surface design confines these ramp-generated vortices, preserving favorable conditions for compression and combustion. The pronounced vorticity near the inlet highlights how minor adjustments in leading-edge geometry or ramp curvature can alter vortex formation and overall engine stability. Collectively, these findings underscore the complexity of hypersonic flow, where shocks and vortices interlink, and where meticulous external and internal geometry design is crucial for high-performance propulsion.

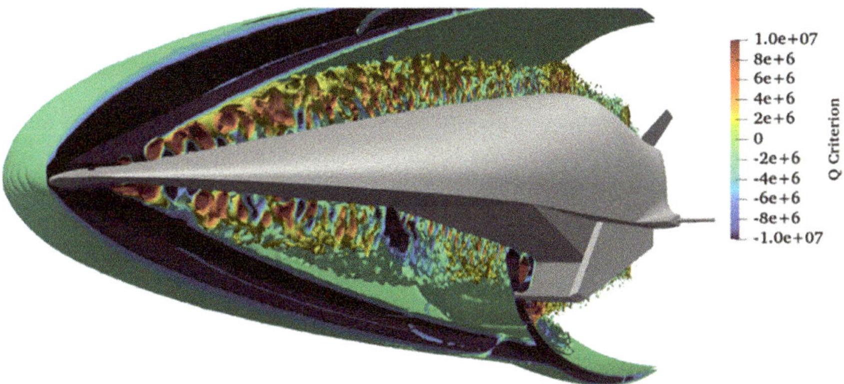

Fig. 6 Three-dimensional instantaneous Q-criterion iso-surfaces. The green–blue conical configuration corresponds to the primary oblique shock, whereas a secondary oblique shock emanates from the forward edge of the engine duct

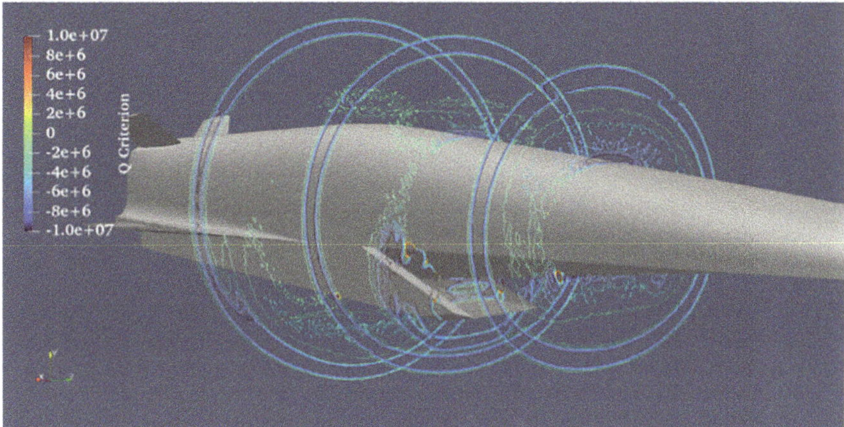

Fig. 7 Q-criterion contours depicted at multiple parallel planes to identify critical flow structures. The circular blue lines represent the locations of oblique shocks, whereas the red regions mark coherent vortex formations

4   Conclusions

This study presents a DSMC-based analysis of hypersonic flowfields around a scramjet-type vehicle, examining both external aerodynamics and internal flow behavior. Mach number contours reveal that oblique shocks at the nose and intake are essential for compressing incoming air to supersonic conditions, as required by scramjet propulsion. Although combustion is not modeled, pressure and density fields confirm effective airflow capture and highlight significant shock–boundary-layer interactions relevant to aerodynamic heating. Within the central cavity, density gradients arise from shock convergence and viscous effects, emphasizing the importance of controlled internal flow. While fuel injection is not simulated, these flow features indicate where injection and mixing would be critical in future reactive models. Q-criterion analysis reveals shock–vortex interactions, showing that ramp-generated vortices can propagate upstream if not managed through careful design. Nozzle expansion downstream produces a drop in density and thermal energy, simulating momentum redistribution consistent with flow acceleration. Due to the absence of combustion, thrust and exit Mach number are not quantified. Nevertheless, the results demonstrate that inlet and nozzle geometry strongly influence shock structure and flow uniformity, with implications for managing thermal and mechanical loads. Future work will incorporate chemical reactions and energy exchange to enable realistic modeling of combustion dynamics, heat transfer, and propulsion performance.

References

1. Fry RS (2004) A century of ramjet propulsion technology evolution. J Propul Power 20(1):27–58
2. Jenkins DR (2007) X-15: extending the frontiers of flight. NASA
3. Heiser WH, Pratt DT (1994) Hypersonic airbreathing propulsion. AIAA Educ Ser
4. Barthelemy R (1989) The national aero-space plane program. In: AIAA maintainability of aerospace systems symposium, paper 5053
5. Voland RT, Huebner LD, McClinton CR (2006) X-43A hypersonic vehicle technology development. Acta Astronaut 59:181–191
6. Oppenheimer MW, Doman DB (2006) A hypersonic vehicle model developed with piston theory. In: AIAA atmospheric flight mechanics conference and exhibit, AIAA, p 6637
7. Rondeau CM, Jorris TR (2013) X-51A Scramjet demonstrator program: waverider ground and flight test. In: SFTE 44th International/SETP southwest flight test symposium, vol 28
8. Qi H, Li X, Yu C, Tong F (2021) Direct numerical simulation of hypersonic boundary layer transition over a lifting-body model HyTRV. Adv Aerodynamics 3(31):1–21
9. Li J, Jiang D, Geng X, Chen J (2021) Kinetic comparative study on aerodynamic characteristics of hypersonic reentry vehicle from near-continuous flow to free molecular flow. Adv Aerodynamics 3(10):1–10
10. Reji RV, Lal SA (2016) Simulations of hypersonic flow past a re-entry capsule using DSMC method. Front Heat Mass Transf 7(1):1–8. https://doi.org/10.5098/hmt.7.27
11. SECRET PROJECTS Homepage, https://secretprojects.co.uk/threads/darpa-boeing-x-51a-waverider.7406/page-11, last accessed 2024/02/28.
12. U.S. Standard Atmosphere, U.S. Government Printing Office, Washington D.C., NOAA-S/T 76-1562 (1976)
13. Plimpton SJ, Moore SG, Borner A, Stagg AK, Koehler TP, Torczynski JR, Gallis MA (2019) Direct simulation Monte Carlo on petaflop supercomputers and beyond. Phys Fluids 31(8)
14. Gallis MA, Torczynski JR (2011) Effect of collision-partner selection schemes on the accuracy and efficiency of the direct simulation Monte Carlo method. Phys Fluids 67(8)
15. Gallis, MA, Bitter N, Koehler T, Torczynski JR, Plimpton SJ, Papadakis G (2010) DSMC simulations of turbulent energy decay in the Taylor-Green vortex flow. In: SAND2017-9109C, Sandia National Lab, Albuquerque
16. Koura K, Matsumoto H (1991) Variable soft sphere molecular model for inverse-power-law or Lennard-Jones potential. Phys Fluids 3(10):2459–2465

17. Klothakis A, Sawant SS, Quintanilha H, Theofilis V, Levin DA (2021) Slip effects on the stability of supersonic laminar flat plate boundary layer. In: AIAA Scitech 2021 Forum, p 1659
18. Klothakis A, Quintanilha Jr H, Sawant SS, Protopapadakis E, Theofilis V, Levin DA, Linear stability analysis of hypersonic boundary layers computed by a kinetic approach: a semi-infinite flat plate at

Numerical Simulation of the Flow Field Around a Sounding Rocket for the PMWE Project

Miklas Schütte, Igor Hörner, Stefan Löhle, Stefanos Fasoulas, and Marcel Pfeiffer

Abstract Atmospheric in-situ measurements with sounding rockets provide high spatial resolution, but face aerodynamic influences. Correction factors from Direct Simulation Monte Carlo (DSMC) are employed to counter disturbances in the atomic oxygen measurements within the PMWE project. For the first time, the resolution of the simulation mesh resolves the sensor geometry and the correction factor is calculated solely from the number density of atomic oxygen. Correction factors vary from 2 (110 km) to 3 (85 km) and are about 80 % of total density correction due to separation effects. They are influenced mainly by free molecular effects and do not depend on the atomic oxygen inflow number density.

Keywords Sounding rocket · PMWE · DSMC · Atomic oxygen density measurement · Aerodynamic influences · Correction factor

1 Introduction

The polar mesosphere winter echoes (PMWE) are reflections of radar waves in the mesosphere/lower thermosphere (MLT) of the atmosphere that extends approximately between 60 and 110 km altitude. The goal of the PMWE-project is to explain their origin. For this purpose, sounding rockets with appropriate instrumentation were launched to conduct measurements in the MLT-region. The Institute of Space Systems (IRS) is involved in the project with the Flux-Φ(Phi)-Probe-Experiment (FIPEX) sensors. These are solid electrolyte sensors for the quantification of atomic oxygen [1–3].

The necessity for high spatial resolution requires in-situ measurement techniques. Sounding rockets are the only suitable carriers for the instruments, since the altitude range of interest is too high for balloons and too low for satellites [4]. However, due to the movement of the rocket at supersonic speed and the presence of the sensor, there are deviations between the measured variables and the value in the undisturbed

M. Schütte (✉) · I. Hörner · S. Löhle · S. Fasoulas · M. Pfeiffer
Institute of Space Systems, University of Stuttgart, Stuttgart, Germany
e-mail: schuettem@irs.uni-stuttgart.de

© The Author(s) 2026

M. Grabe et al. (eds.), *Rarefied Gas Dynamics*, Springer Aerospace Technology,
https://doi.org/10.1007/978-3-032-00094-1_19

atmosphere as a result of aerodynamic influences on the flow field [5, 6]. Therefore, it is of critical importance to understand the effects and to determine their influence.

This paper examines aerodynamic influences and provides correction factors for the measured atomic oxygen density n_O using numerical simulations. Future work will apply these factors to the measurements.

2   Methods

2.1   PMWE 2 Mission

On October 1st 2021 the second measurement campaign (PMWE-2) started from Andøya Space Center in Norway. The apogee of the suborbital flight was at an altitude of about 130 km [2]. Two payloads (FIONA and DUSTIN) equipped with several scientific instruments were used. FIONA carried three FIPEX sensor on its bottom deck and none on the front deck. DUSTIN was equipped with FIPEX sensors on both decks with the bottom deck being identical to the FIONA. Therefore, DUSTIN includes all geometrical information required for this study. The payload is depicted in Fig. 1.   It is basically a long cylinder with a length of ≈ 2.7 m and a diameter of ≈ 36 cm. The measuring decks 1 and 2 are located at the respective ends. Three FIPEX sensors are attached to each deck. A detailed overview of the PMWE-2 mission with a focus on the FIPEX measurements is given in [2]. A correction of the measurement data, to take the aerodynamic influence into account, has not yet been part of the data analysis. The required correction factors are to be provided in the present work.

2.2   Correction Factor

The aerodynamic influences on the flow field are caused by compression and rarefaction effects, composition changes due to mass-dependent separation of species, chemical reactions at the surface and in the shock area, outgassing and desorption, and deviations from local thermodynamic equilibrium at high altitudes [7–9]. Due to theses effects above, the measured number density n_{exp} changes by a factor f_{aero} compared to the value of the undisturbed atmosphere n_∞ [5]:

$$n_{exp} = f_{aero} \cdot n_\infty \tag{1}$$

Fig. 1 Payload with measurement deck 1 (right) and deck 2 (left)

If the factor is known, it is possible to quantify the aerodynamic effects and to draw conclusion about the number density of atomic oxygen of the undisturbed atmosphere $n_{O,\infty}$ from the measurement data. This correction factor, determined through numerical simulation, describes the relationship between the local number density at the sensor position n and the undisturbed number density in the atmosphere n_∞ [10].

$$f_{\text{aero}} = \left(\frac{n}{n_\infty}\right)_{\text{sim}} \tag{2}$$

The factors are determined for several altitudes, because of the change in the aerodynamic conditions. In addition, the factor must be calculated for each sounding rocket project individually. This is due to the dependency of the aerodynamic influences on geometry, speed, angle of attack, atmospheric conditions, measurement principle and position on the payload [9].

The simulative investigation of the aerodynamic influences is complex, since the mean free path increases from a few millimeters at 80 km to several meters at 130 km. Thus, the aerodynamic conditions change from a continuum flow to a free molecular flow. Outside of the continuum regime conventional tools of computational fluid dynamics (CFD) are no longer usable. A common tool for the analysis of rarefied flows, is the direct simulation Monte Carlo (DSMC) method. This microscopic approach describes the movement and interactions of simulation particles, which represents multiple gas molecules or atoms [9]. In this work, the implemented DSMC method in the open source simulation program "PICLas" (Particle-In-Cell Las Vegas), which was and is being developed at IRS, is used for this purpose [11]. PICLas is extensively validated and verified for a wide range of applications [11, 12].

2.3 Calculation of the Correction Factor

The sensor measurements in the ram area are of particular interest for the experimental results, since they were found to reflect the actual variations better than the measurements in the wake area [4]. Therefore, the simulation focuses only on the flow around the measuring deck located in the ram area. The flow around deck 1 is simulated during the upleg only, and deck 2 only during the downleg. This approach significantly reduces the simulation area and minimizes the simulation effort.

For the first time, the simulation geometry includes the sensor itself, which is a small plate (see Fig. 2). The front 4 mm of this plate constitutes the sensitive measurement area. All calculated sensor values in the simulation (e.g. correction factors) are determined as area-averaged values in this area. In addition, the species-specific correction factor for atomic oxygen $n_O/n_{O,\infty}$ is determined. As a result, changes in composition (e.g. separation effects due to mass-dependent reflection) and the increase in number density on the sensor surface can be detected for the first time. Only wall

Fig. 2 Mesh of the FIPEX sensor and measurement deck 2

FIPEX-sensor mesh

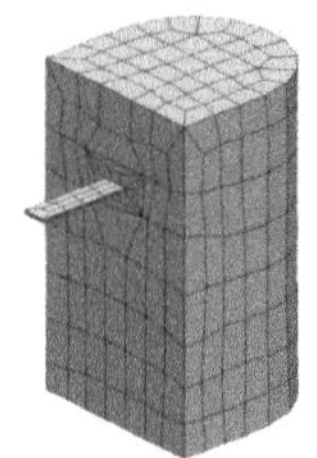

Measurement deck 2 mesh

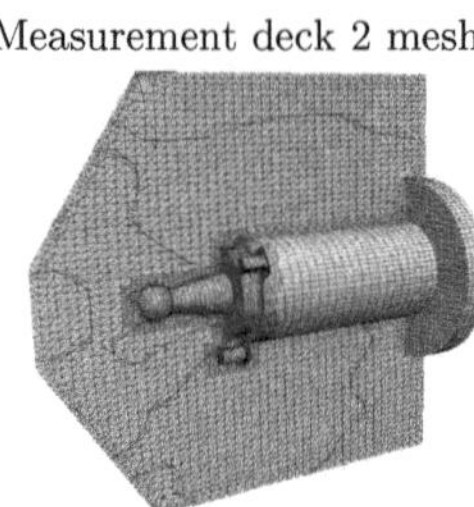

recombination, outgassing and desorption are not taken into account in the simulations. Previous studies only considered the total number density $n_t/n_{t,\infty}$, neglected the sensor geometry, and greatly simplified the overall payload geometry [4].

Simulation settings Since the FIPEX measurments (see [2]) change mainly in the range between 85 and 110 km, the simulation were chosen in the altitudes of 85, 90, 95, 100 and 110 km for the up- and downleg. The simulations employ a five-species model, which includes molecular and atomic nitrogen, molecular and atomic oxygen, and nitric oxide as a potential reaction product. The data for individual number densities and inflow temperatures are taken from the NRLMSISE-00 atmospheric model. For sufficient particle discretization in the DSMC simulation, the mean collision distance must be smaller than the mean free path [11]. Therefore, the particle weighting is between $1 \cdot 10^{10}$ at 110 km altitude and $2 \cdot 10^{11}$ at 90 km altitude. The nearest neighbor pairing algorithm and the octree method are used for particle pairing [13]. The time step was chosen to resolve the collision frequency to ensure a physically realistic simulation [14].

The velocity vector of the inflow $\vec{v}$ is inclined by the angle of attack relative to the center axis of the payload. Therefore, and because of the geometries of the measuring decks, no symmetry conditions can be applied and a complete 3D simulation is necessary. The magnitude of the inflow velocity and the angle of attack results from in-flight measurements. The velocities are between 600 and 1000 m/s. The angle of attack α varies between 5° and 10° for the upleg and between 20° and 30° for the downleg. The spin stabilization of 3 Hz [2] results in a maximum rotational velocity at the sensor position of only 2.17 m/s and is therefore negligible compared to the translational speed of the payload. In the simulation, the inflow direction is selected in such a way that the maximum correction factor occurs at one sensor, because the maximum correction factors for each height are required to correct the measured values. Section 3 summarizes the results for this sensor for the up- and downleg.

In a particle collision process, elastic collisions, relaxation of internal excitations and chemical reactions are considered. The collision cross-sections are calculated with the variable hard sphere model [14]. Chemical reactions are modeled using the total collision energy model. The FIPEX sensors and the rest of the payload surface are modeled as a diffuse reflecting wall. Catalytic surface reactions and electronic excitations are not considered in the simulations. The FIPEX sensors have a constant wall temperature of 1000 K (due to sensor heating) and the remaining payload a temperature of 288.15 K.

Due to the fact that DSMC is a statistic approach, the results are subject to statistical noise [14]. The noise can be reduced by increasing the sample size. This can be done by increasing the number of simulation particles or by sampling over multiple iteration steps/time steps in the case of stationary problems [11]. Eventually, a trade-off between increasing the computing effort (computing time and memory requirements) and reducing the noise must be made. Even though we considered only atomic oxygen, which is only a fraction of the present particles, and the area of interest is small compared to the simulation area, we managed to reduce the noise on the sensor values to just a few percent on the correction factor $n_O/n_{O,\infty}$. This was achieved by meshing the selective sensor area as roughly as possible in order to have as many atomic oxygen simulation particles as possible available for sampling and by increasing the number of iterations. Only in the case of the lowest altitude of 85 km, the noise is larger, due to the small number of atomic oxygen particles in relation to the total number of simulation particles in the whole calculation area. Given the considerable complexity of the geometry and considering the correction factor for atomic oxygen, as well as the limited computational capacity, the noise impact on the factor is very small. The simulation time was between 6 and 13 h on 512 cores at the high performance computing center Stuttgart.

3 Results

3.1 Flow Field

Figure 3 shows the flowfield visualization around the ram oriented bottom deck on the downleg. At the highest altitude the flow is close to free molecular flow. As the payload moves downwards, the density increases and a compression shock forms. Due to the geometry of the payload, the compression shock is very close to the sensors. During the simulations no chemical reactions took place, since the velocities of the payload and the temperatures in the disturbed regions are too low.

3.2 Correction Factor

The maximum correction factors for atomic oxygen $n_O/n_{O,\infty}$ and the maximum correction factors from the total number density $n_t/n_{t,\infty}$ for the sensors are plotted in Fig. 4. The diagram illustrates that the decrease in the correction factor with increasing altitude is extremely small for the upleg and the downleg. The values are between approximately two and three. The same applies for the downleg. Another observation is that the correction factor for atomic oxygen $n_O/n_{O,\infty}$ at 90–110 km is

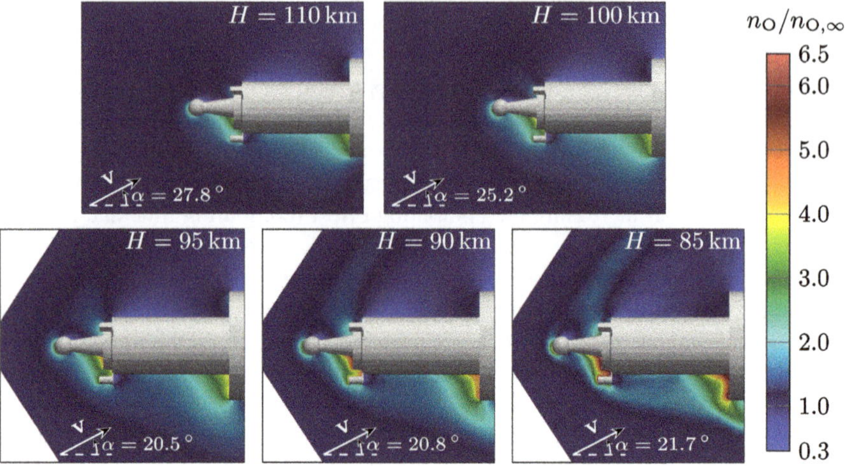

Fig. 3 Flow field around the payload on the downleg at altitudes 85–110 km

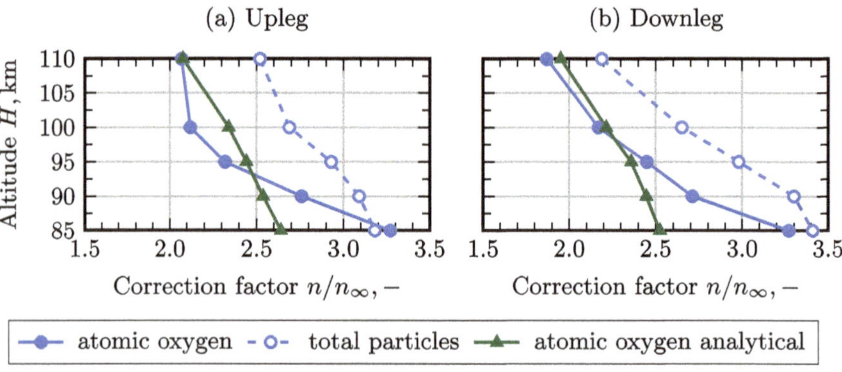

Fig. 4 Simulated correction factors of atomic oxygen $n_O/n_{O,\infty}$ and total particles $n_t/n_{t,\infty}$ as well as analytical determined atomic oxygen factor from Eq. 3

approximately 80 % of the total number density correction factor $n_t/n_{t,\infty}$. This variation is primarily attributed to separation effects, mainly caused by mass-dependent reflection. When comparing the atomic oxygen correction factors in the upleg and downleg, the upleg value at an altitude of 110 km is slightly larger. This difference results from the lower angle of attack in the upleg, which leads to a larger accumulation of particles on the surface and thus to a larger correction factor in this flow regime. At lower altitudes, the up- and downleg values are quite similar. Due to the geometry and angle of attack, the shock is positioned closer to the sensor during the downleg, resulting in a stronger influence on the correction factor compared to the upleg. However, the geometric influence of structures near the sensor increases in the upleg. Consequently, the values are similarly large in both the up- and downleg.

Analytical determination of the correction factor In a free molecular flow, there are no intermolecular collisions. Consequently, the gas reflected from the payload is superimposed on the surrounding gas. The number density for the particles adjacent to the surface n, at complete accommodation and diffuse reflection for a single species, is described as follows [14]:

$$\frac{n}{n_\infty} = \frac{1}{2} \left\{ [1 + \mathrm{erf}\,(s\,\cos\theta)] \cdot \left[1 + \sqrt{\pi}\,s\,\cos\theta\sqrt{\frac{T_\infty}{T_\mathrm{w}}} \right] + \sqrt{\frac{T_\infty}{T_\mathrm{w}}}\,e^{-s^2\cos^2\theta} \right\} \tag{3}$$

The incident and the reflected components are included in n. The parameter n_∞ represents the number density of the inflow of this species, T_∞ is the temperature of the inflow, T_w is the wall temperature, θ describes the angle between the inflow and the surface normal and s is the velocity ratio between the body velocity and the most probable particle velocity of this species for a Maxwell distribution. The ratio n/n_∞ in Eq. 3 describes exactly the aerodynamic correction factor. Figure 4 shows the correction factor for atomic oxygen $n_O/n_{O,\infty}$ at the sensor from the simulation compared to the analytically determined correction factor with Eq. 3. As can be seen, the simulation results agree well with the analytical results, especially at high altitudes. The effects that decisively influence the correction factor are consequently the accumulation of the particles on the surfaces and the separation due to mass-dependent reflection (free molecular effects). The increase in the factor with decreasing altitude is mainly due to increasing velocity and less to an influence from a shock wave. This is the result of the still rarefied flow and the generally low payload velocity. Additionally, due to the generally low angle of attacks, their influence is weak, especially compared to the velocity. The small deviations at high altitudes are due to geometry influences. The differences at low altitudes are due to the increasingly dense flow and the forming compression shock. This leads to a deviation in the Maxwell distribution in the inflow, which violates the assumptions behind Eq. 3. Furthermore, the equation does not account for nearby structures, causing a larger difference in the upleg than in the downleg. Another limitation arises when the sensor is blocked by the payload or impacted by scattered particles, causing the analytical correction factors to be invalid, even in free molecular flow. However, if it is known that the sensor is exposed to the free stream and surrounding geometries have minimal influence without doing a simulation, then the analytical equation is applicable in free molecular flow regime and can provide a good approximation for the correction factor.

4 Conclusion

As part of the PMWE project, the density of atomic oxygen was measured using FIPEX sensors on sounding rockets to study polar mesospheric winter echoes in the MLT region. Due to the intrusive measuring method, the measured values deviate from the undisturbed atmosphere owing to aerodynamic influences. DSMC simulations were conducted at several altitudes in the up- and downleg of the trajectory, focusing on the payload side in the ram area.

The simulations have to be three-dimensional, because of the complex geometry and the angle of attack. The time step and the number of particles were chosen in such a way that the collision frequency and the mean free path are resolved. Additionally, the simulation was refined to the sensor level to capture accumulation and separation effects. Quantitative analysis determined correction factors for atomic oxygen, ranging from two and three. As a result of mass-dependent scattering at the sensor surface, the atomic oxygen correction factor is about 80 % of the total number density correction factor. The factor is primarily determined by free molecular effects and is largely independent of the atomic oxygen inflow number density. Despite the high detail in the simulation, additional effects such as wall recombination, outgassing, and desorption could further influence measurements.

It should be noted that quantitative and qualitative investigations of aerodynamic effects are necessary, as correction factors alone do not capture all aerodynamic influences. As a result, measurements may still be distorted. Qualitative analysis showed, that the upleg measurements are more influenced by geometry, than downleg measurements due to the complexity of the geometry.

Acknowledgements This work is funded by the Deutsche Forschungsgemeinschaft (DFG, German Research Foundation) – Project-ID 516238647 – SFB 1667 (ATLAS - Advancing Technologies of Very Low-Altitude Satellites). The authors also thank the High Performance Computing Center Stuttgart (HLRS) for granting the computational time that allowed the execution of the presented simulations.

References

1. Strelnikov B et al (2021) Sounding rocket project "PMWE" for investigation of polar mesosphere winter echoes. J Atmos Solar-Terrestrial Phys 218
2. Hoerner I et al (2022) atomic oxygen measurments on sounding rockets using solid electrolyte FIPEX sensors: first flight results from the PMWE mission. In: 25th ESA-PAC Symposium, Biarritz, France
3. Hoerner I et al (2025) Fast response solid electrolyte oxygen sensors with porous thin film electrodes. Rev Sci Inst 96(3):015002-1–015002-13
4. Eberhart M et al (2015) Measurement of atomic oxygen in the middle atmosphere using solid electrolyte sensors and catalytic probes. Atmos Meas Tech 8(8):3701–3714
5. Rapp M, Gumbel J, Lübken F-J (2001) Absolute density measurements in the middle atmosphere. Annales Geophysicae 19(5):571–580
6. Allen JB, Perl M, Hauser T (2006) Simulation of aerodynamic influences on rocket-mounted oxygen sensors. J Spacecr Rocket 43(6):1387–1394
7. Bird GA (1988) Aerodynamic effects on atmospheric composition measurements from rocket vehicles in the thermosphere. Planet Space Sci 36(9):921–926
8. Gumbel J, Rapp M, Unckell C (1999) Aerodynamic aspects of rocket-borne in situ studies. In: Kaldeich-Schürmann B (ed) European rocket and balloon programs and related research, vol 437. ESA Special Publication, pp 459 464

9. Gumbel J (2001) Aerodynamic influences on atmospheric in situ measurements from sounding rockets. J Geophys Res Space Phys 106(A6):10553–10563
10. Eberhart M et al (2019) Atomic oxygen number densities in the mesosphere–lower thermosphere region measured by solid electrolyte sensors on WADIS-2. Atmos Meas Tech 12(4):2445–2461
11. Fasoulas S et al (2019) Combining particle-in-cell and direct simulation Monte Carlo for the simulation of reactive plasma flows. Phys Fluids 31(7):072006-1– 072006-19
12. Nizenkov P et al (2016) Verification and validation of a parallel 3D direct simulation Monte Carlo solver for atmospheric entry applications. CEAS Space J 9
13. Pfeiffer M, Mirza A, Fasoulas S (2013) A grid-independent particle pairing strategy for DSMC. J Comput Phys 246:28–36
14. Bird GA (1994) Molecular gas dynamics and the direct simulation of gas flows, 2nd edn. Oxford Engineering Science Series

PICLas-Based Intake Simulation Activities for the Development of an ABEP Specular Intake

Nadine Barth, Jonathan Skalden, Konstantinos Papavramidis, Elizabeth Gutierrez, Franziska Tuttas, Marcel Pfeiffer, Raphael Tietz, Philipp Maier, and Georg Herdrich

Abstract Space missions in very low Earth orbit (VLEO) offer advanced capabilities for Earth observation, telecommunications, and security but are challenged by continuous orbit decay. Atmosphere-Breathing Electric Propulsion (ABEP), which uses atmospheric particles as propellant, provides a potential solution. Within the H2020 DISCOVERER, ESA ram-CLEP and ATLAS projects, IRS is developing a specular intake and helicon plasma thruster to advance ABEP systems. This study uses Direct Simulation Monte Carlo (DSMC) simulations to evaluate specular intake geometries, analyzing particle flux, pressure, mass flow rate, and collection efficiency. The results indicate that reducing the focal length of the parabolic intake significantly improves efficiency and mass flow rate, while increasing the discharge channel diameter proves more effective for achieving these high values as well. As pressure follows an opposing trend to efficiency, and reaching the ignition pressure is crucial, the optimal configuration among the investigated for balancing efficiency and pressure is a discharge channel diameter of 25 mm with a focal length of 3 mm.

Keywords ABEP · Intake · DSMC · Very low earth orbit

1 Introduction

Very Low Earth Orbit (VLEO) missions face significant atmospheric drag, necessitating efficient propulsion systems to maintain orbit and extend mission duration. Atmosphere-Breathing Electric Propulsion (ABEP) addresses this challenge by using residual atmospheric particles as propellant for an electric thruster.

The ABEP system, includes an intake that collects atmospheric particles and directs them into the discharge channel, where they are accelerated to generate thrust. This study focuses on the design and simulation of the ABEP intake system to optimize key parameters such as collection efficiency, pressure, and mass flow rate,

N. Barth · J. Skalden · K. Papavramidis · E. Gutierrez · F. Tuttas · M. Pfeiffer · R. Tietz · P. Maier ·
G. Herdrich (✉)
Institute of Space Systems, Stuttgart, Germany
e-mail: herdrich@irs.uni-stuttgart.de

M. Grabe et al. (eds.), *Rarefied Gas Dynamics*, Springer Aerospace Technology,
https://doi.org/10.1007/978-3-032-00094-1_20

ensuring effective thruster operation while minimizing the intake's size and volume. Additional insights into ABEP and its platform design are available in Herdrich et al. [6, 7]. Related research on VLEO satellite operations is supported by the DFG-funded CRC 1667 ATLAS at the University of Stuttgart.

2　Related Work

In VLEO, the free molecular flow regime is dominated by infrequent collisions, making particle-wall interactions critical for intake design. These interactions, including absorption, scattering, and chemical reactions, are challenging to model due to limited data on surface properties such as finish, cleanliness, and adsorbed gas layers. This is particularly relevant in VLEO, where interactions with gases like atomic oxygen are significant. The Maxwell model, a simplified approach, distinguishes between specular and diffuse reflections [9], forming the basis for intake design classifications.

Romano [12] demonstrated through simulations that specular intake designs outperform diffuse ones, leading to further studies by Romano, Kun, and Herdrich [5, 6, 13]. The specular design uses the optical principles of a parabolic mirror, directing particles to the focal point of the paraboloid. Particles retain their tangential velocity while their normal velocity is reversed, assuming hyperthermal flow, which is less applicable in VLEO [1] but still provides a better performance than diffuse designs.

Figure 1 illustrates the functional principle of a parabolic intake based on specular reflection. This study evaluates three geometric modifications:

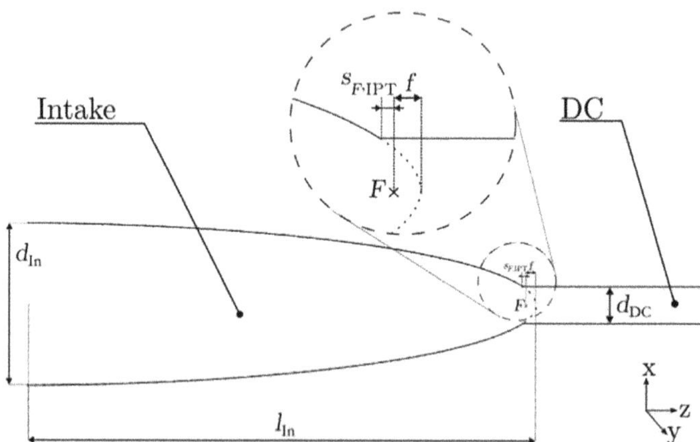

Fig. 1 Functional schematic of the parabolic intake, showing intake diameter d_In, intake length l_In, discharge channel diameter d_DC, and focal point position F inside the discharge channel with the focal length f and vertex-to-focal point distance $s_{F,\mathrm{IPT}}$

1. Intake Length (l_{In}): The intake length is varied between 500 mm, 1000 mm and 1500 mm.
2. DC diameter (d_{DC}): The diameter of the DC is scaled to 15 mm, 25 mm and 37 mm.
3. Focal length f: Focal lengths of 3.0 mm, 3.75 mm, and 5.5 mm are evaluated.

These variations influence intake diameter, intake area, and vertex-to-focal point distance $s_{F,IPT}$.

The location of the focal point F relative to the DC is critical. Romano [12] tested three configurations: one in front of the DC intake vertex, one at the vertex, and one inside the DC. The most efficient setup was with F inside the DC, using a focal length f of 5.5 mm and a d_{DC} of 37 mm, achieving an $s_{F,IPT}$ of 10 mm. Increasing $s_{F,IPT}$ could further improve efficiency but would reduce intake area, limiting particle collection and pressure.

3   Methodology

In rarefied gas conditions, the Navier-Stokes equations are inapplicable, necessitating the use of the Direct Simulation Monte Carlo (DSMC) method for ABEP system simulations instead of continuum methods like Computational Fluid Dynamics (CFD). These simulations are conducted using the three-dimensional open-source particle code PICLas [4], which incorporates the DSMC method [2].

The simulations conducted in this work rely on atmospheric data provided by the NRLMSISE-00 model [10], a widely accepted standard in space research for modeling lower thermospheric conditions. Previous ABEP studies have used this model, including work by Romano [12] and Kun [5]. This study uses atmospheric data under low-activity conditions based on ISO-14222 [8] and ECSS [3] standards (F10.7 = 65, A_P = 0). A fully specular intake material is assumed to focus on geometric and environmental factors. The effect of the specularity assumption of 100% is discussed in detail in [1, 12].

Simulations examine altitudes from 150 to 250 km. The lower limit of 150 km is selected to avoid increased drag, which raises thruster power demands and thermal loads [12]. While there is no strict upper limit, at higher altitudes it becomes increasingly difficult to collect enough particles to sustain thruster operation at a reasonable specific impulse, and the benefits of close proximity to the Earth's atmosphere are lost. Additionally, numerical resources are limited, so 250 km was chosen as the upper limit here. A minimum pressure of 0.1 Pa, sufficient to ignite the ABEP thruster IPT, is assumed based on experimental results from the University of Stuttgart [6]. However, it is possible that a lower threshold pressure may be required.

Rotational symmetry of the intake allows for 2D axisymmetrical simulations in PICLas. Results are averaged across the DC radius at a z-position 164 mm downstream of the intake-DC intersection, considering variations in altitude and geometry.

4 Simulation Results

4.1 Focal Length

The position of the focal point F significantly impacts performance. This study examines three focal lengths inside the DC for a DC diameter of 25 mm: $f = [3\,\text{mm}, 3.75\,\text{mm}, 5.5\,\text{mm}]$, corresponding to $s_{F,\text{IPT}} = [10.02\,\text{mm}, 6.67\,\text{mm}, 1.60\,\text{mm}]$. A focal length of 3 mm was chosen to match earlier IRS research on a DC37 configuration, while 3.75 mm was tested to evaluate the impact of a slightly larger intake area.

Figure 2 compares efficiency Fig. 2a and mean pressure Fig. 2b. Smaller focal lengths lead to higher efficiencies at all altitudes due to better particle focusing, with 5.5 mm performing worse than diffuse intake designs (46 % [11]). Even a 0.75 mm increase of the focal length significantly reduces efficiency. Ignition pressure is reached at similar altitudes for all focal lengths: 220 km for 3.0 mm and 3.75 mm, and 230 km for 5.5 mm. Despite small pressure losses, shorter focal lengths significantly improve efficiency, making 3 mm the optimal focal length for DC25.

4.2 DC Diameter

The comparison of DC diameters used a focal length of 5.5 mm for DC37, 3 mm for DC25, and 3 mm for DC15. A shorter focal length for DC15 was avoided as it reduced the intake area, limiting particle collection.

Figure 3a shows DC25 and DC37 achieving similar efficiencies, reaching up to 94 % for a 500 mm intake length. DC15 performs significantly worse, with its 3 mm focal length resulting in lower efficiencies than DC25's 5.5 mm focal length, even

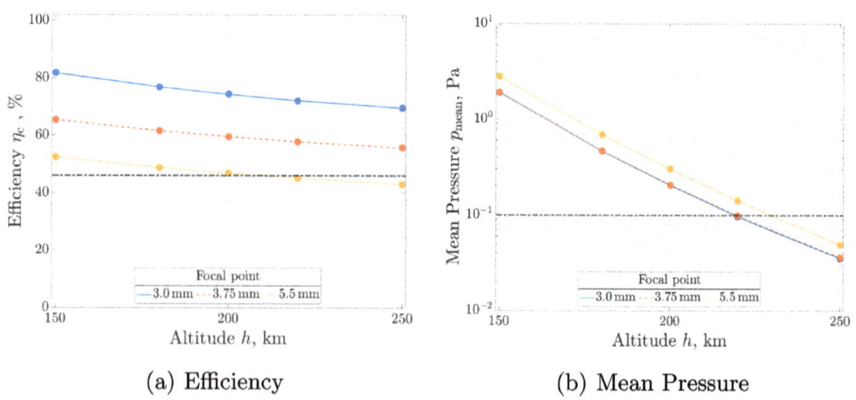

(a) Efficiency   (b) Mean Pressure

Fig. 2 Efficiency and pressure levels for three different focal lengths at a DC diameter of 25 mm and a length of 1000 mm, with the black line showing the current efficiency limit for diffuse intakes (**a**) and the expected minimum pressure required for ignition of the thruster inside the DC (**b**)

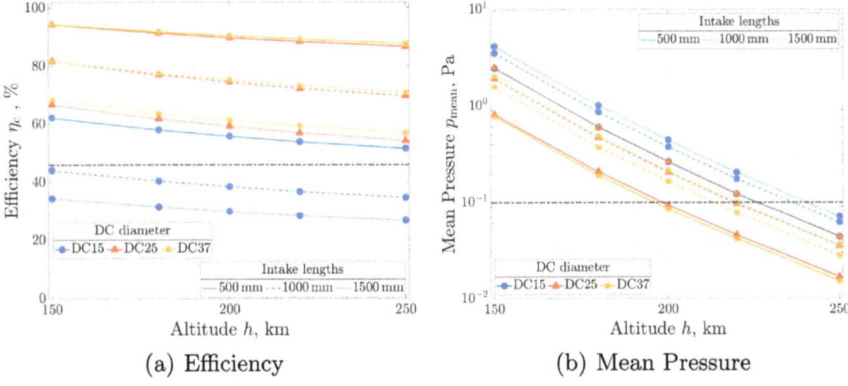

Fig. 3 Efficiency and pressure levels for three different DC diameters and three different intake length over altitude, with the black line showing the current efficiency limit for diffuse intakes (**a**) and the expected minimum pressure required for ignition of the thruster inside the DC (**b**)

below diffuse designs. This highlights the need for shorter focal lengths for smaller DCs to increase $s_{F,\text{IPT}}$ and outperform diffuse designs.

Figure 3b shows DC15 achieving the highest mean pressure across all intake lengths, while DC25 and DC37 show similar pressures, with DC25 slightly higher for longer intakes. As noted in [1], efficiency and pressure depend on the molecular speed ratio, with higher thermal velocities reducing particle focusing for longer intakes. DC37 excels at particle focusing, while DC25 produces more collisions, leading to higher pressure at longer intakes. Differences between DC25 and DC37 are attributed to slight variations in $s_{F,\text{IPT}}$.

4.3 Comparison over Geometry

The previous results indicate a strong relationship between efficiency and $s_{F,\text{IPT}}$ as the results for DC25 and DC37 were similar with comparable $s_{F,\text{IPT}}$. Therefore Fig. 4a presents a comparison of all simulation results based on the vertex-to-focal point distance, $s_{F,\text{IPT}}$, with error bars showing efficiency variations across altitudes. DC diameters are color-coded, and additional DC25 results for a 1000 mm intake length at $f = 3.75$ mm and $f = 5.5$ mm are included. For a given intake length, efficiency shows a linear dependency on $s_{F,\text{IPT}}$, confirming that smaller focal lengths f, which increase $s_{F,\text{IPT}}$, are essential for higher efficiency.

Efficiency is also analyzed as a function of intake area, which is influenced by all geometric factors. Figure 4b illustrates the results, with markers representing intake lengths and colors indicating DC diameters and focal lengths. A fitted curve demonstrates a linear relationship between efficiency and intake area across different configurations. The results indicate that DC37 consistently outperforms DC15

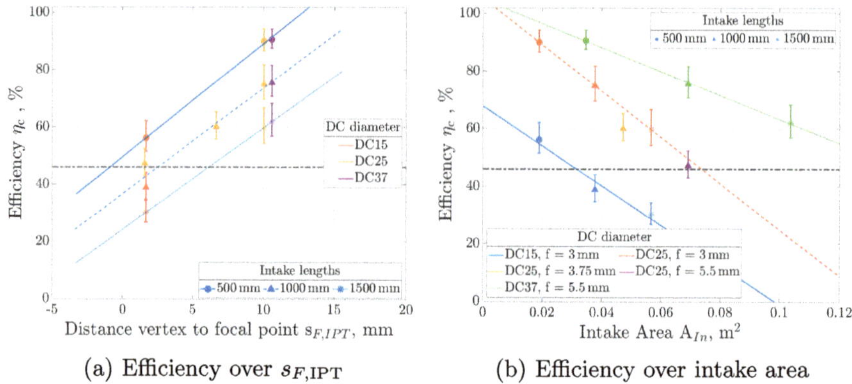

(a) Efficiency over $s_{F,\text{IPT}}$

(b) Efficiency over intake area

Fig. 4 Efficiency for three different DC diameters with different focal lengths and three different intake length over $s_{F,\text{IPT}}$ and intake area, with the black line showing the current efficiency limit for diffuse intakes

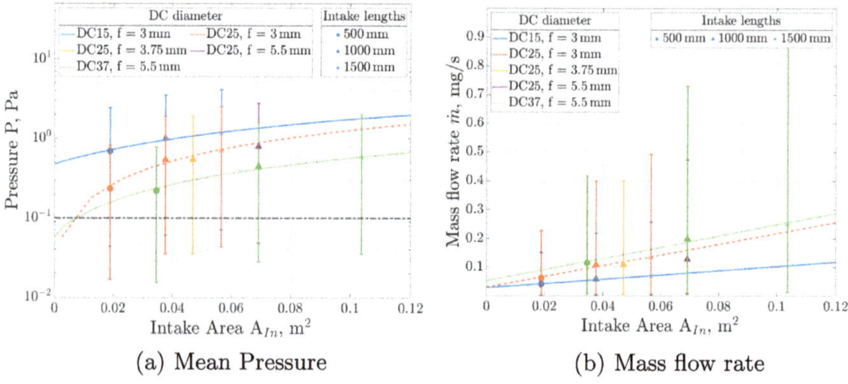

(a) Mean Pressure

(b) Mass flow rate

Fig. 5 Pressure and mass flow rate for three different DC diameters and three different intake length over intake area

and DC25 in terms of efficiency. While DC25 achieves relatively high efficiency at smaller intake areas, its performance declines with longer intake lengths. This decline is expected, as focusing particles into a smaller discharge channel becomes more challenging with increased intake length. This effect is even more pronounced for DC15, where the non-optimal focal length further amplifies the limitations of the smaller discharge channel. Specular intakes outperform diffuse designs for smaller intake areas, but their advantage diminishes as intake area increases.

Efficiency alone, however, is not the sole consideration; sufficient pressure for thruster ignition must also be achieved. The ignition threshold in this study is set at 0.1 Pa, based on experimental results [6], though a lower threshold may be possible. Figure 5a shows the mean pressure in the DC, with a dotted line marking the 0.1 Pa ignition threshold. A linear fit indicates that pressure is highest for DC15 and lowest for DC37. Nonetheless, none of the configurations consistently meet the ignition threshold across all altitudes, as error bars reveal instances where pressure falls below 0.1 Pa. The altitude at which ignition pressure is reached depends on the specific intake system. However, the results suggest that as the DC diameter decreases, the number of collected particles does not decrease proportionally, leading to a higher particle flux and, consequently, greater pressure.

Figure 5b presents the mass flow rate inside the DC, which shows a strong dependence on atmospheric conditions and altitude, as evidenced by the large error bars. Despite this variability, a clear linear trend in the mean mass flow rate across all altitudes is observed, suggesting that, for a given altitude, both mass flow rate and pressure scale linearly with the intake area. DC37 and DC25 have similar slopes for this linear relationship, while DC15 shows a less steep slope. This variation aligns with the previously discussed influence of focal length f and the resulting higher $s_{F,\mathrm{IPT}}$ values for DC25 and DC37, which improves their mass flow rate performance compared to DC15.

5 Conclusion

This simulative study of different ABEP intake geometries examines the impact of intake length, discharge channel diameter, and focal length on collection efficiency, pressure, and mass flow rates. Results indicate a linear relationship between the vertex-to-focal point distance ($s_{F,\mathrm{IPT}}$) and collection efficiency, with larger $s_{F,\mathrm{IPT}}$ resulting in higher efficiencies. For a 25 mm DC, a 3 mm focal length, resulting in $s_{F,\mathrm{IPT}} = 10$ mm, achieves efficiencies up to 94 % for a 500 mm intake length. In contrast, DC15 performs poorly due to its non-optimal larger focal length, highlighting the importance of small focal lengths for high efficiency.

Pressure within the DC is strongly influenced by geometry, with smaller DCs (e.g., DC15) producing higher pressures due to increased particle collisions but lower efficiencies, especially for longer intakes. This trade-off highlights the need to balance collection efficiency and pressure to optimize thruster performance.

Mass flow rate also has a linear dependency on intake area, mirroring the efficiency trend. DC37 and DC25 show similar slopes, while DC15 has a shallower slope due to its suboptimal focal length. Larger DC diameters, combined with optimal focal lengths, improve particle focusing, resulting in higher efficiencies and mass flow rates.

In conclusion, larger DC diameters and shorter focal lengths maximize efficiency and mass flow rates, while higher pressures are achieved with smaller DCs due to increased particle collisions. Among the configurations studied, a DC diameter of 25 mm with a 3 mm focal length offers the best balance between efficiency and pressure.

Acknowledgements This work is funded by the Deutsche Forschungsgemeinschaft project number 516238647-SFB1667/1, titled Advancing Technologies of Very Low Altitude Satellites (ATLAS). Additionally, partial support was provided by the Ram-CLEP project with the name "Technology

Enhancement of Atmosphere-Breathing Cathode-Less Electric Propulsion" funded by ESA under ITT AO/1-10597/20/NL/MG. The views and conclusions presented are solely those of the authors, and neither the Deutsche Forschungsgemeinschaft nor ESA assumes responsibility for any use of the information provided.

References

1. Barth N, Skalden J, Papavramidis K, Hild F, Pfeiffer M, Beyer J, Tietz R, Fasoulas S, Herdrich G (2025) Solar activity dependency of a specular intake for an abep system. J Electric Propulsion
2. Bird GA (1994) Molecular gas dynamics and the direct simulation of gas flows
3. ECSS-E-10-04C: space environment. Tech Rep, ESA Publications Division
4. Fasoulas S, Munz CD, Pfeiffer M, Beyer J, Binder T, Copplestone S, Mirza A, Nizenkov P, Ortwein P, Reschke W (2019) Combining particle-in-cell and direct simulation Monte Carlo for the simulation of reactive plasma flows. Phys Fluids 31(7)
5. Feng K (2023) Investigation of a novel atmosphere-breathing electric propulsion platform and intake
6. Herdrich G, Papavramidis K, Maier P, Skalden J, Hild F, Beyer J, Pfeiffer M, Fugmann M, Klinker S, Fasoulas S, Souhair N (2024) System design study of a VLEO satellite platform using the IRS RF helicon-based plasma thruster. Acta Astronautica 215:245–259
7. Herdrich G, Papavramidis K, Maier P, García-Almiñana D, Rodríguez Donaire S, Sureda Anfres M (2022) Platform and system design study of a VLEO satellite platform using the IRS rf helicon-based plasma thruster. In: Proceedings of the 73rd international astronautical congress, IAC-22-C4.9.1. International Astronautical Federation, pp 1–15
8. ISO-14222 (2013) Space environment (natural and artificial) earth upper atmosphere. Tech Rep
9. Livadiotti S, Crisp NH, Roberts PC, Worrall SD, Oiko VT, Edmondson S, Haigh SJ, Huyton C, Smith KL, Sinpetru LA, Holmes BE, Becedas J, Domínguez RM, Cañas V, Christensen S, Mølgaard A, Nielsen J, Bisgaard M, Chan YA, Herdrich GH, Romano F, Fasoulas S, Traub C, Garcia-Almiñana D, Rodriguez-Donaire S, Sureda M, Kataria D, Belkouchi B, Conte A, Perez JS, Villain R, Outlaw R (2020) A review of gas-surface interaction models for orbital aerodynamics applications. Progress Aerospace Sci 119
10. Picone J, Hedin A, Drob DP, Aikin A (2002) Nrlmsise-00 empirical model of the atmosphere: statistical comparisons and scientific issues. J Geophys Res Space Phys 107(A12): SIA–15
11. Romano F, Binder T, Herdrich G, Fasoulas S, Schönherr T (2015) Air-intake design investigation for an air-breathing electric propulsion system. Int Symp Space Technol Sci
12. Romano F, Espinosa-Orozco J, Pfeiffer M, Herdrich G, Crisp NH, Roberts P, Holmes B, Edmondson S, Haigh S, Livadiotti S, A MR (2021) Intake design for an atmosphere-breathing electric propulsion system (abep). Acta Astronautica 187:225–235
13. Romano F, Herdrich G, Chan YA, Crisp N, Roberts P, Holmes B, Edmondson S, Haigh S, Macario-Rojas A, Oiko V, Sinpetru L (2022) Design of an intake and a thruster for an atmosphere-breathing electric propulsion system. CEAS Space J 14(4):707–715

Effect of Gas-Surface Interactions on Very Low Earth Orbit Mission Operation

Sai Sudha Ramesh, Basman Elhadidi, Boo Cheong Khoo, and Wai Lee Chan

Abstract The recent surge in very low Earth orbit (VLEO) satellite operations necessitates the consideration of gas-surface interactions (GSI) to estimate the aerodynamic drag coefficient. This study utilizes the test particle Monte Carlo (TPMC) method to model the effects of GSI on orbital trajectories. The predominance of atomic oxygen (ATOX) in VLEO increases shear drag that is particularly adverse for VLEO satellite designs with large solar panels for power needs. Mitigation strategies include ATOX-resistant material coatings. This research explores drag reduction impacts on VLEO missions by using different reflective materials, which can be modeled in TPMC using GSI coefficients. Simulations indicate that specularly reflective material coatings on lateral surfaces can achieve approximately 10% thruster fuel savings and 40% longer mission operation windows, with room for improvement through optimization of mission phasing and sequencing.

Keywords Very low earth orbit · Gas-surface interactions · Test particle Monte Carlo · Satellite drag reduction · Satellite fuel efficiency

1 Introduction

The recent interest in very low Earth orbit (VLEO) satellite operations, typically defined by altitudes below 450 km [1], is primarily driven by the potential for improved image resolution, enhanced communication capabilities, reduced power requirements for data downlinking, and a lower risk of collision with space debris.

S. S. Ramesh (✉) · B. C. Khoo
Temasek Laboratories, National University of Singapore, Singapore, Singapore
e-mail: tslssr@nus.edu.sg

B. Elhadidi
School of Engineering and Digital Science, Nazarbayev University, Astana, Kazakhstan

W. L. Chan
Engineering Product Development, Singapore University of Technology and Design, Singapore, Singapore

M. Grabe et al. (eds.), *Rarefied Gas Dynamics*, Springer Aerospace Technology,
https://doi.org/10.1007/978-3-032-00094-1_21

Despite the benefits of operating satellites in VLEO, a major challenge is compensating for the increased atmospheric drag, which can be up to 1000 times greater than in conventional low Earth orbits of 500 km and above. Consequently, accounting for aerodynamic drag effects is crucial in VLEO mission design and analysis.

A more rigorous approach for satellite aerodynamic analysis involves the Monte Carlo methods, which are better suited for the rarefied gas dynamics in VLEO operations. These methods account for the random thermal motion of particles through gas-surface interaction (GSI) models and consider multiple reflections from surfaces, such as antennas, solar panels, and deorbit devices [2]. Most common Monte Carlo methodologies are the test particle Monte Carlo (TPMC) and direct simulation Monte Carlo (DSMC) approaches, for which Jin et al. [2] have compared their computational efficiency and found that TPMC is 10,000 times more efficient than DSMC. This efficiency is due to DSMC's requirement for volume discretization, which TPMC does not need. However, a significant drawback of TPMC is its inability to handle intermolecular collisions, which may be prevalent at altitudes below 120 km [2]. Since typical orbits for VLEO satellites are projected to be sustained at above 200 km where free molecular flow assumption holds, this study chose to use TPMC to model the influence of gas-surface interactions (GSI) on orbital trajectories, thus implicitly opting for the computational efficiency of TPMC at the expense of the reasonable assumption of negligible intermolecular collisions.

In VLEO, the prevalence of atomic oxygen (ATOX) increases the randomness of scattered directions due to diffuse reflection, significantly contributing to shear along surfaces parallel to the incoming flow. This shear drag is particularly adverse for VLEO satellite designs with large solar panels to meet power requirements. The high concentration of atomic oxygen enhances sensitivity to gas-surface interactions, especially on elongated surfaces. A potential mitigation strategy involves applying ATOX-resistant material coatings, which can reduce drag by minimizing random reflections in the tangential direction and promoting more specular reflections. Such specularly reflective materials are suitable for orbit maintenance in VLEO [1, 3].

This study aims to compare the performance of two materials, one diffusive and one specular, using representative GSI coefficients across a range of pitch angles. The sensitivity of the drag coefficient, C_D, was analyzed over various altitudes and satellite orientations. A parametric representation of C_D variation based on satellite attitudes at a nominal altitude was incorporated into the General Mission Analysis Tool (GMAT) orbit propagator [4] to replace the traditional constant C_D value, which may not accurately predict fuel requirements. The advantage of each material type is then quantified in terms of thruster fuel requirements for a target duration in VLEO.

Accordingly, the remainder of the manuscript is organized into three sections. Section 2 presents the performance analysis of satellites using the two types of materials. Section 3 compares these materials in the context of orbit maintenance. The conclusions are presented in Sect. 4.

2 Material Performance in VLEO

The present study is based on the TPMC code that was developed by Los Alamos National Laboratory and openly available on GitHub repository [5]. The original code was modified to: (i) compute aerodynamic forces and moments; and (ii) evaluate the satellite surface pressure. The accuracy of the modified code has been verified in Reference [6]. The satellite model assumed for this study is shown in Fig. 1 at different orientations, based on a prototype of the Extremely Low Earth Orbit Imaging and Technology Explorer (ELITE) [7], a microsatellite developed by the Satellite Research Centre (SaRC) at Nanyang Technological University, Singapore, set to launch in 2026.

In this study, the interactions between impinging gas species and the satellite surface were modeled using the Cercignani–Lampis–Lord (CLL) model. This model can represent diffuse, quasi-specular, or specular reflections through the appropriate selection of two GSI parameters, namely α_n, which denotes normal energy accommodation coefficient, and σ_t, which represents tangential momentum accommodation coefficient. Since C_D depends on GSI coefficients, which determine the nature of particle reflections off the satellite surface and the resultant momentum exchange leading to aerodynamic drag, the assessment of satellite performance with GSI specifications for material coatings across various satellite attitudes is crucial. To this end, the current study considers single-axis pitch rotation to examine the impact of drag reduction on mission operations.

The thermospheric conditions for the performance analysis were derived from the NRLMSISE-00 model, based on data from March 13, 1989, during solar cycle 22 when solar activity was at its peak. Materials M1 and M2, corresponding to GSI parameters of $\{\alpha_n,\ \sigma_t\} = \{0.44,\ 0.25\}$ and $\{0.95,\ 0.78\}$, respectively, were considered. With a smaller σ_t, material M1 will lead to more specular reflections than material M2. Note that the GSI coefficients here were selected such that, under the assumption that $\alpha_n = \alpha_t$, σ_t is close to the surfaces with and without specularly reflective POSS-Novastrat coating in the study of the effects of ATOX scattering on drag reduction by Minton et al. [3]. The rationale for selecting these values is to understand their relative contributions to aerodynamic drag and their subsequent effects in mission analysis and operations.

Figure 2 compares the performances of the two materials M1 and M2 over a range of altitudes from 200–500 km and pitch angles $0°$ to $-90°$. A similar trend was

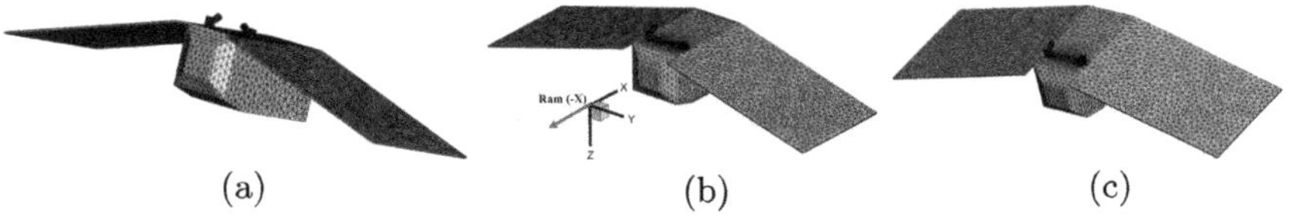

(a)   (b)   (c)

Fig. 1 Illustration of the VLEO satellite in this study modeled after ELITE [7] **a** pitch angle = $-20°$, **b** pitch angle = $0°$ and **c** pitch angle = $20°$. All models have same coordinate system

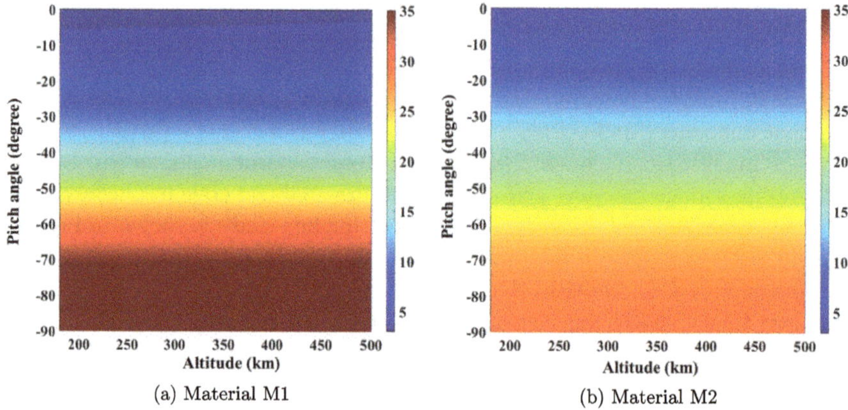

(a) Material M1   (b) Material M2

Fig. 2 C_D over a range of pitch angles and altitudes for **a** Material M1 (more specular) and **b** Material M2 (more diffusive) prescribed by GSI coefficients from Reference [3]

observed for pitch angles from $0°$ to $90°$ and so is not reported for brevity. The purpose of this analysis is to study the sensitivity of C_D to altitude and satellite orientations. Banded distributions can be observed for both materials across altitudes, indicating that C_D is insensitive to altitude variations. Conversely, there is higher sensitivity to pitch angle variations. Additionally, distinct bands are observed for M1 compared to M2, which may be attributed to less randomness in the tangential direction for the former.

Given the greater influence of pitch variations on C_D compared to altitudes as shown in Fig. 2, these variations can be approximated as a polynomial function of pitch angles at a specific altitude range. Doing so allows the incorporation of a varying C_D model with NASA GMAT [4] to assess the performance of the two materials in different mission scenarios. The parametric curves representing C_D variations for a mission scenario at a nominal altitude of 220 km for materials M1 and M2 are given in Eqs. (1) and (2), respectively:

$$C_D = 20.074\alpha^2 + 0.7143\alpha + 3.2005, \tag{1}$$

$$C_D = 258.2\alpha^4 - 263.5\alpha^3 + 106.33\alpha^2 - 1.2843\alpha + 3.6965, \tag{2}$$

where α denotes pitch angle in radians. The above equations will reasonably capture C_D variations within an altitude band of 200 to 265 km.

3 Numerical Simulation of VLEO Maintenance

The performances of the two materials were evaluated in the context of orbit maintenance, with the primary objective of quantifying thruster fuel savings and operational advantages resulting from reduced drag. The objective of this mission scenario is to maintain an average altitude between approximately 200 km and 265 km for a duration of at least 100 days. To simulate a real mission, three operation modes are considered: OM1 for thruster operation; OM2 for nominal operation; and OM3 for attitude maneuver modes. OM1 involves the operation of an electric thruster with a specific impulse of 1200 s and a thrust output of 6 mN [8].

The thruster is activated for a specified duration, based on mission design, to maintain flight within the designated altitude band. OM2 represents the nominal operation mode, where the satellite's attitude is set with yaw, pitch, and roll angles at zero, providing the minimum drag configuration. OM3 involves attitude changes, with the pitch angle adjusting from $0°$ to $20°$ at a rate of $1°/min$ and maintained at $20°$ for a specified duration to perform mission operations.

Referring to Fig. 3, which compares C_D over pitch angles ranging from $0°$ to $20°$, material M1, with its specularly reflective behavior, evidently has a significantly lower C_D than M2 by at least 13%. The reduced drag for M1 is attributed to its smaller σ_t, which results in a lower probability of particles contributing significantly along the flow direction, thereby reducing lateral shear. This behavior is depicted in Fig. 4 which assumes freestream velocity of 7772.5 m/s at an altitude of 220 km. The shear pressure along the solar panels is higher for M2, thus highlighting the effect of lateral shear, which leads to an increase in the drag coefficient. Interestingly, Fig. 4 also indicates that the normal pressure on the ram face of M1 is higher than M2, suggesting that specular reflection may not always be beneficial to drag reduction.

Fig. 3 C_D over a range of pitch angles for materials M1 and M2 for a nominal altitude of 220 km (altitude band between 200 km and 265 km)

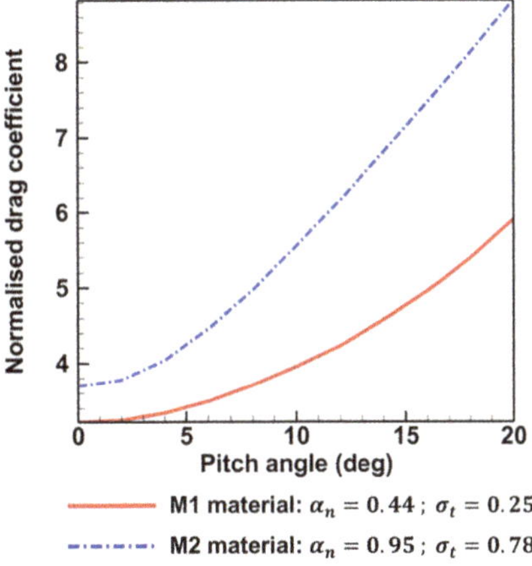

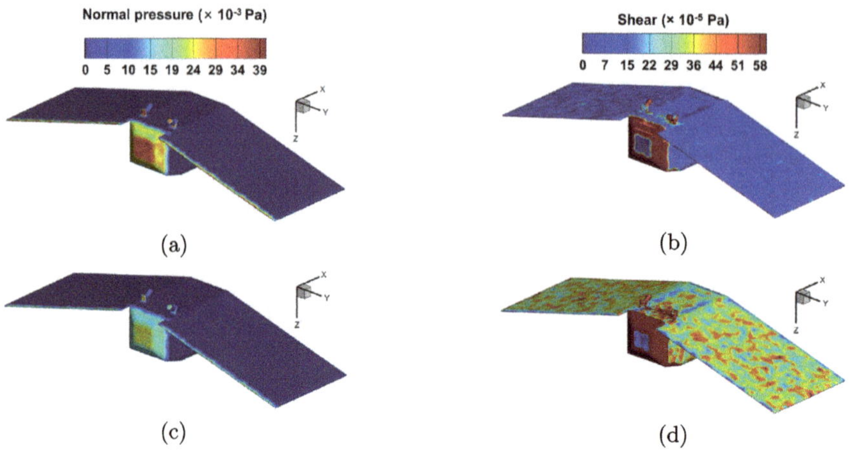

Fig. 4 Pressure distribution on satellite with materials **a,b** M1 and **c, d** M2

Table 1 presents the duration of different phases of mission operations associated with M1 and M2 to maintain VLEO flight between 200 and 265 km. The mission is divided into several phases, each consisting of a sequence of operation modes described earlier. For brevity, only the details for Phases 1 and 9 are presented (see Table 2). According to Table 1, Phase 1, with its mission sequence detailed in Table 2, is operated for 5 days and 2 days for M1 and M2, respectively. Although the current mission sequence is not optimized for M1 and M2, the mission duration and sequence for each phase were incrementally adjusted in small steps to ensure a sustained flight for 120 days. Consequently, this results in variations in the duration and phase sequences of M1 and M2.

Figure 5 compares the orbital trajectories associated with the two materials, with and without the inclusion of thruster firings. While thruster operation is essential for maintaining VLEO flight, it is advantageous to consider specularly reflective materials that reduce drag and fuel consumption compared to materials without AO-resistant coatings.

This is evident in Fig. 6, which shows that fuel depletion for M1 is approximately 10% lower than for M2 over a mission duration of 120 days. Clearly, material M1 is preferred over M2 for VLEO operations, as it aids in orbit maintenance without requiring extended thruster firing for drag compensation. Figure 6b compares the duration of mission operations for each mode, clearly demonstrating that M1 offers about a 27% reduction in thruster operation time compared to M2, ultimately allowing more time (over 40%) for nominal and attitude maneuver operations.

Table 1 Duration of different mission phases

Phase		1	2	3	4	5	6	7	8	9	10	11	12	13
Duration (days)	Material M1	5	5	20	20	8	25	8	3	32	–	–	–	–
	Material M2	2	3	20	15	12	2	10	10	10	7	8	22	12

Table 2 Duration of different operation modes in specific mission phases

Material M1	Phase 1	OM2(2)	OM1(1)	OM3(1)	OM1(1)	OM2(2)	OM3(1)
		OM2(2)	OM1(1)	OM3(1)	OM1(1)	OM2(1)	OM1(1)
	Phase 9	OM2(1)	OM1(3)	OM2(1)	OM1(2)	OM2(1)	OM1(2)
		OM2(1)	OM1(1)	OM2(1)	OM1(2)		
Material M2	Phase 1	OM2(2)	OM1(1)	OM3(1)	OM1(1)	OM2(2)	OM1(1)
		OM2(2)	OM1(1)	OM3(1)	OM1(1)	OM2(1)	OM1(1)
	Phase 9	OM2(2)	OM1(3)	OM2(1)	OM1(3)	OM2(1)	OM1(5)

Note Values inside parentheses indicate the number of orbital periods (approximately 5370 s) during which a particular operation mode is executed

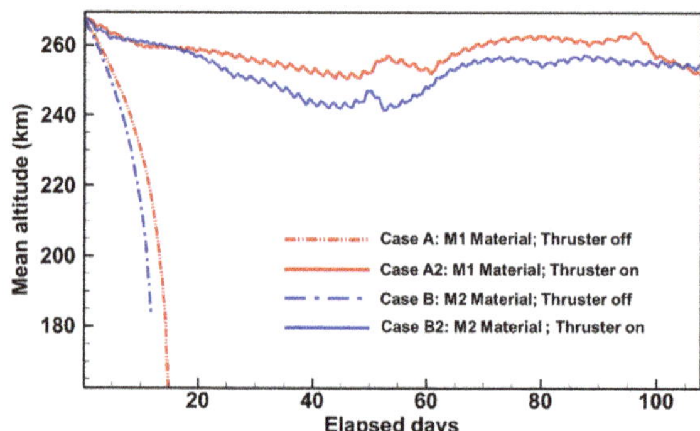

Fig. 5 Orbital trajectories in different scenarios

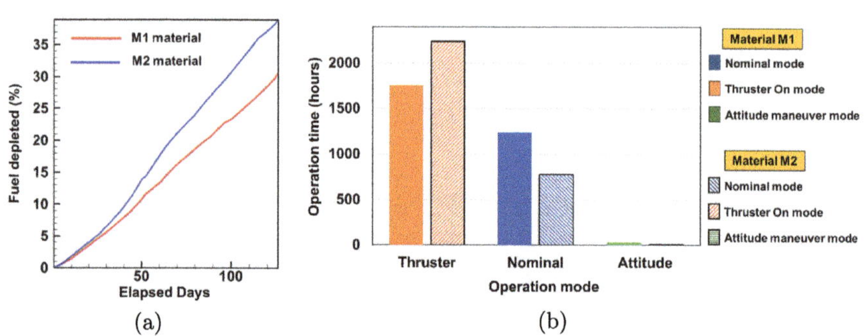

Fig. 6 Comparison of **a** percentage fuel depletion over mission and **b** duration of mission operation modes for materials M1 and M2

4 Conclusions

In this study, materials with different reflective characteristics were studied through prescription of gas-surface interaction coefficients. The reduction of drag for specularly reflective materials is attributed to a decrease in lateral shear that manifests prominently over areas that are parallel to the flow like the solar panels. However, specular reflections were also observed to increase normal pressure drag that acts along the ram face, suggesting that specularly reflective coating may not be suitable for application over all satellite surfaces. The variation of C_D was found to primarily depend on pitch variations, allowing a one-dimensional parametric representation of C_D that can be conveniently integrated with an orbit propagator. Simulations of mission scenarios with varying C_D for VLEO flight maintenance demonstrate that satellites with specularly reflective behavior can reduce drag and decrease thruster operations by 27% compared to those with diffusive materials, resulting in fuel savings of approximately 10%. In turn, this saving provides a 40% larger time window for scientific operations or small attitude maneuvers, thus benefiting the overall mission. Possible future studies include the exploration of selective specification of gas-surface interaction parameters on a satellite design, better characterization of gas-surface interaction coefficients, more robust incorporation of varying C_D model into an orbit propagator, and optimization of VLEO mission phasing and sequencing.

Acknowledgements This study is funded by the Office of Space Technology and Industry, Singapore, under the grant S21-19003-STDP. The computational work for this paper was performed on resources of the National Supercomputing Center, Singapore (https://www.nscc.sg).

References

1. Crisp N, Roberts P, Hanessian V, Sulliotti-Linner V, Herdrich G, García-Almiñana D, Kataria D, Seminari S (2022) A method for the experimental characterisation of novel drag-reducing materials for very low earth orbits using the satellite for orbital aerodynamics research (SOAR) mission. CEAS Space J 14:655–674
2. Jin X, Huang F, Cheng X, Wang Q, Wang B (2019) Monte Carlo simulation for aerodynamic coefficients of satellites in low-Earth orbit. Acta Astronaut 160:222–229
3. Minton T, Schwartzentruber T, Xu C (2021) On the utility of coated POSS-polyimides for vehicles in very low Earth orbit. ACS Appl Mater Interfaces 13:51673–51684
4. https://opensource.gsfc.nasa.gov/projects/GMAT/index.php
5. https://github.com/AndrewCWalker/impact-tpmc
6. Ramesh SS, Elhadidi B, Khoo BC, Chan WL (2025) Aerodynamic force and moment computations for very low Earth orbit satellites using a TPMC model. Acta Astronaut 228:88–100
7. Athreyas KN, Zhang X, Lim WS, Ramesh SS, Khoo BC, Chan WL (2024) Extremely low earth orbit imaging and technology explorer (ELITE): a very low earth orbit mission. In: 38th Small satellite conference, pp SSC24-II-03
8. Potrivitu G-C, Laterza M, Agarwal D, Sun X, Lim JWL (2024) Performance and plume characterization of the multi-stage ignition compact (MUSIC) hall effect thruster. In: 38th IEPC, pp IEPC-2024-586

Internal Flows in Vacuum Systems

Numerical Analysis of Rarefied Gas Flows Induced by Rotor-Stator Blades

Hiroshi Sugimoto, Ayako Miki, and Shogo Sugimoto

Abstract The unsteady gas flow in the two-dimensional model of the turbo-molecular pump is analyzed using the Boltzmann and BGK equations. The deterministic solver enables the visualization of instantaneous behavior in the time-periodic state induced by the rotor-stator system. A more straightforward explanation of the driving mechanism of the turbo-molecular pump is proposed, which leads to the alternative design of the pump. The combination of the traveling finite volume and the spherically designed grid enables the simulation of the full Boltzmann equation to be carried out by a personal computer.

Keywords Turbo-molecular pump · Moving boundary · Finite volume · Spherical design

1 Introduction

The turbo-molecular pump (TMP) may be the most common main pump. It has a high compression ratio and flow speed, and its easy handling is adequate for many laboratories and factories. Gaede proposed its original idea [1], but it made significant advances for practical usage after the improvement by Becker [2] at Pfeiffer Vacuum to have rotors and stators (both of them consist of many blades) stacked alternatively. Because the pressure inside TMP is low, we need the Boltzmann equation to analyze the behavior of the gas for its development and improvements. The problem here is that the direct numerical solution of the Boltzmann system used to be a challenging task due to its complex collision integral. The Direct Simulation Monte Carlo (DSMC) is one of the options [3, 4] to avoid this difficulty. However, the flow in TMP is time-dependent because the rotor blades rotate with a velocity close to

H. Sugimoto (✉) · A. Miki · S. Sugimoto
Department of Aeronautics and Astronautics, Kyoto University, Kyoto, Japan
e-mail: sugimoto.hiroshi.7s@kyoto-u.ac.jp

© The Author(s) 2026

M. Grabe et al. (eds.), *Rarefied Gas Dynamics*, Springer Aerospace Technology,
https://doi.org/10.1007/978-3-032-00094-1_22

sound speed; DSMC results are limited to time averages since present computational resources require some averaging process. Therefore, in the present study, we use our deterministic solver, Traveling Finite Volume method, to clarify the instantaneous flow field in the time-periodic behavior in TMP.

2 Numerical Model

We have to use the simplest model that can reproduce the gas flows in TMP since we do not yet have much information on the time variation of the rarefied gas in the device. Therefore, we throw away everything that is not essential and follow the model proposed by Sharipov [4]. Figure 1 presents our 2D model of TMP. The model consists of two arrays of slanted plates. The red plates, which are a substitute for the rotor of TMP, are slanted with θ_R and move in the direction of the yellow arrow with a constant speed v_R. The black plates at rest are substitutes for the stator. The plates are inversely slanted, and the angle is denoted by θ_S. The temperature of the plates is uniform at T_0. A rarefied gas with average number density n_0 fills the space between the plates. The regions between red plates are identified by part R, and those between black plates by part S. In the present study, two arrays share the same spacing (in the direction of plate motion) between the plates and the height of the slanted plates. They are denoted by, respectively, D and H. We define X_1 and X_2 axis as shown in the figure.

   We numerically analyze the unsteady behavior of the gas under the following assumptions. (i) The gas behavior is periodic with period D in the X_1 direction, and (ii) the behavior is also periodic with period H in the X_2 direction. The latter assumption forces the pressure along the pump axis (i.e., along X_2 direction) periodic. Therefore, the analysis in the present paper targets only the open (or the maximum) flow rate of the device.

3 Governing Equations

The rarefied gas in the model obeys the Boltzmann equation

$$\frac{\partial F}{\partial t} + \boldsymbol{\xi} \cdot \frac{\partial F}{\partial X} = J[F], \tag{1}$$

Fig. 1 2D model of the rotor-stator system. We assume periodicity both in X_1 and X_2 direction

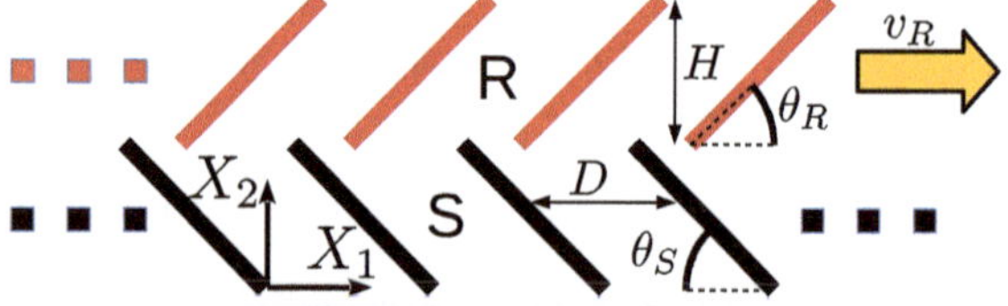

where t is the time, X is the spatial coordinate, ξ is the molecular velocity, F is the velocity distribution function and $J[F]$ is the collision term. For the hard-sphere molecule with molecular mass m and diameter d_m,

$$J_{HS}[F] = \frac{d_m^2}{2} \int_{\zeta \in \mathbb{R}^3, \, \alpha \in \mathbb{S}^2} \left[F(\xi') F(\zeta') - F(\xi) F(\zeta) \right] |V \cdot \alpha| \, d\Omega(\alpha) \, d\zeta, \quad (2)$$

$$V = \zeta - \xi, \ \xi' = \xi + (V \cdot \alpha)\alpha, \ \zeta' = \zeta - (V \cdot \alpha)\alpha,$$

where $d\zeta = d\zeta_1 d\zeta_2 d\zeta_3$ and $d\Omega(\alpha)$ is the solid angle element in the direction of α on a unit sphere $\mathbb{S}^2$. Since the numerical simulation by J_{HS} requires a significant amount of computational resources, we mainly use the BGK model instead:

$$J_{BGK}[F] = A_C n (F_E - F), \ F_E(n, v, T; \xi) = \frac{n}{(2\pi\kappa T/m)^{3/2}} \exp\left(-\frac{|\xi - v|^2}{2\kappa T/m} \right),$$
$$(3)$$

where A_C is a constant to give the collision frequency by $A_C n$, F_E is the Maxwellian belonging to number density n, flow velocity v, and temperature T. The Boltzmann constant is denoted by κ. The values of n, v, etc., are defined by the moments of F as

$$n = \int_{\xi \in \mathbb{R}^3} F \, d\xi \quad nv = \int_{\xi \in \mathbb{R}^3} \xi F \, d\xi, \quad p = \frac{1}{3} \int_{\xi \in \mathbb{R}^3} |\xi - v|^2 F \, d\xi = \kappa n T. \quad (4)$$

The boundary condition on the solid wall (velocity v_w, temperature T_0, and inward normal n_w) is the diffuse reflection:

$$F((\xi - v_w) \cdot n_w > 0) = F_E(\sigma_w, v_w, T_0; \xi) \quad (5)$$

where the number density σ_w is determined by the wall impermeability condition $(v - v_w) \cdot n_w = 0$.

4 Method of Analysis

We use a deterministic numerical scheme that uses the spherical designed grid and the traveling finite volume method as in [5].

4.1 Spherical Designed Grid

The spherical design [6] is a set of points a_k $(1 \le k \le N_\Omega)$ on $\mathbb{S}^2$ arranged so that the average of a smooth function f on them approximates the average on $\mathbb{S}^2$:

$$\frac{1}{4\pi} \int_{\boldsymbol{a} \in \mathbb{S}^2} f(\boldsymbol{a})\, \mathrm{d}\Omega(\boldsymbol{a}) \approx \frac{1}{N_\Omega} \sum_{k=1}^{N_\Omega} f(\boldsymbol{a}_k). \tag{6}$$

Equation (6) achieves the same accuracy of numerical integral of (4), which is essential for the BGK collision term (3), with a several orders small number of grid points $\boldsymbol{\xi}_{\ell,k} = \xi_\ell \boldsymbol{a}_k$ $(1 \le \ell \le N_\xi)$ using the curvilinear coordinate $\mathrm{d}\boldsymbol{\xi} = \xi^2 \mathrm{d}\xi \mathrm{d}\Omega(\boldsymbol{a})$ (cf. [5]). The impact is immense in the hard-sphere model J_{HS} since the four of the quintuple integral in (2), which is the reason for the dense computation, can be replaced by (6). The rest of the work is to carry out the one-dimensional numerical integral with some interpolation process and be carried out by personal computers.

4.2 Traveling Finite Volume Method

The traveling finite volume (TFV) method to solve the Boltzmann Eq. (1) is a finite volume method [7], with the only difference being the application of the characteristic coordinate (i.e., the motion path of the molecules) [5]. The TFV effectively avoids the numerical derivative across the discontinuity of F, a common feature of rarefied gas flows around convex bodies. See [5] for more details.

Before applying TFV, we reshape the parallelogram domains in Fig. 1 into rectangles in Fig. 2a by the appropriate linear transformation $\boldsymbol{Y} = \mathsf{A}\boldsymbol{X}$. The form of A depends on the size of rectangle. Examples for part R are

$$\mathsf{A} = \begin{pmatrix} 1 & -\cot\theta_{\mathsf{R}} & 0 \\ 0 & 1 & 0 \\ 0 & 0 & 1 \end{pmatrix} \text{ (keep height), or } \begin{pmatrix} 1 & -\cot\theta_{\mathsf{R}} & 0 \\ 0 & \csc\theta_{\mathsf{R}} & 0 \\ 0 & 0 & 1 \end{pmatrix} \text{ (keep edge length).}$$

LHS of (1) changes its form to $\dfrac{\partial F}{\partial t} + (\mathsf{A}\boldsymbol{\xi}) \cdot \dfrac{\partial F}{\partial \boldsymbol{Y}}$, but TFV is robust for this change since the molecular motion (i.e., the definition of the characteristics) can incorporate the matrix A in the form of $\mathsf{A}\boldsymbol{\xi}$. Then we introduce the moving frame $\boldsymbol{Y}'$ with

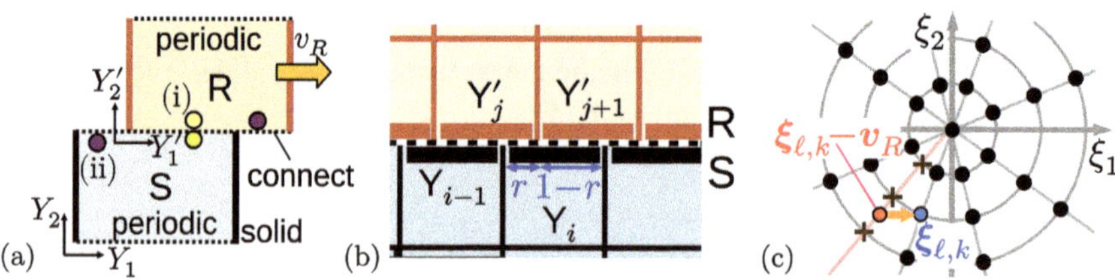

Fig. 2 **a** Moving frame in R, **b** ratio r among cell boundaries in R and S, and **c** a schematic of the interpolation process in $\boldsymbol{\xi}$-space to obtain the value at $\boldsymbol{\xi}_{\ell,k}$ in S, which corresponds $\boldsymbol{\xi}_{\ell,k} - \boldsymbol{v}_R$ in the moving frame in R. The set of grid points $\{\boldsymbol{\xi}_{\ell,k}\}$ in R are indicated by black points

speed $v_R = (v_R, 0, 0)$. Part R rests in the moving frame, as part S does in the original frame. Therefore, we can calculate the gas flows in the model shown in Fig. 1 by investigating the gas in the regions at rest, only by introducing an appropriate coordinate transformation between R and S at their connection surfaces.

Figure 2b shows the boundaries of the finite volume cells between R and S parts. The boundaries in S (indicated by Y_{i-1} and Y_i in the figure) generally differs from those in R (Y'_j and Y'_{j+1}). When Y_i straddles Y'_j and Y'_{j+1} in the ratio $r : 1 - r$, we apply

$$F_i = r F'_j + (1 - r) F'_{j+1} \ (R \to S), \quad F'_j = (1 - r) F_{i-1} + r F_i \ (S \to R). \quad (7)$$

We use the former to obtain the distribution function F_i in Y_i from F'_j and F'_{j+1} for molecules moving $R \to S$ with velocity $\xi_{\ell,k}$, which belong to the grid points in S), and the latter vise versa. Since we use a moving frame in R, $\xi_{\ell,k}$ in S maps to $\xi_{\ell,k} - v_R$ in R, and it does not fall onto the grid points $\xi_{\ell,k}$ in R as shown in Fig. 2c. Thus we need an interpolation process: (i) we approximate F at all radii in the direction of $\xi_{\ell,k} - v_R$ [i.e., the points $+$ in Fig. 2c], and (ii) interpolate the target value at $\xi_{\ell,k} - v_R$ from these points.

The pair of $\left(Y_i, Y'_j\right)$ along with the ratio r vary with time t. It is easy to calculate them if the corresponding cells are in touch as indicated (i) in Fig. 2a. In the case of (ii) there, the cell runs out of the range of the cells of the other region. We can find the corresponding cells even in these cases since we assume the periodicity in the X_1 direction. Since we choose the time step Δ_t evenly divides the wing time-period

$$T_C = D/v_R, \quad (8)$$

we only have to prepare T_C/Δ_t patterns of $\left(Y_i, Y'_j, r\right)$.

5 Results and Discussion

Firstly, we present the time evolution of the BGK solution at Knudsen number $Kn = \ell_0/D = 1$ ($\ell_0 = 2c_0/\sqrt{\pi} A_C n_0$: reference mean free path, $c_0 = \sqrt{2\kappa T_0/m}$: most probable speed), wing speed $v_R = 0.5c_0$, the wing angle $\theta = \theta_R = \theta_S = \pi/4$, and $\sqrt{2}H = D$. Figure 3a shows the time variation of the pump flow speed $-v_2$ of the gas. The black line plots the average $\bar{v}_2(t)$ in the gas defined by

$$\bar{h}(t) = \int_{gas} h(X_1, X_2, t) \, dX_1 dX_2 / \int_{gas} dX_1 dX_2 \quad (9)$$

versus the non-dimensional time t/t_0 ($t_0 = D/c_0$). The gas starts to move in the $-X_2$ direction right after the motion of the rotor at $t = 0$, then, $\bar{v}_2$ approaches a time-periodic state synchronous to the wing time-period T_C in a few periods. We

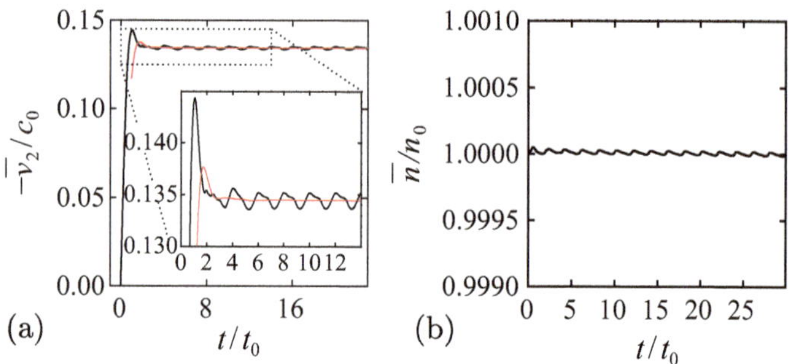

Fig. 3 Time evolution of the average **a** pump flow speed $\bar{v}_2\,(t)$ (black line) and **b** number density $\bar{n}/n_0$. The red line in **a** is the moving average $[\overline{v_2}]\,(t)$

confirm the approach to the quasi-steady state by introducing the moving average

$$\left[\overline{h}\right](t) = \frac{1}{T_C} \int_{t-T_C}^{t} \overline{h}\,(\tau)\, \mathrm{d}\tau. \tag{10}$$

The behavior of $-\left[\overline{v_2}\right](t)$ [red line in Fig. 3a] indicates that several time-periods are sufficient to form the quasi-steady state. From the viewpoint of the numerical procedure, the approach to a steady state can be disturbed by the lack of mass conservation. However, this is not the case for the present result due to the superior conservative feature of the finite volume method; Fig. 3b shows the time variation of the average number density $\overline{n}$, which must be theoretically constant in this closed system. The value of $\overline{n}$ remains close to n_0, although it shows slight vibration resulting from calculating the time-integral of $J\,[F]$.

Figure 4 shows the time-periodic state's pressure (color) and flow velocity (arrows) distributions. The flow velocity distribution relative to the blade position remains almost unchanged despite the motion of the blades. The zero thickness of the blades and the large mean free path by $\mathrm{Kn} = 1$ can be the reason for the stable feature. Slight variation is discernible for the pressure field, but $\partial p/\partial X_1$ remains negative in R and positive in S throughout the period. Its size increases as the wing angle θ, as shown in Fig. 5a, which shows the distribution of p and v for $\theta = \pi/6$ and $\pi/3$. (The

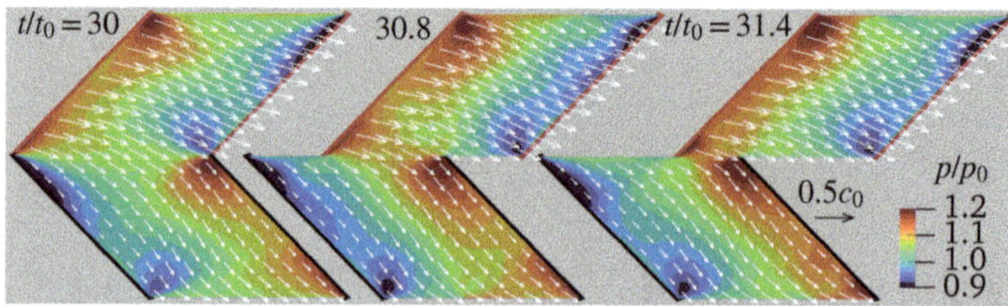

Fig. 4 Flow velocity and pressure field in the time-periodic state. The case for Fig. 3. $p_0 = \kappa n_0 T_0$

value of p for $\theta = \pi/3$ ranges in $0.8 \lesssim p/p_0 \lesssim 1.4$ and exceeds the color range of the figure). The flow speed $- [\overline{v_2}]$ at the quasi-steady state increases as θ, and takes its maximum value around $\theta \sim \pi/3$ (Fig. 5b).

The nonuniform pressure distribution results from the rotor's motion; the molecules entering R from S are more likely to collide with the wall at $-X_1$ side than at $+X_1$ side since walls in R move in the X_1 direction. Therefore, molecules accumulate near the wall at $-X_1$ side. This is the reason for the negative $\partial p/\partial X_1$ in R. On the contrary, the molecules entering S from R mainly collide with the $+X_1$ wall since the wall moves in the $-X_1$ direction relative to R. The positive $\partial p/\partial X_1$ in S shows it. The difference in the number of molecules colliding with the wall between two sides explains why TMP works. As shown in Fig. 6a, the wall surfaces with many molecules invariably face downward (i.e., $-X_2$ direction), and those with lesser molecules always face upward ($+X_2$ direction). As a result, the molecules move downward on average since diffusely reflected molecules must have a velocity normal to the wall, and $\cos\theta$ of the normal is its X_2 component.

The above explanation of the mechanism of TMP is more straightforward than those in the original work in [2]. It can explain the pressure distribution between the blades, giving the reason for the peak θ of the flow speed $- [\overline{v_2}]$ since $\cos\theta \to 0$ as $\theta \to \pi/2$. It also clarifies that the gas flow is a linear phenomenon despite the v_R of the commercial products being close to c_0. Our additional results (omitted here due to the limited space) show that $[\overline{v_2}] \propto v_R$ ($0 < v_R \lesssim c_0$), as the early work [4]

(a)   (b)

Fig. 5 Effect of the angle $\theta = \theta_R = \theta_S$ for $2H = D$, Kn $= 1$, and $v_R/c_0 = 0.5$,**a** te pressure and flow velocity distribution at $t/t_0 = 30.8$ for $\theta = \pi/6$ and $\pi/3$, **b** average flow speed $-v_2'$ at time periodic state versus the angle of the wing $\theta = \theta_R = \theta_S$

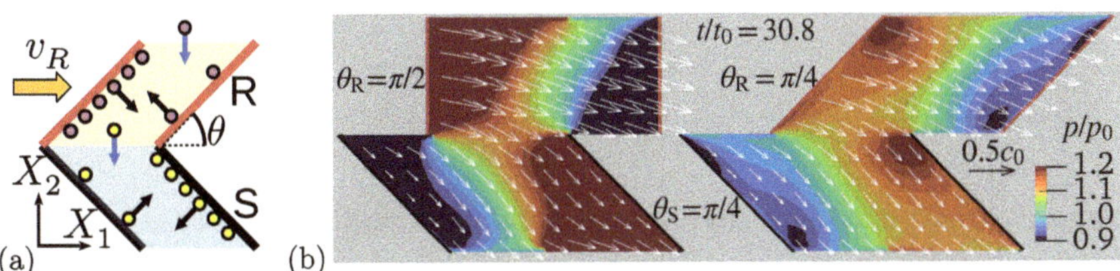

(a)   (b)

Fig. 6 a Driving mechanism of TMP, **b** alternative design of TMP. The distribution of p and v for $2H = D$, Kn $= 1$, and $v_R/c_0 = 0.5$, and $\theta_S = \pi/4$. Left is the case for $\theta_R = \pi/2$ and the right $\theta_R = \theta_S$

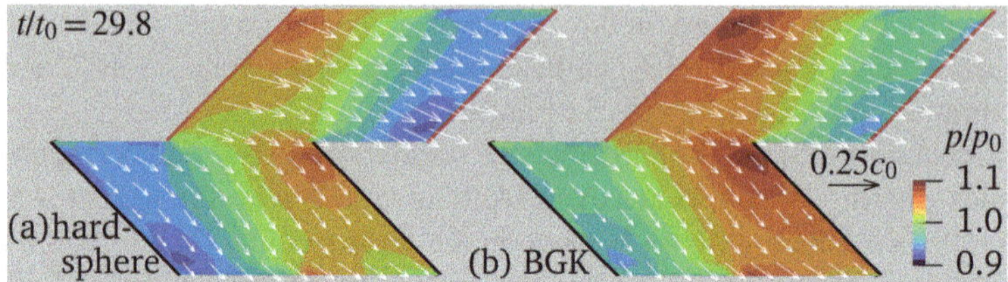

Fig. 7 Comparison between **a** the hard-sphere model and **b** the BGK model. $\theta = \pi/4$, $2H = D$, Kn $= 0.2$, and $v_R/c_0 = 0.25$

reports. Instead, we provide a result for a slightly different design of TMP in Fig. 6b, where the cases for $\theta_R = \pi/2$ and $\pi/4$ are compared. The standing plate of $\theta_R = \pi/2$ induces a greater pressure non-uniformity compared with the standard case $\theta_R = \theta_S$. The significant difference in the pressure (i.e., the number of molecules on the wall surface) results in a slightly stronger flow in the $-X_2$ direction. This result implies that the motion of the wall in the rotor is essential, but the wing-like shape is not.

Finally, we consider the dependency on the collision model. Figure 7 compares the results with J_{HS} in (a) and J_{BGK} in (b). The calculation of J_{HS} is heavy even with the spherical design (6) for unsteady 2D flows. Hence, we use coarser grids in both cases in Fig. 7. That is, each of R and S is divided into 10×5 finite volume cells, time step $\Delta_t/t_0 = 0.05$, and $N_\xi \times N_\Omega = 48 \times 94$, while these values are, respectively, 20×10 cells, time step 0.005, and $N_\xi \times N_\Omega = 48 \times 192$ in Figs 5a and 6b. The two models share the same qualitative feature of v and p distributions. The pressure variation for BGK is slightly larger than that for HS, while the flow speed $[\overline{v_2}]$ differs by only 1.2 %. We do not determine whether the reason is the difference in the collision term or the difference in the mean free path (cf. $\ell_0 = 1/\sqrt{2}\pi d_m^2 n_0$ for HS) at present since our additional calculation shows that $[\overline{v_2}]$ is insensitive to Kn. Reference [4] supports the insensitivity to Kn for the present problem. However, it further reports that the pressure ratio obtained by blocking the flow varies much for small Kn. This implies a possibility of a more significant impact on collision models in pressure-related problems, which should be addressed in future studies. We performed present computations by a computer with a 32-core CPU, and the simulation for period T_C requires 10d for HS and 4 min for BGK.

6 Conclusion

We have visualized the unsteady behavior of the gas in the 2D model of TMP using the deterministic solver of the Boltzmann and BGK equations. The behavior of the gas obtained leads to a simple explanation of the driving mechanism of TMP, which helps explain various characteristics of TMP. Furthermore, the new explanation derives an alternative design of the pump. The qualitative feature of the gas flow is insensitive to the collision model, although the present results are limited to the case of open flow of the pump device. The effect of the pressure load for various Kn, θ_*, … will be reported in the forthcoming paper.

Acknowledgements This work is supported by JSPS grant-in-aid 24K07303.

References

1. Gaede W (1913) Die molekularluftpumpe. Annalen der Physik 346(7):337–380
2. Becker W (1966) The turbomolecular pump, its design, operation and theory; calculation of the pumping speed for various gases and their dependence on the forepump. Vacuum 16(11):625–632
3. Bird GA (2011) Effect of inlet guide vanes and sharp blades on the performance of a turbomolecular pump. J Vac Sci Technol A 29(1)
4. Sharipov F (2010) Numerical simulation of turbomolecular pump over a wide range of gas rarefaction. J Vac Sci Technol 28(6):1312–1315
5. Imazu R, Sugimoto H, Itou M (n.d.) Numerical study on the one-way flow between plates of different temperatures with periodic accommodation distribution. In: Proceedings of the 33rd International symposium on rarefied gas dynamics. Springer Nature, Cham (Submitted in this issue)
6. Womersley RS (2018) Efficient spherical designs with good geometric properties. Springer International Publishing, Cham, pp 1243–1285. https://doi.org/10.1007/978-3-319-72456-0_57
7. LeVeque RJ (2002) Finite volume methods for hyperbolic problems. In: Cambridge Texts in Applied Mathematics. Cambridge University Press, Cambridge

Micro-Tapered-Tubes Flow Analysis via DSMC and Experimental Methods

Christos Tantos, Foteini Litovoli, Franz Schweizer, Lucien Baldas, Juergen J. Brandner, Jan Gerrit Korvink, Thomas Giegerich, and Marcos Rojas-Cárdenas

Abstract A numerical and experimental analysis was performed on pressure-driven flows through micro-tapered tubes, which were fabricated using Two-Photon Polymerization additive manufacturing. Special emphasis was placed on the crucial role of uncertainties associated with the geometric size characterization. The constant volume technique was employed for experimental characterization, and nitrogen was used as the working gas. The DSMC method was utilized for the numerical part. The mass flow rate results show a good agreement between experiments and DSMC, within the numerical and experimental uncertainties. This study highlights the importance of DSMC as an effective tool for flow characterization in 3D microscale structures and of the TPP as a novel manufacturing technique for complex microfluidic devices working with rarefied gas flows.

Keywords Microfluidics · Tapered microchannels · DSMC · Two-photon-polymerization manufacturing

1 Introduction

The correct characterization and modeling of gas flows in microdevices enables the further development of micro-electro-mechanical systems (MEMS). As an extension of constant cross-section tubes and channels, tapered flow configurations are of

C. Tantos (✉) · F. Litovoli · T. Giegerich
Institute for Technical Physics (ITEP), Karlsruhe Institute of Technology (KIT), Karlsruhe, Germany
e-mail: christos.tantos@kit.edu

F. Schweizer · J. J. Brandner · J. G. Korvink
Institute of Microstructure Technology (IMT), Karlsruhe Institute of Technology (KIT), Karlsruhe, Germany

F. Schweizer · L. Baldas · M. Rojas-Cárdenas
Institut Clément Ader (ICA), Université de Toulouse, CNRS, INSA Toulouse, ISAE-SUPAERO, IMT Mines Albi, UT3, Toulouse, France

© The Author(s) 2026

M. Grabe et al. (eds.), *Rarefied Gas Dynamics*, Springer Aerospace Technology,
https://doi.org/10.1007/978-3-032-00094-1_23

increasing interest, because of their potential applications in various technological fields, including aerospace [1] and vacuum technologies [2].

Over the past decade, gas flow through tapered channels has become an increasingly attractive area of study for modelers in the rarefied gas dynamics community. These types of flow configurations have been studied throughout a wide range of gas rarefaction using mainly two different numerical approaches. The first approach, originally proposed in [3] for long tubes of variable radius, is based on mass conservation and its main advantage is that, as a 1D method, it allows fast calculations of the mass flow rate and pressure distribution along the flow direction. Due to its simplicity, this methodology has also been applied to analyze single gas flows through rectangular channels with variable cross-section [4]. Its effectiveness is well-established for channels with large length-to-radius ratios and relatively low variations of radius along the axis of the channel [5, 6]. The second approach is more computationally demanding and involves the complete solution of the flow field within the tapered channel using either Direct Simulation Monte Carlo (DSMC) or Discrete Velocity Method (DVM). Several studies in the literature have used the DSMC or DVM methods to investigate flows through diverging or converging channels, formed by parallel plates placed at a relative angle to each other [6–8]. To the best of our knowledge, most experimental studies on tapered micro-channels focus on planar geometries [1], with limited modeling of tapered tubes [5].

The scope of the present work is to study the computed and measured mass flow rates in pressure-driven flows through micro-tapered tubes manufactured via Two-photon-polymerization additive fabrication technique. Specific attention was given to the analysis of the crucial role of geometric size characterization. The paper is organized as follows. In Sect. 2, the experimental setup and the fabrication of the tapered channel are described. In Sect. 3, the simulated flow configuration and the numerical results are analyzed. In Sect. 4, the experimental and numerical results are presented and discussed, and in Sect. 5 the main conclusions are summarized.

2 Experimental Setup Methodology (Fabrication and the Characterization of the Tapered Channel)

The tapered channel structure was fabricated using Two-Photon-Polymerization (TPP), which is an additive manufacturing technique based on the local polymerization of a liquid photosensitive resin. This fabrication technique was previously utilized for the fabrication of straight microtubes for rarefied gas flows applications [9]. The working principle of the manufacturing technique is illustrated in Fig. 1a. After a droplet of resin (IP-Q, Nanoscribe GmbH) is deposited on a substrate, a femtosecond pulsed laser is focused and directed through the resin with controlled trajectory. The resin solidifies when the threshold of energy intensity is reached. This is only the case in the focus point of the laser, thus generating ellipsoidal voxels. The structure was fabricated by overlapping these ellipsoidal voxels with a slicing distance of 5

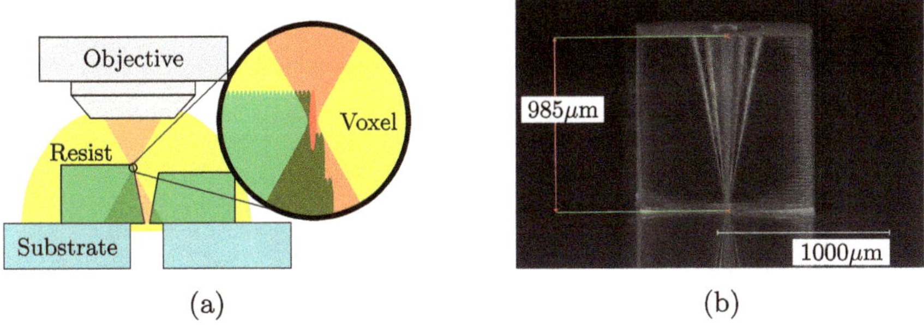

Fig. 1 **a** Schematic of the TPP process; **b** microscope image of the tapered channel

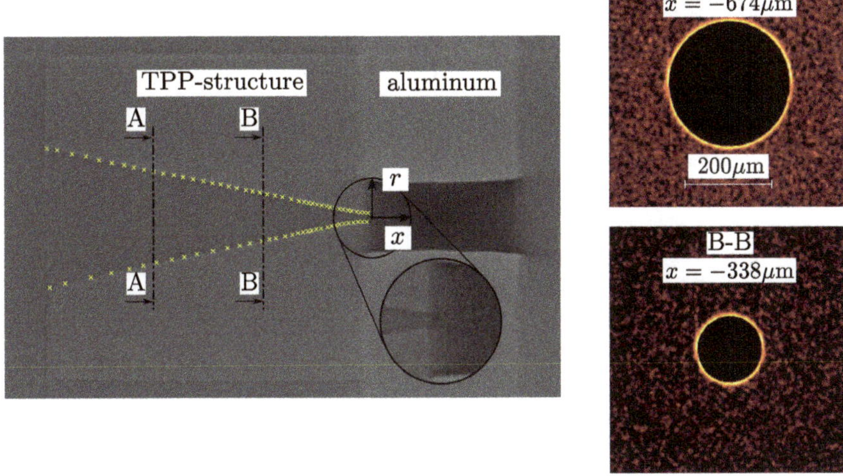

Fig. 2 Tomography image with radius measurement points and cross-sections. Not all measurement points are shown to avoid the overlapping of measurement markers

μm and a lateral hatching distance of 1 μm. After the printing process the sample was submerged in propylene glycol methylether acetate (PGMEA) for a total of 14 h to dissolve the uncured resin. Lastly the structure was rinsed in a solvent and left to dry at 7 °C for 30 min. The fabrication process therefore took approximately 15.5 h, including a printing time of 1 h.

As visible in Fig. 1b the tapered channel was written inside a polymer cube of roughly 1 mm × 1 mm × 1 mm. The structure was deposited on an aluminum substrate of the dimensions 25 mm × 25 mm × 0.5 mm. Furthermore the aluminum substrate encompasses a circular orifice of 200 μm in diameter. This orifice is aligned with the channel to enable the connection of the nozzle of the channel to the experimental setup (Fig. 2). The radius of the channel varies linearly from 215–14 μm over a length of 1012 μm. A high aspect ratio microtube with a circular cross-section,

Table 1 **a** Coefficient values of the polynomials for the radius function $R(x)$; **b** experimental campaign initial conditions

(a)

a_0	13.6 μm
a_1	0.01597
a_2	2.57822e-3 μm^{-1}
a_3	1.868179e-5 μm^{-2}
a_4	9.156089e-8 μm^{-3}
a_5	3.0165235e-10 μm^{-4}
a_6	6.5164167e-13 μm^{-5}
a_7	8.9978361e-16 μm^{-6}
a_8	7.6065986e-19 μm^{-7}
a_9	3.5761114e-22 μm^{-8}
a_{10}	7.150218e-26 μm^{-9}
b_0	111.1 μm
b_1	-6.83e-2
b_2	2.09e-4 μm^{-1}
b_3	-7.9e-08 μm^{-2}

(b)

Case	Converging P_A [Pa]	Converging P_B [Pa]	Diverging P_A [Pa]	Diverging P_B [Pa]
1	210.0	10.0	10.0	206.4
2	303.3	103.2	104.3	304.3
3	409.0	209.0	205.1	401.4
4	505.0	209.0	305.3	505.3
5	604.0	404.0	405.5	599.6
6	705.1	505.1	504.8	700.4
7	800.2	600.2	602.4	802.4
8	902.2	702.2	703.6	901.0
9	1001.7	802.1	802.9	999.7
10	1104.2	904.2	902.6	1097.1
11	1202.4	1002.4	1001.6	1197.8
12	1298.9	1099.1	1101.0	1297.1

whose size varies, is thus obtained. Previous experimental investigations of tapered microchannels have mostly relied on standard planar micro-machining of silicon [1] or micro-milling of metals [10]. Owing to the limitations of these methods only rectangular cross-sections were investigated.

The geometry of the present channel was characterized using a EasyTom XL tomograph (RX Solutions). The resulting tomograms showed an isotropic pixel size of 2.77 μm and were analyzed to extract the internal geometry of the channel. The diameter variation along the length of the channel was measured on the symmetry plane of the channel (Fig. 2, left). The geometry measurement was performed via an edge detection algorithm. Since the nozzle geometry has a critical influence on the flow field, a higher number of measurements were obtained for this region. Furthermore, as it can be seen in Fig. 2 (right), the circularity of the tapered channel cross-section is of excellent quality. To smooth the measurement results, two polynomials were used: the $10th$ order polynomial for the tapered channel section ($R(x) = \sum_{i=0}^{10} a_i x^i$, -1012 μm $\leq x \leq 0$) and a $3rd$ order polynomial for the section inside of the aluminum substrate ($R(x) = \sum_{j=0}^{3} b_j x^j$, $0 \leq x \leq 458$ μm). The coefficients of the polynomials are given in Table 1a.

The flow properties of the channel structure were experimentally investigated utilizing the constant volume technique. This very well established technique relates pressure variation to mass variation in a constant volume [9, 11, 12]. The experimental methodology consisted in setting an initial pressure difference between inlet and outlet of the tapered channel and then allowing the thermodynamic system to relax to equilibrium. The mass flow rate generated through the channel creates a pressure

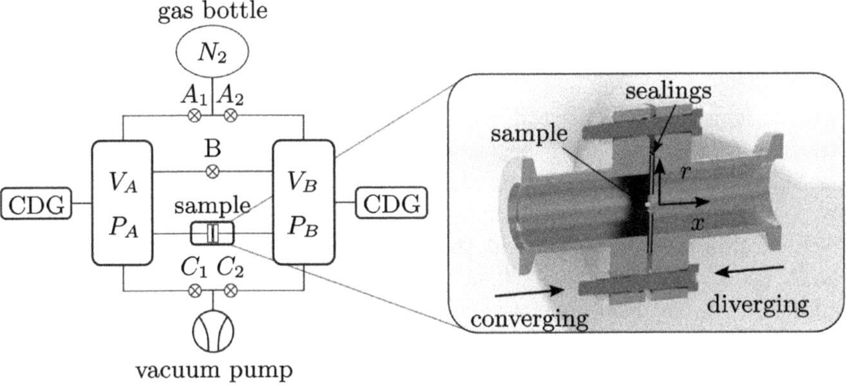

Fig. 3 Constant volume technique experimental setup and sample clamping technique

variation in the inlet and outlet tanks. The initial pressure conditions are specificied in Table 1b. Figure 3 shows a schematic of the used flow setup. The substrate of the flow sample was clamped in a casing using rubber seals. Furthermore the sample was connected to the two volumes V_A and V_B where the pressures P_A and P_B were acquired by Capacitance Diaphragm Gauges (CDG) with full ranges of 1333 Pa and an accuracy of 0.2% of reading. The average pressure and applied pressure difference in the setup could be controlled through the valves $A_{1,2}$ and $C_{1,2}$ which were connected to a pressurized nitrogen bottle and a vacuum pump. This setup allowed initial conditions $P_A > P_B$ for converging flow experiments and $P_A < P_B$ for diverging flow experiments.

The mass flow rate from the volumes and therefore through the sample can be calculated using the ideal gas law assuming isothermal conditions in the volumes $V_{A,B}$ as

$$\dot{m}_{A,B} = \frac{V_{A,B}}{R_g T_0} \frac{\mathrm{d}P_{A,B}}{\mathrm{d}t}, \tag{1}$$

with R_g and T_0 being the specific gas constant and the temperature in the experimental room. The uncertainty of the mass flow rate measurement was estimated to be roughly 3% following the propagation of uncertainty principle taking into account the uncertainty of the volume $V_{A,B}$, the temperature T_0 and pressure $P_{A,B}$ measurements. A more detailed description of the uncertainty estimation is outlined in [11]. The experiments were carried out in a temperature-controlled room ($T = 293.15\mathrm{K} \pm 0.5\mathrm{K}$), justifying the assumption of isothermal conditions in the volumes $V_{A,B}$. Details on the method can be found in [11, 12].

3 Numerical Modeling

The simulated flow setup includes the TPP structure (i.e., the tapered tube) and the substrate area (Fig. 4). The modeling domain corresponds to the physical domain extracted from tomography measurements. In the present model, the open borders (dashed lines) of the channel were maintained at different pressures (P_A and P_B). Two flow directions were considered: the converging flow with $P_A > P_B$ and the diverging flow with $P_A < P_B$, with the corresponding pressure ratio π defined as P_B/P_A and P_A/P_B respectively. In the modeling, the temperature of the solid walls and the gas at the open boundaries were maintained constant at $T_0 = 293.15$ K. Additionally, due to the axisymmetric nature of the flow, only a radial slice of the domain ($r > 0$) was considered. Thus, only two physical dimensions were taken into account, namely the $x-$ and $r-$ axis.

Based on the pressure values shown in Table 1b, the Knudsen number, calculated using the throat radius and the average pressure, ranges from 0.3 to 3.6. This suggests that the gas flow is in the transitional flow regime. In the present work, the Direct Simulation Monte Carlo method proposed by G. A. Bird [13] was applied to describe the flow behavior in the examined microstructure. Over the last few decades, it has evolved into a powerful and well-established numerical tool for studying phenomena over the whole range of the Knudsen number, making it well-suited for the flow problem addressed in this paper. For further details on the DSMC method the reader is referred to G. A. Bird's book [13]. In the present work, the effective collision scheme known as the No-Time-Counter (NTC) method, enhanced by the Variable Hard Sphere (VHS) model introduced by Bird, was used. The VHS model combines the simplicity of a hard-sphere scattering law with the flexibility of a variable collision cross-section, allowing for accurate representation of gas behavior under varying flow conditions. Based on the viscosity data for N_2 [14], the reference VHS diameter d_{ref} and the viscosity index ω for the examined flow conditions are taken as 4.11×10^{-10} m and 0.74 respectively. The gas-surface interaction was modeled based on the Diffuse-Specular model, in which a non-complete accommodation is assumed, while axisymmetric flow was also considered. By analyzing the available data in the literature on the accommodation coefficient α, it can be deduced that for typical technical surfaces without special treatment, a value between 0.9 and 1 is generally

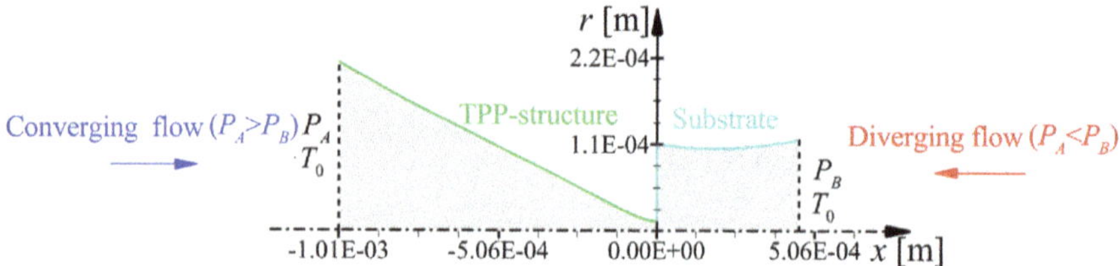

Fig. 4 Schematic view of the flow set-up considered in the present study. The gray area represents the computational domain

expected for N_2 [15, 16]. The computational grid consisted of 128,000 elements. The global cell size was $\Delta x = \Delta r = 1.34 \times 10^{-6}$ m, while in the region R = (0 < $r < 4 \times 10^{-5}$ m, -1.13×10^{-4} m $< x < 7.8 \times 10^{-5}$ m) the cell size was reduced three and six times in the $x-$ and $r-$ directions, respectively. It is noted that, cell size was chosen to be significantly smaller than the mean free path of the gas molecules. The number of DSMC particles varies from 33×10^6 to 40×10^6, with a minimum of at least 30 particles per cell. In order to maintain, as much as possible, uniform distributions of particles at each cell in the whole computational domain, the concept of the weighting factor widely applied in literature has been introduced. The time step Δt has been chosen to be sufficiently smaller (about 1/3) than the cell transversal time, defined as $\min(\Delta r, \Delta x)/u_0$ with u_0 being the most probable speed at reference temperature T_0 and Δr (or Δx) is the cell size in $r-$ (or $x-$) direction. The gas macroscopic quantities are obtained by time averaging over 10^7 time steps after the steady-state has been reached. In this type of pressure driven flows it is often essential to include two large reservoirs at the inlet and outlet of the channel to ensure the proper application of boundary conditions [5, 6, 8, 15]. For this specific setup, simulations using reservoirs up to eight times larger than the corresponding radius at the channel ends demonstrated no significant impact on the mass flow rate (within 1.5%). Consequently, the reservoirs are excluded from the configuration to save computational resources. The applied DSMC code has been successfully compared with results from kinetic models in [17].

4   Results and Discussion

In Fig. 5, a comparison between the DSMC data and the corresponding experimental ones is performed for both converging and diverging flow set-ups. The numerical and experimental data are in very good qualitative agreement for both configurations. As for the reference case, namely, $\alpha = 1$ and reference radius function $R(x)$ (see Table 1a), the comparison shows a consistent error of approximately 20%, which, while not negligible, is within an acceptable range given the geometrical fabrication complexity and the challenges associated with its experimental characterization. It is noted that, the DSMC results consistently underestimate the experimental results. To investigate and understand the sources of this deviation, we first examine the assumption of diffuse boundary conditions. For this purpose, we performed additional simulations assuming incomplete accommodation on the wall, with an accommodation coefficient $\alpha = 0.9$, which stands as the lowest expected value for the considered flow conditions. The corresponding results for $\alpha = 0.9$ are also shown in Fig. 5 for comparison purposes. As it is seen, a decrease in the accommodation coefficient of about 10% causes an increase in the mass flow rate about 10% bringing the DSMC data closer to the experimental ones. This can be considered the maximum uncertainty associated with the imposed pure diffuse boundary conditions.

Another important aspect that we investigated is the potential uncertainty related to the precise definition of the geometry, which remains challenging due to the small

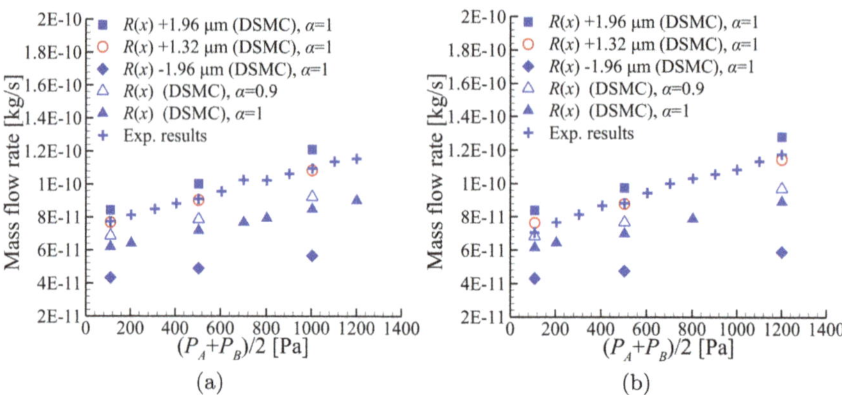

Fig. 5 Comparison of the mass flow rate between the DSMC and the experimental results for converging **a** and diverging **b** flow

scale of the channel. The uncertainty related to this measurement was estimated via a propagation of uncertainty principle. The radius measurement can be understood as a distance between two points. The position of either point has an uncertainty of $\pm dx/2$ (i.e. dx is the voxel size). Assuming both position measurements are not correlated, the uncertainty on the distance measurement is 1.96 μm corresponding to $\pm\sqrt{2}dx/2$. Additional simulations were performed by adding and subtracting 1.96 μm to the radius at each point in the flow direction, corresponding to the estimated uncertainty of the geometry measurement. As it is observed, the geometry uncertainties affect significantly the comparison between DSMC and experimental results. However, the experimental data are always within the two limits imposed by the DSMC simulations, with the experimental results showing better agreement with the larger radius geometry. Based on the above findings we can conclude that an acceptable agreement between the DSMC and experimental data is achieved, given the uncertainties associated with the geometry characterization and gas surface interaction modeling. Based on the mass flow rate data shown in Fig. 5, an attempt was made to estimate the potential radius increase required to obtain acceptable agreement between DSMC and experiments. Interpolation of the data showed that a uniform increase of 1.32 μm resulted in a very good agreement between the experimental and DSMC results with an average error over the entire pressure range of about 2%. This holds true for both examined flow configurations in a very consistent manner. For the sake of clarity and consistency, the corresponding results with $R(x)+1.32$ μm are also shown in Fig. 5. This level of agreement highlights the capability of such a kinetic modeling to complement experimental investigations, particularly in scenarios where the precise definition of the geometry is challenging due to the small scale of the channel.

Figure 6 shows indicative pressure and velocity contours for both converging and diverging flows for the case 1 of Table 1b. This case corresponds to the one with the largest inlet-to-outlet pressure ratio, namely $\pi \approx 0.05$. The velocity contours

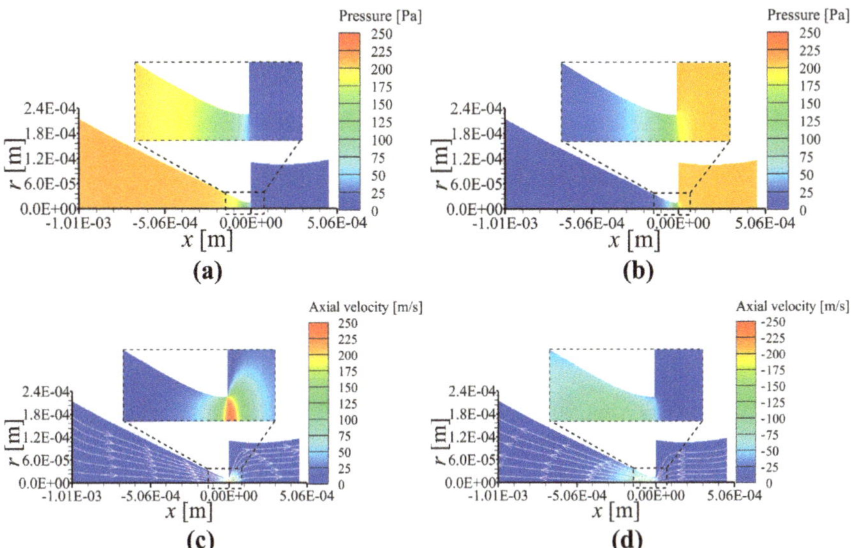

Fig. 6 Contours of pressure **a, b** and velocity **c, d** for converging **a, c** and diverging **b, d** flow, for case 1 ($\pi = 0.05$)

are overlaid by the velocity streamlines. The results show that for both pressure and axial velocity, the greatest variations are observed around the throat at $x = 0$. Similar qualitative behavior was observed for all the examined cases in Table 1b. It is observed that the radial pressure change is not very significant, whereas the radial velocity change is more pronounced, with the maximum values always occurring close to the axis of symmetry $r = 0$, a well-known behavior due to the wall friction effects.

In order to gain a better quantitative understanding of the changes in the macroscopic quantities, the axial distributions of velocity, temperature and pressure are shown in Fig. 7. Two cases are plotted, namely case 1 with $\pi \approx 0.05$ and 8 with $\pi \approx 0.8$, which correspond to high and low inlet-to-outlet pressure ratios, respectively. As can be seen, the velocity remains uniform and noticeably small far from the throat, while it shows strong variations in the throat region. By comparing the velocity profiles of cases 1 and 8, it can be easily deduced that the velocity is an increasing function of the pressure difference. The local Mach number always remains below 0.7 in converging flow direction and below 0.3 in diverging flow direction. In the converging direction, as the gas moves toward the throat, the decreasing area causes it to accelerate (due to the conservation of mass). In the diverging flow, however, the opposite occurs: the gas velocity reaches its maximum after the throat due to expansion effects, and then decreases due to the less resistance to the gas flow as a consequence of the increase in the area. As stated above, the gas flow accelerates as it passes through the narrow area at throat, which leads to an increase in kinetic energy, and part of this energy is extracted from the thermal energy, resulting in

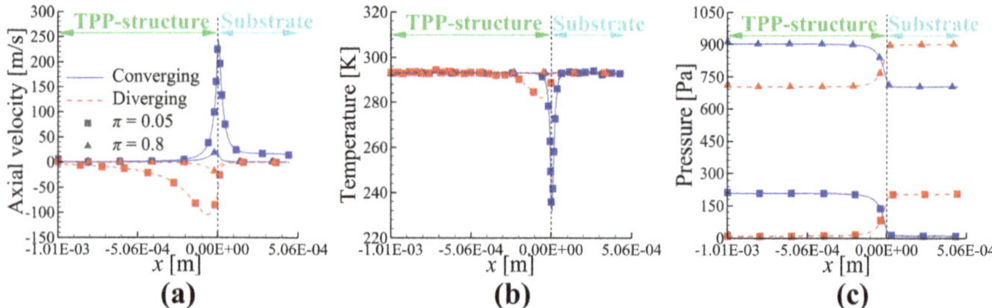

Fig. 7 Axial distributions of the axial velocity **a**, temperature **b** and pressure **c** at $r = 0$

a drop in temperature, as observed from the axial temperature distributions. It is noted that, as the inlet-to-outlet pressure drop decreases, the temperature of the gas flow approaches the wall temperature, and the flow shows isothermal characteristics. Finally, the axial pressure distributions show that the pressure remains uniform and close to its equilibrium values at the two entrances of the flow configuration, while a rapid change in pressure is observed near the throat, coinciding with the aforementioned increase in velocity.

5 Conclusions

This study presents, for the first time, a comparative analysis of experimental and numerical results for pressure-driven flows through micro-tapered tubes manufactured via two-photon-polymerization. The characterization of the flow is performed experimentally via the constant volume technique, as well as numerically via the DSMC method. A good comparison between the experimental and numerical data is observed, given all the uncertainties regarding the geometry characterization and the modeling assumptions which have been quantified in a detailed manner. This study shows that DSMC is a powerful tool for modeling rarefied gas flows in complex microstructures and for characterizing micro-device geometries in meso-scale configurations, where optical or X-ray measurements are challenging. In the future, we plan to extend both the experimental and numerical aspects of this analysis to more complex microstructures with varying cross-sections, relevant to aerospace and vacuum technologies.

Acknowledgements This work was carried out with the support of the Karlsruhe Nano Micro Facility (KNMFi, www.knmf.kit.edu), a Helmholtz Research Infrastructure at Karlsruhe Institute of Technology (KIT). The Research Federation FERMaT (FR 3089, Toulouse - France) is acknowledged for access to its Easy Tom XL X-ray tomograph. This work was supported by computational time granted from the National Infrastructures for Research and Technology S.A. (GRNET S.A.) in the National HPC facility - ARIS - under project ID pr015027-FUSFLOWSIM.

References

1. Alexeenko AA, Levin DA, Gimelshein SF, Collins RJ, Markelov GN (2002) Numerical simulation of high-temperature gas flows in a millimeter-scale thruster. J Thermophys Heat Transf 16. https://doi.org/10.2514/2.6667
2. López Quesada G, Tatsios G, Valougeorgis D, Rojas-Cárdenas M, Baldas L, Barrot C, Colin S (2020) Thermally driven pumps and diodes in multistage assemblies consisting of microchannels with converging, diverging and uniform rectangular cross sections. Microfluid Nanofluid 24:54 . https://doi.org/10.1007/s10404-020-02357-z
3. Sharipov F, Bertoldo G (2005) Rarefied gas flow through a long tube of variable radius. J Vac Sci Technol A 23:531–533. https://doi.org/10.1116/1.1897703
4. Graur I, Ho MT (2014) Rarefied gas flow through a long rectangular channel of variable cross section. Vacuum 101:328–332. https://doi.org/10.1016/j.vacuum.2013.07.047
5. Titarev VA, Shakhov EM, Utyuzhnikov SV (2014) Rarefied gas flow through a diverging conical pipe into vacuum. Vacuum 101:10–17. https://doi.org/10.1016/j.vacuum.2013.07.030
6. Tantos C, Litovoli F, Teichmann T, Sarris I, Day C (2024) Numerical study of rarefied gas flow in diverging channels of finite length at various pressure ratios. Fluids 9:78. https://doi.org/10.3390/fluids9030078
7. Ebrahimi A, Roohi E (2017) DSMC investigation of rarefied gas flow through diverging micro/nanochannels. Microfluid Nanofluid 21:18. https://doi.org/10.1007/s10404-017-1855-1
8. Sazhin O, Sazhin A (2023) Rarefied gas flow into vacuum through linearly diverging and converging channels. Int J Heat Mass Transf 203:123842. https://doi.org/10.1016/j.ijheatmasstransfer.2022.123842
9. Zhang D, López Quesada G, Bergdolt S, Hengsbach S, Bade K, Colin S, Rojas-Cárdenas M (2023) 3D micro-structures for rarefied gas flow applications manufactured via two-photon-polymerization. Vacuum 211:11915. https://doi.org/10.1016/j.vacuum.2023.111915
10. Graur I, Veltzke T, Méolans JG, Ho MT, Thöming J (2015) The gas flow diode effect: theoretical and experimental analysis of moderately rarefied gas flows through a microchannel with varying cross section. Microfluid Nanofluid 18:391–402. https://doi.org/10.1007/s10404-014-1445-4
11. Silva E, Deschamps CJ, Rojas-Cárdenas M, Barrot-Lattes C, Baldas L, Colin S (2018) A time-dependent method for the measurement of mass flow rate of gases in microchannels. Int J Heat Mass Transfer 120:422–434. https://doi.org/10.1016/j.ijheatmasstransfer.2017.11.147
12. Rojas-Cárdenas M, Silva E, Ho MT, Deschamps CJ, Graur I (2017) Time-dependent methodology for non-stationary mass flow rate measurements in a long micro-tube. Microfluid Nanofluid 21:86. https://doi.org/10.1007/s10404-017-1920-9
13. Bird GA (1994) Molecular gas dynamics and the direct simulation of gas flows. Clarendon Press, Oxford, Oxford Engineering Science Series
14. Boushehri A, Bzowski J, Kestin J, Mason EA (1987) Equilibrium and transport properties of eleven polyatomic gases at low density. J Phys Chem Ref Data 16:445-466. https://doi.org/10.1063/1.555800
15. Sharipov F (2016) Rarefied gas dynamics. Wiley-VCH Verlag GmbH & Co, KGaA, Weinheim, Germany
16. Yousefi-Nasab S, Safdari J, Karimi-Sabet J, Norouzi A, Amini E (2019) Determination of momentum accommodation coefficients and velocity distribution function for Noble gas-polymeric surface interactions using molecular dynamics simulation. Appl. Surf. Sci. 493:766–778. https://doi.org/10.1016/j.apsusc.2019.07.033
17. Tantos C, Varoutis S, Day C (2020) Deterministic and stochastic modeling of rarefied gas flows in fusion particle exhaust systems. J Vac Sci Technol B 38:064201. https://doi.org/10.1116/6.0000491

Analysis of Propellant Gas Suppression in Shattered Pellet Injector System of the ITER Experiment

Ákos Gyenge, Richárd László Csiszár, and Balázs Farkas

Abstract Plasma disruptions in tokamak fusion reactors like ITER threaten reactor equipment. To mitigate these events, a Shattered Pellet Injector (SPI) system using cryogenic solid pellets propelled by high-pressure gas was developed at the HUN-REN Centre for Energy Research. A suppressor system, consisting of multiple chambers and separators, was designed to capture the expelled gas. Computational Fluid Dynamics (CFD) simulations and experimental pressure measurements revealed that while CFD accurately predicts pressure peaks in the suppressor compartments, it struggles with gas expansion velocity. Experimental tests show gaps and leaks significantly influence pressure evolution. Findings indicate the suppressor retains some propellant gas but loses efficiency without a pellet. These insights help to develop disruption mitigation strategies for ITER.

Keywords Vacuum-expansion · CFD · Pressure measurement · Fusion technology

1 Introduction

In fusion reactors, such as the International Thermonuclear Experimental Reactor (ITER), plasma disruptions can occur when an instability grows to the point where the stored thermal and magnetic energy is rapidly discharged [3]. A shattered pellet injector (SPI) is planned to mitigate these disruptions and prevent them from damaging the device's plasma-facing wall. With this technique, a 28.5 mm-diameter, cylindrical cryogenic solid pellet is accelerated by a gas pulse and shattered on a plate near the plasma. The resulting fragments dilute and cool the plasma. An SPI injector

Á. Gyenge · R. L. Csiszár
HUN-REN Centre for Energy Research, Institute for Atomic Energy Research, Budapest, Hungary

Á. Gyenge (✉) · B. Farkas
Faculty of Mechanical Engineering, Department of Fluid Mechanics, Budapest University of Technology and Economics, Budapest, Hungary
e-mail: gyenge.akos@ek.hun-ren.hu

M. Grabe et al. (eds.), *Rarefied Gas Dynamics*, Springer Aerospace Technology,
https://doi.org/10.1007/978-3-032-00094-1_24

was developed at the HUN-REN Centre for Energy Research in Budapest (CER) [7]. The solid protium, deuterium or noble gas pellets are propelled by high-pressure gas, which has a higher velocity than the pellet. For magnetohydrodynamic stability, it is disadvantageous if the gas reaches the plasma first. A gas suppressor was designed for the SPI system to prevent the propulsion gas from entering the reactor. To the authors' knowledge, this is the first operational evacuated gas suppressor for such cryogenic pellet injectors. The first half of the SPI test system is depicted in Fig. 1.

During operation, the suppressor volume is evacuated below 0.01 mbar absolute pressure before firing. Due to the low pressure, the gas in the suppressor system is initially considered slightly rarefied (Knudsen number with respect to pellet diameter is around 0.4). In the high-pressure zone, the Knudsen number is significantly lower; therefore, in the expanding gas, the continuum assumption holds. To investigate the developing flow field within the gas retention system and predict the amount of the propellant gas which could potentially reach the plasma, a series of CFD simulations were conducted. This research has two main objectives: first, to present an experimental study of a novel gas-retention system intended for use in fusion reactors; and second, to evaluate the suitability of commercial CFD tools for modelling the expansion of high-pressure gas into a slightly rarefied environment.

1.1 Eddy-Current Actuated Fast Valve

An eddy-current actuated fast valve was designed and built at EK-CER for with two adjustable launch parameters: the initial pressure (up to 100 bar) and the opening voltage (up to 1200 V) [4, 5]. The valve has view ports, where the piston movement is visually observed. The displacement is used as a boundary condition for CFD simulations.

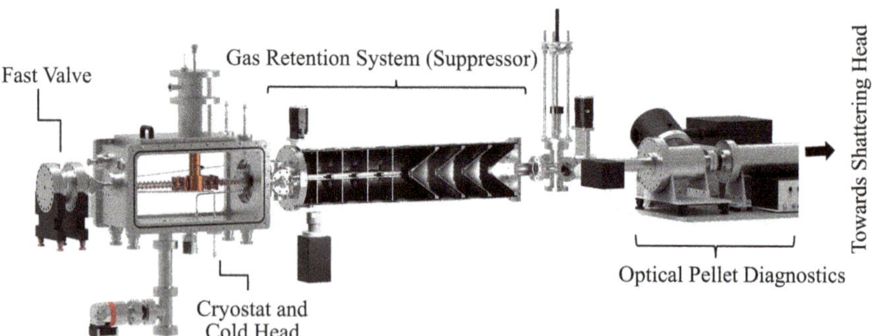

Fig. 1 First half of the Shattered Pellet Injector (SPI) system

1.2   The Suppressor System

In the CER SPI support laboratory, A suppressor model was built and instrumented with pressure sensors and cameras to validate the simulations experimentally. The suppressor is comprised of eight chambers, separated by flat and conical walls for optimal propellant recovery. The first half of the suppressor contains a guiding tube with rectangular cutouts for the gas to escape. The tubular duct serves two purposes: it guides the pellet, and holds the compartment separation plates, which are welded to it. Consequently, a small gap exists between the separator plate edge and the suppressor's inner cylindrical surface. The conical separators are held in place from the outside, allowing an airtight separation at the sides of the chambers.

Each chamber is equipped with a port for Keller M5-HB 0-3 bar fast pressure sensor. Only three pressure sensors are used at a time due to limitations in the data acquisition system. Sensors 1 and 9 are located on the base surfaces, 100 mm from the axis, while sensors 2–8 are placed on the cylindrical surface at 136, 266, 396, 526, 766, 906 and 1046 mm from the base, placing one sensor in each compartment. A close-up view of the suppressor and the pressure ports is presented in Fig. 2.

2   Simulations and Measurements on the Suppressor

2.1   CFD Simulations

A series of simulations was conducted on the SPI system with and without a pellet in ANSYS Fluent software. Although the presence of a slightly rarefied region would typically invoke the use of non-continuum modelling techniques, we chose to use this commercially available CFD solver for multiple reasons. Fluent is the designated tool for fluid dynamics simulations in the ITER project, making it important to assess its suitability for the present study. Secondly, the rarefied regime occurs only locally, just ahead of the expansion front, while the majority of the gas remains in the continuum

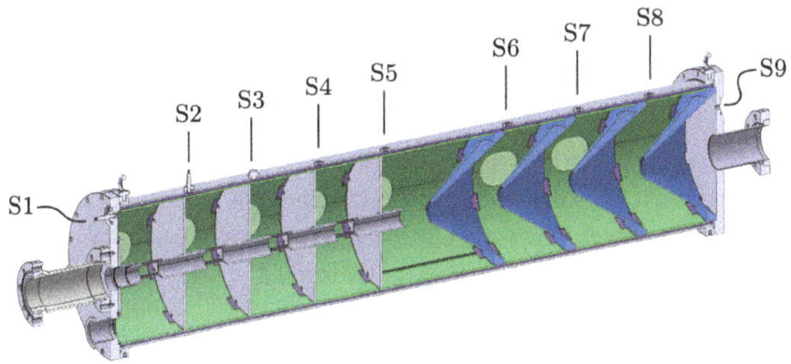

Fig. 2 Cross-section view of the suppressor and sensor locations

regime. If the expansion can be modelled with sufficient accuracy, Fluent can serve as a valid tool for simulating gas retention.

The system is modelled with a 2D axisymmetric geometry. The simulations contain two overset-mesh regions for modelling the valve head and pellet movement [6]. The domain consists of approximately 300 000 cells, all quadrilateral. An initial gap of 65 μ m is introduced at the valve head to maintain numerical stability and preserve a single computational domain. The mesh is refined in both the overset regions and the boundary layer. The simulations use laminar viscosity model. A density-based solver is utilized with implicit time stepping and AUSM+ convective flux, offering advantages in handling compressible flow and shock waves [2]. A time step of 100 ns is used. The zero-time step in the simulations corresponds to the moment when the valve head starts to move.

Due to the continuum assumption's failure at very low operating pressures, the simulations would not converge with the actual background pressure. We set the operating pressure to 1 mbar to overcome this issue. We conducted experiments both at 1 mbar and 1×10^{-3} mbar back pressure, and the pressure evolutions showed minimal deviation, therefore we consider the results of the CFD model applicable for comparison with high-vacuum cases as well.

The test shots were performed in multiple configurations, with or without pellets. For some of the shots, the suppressor internals (flat and conical deflector plates) were removed. The simulations reflected these setups. We conducted the mesh independence analysis using the method from [1]. The suppressor's performance was characterized by the gas retained at pellet exit (approximately 5 ms after launch), used as the representative variable. The analysis employed a fine grid (refinement factor 1.86) and a coarse grid (coarsening factor 1.43). The grid convergence index reached 29% and the total error relative to the fine and medium meshes was 10%. Therefore, we consider the results mesh-independent.

Figure 3 show Mach number and pressure contours for a simulation without pellet initially, right after valve closure, at $t = 1.5$ ms, when the gas starts to fill the suppressor volume and after 5 ms.

2.2 Measurements

A measurement grid, shown in Table 1, was established to ...evaluate the suppressor's performance under different valve pressures and opening voltages. The table lists the ejected gas quantity, expressed in bar-litres, a unit commonly used in fusion technology, for shots without pellets. The ejected gas quantities are even higher when a pellet is present due to shock wave interactions. The incoming gas creates a shock wave upon impact, which travels backward. At high valve pressures and opening voltages, these pressure waves can push the valve head back, causing it to reopen briefly and release additional gas.

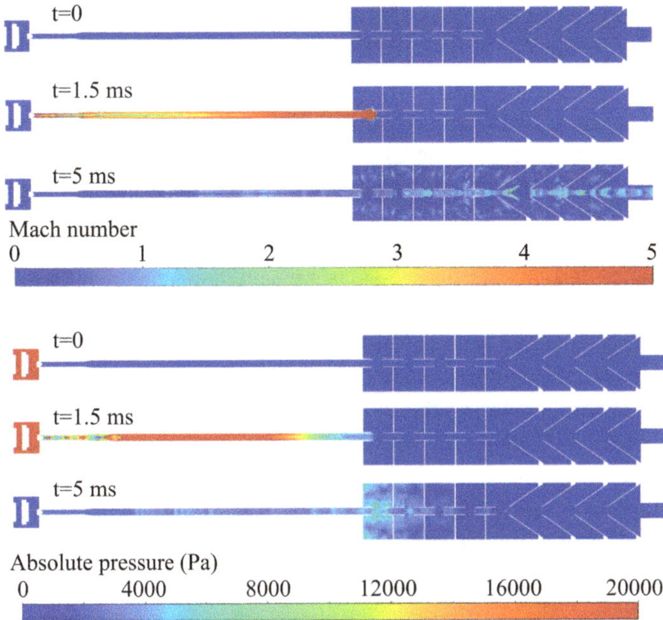

Fig. 3 Mach number and pressure contours initially, at valve closure (1.5 ms) and at 5 ms

Table 1 Test matrix

Valve pressure [bar]	Valve voltage [V]	Ejected gas [bar·litre]
100	900	3.2
100	1000	8.5
100	1100	19.5
80	900	4.8
80	1000	15
80	1100	31.5
60	900	8.7
60	1000	24.8
60	1100	21.8

As mentioned above, the suppressor internals are partially or fully removed for some launches. Only one pellet launch was successful with full suppressor internals, with shot number 1037. This launch was conducted with 80 bar valve pressure and 950 V opening voltage. The results from the whole test matrix are available for tests without pellet or suppressor internals, but those with the same launch parameters were selected for comparison.

3 Comparison of the Results

To compare CFD and measurements, pressure signals in the first, second and last compartments were used. Figure 4 shows the comparison of the aforementioned shot #1037 and the corresponding CFD simulation. In the latter, pressure rises later, especially in more downstream compartments, suggesting that the gas expansion velocity is not properly simulated. The rise and decay of the pressure in the case of the simulation is significantly slower than in the measurements. A probable explanation is the presence of gaps and leaks, which allow a larger mass flow rate between compartments in the actual setup leading to quicker pressure equalization than in the CFD, where the only way is in the central tube.

Figure 5 shows the comparison of simulation and measurements when no pellet is launched. When comparing the pellet-less measurements with simulations, we observe that the CFD simulation captures the pressure rise and peak value in compartment 1, but turbulent dissipation modelling is lacking; large vortices are present in the compartments, causing fluctuations and higher settling pressure in the end. The significant difference seen in Fig. 5 after 2.5 ms in the pressure at the second compartment can also be explained by the appearance of a large (unphysical) vortex in simulation.

In the absence of a pellet, the shape of the measured pressure signal is similar in compartment 1 but has a lower amplitude. The measured pressure in compartment 2 reaches a peak value similar to that of the tests with a pellet. Without a pellet, less gas enters the first few chambers, so we conclude that the gas retention is ineffective without a pellet. This is important since shadowgraph images show that some solid particles and gas (unclear if propellant or sublimated pellet material) precede

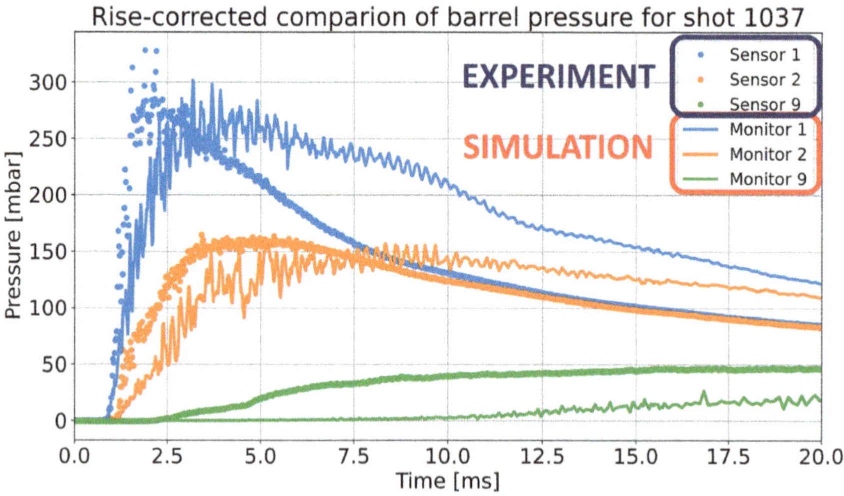

Fig. 4 Comparison of measurements and simulations with full suppressor internals and pellet

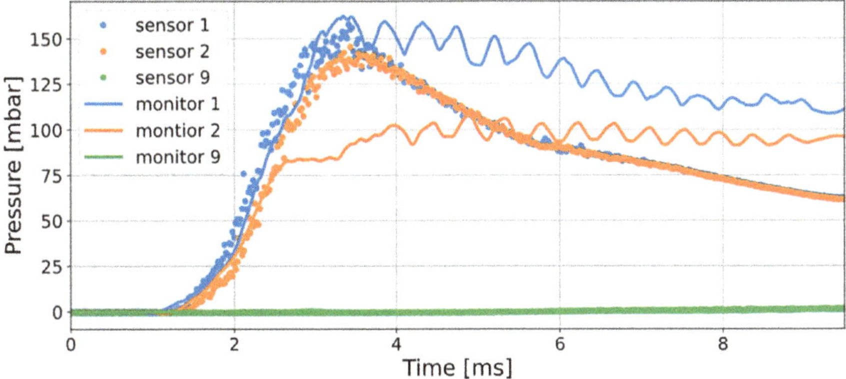

Fig. 5 Comparison of measurements and simulations with full suppressor internals without pellet

the pellet. The gas retention system is ineffective against this gas puff, and it will inevitably reach the reactor plasma before the pellet.

The comparison of measurements against simulation for shots without pellet and without suppressor internals (flat and conical deflector plates) is shown in Fig. 6. Here, the pressure reaches similar peaks as with the internals but lower than the cases that include a pellet. The oscillations seen in the pressure curves are due to the pressure waves travelling back and forth in the empty suppressor. The frequency is consistent with the speed of sound in hydrogen and the suppressor length. The wave amplitude and velocity are reproduced by CFD simulation with small errors.

From the results above, we can conclude that classical CFD tools cannot reproduce the expansion velocity of high-pressure gas puff into low vacuum in all cases. However, the simulation accurately predicts the pressure peaks occurring in the suppressor compartments. Therefore, it can be used to estimate the gas quantity.

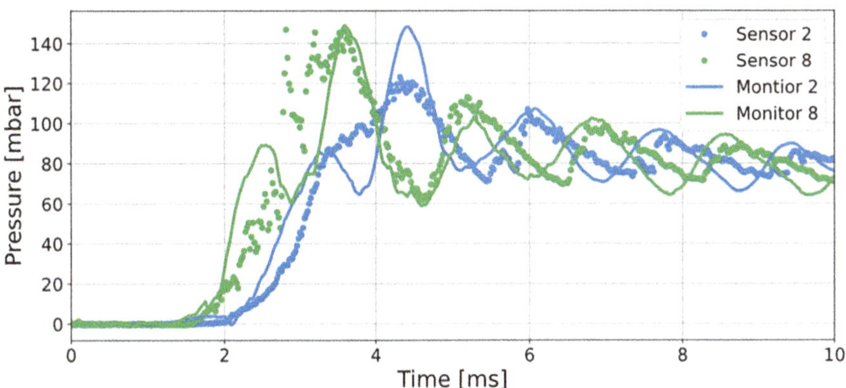

Fig. 6 Comparison of measurements and simulations without suppressor internals or pellet

Another finding of the analysis is that the setup's gaps and leaks significantly affect the pressure evolution. The effect of these gaps in the suppressor separators and around the pellet will be addressed in the future. Different shock-capturing methods can be tested to better model the gas expansion. However, despite the large difference in the operating pressure, the CFD simulations yield acceptable results in modelling the phenomenon. They can serve as preliminary tools for developing vacuum-suppressor systems.

Acknowledgements We acknowledge KIFÜ (Governmental Agency for IT Development, Hungary, https://ror.org/01s0v4q65) for awarding us access to the Komondor HPC facility based in Hungary. Project no. 2024-2.1.2-EKÖP-KDP-2024-00005 has been implemented with the support provided by the Ministry of Culture and Innovation of Hungary from the national research, development and innovation fund, financed under the EKÖP_KDP-24-1-BME-19 funding scheme.

References

1. ASME (2008) Procedure for estimation and reporting of uncertainty due to discretization in CFD applications. J Fluid Eng 130(7):078001. http://FluidsEngineering.asmedigitalcollection. asme.org/article.aspx?articleid=1434171
2. Liou MS (1996) A sequel to AUSM: AUSM+. J Comput Phys 129(2):364–382. https:// linkinghub.elsevier.com/retrieve/pii/S0021999196902569
3. Meitner S, Baylor L, Commaux N, Shiraki D, Combs S, Bjorholm T, Ha T, McGinnis W (2017) Design and commissioning of a three-Barrel Shattered Pellet Injector for DIII-D Disruption Mitigation Studies. In: Fusion science and technology, pp 1–6. https://www.tandfonline.com/ doi/full/10.1080/15361055.2017.1333854
4. Nagy D, Réfy D (2024) Eddy current actuated fast valve development for disruption mitigation applications. Fus Eng Des 202:114400. https://linkinghub.elsevier.com/retrieve/pii/ S0920379624002539
5. Réfy D, Zoletnik S, Walcz E, Nagy D, Szepesi T, Vécsei M, Oravecz D, Katona I, Hegedüs S, Vavrik M, Jachmich S, Kruezi U (2023) Instrumented fast valve for the ITER DMS support laboratory test bench. Fus Eng Des 194:113893. https://linkinghub.elsevier.com/retrieve/pii/ S0920379623004751
6. Steger JL, Benek JA (1987) On the use of composite grid schemes in computational aerodynamics. Comput Methods Appl Mechan Eng 64(1–3):301–320. https://linkinghub.elsevier.com/ retrieve/pii/0045782587900454
7. Zoletnik S, Walcz E, Jachmich S, Kruezi U, Lehnen M, Anda G, Szabolics T, Szepesi T, Bartók G, Cseh G, Boros Z, Dunai D, Gárdonyi G, Hakl J, Hegedüs S, Katona I, Kovacs A, Kocsis G, Lengyel M, Mészáros S, Nagy D, Oravecz D, Poszovecz L, Réfy D, Vad K, Vécsei M (2023) Shattered pellet technology development in the ITER DMS test laboratory. Fus Eng Des 190:113701. https://linkinghub.elsevier.com/retrieve/pii/S0920379623002843

Kinetic Theory

From Kinetic Theory to Hyperbolic-Parabolic Fluid Systems of Equations

Vincent Giovangigli

Abstract Fluid systems of equations can be derived from the kinetic theory by using the Chapman-Enskog asymptotic method. We investigate the deep links between the hyperbolic-parabolic mathematical structure of such systems and the kinetic framework.

Keywords Fluid model · Hyperbolic-parabolic · Kinetic theory

1 Introduction

Fluids models can be derived from the kinetic theory of gases by using the Chapman-Enskog asymptotic method. As a representative example we consider in these notes the case of multicomponent flows with chemistry and transport. Such flows arise in various scientific and engineering applications like atmospheric reentry [1–3], crystal growth [4, 5], or combustion [6, 7].

2 Kinetic Derivation

2.1 Boltzmann Equations

Semiclassical Boltzmann equations in a mixture of polyatomic reactive gases may be written [3, 7–14]

$$\partial_t f_k + \boldsymbol{c}_k \cdot \boldsymbol{\nabla} f_k + \boldsymbol{b}_k \cdot \boldsymbol{\nabla}_{\boldsymbol{c}_k} f_k = \frac{1}{\epsilon} \mathcal{J}_k + \epsilon\, \mathcal{C}_k, \qquad k \in \mathfrak{S}, \tag{1}$$

V. Giovangigli (✉)
CMAP, CNRS, Ecole Polytechnique, Palaiseau Cedex, France
e-mail: vincent.giovangigli@polytechnique.edu
URL: https://cmap.polytechnique.fr/giovangi/

© The Author(s) 2026
M. Grabe et al. (eds.), *Rarefied Gas Dynamics*, Springer Aerospace Technology,
https://doi.org/10.1007/978-3-032-00094-1_25

where ∂_t denotes the time derivative, ∇ the space derivative, ∇_{c_k} the derivative with respect to c_k, $\mathfrak{S} = \{1, \ldots, N\}$ the species indexing set, N the number of species, ϵ the Enskog expansion parameter, and for $k \in \mathfrak{S}$, f_k the distribution function, c_k the particle velocity, b_k the force per unit mass, $\mathcal{J}_k$ and $\mathcal{C}_k$ the scattering and reactive collision operators [3, 7]. The Maxwellian reaction regime with a scaling $\mathcal{O}(\epsilon)$ of chemistry terms is considered for simplicity and we refer to the literature for the tempered reaction regime $\mathcal{O}(\epsilon^0)$ or the kinetic equilibrium regime $\mathcal{O}(\epsilon^{-1})$ [7, 9, 14].

The distribution function of the kth species $f_k(t, x, c_k, K)$ depends on time t, spatial coordinate x, particle velocity $c_k = (c_{k1}, c_{k2}, c_{k3})^t$, quantum number K and $\mathfrak{Q}_k$ denotes the set of quantum energy states of the kth species. We denote by $\xi = (\xi_i)_{i \in \mathfrak{S}}$ any family of functions where $\xi_i = \xi_i(t, x, c_i, I)$ depends on t, x, c_i and I, like the family of distribution functions $f = (f_i)_{i \in \mathfrak{S}}$. The scalar product between two families $\xi = (\xi_i)_{i \in \mathfrak{S}}$ and $\zeta = (\zeta_i)_{i \in \mathfrak{S}}$, where ξ is integrable as f, is defined by

$$\langle\!\langle \xi, \zeta \rangle\!\rangle = \sum_{\substack{i \in \mathfrak{S} \\ I \in \mathfrak{Q}_i}} \int \xi_i(t, x, c_i, I) \zeta_i(t, x, c_i, I) \, dc_i.$$

The collision terms $\mathcal{J}_k$ and $\mathcal{C}_k$ are conveniently written in terms of degeneracy averaged transition probabilities that satisfy reciprocity relations [3, 7–14]. The scattering source term may be written

$$\mathcal{J}_i(f) = \sum_{j \in \mathfrak{S}} \sum_{\substack{I' \in \mathfrak{Q}_i \\ J, J' \in \mathfrak{Q}_j}} \int \left(f_i' f_j' \frac{a_{iI} a_{jJ}}{a_{iI'} a_{jJ'}} - f_i f_j \right) W_{ij}^{IJI'J'} dc_j dc_i' dc_j', \tag{2}$$

where a_{iI} is the degeneracy of the Ith energy state of the ith species, I and J the indices for the quantum states of the ith and jth species before collision, I' and J' the indices after collision, $W_{ij}^{IJI'J'}$ the transition probability for nonreactive collisions, $f_i' = f_i(t, x, c_i', I')$ and $f_j' = f_j(t, x, c_j', J')$ [7, 8, 13]. The transition probabilities satisfy the reciprocity relations $a_{iI} a_{jJ} W_{ij}^{IJI'J'} = a_{iI'} a_{jJ'} W_{ij}^{I'J'IJ}$. The collisional invariants of the scattering operator $\mathcal{J} = (\mathcal{J}_i)_{i \in \mathfrak{S}}$ are given by $\psi^k = (\delta_{ik})_{i \in \mathfrak{S}}$, for $k \in \mathfrak{S} = \{1, \ldots, N\}$, $\psi^{N+l} = (m_i c_{il})_{i \in \mathfrak{S}}$, for $l \in \{1, 2, 3\}$, and $\psi^{N+4} = (\frac{1}{2} m_i |c_i|^2 + E_{iI})_{i \in \mathfrak{S}}$ where δ_{ik} is the Kronecker symbol, m_i the mass of the ith species and E_{iI} the internal energy of the ith species in the Ith state. The corresponding fluid variables are $n_k = \langle\!\langle f, \psi^k \rangle\!\rangle$, $k \in \mathfrak{S}$, $\rho v_l = \langle\!\langle f, \psi^{N+l} \rangle\!\rangle$, $l \in \{1, 2, 3\}$ and $\frac{1}{2} \rho |v|^2 + \mathcal{E} = \langle\!\langle f, \psi^{N+4} \rangle\!\rangle$ where n_k is the number density of the kth species, $v = (v_1, v_2, v_3)^t$ the mixture mass velocity, and $\mathcal{E}$ the mixture internal energy par unit volume.

The chemical reactions are written in the molecular form

$$\sum_{i \in \mathfrak{F}^r} \mathfrak{M}_i \rightleftharpoons \sum_{k \in \mathfrak{B}^r} \mathfrak{M}_k, \qquad r \in \mathfrak{R}, \tag{3}$$

where $\mathfrak{M}_i$ is the symbol of the ith species, $\mathfrak{R} = \{1, \ldots, N^r\}$ the set of reaction indices, N^r the number of reactions, $\mathfrak{F}^r$ and $\mathfrak{B}^r$ the indices for the reactants and products in the rth elementary reaction counted with their order of multiplicity. The letters $\mathfrak{F}^r$ and $\mathfrak{B}^r$ are mnemonics for the forward and backward directions. We denote by ν^f_{ir} and ν^b_{ir} the stoichiometric coefficients of the ith species in the rth reaction, i.e., the order of multiplicity of species i in $\mathfrak{F}^r$ and $\mathfrak{B}^r$, and we also denote by F^r and B^r the quantum states of reactants and products. For a given $i \in \mathfrak{S}$, $\mathfrak{F}^r_i$ denotes the set of reactant indices in which the index for the ith species has been removed once and we introduce a similar notation for $\mathfrak{B}^r_i$, F^r_I and B^r_I. The reactive collision terms may be written $\mathcal{C}_i = \sum_{r \in \mathfrak{R}} \mathcal{C}_{ir}$ with [7]

$$\mathcal{C}_{ir} = \nu^f_{ir} \sum_{F^r_I, B^r} \int \Big(\Big(\prod_{k \in \mathfrak{B}^r} f_k \Big) \frac{\prod_{k \in \mathfrak{B}^r} \beta_{kK}}{\prod_{j \in \mathfrak{F}^r} \beta_{jJ}} - \prod_{j \in \mathfrak{F}^r} f_j \Big) W^{F^r B^r}_{\mathfrak{F}^r \mathfrak{B}^r} \prod_{\substack{j \in \mathfrak{F}^r_i \\ k \in \mathfrak{B}^r}} d\boldsymbol{c}_j d\boldsymbol{c}_k$$

$$+ \nu^b_{ir} \sum_{F^r, B^r_I} \int \Big(\Big(\prod_{j \in \mathfrak{F}^r} f_j \Big) - \frac{\prod_{k \in \mathfrak{B}^r} \beta_{kK}}{\prod_{j \in \mathfrak{F}^r} \beta_{jJ}} \prod_{k \in \mathfrak{B}^r} f_k \Big) W^{F^r B^r}_{\mathfrak{F}^r \mathfrak{B}^r} \prod_{\substack{j \in \mathfrak{F}^r \\ k \in \mathfrak{B}^r_i}} d\boldsymbol{c}_j d\boldsymbol{c}_k, \qquad (4)$$

where $\beta_i = h_P / a_{iI} m_i^3$, h_P is the Planck constant, and $W^{F^r B^r}_{\mathfrak{F}^r \mathfrak{B}^r}$ are transition probabilities with the reciprocity relations $W^{F^r B^r}_{\mathfrak{F}^r \mathfrak{B}^r} \prod_{k \in \mathfrak{B}^r} \beta_{kK} = W^{B^r F^r}_{\mathfrak{B}^r \mathfrak{F}^r} \prod_{j \in \mathfrak{F}^r} \beta_{jJ}$. All transition probabilities, species masses, internal energies, degeneracies, forces per unit mass, and chemical reactions details are assumed to be externally given in the model definition.

2.2 Kinetic Entropy

The kinetic entropy is defined by $\mathcal{S}^{\mathrm{kin}} = -k_B \sum_{i \in \mathfrak{S}, I \in \mathfrak{Q}_i} \int f_i \big(\log(\beta_{iI} f_i) - 1 \big) d\boldsymbol{c}_i$ and satisfies the balance equation

$$\partial_t \mathcal{S}^{\mathrm{kin}} + \nabla \cdot (\mathcal{S}^{\mathrm{kin}} \boldsymbol{v}) + \nabla \cdot \boldsymbol{J}^{\mathrm{kin}} = \mathfrak{v}^{\mathcal{J}} + \mathfrak{v}^{\mathcal{C}}, \qquad (5)$$

where $\boldsymbol{J}^{\mathrm{kin}} = -k_B \sum_{i \in \mathfrak{S}, I \in \mathfrak{Q}_i} \int (\boldsymbol{c}_i - \boldsymbol{v}) f_i \big(\log(\beta_{iI} f_i) - 1 \big) d\boldsymbol{c}_i$ is the diffusive entropy flux, $\mathfrak{v}^{\mathcal{J}} = -k_B \langle\!\langle \mathcal{J}, \log(\beta f) \rangle\!\rangle$ the scattering entropy production rate and $\mathfrak{v}^{\mathcal{C}} = -\sum_{r \in \mathfrak{R}} k_B \langle\!\langle \mathcal{C}_r, \log(\beta f) \rangle\!\rangle$ the chemistry entropy production rate. Both entropy production rates $\mathfrak{v}^{\mathcal{J}}$ and $\mathfrak{v}^{\mathcal{C}}$ are nonnegative with

$$4\frac{\mathfrak{v}^{\mathcal{J}}}{k_{\mathrm{B}}} = \sum_{\substack{i,j\in\mathfrak{S}\\ \mathrm{I},\mathrm{I}'\in\mathfrak{Q}_i\\ \mathrm{J},\mathrm{J}'\in\mathfrak{Q}_j}} \int \Big(f_i'\,f_j'\,\frac{a_{i\mathrm{I}}a_{j\mathrm{J}}}{a_{i\mathrm{I}'}a_{j\mathrm{J}'}} - f_i\,f_j\Big) \log\Big(\frac{f_i'\,f_j'\,a_{i\mathrm{I}}a_{j\mathrm{J}}}{f_i\,f_j\,a_{i\mathrm{I}'}a_{j\mathrm{J}'}}\Big) W_{ij}^{\mathrm{IJI'J'}} d\mathbf{c}_i d\mathbf{c}_j d\mathbf{c}_i' d\mathbf{c}_j',$$

$$\frac{\mathfrak{v}^{\mathcal{C}}}{k_{\mathrm{B}}} = \sum_{\substack{r\in\mathfrak{R}\\ F^r,B^r}} \int \Big(\prod_{k\in\mathfrak{B}^r} f_k\,\frac{\prod_{k\in\mathfrak{B}^r}\beta_{k\mathrm{K}}}{\prod_{j\in\mathfrak{F}^r}\beta_{j\mathrm{J}}} - \prod_{j\in\mathfrak{F}^r} f_j\Big)\log\Big(\frac{\prod_{k\in\mathfrak{B}^r}\beta_{k\mathrm{K}}f_k}{\prod_{j\in\mathfrak{F}^r}\beta_{j\mathrm{J}}f_j}\Big) \mathcal{W}_{\mathfrak{F}^r\mathfrak{B}^r}^{F^r B^r} \prod_{\substack{j\in\mathfrak{F}^r\\ k\in\mathfrak{B}^r}} d\mathbf{c}_j d\mathbf{c}_k.$$

2.3 Enskog Expansion

Application of the Chapman-Enskog method yields the fluid conservation equations, the thermochemistry properties, the transport fluxes, and the transport coefficients [3, 7–14]. The distributions are expanded as $f_k = f_k^0\big(1 + \epsilon\phi_k + \mathcal{O}(\epsilon^2)\big)$ for $k\in\mathfrak{S}$, and must satisfy the Enskog constraints $\langle\!\langle f - f^{(0)}, \psi^l \rangle\!\rangle = 0$ where ψ^l, $1\le l\le \mathrm{N}+4$, are the collisional invariants.

The zeroth order distributions $f^{(0)} = (f_i^{(0)})_{i\in\mathfrak{S}}$ are found to be Maxwellians since $\mathcal{J}(f^{(0)}) = 0$ implies that $\log(\beta f^{(0)})$ is a collisional invariant and thus a linear combination of the ψ^l, $1\le l\le\mathrm{N}+4$

$$f_i^{(0)}(t, \mathbf{x}, \mathrm{I}, \mathbf{c}_i) = \frac{1}{\beta_{i\mathrm{I}}}\frac{n_i}{Z_i}\exp\Big\{-\frac{m_i}{2k_{\mathrm{B}}T}|\mathbf{c}_i - \mathbf{v}|^2 - \frac{E_{i\mathrm{I}}}{k_{\mathrm{B}}T}\Big\}, \tag{6}$$

where $Z_i = Z_i^{\mathrm{tr}}Z_i^{\mathrm{int}}$, $Z_i^{\mathrm{tr}} = \big(\frac{2\pi m_i k_{\mathrm{B}}T}{h_{\mathrm{P}}^2}\big)^{3/2}$, $Z_i^{\mathrm{int}} = \sum_{\mathrm{I}\in\mathfrak{Q}_i} a_{i\mathrm{I}}\exp\big(-\frac{E_{i\mathrm{I}}}{k_{\mathrm{B}}T}\big)$, k_{B} is the Boltzmann constant and T the absolute temperature. The zeroth order equations are the Euler equations with the thermodynamic properties $p = \sum_{i\in\mathfrak{S}} n_i k_{\mathrm{B}}T$ and $\mathcal{E} = \sum_{i\in\mathfrak{S}} n_i\overline{E}_i$ where $\overline{E}_i = \frac{3}{2}k_{\mathrm{B}}T + \overline{E}_i^{\mathrm{int}}$ and $\overline{E}_i^{\mathrm{int}} = \sum_{\mathrm{I}\in\mathfrak{Q}_i}\frac{a_{i\mathrm{I}}E_{i\mathrm{I}}}{Z_i^{\mathrm{int}}}\exp\big(-\frac{E_{i\mathrm{I}}}{k_{\mathrm{B}}T}\big)$.

The linearized Boltzmann operator is in the form

$$\mathcal{I}_i^{\mathcal{J}}(\phi) = \sum_{\substack{j\in\mathfrak{S}\\ \mathrm{I}'\in\mathfrak{Q}_i\\ \mathrm{J},\mathrm{J}'\in\mathfrak{Q}_j}}\int f_j^{(0)}(\phi_i + \phi_j - \phi_i' + \phi_j')W_{ij}^{\mathrm{IJI'J'}} d\mathbf{c}_j d\mathbf{c}_i' d\mathbf{c}_j', \tag{7}$$

and the perturbations $\phi = (\phi_i)_{i\in\mathfrak{S}}$ satisfy the system of integral equations

$$\begin{cases} \mathcal{I}_i^{\mathcal{J}}(\phi) = -\big(\partial_t\log f_i^{(0)} + \mathbf{c}_i\cdot\nabla\log f_i^{(0)}\big) = \Psi_i, & i\in\mathfrak{S} \\ \langle\!\langle f^{(0)}\phi, \psi^l \rangle\!\rangle = 0, & 1\le l\le\mathrm{N}+4. \end{cases} \tag{8}$$

The right hand sides Ψ_i may be decomposed in the form

$$\Psi_i = -\boldsymbol{\Psi}_i^{\eta}:\nabla\mathbf{v} - \frac{1}{3}\boldsymbol{\Psi}_i^{\kappa}\,\nabla\!\cdot\!\mathbf{v} - \sum_{j\in\mathfrak{S}}\boldsymbol{\Psi}_i^{D_j}\!\cdot p\,\mathbf{d}_j - \boldsymbol{\Psi}_i^{\widehat{\lambda}}\!\cdot\nabla\Big(\frac{1}{k_{\mathrm{B}}T}\Big),$$

where $d_j = (\nabla p_j - \rho_j b_j)/p$ is the diffusion driving force of the jth species, p_j the partial pressure of the jth species and the expression of the coefficients Ψ_i^η, Ψ_i^κ, $\Psi_i^{D_j}$, $j \in \mathfrak{S}$, and $\Psi_i^{\widehat{\lambda}}$ may be founded in the literature [7, 13].

The perturbed distribution functions are correspondingly decomposed as

$$\phi_i = -\boldsymbol{\phi}_i^\eta : \nabla \boldsymbol{v} - \tfrac{1}{3}\phi_i^\kappa \, \nabla \cdot \boldsymbol{v} - \sum_{j \in \mathfrak{S}} \boldsymbol{\phi}_i^{D_j} \cdot p \, \boldsymbol{d}_j - \boldsymbol{\phi}_i^{\widehat{\lambda}} \cdot \nabla\left(\frac{1}{k_\mathrm{B} T}\right),$$

and the equation for $\phi^\mu = (\phi_i^\mu)_{i \in \mathfrak{S}}$ is then $\mathcal{I}^{\mathcal{J}}(\phi^\mu) = \Psi^\mu$ with the Enskog constraints $\langle\!\langle f^{(0)}\phi^\mu, \psi^l \rangle\!\rangle = 0$, $1 \leq l \leq N+4$, for $\mu \in \{\kappa, \eta, \widehat{\lambda}\} \cup \{D_1, \ldots, D_N\}$. The quantities $\boldsymbol{\phi}_i^\eta$ and $\boldsymbol{\Psi}^\eta$ are traceless symmetric second order tensors, $\boldsymbol{\phi}_i^{D_j}$ $\boldsymbol{\Psi}_i^{D_j}$, $\boldsymbol{\phi}_i^{\widehat{\lambda}}$ and $\boldsymbol{\Psi}_i^{\widehat{\lambda}}$, are vectors whereas ϕ_i^κ and Ψ^κ are scalars. It is established that $\langle\!\langle f^{(0)}\Psi^\mu, \psi^l \rangle\!\rangle = 0$ and the transport coefficients are typically given by $\mu = [\phi^\mu, \phi^\mu] = \langle\!\langle f^{(0)}\Psi^\mu, \phi^\mu \rangle\!\rangle$. The bracket operator defined by $[\xi, \zeta] = \langle\!\langle f^{(0)}\xi, \mathcal{I}^{\mathcal{J}}\zeta \rangle\!\rangle$ is found to be symmetric $[\xi, \zeta] = [\zeta, \xi]$, positive semi-definite $[\xi, \xi] \geq 0$, and $[\xi, \xi] = 0$ if and only if ξ is a collisional invariant.

Substitution of the gradient decomposition of $\phi = (\phi_i)_{i \in \mathfrak{S}}$ in the expressions of the species diffusion velocities $n_i \boldsymbol{V}_i = \sum_{\mathrm{I} \in \mathfrak{Q}_i} \int (\boldsymbol{c}_i - \boldsymbol{v}) f_i^{(0)} \phi_i d\boldsymbol{c}_i$, viscous tensor $\boldsymbol{\Pi} = \sum_{i \in \mathfrak{S}, \mathrm{I} \in \mathfrak{Q}_i} \int m_i(\boldsymbol{c}_i - \boldsymbol{v}) \otimes (\boldsymbol{c}_i - \boldsymbol{v}) f_i^{(0)} \phi_i d\boldsymbol{c}_i$, and heat flux $\boldsymbol{Q} = \sum_{i \in \mathfrak{S}, \mathrm{I} \in \mathfrak{Q}_i} \int \left(\tfrac{1}{2} m_i |\boldsymbol{c}_i - \boldsymbol{v}|^2 + E_{i\mathrm{I}}\right)(\boldsymbol{c}_i - \boldsymbol{v}) f_i^{(0)} \phi_i d\boldsymbol{c}_i$ then yields the expressions of the fluxes in terms of macroscopic variable gradients as well as the expression of the transport coefficients.

3 Fluid Model

3.1 Governing Equations

The equations for conservation of species mass, momentum and energy derived from the kinetic theory are obtained in the form [7–14]

$$\partial_t \rho_k + \nabla \cdot (\rho_k \boldsymbol{v}) + \nabla \cdot (\rho_k \boldsymbol{V}_k) = m_k \omega_k, \qquad k \in \mathfrak{S}, \tag{9}$$

$$\partial_t (\rho \boldsymbol{v}) + \nabla \cdot (\rho \boldsymbol{v} \otimes \boldsymbol{v} + p \boldsymbol{I}) + \nabla \cdot \boldsymbol{\Pi} = \rho \boldsymbol{b}, \tag{10}$$

$$\partial_t (\mathcal{E} + \tfrac{1}{2}\rho \boldsymbol{v} \cdot \boldsymbol{v}) + \nabla \cdot ((\mathcal{E} + \tfrac{1}{2}\rho \boldsymbol{v} \cdot \boldsymbol{v} + p)\boldsymbol{v}) + \nabla \cdot (\boldsymbol{Q} + \boldsymbol{\Pi} \cdot \boldsymbol{v}) = \rho \boldsymbol{v} \cdot \boldsymbol{b}, \tag{11}$$

where $\rho_k = m_k n_k$ denotes the mass density of the kth species, $\boldsymbol{v}$ the mass average flow velocity, $\boldsymbol{V}_k$ the diffusion velocity of the kth species, m_k the mass of the kth species, ω_k the molecular production rate of the kth species, $\rho = \sum_{k \in \mathfrak{S}} \rho_k$ the total mass density, p the pressure, $\boldsymbol{\Pi}$ the viscous tensor, $\boldsymbol{b}$ the specific force, $\mathcal{E}$ the internal energy per unit volume and $\boldsymbol{Q}$ the heat flux. The force acting on the species $\boldsymbol{b}$ has been assumed to be species independent for simplicity and the thermodynamic properties have been presented in Sect. 2.3.

The transport fluxes $\mathbf{V}_k$, $k \in \mathfrak{S}$, $\mathbf{\Pi}$, and $\mathbf{Q}$ due to fluid variable gradients are obtained in the form [7–13]

$$\mathbf{V}_k = -\sum_{l \in \mathfrak{S}} D_{kl} \mathbf{d}_l - \theta_k \nabla \log T, \qquad k \in \mathfrak{S}, \tag{12}$$

$$\mathbf{\Pi} = -\kappa (\nabla \cdot \mathbf{v}) \mathbf{I} - \eta \big(\nabla \mathbf{v} + \nabla \mathbf{v}^t - \tfrac{2}{3} (\nabla \cdot \mathbf{v}) \mathbf{I} \big), \tag{13}$$

$$\mathbf{Q} = -\widehat{\lambda} \nabla T - p \sum_{k \in \mathfrak{S}} \theta_k \mathbf{d}_k + \sum_{k \in \mathfrak{S}} h_k \rho_k \mathbf{V}_k, \tag{14}$$

where D_{kl}, $k, l \in \mathfrak{S}$, are the multicomponent diffusion coefficients, κ the volume viscosity, η the shear viscosity, $\widehat{\lambda}$ the partial thermal conductivity, θ_k, $k \in \mathfrak{S}$, the thermal diffusion coefficients, $h_k = (\tfrac{5}{2} k_\mathrm{B} T + \overline{E}_k^{\mathrm{int}})/m_k$ the enthalpy per unit mass of the kth species, θ_k, $k \in \mathfrak{S}$, the thermal diffusion coefficients, and ^{t} the transposition operator. The transport coefficients are further obtained in terms of the perturbed distributions

$$D_{ij} = \frac{p k_\mathrm{B} T}{3} [\phi^{D_i}, \phi^{D_j}], \quad \theta_i = -\frac{1}{3} [\phi^{D_i}, \phi^{\widehat{\lambda}}], \quad \widehat{\lambda} = \frac{1}{3 k_\mathrm{B} T^2} [\phi^{\widehat{\lambda}}, \phi^{\widehat{\lambda}}], \tag{15}$$

$$\kappa = \frac{k_\mathrm{B} T}{9} [\phi^\kappa, \phi^\kappa], \quad \eta = \frac{k_\mathrm{B} T}{10} [\phi^\eta, \phi^\eta]. \tag{16}$$

The thermal diffusion ratios $\chi = (\chi_1, \ldots, \chi_\mathrm{N})^t$ introduced by Waldmann [8] are the unique solution of $D\chi = \theta$ with $\langle \chi, \mathbf{I} \rangle = 0$ where $\mathbf{I} = (1, \ldots, 1)^t$. The thermal conductivity λ is then given by $\lambda = \widehat{\lambda} - (p/T) \langle D\chi, \chi \rangle$ and both coefficients λ and χ may also be written in terms of bracket products [13]. The diffusion velocities and the heat flux may then be rewritten as [7, 8, 11–13]

$$\mathbf{V}_k = -\sum_{l \in \mathfrak{S}} D_{kl} (\mathbf{d}_l + \chi_l \nabla \log T), \quad k \in \mathfrak{S}, \tag{17}$$

$$\mathbf{Q} = -\lambda \nabla T + p \sum_{k \in \mathfrak{S}} \chi_k \mathbf{V}_k + \sum_{k \in \mathfrak{S}} h_k \rho_k \mathbf{V}_k, \tag{18}$$

and both formulations (12, 14) and (17, 18) may be used for applications. The mathematical structure and properties of the transport coefficients are then obtained from those of the bracket operator [13]. The diffusion matrix is symmetric $D = D^t$, positive semi-definite with nullspace $N(D) = \mathbb{R}\mathsf{y}$ and the thermal diffusion coefficients $\theta = (\theta_1, \ldots, \theta_\mathrm{N})^t$, are such that $\langle \theta, \mathsf{y} \rangle = 0$ where $\mathsf{y}_k = \rho_k/\rho$, $\mathsf{y} = (\mathsf{y}_1, \ldots, \mathsf{y}_\mathrm{N})^t$, and $\langle , \rangle$ denotes the scalar product. The thermal conductivity λ and the shear viscosity η are positive, and the volume viscosity κ is nonnegative.

The chemical production rates ω_k, $k \in \mathfrak{S}$, are obtained in the form

$$\omega_k = \sum_{r \in \mathfrak{R}} (\nu_{kr}^b - \nu_{kr}^f) \tau_r, \quad \tau_r = \mathcal{K}_r^s \big(\exp \langle \mu, \nu_r^f \rangle - \exp \langle \mu, \nu_r^b \rangle \big), \tag{19}$$

where, for the rth reaction, τ_r is the rate of progress, $\mathcal{K}_r^s$ the symmetric rate constant, $\nu_r^f = (\nu_{1r}^f, \ldots, \nu_{\mathrm{N}r}^f)^t$ and $\nu_r^b = (\nu_{1r}^b, \ldots, \nu_{\mathrm{N}r}^b)^t$ the reaction vectors, and $\mu = (\mu_1, \ldots, \mu_\mathrm{N})^t$, where $\mu_k = \log \frac{n_k}{Z_k}$ are the species chemical potentials. The expressions (19) are obtained from the kinetic reactive collision terms and the rate constants $\mathcal{K}_r^s$ are averaged transition probabilities [7]

$$\mathcal{K}_r^s = \sum_{F^r, B^r} \int \prod_{j \in \mathfrak{F}^r} \widetilde{f}_j^{(0)} \frac{\mathcal{W}_{\mathfrak{F}^r \mathfrak{B}^r}^{F^r B^r}}{\prod_{\substack{j \in \mathfrak{F}^r}} \beta_{jJ}} \prod_{\substack{j \in \mathfrak{F}^r \\ k \in \mathfrak{B}^r}} d\boldsymbol{c}_j d\boldsymbol{c}_k,$$

where $\widetilde{f}_j^{(0)}$ denotes the reduced distribution $\widetilde{f}_j^{(0)} = f_j^{(0)}/n_j$. Decomposing the potentials as $\mu_k = -\log Z_k + \log n_k$, the rates of progress may be written $\tau_r = \mathcal{K}_r^f \prod_{l \in \mathfrak{S}} n_l^{\nu_{lr}^f} - \mathcal{K}_r^b \prod_{l \in \mathfrak{S}} n_l^{\nu_{lr}^b}$ with $\mathcal{K}_r^f = \mathcal{K}_r^s / \prod_{l \in \mathfrak{S}} Z_l^{\nu_{lr}^f}$ and $\mathcal{K}_r^b = \mathcal{K}_r^s / \prod_{l \in \mathfrak{S}} Z_l^{\nu_{lr}^b}$ and are thus compatible with the law of mass action. The equilibrium constant of the rth reaction is $\mathcal{K}_r^e(T) = \prod_{l \in \mathfrak{S}} Z_l^{\nu_{lr}}$ and the reciprocity relations between the forward and backward constants $\mathcal{K}_r^e(T) = \mathcal{K}_r^f(T)/\mathcal{K}_r^b(T)$ may be seen as Onsager relations for chemistry. The forward reaction constants $\mathcal{K}_r^f$, $r \in \mathfrak{R}$, are often approximated with Arrhenius law.

3.2 Fluid Entropy

The fluid entropy $\mathcal{S}$ is the kinetic entropy evaluated with Maxwellian distributions $\mathcal{S} = \mathcal{S}^{\mathrm{kin}(0)}$ and is shown to be second order accurate $\mathcal{S}^{\mathrm{kin}} - \mathcal{S} = \mathcal{O}(\epsilon^2)$. This entropy reads $\mathcal{S} = \sum_{i \in \mathfrak{S}} n_i \overline{S}_i$ where the entropy per molecule of the ith species is $\overline{S}_i = \frac{5}{2} k_\mathrm{B} T + \overline{E}_i^{\mathrm{int}}/T - k_\mathrm{B} \log \frac{n_i}{Z_i}$. The Gibbs function per unit volume is given by $\mathcal{G} = \sum_{i \in \mathfrak{S}} n_i \overline{G}_i$ where $\overline{G}_i = k_\mathrm{B} T \log \frac{n_i}{Z_i} = k_\mathrm{B} T \mu_i$.

From the Gibbs relation $T \, \mathbb{D}\mathcal{S} = \mathbb{D}\mathcal{E} - \sum_{k \in \mathfrak{S}} \overline{G}_k \mathbb{D} n_k$, where $\mathbb{D}$ denotes the total derivative, the governing equation for $\mathcal{S}$ is in the form

$$\partial_t \mathcal{S} + \boldsymbol{\nabla} \cdot (\boldsymbol{v}\mathcal{S}) + \boldsymbol{\nabla} \cdot \left(\frac{\boldsymbol{Q}}{T} - \sum_{k \in \mathfrak{S}} \frac{g_k \rho_k \mathbf{V}_k}{T} \right) = \frac{\kappa}{T}(\boldsymbol{\nabla} \cdot \boldsymbol{v})^2 + \frac{\eta}{2T}|\Sigma|^2 + \frac{\lambda}{T^2}|\boldsymbol{\nabla} T|^2$$

$$+ \frac{p}{T} \sum_{k,l \in \mathfrak{S}} D_{kl}\big(\boldsymbol{d}_k + \chi_k \boldsymbol{\nabla} \log T\big) \cdot \big(\boldsymbol{d}_l + \chi_l \boldsymbol{\nabla} \log T\big) - k_\mathrm{B} \sum_{k \in \mathfrak{S}} \mu_k \omega_k, \tag{20}$$

where $\Sigma = \boldsymbol{\nabla}\boldsymbol{v} + \boldsymbol{\nabla}\boldsymbol{v}^t - \frac{2}{3}(\boldsymbol{\nabla} \cdot \boldsymbol{v})\, \boldsymbol{I}$ denotes of the rate of strain tensor. Since the thermal conductivity λ and the shear viscosity η are positive and the multicomponent diffusion matrix D is positive semidefinite we deduce that entropy production due to macroscopic gradients is nonnegative. On the other hand, the entropy production due to chemistry $-k_\mathrm{B} \sum_{k \in \mathfrak{S}} \mu_k \omega_k$ is such that

$$-\sum_{k \in \mathfrak{S}} \mu_k \omega_k = \sum_{r \in \mathfrak{R}} \mathcal{K}_r^s \big(\langle \mu, \nu_r^f \rangle - \langle \mu, \nu_r^b \rangle\big) \big(\exp\langle \mu, \nu_r^f \rangle - \exp\langle \mu, \nu_r^b \rangle\big), \tag{21}$$

so that the rate of fluid entropy production (20) is a sum of nonnegative terms. Expanding in terms of ϵ the kinetic entropy production due to scattering yields

$\mathfrak{v}^{(-1)\mathcal{J}} = 0$, $\mathfrak{v}^{(0)\mathcal{J}} = 0$ and $\mathfrak{v}^{(1)\mathcal{J}} = k_{\text{B}}[\phi, \phi]$ that coincides with fluid entropy production due to gradients. Similarly, expanding the kinetic entropy production due to reactive collisions yields $\mathfrak{v}^{(0)\mathcal{C}} = 0$ and $\mathfrak{v}^{(1)\mathcal{C}} = -k_{\text{B}}\langle \mu, \omega \rangle$ that coincides with fluid entropy production due to chemistry. Denoting by $\mathcal{R}$ the linear space of chemical reactions $\mathcal{R} = \text{span}\{ v_r; \ r \in \mathfrak{R} \}$ the expression (21) implies notably the following properties of chemical equilibrium

$$\langle \mu, \omega \rangle = 0 \iff \omega_k = 0, \ k \in \mathfrak{S} \iff \tau_r = 0, \ r \in \mathfrak{R} \iff \mu \in \mathcal{R}^{\perp}.$$

3.3 Vector Form

The governing equations for multicomponent fluids may finally be written in vector form $\partial_t \mathsf{u} + \sum_{i \in \mathfrak{D}} \partial_i \mathsf{F}_i + \sum_{i \in \mathfrak{D}} \partial_i \mathsf{F}_i^{\text{diss}} = \Omega$ where u denotes the conservative variable $\mathsf{u} = \left(\rho_1, \ldots, \rho_N, \rho \mathbf{v}, \mathcal{E} + \frac{1}{2}\rho|\mathbf{v}|^2\right)^t$, $\mathfrak{D} = \{1, \ldots, 3\}$ the set of direction indices, ∂_i the derivative in the ith direction, F_i and $\mathsf{F}_i^{\text{diss}}$ the convective and dissipative fluxes in the ith direction, and Ω the source term given by

$$\mathsf{F}_i = \left(\rho_1 v_i, \ldots, \rho_N v_i, \rho \mathbf{v} v_i + p e^i, (\mathcal{E} + p + \tfrac{1}{2}\rho|\mathbf{v}|^2)v_i\right)^t,$$

$$\mathsf{F}_i^{\text{diss}} = \left(\rho_1 \mathcal{V}_{1i}, \ldots, \rho_N \mathcal{V}_{Ni}, \mathbf{\Pi}_i, \ Q_i + \mathbf{\Pi}_i \cdot \mathbf{v}\right)^t,$$

$$\Omega = (m_1 \omega_1, \ldots, m_N \omega_N, \rho \mathbf{b}, \rho \mathbf{v} \cdot \mathbf{b})^t,$$

where e^i, $1 \leq i \leq 3$, denote the standard basis vectors, $\mathbf{\Pi}_i = (\Pi_{1i}, \Pi_{2i}, \Pi_{3i})^t$ and $\mathbf{Q} = (Q_1, Q_2, Q_3)^t$. We may then use the—uniquely defined—dissipation matrices B_{ij} such that $\mathsf{F}_i^{\text{diss}} = -\sum_{j \in \mathfrak{D}} \mathsf{B}_{ij}(\mathsf{u})\partial_j \mathsf{u}$ and the Jacobian matrices $\mathsf{A}_i = \partial_{\mathsf{u}}\mathsf{F}_i$ of the convective fluxes F_i in order to rewrite the governing equation as a second order system of partial differential equations.

4 Hyperbolic-Parabolic Structure

A essential step in order to investigate the mathematical structure and well posedness of a system of conservation laws is symmetrization [7, 15–17]. We discuss in this section the links between kinetic theory and symmetrization.

4.1 Entropy and Symmetrization

Using the conservative variable $\mathsf{u} = \left(\rho_1, \ldots, \rho_N, \rho \mathbf{v}, \mathcal{E} + \frac{1}{2}\rho|\mathbf{v}|^2\right)^t$, the vector notation of Sect. 3.3, the gradient decomposition of the dissipative fluxes $\mathsf{F}_i^{\text{diss}} =$

$-\sum_{j\in\mathfrak{D}} \mathsf{B}_{ij}(\mathsf{u})\partial_j\mathsf{u}$, $i \in \mathfrak{D}$, involving the dissipation matrices B_{ij}, i, $j \in \mathfrak{D}$, the Jacobian $\mathsf{A}_i(\mathsf{u}) = \partial_\mathsf{u}\mathsf{F}_i$ of the convective fluxes F_i, $i \in \mathfrak{D}$, and Ω the source term, the equations governing multicomponent flows read

$$\partial_t\mathsf{u} + \sum_{i\in\mathfrak{D}}\mathsf{A}_i(\mathsf{u})\partial_i\mathsf{u} = \sum_{i,j\in\mathfrak{D}} \partial_i\big(\mathsf{B}_{ij}(\mathsf{u})\partial_j\mathsf{u}\big) + \Omega(\mathsf{u}). \tag{22}$$

The system coefficients $\mathsf{A}_i(\mathsf{u})$, $i \in \mathfrak{D}$, $\mathsf{B}_{ij}(\mathsf{u})$, i, $j \in \mathfrak{D}$, and Ω are smooth functions of the conservative variable u on an open convex set $\mathcal{O}_\mathsf{u}$.

Symmetrization of the quasilinear system (22) may then be achieved by using a mathematical entropy σ compatible with convection, diffusion as well as source terms, that is, a smooth function satisfying (E_1)–(E_7) [7, 17]. Properties $(E_1)(E_2)$ of convective terms originate from the theory of hyperbolic systems [18, 19], properties $(E_3)(E_4)$ of dissipative terms from Kawashima [20], and properties (E_5)-(E_7) of source terms from Chen, Levermore and Liu [21] :

(E_1) $\partial_\mathsf{u}^2\sigma(\mathsf{u})$ is positive definite.

(E_2) $\partial_\mathsf{u}\sigma(\mathsf{u})\mathsf{A}_i(\mathsf{u}) = \partial_\mathsf{u}\mathsf{q}_i(\mathsf{u})$ where q_i is the ith entropy flux.

(E_3) $\big(\mathsf{B}_{ij}(\mathsf{u})\big(\partial_\mathsf{u}^2\sigma(\mathsf{u})\big)^{-1}\big)^t = \mathsf{B}_{ji}(\mathsf{u})\big(\partial_\mathsf{u}^2\sigma(\mathsf{u})\big)^{-1}$.

(E_4) $\sum_{i,j\in\mathfrak{D}}\mathsf{B}_{ij}(\mathsf{u})\big(\partial_\mathsf{u}^2\sigma(\mathsf{u})\big)^{-1}w_i w_j$ is positive semi-definite for $w \in \mathbb{R}^d$.

(E_5) There exists a fixed linear subspace $\mathfrak{E} \subset \mathbb{R}^n$ such that $\Omega(\mathsf{u}) \in \mathfrak{E}^\perp$ and

$$\Omega(\mathsf{u}) = 0 \iff \big(\partial_\mathsf{u}\sigma(\mathsf{u})\big)^t \in \mathfrak{E} \iff \partial_\mathsf{u}\sigma(\mathsf{u})\,\Omega(\mathsf{u}) = 0.$$

(E_6) If $\Omega(\mathsf{u}) = 0$, $\partial_\mathsf{u}\Omega(\mathsf{u})\big(\partial_\mathsf{u}^2\sigma(\mathsf{u})\big)^{-1}$ is symmetric and $N\big(\partial_\mathsf{u}\Omega\,\big(\partial_\mathsf{u}^2\sigma\big)^{-1}\big) = \mathfrak{E}$.

(E_7) We have $\partial_\mathsf{u}\sigma(\mathsf{u})\,\Omega(\mathsf{u}) \leq 0$.

The mathematical entropy balance is then in the form

$$\partial_t\sigma + \sum_{i\in\mathfrak{D}}\partial_i\mathsf{q}_i - \sum_{i,j\in\mathfrak{D}}\partial_i\langle\mathsf{v}, \widetilde{\mathsf{B}}_{ij}\partial_j\mathsf{v}\rangle = -\sum_{i,j\in\mathfrak{D}}\langle\partial_i\mathsf{v}, \widetilde{\mathsf{B}}_{ij}\partial_j\mathsf{v}\rangle + \langle\mathsf{v}, \widetilde{\Omega}\rangle,$$

and the entropic variable is defined by $\mathsf{v} = (\partial_\mathsf{u}\sigma)^t$. For multicomponent flows the mathematical entropy is taken to be $\sigma = -\mathcal{S}/k_\mathrm{B}$ and the entropic variable reads $\mathsf{v} = \frac{1}{k_\mathrm{B}T}\big(g_1 - \frac{1}{2}|\boldsymbol{v}|^2, \ldots, g_\mathrm{N} - \frac{1}{2}|\boldsymbol{v}|^2, \boldsymbol{v}, -1\big)^t$ where $g_k = \overline{G}_k/m_k$, $k \in \mathfrak{S}$, are the species Gibbs functions per unit mass. The map $\mathsf{u} \mapsto \mathsf{v}$ is a smooth diffeomorphism from $\mathcal{O}_\mathsf{u}$ onto an open set $\mathcal{O}_\mathsf{v}$ and letting $\mathsf{u} = \mathsf{u}(\mathsf{v})$ in (22) yields the symmetrized equations

$$\widetilde{\mathsf{A}}_0(\mathsf{v})\partial_t\mathsf{v} + \sum_{i\in\mathfrak{D}}\widetilde{\mathsf{A}}_i(\mathsf{v})\partial_i\mathsf{v} = \sum_{i,j\in\mathfrak{D}}\partial_i\big(\widetilde{\mathsf{B}}_{ij}(\mathsf{v})\partial_j\mathsf{v}\big) + \widetilde{\Omega}(\mathsf{v}), \tag{23}$$

where $\widetilde{\mathsf{A}}_0 = \partial_\mathsf{v}\mathsf{u}$, $\widetilde{\mathsf{A}}_i = \mathsf{A}_i\partial_\mathsf{v}\mathsf{u}$, $\widetilde{\mathsf{B}}_{ij} = \mathsf{B}_{ij}\partial_\mathsf{v}\mathsf{u}$, $\widetilde{\Omega} = \Omega$, and where :

(S_1) $\widetilde{A}_0(\mathsf{v})$ symmetric positive definite.

(S_2) $\widetilde{A}_i(\mathsf{v})$ symmetric.

(S_3) $\widetilde{B}_{ij}^t(\mathsf{v}) = \widetilde{B}_{ji}(\mathsf{v})$.

(S_4) $\sum_{i,j\in\mathfrak{D}} \widetilde{B}_{ij}(\mathsf{v}) w_i w_j$ is positive semi-definite for $w \in \mathbb{R}^d$.

(S_5) There exits a fixed linear subspace $\mathfrak{E} \subset \mathbb{R}^n$ with $\Omega(\mathsf{v}) \in \mathfrak{E}^\perp$ and

$$\Omega(\mathsf{v}) = 0 \iff \mathsf{v} \in \mathfrak{E} \iff \langle \mathsf{v}, \Omega(\mathsf{v}) \rangle = 0.$$

(S_6) If $\Omega(\mathsf{v}) = 0$ then $\partial_{\mathsf{v}}\Omega(\mathsf{v}) = \big(\partial_{\mathsf{v}}\Omega(\mathsf{v})\big)^t$ and $N\big(\partial_{\mathsf{v}}\Omega(\mathsf{v})\big) = \mathfrak{E}$.

(S_7) $\langle \mathsf{v}, \Omega(\mathsf{v}) \rangle \le 0$.

The existence of a mathematical entropy implies symmetrizability and symmetrizability with a change of variable $\mathsf{u} \to \mathsf{v}$ implies the existence of a mathematical entropy such that $\mathsf{v} = (\partial_{\mathsf{u}}\sigma)^t$ [7, 17–21]. The symmetrized system (23) finally reveals in particular the hyperbolic-parabolic structure and may also be used for stabilized finite element methods.

The fundamental links between kinetic theory and symmetrization may then be analyzed. Property (S_1) first results from the expression of Boltzmann entropy and Property (S_2)—or equivalently (E_2)—the existence of the entropy fluxes $q_i = v_i\sigma$, $i \in \mathfrak{D}$, from the conservation of entropy by Euler equations that is itself a consequence of the kinetic entropy equation (5) at zeroth order. Property (S_3), the reciprocity relations $\widetilde{B}_{ij}^t = \widetilde{B}_{ji}$, $i, j \in \mathfrak{D}$, are consequences of the symmetry properties of the transport coefficients and thus of the bracket operator [7, 17]. Property (S_4), the positivity properties of the dissipation matrices, results from that of the bracket operator or equivalently from $\sum_{i,j\in\mathfrak{D}} \langle \partial_i \mathsf{v}, \widetilde{B}_{ij}\partial_j\mathsf{v}\rangle = k_{\text{B}}[\phi, \phi] = \mathfrak{v}^{(1)}\mathcal{J}$. The Properties (S_5) and (S_7) are consequences of the persistence of Boltzmann H theorem for reactive collisions at the fluid level (21) and we also have $\mathfrak{E}^\perp = M\mathcal{R} \times \{0\} \times \{0\}$ and $\langle \mathsf{v}, \widetilde{\Omega}\rangle = -\mathfrak{v}^{(1)}\mathcal{C}$ where $M = \text{diag}(m_1, \ldots, n_{\text{N}})$. Finally Property (S_6) is a consequence of (19), and thus of the expression of kinetic reactive collision terms, that yields at chemical equilibrium the Jacobian $\partial_{\mathsf{v}}\widetilde{\Omega} = -\sum_{r\in\mathfrak{R}} \Lambda_r v_r'\otimes v_r'$ where $v_r' = (Mv_r, 0, 0)^t$, and Λ_r are positive scalars [7]. All the symmetrization properties (S_1)–(S_7) may thus closely be related to the kinetic theory.

4.2   Well Posedness

The symmetrized entropic system may next be written in a normal form where hyperbolic and parabolic variables are split [7, 15–17, 20]. Introducing the partition $I = \{1\}$ and $II = \{2, \ldots, \text{N} + 4\}$, the normal variable w is defined by $\mathsf{w} = (\mathsf{w}_I, \mathsf{w}_{II})^t$ with $\mathsf{w}_I = \rho$ and $\mathsf{w}_{II} = \frac{1}{k_{\text{B}}T}\big(g_2 - g_1, \ldots, g_{\text{N}} - g_1, v, -1\big)^t$. The equations in the w variable, obtained by the change of variable $\mathsf{v} = \mathsf{v}(\mathsf{w})$ and multiplying on the left by $\partial_{\mathsf{w}}\mathsf{v}^t$ in order to maintain symmetry, read

$$\overline{A}_0(\mathsf{w})\partial_t \mathsf{w} + \sum_{i\in\mathfrak{D}}\overline{A}_i(\mathsf{w})\partial_i \mathsf{w} = \sum_{i,j\in\mathfrak{D}} \partial_i\big(\overline{B}_{ij}(\mathsf{w})\partial_j \mathsf{w}\big) + \overline{\Omega}(\mathsf{w}), \qquad (24)$$

with $\overline{A}_0 = (\partial_\mathsf{w}\mathsf{v})^t\widetilde{A}_0\,\partial_\mathsf{w}\mathsf{v}$, $\overline{A}_i = (\partial_\mathsf{w}\mathsf{v})^t\widetilde{A}_i\,\partial_\mathsf{w}\mathsf{v}$, $\overline{B}_{ij} = (\partial_\mathsf{w}\mathsf{v})^t\widetilde{B}_{ij}\,\partial_\mathsf{w}\mathsf{v}$, $\overline{\Omega} = (\partial_\mathsf{w}\mathsf{v})^t\Omega$. The mathematical properties of the normal form (24) are by construction similar to that of the entropic form [7, 17] and we also have the block decompositions $\overline{A}_0 = \mathrm{diag}(\overline{A}_0^{I,I}, \overline{A}_0^{II,II})$ and $\overline{B}_{ij} = \mathrm{diag}(0, \overline{B}_{ij}^{II,II})$, and in addition the matrix $\overline{B}^{II,II} = \sum_{i,j\in\mathfrak{D}} \overline{B}_{ij}^{II,II} w_i w_j$ is positive definite for nonzero w. The resulting system (24) thus appears as a composite system symmetric-hyperbolic in the w_I variable and symmetric-strongly parabolic in the w_{II} variable. Symmetry is then a key point in order to derive a priori estimates that next yield local existence of solutions [15] as well as global solutions around constant equilibrium states and asymptotic stability [7, 16, 17]. There is also an important coupling stability condition between the hyperbolic and the parabolic operators, namely the Kawashima-Shizuta condition that yields strictly dissipative linearized normal form around equilibrium states [7, 17, 20]. A physical characterization of this coupling condition is that all waves associated with multicomponent Euler equations are damped by dissipative processes and this is again a consequence of the underlying physical framework [17].

4.3 Transport Coefficients

A major success of kinetic theory is to provide the transport coefficients (15, 16). A variational space is introduced for each coefficient μ and a Galerkin method is used in order to solve the corresponding system of Boltzmann linearized integral equations with the Enskog constraints [7, 8, 11–13]. This leads to linear systems with constraints and the mathematical properties of the transport linear systems may be extracted from the kinetic framework [7, 13].

Writing the linear system in the form $G\alpha = \beta$, the matrix G is either regular or its nullspace is one dimentional $N(G) = \mathbb{R}\mathcal{N}$ with a constraint $\langle \alpha, \mathcal{Z}\rangle = 0$ where $\langle \mathcal{N}, \mathcal{Z}\rangle \neq 0$. The transport coefficient is then typically given by the scalar product $\langle \alpha, \beta\rangle$. In the singular case, the matrix G is positive semidefinite, the nullspace $\mathbb{R}\mathcal{N}$ is associated with the collisional invariants in the variational approximation space and $\beta \in R(G)$. It is further possible to introduce the sparse transport matrix $db(G)$ constituted by diagonals of blocks of G. Assuming that the variational approximation space is perpendicular to the collisional invariants and orthogonal to constants, it can be shown [13] that $2db(G) - G$ and $db(G)$ are positive definite when $\mathrm{N} \geq 3$.

The solution of the transport linear system can then be obtained either from the symmetric positive definite system $(G + \mathcal{Z}\otimes\mathcal{Z})\alpha = \beta$ or from iterative techniques [22]. The iterative techniques may be generalized conjugate gradients using the very good preconditioning matrix $db(G)$ [13]. Stationary techniques may also be used with a splitting $G = M - L$ with $M = db(G)$, $T = M^{-1}L$ and $P = I - \mathcal{N}\otimes\mathcal{Z}/\langle\mathcal{N}, \mathcal{Z}\rangle$ that yield $\alpha = \sum_{0\leq j<\infty}(PT)^j PM^{-1}P^t\beta$. A key point is then that the matrix $M + L = 2db(G) - G$ is positive definite but this is a consequence from Boltzmann linearized equations [13]. The kinetic theory thus provides the preconditioning matrix $db(G)$ for conjugate gradient methods and insures the convergence of the stationary algorithms [13, 22]

5 Conclusion

The close links between kinetic theory and the hyperbolic-parabolic structure of fluid systems of conservation laws has been discussed for multicomponent flows. A similar analysis may be obtained for other fluid systems like for instance for the relaxation of internal energy [23].

Only classical fluid systems have been intestigated in these notes by using the natural collisional invariants. For moment methods, however, such an ideal mathematical structure is not anymore guaranteed [24, 25]. More complex hyperbolic parabolic structure are also obtained for diffuse interface fluids with antisymmetric diffusion matrices arising from capillarity [26–28].

References

1. Anderson JD Jr (1989) Hypersonics and high temperature gas dynamics. McGraw-Hill Book Company, New-York
2. Capitelli M, Bruno D (2013) Fundamental aspects of plasma physics, transport, springer series on atomic, optical and plasma physics. Springer, New York. https://doi.org/10.1007/978-1-4419-8172-1
3. Nagnibeda E, Kustova E (2009) Non-equilibrium reacting gas flow. Springer, Berlin. https://doi.org/10.1007/978-3-642-01390-4
4. Kee RJ, Coltrin ME, Glarborg P (2003) Chemically reacting flow. Wiley, Hoboken. https://doi.org/10.1002/0471461296
5. Ern A, Giovangigli V, Smooke M (1996) Numerical study of a three-dimensional chemical vapor deposition reactor with detailed chemistry. J Comp Phys 126:21–39. https://doi.org/10.1006/jcph.1996.0117
6. Williams FA (1985) Combustion theory, 2nd edn. Co., Inc., Melo Park, The Benjamin/Cummings Pub. https://doi.org/10.1201/9780429494055
7. Giovangigli V (1999) Multicomponent flow modeling. Birkhaüser, Boston. https://doi.org/10.1007/978-1-4612-1580-6 Erratum at http://cmap.polytechnique.fr/ giovangi
8. Waldmann L (1958) Transporterscheinungen in Gasen von mittlerem Druck. Handbuch der Physik 12:295–514
9. Ludwig G, Heil M (1960) Boundary layer theory with dissociation and ionization, advances in applied mathematics. Academic Press, New York, vol VI, pp 39–118. https://doi.org/10.1016/S0065-2156(08)70110-8
10. Waldmann L, Trübenbacher E (1962) Formale Kinetische Theorie von Gasgemischen aus Anregbaren Molekülen. Zeitschr. Naturforschg. 17a:363–376. https://doi.org/10.1515/zna-1962-0501
11. Chapman S, Cowling TG (1970) The mathematical theory of non-uniform gases. Cambridge University Press, Cambridge

12. Ferziger JH, Kaper HG (1972) Mathematical theory of transport processes in Gases. North Holland Pub. Co., Amsterdam
13. Ern A, Giovangigli V (1994) Multicomponent transport algorithms lectures notes in physics, series monographs m24. Springer, Berlin. https://doi.org/10.1007/978-3-540-48650-3
14. Ern A, Giovangigli V (1998) The kinetic equilibrium regime. Physica-A 260:49–72. https://doi.org/10.1016/S0378-4371(98)00303-3
15. Giovangigli V, Massot M (1998) The local cauchy problem for multicomponent reactive flows in full vibrational nonequilibrium. Math Meth Appl Sci 21:1415–1439. https://doi.org/10.1002/(SICI)1099-1476(199810)21:15<1415::AID-MMA2>3.0.CO;2-D
16. Giovangigli V, Massot M (1998) Asymptotic stability of equilibrium states for multicomponent reactive flows, mathematical models & methods in applied. Science 8:251–297. https://doi.org/10.1142/S0218202598000123
17. Giovangigli V, Matuszewski L (2013) Mathematical modeling of supercritical multicomponent reactive fluids. Math Mod Meth App Sci 23:2193–2251. https://doi.org/10.1142/S0218202513500309
18. Godunov S (1961) An interesting class of quasilinear systems. Sov Math Dokl 2:947–949. https://doi.org/10.1016/j.jcp.2024.113521
19. Friedrichs KO, Lax PD (1971) Systems of conservation laws with a convex extension. Proc Nat Acad Sci USA 68:1686–1688. https://doi.org/10.1073/pnas.68.8.16
20. Kawashima S (1984) Systems of hyperbolic-parabolic composite type, with application to the equations of magnetohydrodynamics, Doctoral Thesis, Kyoto University. https://doi.org/10.14989/doctor.k3193
21. Chen GQ, Levermore CD, Liu TP (1994) Hyperbolic conservation laws with stiff relaxation terms and entropy. Comm Pure Appl Math 47:787–830. https://doi.org/10.1002/cpa.3160470602
22. Ern A, Giovangigli V (1995) Fast and accurate multicomponent property evaluations. J Comp Phys 120:105–116. https://doi.org/10.1006/jcph.1995.1151
23. Bruno D, Giovangigli V (2022) Internal energy relaxation processes and bulk viscosities in fluids. MDPI Fluids 7:356. https://doi.org/10.3390/fluids7110356
24. Struchtrup H (2005) Macroscopic transport equations for rarefied gas flows, interaction of mechanics and mathematics series. Springer, Heidelberg. https://doi.org/10.1007/3-540-32386-4
25. Torrilhon M (2016) Modeling nonequilibrium gas flow based on moment equations. Annu Rev Fluid Mech 48:429–458. https://doi.org/10.1146/annurev-fluid-122414-034259
26. Giovangigli V, Le Calvez Y, Nabet F (2023) Symmetrization and local existence of solutions for diffuse interface fluid models. J Math Fluid Mech 25:82. https://doi.org/10.1007/s00021-023-00825-4
27. Giovangigli V (2020) Kinetic derivation of diffuse-interface fluid models. Phys Rev E 102:012110. https://doi.org/10.1103/PhysRevE.102.012110
28. Giovangigli V (2021) Kinetic derivation of Cahn-Hilliard fluid models. Phys Rev E 104:054109. https://doi.org/10.1103/PhysRevE.104.054109

Relativistic Equilibrium Distribution of Rapidity: A Kinetic Theory Perspective

Guillermo Chacón-Acosta, Ana Laura García-Perciante, and Alma R. Méndez

Abstract The relativistic Boltzmann kinetic equation is reformulated using the rapidity variable, a Lorentz-invariant quantity with additive properties under boosts. From this, the equilibrium distribution function is obtained and its properties are explored, highlighting its applications and comparing its behavior with the usual Jüttner distribution function.

Keywords Relativistic kinetic theory · Jüttner distribution · Rapidity

1 Introduction

Historically, the Jüttner distribution function has been widely accepted as the equilibrium distribution for a relativistic gas given its compatibility with the principles of relativity, and statistical mechanics. However, the debate regarding whether this function corresponds to the correct relativistic generalization of the Maxwell-Boltzmann distribution has resurfaced, principally driven by a lack of detailed simulations and experimental evidence. Moreover, the so-called modified Jüttner distribution, which is characterized by a lower particle population in the high energy tail, has been proposed. This led to further questioning on the standard Jüttner distribution being the best description in some specific contexts.

In response to this debate, significant efforts have been made to validate either of the proposed distributions. Numerical simulations, in one [5], two [14] and three dimensions [9] strongly support Jüttner's distribution even under reference frame boosts. Also, theoretical studies confirm the Jüttner function as the equilibrium solution to the relativistic Boltzmann equation in collisionless systems, consistently with the H-theorem and thus ensuring a non-negative entropy production [4, 10]. However, additional studies revealed that the observed distribution could vary depending on specific modeling choices. For instance, the use of proper time as the temporal

G. Chacón-Acosta (✉) · A. L. García-Perciante · Alma R. Méndez
Departamento de Matemáticas Aplicadas y Sistemas, Universidad Autónoma Metropolitana Cuajimalpa, Cuajimalpa de Morelos, Ciudad de México, Mexico
e-mail: gchacon@cua.uam.mx

© The Author(s) 2026

M. Grabe et al. (eds.), *Rarefied Gas Dynamics*, Springer Aerospace Technology,
https://doi.org/10.1007/978-3-032-00094-1_26

variable leads to the modified Jüttner distribution, as shown in Refs. [8, 11]. The thermostatistical properties of this function, extended to any exponent of the relativistic energy, were studied in Ref. [1]. Also, a generalized distribution was found when the effects of a gravitational field are taken into account [16]. In this case a transition is encountered between a Jüttner function in the weak-field limit to a modified Jüttner in the strong field case. Recently, an alternative Lorentz invariant distribution has been derived by using the central limit theorem for the sum of the hyperbolic angles associated with Lorentz transformations [6, 7]. This variable, called rapidity, has the property of being additive under Lorentz boosts, as opposed to velocity. Rapidity is widely used to model high-energy experiments where relativistic effects primarily occur along the collision axis [17]. This approach offers intriguing perspectives but still lacks a direct kinetic justification.

The results mentioned above highlight the need for a more thorough theoretical understanding of statistical mechanics and kinetic theory in the relativistic regime. In this brief contribution, we address the kinetic theory derivation of the equilibrium distribution function in the rapidity variable. For this, we rewrite the relativistic Boltzmann equation in such variable and establish the equilibrium Jüttner distribution as the corresponding homogeneous solution. A qualitative analysis is carried out, aiming to provide new insights into its structure and behavior with increasing temperature. In particular, we seek to characterize the counterpart, in the rapidity framework, of the morphological transition to bimodality encountered in the Jüttner distribution [3, 13, 15]. This transition, which is interpreted as the threshold between non-relativistic and relativistic scenarios, reflects the increasing contribution of higher energy states as the system enters the relativistic domain. The rapidity distribution function is here found to retain its bell-shape throughout the transition with the hight of the single peak changing from increasing to decreasing at a critical temperature. This behavior may give additional insights into the transient regime.

The paper is organized as follows. Section 2 introduces the rapidity variable, discussing its definition and fundamental properties. In Sect. 3, we rewrite the relativistic Boltzmann equation in terms of the rapidity and derive the corresponding Jüttner equilibrium distribution. In Sect. 4, we analyze the implications of the bell-shaped rapidity distribution in the region where relativistic effects become relevant. Finally, in Sect. 5, we summarize and discuss our results and outline potential future research lines.

2   Rapidity as Relativistic Realization of Velocity

The one-dimensional Lorentz transformation of a particle's velocity v to a moving frame with relative velocity V is well known to be given by

$$v' = \frac{v + V}{1 + vV},$$

(1)

where here and in what follows we consider natural units with $c = 1$. From this expression one can notice that $(v' - 1)/(v' + 1)$ does not depend on vV. Moreover, one can define a new variable, usually referred to as rapidity [17], as follows

$$y := \frac{1}{2} \ln \left(\frac{1 + v}{1 - v} \right), \tag{2}$$

such that y transforms in an additive fashion under Lorentz boosts. That is, if Y is the rapidity corresponding to the velocity V of the boost, one obtains $y' = y + Y$, which resembles a Galilean transformation. Moreover, Lorentz transformations can be parametrized in terms of rapidity by observing that $\cosh y$ reduces to the Lorentz factor γ and $\sinh y = \gamma V$. Consequently, the rapidity can be also written as $y = \operatorname{arctanh} V$ and is thus sometimes referred to as the hyperbolic angle in spacetime diagrams.

Rapidity is a fundamental concept in accelerator physics, mainly because collisions occur along a given axis in high-energy experiments [17]. For instance, the rapidity and angular differences between two events remain invariant under Lorentz boosts parallel to the beam axis, preserving the intrinsic symmetries of the system. This invariance simplifies the analysis and interpretation of collision data since it guarantees consistency regardless of the observer's reference frame. Experimental results are typically represented by histograms of rapidity distributions, which are often modeled with Gaussian profiles to provide the best fit to the observed data. These distributions are particularly helpful because they remain non-deformed with respect to the center-of-mass frame by boosts parallel to the collision axis.

Another interesting feature of rapidity is its behavior in the limiting cases. In the non-relativistic limit, where velocities are much smaller than the speed of light, one obtains $y \sim v$. For speeds close to the speed of light, it diverges as $y \sim \sqrt{2(\gamma - 1)}$, which coincides with the first approximation of the velocity in terms of γ. This divergence emphasizes its role in relativistic kinematics, where it provides a continuous and additive measure for relative motion. Therefore, the rapidity of a particle can be considered a relativistic realization of its velocity, which has advantages in describing high-speed dynamics.

Rapidity can also be expressed as a function of the four-momentum through the velocity in the boost direction, defined as $V = p_x/E$, as follows:

$$y = \frac{1}{2} \ln \left(\frac{E + p_x}{E - p_x} \right) = \ln \left(\frac{E + p_x}{m_T} \right). \tag{3}$$

Here E is the energy, $m_T^2 = E^2 - p_x^2 = m^2 + \mathbf{p}_T^2$ is the so-called transverse mass [17] and $\mathbf{p}_T$ is the momentum transverse to the direction of the boost. In these variables, momentum can be decomposed as $\mathbf{p} = p_x \hat{e}_x + \mathbf{p}_T$, with p_x being the longitudinal component of $\mathbf{p}$. For a free particle on the mass shell, i.e. satisfying $E^2 = \mathbf{p} \cdot \mathbf{p} + m^2$, the four-momentum has only three independent degrees of freedom. These can be conveniently parametrized by $(y, \mathbf{p}_T)$, where y replaces the longitudinal momentum

p_x. In this representation, E and p_x are directly expressed in terms of rapidity and transverse mass as:

$$E = m_T \cosh y, \quad p_x = m_T \sinh y. \tag{4}$$

This parametrization offers significant advantages in high-energy physics, particularly in collider experiments, where rapidity provides a natural measure for describing particle distributions due to its additive property under successive Lorentz transformations. Furthermore, using m_T and y highlights the separation between transverse and longitudinal dynamics, simplifying the analysis of processes involving boosts along a preferred direction.

3 Distribution Function of Rapidity from Relativistic Kinetic Theory

The covariant Boltzmann equation for the momentum distribution function $f(p)$ of a system of neutral relativistic particles in flat spacetime is given by [2]

$$p^\mu \frac{\partial f}{\partial x^\mu} = \int \frac{d^3 p_1}{E_1} \, d\Omega \, F \, \sigma \left(f_1' f' - f_1 f \right). \tag{5}$$

The integral on the right-hand side of Eq. (5) is the well-known collision term, accounting for interactions with particles of the system labeled with a subscript 1. Here $d^3 p_1/E_1$ is the invariant line element, σ the interaction cross-section, $d\Omega$ the solid angle and $F = \sqrt{(p_1^\mu p_\mu)^2 - m^4}$ denotes the so-called invariant flux. Using Eq. (4) and the definitions of transverse mass and momentum, we can reformulate the Boltzmann equation (5) as a function of rapidity, as shown below

$$m_T \cosh y \frac{\partial f}{\partial t} + m_T \sinh y \frac{\partial f}{\partial x} + \mathbf{p}_T \cdot \nabla_T f = \int dy \, d^2 p_T \, d\Omega \, F \, \sigma \left(f_1' f' - f_1 f \right), \tag{6}$$

where ∇_T stands for the gradient in the direction orthogonal to the boost.

Following [2, 4], one can obtain the equilibrium distribution function f_{eq} by imposing the condition of zero entropy production. This leads to the fact that f_{eq} must depend only on the collisional invariants $\{1, p^\mu\}$ and remain invariant under Lorentz transformations. This, in turn, allows for evaluation in a convenient rest frame where $B^\mu \rightarrow (B, \mathbf{0})$, such that f_{eq} can be written as

$$f_{eq} = A \, e^{-B \, m_T \cosh y}, \tag{7}$$

where the coefficients A and B can be related to the physical moments of the distribution. Indeed, if N^μ and $T^{\mu\nu}$ are the particle four-flux and energy momentum tensor respectively, the following relations hold:

$$N^\mu = -A\frac{\partial \mathcal{I}}{\partial B_\mu}, \quad T^{\mu\nu} = A\frac{\partial^2 \mathcal{I}}{\partial B_\mu \partial B_\nu}, \tag{8}$$

where

$$\mathcal{I} = \int \frac{d^3 p}{E} e^{-B^\mu p_\mu}.$$

In the frame here considered, $\mathcal{I}$ can be written in terms of $B = \sqrt{-B_\mu B^\mu}$ as follows

$$\mathcal{I} = \int dy\, d^2 p_T e^{-B m_T \cosh y} = \frac{4\pi m}{B} K_1(m B), \tag{9}$$

where and $K_i(z)$ is the modified Bessel function of the second kind. Aligning the rest frame with the fluid frame and using the definition of entropy flux, together with the Gibbs relation to express the energy density in terms of temperature [2, 4], leads to

$$A = \frac{nm B}{4\pi m^3 K_2(m B)}, \quad B = \frac{1}{T}, \tag{10}$$

where we also consider units in which the Boltzmann constant is $k_B = 1$. Accordingly, the equilibrium distribution function can be written as

$$f_{eq} = f_0 \frac{z}{K_2(z)} \exp\left(-z\mu \cosh y\right), \tag{11}$$

where $f_0 = n/4\pi m^3$. In Eq. (11) we have also introduced the relativistic factor $z = m/T$ and the fraction of the transverse mass $\mu = m_T/m$, such that the non-relativistic limit is characterized by $z \gg 1$ and $\mu \to 1$, additional to $y \ll 1$.

It is worth noticing that, by using the known physical moments given in Eq. (8), the resulting distribution should match the well-known Jüttner distribution, and thus Eq. (11) is indeed, the Jüttner distribution in terms of rapidity. When analyzed as a function of rapidity, transverse mass, and temperature, this distribution exhibits several interesting features. First, it is symmetric with respect to y, with a maximum at $y = 0$ for any positive value of z and μ. Moreover, as shown in Fig. 1, it resembles a gaussian-like bell-shaped curve in contrast with the unimodal to bimodal transition observed when expressed in terms of velocity. Furthermore, when analyzed as a function of temperature, the distribution exhibits a peak at a critical point, z_m, whose magnitude depends on the transverse mass, suggesting a possible correlation with the relativistic morphological transition between limiting cases. This feature is further discussed in the next section.

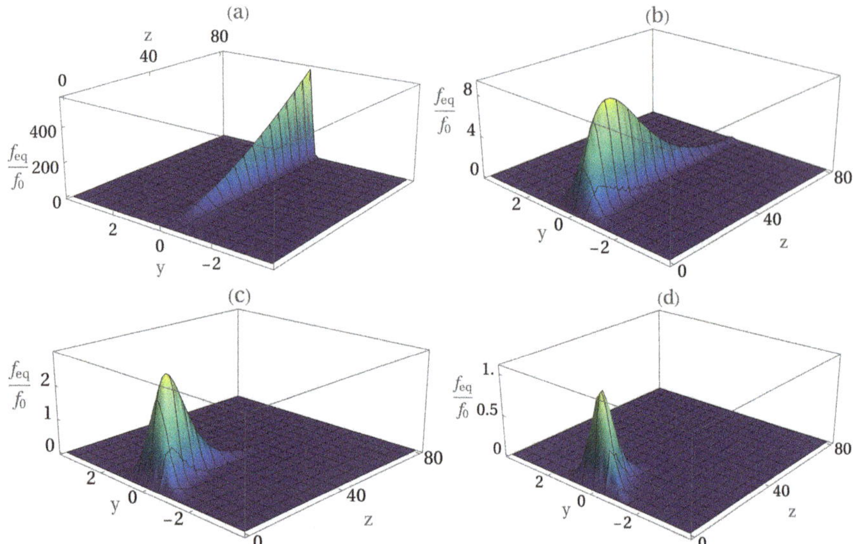

Fig. 1 Relativistic distributions of the rapidity for different values of the transverse mass fraction: **a** $\mu = 1.0$, **b** $\mu = 1.1$, **c** $\mu = 1.2$, **d** $\mu = 1.35$. We observe how z decreases with μ

4   Relation to the Relativistic Morphological Transition

As previously mentioned, the relativistic distribution exhibits a transition from uni-modal to bimodal as the temperature increases. Moreover, relativistic effects are considered to become significant precisely at the critical temperature T_c of the transition, which depends on dimensionality and statistics [3, 13, 15].

In contrast, the distribution given in Eq. (11) does not show such a morphological transition. Instead, it has a single global maximum at $y = 0$, whose height depends on the temperature through the parameter z. As shown in Fig. 1, the behavior of $f_{eq}(y = 0)$ changes from increasing to decreasing at a critical value $z = z_m$ for each $\mu > 1$. However, in the non-relativistic regime, the distribution simplifies to $f_{eq} \sim \sqrt{2/\pi}\, z^{3/2} e^{-(y^2 z)/2}$ and thus diverges at $y = 0$, indicating the absence of a maximum temperature in this limit (see Fig. 1a). Thus the existence of the maximum temperature given by z_m appears to be an exclusive property for transverse masses $\mu \gtrsim 1$.

The behavior of the temperature corresponding to a maximum in $f_{eq}(y = 0)$ for $\mu > 1$ can be further assessed by inspection of Fig. 2, where z_m is shown for different values of μ. The red points correspond to the numerical values obtained by maximizing the distribution as a function of z for various values of μ, with the black curve corresponding to the interpolation of these points. The gray lines indicate the distribution contour lines as a function of (z_m, μ) for $y = 0$. Remarkably, for $\mu \simeq 1.36$ the inverse of the maximum temperature becomes close to the critical value of the relativistic transition, which in three dimensions corresponds to $z_c = 5$.

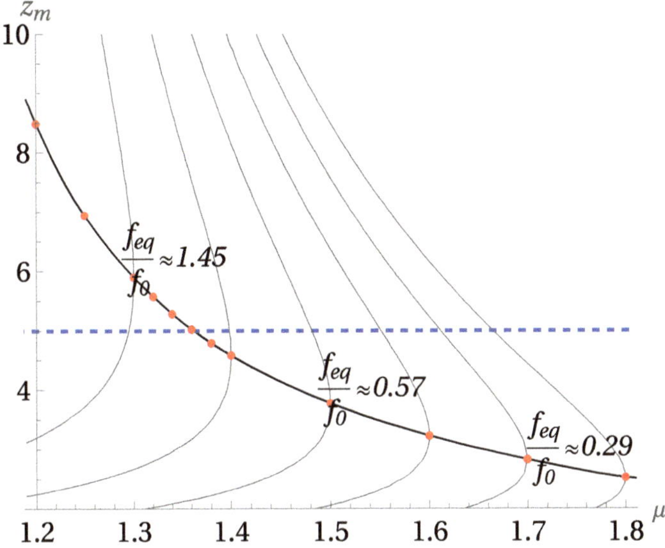

Fig. 2 Graph of z_m as a function of traverse mass fraction. The red dots are numerical values from the maximization of the distribution, the black curve is the interpolation, and the gray lines are the distribution contour lines for $y = 0$. The dashed blue line indicates the critical factor $z_c = 5$

Moreover, one finds that the level curve corresponding to $f_{eq}(y = 0)/f_0 \simeq 1$ in Fig. 2 passes through this region, suggesting that relativistic effects become important in this representation from these coincident values onwards.

5 Summary and Discussion

In this contribution, we established the rapidity distribution in an equilibrium configuration from the kinetic theory point of view and qualitatively analyzed its properties, principally in relation to the relativistic transition. The choice of the rapidity variable is mostly motivated by its additive property under Lorentz boosts, which results in its wide use in high-energy experiments [6, 7, 17]. To accomplish this task we reformulated the relativistic Boltzmann equation in terms of rapidity and transverse momentum, and solved it for the equilibrium case.

The resulting solution corresponds to the Jüttner distribution (Eq. 11) depending on rapidity, transverse mass fraction, and temperature. This distribution exhibits a bell-shaped profile for all temperatures, flattening in the ultra-relativistic regime and approximating a Gaussian in the non-relativistic limit. Interestingly, the peak of the distribution reaches a maximum at z_m depending on μ. This observation suggests a relation with the relativistic transition temperature, which has been previously studied. It is important point out that in this contribution we focus on the non-quantum three-dimensional case in the comoving frame. However, it is very likely that the maximum temperature will also depend on frame velocity, statistics and dimensionality, similar to how the critical temperature of the relativistic transition described above behaves [3, 13, 15].

We found that the value of z_m for which the peak of the distribution reaches a maximum value is close to the critical value $z_c = 5$ of the relativistic transition [3, 13, 15], at approximately $\mu \simeq 1.36$. This suggests that z_m is the threshold between the ultra-relativistic (for $z < z_m$) and non-relativistic (for $z > z_m$) regimes. Thus, this approach offers an alternative perspective for studying the relativistic transition, in terms of the transverse mass and the maximum temperature, that may influence morphological transitions when analyzed in the original variables.

Future work could explore moments of rapidity to provide deeper insight into the relativistic transition. For instance, kurtosis may be relevant in the ultra-relativistic regime, where the distribution flattens. This analysis will follow from the transfer equation by integrating Eq. (5). Additionally, alternative distributions, like the modified Jüttner [1] or those derived from nonlinear Fokker-Planck equations in high-energy processes [12, 18], deserve further study as they differ from Jüttner distribution by energy functions, offering potentially valuable theoretical and experimental insights.

Acknowledgements We thank the organizers of the 33rd International Symposium on Rarefied Gas Dynamics for providing an excellent forum for scientific interaction.

References

1. Aragón-Muñoz L, Chacón-Acosta G (2018) J Phys: Conf Ser 1030:012004
2. Cercignani C, Kremer GM (2002) The relativistic Boltzmann equation: theory and applications. Birkhäuser Verlag
3. Chacón-Acosta G (2016) AIP Conf Proc 1786:070016
4. Chacón-Acosta G, Dagdug L, Morales-Técotl H (2010) Phys Rev E 81:021126
5. Cubero D, Casado-Pascual J, Dunkel J, Talkner P, Hänggi P (2007) Thermal equilibrium and statistical thermometers in special relativity. Phys Rev Lett 99:170601
6. Curado EMF, Cedeño CE, Soares ID, Tsallis C (2022) Chaos 32:103110
7. Curado EMF, Germani FTL, Soares ID (2016) Phys A 444:963–969
8. Dunkel J, Hänggi P (2009) Phys Rep 471:1–73
9. Dunkel J, Hänggi P, Hilbert S (2009) Nature Phys 5:741–747
10. García-Perciante AL, Méndez AR, Escobar-Aguilar E (2017) J Stat Phys 167:123–134
11. Ghodrat M, Montakhab A (2010) Phys Rev E 82:011110
12. Lavagnno A (2002) Phys A 305:2038–241
13. Mendoza M, Araújo NAM, Succi S, Herrmann HJ (2012) Sci Rep 2:611
14. Montakhab A, Ghodrat M, Barati M (2009) Statistical thermodynamics of a two-dimensional relativistic gas. Phys Rev E 79:031124
15. Montakhab A, Shahsavar L, Ghodrat M (2014) Phys A 412:32–38
16. Sánchez-Rey B, Chacón-Acosta G, Dagdug L, Cubero D (2013) Phys Rev E 87:052121
17. Torres-Rincon JM (2014) Hadronic transport coefficients from effective field theories. Springer
18. Wolschin G (2004) Phys Rev E 69:024906

Classical Trajectories Estimation of Transport Properties of N_2-O_2 Mixtures

Domenico Bruno and Aldo Frezzotti

Abstract Estimations of transport coefficients of N_2-O_2 mixtures are obtained by describing N_2-N_2, O_2-O_2 and N_2-O_2 binary collisions with three distinct Potential Energy Surfaces. Classical Trajectories are then used in combination with Monte Carlo and with non-equilibrium Direct Simulation Monte Carlo simulations to obtain transport coefficients of N_2-O_2 mixtures.

Keywords Air transport coefficients · Classical trajectories

1 Introduction

Elastic light scattering, or Rayleigh-Brillouin scattering (RBS), is currently employed for remote sensing of the atmosphere [29]. The current practice is to use the Tenti S6 model [7, 26] as line-shape model to interpret the experiments and extract the required gas temperature. It has been shown, however, that the Tenti S6 model fails to reproduce the RBS spectrum when the relaxation of internal degrees of freedom is slow enough that non-equilibrium models are required [10]. It is thus worth exploring more accurate theoretical tools to analyse Rayleigh-Brillouin scattering (RBS) experiments on air species. Direct Simulation Monte Carlo (DSMC) [6] is a natural choice since it solves the Boltzmann equation with no approximations. Phenomenological binary collision models [6] are replaced with Classical Trajectories (CT) calculations [15]. Simulations based on CT-DSMC techniques [19] benefit from the availability of accurate Potential Energy Surfaces (PES), describing the interaction of two colliding molecules. Here, attention is limited to PES's, describing N_2-N_2, O_2-O_2, and N_2-O_2 pairwise interactions between rigid, dumbbell-like molecules. The effects of vibrational excitations can, in fact, be neglected under the conditions

D. Bruno (✉)
National Research Council, Institute for Plasma Science and Technology, Bari, Italy
e-mail: domenico.bruno@cnr.it

A. Frezzotti
Department of Aerospace Science & Technology, Politecnico di Milano, Milan, Italy

© The Author(s) 2026

M. Grabe et al. (eds.), *Rarefied Gas Dynamics*, Springer Aerospace Technology,
https://doi.org/10.1007/978-3-032-00094-1_27

of RBS experiments conducted near room temperature. The considered PES's have been analyzed by computing transport coefficients by different methods, including Monte Carlo quadrature (CT-MC) applied to explicit expressions of transport coefficients [21] and non-equilibrium CT-DSMC simulations. The latter are, in general, more expensive than CT-MC estimations but provide values of transport coefficients that do not require higher order corrections [12]. Moreover, they can be used for complex gas mixtures when expressions of transport coefficients are not immediately available. The general form of the N_2-N_2, O_2-O_2, and N_2-O_2 PES's here used to estimate air transport coefficients is described in Sect. 2. Individual PES are described and discussed in Sects. 2.1–2.3. The results on air shear and bulk viscosity are given in Sect. 3.

2 Potential Energy Surfaces

The intermolecular potentials considered have the following general structure [2–4]:

$$V(R_{12}, \theta_1, \theta_2, \phi) = \sum_{L_1, L_2, L} V^{L_1, L_2, L}(R_{12}) \mathcal{Y}^{L,0}_{L_1, L_2}(\theta_1, \theta_2, \phi) \tag{1}$$

$$L_1, \ L_2 = 0, 1, 2, \ldots, \quad |L_1 - L_2| < L < L_1 + L_2$$

where R_{12} is the distance between the molecular centers of mass, θ_1, θ_2 and ϕ are three angles that specify the relative orientation of molecular pairs (see Fig. 1). Potentials are expressed as superpositions of a finite number of terms, each resulting from the product of a radial function $V^{L_1, L_2, L}(R_{12})$ and a bipolar spherical harmonic function $\mathcal{Y}^{L,0}_{L_1, L_2}(\theta_1, \theta_2, \phi)$ [4]. In the expansion (1), the first term $V^{0,0,0}(R_{12})$ represents the isotropic contribution to the potential. The number of terms depends on interacting species, varying from 4 in the O_2-O_2 PES, to 6 in the N_2-O_2 PES. The form of the radial coefficients is described in Refs. [2–4] and it will not be given here, for brevity. However, it is important to mention that the adjustable parameters present in the radial functions have been determined from molecular beam scattering experimental data. Accordingly, parameters uncertainties are also provided in the case of O_2-O_2 and N_2-N_2 PES's, whereas no error intervals are given for the parameters that characterize N_2-O_2 PES.

The interpretation of parameters uncertainties requires some caution. Actually, an uncertainty quantification analysis, conducted within the present work, has shown that a considerable fraction of the PES parameters combinations, although compatible with the given error bounds, lead to strongly distorted and unrealistic PES's. Nevertheless, sensible variations, involving one or two parameters at a time, can be exploited to improve the predictions of transport coefficients, as shown in the following.

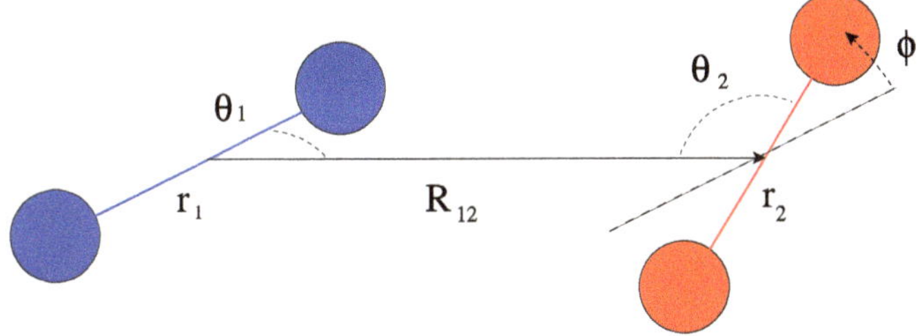

Fig. 1 Coordinates system describing the relative position and orientation of two interacting linear, rigid molecules

2.1 N_2-N_2 PES

The quality of N_2-N_2 PES [3] has been assessed by computing N_2 transport coefficients in the temperature range [250–600 K], by CT-MC simulations based on the expressions given in [21]. CT-MC estimations are compared with the reference experimental data from [20]. Figure 2 shows that the nominal PES, based on the central values of the confidence intervals of the parameters, underestimates both shear viscosity η_s and thermal conductivity λ. The agreement with experiments is improved by reducing the range of repulsive forces in the isotropic contribution of PES, $V^{0,0,0}(R_{12})$. Accordingly, the nominal value of the scale parameter R_m [4], 4.11Å, has been reduced to 4.065 Å, about half standard deviation below the central value. The correction makes the molecule "smaller" and increases η_s and λ. The accuracy of the corrected PES is comparable with that proposed in [15], which is probably the most accurate N_2-N_2 PES for low temperature applications.

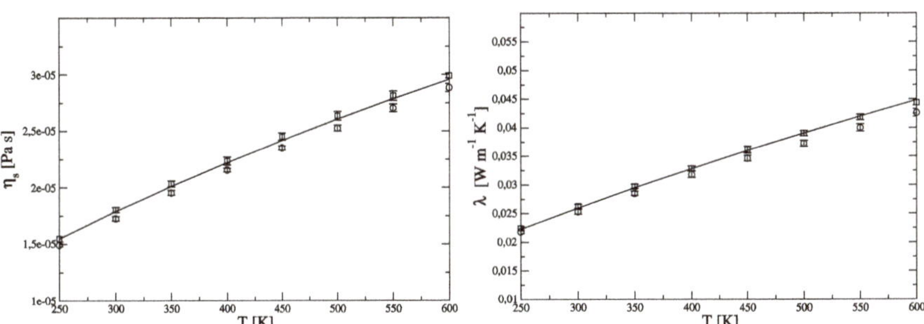

Fig. 2 *Left*—N_2 shear viscosity. *Right*—N_2 thermal conductivity. Solid lines: NIST data [20]; o:N_2-N_2 PES with nominal parametes; □: modified N_2-N_2 PES

2.2 O_2-O_2 PES

A similar evaluation has been conducted for O_2-O_2 PES [4], extending a previous investigation [9] to include the effects of spin on molecular interaction. As shown in [4], spin coupling can be accounted for by introducing three separate PES's, describing singlet ($S = 0$), triplet ($S = 1$) and quintet ($S = 2$) states, being S the total spin. The weights associated with the three PES's are $\frac{1}{9}, \frac{3}{9}, \frac{5}{9}$, respectively [4]. Figure 3, presents the results of the shear viscosity and thermal conductivity obtained from the CT-MC simulations. As in the case of N_2, the nominal O_2-O_2 PES underestimates experimental η_s and λ. Moreover, spin corrections seem to have no effect on η_s and marginal effects on λ.

A closer inspection revealed that the transport properties are different for the three individual PES. However, their weighting compensates for the differences and returns values close to those obtained by the O_2-O_2 PES without spin corrections. As for N_2, the underestimation of the transport coefficients can be mitigated by reducing the effective range of molecular repulsive forces, in the isotropic term $V^{0,0,0}(R_{12})$. Since the deviation from the experimental data is greater than in the case of N_2-N_2 PES, the nominal value of the potential parameter R_m, 3.9 Å, has been reduced to 3.82 Å, about one standard deviation below the nominal value. As shown in Fig. 3, reducing R_m considerably improves the agreement with the experimental η_s, but it is not sufficient to reduce the deviation from the experimental λ by a comparable amount.

2.3 N_2-O_2 PES

The N_2-O_2 PES for the interaction between the N_2 and O_2 molecules is described in [3]. It shares its structure with N_2-N_2 and O_2-O_2 PES's, but no error intervals

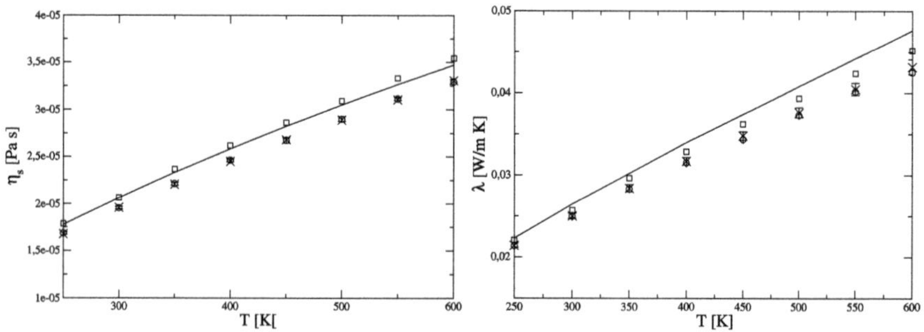

Fig. 3 *Left—O_2* shear viscosity. *Right—O_2* thermal conductivity. Solid lines: NIST data [20]; o:O_2-O_2 PES with nominal parametes, no spin correction; X:PES with nominal parametes and spin correction; □: modified O_2-O_2 PES, with spin correction

are given for the parameters appearing in the radial function $V^{L_1,L_2,L}(R_{12})$. To the authors' knowledge, this particular PES has not yet been evaluated in regard to predictions of transport properties of N_2-O_2 mixtures. In addition to the shear and bulk viscosity estimations, described in Sect. 3, the quality of the PES has been evaluated by comparing data from binary diffusion coefficient data for equimolar mixtures N_2-O_2 [11] with CT-MC and CT-DSMC estimations. The latter have been obtained from equilibrium CT-DSMC simulations, in which the variance of molecular displacement from initial position, $< \Delta r^2 > (t)$, is recorded. The diffusion coefficient, D_{12}, is obtained by applying the Einstein relation [24]

$$D_{12} = \frac{1}{6}\frac{d}{dt} < \Delta r^2 > \tag{2}$$

after the slope of $< \Delta r^2 > (t)$ becomes constant (see Fig. 4). As shown in Fig. 4a, the agreement between the reference data, the estimations obtained by equilibrium CT-DSMC simulations, and the CT-MC direct evaluation of the orientation-averaged integral $\Omega^{(1,1)}_{(1,2)}$ [14] is very good, indicating that the PES describes the N_2-O_2 interactions well.

3 Estimation of Air Shear and Bulk Viscosities

Following Ref. [9], the shear viscosity η_s of N_2-O_2 mixtures was calculated by nonequilibrium CT-DSMC simulations of isothermal, uniform shear flows (USF). These are amenable to a spatially 0-dimensional problem in which the gas is subject to a velocity-dependent force field whose intensity is proportional to the imposed

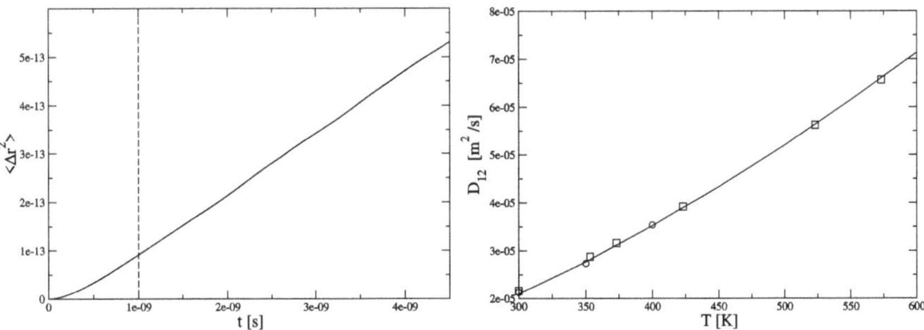

Fig. 4 *Left*—Time evolution of the variance of molecular displacements in an equimolar N_2-O_2 mixture at 1 bar and 300 K. The vertical dashed line, separates the initial, nearly collisionless regime, from the linear one, where Eq. (2) can be used to obtain D_{12}. *Right*—Temperature dependence of D_{12} in an equimolar N_2-O_2 mixture at 1 bar. Solid line: CT-MC estimation from $\Omega^{(1,1)}_{(1,2)}$; o: CT-DSMC estimation from Eq. (2); □: D_{12} data from Ref. [11]

shear rate α. A thermostat is also applied to contrast viscous heating and keep the desired temperature value.

The shear viscosity has been obtained by linear regression of shear stress data, computed for small values of α [9]. Since using the nominal PES's would have led to underestimation of η_s and masked the performance of the N_2-O_2 PES, the modified N_2-N_2 and O_2-O_2 PES's, discussed in Sects. 2.1 and 2.2, have been adopted in the comparisons with reference data.

The left panel of Fig. 5 shows the computed η_s as a function of the mixture temperature T, for two mixture compositions: $X_{O_2} = 0.21$ (air) and $X_{O_2} = 0.5$ (equimolar). In the first case η_s data lie slightly above the pure N_2 viscosity, computed by CT-MC. The point at $T = 300$ K is very close to NIST corresponding point. In the case of the equimolar mixture, the computed η_s data are slightly below the viscosity data from Ref. [11] but the overall agreement is good.

Bulk viscosity arises from the finite relaxation time of internal degrees of freedom [14]:

$$\eta_v = \frac{2c^{(i)}}{3c_V} p\tau^{(i)} \tag{3}$$

where $c^{(i)}$, c_V are the internal and total specific heats (adimensional), p is the pressure and $\tau^{(i)}$ the relaxation time of the internal degrees of freedom:

$$\frac{dE^{(i)}}{dt} = \frac{E^{(i)}_\infty - E^{(i)}}{\tau^{(i)}} \tag{4}$$

where $E^{(i)}$ is the internal energy and $E^{(i)}_\infty$ is the internal energy at equilibrium.

For each gas mixture, the rotational relaxation time at temperature T_0 has been measured from the results of homogeneous (i.e. adiabatic) relaxation simulations where the starting conditions for translational and rotational temperatures are:

$$T_0^t = T_0 + \frac{\Delta T}{3}, \quad T_0^r = T_0 - \frac{\Delta T}{2} \tag{5}$$

The simulated relaxation of rotational energy is fitted with a simple exponential, as suggested by Eq. (4). Then an estimation of the true relaxation time is obtained using small ΔT's, of the order of 10–20 K. The results shown in Figs. 5 and 6 have been obtained using 2×10^6 molecules per simulation. The error bars are indicative of the run-to-run variation of the estimated relaxation times. The accuracy is not high, but sufficient to judge about the behavior of the quantities of interest. In variance with shear viscosity simulations, nominal PES parameters have been used for bulk viscosity estimations. In fact, the uncertainties of the anisotropic contributions of the PES are too large to obtain a reasonably precise estimate of η_v.

Figure 5 shows the results for air. The results of homogeneous relaxation are compared with values obtained from simulation of RBS spectra [10] with DSMC and to the values obtained by fitting the Tenti S6 model to experimental RBS spectra produced with two different driving laser wavelengths. We note that the experiment

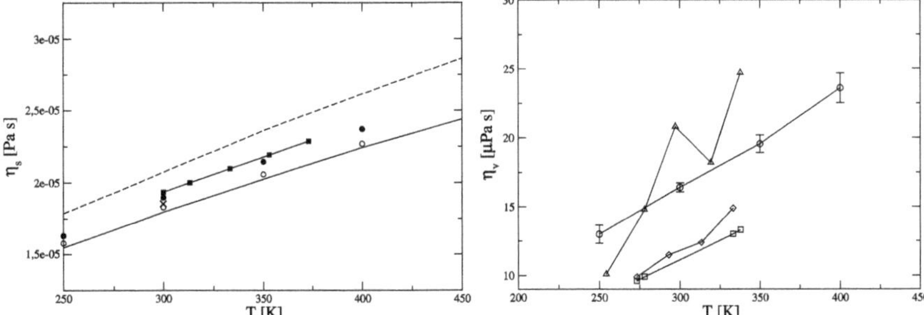

Fig. 5 *Left*—Shear viscosity of N_2-O_2 mixtures. Solid line: pure N_2; Dashed line: Pure O_2; ∘: $X_{O_2} = 0.21$ from CT-DSMC simulations; •: $X_{O_2} = 0.50$ from CT-DSMC simulations; cross: air viscosity from NIST [20]; ■: equimolar N_2-O_2 mixture viscosity data from [11]. *Right*—Bulk viscosity of air. ∘: $X_{O_2} = 0.21$ from CT-DSMC simulations; □: DSMC estimations from RBS spectra [10]; ◇: Tenti S6 model estimations from RBS spectra (green light) [10]; △: Tenti S6 model estimations from RBS spectra (blue light) [10];

conducted with blue light is done under more rarefied conditions, where the accuracy of the Tenti S6 model is worse. The CT-DSMC estimations of η_v lie above the points obtained by fitting experimental RBS spectra to DSMC simulations of N_2-O_2 mixtures, based on VSS collision cross sections and Larsen-Borgnakke model for rotational-translational coupling [6]. Data obtained from Tenti S6 model and the same experimental RBS spectra are somehow closer to CT-DSMC estimations but the agreement is only qualitative. It is of some interest to consider bulk viscosity CT-DSMC estimations for pure N_2 and O_2, for which results from the literature are available. In Fig. 6 CT-DSMC results, obtained from homogeneous relaxation, are compared with the recommended values [27], to semiclassical calculations [5, 22, 28], and to experimental results [1, 17, 18, 23, 25]. Experimental results are mostly based on (old) ultrasound absorption measurements, they are available only at room temperature and show a large spread; semiclassical calculations report results for the first Chapman-Enskog approximation and therefore represent a lower bound; the recommendation in [27] is based on correlations obtained from the experimental measurements fitted to the temperature dependence obtained in the theoretical treatment of rotational relaxation of classical rigid rotors [8]. In the case of N_2, CT-DSMC points lie above all other data sets, suggesting that air bulk viscosity is also overestimated by the considered PES's. Oxygen CT-DSMC points are slightly below the recommended data from [27] but overestimate all other values, with the exception of Connor's measurement at $T = 300$ K.

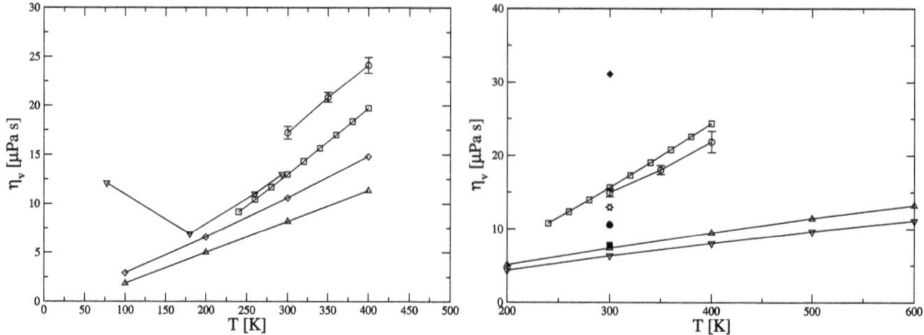

Fig. 6 *Left*—Bulk viscosity of pure N_2. ○: $X_{O_2} = 0.21$ from CT-DSMC simulations; □: from [27]; ◇: from [22]; △: from [5]; ▽: from [18]. *Right*—Bulk viscosity of pure O_2. ○: $X_{O_2} = 0.21$ from CT-DSMC simulations; □:from [27]; △, ▽: from [28]; ●: from [17]; ■: from [23]; ◆: from [13]; ▶: from [25]; ◀: from [1]

4 Conclusions

Shear and bulk viscosity of N_2-O_2 mixtures have been computed assuming that N_2 and O_2 molecules behave as rigid dumbbell, interacting by three PES's, including isotropic and anisotropic contributions. Results, obtained by combining Classical Trajectories integration with Monte Carlo and Direct Simulation Monte Carlo, show that the nominal N_2 and O_2 PES's underestimate experimental pure gas shear viscosity and thermal conductivity. Modifying N_2 PES parameters within the given confidence intervals allows improving the agreement with experimental data but the modification on O_2 PES is not as effective. The N_2-O_2 PES seems to provide a good description of N_2 and O_2 coupling, based on CT-DSMC estimations of the binary diffusion coefficient. CT-DSMC estimations of the bulk viscosity of pure N_2, O_2 and air suggest that the nominal PES's overestimate experimental data.

The above considerations point at the necessity of replacing the N_2-N_2 and O_2-O_2 PES's analyzed here with the more accurate ones, respectively proposed in [15] for N_2, and in [16] for O_2.

References

1. Anderson W, Hornig D (1958) Technical Report 8. Brown University, Metcalf Research Laboratory
2. Aquilanti V, Bartolomei M, Cappelletti D, Carmona-Novillo E, Pirani F (2002) The N2–N2 system: an experimental potential energy surface and calculated rotovibrational levels of the molecular nitrogen dimer. J Chem Phys 117(2):615–627

3. Aquilanti V, Bartolomei M, Carmona-Novillo E, Pirani F (2003) The asymmetric dimer N2–O2: characterization of the potential energy surface and quantum mechanical calculation of rotovibrational levels. J Chem Phys 118(5):2214–2222

4. Aquilanti V, Ascenzi D, Bartolomei M, Cappelletti D, Cavalli S, de Castro Vitores M, Pirani F (1999) Molecular beam scattering of aligned oxygen molecules. The nature of the bond in the O2-O2 dimer. J Am Chem Soc 121(46):10794–10802

5. Billing G, Wang L (1992) Semiclassical calculations of transport coefficients and rotational relaxation of nitrogen at high temperatures. J Phys Chem 96:2572

6. Bird GA (1994) Molecular gas dynamics and the direct simulation of gas flows. Clarendon Press, Oxford

7. Boley C, Desai R, Tenti G (1972) Kinetic models and brillouin scattering in a molecular gas. Can J Phys 50:2158

8. Brau C, Jonkman R (1970) Classical theory of rotational relaxation in diatomic gases. J Chem Phys 52:477

9. Bruno D, Frezzotti A, Ghiroldi G (2015) Oxygen transport properties estimation by classical trajectory-direct simulation Monte Carlo. Phys Fluids 27(5)

10. Bruno D, Frezzotti A, Jamali SH, Van De Water W (2023) Finding the bulk viscosity of air from Rayleigh-Brillouin light scattering spectra. J Chem Phys 158(3):031101

11. Bzowski J, Kestin J, Mason EA, Uribe FJ (1990) Equilibrium and transport properties of gas mixtures at low density: eleven polyatomic gases and five noble gases. J Phys Chem Ref Data 19(5):1179–1232

12. Chapman S, Cowling TG (1990) The mathematical theory of non-uniform gases. Cambridge University Press, Cambridge

13. Connor J (1958) Ultrasonic dispersion in oxygen. J Acoust Soc 30:297

14. Ferziger JH, Kaper HG (1972) Mathematical theory of transport processes in gases. North Holland, Amsterdam

15. Hellmann R (2013) Ab initio potential energy surface for the nitrogen molecule pair and thermophysical properties of nitrogen gas. Molecul Phys 111(3):387–401

16. Hellmann R (2023) Ab initio potential energy surfaces for the O2–O2 system and derived thermophysical properties. J Chem Phys 159(10):104303

17. Holmes R, Jones G, Pusat N, Tempest W (1962) Rotational relaxation in helium + oxygen and helium + nitrogen mixtures. Trans Faraday Soc 58:2342–2347

18. van Houten H, Hermans L, Beenakker J (1985) A survey of experimental data related to the non-spherical interaction for simple classical linear molecules and their mixtures with noble gases. Physica 131A:64–103

19. Koura K (1997) 4 Carlo direct simulation of rotational relaxation of diatomic molecules using classical trajectory calculations: nitrogen shock wave. Phys Fluids 9(11):3543–3549

20. Lemmon EW, Jacobsen RT (2004) Viscosity and thermal conductivity equations for nitrogen, oxygen, argon, and air. Int J Thermophys 25(1):21–69

21. Taxman N (1958) Classical theory of transport phenomena in dilute polyatomic gases. Phys Rev 110(6):1235–1239

22. Nyeland C, Billing G (1988) Transport coefficients of diatomic gases: Internal-state analysis for rotational and vibrational degrees of freedom. J Phys Chem 92:1752

23. Parker J, Adams C, Stavseth R (1953) Absorption of sound in argon, nitrogen, and oxygen at low pressure. J Acoust Soc Am 25:263

24. Resibois P, DeLeener M (1977) Classical kinetic theory of fluids. Wiley, New York

25. Tempest W, Parbrook H (1957) The absorption of sound in argon, nitrogen and oxygen. Acta Acust 7:354–362

26. Tenti G, Boley C, Desai R (1974) On the kinetic model description of rayleigh-brillouin scattering from molecular gases. Can J Phys 52:285

27. Uribe F, Mason E, Kestin J (1989) A correlation scheme for the thermal conductivity of polyatomic gases at low density. Physica A 156:467–491

28. Wang L, Billing G (1993) Rotational relaxation and transport properties of oxygen by using the importance sampling method. J Phys Chem 97:2523

29. Witschas B, Lemmerz C, Reitebuch O (2014) Daytime measurements of atmospheric temperature profiles (2–15 km) by lidar utilizing rayleigh-brillouin scattering. Opt Lett 39:1972

Micro- and Nanoscale Flows

Uniform Flow Past a Circular Disk: Numerical Analysis and Modeling of Thermal Polarization

Takuma Tomita, Satoshi Taguchi, and Tetsuro Tsuji

Abstract In this paper, we investigate a slow rarefied gas flow past a circular disk with an emphasis on the thermal polarization effect. First, the behavior of the gas is numerically studied based on the Bhatnagar–Gross–Krook model of the Boltzmann equation and the diffuse reflection boundary condition, under the assumption that the system can be linearized. A key highlight of the numerical results is the localization of temperature variations near the edge of the disk in the near-continuum regime. Consistent with previous studies, we find that the temperature near the edge displays distinct behavior, differing from the classical kinetic boundary layer typically observed near smooth boundaries. Next, we explore the role of the free motion of gas molecules and construct a simplified toy model to interpret the above numerical results. The model demonstrates qualitative agreement with the numerical results, providing valuable insights into the observed phenomena.

Keywords Thermal polarization · Kinetic model · Edge effects

1 Introduction

When the mean free path of gas molecules is comparable to the characteristic length of a system, the continuum hypothesis and the assumption of local thermal equilibrium break down. Under such rarefied conditions, it is well-known that a non-uniform temperature distribution on a surface induces gas flow, even in the absence of external forces. Complementary to this phenomenon is thermal polarization [2], in which the gas temperature varies around a body in a flow, even when the surface of the body is maintained at the same temperature as the surrounding uniform flow. Thermal polarization is fundamentally a non-equilibrium effect and remains relevant even when the flow speed is sufficiently low for the system to be linearized, that is, when nonlinear effects such as viscous heating are negligible.

T. Tomita · S. Taguchi (✉) · T. Tsuji
Graduate School of Informatics, Kyoto University, Kyoto, Japan
e-mail: taguchi.satoshi.5a@kyoto-u.ac.jp

© The Author(s) 2026

M. Grabe et al. (eds.), *Rarefied Gas Dynamics*, Springer Aerospace Technology,
https://doi.org/10.1007/978-3-032-00094-1_28

From a theoretical perspective, these phenomena originate from the non-equilibrium velocity distribution of gas molecules. An important challenge in rarefied gas dynamics is bridging kinetic descriptions with macroscopic behavior. To this end, asymptotic expansion techniques have been widely used in the literature. In particular, with the aid of a systematic analysis of the kinetic boundary layer (Knudsen layer), Sone [4] classified various kinetic effects according to the orders of the Knudsen number Kn, which is defined as the ratio of the mean free path of gas molecules to the characteristic length of a system. For slow flows, which are of concern here, the zeroth order of Kn corresponds to the continuum theory, while the first, second, and higher orders account for non-equilibrium effects. This framework is known as generalized slip-flow theory.

However, the generalized slip-flow theory assumes that both boundary shapes and boundary conditions are smooth, making it inapplicable to flows around edge-shaped bodies. As reported in [6], a numerical study on thermal edge flow and radiometric flow in a two-dimensional setting revealed that macroscopic quantities exhibit anomalous behavior near an edge. Specifically, a fractional order of Kn was observed, which cannot be predicted by conventional slip-flow theory. It is worth noting that such non-smooth boundaries can be commonly encountered in modern applications, including micro gas pumps (Knudsen pumps) [8] and self-propelled particles (Janus particles) [1].

In this study, we extend the previous work [6] to a more realistic three-dimensional case by examining the thermal polarization effect induced by a slow uniform flow past a circular disk. The paper is divided into two main parts. In the first part, we numerically investigate the behavior of the gas based on a kinetic equation. In the second part, we construct a simplified toy model to qualitatively reproduce the numerical results and obtain an insight into the thermal polarization effect around the disk.

2   Numerical Analysis Based on a Kinetic Formulation

2.1 *Problem and Formulation*

Let us consider a situation where an infinitely thin circular disk is placed in a uniform flow at speed U. We take the radius of the disk L as the characteristic length of the system and introduce Cartesian coordinates Lx_i $(i = 1, 2, 3)$ in such a way that the origin is centered on the disk and the x_1-axis is perpendicular to the disk. The uniform flow is perpendicular to the disk, and the disk surface is maintained at a uniform temperature T_0. Far from the disk, the gas is in the equilibrium state with density ρ_0, flow velocity $(U, 0, 0)$, and temperature T_0. There is no external force acting on the gas. We investigate the steady behavior of the gas under the following assumptions: (i) The behavior of the gas is governed by the Bhatnagar–Gross–Krook (BGK) model of the Boltzmann equation; (ii) A gas molecule is diffusely reflected on the disk surface; (iii) The flow speed U is so small compared to the thermal speed

$(2RT_0)^{1/2}$ that the system can be linearized around the reference equilibrium state at rest with density ρ_0 and temperature T_0, where R is the specific gas constant.

Next, we formulate the problem. Let $(2RT_0)^{1/2}\boldsymbol{\zeta}$ be the molecular velocity and $\rho_0(2RT_0)^{3/2}(1 + \phi(\boldsymbol{x}, \boldsymbol{\zeta})E)$ be the velocity distribution function (VDF), where $E = E(\boldsymbol{\zeta}) := \pi^{-3/2}\exp(-\zeta^2)$. In addition, we denote the mass density of the gas by $\rho_0(1 + \omega(\boldsymbol{x}))$, the flow velocity by $(2RT_0)^{1/2}\boldsymbol{u}(\boldsymbol{x})$, and the temperature of the gas by $T_0(1 + \tau(\boldsymbol{x}))$. These macroscopic quantities are related to $\phi(\boldsymbol{x}, \boldsymbol{\zeta})$ through its moments with respect to $\boldsymbol{\zeta}$, as defined in (1c) below. The linearized BGK equation and its boundary conditions are summarized as follows (see §1.11 of [4]):

$$\boldsymbol{\zeta} \cdot \nabla_x \phi = \frac{1}{\kappa}(\phi_e - \phi), \tag{1a}$$

$$\phi_e = \omega + 2\boldsymbol{\zeta} \cdot \boldsymbol{u} + (\zeta^2 - \frac{3}{2})\tau, \tag{1b}$$

$$\omega = \int_{\mathbb{R}^3} \phi E d\boldsymbol{\zeta}, \quad \boldsymbol{u} = \int_{\mathbb{R}^3} \boldsymbol{\zeta}\phi E d\boldsymbol{\zeta}, \quad \tau = \frac{2}{3}\int_{\mathbb{R}^3} (\zeta^2 - \frac{3}{2})\phi E d\boldsymbol{\zeta}, \tag{1c}$$

$$\text{b.c.} \quad \phi \to \sigma_w^\pm(\boldsymbol{x}), \quad \sigma_w^\pm(\boldsymbol{x}) = 2\sqrt{\pi} \int_0^{\mp\infty} \zeta_1 \left(\int_{-\infty}^{\infty} \int_{-\infty}^{\infty} \phi E d\zeta_2 d\zeta_3 \right) d\zeta_1$$

$$(x_1 \to \pm 0, \; x_2^2 + x_3^2 < 1, \; \zeta_1 \gtrless 0), \tag{1d}$$

$$\text{b.c.} \quad \phi \to 2\zeta_1 u_\infty \quad (|\boldsymbol{x}| \to \infty). \tag{1e}$$

Here, κ in (1a) is the sole parameter of the system and is defined as $\kappa = (\sqrt{\pi}/2)\mathrm{Kn} = (\sqrt{\pi}/2)(\ell_0/L)$, where ℓ_0 represents the mean free path of the gas molecules in the reference equilibrium state at rest, while $\mathrm{Kn} = \ell_0/L$ is the Knudsen number of the system. The diffuse reflection boundary condition (1d) states that gas molecules are reflected on the surface according to the corresponding part of a stationary Maxwellian distribution, determined by the temperature of the disk surface and by the condition of no net flux across the boundary. This is also referred to as the complete accommodation condition. While more general boundary conditions incorporating accommodation coefficients are also used in the literature, we focus on the diffuse reflection condition for the sake of simplicity. Equation (1e) implies that ϕ approaches the uniform equilibrium state at infinity, where u_∞ denotes the non-dimensional speed of the uniform flow, defined by $u_\infty = U(2RT_0)^{-1/2}$. Note that $u_\infty \ll 1$ due to the above assumption (iii), and that ϕ, ω, $\boldsymbol{u}$, and τ linearly depend on u_∞.

In this study, we employ the BGK equation to simplify the analysis, although other models that yield more realistic Prandtl numbers could be used with additional computational effort.

2.2 Integral Equations

To balance accuracy and feasibility, our numerical analysis is fully based on the integral formulation of the BGK equation. The integral equation is briefly explained in this section, with further details deferred to a forthcoming paper.

The BGK equation (1a) is a transport equation, as its left-hand side represents the transport effect in the direction of the molecular velocity ζ. For a given (x, ζ), the characteristic curve is the line passing through the point x and parallel to ζ. To derive the integral forms of the BGK equation (1a), we consider the following two cases separately. The first case arises when the disk is encountered while tracing the characteristic curve backward in the direction of $-\zeta$ from the point x. In this case, integrating (1a) along the characteristic curve gives

$$\phi(x, \zeta) = \frac{1}{\kappa|\zeta|}\left[\int_0^{s_w} \exp\left(-\frac{s}{\kappa|\zeta|}\right)\phi_e\left(x - \frac{\zeta}{|\zeta|}s, \zeta\right)ds + \exp\left(-\frac{s_w}{\kappa|\zeta|}\right)\sigma_w^{\pm}(x_*)\right]. \quad (2)$$

Here, the integral variable s represents the backward distance along the characteristic curve, s_w denotes the distance from x to the disk along this curve, and ϕ_e and $\sigma_w^{\pm}$ are given by (1b) (with (1c)) and (1d), respectively. The point $x_* = x - (\zeta/|\zeta|)s_w$ is the intersection of the characteristic curve with the disk. The plus-minus sign above σ_w indicates $+$ for $x_1 > 0$ and $-$ for $x_1 < 0$. Note that ϕ_e depends on ϕ through ω, u and τ, and that $\sigma_w^{\pm}$ also depends on ϕ. A similar comment applies to (3) below. The second case arises when the backward characteristic curve extends to infinity without intersecting the disk. In this case, the integral equation takes the form

$$\phi(x, \zeta) = \frac{1}{\kappa|\zeta|}\int_0^{\infty} \exp\left(-\frac{s}{\kappa|\zeta|}\right)\phi_e\left(x - \frac{\zeta}{|\zeta|}s, \zeta\right)ds, \quad (3)$$

where ϕ_e is given by (1b) with (1c).

Equations (2) and (3) are solved by an iterative method similar to the one used in [7]. The discrepancy between (2) and (3), arising from distinct molecular trajectories, leads to discontinuities in ϕ (see, e.g., [5]). For an accurate analysis, these discontinuities were carefully taken into account in our numerical computations.

2.3 Numerical Results

First, let us examine the overall flow and temperature fields. In Fig. 1a, the flow velocity and the temperature fields around the disk are presented in the case of $\kappa = 0.5$. This figure shows the vectors of u/u_∞ and contours of τ/u_∞ in the region $-2 \leq x_1 \leq 2$, $-2 \leq x_2 \leq 2$ on the plane $x_3 = 0$. The cross-section of the disk, located at $x_1 = 0$,

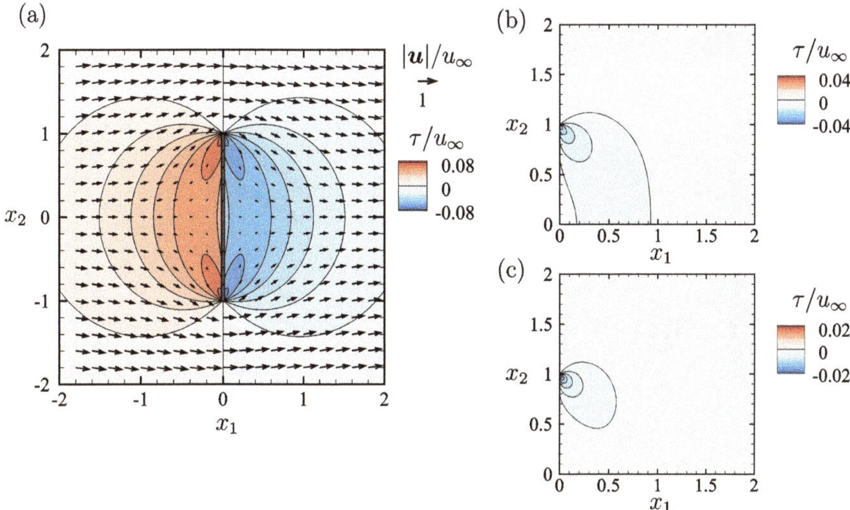

Fig. 1 **a** Flow velocity u/u_∞ (vector field) and temperature τ/u_∞ (contours) in the region $-2 \leq x_1 \leq 2$, $-2 \leq x_2 \leq 2$ on the plane $x_3 = 0$ for $\kappa = 0.5$. The thick line at the center of the figure indicates the cross-section of the disk. **b, c** Temperature distributions τ/u_∞ in the region $0 \leq x_1 \leq 2$, $0 \leq x_2 \leq 2$ on the plane $x_3 = 0$. **b** $\kappa = 0.1$, **c** 0.05

$-1 < x_2 < 1$, is also included and depicted as a thick segment for visibility. Owing to the axisymmetry of the system, the flow and temperature fields are symmetric with respect to the x_1 axis. In addition, τ is antisymmetric with respect to the x_2 axis in the figure. As shown, the flow is obstructed by the disk, and the gas temperature is higher than T_0 on the upstream side and lower on the downstream side, despite the fact that the disk surface is maintained at a uniform temperature T_0. This behavior exemplifies the thermal polarization effect.

Next, we examine how the temperature field depends on κ. In Fig. 1b, c, we present the contour plots of τ/u_∞ for (b) $\kappa = 0.1$ and (c) $\kappa = 0.05$. Considering the symmetric properties discussed earlier, only the region $0 \leq x_1 \leq 2$, $0 \leq x_2 \leq 2$ on the plane $x_3 = 0$ is shown. It is observed that the isothermal lines become increasingly concentrated near the edge of the disk located at $(x_1, x_2) = (0, 1)$ as κ decreases. This indicates that the thermal polarization effect becomes more localized near the edge with smaller values of κ.

To gain further insights, we analyze the relationship between τ and κ as follows. On the plane $x_3 = 0$, we select three points κ away from the edge: $(x_1, x_2) = (\kappa/\sqrt{2}, 1 + \kappa/\sqrt{2})$, $(\kappa, 1)$, $(\kappa/\sqrt{2}, 1 - \kappa/\sqrt{2})$, referred to as P, Q, and R, respectively (Fig. 2a). This scaling reflects the use of the mean free path ℓ_0 as the reference length, rather than L, in the region near the edge. Figure 2b shows the log-log plot of $|\tau/u_\infty|$ versus κ at these three points (red square for P, green triangle for Q, blue circle for R). The dashed line represents a slope of $\kappa^{1/2}$. The figure clearly shows that the magnitude of the perturbed temperature, $|\tau/u_\infty|$, decreases proportionally to

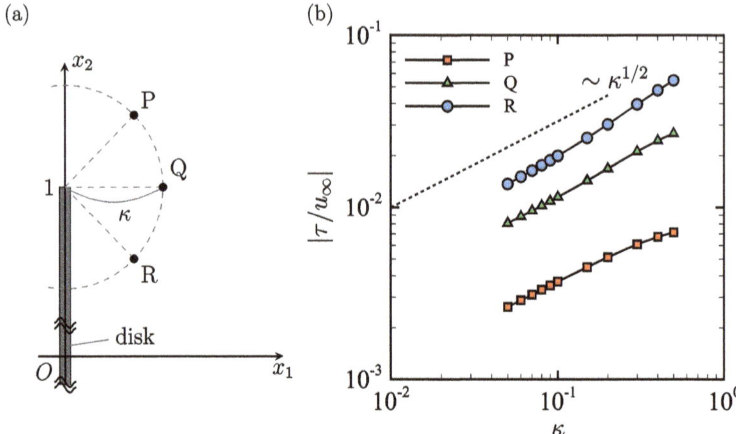

Fig. 2 **a** Three points located κ away from the edge: P($\kappa/\sqrt{2}$, $1 + \kappa/\sqrt{2}$), Q(κ, 1), R($\kappa/\sqrt{2}$, $1 - \kappa/\sqrt{2}$). **b** Log-log plots of $|\tau/u_\infty|$ as a function of κ at these points: red squares for P, green triangles for Q, and blue circles for R. The dashed line indicates a slope of $\kappa^{1/2}$

$\kappa^{1/2}$ as $\kappa \to 0$ at all three points. As previously mentioned, such a fractional power behavior cannot be captured in the conventional framework of the generalized slip-flow theory, in which all quantities are expanded as a power series in the Knudsen numbers (i.e., the Hilbert expansion) [4].

3 Modeling of Thermal Polarization

In this section, we develop a toy model to qualitatively explain the thermal polarization effect around the disk. To this end, we focus on the role of the free motion of gas molecules. To better understand the significance of free motion, let us review the concept of thermal creep flow, which occurs near a wall with a non-uniform temperature distribution under rarefied conditions. It is well-known that the flow arises due to differences in the momentum of gas molecules reaching the wall. Specifically, molecules originating from the hotter side carry, on average, greater momentum than those from the colder side, while molecules are isotropically reflected off the wall due to diffuse reflection. Consequently, gas molecules impart a net momentum to the wall from the hotter side toward the colder side, inducing a counteracting gas flow from the colder side toward the hotter side. The critical point here is that this momentum exchange is enabled by the finite distance gas molecules travel after a molecular collision. (As the Knudsen number, $\mathrm{Kn} = \ell_0/L$, approaches zero, the free motion of molecules shrinks to a single point on the wall, where no temperature difference exists.) Through the construction of our toy model, we will see that the thermal polarization effect is closely connected to the free motion of molecules and that its mechanism is analogous to that of the thermal creep flow.

3.1 Modeling Method

We begin by interpreting the integral equations (2) and (3) from a physical perspective. If we make the change of integral variable $s = |\zeta|t$ in (3), we obtain

$$\phi(\boldsymbol{x}, \zeta) = \int_0^\infty \phi_e(\boldsymbol{x} - \zeta t, \zeta) f(t)\,\mathrm{d}t, \quad f(t) = \frac{1}{\kappa}\exp\left(-\frac{t}{\kappa}\right). \tag{4}$$

The new integral variable t represents the backward travel time of a molecule. Consequently, $\boldsymbol{x} - \zeta t$ is the position of the molecule at time t before the present, and $\phi_e(\boldsymbol{x} - \zeta t, \zeta)$ is the (perturbed) local Maxwellian at that location. Furthermore, $f(t)$ can be interpreted as the probability density function of an exponential distribution with the parameter $1/\kappa$. Therefore, the right-hand side of (4) represents the mean value of ϕ_e along the path of a molecule, weighted by an exponential distribution. In other words, ϕ_e encodes information about the molecular velocity at the moment of a molecular collision, and this information is transmitted via the free motion of the molecule. According to (4), the elapsed time since the most recent molecular collision follows an exponential distribution. Regarding (2), after changing the integral variable to $s = |\zeta|t$ and performing a slight transformation of the second term, we obtain

$$\phi(\boldsymbol{x}, \zeta) = \int_0^{s_{\mathrm{w}}/|\zeta|} \phi_e(\boldsymbol{x} - \zeta t, \zeta) f(t)\,\mathrm{d}t + \int_{s_{\mathrm{w}}/|\zeta|}^\infty \sigma_{\mathrm{w}}^{\pm}(\boldsymbol{x}_*) f(t)\,\mathrm{d}t. \tag{5}$$

In other words, if the path of the molecule intersects the disk during the backward tracing, the remaining contribution of ϕ_e must be replaced by $\sigma_{\mathrm{w}}^{\pm}(\boldsymbol{x}_*)$.

The set of equations (4) and (5) forms an implicit equation of ϕ, as the right-hand side also depends on ϕ through ϕ_e and $\sigma_{\mathrm{w}}^{\pm}$. To eliminate this dependence, we approximate ϕ_e and $\sigma_{\mathrm{w}}^{\pm}$ using a function independent of ϕ. Assuming a given density field $\omega_0(\boldsymbol{x})$, flow velocity field $\boldsymbol{u}_0(\boldsymbol{x})$, and temperature temperature field $\tau_0(\boldsymbol{x})$, the local Maxwellian $\phi_0(\boldsymbol{x}, \zeta) = \omega_0(\boldsymbol{x}) + 2\zeta \cdot \boldsymbol{u}_0(\boldsymbol{x}) + (\zeta^2 - \frac{3}{2})\tau_0(\boldsymbol{x})$ can be used as a crude approximation of ϕ_e. Recall that molecular collisions dominate in the continuum regime. Taking this into account, we assume that $\omega_0(\boldsymbol{x}) = 0$, and $\boldsymbol{u}_0(\boldsymbol{x})$ and $\tau_0(\boldsymbol{x})$ satisfy the equations of continuum theory. The solution to the Stokes equation with no-slip boundary conditions can be found in [3]. In our problem setting, $\tau_0(\boldsymbol{x}) = 0$ serves as the trivial solution to the Laplace equation. In addition, the simplest way to approximate $\sigma_{\mathrm{w}}^{\pm}$ is to substitute ϕ_0 for ϕ in the definition of $\sigma_{\mathrm{w}}^{\pm}$ given in (1d). As a result, the integral on the right-hand side of (1d) becomes zero, causing the second term on the right-hand side of (5) to vanish.

After these manipulations, we obtain

$$\phi(\boldsymbol{x}, \boldsymbol{\zeta}) = \int_0^{s_B} \phi_0(\boldsymbol{x} - \boldsymbol{\zeta} t, \boldsymbol{\zeta}) f(t) \mathrm{d}t, \tag{6}$$

where $s_B = s_w/|\boldsymbol{\zeta}|$ if the backward molecular path intersects the disk; otherwise, $s_B = \infty$. Since the right-hand side of (6) no longer depends on ϕ, the formula is significantly easier to compute than (4) and (5). Despite its simplicity, it retains the mathematical structure required to capture the effects of free molecular motion.

3.2 Computation of Thermal Polarization Using the Model

We are now ready to estimate the thermal polarization effect around the disk using the model. At this stage, analytical evaluation of the integrals in (6) and the third equation of (1c) is no longer feasible. Consequently, the results presented below are obtained through numerical computation. It is worth noting that the model (6) requires only a single set of numerical integrations without any iterative procedures. As a result, the computational costs are significantly lower compared to the numerical analysis discussed in Sect. 2.

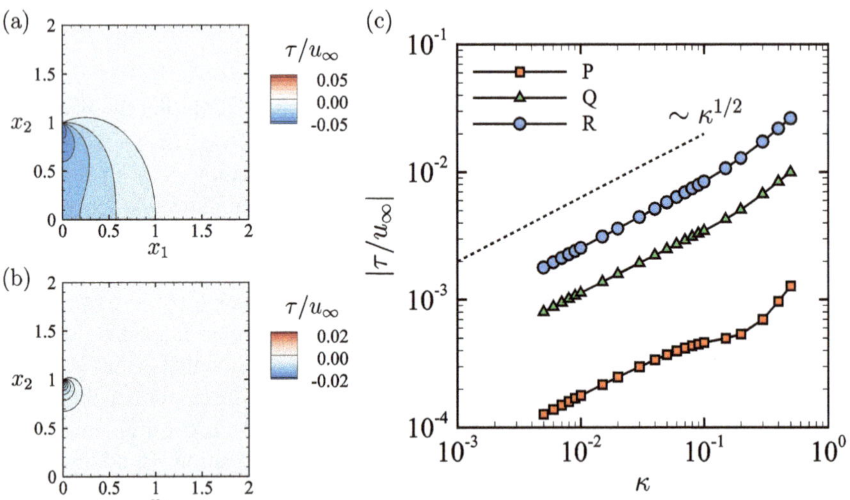

Fig. 3 **a, b** Temperature distribution obtained using (6). The values of τ/u_∞ are shown in the region $0 \le x_1 \le 2, 0 \le x_2 \le 2$ on the plane $x_3 = 0$. **a** $\kappa = 0.5$, **b** 0.1. **c** Log-log plots of $|\tau/u_\infty|$ as a function of κ at points P, Q, and R (Refer to the last paragraph in Sect. 2 and Fig. 2a). Red squares for P, green triangles for Q, and blue circles for R. The dashed line represents a slope of $\kappa^{1/2}$

In Fig. 3a, b, we present the spatial distributions of τ/u_∞ obtained using the model for (a) $\kappa = 0.5$ and (b) $\kappa = 0.1$. Only the region $0 \leq x_1 \leq 2, 0 \leq x_2 \leq 2$ on the plane $x_3 = 0$ is presented, consistent with Fig. 1b, c. From these figures, we can identify the localization of temperature variation as κ decreases. Furthermore, Fig. 3c shows the log-log plot of $|\tau/u_\infty|$ versus κ at the points P, Q, and R (refer to the last paragraph of Sect. 2.3 and Fig. 2a). Remarkably, Figs. 3c and 2b share a common feature: in both the results derived from the BGK equation and those from our toy model, $|\tau/u_\infty|$ decreases proportionally to $\kappa^{1/2}$ as κ approaches zero.

4 Conclusion

In this study, we investigated the steady behavior of a slow rarefied gas flow past a circular disk based on the linearized BGK equation and the diffuse reflection boundary condition. Our precise numerical analysis, in which the discontinuities in the VDF were fully accounted for, revealed the following key findings: (i) The temperature around the disk is higher on the upstream side and lower on the downstream side, demonstrating the thermal polarization effect; (ii) As κ decreases, temperature variation becomes increasingly localized near the edge of the disk; (iii) At points located κ away from the edge, the magnitude of the temperature decreases proportionally to $\kappa^{1/2}$ as $\kappa \to 0$.

Next, we proposed the formula (6) as a toy model. The temperature fields computed using this model showed good qualitative agreement with the numerical results based on the BGK equation, particularly regarding the three features discussed above. This suggests that the free motion of molecules plays a crucial role in the thermal polarization effect. Moreover, the computational cost of the model is significantly lower, making it an efficient tool for quickly exploring various kinetic effects.

Acknowledgements This work was supported by JSPS KAKENHI Grant Numbers JP20H02067, JP22K03924, JP22K18770, and JST SPRING Grant Number JPMJSP2110. Numerical computations were carried out using the supercomputer of ACCMS, Kyoto University.

References

1. Baier T, Tiwari S, Shrestha S, Klar A, Hardt S (2018) Thermophoresis of Janus particles at large Knudsen numbers. Phys Rev Fluids 3(9):094202
2. Bakanov SP, Vysotskij VV, Derjaguin BV, Roldughin VI (1983) Thermal polarization of bodies in the rarefied gas flow. J Non-Equilib Thermodyn 8(1):75–83
3. Happel J, Brenner H (1983) Low Reynolds number hydrodynamics: with special applications to particulate media. Springer, Netherlands
4. Sone Y (2007) Molecular gas dynamics: theory, techniques, and applications. Birkhäuser Boston

5. Sone Y, Takata S (1992) Discontinuity of the velocity distribution function in a rarefied gas around a convex body and the S layer at the bottom of the Knudsen layer. Trans Theory Statist Phys 21(4–6):501–530
6. Taguchi S, Aoki K (2012) Rarefied gas flow around a sharp edge induced by a temperature field. J Fluid Mechan 694:191–224
7. Taguchi S, Tsuji T, Kotera M (2020) Transient behaviour of a rarefied gas around a sphere caused by impulsive rotation. J Fluid Mechan 909:A6
8. Wang X, Su T, Zhang W, Zhang Z, Zhang S (2020) Knudsen pumps: a review. Microsyst Nanoeng 6(1):26

Modeling Transient Slip Behavior from Laminar to Molecular Flow

Rodion Groll [ID]

Abstract One of the greatest challenges of microfluid dynamics remains to describe the physical behavior in the transition region between the continuum and the free molecular regime. In this work, the non-dimensionalized relationship between mass flow and pressure drop in microchannels of different cross-sections is investigated. The flow resistance in microchannels is affected by increasing Knudsen numbers because the molecular wall collisions have an increasing influence. The reduced collision probabilities between the particles correspondingly reduce the influence of the effective viscosity of the fluid. In this work, approaches are investigated which have been able to predict the effective pressure difference specific permeation of a fluid as a function of the Knudsen number of an increasingly thinned gas flow. A further challenge is posed by the question of the origin of the local Knudsen minimum. This minimum of the non-dimensioned mass flow in the range $Kn \approx 1$ is cancelled out in the range of free molecular flows by the increasing permeation. The model presented combines the effects of different scale ranges by superposition. The model data are compared with measured data and BGK simulation results in both the transient slip regime and the free molecular convection regime and show excellent agreement for both plane channel flows and pipe flows.

Keywords Scale-transitioning gas flow · Analytical modeling · Slip model · Knudsen flow

1 Introduction

This relationship of mass flow and pressure loss in microchannel flows is controlled on macroscopic scales by temperature or viscosity. When the Knudsen range is reached, the influence of molecular wall collisions increases significantly compared to inter-particle collisions, so that even at constant temperature the pressure loss varies with decreasing molecular number density. The hydraulic diameter of a plane

R. Groll (✉)
Center of Applied Space Technology and Microgravity, University of Bremen, Bremen, Germany
e-mail: groll@zarm.uni-bremen.de

© The Author(s) 2026

M. Grabe et al. (eds.), *Rarefied Gas Dynamics*, Springer Aerospace Technology,
https://doi.org/10.1007/978-3-032-00094-1_29

channel flow is described by the limit value of a channel flow with a rectangular cross-section, where the width w is considered to be much greater than the height h and is regarded as infinity. The hydraulic diameter of a flat channel flow corresponds to twice the channel height. In highly diluted gas flows, the no-slip condition known from laminar flows in fluid mechanics can no longer be fulfilled for higher Knudsen numbers. As a consequence, slip conditions are introduced for $Kn > 0.01$, for which it is known that they are no longer fulfilled at $Kn < 1$ at the latest. The laminar channel flow is described by a parabolic velocity profile. So, pressure gradient and mass flow are proportional as follows:

$$u_0(y) = \frac{1}{2\mu}\frac{dp}{dx}(y^2 - hy) \Rightarrow \dot{m}_0 = \int_0^h u_0 dy = \rho\frac{h^2}{12\mu}A\frac{dp}{dx};$$

$$\mu = \frac{5\pi}{32}\rho\lambda\bar{c}$$

Corresponding to the following relation the definition of the viscosity μ and the hydraulic diameter D_h (ratio between four-times the cross-sectional area A and the circumference/perimeter Γ) non-dimensional mass flow G is inversely proportional to the hydraulic Knudsen number $Kn_h = \lambda/D_h$ [16]:

$$G = -\frac{3}{D_h}\frac{\dot{m}_0}{A\frac{dp}{dx}}\frac{\sqrt{\pi}}{4}\bar{c} = \frac{1}{10\sqrt{\pi}}\frac{D_h}{\lambda} = \frac{Kn_h^{-1}}{10\sqrt{\pi}} \tag{1}$$

Thus, in this laminar description, a hyperbolic G-profile results, which deviates from the reference data in the said Knudsen range with increasing Knudsen numbers.

2 Slip Flow

First-Order Slip Model. *The higher mass flow in rarefied follows the. increasing velocity near the wall—so called velocity slip us.*

With the parabolic velocity profile the slip velocity generates an effective offset and mass flow follows (see Fig. 1 (right)):

$$A = wh \Rightarrow \dot{m} = \dot{m}_0 + \dot{m}_s = \rho\frac{wh^3}{12\mu}\frac{dp}{dx} + \rho u_s wh$$

The reality is anywhere between the ideal elastic and the purely diffusive reflection (see Fig. 1 (left and middle)). Based on the definitions of slip velocity $u_s = (u_\lambda + u_r)/2$ and rebound velocity $u_r = (1 - \sigma_v)u_\lambda + \sigma_v u_w$ the relative slip velocity $u_s - u_w$ is the average of impact u_λ and rebound velocity u_r. The rebound velocity depends on the weighting accommodation coefficient σ_v [14]. So, the effective model depends

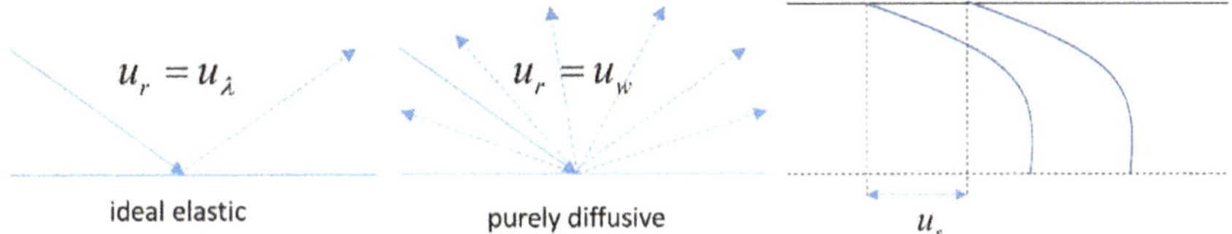

Fig. 1 Relation between averaged rebound velocity u_r, the averaging velocity from mean-free path-length distance u_λ and the velocity of the wall u_w (left and middle), Slip velocity shifts the velocity profile in channel flow and increases the effective volume flow (right)

on the velocity gradient at the wall. With the parabolic velocity profile slip velocity and mass flow follows neglecting wall velocity $u_w = 0$:

$$u_s = \frac{2 - \sigma_v}{\sigma_v} \lambda \frac{du}{dy}\bigg|_{y=0} = -\frac{2 - \sigma_v}{\sigma_v} \frac{\lambda h}{2\mu} \frac{dp}{dx} \Rightarrow \frac{\dot{m}}{\dot{m}_0}$$

$$= \frac{\dot{m}_0 + \dot{m}_s}{\dot{m}_0} = 1 + 6\frac{2 - \sigma_v}{\sigma_v} \frac{\lambda}{h}$$

With this linear slip velocity approximation, the corresponding non-dimensional parameter G is shifted by a constant depending on the accommodation coefficient σ_v.

$$-\frac{\dot{m}}{A\frac{dp}{dx}} = \left(1 + 6\frac{2 - \sigma_v}{\sigma_v} \frac{\lambda}{h}\right) \rho \frac{h^2}{12\mu}; \mu = \frac{5\pi}{32}\rho\lambda\bar{c}$$

$$G = -\frac{3}{D_h} \frac{\dot{m}}{A\frac{dp}{dx}} \frac{\sqrt{\pi}}{4}\bar{c} = \frac{1}{10\sqrt{\pi}} \frac{D_h}{\lambda} + \frac{2 - \sigma_v}{\sigma_v} \frac{6}{5\sqrt{\pi}} \tag{2}$$

Following this relation, the permeation is increasing with decreasing σ_v values. In the present example (Eq. 2), a superposition principle can be read off directly, whereby the effective permeation is represented as a summation of mass flows added to the laminar solution (Eq. 1), each of which is generated by different decoupled effects, as shown in Fig. 1.

Second-Order Slip Model. The second-order model describing the slip velocity (e.g., [1, 10]):

$$u_s = \frac{2 - \sigma_v}{\sigma_v}\left(\lambda \frac{du}{dy}\bigg|_{y=0} + \lambda^2 \frac{d^2u}{dy^2}\bigg|_{y=0}\right) = -\frac{2 - \sigma_v}{\sigma_v}\left(\frac{\lambda h}{2\mu} - \frac{\lambda^2}{2\mu}\right)\frac{dp}{dx}$$

and the effective mass flow increases with this non-linear form:

$$\frac{\dot{m}}{\dot{m}_0} = \frac{\dot{m}_0 + \dot{m}_s}{\dot{m}_0} = 1 + 6\frac{2 - \sigma_v}{\sigma_v} \frac{\lambda}{h}\left(1 - \frac{\lambda}{h}\right) = 1 + 12\frac{2 - \sigma_v}{\sigma_v} \frac{\lambda}{D_h}\left(1 - 2\frac{\lambda}{D_h}\right)$$

With this second-order slip velocity approximation the corresponding non-dimensional parameter G is shifted by a linear function depending on the coefficient σ_v.

$$-\frac{\dot{m}}{A\frac{dp}{dx}} = \frac{\rho D_h^2}{48\mu} + \rho\frac{2-\sigma_v}{\sigma_v}\frac{\lambda D_h}{4\mu}\left(1 - 2\frac{\lambda}{D_h}\right)$$

$$\Rightarrow G = \frac{D_h}{10\sqrt{\pi}\lambda} + \frac{2-\sigma_v}{\sigma_v}\frac{6}{5\sqrt{\pi}}\left(1 - 2\frac{\lambda}{D_h}\right)$$

However, the second-order equation is also a superposition with a non-linear addition to the laminar solution. This non-linearity increases the error near in the region close to Kn = 1, the region where the mean-free path of molecules is similar to the channel height.

Slip Correction for higher Knudsen numbers. Analog algebraic approaches do not use accommodation coefficients. The non-dimensional mass flow is developed in a way analogue to second order with h = D_h/2. The results with the estimated coefficients of Cercignani and Daneri [3] (C_1 = 1.1466 and C_2 = 0.2439) as of Deissler [5] (C_1 = 1 and C_2 = 9/32) are shown in Fig. 2 (left). With accommodation coefficient depending "constants" the developed model can be approximated by the algebraic form C_1 = 2/σ_v − 1 and C_2 = 1/2 − 1/σ_v (e.g., [1, 10]). So the non-dimensional mass flow G become independent from optimized constants:

$$G = \frac{1}{10\sqrt{\pi}}\frac{D_h}{\lambda} + \frac{6}{5\sqrt{\pi}}\left(C_1 + 4C_2\frac{\lambda}{D_h}\right) = \frac{Kn_h^{-1}}{10\sqrt{\pi}} + \frac{6}{5\sqrt{\pi}}\frac{2-\sigma_v}{\sigma_v}(1 - 2Kn_h)$$

Because of the increasing influence of wall collisions for $\lambda > 0.1$ h the effective mean-free path of molecules for characterizing the slip behavior is smaller than λ.

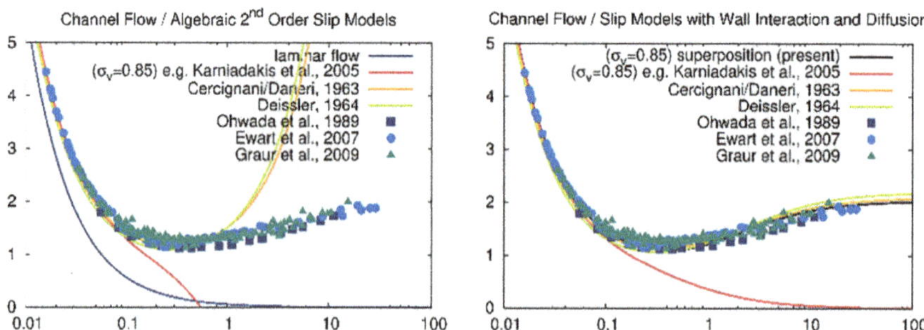

Fig. 2 Left: Non-dimensional mass flow [G] over hydraulic Knudsen number [Kn_h] increases unphysically by second-order optimized algebraic approaches for $Kn_h > 1$. [3, 5, 6, 8, 10, 12], right: Channel flow modeling non-dimensional mass flow [G] over hydraulic Knudsen number [Kn_h] including slip, wall-collision and free-molecular correction terms (Data: [3, 5, 6, 8, 10, 12])

With a model of the summarization of the wave numbers $1/\lambda_{\text{eff}} = 1/\lambda + 1/h$ (see e.g., [2, 4]). The effective Knudsen is decreased too.

$$Kn_{h,eff} = \frac{\lambda_{eff}}{D_h} = \frac{1}{\frac{D_h}{\lambda} + \frac{D_h}{h}} = \frac{1}{Kn_h^{-1} + 2} = \frac{Kn_h}{1 + 2Kn_h}$$

With the new-defined Knudsen number $Kn_{h,eff}$ the corrected transport results as follows:

$$\begin{aligned}
G_{eff} &= \frac{Kn_h^{-1}}{10\sqrt{\pi}} + \frac{6}{5\sqrt{\pi}} \frac{2 - \sigma_v}{\sigma_v} \left(1 - 2Kn_{h,eff}\right) \\
&= \frac{Kn_h^{-1}}{10\sqrt{\pi}} + \frac{6}{5\sqrt{\pi}} \frac{2 - \sigma_v}{\sigma_v} \left(1 - \frac{2Kn_h}{1 + 2Kn_h}\right)
\end{aligned}$$

3 Free-Molecular Flow

Depending on momentum conservation the free molecular mass flow is calculated by the definition of viscosity as follows (e.g., [9]):

$$\frac{\dot{m}_{FM}}{A} = \frac{\mu}{n\bar{c}} \frac{d}{dx}(n\bar{c}); \ \mu = \frac{5\pi}{32}\rho\lambda\bar{c} \Rightarrow -\frac{\dot{m}_{FM}}{A\frac{dp}{dx}} = \frac{5\lambda}{8\bar{c}} \frac{\gamma + 1}{\gamma}$$

Inserted to the definition of non-dimensional mass flow G the free-molecular approach results as follows.

$$G_{FM} = \frac{3}{D_h} \frac{\dot{m}_{FM}}{A\frac{dp}{dx}} \frac{\sqrt{\pi}}{4}\bar{c} = \frac{15\sqrt{\pi}}{32} \frac{\gamma + 1}{\gamma} \frac{\lambda}{D_h} = \frac{15\sqrt{\pi}}{32} \frac{\gamma + 1}{\gamma} Kn_h$$

With an effective, reduced mean-free path depending on characteristic geometric length proportional to hydraulic diameter $l_\alpha = \alpha D_h$ the summarization of wave numbers leads to $1/\lambda_{FM,eff} = 1/\lambda + \alpha/D_h$. The Knudsen number definition follows the effective Knudsen number describing free-molecular flows:

$$Kn_{FM,eff} = \frac{\lambda_{FM,eff}}{D_h} = \frac{1}{D_h\left(\frac{1}{\lambda} + \frac{\alpha}{D_h}\right)} = \frac{\frac{\lambda}{D_h}}{1 + \alpha\frac{\lambda}{D_h}} = \frac{Kn_h}{1 + \alpha Kn_h}$$

And the effective mass flow of free-molecular flows predicts the free molecular convection decreasing with smaller Knudsen numbers:

$$G_{FM,eff} = \frac{15\sqrt{\pi}}{32}\frac{\gamma+1}{\gamma}Kn_{h,eff} = \frac{15\sqrt{\pi}}{32}\frac{\gamma+1}{\gamma}\frac{Kn_h}{1+\alpha Kn_h}$$

Channel coefficient. Based on comparison of the limiting coefficients of (see [13, 15]):

$$\lim_{Kn_h\to\infty} G_{FM,eff} = \frac{2-\sigma_v}{\sigma_v}\frac{8}{3\sqrt{\pi}}$$

the equivalence approximation with the present study results the coefficient α and the effective free-molecular gas flow therefore (see [9]):

$$\frac{15\sqrt{\pi}}{32}\frac{\gamma+1}{\gamma}\alpha^{-1} = \frac{2-\sigma_v}{\sigma_v}\frac{8}{3\sqrt{\pi}} \Rightarrow \alpha = \frac{\sigma_v}{2-\sigma_v}\frac{45\pi}{256}\left(1+\frac{1}{\gamma}\right)$$

The effective molecular mass flow $G_{FM,eff}$ results with the implemented constant α as follows (see Fig. 2 (right)):

$$G_{FM,eff} = \frac{15\sqrt{\pi}}{32}\frac{\gamma+1}{\gamma}\frac{Kn_h}{1+\frac{\sigma_v}{2-\sigma_v}\frac{45\pi}{256}\left(1+\frac{1}{\gamma}\right)Kn_h}$$

4   Results

4.1   *Channel Flow*

Summarizing molecular free-flow solution to the slip solution $G_{tot} = G_{eff} + G_{FM,eff}$ the transition from laminar to free-molecular flow is modeled.

The closing result modeling this transient behavior of channel flows ($\sigma_v = 0.85$) is this explicit formulation describing the transient mass flow development by summarization (Eq. 3) of the components: the laminar, the slip and the free-molecular component (see Fig. 3 (left)):

$$G_{tot} = \frac{Kn_h^{-1}}{10\sqrt{\pi}} + \frac{2-\sigma_v}{\sigma_v}\frac{6}{5\sqrt{\pi}}\left(1-2\frac{Kn_h}{1+2Kn_h}\right) + \frac{\frac{15\sqrt{\pi}}{32}\frac{\gamma+1}{\gamma}Kn_h}{1+\frac{\sigma_v}{2-\sigma_v}\frac{45\pi}{256}\left(1+\frac{1}{\gamma}\right)Kn_h}$$

$$(3)$$

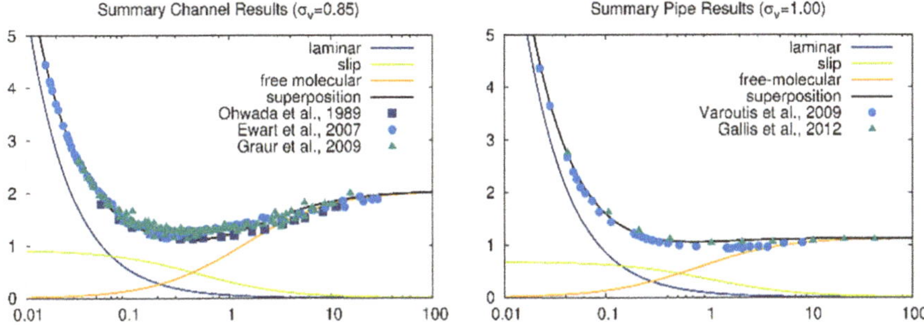

Fig. 3 Left: Channel flow modeling non-dimensional mass flow G depending on hydraulic Knudsen number (Kn_h) as superposition of laminar, slip and free-molecular component. [6, 8, 12], right: Pipe flow modeling non-dimensional mass flow G depending on hydraulic Knudsen number (Kn_h) as superposition of laminar, slip and free-molecular component [7, 17]

4.2 Pipe Flow

With the different form of the cross-sectional area of a pipe flow the definition of the circle area produces an equivalence of hydraulic diameter D_h and diameter d of the pipe. The solution of a laminar Hagen-Poiseuille flow produces a hyperbolic relation G_0 and the first-order slip model shifts the curve with a constant to G:

$$G_0 = \frac{3}{20\sqrt{\pi}}Kn_h^{-1} \Rightarrow G = \frac{3}{20\sqrt{\pi}}Kn_h^{-1} + \frac{2-\sigma_v}{\sigma_v}\frac{6}{5\sqrt{\pi}}$$

The second-order model transforms the curve with large differences in the Knudsen regime:

$$G = \frac{3}{20\sqrt{\pi}}Kn_h^{-1} + \frac{2-\sigma_v}{\sigma_v}\frac{6}{5\sqrt{\pi}}(1 - Kn_h)$$

For circular cross sections with an accommodation coefficient of $\sigma_v = 1.0$, the curves show a better fit in the slip regime. (Kn < 1) (e.g., [11]). With the Knudsen number correction of the slip component show as estimated a much better first-order approximation of the mass flow behavior in in slip regime for $\sigma_v = 1.0$. With substituted accommodation coefficients the wall-collision approximation of the second-order slip model produces analog with channel flow results also a decreasing curve with a limit G = 0 at infimum.

$$G_{eff} = \frac{3}{20\sqrt{\pi}}Kn_h^{-1} + \frac{6}{5\sqrt{\pi}}\left(1 - \frac{Kn_h}{1 + Kn_h}\right)$$

Also, for tubes the new model is the superposition (Eq. 4) of laminar flow, slip flow and molecular flow/diffusion.

$$G_{tot} = \frac{3}{20\sqrt{\pi}} Kn_h^{-1} + \frac{6}{5\sqrt{\pi}}\left(1 - \frac{Kn_h}{1 + Kn_h}\right)$$

$$+ \frac{15\sqrt{\pi}}{32}\left(1 - \frac{\left(1 + \frac{1}{\gamma}\right)Kn_h}{1 + \frac{15\pi}{64}\left(1 + \frac{1}{\gamma}\right)Kn_h}\right) \tag{4}$$

And also, this approximation matches with reference data in an excellent way as shown in Fig. 3 (right).

5   Conclusion

The results of this superposition model provide excellent results for the channel and pipe flow based on the hydraulic diameter and the corresponding Knudsen number. Except for minor adjustments to the accommodation coefficient in the slip regime, it can be concluded from the investigations presented that this method can be used to predict mass flows in microchannels independently of the shape of the cross-section in a Knudsen number-dependent manner.

Based on the modeling by three complementary causes, the previously only measured local Knudsen minimum can be quantitatively classified as a slip transition between continuum mechanical description and free molecular flow. In detail, the approaches can be summarized as follows. The slip and free-molecular flow effects follow the reflective and diffusive molecular rebound behavior and are modeled as an additional mass flow corresponding to an additional velocity profile. As shown many algebraic models show good results for $0.1 < Kn < 1$, but they cannot model the behavior for $Kn > 1$. For the wall collision effects, the influence of the slip flow decreases for higher Knudsen numbers ($Kn > 1$). So, three mass flows are superimposed, two of which disappear for large Knudsen numbers and the third is so small even for large Knudsen numbers that it is negligible in other areas compared to the other values.

The strictly hyperbolic form of the laminar flow is supplemented by slip diffusion, which disappears for large Knudsen numbers and converges to a constant value for small Knudsen numbers. Relative to the laminar solution, this naturally disappears for small Knudsen numbers. The significance of the shape of the cross-sectional area on the model parameters in the slip regime could not be clarified conclusively. However, it is clear that concavely curved surfaces favor multiple collisions, which drastically reduces the effective free path length. This could be the starting point for further research. While the first two model components disappear for large Knudsen numbers, the third term, the free molecular diffusion, becomes dominant for the limiting range at $Kn \gg 1$.

References

1. Beskok A, Karniadakis GE (1999) A model for flows in channels, pipes and ducts at micro and nano scales. Microscale Thermophys Eng 3:43
2. Bird GA (1994) Molecular gas dynamics and the direct simulation of gas flows. Clarendon, Oxford
3. Cercignani C, Daneri A (1963) Flow of a rarefied gas between two parallel plates. J Appl Phys 34:3509
4. Chapman S, Cowling TG (1970) The mathematical theory of non-uniform gases. Cambridge University Press, Cambridge
5. Deissler RG (1964) An analysis of second-order slip flow and temperature-jump boundary conditions for rarefied gases. Int J Heat Mass Transf 7(6):681
6. Ewart T, Perrier P, Graur IA, Méolans JG (2007) Mass flow rate measurements in a microchannel, from hydrodynamic to near free molecular regimes. J Fluid Mech 584:337
7. Gallis MA, Torczynski JR (2012) Direct simulation Monte Carlo-based expressions for the gas mass flow rate and pressure profile in a microscale tube. Phys Fluids 24:012005
8. Graur IA, Perrier P, Ghozlani W, Méolans JG (2009) Measurements of tangential momentum accommodation coefficient for various gases in plane microchannel. Phys Fluids 21:102004
9. Groll R, Kunze S, Besser B (2020) Correction of second-order slip condition for higher Knudsen numbers by approximation of free-molecular diffusion. Phys Fluids 32:092008
10. Karniadakis GE, Beskok A, Aluru N (2005) Microflows and nanoflows: fundamentals and simulations. Springer, New York
11. Kunze S, Groll R, Besser B, Thöming J (2022) Molecular diameters of rarefied gases. Sci Rep 12:2057
12. Ohwada T, Sone Y, Aoki K (1989) Numerical analysis of the Poiseuille and thermal transpiration flows between two parallel plates on the basis of the Boltzmann equation for hard-sphere molecules. Phys Fluids A 1:2042
13. Perrier P, Graur IA, Ewart T, Meolans JG (2011) Mass flow rate measurements in microtubes: from hydrodynamic to near free molecular regime. Phys Fluids 23:042004
14. Schaaf SA, Chambre PL (1961) Rarefied gas dynamics. Princeton University Press, Princeton
15. Sharipov F, Seleznev V (1998) Data on internal rarefied gas flows. J Phys Chem Ref Data 27:657
16. Veltzke T (2013) On gaseous microflows under isothermal conditions. Universität Bremen. https://media.suub.uni-bremen.de/handle/elib/470
17. Varoutis S, Naris S, Hauer V, Day C, Valougeorgis D (2009) Computational and experimental study of gas flows through long channels of various cross sections in the whole range of the Knudsen number. J Vac Sci Technol A 27:89

A Meshless Approach to Study Rarefied Gas Flows in Lid-Driven Square Cavities

Himanshi, Anirudh Singh Rana, and Vinay Kumar Gupta

Abstract This paper investigates rarefied gas flows in a lid-driven square cavity using the method of fundamental solutions (MFS). Unlike traditional techniques requiring extensive computational resources and meshing for simulating rarefied gas flows in complex geometries, the MFS bypasses the need for mesh generation and offers a promising alternative. This research focuses on rarefied gas flows confined within an isothermal square cavity with single-sided and two-sided lid-driven configurations. The coupled constitutive relations (CCR) have been adopted in the present work to capture rarefaction effects.

1 Introduction

Gas flows in microchannels have gained importance with the miniaturization of devices like micro heat-exchangers, inkjet printheads, and micro pumps. In these systems, the Knudsen number (ratio of the molecular mean-free path to a characteristic length scale) becomes large and rarefaction effects come into play. While the Navier–Stokes–Fourier (NSF) equations are accurate for low Knudsen numbers, they fail at higher values, where effects like velocity slip and temperature jump dominate. The Boltzmann equation provides a comprehensive description in such cases but is computationally demanding. To bridge this gap, extended macroscopic transport equations have been derived from the Boltzmann equation such as those from the Chapman–Enskog technique [2] and moment methods like, Grad's moments approach [3] and regularized moment methods [4, 13]. More recently, Rana et al. [12] introduced a model to extend classical continuum theories which closes the mass, momentum, and energy balance equations with the coupled constitutive relations (CCR), and is therefore known as the CCR model.

Himanshi (✉) · V. K. Gupta
Department of Mathematics, Indian Institute of Technology Indore, Indore, India
e-mail: hkhungar@gmail.com

A. S. Rana
Department of Mathematics, Birla Institute of Technology and Science Pilani, Rajasthan, India

M. Grabe et al. (eds.), *Rarefied Gas Dynamics*, Springer Aerospace Technology,
https://doi.org/10.1007/978-3-032-00094-1_30

315

Traditional methods like finite element and finite volume approaches, while precise, require substantial computational resources and complex meshing for 3D problems. In contrast, meshless methods, such as the method of fundamental solutions (MFS) offer a flexible and efficient alternative by superimposing Green's functions to approximate solutions to linear boundary value problems. The MFS has been successfully applied to rarefied gas flows [6–9, 11] due to its computational efficiency and straightforward implementation. Building on our earlier work [6], we extend the MFS framework based on the fundamental solutions of the CCR model in two dimensions to study the classic lid-driven cavity problem. The results are validated against the regularized 13-moment (R13) equations and direct simulation Monte Carlo (DSMC) data [10], and we further explore flow scenarios driven by the motion of two walls in both the same and opposite directions.

2   Problem Description and Mathematical Formulation

2.1   Problem Statement

We consider a monatomic rarefied gas contained within an isothermal square cavity of unit side length ($L = 1$). Three configurations are studied:

1. Single-sided lid-driven cavity, where the top wall moves at a constant (dimensionless) velocity v_w in the positive x-direction, while the other walls remain stationary (Fig. 1(a)).
2. Two-sided cavity with both top and bottom walls moving in the positive x-direction with the same velocity v_w (Fig. 1(b)).
3. Two-sided cavity with the top and bottom walls moving in opposite directions, each at speed v_w (Fig. 1(c)).

2.2   The Model and Numerical Technique

The complete CCR model includes the mass, momentum, and energy balance equations along with the coupled constitutive relations [12]. In the present work, we directly present the linearized CCR model in the steady state in order to avoid repetition. For details, the reader is referred to [6]. The steady-state linear dimensionless CCR model consists of the conservation laws

$$\nabla \cdot v = 0, \quad \nabla p + \nabla \cdot \sigma = 0 \quad \text{and} \quad \nabla \cdot q = 0 \tag{1}$$

closed with the relations

$$\sigma = 2\mathrm{Kn}(\overline{\nabla v} + \alpha_0 \overline{\nabla q}) \quad \text{and} \quad q = \frac{5\mathrm{Kn}}{2\mathrm{Pr}}(\nabla T + \alpha_0 \nabla \cdot \sigma). \tag{2}$$

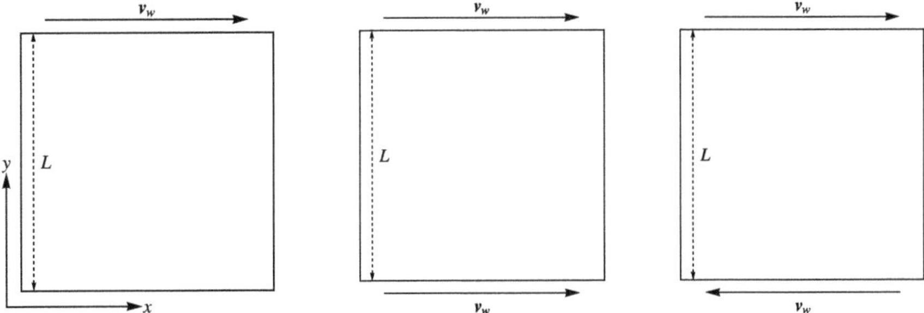

Fig. 1 Schematics of (**a**) single-sided lid-driven cavity, and two-sided lid-driven cavities with top and bottom walls moving in (**b**) same and (**c**) opposite directions

The variables v, p, q, T and σ denote the (dimensionless) velocity, pressure, heat flux, temperature and stress, respectively. Here, α_0 is the coupling coefficient, Kn is the Knudsen number, and Pr is the Prandtl number. The trace-free symmetric components of the velocity and heat flux gradients are represented by $\overline{\nabla v}$ and $\overline{\nabla q}$, respectively. In this study, $\alpha_0 = 2/5$ and Pr $= 2/3$, corresponding to Maxwell molecules [12]. The thermodynamically-consistent boundary conditions associated with the CCR model read [6, 11]

$$v \cdot n - v_w \cdot n = 0, \tag{3}$$

$$q \cdot n = -1.3568(T + \alpha_0 n \cdot \sigma \cdot n), \tag{4}$$

$$n \cdot \sigma \cdot t = -0.7019(v \cdot t - v_w \cdot t + \alpha_0 q \cdot t), \tag{5}$$

where v_w is the wall velocity, n and t are the unit normal and tangent vectors at the boundary, respectively.

The fundamental solutions of the CCR model (1)–(2) in 2D have been determined in [6]. Implementation of the MFS for current problem involves the placement of source points outside the flow domain on a circular fictitious boundary, a schematic of which is illustrated in Fig. 2. A linear system is formed by evaluating the boundary conditions at each boundary node to calculate the unknowns (point force $f_i = (f_{1_i}, f_{2_i})^{\mathsf{T}}$, and point heat source g_i) corresponding to i^{th} source point x_i^s; see Refs. [6, 8] for details. The overall solution at a point in the domain having position x is given by superposition of the fundamental solutions which read [6]

$$v = \sum_{i=1}^{N_s} \left[\frac{f_i}{8\pi\,\text{Kn}} \cdot \left(\frac{2r_i r_i}{r_i^2} - (2\ln r_i - 1)I \right) + \frac{1}{4\pi}\frac{5\text{Kn}}{\text{Pr}}\alpha_0^2 f_i \cdot \left(\frac{2r_i r_i}{r_i^4} - \frac{I}{r_i^2} \right) \right], \tag{6}$$

$$p = \sum_{i=1}^{N_s} \frac{f_i \cdot r_i}{2\pi r_i^2}, \tag{7}$$

Fig. 2 Schematic of the distribution of collocation points (or boundary nodes) on the boundaries and source points outside the flow region for the problem described in Sect. 2.1. The magenta and blue arrows demonstrate the tangential and normal directions at each boundary node, respectively

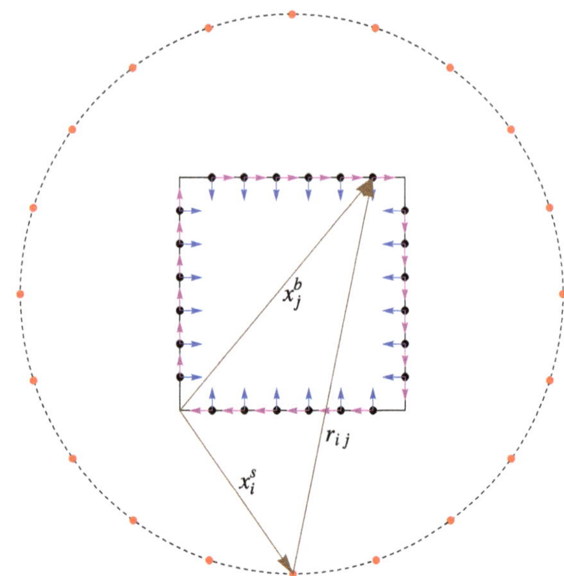

$$\sigma = \sum_{i=1}^{N_s} \frac{\boldsymbol{f}_i \cdot \boldsymbol{r}_i + 2\text{Kn}\, g_i\, \alpha_0}{2\pi} \left(\frac{2\boldsymbol{r}_i \boldsymbol{r}_i}{r_i^4} - \frac{\boldsymbol{I}}{r_i^2} \right), \tag{8}$$

$$T = -\sum_{i=1}^{N_s} \frac{\text{Pr}}{5\text{Kn}} \frac{g_i\, \ln r_i}{\pi}, \tag{9}$$

$$q = \sum_{i=1}^{N_s} \left[\frac{g_i}{2\pi} \frac{\boldsymbol{r}_i}{r_i^2} - \frac{1}{4\pi} \frac{5\text{Kn}}{\text{Pr}} \alpha_0 \boldsymbol{f}_i \cdot \left(\frac{2\boldsymbol{r}_i \boldsymbol{r}_i}{r_i^4} - \frac{\boldsymbol{I}}{r_i^2} \right) \right], \tag{10}$$

where $\boldsymbol{r}_i$ denotes the relative position vector of the ith source point (having position x_i^s) from any position x in the domain.

3 Results and Discussion

The results have been obtained by fixing an equal number of boundary nodes and source points, i.e., $N_b = N_s = 200$ and the cavity region $\{(x, y) : 0 \le x \le 1, 0 \le y \le 1\}$. The fictitious circular boundary on which the source points are placed is centered at $(0.5, 0.5)$ with radius $R_s = 2$.

3.1   Results for a Single-Sided Lid-Driven Cavity

The dimensionless velocity of the lid is fixed at $(v_x, v_y) = (1, 0)$. The results obtained from the MFS applied to the CCR model are validated against data from the DSMC method and R13 equations reported in [10]. The left panel of Fig. 3 illustrates the variation of the vertical velocity along the x-direction at fixed $y = 0.5$, i.e., the variation of $v_y(x, 0.5)$ along the horizontal centerline of the cavity for Kn $= 0.08$. Analogously, the right panel of Fig. 3 depicts the variation of the horizontal velocity $v_x(0.5, y)$ along the vertical centerline of the cavity for Kn $= 0.08$. It is evident that there is a good agreement among the results from the CCR model, DSMC method and R13 model in both panels. Small deviations near the corners are observed, likely due to the inability of the CCR model to capture Knudsen layers, which are more pronounced near solid boundaries.

Figure 4 further illustrates the variation of v_y along horizontal lines at $y = 0.1, 0.4$, and 0.8 (left panel) and v_x along vertical lines at $x = 0.1, 0.4$, and 0.8 (right panel). The increasing tendency of horizontal velocity component v_x with increase in y is due to the maximum horizontal velocity at lid, whereas the wave-like nature of the vertical velocity component v_y with variation in x is due to the formation of vortices which could be clearly depicted via velocity streamlines.

The left panel of Fig. 5 shows velocity streamlines over shear stress σ_{xy} contours for Kn $= 0.08$, indicating the clockwise vortex structure driven by the moving lid. The contours represent the distribution of shear stress σ_{xy} within the cavity induced due to velocity gradient. The right panel in Fig. 5 depicts the heat flux lines plotted over the temperature contours for Kn $= 0.08$. The second-order temperature-jump condition (4) causes cold and hot regions at the left and right corners near the moving

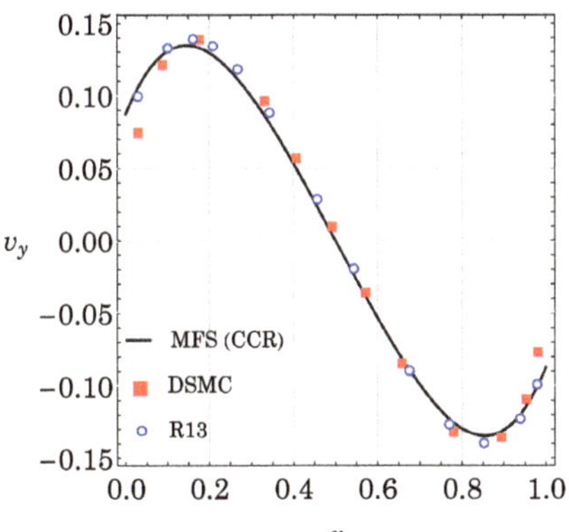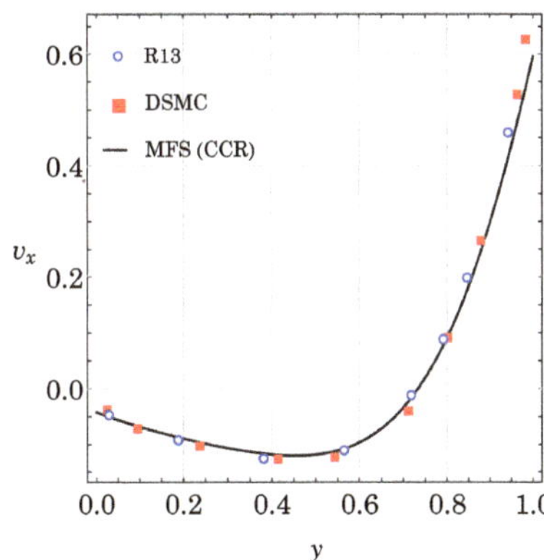

Fig. 3 Variation of v_y along the horizontal centerline (left) and variation of v_x along the vertical centerline (right) for Kn $= 0.08$. The solid black curve represents the results for the MFS applied to the CCR model, the red (square) and blue (circle) symbols denote the data from the DSMC method and R13 model, respectively, taken from Ref. [10]

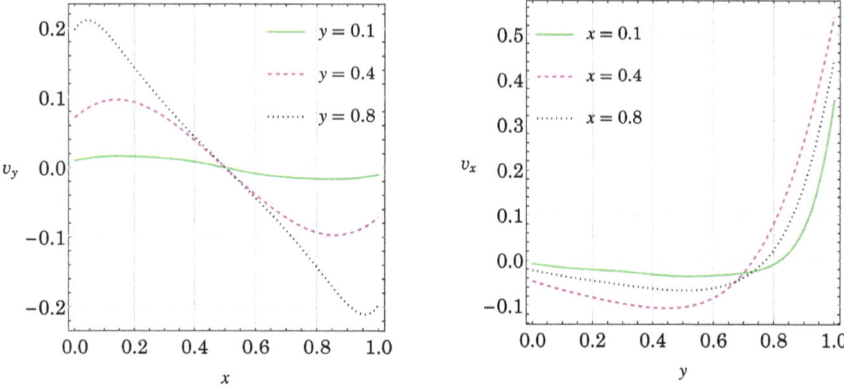

Fig. 4 Variation in v_y along different horizontal lines $y = 0.1, 0.4$ and 0.8 inside the cavity (left) and the variation of v_x along different vertical lines $x = 0.1, 0.4$ and 0.8 inside the cavity (right)

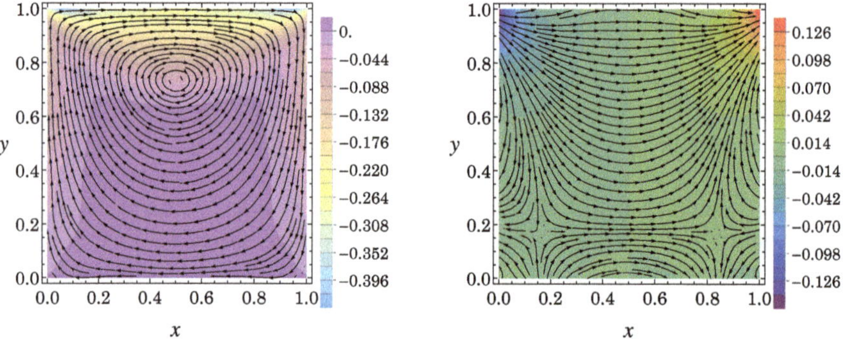

Fig. 5 Velocity streamlines plotted over shear stress contours (left) and heat flux lines plotted over temperature contours (right) for the case when top wall is moving in the positive x-direction

lid, respectively, as wall temperature is influenced by stress. Moreover, the coupling between stress and heat flux (substituting $(1)_2$ into $(2)_2$), gives

$$q = -\frac{5\mathrm{Kn}}{2\mathrm{Pr}}(\nabla T - \alpha_0 \nabla p),\tag{11}$$

highlighting that heat flux depends on both temperature and pressure gradients. This relation introduces a competition between these gradients in determining the heat flow direction. In the present scenario, the pressure gradient dominates, leading to an anti-Fourier heat flow, as evidenced by the heat flux lines in the right panel of Fig. 5. This non-classical phenomenon, observed in microscale and rarefied gas flows, occurs when heat flows from cooler to warmer regions, opposite to the conventional Fourier law of heat conduction [1]. This phenomenon can be predicted well by the CCR model due to the relation (11). Although not shown here, the NSF model even with second-order slip and jump conditions is unable to predict the anti-Fourier heat flow.

We also note that although the bulk flow and centerline velocity profiles agree closely with those in Ref. [10], significant differences appear in the temperature and heat-flux contours. The tilting and asymmetry reported in Ref. [10] arises due to fully non-linear equations, whereas the MFS framework employs the *linearized* CCR equations. The omission of non-linear convective enforces strictly symmetric fields. Such symmetric profiles are also seen in lid-driven cavity simulations for gas mixtures using linearized equations [5].

3.2 Results for the Two-Sided Lid-Driven Cavity with Top and Bottom Walls Moving in the Same Direction

In this subsection, we showcase the flow characteristics when the top and bottom walls are moving in the same direction with the same horizontal velocity v_x which is fixed as $v_x = 1$ for computational purpose. The left panel of Fig. 6 shows velocity streamlines superimposed on shear stress σ_{xy} contours for Kn $= 0.1$. The flow pattern reveals two primary vortices, each occupying the upper and lower halves of the cavity, rotating in the same direction as the moving walls. The contours indicate that the shear stress magnitude is highest near the moving walls, with positive and negative values corresponding to the direction of the local velocity gradients. The heat flux lines plotted over the temperature contours for Kn $= 0.1$ are illustrated in the right panel of Fig. 6. The temperature contours depict hot and cold regions near the corners associated with both top and bottom walls. As evident from the right panel, the heat flows from colder to hotter regions again depicting the anti-Fourier effect produced due to pressure gradients inside the cavity.

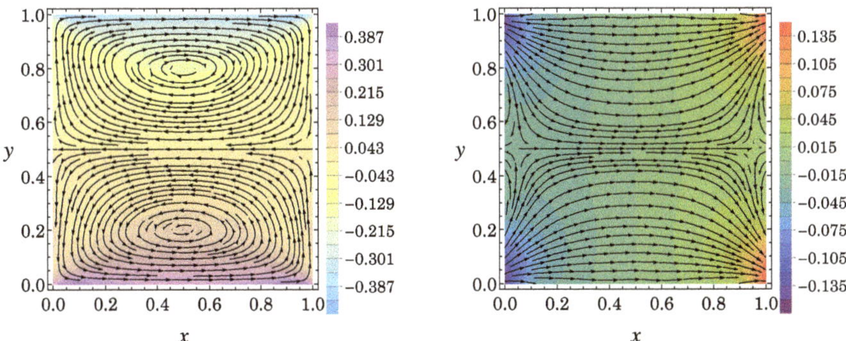

Fig. 6 Same as Fig. 5 but for the case when top and bottom walls are moving in same directions with the same speeds

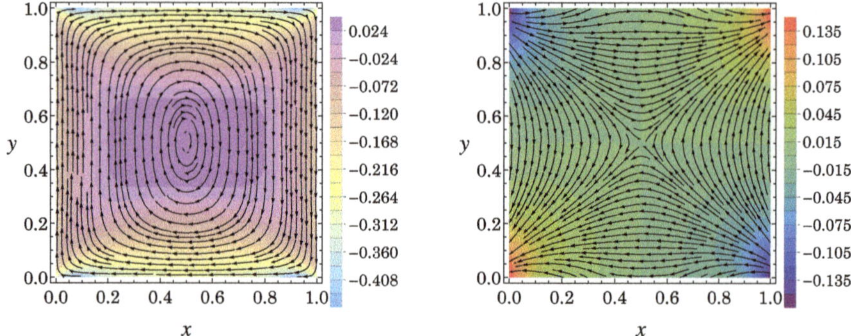

Fig. 7 Same as Fig. 5 but for the case when top and bottom walls are moving in opposite directions with the same speeds

3.3 Results for the Two-Sided Lid-Driven Cavity with Top and Bottom Walls Moving in Opposite Directions

In this case, the horizontal velocity is fixed as $v_x = 1$ and $v_x = -1$ for top and bottom walls, respectively, and the movement of walls leads to formation of a large vortex covering the entire cavity. As evident from the left panel of Fig. 7, the absolute value of shear stress is greatest near the corners of both top and bottom walls. In the present scenario, the hot and cold regions along the bottom walls are opposite as compared to the previous case (Sect. 3.2). However, the heat flux lines again depict the anti-Fourier effect.

References

1. Akhlaghi H, Roohi E, Stefanov S (2018) Ballistic and collisional flow contributions to anti-Fourier heat transfer in rarefied cavity flow. Sci Rep 8:13533
2. Chapman S, Cowling TG (1970) The mathematical theory of non-uniform gases. Cambridge University Press, Cambridge
3. Grad H (1949) On the kinetic theory of rarefied gases. Commun Pure Appl Math 2:331–407
4. Gu XJ, Emerson DR (2009) A high-order moment approach for capturing non-equilibrium phenomena in the transition regime. J Fluid Mech 636:177–216
5. Gupta VK (2015) Mathematical modeling of rarefied gas mixtures. Ph.D. thesis, RWTH Aachen University, Germany
6. Himanshi, Rana AS, Gupta VK (2023) Fundamental solutions of an extended hydrodynamic model in two dimensions: derivation, theory, and applications. Phys Rev E 108:015306
7. Himanshi, Rana AS, Gupta VK (2024) A viewpoint on thermally-induced transport in rarefied gases through the method of fundamental solutions. J Comput Theor Transp 53:279–301
8. Himanshi, Rana AS, Gupta VK (2025) Exploring external rarefied gas flows through the method of fundamental solutions. Phys Rev E 111:015101
9. Lockerby DA, Collyer B (2016) Fundamental solutions to moment equations for the simulation of microscale gas flows. J Fluid Mech 806:413–436

10. Rana A, Torrilhon M, Struchtrup H (2013) A robust numerical method for the R13 equations of rarefied gas dynamics: application to lid driven cavity. J Comput Phys 236:169–186
11. Rana AS, Saini S, Chakraborty S, Lockerby DA, Sprittles JE (2021) Efficient simulation of non-classical liquid-vapour phase-transition flows: a method of fundamental solutions. J Fluid Mech 919:A35
12. Rana AS, Gupta VK, Struchtrup H (2018) Coupled constitutive relations: a second law based higher-order closure for hydrodynamics. Proc Roy Soc A 474:20180323
13. Struchtrup H, Torrilhon M (2003) Regularization of Grad's 13 moment equations: derivation and linear analysis. Phys Fluids 15:2668–2680

Modelling of the Flow of a Thermal Binary Gas Mixture in a Microchannel with a Sinusoidal Varying Thickness

Cédric Croizet

Abstract Interest is significant in the ability to numerically predict the gas flows in microchannels. One of the reference methods used for this purpose is the DSMC method by Bird (Aktas et al., J Microelectromech S 10:538–549 (2001) [1]; Bird, Gas dynamics and the direct simulation of gas flows. Clarendon Press (1994) [2]). Although effective, this statistical method requires significant computational resources. A more economical alternative is to use an asymptotic model. We use this approach in the case of a planar microchannel whose thickness varies sinusoidally and whose wall temperature is a function of the longitudinal spatial variable. We are interested in the case of flows of a mixture of two inert gases with a low Mach number and a Knudsen number of the order of 10^{-1}. These flows correspond to the slip flow regime. As a consequence, the Navier-Stokes-Fourier equations remain valid, provided they are supplemented with slip conditions for velocity and temperature. The aim of this contribution is to study the effects of the channel cross-section function on the flow parameters.

1 Introduction

Miniaturization is a significant challenge in many industrial sectors. As a consequence, numerous applications and many devices involve channels whose size is typically in the micrometer range. For instance, microchannels are found in microturbines, micropumps, microheat exchangers, or microcooling systems. The existence of numerous and varied applications makes the study of gas flows in microchannels very practical and interesting. Several studies have been published discussing the theoretical, numerical, and experimental aspects of such flows [2, 5, 10–12, 15]. From a numerical standpoint, Bird's DSMC is relevant for simulating these flows but requires significant computational resources. Our approach is based on an asymptotic description of the flow that takes into account the main physical phenomena involved. This approach has already been applied in various cases: for a single gas [9,

C. Croizet (✉)
Sorbonne Université, CNRS, Institut Jean Le Rond d'Alembert, Paris, France
e-mail: cedric.croizet@sorbonne-universite.fr

325

M. Grabe et al. (eds.), *Rarefied Gas Dynamics*, Springer Aerospace Technology,
https://doi.org/10.1007/978-3-032-00094-1_31

12], for mixtures in planar channels [4, 8, 10], and in axisymmetric channels [5, 6]. The Knudsen number, Kn, is equal to the ratio of the mean free path to the channel width. At standard conditions for temperature and pressure, the Knudsen number of gas flows in microchannels is of the order 10^{-1} [11, 12]. The Reynolds and Mach numbers are low. From a phenomenological standpoint, these values characterize flows in the weakly rarefied slip regime. As a consequence, the usual mass, momentum and energy balance equations (Navier-Stokes-Fourier equations) remain valid [3–5, 7, 10]. However, the usual no-slip boundary conditions are no longer valid and must be replaced with slip conditions involving coefficients determined from experimental measurement [11]. In this work, we set first order jump conditions for the temperatures and first order slip conditions for the velocities [12]. In order to take into account the interactions between the two gases, coupling terms are added to the momentum and energy equation.

In this contribution, we are interested in flows of a mixture of two compressible ideal gases in a two-dimensional microchannel. Our goal is to take into account small variations in the channel width and to observe their consequences on the flow. In a previous contribution [4], the case of weakly converging or diverging channels with straight walls was studied. In this contribution, the case of channels where the wall equation is a sinusoid is investigated (see Fig. 1). The gas flow can be induced by both a pressure gradient and a temperature gradient at the wall. Regarding the geometry of the channel, we formulate two hypotheses. Firstly, the aspect ratio of the channel, ϵ_1, is small: The channel length (L) is much greater than its width ($\bar{h}$). Secondly, we assume that the width of the channel varies slightly around its average value $\bar{h}$. Consequently, a second small parameter ϵ_2 is introduced to describe the local deviation of the channel width from its mean value $\bar{h}$. In addition, the walls are at rest and their temperature $\bar{T}_w$ depends on the longitudinal space variable $\bar{x}$. Finally, we assume that the flow is steady and symmetric with respect to the channel axis. The pressure and velocity of each gas are specific at the channel inlet. After a brief presentation, the model will be applied to a channel with sinusoidal walls.

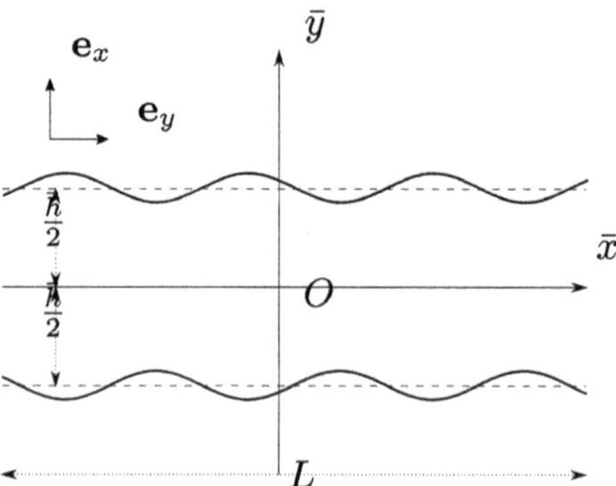

Fig. 1 Channel geometry

To focus our analysis on the effects of channel geometry, we consider an isothermal case. The solution of the first-order approximation is investigated.

2 Model Equations

We consider a two-dimensional channel, symmetric with respect to its longitudinal axis, of length $L = 20\,\mu$m. The $\bar{x}$-axis is parallel to the channel axis (see Fig. 1). Quantities with dimensions (except when they are constants) are denoted with a bar which is removed when referring to the corresponding dimensionless quantities. The walls equation is given by $\bar{y} = \pm \bar{H}(\bar{x})$.

The equations of motion are the Navier-Stokes-Fourier equations for a mixture of two ideal gases. The two gases (denoted with the subscripts "a" and "b") have constant heat capacities at constant pressure (c_{pa} and c_{pb}) and at constant volume (c_{va} and c_{vb}). We assume that the shear viscosities, μ_a and μ_b, and the thermal conductivities, κ_a and κ_b, depend only on the temperature and are such that the Prandtl numbers are constant. For the sake of conciseness, the model is not detailed here; a precise description can be found in [4, 10]. The velocity of the gas θ ($\theta = a, b$) is denoted: $\bar{\mathbf{u}}_\theta = \bar{u}_\theta(\bar{x}, \bar{y}, t)\,\mathbf{e}_x + \bar{v}_\theta(\bar{x}, \bar{y}, t)\,\mathbf{e}_y$, ρ_θ is the volumetric mass of the gas θ and r_θ its specific ideal gas constant. The boundary conditions are first-order slip conditions for the velocities and first order jump conditions for the temperature. For the gas θ at the upper wall $\bar{y} = \bar{H}(\bar{x})$, we have [11, 12]:

$$\bar{\mathbf{u}}_\theta(\bar{x}, \bar{y} = \bar{H}(\bar{x})) \cdot \tau = \left[\frac{\sigma_{p\theta}\,\bar{\mu}_\theta}{\bar{p}_\theta} \sqrt{2\,r_\theta\,\bar{T}_\theta}\,\bar{\nabla}(\bar{\mathbf{u}}_\theta \cdot \tau) \cdot \mathbf{n} \right.$$

$$\left. + \frac{\sigma_{Ta}r_\theta\bar{\mu}_\theta}{\bar{p}_\theta}\,\bar{\nabla}(\bar{T}_\theta \cdot \tau) \right]_{\bar{y}=\bar{H}(\bar{x})} , \tag{1}$$

$$\bar{\mathbf{u}}_\theta(\bar{x}, \bar{y} = \bar{H}(\bar{x})) \cdot \mathbf{n} = 0 , \tag{2}$$

$$\bar{T}_\theta(\bar{x}, \bar{y} = \bar{H}(\bar{x})) = \bar{T}_w + \left[\xi_{T\theta}\sqrt{2\,r_\theta\,\bar{T}_\theta}\,\bar{\nabla}(\bar{T}_\theta) \cdot \mathbf{n} \right]_{\bar{y}=\bar{H}(\bar{x})} , \tag{3}$$

where $\mathbf{n}$ and τ are the unit vectors respectively normal and tangential to the wall. The constant coefficients $\sigma_{p\theta}$, $\sigma_{T\theta}$ and $\xi_{T\theta}$ characterize the viscous and thermal slip of the species θ on the walls [13]. At the wall at $\bar{y} = -\bar{H}(\bar{x})$, analogous boundary conditions are set. Boundary conditions at the channel inlet and outlet will be specified further. To simplify the problem, we will analyze and compare the orders of magnitude of the different terms in the problem's equations in a dimensionless form. The length of the channel, L, is set as the longitudinal length scale. The channel width is characterized by the transversal length scale $\bar{h}$. Two small parameters are introduced: the aspect ratio of the channel, $\epsilon_1 = \bar{h}/L \ll 1$, and a second small parameter, ϵ_2, that characterizes the deviation of the local width of the channel from $\bar{h}$. We set $\bar{H}(\bar{x}) = \bar{h}\left[\frac{1}{2} + \epsilon_2 h(x)\right]$ with $h(x) = \sin\left(\frac{n\pi x}{L} + \lambda\right)$. Finally, U and V, T_c

(for the two species), $p_{\theta c}$ and $\rho_{\theta c}$ (with $p_{\theta c} = r_\theta \rho_{\theta c} T_c, \theta = a, b$) are set for the characteristic values of the axial and transversal velocities, the temperature, the pressure, and the volumetric mass, respectively. With these characteristic scales, we build the following dimensionless numbers:

$$\gamma_\theta = \frac{c_{p\theta}}{c_{v\theta}}, \quad Pr_\theta = \frac{\mu_\theta c_{p\theta}}{\kappa_\theta}, \quad Re_\theta = \frac{\rho_{\theta c} U \bar{h}}{\mu_{\theta c}},$$

$$Ma_\theta = \frac{U}{\sqrt{\gamma_\theta r_\theta T_c}}, \quad Kn_\theta = \sqrt{\frac{\gamma_\theta \pi}{2}} \frac{Ma_\theta}{Re_\theta},$$

where μ_θ and κ_θ are proportional to $T_\theta^{\omega_\theta}$ since Pr_θ is constant. The Principle of Least Degeneracy applied to the mass balance equation implies that $U = \epsilon_1 V$. In the momentum balance projected in the main direction of the flow $(O, \mathbf{e_x})$, the pressure gradient appears as the driving force. We also recall that we are interested in flows at Mach and Reynolds numbers small or of order unity. To determine the degeneracies that correspond to the physics of our problem, we set $Re_a = R_a \epsilon_1^\alpha$ and $Ma_a = M_a \epsilon_1^\beta$, where R_a and M_a are of order one. From the dimensionless balance of momentum and energy, we have $\alpha = \beta = 1$. This case corresponds to Mach and Reynolds numbers of order ϵ_1. As a consequence we have $Kn_\theta = O(1), \theta = a, b$. At the first approximation order in ϵ_1, we have for the gas a:

$$\frac{\partial}{\partial x}(\rho_a u_a) + \frac{\partial}{\partial y}(\rho_a v_a) = 0 \tag{4}$$

$$-\frac{1}{\gamma_a M_a^2} \frac{\partial p_a}{\partial y} = 0 \tag{5}$$

$$\frac{1}{\gamma_a M_a^2} \frac{\partial p_a}{\partial x} = \frac{1}{R_a} \frac{\partial}{\partial y}\left(\mu_a \frac{\partial u_a}{\partial y}\right)$$

$$+ F_{ab} \frac{\rho_{bc}}{\rho_{ac}} \left(\frac{\mu_{ac}}{\mu_{bc}}\right)^{\frac{1}{2}} \frac{R_a}{\gamma_a M_a^2} \frac{p_a p_b}{\sqrt{\mu_a \mu_b}} \left(\frac{T_b}{T_a}\right)^{\frac{1}{2}} (u_b - u_a) \tag{6}$$

$$\frac{\gamma_a}{R_a Pr_a} \frac{\partial}{\partial y}\left(\kappa_a \frac{\partial T_a}{\partial y}\right) + G_{ab} \frac{R_a}{\gamma_a M_a^2} \frac{\rho_b p_a}{\sqrt{\mu_a \mu_b}} \sqrt{\frac{T_b}{T_a}} (T_b - T_a) = 0, \tag{7}$$

where F_{ab} and G_{ab} depend on the gas properties and are explicitly given in [4]. From Eq. (5) and the analogous equation for gas b, it is obvious that p_a and p_b depend only on x. The equations for the gas b are obtained by exchanging a and b in Eqs. (4) to (7). At the same approximation order, the temperature jump at the upper wall is written as:

$$T_\theta\left(x, \frac{1}{2} + \epsilon_2 h(x)\right) = T_w(x) - \frac{\sqrt{2 \gamma_\theta} M_\theta}{R_\theta} \xi_{\theta T} \frac{\mu_\theta \sqrt{T_\theta}}{p_\theta} \frac{\partial T_\theta}{\partial y}\left(x, \frac{1}{2} + \epsilon_2 h(x)\right) \tag{8}$$

If we assume that the flow is in thermal equilibrium, meaning both gases are at the same temperature, Eq. (7) with the wall condition (8) implies that the two gases are at the wall temperature : $T_a(x, y) = T_w(x) = T_b(x, y) = T(x)$. Consequently, the dimensionless velocity conditions at the upper wall are:

$$u_\theta\left(x, \frac{1}{2} + \epsilon_2 h(x)\right) = -\frac{K_\theta\, [T]^{\omega_\theta + \frac{1}{2}}}{p_\theta}\, \frac{\partial u_\theta}{\partial y}\left(x, \frac{1}{2} + \epsilon_2 h(x)\right) + \frac{\sigma_{\theta T}}{R_\theta}\, \frac{[T]^{\omega_\theta}}{p_\theta}\, \frac{dT}{dx} \tag{9}$$

$$v_\theta\left(x, \frac{1}{2} + \epsilon_2 h(x)\right) - \epsilon_2 \frac{dh}{dx}(x)\, u_\theta\left(x, \frac{1}{2} + \epsilon_2 h(x)\right) = 0 \tag{10}$$

where: $K_\theta = \sigma_{\theta\, p}\, \sqrt{2\gamma_\theta}\, M_\theta / R_\theta$. At this approximation order, we have a set of six equations derived from the mass and momentum balance of the two gases. With the ideal gas law for the two species, we have: $p_a(x) = \rho_a(x)\, T(x)$ and $p_b(x) = \rho_b(x)\, T(x)$ and the problem is reduced to:

$$\frac{\partial}{\partial x}(\rho_a\, u_a) + \frac{\partial}{\partial y}(\rho_a v_a) = 0, \quad \frac{\partial}{\partial x}(\rho_b\, u_b) + \frac{\partial}{\partial y}(\rho_b v_b) = 0, \tag{11}$$

$$\begin{aligned}
\frac{\partial^2 u_a}{\partial y^2} - \frac{B_a}{A_a}\, p_a\, p_b(u_a - u_b) &= \frac{1}{A_a}\, \frac{dp_a}{dx}, \\
\frac{\partial^2 u_b}{\partial y^2} + \frac{B_b}{A_b}\, p_a\, p_b(u_a - u_b) &= \frac{1}{A_b}\, \frac{dp_b}{dx},
\end{aligned} \tag{12}$$

where $A_\theta(x)$ and $B_\theta(x)$ $(\theta = a, b)$ are given in [4]. From the longitudinal momentum equations (12), an explicit expression of the longitudinal component of the velocities at the first approximation order is obtained. So, by using the conservation of mass in its global form and introducing the mass flow rates of each gas, it is straightforward to obtain two coupled ordinary differential equations of the first order for the pressures [4]:

$$X_a \frac{dp_a}{dx} + X_b \frac{dp_b}{dx} = S_a, \qquad Y_a \frac{dp_a}{dx} + Y_b \frac{dp_b}{dx} = S_b. \tag{13}$$

For the sake of conciseness, the detailed expressions of X_θ and Y_θ $(\theta = a, b)$, are not explicitly given here.

3 Numerical Solution

In order to solve Eq. (13), we need to impose the wall temperature and boundary conditions at the inlet and at the outlet of the channel. Given the form of the equations, we set the pressures at the inlet for each gas, as well as the constant mass flow rates $\bar{Q}_{ma}$ and $\bar{Q}_{mb}$, of each gas. The two Eq. (13) are solved with MATLAB ; a fourth order Runge-Kutta method with variable stepsizes is used.

Table 1 Physical properties of the gases [2]

	Argon	Neon
Molecular mass (kg)	$66.3\ 10^{-27}$	$33.5\ 10^{-27}$
Molecular diameter (m)	$4.11\ 10^{-10}$	$2.72\ 10^{-10}$
γ	5/3	5/3
μ_{VSS} (Pa.s)	$2.117\ 10^{-5}$	$2.975\ 10^{-5}$
α	1.40	1.31
ω	0.81	0.66

Although the model is capable of addressing problems involving arbitrary temperature distributions along the walls, an isothermal case ($T = 300$ K) is considered here in order to isolate and analyze the effect of the channel geometry. We consider the case of a mixture of Argon (a) and Neon (b) (Table 1). From a geometric point of view, the walls are sinusoidal curves chosen such that the length of the channel is equal to a quarter of the sinusoid period. We set: $\epsilon_2 = 10^{-1}$, $h_2(x) = \sin(\frac{\pi x}{2})$, and $h_2(x) = \sin(\frac{\pi x}{2} + \pi)$. In the first case, a converging channel is built with an inlet half-width of 0.5 μm and an outlet half-width of 0.4 μm. In the second case, a diverging channel is constructed with an inlet half-width of 0.5 μm and an outlet half-width of 0.6 μm. A third geometry is considered: a channel with straight walls and a constant half-width of 0.5 μm. In addition, we set: $\bar{p}_a(-L/2) = 3.533\ 10^4$ Pa, $\bar{p}_b(-L/2) = 1.514\ 10^4$ Pa, $\bar{Q}_{ma} = 2.248\ 10^{-6}$ kg m^{-1} s^{-1} and $\bar{Q}_{mb} = 0.485\ 10^{-6}$ kg m^{-1} s^{-1}. For each of the three geometries, the evolution of the pressures along the channel axis is plotted in Fig. 2a. We observe that the results are very similar to those obtained in the case of channels with straight walls [4], with inlet and outlet half-widths identical to those of the channels studied in this contribution and with identical boundary conditions. The channel geometry influences the pressure of the two gases: When the channel width at the outlet increases, the outlet pressures increase. The profiles of the longitudinal component of the velocity at the channel outlet are plotted in Fig. 2b. At the channel outlet, the slip velocity of Argon is smaller than the slip

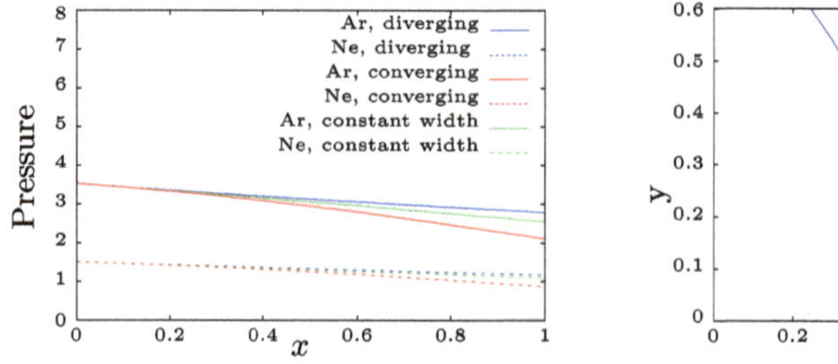

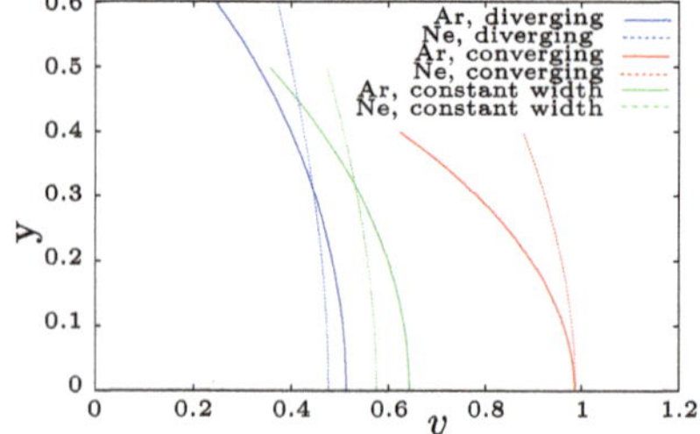

Fig. 2 Dimensionless pressures along the channel axis for $T = 300$ K and different channel shapes: converging channel $\bar{h} = 0.4$ μm, straight channel $\bar{h} = 0.5$ μm, diverging channel $\bar{h} = 0.6$ μm (**a**), dimensionless axial velocities (**b**)

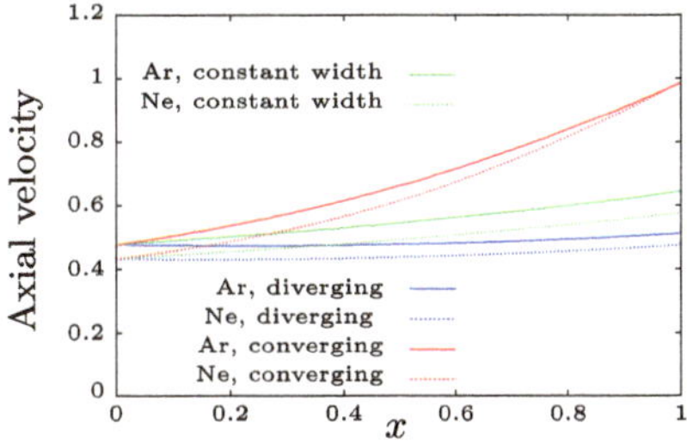
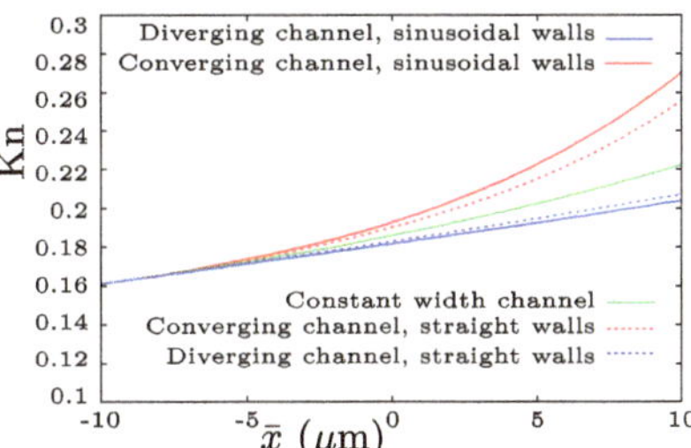

Fig. 3 Dimensionless axial velocity along the channel axis for $T = 300$ K and different channel shapes: converging channel $\bar{h} = 0.4$ μm, constant width channel $\bar{h} = 0.5$ μm, diverging channel $\bar{h} = 0.6$ μm (**a**), Knudsen number (**b**)

velocity of Neon. On the other hand, when the channel width at the outlet increases the axial velocity at the channel outlet decreases.

For the three geometries, the evolution of the value of the longitudinal component of the velocity along the channel axis is plotted in Fig. 3a. For the two gases, we observe that the axial velocity increases along the channel axis. Its value for Neon is smaller than for Argon. Moreover, when the channel width at the outlet increases, the axial velocity decreases. These observations are consistent with what has been observed in the case of straight walls [4]. In Fig. 3b the evolution of the Knudsen number of the mixture along the channel axis is plotted for the constant width channel, the diverging channels with straight and sinusoidal walls and the converging channels with straight and sinusoidal walls.

The expression of the Knudsen number of the mixture is given in [5, 14] for instance. Whether for sinusoidal or straight walls, the results are quite similar: the Knudsen number increases along the channel axis and, at the outlet, it increases as the channel thickness decreases. This behavior seems to be amplified by the sinusoidal walls: in this case, the difference between the Knudsen number values at the outlet is greater than in the case of straight walls. In Fig. 4a, we plot the molar fraction of Argon along the channel axis in the five cases. It is slightly affected by the shape

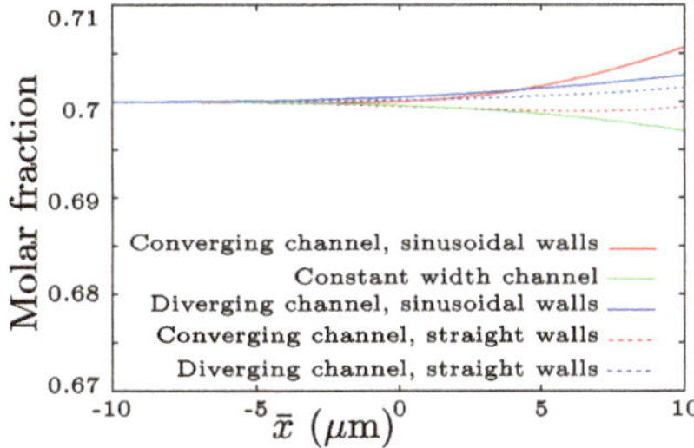
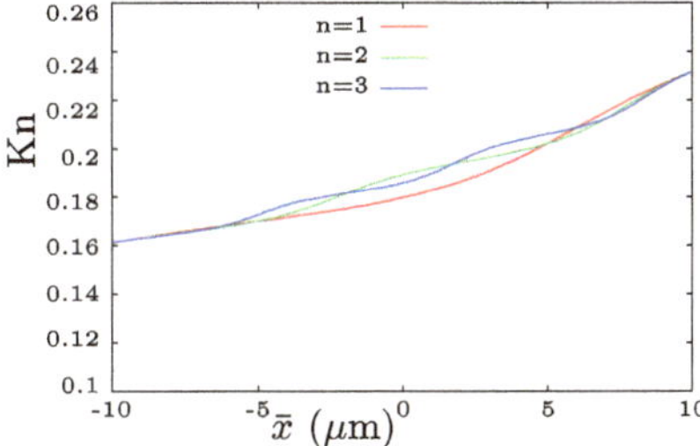

Fig. 4 Molar fraction of Argon along the channel axis: converging channel $\bar{h} = 0.4$ μm, straight channel $\bar{h} = 0.5$ μm, diverging channel $\bar{h} = 0.6$ μm (**a**), molar fraction of Argon along the channel axis for $n = 1$, $n = 2$ and $n = 3$ (**b**)

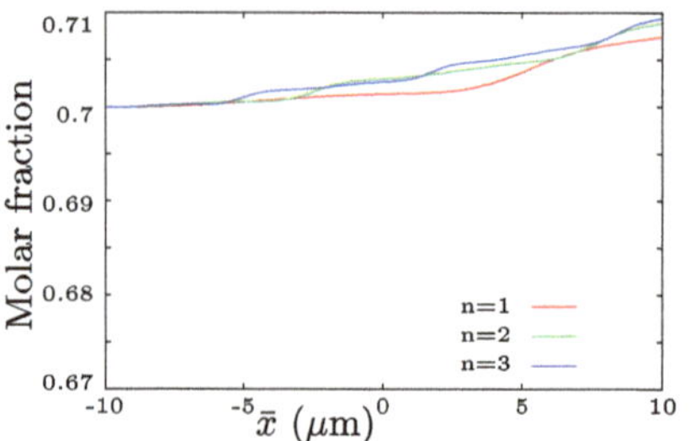 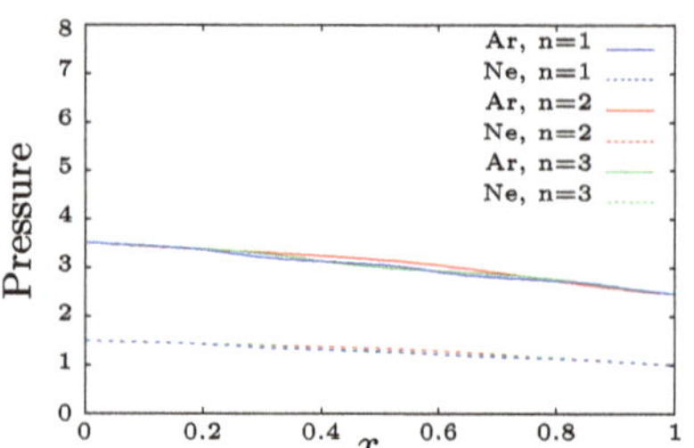

Fig. 5 Molar fraction of Argon along the channel axis for different channel shapes: $n = 1$, $n = 2$ and $n = 3$ (**a**), dimensionless pressures along the channel axis for different channel shapes: $n = 1$, $n = 2$ and $n = 3$ (**b**)

of the wall as it varies by less than one percent; however, this variation seems more noticeable in the case of sinusoidal walls. It follows that the molar diffusion is weak, and thus the flow is dominated by the pressure gradient. It follows that the flow is close to Poiseuille flow with slip boundary conditions; the pressure decreases along the channel (Fig. 2a) as a consequence of the viscous dissipation.

We now consider the case where the wall equation is of the form $h_2(x) = \sin(2n\pi x)$ with $\epsilon_2 = 10^{-1}$ and n an integer. In Figs. 4b, 5a, b, we depict the Knudsen number, the molar fraction of Argon, and the pressures along the channel axis, respectively, for $n = 1, 2$, and 3. It is observed that in this isothermal case, the Knudsen number increases along the channel axis. The effects of the variation in the channel thickness are local and the Knudsen number at the outlet is approximately equal in the three considered cases. The geometry has a rather weak influence on the molar fraction of Argon. Its variation is lower than 1.5 %, but its value at the outlet increases when n increases. The pressures decrease along the channel axis and the outlet pressures are very close in the three considered cases.

4 Conclusion

This contribution deals with a microchannel with a sinusoidal-shaped wall. From the Navier-Stokes-Fourier equations for each gas, we derived an asymptotic model. The case of converging and diverging channels has been investigated. At the first approximation order, the channel geometry has an influence on the flow: quantitative differences are observed between the case of a channel with straight walls and that of a channel with sinusoidal walls. In the case of channels with identical width at the inlet and at the outlet, the molar fraction of Argon at the outlet increases with the number of period n of the sine function.

References

1. Aktas O, Aluru NR, Ravaioli U (2001) Application of a parallel DSMC technique to predict flow characteristics in microfluidic filters. J Microelectromech S 10:538–549
2. Bird GA (1994) Gas dynamics and the direct simulation of gas flows. Clarendon Press
3. Chapman S, Cowling TG (1952) The mathematical theory of non-uniform gases. C. U. P
4. Croizet C, Gatignol R (2024) Asymptotic modelling of the flow of a thermal binary gas mixture in a microchannel with variable width. AIP Conf Proc 2996(1):100002. https://doi.org/10.1063/5.0187532
5. Croizet C, Gatignol R (2019) Asymptotic modelling of thermal binary gas flows in circular micro-channels. Microfluid. Nanofluidics 23(7):94. https://doi.org/10.1007/s10404-019-2260-8
6. Croizet C, Gatignol R (2019) Molecular mass effect on the flow of a thermal binary gas mixture in a circular micro channel. AIP Conf Proc 2132(1):160002. https://aip.scitation.org/doi/abs/10.1063/1.5119649
7. Elizarova TG, Graur IA, Lengrand JC (2001) Two-fluid computational model for a binary gas mixture. Eur J Mech B-Fluids 3:351–369
8. Gatignol R, Croizet C (2014) Asymptotic modelling of the flow of a thermal binary gas mixture in a microchannel. AIP Conf Proc 1628(1):807–814. https://aip.scitation.org/doi/abs/10.1063/1.4902676
9. Gatignol R, Croizet C (2011) Asymptotic modelling of the flows in micro-channel by using macroscopic balance equation. AIP Conf Proc 1333(1):730–735. https://aip.scitation.org/doi/abs/10.1063/1.3562733
10. Gatignol R, Croizet C (2017) Asymptotic modeling of thermal binary monatomic gas flows in plane microchannels—comparison with DSMC simulations. Phys Fluids 29(4):042001. http://dx.doi.org/10.1063/1.4979683
11. Kandlikar S, Garimella S, Li D, Colin S, King M (2005) Transfer and fluid flow in minichannels and microchannels. Elsevier
12. Karniadakis G, Beskok A (2006) Microflows–fundamentals and simulation. Springer
13. Sharipov F, Seleznev V (1998) Data on internal rarefied gas flows. J Phys Chem Ref Data 27:657–706
14. Sharipov F, Kalempa D (2002) Gaseous mixture flow through a long tube at arbitrary knudsen numbers. J Vac Sci Technol A 20(3):814–822
15. Silva E, Deschamps CJ, Rojas-Cárdenas M, Barrot-Lattes C, Baldas L, Colin S (2018) A time-dependent method for the measurement of mass flow rate of gases in microchannels. Int J Heat Mass Transf 120:422–434

Moment Methods

Spatially Adaptive Moment Model Using Padded Buffer Cell for Linear Hierarchical Moment Equations

Rik Verbiest and Julian Koellermeier

Abstract The need for spatial adaptivity in simulations of rarefied gases arises because different subdomains within a rarefied gas domain often require varying levels of modeling complexity. Different moment models are effective at describing rarefied flows with respective degrees of complexity in each subdomain. However, currently there is no operational spatially adaptive method to efficiently couple moment models in a single simulation. This paper presents a Spatially Adaptive Moment Model (SAMM) for simulating linear hierarchical moment models. A padded buffer cell approach is proposed to couple these varying-order moment models. Numerical shock tube simulations demonstrate that the proposed model achieves accurate results comparable to a high-order model while providing a computational speedup corresponding to the reduction of variables.

Keywords Moment models · Rarefied gases · Adaptive simulation

1 Introduction

Many complex gas flows are characterized by the presence of both rarefied and continuum regimes. Examples are high-altitude flights and atmospheric reentry flights [8, 12, 16], in which the mean free path varies in time and space. While in domains of small Knudsen number, standard fluid dynamics models yield sufficient accuracy, the nonequilibrium effects in domains of large Knudsen number need models beyond the standard fluid solvers [14]. Particle simulations such as DSMC [1] have become the main computational tool for simulating large Knudsen number rarefied gas flows, but result in prohibitively large computational effort for more moderate Knudsen numbers [7].

R. Verbiest (✉) · J. Koellermeier
Bernoulli Institute, University of Groningen, Groningen, The Netherlands
e-mail: r.verbiest@rug.nl

J. Koellermeier
Department of Mathematics, Computer Science and Statistics, Ghent University, Gent, Belgium

© The Author(s) 2026

M. Grabe et al. (eds.), *Rarefied Gas Dynamics*, Springer Aerospace Technology,
https://doi.org/10.1007/978-3-032-00094-1_32

Moment models for kinetic equations have proven to be a powerful modeling technique for the simulation of rarefied gases with moderate Knudsen numbers [3, 15]. Moment models achieve increasing accuracy by successively including more variables and evolution equations. A moment model is characterized by its order. A higher order moment model is often more accurate but computationally more expensive than a lower order moment model as more variables are included in the model. Moment models are ideally suited for spatial adaptivity because they are hierarchical, which simplifies the coupling considerably. However, there is no spatially adaptive scheme for moment models, yet.

The goal of this paper is to construct a spatially adaptive numerical scheme for the simulation of linear hierarchical moment models. We introduce the first spatially adaptive moment model simulations using a domain decomposition approach with padded buffer cells at the interfaces between two models with a different number of moments. The proposed model is called the Spatially Adaptive Moment Model using Padded Buffer Cell (SAMM-PBC). We investigated the precision using a shock tube test case.

2   Moment Equations: Hermite Spectral Method

For simplicity, we consider the 1D spatial space and leave the extension to higher dimensions to future work. The motion of gas particles is described by the 1D Boltzmann transport equation

$$\partial_t f(t, x, c) + c\partial_x f(t, x, c) = S(f) ,\tag{1}$$

with particle distribution $f(t, x, c)$, spatial position $x \in \Omega \subseteq \mathbb{R}$, microscopic velocity $c \in \mathbb{R}$ and collision operator $S(f)$. The non-linear Hyperbolic Moment Equations (HME) [2] are obtained by expanding $f(t, x, c)$ in a truncated expansion of order $M \in \mathbb{N}$ in the Hermite polynomials $\Phi_i^{u,T}\left(\frac{c-u}{\sqrt{T}}\right)$ [6], i.e.,

$$f(t, x, c) = \sum_{i=0}^{M} m_i(t, x)\Phi_i^{u,T}\left(\frac{c - u}{\sqrt{T}}\right) ,\tag{2}$$

with velocity u, temperature T and density ϱ. The moment equations are then obtained by projecting (1) onto test functions up to a desired order. We consider the linearized version of the HME model (that is, linearized around $u_0 \approx 0$ and $T_0 \approx 1$), called the Hermite Spectral Method (HSM) [5, 10]. In the HSM, the evolution of the expansion coefficients $\mathbf{w}_M = (m_0, m_1, \ldots, m_M)^T \in \mathbb{R}^{M+1}$ is governed by the linear hyperbolic PDE

$$\frac{\partial \mathbf{w}_M}{\partial t} + A_M \frac{\partial \mathbf{w}_M}{\partial x} = S(\mathbf{w}_M) ,\tag{3}$$

with constant system matrix

$$
A_M = \begin{pmatrix} & 1 & & & \\ 1 & & \ddots & & \\ & & & & M \\ & \ddots & & & \\ & & 1 & & \end{pmatrix} \in \mathbb{R}^{(M+1)\times(M+1)} \; ,
$$

and where the right-hand side collision term $S(\mathbf{w}_M)$ is given by the simple BGK model $S(\mathbf{w}_M) = -\frac{1}{\tau}(0, 0, 0, m_3, \dots, m_M)^T \in \mathbb{R}^{M+1}$ with constant relaxation time $\tau \in \mathbb{R}^+$ in this paper.

3 Spatially Adaptive Simulation of Moment Models

Equation (3) is discretized in Ω with grid size Δx, producing semi-discretized cell-averaged moment vectors $w_i \in \mathbb{R}^{M+1}$ in cells $C_i = [x_{i-\frac{1}{2}}, x_{i+\frac{1}{2}}]$ with equidistant cell centers x_i and cell interfaces $x_{i\pm\frac{1}{2}} = x_i \pm \frac{\Delta x}{2}$, $i = 1, 2, \dots, N_x$, and discretized in time with time step Δt, yielding the fully discretized cell-averaged moment vectors $w_i^n \in \mathbb{R}^{M+1}$ at discrete times t_n, $n = 0, 1, \dots, N_t$. For the numerical simulation of (3), we use a finite volume scheme with the Polynomial Viscosity Method (PVM) [4] notation due to its generality and due to its applicability to nonconservative systems. The PVM reads

$$
w_i^{n+1} = w_i^n - \frac{\Delta t}{\Delta x}\left(D_{i-\frac{1}{2}}^+ + D_{i+\frac{1}{2}}^-\right) - \Delta t\, S(w_i^n) \; , \tag{4}
$$

with fluctuations $D_{i+1/2}^{\pm} = A_{\Phi}^{\pm}(w_i, w_{i+1})$ given by

$$
A_{\Phi}^{\pm}(w_L, w_R) = \frac{1}{2}\left(A_{\Phi}(w_L, w_R)\cdot(w_R - w_L) \pm Q_{\Phi}\cdot(w_R - w_L)\right) \; , \tag{5}
$$

with generalized Roe linearization $A_{\Phi} = A_{\Phi}(w_L, w_R)$ given by

$$
A_{\Phi}(w_L, w_R)\cdot(w_R - w_L) = \int_0^1 A(\Phi(s; w_L, w_R))\frac{\partial \Phi}{\partial s}(s; w_L, w_R)ds \; , \tag{6}
$$

and where the numerical viscosity matrix $Q_{\Phi} = Q_{\Phi}(w_L, w_R) = P(A_{\Phi}(w_L, w_R))$ is a function $P(\cdot)$ of the generalized Roe matrix $A_{\Phi}(w_L, w_R)$. The Lax-Friedrichs flux [11] is chosen, which is equivalent to $Q_{\Phi} = \frac{\Delta x}{\Delta t}I$, where I is the identity matrix. Here, $\Phi(s; w_L, w_R)$ denotes a path connecting the left and right states at the cell interface, such that $\Phi(0; w_L, w_R) = w_L$ and $\Phi(1; w_L, w_R) = w_R$. Note that the expression (4) is written in nonconservative form for generality.

3.1 Domain Decomposition

The proposed spatially adaptive simulation of moment models consists of two steps:
(1) domain decomposition into subdomains, each modeled by a moment model of
a certain order, and (2) coupling of the different order moment models at their
interfaces. We consider a decomposition into two domains that is performed in
some time step during the computation. For now, the time index will be dropped
to simplify notation. Assume that the domain interface is given by $x_{I+\frac{1}{2}} \in \Omega$, so
that for $x < x_{I+\frac{1}{2}}$, an Mth order moment model is used, while for $x \geq x_{I+\frac{1}{2}}$, an
$(M+1)$th order moment model is used. The situation is sketched in Fig. 1. The
spatial domain Ω is divided into two subdomains $\Omega_M = \{x : x \in \Omega, x < x_{I+\frac{1}{2}}\}$ and
$\Omega_{M+1} = \{x : x \in \Omega, x \geq x_{I+\frac{1}{2}}\}$. In each cell $C_l \subset \Omega_M$ of the left subdomain, the
moment vector $w_l \in \mathbb{R}^{M+1}$ contains the moments $m_{l,0}$ to $m_{l,M}$, while in each cell
$C_r \subset \Omega_{M+1}$ of the right subdomain, the moment vector $w_r \in \mathbb{R}^{M+2}$ contains the
moments $m_{r,0}$ to $m_{r,M+1}$:

$$w_l = (m_{l,0}, \ldots, m_{l,M})^T, \; C_l \subset \Omega_M \;, \; \text{and} \; w_r = (m_{r,0}, \ldots, m_{r,M+1})^T, \; C_r \subset \Omega_{M+1} \;.$$

The generalization towards more domains or an arbitrary difference in the number
of moments is straightforward and follows directly from the presented analysis.

3.2 Spatially Adaptive Moment Model Using Padded Buffer Cell (SAMM-PBC)

We have $C_I \subset \Omega_M$, so that $w_I \in \mathbb{R}^{M+1}$, and $C_{I+1} \subset \Omega_{M+1}$, so that $w_{I+1} \in \mathbb{R}^{M+2}$.
The difficulty is the computation of the flux at the boundary interface between the
cells C_I and C_{I+1}. We propose padded buffer cells to couple different order moment
models at the boundary interface. The padded buffer cell locally uses the cell C_I
as a buffer cell and pads it with certain values for the missing moments. Padding
can be understood as increasing the size of the shorter moment vector by adding
the additional last moments. In this example, this leads to the situation illustrated in
Fig. 2. The moment vector w_I is padded with a last moment $\tilde{m}_{I,M+1}$. The choice of
$\tilde{m}_{I,M+1}$ determines how the two moment models are coupled and should represent
the physics of the system. We will define the new padded moment vector as $\tilde{w}_I :=$

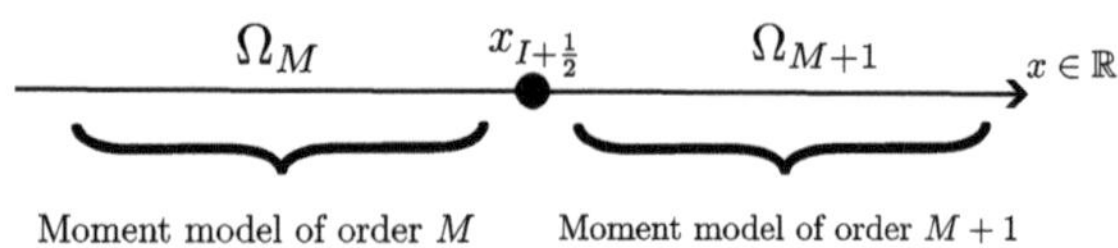

Fig. 1 The domain is split into two subdomains Ω_M and Ω_{M+1} in which a moment model of order
M and $M+1$ is used, respectively

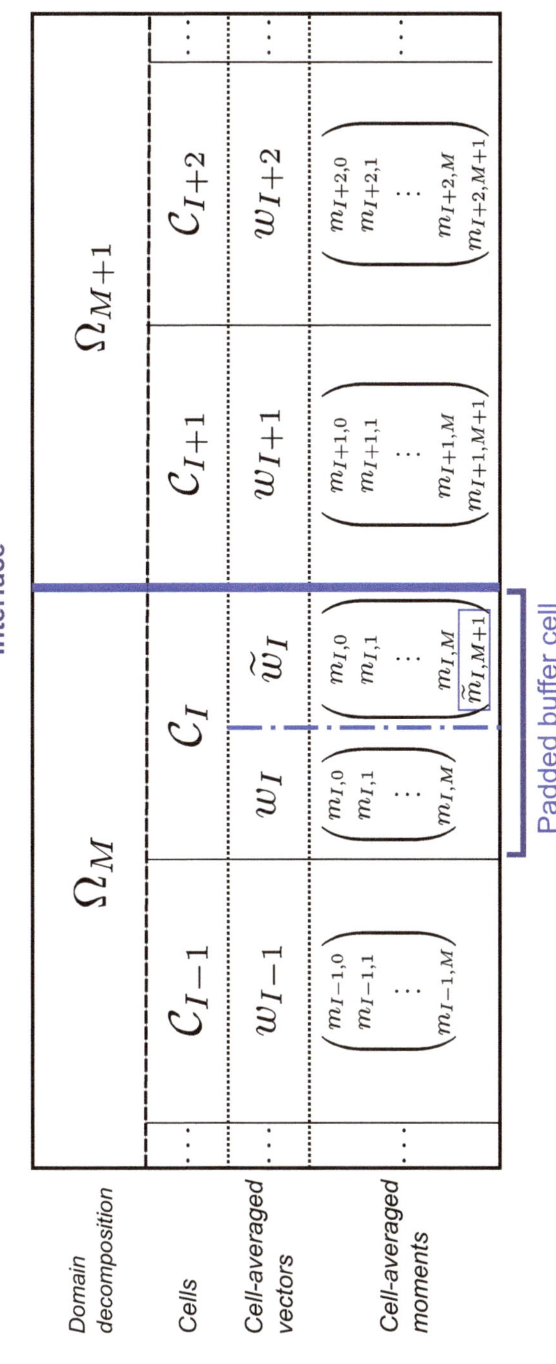

Fig. 2 Padded buffer cell with added last moment $\tilde{m}_{I,M+1}$ for the coupling of Ω_M and Ω_{M+1}.

$(m_{I,0}, m_{I,1}, \ldots, m_{I,M}, \widetilde{m}_{I,M+1})^T \in \mathbb{R}^{M+2}$. The still to be defined padded moment $\widetilde{m}_{I,M+1}$ is solely used to compute the flux at the boundary interface between the cells C_I and C_{I+1}; all other fluxes are between cells with the same number of moments, which predefines the moment model for the flux computation at those cells. The new padded moment vector in the padded buffer cell can then be inserted in the relevant PVM equations (4). We will call the resulting finite volume method the Spatially Adaptive Moment Model using Padded Buffer Cell (SAMM-PBC).

The SAMM-PBC introduced here takes the same fluctuation-differencing form as (4). We assume that the domain decomposition is performed after collecting the cell averages at time t_n but before computing the cell averages at time t_{n+1} using the PVM formula (4). The Mth order and the $(M + 1)$th order HSM moment equations (3) read

$$\frac{\partial \mathbf{w}_M}{\partial t} + A_M \frac{\partial \mathbf{w}_M}{\partial x} = S(\mathbf{w}_M) , \quad \text{and} \quad \frac{\partial \mathbf{w}_{M+1}}{\partial t} + A_{M+1} \frac{\partial \mathbf{w}_{M+1}}{\partial x} = S(\mathbf{w}_{M+1}) , \quad (7)$$

respectively, with $\mathbf{w}_M \in \mathbb{R}^{M+1}$, $A_M \in \mathbb{R}^{(M+1)\times(M+1)}$, $\mathbf{w}_{M+1} \in \mathbb{R}^{M+2}$ and $A_{M+1} \in \mathbb{R}^{(M+2)\times(M+2)}$. To compute the flux across the interface between C_I and C_{I+1}, the generalized Roe linearization (6) needs to be computed:

$$A_\Phi(\widetilde{w}_I^n, w_{I+1}^n) \cdot (w_{I+1}^n - \widetilde{w}_I^n) = \int_0^1 \widetilde{A}(\Phi(s; \widetilde{w}_I^n, w_{I+1}^n)) \frac{\partial \Phi}{\partial s}(s; \widetilde{w}_I^n, w_{I+1}^n)ds , \quad (8)$$

with padded moment vector $\widetilde{w}_I^n := (m_{I,0}^n, m_{I,1}^n, \ldots, m_{I,M}^n, \widetilde{m}_{I,M+1}^n)^T \in \mathbb{R}^{M+2}$, padded with the still to be defined last moment $\widetilde{m}_{I,M+1}^n$. The interface transport matrix $\widetilde{A}(\Phi(s; \widetilde{w}_I^n, w_{I+1}^n)) \in \mathbb{R}^{(M+2)\times(M+2)}$ models propagation of information between $\widetilde{w}_I^n$ and w_{I+1}^n along the path $\Phi(s; \widetilde{w}_I^n, w_{I+1}^n)$. Using the constant matrix $\widetilde{A} = A_{M+1}$ to be consistent with the higher order model, the integral along the path does not need to be evaluated. The fluctuation $D_{I-1/2}^+ \in \mathbb{R}^{M+2}$ needs to implement a boundary condition for the added moment $\widetilde{m}_{I,M+1}^n$, while the fluctuation $D_{I-1/2}^- \in \mathbb{R}^{M+1}$ uses w_I^n instead of the padded moment vector $\widetilde{w}_I^n$, i.e.,

$$D_{I-\frac{1}{2}}^- = A_\Phi^-(w_{I-1}^n, w_I^n) , \qquad D_{I-\frac{1}{2}}^+ = \begin{pmatrix} A_\Phi^+(w_{I-1}^n, w_I^n) \\ a_{\text{bound}} \end{pmatrix} , \quad \text{with} \qquad (9)$$

$$A_\Phi^\pm(w_{I-1}^n, w_I^n) = \frac{1}{2}(A_M(w_I^n - w_{I-1}^n) \pm Q_M(w_I^n - w_{I-1}^n)) , \qquad (10)$$

and where a_{bound} is some interface boundary condition for the padded moment $\widetilde{m}_{I,M+1}^n$, which originates from the fact that there is an artificial boundary for the padded moment. We propose $a_{\text{bound}} = a_{M+1} \cdot (\widetilde{w}_I^n - \widetilde{w}_{I-1}^n)$, where we defined $\widetilde{w}_{I-1}^n := (m_{I-1,0}, m_{I-1,1}, \ldots, m_{I-1,M}, \widetilde{m}_{I,M+1})^T \in \mathbb{R}^{M+2}$ and where $a_{M+1} \in \mathbb{R}^{1\times(M+2)}$ is the last row of A_{M+1}. The Roe linearizations at the remaining interfaces are simply given by

$$A_\Phi(w_i, w_{i+1})(w_{i+1} - w_i) = A_M(w_{i+1} - w_i) , \quad i = 1, \ldots, I - 2 , \tag{11}$$

$$A_\Phi(w_i, w_{i+1})(w_{i+1} - w_i) = A_{M+1}(w_{i+1} - w_i) , \quad i = I + 1, \ldots, N_x - 1. \tag{12}$$

The numerical viscosity matrix Q_Φ can then be computed by using $Q_\Phi(w_L, w_R) = P(A_\Phi(w_L, w_R))$ at all interfaces. Expressions (8), (9) (11) and (12) can be readily inserted into the PVM equations (4), yielding the SAMM-PBC.

4 Numerical Simulation

We consider a 1D shock tube test case with initial conditions

$$(\varrho, u, T)^T (0, x) = \begin{cases} (3 + 0.2x, 0, 1)^T, & x \leq 0, \\ (1, 0, 1)^T, & x > 0, \end{cases} , \quad m_i(0, x) = 0 , \quad i > 2 ,$$

requiring high accuracy because of the density discontinuity at $x = 0$ [9, 13], and a small linear density profile in the left subdomain, modelled with sufficient accuracy by a lower order model. The relaxation time is $\tau = 1$. The 1D HSM (3) is simulated on the interval $\Omega = [-2, 1]$, that is decomposed into $\Omega_M = [-2, -1]$ and $\Omega_{M+2} = [-1, 1]$, using three different models:

1. **SAMM-PBC46**: The SAMM-PBC proposed in this paper, using $M = 4$ moments in Ω_M and $M + 2 = 6$ moments in Ω_{M+2}.
2. **HSM4**: the lower order moment model with $M = 4$ moments in Ω.
3. **HSM6**: the higher order moment model with $M + 2 = 6$ moments in Ω.

The order of the moment models was chosen based on the observation that the expansions had converged sufficiently well for these orders in the corresponding subdomains. The derivation of suitable domain decomposition criteria is ongoing work. Note that we use a difference of two moments between the lower order model HSM4 and the higher order model HSM6 instead of only one moment difference that was used in Sect. 3 to simplify the notation in the description of the method. The initial values for the higher order moments are zero, such that the only logical initial choice for the two padded moments in the SAMM-PBC46 is to set them to zero. The goal of this test case is to verify whether the SAMM-PBC46 yields comparable results to the higher order model HSM6 while realizing a computational speedup. Note that the expected speedup cannot exceed some theoretical bound based on the domain decomposition and number of moments.

The spatial grid is composed of $N_x = 300$ equally sized cells. The end time of the simulation is $t_{end} = 0.3$ with a fixed time step of $\Delta t = 6.7 \times 10^{-4}$ based on the CFL condition and the stable integration of the relaxation term. Outflow boundary conditions are implemented by zero-order extrapolation in the ghost cells.

The numerical results are shown in Fig. 3. For each of the variables ϱ (Fig. 3a), u (Fig. 3b) and T (Fig. 3c), the results for SAMM-PBC46, HSM4, and HSM6 are shown. Our new SAMM-PBC46 and the higher order model HSM6 yield nearly identical results and cannot be visually distinguished, apart from a small bump in velocity and temperature at the boundary interface $x = -1$. This small bump is a result of the coupling of two different order models and should be reduced as much as possible by using appropriate padded values in the buffer cell. The difference between the lower order model HSM4 and the other two models is significant. These observations are confirmed in Fig. 3d, which shows the relative 2-norm difference of HSM4 and SAMM-PBC46 with respect to HSM6 for different relaxation times. The greater the non-equilibrium in the gas, the greater the gain achieved by using the adaptive scheme compared to the low-order scheme. The speedup of SAMM-PBC46 compared to HSM6 was $\approx 19\%$. We can conclude that for this specific test case, SAMM-PBC46 yields nearly identical accuracy as HSM6, while realizing a speedup of 19 percent. We note that only a small part of the domain ($\approx 33\%$) uses HSM4, for which this speedup is significant.

5 Conclusion

We introduced a proof-of-concept for the spatially adaptive simulation of a 1D linear moment model with space-dependent order. A lower order and a higher order moment model are coupled at their boundary interface using a buffer cell in which the moment vector is padded with additional moments. Due to the structure of hierarchical moment models, an efficient implementation is possible. Numerical simulations of a shock tube with a small linear density profile in a subdomain using the spatially adaptive model yielded accurate results, compared with a high order model, while realizing a significant speedup. This paper serves as an introduction to adaptive simulation of moment models. Future extensions could include time adaptive domain decomposition, multiple domains, and the extension to non-linear systems.

References

1. Bird GA (1994) Molecular gas dynamics and the direct simulation of gas flows. Oxford University Press, Oxford. https://doi.org/10.1093/oso/9780198561958.001.0001
2. Cai Z, Fan Y, Li R (2013) Globally hyperbolic regularization of grad's moment system in one-dimensional space. Commun Math Sci 11:547–571
3. Cai Z, Fan Y, Li R (2015) A framework on moment model reduction for kinetic equation. SIAM J Appl Math 75(5):2001–2023. https://doi.org/10.1137/14100110X
4. Castro Díaz MJ, Fernández-Nieto E (2012) A class of computationally fast first order finite volume solvers: PVM methods. SIAM J Sci Comput 34(4):A2173–A2196. https://doi.org/10.1137/100795280

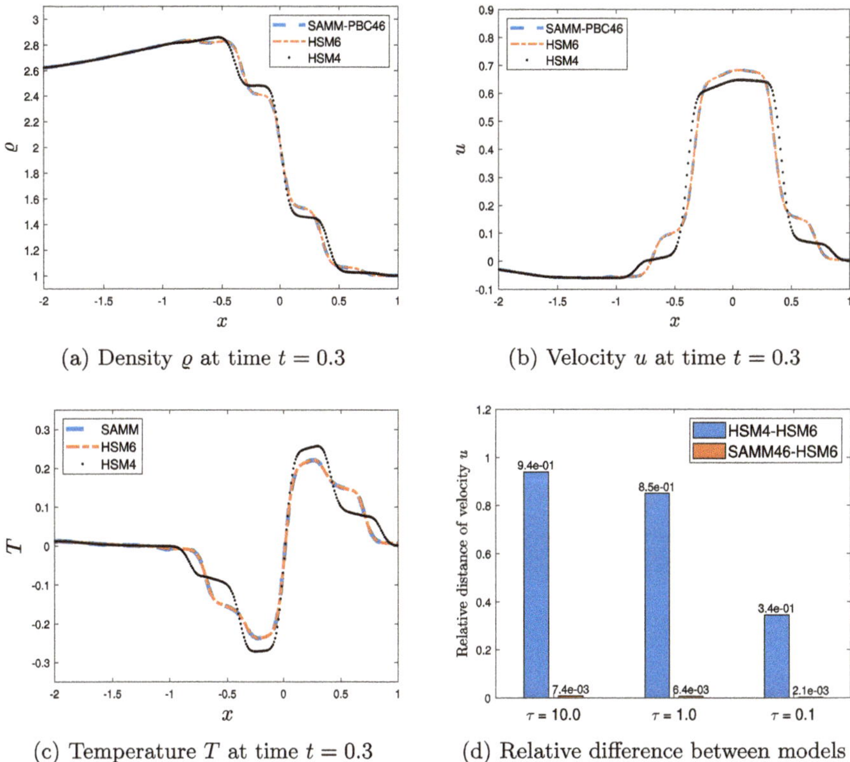

(a) Density ϱ at time $t = 0.3$

(b) Velocity u at time $t = 0.3$

(c) Temperature T at time $t = 0.3$

(d) Relative difference between models

Fig. 3 Numerical results for 1D shock tube test case for HSM4 (black dotted), HSM6 (orange dash-dotted) and SAMM-PBC46 (blue dashed). (**a**) density ϱ. (**b**) velocity u. (**c**) Temperature T. (**d**) relative difference of velocity u of SAMM-PBC46 and HSM4 with respect to HSM6 for varying Knudsen number τ.

5. Fan Y, Koellermeier J (2020) Accelerating the convergence of the moment method for the Boltzmann equation using filters. J Sci Comput 84(1):1. https://doi.org/10.1007/s10915-020-01251-8

6. Grad H (1949) On the kinetic theory of rarefied gases. Commun Pure Appl Math 2(4):331–407. https://doi.org/10.1002/cpa.3160020403

7. Hash D, Hassan H (1997) Two-dimensional coupling issues of hybrid DSMC/Navier-Stokes solvers. In: 32nd thermophysics conference. Fluid dynamics and co-located conferences, AIAA. https://doi.org/10.2514/6.1997-2507

8. Ivanov MS, Gimelshein SF (1998) Computational hypersonic rarefied flows. Annu Rev Fluid Mech 30:469–505. https://www.annualreviews.org/content/journals/10.1146/annurev.fluid.30.1.469

9. Koellermeier J (2019) Error estimators for adaptive simulation of rarefied gases using hyperbolic moment models. AIP Conf Proc 2132(1):120004. https://doi.org/10.1063/1.5119617

10. Koellermeier J, Samaey G (2021) Projective integration schemes for hyperbolic moment equations. Kinet Relat Models 14(2):353–387. https://www.aimsciences.org/article/doi/10.3934/krm.2021008

11. Lax PD (1954) Weak solutions of nonlinear hyperbolic equations and their numerical computation. Commun Pure Appl Math 7(1):159–193. https://doi.org/10.1002/cpa.3160070112

12. Li ZH, Zhang HX (2009) Gas-kinetic numerical studies of three-dimensional complex flows on spacecraft re-entry. J Comput Phys 228(4):1116–1138. https://www.sciencedirect.com/science/article/pii/S0021999108005408
13. Rovenskaya O, Croce G (2014) Application a hybrid solver to gas flow through a slit at arbitrary pressure ratio. Vacuum 109:266–274
14. Struchtrup H (2005) Macroscopic transport equations for rarefied gas flows-approximation methods in kinetic theory. Springer, Berlin
15. Torrilhon M (2016) Modeling nonequilibrium gas flow based on moment equations. Annu Rev Fluid Mech 48(48):429–458. https://www.annualreviews.org/content/journals/10.1146/annurev-fluid-122414-034259
16. Tsien HS (1946) Superaerodynamics, mechanics of rarefied gases. J Astronaut Sci 13(12):653–664. https://doi.org/10.2514/8.11476

Application of Regularized 13-Moment Equations to Continuum–Rarefied Gas Flow Over Aerospike Blunt Body

Satyvir Singh and Manuel Torrilhon

Abstract The accurate modeling of continuum-rarefied gas flows is crucial for understanding non-equilibrium phenomena in aerospace applications, particularly for high-speed vehicles encountering rarefied flow conditions. Traditional Navier–Stokes-Fourier equations fail to capture these effects, while kinetic approaches like the Boltzmann equation are computationally prohibitive. The regularized 13-moment (R13) equations offer a promising alternative by bridging the gap between continuum and rarefied regimes with improved accuracy and computational efficiency. This study investigates the application of the R13 equations to simulate gas flow over aerospike blunt bodies, a geometry known for its aerodynamic and thermal efficiency in high-speed conditions. The R13 framework effectively captures non-equilibrium phenomena, which are critical in transitional flow regimes. For numerical simulations, a two-dimensional modal discontinuous Galerkin method is developed to solve the R13 equations. Key flow characteristics, such as shock wave interactions, rarefaction effects, and aerodynamics coefficients, are analysed to assess the performance of the R13 model. The findings demonstrate the capability of the R13 equations to accurately predict flow behavior across a wide range of Knudsen numbers while maintaining computational efficiency.

Keywords R13 equations · Continuum-rarefied · Discontinuous galerkin method · Aerospike body

1 Introduction

Aerospace applications, particularly in hypersonic flight and spacecraft re-entry, often involve gas flows that span multiple regimes, from continuum to rarefied [1]. Understanding and accurately modeling these flows is critical for aerodynamic performance and thermal protection system design [2]. Hypersonic flows over blunt bodies generate strong shocks and significant non-equilibrium effects, such as velocity

S. Singh (✉) · M. Torrilhon
Applied and Computational Mathematics, RWTH Aachen University, Aachen, Germany
e-mail: singh@acom.rwth-aachen.de

M. Grabe et al. (eds.), *Rarefied Gas Dynamics*, Springer Aerospace Technology,
https://doi.org/10.1007/978-3-032-00094-1_33

slip and temperature jump, which traditional continuum-based models like Navier-Stokes equations cannot resolve adequately [3]. These effects become pronounced in transitional flow regimes, where the continuum assumption breaks down. Aerospike blunt bodies are an advanced aerodynamic design aimed at reducing drag and heat flux by modifying shock-wave behavior and redistributing pressure loads [4]. Their potential to enhance efficiency in high-speed vehicles makes them a key area of research. However, simulating the flow physics around these configurations remains complex due to the interplay of continuum and rarefied dynamics.

Traditional macroscopic models struggle in non-equilibrium regimes, necessitating approaches rooted in kinetic theory. The moment-based methods derive macroscopic equations by averaging the Boltzmann equation [5], enabling better representation of rarefied flows. Compared to direct simulation Monte Carlo (DSMC) methods [1], which are computationally expensive for near-continuum flows, moment methods provide a more efficient and deterministic alternative. A versatile framework for deriving evolution equations for macroscopic quantities from the Boltzmann equations is offered by moment approaches. By anticipating higher-order rarefaction effects, these methods generally close the gap between fluid dynamics and kinetic gas theory in extreme non-equilibrium situations.

In 1949, Grad brought moment method into kinetic theory using a Hilbert expansion of the distribution function in Hermite polynomials [6]. The field equations based on Grad's 13-moment methods use an extended set of variables to describe the state of a gas, starting with the fluid dynamic quantities-density, velocity, and temperature including higher-order moments of the distribution function. Interestingly, Grad's 13-moment equations were an initial step, but they suffered from instabilities, especially in strong non-equilibrium conditions [7]. The regularized 13-moment (R13), derived from the Boltzmann equation using moment closure techniques, extend the classical Grad-13 moment equations by incorporating regularization terms to ensure hyperbolicity and stability [8]. These equations are well-suited for capturing non-equilibrium effects, such as heat flux and stress anisotropy, which are critical in rarefied gas dynamics. A comprehensive summary of the applied numerical methods for solving Grad's system can be found in the study of Torrilhon [9]. The R13 model has been successfully applied to various benchmark problems, but its potential in addressing the aerodynamic challenges of high-speed vehicles, particularly those with complex geometries like aerospike blunt bodies, remains underexplored.

Aerospike blunt bodies are known for their drag reduction and enhanced heat management, critical for high-speed and space applications [10]. Their unique flow features, including detached shock waves and complex interactions in the wake region, require advanced modeling techniques. Few studies leverage moment-based methods to explore the continuum-rarefied regime, leaving a gap this study aims to address [11]. This study aims to investigate the application of the R13 equations to simulate continuum-rarefied gas flows over aerospike blunt bodies. The focus is on assessing the model's capability to predict flow phenomena across different regimes, and exploring its computational efficiency for practical engineering applications. By leveraging the R13 equations, this work seeks to enhance the understanding of non-equilibrium effects in supersonic flows.

The remainder of the study is structured as follows: Sect. 2 introduces the mathematical formulation for two-dimensional R13 moment equations. Section 3 discusses the employed numerical solver. Section 4 presents simulation results and provides a detailed description of continuum-rarefied gas flow over an aerospace blunt body. Finally, Sect. 5 gives the concluding remarks and outlook for the future work.

2 Two-Dimensional Regularized 13-Moment Equations

In this study, we performed the numerical simulations for the two-dimensional regularized 13-moment (R13) equations in continuum-rarefied gas flows. The two-dimensional R13 equations are derived by Torrilhon [12] in conservative form as

$$\frac{\partial \mathbf{U}(W)}{\partial t} + \nabla \cdot \mathbf{F}(\mathbf{U}) = \mathbf{P}(\mathbf{U}), \tag{1}$$

with nine independent variables $W = \{\rho, v_x, v_y, p, p_x, p_y, \sigma, q_x, q_y\}$. Here, ρ is mass density; v_x, v_y are velocity components; p is the pressure; p_x, p_y are the directional pressure; σ is the shear stress; and q_x, q_y are the components of heat flux vector in $x-$ and $y-$ directions. Pressure tensor (p_{ij}) is related with stress tensor σ_{ij} by the mathematical expression $p_{ij} = p\delta_{ij} + \sigma_{ij}$. In Eq. (1), $\mathbf{U}$ is the vector of conservative variables, $\mathbf{F} = (F, G)$ is the vector of flux functions, and $\mathbf{P}$ is the vector of production terms, which are defined as follows:

$$\mathbf{U} = \begin{bmatrix} \rho \\ \rho v_x \\ \rho v_y \\ \rho v^2 + 3p \\ \rho v_x^2 + p_x \\ \rho v_y^2 + p_y \\ \rho v_x v_y + \sigma \\ \rho v_x v_x^2 + 3\rho v_x + 2(p_x v_x + \sigma v_y) + 2q_x \\ \rho v_y v_y^2 + 3\rho v_y + 2(p_y v_y + \sigma v_x) + 2q_y \end{bmatrix}, \quad \mathbf{P} = -\frac{p}{\mu} \begin{bmatrix} 0 \\ 0 \\ 0 \\ 0 \\ p_x - p \\ p_y - p \\ \sigma \\ \sigma v_y + (p_x - p)v_x + q_x \\ \sigma v_x + (p_y - p)v_y + q_y \end{bmatrix}$$

$$F = \begin{bmatrix} \rho v_x \\ \rho v_x^2 + p_x \\ \rho v_x v_y + \sigma \\ \rho v_x v^2 + 2(p_x v_x + 2\sigma v_y) + 3p v_x + 2q_x \\ \rho v_x^3 + 3p_x v_x + \frac{6}{5}q_x + m_{xxx} \\ \rho v_x v_y^2 + p_y v_x + 2\sigma v_y + \frac{2}{5}q_x + m_{xyy} \\ \rho v_y v_x^2 + p_x v_y + 2\sigma v_x + \frac{2}{5}q_y + m_{xxy} \\ (pv^2 + 3p + 4p_x)v_x^2 + (7\theta + v^2)p_x + 4\sigma v_x v_y + \frac{32}{5}q_x v_x + \frac{4}{5}q_y v_y - 2\theta p + \hat{R}_{xx} \\ (\rho v^2 + 3p + 2(p_x + p_y))v_x v_y + (7\theta + 3v^2)\sigma + \frac{14}{5}(q_x v_y + q_y v_x) + \hat{R}_{xy} \end{bmatrix}$$

$$
G = \begin{bmatrix}
\rho v_y \\
\rho v_x v_y + \sigma \\
\rho v_y^2 + p_y \\
\rho v_y v^2 + 2(p_y v_y + 2\sigma v_x) + 3 p v_y + 2 q_y \\
\rho v_y v_x^2 + p_x v_y + 2\sigma v_x + \frac{2}{5} q_x + m_{yxx} \\
\rho v_y^3 + 3 p_y v_y + \frac{6}{5} q_y + m_{yyy} \\
\rho v_x v_y^2 + p_y v_x + 2\sigma v_y + \frac{2}{5} q_y + m_{xyy} \\
(\rho v^2 + 3p + 2(p_x + p_y)) v_x v_y + (7\theta + 3v^2)\sigma + \frac{14}{5}(q_x v_y + q_y v_x) + \hat{R}_{xy} \\
(p v^2 + 3p + 4 p_y) v_y^2 + (7\theta + v^2) p_y + 4\sigma v_x v_y + \frac{32}{5} q_y v_y + \frac{4}{5} q_x v_x - 2\theta p + \hat{R}_{yy}
\end{bmatrix}
$$

Here, $p = \rho\theta$, and μ is the shear viscosity. The terms m_{ijk} and R_{ij} represent higher-order moments introduced to regularize Grad's original equations. Their definitions are derived from a Chapman–Enskog expansion around the Grad 13-moment approximation, as established by Struchtrup and Torrilhon [8, 12]. These regularization terms enhance the stability of the stress and heat flux equations and improve the model's accuracy across the continuum to rarefied flow regimes. These higher moments are connected by the relationship: $\hat{R}_{ij} = m_{ijk} v_k + R_{ij}$. For two-dimensional R13 equations, m_{ijk} and R_{ij} are defined by

$$
\begin{bmatrix} m_{xxx} \\ m_{xxy} \\ m_{xyy} \\ m_{yyy} \end{bmatrix}
= -2\frac{\mu}{p}\theta
\begin{bmatrix}
\frac{3}{5}\left(\frac{\partial p_x}{\partial x} - \frac{\partial p}{\partial x}\right) - \frac{2}{5}\frac{\partial \sigma}{\partial y} \\
\frac{1}{3}\frac{\partial p_x}{\partial y} - \frac{1}{5}\frac{\partial p}{\partial y} + \frac{8}{15}\frac{\partial \sigma}{\partial x} - \frac{2}{15}\frac{\partial p_y}{\partial y} \\
\frac{1}{3}\frac{\partial p_y}{\partial x} - \frac{1}{5}\frac{\partial p}{\partial x} + \frac{8}{15}\frac{\partial \sigma}{\partial y} - \frac{2}{15}\frac{\partial p_x}{\partial x} \\
\frac{3}{5}\left(\frac{\partial p_y}{\partial y} - \frac{\partial p}{\partial y}\right) - \frac{2}{5}\frac{\partial \sigma}{\partial x}
\end{bmatrix},
$$

$$
\begin{bmatrix} R_{xx} \\ R_{xy} \\ R_{yy} \end{bmatrix}
= -2\frac{\mu}{p}\theta
\begin{bmatrix}
\frac{5}{3}\frac{\partial q_x}{\partial x} + \frac{2}{3}\frac{\partial q_y}{\partial y} \\
\frac{1}{2}\left(\frac{\partial q_y}{\partial x} + \frac{\partial q_x}{\partial y}\right) \\
\frac{5}{3}\frac{\partial q_y}{\partial y} + \frac{2}{3}\frac{\partial q_x}{\partial x}
\end{bmatrix}.
$$

3   Numerical Solver and Validation

In this study, a modal discontinuous Galerkin method is developed to solve the two-dimensional R13 equations in the continuum-rarefied regime. The domain is discretized using non-overlapping triangular elements with Dubiner modal basis functions [3]. Gauss-Legendre quadrature is used for volume and flux integration, and the local Lax–Friedrichs (LLF) scheme computes the interface fluxes. Time advancement is performed using a third-order strong stability-preserving Runge-Kutta method. To ensure physical consistency, a positivity-preserving limiter [3] is applied to avoid non-physical values such as negative density or pressure.

To validate the R13-based computational model, we compare it with classical NSF results and experimental data for a one-dimensional normal shock in argon gas. Figure 1a shows that at $Ma = 3.8$, the R13 model matches the experimental density profile from Alsmeyer [13] more closely than the NSF model, accurately capturing the shock structure. Figure 1b presents the inverse density thickness $1/\delta$ over a range of Mach numbers, where R13 results show good agreement with experiments from Alsmeyer and Camac [14], while NSF consistently overpredicts the thickness. These comparisons demonstrate the superior accuracy of R13 in capturing non-equilibrium effects in the transitional regime (Fig. 2).

4　Results and Discussion

In this section, we present the numerical results of the R13 equations for supersonic gas flow over an aerospike blunt body in the continuum–rarefied regime. The computational domain is discretized using 30,739 non-overlapping triangular elements, as shown in Fig. 1, with mesh refinement near the wall to accurately resolve the boundary layer and shock structures. For the aerospike configuration, a geometric ratio of $d/D = 0.2$ and $L/D = 1.5$ is employed, where d is the aero-disk diameter, D is the diameter of the blunt body (used as the reference length), and L is the spike length. The outer boundary of the domain is treated as a far-field inflow, while the inner boundaries representing the solid surface are imposed with no-slip and isothermal wall conditions. The free-stream conditions are specified as Mach number $Ma = 2.0$, angle of attack $AoA = 0°$, temperature $T_\infty = 26.6$. The wall temperature is fixed as $T_w = 26.6$. Simulations are conducted for three Knudsen numbers: $Kn = 0.0001$, 0.001, and 0.01, capturing the transition from near-continuum to rarefied regimes. The corresponding Reynolds numbers, based

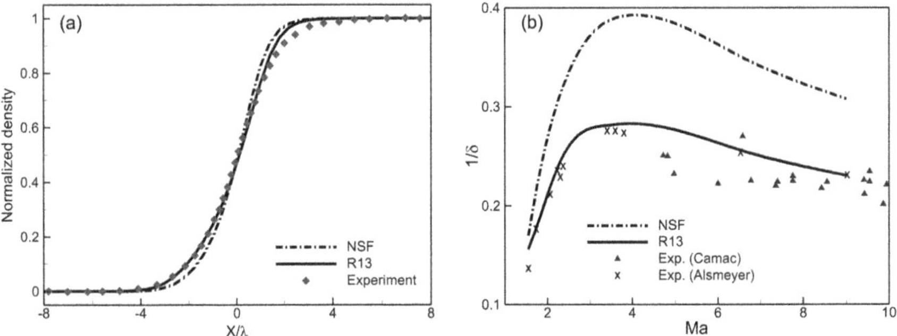

Fig. 1　Comparison of **a** normalized density profiles at $Ma = 3.8$, and **b** inverse density thickness versus Mach number for NSF, R13, and experimental data. Results are validated against experimental data from Alsmeyer [13] and Camac [14] for argon gas in a one-dimensional shock structure problem

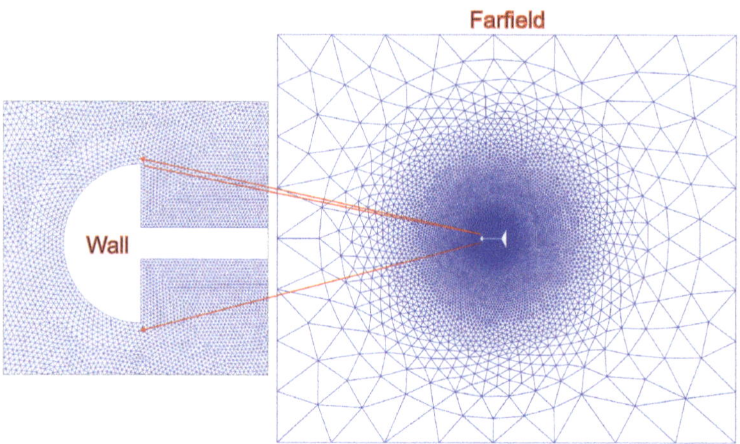

Fig. 2 Computational domain and boundary conditions

on the reference length D, are $Re = 19412, 1941.2$, and 194.12, respectively. The working gas is monatomic argon, selected for its well-characterized thermophysical properties, making it suitable for rarefied flow studies.

Figure 3 presents the contours of density, Mach number, and temperature for supersonic flow over an aerospike blunt body at three different Knudsen numbers ($Kn = 0.0001, 0.001, 0.01$), illustrating the transition from continuum to rarefied regimes. At $Kn = 0.0001$, the flow exhibits continuum-like behavior with a well-defined bow shock, resulting in sharp increases in density and temperature and a distinct deceleration in Mach number near the spike. As Kn increases, the shock structure becomes increasingly diffused due to non-equilibrium effects, leading to lower post-shock density and temperature values, and smoother Mach number transitions. At $Kn = 0.01$, significant rarefaction effects emerge, weakening the compression and thermal gradients, and revealing broader, non-equilibrium shock layers. The temperature fields particularly demonstrate reduced heat conduction and less pronounced heating in the stagnation region. Across all cases, the aerospike effectively modifies the flow structure by increasing the shock stand-off distance and redistributing the compression away from the main body, thereby reducing aerodynamic heating and drag.

Figure 4 shows the effect of the Knudsen number (Kn) on the computed drag coefficient for supersonic gas flow over an aerospike blunt body, highlighting the transition from continuum to rarefied flow regimes. At low Kn values (Kn < 0.001), corresponding to the continuum regime, the drag coefficient is highest due to significant collisional interactions and strong shock waves, resulting in pronounced viscous and pressure drag. As Kn increases into the transitional regime ($0.001 \leq Kn \leq 0.01$), non-equilibrium effects such as velocity and temperature slip reduce momentum transfer and weaken shock structures, leading to a sharp decline in drag. In the rarefied

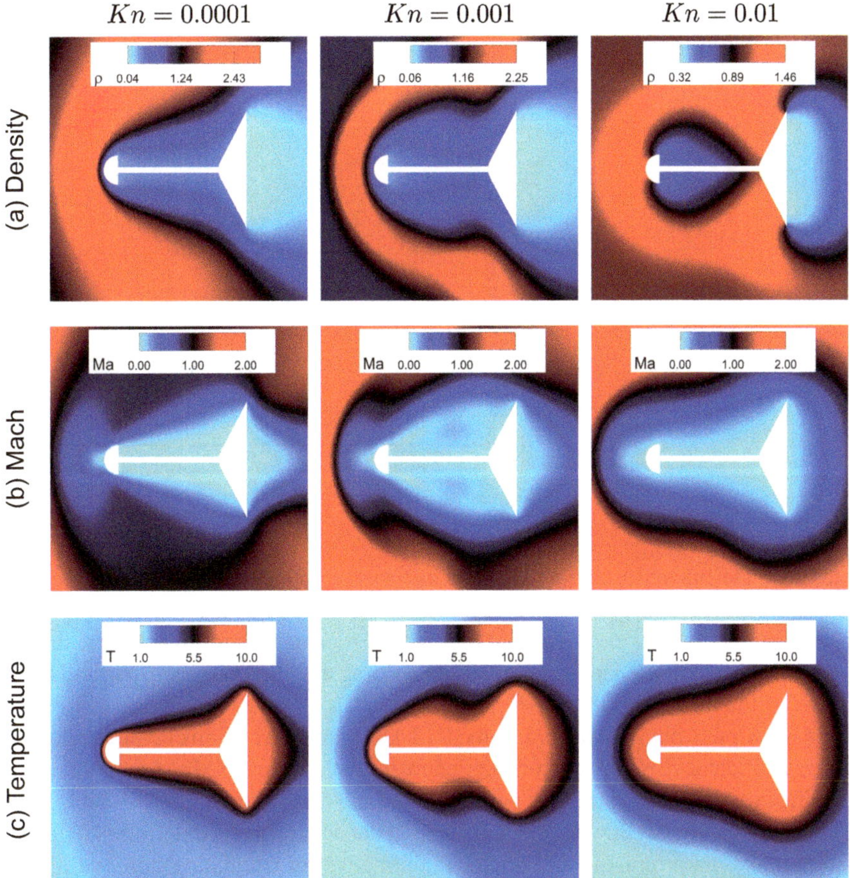

Fig. 3 Contours of **a** density, **b** Mach, and **c** temperature obtained for the supersonic gas flow over an aerospike blunt body with three Knudsen numbers ($Kn = 0.0001, 0.001,$ and 0.01)

regime (Kn > 0.01), the drag coefficient stabilizes at a lower value as free-molecular effects dominate, and collisional drag mechanisms become negligible.

5 Concluding Remarks

This study demonstrates the capability of the R13 equations to effectively model continuum-rarefied gas flows over an aerospike blunt body. By employing a two-dimensional modal discontinuous Galerkin method, the R13 framework accurately captures key non-equilibrium phenomena-such as stress anisotropy, heat flux effects, and rarefaction-induced deviations from classical behavior-across a range of Knudsen

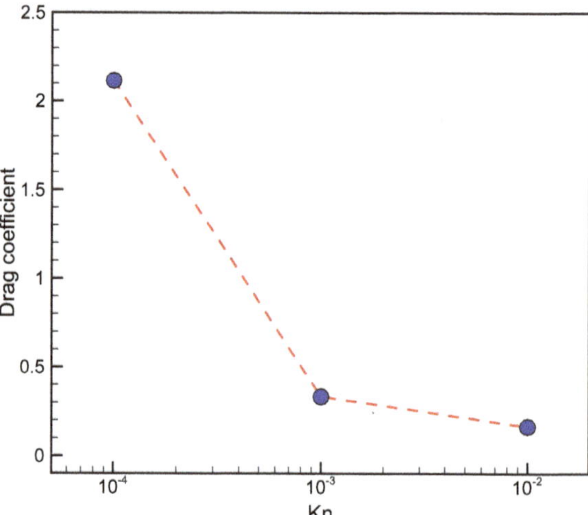

Fig. 4 Knudsen number effect on computed drag coefficient for the supersonic gas flow over an aerospike blunt body

numbers. Despite enforcing no-slip and isothermal boundary conditions at the wall, the model successfully reveals critical transitions in flow structure and aerodynamic performance from the continuum to the rarefied regime. The results confirm that the aerospike configuration substantially reduces drag and thermal loads, particularly in near-continuum conditions.

In addition to physical accuracy, the R13-DG approach exhibits favorable computational efficiency. While it incurs higher costs than classical NSF solvers, it remains significantly more efficient than particle-based methods such as DSMC and Fokker-Planck in transitional flow regimes. Its high-order accuracy and deterministic formulation make it a promising and practical alternative for simulating non-equilibrium gas dynamics in complex aerospace configurations.

References

1. Bird GA (1994) Molecular gas dynamics and the direct simulation of gas flows. Oxford University Press, Oxford
2. Reese JM, Gallis, Michael A, Lockerby, Duncan A (2003) New directions in fluid dynamics: non-equilibrium aerodynamic and microsystem flows. Philos Trans R Soc A 361:2967–298
3. Singh S, Karchani A, Chourushi T, Myong RS (2022) A three-dimensional modal discontinuous Galerkin method for the second-order Boltzmann-Curtiss-based constitutive model of rarefied and microscale gas flows. J Comput Phys 457:111052
4. Chinnappan AK, Malaikannan G, Kumar R (2017) Insights into flow and heat transfer aspects of hypersonic rarefied flow over a blunt body with aerospike using direct simulation Monte-Carlo approach. Aerosp Sci Technol 66:119–128
5. Boltzmann L (1872) Weitere Studien uber das Wa rmegleichgewicht unter Gasmolekulen. Wien
6. Grad H (1949) On the kinetic theory of rarefied gases. Commun Pure Appl Math 2:331–407

7. Singh S, Song H, Torrilhon M (2024) Modal discontinuous Galerkin simulations for Grad's 13 moment equations: application to Riemann problem in continuum-rarefied flow regime. J Comput Theor Transp 398–422
8. Struchtrup H, Torrilhon M (2003) Regularization of Grad's 13 moment equations: derivation and linear analysis. Phys Fluids 15:2668–2680
9. Torrilhon M (2016) Modeling nonequilibrium gas flow based on moment equations. Annu Rev Fluid Mech 48:429–458
10. Wood CJ (1962) Hypersonic flow over spiked cones. J Fluid Mech 12(48):614–624
11. Chourushi T, Singh S, Vishnu AS, Myong RS (2024) Numerical simulations of rarefied gas flow over an aero-spiked hypersonic blunt body using the second-order Boltzmann-Curtiss constitutive model. AIP Conf Proc 2996:140001
12. Torrilhon M (2006) Two-dimensional bulk microflow simulations based on regularized Grad's 13-moment equations. Multiscale Model Sim 5:695–728
13. Alsmeyer H (1976) Density profiles in argon and nitrogen shock waves measured by the absorption of an electron beam. J Fluid Mech 74:497–513
14. Camac M (1965) Argon shock structure. In: Proceedings of the fourth international symposium on rarefied gas dynamics, vol 1, p 240

Suitability of Moment Methods to Model the Ion Dynamics in a Low-Temperature Plasma

Anatole Berger, Nicolas Lequette, Thierry Magin, Anne Bourdon, and Alejandro Alvarez Laguna

Abstract We present a study of the suitability of several moment methods to capture the ion dynamics in a bounded non-equilibrium low-temperature plasma between two electrically-floating walls under different pressure regimes. We use a set of kinetic simulations in order to study the properties of one-dimensional five-moment (1D 5M) closures. We discuss on the advantages and disadvantages of some of the most common moment methods: Grad's closure, the quadrature method of moments (QMOM), the extended QMOM (EQMOM) and maximum entropy. For this study, we analyse the capability of each closure to capture the dynamics of the closing flux and its ability to retrieve the distribution function. The results show that the accuracy of each model largely depends on the pressure regime. In general, all the considered methods behave rather fine at high pressure and exhibit different challenges in order to represent the low-pressure regime.

Keywords Plasma discharge · Ions · Moment methods · Grad · QMOM · EQMOM · Maximum entropy · Particle-in-cell

1 Introduction

Low-temperature plasmas are fundamental to a wide number of industrial applications, from plasma etching for microelectronics [6, 20] and electric propulsion [11] to biomedical applications [9]. These plasmas are characterized by a large temperature difference between the heavy species (ions and neutral atoms or molecules) and electrons. The former are close to room temperature, while the latter are usually several eVs (or tens of thousands of Kelvin). Low-temperature plasmas are weakly

A. Berger (✉) · N. Lequette · A. Bourdon · A. Alvarez Laguna
Laboratoire de Physique des Plasmas, Centre National de la Recherche Scientifique, Sorbonne Université, École Polytechnique, Institut Polytechnique de Paris, Palaiseau, France
e-mail: anatole.berger@lpp.polytechnique.fr

T. Magin
Aeronautics and Aerospace Department, von Karman Institute for Fluid Dynamics, Sint-Genesius-Rode, Belgium

M. Grabe et al. (eds.), *Rarefied Gas Dynamics*, Springer Aerospace Technology,
https://doi.org/10.1007/978-3-032-00094-1_34

ionized, with an ionization degree often much below the percent (usually around 10^{-4}). Furthermore, depending on the application, we deal with pressures ranging from a few tenths of Pascals up to atmospheric or higher pressures.

The charged species of low-temperature plasmas are often out of equilibrium [13] due to the presence of strong electric fields, gradients, and collisions with the neutral species. In this non-equilibrium regime, kinetic simulations are very computationally expensive (especially at higher pressures), whereas fluid models are less accurate because of the conditions far from the continuum regime. As a result, different moment methods (see e.g. [22] for a recent review on general applications of moment models) have been proposed in the low-temperature plasma literature. These include Grad [2, 16, 26], maximum entropy [3, 4], anisotropic Gaussian closures [14, 15], and the quadrature-based methods of moments (QMOM) [24]. Despite these efforts, the comparison with kinetic simulations that are representative of a bounded low-temperature plasma is still a challenge due to the presence of strong electric fields in the plasma sheath [1, 23]. In this paper, we study some of the most common moment closures in order to represent the ion dynamics and distribution functions of low-temperature plasmas, with realistic collision cross sections in a wide range of pressures and collisional regimes. In this paper, we assess the suitability of these models whereas in a future work, we will present non-linear numerical solutions of the moment models.

2 Kinetic Simulations of a Low-Temperature Bounded Plasma at Different Pressures

We study a one-dimensional bounded argon plasma between two electrically-floating walls, similar to the numerical set-up explained by Tavant et al. [25]. We consider a one-dimensional domain of 10 cm and neutral gas pressures that vary from 0.05 to 100 mTorr (0.0067–13 Pa). We use an in-house Particle-In-Cell (PIC) code with Monte Carlo collisions for the charged species, i.e., singly-charged ions and electrons [18]. The cross sections are taken from the LxCat data-base [21], considering ion isotropic elastic scattering and charge exchange. The neutral gas background is considered to be at rest with a Maxwellian distribution at room temperature (300 K). Therefore, in these simulations, only the charged particles (ions and electrons) are simulated, and their dynamics are coupled with Poisson's equation. We impose the ionization, proportional to the local plasma density, as proposed by Tavant et al. [25], in order to have a steady-state solution with an average ion density of 10^{15} m^{-3}. The ions are injected with the neutral distribution whereas the electrons are injected with a Maxwellian distribution at 5 eV. Note that although the simulation domain is one-dimensional, the three velocity directions are considered in the kinetic model. In this work, we focus on the ion dynamics and therefore only ion data will be presented.

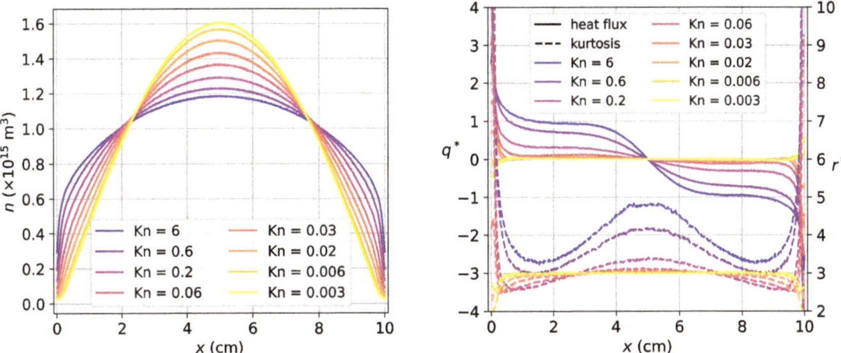

Fig. 1 Ion density profile (left), and normalized heat-flux and kurtosis profiles (right) for different pressure regimes. The right plot is zoomed in, but the moments reach higher values at the wall (namely, 6 for the heat-flux and 60 for the kurtosis in the most extreme cases). The Knudsen number refers to the ion-neutral collisions

We define the characteristic ion-neutral Knudsen number as

$$\mathrm{Kn} = \frac{l_{\mathrm{ig}}}{L},\tag{1}$$

with the domain length L and the ion-neutral mean-free-path $l_{\mathrm{ig}} = (\sigma_{\mathrm{ig}}^0 n_{\mathrm{g}})^{-1}$ where $\sigma_{\mathrm{ig}}^0 = 10^{-18}$ m^2 is the ion-neutral characteristic cross section, and n_{g} the atomic argon density, which is constant in this study and varies depending on the pressure, following an ideal perfect gas law. Note that in these kinetic simulations, the ion-ion and ion-electron collisions are neglected as they are much less frequent than the ion-neutral collisions. Therefore, the ion-neutral Knudsen of Eq. (1) does not necessarily quantify the tendency to thermodynamic equilibrium of the ion population.

The considered system is one-dimensional in space and, therefore, we are interested in the ion dynamics in the x direction. Nevertheless, the system is simulated with a kinetic model that simulates the three velocity components. Because of the studied geometry, the impact of the transverse velocity of the ions on the dynamics is very reduced: e.g., the transverse mean velocity is zero and the transverse temperature is nearly the constant gas temperature in the whole domain at all considered pressures. As a result, we can simplify the kinetic problem by considering that the distribution function along the transverse direction is nearly at equilibrium and perform the integration of the distribution function in the perpendicular direction in order to reduce the problem to 1D-1V, as follows,

$$f_x^{1\mathrm{D}}(v_x) = \int_{\mathbb{R}^2} f^{3\mathrm{D}}(\vec{v}) \, \mathrm{d}v_y \, \mathrm{d}v_z .\tag{2}$$

Similarly, the considered moments will be only along the x direction, as follows,

$$M_x^{(k)} = \int_{\mathbb{R}^3} \psi_k(v_x) f^{3D}(\vec{v}) \, d\vec{v} = \int_{\mathbb{R}} \psi_k(v_x) f_x^{1D}(v_x) \, dv_x \,. \tag{3}$$

where $\psi_k(v_x)$ is a monomial in the x component of the velocity, that depends on the moment considered.

In the rest of the article, we will denote f the 1D VDF and drop the x subindex for the velocity and VDF. We define the following moments that are relevant to the discussion of the 1D 5-moment model, referred to as number density, drift velocity, pressure, heat flux, kurtosis, and kurtosis flux, as follows,

$$n = \int_{\mathbb{R}^3} f^{3D}(\vec{v}) \, d^3 v = \int_{-\infty}^{\infty} f(v) \, dv \,, \quad nu = \int_{-\infty}^{\infty} vf \, dv, \quad p = m \int_{-\infty}^{\infty} (v - u)^2 f \, dv \,,$$

$$q = m \int_{-\infty}^{\infty} (v - u)^3 f \, dv, \quad r = m \int_{-\infty}^{\infty} (v - u)^4 f \, dv \,, \quad s = m \int_{-\infty}^{\infty} (v - u)^5 f \, dv \,,$$

where m is the ion mass. Note that, as explained above, the moments are along the x direction, and therefore the high order moments written in the tensorial notation correspond to: $u \equiv u_x$, $p \equiv P_{xx}$, $q \equiv Q_{xxx}$, etc. In the following, we will use the normalized higher-order centered moments, defined as,

$$q^* = \frac{q}{\rho v_{\text{th}}^3}, \qquad r^* = \frac{r}{\rho v_{\text{th}}^4} \quad \text{and} \quad s^* = \frac{s}{\rho v_{\text{th}}^5} \,, \tag{4}$$

where the mass density is $\rho = mn$ and the thermal speed is $v_{\text{th}} = \sqrt{p/\rho}$.

The number density profile of the particle-in-cell simulations at different pressures is shown on the left panel of Fig. 1. In all cases, it has a dome-like shape with a maximum in the center and approaches zero at the wall since the charged particles are created in the bulk by ionization and then lost at the floating wall. The density profile flattens for decreasing pressures. This figure clearly illustrates the effect of the ion-neutral collisions in the transport of charged species of a bounded plasma.

The right panel of Fig. 1 shows the profiles of the normalized heat flux and kurtosis, as computed with the results of the particle-in-cell simulations. At high-pressure, both the heat-flux and the kurtosis are mostly at equilibrium in the whole domain, except in the sheath (i.e., close to the boundary). At low pressure, they are out-of-equilibrium in all the domain and increase sharply in the sheath, to reach very high values.

In the left panel of Fig. 2, we present the heat flux and kurtosis profiles in the realizability phase space. In this plot, we draw the boundary of the realizability domain (the black solid line in figure) that separates realizable and unrealizable velocity distribution functions (VDFs). Any physical VDF (positive definite integrable function) lies above this curve. Similarly, we present the Junk line, that is relevant for the

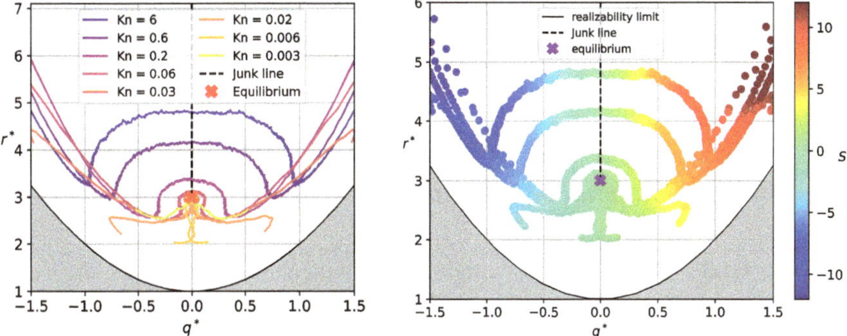

Fig. 2 (Left) high-order moment evolution in the realizability space for different pressure level. (Right) variation of the kurtosis-flux in this same space. The plots are zoomed in around the equilibrium. The grey area is the non-realizable zone in which no physical VDF exists with the corresponding heat-flux and kurtosis

discussion of maximum entropy and EQMOM closures. Note that the low-pressure simulations cross the Junk line.

We can then plot the variation of the normalized kurtosis-flux (the fifth order moment) in this space. This is shown on the right plot of Fig. 2. This plot shows an interesting trend. First, the behaviour of the fifth order moment seems to be similar in all the studied pressure regimes. As seen in the right panel of the figure, the fifth order moment seems to have a dependence on the normalized heat flux that is nearly linear and that it seems to be the same and independent of the background pressure. This is an interesting finding as it can provide guidelines in order to choose a closure model, as it will be shown in the next section. In addition, as the fifth order moment seems to have a behavior that can be determined by the previous moments (mainly on the heat-flux), it strongly motivates one to consider the kurtosis-flux as the closing flux and hence to consider five-moment (5M) models.

3 Comparison of Different Moment Closures for Ions in Low-Temperature Plasmas

In this section, we will first review some of the most common one-dimensional 5M closures: Grad, QMOM, EQMOM and maximum entropy. A discussion on these methods for 1D shocks can be found in Ref. [17]. Secondly, we will assess their suitability for capturing the ion properties as observed in the kinetic simulations. We base our assessment on a comparison of the closing flux (including their realizability) and the reconstructed VDF. We underline that in this work we do not implement a numerical solver of the moment closures, which is left to a future work. Instead, we take the first five moments (up to order four) of the kinetic simulation, as show in the previous section, and we use them to study the above-mentioned properties of the closures.

3.1 One-Dimensional 5M-Closures

We will consider four of the most common closures, namely, Grad [12], the hyperbolic quadrature method of moment (HyQMOM) [10], the extended quadrature method of moment (EQMOM) or bi-Maxwellian closure [7], and maximum entropy (Max. Entr.) [8, 19] , with five moments. The expression of the VDF for each closure is

$$
f_{\text{Grad}}^{\text{1D}}(c) = \frac{\rho \, e^{-c^2/2}}{\sqrt{2\pi k_{\text{B}} T}} \left(1 - \frac{q^*}{6} \left(3c - c^3 \right) + \frac{r^* - 3}{6} \left(3 - 6c^2 + c^4 \right) \right) ,
$$

$$
f_{\text{HyQMOM}}^{\text{1D}}(c) = \rho_0 \delta(c) + \rho_1 \delta(c - c_1) + \rho_2 \delta(c - c_2) ,
$$

$$
f_{\text{EQMOM}}^{\text{1D}}(c) = \frac{\rho_1 \, e^{-(c-c_1)^2/2}}{\sqrt{2\pi k_{\text{B}} T_0}} + \frac{\rho_2 \, e^{-(c-c_2)^2/2}}{\sqrt{2\pi k_{\text{B}} T_0}} ,
$$

$$
f_{\text{Max.Ent.}}^{\text{1D}}(c) = \exp \left(k_0 + k_1 c + k_2 c^2 + k_3 c^3 + k_4 c^4 \right) .
$$

where c is the peculiar velocity along the x direction normalized to the thermal speed, i.e., $c \equiv v_x / v_{\text{th}}$ (the x index is dropped for the sake of simplicity), normalized to the thermal speed, ρ, ρ_1 and ρ_2 are densities, $T = p/(n k_B)$ and T_0 are temperatures, k_{B} is the Boltzmann constant, $k_0, \cdots , k_4$ are some non-dimensional coefficients that are function of the accessible moments and are computed numerically by solving an optimization problem, and $\delta(c)$ are Dirac delta distributions.

3.2 Closing Flux

The closure of a 1D 5M model is completely characterized by the dependency of the kurtosis-flux on the heat-flux and the kurtosis. In Fig. 3, we present s^* in the realizability space for the four considered closures. This can be compare to Fig. 2, where the kurtosis-flux from the kinetic simulations was presented. At high pressures (Kn $<$ 0.2) the kinetic data are close to the equilibrium and the four closures resemble the PIC results. Alternatively, at low-pressures (Kn $\geq$ 0.2), the kinetic simulations cross the Junk line at the center of the domain. This line is singular in the EQMOM and maximum entropy closures so this may be an issue for these closures. Furthermore, the kurtosis-flux does not have the right trend as compared to the kinetic solutions around this line (it diverges while it is zero in the kinetic data). Grad's closure and HyQMOM have, on the other hand, a more similar behavior. In particular, HyQMOM

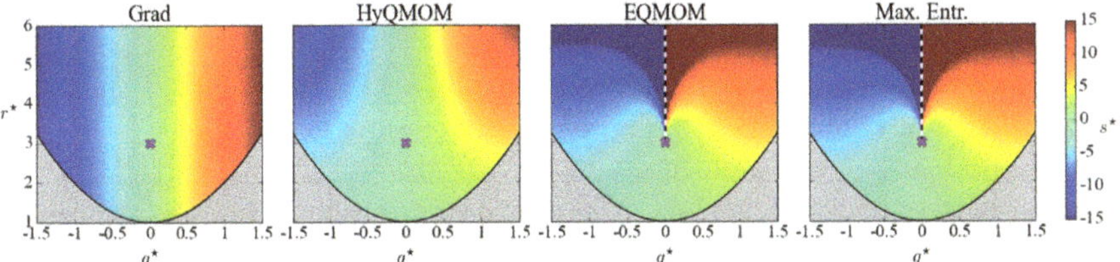

Fig. 3 Value of the kurtosis-flux in the realizability space for the different closures considered. The violet cross represent the equilibrium, the grey area is the non-realizable zone and the dashed line is the Junk line. The colorbar is cut-off at 25 because the kurtosis-flux diverges around the Junk line for EQMOM and Maximum entropy

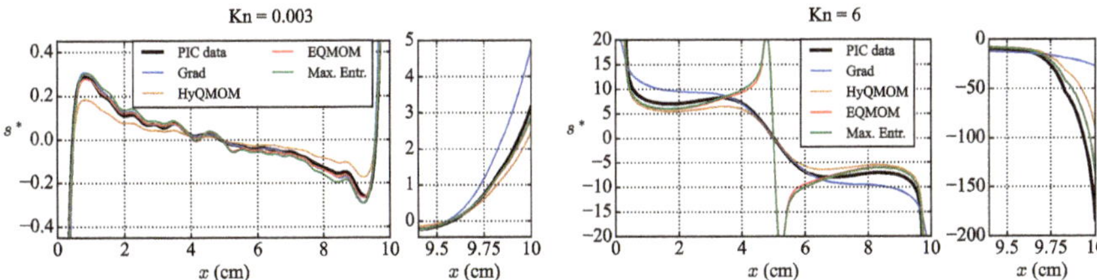

Fig. 4 Kurtosis-flux profile as directly computed from the VDF of PIC data and from the previous moments using the 4 closures considered in this study at two different pressures. The right panel in each subfigure show a zoom of the region close to the wall, i.e., the sheath region

closure seems to match the best the behavior at low pressure, as the closing flux takes into account the dependency on the kurtosis (as seen in the kinetic solutions) while Grad's closing depends only on the heat-flux.

Figure 4 shows kurtosis-flux profiles for two pressures, high pressure (Kn = 0.003) and low pressure (Kn = 6). We plot the kurtosis-flux directly taken from the PIC data, as well as the kurtosis-flux computed from the heat-flux and kurtosis of the PIC data using the 4 closures introduced in this section. At high pressure, in the bulk (and pre-sheath, i.e., $x \in [0.5, 9.5]$ cm) all closures but HyQMOM have similar and very good accuracy. The latter is less accurate but still follows the right trend. The sheath region is shown in the right panel of each subfigure of Fig. 4. In this region, Maximum entropy and EQMOM capture the kinetic solution and HyQMOM provides a less accurate approximation, while Grad becomes far from the kinetic solution. At low pressure (Kn = 0.6), HyQMOM becomes the best approximation of the flux in the all domain. Maximum entropy and EQMOM are accurate in the sheath and in a large part of the bulk but diverge in the center (because of the crossing of the Junk line). Grad is the less accurate, in particular in the sheath region. Overall, each closure has its own domains and regimes of accuracy, maximum entropy and EQMOM being extremely similar.

3.3 VDF Reconstruction from the Moments

We will now study the accuracy of the considered closures to recover the VDF of the kinetic solutions from its moments. Grad's VDF reconstruction is straightforward since its expression as a function of the moment is analytical. For HyQMOM and EQMOM, the link between the moments and the parameters of the VDF can be reduced to a single polynomial equation, that is solved using Brent algorithm [5]. For maximum entropy, there is no known simplification, and the total system of five equations is solved using Newton-Raphson algorithm.

At high pressure in the bulk, the kinetic VDF is mostly Maxwellian, so the reconstruction from the moments is almost exact and equivalent for all closures but HyQ-MOM. HyQMOM VDF being a sum of Dirac delta distributions, it will not be a continuous function, which is a major drawback for the physical consistency and for the computation of the collisional processes.

In Fig. 5, we present some cases with non-Maxwellian VDF: at high pressure in the sheath (left plot), at low pressure in the center (center plot) and at low pressure in the sheath (right plot). For each case, a Maxwellian with the same first three moments is also shown as a reference. Despite having an almost identical closing flux, maximum entropy and EQMOM have very different VDF reconstructions. The former is extremely accurate in the high-pressure regime and manages to recover an almost exact VDF in all the domain, even in the sheath as can be seen in the left plot. It is also the most accurate (and smoothest) in most of the domain in the low pressure regime except around the center of the domain (close to the Junk line), otherwise it only struggles to catch the stiffness of the PIC VDF in the sheath at very low pressure. EQMOM being a sum of two Gaussians, its VDF will be less smooth. Indeed, as can be seen in the right plot, its only way to create a large skewness is by increasing the space between the two Gaussians, leading to two distinct populations in extreme cases, which is not representative of what happens to the kinetic VDF. Finally, Grad is a small improvement from a Maxwellian for small perturbations but becomes unpractical for large deviations as its tail can become negative (as can be seen on the right plot).

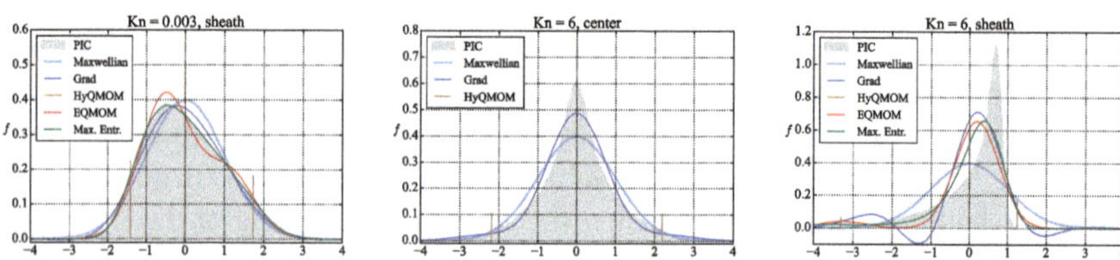

Fig. 5 Comparison of the VDF reconstructed from the first five moments with different closures. The HyQMOM VDF being a sum of Dirac delta distributions, it is represented with vertical lines with heights corresponding to the weights of each Dirac. In the center plot, there is no fit for EQMOM and Maximum entropy since the moment set is on the Junk line

4   Conclusion and Discussion

We studied the suitability of several one-dimensional moment models for ions in a one-dimensional bounded low-temperature plasma. We have used for this study kinetic simulations under a wide pressure range, considering realistic collisional processes, including elastic isotropic scattering and charge-exchange collisions. We used the first five moments of the kinetic simulations in order to analyze the behavior of the closures with these accessible moments. We summarize here our conclusions:

- The study of the fifth-order moment, as seen from the kinetic data, in the 5M realizability space, suggests that there is a mapping between this moment and the normalized heat flux and kurtosis. This mapping seems to be unique, independent of the pressure conditions. This leads us to believe that a 5M closure is a sound choice for this problem.
- When compared to the different considered closures, we observe that the maximum-entropy and EQMOM closures are able to capture the closing flux with a high degree of accuracy at high pressure. At low-pressure, the kinetic solutions lie on the Junk-line in the center of the domain, which hinders the suitability of maximum-entropy and EQMOM. Nevertheless, maximum-entropy, EQMOM, and HyQMOM capture the closing flux in the sheath with remarkable accuracy.
- The reconstructed VDFs at the wall are particularly important in order to study wall physical processes, e.g., etching rates for microelectronics applications. Maximum-entropy and EQMOM are able to capture this VDF at the high-pressure regime. However, none of the considered closures provide a satisfactory approximation at low-pressure.

This study will be continued with the numerical resolution of the moment equations, which will be presented in a separate paper. The current analysis shows that a unique model able to capture all the pressure regimes is a challenge. Also, the model should be simple enough in order to include the collisional processes in a realistic manner. One possible solution could be the consideration of a hierarchy with higher number of moments, which can help to solve some of the shown limitations.

One example of these limitations is the singular behaviour of EQMOM along the Junk line, which might be solved by considering a larger number of moments. The usual EQMOM considers a sum of Gaussians with a single temperature because the moment inversion algorithm (often based on the Wheeler algorithm) requires this assumption in order to invert the problem in a computationally efficiently manner. A possible alternative that would solve this singular behaviour could be to consider two different temperatures for the Gaussians, which would require a 6M model. However, this would come with an increase of the computational cost.

Alternatively, the limitations associated to the Junk line for maximum-entropy models in the center of the domain at low pressure might be more challenging to solve by increasing the number of moments. By observing the VDF at the center of the simulation at low pressure, we can note that it decays as an exponential with $-|v|$ in the exponent and hence it has symmetric heavy tails (positive excess kurtosis).

However, the symmetric maximum entropy leading-order terms at $v \to \infty$ are by construction of order $-v^{2n}$ inside the exponential (for $n \geq 1$). As a result, even with an increased number of moments, the maximum-entropy will have problems to capture the decay as seen in the simulations and therefore the Junk line will still be a problem in the center of the domain at low pressures.

Acknowledgements This work is part of a PhD funded by the Institut Polytechnique de Paris (IPP) and the EUR PlasmaScience. The authors would like to thank Clinton Groth, whose course on the moment methods and discussions greatly helped realizing this study.

References

1. Alvarez Laguna A, Pichard T, Magin T, Chabert P, Bourdon A, Massot M (2020) An asymptotic preserving well-balanced scheme for the isothermal fluid equations in low-temperature plasmas at low-pressure. J Comput Phys 419:109634
2. Alvarez Laguna A, Esteves B, Bourdon A, Chabert P (2022) A regularized high-order moment model to capture non-Maxwellian electron energy distribution function effects in partially ionized plasmas. Phys Plasmas 29(8):083507. https://doi.org/10.1063/5.0095019
3. Boccelli S, Giroux F, Magin TE, Groth CPT, McDonald JG (2020) A 14-moment maximum-entropy description of electrons in crossed electric and magnetic fields. Phys Plasmas 27(12):123506. https://doi.org/10.1063/5.0025651
4. Boccelli S, McDonald JG, Magin TE (2022) 14-moment maximum-entropy modeling of collisionless ions for Hall thruster discharges. Phys Plasmas 29(8):083903. https://doi.org/10.1063/5.0100092
5. Brent RP (1973) Algorithms for minimization without derivatives. Prentice-Hall
6. Chabert P, Braithwaite N (2011) Physics of radio-frequency plasmas. Cambridge University Press
7. Chalons C, Fox R, Massot M (2010) A multi-gaussian quadrature method of moments for gas-particle flows in a les framework. In: Proceedings of the summer program, pp 347–358
8. Dreyer W (1987) Maximisation of the entropy in non-equilibrium. J Phys A: Math General 20(18):6505. https://dx.doi.org/10.1088/0305-4470/20/18/047
9. Duchesne C, Banzet S, Lataillade J, Rousseau A, Frescaline N (2019) Cold atmospheric plasma modulates endothelial nitric oxide synthase signalling and enhances burn wound neovascularisation. J Pathol 249
10. Fox RO, Laurent F, Vié A (2018) Conditional hyperbolic quadrature method of moments for kinetic equations. J Comput Phys 365:269–293. https://www.sciencedirect.com/science/article/pii/S0021999118301852
11. Goebel D, Katz I (2008) Fundamentals of electric propulsion: ion and hall thrusters. In: JPL space science and technology series. Wiley. https://books.google.fr/books?id=P5OGFXcBKcwC
12. Grad H (1949) On the kinetic theory of rarefied gases. Commun Pure Appl Math 2(4):331–407. https://onlinelibrary.wiley.com/doi/abs/10.1002/cpa.3160020403
13. Kolobov V, Godyak V (2019) Electron kinetics in low-temperature plasmas. Phys Plasmas 26(6):060601. https://doi.org/10.1063/1.5093199
14. Kuldinow DA, Yamashita Y, Mansour AR, Hara K (2024) Ten-moment fluid model with heat flux closure for gasdynamic flows. J Comput Phys 508:113030. https://www.sciencedirect.com/science/article/pii/S0021999124002791
15. Kuldinow D, Yamashita Y, Hara K (2024) Ten-moment fluid model for low-temperature magnetized plasmas. Phys Plasmas 31:3

16. Laguna AA, Esteves B, Raimbault JL, Bourdon A, Chabert P (2023) Discussion on the transport processes in electrons with non-maxwellian energy distribution function in partially-ionized plasmas. Plasma Phys Controlled Fusion 65(5):054002
17. Laplante J, Groth CPT (2016) Comparison of maximum entropy and quadrature-based moment closures for shock transitions prediction in one-dimensional gaskinetic theory. AIP Conf Proc 1786(1):140010. https://doi.org/10.1063/1.4967641
18. Lequette N, Alvarez Laguna A, Chabert P, Bourdon A, Esteves B (2024) Pic fluid hybrid simulation of a discharge for electric propulsion. In: Proceedings of the 38th international electric propulsion conference, Toulouse, France, pp 2024–686
19. Levermore CD (1996) Moment closure hierarchies for kinetic theories. J Statist Phys 83(5–6):1021–1065
20. Lieberman MA, Lichtenberg A (2005) Principles of plasma discharges and materials processing, 2nd edn. Wiley-Interscience
21. Phelps Database (2021) www.lxcat.net, retrieved on 14 Apr 2021
22. Pichard T (2023) Some recent advances on the method of moments in kinetic theory. ESAIM ProcS 75:86–95
23. Sahu R, Mansour AR, Hara K (2020) Full fluid moment model for low temperature magnetized plasmas. Phys Plasmas 27(11):113505
24. Taunay PYC, Mueller ME (2023) Quadrature-based moment methods for kinetic plasma simulations. J Comput Phys 473:111700. https://www.sciencedirect.com/science/article/pii/S002199912200763X
25. Tavant A, Lucken R, Bourdon A, Chabert P (2019) Non-isothermal sheath model for low pressure plasmas. Plasma Sources Sci Technol 28(7):075007. https://dx.doi.org/10.1088/1361-6595/ab279b
26. Zhdanov V, Tirskii G (2003) The use of the moment method to derive the gas and plasma transport equations with transport coefficients in higher-order approximations. J Appl Math Mechan 67(3):365–388. https://www.sciencedirect.com/science/article/pii/S0021892803900219

A Study on Gramian Closures in Bimodal Phenomena

Eda Yilmaz, Georgii Oblapenko, and Manuel Torrilhon

Abstract This work analyzes the prediction quality of Gramian closures using a strongly bimodal distribution modeled as the superposition of two Gaussian functions. In [1], a new class of hyperbolic closures based on orthogonal polynomials named Gramian and extended Gramian closures has been introduced. These new closures are compared with Grad and the maximum entropy closures in terms of approximation quality. Numerical experiments show that the extended Gramian closure achieves competitive accuracy with the maximum entropy approach, particularly for even-order moments.

Keywords Kinetic theory of gases · Moment methods · Non-linear closures

1 Introduction

Capturing the dynamics of non-equilibrium gas flows requires models that go beyond classical continuum formulations. In such regimes, traditional continuum laws such as Navier-Stokes and Fourier fail to provide accurate results. As continuum assumptions break down, it becomes necessary to model gas behavior at the kinetic level, where individual molecular motion is considered. Kinetic theory provides a microscopic description of gas behavior by modeling the time evolution of the particle distribution function governed by the Boltzmann equation.

Despite providing a detailed description of microscopic gas dynamics, the Boltzmann equation remains computationally taxing due to its high dimensionality and the complexity of the collision operator. A widely used simplification technique is the method of moments, which projects the Boltzmann equation onto a finite set of macroscopic quantities such as density, momentum, and energy. However, this process generates an infinite hierarchy of moment equations, as the evolution of each moment depends on higher-order ones. In order to close this set of equations,

E. Yilmaz (✉) · G. Oblapenko · M. Torrilhon
Applied and Computational Mathematics (ACoM), RWTH Aachen University, Aachen, Germany
e-mail: yilmaz@acom.rwth-aachen.de

© The Author(s) 2026

M. Grabe et al. (eds.), *Rarefied Gas Dynamics*, Springer Aerospace Technology,
https://doi.org/10.1007/978-3-032-00094-1_35

information of the next moment needs to be defined based on the given set of lower moments, which is called a closure relation.

Over the past decades, many closure strategies have been proposed. One well-known classical technique is constructed by Grad [2] for closing the moment system (2) by expanding the distribution function as a series. In this approach, the distribution function is expressed as a weighted sum of Hermite polynomials based on equilibrium. The coefficients of the series represent the parameters of the ansatz. Grad's closure is a well-known and widely used classical method, but it loses hyperbolicity under non-equilibrium conditions. Another approach is centered on entropy and provides a nonlinear closure that maximizes entropy under the constraints of given moments [3, 4]. This method guarantees hyperbolicity and leads to a system of equations in conservative form, making it attractive for modeling non-equilibrium flows. However, the maximum entropy closure may not produce a valid distribution for all moment sets, and the resulting fluxes often lack closed form expressions, making the method computationally intensive and limited to lower order systems.

Quadrature-based methods offer another direction. McGraw [5] first introduced such an approach for modeling aerosol dynamics, and it was later extended in various contexts [6–9]. However, classical quadrature-based closures often result in non-hyperbolic systems, which can lead to nonphysical shock formations. More recently, the Hyperbolic Quadrature Method of Moments (HyQMOM) [6] has been proposed. This method relies on orthogonal polynomials and provides a solution to the moment closure problem without reconstructing the underlying probability distribution function.

This work builds upon [1] by introducing a new benchmark test problem. Specifically, we examine a strongly bimodal distribution constructed as a superposition of two Gaussians with distinct velocities. We evaluate the performance of Gramian and extended Gramian closures in predicting the $(M + 1)^{\text{th}}$ moment based on n lower-order moments, and compare their accuracy against those of Grad's and maximum entropy closures.

The remainder of the paper starts with a brief overview of moment closure problem, including the one-dimensional kinetic transport equation and its associated moment system. In addition, the relation of orthogonal polynomial with Gramian matrix is also introduced. Section 3 summarizes the various closure strategies used in this work. Then, in Sect. 4, details of the numerical procedure used to evaluate the accuracy of each closure, based on reference moments computed from a bimodal distribution are presented. Lastly, the bimodal test case is defined and numerical results comparing the relative errors of each closure method across a range of velocity values and moment orders are drawn.

2 Moment Closures

We focus on the solution of the one-dimensional kinetic transport equation

$$\partial_t f + c \partial_x f = S(f) \qquad (x, t, c) \in \Omega \times [0, T] \times \mathbb{R}, \tag{1}$$

where $f(x, t, c)$ represents the particle distribution over space $\mathbb{R}$ and time $[0, T]$. The variable c represents particle velocity at each point (x, t), and $S(f)$ denotes the collision operator.

The distribution function $f : \mathbb{R} \to \mathbb{R}^+$, defined by $c \mapsto f(c)$, is a key component in the formulation of moment equations. The moments of a distribution on the real line are defined as $u_k = \int_{\mathbb{R}} c^k f(c)dc$, where $f(c)$ satisfies the usual requirements of, e.g., positivity and integrability. The set of moments $u := \{u_k\}_{k=0,1,...,M}$ up to a given order M are the only information available about the distribution, while $f(c)$ is generally unknown. When the distribution function depends on space and time, moments also vary respectively. The process of multiplying the kinetic transport equation by successive powers of c and integrating over the velocity space generates a system of partial differential equations. The system of moment equations has the following form

$$\begin{aligned}
\partial_t u_0 + \partial_x u_1 &= s_0 \\
\partial_t u_1 + \partial_x u_2 &= s_1 \\
&\;\;\vdots \\
\partial_t u_M + \partial_x u_{M+1} &= s_M
\end{aligned} \tag{2}$$

where the s_k are the moments of the collision operator $S(f)$. It is important to highlight that this system follows a cascading structure, where the equation for the nth-order moment depends on the next higher-order moment. The closure relation for u_{M+1} must be defined to complete the system.

Definition 1 A prediction of the next moment based on a set of given moments in the form $u_{M+1} = C(u_0, u_1, \cdots, u_M)$, where $C : \mathcal{R} \to \mathbb{R}$ is a smooth function, is called a *moment closure*.

Orthogonal Polynomials
A scalar product defined using a distribution function $f(c)$ is given by $\langle v, w \rangle_f = \int_{\mathbb{R}} v(c)w(c)f(c)dc$. This inner product can be used to define orthogonal polynomials $p_k \in \mathbb{P}_k$, which satisfy the orthogonality condition $\langle p_k, p_l \rangle_f = 0$ for $k \neq l$. Here, $\mathbb{P}_k$ represents the space of univariate polynomials with degree at most k, and monic polynomials are considered. Recent work replaces Gram-Schmidt orthogonalization with a Gramian matrix to determine the orthogonal polynomials $\{p_k\}$.

Definition 2 Let the moments $\{u_k\}_{k=0,\dots,2n} \subset \mathcal{R}$ of a distribution be given and finite. Then the matrix $G_n = \{\langle c^i, c^j \rangle_f\}_{i,j=0,\dots,n}$ is called the monomial-based *Gramian matrix* of order $n \in \mathbb{N}$.

In the classical book [10, Sect. 2.2], the explicit representation of the orthogonal polynomial p_n is defined by

$$p_n(c) = \frac{1}{\det G_{n-1}} \det \begin{pmatrix} & & u_n \\ & G_{n-1} & \vdots \\ & & u_{2n-1} \\ 1, \, c, \, \cdots, \, c^{n-1} & & c^n \end{pmatrix}. \tag{3}$$

For simplicity, the determinant of Schur complement is used and Eq. (3) is written in the following form

$$p_n(c) = c^n - (1, c, \cdots, c^{n-1}) G_{n-1}^{-1} u_{n,2n-1} \tag{4}$$

where we used the abbreviation $u_{k,l} := (u_k, u_{k+1}, \dots, u_l)^{\mathsf{T}} \in \mathbb{R}^{l-k+1}$ for $k \leq l$ for the vector $u_{n,2n-1} \in \mathbb{R}^n$.

3 Closure Techniques

The choice of closure plays a key role in the accuracy and stability of moment methods. In this study, we investigate two recently developed closures, Gramian and extended Gramian closures, and compare them with classical approaches such as Grad's method and the maximum entropy closure. Gramian closures are based on orthogonal polynomials derived from the Gramian matrix and provide explicit expressions for the next higher moment. The extended version introduces an additional parameter χ, which allows adjusting certain mathematical properties of the closure.

The mathematical properties of these closures vary depending on whether the number of moments M is even or odd. For even values of M, the extended Gramian closure satisfies strict hyperbolicity over a broad range of parameters, ensures gauge invariance for a specific choice, and preserves equilibrium. In contrast, for odd M, equilibrium preservation is no longer possible. The standard Gramian closure in this case is only weakly hyperbolic, while the extended version offers improved behavior near equilibrium and achieves gauge invariance under certain conditions.

Even Gramian Closure

Let $M \in \mathbb{N}$ be even and $n = \frac{M}{2}$. The condition $\int_{\mathbb{R}} p_n(c) c^{n+1} f(c) dc = 0$ defines the relation $\mathbf{u_{2n+1}} = \mathbf{u}_{n+1,2n}^{\mathsf{T}} \mathbf{G}_{n-1}^{-1} \mathbf{u_{n,2n-1}}$ between $u_{M+1} = u_{2n+1}$ and $\{u_k\}_{k=0,1,\dots,M}$, which we call *Gramian moment closure*. The closure relation arises directly by inserting the polynomial in (4) into the integral condition.

Extended Even Gramian Closure

Let $M \in \mathbb{N}$ be even and $n = \frac{M}{2}$. For an arbitrary $\chi \in \mathbb{R}$, the relation

$$u_{2n+1} = u_{n+1,2n}^T G_{n-1}^{-1} u_{n,2n-1}$$
$$+ \chi \frac{n+1}{n} \frac{u_{2n} - u_{n,2n-1}^T G_{n-1}^{-1} u_{n,2n-1}}{u_{2n-2} - u_{n-1,2n-3}^T G_{n-2}^{-1} u_{n-1,2n-3}} \left(u_{2n-1} - u_{n,2n-2}^T G_{n-2}^{-1} u_{n-1,2n-3} \right)$$

between $u_{M+1} = u_{2n+1}$ and $\{u_k\}_{k=0,1,\ldots,M}$, which we call *extended Gramian moment closure* for even M.

Odd Gramian Closure

Let $M \in \mathbb{N}$ be odd and $n = \frac{M+1}{2}$. The condition $\displaystyle\int_{\mathbb{R}} p_n(c) c^n f(c) dc = 0$ defines the relation $u_{2n} = u_{n,2n-1}^T G_{n-1}^{-1} u_{n,2n-1}$ between $u_{M+1} = u_{2n}$ and $\{u_k\}_{k=0,1,\ldots,M}$, which we call *Gramian moment closure* for odd M.

Extended odd Gramian Closure

Let $M \in \mathbb{N}$ be odd and $n = \frac{M+1}{2}$. The relation

$$u_{2n} = u_{n+1,2n-1}^T G_{n-2}^{-1} u_{n-1,2n-3} + \chi \frac{\left(u_{2n-1} - u_{n,2n-2}^T G_{n-2}^{-1} u_{n-1,2n-3} \right)^2}{u_{2n-2} - u_{n-1,2n-3}^T G_{n-2}^{-1} u_{n-1,2n-3}}$$

between $u_{M+1} = u_{2n}$ and $\{u_k\}_{k=0,1,\ldots,M}$, which we call *extended Gramian moment closure* for odd M.

4 Computation Procedure

In this work, dimensionless moments scaled by density, temperature, and shifted by velocity are considered. First, we evaluate reference moment values $\{u_k^{(\text{reference})}\}_{k=0,1,\ldots,M+1}$ with the bimodal distribution function as defined below. Using the first M reference moment values, the next higher moment is calculated by different closure methods. This can be achieved by applying a reconstructed distribution, as is done in the Grad and maximum entropy closures, or through a direct approach, which is the case for all Gramian closures. For the maximum entropy closure, we compute point values of $f \in \mathbb{R}^{N_c}$ on a discrete velocity grid $\{c_j\}_{j=1,\ldots,N_c}$ on a finite domain $[c_{\min}, c_{\max}]$. To improve the convergence behavior, the logarithm is modified by writing $-f \log(f + \varepsilon)$ where $\varepsilon = 10^{-8}$. It is important to mention that this method allows handling odd cases when the maximum entropy distribution fails to be integrable. We define the relative error as the measure of the difference between the reference moment value $u_{M+1}^{(\text{reference})}$ and the predicted moment u_{closure} obtained by the closure method. It is formulated as

$$e_{\text{rel}}(u_{\text{closure}}) = \left| \frac{u_{\text{closure}} - u_{M+1}^{(\text{reference})}}{u_{M+1}^{(\text{reference})}} \right|.$$

5 Bimodal Distribution

The following superposition of Maxwellian type distribution functions

$$f_{\mathrm{BM}}(c) = f^{(\mathrm{eq})}(\rho_1, v_1, \theta_1, c) + f^{(\mathrm{eq})}(\rho_2, v_2, \theta_2, c),$$

where $f^{(\mathrm{eq})}$ represents the equilibrium distribution characterized by densities (ρ_i), velocities (v_i), and temperatures (θ_i) for $i = 1, 2$, is considered as a bimodal test problem. We fix the parameters $\rho_1 = 0.4$, $v_1 = 0$, $\theta_1 = 1 = \theta_2$, $\rho_2 = 0.6$, and using the distribution function f_{BM} we calculate the first M moments for each v_2 value between 0.5 and 4 with step size 0.05. For the maximum entropy closure, the computational domain is taken with $c \in [-4, 6]$ with 1000 grid points. The resulting distributions are shown in Fig. 1. The distribution function for the velocity $v_2 = 0.5$ is actually very close to equilibrium, but it goes into non-equilibrium states as v_2 increases and two peaks slowly begin to form.

The relative error of the next moment is plotted in Fig. 2. As expected, close to equilibrium the relative error of Grad's closure results goes down to zero. Furthermore, around equilibrium, compared to other methods Grad's closure mostly gives better results, yet maximum entropy is also competitive. The disadvantage of Grad's closure is loss of hyperbolicity in non-equilibrium processes, which limits its use even though the relative error may be performing well.

The error curve of the maximum entropy closure shows a strongly nonlinear behavior including singularities where the error actually vanishes. Similar behavior is visible in the other examples below. A general explanation of this behavior is difficult and out of the scope of this paper. Close to equilibrium, for small v_2 the maximum

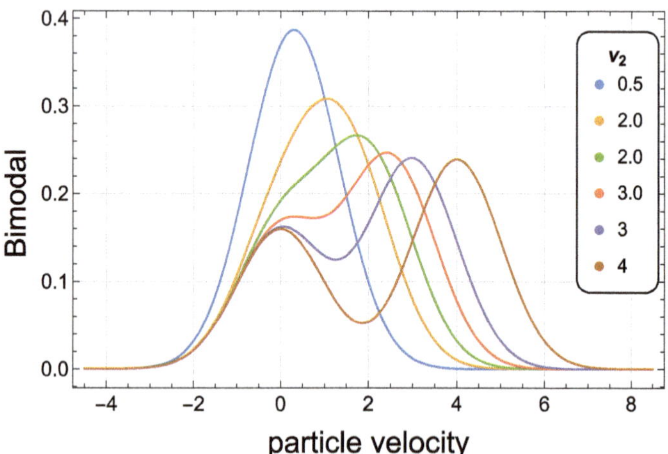

Fig. 1 The bimodal distribution $f_{BM}(c)$ is shown with varying values of v_2, while other parameters are kept fixed

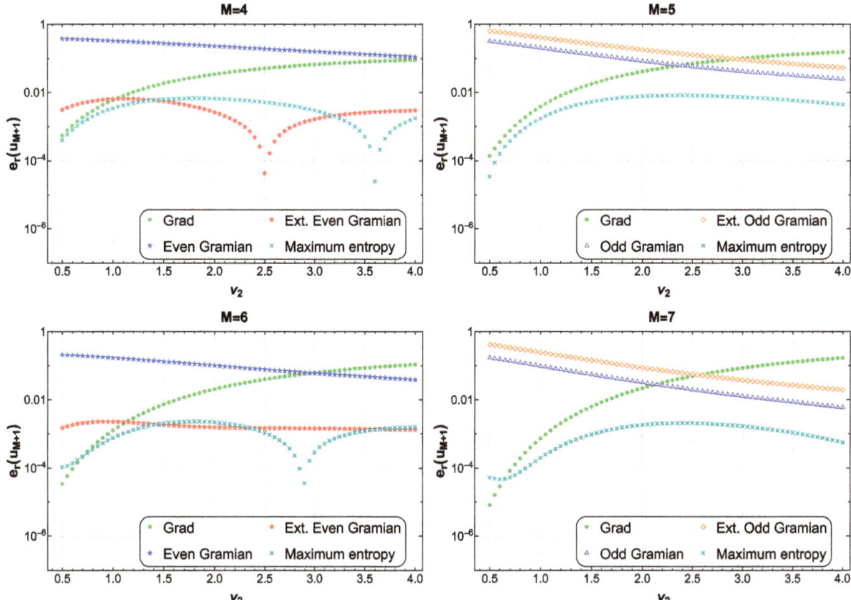

Fig. 2 Relative error $e_r(u_{M+1})$ of the next higher moment for a range velocity shifts in the bimodal test case. The left part of the figure shows the relative error for even number of moments $M = 4, 6$ while the right one shows odd cases $M = 5, 7$. Different plot markers with colors represent the closure. Note, that the Gramian and extended Gramian closure are defined in the even and odd case differently

entropy closure should show vanishing errors. For larger moment orders $M = 6, 7$ the calculation of the maximum entropy distribution becomes significantly more difficult and numerical errors prevent the approximation from becoming better in proximity to equilibrium. Some of the error curves show sharp spikes at different points due to the error briefly passing through zero as parameters change, which is difficult to explain. These rare cases are considered outliers, and the main focus is placed on the overall behavior of the error. However, overall the maximum entropy approach mostly provides better relative errors compared to the other closure methods. For the new closures the extended Gramian closure in the even case performs best, even comparable to maximum entropy. Surprisingly, that closure even mimics the nonlinear behavior of the maximum entropy error. In the even cases, the impact of gauge-invariance between the extended Gramian and simple Gramian closures can be observed. For the odd case, surprisingly, the simple Gramian closure is slightly better than the extended version and it is not clear what causes this. The closures for odd values of M demonstrate relatively large errors, yet still show decay, despite the absence of equilibrium preservation. Although the significant errors in the odd-order case can be attributed to the absence of equilibrium preservation, additional research is required, and this will be explored in future studies. Odd Gramian and extended odd Gramian relative errors are not as good as maximum entropy or Grad close to equilibrium states, but when two peaks begin to form in the distribution function their relative errors become clearly better than Grad's closure.

6　Conclusion

In this work, we examined the predictive accuracy of various moment closure techniques, namely Grad, maximum entropy, Gramian, and extended Gramian closures, using a highly non-equilibrium bimodal distribution as a benchmark test case. The outcomes of our numerical investigation demonstrate that the extended Gramian closure exhibits performance comparable to the maximum entropy closure, especially in predicting even-order moments. Moreover, unlike maximum entropy methods, Gramian-based closures provide explicit and computationally efficient expressions for the higher-order moments without requiring iterative optimization or function reconstruction. This highlights their potential for scalable implementation in complex flow simulations where computational efficiency and robustness are critical. Future work will explore extensions to multidimensional settings and incorporate realistic collision models to further assess the applicability of these closures in kinetic theory.

References

1. Yilmaz E, Oblapenko G, Torrilhon M (2024) On nonlinear closures for moment equations based on orthogonal polynomials. arXiv preprint, arXiv:2407.05894v2. https://doi.org/10.48550/arXiv.2407.05894
2. Grad H (1949) On the kinetic theory of rarefied gases. Commun Pure Appl Math 2(4):331–407. Wiley Online Library
3. Levermore CD (1997) Entropy-based moment closures for kinetic equations. Transp Theor Statist Phys 26(4–5):591–606
4. Dreyer W (1987) Maximisation of the entropy in non-equilibrium. J Phys A: Math Gener 20(18):6505. https://doi.org/10.1088/0305-4470/20/18/047
5. McGraw R (1997) Description of aerosol dynamics by the quadrature method of moments. Aerosol Sci Technol 27(2):255–265
6. Fox RO, Laurent F (2022) Hyperbolic quadrature method of moments for the one-dimensional kinetic equation. SIAM J Appl Math 82:750–771
7. Kah D, Chalons C, ve Massot M (2010) Beyond pressureless gas dynamics: quadrature-based velocity moment models. Commun Math Sci (submitted)
8. Fox RO (2008) A quadrature-based third-order moment method for dilute gas-particle flows. J Comput Phys 227(12):6313–6350
9. Huang Q, Li S, Yong W-A (2020) Stability analysis of quadrature-based moment methods for kinetic equations. SIAM J Appl Math 80(1):206–231
10. Szegö G (1939) Orthogonal polynomials. AMS, Providence
11. Struchtrup H (2005) Macroscopic transport equations for rarefied gas flows. Springer
12. Torrilhon M (2016) Modeling nonequilibrium gas flow based on moment equations. Ann Rev Fluid Mechan 48:429–458
13. Aheizer N, Krein M (1962) Some questions in the theory of moments. Translations of Mathematical Monographs, American Mathematical Society, vol 2

14. Gautschi W (1996) Orthogonal polynomials: applications and computation. Acta Numerica 5:45–119
15. Hamburger HL (1944) Hermitian transformations of deficiency-index (1,1), Jacobi matrices and undetermined moment problems. Am J Math 66(4):489–522

Numerical Study of Grad-14 and 17 Moment Equations for Rarefied Polyatomic Gases

Hang Song, Satyvir Singh, and Manuel Torrilhon

Abstract The Grad-type moment equations provide an extended hydrodynamic framework for describing the non-equilibrium behavior of rarefied polyatomic gases beyond the classical Navier-Stokes-Fourier equations. In this study, we present a numerical investigation of Grad-14 and Grad-17 moment equations to analyze their accuracy and applicability in capturing rarefaction effects, translational-rotational energy exchange, and non-equilibrium transport phenomena. The numerical scheme employs a high-order nodal discontinuous Galerkin method to solve the Grad-type equations for rarefied polyatomic flows. The results are compared with solutions from the direct simulation Monte Carlo (DSMC) method and experimental data to assess the predictive capability of the moment approach. Our analysis demonstrates that the Grad formulation offers superior accuracy over the Navier-Stokes-Fourier equations in capturing non-equilibrium phenomena at Mach numbers up to 1.6, particularly under strong velocity/temperature gradient conditions. Nevertheless, persistent non-physical sub-shock formations emerge as a critical limitation at Mach 2.0. Additionally, the study highlights the limitations of moment closure assumptions and their implications for extending higher-order moment theories to polyatomic gases.

Keywords Polyatomic gas · Moment equations · Grad's closure · Shock wave

1 Introduction

Gaseous systems exhibiting non-equilibrium behavior hold critical significance across diverse technical disciplines, particularly when encountered in rarefied environments or under extreme velocity and thermal gradients. Such deviations from thermodynamic equilibrium emerge when molecular velocity distributions substantially depart from Maxwellian statistics, a characteristic feature observed in microscale fluid dynamics and vacuum-based gas transport applications. For polyatomic gas molecules, additional complexities arise from their internal degrees of freedom

H. Song · S. Singh (✉) · M. Torrilhon
Applied and Computational Mathematics, RWTH Aachen University, Aachen, Germany
e-mail: singh@acom.rwth-aachen.de

© The Author(s) 2026

M. Grabe et al. (eds.), *Rarefied Gas Dynamics*, Springer Aerospace Technology,
https://doi.org/10.1007/978-3-032-00094-1_36

379

including rotational and vibrational excitations, which significantly influence energy redistribution mechanisms and relaxation dynamics during molecular collisions [1].

In non-equilibrium gas flows, the Knudsen number (Kn) is an essential parameter for evaluating the influence of non-equilibrium effects on flow dynamics [2]. This dimensionless quantity is defined as the ratio of the mean free path of gas molecules (λ) to a characteristic length scale (L) of the system: $Kn = \lambda/L$. Based on the Kn number, gas flows are divided into four regimes: continuum ($Kn \lesssim 0.001$), slip ($0.001 \lesssim Kn \lesssim 0.01$), transition ($0.01 \lesssim Kn \lesssim 10$), and free molecular ($Kn \gtrsim 10$) [3]. At higher Knudsen numbers, the lack of sufficient particle collisions becomes the primary factor for significant non-equilibrium effects, which are further exacerbated as the Mach number increases. This results in considerable challenges in modeling such gas flows both theoretically and computationally.

The Boltzmann transport equation (BTE) remains fundamental for modeling non-equilibrium gas flows where continuum assumptions fail. It characterizes gas behavior through a velocity distribution function tracking particle motions in space-time. While macroscopic models like the Euler or Navier-Stokes-Fourier (NSF) equations offer computational efficiency in transition regimes, their accuracy diminishes with increasing Knudsen number. Grad's pioneering work introduced moment-based formulations into kinetic theory through Hermite polynomial expansions of the distribution function, subsequently enabling the gas state description via extended variable sets in field equations [4]. This framework extends beyond traditional fluid dynamic variables to systematically incorporate higher-order moments of the distribution function. The Grad-14 (G14) moment equations with a Boltzmann collision operator in the continuous internal energy setting were developed to describe polyatomic gases with temperature-dependent specific heats or thermally perfect (non-polytropic) gases [5]. A macroscopic system of Grad-17 (G17) moment equations and its transport coefficients has been derived through analysis of the BTE based on a continuous internal energy variable for modeling polyatomic gases with constant specific heats [6].

This study investigates non-equilibrium flow simulations of polyatomic gases through the G14 and G17 moment equations, focusing on elucidating complex shock phenomena under low non-equilibrium conditions (Ma $\leq$ 2.0), where both translational and rotational energies are explicitly considered while vibrational energy remains excluded from the analysis. The objective is to evaluate the effect of Grad models when predicting non-equilibrium phenomena in high-speed flows. By analyzing one-dimensional shock structures, the study will validate the Grad models through comparison with DSMC results. The remainder of the paper is structured as follows: Sect. 2 introduces the mathematical formulation for polyatomic gases, including various Grad-type moment equations. Section 3 discusses the employed numerical solver. Section 4 presents numerical results and provides a detailed description of non-equilibrium flow simulations in polyatomic gases. Finally, Sect. 5 gives the concluding remarks and outlook for the future work.

2 Mathematical Formulation

The BTE for polyatomic gases extends the classical kinetic equation to incorporate additional internal energy modes inherent in molecular structures with multiple atoms. Unlike monatomic gases (with purely translational motion), polyatomic systems exhibit rotational and vibrational energy states due to intramolecular configurations. A continuous parameter $I \in \mathbb{R}_+$ is employed, termed "microscopic internal energy", to collectively model non-translational energy states, acting as a lumped descriptor for internal degrees of freedom. The present study exclusively considers rotational energy in the analysis.

The macroscopic state of a rarefied polyatomic gas is statistically characterized by a non-negative distribution function: $f(\mathbf{x}, \mathbf{v}, I, t) \geq 0$, in which spatial coordinates $\mathbf{x} \in \mathbb{R}^3$, molecular velocity $\mathbf{v} \in \mathbb{R}^3$, internal energy $I \in \mathbb{R}_+$, and time $t \in \mathbb{R}_+$. The BTE governs its spatiotemporal evolution as Eq. 1. The bilinear collision operator $R[f]$ acts exclusively on $\mathbf{v}$ and I, preserving conservation laws through its integral structure.

$$\left(\frac{\partial}{\partial t} + \mathbf{v} \cdot \nabla_\mathbf{x}\right) f = R[f](\mathbf{v}, I). \tag{1}$$

2.1 Moment Methods

The moment equations governing polyatomic gases originate from the BTE, establishing a system of conservation laws that describe macroscopic quantities and higher-order moments. For polyatomic gases, these conservation relations exhibit increased complexity due to the necessity of incorporating internal degrees of freedom beyond translational motion. In kinetic theory, the moments are mathematically constructed through phase space integration of the distribution function. Specifically, a seris of vaiables are defined: mass density (ρ), flow velocity (u_i), dynamic pressure (p), average pressure $(\overline{p})$, pressure difference $(p_{\text{diff}} = p - \overline{p})$, pressure tensor (p_{ij}), stress tensor (σ_{ij}), translational heat flux (q_i), rotational heat flux (s_i), and total heat flux $(q_i^{\text{tot}} = q_i + s_i)$.

$$\begin{pmatrix} \rho \\ \rho u_i \\ p \\ \overline{p} \\ p_{\text{diff}} \\ p_{ij} \\ \sigma_{ij} \\ q_i \\ s_i \\ q_i^{\text{tot}} \end{pmatrix} = \int_{\mathbb{R}^3 \times \mathbb{R}_+} \begin{pmatrix} m \\ m v_i \\ \frac{1}{3} m c^2 \\ \frac{2}{\delta+3}\left(\frac{1}{2} m c^2 + I\right) \\ \frac{1}{3} m c^2 - \frac{2}{\delta+3}\left(\frac{1}{2} m c^2 + I\right) \\ m c_i c_j \\ m c_i c_j - \frac{1}{3} m c^2 \delta_{ij} \\ \frac{1}{2} m c^2 c_i \\ I c_i \\ \frac{1}{2} m c^2 c_i + I c_i \end{pmatrix} f \, d\mathbf{v} \, dI, \quad i, j \in \{1, 2, 3\}.$$

Here, $\mathbf{c} = \mathbf{v} - \mathbf{u}$ is the thermal velocity of the gas molecule, and $\delta > 0$ represents the internal degrees of freedom, δ_{ij} is the Kronecker delta. Proceeding with the methodology, the BTE (1) for polyatomic gases is integrated using the moments specified in the list, yielding the fundamental conservation principles for the associated densities. The system can be formulated in the conservative form Eq. (2) as

$$\frac{\partial \mathbf{U}}{\partial t} + \nabla \cdot \mathbf{F}(\mathbf{U}) = \mathbf{S}(\mathbf{U}), \tag{2}$$

where $\mathbf{U}$ represents the vector of conserved quantities, $\mathbf{F}(\mathbf{U})$ denotes the flux tensor, and $\mathbf{S}(\mathbf{U})$ characterizes the source terms.

2.2 Grad-17 Moment Equations

Based on the theory of Grad moment equations, the heat flux of translation energy (q) and rotation energy (s) are analyzed separately in G17 equations. For polyatomic gas, the difference between kinetic pressure (p) and average pressure $(\overline{p})$ cannot be neglected. The equation of kinetic pressure and heat flux are added to the system, the set of primitive variables $W = \{\rho, u_i, \overline{p}, p_{ij}, q_i, s_i\}^{\mathsf{T}}$ is applied. In three-dimensional simulations, this system contains 17 independent variables: ρ (1 scalar), u_i (3 vector components), $\overline{p}$ (1 scalar), p_{ij} (6 independent components of symmetric tensor), q_i (3 vector components), and s_i (3 vector components). G17 equations are shown as Eq. (3) [6].

$$\mathbf{U} = \begin{pmatrix} \rho \\ \rho u_i \\ \rho|u|^2 + (\delta + 3)\overline{p} \\ \rho u_i u_j + p_{ij} \\ (\rho|u|^2 + 3p)u_i + 2p_{ij}u_j + 2q_i \\ 2s_i + (\delta + 3)\overline{p}u_i - 3pu_i \end{pmatrix},$$

$$\mathbf{F}(\mathbf{U}) = \begin{pmatrix} \rho u_k \\ \rho u_i u_k + p_{ik} \\ (\rho|u|^2 + (\delta + 3)\overline{p})u_k + 2p_{jk}u_j + 2q_k + 2s_k \\ \rho u_i u_j u_k + u_i p_{jk} + u_j p_{ki} + u_k p_{ij} + \frac{2}{5}(q_i \delta_{jk} + q_j \delta_{ik} + q_k \delta_{ij}) \\ (\rho|u|^2 + 3p)u_i u_k + 2u_i u_j p_{jk} + 2u_j u_k p_{ij} + |u|^2 p_{ik} + 2u_i q_k + 2u_k q_i \\ \quad + \frac{4}{5}(u_i q_k + u_k q_i + u_j q_j \delta_{ik}) + 7\frac{\overline{p}}{\rho}p_{ik} - 5\frac{\overline{p}^2}{\rho}\delta_{ik} + 3\frac{\overline{p}}{\rho}p\delta_{ik} \\ \frac{\overline{p}}{\rho}(\delta p_{ik} - \delta p + \delta p\delta_{ik} - 3p\delta_{ik} + 3\overline{p}\delta_{ik}) \\ \quad + 2u_i s_k + 2u_k s_i + u_i u_k((\delta + 3)\overline{p} - 3p) \end{pmatrix}, \tag{3}$$

$$\mathbf{S}(\mathbf{U}) = \begin{pmatrix} 0 \\ 0 \\ 0 \\ -\frac{\overline{p}}{\mu}(p_{ij} - p\delta_{ij}) - \frac{\overline{p}}{\mu_b}\frac{2\delta}{3\delta+9}(p - \overline{p})\delta_{ij} \\ -\frac{\overline{p}}{\mu}(2u_j p_{ij} - 2pu_i) - \frac{\overline{p}}{\mu_b}\frac{2\delta}{3\delta+9}(p - \overline{p})(5u_i) - \frac{\overline{p}}{\mu}(2P_q^{(0)}q_i + 2P_q^{(1)}s_i) \\ -\frac{\overline{p}}{\mu}(2P_s^{(0)}s_i + 2P_s^{(1)}q_i) + \frac{\overline{p}}{\mu_b}\frac{2\delta}{3\delta+9}(p - \overline{p})(3u_i) \end{pmatrix}.$$

The coefficients $P_q^{(0)}$, $P_q^{(1)}$, $P_s^{(0)}$, $P_s^{(1)}$ of source terms depend on the collision model applied in the simulation. Their relationship with Pr number is shown as Eq. (4).

$$\mathrm{Pr} = \frac{\delta + 5}{2} \frac{2(P_q^{(0)} P_s^{(0)} - P_q^{(1)} P_s^{(1)})}{5(P_s^{(0)} - P_s^{(1)}) + \delta(P_q^{(0)} - P_q^{(1)})} \tag{4}$$

In this study, the default values are $P_q^{(0)} = \frac{2}{3}$, $P_s^{(0)} = 1$, $P_q^{(1)} = P_s^{(1)} = 0$ which are obtained through BGK model. The default parameters yield $\delta = 2$ for diatomic gases and $\mathrm{Pr} = 14/19$, demonstrating close agreement with both DSMC simulations and experimental measurements of nitrogen.

2.3 Grad-14 Moment Equations and NSF Equations

It is also possible to avoid using an independent equation of heat flux of rotational energy. However, the difference of average pressure and kinetic pressure is still significant. An additional equation is contained in the system compared with Grad-13 moment equations. The set of primitive variables is $W = \{\rho, u_i, \overline{p}, p_{ij}, q_i^{\mathrm{tot}}\}^T$. The G14 equations are shown in [7]. The mathematical expression for these heat fluxes are then combined into a single equation representing the total energy heat flux.

The NSF equations emerge as a limiting regime of the G14/G17 moment equations under near-equilibrium conditions, achieved through appropriate closure relations. In this framework, the gas resides near thermodynamic equilibrium, resulting in small deviations of higher-order moments (e.g., stress tensor and heat flux) that scale linearly with the gradients of conserved variables (velocity and temperature). For polyatomic gases, the shear stress tensor and heat flux can be respectively expressed as $\tau_{ij} = \mu_b \frac{\partial u_j}{\partial x_j} \delta_{ik} + \mu \left(\frac{\partial u_i}{\partial x_k} + \frac{\partial u_k}{\partial x_i} - \frac{2}{3} \frac{\partial u_j}{\partial x_j} \delta_{ik} \right)$ and $q_i = \lambda \frac{\partial T}{\partial x_i}$.

3 Employed Numerical Solver

The discontinuous Galerkin (DG) method is a variant of the finite element approach, which allows it to efficiently handle complex geometries and irregular meshes [8]. The DG method excels in solving problems with discontinuous solutions, making it particularly well-suited for fluid dynamics and other fields requiring high-precision numerical simulations. The numerical method employed in this research is the nodal DG spectral element method (DGSEM), which is a combination of the DG method and the spectral element method. DGSEM provides higher accuracy, fewer degrees of freedom, and better numerical stability, making it particularly suitable for complex flows and situations requiring high-order solutions. The foundation of this work is Trixi.jl, a numerical framework centered on conservation laws that incorporates error-based and CFL-based adaptive step size control [9]. The approximation of the function on each element is based on the third-order Lobatto-Legendre basis.

4   Numerical Results and Discussion

The one-dimensional steady shock problem is a fundamental issue in fluid dynamics, involving the steady propagation of a shock wave in one-dimensional flow. In this scenario, the properties of the gas, contains density, pressure, velocity, and temperature, undergo discontinuous changes across the shock, while the shock wave propagates at a constant speed. The left inflow boundary condition and initial state for $x < 0$ are rigorously prescribed to replicate the pre-shock gas properties governed by the Rankine-Hugoniot relations, while the right outflow boundary condition and initial state for $x > 0$ are systematically configured to enforce the post-shock thermodynamic state through strict conservation of Rankine-Hugoniot jump conditions.

In this section, the research gas is nitrogen, of which the rotational energy cannot be ignored. The mean free path of pre-shock region (left side) is 1 mm and the temperature is 273.15 K, and the reference values are provided by Bird [10]. Three inflow Mach numbers Ma = 1.2, 1.6, and 2.0 were discussed. Except for the Euler, NSF, G14 and G17 equations, the particle method DSMC is also applied to verify the effect. Two popular collision models of DSMC, the variable hard sphere (VHS) and the variable soft sphere (VSS) are employed in the validation.

As shown in Fig. 1, the G14 and G17 equations provide a more accurate description of density, velocity, and average pressure compared to the NSF equations at inflow Ma $= 1.2, 1.6$. Nonlinear processes become more significant as Mach number of inflow increasing, causing the constitutive relations between fluxes and spatial derivatives to break down. It is evident that the NSF equations deviate more from the DSMC results in the regions $-8 < x/\lambda < -6$ and $2 < x/\lambda < 4$ compared to the G14 and G17 equations at inflow Ma $= 1.6$. For Ma $= 2.0$, a small sub-shock appears near the pre-shock region ($x/\lambda \approx -5$) because of the symmetric hyperbolicity, leading to inaccuracies in the G14 and G17 equation simulations. The largest deviation of propagation speeds of G14 and G17 equations from bulk velocity are $2.032\sqrt{p/\rho}$ and $2.130\sqrt{p/\rho}$, which are 1.72 and 1.8 times that of the sound speed of polyatomic gas [11]. Mathematically, the polyatomic gas flow at Ma > 1.8 cannot be simulated accurately through G14 and G17 equations. The discrepancies between the NSF equations and DSMC are also significant, highlighting that the strong non-equilibrium effects caused by inflow at Ma > 2.0 cannot be precisely captured by either the NSF equations or the Grad moment equations.

In Fig. 2, both the G14 and G17 equations demonstrate good agreement with the DSMC results regarding pressure difference and stress at Mach numbers of 1.2 and 1.6, whereas the NSF equations are inadequate. This discrepancy arises because, when the high-speed airflow passes through a shock wave, the molecular translational energy increases rapidly, leading to a sharp rise in kinetic pressure.

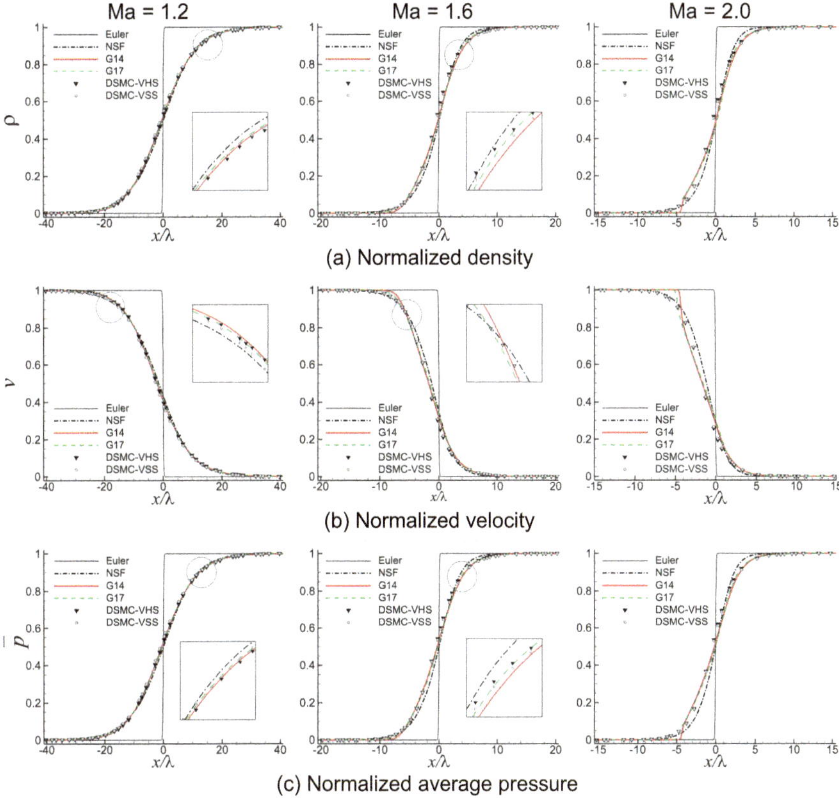

Fig. 1 1D shock structure problem: profiles of normalized density (ρ), normalized velocity (v), normalized average pressure ($\overline{p}$) at Ma = 1.2, 1.6 and 2.0

However, the response of rotational energy is slower, resulting in a difference between kinetic pressure (p) and average pressure ($\overline{p}$). Except for the region of sub-shock at Ma $= 2.0$, the agreements are also acceptable. The simulations of heat flux by both Grad moment equations and the NSF equations are not ideal at Ma $= 1.6, 2.0$, which means the calculation of third-order terms are not satisfactory. It is necessary to adjust the coefficients in the source terms of G17 equations to quantify model-form errors and establish baseline performance before pursuing more complex moment system extensions. This parametric study will inform the development of higher-order Grad models while maintaining computational tractability, which will be addressed in subsequent research.

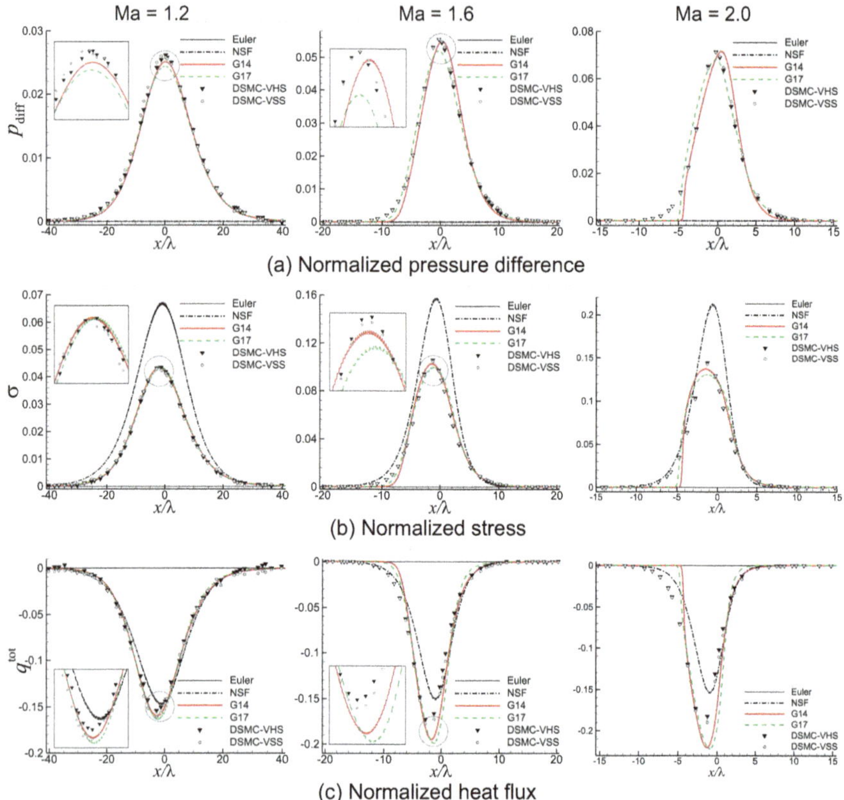

Fig. 2 1D shock structure problem: profiles of normalized pressure difference (p_{diff}), normalized stress (σ), normalized heat flux of total energy (q^{tot}) at Ma = 1.2, 1.6 and 2.0

5 Conclusion

In this research, the non-equilibrium processes are simulated using polyatomic G14 and G17 moment equations through a one-dimensional shock wave. A solver based on the DGSEM is employed, forming the foundation for high-order Grad moment equations and regularized equation systems. It was found that the Grad models, particularly the G17 model, offer superior accuracy in predicting non-equilibrium effects compared to traditional Euler and NSF models. These results were validated through comparisons with DSMC data, highlighting the Grad models' ability to resolve shock thickness, stress, and heat flux in high-speed flows, where non-equilibrium phenomena are significant. Nevertheless, the persistence of hyperbolicity-induced sub-shocks remains an unresolved challenge in the current framework, necessitating the systematic investigation through regularization techniques in forthcoming research initiatives.

References

1. Vincenti WG, Kruger CH, Teichmann T (1966) Introduction to physical gas dynamics. Phys Today 19(10):95
2. Struchtrup H (2005) Macroscopic transport equations for rarefied gas flows. In: Macroscopic transport equations for rarefied gas flows. Interaction of mechanics and mathematics. Springer, Berlin, pp 159–186
3. Torrilhon M (2016) Modeling nonequilibrium gas flow based on moment equations. Ann Rev Fluid Mechan 48(1):429–458
4. Grad H (1952) The profile of a steady plane shock wave. Commun Pure Appl Math 5:257
5. Pavi'c-Čoli'c M, Simi'c S (2022) Kinetic description of polyatomic gases with temperature-dependent specific heats. Phys Rev Fluids 7:083401
6. Djordji'c V, Oblapenko G, Pavi'c-Čoli'c M, Torrilhon M (2023) Boltzmann collision operator for polyatomic gases in agreement with experimental data and DSMC method. Continuum Mech Thermodyn 35:103–119
7. Pavić-Čolić M, Simić S (2014) Moment equations for polyatomic gases. Acta Appl Math 132:469–482
8. Singh S, Karchani A, Myong RS (2022) A three-dimensional modal discontinuous Galerkin method for the second-order Boltzmann-Curtiss-based constitutive model of rarefied and microscale gas flows. J Comput Phys 457:111052
9. Schlottke-Lakemper M et al (2021) A purely hyperbolic discontinuous Galerkin approach for self-gravitating gas dynamics. J Comput Phys 442:110467
10. Bird G (1970) Direct simulation and the Boltzmann equation. Phys Fluids 13:2676–2681
11. Song H, Singh S, Torrilhon M (2025) Non-equilibrium flow simulations based on Grad-14 and Grad-17 moment equations for polytropic gases. Phys Fluids 37(3)

Numerical Methods for Kinetic Equations

Adaptation of the Unified Gas-Kinetic Scheme to ES-BGK Models

Céline Baranger, Alexis Coëpeau, and Luc Mieussens

Abstract The Unified Gas-Kinetic Scheme stands out from traditional deterministic numerical approaches to gas dynamics by enabling the resolution of flow regimes ranging from rarefied to continuum, within a simulation time independent of the regime. Its effectiveness in solving complex flows has primarily been demonstrated with Shakhov or Rykov models. Here, we propose to examine in detail the applicability of UGKS to ES-BGK models, extending its application from the monoatomic framework to the polyatomic framework by using recent ES-BGK models that take into account rotational and vibrational energy modes.

Keywords Gas dynamics · UGKS · ES-BGK models

1 Introduction

The Ellipsoidal-Statistical BGK (ES-BGK) model [5] is an alternative modelling of the Boltzmann equation, developed to enable faster numerical simulations of transport phenomena within a monoatomic gas in a transitional state between rarefied and continuous regimes. This model retains the same fundamental properties as the Boltzmann equation since it ensures the positivity of the mass distribution, has the same collisions invariants, satisfies the H-Theorem [1], and is able to recover a Prandtl number of 2/3 for monoatomic gases. It has also been extended to polyatomic gases [1] and has undergone various modifications, such as capturing rotational and vibrational energy relaxation processes [4, 7].

For a long time, these kinetic models were only solved numerically using methods that were extremely time-consuming as the density of the resolved flow increased. However, since 2010, Kun Xu et al. have been developing a unified method known

C. Baranger · A. Coëpeau
CEA-CESTA, Le Barp Cedex, France

A. Coëpeau (✉) · L. Mieussens
University of Bordeaux, CNRS, Bordeaux INP, IMB, Talence, France
e-mail: alexis.coepeau@u-bordeaux.fr

© The Author(s) 2026

M. Grabe et al. (eds.), *Rarefied Gas Dynamics*, Springer Aerospace Technology,
https://doi.org/10.1007/978-3-032-00094-1_37

as the "Unified Gas-Kinetic Scheme" [10]. This singular method allows the accurate resolution of all ranges of flows with a computational cost independent of the rarefaction of the gas.

The UGKS scheme has only been applied to the monoatomic ES-BGK model [3, 6]. Here, we propose to extend it to phenomena typical of diatomic models, such as internal energy storage, as described by Andriès et al. [1].

2 The Ellipsoidal-Statistical BGK Model for Polyatomic Gases

To describe the dynamics of a polyatomic gas, it is customary to use a microscopic distribution of mass F defined over the phase space $(t, \mathbf{x}, \mathbf{v}, I) \in \mathbb{R}^+ \times \mathbb{R}^3 \times \mathbb{R}^3 \times \mathbb{R}^+$ such that, at any time $t \in \mathbb{R}^+$, $F(t, \mathbf{x}, \mathbf{v}, I)\, d\mathbf{x}\, d\mathbf{v} dI$ represents the mass of gas in the volume $d\mathbf{x}\, d\mathbf{v} dI$ centred at the spatial point $\mathbf{x}$, the particle velocity $\mathbf{v}$ and the internal energy $\epsilon(I) = I^{2/\delta}$. The number of internal degrees of freedom δ is typically set to 2 for diatomic molecules at sufficiently low temperature to enable only the rotational mode of energy.

The densities of mass ρ, momentum $\rho\mathbf{u}$ (where $\mathbf{u}$ is the macroscopic flow velocity), and total energy E, which are macroscopic quantities depending on space and time only, are recovered as velocity and internal energy moments of the microscopic distribution:

$$\rho = \langle\langle F \rangle\rangle, \quad \rho\mathbf{u} = \langle\langle \mathbf{v} F \rangle\rangle, \quad E = \left\langle\left\langle \left(\frac{1}{2}|\mathbf{v}|^2 F + \epsilon(I)F\right)\right\rangle\right\rangle, \tag{1}$$

with $\langle\langle \phi \rangle\rangle = \int_{\mathbb{R}^3} \int_{\mathbb{R}^+} \phi dI d\mathbf{v}$ for any distribution $\phi(\mathbf{v}, I)$. We also define for further needs: $\langle \psi \rangle = \int_{\mathbb{R}^3} \psi d\mathbf{v}$ for any distribution $\psi(\mathbf{v})$. Moreover, additional quantities, such as the modal temperatures, can be derived as follows:

$$E_c = \frac{1}{2}\rho|\mathbf{u}|^2, \quad E_{tr} = \frac{3}{2}\rho R_s T_{tr}, \quad E_{int} = \frac{\delta}{2}\rho R_s T_{int} . \tag{2}$$

In these expressions, R_s is the specific constant of the gas. The subscripts tr and int refer to the translational and internal modes of energy. Finally, T_{eq} is the equilibrium temperature associated with both translational and internal modes.

The distribution F is governed by the following equation [1]:

$$\partial_t F + \mathbf{v} \cdot \nabla_\mathbf{x} F = \frac{1}{\tau}(G[F] - F), \tag{3}$$

where the anisotropic equilibrium state G is defined as the product of G_{tr} and G_{int}, expressed as follows:

$$G_{\mathrm{tr}}(\mathbf{v}) = \frac{\rho}{\sqrt{\det(2\pi\,\mathsf{T})}}\exp\left(-\frac{1}{2}(\mathbf{v}-\mathbf{u})\cdot\mathsf{T}^{-1}\cdot(\mathbf{v}-\mathbf{u})\right),\tag{4}$$

$$G_{\mathrm{int}}(I) = \frac{\Lambda_\delta}{(R_s T_{\mathrm{int}}^{\mathrm{rel}})^{\delta/2}}\exp\left(-\frac{\epsilon(I)}{R_s T_{\mathrm{int}}^{\mathrm{rel}}}\right).\tag{5}$$

Here, $\Lambda_\delta^{-1} = \int_{\mathbb{R}^+} e^{-\epsilon(I)}dI$ is constant and $T_{\mathrm{int}}^{\mathrm{rel}}$ is the relaxation internal temperature. The relaxation temperature tensor T/R_s is related to the anisotropic tensor of temperature Θ/R_s, and the relaxation translational temperature $T_{\mathrm{tr}}^{\mathrm{rel}}$ by the following expressions:

$$\mathsf{T} = R_s T_{\mathrm{tr}}^{\mathrm{rel}} + \left(1 - \frac{1}{\mathrm{Pr}}\right)[\Theta - R_s T_{\mathrm{tr}}],$$

$$\Theta = \frac{1}{\rho}\langle\langle(\mathbf{v}-\mathbf{u})\otimes(\mathbf{v}-\mathbf{u})F\rangle\rangle,$$

$$T_{\mathrm{tr}}^{\mathrm{rel}} = T_{\mathrm{tr}} + \frac{1}{Z_{\mathrm{int}}}(T_{\mathrm{eq}} - T_{\mathrm{tr}}),\quad T_{\mathrm{int}}^{\mathrm{rel}} = T_{\mathrm{int}} + \frac{1}{Z_{\mathrm{int}}}(T_{\mathrm{eq}} - T_{\mathrm{int}}).$$

In these relations, Pr denotes the Prandtl number, and Z_{int} corresponds to the average number of collisions required to realise an energy exchange between equilibrium and internal modes. Finally, the characteristic relaxation time τ is related to the fluid viscosity μ and the pressure $p = \rho R_s T_{\mathrm{tr}}$ by $\tau = \frac{\mu}{\mathrm{Pr}\,p}$.

With parameters $2/3 \le \mathrm{Pr}$ and $1 < Z_{\mathrm{int}}$, this model has been proven in [1] to conserve mass, momentum, and total energy, satisfy the H-theorem, give the correct Prandtl number, and admit the Maxwell-Boltzmann distribution $M(\mathbf{v}, I)$ below as full equilibrium:

$$M(\mathbf{v}, I) = \frac{\rho}{(2\pi R_s T_{\mathrm{eq}})^{3/2}}\exp\left(-\frac{|\mathbf{v}-\mathbf{u}|^2}{2R_s T_{\mathrm{eq}}}\right)\frac{\Lambda_\delta}{(R_s T_{\mathrm{eq}})^{\delta/2}}\exp\left(-\frac{\epsilon(I)}{R_s T_{\mathrm{eq}}}\right).\tag{6}$$

3 The Unified-Gas Kinetic Scheme (UGKS)

The numerical method will be presented in a 1D spatial framework for simplicity, although it can be extended to 2D or 3D. We begin by outlining the framework before constructing the UGKS for the ES-BGK model introduced by [1].

3.1 A Discrete-Velocity-Model Finite-Volume Framework

The ES-BGK model is an integro-differential equation expressed in a advection-relaxation form, which makes the finite volume framework intrinsically well suited.

In this context, all phases of the equation will be discretised. In that sense, the time space $\mathbb{R}^+$ and the physical space $\mathbb{R}$ are divided into intervals $([t^n, t^{n+1}])_{n \in \mathbb{N}}$ and $([x_{i-\frac{1}{2}}, x_{i+\frac{1}{2}}])_{i \in \mathbb{Z}}$, respectively. For simplicity, the spatial interval length will be constant, denoted as $\Delta x = x_{i+\frac{1}{2}} - x_{i-\frac{1}{2}}$. Finally, following the methodology of Discrete Velocity Models (DVM), we consider a finite velocity set $\mathcal{V} \subset \mathbb{R}^3$ and a quadrature rule $\langle \rangle_{\mathcal{V}}$ that enables the computation of the moments of the set of microscopic distributions.

We introduce the distribution families $(F_{i,k}^n(\cdot))_{n \in \mathbb{N}, i \in \mathbb{Z}, k \leq |\mathcal{V}|}$ and $(G_{i,k}^n(\cdot))_{n \in \mathbb{N}, i \in \mathbb{Z}, k \leq |\mathcal{V}|}$ defined as the mean value of F and G on a spatial cell $[x_{i-\frac{1}{2}}, x_{i+\frac{1}{2}}]$ at time t^n, and velocity near $\mathbf{v}_k$. All of these considerations lead to the classical finite volume formulation:

$$F_{i,k}^{n+1}(I) - F_{i,k}^n(I) = -\frac{\Delta t}{\Delta x} \left[\phi_{i+\frac{1}{2},k}^n - \phi_{i-\frac{1}{2},k}^n \right](I)$$

$$+ \int_{t^n}^{t^{n+1}} \int_{x_{i-\frac{1}{2}}}^{x_{i+\frac{1}{2}}} \frac{G - F}{\tau}(t, x, \mathbf{v}_k, I)\,dx\,dt, \tag{7}$$

$$\phi_{i+\frac{1}{2},k}^n(I) = \frac{1}{\Delta t} \int_{t^n}^{t^{n+1}} \mathbf{v}_k F(t, x_{i+\frac{1}{2}}, \mathbf{v}_k, I)\,dt\ . \tag{8}$$

Finally, for further needs, we also introduce the moments $(\mathbf{W}_i^n)_{n \in \mathbb{N}, i \in \mathbb{Z}}$ of $(F_{i,k}^n(\cdot))_{n \in \mathbb{N}, i \in \mathbb{Z}, k \leq |\mathcal{V}|}$, where $\langle\langle \cdot \rangle\rangle_{\mathcal{V}}$ combines a continuous integration over the internal energy phase and the chosen quadrature rule on the velocity phase:

$$\mathbf{W}_i^n = \langle\langle \mathbf{m}_k(I) F_{i,k}^n(I) \rangle\rangle_{\mathcal{V}}, \quad \mathbf{m}_k(I) = (1 \ \ \mathbf{v}_k \ \ \mathbf{v}_k \otimes \mathbf{v}_k \ \ \epsilon(I))^T\ . \tag{9}$$

3.2 The Construction of the UGKS

A Second Order Multi-scale Formulation of the Flux Part of the Numerical Scheme The key idea of the UGKS [10] is to use the characteristics method on the entire model Eq. (3) to express the evolution of the distribution F at cell interface position. Assuming τ is constant over $[t^n, t]$, we get:

$$F(t, \mathbf{x}_{i+\frac{1}{2}}, \mathbf{v}_k, I) = \ \exp\left(-\frac{t - t^n}{\tau}\right) F(t^n, \mathbf{x}_{i+\frac{1}{2}} - \mathbf{v}_k(t - t^n), \mathbf{v}_k, I) \tag{10}$$

$$+ \int_{t^n}^{t} \exp\left(-\frac{t - s}{\tau}\right) \frac{1}{\tau} G(s, \mathbf{x}_{i+\frac{1}{2}} - \mathbf{v}_k(t - s), \mathbf{v}_k, I)\,ds\ .$$

For accuracy, the continuous distributions F and G are approximated using linear reconstructions based on discretised distributions $(F_{i,k}^n(\cdot))$ and $(G_{i,k}^n(\cdot))$. Classically [10], reconstructions for F and G are performed at each cell interface involving discrete approximations of partial derivatives $\overline{\partial_x F}, \overline{G}, \overline{\partial_x G},$ and $\overline{\partial_t G}$. The construction of $\overline{\partial_x F}$, moments $\mathbf{W}^n_{i+\frac{1}{2}}$, and the term $\overline{G}$ is standard [10].

A Major Replacement in the Discrete Derivative Terms The key idea employed here to ensure the proper behaviour of the scheme is to replace the Gaussian derivatives with those of the corresponding Maxwellian distribution.

First, note that the derivatives of G and M are close to within a error of the order of τ. This is a classical result from Chapman-Enskog analysis (see, e.g., [1]). Thus, for a large time step $\Delta t \gg \tau$, the average time derivative of G closely matches that of M, with an error depending on $\tau \ll \Delta t$.

Second, the discrete time derivative contribution in the numerical fluxes is mainly important in the continuum regime. An approximation of $O(\tau)$ on the derivative of equilibrium would not break down the Navier-Stokes asymptotic since it is a convenient replacement in the Chapmann-Enskog expansion.

The last and deeper reason is stability. In the continuum regime, numerical fluxes achieve a second-order time scheme by approximating indirectly the equilibrium at $t^n + \Delta t/2$. This approach resembles a forward Euler method based on the time derivative at t^n. However, since the relaxation time of the Gaussian state G toward the Maxwellian M scales with τ, using a time step $\Delta t/2 \gg \tau$ may over-relax G, potentially leading to negative distribution values

Finally, this approach was used for the monoatomic ES-BGK model in [3] and for the Shakhov model in [11]. Since the terms $\overline{\partial_x M}$ an $\overline{\partial_t M}$ are classic in the UGKS framework [10], their constructions are not presented here.

Conclusion on the Flux Part of the Scheme Injecting the reconstructions in the numerical flux definition (8) leads to expression of the numerical flux ϕ :

$$\phi^n_{i+\frac{1}{2},k}(I) = q_1 \overline{G}^n_{i+\frac{1}{2},k}(I) + q_2 v_{kx} \overline{\partial_x M}^n_{i+\frac{1}{2},k}(I) + q_3 \overline{\partial_t M}^n_{i+\frac{1}{2},k}(I)$$

$$+ q_4 \overline{F}^n_{i+\frac{1}{2},k}(I) + q_5 v_{kx} \overline{\partial_x F}^n_{i+\frac{1}{2},k}(I) . \tag{11}$$

The coefficients $(q_i)_{1 \le i \le 5}$ determine the behaviour of the scheme and depend locally on Δt and $\tau^n_{i+\frac{1}{2}}$ [10].

Relaxation Part of the Scheme A Crank-Nicolson based approximation is used for the relaxation part [10]. The numerical scheme is therefore:

$$F^{n+1}_{i,k}(I) = F^n_{i,k}(I) - \frac{\Delta t}{\Delta x} \left[\phi^n_{i+\frac{1}{2},k} - \phi^n_{i-\frac{1}{2},k} \right](I)$$

$$+ \frac{\Delta t}{2} \left[\left(\frac{G-F}{\tau} \right)^n_{i,k} + \left(\frac{G-F}{\tau} \right)^{n+1}_{i,k} \right](I) . \tag{12}$$

Classically, the equilibrium state at time t^{n+1} is computed based on the moments of F, which are obtained from the moments of the microscopic scheme (12) [6, 10].

This accurate technique could lead to a non-symmetric positive-definite tensor Θ and could compromise the calculation of an equilibrium G^{n+1}. This is why we instead use a backward Euler relaxation term for the macroscopic scheme:

$$\mathbf{W}_i^{n+1} = \mathbf{W}_i^n - \frac{\Delta t}{\Delta x}\left[\Phi_{i+\frac{1}{2}}^n - \Phi_{i-\frac{1}{2}}^n\right] + \Delta t \left(\frac{\mathbf{V} - \mathbf{W}}{\tau}\right)_i^{n+1}, \tag{13}$$

$$\Phi = \langle\langle \mathbf{m}\phi \rangle\rangle_{\mathcal{V}}, \quad \mathbf{V} = \langle\langle \mathbf{m}G \rangle\rangle_{\mathcal{V}}. \tag{14}$$

4 Numerical Results

To assess the effectiveness of incorporating the collision process into the characteristic method (10) used to define numerical fluxes, we introduce the KO2 scheme, which is formulated without these relaxation effects. This scheme is obtained from (11) by fixing $q_{1,2,3}$ to 0, and q_4 and q_5 to 1 and $-\Delta t/2$, respectively.

For faster simulations, these numerical schemes are combined with the reduced distribution technique: by integrating with respect to the energy variable I as well as the velocity components v_z and/or v_y in 1D and 2D problems, it permits to reduce the computational cost, without any approximation (see [1]).

4.1 Couette Flow

In the Couette configuration, the gas flows between two parallel, isothermal, infinite plates separated by a distance L. One plate is stationary, while the other moves with a finite velocity u_{w} in the y-direction. Under the continuum approximation, with no slip boundary conditions, the compressible Navier-Stokes-Fourier equations (CNS) yield to:

$$u_y(x) = \frac{x}{L}u_{\mathrm{w}}, \quad T(x) = T_{\mathrm{w}} + \frac{1}{2}\frac{\mathrm{Pr}}{c_{\mathrm{p}}}\frac{x(L-x)}{L^2}u_{\mathrm{w}}^2, \quad p(x) = cst. \tag{15}$$

The Knudsen number of the flow is here defined as: $\mathrm{Kn} = \lambda/L$, where λ is the mean free path, that can be defined as $\lambda = \tau v_{\mathrm{th}}$ at the boundary, where $v_{\mathrm{th}} = \sqrt{R_{\mathrm{s}}T_{\mathrm{w}}}$ is the thermal velocity at the wall temperature T_{w}. Simulations are conducted in near continuum regime, at different Knudsen numbers by varying the initial mass density of the flow.

The parameters defining the gas are taken constant as follows: the specific constant R_{s} is $296.8\,\mathrm{J\,kg^{-1}\,K^{-1}}$, the viscosity μ is $1.656 \times 10^{-5}\,\mathrm{Pa\,s}$, the Prandtl number Pr is 0.71, the degree of freedom number δ is 2.0, and the average number Z_{int} is set to 5.0. Furthermore, the boundary walls are separated by a distance $L = 1\,\text{m}$, are maintained at $273\,\text{K}$ and the wall velocity u_w is set to $300\,\text{m s}^{-1}$.

The spatial domain is discretised exclusively in the x-direction using 25 uniform cells. The velocity grid is Cartesian, defined over the range $[\pm 1200] \times [-1200, +1500]\,\text{m s}^{-1}$, with 50 velocity points in both the x- and y-directions.

Figure 1 presents comparisons of the UGKS and KO2 schemes with the CNS solutions at different Knudsen numbers. Simulations with the KO2 scheme and a finer mesh have been conducted to highlight the deviation of the CNS solution from the rarefied flow solutions caused by the boundary conditions.

First, Fig. 1 illustrates that both UGKS and KO2 can capture near-continuum flows. However, UGKS achieves higher accuracy on a same mesh, while KO2 requires a significantly finer mesh to match UGKS accuracy. Additionally, as the Knudsen number decreases, predictions deviate only for KO2.

Second, using the temperature Eq. (15), the Prandtl number can be roughly evaluated through a parabolic regression using the least squares method. The estimations on Fig. 1 shows the excellent capability of UGKS to simulate a correct Prandtl number for a given mesh, compared to the KO2 scheme.

4.2 Flow Passing a Flat Plate

This second case reproduces the experiment conducted by [8], where a flat plate is immersed in a supersonic near-continuum nitrogen cold flow. The internal energy of the gas is entirely associated with the rotational mode. Therefore, the ES-BGK model from [1] with parameter $\delta = 2$ is appropriate.

The gas is modelled according to the VSS model, with parameters taken from [2]. The spatial domain is meshed using a 2D structured mesh all around the plate. The velocity grid is bounded within $[-1200, 2200] \times [-1700, 1500]\,\text{m s}^{-1}$ and discretised with 25 points in both x- and y-directions.

Comparisons between the simulation and experimental measurements of rotational temperature, taken at 5 and 20 mm from the stagnation point on vertical lines, are presented in Fig. 2. The shapes of the rotational temperature are well reproduced, although some discrepancies arise due to uncertainties in modelling the real relaxation processes involved. To align the relaxation processes used by Bird [2] or Mathiaud et al. [7], a simple adjustment of Z_int can be achieved as $Z'_\text{int} = \tau_C Z_\text{int}/\tau$ or $Z''_\text{int} = \frac{3}{5}\tau_C Z_\text{int}/\tau$, respectively, where τ_C is the mean collision time defined in the VSS model [2]. These modifications significantly influence the profiles of rotational temperature, as illustrated in Fig. 2, even though the results are in a general good agreement with the experimental measurements.

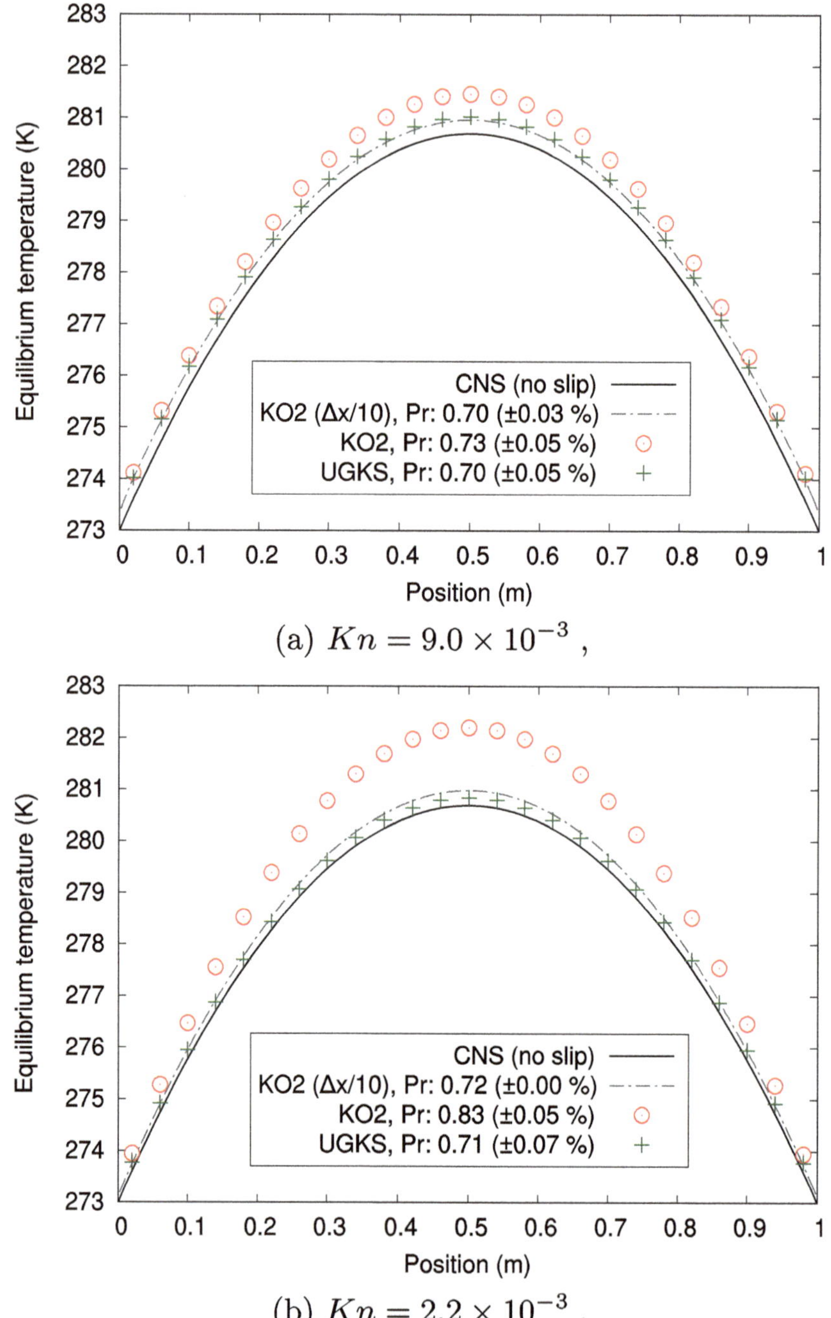

(a) $Kn = 9.0 \times 10^{-3}$,

(b) $Kn = 2.2 \times 10^{-3}$,

Fig. 1 Comparison of UGKS and KO2 schemes with CNS solution in 1D Couette flow

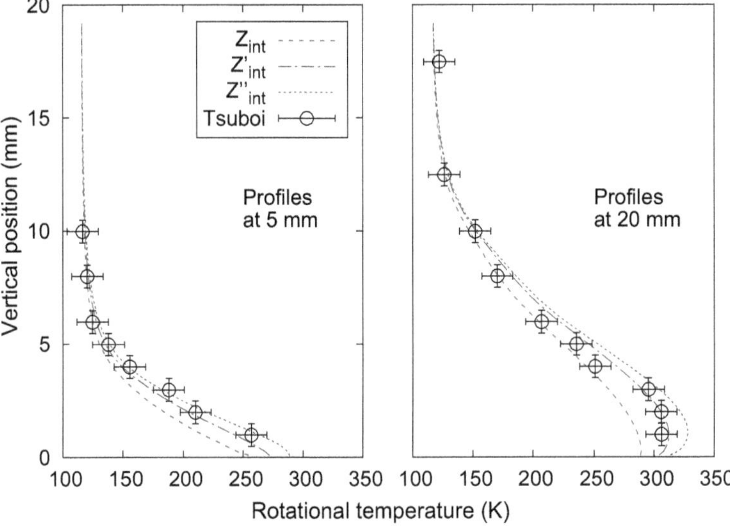

Fig. 2 Comparison of rotational temperature predictions between ES-BGK models at 5 and 20 mm after the stagnation point (*Diffuse wall reflection*)

5 Conclusion

We proposed an extension of the Unified Gas-Kinetic Scheme (UGKS) to an ES-BGK model that accounts for internal energy modes. This adaptation extends the UGKS-ES-BGK framework from monoatomic to diatomic gases using techniques similar to those applied in extending UGKS-BGK to Shakhov and Rykov models. The method shows good agreements both for a 1D viscous-driven flow and an experimental 2D hypersonic flow. Consequently, future work will address the incorporation of vibrational effects within the framework of ES-BGK models.

References

1. Andriès P, Le Tallec P, Perlat J-P, Perthame B (2000) The Gaussian-BGK model of Boltzmann equation with small Prandtl number. Eur J Mechan B/Fluids 19(6):813–830
2. Bird GA (1994) Molecular gas dynamics and the direct simulation of gas flows. Oxford Science Publications, Oxford
3. Chen S, Xu K, Cai Q (2015) A comparison and unification of ellipsoidal statistical and Shakhov BGK models. Adv Appl Math Mechan 7(2):245–266
4. Dauvois Y, Mathiaud J, Mieussens L (2021) An ES-BGK model for polyatomic gases in rotational and vibrational nonequilibrium. Eur J Mechan B/Fluids 88:1–16
5. Holway LH (1965) Kinetic theory of shock structure using an ellipsoidal distribution function. Rarefied Gas Dyn 1:193–215

6. Liu S, Zhong C (2014) Investigation of the kinetic model equations. Phys Rev E 89(3):033306–033322
7. Mathiaud J, Mieussens L, Pfeiffer M (2022) An ES-BGK model for diatomic gases with correct relaxation rates for internal energies. Eur J Mechan B/Fluids 96:65–77
8. Tsuboi N, Matsumoto Y (2005) Experimental and numerical study of hypersonic rarefied gas flow over flat plates. AIAA J 43(6):1243–1255
9. Xu K (2021) A Unified Computational Fluid Dynamics Framework from Rarefied to Continuum Regimes. Cambridge University Press, Cambridge
10. Xu K, Huang J-C (2010) A unified gas-kinetic scheme for continuum and rarefied flows. J Comput Phys 229(20):7747–7764
11. Xu K, Huang J-C (2011) An improved unified gas-kinetic scheme and the study of shock structures. IMA J Appl Math 76(5):698–711

Spatially-Parallel Collision Scheme for Implementing Direct Simulation Monte Carlo in Field-Programmable Gate Arrays

Saleen Bhattarai, Sean O' Byrne, Edwin Peters, and David Petty

Abstract Direct simulation Monte Carlo is a numerical method to simulate gas flows, leveraging some embarrassingly parallel steps like random number generation, advection, and sampling. Field-programmable gate arrays—devices with reconfigurable logic—are well suited for these tasks. Efficiently parallelising the commonly used no-time-counter collision evaluation scheme on these devices, however, introduces challenges, particularly repeated particle selections and race conditions during simultaneous post-collision velocity updates. These issues reduce the collision frequency and compromise accuracy. We address this with a modified scheme that splits and shuffles particle indices within a cell, enabling fine-grained, highly parallel collision evaluations without repeated particles or pairs and a preservation of collision frequency while still realising coarse-grained cell-based spatial parallelism within the programmable logic of a single device. We demonstrate the method on a one-dimensional heat transfer problem, achieving accurate results comparable to standard no-time-counter-based implementations.

Keywords DSMC · Spatial parallelism · FPGA acceleration

S. Bhattarai (✉) · E. Peters · D. Petty
The University of New South Wales Canberra, Campbell, ACT, Australia
e-mail: saleen.bhattarai@unsw.edu.au

E. Peters
e-mail: edwin.peters@unsw.edu.au

D. Petty
e-mail: d.petty@unsw.edu.au

S. O' Byrne
The Australian National University, Acton, ACT, Australia
e-mail: sean.obyrne@anu.edu.au

M. Grabe et al. (eds.), *Rarefied Gas Dynamics*, Springer Aerospace Technology,
https://doi.org/10.1007/978-3-032-00094-1_38

1   Introduction

Direct simulation Monte Carlo (DSMC) [4] is a stochastic particle-based algorithm for simulating rarefied gas flows. It has a wide range of applications from hypersonics [10] to micro-scale flows [9]. Advancements in hardware have enabled large-scale DSMC simulations, often leveraging heterogeneous architectures that combine CPUs and GPUs for efficient parallelisation [7], but at the cost of using more computational resources than continuum methods [10].

In contrast, field-programmable gate arrays (FPGAs) offer reconfigurable logic that can be tailored to specific tasks, allowing custom-width data pathways, continuous data flow every clock cycle without the overhead of operating systems and fixed-width buses, and potential for parallelism. This degree of control can provide higher power efficiency and better performance than CPUs or GPUs [11]. The limited size of current FPGAs may make them impractical for large-scale DSMC simulations. However, micro-scale flows, where the bulk velocity is low and only a few particles and collision cells are needed – yet demand significant ensemble averaging – are therefore more suitable for FPGA implementation.

A critical step in DSMC involves simulating binary collisions. While CPUs or GPUs manage collision schemes through serialised steps, attempting parallel collision evaluation on FPGAs can lead to race conditions. This paper presents a novel method to eliminate these issues, ensuring accurate and efficient FPGA-based parallel collision evaluations. Traditional DSMC codes use message passing interface (MPI) for spatial parallelism, partitioning the domain into large subdomains across ranks that each still perform collision tests in a loop (possibly via threads). In contrast, our FPGA scheme instantiates one collision kernel per cell (multiple of which can fit inside a single FPGA chip) and further replicates pair-test/velocity-update logic within each kernel so that all collisions in a cell execute truly concurrently on-chip.

2   Spatial Parallelism and the DSMC Algorithm

FPGAs excel as computational accelerators due to inherent parallelism, offering both task and spatial parallelism. Task parallelism allows different algorithm stages to be executed in a pipelined manner, reducing the overall runtime. Spatial parallelism enables replication of a task multiple times at different locations within the FPGA, maximising computational throughput. These methods rely on FPGA resources such as look-up tables (LUT), flip flops (FF), block RAMs (BRAM), and digital signal processor (DSP) blocks.

The DSMC algorithm uses collision cells to locally evaluate binary elastic collisions. Thus, an ideal FPGA implementation of the DSMC algorithm would include a collision kernel for each cell that is spatially replicated within the resource constraints of the FPGA logic. This would be equivalent to performing a domain decomposition at the collision cell level. We aim to go a step further by proposing a spatially parallel evaluation of all collisions within a single collision cell every time step. Details on the FPGA implementation can be found in [3].

The DSMC algorithm [4] has three major phases: advection of particles, simulating particle collisions, and sampling of macroscopic variables from the particle velocity distribution within a cell. The advection and sampling phases are embarrassingly parallel over all particles due to the particle-specific independence in the nature of these calculations. On the contrary, the collision phase is evaluated stochastically in DSMC and is not embarrassingly parallel. The process has two stages: first, potential collision partners within a collision cell are identified using the no-time-counter (NTC) method [4], followed by collision testing and updating the post-collision velocities for the accepted pairs.

3 Limitations Preventing Spatial Parallelism on FPGAs

In many parallel implementations of the NTC method, potential collision pairs are generated randomly. This means that either the same pair (type A) or the same particle (type B) can be selected multiple times. Previous studies [1] have tried to quantify the effects of successive type A pairs in DSMC simulations showing that their presence tends to reduce the collision frequency affecting the distribution of properties of the particles in the collision cells. Alternative methods [8] have been proposed to avoid the generation of such pairs.

Type B pairs are permissible in conventional hardware (CPUs or GPUs) with sequential calculation of collisions as subsequent interactions will use the correct intermediate particle dynamics. This is because even if the same particle is selected twice and accepted both times, the post-collision velocity updates happen at different times. Spatially parallel evaluation of these collisions (involving both type A and B pairs) presents a race condition where multiple post-collision velocity updates may occur simultaneously. This prevents the implementation of a completely spatially parallel version of the collision kernel using the NTC method on an FPGA. An earlier attempt [3] at spatially parallelising the collisions achieved this by generating unique pairs. However, until now, no attempt has been made to discuss the validity of results obtained from computing DSMC with spatial parallelism that involves an absence of type B pairs.

4 Methodology for Resolving the Limitations

An early collision scheme for vectorised DSMC implemented on the Cray supercomputer [2] tackled a similar problem. The scheme avoided race conditions by employing fixed collision pairs followed by a lengthening of the local particle index array with a simple pairing strategy based on a cyclically selected first particle paired with another offset by a fixed gap. Our approach to resolve the limitations highlighted in the previous section is also to deterministically select collision pairs, particularly in a way that prevents the occurrence of successive type A pairs. However, unlike a simple cyclical pairing strategy of [2], pseudo-randomness is introduced into the simulation in a post-collision step by shuffling the particles referenced by the local particle indices. Lastly, batched repetitions of the entire set of collisions are performed to introduce type B pairs into the simulation. The algorithm (termed as half-split-shuffle or HSS) is as follows:

1. Generate a local particle index list for each collision cell every time step.
2. Pair the particles referenced by the first half of the index list with the second half. All pairings are unique and without any type A or type B pairs. Thus, the number of potential collision pairs per batch is given by:

$$N_{\text{sel}} \leq 0.5 N_{\text{p}} \tag{1}$$

3. Collide all N_{sel} pairs at once using the following probability:

$$P_{\text{col, } ab} = \frac{1}{N_{\text{batch}}} N_{\text{x}} N_{\text{eff}} (\sigma c_r)_{ab} \frac{\Delta t}{V_{\text{cell}}} \tag{2}$$

$$N_{\text{x}} = \begin{cases} N_{\text{p}}, & \text{when } N_{\text{p}} \text{ is odd} \\ N_{\text{p}} - 1, & \text{when } N_{\text{p}} \text{ is even} \end{cases} \tag{3}$$

where N_{batch} = number of batched repetitions, N_{eff} = number of molecules represented by a simulated particle, $(\sigma c_r)_{ab}$ = product between collision cross section (σ) and relative velocity (c_r) of particles (a, b), Δt = time step, and V_{cell} = volume of the cell. Collision occurs if $P_{\text{col, } ab} \geq R$ where R is a uniformly distributed random number between 0 and 1.

4. Calculate and update the post-collision velocities (V'_a, V'_b) of all colliding pairs in parallel using the following equations in a step termed a batch:

$$V'_{r\,ab} = V_{r\,ab} [\cos \chi, \quad \sin \chi \cos \phi, \quad \sin \chi \sin \phi], \tag{4}$$

$$V'_a \text{ or } V'_b = V_{\text{CM}} \pm V'_{r\,ab}/2, \tag{5}$$

where V_{CM} = centre-of-mass velocity of the pairs and (χ, ϕ) = scattering angles obtained using different Rs.

5. Shuffle the local particle index list using a Fisher-Yates algorithm [6].
6. Repeat steps 3 and 4 for a specified number of batches (N_{batch}).

Compared to the earlier scheme, the proposed HSS method applies a more robust shuffle to ensure randomness across batches. By design, it eliminates type A pairs (equivalent to generating unique pairs i.e. selecting two without replacement from a pool of N_{p} particles) and remains highly spatially parallelisable, with minimal FPGA resource usage because it does not require any complex logic to keep track of type A and type B pairs. The proposed method also ensures that type B pairs, while being eliminated from a single batch to allow effective spatial parallelism, are still introduced through batched repetitions. A consequence of this method is that there is a requirement to ensure that $P_{\text{col, }ab} \leq 1$ by appropriately choosing Δt and V_{cell}.

To analyse the proposed method, we compare its results against an implementation of the standard NTC algorithm for a benchmark Fourier heat transfer problem as outlined in [1]. We consider Argon gas (molecule mass $m = 6.63 \times 10^{-26}$ kg, molecule diameter $d = 3.66 \times 10^{-10}$ m, initial temperature $T = 273$ K, and initial density $\rho = 1.78$ kg/m^3) with a hard-sphere collision model. The length of the domain is 10^{-3} m in the y-direction bounded by two thermal walls with perfect accommodation: a hot wall at the top and a cold wall at the bottom with a temperature difference of 100 K each with an area A. The walls are separated (Knudsen number 0.024) by a varying number of collision cells N_c. We observe the distribution of the temperature and the number of accepted collisions in the domain. We also calculate the wall heat flux q by tracking the change in kinetic energy of molecules impacting the wall as follows:

$$q = \frac{m N_{\text{eff}}}{A \Delta t} \sum \left(v_i^2 - v_r^2 \right), \ \text{W/m}^2 \tag{6}$$

In addition, we consider the effects of the proposed HSS technique compared to the NTC method on more subtle parameters like the probability distribution of the relative velocity of accepted collisions. Finally, we observe the effects of the number of batched repetitions on our simulation results.

The method was implemented on a Versal VPK120 FPGA (containing approximately 10^6 LUTs, 2×10^6 FFs, 4000 DSPs, and 1350 BRAMs) using the Vitis HLS 2023.4 tool. Each collision kernel consumed 7093 LUTs (0.79 %), 3601 FFs (0.20 %), 49 DSPs (1.23 %), and 7 BRAMs (0.52 %). It is evident that DSPs are the constraining resource which limits theoretical scalability to 81 collision kernels. In order to prevent routing congestion in the FPGA, a total of 50 collision kernels were implemented. All results obtained using the current method were obtained from this implementation. It should be noted that for obtaining the results for cases with $N_c > 50$, the computation was carried out in groups of 50 cells at a time.

5 Results and Discussion

A detailed comparison between the results obtained for the Fourier heat transfer problem using the NTC and the current method is shown in Fig. 1. Qualitative agreement between the temperature profiles obtained using the two methods is shown in Fig. 1a. The degree of agreement between the two results is supported by the maximum normalised difference in the temperature profile being less than 0.005 as shown in Fig. 1b, implying that these different collision evaluation techniques yield nominally equivalent macroscopic temperature profiles.

Moving from temperature to collision rate, Fig. 1c depicts the distribution of accepted collisions normalised by half the average number of particles in the cell (see Eq. 1). The figure shows that despite testing a larger but deterministic number

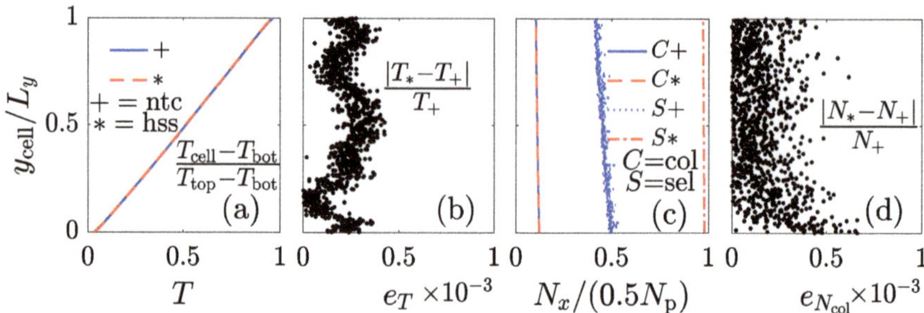

Fig. 1 Results: Fourier heat transfer problem for N_p=20, N_c=1000, $\Delta t/\tau$=10: **a** temperature, **b** difference in the temperature, **c** number of pair selections (sel) and accepted collisions (col), **d** difference in the accepted collisions

of potential collision pairs than the NTC method, the HSS method is able to replicate the number of accepted collisions, suggesting a preservation of collision frequency. Fig. 1d shows the normalised difference in the accepted collision distribution between the two methods with the maximum being less than 0.007, once again, showing that both methods are in close agreement with each other.

Fig. 2a illustrates the probability distribution of relative velocities of accepted collision pairs across several batches of simulations compared with the theoretical values and the results obtained using the NTC method. Each theoretical curve is constructed at a specific y-location using the value of temperature obtained at that location from Fig. 1a using the following relation [5]:

$$P_{\text{col, accept}}(c_r) = 2 \left(\frac{m_r}{2K_BT} \right)^2 c_r^3 \exp\left(-\frac{m_r c_r^2}{2K_BT} \right) \tag{7}$$

where m_r = the reduced mass of the pair and K_B = Boltzmann's constant. Each curve represents the probability distribution function $P(c_r)$ of relative velocities of accepted collision pairs in different y-locations (different colours) and for different numbers of batched evaluations (different line styles). The figure shows that as N_{batch} increases, the obtained distributions tend to converge towards the theoretical curve, highlighting the accuracy of the batched repetition approach. Fig. 2b shows the normalised mean error (with respect to the theoretical curve) as a function of the batch number. The plot indicates asymptotic convergence towards the theoretical result with increasing N_{batch}. The results from Figs. 1, 2a, b collectively highlight the reliability of using batched repetitions to capture both macroscopic and microscopic properties of the problem.

Figure 2c shows how the value of the wall heat flux (normalised by the values obtained in similar conditions using the NTC method) converges as N_{batch} is increased for different choices of N_c. The value of wall heat flux obtained from the HSS method converges to within a single percentage point of the value obtained from NTC after performing just two batched repetitions. It is interesting to note that

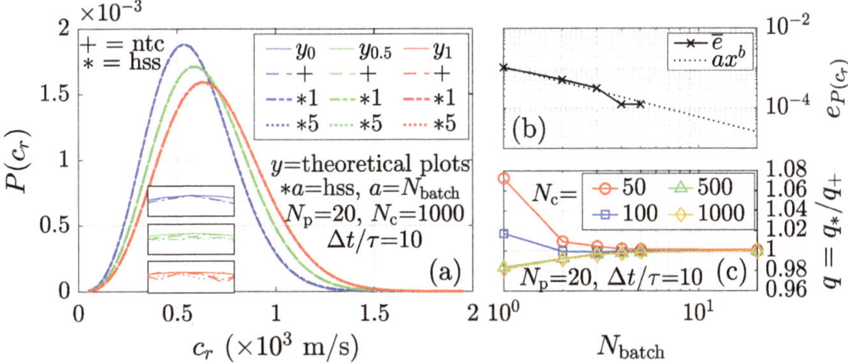

Fig. 2 **a** Distribution of the relative velocities of accepted collision pairs with batched repetitions, **b** error as a function of the batch number, **c** convergence in the value of wall heat flux with increasing batches

there is a large deviation in the results when performing only a single batch. This is explained by the deviation in the probability distributions for $N_{\text{batch}} = 1$ compared to the theoretical and NTC curves in Fig. 2a. The wall heat flux is a computationally sensitive quantity that directly depends on the number of particles, and their associated energy, that reach the wall. A small deviation in the distribution of the relative velocities of accepted collisions implies a small bias towards accepting pairs whose relative velocities only lie within a specific range. These biases are amplified when updating the post-collision velocities and materially alter the number of particles, and their associated energy, which interact with the bounding walls. As a consequence, the value of the wall heat flux is affected to a higher degree than the calculated temperature profile.

The reason why the probability distribution is different for a single batched HSS result compared to the same from NTC is because, by design, a single batch can never contain type B pairs to prevent race conditions as discussed earlier. This can be illustrated by keeping track of potential collision pairs as N_{batch} is increased and comparing that with the same from the NTC method. This has been demonstrated in Fig. 3. The graphs represent the distribution of the probability that a pair (a, b) is selected. Each box represents a pair chosen once or more during the simulation, while its colour represents the probability of that pair being selected. The lower plots represent the method of choosing pairs randomly with no restriction on either repeated particles or repeated pairs (equivalent to NTC). The upper plots represent the method of selecting unique pairs (equivalent to HSS). For a single repetition ($N_{\text{batch}} = 1$), we find that the lower plot will likely have boxes adjacent to each other or boxes sharing rows or columns, which represent type B pairs. The upper plots, for a single repetition, will never have such occurrences. This difference is enough to bias the pair selection such that it can affect the wall heat flux. The heat flux value obtained by HSS converges to the NTC value by introducing type B pairs with several repetitions of HSS. A key point to note here is that after the second set of repetitions,

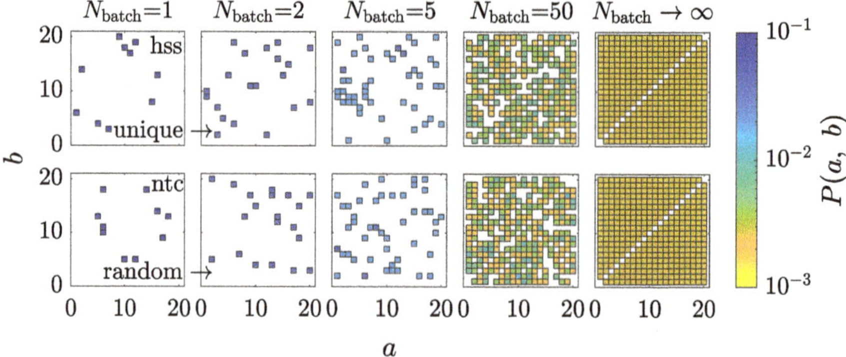

Fig. 3 Comparison of collision pair selection schemes ($N_\mathrm{p} = 20$, $N_\mathrm{sel} = 0.5N_\mathrm{p}$)

the behaviour of both methods (in regard to the pair selection pattern) is no different from one another. The number of batches required to achieve this convergence is much less that the number of accepted collisions that must be sequentially calculated using NTC. The probability of a pair being selected by the HSS method becomes equivalent to any other pair in the limit of infinite batching.

6 Conclusion

This paper has presented a spatially-parallel collision scheme for implementing DSMC on FPGAs. The proposed HSS method addresses the limitations preventing parallel evaluation of collisions on an FPGA by eliminating type A pairs and introducing type B pairs through batched repetitions. The scheme preserves macroscopic and microscopic properties of interest in DSMC simulations as verified through a benchmark Fourier heat transfer problem. Results indicate that the temperature distribution and the number of accepted collisions obtained using this method agree closely with those obtained using the NTC method. The distribution of the relative velocities of accepted collision pairs converges towards theoretical values as the number of batches increases. While performing only a single batch results in significant deviations in the wall heat flux due to the absence of type B pairs, convergence to within 1% of the NTC values is achieved after just two batched repetitions. This is supported by an analysis of the probability distribution of potential collision pair selection which shows that the behaviour of both methods, with the pattern in which particle pairs are selected, becomes indistinguishable after the second set of repetitions. By combining an intra-cell fine-grained FPGA parallelism with a coarse-grained cell-based collision kernel hardware replication, this method opens new opportunities for accelerating micro-scale flow simulations with a small number of particles and collision cells. Future work will focus on quantifying the performance and energy efficiency benefits of this implementation against CPUs and GPUs.

Acknowledgements This work was supported by the Air Force Office of Scientific Research (23IOA106). We would like to thank Prof. Andrew Lambert and Dr. Jong-Leng Liow for their contributions to the discussion phase of this work.

References

1. Akhlaghi H, Roohi E, Stefanov S (2018) On the consequences of successively repeated collisions in no-time-counter collision scheme in DSMC. Comput Fluids 161:23–32
2. Baganoff D, McDonald JD (1990) A collision-selection rule for a particle simulation method suited to vector computers. Phys Fluids 2(7):1248–1259
3. Bhattarai S, Petty D, O'Byrne S, Peters E (2024) Parallel implementation of direct simulation monte carlo using field-programmable gate arrays. In: Australasian fluid mechanics conference. AFMS, Canberra
4. Bird GA (2013) The DSMC method, 2nd edn. CreateSpace, US
5. Boyd ID, Schwartzentruber TE (2017) Nonequilibrium gas dynamics and molecular simulation. Cambridge University Press, pp 323–328
6. Fisher RA, Yates F (1948) Statistical tables for biological, agricultural and medical research. Oliver & Boyd, 3 edn, pp 26–27
7. Hagleitner C, Diamantopoulos D, Ringlein B, Evangelinos C, Johns C, Chang RN, D'Amora B, Kahle JA, Sexton J, Johnston M, Pyzer-Knapp E, Ward C (2021) Heterogeneous computing systems for complex scientific discovery workflows. In: Design, automation and test in Europe conference and Exhibition. IEEE, Grenoble
8. Javani M, Roohi E, Stefanov S (2024) Symmetrized generalized and simplified Bernoulli-trials collision schemes in DSMC. Comput Fluids 272:106188
9. Mirnezhad N, Amiri-Jaghargh A, Qaderi A (2022) DSMC analysis of flow in Knudsen pumps with square-shaped wall roughness. AUT J Mech Eng 6(1):77–93
10. Plimpton SJ, Moore SG, Borner A, Stagg AK, Koehler TP, Torczynski JR, Gallis MA (2019) Direct simulation Monte Carlo on petaflop supercomputers and beyond. Phys Fluids 31(8):086101
11. Thomas DB, Howes L, Luk W (2009) A comparison of CPUs, GPUs, FPGAs, and massively parallel processor arrays for random number generation. International symposium on field programmable gate arrays. ACM, Monterey, pp 63–72

Higher-Order Meshfree Methods for BGK Equations with Moving Boundaries

Klaas Willems, Sudarshan Tiwari, and Axel Klar

Abstract We present a numerical method for simulating mixtures of rarefied gases that interact with moving boundaries and rigid bodies. The gas mixture is described by the general consistent BGK model for gas mixtures (Bobylev et al. in Kinetic Related Models 11(6):1377–1393, 2018) [1]. The BGK equations are solved using an Arbitrary Lagrangian-Eulerian method, in which grid points move with the local mean velocity of the gas (Tiwari et al. in J Comput Phys 458, 2022) [2]. The main advantage of the moving grid is that the algorithm can deal well with cases where the domain boundaries are time-dependent and the simulation domain contains rigid objects. Due to the irregular nature of the grid, we use a novel MUSCL-like Moving Least Squares Method (MLS) spatial discretization coupled with a second-order IMEX method. To avoid spurious oscillations at discontinuities, we use the so-called MOOD algorithm (Clain et al. in J Comput Phys 230(10):4028–4050, 2011) [3]. As a proof-of-concept, we apply the algorithm to Sod's shock tube and a moving boundary problem.

Keywords Multi-species BGK · Meshfree methods · MOOD

1 Introduction

In recent years, moving boundary problems for rarefied gas dynamics have been extensively investigated in connection with Micro-Electro-Mechanical-Systems (MEMS) [4]. In these applications, a low-Mach flow interacts with a small electronic sensor such as an accelerometer or micro engine. Because the mean free path of the gas is on the order of magnitude of the size of the MEMS, a kinetic description of the gas is required. Historically, simulations of rarefied flows were performed

K. Willems (✉) · S. Tiwari · A. Klar
Department of Mathematics, Rheinland-Pfälzische Universität Kaiserslautern-Landau, Kaiserslautern, Germany
e-mail: klaas.willems@math.rptu.de
URL: https://math.rptu.de/en/wgs/techno

M. Grabe et al. (eds.), *Rarefied Gas Dynamics*, Springer Aerospace Technology,
https://doi.org/10.1007/978-3-032-00094-1_39

411

with Direct Simulation Monte Carlo (DSMC). However, due to the transient, low-Mach nature of the flow, DSMC is not well-suited for these problems [2]. Another approach to modelling these problems is through deterministic methods in which the Boltzmann equation is replaced by the more simple BGK model. Several numerical methods for the BGK-equation adapted to moving boundaries have been presented, such as a semi-Lagrangian method [5] and a finite volume method [6]. See latter reference for a overview. In this text, we present a Lagrangian method in which the grid points move with the mean velocity of the gas. This method is easily extendable to higher dimensions and can keep track of the gas-obstacle boundary exactly.

This text is organised as follows. In Sect. 2, we introduce the model. In Sect. 3, we discuss the space, velocity and time discretization. Finally, in Sect. 4, we validate the method on two numerical examples.

2 The Multi-Species BGK Model

We consider a mixture of rarefied gases described by the general consistent BGK model from [1]. Each kinetic equation in said model describes the evolution of the population $f_s(x, v, t)$ of gas species s as a function of space $x \in \mathbb{R}^3$, microscopic velocity $v \in \mathbb{R}^3$, and time $t \in \mathbb{R}^+$

$$\frac{\partial f_s}{\partial t} + v \cdot \nabla_x f_s = \frac{1}{\tau} \sum_{k=1}^{L} v_{sk}(n_s M_{sk} - f_s), \quad s = 1 \ldots L . \tag{1}$$

The parameters $v_{sk} = \lambda_{sk} n_k$ are collision frequencies, and τ is the mean free path. The Maxwellians M_{sk} are defined as follows,

$$M_{sk} = \frac{1}{(2\pi R T_{sk})^{3/2}} \exp \frac{|v - u_{sk}|^2}{2 R T_{sk}}, \tag{2}$$

in which the pseudo-velocities u_{sk} and temperatures T_{sk} are defined such that the BGK relaxation operators have the same collision invariants as the associated multi-species Boltzmann collision operator. We define the number density n_s, mean velocity u_s and temperature T_s for species s as

$$n_s = \langle 1, f_s \rangle, \quad n_s u_s = \langle v, f_s \rangle, \quad 3 n_s R T_s = \langle |v - u_s|^2, f_s \rangle. \tag{3}$$

The macroscopic values for the whole mixture can then be obtained from

$$n = \sum_{s=1}^{L} n_s, \quad u = \frac{1}{\rho} \sum_{s=1}^{L} \rho_s u_s, \quad 3 n R T = 3 \sum_{s=1}^{L} n_s T_s + \sum_{s=1}^{L} n_s |u_s - u|^2. \tag{4}$$

The model features the same structure as the corresponding Boltzmann equations and fulfils all consistency requirements concerning conservation laws, equilibria and H-theorem. The hydrodynamical limits of this BGK model were formally derived in [7].

We intend to solve (1) in complex geometries with moving boundaries, while avoiding cumbersome algorithms such as a cut-cell method [6]. To this end, we write (1) in 'Lagrangian' form

$$\frac{\mathrm{d}x}{\mathrm{d}t} = u \tag{5}$$

$$\frac{\mathrm{d}f_s}{\mathrm{d}t} = -(v - u) \cdot \nabla_x f_s + \frac{1}{\tau} \sum_{k=1}^{L} v_{sk}(n_s M_{sk} - f_s), \quad s = 1 \ldots L. \tag{6}$$

In other words, we move grid points with the mean macroscopic velocity of the gas mixture. To account for this motion, the population density is advected with velocity $v - u$. In the following section, we discretize the equation on a moving irregular grid. For the purpose of this paper, we limit ourselves to 1D examples. To avoid the large time complexity associated with the full three-dimensional velocity field, we reduce the dimension of the equation using a Chu-reduction [8].

Remark 1 By moving the grid points with the mean velocity of the gas, the grid may become distorted throughout the simulation, leading to large discretization errors. A particle management algorithm is run periodically to avoid gaps in the grid or clustering of points. This algorithm uses a background grid to remove points where the point density is higher and adds points where the the point density is low [2].

Remark 2 This method only concerns the advection term in the BGK equation and is thus applicable to any BGK-type equation. For an overview of multi-species BGK-type equations, see for example [9].

3 Discretization Scheme

3.1 MUSCL-Like Spatial Discretization

The main difficulty associated with the Lagrangian form of the multi-species BGK equations is that the grid moves. One option to solve the equations is to use a finite volume method for moving unstructured grids [10]. However, for simplicity, we restrict ourselves to finite difference methods. Due to the unstructured nature of the grid, we cannot rely on classical finite difference schemes. Instead, we use the so-called Generalised Finite Difference Method (GFDM), which relies on a Moving Least Squares Approximation (MLS) [11]. MLS uses a Taylor expansion around a central point $f_s(x_i, v, t) \approx f_i$

$$f_j = f_i + \Delta x_{ij}\frac{\partial f_i}{\partial x} + \frac{\Delta x_{ij}^2}{2}\frac{\partial^2 f_i}{\partial x^2} + e_{ij}, \quad j \in \mathcal{C}_i, \tag{7}$$

where $\Delta x_{ij} = x_j - x_i$, and $\mathcal{C}_i$ is the set of all points that are within some distance of x_i. The spatial derivatives are then obtained by minimizing the $\mathcal{L}^2$ norm of the errors e_{ij} with respect to a weight function. One then obtains a result of the form

$$\frac{\partial f_i}{\partial x} \approx \sum_{j \in \mathcal{C}_i} \alpha_{ij}(f_j - f_i), \tag{8}$$

$$\frac{\partial^2 f_i}{\partial x^2} \approx \sum_{j \in \mathcal{C}_i} \beta_{ij}(f_j - f_i), \tag{9}$$

which can then be substituted into equation (6). To obtain a stable discretization, upwinding is required. Thus, the choice of stencil C_{ij} should depend on the velocity v. On the other hand, a central stencil is preferred due to significantly lower cost: the coefficients α_{ij} can only be computed once and can be reused for each velocity v. The novel MUSCL-like method achieves a stable discretization using central stencils and is based on the first-order positive method from [12]. We write the spatial derivative as

$$\frac{\partial f_i}{\partial x} \approx 2 \sum_{j \in \mathcal{C}_i} \alpha_{ij}(f_{ij} - f_i), \tag{10}$$

where f_{ij} is an approximation for the midpoint between x_i and x_j. The midpoint f_{ij} is reconstructed using a Taylor expansion from point x_i or x_j depending of the velocity

$$f_{ij} = \begin{cases} f_i + \frac{\Delta x_{ij}}{2}\sum_{k \in C_i} \alpha_{ik}(f_k - f_i) + \frac{\Delta x_{ij}^2}{8}\sum_{k \in C_i} \beta_{ik}(f_k - f_i) & \text{if } (v-u)\Delta x_{ij} > 0, \\ f_j - \frac{\Delta x_{ij}}{2}\sum_{k \in C_j} \alpha_{jk}(f_k - f_j) + \frac{\Delta x_{ij}^2}{8}\sum_{k \in C_j} \beta_{jk}(f_k - f_j) & \text{else.} \end{cases} \tag{11}$$

Upwinding is thus obtained by selecting the midpoint; the stencil is purely central. The scheme above is of second-order, or one order lower than the order of the Taylor expansion. Note that the extension to higher orders is straightforward. First, one should solve the least squares problem to a higher-order Taylor expansion, analogous to (7). Using the coefficients from the least squares problem, one can then do a higher-order reconstruction of the midpoint.

3.2 Velocity Discretization

For simplicity, we use a uniform zero-centered velocity discretization $[-v_{max}, v_{max}]$. All continuous inner products are approximated using the midpoint rule. Schemes that use this type of velocity discretization are referred to as discrete velocity models.

Remark 3 Due to the finite-difference nature of the spatial discretization, the discrete velocity model, and the non-conservative flux in (6), this algorithm is not conservative.

3.3 Time Discretization

We use the ARS222 IMEX Runge-Kutta scheme from [] for time integration of (6). The advection term is time-integrated using the explicit RK method. BGK-collision terms must be treated implicitly as in [13]. This avoids a time restriction due to the (possibly) stiff term, and ensures that the scheme is asymptotic preserving, i.e., in the limit for $\tau \to 0$, the numerical solution tends to the solution of the hydrodynamical limit. The explicit scheme of the IMEX method is used to time integrate the evolution equation for the position of the grid points (5).

Although the combination of an IMEX scheme and the MUSCL-like discretization is stable, it is possible to show that it is not positive. As a result, the algorithm will yield spurious oscillations at discontinuities. To avoid this, we use the so-called Multidimensional Optimal Order Detection (MOOD) method [3]. In the most basic version of MOOD, one checks if the numerical solution satisfies the local discrete maximum property (DMP)

$$\min_{j \in C_i}(f_i, f_j) \leq f_i^* \leq \max_{j \in C_i}(f_i, f_j). \tag{12}$$

If the proposed numerical solution does not satisfy the DMP property, the solution is recomputed using a lower-order method. In our case, we use the first-order positive method from [12], combined with a forward Euler time integration scheme. We apply MOOD on the numerical solution after every explicit RK stage f_i^* to avoid spurious oscillations due to the linear transport step. To avoid loss of accuracy at extrema, we relax criterion (12) with the so-called u2 detection criterion from [14].

4 Numerical Examples

4.1 Multi-Species Shock Tube

We consider the multi-species shock tube from Section 6.3 in [13]. We assume a spatial domain $[-1, 1]$ with diffusive reflective boundary conditions. The velocity domain is truncated to $[-15, 15]$. The numerical solutions are computed with $N_x = 200$ spatial grid points and $N_v = 50$ velocity grid points until the final time $t_f = 0.2$. We set the CFL number to 0.5 and set $\tau = 10^{-5}$. We compare three spatial discretisation methods: the second-order MUSCL-like scheme from 3.1 with and without MOOD, and a second-order meshless WENO method from [2]. All simulations are performed with the second-order IMEX ARS(2, 2, 2) time integration scheme. The density and temperature are shown in Fig. 1. The meshless MUSCL scheme without MOOD yields significant oscillations. Adding MOOD at the kinetic level removes the oscillations and results in sharp discontinuities. The meshless WENO scheme is a bit more diffusive than the MUSCL scheme.

4.2 *Single Species Moving Plate Problem*

As a final test, we consider the single species slider problem from [2]. This test illustrates the flexibility of meshless schemes when it comes to moving boundaries. A schematic of the test problem is provided in Fig. 2. A moving plate separates two chambers filled with Argon gas ($R_s = 208$, $\tau = 5.398 \times 10^{-4}$). Initially, the plate is at rest in the centre of the domain. We take as initial data of the gas in the left and right chamber Maxwellians with density $\rho_0 = 6.87 \times 10^{-7}$, velocity $U_0 = 0$, and temperature $T_0 = 270$. The density of the plate is ten times the density of the gas. The temperature of the walls of the left and right gas chambers are set to T_0 and $T_w = 330$ respectively. Due to the higher temperature of the walls of the right gas chamber, the pressure will increase, pushing the moving plate to a new equilibrium position. The motion of the plate is described by Newton's equation of motion. In the one-dimensional setting, the force that the gas applies to the moving plate is given by

$$F = \left(\psi_{left} - \psi_{right}\right) A$$
$$\psi_i = \int (v - U_w)^2 f \, dv,$$

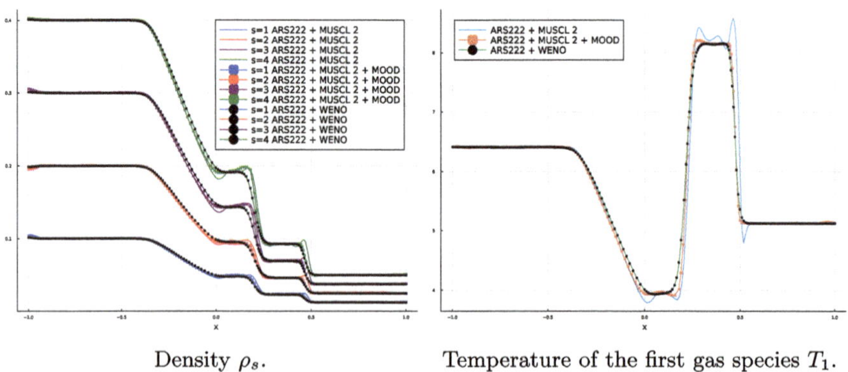

Density ρ_s. Temperature of the first gas species T_1.

Fig. 1 Density and temperature for the multi-species shock tube

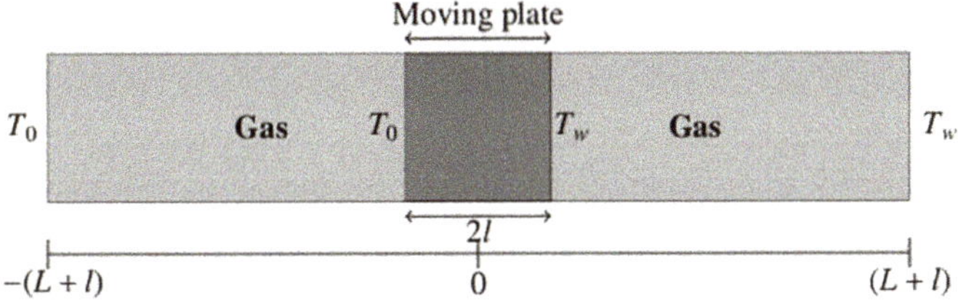

Fig. 2 Schematic of moving plate problem with $L = 1$ and $l = 0.1$ [2]

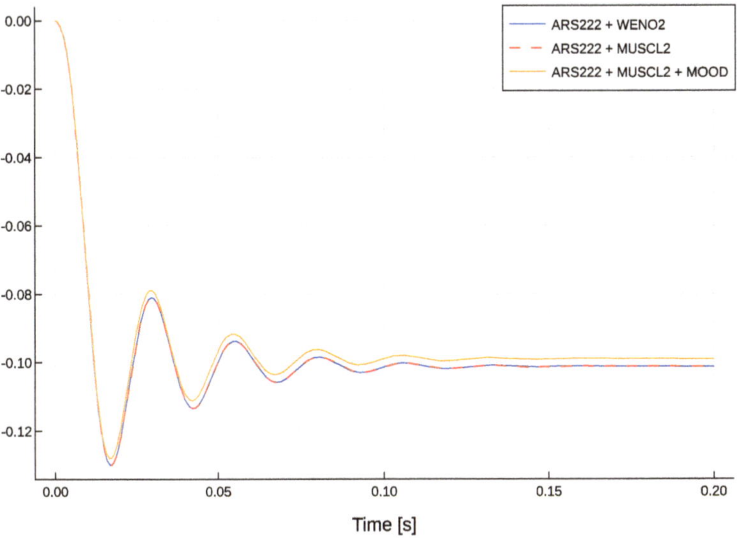

Fig. 3 The position of the centre of mass of the moving plate

where $A = 1$ is the surface area of the plate and U_w the velocity of the plate. This problem is often used as a benchmark because the equilibrium position of the slider can be computed explicitly from

$$x_{eq} = L\frac{T_0 - T_w}{T_0 + T_w} = -0.1.$$

The spatial domain and velocity domain $[-1600, 1600]$ are discretized using $N_x = 200$ and $N_v = 30$ points respectively. We perform the simulation up to time $t_f = 0.2$ with CFL $= 0.5$ using the same algorithms as in Sect. 4.1. The position of the centre of mass of the moving plate is plotted as a function of time in Fig. 3. The meshless WENO and MUSCL methods yield very comparable results. The main takeaway from this test is that no additional algorithms are required to deal with the moving boundary compared to the shock tube simulation. Due to the moving points and the particle management algorithm (see remark 1), the grid automatically adapts to the boundary.

5 Conclusion

We have presented a meshless scheme for linear hyperbolic equations on unstructured grids and applied it to the Lagrangian form of the multi-species BGK model. The numerical scheme uses a central stencil, which allows for the reuse of the stencil coefficients for each kinetic velocity. The MUSCL-like reconstruction at the stencil midpoints dynamically selects the upwind direction for each kinetic velocity and ensures the stability of the scheme. The meshless scheme is coupled with an IMEX time discretization scheme and a MOOD algorithm to avoid spurious oscillations at discontinuities. As a proof-of-concept, we have applied the algorithm to two test cases: a multi-species shock tube and the moving slider problem. In the former test case, the new MUSCL scheme with MOOD yields sharper discontinuities and no oscillations than the meshless WENO scheme from [2]. The latter test case illustrates the flexibility of meshless schemes when the domain contains moving boundaries. In future work, we intend to extend the scheme to fourth order and investigate the stability and conservation properties. In addition, we will extend the scheme to 2D and attempt simulations of more complex geometries.

References

1. Bobylev AV, Bisi M, Groppi M, Spiga G, Potapenko IF (2018) A general consistent BGK model for gas mixtures. Kinetic Related Models 11(6):1377–1393. https://doi.org/10.3934/krm.2018054
2. Tiwari S, Klar A, Russo G (2022) A meshfree arbitrary Lagrangian-Eulerian method for the BGK model of the Boltzmann equation with moving boundaries. J Computat Phys 458. https://doi.org/10.1016/j.jcp.2022.111088
3. Clain S, Diot S, Loubère R (201) A high-order finite volume method for systems of conservation laws—Multi-dimensional Optimal Order Detection (MOOD). J Comput Phys 230(10):4028–4050. https://doi.org/10.1016/j.jcp.2011.02.026
4. Karniadakis G, Beskok A, Aluru N (2005) Microflows and nano-flows: fundamentals ans simulations, Springer, New York. https://link.springer.com/book/10.1007/0-387-28676-4
5. Russo G, Filbet F (2009) Semilagrangian schemes applied to moving boundary problems for the BGK model of rarefied gas dynamics. Kinetic and Related Models 2:231–250. http://aimsciences.org//article/doi/10.3934/krm.2009.2.231
6. Déchristé G, Mieussens L (2016) A Cartesian cut cell method for rarefied flow simulations around moving obstacles. J Comput Phys 314:012030. https://doi.org/10.1016/j.jcp.2016.03.024
7. Bisi M, Groppi M, Martalò G (2021) Macroscopic equations for inert gas mixtures in different hydrodynamic regimes. J Phys A: Math Theoret 54(8):085201. https://iopscience.iop.org/article/10.1088/1751-8121/abbd1b
8. Chu CK Kinetic-theoretic description of the formation of a shock wave. In: The physics of fluids https://pubs.aip.org/aip/pfl/article-abstract/8/1/12/957838/Kinetic-Theoretic-Description-of-the-Formation-of?redirectedFrom=fulltext
9. Brull S, Pavan V, Schneider J (2012) Derivation of BGK model for mixtures. Eur J Mech B&Fluids, 33:74–86. https://linkinghub.elsevier.com/retrieve/pii/S0997754611001324
10. Boscheri W (2017) High order direct arbitrary-lagrangian–eulerian (ALE) finite volume schemes for hyperbolic systems on unstructured meshes. Arch Comput Methods Eng 751–801. https://doi.org/10.1007/s11831-016-9188-x

11. Lancaster P, Salkauskas K (1981) Surfaces generated by moving least squares methods. Math Comput 37:141–158. https://www.ams.org/journals/mcom/1981-37-155/S0025-5718-1981-0616367-1/
12. Chandrashekar P A positive meshless method for hyperbolic equations. https://www.researchgate.net/publication/277759856_A_positive_meshless_method_for_hyperbolic_equations
13. Cho SY, Boscarino S, Groppi M, Russo G Conservative semi-lagrangian schemes for a general consistent BGK model for inert gas mixtures. https://dx.doi.org/10.4310/CMS.2022.v20.n3.a4
14. Diot S, Clain S, Loubère R (2012) Improved detection criteria for the multi-dimensional optimal order detection (MOOD) on unstructured meshes with very high-order polynomials. Comput Fluids 64:43–63. https://doi.org/10.1016/j.compfluid.2012.05.004
15. Déchristé G, Mieussens L (2012) Numerical simulation of micro flows with moving obstacles. J Phys: Conf Ser 362:465–488. https://iopscience.iop.org/article/10.1088/1742-6596/362/1/012030
16. Ascher UM, Ruuth SJ, Spiteri RJ (1997) Implicit-explicit Runge-Kutta methods for time-dependent partial differential equations. Appl Numer Math 25(2–3):151–167. https://doi.org/10.1016/S0168-9274(97)00056-1

Comparison of Stochastic BGK and FP Methods for the Simulation of Non-Equilibrium Multi-species Molecular Gas Flows

Franziska Tuttas and Marcel Pfeiffer

Abstract Due to limited possibilities of experimental investigations for non-equilibrium gas flows, numerical results are of highest interest. Although the well-established Direct Simulation Monte Carlo (DSMC) method achieves highly accurate solutions, the computational requirements increase excessively for lower Knudsen regimes. Computationally more efficient simulations can be achieved with stochastic continuum-based methods using either the Bhatnagar-Gross-Krook (BGK) or the Fokker-Planck (FP) approximations where, instead of particle collisions, particle relaxation processes are considered. This paper explains the implementation of different stochastic BGK and FP methods in the open-source particle code PICLas for multi-species molecular gas flows. For verification, the results of different test cases are compared.

Keywords DSMC · BGK · Fokker-Planck · Multi-species · Mixture

1 Introduction

Looking at multi-scale flow problems across a wide range of Knudsen numbers, the well-established Direct Simulation Monte Carlo (DSMC) method [1] achieves highly accurate solutions. However, the computational requirements increase excessively for lower Knudsen regimes, making it computationally infeasible for many practical applications. A possible solution is a coupling of the DSMC method with continuum-based methods, leading to reduced computational costs. Using the same particle approach for simplicity of the coupling, different stochastic particle-based methods such as the Bhatnagar-Gross-Krook (BGK) [2–4] and the Fokker-Planck (FP) [5–10] approximations for the Boltzmann collision integral can be used for this purpose. The advantage of these methods in the computational efficiency in transition and continuum regimes, as they consider relaxation processes of the particles instead of binary particle collisions.

F. Tuttas (✉) · M. Pfeiffer
Institute of Space Systems, University of Stuttgart, Stuttgart, Germany
e-mail: tuttasf@irs.uni-stuttgart.de

© The Author(s) 2026

M. Grabe et al. (eds.), *Rarefied Gas Dynamics*, Springer Aerospace Technology,
https://doi.org/10.1007/978-3-032-00094-1_40

In this paper, the implementation of different stochastic BGK and FP methods for multi-species molecular gas flows in the open-source particle code PICLas [11, 12] is explained. First, the different models are presented, including an Ellipsoidal Statistical BGK (ESBGK) model [3], a Shakhov BGK model (SBGK) [4], and an Ellipsoidal Statistical Fokker-Planck (ESFP) model [7]. The focus of this paper lies in the differences between the models and their implementation in PICLas, mainly regarding the relaxation and sampling of the particle velocities. The relaxation of internal energies are described through Landau-Teller equations as shown in Hild et al. [13], which is the same for all models and thus is not presented any further here. The energy and momentum conservation scheme is discussed briefly, which is crucial for the stability of the simulations. To assess the accuracy of the different models, the results of two supersonic Couette flow test cases are compared to DSMC reference solutions.

2 BGK Method

The BGK operator approximates the Boltzmann collision integral by a relaxation process of a particle distribution function f_s $(\mathbf{x}, \mathbf{v}, t)$ of a species s at position $\mathbf{x}$ and with velocity $\mathbf{v}$ towards a target distribution f_s^t [2], using the relaxation frequency ν:

$$\partial_t f_s + \mathbf{v}\partial_{\mathbf{x}} f_s = \nu(f_s^t - f_s). \tag{1}$$

For atomic species, the target distribution function only consists of a translational part $f_s^{t,\mathrm{tr}}(\mathbf{v})$, whereas for molecular species, the distribution function can be separated into a translational, a rotational, and a vibrational part as demonstrated in [14, 15]. Both the ESBGK [3] and the SBGK [4] models use one relaxation term per species, produce the Maxwellian distribution in the equilibrium state, and fulfill the indifferentiability principle, yielding same outcomes for one species or two identical species. While there the H theorem is proven for the ESBGK model, there is no general proof for the SBGK model. Furthermore, the positivity of SBGK target distribution function $f_s^{\mathrm{SBGK,tr}}$ cannot be guaranteed.

2.1 ESBGK Mixture Model

The ESBGK model presented here combines different models of Mathiaud et al. [14], Pfeiffer [16] and Brull [17, 18] into a mixture model that allows for non-equilibrium effects in the internal degrees of freedom. It is already discussed in greater detail in [13]. Instead of a Maxwellian target distribution, the correct Prandtl number is achieved by using an anisotropic Gaussian distribution [3]:

$$f_s^{\text{ESBGK,tr}} = \frac{n_s}{\sqrt{\det 2\pi \mathbf{A}_s}} \exp\left[-\frac{1}{2}\mathbf{c}^{\mathrm{T}}\mathbf{A}_s^{-1}\mathbf{c}\right]. \tag{2}$$

The anisotropic matrix $\mathbf{A}_s$ is defined as

$$\mathbf{A}_s = \frac{k_{\mathrm{B}}T_{\text{tr,rel}}}{m_s}\mathbf{I} - \frac{1-\alpha Pr}{\alpha Pr}\left(\frac{T_{\text{tr,rel}}}{T_{\text{tr}}}\Theta - \frac{k_{\mathrm{B}}}{m_s}T_{\text{tr,rel}}\mathbf{I}\right) \tag{3}$$

with the identity matrix $\mathbf{I}$, and the symmetric stress tensor Θ, calculated with the pressure tensor $\mathbf{P}$, and the mass density ρ:

$$\Theta = \frac{\mathbf{P}}{\rho} = \frac{1}{\rho}\sum_{s=1}^{M} m_s \int \mathbf{c}\mathbf{c}^{\mathrm{T}} f_s \, d\mathbf{v} \tag{4}$$

$T_{\text{tr,rel}}$ is the translational relaxation temperature, defined according to [13, 14]. The parameter α is a model variable that depends on mass fraction, density fraction and internal degrees of freedom [18]:

$$\alpha = m \frac{\sum_{s=1}^{M} \frac{n_s}{m_s}\left(5 + \xi_{\text{vib},s} + \xi_{\text{rot},s}\right)}{\sum_{s=1}^{M} n_s \left(5 + \xi_{\text{vib},s} + \xi_{\text{rot},s}\right)} \tag{5}$$

The relaxation frequency ν_{ESBGK} of the model is defined as

$$\nu_{\text{ESBGK}} = \frac{nk_{\mathrm{B}}T_{\text{tr}}}{\mu}\alpha Pr \tag{6}$$

with the viscosity of the mixture μ and the translational temperature T_{tr}. Pr is the targeted Prandtl number of the gas mixture, which is calculated using collision integrals [19]. For molecular species, the rotational and vibrational components of the distribution function are taken directly from [14]. The relaxation temperatures are chosen to match the Landau-Teller relaxation. For more details, the reader is referred to [13].

2.2 Shakhov BGK Mixture Model

In the SBGK model, the heat flux is modified to produce the correct Prandtl number [4]. Thus, the target distribution function for a mixture model reads

$$f_s^{\text{SBGK,tr}} = f_s^{\mathrm{M}}\left[1 + (1 - \alpha Pr)\frac{\mathbf{c}^{\mathrm{T}}\mathbf{q}/m_s}{5n\left(k_{\mathrm{B}}T_{\text{tr}}/m_s\right)^2}\left(\frac{|\mathbf{c}|^2}{2k_{\mathrm{B}}T_{\text{tr}}/m_s} - \frac{5}{2}\right)\right] \tag{7}$$

with the heat flux vector

$$q = \sum_{s=1}^{M} m_s \int \mathbf{c} \, |\mathbf{c}|^2 \, f_s \, d\mathbf{v} \tag{8}$$

and the Maxwellian distribution f_s^M. The relaxation frequency ν_{SBGK} of the model is defined as

$$\nu_{\mathrm{SBGK}} = \frac{n k_B T_{\mathrm{tr}}}{\mu}. \tag{9}$$

For molecular species, the rotational and vibrational components of the distribution function are defined similar as for the ESBGK model (see Sect. 2.1).

3 FP Method

The FP operator approximates the Boltzmann collision term by using a drift vector $\mathbf{a} = a_i$ and a diffusion matrix $\mathbf{D}_s = D_{ij,s}$ [5]:

$$\partial_t f_s + \mathbf{v} \partial_\mathbf{x} f_s = -\sum_{i=1}^{3} \frac{\partial}{\partial v_i} (a_i f_s) + \sum_{i=1}^{3} \sum_{j=1}^{3} \frac{\partial^2}{\partial v_i \partial v_j} (D_{ij,s} f_s). \tag{10}$$

For molecular species, the internal energies are handled following the BGK approach presented in Sect. 2. In this paper, the ESFP model [7] for molecular gas mixtures is presented and compared to the BGK models. Same as the ESBGK model, it fulfills the H theorem.

3.1 *ESFP Mixture Model*

The ESFP mixture model modifies the diffusion matrix $\mathbf{D}_s$ to correct the Prandtl number of the Standard FP model [7]:

$$\mathbf{D}_s^{\mathrm{ESFP}} = \nu_{\mathrm{ESFP}} \left((1 - \chi_s) \frac{k_B T_{\mathrm{tr}}}{m_s} \mathbf{I} + \chi_s \mathbf{\Theta} \right). \tag{11}$$

$\mathbf{D}_s$ is a convex combination of the stress tensor $\mathbf{\Theta}$ and its equilibrium value $(k_B T_{\mathrm{tr}} / m_s) \mathbf{I}$ for a species s with a parameter χ_s including $\lambda_{\max}$ as the (positive) maximum eigenvalue of $\mathbf{\Theta}$, so that $\mathbf{D}_s$ remains strictly definite positive:

$$\chi_s = \max \left(1 - \frac{3}{2Pr}, -\frac{k_B T_{\mathrm{tr}}/m_s}{\lambda_{\max} - k_B T_{\mathrm{tr}}/m_s} \right). \tag{12}$$

The drift vector is defined similar to the Standard FP model:

$$\mathbf{a}^{\text{ESFP}} = -\nu_{\text{ESFP}}\mathbf{c}. \tag{13}$$

The relaxation frequency ν_{ESFP} of the model is defined as

$$\nu_{\text{ESFP}} = \frac{nk_{\text{B}}T_{\text{tr}}}{3\mu}Pr. \tag{14}$$

4   Implementation

All the particle methods presented in this paper are implemented in the open-source code PICLas [11]. The different implementations of the models for the translational relaxation of the particles and the sampling of the new velocities are discussed. In addition, the basic energy and momentum conservation scheme is described (for details see [13]). For details on the rotational and vibrational relaxations, the reader is also referred to [13].

4.1   Relaxation and Sampling of Particle Velocities

For the BGK models, all particles in a cell relax with a probability P in each time step Δt towards the specified target distribution function $f_s^{\text{ESBGK,tr}}$ or $f_s^{\text{SBGK,tr}}$:

$$P = 1 - \exp\left[-\nu\Delta t\right], \quad \nu = \nu_{\text{ESBGK}}, \nu_{\text{SBGK}}. \tag{15}$$

The relaxation frequency ν is calculated in each cell and time step using collision integrals [19], as described in [13]. If a uniform random number is smaller than the probability, the particle considered relaxes and its new velocity is sampled from the target distribution function of the used method.

For the ESBGK method, the transformation matrix $\mathbf{S}_s$ [20] with $\mathbf{A}_s = \mathbf{S}_s\mathbf{S}_s^{\text{T}}$ is approximated and used to transform a random Maxwellian vector $\mathbf{r}_p$ for a particle p to a velocity vector from the ESBGK target distribution:

$$\mathbf{v}_{p,\text{ESBGK}}^* = \mathbf{u} + \mathbf{S}_s\mathbf{r}_p, \tag{16}$$

$$\mathbf{S}_s = \sqrt{\frac{k_{\text{B}}T_{\text{tr,rel}}}{m_s}}\left[\mathbf{I} - \frac{1-\alpha Pr}{2\alpha Pr}\left(\frac{m}{k_{\text{B}}T_{\text{tr}}}\Theta - \mathbf{I}\right)\right]. \tag{17}$$

The superscript * marks a parameter after relaxation, but before energy and momentum conservation.

For the SBGK method, however, an analytical expression for a similar vector transformation is not available, which is why an acceptance-rejection (AR) algorithm [21] is used. Hereby, an envelope function $g_s(\mathbf{r}_p)$ is defined. For the presented particle SBGK method, good results are found by choosing

$$g_s(\mathbf{r}_p) = A_s f_s^{\text{SBGK,tr}}(\mathbf{r}_p) \tag{18}$$

$$A_s = 1 + 10\frac{\max(\mathbf{q}/m_s)}{(k_{\mathrm{B}}T_{\mathrm{tr}}/m_s)^{3/2}} \tag{19}$$

as envelope function and

$$P_s^{\text{SBGK,AR}}(\mathbf{r}_p) = 1 + (1 - Pr)\frac{\mathbf{r}_p^{\mathrm{T}}\mathbf{q}/m_s}{5n(k_{\mathrm{B}}T_{\mathrm{tr}}/m_s)^{3/2}}\left(\frac{|\mathbf{r}_p|^2}{2} - \frac{5}{2}\right) \tag{20}$$

as acceptance probability. A normal distributed vector $\mathbf{r}_p$ is accepted if a uniform random number $r < P_s^{\text{SBGK,AR}}(\mathbf{r}_p)/A_s$. The new velocity is then calculated with

$$\mathbf{v}_{p,\text{SBGK}}^* = \mathbf{r}_p\sqrt{\frac{k_{\mathrm{B}}T_{\mathrm{tr,rel}}}{m_s}}. \tag{21}$$

For a detailed description of the different BGK sampling options, the reader is referred to [12], in which also alternative methods for sampling are described.

The ESFP model relaxes each particle according to the Ornstein-Uhlenbeck process [22] with the following stochastic differential equation to solve the corresponding FP Eq. (10):

$$\mathrm{d}\mathbf{c} = \mathbf{a}\,\mathrm{d}t + \sqrt{2}\mathbf{D}^{1/2}\,\mathrm{d}\mathbf{w} \tag{22}$$

$\mathrm{d}\mathbf{w}$ is the standard three-dimensional Wiener process [7]. Using an exact time integration [5], the new particle velocity for the ESFP model is calculated by

$$\mathbf{v}_{p,\text{ESFP}}^* = \mathbf{u} + \mathbf{c}_p \exp\left(-\nu_{\text{ESFP}}\Delta t\right) + \frac{\left(\mathbf{D}_s^{\text{ESFP}}\right)^{1/2}}{\sqrt{\nu_{\text{ESFP}}}}\sqrt{1 - \exp\left(-2\nu_{\text{ESFP}}\Delta t\right)} \cdot \mathbf{r}_p \tag{23}$$

with $\mathbf{c}_p$ being the thermal particle velocity before relaxation.

4.2 Energy and Momentum Conservation

Due to the stochastic approach with particles, an energy and momentum conservation scheme needs to be established even though the presented BGK and FP models are in general both momentum and energy conserving. The reason for this lies in the random choice of the new states of the particles from the distribution functions as described in Sect. 4.1, whereby the conservation of energy and momentum is not automatically given. Thus, a large number of particles would be needed for stable simulation results if no additional energy and momentum conservation step would be implemented. In general, the chosen conservation scheme is based on the work in [16, 23] and extended to polyatomic molecules and mixtures of different gas species in [13]. The main idea is to distribute the energy difference between the old energy and the new energy from the sampling evenly according to the respective degrees of freedom to the translational energy of all particles in the cell and the internal energies of the respective relaxing particles within the cell [16]. More details on the implementation can be found in [13].

5   Simulation Results for Supersonic Couette Flows

The different multi-species BGK and FP implementations are verified and compared with a test case of two one-dimensional supersonic Couette flows. Simulating a 50%-50% N_2-He mixture first, the initial particle density is $n_0 = 1.3 \times 10^{20} m^{-3}$, leading to a Knudsen number of $Kn_{VHS} = 0.0112$ with a characteristic length of $L = 1m$. As second case, an air mixture with N, O, N_2, O_2, and NO, with $n_0 = 1.25 \times 10^{20} m^{-3}$ and a corresponding Knudsen number of $Kn_{VHS} = 0.0126$ is simulated in order to test the models with a more complicated case of multiple molecular species in a mixture. The velocity and temperature of the gas are initialized at $v_0 = 0 ms^{-1}$ and $T_0 = 273K$ in both cases. The boundaries in y direction have a velocity of $v_{wall,1} = 350 ms^{-1}$ and $v_{wall,2} = -350 ms^{-1}$, respectively, assuming diffuse reflection and complete thermal accommodation at a constant wall temperature of $T_{wall} = 273K$. The time steps and particle weighting factors are chosen so that the mean free path and the collision frequency are resolved. The VHS model is used [1], and the parameters for a collision pair of unlike species are determined as an average (referred to as collision-averaged).

The translational temperature results in comparison for the different methods are shown in Fig. 1. While there is very good agreement for the ESBGK and SBGK

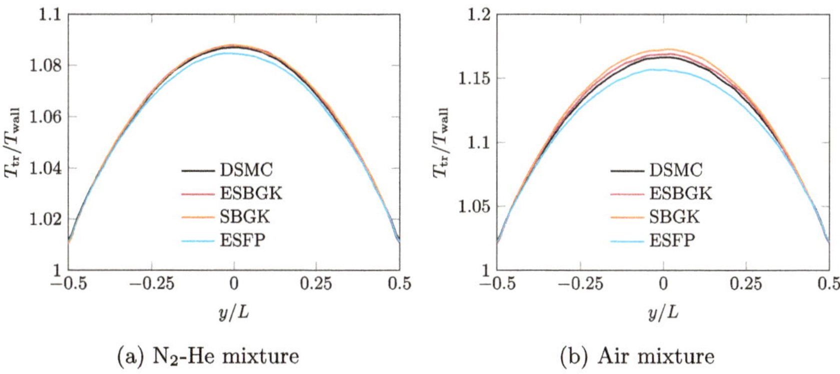

(a) N$_2$-He mixture   (b) Air mixture

Fig. 1   Translational temperature for the ESBGK (purple), SBGK (orange), and ESFP (blue) models as well as the DSMC result (black) as reference

models compared to the DSMC reference solution, the ESFP model shows deviations in the peak temperature. Similar results for a comparable Knudsen number are shown in [24] for a single-species ESFP model, which is why the deviations are attributed to the model itself rather than the mixture modeling. The results for the rotational temperatures are similar due to the thermal equilibrium in the Couette flow.

6 Conclusion

Different continuum-based particle methods, namely an ESBGK model, a Shakhov BGK model, and an ESFP model are proposed. Aiming for solutions of multi-scale problems, the handling of molecular gas mixture flows is implemented in the open-source particle code PICLas. Two different supersonic Couette flows with molecular gas mixtures indicate overall very good agreement of the results between both the BGK models and the DSMC method, whereas small deviations occur for the ESFP model. In future work, the implementation of chemical reactions into all the different models is envisioned.

Acknowledgements This project has received funding from the European Research Council (ERC) under the European Union's Horizon 2020 research and innovation programme (grant agreement No. 899981 MEDUSA).

References

1. Bird GA (1994) Molecular gas dynamics and the direct simulation of gas flows. Oxford University Press, New York
2. Bhatnagar PL, Gross EP, Krook M (1954) A model for collision processes in gases. I. Small amplitude processes in charged and neutral one-component systems. Phys Rev 94(3):511–525
3. Holway HL Jr (1966) New statistical models for kinetic theory: methods of construction. Phys Fluids 9:1658–1673
4. Shakhov EM (1968) Generalization of the Krook kinetic relaxation equation. Fluid Dyn 33:95–96
5. Jenny P, Torrilhon M, Heinz S (2010) A solution algorithm for the fluid dynamic equations based on a stochastic model for molecular motion. J Comput Phys 229(4):1077–1098
6. Pfeiffer M, Gorji MH (2017) Adaptive particle-cell algorithm for Fokker-Planck based rarefied gas flow simulations. Comput Phys Commun 213:1–8
7. Mathiaud J, Mieussens L (2016) A Fokker-Planck model of the Boltzmann equation with correct Prandtl number. J Stat Phys 162:397–414
8. Gorji MH, Jenny P (2010) An efficient particle Fokker-Planck algorithm for rarefied gas flows. J Comput Phys 262:325–343
9. Kim S, Jun E (2025) A particle Fokker-Planck method for rarefied gas flows of monatomic mixtures. Phys Fluids 37:016104
10. Hepp C, Grabe M, Hannemann K (2020) A kinetic Fokker-Planck approach for modeling variable hard-sphere gas mixtures. AIP Advances 10:085219

11. Fasoulas S, Munz C-D, Pfeiffer M, Beyer J, Binder T, Copplestone S, Mirza A, Nizenkov P, Ortwein P, Reschke W (2019) Combining particle-in-cell and direct simulation Monte Carlo for the simulation of reactive plasma flows. Phys Fluids 31:072006
12. Pfeiffer M (2018) Particle-based fluid dynamics: comparison of different Bhatnagar-Gross-Krook models and the direct simulation Monte Carlo method for hypersonic flows. Phys Fluids 30:106106
13. Hild F, Pfeiffer M (2024) Multi-species modeling in the particle-based ellipsoidal statistical Bhatnagar-Gross-Krook method including internal degrees of freedom. J Comput Phys 514:113226
14. Mathiaud J, Mieussens L, Pfeiffer M (2022) An ES-BGK model for diatomic gases with correct relaxation rates for internal energies. Eur J Mech B/Fluids 96:65–77
15. Dauvois Y, Mathiaud J, Mieussens L (2021) An ES-BGK model for polyatomic gases in rotational and vibrational nonequilibrium. Eur J of Mech B/Fluids 88:1–16
16. Pfeiffer M (2018) Extending the particle ellipsoidal statistical Bhatnagar-Gross-Krook method to diatomic molecules including quantized vibrational energies. Phys Fluids 30:116103
17. Brull S (2014) An ellipsoidal statistical model for gas mixtures. Commun Math Sci 13:1–13
18. Brull S (2021) An ellipsoidal statistical model for a monoatomic and a polyatomic gas mixture. Commun Math Sci 19(8):2177–2194
19. Hirschfelder JO, Curtiss CF, Bird RB (1964) The molecular theory of gases and liquids. Wiley-Interscience
20. Gallis MA, Torczynski JR (2000) The application of the BGK model in particle simulations. In: 34th thermophysics conference, American institute of aeronautics and astronautics, Denver, Colorado, USA
21. Garcia AL, Aldr BJ (1998) Generation of the Chapman-Enskog distribution. J Comput Phys 140:66–70
22. Uhlenbeck GE, Ornstein LS (1930) On the theory of the Brownian motion. Phys Rev 36:823–841
23. Burt JM, Boyd ID (2006) Evaluation of a particle method for the ellipsoidal statistical bhatnagar-gross-krook equation. In: 44th AIAA aerospace sciences meeting and exhibit
24. Jun E, Pfeiffer M, Mieussens L, Gorji MH (2019) Comparative study between cubic and ellipsoidal Fokker-Planck kinetic models. AIAAJ 57(6):2524–2533

Molecular-Level Simulations of Turbulent Couette Flow Over Rough Surfaces

Ryan M. McMullen, Timothy P. Koehler, and Michael A. Gallis

Abstract We report flow statistics from simulations using the direct simulation Monte Carlo method for turbulent Couette flow where the lower wall is replaced by an idealized rough surface. We focus on conditions yielding a modest global Reynolds number, supersonic Mach number, and a global Knudsen number in the near-continuum regime, allowing for comparison with direct numerical simulation of the Navier-Stokes equations for the same conditions. While we observe discrepancies on the order of a few percent, we find the overall level of agreement between macroscopic and molecular-level descriptions of turbulent flow encouraging.

Keywords DSMC · Turbulence · Rough surfaces · TPS materials

1 Introduction

Thermal-protection-system (TPS) materials used to coat the surfaces of reentry vehicles are often rough and permeable. This surface layer has the potential to significantly modify the turbulent boundary layer over the vehicle and thereby alter the aerothermal loading, which could in turn compromise vehicle performance [1]. Consequently, the ability to make accurate predictions about turbulent flows over TPS materials is of great interest. However, the length scales of common TPS materials can be on the order of the gas-molecular mean free path [2], which itself may not be small compared to other characteristic flow length scales. Therefore, the validity of predictions from conventional continuum computational fluid dynamics needs to be assessed in the context of flow over reentry vehicles coated with TPS materials.

On the other hand, the direct simulation Monte Carlo (DSMC) method of molecular gas dynamics is suitable for a wider range of flow physics, including situations in which noncontinuum and nonequilibrium effects are important [3]. DSMC simulates gas flows by tracking large numbers of computational particles (each of which in turn represents a large number of real gas molecules) that move, collide with

R. M. McMullen (✉) · T. P. Koehler · M. A. Gallis
Engineering Sciences Center, Sandia National Laboratories, Albuquerque, NM, USA
e-mail: rmmcmul@sandia.gov

M. Grabe et al. (eds.), *Rarefied Gas Dynamics*, Springer Aerospace Technology,
https://doi.org/10.1007/978-3-032-00094-1_41

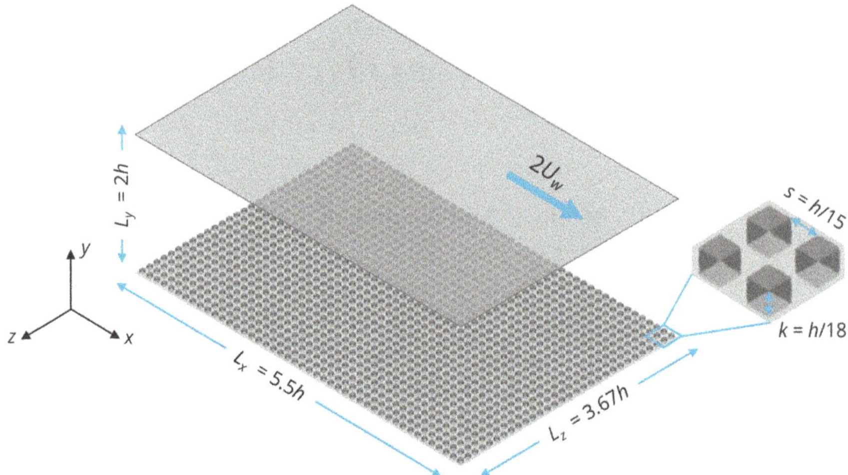

Fig. 1 Configuration for turbulent Couette flow over a rough wall with cubical roughness elements

each other, and reflect from solid surfaces just like real gas molecules do. Although DSMC is a molecular-level method and thus computationally intense, algorithmic and computing-power advances have recently made DSMC simulations of near-continuum turbulent flows computationally feasible. DSMC has now been shown to be able to simulate turbulent flows accurately, reproducing known features of decaying turbulence [4] and self-sustaining wall-bounded turbulence [5]. Furthermore, DSMC inherently includes noncontinuum effects, such as Knudsen layers adjacent to solid surfaces and molecular thermal fluctuations, which become significant in the dissipation range of the turbulent energy cascade [6–8].

With these capabilities, DSMC should be well suited for performing high-fidelity simulations of turbulent flows over complex TPS-like rough and permeable surfaces. However, the ability of DSMC to simulate turbulent flow over rough surfaces has not yet been verified. Therefore, we herein compare DSMC simulations of turbulent Couette flow over an idealized and canonical rough surface consisting of regularly spaced cubical roughness elements to direct numerical simulation of the Navier-Stokes (NS) equations. The flow conditions and surface geometry have been chosen such that good agreement between the two methods is anticipated.

2 Rough-Wall Simulations

2.1 Flow Configuration

Figure 1 shows a schematic diagram of the flow configuration. x, y, and z are the streamwise, wall-normal, and spanwise directions, respectively. The velocity components corresponding to these directions are denoted u, v, w, respectively. The domain extent is $(L_x, L_y, L_z) = (5.5h, 2h, 3.67h)$, where $h = 500\,\mu\text{m}$ is the channel half-height. The boundary conditions are periodic in the x and z directions, and all solid surfaces are fully accommodating. The lower wall is roughened with cubical roughness elements of height $k = h/18$, which are separated by a distance $s = h/15$.

The global Reynolds number is $\text{Re} = \rho_0 U_w h / \mu_w = 2790$, where $\rho_0 = 0.345$ kg m^{-3} is the volume-averaged density, $U_w = 461.85$ m s^{-1} is half the velocity difference between the walls, and $\mu_w = 2.86 \times 10^{-5}$ Pa s is the value of the dynamic viscosity at the constant wall temperature $T_w = 273.15$ K. The global Mach number is $\text{Ma} = U_w/a_w = 1.5$, where $a_w = \sqrt{\gamma k_B T_w/m} = 307.9$ m s^{-1} is the sound speed at T_w, $\gamma = 5/3$ is the specific heat ratio, k_B is the Boltzmann constant, and $m = 6.63 \times 10^{-26}$ kg is the molecular mass. The global Knudsen number is $\text{Kn} = \lambda_0/h = \sqrt{\pi\gamma/2}\,\text{Ma}/\text{Re} = 8.7 \times 10^{-4}$, where λ_0 is the molecular mean-free path at ρ_0 and T_w. Velocities and lengths nondimensionalized by U_w and h, respectively, are referred to as being in outer units or as outer-scaled. Velocities and lengths nondimensionalized by the friction velocity $u_\tau = \sqrt{\tau_w/\rho_w}$ and viscous length scale $\mu_w/\rho_w u_\tau$, respectively, are referred to as being in inner units or as inner-scaled and are denoted with a superscript "+".

2.2 Numerical Details

The DSMC simulations are performed using Sandia's open-source massively parallel code SPARTA [9]. The domain is divided into $N_x \times N_y \times N_z = 2141 \times 823 \times 1475$ uniform cells. For the conditions considered herein, this cell size yields a mean collision separation comparable to λ_0, such that any potential shocks are well resolved. On average, each cell contains 30 particles, giving approximately 73 billion particles total. The time step is 9.12 ps, which is less than 1% of the mean collision time. Particle-particle collisions are performed using hard-sphere interactions, which produces a viscosity temperature dependence of the form $\mu(T) = \mu_w\sqrt{T/T_w}$ [3]. The simulation was performed on the Los Alamos National Laboratories Crossroads supercomputer and was run to approximately 77×10^6 time steps, corresponding to a dimensionless stopping time of $tU_w/h \approx 650$.

A cut-cell method is used to conform the mesh to the rough lower surface [9]. In this approach, a closed three-dimensional structure represented by triangular surface elements is embedded in the Cartesian grid. The surface elements intersecting a given grid cell cut the portion of its volume accessible to particles, and the reduced volume for each cut cell is then calculated and taken into account when performing particle-particle collisions.

Macroscopic quantities of interest are obtained by averaging over the particles contained within each cell. However, since each simulated particle represents $O(10^5)$ real molecules, the statistical variations arising from thermal molecular fluctuations are drastically overestimated [3], often resulting in an unacceptably low signal-to-noise ratio. A common technique to mitigate this effect is to additionally perform temporal averages within the cells. Ideally, an averaging time that gives acceptable noise reduction is still much smaller than any macroscopic flow timescale, though this is not always possible. Here, we found that averaging over 10^3 consecutive time steps, which is less than 1% of the turnover time h/U_w and less than 10% of the viscous time μ_w/τ_w, reduced noise acceptably without significantly smearing out the turbulent structures.

The initial condition for the simulation is the incompressible laminar smooth-wall solution $u = U_w(y/h + 1)$ plus a sinusoidal perturbation with amplitude sufficient to trigger transition to turbulence. Turbulence statistics are collected for $tU_w/h \geq 200$ to allow the flow to reach statistically steady state, which is confirmed by examination of the skin friction shown in Fig. 2. Subsequently, DSMC realizations were collected every 10^6 time steps (every $8.4\,tU_w/h$).

Companion direct numerical simulation (DNS) of the NS equations for the same flow conditions was performed using the Sandia-developed finite-volume code SPARC [10]. The DNS is performed using a grid of $N_x \times N_y \times N_z = 495 \times 278 \times 330$. Uniform grid spacing is used in the homogeneous wall-parallel directions, resulting in $\Delta x^+ = \Delta z^+ = 2.3$. The grid in the wall-normal direction is uniform within the cubical roughness elements, with $\Delta y_r^+ = 0.42$. Above the roughness elements, the grid is stretched using a hyperbolic tangent distribution, such that the resolution at the channel center is $\Delta y_c^+ = 3.8$. Likewise, the grid is stretched moving from the top wall toward the channel center, with a near-wall resolution of $\Delta y_w^+ = 0.45$. Time is advanced using a fourth-order explicit Runge-Kutta scheme with a Courant-Friedrichs-Lewy number of 0.5 to determine the time step. The present simulation thus has resolutions comparable to other DNS studies of compressible turbulent Couette flow [11]. The initial condition for the NS simulation is identical to the one used for the DSMC simulation. For $tU_w/h \geq 200$, statistics were collected every turnover time h/U_w up to the simulation stopping time of $tU_w/h \approx 3700$.

2.3 Results

First, we compare the DSMC and NS wall skin friction τ_w (drag per unit plan area). Figure 2 shows the DSMC and NS values of the top- and bottom-wall skin friction nondimensionalized by the laminar value $\mu_w U_w/h$. Both agree almost exactly for early times $tU_w/h \lesssim 50$ corresponding to breakdown of the deterministic initial condition into turbulence. The flow reaches statistically steady state at $tU_w/h \approx 200$, after which the DSMC and NS wall shear stress have very similar mean values (10.6

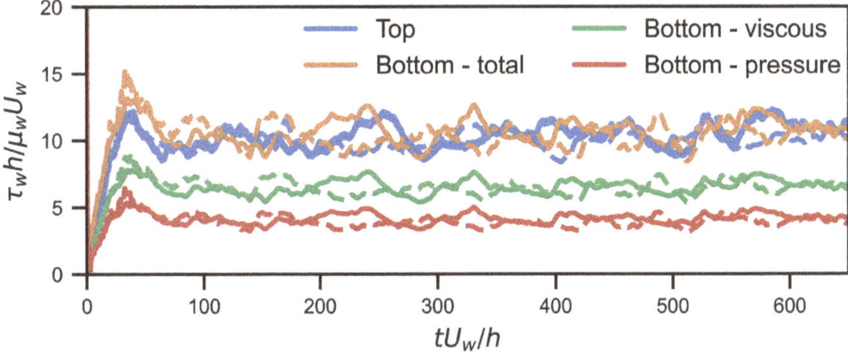

Fig. 2 Comparison of top- and bottom-wall skin friction for DSMC (solid lines) and NS (dashed lines). Also shown is the decomposition of bottom-wall skin friction into viscous- and pressure-drag contributions

and 10.1, respectively) but different time histories due to the chaotic nature of turbulent flows. The slight discrepancy of $\sim 5\%$ between the mean values can possibly be attributed to the relatively short record length of DSMC simulation compared to that of the NS simulation.

Note that both the top (smooth) and bottom (rough) walls are guaranteed to have the same mean skin friction since the mean momentum balance for Couette flow states that the total (viscous plus turbulent) shear stress is a constant (equal to τ_w) across the entire channel. However, this does not mean that the rough wall has no effect on the drag. Indeed, NS simulations for the same Re and Ma in which both walls are smooth (not shown in Fig. 2) yield a mean dimensionless skin friction value of 9.23, so that the rough wall increases drag by $\sim 10\%$.

While the top-wall skin friction is entirely due to viscous drag (shear stress), the skin friction on the roughened bottom wall is comprised of both pressure drag (normal stress) and viscous drag on the cubical roughness elements. Each of these contributions to the bottom-wall skin friction is also shown in Fig. 2, where once again the DSMC and NS results are found to be in good agreement. Pressure drag comprises 39% and 38% of the total bottom-wall skin friction for the DSMC and NS simulations, respectively. This implies that the flow falls into the so-called "transitionally rough" regime [12], in which both viscous and pressure drag are both significant.

Next, we compare the DSMC and NS mean density and velocity profiles, which are shown in Fig. 3. For reference, profiles for a smooth-wall NS simulation with the same Re and Ma are also shown (black lines). The location of the tops of the cubical roughness elements is indicated by the vertical gray lines. The agreement between the rough-wall DSMC and NS density profiles shown in the left panel of Fig. 3 is excellent, and both are in good agreement with the smooth-wall density profile, except in the vicinity of the rough wall, which causes the density to rise more abruptly. This level of agreement between the rough-wall DSMC and NS density

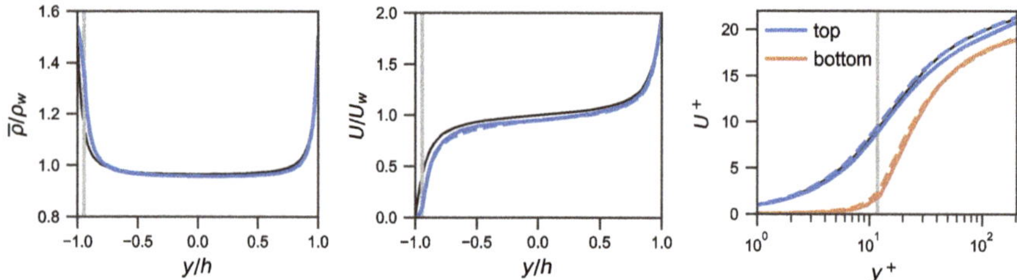

Fig. 3 Comparison of DSMC (solid lines) and NS (dashed lines) mean density (left), outer-scaled mean velocity (middle), and inner-scaled velocity (right) profiles. The vertical gray line indicates the location of the top of the roughness elements. The thin black line is a smooth-wall NS simulation for the same Re and Ma

profiles is unsurprising given that the density fluctuations for the present Ma are relatively small.

The outer-scaled mean velocity profiles are plotted in the middle panel of Fig. 3. Overall the DSMC and NS velocity profiles agree fairly well, and both deviate significantly from the smooth-wall velocity profile across the majority of the channel. The minor discrepancies between the rough-wall DSMC and NS profiles can again likely be attributed to the limited number of DSMC samples. More significant discrepancies are apparent in the upper-half inner-scaled profiles shown in the right panel of Fig. 3. However, most of this discrepancy arises from the aforementioned difference in the mean value of the skin friction τ_w, which in turn results in a different value of the friction velocity u_τ used to normalize the inner-scaled velocity profiles. Note that the NS upper-half inner-scaled velocity profile collapses onto the smooth-wall velocity profile, indicating that significant roughness effects are confined to the lower half of the channel. In other words, the primary effect of the rough bottom wall on the flow near the smooth upper wall is a modification of the boundary condition by adjusting τ_w (and hence u_τ).

Thus far we have examined spatially averaged quantities (skin friction, mean velocity profiles). Now, we turn our attention to comparison of the structure of the turbulent fluctuations, quantified via the (density-weighted) premultiplied velocity spectra, e.g., $k_x k_z E_{\rho uu}$, which describe how the kinetic energy associated with turbulent fluctuations is distributed in scale (Fourier) space, and where k_x and k_z are the streamwise and spanwise wavenumbers, respectively. Premultiplication by $k_x k_z$ ensures that equal volumes in scale space correspond to equal energy when plotted on logarithmic axes. The velocity spectra plotted as functions of streamwise and spanwise wavelengths $\lambda_x = 2\pi/k_x$ and $\lambda_z = 2\pi/k_z$, respectively, are shown in Fig. 4 for the plane immediately above the roughness elements, $y/h = -0.94$. Because the DSMC simulation has a higher spatial resolution than the NS simulation, smaller wavelengths are resolved in the DSMC spectra. Accordingly, the region containing the wavelengths in the NS simulation are indicated by the cyan rectangle. Also, small wavelengths in the DSMC simulation are dominated by thermal fluctuations, which are absent from the NS simulation [8]. Otherwise, the DSMC and NS spectra agree

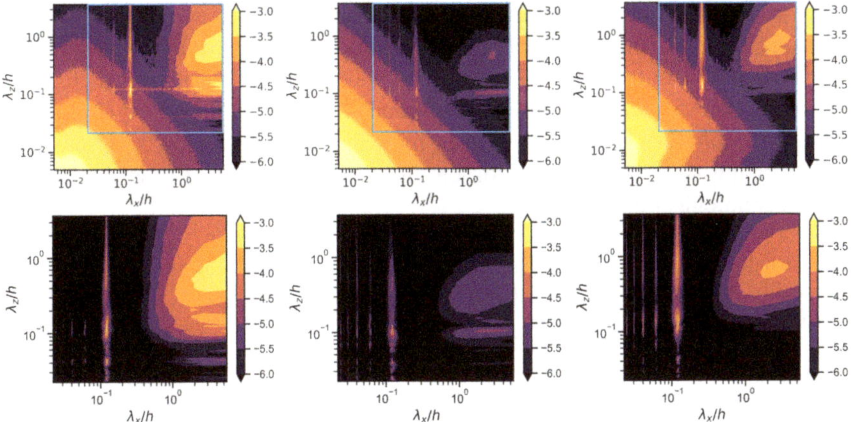

Fig. 4 Comparison of DSMC (top) and NS (bottom) velocity spectra in the plane $y/h = -0.94$ (immediately above the roughness elements). Left column: $k_x k_z E_{\rho uu}$. Center column: $k_x k_z E_{\rho vv}$. Right column: $k_x k_z E_{\rho ww}$. The blue interior of the cyan rectangles indicate the wavenumbers included in the NS simulations

very well in both shape and magnitude, indicating that both simulations capture the same turbulence structures.

For both simulations, the most energetic structures in Fig. 4 correspond to long streamwise wavelengths, $\lambda_x > h$, and spanwise wavelengths $\lambda_z \sim h$, consistent with large streaky structures commonly observed in turbulent Couette flow [11]. The other most prominent feature in the spectra is the strong peak located at $\lambda_x = \lambda_z = 0.12h$, which corresponds to the periodicity of the roughness elements, $k + s$. Several harmonics and subharmonics of this wavelength are also visible. Note also that the roughness introduces "side bands" of energetic streaky structures in the u and v spectra with $\lambda_x > h$ and $\lambda_z \approx k + s$, which are the spectral signature of intense narrow streaks corresponding to fluid being accelerated between columns of roughness elements.

3 Summary and Conclusions

We compared statistical quantities from DSMC and NS simulations of turbulent Couette flow with a rough wall. The simulations agree well for the skin friction, mean density and velocity profiles, and velocity spectra. The minor discrepancies (on the order of a few percent) in these quantities can reasonably be attributed to the relatively limited number of samples available for calculating statistics from the DSMC simulation. Nonetheless, we find the overall level of agreement between macroscopic (NS) and molecular-level (DSMC) descriptions of turbulent flow encouraging.

This successful demonstration of DSMC's capability to reproduce rough-wall turbulence effects in the near-continuum regime sets the stage for DSMC simulations of turbulent flow over realistic TPS materials having mean-free-path-scale features, for which the NS equations may not be accurate. Additionally, recent efforts have aimed at improving SPARTA's ablation modeling capability [13], so future work incorporating the effects of ablation on the overlying turbulent flow would be of great technological interest.

Acknowledgements Sandia National Laboratories is a multi-mission laboratory managed and operated by National Technology & Engineering Solutions of Sandia, LLC (NTESS), a wholly owned subsidiary of Honeywell International Inc., for the U.S. Department of Energy's National Nuclear Security Administration (DOE/NNSA) under contract DE-NA0003525. This written work is authored by an employee of NTESS. The employee, not NTESS, owns the right, title and interest in and to the written work and is responsible for its contents. Any subjective views or opinions that might be expressed in the written work do not necessarily represent the views of the U.S. Government. The publisher acknowledges that the U.S. Government retains a nonexclusive, paid-up, irrevocable, world-wide license to publish or reproduce the published form of this written work or allow others to do so, for U.S. Government purposes. The DOE will provide public access to results of federally sponsored research in accordance with the DOE Public Access Plan [14]. The authors thank L. J. DeChant for helpful discussions about this work.

References

1. NASA identifies cause of artemis i orion heat shield char loss. https://www.nasa.gov/missions/artemis/nasa-identifies-cause-of-artemis-i-orion-heat-shield-char-loss/, Accessed: 19 Dec 2024
2. Borner A, Panerai F, Mansour NN (2017) High temperature permeability of fibrous materials using direct simulation monte carlo. Int J Heat Mass Transfer 106:1318–1326
3. Bird GA (1994) Molecular gas dynamics and the direct simulation of gas flows. Oxford University Press, Oxford
4. Gallis M, Bitter N, Koehler T, Torczynski J, Plimpton S, Papadakis G (2017) Molecular-level simulations of turbulence and its decay. Phys Rev Lett 118(6):064501
5. Gallis MA, Torczynski JR, Bitter NP, Koehler TP, Plimpton SJ, Papadakis G (2018) Gas-kinetic simulation of sustained turbulence in minimal Couette flow. Phys Rev Fluids 3(7):071402
6. Bandak D, Goldenfeld N, Mailybaev AA, Eyink G (2022) Dissipation-range fluid turbulence and thermal noise. Phys Rev E 105(6):065113
7. Bell JB, Nonaka A, Garcia AL, Eyink G (2022) Thermal fluctuations in the dissipation range of homogeneous isotropic turbulence. J Fluid Mech 939:A12
8. McMullen RM, Krygier MC, Torczynski JR, Gallis MA (2022) Navier-Stokes equations do not describe the smallest scales of turbulence in gases. Phys Rev Lett 128(11):114501
9. Plimpton S, Moore S, Borner A, Stagg A, Koehler T, Torczynski J, Gallis M (2019) Direct simulation Monte Carlo on petaflop supercomputers and beyond. Phys Fluids 31(8)
10. Howard M, Bradley A, Bova SW, Overfelt J, Wagnild R, Dinzl D, Hoemmen M, Klinvex A (2017) Towards performance portability in a compressible CFD code. In: 23rd AIAA computational fluid dynamics conference, p 4407
11. Yao J, Hussain F (2023) Study of compressible turbulent plane Couette flows via direct numerical simulation. J Fluid Mech 964:A29
12. Jiménez J (2004) Turbulent flows over rough walls. Annu Rev Fluid Mech 36(1):173–196

13. Hong AY, Gallis MA, Moore SG, Plimpton SJ (2024) Toward physically realistic ablation modeling in direct simulation Monte Carlo. Phys Fluids 36(12):126132
14. https://www.energy.gov/doe-public-access-plan

Simulation of Multi-species Kinetic Instabilities with the Numerical Flow Iteration

Rostislav-Paul Wilhelm and Manuel Torrilhon

Abstract Kinetic instabilities are one of the most challenging aspects in computational plasma physics. Accurately capturing their onset and evolution requires fine resolution of the high-dimensional distribution functions of each relevant species, which quickly becomes computationally prohibitively expensive. Additionally, plasma dynamics is an inherently multi-scale phenomenon due to the vast separation of scales between heavy ions and light-weight electrons. In previous work the *Numerical Flow Iteration* (NuFI) was suggested as a high-fidelity alternative with reduced memory complexity making it an interesting candidate to simulate complicated kinetic instabilities. In this work we extend NuFI to non-periodic boundary conditions and demonstrate how it is possible to reduce the computational complexity to allow for longer simulation periods.

Keywords Computational plasma physics · Kinetic theory · Numerical methods

1 Introduction

In many applications such as ion thrusters or nuclear fusion devices it is relevant to fundamentally understand how instabilities in a plasma are triggered and how they evolve over time. This often requires modelling the plasma evolution through the (collisionless) Vlasov equation [1]

$$\partial_t f^\alpha + v \cdot \nabla_x f^\alpha + F^\alpha \cdot \nabla_v f^\alpha = 0. \tag{1}$$

R.-P. Wilhelm (✉)
Centre for Mathematical Plasma Astrophysics, KU Leuven, Leuven, Belgium
e-mail: paul.wilhelm@kuleuven.be

R.-P. Wilhelm · M. Torrilhon
Institute for Applied and Computational Mathematics, RWTH Aachen University, Aachen, Germany

© The Author(s) 2026

441

M. Grabe et al. (eds.), *Rarefied Gas Dynamics*, Springer Aerospace Technology,
https://doi.org/10.1007/978-3-032-00094-1_42

where f^α is the probability distribution of the species α in the up to six-dimensional phase-space. To discuss the fundamental methodical issues we restrict ourselves to the electro-static case, i.e., where the forces reduce to the self-induced electric field force $F^\alpha = \frac{q_\alpha}{m_\alpha} E$ computed through

$$- \Delta_x \varphi = \rho = \sum_\alpha q_\alpha \int_{\mathbb{R}^3} f_\alpha dv, \tag{2}$$

$$E = -\nabla_x \varphi. \tag{3}$$

This coupling yields the (non-linear) Vlasov–Poisson system.[1]

Due to the high-dimensionality of the problem it is already challenging to handle by itself even on modern large-scale supercomputers. In addition f is notoriously known for its turbulent behaviour and development of fine scale structures, which are relevant for the plasma dynamics and, in particular, are crucial to capture the onset of kinetic instabilities. Additionally, the problem is naturally multi-scale due to the large difference in mass between electrons and ion-species, however, in-situ and numerical measurements suggest that each stage of the cascade can have a non-negligible effect on the global dynamics [2, 3].

We argue that many problems of "classical Vlasov solvers" stem from trying to directly discretize the complicated and high-dimensional distribution functions. Thus NuFI arises from a change of paradigm: Instead of a direct discretization of f^α we approximate the characteristics. This is possible in an efficient way as they can be reconstructed on-the-fly using operator-splitting and only requires knowledge of the past electric fields at the cost of increasing the computational complexity from linear to quadratic in the number of time-steps [4]. Previous publications discussed the efficient implementation of NuFI for single- and multi-species simulations in the electro-static limit with periodic boundary conditions [4–6]. This work demonstrates how NuFI can be extended to non-periodic boundary conditions and showcases a prototype implementation of a restart to reduce the computational complexity. In Sect. 2 we give a brief introduction to NuFI followed, including multi- species simulations and handling of non-periodic boundary conditions. In Sect. 3 we discuss a restart procedure for NuFI and discuss its impact on accuracy and computational time. In Sect. 4 we showcase how NuFI can be applied by simulating an ion-acoustic shock with non-periodic boundaries.

2 Simulation of Kinetic Plasma Dynamics with NuFI

Before moving on to the main focus of this work, we want to briefly recap the ideas of the Numerical Flow Iteration (NuFI). The underlying idea of NuFI is to approximate the flow-map of the Vlasov–Poisson equation instead of directly solving for the

---

[1] The normalization of units is discussed in Appendix 5.

distribution function f^α. The solution of (1) can be written as

$$f^\alpha(t, x, v) = f_0^\alpha(\Phi_t^{0,\alpha}(x, v)),\tag{4}$$

where the *backward flow* $s \mapsto \Phi_t^{s,\alpha}(x, v) = (\hat{x}_\alpha(s), \hat{v}_\alpha(s))$ is the solution to

$$\begin{aligned}
\tfrac{\mathrm{d}}{\mathrm{d}s}\hat{x}^\alpha(s) &= -\hat{v}^\alpha(s), & \hat{x}^\alpha(t) &= x, \\
\tfrac{\mathrm{d}}{\mathrm{d}s}\hat{v}^\alpha(s) &= -\tfrac{q_\alpha}{m_\alpha} E\left(s, \hat{x}^\alpha(s)\right), & \hat{v}^\alpha(t) &= v.
\end{aligned}\tag{5}$$

To solve (5) numerically we use the Störmer–Verlet time-integration method: Starting from $\hat{x}_n^h = x$ and $\hat{v}_n^h = v$ at the time-step $t = t_n$ compute

$$\hat{v}_{i-1/2}^{h,\alpha} = \hat{v}_i^{h,\alpha} - \tfrac{\Delta t}{2}\tfrac{q_\alpha}{m_\alpha} E(t_i, \hat{x}_i^{h,\alpha}),\tag{6}$$

$$\hat{x}_{i-1}^{h,\alpha} = \hat{x}_i^{h,\alpha} - \hat{v}_{i-1/2}^{h,\alpha},\tag{7}$$

$$\hat{v}_{i-1}^{h,\alpha} = \hat{v}_{i-1/2}^{h,\alpha} - \tfrac{\Delta t}{2}\tfrac{q_\alpha}{m_\alpha} E(t_{i-1}, \hat{x}_{i-1}^{h,\alpha})\tag{8}$$

for $i = 0, \ldots, n$. Now

$$f^\alpha(t, x, v) = f_0^\alpha(\hat{x}_0^{h,\alpha}, \hat{v}_0^{h,\alpha}) + \mathcal{O}\left(\Delta t^2\right).\tag{9}$$

Note that to compute ρ via (2) we can skip the first half-step (6) as it can be recast as a transformation of the respective integral with functional determinant 1. With this we can state the full NuFI algorithm. Note that we only write it out for $d = 1$ but extending to $d = 2, 3$ is trivial [4].

Algorithm 1 Solving the Vlasov–Poisson system in $d = 1$.

---

function NuFI(f_0^e, f_0^i, N_t, Δt, N_x, N_v)
    Allocate a array C for the coefficients of φ ($N_t N_x$ floats).
    for $n = 0, \ldots, N_t$ **do**
        for $k = 0, \ldots, N_x$ **do**
        $\rho_k^e, \rho_k^i = 0$.
        for $l = 0, \ldots, N_v$ **do**
            Evaluate $\tilde{f}_e = f^e(t_n, x_k, v_l)$ and $\tilde{f}_i = f^i(t_n, x_k, v_l)$ using (9).
            $\rho_k^e \mathrel{+}= h_v \tilde{f}_e$ and $\rho_k^i \mathrel{+}= h_v \tilde{f}_i$.
        end for
        $\rho_k = q_i \rho_k^i + q_e \rho_k^e$.
        end for
        Solve the Poisson's equation via FFT to obtain $\varphi_0, \ldots, \varphi_{N_x}$ from $\rho_0, \ldots, \rho_{N_x}$.
        Interpolate $\varphi_0, \ldots, \varphi_{N_x}$ and store the coefficients of $\varphi_{\Delta t, h}$ in the B-Spline basis.
    end for
end function

A detailed derivation and discussion of the algorithm can be found in Kirchhart and Wilhelm's previous work [4]. The extension to multiple species and adaptive integration (instead of the mid-point integration rule used above) was presented in Wilhelm et al. [5]. A convergence analysis and accuracy comparisons to Particle-In-Cell (PIC) and semi-Lagrangian approaches was also carried out in previous works [4–6].

2.1 Handling of Boundary Conditions

In the following we list how common boundary conditions can be handled with the NuFI

- **Periodic boundary**: When the characteristic hits a periodic boundary, let it re-enter on the opposite side and continue the backwards-tracing as before.
- **In-Flow boundary**: When the characteristic hits a in-flow boundary stop the back-tracing procedure and return the value of f prescribed at the respective boundary (instead of f_0).
- **Open boundary**: No additional implementation is required for this case as we move backwards along the characteristics and thus we are not interested in characteristics leaving the domain, i. e., they are naturally accounted for.
- **Reflective boundary**: For perfect reflection it is sufficient to reflect the characteristic analogous to a reflection of a particle. If a temperature is prescribed at a wall, this is analogous to prescribing a Maxwellian as boundary value.

3 Restarting NuFI

The improved accuracy and low-memory consumption of NuFI comes at cost of having to evaluate the entire characteristic map backwards until the initial data is reached, thus the computational complexity of NuFI is quadratic, not linear, in the total number of time-steps [4, 5]. To elevate this limiting factor, in particular, considering that we are interested in kinetic instabilities and turbulence which can take substantial time periods to develop it is of interest to restart NuFI after a fixed number of time-steps throughout a longer simulation.

A simple, yet effective approach is to store values of $f^\alpha(t, \cdot, \cdot)$ on a (regular) grid with $N_x^r \times N_v^r$ degrees of freedom in phase-space every $n_r \gg 1$ time-steps and evaluating in between the grid-points using linear interpolation.

Remark 1 This naive approach only works for lower-dimensional simulations as otherwise the memory-requirement would grow too large same as for any classical approach directly approximating the distribution function. A more efficient approach would be adaptive or sparse meshes such as they are discussed in e.g. Gerhard et al. [7]. However, this is a prototype implementation to demonstrate the feasibility

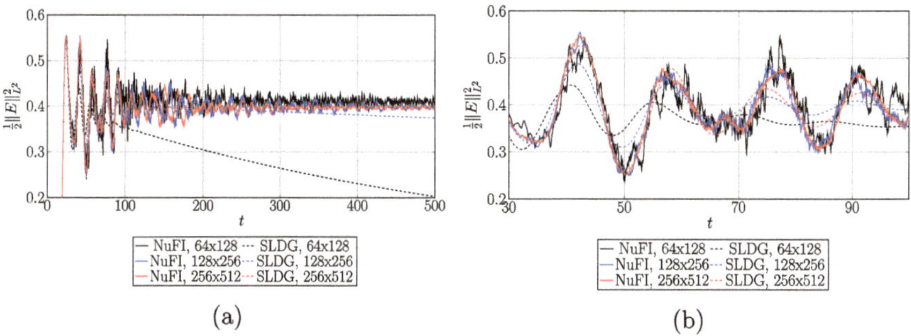

Fig. 1 A comparison of the electric energy in a two stream instability simulation between NuFI and SLDG. NuFI uses the restart procedure discussed in Sect. 3 with $n_r = 200$ and $N_x^r = N_v^r = 1024$

of restarting NuFI and therefore we kept it simple. Additionally, this type of storage is easy to compress using low-rank tensor-based approaches as was discussed in Kormann's work [8].

In the following we compare simulation results of NuFI with SLDG (Semi-Lagrangian Discontinuous Galerkin) [9]. To this end we consider the *two stream instability* benchmark, which is known as a notoriously complicated kinetic instability [4]. In Fig. 1 we compare the evolution of the electric energy over time until $T = 500$. Figure 1b zooms in on the electric energy after the kinetic instability saturates at which point the electric energy starts periodically oscillating around a fixed level. While NuFI captures this effect even with low resolution until $t = 100$, but SLDG requires a four times finer resolution to capture the oscillation even at early times. At later stages the electric energy essentially "flat-lines" and starts decreasing (the faster the lower the resolution), which is a consequence of numerical diffusion.

We also checked how restarting NuFI with above method influences its conservation properties, see Fig. 2. As expected some of the conservation properties are no longer fulfilled exactly by the restarted NuFI approach: There is a relative deviation of up to 1% for the entropy and of up to 4% for the L^2-norm until $T = 500$. The conservation of total energy seems to be broken during the first restart, when the total energy increases by roughly 1% but remains (essentially) constant afterwards. However, these drifts are still substantially lower than for SLDG with the same resolution. The error in entropy conservation is at 7% and the error in the L^2-norm exceeds 14%. Therefore we can conclude that even if NuFI looses its exact conservation properties through a this restart procedure the overall accuracy and conservation properties still substantially exceed those of SLDG.

Finally let us remark that the restart procedure indeed reduced the computational complexity back to linear for NuFI. Therefore in the case of the simulations run for Fig. 2 we observed a speed-up of a factor 35, i.e., a simulation until $T = 500$ would take 2240 s without restart, while with restart it only takes 64 s. SLDG requires for a simulation with same grid resolution roughly 20 s, i.e., is roughly 3 times faster.

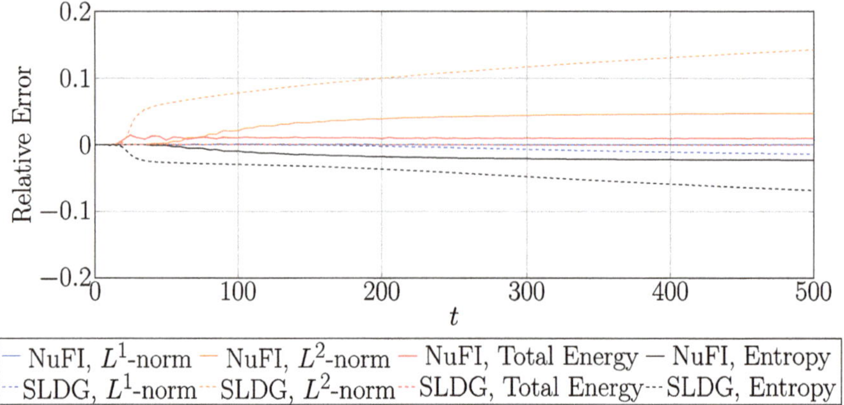

Fig. 2 Compared are the relative errors in L^1-, L^2-norm as well as total energy and entropy between NuFI and SLDG. Both approaches used a resolution of $N_x = N_v = 64$ and $\Delta t = \frac{1}{16}$ and NuFI was restarted every $n_r = 200$ time-steps on a uniform grid with $N_x^r = N_v^r = 1024$ grid points

However, when accounting for the better accuracy and conservation properties of NuFI we argue that it is worthwhile to use NuFI.[2]

4 Ion-Acoustic Shock with Reflecting Wall

After establishing in theory how NuFI can be used run simulations of the Vlasov–Poisson system with a range of boundary conditions and how it is possible to improve the computational complexity enabling longer simulations, we want to showcase some results obtained in a setting with non-periodic boundary conditions. The simulation setup is inspired by the setup used by Liseykina et al. in their work, however, with some slight modifications[3] [10].

We again restrict ourselves to $d = 1$ with only electro-static forces. This time we consider both electrons and ions (Hydrogen-ions), where we set the realistic mass ratio of $M_r = 1836$, i.e., $m_e = 1$ and $m_i = 1836$. The physical domain is from $x_{\min} = -L$ to $x_{\max} = 0$ with $L \gg 1$. The ions and electrons are initialized with unperturbed Maxwellian velocity distributions, where the ion distribution is centred around a drift speed $u_s > 0$. The right boundary is set to be a perfectly reflecting wall for both electrons and ions. The left boundary is an open boundary for particles leaving the domain and for particles entering the domain prescribes an

---

[2] All of the above simulations were carried out on a local workstation using a Intel(R) Xeon(R) E-2276M CPU @ 2.80GHz with 6 physical cores.

[3] Our initial goal to reproduce the results from Liseykina's work failed due to a unclear parameter choices in their work. Still we believe that the setup presented there is an interesting setup to verify a Vlasov solver in non-periodic boundary conditions.

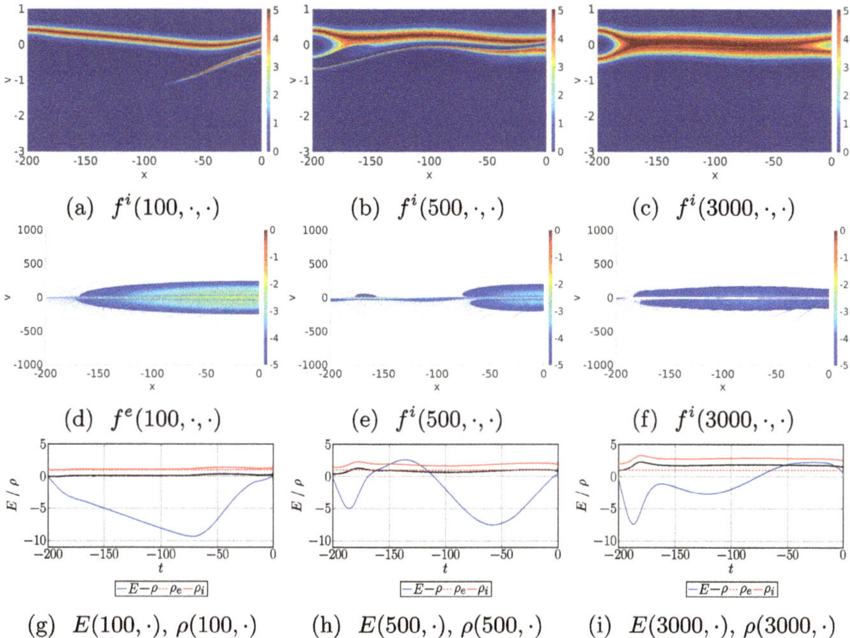

Fig. 3 Simulation of setup described in Sect. 4. Top row are the ion distribution functions at $t = 100, 500, 3000$. The middle row shows the absolute distance between electron distribution function and the Maxwellian distribution. The bottom row shows the electric field, the total charge density as well as the charge densities of the electrons and ions

in-flow with the same Maxwellian as the initial condition for the particle species. Both boundaries are set to be perfectly conducting, i.e., for the Poisson equation we set zero Neumann-boundary conditions.

Following the suggestions of Liseykina et al. in the $u_s = 0.4$ case we chose $T_e = 11475 \gg T_i = 10$ and $L = 200$. We simulate with NuFI choosing $N_x = 1024$, adaptive integration in velocity with at least $N_v = 64$ cells and a time-integration step of $\Delta t = \frac{1}{10}$. To restart NuFI we use $n_r = 200$ and $N_x^r = N_v^r = 1024$.

In Fig. 3 we display the distribution functions as well as the electric field and charge densities at the times $t = 100, 500, 3000$. To be better able to identify the shock in the electron distribution function we show the (absolute) distance between f^e and the Maxwellian (with T_e and m_e from the initial data) in a logarithmic scale. The reflected ion population moving from right to left induces a shock in the electrons, which initially also moves from right to left. When the reflected ion population reaches the left boundary a vortex starts forming.[4]

Another big vortex forms at the right boundary and grows with time. Additionally smaller vortices also form in the ion distribution in the middle and move to the left finally uniting with the stationary vortex at the left boundary.

---

[4] The vortex at the left boundary is a boundary effect and could be avoided by choosing L even larger. As we are only interested in triggering the shock this setup is sufficient.

5   Conclusion

In this work we discussed how the Numerical Flow Iteration (NuFI) can be extended
to handle multi-species Vlasov simulations with non-periodic boundary conditions
and additionally we showed that NuFI can be restarted while still remaining more
accurate than other semi-Lagrangian approaches with the same resolution. Finally
we successfully applied the resulting algorithm to a simulation of an ion-acoustic
shock caused by an ion-population being reflected at a wall.

Acknowledgements Special thanks goes to Jan Eifert who assisted in the initial implementation of the Dirichlet extension to NuFI's codebase. Furthermore we want to acknowledge the
support of from the German national high performance computing organisation (NHR) funded
by the Federal Ministry of Education and Research as well as the state governments and by the
Deutsche Forschungsgemeinschaft (DFG, German Research Foundation)—333849990/GRK2379
(IRTG Hierarchical and Hybrid Approaches in Modern Inverse Problems).

Normalization of Units

To normalize the quantities in our work we follow Arber and Vann: The spatial length
are given in Debye length $\lambda_D = \sqrt{(\epsilon_0 k_B T_e)/(n_0 e^2)}$ (n_0 is the equilibrium number
density), velocities are given in the thermal speed $v_{th} = \sqrt{(k_B T_e)/(m_e)}$, time is given
in ω_p^{-1} and the electric field is in[11] $(m_e v_{th}^2)/(e\lambda_D)$.

References

1. Chen F (2015) Introduction to plasma physics and controlled fusion. Springer Cham. https://
   doi.org/10.1007/978-3-319-22309-4
2. Lapenta G et al (2022) Do we need to consider electrons' kinetic effects to properly model a
   planetary magnetosphere: the case of mercury. Space Phys, J Geophys Res. https://doi.org/10.
   1029/2021JA030241
3. Verscharen D et al (2019) The multi-scale nature of the solar wind. Living Rev Sol Phys. https://
   doi.org/10.1007/s41116-019-0021-0
4. Kirchhart M et al (2024) The Numerical flow iteration for the Vlasov-Poisson equation. SIAM
   J Sci Comp. https://doi.org/10.1137/23M154710X
5. Wilhelm R-P, Eifert J et al (2024) High fidelity simulations of the multi- species Vlasov equation
   in the electro-static, collisional-less limit. J Plasma Phys Contr Fusion. https://doi.org/10.1088/
   1361-6587/ad9fdb
6. Wilhelm R-P et al (2023) Introduction to the numerical flow iteration for the Vlasov-Poisson
   equation. Pro Appl Math Mech. https://doi.org/10.1002/pamm.202300162
7. Gerhard N et al (2022) A wavelet-free approach for multiresolution-based grid adaptation for
   conservation laws, communications on applied mathematics and computation. https://doi.org/
   10.1007/s42967-020-00101-6
8. Kormann K (2015) A Semi-Lagrangian Vlasov Solver in Tensor Train Format. SIAM Journal
   on Scientific Computing. https://doi.org/10.1137/140971270

9. Einkemmer L (2016) High performance computing aspects of a dimension independent semi-Lagrangian discontinuous Galerkin code. Comput Phys Commun. https://doi.org/10.1016/j.cpc.2016.01.012
10. Liseykina TV et al (2015) Ion-acoustic shocks with reflected ions: modelling and particle-in-cell simulations. J Plasma Phys. https://doi.org/10.1017/S002237781500077X
11. Arber T et al (2002) A critical comparison of Eulerian-grid-based Vlasov solvers. J Comput Phys. https://doi.org/10.1006/jcph.2002.7098

A Novel Splitting Method for Vlasov-Ampère

James A. Rossmanith and Christine Vaughan

Abstract Vlasov equations model the dynamics of plasma in the collisionless regime. A standard approach for numerically solving the Vlasov equation is to operator split the spatial and velocity derivative terms, allowing simpler time-stepping schemes to be applied to each piece separately (known as the Cheng-Knorr method). One disadvantage of such an operator split method is that the order of accuracy of fluid moments (e.g., mass, momentum, and energy) is restricted by the order of the operator splitting (second-order accuracy in the Cheng-Knorr case). In this work, we develop a novel approach that first represents the particle density function on a velocity mesh with a local fluid approximation in each discrete velocity band and then introduces an operator splitting that splits the inter-velocity band coupling terms from the dynamics within the discrete velocity band. The advantage is that the inter-velocity band coupling terms are only needed to achieve consistency of the full distribution functions, but the local fluid models within each band are sufficient to achieve high-order accuracy on global moments such as mass, momentum, and energy. The resulting scheme is verified on several standard Vlasov-Poisson test cases.

1 Introduction

Plasma is a highly energetic state of matter where electrons have dissociated from their nuclei, creating a mixture of charged particles that interact through electromagnetic forces and particle-particle collisions. In the collisionless approximation, the plasma can be modeled via the celebrated Vlasov equation. In particular, we are concerned here with the (non-dimensionalized) single-species (i.e., streaming electrons with stationary neutralizing background ions) Vlasov-Ampère system [5] in 1D1V:

J. A. Rossmanith (✉)
Department of Mathematics, Iowa State University, Ames, IA, USA
e-mail: rossmani@iastate.edu

C. Vaughan
Physicians Health Plan, Lansing, MI, USA

M. Grabe et al. (eds.), *Rarefied Gas Dynamics*, Springer Aerospace Technology,
https://doi.org/10.1007/978-3-032-00094-1_43

$$f_{,t} + v f_{,x} - E(t, x) f_{,v} = 0, \quad E_{,t} = \mathbb{M}_1(t, x) - J_0, \tag{1}$$

$$\mathbb{M}_1(t, x) = \int_{-\infty}^{\infty} v f(t, x, v)\, dv, \quad J_0 = \frac{1}{b - a} \int_a^b \mathbb{M}_1(t = 0, x)\, dx, \tag{2}$$

where $t \in \mathbb{R}_{\geq 0}$ is time, $x \in [a, b]$ is the spatial coordinate, $v \in \mathbb{R}$ is the velocity coordinate, $f(t, x, v) : \mathbb{R}_{\geq 0} \times [a, b] \times \mathbb{R} \mapsto \mathbb{R}_{\geq 0}$ is the particle density function (PDF) describing the distribution of electrons in phase space, $E(t, x) : \mathbb{R}_{\geq 0} \times [a, b] \mapsto \mathbb{R}$ is the electric field, $\mathbb{M}_1(t, x) : \mathbb{R}_{\geq 0} \times [a, b] \mapsto \mathbb{R}$ is the electron momentum density, and $J_0 \in \mathbb{R}$ is the stationary background ion momentum density. The initial electric field can be computed from the Gauss' law:

$$E_{,x} = \rho_0 - \rho(t, x), \; \rho(t, x) = \int_{-\infty}^{\infty} f(t, x, v)\, dv, \; \rho_0 = \frac{1}{b - a} \int_a^b \rho(t = 0, x)\, dx, \tag{3}$$

where $\rho(t, x) : \mathbb{R}_{\geq 0} \times [a, b] \mapsto \mathbb{R}_{\geq 0}$ is the electron mass density and $\rho_0 \in \mathbb{R}$ is the stationary background ion mass density.

A standard approach for numerically solving the Vlasov equation is to operator split the spatial and velocity derivative terms, allowing simpler time-stepping schemes to be applied to each piece separately (known as the Cheng-Knorr method [1]):

$$\textbf{Problem A:} \quad f_{,t} + v f_{,x} = 0, \quad E_{,t} = \mathbb{M}_1(t, x) - J_0; \tag{4}$$

$$\textbf{Problem B:} \quad f_{,t} - E f_{,v} = 0. \tag{5}$$

One disadvantage of such an operator split method is that the order of accuracy of fluid moments (e.g., mass, momentum, and energy) is restricted by the order of the operator splitting (second-order accuracy in the Cheng-Knorr [1] case, which is based on Strang operator splitting [10].)

2   A Novel Operator Splitting Method

We develop in this work a novel operator splitting approach for the Vlasov-Ampère system (1)–(2). We begin by discretizing the truncated velocity space, $v \in [V_{\min}, V_{\max}]$, into non-overlapping *velocity bands* each of width and center:

$$\Delta v = (V_{\max} - V_{\min}) / N_v, \quad v_j = V_{\min} + \left(j - \frac{1}{2} \right) \Delta v \quad \text{for} \quad j = 1, \ldots, N_v. \tag{6}$$

where $V_{\max}$ is the maximum of velocity domain, $V_{\min}$ is the minimum velocity, and N_v is the number of velocity layers. Within each velocity layer, we define the *local fluid moments* as,

$$\mathbb{M}_\ell^{(j)} = \int_{v_{j-\frac{1}{2}}}^{v_{j+\frac{1}{2}}} v^\ell f(t, x, v)\, dv, \quad \ell = 0, 1, 2, 3, 4, \tag{7}$$

where $v_{j\pm\frac{1}{2}} = v_j \pm \Delta v/2$. Theoretically, we can take as many local fluid moments as desired; in the current work, we restrict ourselves to five (e.g., $\ell = 0, 1, 2, 3, 4$). In practice, at least the first three moments should be included (mass, momentum, energy) to guarantee high-order accuracy on the lowest-order statistics. The more moments included, the more computationally complex the system in each velocity band will be. Five moments were chosen to balance accuracy and computational complexity. We then take Vlasov equation (1), multiply by v^ℓ, and integrate over each discrete velocity band, which, after integration by parts in velocity, becomes the following expression:

$$\mathbb{M}_{\ell,t}^{(j)} + \mathbb{M}_{\ell+1,x}^{(j)} - \left[E v^\ell f \right]_{v_{j-\frac{1}{2}}}^{v_{j+\frac{1}{2}}} = -\ell \mathbb{M}_{\ell-1}^{(j)} E, \tag{8}$$

for $j = 1, \ldots, N_v$ and $\ell = 0, 1, 2, 3, 4$. We note that the global moments (e.g., mass, momentum, and energy) can be directly computed from the local moments:

$$\mathbb{M}_\ell = \sum_{j=1}^{N_v} \mathbb{M}_\ell^{(j)}, \quad \text{for } \ell = 0, 1, 2, 3, 4. \tag{9}$$

In each velocity band, we arrive at the following system of local moments:

$$\underline{\mathbb{M}}_{,t}^{(j)} + \underline{\underline{A}}^{(j)}\, \underline{\mathbb{M}}_{,x}^{(j)} = -E \underline{S}^{(j)} + E \underline{S}_+^{(j)} f\big|_{v_{j+\frac{1}{2}}} - E \underline{S}_-^{(j)} f\big|_{v_{j-\frac{1}{2}}}, \tag{10}$$

where

$$\underline{\mathbb{M}}^{(j)} = \begin{bmatrix} \mathbb{M}_0^{(j)} \\ \mathbb{M}_1^{(j)} \\ \mathbb{M}_2^{(j)} \\ \mathbb{M}_3^{(j)} \\ \mathbb{M}_4^{(j)} \end{bmatrix}, \quad \underline{\underline{A}}^{(j)} = \begin{bmatrix} 0 & 1 & 0 & 0 & 0 \\ 0 & 0 & 1 & 0 & 0 \\ 0 & 0 & 0 & 1 & 0 \\ 0 & 0 & 0 & 0 & 1 \\ A_{51} & A_{52} & A_{53} & A_{54} & A_{55} \end{bmatrix}, \quad \underline{S}^{(j)} = \begin{bmatrix} 0 \\ \mathbb{M}_0^{(j)} \\ 2\mathbb{M}_1^{(j)} \\ 3\mathbb{M}_2^{(j)} \\ 4\mathbb{M}_3^{(j)} \end{bmatrix}, \quad \underline{S}_\pm^{(j)} = \begin{bmatrix} 1 \\ v_{j\pm\frac{1}{2}} \\ v_{j\pm\frac{1}{2}}^2 \\ v_{j\pm\frac{1}{2}}^3 \\ v_{j\pm\frac{1}{2}}^4 \end{bmatrix}, \tag{11}$$

where

$$A_{51} = \frac{5\Delta v^4 v_j}{336} - \frac{5\Delta v^2 v_j^3}{18} + v_j^5, \quad A_{52} = \frac{-5\Delta v^4}{336} + \frac{5\Delta v^2 v_j^2}{6} - 5v_j^4,$$

$$A_{53} = \frac{-5\Delta v^2 v_j}{6} + 10v_j^3, \quad A_{54} = \left(\frac{5}{18}\right)\left(\Delta v^2 - 36v_j^2\right), \quad A_{55} = 5v_j. \tag{12}$$

It can readily be shown that $\underline{\underline{A}}^{(j)}$ is fully diagonalizable with only real eigenvalues. For a full derivation, please see Vaughan [11].

We note that the first three terms in equation (10) are entirely local to the jth velocity band, and only the final two terms in (10) are responsible for coupling the jth band to the $j+1$ and $j-1$ bands. This suggests a new operator-splitting approach:

Problem A: $\quad \underline{M}_{,t}^{(j)} = E\underline{S}_+^{(j)} f\big|_{v_{j+\frac{1}{2}}} - E\underline{S}_-^{(j)} f\big|_{v_{j-\frac{1}{2}}};$ $\qquad\qquad$ (13)

Problem B: $\quad \underline{M}_{,t}^{(j)} + \underline{\underline{A}}^{(j)}\underline{M}_{,x}^{(j)} = -E\underline{S}^{(j)}, \quad E_{,t} = \mathbb{M}_1(t,x) - J_0.$ $\qquad$ (14)

Just as in Cheng and Knorr [1], we introduce Strang operator splitting [10]: we will solve **Problem A** on a half time step, $\frac{\Delta t}{2}$, followed by solving **Problem B** on a full time step, Δt, and then finally again solving **Problem A** on a half time.

Proposition 1 *Under the assumption that the distribution function, f, is zero at $v = V_{min}$ and $v = V_{max}$, the operator splitting technique shown in (13) and (14) clearly introduces an $\mathcal{O}\left(\Delta t^2\right)$ splitting error into the accuracy of the PDF, f, but there is no operator splitting error in the global moments: $\mathbb{M}_0, \dots, \mathbb{M}_4$.*

Proof The global moments, $\mathbb{M}_0, \mathbb{M}_1, \mathbb{M}_2, \mathbb{M}_3, \mathbb{M}_4$, as defined by (9) remain constant during **Problem A**, since $\underline{S}_+^{(j)} = \underline{S}_-^{(j+1)}$ for all j, and summing over j in (13) results in a telescoping series such that the only remaining terms involve f at $v = V_{min}$ and $v = V_{max}$, which we assume are zero. $\qquad\qquad\square$

3　Numerical Method

We discretize (13)–(14) using the Lax-Wendroff discontinuous Galerkin (LxW-DG) as described in Johnson et al. [4], which is based on previous versions of the LxW-DG work of Qiu et al. [7] and Gassner et al. [2]. The method allows us to discretize in space and time with piecewise polynomials of degree M_O (i.e., $\mathcal{P}^{M_O}$), which in the case of smooth solutions results in errors of size $\mathcal{O}\left(\Delta t^{M_O+1} + \Delta x^{M_O+1}\right)$. We omit the details of this here and instead refer the reader to Johnson et al. [4].

The novel aspect of this work is the introduction of **Problem A** (13); and, as such, we include details here on how this equation is solved numerically. If $E \geq 0$, we construct the following Taylor expansion comparing moments at $t = t^n$ and $t = t^n + \Delta t$ (see Vaughan [11]):

$$\underline{M}^{(j)}(t^n + \Delta t) = \underline{M}^{(j)} - \Delta t \left(\underline{S}^{(j)}_+ E \beta^{(j)}_+ - \underline{S}^{(j)}_- E \beta^{(j-1)}_+ \right) + \mathcal{O}(\Delta t^5), \qquad (15)$$

where

$$\begin{aligned}
\beta^{(j)}_+ &= \left(1 - \frac{5\kappa}{2} + \frac{25\kappa^2}{6} - \frac{125\kappa^3}{24}\right) \alpha^{(j)}_+ + \left(\frac{\kappa}{2} - \frac{5\kappa^2}{3} + \frac{25\kappa^3}{8}\right) \alpha^{(j-1)}_+ \\
&\quad + \left(\frac{\kappa^2}{6} - \frac{5\kappa^3}{8}\right) \alpha^{(j-2)}_+ + \left(\frac{\kappa^3}{24}\right) \alpha^{(j-3)}_+, \quad \kappa = \frac{5E\Delta t}{\Delta v}, \quad \alpha^{(j)}_+ = \underline{N}^{(j)}_+ \cdot \underline{M}^{(j)},
\end{aligned} \qquad (16)$$

where $\underline{N}^{(j)}_+$ is an interpolated vector describing the distribution function at the upper interface of the velocity band j. There is a similar formula for $E \leq 0$ [11].

We define on each element, Gauss-Legendre quadrature points, ξ_m, for $m = 1, 2, \ldots, M_O$. We also define the Legendre polynomials $\psi_k(\xi)$ for $k = 1, 2, \ldots, M_O$. Putting all the pieces together, we find that

$$\underline{M}^{(j)}_k(t^n + \Delta t) = \underline{M}^{(j)}_k(t^n) - \Delta t \left(\underline{\gamma}^{(j+\frac{1}{2})}_k - \underline{\gamma}^{(j-\frac{1}{2})}_k \right) + \mathcal{O}(\Delta t^5), \qquad (17)$$

where

$$\underline{\gamma}^{(j\pm\frac{1}{2})}_k = \underline{S}^{(j)}_\pm \left[\frac{1}{2} \sum_{m=1}^{M_O} \omega_m \psi_\ell(\xi_m) \beta^{(j\pm\frac{1}{2})}_m \right], \qquad (18)$$

where

$$\beta^{(j-\frac{1}{2})}_m = \min(E(\xi_m), 0) \, \beta^{(j-1)}_- + \max(E(\xi_m), 0) \, \beta^{(j)}_+, \qquad (19)$$

$$\beta^{(j+\frac{1}{2})}_m = \min(E(\xi_m), 0) \, \beta^{(j)}_- + \max(E(\xi_m), 0) \, \beta^{(j+1)}_+. \qquad (20)$$

4 Numerical Examples

In this section we discuss the examples used to test the velocity-banded quadrature-based moment closure. We begin with a forced problem, using the method of manufactured solutions. We used this example to perform a convergence test to show the method converges up to fourth order in space and time. We also show that error in space and time as well as round-off error dominate the error over that produced by the discretization of the velocity space. Next, we show the results from the weak Landau damping problem, the strong Landau damping problem, and two-stream instability. Finally, we show the results from a plasma sheath example.

Table 1 Relative L^2 errors for the forced manufactured problem with 200 bands for the Vlasov-Ampère equation with the $\mathcal{P}^1$ (Columns 2 and 3), $\mathcal{P}^2$ (Columns 4 and 5), and $\mathcal{P}^3$ (Columns 6 and 7) Lax-Wendroff DG schemes

N	e_N ($M_O = 1$)	$\log_2 \frac{e_{N/2}}{e_N}$	e_N ($M_O = 2$)	$\log_2 \frac{e_{N/2}}{e_N}$	e_N ($M_O = 3$)	$\log_2 \frac{e_{N/2}}{e_N}$
20	5.203e−02	–	2.995e−03	–	1.144e−04	–
40	1.454e−02	1.840	4.075e−04	2.878	7.829e−06	3.869
80	3.890e−03	1.902	4.964e−05	3.037	4.927e−07	3.990
160	9.943e−04	1.968	6.170e−06	3.008	3.098e−08	3.991
320	2.502e−04	1.991	7.715e−07	3.000	2.335e−09	3.730

Here N refers to the number of spatial elements, $\Delta x = 2\pi/N$, and $\Delta t = \text{CFL} \times \Delta x / V_{\max}$, where $V_{\max} = 4$ and $\text{CFL} = 0.09$

4.1 Manufactured Solution

Consider the forced Vlasov-Poisson equation:

$$f_{,t} + v f_{,x} + E f_{,v} = \psi(t, x, v), \quad E_{,x} = \sqrt{\pi} - \int_{-\infty}^{\infty} f(t, x, v)\, dv, \qquad (21)$$

on the domain $(t, x, v) \in [0, 0.1] \times [-\pi, \pi] \times [-4, 4]$ with periodic boundary conditions in x. The forced exact solution is given by [8]:

$$f(t, x, v) = (2 - \cos(2x - 2\pi t))\, e^{-\frac{(4v-1)^2}{4}}, \quad E(t, x) = \frac{\sqrt{\pi}}{4} \sin(2x - 2\pi t), \tag{22}$$

$$\psi(t, x, v) = \frac{1}{2} S e^{-\frac{1}{4}(4v-1)^2} \left[\left(2\sqrt{\pi} + 1\right)\left(4v - 2\sqrt{\pi}\right) - \sqrt{\pi}\,(4v - 1)\, C \right], \tag{23}$$

where $S := \sin(2x - 2\pi t)$ and $C := \cos(2x - 2\pi t)$. A convergence table for 3 different polynomial orders are shown in Table 1. In the convergence study, we are decreasing Δx and Δt (both are proportional to $1/N$), but fixing Δv (i.e., the number of velocity bands).

4.2 Weak and Strong Landau Damping

Landau damping is a well-studied problem both numerically and analytically [1, 3], and the damping effect of the L_2 norm of the electric field is well-established and has been rigorously studied by Mouhot and Villani [6]. This work earned Villani the 2010 Fields Medal. Given that the Landau damping problem is so well-studied, it is a

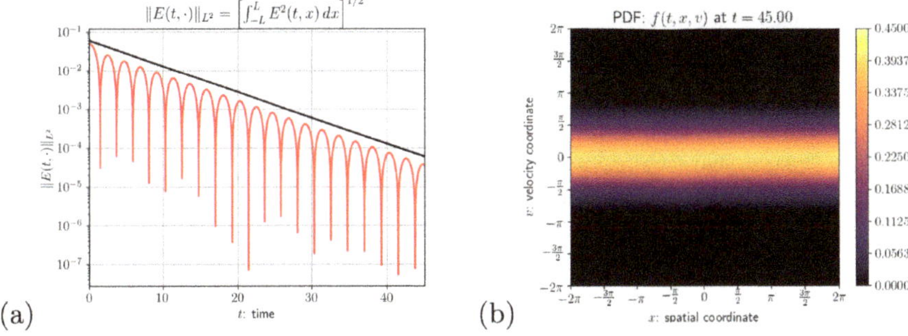

Fig. 1 Weak Landau damping problem. Panel **a** shows the exponential decay of the L_2-norm of the electric field and the line $0.06e^{\gamma t}$ with the numerically computed decay rate $\gamma = -0.1536$, and Panel **b** shows the reconstructed PDF

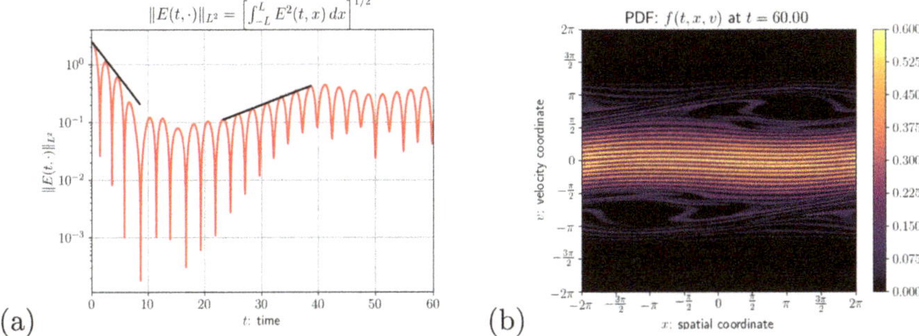

Fig. 2 Strong Landau damping problem. Panel (a) shows the initial exponential decay of the L_2-norm of the electric field ($\gamma_1 = -0.2918$) followed by exponential growth due to nonlinear effects ($\gamma_2 = 0.08584$), and Panel (b) shows the reconstructed PDF

favorite test example to study methods for plasma dynamics. The initial distribution function for the weak Landau damping problem is,

$$f(t = 0, x, v) = (1 + \alpha \cos(kx)) \frac{1}{\sqrt{2\pi}} e^{-\frac{v^2}{2}}, \tag{24}$$

with $\alpha = 0.01$ (weak) or $\alpha = 0.5$ (strong) and $k = 0.5$ on the domain $(x, v) \in [-2\pi, 2\pi] \times [-2\pi, 2\pi]$.

The weak Landau simulation was run to $t = 45$ with 40 cells in space and 80 bands in velocity. The simulation results are presented in Fig. 1. Panel (a) shows the damping of the L_2 norm of the electric field over time. Panel (b) shows the reconstructed distribution function at the final time. These results closely match those from the literature.

The strong Landau simulation was run to $t = 60$ with 40 cells in space and 80 bands in velocity. The simulation results are presented in Fig. 2. Panel (a) shows the

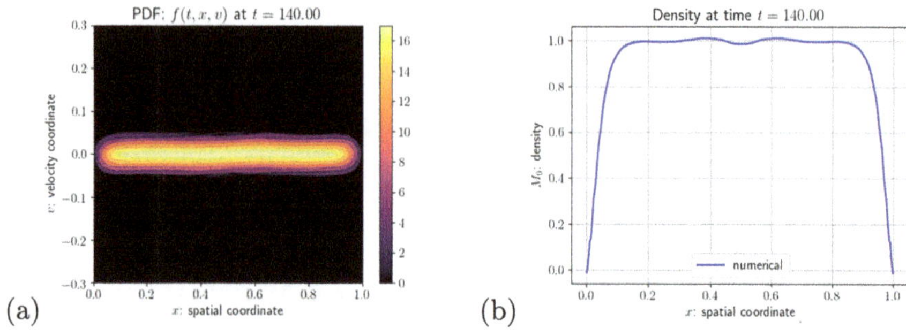

Fig. 3 Plasma sheath problem. Panel **a** shows the PDF, and Panel **b** shows the density

damping and growth of the L_2 norm of the electric field over time. Panel (b) shows the reconstructed distribution function at the final time. These results closely match those from the literature.

4.3 Plasma Sheath

The plasma sheath problem is a classic example where the distribution function represents the electrons inside the region, $0 \leq x \leq L$. We use a zero-inflow boundary condition, assuming no electrons can enter the domain. The initial condition is given by

$$f(0, x, v) = \frac{\rho_0}{\sqrt{2\pi\theta_0}} e^{-\frac{v^2}{2\theta_0}}. \tag{25}$$

The problem is solved on the domain $(t, x, v) \in [0, 140] \times [0, 1] \times [-0.2, 0.2]$. The setup shown here is identical to that in Seal [9].

Figure 3 displays the results of the simulation with 80 grid cells in space and 80 bands in velocity. Panel (a) shows the PDF and Panel (b) shows the density at $t = 140$, where a plasma sheath has formed due to the net positive charge, keeping the electrons from leaving the domain. These results match those found in literature (e.g., Figure 20(b) of Seal [9]).

5 Conclusions

In this work, a novel discontinuous Galerkin scheme for the Vlasov-Ampère system was developed. The scheme uses a spectral element moment-closure that results in a series of fluid models in each velocity band. Strang operator splitting is introduced to decouple these local fluid models and their inter-velocity coupling term. The coupling terms do not directly affect the global moments (e.g., mass, momentum, and energy);

thus, high-order accuracy is achieved on the global moments. The scheme was tested on several numerical examples.

Acknowledgements We would like to thank the anonymous referees for their thoughtful comments and suggestions that helped to improve this paper. This research was partially funded by NSF Grants DMS–2012699 and DMS–2410538.

References

1. Cheng C, Knorr G (1976) The integration of the Vlasov equation in configuration space. J Comput Phys 22:330–351
2. Gassner G, Dumbser M, Hindenlang F, Munz C-D (2011) Explicit one-step time discretizations for discontinuous Galerkin and finite volume schemes based on local predictors. J Comput Phys 230:4232–4247
3. Heath RE, Gamba IM, Morrison PJ, Michler C (2012) A discontinuous Galerkin method for the Vlasov-Poisson system. J Comput Phys 231:1140–1174
4. Johnson ER, Rossmanith JA, Vaughan C (2023) Positivity-preserving Lax-Wendroff discontinuous Galerkin schemes for quadrature-based moment-closure approximations of kinetic models. J Sci Comput 95
5. Landau L (1946) On the vibrations of the electronic plasma. J Phys (USSR) 10:25–34
6. Mouhot C, Villani C (2010) Landau damping. J Math Phys 51
7. Qiu J, Dumbser M, Shu C-W (2005) The discontinuous Galerkin method with Lax-Wendroff type time discretizations. Comput Methods Appl Mech Engr 194:4528–4543
8. Rossmanith JA, Seal DC (2011) A positivity-preserving high-order semi-Lagrangian discontinuous Galerkin scheme for the Vlasov-Poisson equations. J Comput Phys 230:6203–6232
9. Seal DC (2012) Discontinuous Galerkin methods for vlasov models of plasma. PhD thesis, University of Wisconsin. https://www.proquest.com/pqdtglobal/docview/1022644798/
10. Strang G (1968) On the construction and comparison of difference schemes. SIAM J Numer Anal 5:506–517
11. Vaughan C (2021) Spectral element moment-closure methods for kinetic Vlasov models of plasma. PhD thesis, Iowa State University. https://www.proquest.com/pqdtglobal/docview/2572602286/

Modeling Gap Closure Within Vaporizing Transmission Lines of Pulsed Power Machines Using Particle-in-Cell Direct Simulation Monte Carlo (PIC-DSMC)

Grace Kirk, Georgii Oblapenko, David Sirajuddin, Christopher Moore, David Goldstein, and Philip Varghese

Abstract We use a hybrid PIC-DSMC approach for modeling gap closure and examine the effectiveness of the novel variable weight DSMC method Event Splitting for this problem. Event Splitting has been shown to decrease computational costs and simulation noise in 0-D and 1-D electrical breakdown simulations by modeling all collision outcomes with the probability of each event reflected in post-collision macroparticle weights. Gap closure occurs in high-energy vacuum diodes where vaporized diode material creates a quasineutral plasma capable of disrupting the current. We examine the framework of our reduced surrogate simulation for reaching a fully closed gap and provide a discussion of the physics present during wave propagation across the domain as the quasineutral plasma evolves. Our results show that due to a lack of a more robust merging scheme, event splitting provides little benefit for the gap closure simulations considered in the present work.

Keywords PIC · DSMC · Event splitting · Gap closure

G. Kirk (✉) · D. Goldstein · P. Varghese
ASE-EM Department, The University of Texas at Austin, Austin, TX, USA
e-mail: grace.kirk@utexas.edu

G. Oblapenko
Applied and Computational Mathematics, RWTH Aachen, Aachen, Germany

G. Kirk · D. Goldstein · P. Varghese
Oden Institute for Computational Engineering and Sciences, The University of Texas at Austin, Austin, TX, USA

D. Sirajuddin · C. Moore
Sandia National Laboratories, Albuquerque, NM, USA

M. Grabe et al. (eds.), *Rarefied Gas Dynamics*, Springer Aerospace Technology,
https://doi.org/10.1007/978-3-032-00094-1_44

1 Introduction

1.1 *Gap Closure*

Gap closure is a phenomenon associated with the high-energy physics of vacuum diodes. "Closing the gap" describes the formation and expansion of a quasineutral plasma between the anode and the cathode [6]. In a terawatt pulsed-power accelerator, the combination of resistive heating from mega-ampere currents and the energy deposition from relativistic electrons impacting the anode lead to vaporizing diode surfaces during operation. Vaporized particles diffuse into the vacuum, where they are ionized by collisions with electrons. This causes the number density of ions to increase, which slows the electrons, resulting in a build-up of charge density capable of screening the externally applied electric field to near zero. Once the expanding plasma fills the entire gap, the lack of gradients in the electric potential will stop electron acceleration, effectively terminating the discharge [12]. Increasing simulation fidelity of gap closure physics is needed to develop predictive modeling capabilities for pulsed power engineering needs.

1.2 *Modeling Methods*

Although not the usual application of these methods, modeling gap closure requires the hybrid coupling of PIC and DSMC to capture proper ionization rates, particle accelerations, and electrostatic field variations while still maintaining the coupling between the collision and field solver.

Particle-in-Cell (PIC) is a particle approach focused on modeling ideal plasma dynamics with either electrostatic or electromagnetic fields. For our code the domain is discretized using an Eulerian focus that allows electrostatic information to be interpolated from the grid nodes to the particles, accelerating the computational macroparticles and then interpolating charges back to the grid to solve Poisson's equation for the new electric field [3].

$$\nabla^2 \Phi = \frac{\rho}{\varepsilon_0} = \frac{e(n_i - n_e)}{\varepsilon_0} \tag{1}$$

Here Φ is the electric potential, ρ is the charge density, ε_0 is the permittivity of free space, and n_i and n_e are the number densities of ions and electrons, respectively.

Direct Simulation Monte Carlo (DSMC) is a kinetic method that leverages first principle equations of motion and statistical collision ensembles to solve the Boltzmann equation [2] that describes the evolution of the particle velocity distribution function f in physical and velocity space:

$$\frac{\partial f}{\partial t} + \boldsymbol{v} \cdot \nabla_x f + \boldsymbol{a} \cdot \nabla_v f = \left(\frac{\partial f}{\partial t}\right)_{coll} \tag{2}$$

where $\boldsymbol{v}$ is the velocity, $\boldsymbol{a}$ is the acceleration, and the effect of collisions is represented by $\left(\frac{\partial f}{\partial t}\right)_{coll}$.

Event Splitting (ES) is a novel numerical method that realizes all collisional outcomes with macroparticle weights commensurate with the event probabilities – essentially every possible outcome is modeled. This method has been proven to reduce noise with fewer macroparticles than standard DSMC when applied to plasma flows. This can result in lower computational costs since rare events are simulated more efficiently [8–10]. However, a consequence of this method is an explosion of macroparticle number due to the frequent particle splitting: every collision pair produces additional particles whereas standard DSMC follows the accept-reject method. In this work, we mitigate costly exponential growth using the octree merging method, which periodically collapses the particle distribution function into a much smaller set of particles while preserving mass, momentum, angular momentum, and energy [7].

2 Simulation Setup

Our goal is not to create a self-consistent diode gap closure simulation but rather to exercise novel approaches (e.g. ES) to see if they could benefit such simulations. Towards that end, we design a surrogate simulation that produces gap closure so that we can grade these techniques on our problem of interest. Our "bare bones" simulation has a monoenergetic beam of neutrals and electrons fluxing into a vacuum while an external electric field is applied.

The focus of our study was on gap closure within the pulsed-power machine Saturn at Sandia National Labs. Specifically, we study its last millimeter of circuitry where there is a field of 4×10^8 V/m [11]. In these conditions, electrons are shot across the gap with relativistic velocities, while neutrals (accumulating to densities on the order of 10^{25} particles/m^3) passively diffuse into the vacuum, resulting in extremely steep density gradients. To reduce the high computational costs in terms of spatial discretization and time steps required as well as avoid relativistic effects, we scaled the problem down to a smaller domain (gap length L) and lower applied voltage while maintaining the original environment through a base Knudsen number ($Kn = \lambda_{MFP}/L = 0.02$) and electric field ($\nabla\Phi = V/L = 400$ MV/m) [11].

Within our simulation, we control these two properties – loosely, as this is a very dynamic environment – with the fluxes of neutrals and electrons and the chosen gap length. We further simplify the problem by using argon as our neutral of choice for its monoatomic structure (breakdown of a molecule would result in several species) and well documented cross-sections. Its flux ($\Gamma_{Ar} = n_{Ar} v_{Ar}$) was chosen such that the collisionality was high enough for breakdown of the injected material to occur. To maintain a realistic electron flux we used the Child-Langmuir space-charge limit law.

$$J = \frac{4\varepsilon_0}{9}\sqrt{\frac{2e}{m_e}}\frac{V^{\frac{3}{2}}}{L^2} \tag{3}$$

Here, e is the electron charge, m_e is the electron mass, V is the applied voltage, and L the gap length [5]. A second-order finite difference scheme was used for the field solver, with linear basis functions for interpolation of the charge density and electrostatic forces. The velocity Verlet time integrator was used.

The necessary resolution for the temporal and spatial scales are determined by the plasma frequency ω_{pe} and Debye length λ_D for electrons and the mean free path λ_{MFP} and mean time between collisions τ for argon.

$$\Delta t_e \approx \frac{1}{\omega_{pe}} = \sqrt{\frac{m_e\varepsilon_0}{n_e e^2}}, \tag{4}$$

$$\Delta t_{Ar} \approx \tau = \frac{1}{n_{Ar}\sigma_i v_{Ar}} \approx \frac{1}{\Gamma_{Ar}\sigma_i} \tag{5}$$

$$\Delta\lambda_e \approx \lambda_D = \sqrt{\frac{\varepsilon_0 k_B T_e}{n_e e^2}}, \tag{6}$$

$$\Delta\lambda_{Ar} \approx \lambda_{MFP} = \frac{1}{n_{Ar}\sigma_{iz}} \tag{7}$$

The number densities of electrons and argon are denoted by n_e and n_{Ar}, respectively. T_e is the temperature of electrons, k_B is Boltzmann's constant, and σ_{iz} represents the ionization cross-section. Cross-sections for Ar-Ar and Ar-e collisions were determined using the Variable Hard Sphere (VHS) model and the IST-LISBON database, respectively [1, 2].

When selecting flux values, it was critical to assess their influence on the required spatial and temporal resolution and whether such demands were computationally feasible. In typical gap closure scenarios, it is impractical to fully resolve the smallest Debye length and mean free path (nm scale), particularly near the cathode surface. Fortunately, the most extreme gradients are highly localized near the cathode wall, where vaporized material is rapidly ablated and diffused into the domain. In this small region, the Debye length and mean free path were under-resolved with our uniform mesh by factors of approximately 370% and 850%, respectively. Despite this, we maintain stable and physically accurate solutions (e.g., consistent expansion velocities). In the bulk plasma away from the cathode, these scales were substantially better resolved, with the Debye length and mean free path orders of magnitude larger. Any numerical grid heating that arose was confined to a narrow region near the cathode, and affected particles quickly migrated to areas with higher spatial resolution. A more detailed discussion can be found in the thesis by Grace Kirk [4].

Ultimately we identified a "Goldilocks zone" of gap length and simulation parameters (Table 1) that balanced computational constraints with the physical fidelity required for studying gap closure. Achieving this involved applying standard

Table 1 Final parameters used for simulating gap closure

Parameter	Value	Units
Neutral flux	1.5×10^{29}	$m^{-2}\,s^{-1}$
Electron flux	1.0×10^{29}	$m^{-2}\,s^{-1}$
Macroparticles per cell	250:500	–
Wall temperature	3000	K
Neutral initial velocity	9.82×10^3	$m\,s^{-1}$
Electron initial velocity	0	$m\,s^{-1}$
Voltage	2000	V
Domain length	5.0×10^{-6}	m
Cells	500	–
Time step size	1.0×10^{-16}	s
Number of time steps	30×10^6	–

PIC-DSMC resolution Eqs. (4–7) alongside iterative trial-and-error adjustments to mitigate numerical grid heating and preserve accuracy.

3 Results and Discussion

The following results are broken down into several significant phases (Table 2) and are graphically represented by the number density of individual species and the electric field (Fig. 1). Attaining these results of full gap closure took about three days using sequential computing on a workstation.

As seen in the top left pane of Fig. 1, the electrons fill the entire gap with a wide density distribution, whereas argon lags significantly behind. As the collisionality increases near the right side from the continuous argon injection, the electron density spread narrows (from having a range of $10^{20} : 10^{23}$ to $5 \times 10^{21} : 5 \times 10^{22}$ m^{-3}) and the number of ions increases. The electrons begin to be slowed by increased collisions and a growing attraction to the denser ion region. The slowed electrons

Table 2 Gap closure phases and their descriptions

Phase	Description
Pre-ionization ($T \approx 9$ ps)	Flux of neutrals and electrons off one of the diode surfaces
Initial ionization ($T \approx 1$ ns)	Significant ionization of neutral argon leading to ion creation
Steady quasineutral expansion ($T \approx 2$ ns)	Plasma expansion continues
Gap closure ($T \approx 2.7$ ns)	The plasma fills the gap

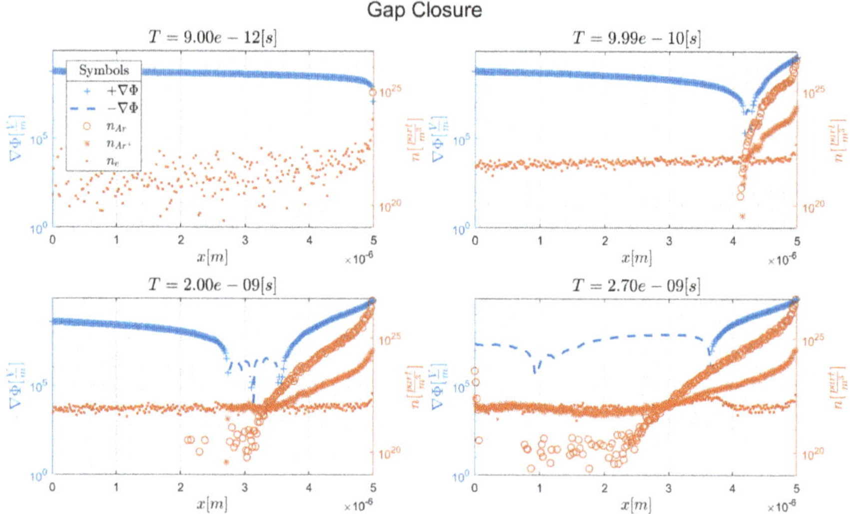

Fig. 1 Gap Closure Phases: Pre-Ionization (Top Left): steady electric field with an electron beam; Significant Ionization (Top Right): electric field varies from increase in charged particles; Steady Quasineutral Expansion (Bottom Left): larger regions of electric field screening by plasma growth; Closure (Bottom Right): field reversal with the plasma filling entire gap. Note log scales used for y-axis where the $-$ and $+$ signs denote positive and negative electric field

will then neutralize or even reverse the electric field (depending on local electron densities). The continual diffusion and ionization of neutrals acts as fuel to the plasma, expanding it further across the gap until it fills the full gap, reversing most of the electric field. This then quickly leads to an interruption of the current as there is no continuous direction of the field for electrons to follow.

Our simulations show in full kinetic detail that while plasma expansion has steady growth (a representative expansion velocity can be calculated), its advancing front is complicated, exhibiting "sloshing and spillage". For our case this was between 0.35 and 0.99 ns; in a larger scale simulation this would occur very quickly, possibly numerous times, and would go relatively unnoticed. Essentially some ions travel too far ahead of the quasineutral region during their creation and expansion and are then aggressively drawn back to the cathode by a very strong positive field. This can be imagined as a glass of water jerked too quickly where the spilled water represents ions, and gravity plays the role of the electric field. While this instability lasts briefly, it can irreparably destabilize the simulation as seen in the graphs titled "ES on" in Fig. 2. This highlights one of many challenges in faithfully simulating a system having both complicated dynamics and disparate scales such as a closing diode.

When testing Event Splitting for Gap Closure, we saw that it failed to maintain a stable solution. In the above figure, we can see, as denoted by the black arrows, the sloshing motion of the ions clearly shown in the "ES Off" results. When ES is on, we see a blow-out effect of the ions where they reach unreasonable velocities,

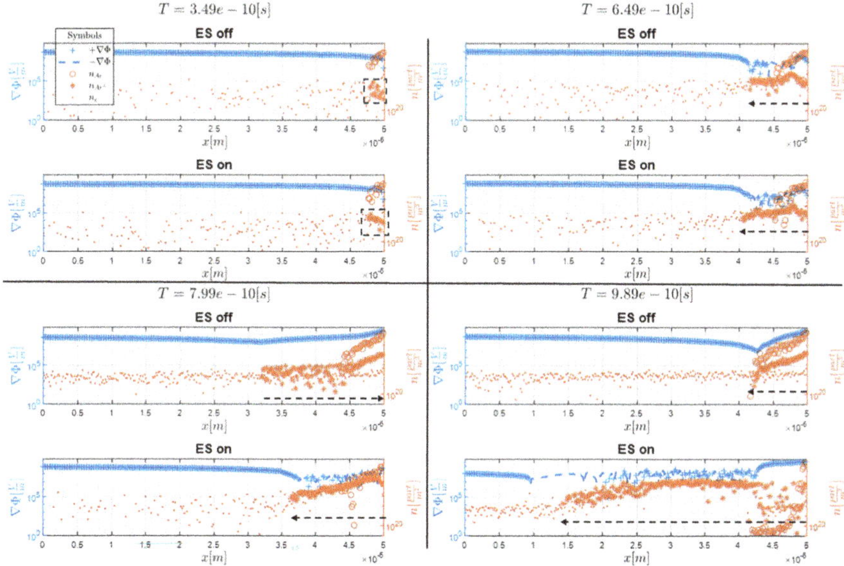

Fig. 2 Ion instability: comparison of results with and without event splitting (ES) where dashed lines/arrows denote regions of interest and general ion motion

beyond what would be expected from predicted field effects. We believe that they could be artificially accelerated by numerical grid heating arising from distortion of the distribution functions by excessive merging, which pushes an unsteady and barely resolved solution into instability. However, to Event Splitting's credit, we do see reduced noise in early formation of ions when comparing the ES on vs ES off solutions highlighted by the dashed box in the top left panel of Fig. 2.

4 Conclusions

The purpose of this paper was to describe our creation of a model capable of simulating full gap closure as well as examine the feasibility of a novel numerical method for reducing computational costs. This was done using a 1-D PIC-DSMC code with numerous simplifying assumptions, as the goal was to simply test these methods under representative conditions.

The extreme gradients in particle density and electrostatic properties associated with gap closure make it far too computationally expensive to fully resolve the maximal spatial and temporal requirements across the full domain. To reach gap closure, we required a reduced domain that maintained the desired physical properties by preservation of the Knudsen number and electric field.

Event Splitting improves the noise levels when compared to standard DSMC for the same macroparticle counts given a fully resolved simulation domain, i.e. adequate temporal and spatial resolutions. For an extremely unsteady and marginally unresolved simulation, however, event splitting creates so many particles, which then need to be merged so frequently, that the distribution function is greatly distorted. This then creates a heightened sensitivity to DSMC noise fluctuations which leads to numerical instabilities. For the simulation conditions of the gap closure problem examined in this work we found no benefit from the use of the Event Splitting.

Future work could include developing merge schemes that preserve more details of the distribution function. This would enable more robust use of ES and on a broader set of problems including the gap closure problem in this work. Additionally, mixed strategies could be considered, e.g. employing ES for some collisions, while using conventional DSMC for others, or using an adaptive spatial grid for the DSMC and PIC solvers. These pose their own challenges of particle tracking and preserving higher moments of the distribution functions while merging particles (merging will be required regardless of the method). Ultimately, finding a balance between simulation fidelity and computational cost will be required to simulate these detailed physics processes on real-world systems (e.g., simulations of the Saturn accelerator). Strategies such as those explored in this work are among a growing set of techniques that can play such a role.

Acknowledgements This work was supported in part by Sandia National Laboratories. Sandia National Laboratories is a multimission laboratory managed and operated by National Technology and Engineering Solutions of Sandia, LLC., a wholly owned subsidiary of Honeywell International, Inc., for the U.S. Department of Energy's National Nuclear Security Administration under contract DE-NA0003525.

References

1. Alves LL (2014) The IST-LISBON database on LXCat. J Phys: Conf Ser 565(1)
2. Bird GA (1994) Molecular gas dynamics and the direct simulation of gas flows. Oxford University Press. https://doi.org/10.1093/oso/9780198561958.001.0001
3. Birdsall CK, Langdon AB (2018) Plasma physics via computer simulation. CRC Press. https://doi.org/10.1201/9781315275048
4. Kirk G, Oblapenko G, Sirajuddin D, Moore C, Goldstein D, Varghese P (2023) Modeling gap closure within vaporizing transmission lines of pulsed power machines using particle-in-cell direct simulation Monte Carlo (PIC-DSMC). Master's thesis, University of Texas at Austin, Austin, TX. https://repositories.lib.utexas.edu/items/3c50abed-c1a3-4218-bb5e-cee6630df771/full
5. Langmuir I (1913) The effect of space charge and residual gases on thermionic currents in high vacuum. Phys. Rev. 2(6):450. https://doi.org/10.1103/PhysRev.2.450
6. Latham RV (1995) High voltage vacuum insulation: basic concepts and technological practice. Elsevier
7. Martin RS, Cambier JL (2016) Octree particle management for DSMC and PIC simulations. J Comput Phys 327:943. https://doi.org/10.1016/j.jcp.2016.01.020

8. Oblapenko G, Goldstein D, Varghese P, Moore C (2022) Hedging direct simulation monte carlo bets via event splitting. J Comput Phys 466:111390. https://doi.org/10.1016/j.jcp.2022.111390

9. Oblapenko G, Goldstein DB, Varghese P, Moore C (2022) Improving PIC-DSMC simulations of electrical breakdown via event splitting. AIAA SciTech 2022 Forum. https://doi.org/10.2514/6.2022-2152

10. Oblapenko G, Goldstein D, Varghese P, Moore C (2022) Improving PIC-DSMC simulations of RF plasmas via event splitting. 9th European conference for aero-nautics and aerospace sciences (EUCASS) 2022

11. Renk T, Struve K, Hustel B, Ulmen B (2023) Development of saturn transmission line circuit modeling: from bertha origins to CASTLE modeling of original and replacement hardware performance as part of saturn recap program. SAND2023-10990

12. Roy A, Menon R, Mitra S, Kumar S, Sharma V, Nagesh KV, Mittal KC, Chakravarthy DP (2009) Plasma expansion and fast gap closure in a high power electron beam diode. Phys. Plasmas 16(5):053103. https://doi.org/10.1063/1.3129802

Evaporation Simulations in High Pressure Environments with a Multi-species Enskog-Vlasov Solver

Raphael Tietz, Rolf Stierle, Stefanos Fasoulas, and Marcel Pfeiffer

Abstract This paper presents a novel multi-species Enskog-Vlasov solver designed to simulate mixtures with different particle diameters and masses and their evaporation coefficients. Traditional combustion engines face efficiency deficits due to suboptimal two-phase mixing of liquid fuel droplets and air. Recent approaches suggest heating fuel droplets above their critical temperature to achieve single-phase mixing, but accurate evaporation coefficients for high-pressure and high-temperature conditions are lacking. The proposed solver closes the gap to other EV solvers which are only capable of solving single-species simulations or multi-species simulations with equal particle diameter and masses. The solver is validated against existing single-species solvers and SAFT-VRQ Mie model. The results show good agreement with reference data, demonstrating the solver's capability to replicate multi-species simulations and determine evaporation and condensation coefficients.

Keywords Enskog-Vlasov equation · DSMC · Evaporation · Multi-species · Combustion engines · SAFT-VRQ mie

1 Introduction

The ongoing climate crisis requires a transition from traditional combustion engines to electric alternatives. However, this transition is not always feasible like in aviation or nautical applications. A mayor problem with combustion engines is their efficiency deficit, in part due to suboptimal two-phase mixing of liquid fuel droplets and air.

R. Tietz (✉) · S. Fasoulas · M. Pfeiffer
University of Stuttgart, Institute of Space Systems, Stuttgart, Germany
e-mail: rtietz@irs.uni-stuttgart.de
URL: https://www.irs.uni-stuttgart.de/en/research/space-transport-technology/numerical-modeling-and-simulations/mehrphasen/

R. Stierle
Institute of Thermodynamics and Thermal Process Engineering, University of Stuttgart, Stuttgart, Germany

© The Author(s) 2026
M. Grabe et al. (eds.), *Rarefied Gas Dynamics*, Springer Aerospace Technology,
https://doi.org/10.1007/978-3-032-00094-1_45

Recent research by Lamanna et al. [1] proposes an innovative approach to address this inefficiency by heating fuel droplets above their critical temperature so that the liquid phase disappears, allowing for more efficient and simpler single-phase mixing. However, their model relies on evaporation coefficients to model the heating process of the cold fuel droplet, until it reaches super-critical state. But they are not available for high pressure and high temperature conditions as they appear in combustion engines. This paper presents a novel multi-species Enskog-Vlasov (EV) solver and simulations designed to obtain accurate evaporation coefficients at distinct temperatures and pressures. The composition of the liquid and vapor phase of these simulations are then compared with results of the SAFT-VRQ Mie model to validate their accuracy and potential application in improving the efficiency of combustion engines. This model and evaporation studies may also be of interest in other disciplines, such as the development of new combustors for rocket engines, where higher pressures lead to higher thrust.

To date, only single-species EV solvers or multi-species EV solvers with equal particle diameters and masses of all species have been developed [2, 3]. This work goes a step further and presents a multi-species EV solver capable of simulating mixtures with different particle diameters and masses.

2 Numerical Modeling

In this work the Enskog-Vlasov equation is solved for a multi-species gas mixture. The Enskog-Vlasov equation [4] is an extension of Boltzmann's equation that describes the evolution of the distribution function $f(\underline{x}, \underline{v}, t)$ in dense gases and fluids and is given by:

$$\frac{\partial f}{\partial t} + \left\langle \underline{v} \cdot \frac{\partial f}{\partial \underline{x}} \right\rangle + \left\langle \frac{\underline{F}}{m} \cdot \frac{\partial f}{\partial \underline{v}} \right\rangle = C_E(f, f), \tag{1}$$

where t is the time, $\underline{x}$ and $\underline{v}$ are the position and velocity vectors, respectively, m is the particle mass, $\underline{F}$ is the mean-field force, and $C_E(f, f)$ is the Enskog collision integral [5].

The interaction potential of the particles is described by the Sutherland potential $\phi(r)$, which is split at its most negative position. The inner part, the hard sphere part, is modeled by the collision operator $C_E(f, f)$, while the attractive tail is modeled by the Vlasov integral and coupled in by the mean-field force $\underline{F}$.

$$\phi(r) = \begin{cases} +\infty & r < d \\ -\phi_d \left(\frac{r}{d}\right)^{-\gamma} & r \geq d \end{cases}, \tag{2}$$

$$\underline{F} = \int_{|\underline{x}_1 - \underline{x}| > d} \frac{d\phi}{dr} \frac{\underline{x}_1 - \underline{x}}{|\underline{x}_1 - \underline{x}|} n(\underline{x}_1) d\underline{x}_1, \tag{3}$$

where r is the radius, ϕ_d is the depth of the potential well, d is the particle diameter, and γ is the power of the potential. The mean-field force $\underline{F}$ is calculated by integrating the gradient of the potential over the particle distribution $n(\underline{x}_1)$, if the distance between $\underline{x}_1$ and $\underline{x}$ is greater than d.

The collision integral $C_E(f, f)$ reads as follows:

$$C_E(f, f) = d^2 \int\limits_{S} \int\limits_{\mathbb{R}^3} R(\langle \underline{c}_r \cdot \underline{k} \rangle)$$

$$\cdot \left[\begin{array}{l} Y(n(\underline{x} - \frac{d\underline{k}}{2})) f(\underline{x}, \underline{v}') f(\underline{x} - d\underline{k}, \underline{v}'_*)\dots \\ -Y(n(\underline{x} + \frac{d\underline{k}}{2})) f(\underline{x}, \underline{v}) f(\underline{x} + d\underline{k}, \underline{v}_*) \end{array} \right] d^3\underline{v}_* \, d^2\underline{k}, \qquad (4)$$

where c_r is the relative velocity of the colliding particles and $\underline{k}$ is the unit vector through the position of both particles. $\underline{v}$ and $\underline{v}_*$ are the pre-collision velocities of the first and second particles, and $\underline{v}'$ and $\underline{v}'_*$ are the post-collision velocities of them. Y is the pair correlation function at contact.

The Enskog collision integral includes several additional effects compared to the Boltzmann integral. First, colliding particles are only allowed to collide if they are moving toward another due to the Ramp function $R(\langle \underline{c}_r \cdot \underline{k} \rangle) = \mathbf{max}\,(\langle \underline{c}_r \cdot \underline{k} \rangle, 0)$ to model the correct virial. Second, the collision distance, which is equal to the particle diameter, is not neglected, which leads to a non-locality of the collisions. Third, the collision frequency is increased by the pair correlation function at contact Y due to higher number densities near the particles in dense media, where usually the Carnahan and Starling equation of state (EOS) for a single-species Sutherland fluid [6] is used to derive it.

3 Multi-species Enskog-Vlasov DSMC

The multi-species EV solver is based on the single-species solvers of Montanero et al. [7] and Frezzotti et al. [8]. They developed a single-species DSMC-like solver for the EV equation. The main idea is to divide the collision procedure into several passes, one for each species of the second collision particle, since this must be known before pairing. The pair correlation function Y is changed to one based on the BMCSL EOS [9], which is a multi-species extension of the Carnahan and Starling EOS. Finally, the calculation of post-collision velocities and sampling must be adjusted to allow for different particle masses and diameters.

4   Results

The solver was tested against the simulations of Ohashi et al. [2] to demonstrate
the capability of the solver to replicate state-of-the-art results. They performed similar simulations as required for this work, but the only difference between the two
species is their respective ϕ_d. The interspecies attraction is calculated via the Lorentz-
Berthelot combining rule. The simulations are conducted with an argon-neon mixture
with equal particle diameters d and masses m. The critical temperature of neon $T_{c,\mathrm{Ne}}$
is lower than in reality by a factor of about 4 to minimize perturbations of the liquid
phase by dissolved neon. The simulations are performed with varying amounts of
neon in the simulation domain, resulting in varying pressures and a constant temperature of $0.633 T_{c,\mathrm{Ar}}$. The number density profiles are shown in Fig. 1 and follow the
expected trend.

The evaporation and condensation coefficients (σ_e and σ_c) are calculated using
the following formulas:

$$\sigma_e = \frac{J_{\mathrm{evap}}}{J_{\mathrm{evap}} + J_{\mathrm{refl}}} \quad \text{and} \quad \sigma_c = \frac{J_{\mathrm{cond}}}{J_{\mathrm{cond}} + J_{\mathrm{refl}}}, \tag{5}$$

where J_{evap} and J_{cond} are the particle flux from the liquid to the gas phase and vice
versa, and J_{refl} is the particle flux of particles reflected at the interface according
to [10]. The evaporation and condensation coefficients shown in Fig. 2 are in good
agreement with the reference. This shows the ability of the solver to reproduce
state-of-the-art multi-species simulations with equal masses and diameters of all
species.

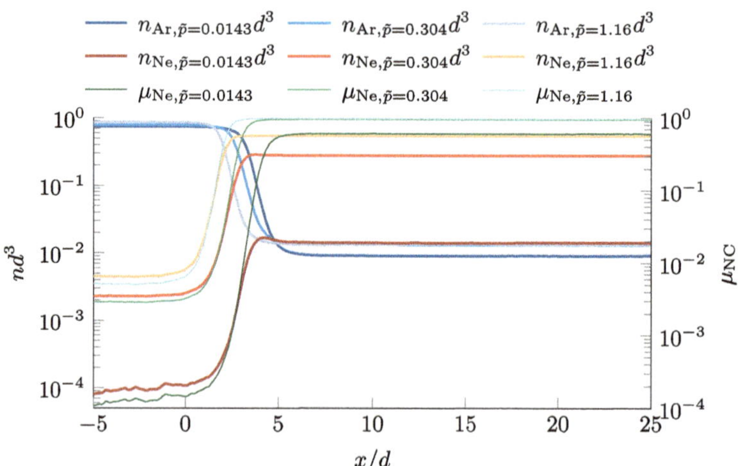

Fig. 1   Argon and neon number density profiles and neon mole fractions of cases 1, 5 and 8 of
Ohashi et al. [2]

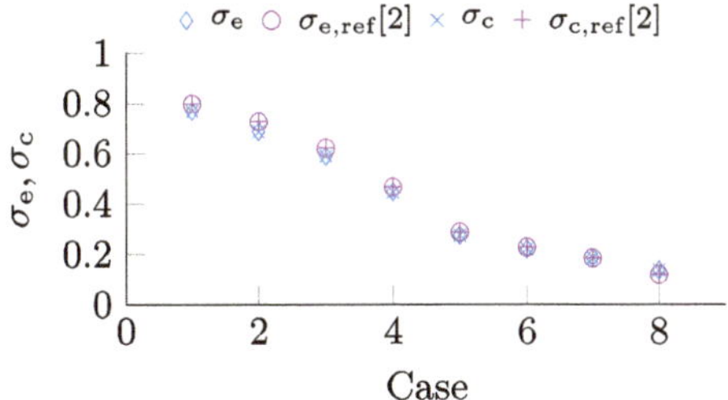

Fig. 2 Evaporation and condensation coefficients compared to Ohashi et al. [2]

Table 1 Particle Diameters of the subsequent simulations

T/T_c	$d_{Ar}/\text{Å}$	$d_{Ne}/\text{Å}$
0.633	3.3322	2.5898
0.767	3.2270	2.5663
0.900	3.1418	2.5489

To demonstrate the full multi-species capability of the solver, vapor and liquid phase densities are compared with SAFT-VRQ Mie model [11, 12]. This is a sophisticated EOS capable of describing the vapor-liquid equilibrium of mixtures and their composition by simplified Mie interaction potentials. For this comparison, the critical temperature of neon is raised to its correct value and the particle diameters are fitted to external data: The argon diameter is fitted to single species results of the SAFT-VRQ Mie EOS by replicating the correct liquid density, and the neon diameter is fitted to NIST pressure data [13]. The diameters are listed in Table 1.

The simulations are executed at three different temperatures 0.633, 0.767, and $0.900 T_{c,Ar}$ and the pressure is varied by different amounts of neon in the domain.

The simulation domain was initialized with pure argon in the liquid phase with a number density of $2 \cdot 10^{28}\,\frac{1}{\text{m}^3}$ and a length of 25.538Å. On the right boundary a surface flux of neon was applied with a neon pressure of 0, 1, 10, 50 and 100bar. The part of the computational domain what is still empty, was initialized with neon at the same pressure. The left boundary is a symmetric boundary with a thermostatted region next to it to counteract evaporation heat losses tuned to the temperature of the simulation.

The results are depicted in Fig. 3 and show that more neon is dissolved in the liquid phase due to the higher neon attraction, smaller particle diameter, and higher simulated temperatures. Especially at the higher temperatures the amount of neon disturbs the liquid phase due to a high neon mole fraction in it. For the highest pressure case at the highest temperature no stable liquid phase formed. The agglomeration of neon atoms at the gas-liquid interface region can also be observed in the results of other work.

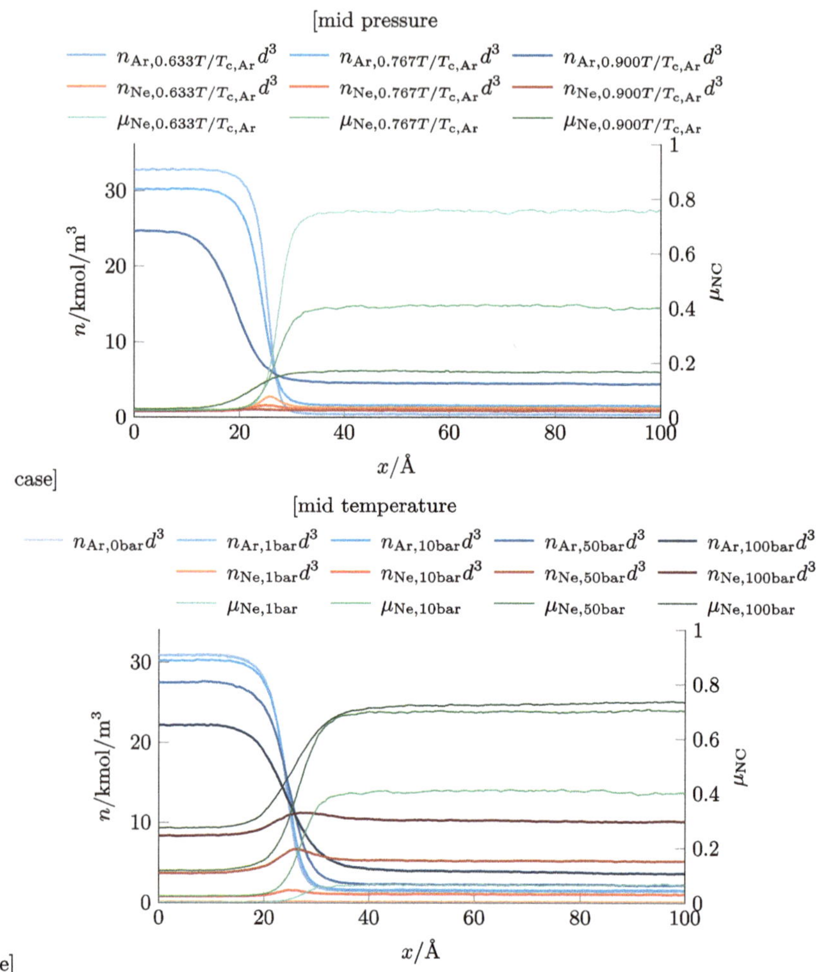

Fig. 3 Number density profiles and neon mole fractions with different particle diameters and masses of the species

The equilibrium compositions of the liquid and gaseous phases are compared with results of the SAFT-VRQ Mie model [11, 12] in Fig. 4. The results of both methods are in reasonable agreement to each other. The deviations may be due to the different interaction potentials used – Sutherland vs. Mie potential. The higher the temperature or pressure the higher are the deviations. Therefore, the particles come closer in those circumstances and this is the point where the used interaction potentials have their highest differences. The results indicate that the presented multi-species EV solver is able to simulate mixtures with different particle diameters and masses.

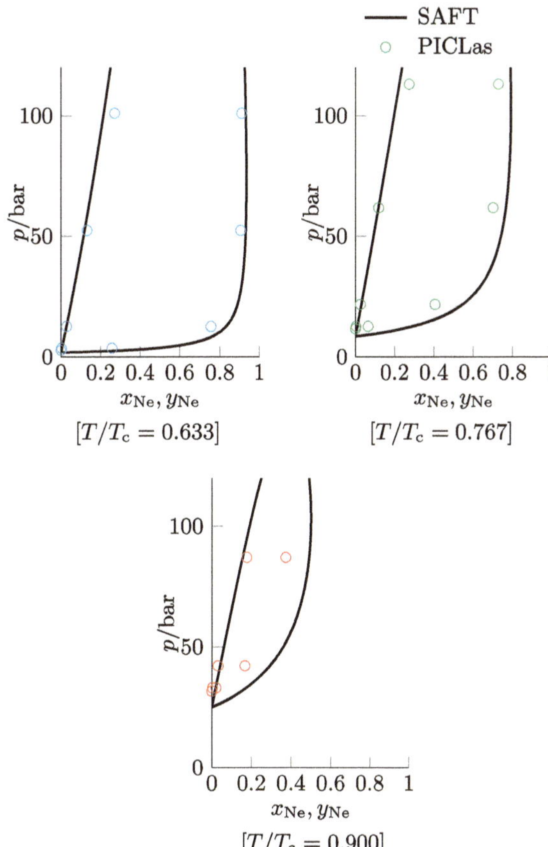

Fig. 4 Enskog-Vlasov result compared to SAFT-VRQ Mie [11, 12]

With this knowledge the solver can be used to sample the evaporation coefficients. The same sampling technique for the evaporation coefficients as in the simulations before was used and are shown in Fig. 5. The coefficients follow the expected trend. For higher pressures Ohashi et al. [2] observed a decrease of the evaporation coefficient. This is due to the higher number density of the gas phase leading to a higher number of collisions and therefore a lower probability of evaporation. Frezzotti et al. [8] observed a decrease in the evaporation coefficient with increasing temperature. This trend continues with higher temperatures.

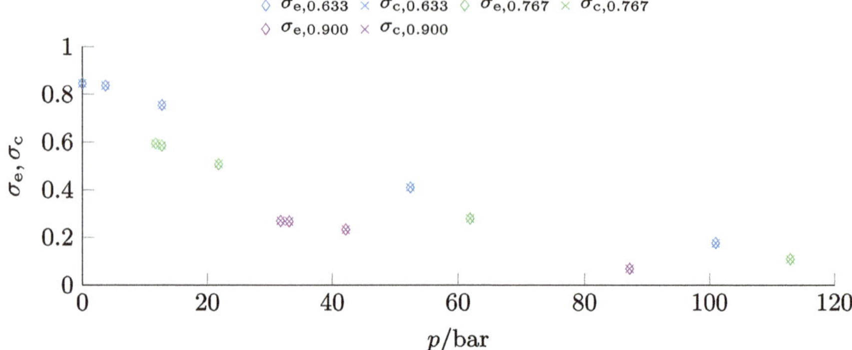

Fig. 5 Evaporation and condensation coefficients

5 Conclusion

In this paper, we present a novel multi-species EV solver designed to simulate evaporation in high-pressure environments. This solver simulates mixtures with different particle diameters and masses, filling the gap in existing models. The solver has been validated against single-species EV solvers and SAFT-VRQ Mie model, showing good agreement with reference data of Ohashi et al. [2]. The results demonstrate the ability of the solver to reproduce multi-species simulations and to determine evaporation and condensation coefficients. In the future this novel EV solver can be used in non-equilibrium simulations to demonstrate its capabilities in such scenarios.

These results have significant implications for improving the efficiency of internal combustion engines and reducing emissions. Future work will focus on further validation and application of the solver to real-world scenarios, as well as exploring its potential in other areas requiring accurate evaporation modeling.

References

1. Lamanna G, Steinhausen C, Weigand B (2023) On the role of trancritical evaporation in controlling the transition from two-phase to single-phase mixing. In: ILASS Europe 2023, 32nd European conference on liquid atomization and spray systems
2. Ohashi K, Kobayashi K, Fujii H, Watanabe M (2020) Evaporation coefficient and condensation coefficient of vapor under high gas pressure conditions. Sci Rep 10(1):8143. https://doi.org/10.1038/s41598-020-64905-5
3. Frezzotti A, Gibelli L, Lockerby DA, Sprittles JE (2018) Mean-field kinetic theory approach to evaporation of a binary liquid into vacuum. Phys Rev Fluids 3(5):054,001. https://doi.org/10.1103/PhysRevFluids.3.054001
4. Grmela M (1971) Kinetic equation approach to phase transitions. J Statist Phys 3(3):347–364. https://doi.org/10.1007/BF01011389
5. Enskog D (1922) Kinetische Theorie der Wärmeleitung: reibung und Selbst-diffusion in Gewissen verdichteten gasen und flüssigkeiten. Almqvist & Wiksells boktryckeri-a.-b

6. Carnahan NF, Starling KE (1969) Equation of state for nonattracting rigid spheres. J Chem Phys 51(2):635–636. https://doi.org/10.1063/1.1672048
7. Montanero JM, Santos A (1997) Simulation of the Enskog equation à la Bird. Phys Fluids 9(7):2057–2060. https://doi.org/10.1063/1.869325
8. Frezzotti A, Gibelli L, Lorenzani S (2005) Mean field kinetic theory description of evaporation of a fluid into vacuum. Phys Fluids **17**(1), 012,102. https://doi.org/10.1063/1.1824111
9. Yau DHL, Chan KY, Henderson D (1997) Pair correlation functions for a hard sphere mixture in the colloidal limit. Molecular Physics 91(6):1137–1142. https://doi.org/10.1080/002689797170860
10. Ishiyama T, Yano T, Fujikawa S (2004) Molecular dynamics study of kinetic boundary condition at an interface between argon vapor and its condensed phase. Phys Fluids 16(8):2899–2906. https://doi.org/10.1063/1.1763936
11. Aasen A, Hammer M, Ervik Å, Müller EA, Wilhelmsen Ø (2019) Equation of state and force fields for feynman–hibbs-corrected mie fluids. i. application to pure helium, neon, hydrogen, and deuterium. J Chem Phys 151(6):064,508 https://doi.org/10.1063/1.5111364
12. Aasen A, Hammer M, Müller EA, Wilhelmsen Ø (2020) Equation of state and force fields for feynman–hibbs-corrected mie fluids. ii. application to mixtures of helium, neon, hydrogen, and deuterium. J Chem Phys 152(7):074,507. https://doi.org/10.1063/1.5136079
13. Acree Jr, WE, Chickos JS (2000) NIST chemistry webbook: NIST standard reference database number 69. Nat Inst Stand Technol Gaithersburg MD, 20899 (2000). https://doi.org/10.18434/T4D303

Combining Stochastic Particle BGK and Discrete Velocity Method for Efficient Multiscale Simulations

Félix Garmirian and Marcel Pfeiffer

Abstract The Bhatnagar-Gross-Krook (BGK) equation is widely used as an efficient model for flow simulations in transitional Knudsen number regimes. The two main types of solvers based on the BGK model, the deterministic discrete velocity methods (DVM) and the stochastic particle-based BGK solvers, each come with specific disadvantages such as six-dimensional discretization or statistical noise. This work introduces a coupling in velocity space between both approaches, creating a flexible noise-reduced method capable of accelerating the resolution of multiscale problems. The method is verified on test cases, showing improved efficiency compared to the pure DVM or the pure particle-based solver.

Keywords Stochastic particle BGK · Discrete velocity method · Velocity-space coupling · Multiscale methods

While the discrete simulation Monte Carlo (DSMC) method [4] has become the standard for rarefied gas simulations, it still suffers from a decreasing efficiency in the transition regime, where the Knudsen number is still too high to enable valid Navier-Stokes simulations. In such conditions, where the low mean-free path and large collision frequency would require computationally expensive DSMC simulations, the use of the BGK model [3] has proven to be efficient [10, 17].

Indeed, BGK solvers can avoid the physics-based restrictions of DSMC, and with the right integration method, even the relaxation frequency of this model does not need to be resolved. Several methods that capitalize on these advantages have been developed, based either on the discrete velocity method (DVM) [9, 10, 14, 19], or on a stochastic particle-based approach (SPBGK) similar to DSMC [7, 18], all able to reach at least second-order accuracy in time and space. While the particle-based solvers retain some of the flexibility of DSMC when it comes to boundary interaction, chemical reactions or supersonic flows, the DVMs have the benefit of noise-free results due to their fully deterministic nature, making them particularly useful for unsteady low-Mach number flows where DSMC results with a high signal-to-noise ratio are hardly achievable.

F. Garmirian (✉) · M. Pfeiffer
Institute of Space Systems, University of Stuttgart, Stuttgart, Germany
e-mail: garmirianf@irs.uni-stuttgart.de

© The Author(s) 2026

M. Grabe et al. (eds.), *Rarefied Gas Dynamics*, Springer Aerospace Technology,
https://doi.org/10.1007/978-3-032-00094-1_46

In order to develop a method that would profit from the advantages of the BGK model in all flow regimes, both approaches, particle-based and DVM, have been implemented in the open-source kinetic suite PICLas [6]. The stochastic and deterministic solvers are then coupled in velocity space, with the goal of using a limited range and resolution for the DVM velocity discretization while stochastic particles make up for the rest of the distribution function. A similar coupling in the context of DSMC has been investigated in the past [15, 16], but with the opposite decomposition, using a deterministic solver to better resolve the distribution tails. Another type of stochastic/deterministic hybrid BGK method has also been developed [5], in which the velocities of particles representing the deviation from equilibrium have to lie on DVM grid points.

With the coupling described in this paper, the aim is to reduce the statistical noise produced by the stochastic particle-based approach, while avoiding the increase in memory and CPU cost for 3D supersonic flows associated with DVM. This first-order SPBGK/DVM coupling could also be extended to a second-order method by using the same second-order integration for both components, as the boundary between stochastic particles and the deterministic solver is fixed throughout the simulation, what could be an advantage over other noise-reduced particle methods [1, 12, 13].

1   The Bhatnagar-Gross-Krook Model

The BGK equation [3] is an approximation of the Boltzmann equation, where the collision integral is replaced by a relaxation of the velocity distribution function $f(\mathbf{x}, \mathbf{v}, t)$ towards a target distribution f^t at the frequency $\nu = 1/\tau$:

$$\frac{\partial f}{\partial t} + \mathbf{v} \cdot \frac{\partial f}{\partial \mathbf{x}} = -\nu(f - f^t), \tag{1}$$

with position $\mathbf{x}$, velocity $\mathbf{v}$ and time t. Using the equilibrium Maxwell distribution as target results in a fixed Prandtl number $\mathrm{Pr} = 1$, several corrected models have therefore been developed to account for thermal and viscous effects at the same time [11, 17]. In this paper, the ellipsoidal statistical BGK (ESBGK) model is preferred for its positivity because it satisfies Boltzmann's H-theorem [2, 11].

In the ESBGK model, the relaxation frequency and target distribution are defined as

$$\nu = \mathrm{Pr}\frac{nk_\mathrm{B}T}{\mu}, \quad f^{\mathrm{ES}} = \frac{n}{\sqrt{\det(2\pi\mathsf{A})}} \exp\left[-\frac{\mathbf{c}^T\mathsf{A}^{-1}\mathbf{c}}{2}\right], \tag{2}$$

with particle density n, particle mass m, temperature T, viscosity μ and Boltzmann constant k_B. The matrix A is constructed from the pressure tensor P:

$$\mathsf{A} = \frac{k_\mathrm{B}T}{m\mathrm{Pr}}\mathsf{Id} + \left(1 - \frac{1}{\mathrm{Pr}}\right)\frac{\mathsf{P}}{nm}, \tag{3}$$

where Id is the identity matrix and $\mathsf{P} = \int m\mathbf{c}\mathbf{c}^T f d\mathbf{v}$.

Most numerical methods that use the BGK model to simulate rarefied gas flows are based either on the stochastic particle-based approach [8] or on the deterministic DVM [14]. The first one uses the same representation of the distribution function by simulation particles as DSMC. It also shares with DSMC the particle movement step, differing only in the collision step, replaced by a resampling of some particles from the chosen target distribution. The DVM approach, on the other hand, consists in a deterministic discretization of the velocity space into an appropriate grid. The BGK equation is then solved for each phase space point, usually with a finite volume scheme.

2 Stochastic Particle/DVM Coupling

One major disadvantage of discrete velocity methods for practical use is that, along the usual space mesh and time steps, a velocity grid must be constructed. The dimension of this grid depends of course on the problem considered, exponentially increasing the computational cost when going from 1D to 3D. Besides, even though many types of adapted grids have been developed to reduce the number of velocities, selecting the resolution and the bounds of the velocity grid for maximum accuracy still remains a challenge, especially in strong non-equilibrium conditions where the shape of the distribution function is unknown.

On the other hand, particle-based methods are more flexible, as their Monte-Carlo approach leads to cost increasing only as the square root of the dimension. The particle velocities are only limited by computer precision, enabling in theory a more accurate representation of non-equilibrium distribution tails. However, these stochastic methods introduce statistical noise, particularly problematic in the low-Mach number regime where the signal-to-noise ratio is low.

We therefore choose to couple DVM and SPBGK in velocity space, using DVM for the velocities closer to the mean velocity, where most of the density is located, and SPBGK for high-velocity particles making up the distribution tails. The cost of SPBGK is thus concentrated in the parts of the distribution that determine the non-equilibrium effects hard to capture with DVM, and the central DVM grid greatly reduces the number of particles needed to achieve low-noise results.

2.1 Velocity Space Decomposition

The coupling is done by splitting the velocity space into two regions, $\mathcal{V}_{\text{DVM}}$ and $\mathcal{V}_{\text{part}}$, delimited by velocity bounds $\mathbf{v}_{\min}$ and $\mathbf{v}_{\max}$ between which a DVM grid with evenly spaced points is constructed. Such a decomposition can be seen in two dimensions on Fig. 1. The distribution function is then separated according to this decomposition:

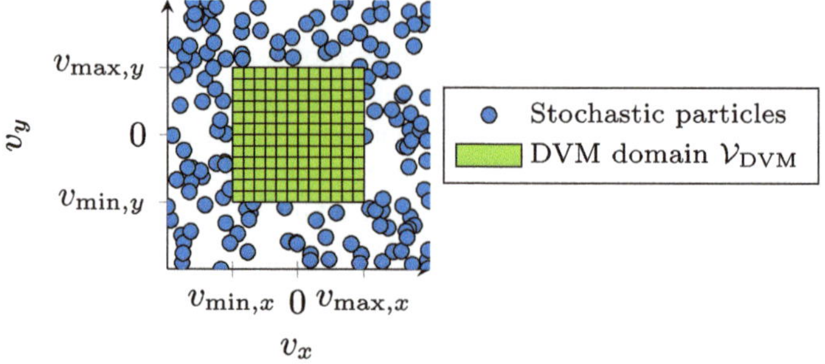

Fig. 1 Example of a decomposition of the 2D velocity space in DVM and SPBGK regions

$$f(\mathbf{v}) = f^{\mathrm{DVM}}(\mathbf{v}) + f^{\mathrm{part}}(\mathbf{v}) \quad \text{with} \quad \mathrm{supp}(f^{\mathrm{DVM}}) = \mathcal{V}_{\mathrm{DVM}} = [\mathbf{v}_{\min}, \mathbf{v}_{\max}], \quad (4)$$

meaning that the DVM representation f^{DVM} is limited to the central velocity grid while stochastic particles f^{part} can exist in the whole domain. The relaxation algorithm is however constructed so that the density represented by particles that fall inside the DVM domain will automatically pass from f^{part} to f^{DVM} for the next time step.

In this first approach, the BGK equation is solved using a first-order splitting of the advection and relaxation terms. For the advection, both parts of the distribution are evolved using their corresponding methods, i.e. particles are moved according to their velocity:

$$f^{\mathrm{part}}(\mathbf{x}, \mathbf{v}, t + \Delta t) = f^{\mathrm{part}}(\mathbf{x} - \mathbf{v}\Delta t, \mathbf{v}, t), \quad (5)$$

while, for the DVM part, a simple first-order upwind finite volume scheme is applied to solve:

$$\frac{\partial f^{\mathrm{DVM}}}{\partial t} + \mathbf{v} \cdot \frac{\partial f^{\mathrm{DVM}}}{\partial \mathbf{x}} = 0. \quad (6)$$

For the relaxation, the moments of the total distribution $f = f^{\mathrm{DVM}} + f^{\mathrm{part}}$ are then computed and the target distribution $f^{\mathrm{ES}}(\mathbf{x}, \mathbf{v})$ is constructed at each point of the DVM grid ($\mathbf{v} \in \mathcal{V}_{\mathrm{DVM}}$). The relaxation term can be integrated exactly:

$$\frac{\partial f}{\partial t} = \nu \left(f^{\mathrm{ES}} - f \right) \implies f(t + \Delta t) = \gamma f(t) + (1 - \gamma) f^{\mathrm{ES}}, \qquad \gamma = \exp\left(-\nu \Delta t\right). \quad (7)$$

The relaxation step is first applied to the limited DVM domain:

$$\forall \mathbf{v} \in \mathcal{V}_{\mathrm{DVM}}, \quad f^{\mathrm{DVM}}(\mathbf{x}, \mathbf{v}, t + \Delta t) = \gamma f^{\mathrm{DVM}}(\mathbf{x}, \mathbf{v}, t) + (1 - \gamma) f^{\mathrm{ES}}(\mathbf{x}, \mathbf{v}). \quad (8)$$

Note that, although the BGK collision operator is conservative, this DVM step is not. Indeed, f^{ES} is calculated from the total moments, and can therefore already include the density represented by particles with an initial velocity outside of the $\mathcal{V}_{\mathrm{DVM}}$ but relaxing into it.

Finally, some particles are randomly picked to relax with probability $1 - \gamma$, their velocity is resampled from f^{ES}, but they are then deleted if their new velocity falls inside $\mathcal{V}_{\mathrm{DVM}}$, as this part of the distribution has already been handled with (8). With $\mathbb{1}$ the indicator function, this can be written as:

$$f^{\mathrm{part}}(\mathbf{x}, \mathbf{v}, t + \Delta t) = \gamma\, f^{\mathrm{part}}(\mathbf{x}, \mathbf{v}, t) + (1 - \gamma)\, \mathbb{1}_{\mathbb{R}^3 \setminus \mathcal{V}_{\mathrm{DVM}}}\, f^{\mathrm{ES}}(\mathbf{x}, \mathbf{v}), \qquad (9)$$

which, combined with (8), corresponds to the full relaxation step (7).

2.2 Conservation of Moments

For the output of results, and for the calculation of f^{ES} as previously mentioned, the moments of the total distribution function have to be computed. To this end, the conservative moments of f^{part} and f^{DVM} are simply summed together, accounting for all discretization points, i.e. particles and DVM grid points.

However, as the selection and resampling process for the relaxation of particles involves random numbers, the density of deleted particles can differ from the one added to the DVM domain. To enforce mass conservation, the density of the particles resampled from f^{ES} that do not fall inside $\mathcal{V}_{\mathrm{DVM}}$ is adjusted to:

$$n_{\mathrm{resampled}} = n(f_{\mathrm{old}}) - n(f_{\mathrm{new}}^{\mathrm{DVM}}) - n(\gamma f_{\mathrm{old}}^{\mathrm{part}}), \qquad (10)$$

where old and new refer to the state before and after the relaxation step. Besides, as this calculated density $n_{\mathrm{resampled}}$ can sometimes be negative due to the numerical error associated with a very low density outside $\mathcal{V}_{\mathrm{DVM}}$ or a too low number of relaxing particles, it is set to zero in such cases. Mass conservation is instead enforced by slightly rescaling the whole DVM distribution.

For momentum and temperature, conservation is enforced by computing their deviation on the total distribution, and applying a similar velocity rescaling as Gallis and Torczynski [8] to the particles only.

2.3 Boundary Conditions

While immobile specular boundaries do not require any special treatment, as reflected velocities stay in the same domain as incoming ones, the coupling of velocities through a diffusive wall's Maxwell distribution needs to be addressed. First, the outgoing DVM flux at each diffusive boundary is computed and summed with the flux of particles hitting the same boundary. Those particles are deleted, and new ones are sampled from the wall distribution function, with a density corresponding to the outgoing flux. To prevent an always increasing number of particles in the simulation domain, new incoming particles with a velocity inside $\mathcal{V}_{\mathrm{DVM}}$ are deleted, and replaced by an incoming DVM flux with the same density.

3 Results

3.1 Sod Shock Tube

The first test case is the Sod shock tube, with argon at an initial density of $n_1 = 1.294 \times 10^{21}$ m^{-3} on the left part of the domain and $n_r = 1.618 \times 10^{20}$ m^{-3} on the right, with temperatures $T_1 = 273$ K and $T_r = 218.4$ K. The flow is thus in the transition regime ($10^{-3} < \mathrm{Kn} < 10^{-2}$). The simulation is run on a 500 cells mesh, with a time step of $\Delta t = 10^{-6}$ s. Figure 2 shows the temperature profiles at the beginning and at the end of the simulation, for a noisy pure SPBGK simulation and for a coupled SPBGK/DVM simulation. Results are compared to a DSMC reference that used around 10^6 particles per cell to lower the noise level. Both the pure SPBGK and the coupled solver are initialized using around 300 000 particles to represent the total density. The DVM domain for the SPBGK/DVM coupling consists of 15 Gauss-Hermite points between -1500 and 1500 m/s. As the coupled simulation runs, the number of particles decreases down to zero in a few hundred time steps, because all the density they represent is passed to the DVM domain that encompasses most of the distribution function. The statistical noise is thus clearly reduced while maintaining very good accuracy.

3.2 Couette Flow

In order to evaluate the efficiency of the coupled method, two Couette flows are simulated, one with a Knudsen number of $\mathrm{Kn} = 0.01$ between plates moving in

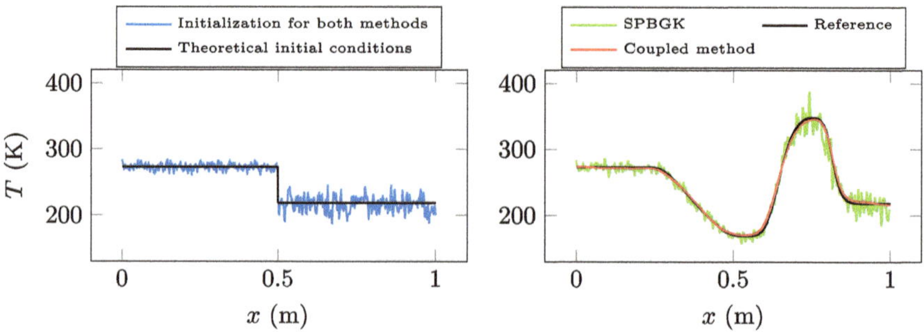

Fig. 2 Temperature profiles of the Sod shock tube test case. Left: initial conditions. Right: Profile after $t = 7 \times 10^{-4}$ s

opposite directions at ± 50 m/s, the other with Kn $= 0.1$ and wall velocities of ± 500 m/s resulting in a supersonic flow with Mach number Ma $\simeq 1.6$. Both cases are run using argon gas with a time step of $\Delta t = 10^{-6}$ s.

For the first case, shown on the left of Fig. 3, a 2D DVM velocity grid of 15×15 equally spaced points between -1500 and 1500 m/s is constructed. The coupled simulation is initialized with all the density as stochastic particles, which, as the simulation runs, are deleted as they fall inside the DVM domain. The simulation runs until the relative variation of the maximum temperature stays under 10^{-6}. With 440 initial particles per cell, this resulted in the coupled simulation taking 158 min on 5 CPU cores. As the limited velocity grid alone already allows for accurate results in this case, the pure DVM simulation only lasted 52 min. However, this was much more efficient than the pure SPBGK simulation initialized with the same number of particles, which needed 198 min to reach steady state, plus another 75 min to produce time-averaged results, still much noisier than the instantaneous results of the coupled simulation.

The second case (right part of Fig. 3) is less costly for pure particle methods as the high gradients in the flow macroscopic values lead to a higher signal-to-noise ratio. Using a converged SPBGK simulation as reference, the coupled solver with a 25×25 grid limited to $\mathcal{V}_{\mathrm{DVM}} = [-900, 900]^2$ m/s, along with 10 to 100 particles per cell at steady state, results in a L^2 error of less than 1.5 K in the temperature profile. However, as the velocities span a larger range and because the non-equilibrium effects are more pronounced, a pure DVM simulation with the same 25×25 grid is inaccurate, and a 51×51 velocity grid between -3000 and 3000 m/s was required to reach the same accuracy. The coupled representation compensates the DVM limitations with a few high-velocity particles, thus greatly reducing the memory usage when compared to pure DVM.

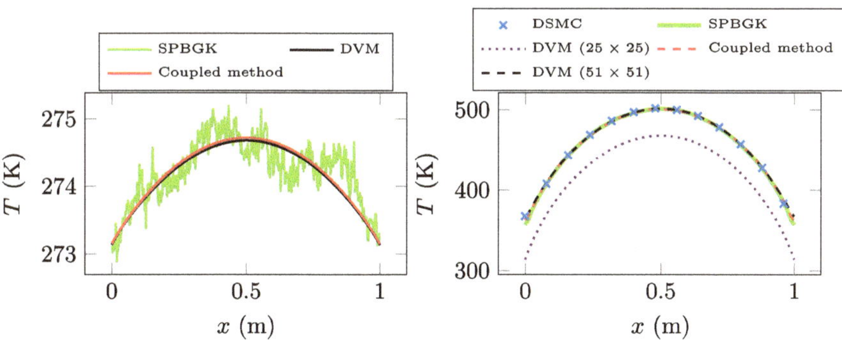

Fig. 3 Temperature profile of the Couette flows. Left: low-Mach number flow. Right: supersonic flow

4 Conclusion

A coupling of SPBGK and DVM in velocity space has been developed and evaluated on two common test cases, showing improved efficiency on one or the other parent method depending on the flow regime. This coupled method would therefore be interesting for multiscale simulations, where the flow goes from one regime to another, in terms of rarefaction or Mach number. Future work would include the extension of the method to second order in time and space, and the development of a better strategy for velocity domain decomposition, perhaps using a spherical boundary for the DVM domain. It would then be interesting to compare the resulting method to other noise-reduced particle methods.

Acknowledgements This project has received funding from the European Research Council (ERC) under the European Union's Horizon 2020 research and innovation programme (grant agreement No. 899981 MEDUSA).

References

1. Al-Mohssen HA, Hadjiconstantinou NG (2010) Low-variance direct Monte Carlo simulations using importance weights. ESAIM: Mathematical Modelling Numer Anal 44(5):1069–1083
2. Andries P, Le Tallec P, Perlat JP, Perthame B (2000) The Gaussian-BGK model of Boltzmann equation with small Prandtl number. Eur J Mech - B/Fluids 19(6):813–830
3. Bhatnagar PL, Gross EP, Krook M (1954) A model for collision processes in gases. I. Small amplitude processes in charged and neutral one-component systems. Phys Rev 94(3):511–525
4. Bird GA (1994) Molecular gas dynamics and the direct simulation of gas flows. Oxford University Press, Oxford, New York, Oxford Engineering Science Series
5. Dimarco G, Pareschi L (2008) Hybrid multiscale methods II. Kinetic equations. Multiscale Modeling Simulat 6(4):1169–1197
6. Fasoulas S, Munz CD, Pfeiffer M, Beyer J, Binder T, Copplestone S, Mirza A, Nizenkov P, Ortwein P, Reschke W (2019) Combining particle-in-cell and direct simulation Monte Carlo for the simulation of reactive plasma flows. Phys Fluids 31(7):072006
7. Fei F, Zhang J, Li J, Liu Z (2020) A unified stochastic particle Bhatnagar-Gross-Krook method for multiscale gas flows. J Comput Phys 400:108972
8. Gallis MA, Torczynski JR (2011) Investigation of the ellipsoidal-statistical Bhatnagar-Gross-Krook kinetic model applied to gas-phase transport of heat and tangential momentum between parallel walls. Phys Fluids 23(3):030601
9. Garmirian F, Pfeiffer M (2025) Implementation of asymptotic preserving discrete velocity methods into the simulation code PICLas. Comput Phys Commun p 109648
10. Guo Z, Xu K (2021) Progress of discrete unified gas-kinetic scheme for multiscale flows. Adv Aerodynam 3(1):6
11. Holway LH Jr (1966) New statistical models for kinetic theory: methods of construction. Phys Fluids 9(9):1658–1673
12. Homolle TMM, Hadjiconstantinou NG (2007) A low-variance deviational simulation Monte Carlo for the Boltzmann equation. J Comput Phys 226(2):2341–2358
13. Liu C, Zhu Y, Xu K (2020) Unified gas-kinetic wave-particle methods I: continuum and rarefied gas flow. J Comput Phys 401:108977
14. Mieussens L (2000) Discrete velocity model and implicit scheme for the BGK equation of rarefied gas dynamics. Math Models Methods in Appl Sci 10(08):1121–1149

15. Oblapenko G, Goldstein D, Varghese P, Moore C (2020) A velocity space hybridization-based Boltzmann equation solver. J Comput Phys 408:109302
16. Pan TJ, Stephani KA (2016) Investigation of velocity-space coupling approach in DSMC for tail-driven processes. In: 30th international symposium on rarefied gas dynamics. Victoria, BC, Canada, p 050017
17. Pfeiffer M (2018) Particle-based fluid dynamics: comparison of different Bhatnagar-Gross-Krook models and the direct simulation Monte Carlo method for hypersonic flows. Phys Fluids 30(10):106106
18. Pfeiffer M, Garmirian F, Gorji MH (2022) Exponential Bhatnagar-Gross-Krook integrator for multiscale particle-based kinetic simulations. Phys Rev E 106(2):025303
19. Xu K, Huang JC (2010) A unified gas-kinetic scheme for continuum and rarefied flows. J Comput Phys 229(20):7747–7764

An Improved Particle Scheme for Solving Fokker-Planck Models of the Boltzmann Equation

Tobias Ott, Hossein Gorji, and Marcel Pfeiffer

Abstract This paper presents an improved particle scheme for solving Fokker-Planck models of the Boltzmann equation. The proposed method employs a second-order Strang splitting scheme for accurate time integration. Element-wise basis functions are chosen for spatial discretization, and the moments of the distribution function are calculated by projection onto these basis functions. The method is validated using a 1D Couette flow test case, demonstrating its temporal and spatial convergence properties. The results show that the proposed scheme provides accurate solutions for the Boltzmann equation with the Ellipsoidal-Statistical Fokker-Planck collision operator.

Keywords Particle methods · Boltzmann equation · Fokker-planck

1 Introduction

Numerical simulation proves to be especially important for non-equilibrium flows, where conducting experiments is often prohibitively expensive or not feasible at all. Traditionally, the Direct Simulation Monte Carlo (DSMC) method [1] has been the standard approach for the numerical solution of problems in this flow regime. However, a significant limitation of the DSMC method is its requirement to resolve the mean collision frequency and the mean free path, constraints that become particularly restrictive in denser flow regimes. To circumvent this limitation, the Boltzmann collision integral is often replaced by approximations such as the Fokker-Planck (FP) equation. Recently, more advanced FP models, such as the ellipsoidal statistical FP model [2] or the cubic FP model [3], have been developed that reproduce the correct

T. Ott (✉) · M. Pfeiffer
Institute of Space Systems, University of Stuttgart, Stuttgart, Germany
e-mail: ottt@irs.uni-stuttgart.de

H. Gorji
Laboratory for Computational Engineering, Empa, Swiss Federal Laboratories for Materials Science and Technology, Dübendorf, Switzerland

© The Author(s) 2026

M. Grabe et al. (eds.), *Rarefied Gas Dynamics*, Springer Aerospace Technology,
https://doi.org/10.1007/978-3-032-00094-1_47

relaxation rates for the pressure tensor and the heat flow and thus reproduce the correct Prandtl number. In theory, these advanced models are applicable to a wide range of flow phenomena from rarefied flows to flows near thermal equilibrium. However, due to the accuracy of the numerical methods, fine temporal and spatial discretization is required. In this work, we introduce a new particle scheme for solving the Boltzmann equation with the Ellipsoidal Statistical Fokker-Planck (ESFP) collision operator. First, we present the theoretical background of the ESFP model and the particle-based method. Then, we describe the time integration and spatial discretization methods used in the scheme. Finally, we validate the method using a 1D Couette flow test case and compare its accuracy and convergence properties with standard methods.

2 Theory

The behavior of monatomic gases can be described by the Boltzmann equation

$$\frac{\partial f}{\partial t} + v_i \cdot \frac{\partial f}{\partial x_i} = \left.\frac{\partial f}{\partial t}\right|_{\text{coll}}, \tag{1}$$

where $f = f(\mathbf{x}, \mathbf{v}, t)$ is the distribution function in phase space spanned by the position $\mathbf{x}$ and the velocity $\mathbf{v}$ at time t. The collision term $\left.\frac{\partial f}{\partial t}\right|_{\text{coll}}$ describes the interaction between particles and can be approximated by the Fokker-Planck operator:

$$\left.\frac{\partial f}{\partial t}\right|_{\text{coll}} = \frac{\partial}{\partial v_i}\left(a_i(\mathbf{x}, t) f(\mathbf{x}, \mathbf{v}, t)\right) + \frac{\partial^2}{\partial v_i \partial v_j}\left(D_{ij}(\mathbf{x}, t) f(\mathbf{x}, \mathbf{v}, t)\right). \tag{2}$$

In this paper, we will focus on the ESFP model by Mathiaud and Mieussens [2], where the coefficients a_i and D_{ij} are given by

$$a_i = \frac{1}{\tau}(v_i - u_i), \quad D_{ij} = \frac{1}{\tau}\left((1 - v)\frac{\text{Tr}(\Sigma)}{3}\delta_{ij} + v\Sigma_{ij}\right). \tag{3}$$

Here, τ is the relaxation time and v is the anisotropy parameter given as a function of the Prandtl number Pr by

$$\tau = \frac{3}{Pr}\frac{\mu}{p}, \quad v = 1 - \frac{3}{2}\frac{1}{Pr}. \tag{4}$$

Additionally, we require the zeroth, first, and second moments of the distribution function which are the number density n, the velocity $\mathbf{u}$, and the covariance matrix Σ respectively:

$$n(\mathbf{x}, t) = \int f \, d\mathbf{v}, \quad \mathbf{u}(\mathbf{x}, t) = \frac{1}{n} \int \mathbf{v} f \, d\mathbf{v}, \quad \Sigma(\mathbf{x}, t) = \frac{1}{n} \int (\mathbf{v} - \mathbf{u})(\mathbf{v} - \mathbf{u})^T f \, d\mathbf{v}.$$

(5)

Using the mass density $\rho = m_s n$ with the mass of the species m_s and the Boltzmann constant k_B, the temperature and pressure can be calculated as

$$T = \frac{m_s}{k_B} \frac{\mathrm{Tr}(\Sigma)}{3}, \quad p = n k_B T.$$

(6)

The dynamic viscosity μ needed to calculate the relaxation time is assumed to follow the power law:

$$\mu = \mu_{\mathrm{ref,s}} \left(\frac{T}{T_{\mathrm{ref,s}}} \right)^{\omega_s},$$

(7)

with the reference viscosity given as a function of the reference temperature $T_{\mathrm{ref,s}}$, the power law exponent ω_s and the mass m_s and the diameter of the species d_s by [4]

$$\mu_{\mathrm{ref,s}} = \frac{30\sqrt{m_s k_B T_{\mathrm{ref,s}}}}{4\sqrt{\pi}(5 - 2\omega_s)(7 - 2\omega_s)d_s^2}.$$

(8)

3 Implementation

We use a particle-based method to solve the Boltzmann equation with the ESFP collision operator. For this, we can write

$$f(\mathbf{x}, \mathbf{v}, t) = \int \int f(\mathbf{x}_p, \mathbf{v}_p, t)\delta(\mathbf{x} - \mathbf{X}_k)\delta(\mathbf{v} - \mathbf{V}_k) \, d\mathbf{X}_k d\mathbf{V}_k.$$

(9)

This integral can be solved approximately by Monte Carlo integration. If we assume that the samples are distributed according to the distribution function f, the approximate integral is given as a sum over the particles:

$$f(\mathbf{x}, \mathbf{v}, t) \approx \frac{N}{N_p} \sum_{k=1}^{N_p} \delta(\mathbf{x} - \mathbf{X}_k)\delta(\mathbf{v} - \mathbf{V}_k),$$

(10)

where $N = \int \int f \, d\mathbf{x}d\mathbf{v}$ is the number of physical particles, N_p is the number of simulation particles, and $\mathbf{x}_p$ and $\mathbf{v}_p$ are the position and velocity of particle p, respectively.

3.1 Time Integration

The solution of the Boltzmann Equation (1) can be described by the *flow* φ given as

$$\varphi(t, f_0(\mathbf{x}, \mathbf{v})) = f(\mathbf{x}, \mathbf{v}, t), \quad \text{with} \quad f(\mathbf{x}, \mathbf{v}, 0) = f_0(\mathbf{x}, \mathbf{v}). \tag{11}$$

We can split the Boltzmann equation into two equations, one describing the movement of particles and one describing the collision term:

$$\frac{\partial f}{\partial t} = -v_i \frac{\partial f}{\partial x_i} \quad \text{and} \quad \frac{\partial f}{\partial t} = \frac{\partial f}{\partial t}\bigg|_{\text{coll}}. \tag{12}$$

The solutions of these equations can also be described by their flows $\varphi_t^M(f_0) = \varphi^M(t, f_0)$ and $\varphi_t^C(f_0) = \varphi^C(t, f_0)$, respectively. Using a composition of these flows, the second order Strang splitting scheme [5] can be written as:

$$\varphi(\Delta t, f_0) = \left(\varphi_{\Delta t/2}^M \circ \varphi_{\Delta t}^C \circ \varphi_{\Delta t/2}^M\right)(f_0(\mathbf{x}, \mathbf{v})). \tag{13}$$

The movement flow φ_t^M can be solved exactly by simply pushing the particles along their velocity vector:

$$f(\mathbf{x}, \mathbf{v}, \Delta t) = f(\mathbf{x} - \mathbf{v}\Delta t, \mathbf{v}, 0), \quad \rightarrow \quad \mathbf{X}_k(\Delta t) = \mathbf{X}_k(0) + \mathbf{V}_k\Delta t. \tag{14}$$

For the collision flow φ_t^C, we must solve the ESFP operator (2, 3). To do this, we can first rewrite the ESFP operator as a stochastic differential equation:

$$d\mathbf{V}_k = \mathbf{a}(\mathbf{X}_k, t)dt + \sqrt{2}\mathbf{D}^{1/2}(\mathbf{X}_k, t)d\mathbf{W}, \tag{15}$$

where $\mathbf{D}^{1/2}(\mathbf{X}_k, t)$ is the Cholesky decomposition of the diffusion matrix $\mathbf{D}(\mathbf{X}_k, t)$ and $\mathbf{W}_k$ is a Wiener process. A standard approach to solve this equation is to assume that the stochastic term is constant [6, 7] which can then be easily solved to get

$$\mathbf{V}_k(\Delta t) = (\mathbf{V}_k(0) - \mathbf{u})e^{-\Delta t/\tau} + \sqrt{1 - e^{-2\Delta t/\tau}}\mathbf{D}^{1/2}(\mathbf{X}_p, 0)\xi. \tag{16}$$

Here, ξ is a random normal vector with zero mean and unit variance. For the ESFP operator, however, the equation can be solved exactly by [6]

$$\mathbf{V}_k(\Delta t) = (\mathbf{V}_k(0) - \mathbf{u})e^{-\Delta t/\tau}$$
$$+ \left(\left(1 - e^{-3\Delta t/(\tau Pr)}\right)\frac{\text{Tr}(\Sigma)}{3}\mathbf{I} + \left(e^{-3\Delta t/(\tau Pr)} - e^{-2\Delta t/\tau}\right)\Sigma\right)^{1/2}\xi. \tag{17}$$

All macroscopic values must be evaluated at the position of the particle $\mathbf{X}_i$. For example, the velocity and covariance matrix are given by $\mathbf{u}(\mathbf{X}_i, 0)$ and $\Sigma(\mathbf{X}_i, 0)$, respectively.

3.2 Deposition and Interpolation

In each time step, the particles must be deposited onto the grid to calculate the moments. In this work, we first choose an element-wise basis to represent the moments M,

$$M(\mathbf{x}, t) \approx M^E(\mathbf{x}, t) = \sum_j \hat{M}_i^E(t)\phi_j^E(\mathbf{x}), \tag{18}$$

with the degrees of freedom $\hat{M}_j^E(t)$ and the basis functions $\phi_j^E(\mathbf{x})$. This choice allows for element-local deposition at the cost of allowing for discontinuities at the element interfaces. The unknown degrees of freedom $\hat{M}_j^E(t)$ can be calculated by projecting the moments onto each basis function ϕ_i^E, specifically by solving

$$\int \phi_i^E(\mathbf{x})M(\mathbf{x}, t)\mathrm{d}\mathbf{x} \stackrel{!}{=} \int \phi_i^E(\mathbf{x})M^E(\mathbf{x}, t)\mathrm{d}\mathbf{x} = \sum_j \hat{M}_j^E(t) \int \phi_i^E(\mathbf{x})\phi_j^E(\mathbf{x})\mathrm{d}\mathbf{x}. \tag{19}$$

By using orthogonal basis functions, the projection matrix becomes diagonal and the degrees of freedom can be calculated without the need to solve a system of equations. Although any polynomial degree can be used, in this work we limit ourselves to constant and linear basis functions.

Another approach to achieve a higher accuracy when using the constant basis is the stochastic interpolation scheme proposed by Fei et al. [8], where the moments for a particle i are evaluated at a random position $\mathbf{x}_i^*$ within half the element size of the particle position $\mathbf{X}_i$:

$$x_{i,j}^* = X_{i,j} + R_j\Delta x_j, \quad R_j \sim \mathcal{U}(-0.5, 0.5). \tag{20}$$

3.3 Momentum and Energy Conservation

After the relaxation, the particle velocities are updated to enforce conservation of momentum and energy:

$$\mathbf{V}_i = \mathbf{u}_{\text{old}} + \sqrt{\frac{\text{Tr}(\Sigma_{\text{old}})}{\text{Tr}(\Sigma_{\text{new}})}}(\mathbf{V}_{i,\text{new}} - \mathbf{u}_{\text{new}}). \tag{21}$$

In this equation, the subscripts old and new denote the values before and after the relaxation step, respectively. Note that this conservation step is always done on the constant basis, as the linear basis would require the solution of a coupled system of all particles in each element.

4 Results

The time integration methods and spatial discretization methods are compared using a simple 1D Couette flow. In this test case, two parallel plates are placed at a distance of $L = 1\,\mathrm{m}$. The two plates are moving in opposite directions with a velocity of $|v_y| = 500\,\mathrm{m\,s^{-1}}$ in the y direction. The interaction between the particles and the plates is assumed to be fully diffuse with a constant plate temperature of $T_w = 273\,\mathrm{K}$. Argon gas, with properties listed in Table 1, is used as the working fluid. The simulations are initialized with a uniform number density of $n = 1.37 \times 10^{20}\,\mathrm{m^{-3}}$ leading to a Knudsen number of approximately $\mathrm{Kn} \approx 0.01$. In all simulations, a total of $N_p = 137000$ particles are simulated.

4.1 Temporal Convergence

In order to test the temporal convergence of the schemes, the simulation is performed with different time steps, varying from $\Delta t = 2.5 \times 10^{-6}\mathrm{s}$ to $\Delta t = 1.6 \times 10^{-4}\,\mathrm{s}$. In all temporal convergence tests, the spatial discretization is kept constant with $N_e = 64$ elements and the linear basis. Figure 1a shows the simulated temperature profile of the two different time integration schemes simulated with $\Delta t = 2 \times 10^{-5}\,\mathrm{s}$. As a reference solution, the simulation results with a very small time step, $\Delta t = 1 \times 10^{-6}\,\mathrm{s}$, are used. We see that the result of the method using the exact integration of the collision term is in much better agreement with the reference solution than that of the standard approach.

In order to further investigate the temporal convergence of the two methods, the L_2-error can be plotted over the different time steps. The L_2-error is given by

$$L_2 = \sqrt{\frac{1}{N_e} \sum_{i=1}^{N_e} \left(T_i - T_{\mathrm{ref},i}\right)^2}, \tag{22}$$

Table 1 Species-specific parameters used for all simulations

Species	$d_s\,/\,\mathrm{m}$	$T_{\mathrm{ref},s}\,/\,\mathrm{K}$	ω_s	$m_s\,/\,\mathrm{kg}$
Ar	4.05×10^{-10}	273	0.77	6.63×10^{-26}

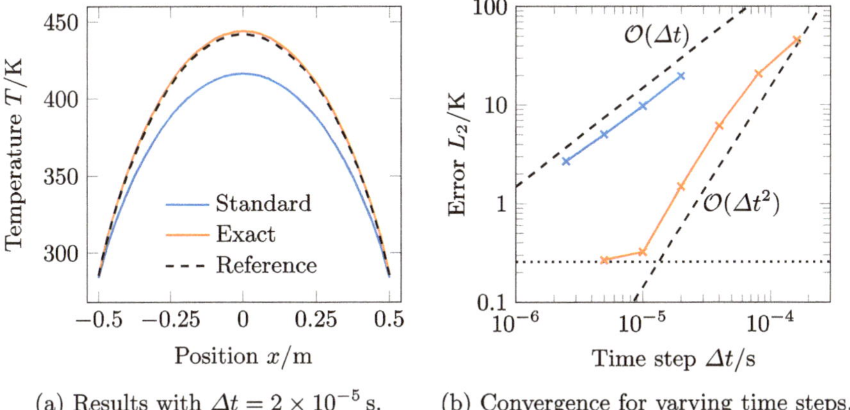

(a) Results with $\Delta t = 2 \times 10^{-5}$ s. (b) Convergence for varying time steps.

Fig. 1 Comparison of the exact integration of the collision operator against the standard approach that assumes a constant diffusion matrix. The dotted line on the right plot shows the estimated bias of the temperature

where T_i and $T_{ref,i}$ are the temperatures of the current simulation and the reference solution evaluated at the center of element i respectively. Figure 1b shows the error L_2 for the two different time integration schemes. It can be seen that, while both methods theoretically use a second-order splitting scheme, only the method using the exact integration of the collision term shows a second-order convergence rate. The standard approach only shows a first-order convergence rate because of the first-order time integration of the collision term. For the smallest time step used in the exact integration of the collision term, the method does not exhibit second-order convergence. This is because the numerical error caused by the finite number of particles dominates at this time step, making the time integration error negligible in comparison. The approach used to calculate the moments on the grid gives a biased estimate of the covariance matrix, see Eq. (5), and thus leads to a biased estimate of the temperature. An estimate of the bias is drawn in Fig. 1b as a dotted line. The error is approximated by assuming a constant temperature T_{ref} and a constant number density, which leads to $N_{\text{DOF}} = N_p/(2N_e)$ particles on average per degree of freedom. Then, the bias can be calculated by using Bessel's correction:

$$\Delta T_{bias} \approx T_{\text{ref}} \left(\frac{N_{\text{DOF}}}{N_{\text{DOF}} - 1} - 1 \right). \tag{23}$$

4.2 Spatial Convergence

The convergence behavior for the different spatial interpolation schemes is tested by varying the number of elements from $N_e = 2$ to $N_e = 128$. During these tests, the time step is kept constant at $\Delta t = 5 \times 10^{-6}$ s and the exact integration of the collision term is used. In Fig. 2, the simulation results of the spatial discretization methods are plotted when simulated using 16 elements. As with the previous example, a very fine discretization is chosen as the reference solution. In this case, the reference solution is simulated with 128 elements and a linear basis. While the constant basis exhibits a large temperature error especially in the center of the channel, the linear basis shows a near perfect agreement with the reference solution.

As for the temporal convergence, the L_2-error can be plotted over the different discretization levels. Figure 2b shows the L_2-error over the number of elements for the different discretization methods. We see that, at a given number of elements, the absolute error of the linear basis is lowest and the error of the constant basis without stochastic interpolation is highest. Interestingly, all methods appear to converge with a second-order convergence rate when the number of elements is greater than $N_e = 8$. Unfortunately, when using the linear basis, the spatial discretization error is found to be of the order of error due to the biased temperature estimate, even when only 32 elements are used. Thus, the convergence rate of the linear basis is limited by the biased estimate of the temperature. The estimate bias shown as a dotted line in Fig. 2b is calculated in the same way as for the temporal convergence. Since the number of degrees of freedom depends on the the number of elements, this bias is not constant. Also, no results could be obtained for the linear basis with fewer than 8 elements, as the stochastic noise would lead to negative pressures. The reason for this is that the orthogonal linear basis functions are not strictly positive. This can lead to negative densities as well as negative eigenvalues of the covariance matrix, which in turn leads to negative pressures.

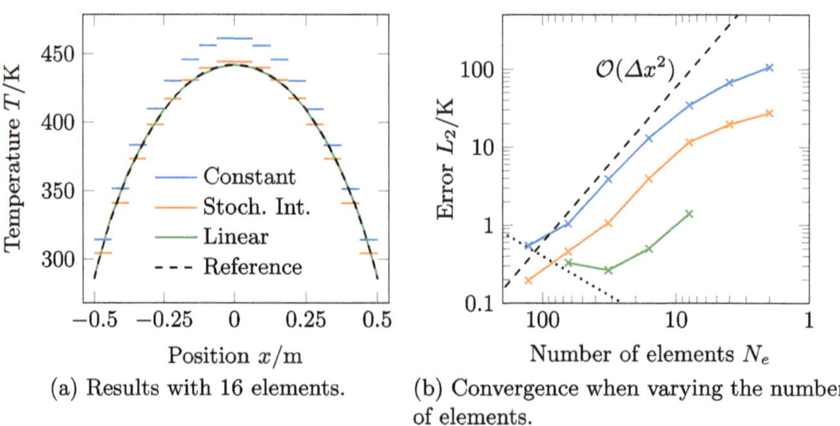

(a) Results with 16 elements.   (b) Convergence when varying the number of elements.

Fig. 2 Comparison of the deposition and interpolation schemes using a constant basis, a constant basis with stochastic interpolation and the linear basis. The dotted line on the right plot shows the estimated bias of the temperature for the linear basis

5 Conclusion

This paper presented an improved particle scheme for solving Fokker-Planck models of the Boltzmann equation. The simulation results of a 1D Couette flow showed that the Strang splitting as well as using linear basis functions provide accurate solutions for the Boltzmann equation with the Ellipsoidal-Statistical Fokker-Planck collision operator. Future work will focus on the possibility to limit the linear basis functions to avoid negative pressures. Additionally, further investigation will be done on extending the method to other Fokker-Planck operators and exploring higher-order spatial and temporal discretizations.

Acknowledgements This project has received funding from the European Research Council (ERC) under the European Union's Horizon 2020 research and innovation program (grant agreement No. 899981 MEDUSA).

References

1. Bird GA (1994) Molecular gas dynamics and the direct simulation of gas flows. Oxford University Press, New York
2. Mathiaud J, Mieussens L (2016) A Fokker–Planck model of the Boltzmann equation with correct Prandtl number. J Statist Phys 162(2):397–414
3. Gorji MH, Jenny P (2014) An efficient particle Fokker-Planck algorithm for rarefied gas flows. J Computat Phys 262:325–342
4. Burt JM, Boyd ID (2006) Evaluation of a particle method for the ellipsoidal statistical Bhatnagar-gross-Krook equation. In: 44th AIAA aerospace sciences meeting and exhibit
5. Strang G (1968) On the construction and comparison of difference schemes. J Numer Anal 5(3):506–517
6. Jiang Y, Gao Z, Lee C-H (2018) Particle simulation of nonequilibrium gas flows based on ellipsoidal statistical Fokker-Planck model. Comput Fluids 170:106–120
7. Jun E, Pfeiffer M, Mieussens L, Gorji MH (2019) Comparative study between cubic and ellipsoidal Fokker-Planck kinetic models. AIAA J 57:2524–2533
8. Fei F, Ma Y, Wu J, Zang J (2021) An efficient algorithm of the unified stochastic particle Bhatnagar-Gross-Krook method for the simulation of multi-scale gas flows. Adv Aerodynam 1–16

Impact of Entropy in Shock Recovery Through Fokker-Planck Kinetics

Veronica Montanaro, Lukas Netterdon, Manuel Torrilhon, and Hossein Gorji

Abstract While numerical methods based on the Boltzmann equation are often computationally demanding in near-continuum regime, Fokker-Planck approximations can offer efficient alternative. However, existing Fokker-Planck models remain largely heuristic and often lack global accuracy across diverse kinetic regimes, from micro-nano fluidics to the hypersonic setups. We propose a systematic recipe to construct Fokker-Planck models with special emphasis on the role of entropy and its evolution. We make the case that the Fisher entropic constraint is a key to build viable Fokker-Planck models, particularly for resolving physically accurate shock profiles. Our arguments are supported by numerical tests on a two-dimensional shock problem.

Keywords Fokker-Planck equation · Entropy · Langevin dynamics

1 Introduction

Efficient and accurate simulations of rarefied gas flows, which occur in micro-nano geometries and/or atmospheric reentry flights, remain an active area of research. This is mainly due to the limitations of conventional hydrodynamic equations, such as the Navier-Stokes-Fourier system, which often fail to provide a satisfactory macroscopic description. The Boltzmann equation is, on the other hand, the cornerstone of gas kinetics across the whole range of rarefactions [1]. Despite its mathematically rich structure and physically accurate descriptions, the numerical solution of the Boltzmann equation remains by far computationally expensive. A particle based variant

V. Montanaro · H. Gorji (✉)
Computational Engineering Laboratory, Empa, Switzerland
e-mail: mohammadhossein.gorji@empa.ch

V. Montanaro
Institute of Mathematics, EPFL, Switzerland

L. Netterdon · M. Torrilhon
Applied and Computational Mathematics, RWTH Aachen University, Aachen, Germany

© The Author(s) 2026

501

M. Grabe et al. (eds.), *Rarefied Gas Dynamics*, Springer Aerospace Technology,
https://doi.org/10.1007/978-3-032-00094-1_48

of the Boltzmann equation, provided by Direct Simulation Monte-Carlo (DSMC) [2] is often the method of choice, especially when practical rarefied gas settings are considered. Despite the Monte-Carlo basis of DSMC, which makes it efficient in terms of scaling with respect to the dimension, it may suffer from high computational costs in transitional regime. This is mainly due to the sharp increase in the time complexity entailed in resolving binary collisions, besides sample complexity arising in low signal-to-noise ratio regime [3, 4].

Simplified kinetic models surged as a viable tool to study different aspects of the Boltzmann equation and corresponding physical processes. Simplified collision operators such as linearized Boltzmann [5], BGK [6], and Fokker-Planck [7, 8] were introduced and studied as an alternative to solution algorithms focusing on the Boltzmann equation. Development of such models can also be of interest in broader communities concerned with micro-macro couplings [9].

Kinetic equations are inherently high-dimensional and therefore their computational complexity and memory requirements can become a limiting factor as richer processes are considered. However, they can be reset in the Monte-Carlo framework, in the same fashion as DSMC, to leverage favourable scaling while avoiding the stringent requirement of resolving collisional scales. This was the main motivation behind Fokker-Planck (FP) based particle algorithms targetting rarefied gas flow simulations. Nevertheless, one should note that these approaches do not provide a direct remedy to the sample complexity problem encountered in the low Mach regime [10, 11].

Historically, the FP equation has been studied as diffusive limit of Boltzmann kinetics [8]. Modern approaches were built on and enriched this asymptotic result, in order to improve the accuracy of FP based rarefied gas simulations. Most notably, Linear-FP [12], Cubic-FP [13], ES-FP [14], and EFP [15] were proposed. To a large degree, the modeling rational behind these approaches remain heuristic. Having these models in hand, the aim of the current study is to propose a methodology which can bring together flexibility of nonlinear-drift models and favourable entropy aspect of the Linear-FP model. The focus of the work remains at the modeling front, and we leave the more advanced algorithmic aspects including high-order time integration to separate studies.

2 Background on Fokker-Planck Models

Physically motivated derivation of FP has been pursued from different perspectives. For example, the Hamiltonian dynamics can be projected onto a resolved particle leading to the Generalised Langevin Equation, as performed in the Mori-Zwanzig formalism [16]. Alternatively, in the Kramers-Moyal expansion, the FP equation can be derived after keeping the first and second order terms [17]. No matter which path is taken, the appeal of the FP equation seems to be in its expressive power to describe complex distributions.

For our purpose, the FP equation remains an approximation of the Boltzmann equation and therefore we do not intend to pursue a first-principled derivation of FP. Instead, we postulate a flexible FP model and impose the desired properties on its structure. Yet note that the resulting model only weakly correspond to the collision dynamic underlying the Boltzmann equation, as it approximates the latter in the statistical sense. In other words, we do not attempt to recover the path of a sample particle in the FP model. This weak approximation enables us to replace collisions with stochastic paths governed by Langevin dynamics.

The FP model should be designed to ensure that (certain) non-equilibrium moments relax at rates similar to those in the Boltzmann equation. This design choice of the model is motivated by the fact that the quantities of interest in rarefied gas flow simulations can be described as macroscopic moments. Furthermore, the model should obey the entropy law, specifically with the same H-functional as in the Boltzmann equation. Besides the theoretical appeal of an entropy law (rooted in stability of the governing equation), it has been suggested that imposing certain dynamics on the entropy offers physically more accurate results, especially in shock related problems [15, 18, 19].

The main objective behind FP models is to approximate the collision operator, $S^{Boltz}[f]$, governed by the Boltzmann equation

$$\partial_t f + \nabla_{x_i}(v_i f) = S^{Boltz}[f], \tag{1}$$

with a FP one

$$S^{FP}[f] = -\nabla_{v_i}(A_i[f]f) + \Delta_{v_i v_j}(D_{ij}[f]f). \tag{2}$$

Note that $f(v; x, t) : \mathbb{R}^6 \times \mathbb{R}^+ \to \mathbb{R}^{\geq 0}$ is the (normalized) density of probability of observing a particle with velocity near $v \in \mathbb{R}^3$, at position $x \in \mathbb{R}^3$ and time $t \in \mathbb{R}^+$. The drift, $A : f \mapsto A[f] \in \mathbb{R}^3$, and diffusion, $D : f \mapsto D[f] \in S^3_+$ with S^3_+ being the set of 3×3 positive semi-definite matrices with component values in $\mathbb{R}^3$, are model unknowns. Note that throughout this manuscript, we employ the index summation convention.

The path taken by the Cubic-FP model was to introduce enough degrees of freedom in the drift functional, such that the FP operator fulfills

$$\langle S^{FP}[f] H_\alpha(v) \rangle_f \overset{!}{=} \langle S^{Boltz}[f] H_\alpha(v) \rangle_f \tag{3}$$

for $H_\alpha(v) \in \{1, v, v \otimes v, v\|v\|^2\}$, where $\langle . \rangle_f$ is shorthand for the integration, i.e., $\langle g \rangle_f = \int_{\mathbb{R}^3} gf\ dv$. Despite the fact that introducing higher order terms in $A[f]$ leads to fulfillment of the constraint (3), it turns out the specific form of the entropy relaxation is crucial for accuracy of the model in shock structure problems [15]. Upon close inspection of the Linear-FP model, we observe that the entropy functional

$$\mathcal{H}[f] = \langle \log f - 1 \rangle_f \tag{4}$$

follows specific dynamics. In particular, we get

$$\partial_t \mathcal{H}[f] = -D[f]\overbrace{\left\langle \nabla_{v_i} \log \frac{f}{f_{eq}} \nabla_{v_i} \log \frac{f}{f_{eq}} \right\rangle_f}^{=:I[f|f_{eq}]}, \tag{5}$$

where f_{eq} is the equilibrium density and $D_{ij} = D\delta_{ij}$. Note that the term $I[f|f_{eq}]$ is the Fisher information, which has rich physical interpretations [20], and we refer to Eq. (5) as the Fisher entropic constraint.

3 Problem Statement and Main Idea

Motivated by flexibility of the Cubic-FP model, we propose a modeling approach where moment constraints (3) along with the Fisher entropic constraint (5) are fulfilled. We construct the model as an expansion around the linear one, with the time scale $\tau > 0$. Therefore, we set

$$A[f] = -\frac{1}{\tau}v_i' + A^{ho}[f] \tag{6}$$

where $v' = v - U$, $U = \langle v \rangle_f$, and

$$D_{ij}[f] = \frac{1}{3\tau}\langle v_l'v_l' \rangle_f \delta_{ij} =: D\delta_{ij} , \tag{7}$$

with δ_{ij} as the Kronecker delta. Direct computation shows that the Fisher entropic constraint can be translated into

$$\langle \nabla_{v_i} A_i^{ho}[f] \rangle_f = 0 . \tag{8}$$

This is an interesting result since it allows us to embed the entropy constraint given by Eq. (5) into the high-order terms of the drift. Considering the moment constraints for momentum, pressure tensor, and heat-fluxes, besides Eq. (8), we follow a polynomial expansion of the drift

$$A_i^{ho}[f](v) = c_i^{(0)} + c_{ij}^{(1)} v_j' + 2c_j^{(2)} v_i'v_j' + c_i^{(2)} \|v'\|^2 + c^{(3)} v_i'\|v\|^2 + c^{(4)} v_i'\|v'\|^4 , \tag{9}$$

where the 11 coefficients $\{c^{(0)}, c^{(1)}, c^{(2)}, c^{(3)}\}$ depend on f (through its moments), ensuring the moment consistency relations together with the entropy constraint. The coefficient $c^{(4)} < 0$, $|c^{(4)}| \ll 1$ is introduced for the stability of the drift, guaranteeing that $A_i \to -\infty$ as $\|v\| \to \infty$. Besides $c^{(0)}$, that can be found based on momentum conservation and $c^{(4)}$, which is set as a stabilizer, the 10 remaining coefficients (six components in $c^{(1)}$, three components in $c^{(2)}$ and one in $c^{(3)}$) can be computed from a 10×10 linear system via plugging the ansatz (9) into the constraints (3) and (8). We refer to the model with coefficients described above as Fisher Entropic FP (FE-FP) model. Some remarks are in order.

1. The model coefficients can be found from the linear system where the explicit knowledge of f is not required. The system is comprised of the moments which can be computed from particle ensemble.
2. In comparison to EFP [15], which only ensures $\partial_t \mathcal{H} \leq 0$, the introduced model fulfills a stronger constraint on the entropy evolution given by Eq. (5).
3. The time scale τ which controls the rate of entropy decay via

$$\partial_t \mathcal{H} = -\frac{1}{3\tau} \langle v_i' v_i' \rangle_f \left\langle \nabla_{v_j} \log \frac{f}{f_{eq}} \nabla_{v_j} \log \frac{f}{f_{eq}} \right\rangle_f , \tag{10}$$

remains a free parameter and can be adjusted either from physical considerations or empirically based on model performance. Following the first path, it is straight-forward to find $\tau = 2\mu/p$ (where μ is the viscosity and p is the pressure) leads to entropy relaxation rate of the FE-FP consistent with the Botlzmann equation, for anisotropic Gaussian initial distribution. f_G. Note that this time scale is similar to the one obtained in the linear drift model.

4 Overview of Numerics

Similar to DSMC, a particle-based Monte-Carlo approach is employed to solve the FP model. However, instead of free flight and collisions, the particle trajectories follow the Langevin equations

$$dM_i = A_i[f](M)\,dt + \sqrt{2D[f]}\,dW_i \quad \text{and} \quad dX_i = M_i\,dt, \tag{11}$$

where $M(t)$, $X(t) \in \mathbb{R}^3$ are random variables corresponding to the particle velocity and position, respectively. The term dW denotes Wiener process with zero mean and variance dt. It follows from classic arguments that the density of the random variables, which evolve according to the Langevin equations given above, coincides with the solution of the FP equation with drift A and diffusion D [17].

Time integration of the Langevin system of the form given above requires a spatio-temporal discretization, since the coefficients depend on the macroscopic quantities through f. For simplicity, we focus on the first-order Euler-Maruyama scheme, similar to [15]. The algorithm then proceeds for each time steps, according to the following steps.

1. Estimate moments from the particle ensemble and solve for $\{c^{(0)}, c^{(1)}, ..., c^{(4)}\}$.
2. Compute the drift and diffusion, Eqs. (9) and (7).
3. Update particles velocities and positions, Eq.(11).

4. Apply the boundary conditions (similar to DSMC).
5. Output macroscopic quantities of interest.

5   Shock Results

We test the FE-FP model in supersonic flow of argon (with hard sphere molecular model and the mass $m = 66.3 \times 10^{-27}$ kg) over an infinitely thin vertical plate of length $l = 0.1828$ m. The chosen setup is the one appearing in [15], in order to highlight the deficiency of previous FP models in capturing shock structure. The vertical plate is placed at the center of a rectangular domain of $2.5l \times 1.5l$. The number density $n = 10^{20}$ $1/m^3$, and temperature $T_0 = 273$ K (similar to the isothermal plate). The flow enters the domain from the left boundary. The simulation employs 3×10^6 number of particles, 120×120 computational cells, and time-step size $\Delta t = 1.5978 \times 10^{-6}$ s. After $2'000$ initial time-steps, $5'000$ steps are used for time averaging.

We run the simulation with different value of the Mach number of the upstream flow Ma $= [4, 5, 6, 7]$, using the DSMC results as the benchmark. The Mach contour is compared with respect to both DSMC and Cubic-FP results, as shown in Fig. 1. Excellent agreement is observed between FE-FP and DSMC for all the range of Mach numbers, while the Cubic-FP exhibits diffused shock profile increasing proportionally with Ma. To gain further insight into the deviations between considered models, temperature profiles at $y = 1.875l$ are depicted as shown in Fig. 2. The agreement with the DSMC solution is further confirmed there.

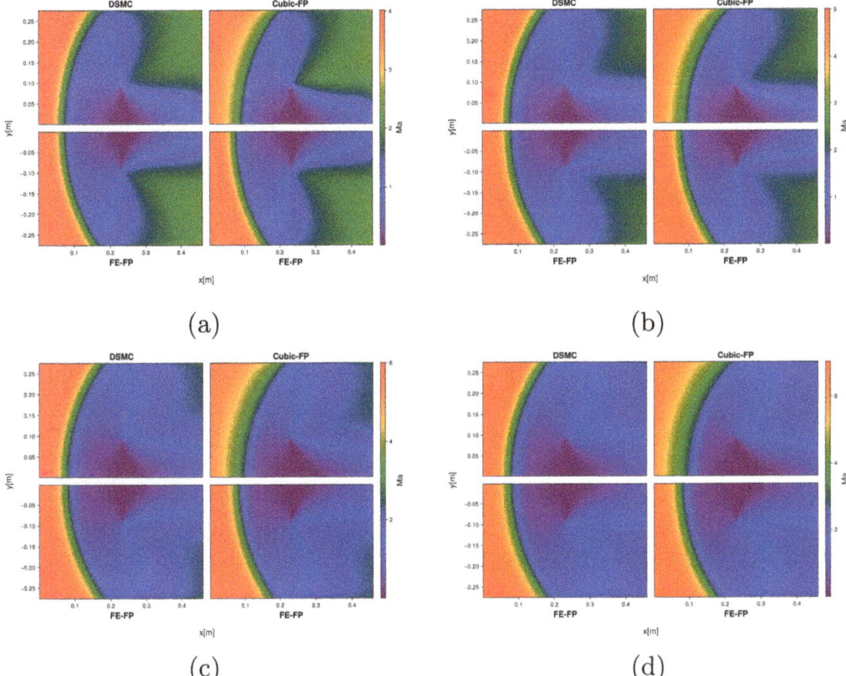

Fig. 1 Mach contours for a flow of argon over a vertical plate; Results for **a** Ma = 4, **b** Ma = 5, **c** Ma = 6, and **d** Ma = 7; DSMC and Cubic-FP results are shown on top and compared with respect to the FE-FP results at the bottom

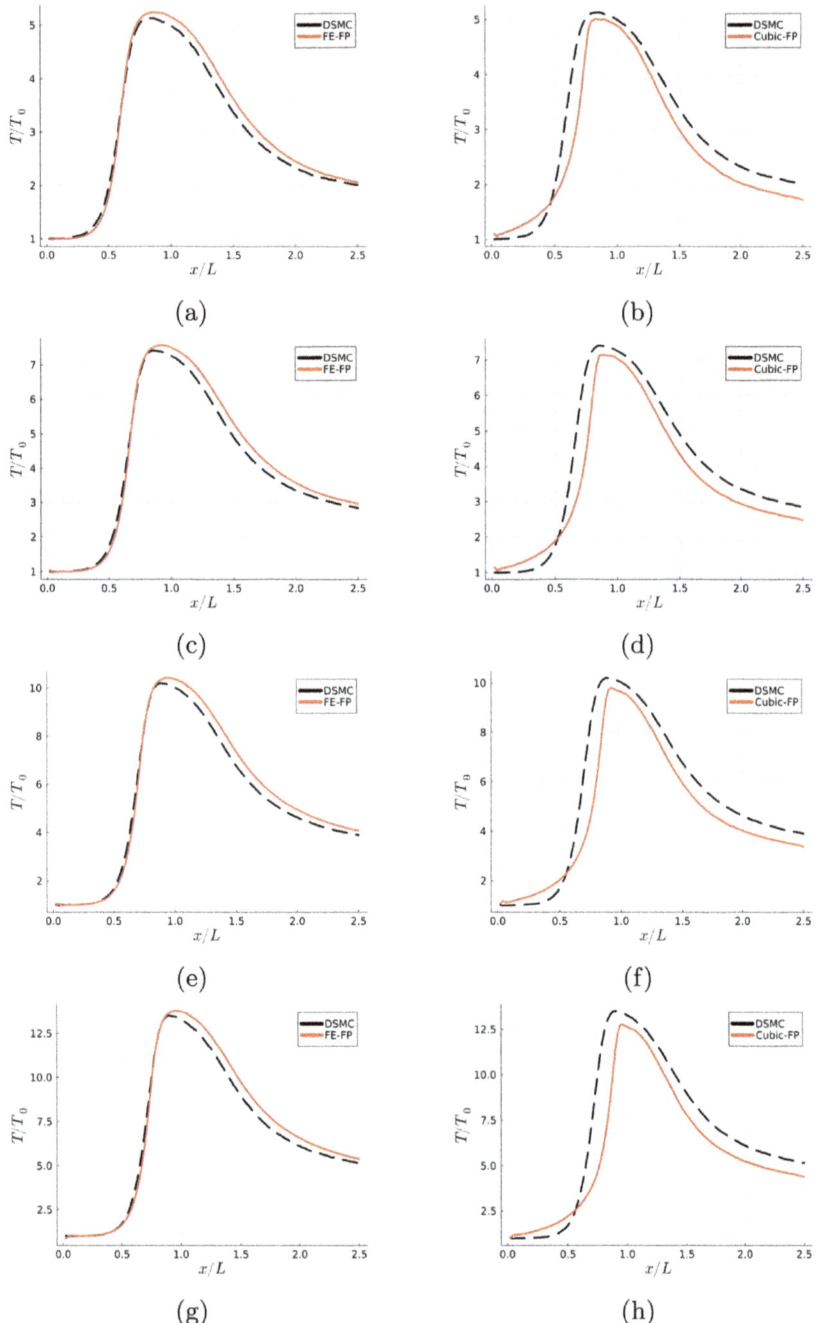

Fig. 2 Normalized temperature profiles along $y = 1.875l$; **a** and **b** Ma $= 4$, **c** and **d** Ma $= 5$, **e** and **f** Ma $= 6$, **g** and **h** Ma $= 7$; The left column compares the DSMC reference solution with the FE-FP model, while the right column shows comparisons with the Cubic-FP model

6 Conclusion

In this study, we developed a systematic approach to derive Fokker-Planck (FP) models that preserve the Fisher-entropic property. Our framework integrates the flexibility of the Cubic-FP model with the favorable entropy treatment of the Linear-FP model. This combination led to the formulation of the FE-FP model, which accurately captures shock structures given an appropriate choice of the time scale governing the entropy rate. Future works will include more extensive investigation of the FE-FP model in different benchmark problems, along with a further characterization of the time scale's role in model accuracy.

Acknowledgements This work was supported by Swiss National Science Foundation under grant number 217408, and by the Deutsche Forschungsgemeinschaft (DFG, German Research Foundation)—SPP 2410 "Hyperbolic Balance Laws in Fluid Mechanics: Complexity, Scales, Randomness (CoScaRa)" within the Project 525660607 "Model Cascades for Stochastic Particle Simulations of Rarefied Polyatomic Gases".

References

1. Carlo C (1988) The boltzmann equation. Springer, New York
2. Bird GA (1994) Molecular gas dynamics and the direct simulation of gas flows. Oxford University Press
3. Baker LL, Hadjiconstantinou NG (2005) Variance reduction for Monte Carlo solutions of the Boltzmann equation. Phys Fluids 17(5)
4. Garcia Alejandro L, Wolfgang W (2000) Time step truncation error in direct simulation Monte Carlo. Phys Fluids 12(10):2621–2633
5. Gross EP, Atlee Jackson E (1959) Kinetic models and the linearized Boltzmann equation. Phys Fluids 2(4):432–441
6. Shakhov EM (1968) Generalization of the Krook kinetic relaxation equation. Fluid Dynam 3(5):95–96
7. Pawula RF (1967) Approximation of the linear Boltzmann equation by the Fokker-Planck equation. Phys Rev 162(1):186
8. Lebowitz JL, Frisch HL, Helfand E (1960) Nonequilibrium distribution functions in a fluid. Phys Fluids 3(3):325–338
9. Sadr M, Hadjiconstantinou NG, Hossein Gorji M (2024) Wasserstein-penalized Entropy closure: a use case for stochastic particle methods. J Comput Phys 511(2024):113066
10. Gorji MH, Andric N, Jenny P (2015) Variance reduction for Fokker-Planck based particle Monte Carlo schemes. J Comput Phys 295:644–664
11. Collyer Benjamin S, Colm C, Lockerby Duncan A (2016) Importance sampling variance reduction for the Fokker-Planck rarefied gas particle method. J Comput Phys 325:116–128
12. Patrick J, Manuel T, Stefan H (2010) A solution algorithm for the fluid dynamic equations based on a stochastic model for molecular motion. J Comput Phys 229(4):1077–1098
13. Gorji MH, Torrilhon M, Jenny P (2011) Fokker-Planck model for computational studies of monatomic rarefied gas flows. J Fluid Mech 680:574–601
14. Mathiaud J, Mieussens L (2016) A Fokker-Planck model of the Boltzmann equation with correct Prandtl number. J Stat Phys 162:397–414
15. Gorji MH, Torrilhon M (2021) Entropic Fokker-Planck kinetic model. J Comput Phys 430:110034

16. Chorin Alexandre J, Hald Ole H, Raz Kupferman (2000) Optimal prediction and the Mori-Zwanzig representation of irreversible processes. Proc the Nat Acad Sci 97(7):2968–2973
17. Risken H (1996) The Fokker-Planck equation
18. Jun E, Pfeiffer M, Mieussens L, Hossein Gorji M (2019) Comparative study between cubic and ellipsoidal Fokker-Planck kinetic models. AIAA J 57(6):2524–2533
19. Kim S, Gorji H, Jun E (2023) Critical assessment of various particle Fokker–Planck models for monatomic rarefied gas flows. Phys Fluids 35(4)
20. Frieden BR (1990) Fisher information, disorder, and the equilibrium distributions of physics. Phys Rev A 41(8):4265

A Coupled Semi-Lagrangian/Finite Volume Scheme to Deal with Close-Packing Limit in Thick Sprays

Christophe Buet, Bruno Després, and Victor Fournet

Abstract This work is dedicated to the numerical discretization of a thick spray model. We present a coupled method combining a Finite Volume scheme for the fluid phase, and a semi-Lagrangian scheme for the dispersed phase. We also present two methods for taking into account the close-packing limit of the particles in the model.

Keywords Fluid-kinetic system · Numerical method · Vlasov equation

1 Introduction

In this work, we are interested in the numerical simulation of dense suspensions of particles within a carrying fluid. The starting point of this work is the following fluid kinetic system

C. Buet · V. Fournet (✉)
CEA, DAM, DIF, Arpajon, France
e-mail: victor.fournet@sorbonne-universite.fr

C. Buet
e-mail: christophe.buet@cea.fr

C. Buet
Universite Paris-Saclay, CEA DAM DIF, Laboratoire en Informatique Haute Performance pour le Calcul et la simulation, Arpajon, France

B. Després · V. Fournet
Laboratoire Jacques Louis Lions, Sorbonne Universite, Paris, France
e-mail: despres@ann.jussieu.fr

M. Grabe et al. (eds.), *Rarefied Gas Dynamics*, Springer Aerospace Technology,
https://doi.org/10.1007/978-3-032-00094-1_49

511

$$\begin{cases} \partial_t(\alpha\varrho) + \nabla_x \cdot (\alpha\varrho\mathbf{u}) = 0 \\[4pt] \partial_t(\alpha\varrho\mathbf{u}) + \nabla_x \cdot (\alpha\varrho\mathbf{u} \otimes \mathbf{u}) + \alpha\nabla_x p = D_\star \int_{\mathbb{R}^3} (\mathbf{v} - \mathbf{u})f \, dv \\[4pt] \partial_t f + \mathbf{v} \cdot \nabla_x f + \nabla_v \cdot (\Gamma f) = 0 \\[4pt] \alpha = 1 - m_\star \int_{\mathbb{R}^3} f \, dv \\[4pt] m_\star \Gamma = -m_\star \nabla_x p - D_\star(\mathbf{v} - \mathbf{u}). \end{cases} \tag{1}$$

The first two equations are modified isentropic Euler equations, where the fluid is described by its volume fraction $\alpha(t, \mathbf{x})$, its density $\varrho(t, \mathbf{x})$ and its velocity $\mathbf{u}(t, \mathbf{x})$. We moreover assume that this fluid is ideal and barotropic, with a pressure law $p(\rho)$. The particle are described by a distribution function $f(t, \mathbf{x}, \mathbf{v})$ which evolves according to the third equation of (1). The coupling between the two phases is done not only through a linear drag force, where $D_\star > 0$ is a (constant) drag coefficient, but also through the volume fraction of the fluid α in the fluid equations and through the presence of the pressure gradient $-\nabla_x p$ in the force field of the Vlasov equation. The system (1) is a prototype for describing dense flow of particles within a carrying gas, it is sometimes refered to as a thick spray model [9]. In this work, we only consider the case where $p(\varrho) = \varrho^\gamma$, where $\gamma > 1$ is the adiabatic index. We place ourselves in a system of unit where the density of particles is 1, so that the coefficient $m_\star$ is an adimensional volume. Traditionaly, the numerical discretization of this kind of fluid-particle coupling is done with a Particle-In-Cell (PIC) method for the Vlasov equation [1, 7]. The PIC method has the primary advantage of only discretizing the physical space x and not the velocity space v. This drastically reduces the computational cost of a direct discretization of the Vlasov equation. Indeed, the primary computational challenge comes from the fact that the distribution function f lives in an up to six-dimensional phase space. However, particle methods are known to suffer from numerical noise that decreases very slowly, as the square root of the number of particles. They are also known to be highly sensitive to sampling errors. Recently, especially in the plasma physics community, Eulerian methods for the Vlasov equation have been developed with remarkable success [4, 5, 10–12, 14, 19]. Due to the substantial increase in computational power, 4D simulation can now be performed routinely on high performance computing (HPC) systems.

In this work, we present a numerical method for the discretization of thick sprays, based on semi-Lagrangian method for the Vlasov equation, and a Finite Volume method for the fluid phase [2, 3]. A point of interest in this work is to take into account the *close-packing limit* in the system, characterized by a state where the particle density reaches a maximum.

1.1 Close-Packing Limit

By close-packing limit, we refer to the configuration where the maximal density for the particles is reached, which typically corresponds to the case where the particles are packed on top of each other. The close-packing of particles is a part of a larger family of congestion phenomena in fluid equations. The study of such phenomena represents a prolific area of the literature, see [6, 8, 15, 17, 18] and the references therein. The system (1) does not take into account the modelling of close-packing limit. We would like to modify the model (1) so that, given a lower bound $0 \leq \alpha_\star < 1$ and assuming that the initial data verifies $\alpha(t = 0, \cdot) > \alpha_\star$, the solutions preserve the bound $\alpha(t, \cdot) > \alpha_\star$ for all $t > 0$. A priori, the system (1) does not verify this property. In the literature, multiple ways to model the close-packing limit can be found. For instance, a possible way [1, 3] to modify the Vlasov equation is to add a pressure term in the force field: $m_\star \Gamma = D_\star(\mathbf{u} - \mathbf{v}) - m_\star \nabla_x p - \frac{1}{1-\alpha} \nabla_x \pi(\alpha)$, The function $\pi : [0, \alpha_\star) \to [0, +\infty)$ is required to satisfy $\pi(0) = 0$ and $\lim_{\alpha \to \alpha_\star} \pi(\alpha) = +\infty$. Another possibility, found for instance in [13] is to modify the free transport operator directly: $\partial_t f + \nabla_x \cdot (\mathbf{v} f - \pi(\alpha) \nabla_v f) + \nabla_v \cdot (\Gamma f) = 0$. Our first proposition to implement close-packing effects in the model (1) is to modify Vlasov equation $\partial_t f + \nabla_x \cdot (\mathbf{v} f - \nabla_x \pi f) + \nabla_v \cdot (\Gamma f) = 0$.

In the literature [8], it is common to give a formula for π, which is mainly of empirical nature. A typical formula is

$$\pi(\alpha) = \varepsilon \left(\alpha_f\right)^\beta (\alpha_\star - \alpha_f)^{-\beta}, \quad \beta > 1. \tag{2}$$

The coefficient $\varepsilon > 0$ is usually taken small. Its role is to make π negligible when α_f is not very close to $\alpha_\star$. The coefficient $\beta > 1$ is given empirically. This kind of expression for π corresponds to so-called *soft congestion models* [8]. The idea is that the singular pressure π acts as a barrier that prevents the particle volume fraction from becoming too big. When $\varepsilon \to 0$, the singular pressure π tends to a limit pressure π_∞ such that $(\alpha - \alpha_\star)\pi_\infty = 0$, $\alpha \geq \alpha_\star$, $\pi_\infty \geq 0$, and the corresponding models that use this type of pressure are known as *hard congestion models* [8]. The link between the hard and soft models is interesting from a numerical point of view. The hard limit is difficult to handle because of the sharp unknown interface between the free domain (where $\pi_\infty = 0$) and the congested domain. One can instead use the soft model with small ε to approximate the hard model [6]. In this work, we only consider the case $\varepsilon > 0$. Our second proposition to take into account the close packing limit is purely of numerical nature. The idea is to introduce a correction procedure in the numerical scheme that imposes the bound $\alpha \geq \alpha_\star$. The correction procedure we propose is inspired by works done in [15, 16, 18] on the modeling of crowd motion. To finish this section, let us write the two systems that we will study. We rewrite the fluid phase in terms of the conserved variables $n := \alpha \varrho$, $\mathbf{q} := \alpha \varrho \mathbf{u} + m_\star \int_{\mathbf{R}^3} \mathbf{v} f \, dv$,. The original system without packing pressure π is

$$\begin{cases} \partial_t n + \nabla_x \cdot \left(\mathbf{q} - m_\star \int \mathbf{v} f \, dv \right) = 0 \\ \partial_t \mathbf{q} + \nabla_x \cdot \left(\dfrac{\left(\mathbf{q} - m_\star \int \mathbf{v} f \, dv \right)^{\otimes 2}}{n} + m_\star \int \mathbf{v} \otimes \mathbf{v} f \, dv \right) + \nabla_x p \left(\dfrac{n}{\alpha} \right) = 0 \\ \partial_t f + \nabla_x \cdot (\mathbf{v} f) + \nabla_v \cdot (\Gamma f) = 0 \\ \alpha = 1 - m_\star \displaystyle\int_{\mathbf{R}^3} f \, dv \\ m_\star \Gamma = -m_\star \nabla_x p - D_\star (\mathbf{v} - \mathbf{u}), \end{cases} \tag{3}$$

that we will refer to as the "original model". Then, we have the system with added packing pressure

$$\begin{cases} \partial_t n + \nabla_x \cdot \left(\mathbf{q} - m_\star \int \mathbf{v} f \, dv \right) = 0 \\ \partial_t \mathbf{q} + \nabla_x \cdot \left(\dfrac{\left(\mathbf{q} - m_\star \int \mathbf{v} f \, dv \right)^{\otimes 2}}{n} + m_\star \int \mathbf{v} \otimes \mathbf{v} f \, dv - m_\star \partial_x \pi \int f v \, dv \right) + \nabla_x p \left(\dfrac{n}{\alpha} \right) = 0 \\ \partial_t f + \nabla_x \cdot [(\mathbf{v} - \nabla_x \pi(\alpha)) f] + \nabla_v \cdot (\Gamma f) = 0 \\ \alpha = 1 - m_\star \displaystyle\int_{\mathbf{R}^3} f \, dv \\ m_\star \Gamma = -m_\star \nabla_x p - D_\star (\mathbf{v} - \mathbf{u}) \\ \pi(\alpha) = \varepsilon \left(\dfrac{\alpha_f}{\alpha_\star - \alpha_f} \right)^\beta, \quad \beta > 1, \end{cases} \tag{4}$$

and we will refer to it as the "soft model".

2 Numerical Methods

In this work, the systems (3) and (4) are discretized by combining classical methods. We use an appropriate time-splitting where the fluid phase is discretized using a Finite Volume scheme with Rusanov flux. The Vlasov equation is discretized using a conservative semi-Lagrangian scheme, the PFC scheme [11].

Correction procedure. The spirit of the correction procedure is to compare α_i^k with $\alpha_\star$ and check if $\alpha_i^k \geq \alpha_\star$. In this work, $\alpha_\star$ is the same in all the cells, but one could easily extend this work to a non-homogenous packing limit. Assume that $\alpha_i^k < \alpha_\star$, this means that the particle volume fraction in the cell i, $m_\star \Delta v \sum_j f_{i,j}^k$ is too big, and one has to redistribute the mass in the cells around the cell i. Taking inspiration from [15], we propose to redistribute the mass with a stochastic algorithm. Starting from the saturated cell, we initiate a random walk. When the walk encounters a cell which is not saturated, it get rids of as much mass as it can. Then the random walk continues, until the exceeding mass has been completely redistributed.

3 Numerical Results

In this section, we present several numerical results to validate our methods.

3.1 Collision of Two Pack of Particles

In this test, we consider the case with a gas initially at rest, with two clouds of particles going into the middle of the domain. The initial condition for the gas is a gas at rest with constant density and pressure $(\varrho_0, u_0, p_0)(x) = (1, 0, 1)$, with $\gamma = 1.4$. The initial condition for the particles is

$$f_0(x, v) = \frac{1}{\sqrt{2\pi}\sigma_v} e^{-\frac{(v-6)^2}{2\sigma_v^2}} \frac{1}{\sqrt{2\pi}\sigma_x} e^{-\frac{(x-2.6)^2}{2\sigma_x^2}} \mathbf{1}_{v>0}(v)$$
$$+ \frac{1}{\sqrt{2\pi}\sigma_v} e^{-\frac{(v+6)^2}{2\sigma_v^2}} \frac{1}{\sqrt{2\pi}\sigma_x} e^{-\frac{(x-3.4)^2}{2\sigma_x^2}} \mathbf{1}_{v<0}(v). \tag{5}$$

with $\sigma_x = \sigma_v = 0.1$. It corresponds to two clouds of particles with opposite velocities. The drag coefficient is $D_\star = 1$, and $m_\star = 0.32$, $N_x = N_v = 200$. Moreover, $\alpha_\star = 0.4$. The computational domain is $[0, 6] \times [-10, 10]$ with periodic boundary conditions on both dimensions, and discretization parameters $N_x = N_v = 200$.

The results are visible in Fig. 1 with the system (3) and in Fig. 2 with the system (4).

Initially, the two particle clouds move toward each other, the particle volume fraction increases until reaching the close-packing limit, $\alpha_\star$. This collision compresses the gas, causing a sharp rise in fluid density in the congested zone. Over time, some particles begin to penetrate this dense layer, reducing both the particle concentration and gas density as the congestion dissipates.

3.2 Shock Wave Passing Through a Layer of Particles

In this test, we examine the interaction between a shock wave and a stationary particle cloud. The gas is initialized as:

$$(\varrho_0, u_0, p_0)(x) = \begin{cases} (0.54, 1.54, 1.54^\gamma) & \text{for } x < 0.2 \\ (1, 0, 1) & \text{for } x > 0.2. \end{cases}$$

where $\gamma = 1.4$. Immediately after the shock front, there is a cloud of particles at equilibrium, with an initial distribution function given by

$$f_0(x, v) = \frac{1}{\sqrt{2\pi}\sigma_x} e^{-\frac{(x-3)^2}{\sigma_x}} \frac{1}{\sqrt{2\pi}\sigma_v} e^{-\frac{v^2}{\sigma_v}}, \tag{6}$$

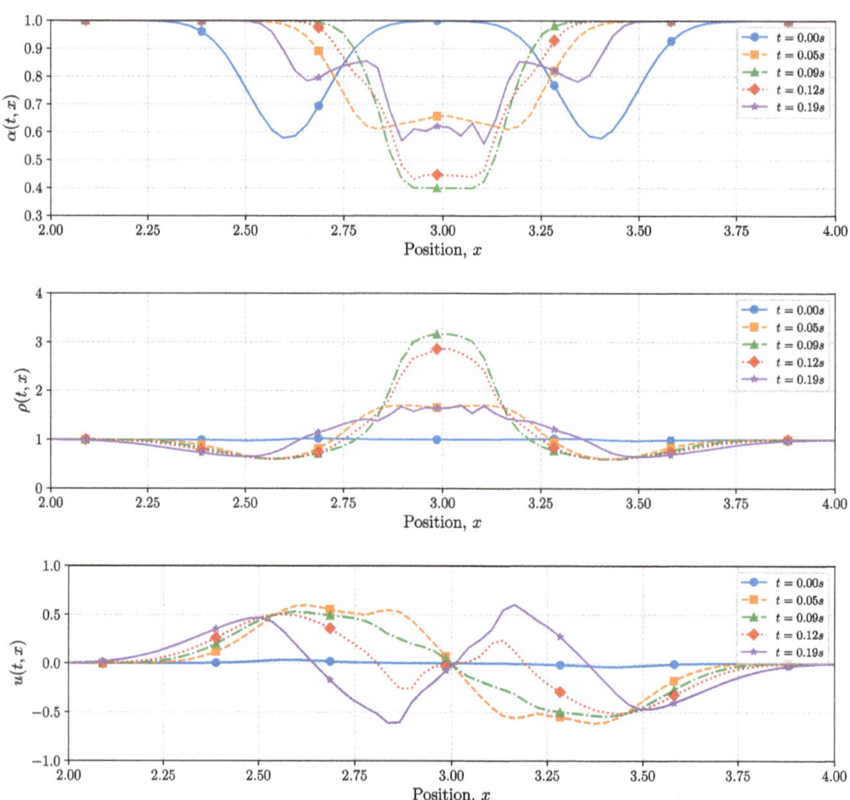

Fig. 1 Collision with the original model (3) with the PFC scheme

where $\sigma_x = \sigma_v = 0.1$. The computational domain is $[0, 6] \times [-10, 10]$. Other parameters include a drag coefficient $D_\star = 1$, $m_\star = 0.32$, $\alpha_\star = 0.4$, and discretization parameters $N_x = N_v = 200$. The results for the PFC scheme are shown in Fig. 3 for system (3) and in Fig. 4 for system (4). Initially, we observe that the shock wave compresses and pushes the particle cloud forward. This interaction causes a noticeable increase in the gas density on the left side of the particle cloud, as particles concentrate and reach the close-packing limit. At $t = 0.38$, the fluid volume fraction reaches the close-packing limit $\alpha_\star$, marking the peak compression phase in the congested region. Due to drag effects and the pressure of the gas, the gas shock transfers momentum to the particles, accelerating them in the shock's direction. This momentum transfer results in particle movement and spreading of the initial particle distribution as the shock front propagates. As particles disperse following the shock passage, the gas density and particle volume fraction start to decline, gradually returning to equilibrium. The density profile residual perturbations as the particles and gas adjust post-shock.

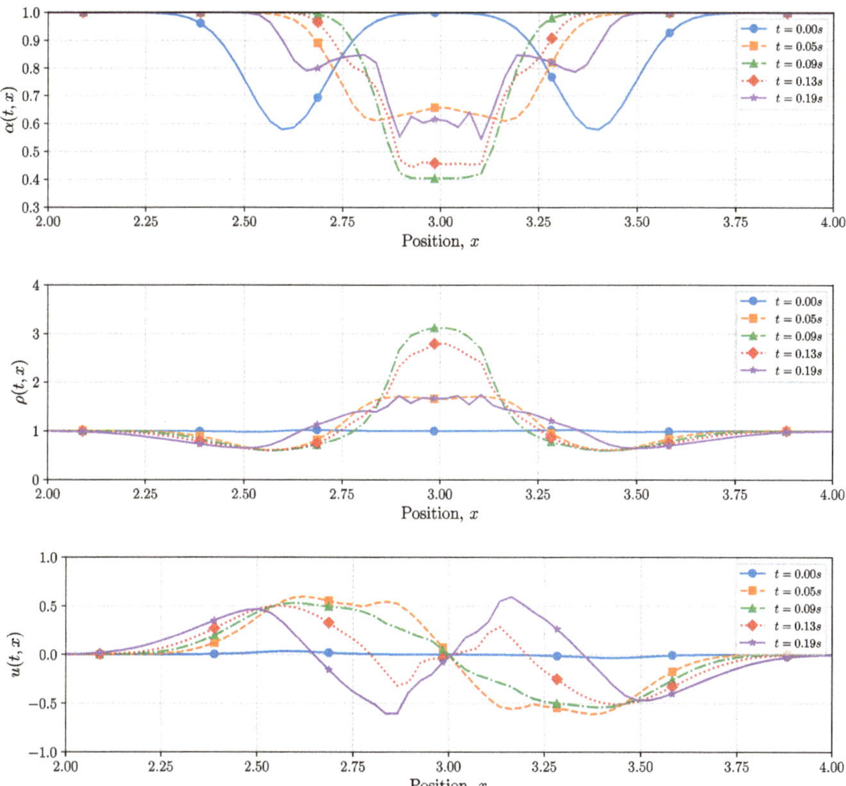

Fig. 2 Collision with the soft-congested model (4) with $\varepsilon = 10^{-6}$ with the PFC scheme

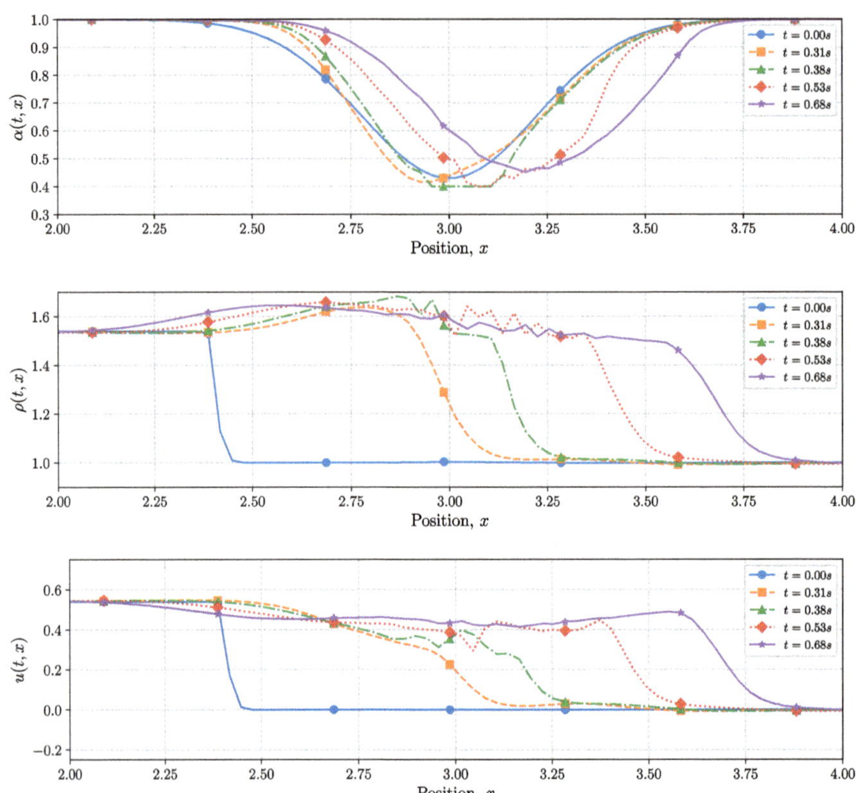

Fig. 3 Shock with the original model with the PFC scheme

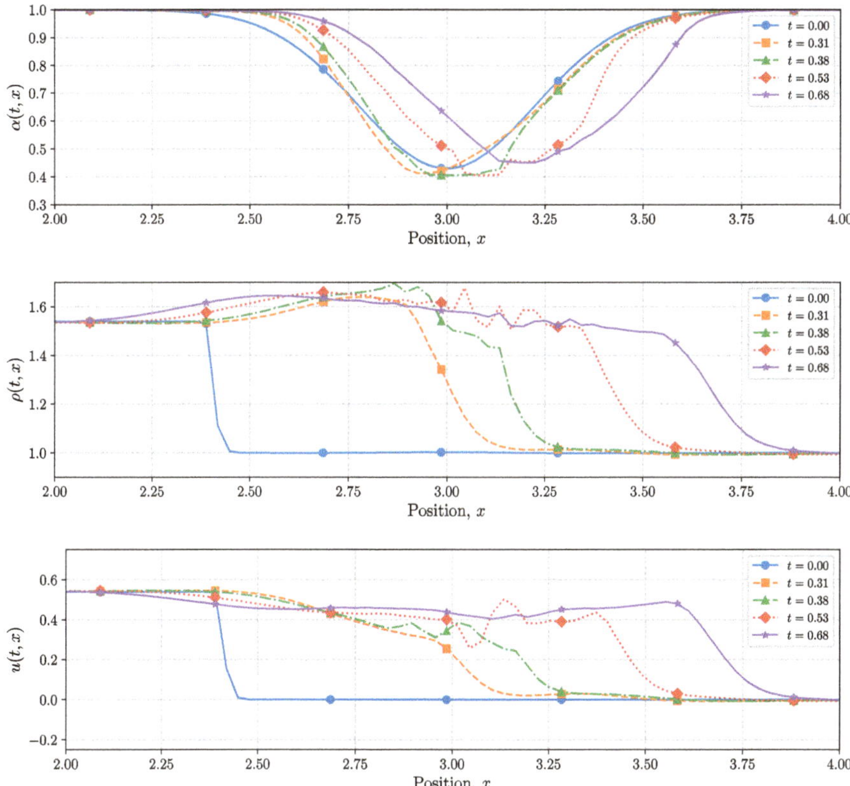

Fig. 4 Shock with the soft-congested model with $\varepsilon = 10^{-6}$ with the PFC scheme

References

1. Andrews M, O'Rourke P (1996) The multiphase particle-in-cell (mp-pic) method for dense particulate flows. Int J Multiphase Flow 22(2):379–402. https://www.sciencedirect.com/science/article/pii/0301932295000720
2. Bernard-Champmartin A, Braeunig J, Fochesato C, Goudon T (2016) A semi-lagrangian approach for dilute non-collisional fluid-particle flows. Commun Comput Phys 19(3):801–840
3. Berthelin F, Goudon T, Minjeaud S (2016) Multifluid flows: a kinetic approach. J Sci Comput 66(2):792–824
4. Besse N, Sonnendrücker E (2003) Semi-lagrangian schemes for the vlasov equation on an unstructured mesh of phase space. J Comput Phys 191:341–376
5. Cheng CZ, Knorr G (1976) The integration of the vlasov equation in configuration space. J Comput Phys 22(3):330–351
6. Degond P, Hua J, Navoret L (2010) Numerical simulations of the euler system with congestion constraint. J Comput Phys
7. Dukowicz JK (1980) A particle-fluid numerical model for liquid sprays. J Comput Phys 35(2):229–253
8. EDP J (ed) (2018) An overview on congestion phenomena in fluid equations
9. Ertzbischoff L, Han-Kwan D (2023) On well-posedness for thick spray equations, arxiv

10. Filbet F, Russo G (2003) High order numerical methods for the space non-homogeneous Boltzmann equation. J Comput Phys 186:457–480
11. Filbet F, Sonnendrücker E, Bertrand P (2001) Conservative numerical schemes for the vlasov equation. J Comput Phys 172:166–187
12. Filbet F, Sonnendrücker E, Bertrand P (2001) Conservative numerical schemes for the vlasov equation. J Comput Phys 172(1):166–187
13. Fox RO, Posey JW, Houim RW, Laurent F (2024) A kinetic-based model for polydisperse, high-speed, fluid-particle flows. Int J Multiphase Flow 171:104–698
14. Mangeney A, Califano F, Cavazzoni C, Travnicek P (2002) A numerical scheme for the integrations of the vlasov-maxwell system of equations. J Comput Phys 179:495–538
15. Maury B, Roudneff-Chupin A, Santambrogio F, Venel J (2011) Handling congestion in crowd motion modeling. Netw Heterogeneous Media
16. Maury B, Roudneff-Chupin A, Santambrogio F (2010) A macroscopic crowd motion model of gradient flow type. Math Models Methods Appl Sci
17. Perrin C, Zatorska E (2015) Free/congested two-phase model from weak solutions to multi-dimensional compressible navier-stokes equations. Commun Partial Diff Eq (2015)
18. Santambrogio F (2018) Crowd motion and evolution pdes under density constraints. ESAIM: Proceedings Surveys 64:137–157
19. Umeda T (2008) A conservative and non-oscillatory scheme for vlasov code simulations. Earth Planet SP 60:773–779

Numerical Methods for Non-equilibrium Flows

Effect of Scaled Angle on Unsteadiness Characteristics of Large Separation Bubbles in High Speed Flows

Irmak T. Karpuzcu, Deborah A. Levin, and Vassilis Theofilis

Abstract In this work, high-speed flows at moderate Reynolds numbers over compression ramps with large scaled angles were computed using the Direct Simulation Monte Carlo (DSMC) method. The free-stream conditions and ramp angles were selected to ensure that the resulting scaled angles were large, leading to correspondingly large separation bubbles. The study revealed that once the separation bubble reached a certain size, further decreasing the wall temperature and increasing the free-stream speed and physical ramp angle destabilized the flow, and resulted in the formation of secondary recirculation regions. Moreover, configurations that had secondary recirculation regions were found to exhibit unsteady behavior. This unsteadiness in the near-reattachment region was particularly pronounced across all cases.

Keywords Direct simulation monte carlo · High speed flows · Unsteadiness · Shock boundary layer interactions

1 Problem Definition and Motivation

Compression ramps are representative of many components widely used in high-speed aircraft, such as inlets, swept wings, and control surfaces. At supersonic and hypersonic speeds, these components generate complex flow structures, beginning with a leading-edge shock. As the flow moves downstream, the compression corner produces a compression shock, and the interaction of this shock with the incoming boundary layer typically results in boundary layer separation. This separation gives rise to a separation bubble, which is enclosed by the separated shear layer, the separation shock, and, farther downstream, a reattachment shock. The interactions among

I. T. Karpuzcu (✉) · D. A. Levin
University of Illinois, Urbana-Champaign, IL, USA
e-mail: itk3@illinois.edu

V. Theofilis
Center for High-Speed Flow, Faculty of Aerospace Engineering, Technion—Israel Institute of Technology, Haifa, Israel

© The Author(s) 2026

M. Grabe et al. (eds.), *Rarefied Gas Dynamics*, Springer Aerospace Technology,
https://doi.org/10.1007/978-3-032-00094-1_50

these flow structures can be classified into several types of Edney interactions [5], where each different type of interaction can cause different flow responses in terms of separation length, unsteadiness, and surface properties. It has been established that these interactions are also prone to generating unsteadiness [3, 4, 7].

One of the strong theoretical approaches to simulate such flows over compression ramps is triple deck theory [2, 12]. It assumes three decks or length scales to describe the flow where in each deck the importance of viscous and inviscid forces interchange. Three decks of the triple deck theory consists of a lower, main and an upper deck characterized by the dominant flow mechanisms such as viscous forces in the lower deck, a combination of viscous and inviscid forces in the main deck and the inviscid and irrotational forces in the upper deck. The lengths of these decks are determined by the scalings given Refs. [2, 12]. These lengths scale with $\mathrm{Re}^{-\frac{5}{8}}$ in the lower, $\mathrm{Re}^{-\frac{1}{2}}$ in the main and $\mathrm{Re}^{-\frac{3}{8}}$ in the upper deck, which also relates to the classical boundary layer theory. Triple deck theory has been used to predict surface parameters such as skin friction and pressure distributions over a flat plate and compression ramp walls. Moreover, a scaled angle parameter that uses the ratio of momentum viscous and inviscid forces is defined as [11, 14];

$$\alpha = \alpha^* \frac{Re^{\frac{1}{4}}}{0.332^{\frac{1}{2}} C^{\frac{1}{4}} (M_\infty^2 - 1)^{\frac{1}{4}}}, \qquad C = \frac{\mu_w T_\infty}{\mu_\infty T_w} \tag{1}$$

where α^* is the angle of the ramp in radians, subscript w denotes values at the wall, and ∞ at the freestream. The experiments of Egorov et al. [6] have been compared with triple deck theory to explain many flow characteristics such as the appearance of the secondary recirculation regions and the three dimensionality of the flow.

Our previous work on a 42° angle configuration demonstrated that the base flow is steady, with a large single recirculation bubble present in the flow [8]. A BiGlobal linear stability analysis revealed that this flow exhibits a growing mode in the spanwise direction as was confirmed by a three-dimensional spanwise-periodic DSMC computations. Interestingly, this unstable mode originates at the leading edge, rather than from the conventional modes typically associated with the separation bubble. Recognizing that this 42° angle case is critically stable, we modified one parameter from this case—hereafter referred to as the base configuration—and investigated the effects of wall temperature, free-stream velocity, and physical ramp angle. In our follow-up study, we established the relationship between the scaled angle and the emergence of secondary recirculation regions, as well as the dependence of separation length and recirculation strength on these parameters [10].

In this work, we further investigate the relation between the scaled angle and the unsteadiness of the flowfields. We use direct simulation Monte Carlo (DSMC) to compute the flowfields [1]. DSMC provides high-fidelity simulations for shock-dominated flowfields, such as the compression ramp flow cases considered in this study. It effectively resolves high-gradient regions, including shocks and shear layers, where continuum breakdown occurs, without introducing numerical viscosity. Furthermore, DSMC's inherent time accuracy is crucial for capturing the unsteady

Table 1 Free stream conditions for each configuration and corresponding scaled angles

	Base	Tw175	M6	CR45
Ramp angle°	42	42	42	45
Free stream velocity (m/s)	432	432	864	432
Reynolds number	11.2×10^3	11.2×10^3	22.4×10^3	11.2×10^3
Mach number	3.0	3.0	6.0	3.0
Wall temperature (K)	275	175	275	275
Scaled angle	8.60	8.45	7.07	9.22

flow characteristics typical of these flowfields. Moreover, even though the free stream conditions are continuum-like, continuum breakdown[1] occurs within the shock and shear layers of the considered flows. Therefore a kinetic method must to be used.

2 Relation Between Unsteadiness and Scaled Angle

We utilized our in-house code SUGAR [13] in this work using the free stream conditions given in Table 1. Note that the base configuration of 42° angle given in column 1 is stable and steady in 2D. In all configurations isothermal walls are used with the given wall temperatures. The free stream flow consists of molecular nitrogen (N_2) and to capture viscosity temperature dependence Variable Hard Sphere (VHS) model is used with the molecular diameter $4.17 \times 10^{-10} m$ and viscosity coefficient of 0.255. Internal modes of N_2 molecule are modeled with rigid rotor and simple harmonic oscillator models however due to the low temperature there was no effect of the internal modes on the flow conditions considered in this work. For each case, 1.1 million sampling cells are used with at least 150 simulation particles in each sampling cell giving total number of simulation particles of 200 million. SUGAR employs adaptive mesh refinement and ensures that at least 20 particles are present in each collision cell per time step.

Streamwise velocity contours with streamlines for each configuration are shown in Fig. 1 to describe the overall flowfield characteristics. As mentioned before, the base configuration is steady with a single recirculation region occupying most of the flat plate and ramp surface. Increasing the free stream velocity results in three recirculation regions in the separation bubble as well as a better defined reattachment shock. A similar result is observed when decreasing the wall temperature to 175K however there are two recirculation regions for this case, which means the effect is weaker compared to the free stream velocity. Lastly, increasing the ramp angle, the CR45 case, also results in a secondary recirculation region, however it is very small and contained at the corner location of the flow.

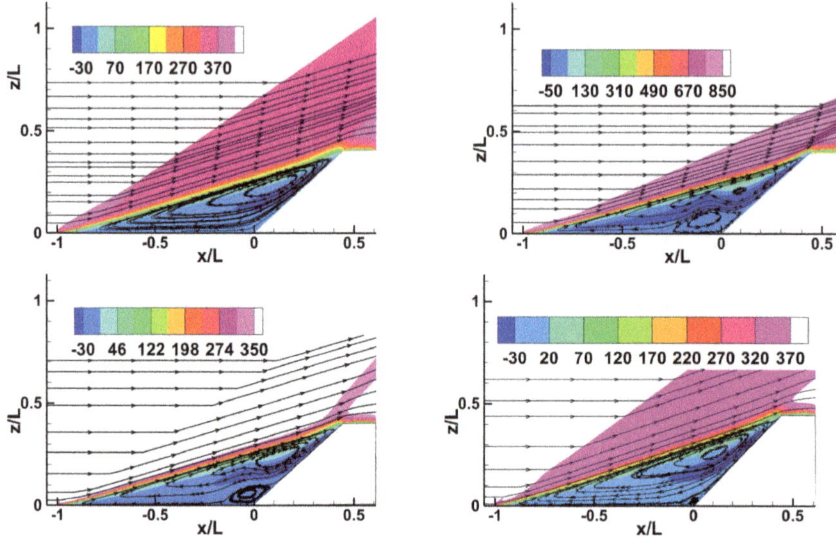

Fig. 1 Streamwise velocity, V_x (m/s), contours with streamlines. Top left: base case, top right: M6 case, bottom left: Tw175 case and bottom right: CR45 case

Next, we investigate the correlation between recirculation strength Δv [15]—defined as the ratio of the maximum reverse streamwise velocity in the separation bubble to the free-stream velocity—scaled angle, and unsteadiness characteristics. Figure 2 shows that, as the scaled angle increases, recirculation strength generally follows an increasing trend, with portions of the case results taken from earlier work [9, 10]. However, for similar scaled angle values, variations in recirculation strength are observed.

More notably, configurations with secondary recirculation regions are consistently unsteady, as indicated by the rectangular symbols in Fig. 2. These findings suggest that scaled angle alone may not fully capture the characteristics of separation bubbles at large scaled angle values. One possible reason is that the scaled angle formulation assumes a developed boundary layer, whereas the configurations in this study experience separation near the leading edge. In such cases, traditional length scalings, such as the full flat plate length, may not be appropriate and a physics-informed length, such as the boundary layer thickness prior to separation, might be more suitable.

To further investigate the unsteadiness in the flowfields, numerical probe data instantaneous pressure was taken from several locations of the separation bubble and important flow structures such as the separation and reattachment shock. It was seen that the strongest unsteadiness occurs near the reattachment location where the reattachment shock interacts with the shear layer and an immediate expansion occurs right after the reattachment. Power spectral densities (PSDs) of the mean subtracted pressure taken from the probes located near the reattachment shock foot for each case are shown in Fig. 3. It can be seen that a low frequency broadband is present for

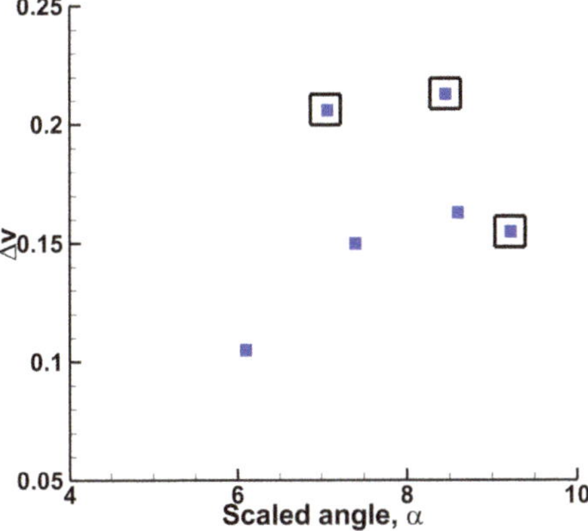

Fig. 2 Recirculation strength comparison with scaled angle. Squares indicate unsteady cases

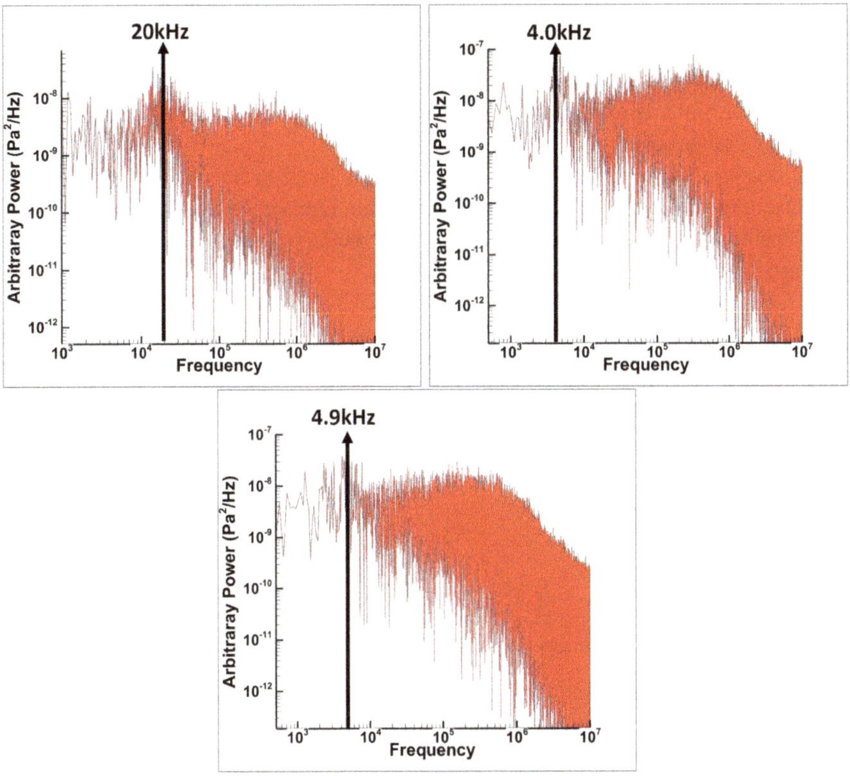

Fig. 3 Power spectral density calculated for the probe located near the reattachment for all cases. Top left: M6 case, top right: Tw175 case, and bottom: CR45 case. The arrows indicate the center of the low frequency broadband regions of the dominant frequencies

all configurations with considerably higher power peaks than the expected thermal fluctuations in the MHz range. Similar peak frequencies around 4–5 kHz are observed for Tw175 and CR45 configurations whereas a higher 20kHz band is present for the M6 configuration. This might be partly related to the fact that M6 case has twice the free stream speeds of the other cases as well as having a higher free stream Mach number that creates more complex shock interactions. These results show that the flowfields reported in this work apart from the baseline case have unsteady flow features, that needs to be investigated in more detail in a future work.

3 Summary

In this study, compression ramps with large scaled angles were simulated using DSMC in supersonic and hypersonic flows. It was shown that varying parameters such as free-stream speed, wall temperature, and ramp angle—despite producing similar scaled angle values—resulted in vastly different flowfields. Using the base configuration from Karpuzcu et al. [8], the two-dimensional base flows were found to exhibit multiple recirculation regions and unsteadiness characterized by different frequency components.

Future work will employ data-driven methods, such as Spectral Proper Orthogonal Decomposition (SPOD) and Dynamic Mode Decomposition (DMD), to investigate the unsteadiness in greater detail and extract coherent structures within these stochastic flowfields.

Acknowledgements The research conducted in this work is supported by the Office of Naval Research under Grant No. N000141202195 titled "Multi-scale modelling of unsteady shock-boundary layer hypersonic flow instabilities," with Dr. Eric Marineau as the Program Officer. This research is also supported by NSF TACC Frontera supercomputer with the project number CTS23002.

References

1. Bird GA (1994) Molecular gas dynamics and the direct simulation of gas flows. Clarendon, Oxford, England, U.K
2. Cassel KW, Ruban AI, Walker JDA (1995) An instability in supersonic boundary-layer flow over a compression ramp. J Fluid Mech 300:265–285
3. Clemens NT, Narayanaswamy V (2014) Low-frequency unsteadiness of shock wave/turbulent boundary layer interactions. Ann Rev Fluid Mech 46(1):469–492
4. Dolling DS (2001) Fifty years of shock-wave/boundary-layer interaction research: what next? AIAA J 39(8):1517–1531
5. Edney B (1968) Anomalous heat transfer and pressure distributions on blunt bodies at hypersonic speeds in the presence of an impinging shock. Aeronautical Research Institute of Sweden, Report 115 (1968)

6. Egorov I, Neiland V, Shredchenko V (2011) Three-dimensional flow structures at supersonic flow over the compression ramp. In: 49th AIAA aerospace sciences meeting including the new horizons forum and aerospace exposition. No. AIAA 2011-730
7. Gaitonde DV (2015) Progress in shock wave/boundary layer interactions. Progress in Aerospace Sciences 72: 80–99. Celebrating 60 Years of the Air Force Office of Scientific Research (AFOSR): A Review of Hypersonic Aerothermodynamics
8. Karpuzcu IT, Theofilis V, Levin DA (2024) On linear stability of supersonic flow over a short compression corner at large ramp angles. https://doi.org/10.48550/arXiv.2405.06775
9. Karpuzcu IT, Levin DA, Cerulus N, Theofilis V (2023) On the unsteadiness and three dimensionality of a laminar separation bubble for a supersonic flow over a compression corner. AIAA Paper 2023-0679
10. Karpuzcu IT, Levin DA, Theofilis V (2024) Effect of the scaled angle on the stability of a large separation bubble. AIAA Paper 2024-1359
11. Neiland VY (1969) Theory of laminar boundary layer separation in supersonic flow. Fluid Dynamics 4(4):33–35
12. Rizzetta D, Burggraf O, Jenson R (1978) Triple-deck solutions for viscous supersonic and hypersonic flow past corners. J Fluid Mech 89(3):535–552
13. Sawant SS, Tumuklu O, Jambunathan R, Levin DA (2018) Application of adaptively refined unstructured grids in DSMC to shock wave simulations. Comput Fluids 170:197–212
14. Stewartson K (1970) On laminar boundary layers near corners. Quarterly J Mech Appl Math 23(2):137–152
15. Theofilis V, Hein S, Dallmann U (2000) On the origins of unsteadiness and three-dimensionality in a laminar separation bubble. Philosophical Trans Royal Soc London 358:3229–3324

Vibronic Modelling in Direct Simulation Monte Carlo

Clément H. B. Civrais, Marcel Pfeiffer, Craig White, and René Steijl

Abstract The traditional methods for modelling the electronic excitation in direct simulation Monte Carlo (DSMC) simulations treat each internal mode individually and disregard any interaction between modes. To address this, a vibronic model was developed by Civrais et al. (Phys. Fluids 36:086112, 2024) allowing each electronic excited state to excite its unique vibrational quantum levels. This work expands on previous work by providing complementary physical insights into the modelling of the vibrational quantum levels of electronic states. The fundamental differences between the uncoupled and coupled approaches in terms of the internal energies and distribution functions are outlined. Then, the scalability of the vibronic model against the traditional approach is demonstrated for elementary operations. It is shown that the coupled approach demonstrates a comparable runtime as the uncoupled approach. Finally, the traditional uncoupled and coupled approaches are applied to the modelling of the electronic excited states on the forebody of the Space Shuttle at an altitude of 99.49 km, demonstrating the inaccuracy of the traditional assumption of the electronic excited states being distributed according to the Boltzmann statistics. It is also shown that the population of the electronic excited states and their vibrational quantum levels strongly deviate between the two approaches, suggesting a significant impact on the line intensity of molecular species' absorption spectrum.

Keywords Electronic excited states · Scalability · Space shuttle · DSMC

C. H. B. Civrais
Present institution: Center for Hypersonics and Entry Systems Studies, University of Illinois at Urbana-Champaign, Urbana, USA

C. H. B. Civrais (✉) · C. White · R. Steijl
James Watt School of Engineering University of Glasgow, Glasgow, UK
e-mail: ccivrais@uci.edu

M. Pfeiffer
Institute of Space Systems University of Stuttgart, Stuttgart, Germany

© The Author(s) 2026

M. Grabe et al. (eds.), *Rarefied Gas Dynamics*, Springer Aerospace Technology,
https://doi.org/10.1007/978-3-032-00094-1_51

1   Introduction

Modelling the electronic excited states remains a critical challenge for numerous aerothermodynamics applications. For instance, in hypersonic reentries [14], ionisation reactions and radiation are predominantly governed by the excitation of the electronic states. Furthermore, state-of-the-art chemistry calculations have demonstrated that chemical reactions not only occur from ground electronic states but also electronic excited states [12, 13]. In the context of the development of versatile numerical platforms [18, 21], it is important to allow for consistent modelling of thermodynamic and transport properties to ensure accuracy and global stability.

Over the past few decades, the direct simulation Monte Carlo (DSMC) method has emerged as a standard approach for simulating rarefied non-equilibrium flows [3]. One reason for this is the relatively simple implementation of the method, along with the various physical and chemical models that can be implemented. This adaptability enables the DSMC method to simulate complex thermal and chemical non-equilibrium effects in a flow [6, 9, 10]. Beyond its successful application in reproducing experimental measurements of typical hypersonic flows [11], recent studies have also demonstrated the successful applications of the DSMC method to combustion applications [20].

In the DSMC community, three methods have been developed to model the electronic excitation in flows [4, 14, 15]. A common aspect of all methods is that each internal mode, i.e. rotational, vibrational, and electronic, is regarded separately. However, real-gas effects involve adiabatic and non-adiabatic interactions between the internal modes. This becomes particularly significant in the analysis of molecular radiation, where achieving the correct distribution of vibrational and electronic excitation energy is crucial. In recent years, coupling of the DSMC method with radiation transport solvers has emerged for such investigations [1, 2, 18]. While in the DSMC method, the internal modes are treated separately and any interaction between modes is neglected, radiation solvers adopt a more detailed description of the internal modes in which all types of interactions between the internal modes are modelled. To perform such coupling, assumptions are made to transfer uncoupled information from DSMC into the coupled models of radiation solvers [1, 2, 14, 18].

This issue was partially addressed by Civrais et al. [5] with the development of a model for coupling the electronic and vibrational modes of molecular species in DSMC. The derivation and implementation of the vibronic model were verified for a series of adiabatic reactor simulations. In the present article, complementary studies are conducted to provide further physical insights into the modelling of the vibrational quantum levels of the electronic excited states. This work intends to demonstrate the scalability of the vibronic model against the traditional approach for elementary operations such as the initialisation of the particles in the domain and the collision routine. Additionally, it aims to compare the two approaches to the modelling of the electronic excited states of N_2 for a non-reactive air mixture about the forebody of the Space Shuttle at an altitude of 99.49 km [17].

2 Mathematical Description

2.1 Uncoupled Approach

The traditional approach [19] models the vibrational excitation using the harmonic oscillator (HO) model, where the vibrational energy is expressed as

$$\epsilon_j = hc\omega_{e,0}\left(j + \frac{1}{2}\right), \tag{1}$$

where h is the Planck constant, c the speed of light in vacuum, and $\omega_{e,0}$ is the harmonic oscillator spectroscopic constant of the ground electronic state. All spectroscopic constants presented in this article are extracted from the National Institute of Standards and Technology (NIST) database [16].

The harmonic oscillator model allows an infinite number of vibrational quantum levels, with each level being non-degenerate, i.e. $g_v = 1$. While more detailed approaches for modelling the vibrational energy of molecular species exist [7, 8], this article focuses on evaluating the coupled approach against the traditional approach for modelling the vibrational mode, i.e. Equation (1). Consequently, these more detailed approaches are not considered in this study.

Additionally, the energies associated to each electronic state are calculated as,

$$\epsilon_i = hc\kappa_{e,i}, \tag{2}$$

where $\kappa_{e,i}$ is the electronic transition of state i. Note that all electronic states have a distinct degeneracy corresponding to their electronic configuration [16].

In the uncoupled approach, each internal mode is treated separately and any interaction between modes is neglected. Therefore, each electronic excited state can only excite the same set of vibrational levels as the ground electronic state. The joint distribution function for the vibrational-electronic mode is given by the product of each individual distribution function which yields,

$$f_{i,j} = f_i \times f_j = \frac{g_i e^{-\frac{\varepsilon_i}{kT}}}{\sum_{i \in \mathscr{I}} g_i e^{-\frac{\varepsilon_i}{kT}}} \times \frac{e^{-\frac{\varepsilon_j}{kT}}}{\sum_{j \in \mathscr{J}} e^{-\frac{\varepsilon_j}{kT}}} = \frac{g_i e^{-\frac{\varepsilon_i}{kT}}}{Q_e} \times \frac{e^{-\frac{\varepsilon_j}{kT}}}{Q_v}, \tag{3}$$

where g_i is the degeneracy of the electronic state, $\mathscr{I}$ is the ensemble of electronic quantum levels, $\mathscr{J}$ is the ensemble of vibrational quantum levels of the ground electronic state, T is the temperature, Q_v is the vibrational partition function and Q_e is the electronic partition function.

2.2 Coupled Approach

The coupled approach assumes a coupling between the vibrational and electronic modes. Specifically, each electronic excited state has a unique set of vibrational levels described by

$$\epsilon_{i,j} = hc\kappa_{e,i} + hc\omega_{e,i}\left(j + \frac{1}{2}\right) - hc\omega_{e,i}\chi_{e,i}\left(j + \frac{1}{2}\right)^2, \tag{4}$$

where $\kappa_{e,i}$, $\omega_{e,i}$ and $\omega_{e,i}\chi_{e,i}$ are the spectroscopic constants of electronic state i.

Considering an electronic state i, the maximum vibrational quantum level permissible is determined as the vibrational quantum immediately before the dissociation energy of electronic state i, i.e. $\epsilon_{i,j} < \epsilon_{d,i}$. For some electronic configurations, all the vibrational quantum levels lie below the corresponding dissociation energy. In this particular scenario, the maximum vibrational quantum level is determined as the last vibrational quantum level before the gradient $\partial\epsilon_{i,j}/\partial j$ changes sign.

Since the electronic excited states allow for vibrational excitation, the corresponding distribution function for the vibronic mode consists of a two-dimensional Boltzmann distribution such that,

$$f_{i,j} = \frac{g_i e^{-\frac{\epsilon_{i,j}}{kT}}}{\displaystyle\sum_{i\in\mathscr{I}}\sum_{j\in\mathscr{J}_i} g_i e^{-\frac{\epsilon_{i,j}}{kT}}} = \frac{g_i e^{-\frac{\epsilon_{i,j}}{kT}}}{Q_{ve}}, \tag{5}$$

where $\mathscr{J}_i$ is the set of vibrational quantum levels of electronic state i and Q_{ve} is the partition function of the vibronic mode.

As shown by Eqs. (3) and (5), the two approaches yield distinct distribution functions. In the uncoupled approach, the vibrational and electronic quantum levels are distributed through two one-dimensional distribution functions. In contrast, the coupled approach distributes the two modes on a two-dimensional distribution function that accounts for the coupling between the vibrational and electronic modes. As a result, the coupled approach defines a unique partition function for the vibronic mode which imposes the definition of a vibronic temperature, i.e. T_{ve}.

This contrasts with the traditional uncoupled approach which requires separate internal temperatures for each mode, i.e. T_v and T_e. The vibronic temperature must be used with caution to avoid misleading interpretations. Specifically, it describes a mean excitation of the vibrational-electronic mode. Since the electronic mode admits a set of vibrational levels the vibrational-electronic temperature cannot reflect the vibrational excitation of each electronic excited state but, rather, a mean excitation of the electronic excited states and their corresponding vibrational quantum levels. While it may serve the purpose of a verification exercise to verify the implementation of the model in thermal equilibrium conditions, for any non-equilibrium conditions, the vibronic temperature carries minimal physical insights. It is recommended that the local distribution functions rather than the vibronic temperature be evaluated.

3 Scalability

In the present section, the two approaches are compared on the scalability of the initialisation and collision routines. The test case involves an adiabatic reactor represented by a single cubic cell with a side length of 1.88×10^{-4} m filled with a number of DSMC simulator particles ranging between 10^4 and 2×10^6 with periodic boundaries at an equilibrium temperature of $20,000$ K. The scalability of the two approaches is monitored and depicted in Fig. 1. The ratio between the two approaches is included for reference.

Figure 1a shows that the coupled approach requires a slightly larger runtime of the initialisation routine in comparison to the uncoupled approach. For the uncoupled approach, the distribution function simplifies to an analytical expression from which the vibrational quantum level is sampled. In contrast, the electronic quantum level is sampled from a Boltzmann distribution using an acceptance-rejection scheme. The coupled approach requires sampling the vibrational and electronic quantum levels from a two-dimensional distribution function using an acceptance-rejection scheme [5]. This difference in the sampling schemes for the initialisation routine is responsible for the slight increase in the computational expense depicted in Fig. 1a.

Figure 1b shows that the coupled approach requires a similar runtime of $100\Delta t$ to the uncoupled approach as shown with a quasi-constant ratio of 1 for the range of particles considered. For the two approaches, the post-collision quantum levels are sampled from a combined distribution function of the translational and an internal mode of the colliding particle. Thus, the two approaches utilise similar sampling techniques and only differ in the description of the distribution function.

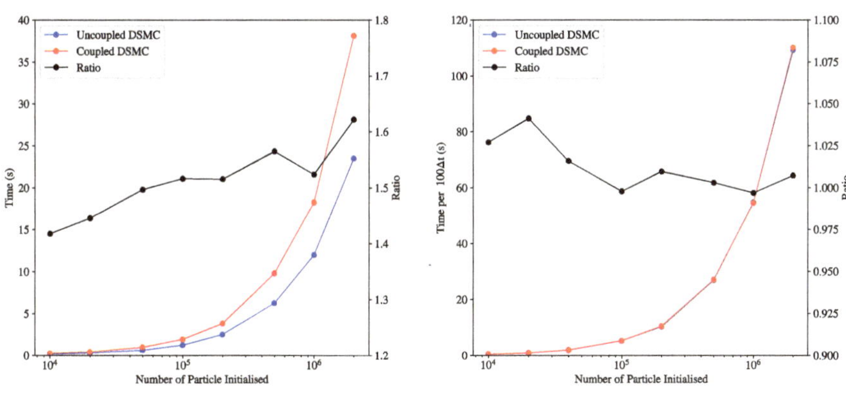

(a) Initialisation routine.   (b) Runtime for $100\Delta t$.

Fig. 1 Scalability of the coupled approach

4 Space Shuttle Forebody at 99.49 Km

In the present section, the uncoupled and coupled approaches are compared on the modelling of the electronic excited states of N_2 for a non-reactive air mixture about the forebody of the Space Shuttle at an altitude of 99.49 km. The freestream conditions are extracted from Moss and Bird [17].

The main objective of this section is to compare the two approaches for the distribution of the internal quantum levels measured by the DSMC method and to assess the assumption that the internal quantum levels are distributed according to the Boltzmann distribution. For both approaches, the vibrational and electronic density functions are calculated with the local internal temperatures, namely $f = f_i(T_e)f_j(T_v)$ for the uncoupled approach and $f = f_{i,j}(T_{ve})$ for the coupled approach.

The computational domain extends to a length of 1.5 m upstream of the stagnation point. The freestream mean free path is approximately $\lambda_\infty = 0.133$ m resulting in a Knudsen number based on the nose radius of about $Kn = 0.098$. The mesh is refined near the stagnation point to ensure the cell size, Δx, remains smaller than the local mean free path, λ, resulting in a total number of 15, 625 cells. The time step is carefully chosen to resolve the collision frequency which results in $\Delta t = 1 \times 10^{-8}$ s. Intermolecular collisions are computed with the VHS model for a reference temperature of $T_{ref} = 273$ K. For the uncoupled approach, a finite probability of $P_r = 0.2$, $P_v = 0.02$, and $P_e = 0.002$ are, respectively, considered for rotational, vibrational, and electronic relaxation. To reproduce the typical relaxation behaviour of the uncoupled approach, a finite probability of $P_r = 0.2$, $P_1 = 0.02$, and $P_2 = 0.1$ for a particle to undergo rotational relaxation, pure vibrational excitation/de-excitation, and electronic and vibrational excitation/de-excitation, respectively, is applied [5]. The gas-surface interactions are fully diffusive with a wall temperature $T_w = 800$ K. To maintain a minimal number of 100 particles per cell throughout, the numerical mesh is populated with a total number of 1.53×10^7 DSMC particles at steady state.

Figure 2 shows the distribution functions probed at four stations along the stagnation streamline. It is shown that the DSMC measurements of the distribution function at about 1 m to the stagnation point are similar to that of the distribution function located at 0.25 m to the stagnation point. This suggests that post-shock information is carried upstream of the shock wave. This phenomenon is attributed to high Knudsen number regimes in which the degree of rarefaction is sufficient for a particle to collide with the surface of the body and travel backwards without undergoing sufficient collisions with any other particles to de-excite and return to an equilibrium distribution. This non-equilibrium effect is characterised by a strong deviation from a Maxwell velocity distribution [1, 14, 15].

Additionally, Fig. 2 indicates a strong deviation between the two approaches in the DSMC measurements. For all locations, the two approaches appear to present similar distributions for the first few quantum levels where the two approaches differ in the tail. This deviation is the result of the difference in the description of the vibrational quantum between the two approaches. Specifically, the traditional approach assumes

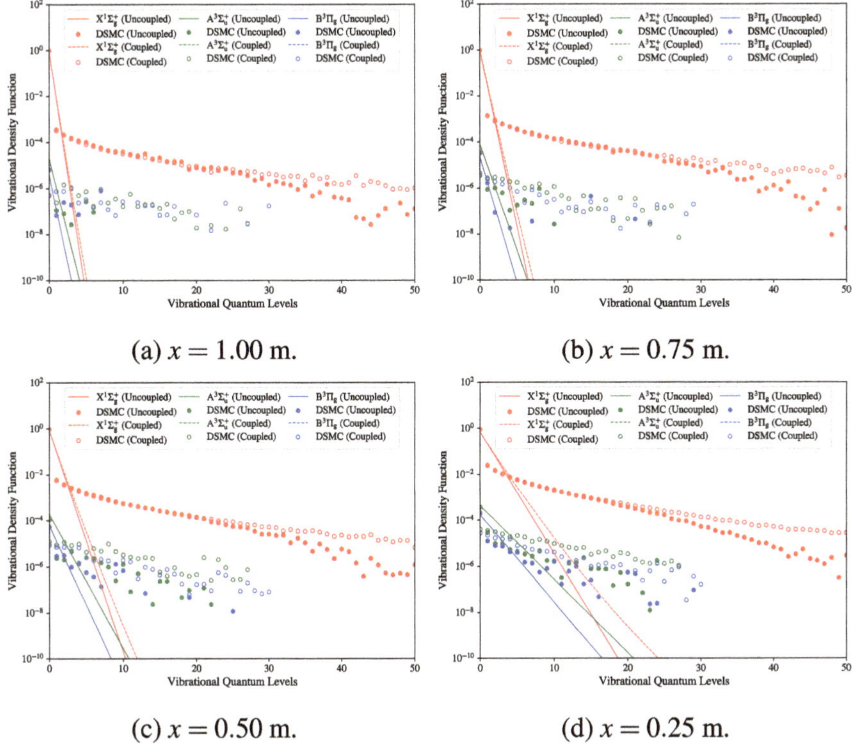

(a) $x = 1.00$ m.

(b) $x = 0.75$ m.

(c) $x = 0.50$ m.

(d) $x = 0.25$ m.

Fig. 2 Vibrational density function of the first three electronic states of N_2 at four locations along the stagnation streamline

that each electronic state excites the same set of infinite vibrational quantum levels as the ground state, whereas the coupled approach models the vibrational excitation of the electronic states with a state-specific set of vibrational quantum levels.

Furthermore, Fig. 2 shows that the DSMC measurements largely deviate from the traditional assumption of the quantum levels being distributed according to the Boltzmann distribution. The derivation of the Boltzmann distribution imposes the system to be in thermal equilibrium which is not applicable in such an application. In light of these results, when coupling a DSMC and a radiation solver, the authors recommend transferring the local distribution function rather than the mean internal temperatures. While this recommendation does not fully address the limitations in modelling all the interactions between internal modes, it can at least carry more accurate and meaningful information between solvers.

5 Conclusion

The merits of the recently developed coupled approach for modelling the vibrational quantum levels of the electronic states are discussed. The scalability of the initialisation and collision routines of the coupled approach is assessed against the uncoupled approach. Additionally, the coupled approach is applied to a non-reactive air mixture past the Space Shuttle at an altitude of about 99.49 km to assess the traditional assumption made to transfer uncoupled information from the DSMC method to radiation solvers.

The scalability study indicates that the coupled approach has similar computational requirements to the uncoupled approach. The initialisation routine presents a slight increase in the computation expenses due to the sampling on a two-dimensional distribution function. However, the coupled approach demonstrates comparable runtime per 100 iterations to the uncoupled approach.

The STS-II study demonstrates that the traditional assumption of the vibrational and electronic quantum levels being Boltzmann distributed is invalid. This discrepancy is attributed to non-equilibrium effects in high Knudsen number regimes. The authors recommend taking advantage of the DSMC method by measuring the local distribution functions rather than the internal temperatures and transferring this measurement to radiation solvers.

In the present article, chemical reactions have been neglected; however, it is evident that they play a significant part in aerothermodynamic applications. Therefore, further investigations must be carried out to integrate the contribution of the electronic excited states to the reaction rates. Given the limited data available on chemical reactions involving electronic excited states, one may benefit from the quantum-kinetic (QK) chemistry model which solely relies on the collision properties of the colliding partners.

Acknowledgements This work was initiated during a secondment at the University of Stuttgart. The lead author gratefully acknowledges scholarship funding from the James Watt School of Engineering through the Jim Gatheral Travel Grant 2023.

References

1. Beyer J, Nizenkov P, Fasoulas S, Pfeiffer M (2024) Simulation of radiating non-equilibrium flows around a capsule entering Titan's atmosphere. In: AIP conference proceedings, vol 2996, issue 1, p 200002
2. Beyer J, Pfeiffer M, Fasoulas S (2022) Non-equilibrium radiation modeling in a gas kinetic simulation code. J Quantitative Spectrosc Radiative Transfer 280:108083
3. Bird GA (1994) Molecular gas dynamics and the direct simulation of gas flows. Oxford University Press Inc
4. Burt JM, Josyula E (2017) Direct simulation Monte Carlo modeling of gas electronic excitation for hypersonic sensing. J Thermophysics Heat Transfer 31(4):858–870

5. Civrais CHB, Pfeiffer M, White C, Steijl R (2024) Modelling of the electronic excited states in high-temperature flows. Phys Fluids 36(8):086112
6. Civrais CHB, Pfeiffer M, White C, Steijl R (2025) Development and validation of the vibronic QK model for oxygen dissociation reactions. In: 3rd international conference on flight vehicles, aerothermodynamics and re-entry (FAR), Arcachon
7. Civrais CHB, White C, Steijl R (2022) Vibrational modeling with an anharmonic oscillator model in direct simulation Monte Carlo. J Thermophysics Heat Transfer 37(3):534–548
8. Civrais CHB, White C, Steijl R (2023) Extension of the normal shock wave relations for calorically imperfect gas. Shock Waves 33:533–551
9. Civrais CHB, White C, Steijl R (2024) Quantum kinetics chemistry models with an anharmonic oscillator model. Model derivation and limitations. Phys Fluids 36(8):086120
10. Civrais CHB, White C, Steijl R (2024) Quantum kinetics chemistry models with an anharmonic oscillator model. Model extension and validation. Phys Fluids 36(11):116128
11. Gimelshein SF, Wysong IJ (2019) Nonequilibrium air flow predictions with a high-fidelity direct simulation Monte Carlo approach. Phys Rev Fluids 4(3):033405
12. Grover MS, Bisek NJ, Valentini P (2025) Direct simulation Monte Carlo of hypervelocity experiments. In: AIAA SCITECH 2025 Forum. Orlando, FL, AIAA Paper 2025-0209
13. Jo SM, Venturi S, Kim JG, Panesi M (2023) Rovibrational internal energy transfer and dissociation of high-temperature oxygen mixture. J Chem Phys 158(6):064305
14. Li Z, Ozawa T, Sohn I, Levin DA (2011) Modeling of electronic excitation and radiation in non-continuum hypersonic reentry flows. Phys Fluids 23(6):066102
15. Liechty DS, Lewis MJ (2014) Extension of the quantum-kinetic model to lunar and Mars return physics. Phys Fluids 26(2):027106
16. Linstrom PJ, Mallard WG (2024) NIST Chemistry WebBook, NIST Standard Reference Database Number 69. National Institute of Standards and Technology. Retrieved 30 Jan 2024
17. Moss JN, Bird GA (2003) Direct simulation of transitional flow for hypersonic reentry conditions. J Spacecraft Rockets 40(5):830–843
18. Pfeiffer M, Beyer J, Vaubaillon J, Matlovič P, Tóth J, Fasoulas S, Löhle S (2024) Numerical simulation of an iron meteoroid entering into Earth's atmosphere using DSMC and a radiation solver with comparison to ground testing data. Icarus 407:115768
19. Pfeiffer M, Nizenkov P, Mirza A, Fasoulas S (2016) Direct simulation Monte Carlo modeling of relaxation processes in polyatomic gases. Phys Fluids 28(2):027103
20. Trivedi S, Cant RS, Harvey JK (2021) Molecular level simulations of combustion processes using the DSMC method. Combustion Theory Modelling 25(2):351–363
21. Vasileiadis N, White C (2024) hybridDCFoam: a coupled DSMC/Navier–Stokes–Fourier solver for steady-state multiscale rarefied gas flows. Adv Eng Softw 193:103669

A Non-negative Least Squares-Based Approach for Moment-Preserving Particle Merging

Georgii Oblapenko

Abstract In the present work, a novel particle merging scheme is proposed for PIC-DSMC simulations, based on the solution of a Non-negative Least Squares problem. The merging algorithm conserves arbitrary moments of the velocity distribution function, and a collision rate-conserving version of the algorithm is presented as well. Numerical simulations show excellent performance of the merging algorithm in terms of accuracy.

Keywords DSMC · Particle merging · Non-negative least squares

1 Introduction

Rarefied gas flows arise in a variety of engineering applications, such as aerospace engineering, semiconductor manufacturing, and nanotechnology applications [17]. Their non-equilibrium nature precludes the use of traditional continuum-based computational fluid dynamics approaches for their simulation and necessitates the use of either higher-order moment equations [18], or of kinetic solvers, such as the Direct Simulation Monte Carlo (DSMC) method [3]. The DSMC method has established itself as one of the main tools for rarefied gas dynamics simulations. In its standard formulation, the DSMC approach assumes that each computational particle represents a fixed number of actual particles. This causes issues in simulating flows with large density gradients and/or trace chemical species, as resolving the low populations requires a large number of simulation particles.

To resolve this issue, variable-weight DSMC approaches have been suggested, where the weight of the computational particles are not fixed [2, 5, 16]. The drawback is that, during collisions, particles have to be split, leading to an exponential growth in the number of particles if left unchecked. The growth becomes even more rapid if modern collision schemes with improved resolution of low-probability events are used [2, 14]. Therefore, particle merging is required, i.e. a procedure that reduces

G. Oblapenko (✉)
Applied and Computational Mathematics, RWTH Aachen, Schinkelstrasse 2, Aachen, Germany
e-mail: oblapenko@acom.rwth-aachen.de

© The Author(s) 2026

M. Grabe et al. (eds.), *Rarefied Gas Dynamics*, Springer Aerospace Technology,
https://doi.org/10.1007/978-3-032-00094-1_52

the number of particles in a simulation. This inherently incurs an information loss and may lead to significant numerical errors. Multiple merging algorithms have been developed [6, 12, 15, 19] in an attempt to reduce this error and improve the efficiency of the merging procedure. They frequently rely on a "divide-and-conquer" strategy: the particles chosen for merging are grouped into smaller sets in velocity space, and in each of these subsets all the particles are replaced by a very small number (usually 1 or 2) of particles. The advantage of such an approach is that it is possible to compute the weights, positions, and velocities of the post-merge particles in each subset analytically. However, the analytical merging of particles is restricted to conservation of lower-order moments of particle distribution, as conservation of higher-order moments requires solving of systems of non-linear equations. In addition, it is in general not possible to guarantee that post-merge particles will not end up outside of the physical domain or that their velocities will not exceed the velocities of the original set of particles. Therefore, it is of interest to design merging schemes that can ensure conservation of higher-order moments and constraint satisfaction [10]. The recent scheme of Gonoskov [7] is an example of a moment-preserving merging scheme that has also been successfully coupled with the octree grouping approach [8]. In this work we investigate an alternative moment-preserving merging scheme based on solution of a non-negative least squares problem and apply it to simulation of two model spatially homogeneous problems. In addition, a rate-preserving version of the scheme is formulated for plasma simulations. We compare the results with those obtained by the octree merging scheme and show that the new scheme leads to lower bias in the numerical solution.

2   Particle Merging

We consider a set of particles $\mathcal{P}_N = \{(w^{(i)}, v_x^{(i)}, v_y^{(i)}, v_z^{(i)})\}_{i=1}^{N}$, where each particle is characterized by its non-negative computational weight $w^{(i)} \geq 0$ and the 3 velocity components $v_x^{(i)}, v_y^{(i)}, v_z^{(i)}$. In the present work we disregard the particle positions and focus only on preservation of velocity moments. Merging of particles corresponds to computation of a new set of particles $\hat{\mathcal{P}}_M = \{(\hat{w}^{(i)}, \hat{v}_x^{(i)}, \hat{v}_y^{(i)}, \hat{v}_z^{(i)})\}_{i=1}^{M}$, $M < N$.

We are interested in conserving central moments, but since we assume Galilean invariance, once can always compute the mean velocities $\bar{v}_x, \bar{v}_y, \bar{v}_z$, subtract them from the velocities of $\mathcal{P}_N$, merge down the particles, and add the mean velocity components to the post-merge particles $\hat{\mathcal{P}}_M$. Therefore, we assume that $\bar{v}_x = \bar{v}_y = \bar{v}_z = 0$ and focus on the non-central moments

$$M_{jkl}(\mathcal{P}) = \frac{1}{\sum_i w^{(i)}} \sum_i w^{(i)} \left(v_x^{(i)}\right)^j \left(v_y^{(i)}\right)^k \left(v_z^{(i)}\right)^l . \tag{1}$$

We are interested in developing a merging approach that can conserve not only the basic invariants of mass, momentum, and energy, but also certain higher-order moments M_{jkl} of prescribed order j, k, l.

2.1 *Non-negative Least Squares Merging*

Now we consider an alternative merging algorithm, based on the non-negative least squares approach. Let us consider the definition (1) for a set of moments $(M_{j_1 k_1 l_1}, \ldots, M_{j_L k_L l_L})$ in matrix form:

$$\mathbf{Vw} = \mathbf{M}, \tag{2}$$

where the matrices and vectors $\mathbf{V} \in \mathbb{R}^{L \times N}$, $\mathbf{w} \in \mathbb{R}^N$, $\mathbf{M} \in \mathbb{R}^L$ are defined as

$$\mathbf{V} = \begin{pmatrix} \left(v_x^{(1)}\right)^{j_1} \left(v_y^{(1)}\right)^{k_1} \left(v_z^{(1)}\right)^{l_1} & \cdots & \left(v_x^{(N)}\right)^{j_1} \left(v_y^{(N)}\right)^{k_1} \left(v_z^{(N)}\right)^{l_1} \\ & \ddots & \\ \left(v_x^{(1)}\right)^{j_L} \left(v_y^{(1)}\right)^{k_L} \left(v_z^{(1)}\right)^{l_L} & \cdots & \left(v_x^{(N)}\right)^{j_L} \left(v_y^{(N)}\right)^{k_L} \left(v_z^{(N)}\right)^{l_L} \end{pmatrix}, \tag{3}$$

$$\mathbf{w} = \left(w^{(1)} \ldots w^{(N)}\right)^{\mathrm{T}}, \quad \mathbf{M} = \left(M_{j_1 k_1 l_1} \ldots M_{j_L k_L l_L}\right)^{\mathrm{T}}. \tag{4}$$

We can switch our point of view and consider (2) not as a definition (i.e. given particle velocities and weights, one can compute the moments), but as a system of linear equations for $\mathbf{w}$, i.e. the particle weights: given a set of moments and particle velocities, one can compute the weights. Depending on the number of particle velocities and moments, this system is be either under- ($N > L$), well- ($N = L$, or over-determined ($N < L$). We consider the under-determined case ($N > L$), that is, we have more unknown particle weights than moments. However, we augment the system with the requirement that $w_i \geq 0$. This leads to a non-negative least squares (NNLS) problem [11]. One property of solutions of under-determined NNLS problems is their sparsity: it is highly likely that some entries of the solution vector $\mathbf{w}$ will be 0. We can leverage this sparsity property of NNLS solutions in order to use the NNLS algorithm as a merging method:

1. Given a set of particles $\mathcal{P}_N$, we compute the matrix $\mathbf{V}$ and vector $\mathbf{M}$, as defined by (3)–(4)
2. Next, we solve system (2) for $\mathbf{w}'$ using the NNLS method
3. We replace the original set of particles $\mathcal{P}_N$ with new particles, by taking the velocities from the original set of velocities used to construct $\mathbf{V}$ and new weights from the solution vector $\mathbf{w}$, skipping creation of particles whose weights are equal to 0.

Thus one can can conserve any velocity and spatial moments (not considered in the present work, but incorporating them only means adding additional rows to $\mathbf{V}$ and

M), and also has full control over the post-merge velocities and spatial locations of the particles, since they are explicitly used to construct **V**. However, the existence of a solution is not guaranteed. In addition, we have no control of the number of non-zero weights, although in practice it is usually equal to L. The algorithm conserves mass, momentum, and energy, as long as the moments M_{000}, M_{100}, M_{010}, M_{001}, M_{200}, M_{020}, M_{002} are conserved; therefore, these are always included in the set of conserved moments.

To increase the chances of finding a solution, we compute the velocities defined by the second-order moments: $v_{x,M_2} = \sqrt{M_{200}}$, $v_{y,M_2} = \sqrt{M_{020}}$, $v_{z,M_2} = \sqrt{M_{020}}$. We then augment the matrix **V** with additional columns computed using new velocities with a magnitude of $\alpha v_{i,M_2}$, $i = x, y, z$, where $\alpha \in [0, 1]$ is a user-defined multiplier, with the velocity signs chosen so that these additional velocities cover all octants in velocity space. In the present work, 16 additional columns are added to **V** in this manner: 8 with $\alpha = 1$ and 8 with $\alpha = 0.5$. An additional constraint is added that the velocities of these points do not exceed the minimum and maximum velocities in the corresponding directions. This has been found to improve the stability of the algorithm, but a more detailed investigation of the optimal choice of additional velocities is left for future work. If on a given timestep the NNLS merging approach fails to produce a sparse solution to the linear system (2), one can switch to an alternative merging approach.

We also consider conservation of inter-species collision rates. This is complicated by the fact that collision rates depend on the relative velocity of the colliding particles g, i.e. on both species' VDFs. However, for the specific case of electron-neutral collisions in a plasma, one can use the common assumption that the electron velocities are significantly higher than those of the neutrals:

$$v_e \gg v_n, \tag{5}$$

where v_e is the velocity of electrons, and v_n is the velocity of the neutrals. Using this assumption, one can approximate an electron-neutral collision rate k as

$$k(g) = \frac{1}{w_e w_n} \sum_i \sum_j w_e^{(i)} w_n^{(j)} |v_e^{(i)} - v_n^{(j)}| \sigma(|v_e^{(i)} - v_n^{(j)}|) \approx \frac{1}{w_e} \sum_i w_e^{(i)} |v_e^{(i)}| \sigma(|v_e^{(i)}|). \tag{6}$$

Here $\sigma(g)$ is the cross-section of the process for which the rate is being calculated. With this assumption, the approximate rate becomes a function of the electron VDF only, and can thus be conserved when merging the electron particles by adding the following row to matrix **V**:

$$\left(|v_e^{(i)}| \sigma(|v_e^{(i)}|) \ldots |v_e^{(N)}| \sigma(|v_e^{(N)}|) \right) \tag{7}$$

and the corresponding entry to the right-hand side vector **M**:

$$\sum_i w_{\mathfrak{e}}^{(i)} |v_{\mathfrak{e}}^{(i)}| \sigma(|v_{\mathfrak{e}}^{(i)}|). \tag{8}$$

This procedure can be carried out for multiple different processes, such as elastic scattering, electron-impact electronic excitation, electron-impact ionization, etc.

3 Numerical Results

The octree merging and NNLS-based merging approaches were implemented in Merzbild.jl, a DSMC code implemented in the Julia programming language. The code and the input files used to produce the results for the present work are publicly available on Github [13]. As a baseline merging approach, the octree binning algorithm [12] was used.

3.1 BKW Relaxation

We first consider the BKW relaxation problem [4, 9], as we can compare the numerical solutions to the analytical solution. The problem is initialized by evaluating the velocity distribution function on a 32^3 velocity grid and converting the points on the grid with their associated VDF values to variable-weight particles. For NNLS merging, different numbers of conserved central moments were considered, with **all** mixed moments of order up to $L_{m,1}$ conserved (that is, all moments $M_{j_i k_i l_i}$ such that $j_i + k_i + l_i \leq L_{m,1}$), and additional conservation of moments $M_{j_i 00}$, $M_{0 j_i 0}$, $M_{0 j_i 0}$, where $L_{m,1} < j_i \leq L_{m,2}$. The target number of particles for the octree merging algorithm was chosen to be similar to that of the post-merge particles obtained via the NNLS merging. 400 simulations were performed for 500 timesteps each for each set of merging settings and then averaged, to reduce the impact of stochastic noise on the analysis of the results.

As an error metric, we consider the bias of the ensemble-averaged solution w.r.t. the analytical solution:

$$\mathcal{B}\left(\hat{M}^{2l}\right) = \sqrt{\frac{1}{N_t} \sum_{t_i} \left(\overline{\hat{M}^{2l}}(t_i) - \hat{M}^{2l}_{an}(t_i)\right)^2}. \tag{9}$$

Here $\hat{M}^{2l}$ is the total moment of order $2l$ (the moment of the particle speed of order $2l$), N_t is the number of timesteps over which the bias is computed, t_i is the time at timestep i, $\overline{\hat{M}^{2l}}$ is the ensemble average of the computed moments, and $\hat{M}^{2l}_{an}$ is the analytical value of the moment.

Figure 1 shows the ensemble-averaged evolution of the 4-th total moment of the BKW distribution for different merging algorithms and different numbers of particles

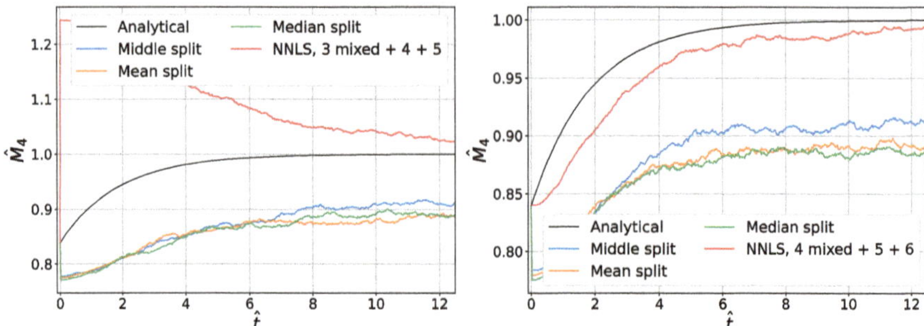

Fig. 1 Evolution of the 4-th total moment of the BKW distribution, 50:30 merge (left) and 75:50 merge (right)

considered. The $N:M$ numbers on the legend denote the threshold number of particles N (if it is exceeded, particle merging is performed) and target post-merge number of particles M. The "middle split", "mean split", and "median split" are different versions of the octree merging algorithm with different splitting of the bins for refinement: either along the middle of the bean (regardless of the particles' velocities), along the mean velocity of the particles in the bin, or along the median velocity of the particles in the bin. On the left subplot, we see significant deviations from the analytical solution, and both the octree merging and NNLS-based merging algorithms exhibit a large jump at $t = 0$, as the very finely sampled distribution is merged down to only a few dozen particles. On the right subplot, the octree merging algorithm exhibits similar behaviour, although the error is somewhat lower, as the number of particles is increased. The NNLS merging algorithm however now conserves a sufficient number of moments to be able to preserve the 4-th total moment during merging—no jump can be seen at $t = 0$, and the solution in general is much closer to the analytical one.

Figure 2 shows the bias in the 4-th total moment of the BKW distribution as a function of the time-averaged number of particles $\overline{N}_p$. We see that all versions of the octree merging exhibit a significantly higher bias than the proposed moment-preserving approach, requiring at least 5 times as many particles to achieve a similar level of solution bias. Out of all the octree bin splitting strategies, the "middle split" performs the best, due to it avoiding any bias induced by the distribution of particles in phase space.

3.2 0-D Ionization

Next, we consider a spatially homogeneous ionization in an argon plasma driven by a constant electric field of 400 Tn. Only electron-neutral elastic scattering and electron-impact ionization were accounted for, with isotropic scattering, and the collision cross-sections taken from the IST-Lisbon database [1]. After an initial burn-in time,

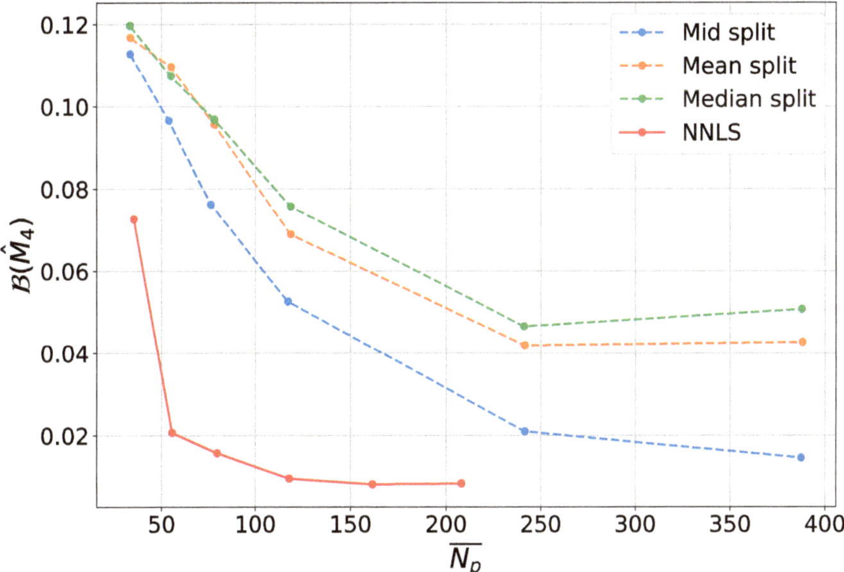

Fig. 2 Bias in the 4-th total moment of the BKW distribution as a function of the average number of particles

the system reaches a quasi-steady state, characterized by a constant ionization rate coefficient. The event-splitting scheme was used for collisions [14]. We analyze the bias introduced into the ionization rate coefficient in the first 100 timesteps after merging of the electron VDF is performed (averaged over all merging events). Since the system is driven by an external force, it has time to "recover" between merging events, and thus the bias immediately after merging has been performed is a better metric to compare the error introduced by the different merging approaches.

Figure 3 shows the average bias in the ionization rate coefficient immediately after merging events. Two versions of the NNLS merging approach were considered: with and without preservation of approximate electron-neutral collision rates (red and black lines, correspondingly). We see that for low numbers of particles, the NNLS-based approaches outperform the octree merging approach. The addition of the approximate rate preservation constraints also reduces the bias; however, as more and more moments are conserved in the NNLS merging algorithm (corresponding to a larger average number of particles $\overline{N_p}$), the difference between the rate preserving and non-rate preserving formulations becomes negligible, as the high-velocity tails of the electron VDF are sufficiently well-preserved during merging when $\overline{N_p}$ is sufficiently large.

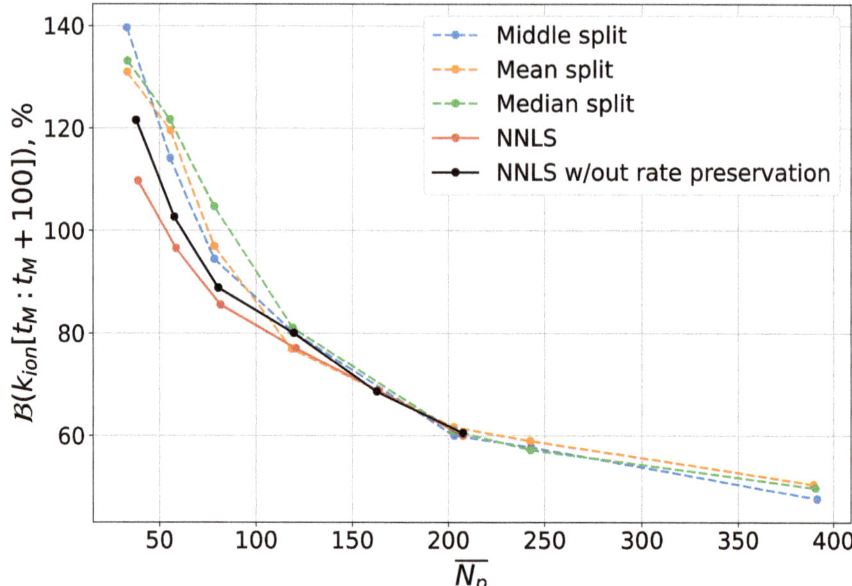

Fig. 3 Average Bias in the ionization rate coefficient in the first 100 timesteps after merging events as a function of the average number of particles

4 Conclusions

A novel moment-preserving particle merging approach based on the solution of a non-negative least-squares problem has been developed, along with an approximate rate-conserving version for plasmas. It has been shown to noticeably reduce the merging-induced the bias in higher-order moments of the velocity distribution function, as well as provide a better representation of the ionization rate coefficient for low numbers of computational particles.

Future planned extensions to the work include coupling the NNLS approach with the octree merging algorithm [12] for a greater degree of adaptivity, and consideration of spatial moments for inhomogeneous problems.

Acknowledgements This work has been supported by the German Research Foundation within the research unit SFB 1481.

References

1. Alves L (2014) The IST-LISBON database on LXCat. J Phys Conf Ser 565:012007 (IOP Publishing)
2. Araki SJ, Martin RS (2020) Interspecies fractional collisions. Phys Plasmas 27(3)

3. Bird GA (1994) Molecular gas dynamics and the direct simulation of gas flows. Clarendon, Oxford, England, UK
4. Bobylev AV (1976) One class of invariant solutions of the Boltzmann equation. In: Akademiia Nauk SSSR Doklady. vol 231, pp 571–574
5. Boyd ID (1996) Conservative species weighting scheme for the direct simulation Monte Carlo method. J Thermophysics Heat Transfer 10(4):579–585
6. Chen Z, Chen Z, Wang Y, Xu J, Chen Z, Jiang W, Wang H, Zhang Y (202) Stochastic weighted particle control for electrostatic particle-in-cell Monte Carlo collision simulations in an axisymmetric coordinate system. In: Computer physics communications, p 109390
7. Gonoskov A (2022) Agnostic conservative down-sampling for optimizing statistical representations and PIC simulations. In: Computer physics communications, vol 271, p 108200
8. Huerta C, Eckhardt D, Martin R (2024) In-situ, conservative particle merging with octree sorting. In: AIP conference proceedings, vol 2996. AIP Publishing
9. Krook M, Wu TT (1977) Exact solutions of the Boltzmann equation. Phys Fluids 20(10):1589–1595
10. Lama S, Zweck J, Goeckner M (2020) A higher order moment preserving reduction scheme for the stochastic weighted particle method. SIAM J Sci Comput 42(5):A2889–A2909
11. Lawson CL, Hanson RJ (1995) Solving least squares problems. SIAM
12. Martin RS, Cambier JL (2016) Octree particle management for DSMC and PIC simulations. J Comput Phys 327:943–966
13. Oblapenko G (2024) Merzbild.jl: A Julia DSMC code. https://github.com/merzbild/Merzbild.jl
14. Oblapenko G, Goldstein D, Varghese P, Moore C (2022) Hedging direct simulation Monte Carlo bets via event splitting. J Comput Phys 466:111390
15. Rjasanow S, Schreiber T, Wagner W (1998) Reduction of the number of particles in the stochastic weighted particle method for the Boltzmann equation. J Comput Phys 145(1):382–405
16. Rjasanow S, Wagner W (1996) A stochastic weighted particle method for the Boltzmann equation. J Comput Phys 124(2):243–253
17. Shen C (2006) Rarefied gas dynamics: fundamentals, simulations and micro flows. Springer Science & Business Media
18. Torrilhon M (2016) Modeling nonequilibrium gas flow based on moment equations. Annu Rev Fluid Mech 48:429–458
19. Vranic M, Grismayer T, Martins JL, Fonseca RA, Silva LO (2015) Particle merging algorithm for PIC codes. Comput Phys Commun 191:65–73

Data-Driven Extraction of 3D Orthogonal Velocity Distribution Modes from 1D DSMC Shocks

Robert S. Martin

Abstract This work investigates the structure of 3D particle velocity distribution functions (VDFs) within 1D normal shock waves. To explore the potential to extend continuum Boltzmann reduced order models previously studied only for a subset of homogeneous relaxation problems, low-rank modal structures are extracted from normalized Mach 2 and Mach 8 shock structure solutions. Because full 1D shock structure solutions are not readily available, cell-based continuum VDFs are instead sampled from highly resolved Direct Simulation Monte Carlo (DSMC) results. Normalized heat fluxes for Mach 2 and Mach 8 shocks are compared with expansion in terms of the Sonine polynomials traditionally used in Chapman-Enskog expansions. The viability of adapting the resulting data-driven modes for development of spatially inhomogeneous Boltzmann-ROMs and potential for other future algorithmic hybridization are also discussed.

Keywords Direct Simulation Monte Carlo · Proper Orthogonal Decomposition (POD) · Data-driven · Boltzmann Equation · Shock structure

1 Introduction

In Ref. [1], three orders of magnitude speedup relative to a nodal discontinuous Galerkin method for a class of 0D homogeneous relaxation solutions to the Boltzmann equation using a projective Reduced Order Model (ROM) was achieved. However, since the modes used to construct the ROM were extracted using singular values of solution vectors from randomly sampled trajectories of nearby initial conditions, extending this approach to spatially inhomogeneous flows would require a large database of spatially inhomogeneous solution trajectories. Such solution databases would quickly become prohibitively large and expensive to compute with direct

DISTRIBUTION STATEMENT A. Approved for public release. Distribution is unlimited.

R. S. Martin (✉)
DEVCOM Army Research Laboratory, Army Research Office, Research Triangle Park, NC, USA
e-mail: robert.s.martin163.civ@army.mil

© The Author(s) 2026

M. Grabe et al. (eds.), *Rarefied Gas Dynamics*, Springer Aerospace Technology,
https://doi.org/10.1007/978-3-032-00094-1_53

Boltzmann solvers, especially for higher spatial dimension problems. Instead, this work explores extracting such mode structures from 1D DSMC solutions to the Boltzmann equation. The work of Ref. [10] was extended in the 32nd symposium with streaming proper orthogonal decomposition (POD) based on Ref. [7] to extract dominant spatial modes of cell based density fields from transient particle simulations. In theory, such streaming POD could be used to similarly extract other observables of the flow such as VDF densities. In an operator split implementation of a Boltzmann solver, the multi-dimensional advection schemes could be coupled with a precomputed ROM for homogeneous collisional relaxation in each cell if a low-rank structure can be identified in the VDF space. This is not particularly distinct from the mindset of perturbation solutions used in moment-methods. The low-rank assumption in this case is equivalent to an expectation that the flow is sufficiently collisional as to be dominated by the equilibrium distribution and a sequence of perturbations. In this case, that sequence is simply data-driven and extracted from stochastic particle simulation rather than functional forms preselected for analytical convenience.

As with spatial cells, there is an inherent tradeoff between cell size and statistical noise. Because such high-dimensional noise is generally orthogonal to itself, the noise rejection properties of streaming POD are particularly useful. However, as VDFs exponentially decay in unbounded velocity domains, the large dynamic range in density makes retention of statistically significant correlations in distribution tails challenging. This work represents a first cut at exploring these issues in the relatively simplified restricted configuration of 1D shock structures.

2 Method

The structure of 1D normal shock-waves, as in Ref. [3], was an early application of the DSMC method. Following the basic DSMC formulation presented in [4], Mach 2 and Mach 8 stationary Xe gas shock-wave cases were setup with the parameters defined in Table 2 of the Appendix within the Thermophysics Universal Research Framework (TURF) [2, 8] previously shown to compare favorably to DS1V results.[1] Additional details of the setup and numerical parameters are provided in Appendix A.1. The solution was allowed to evolve $50\,\mu s$ with a time-step of $10\,\text{ns}$ to establish a steady state. No shock stabilization technique was applied though the drift is expected to be negligible with high particle counts of $\tilde{1}6.4$ and 24.0 million computational particles for the Mach 2 and Mach 8 cases respectively. After an initial $50\,\mu s$ transient, VDF sampling began. The solution was run for an additional $50\,\mu s$ with particle distributions sampled to the mesh every $1\,\mu s$ (100 iterations) though full output of all VDFs was only performed at 16, 33, $50\,\mu s$ after restart due to the relatively large output files. The steps between sampling were selected to ensure that samples of the low

---

[1] Note equivalent results would be expected using DS1V. TURF was only used for convenience due to preexisting data structures and I/O functionality for continuum VDF data requiring only minimal modification of sampling procedures.

speed particles in the stationary reference frame were not overly correlated due to the same slow particle residing in the same spatial cell for multiple time-steps, although any potential impact of such correlations are outside the scope of this work. Samples were accumulated for the full time from restart with time averaging performed in post processing.

Accumulating VDF solutions to a static mesh induces a few challenges. First, the change in mean velocity and temperature through the shock generally demand a relatively large velocity space mesh. With the idea that the extracted basis functions might serve as modes for a future POD Boltzmann ROM, the need to minimize the VDF mesh dimension further directly impacts the viability of preprocessing the ROM collision integrals. For this reason, the continuum VDF mesh followed the normalization used in Ref. [1]. To ensure that all of the initial conditions converged to the same equilibrium VDF, the VDF initial conditions were normalized to a standard, zero-drift Maxwellian with non-dimensional temperature of 0.1. Prior to sampling in the particle VDF in each spatial cell, the instantaneous particle temperature and mean velocity were used to similarly transform the particle VDF. While the initial results used the same 41^3 of ± 3 dimensionless velocity mesh as the 0D cases, this mesh was shifted in the shock normal direction to -4.317 to 1.683 normalized velocity units to account for the heavy backscattered VDF tail observed in the Mach 8 case, similar to tail distributions noted in Figs. 12.40 & 12.44 of Ref. [4].

To construct data-driven reduced order models (ROMs) as described in Ref. [5], a common approach is to first build a more efficient linear basis for representing the solutions. In the case of collision integrals, $\partial f / \partial t |_{\text{collision}}$, this corresponds to a truncated singular value decomposition (SVD) based compressed representation referred to as POD. Whereas POD is most commonly used to represent spatial "snapshot" vectors across time sequences, the same philosophy can be ascribed to a velocity space basis across spatial cells for a steady-state problem like the shock structure calculation. A modified "streaming" version of POD derived as a subset of the dynamic mode decomposition algorithm of Ref. [7] was used to extract the mode structures. Additional algorithmic details are provided in Appendix A.2. For the results presented, a maximum rank of 35 with periodic re-compression to 25 was used. The modes were also processed sequentially from upstream to downstream. These sizes and order were selected arbitrarily to appear sufficient capture a dominant low-rank structure and subsequent noise floor in combination with the sampling procedure described. Further optimization and the impact of truncation making long range correlations undetectable are left to future work.

Though full distribution POD modes were also computed, this paper focuses only on the δf results corresponding to a modal expansion around the equilibrium solution $\delta f = f - f_0$. The equilibrium Maxwellian is subtracted from each VDF prior to constructing the modes. This corresponds to the more stable solution in the previous 0D ROM results and also mitigates the issue that the dominant 0-mode was the spatial mean VDF rather than Maxwellian. In particular, the non-Maxwellian 0-mode caused a heavy upstream tail in the 0-mode and significant residual downstream mode amplitudes in the Mach 8 case. By focusing on δf-modes only, the amplitudes also correspond to the degree of non-equilibrium captured in the low-rank approximation.

Table 1 Mode energy preserved relative δf^2 of meshed VDF. Top-25 mode and further truncated fraction provided

Sampling Duration	16 μs	33 μs	50 μs
Mach 2 (Top-3)	48.9% (40.4%)	64.1% (58.8%)	72.2% (67.6%)
Mach 8 (Top-7)	99.27% (99.15%)	99.64% (99.58%)	99.76% (99.72%)

The non-equilibrium portion of the shock structure solutions are then approximated by parametric loops starting and ending at the origin of the r-dimensional space.

3 Results

Figure 1 shows results for the δf VDF modes for the Mach 2 and Mach 8 Xenon test cases respectively. The figures provide 2D summed distributions of 3D orthonormal POD modes in one of the two equivalent transverse velocity directions. The figures also provide projection of the 1D shock structure solutions into the i-th discrete POD mode amplitudes, $a_i(x_j) = \phi_i \cdot f(x_j)$. The corresponding VDF energy spectra, and a comparison of the non-dimensionalized heat-flux constructed by discretely summing the VDF representation with $C^2 C_x$ normalized velocity in the transformed coordinate mesh. Table 1 shows the fraction of non-equilibrium δf^2 energy retained by the POD expansions.

Though a larger fraction of the energy is retained in the Mach 8 case compared to the Mach 2 case, this is because the non-equilibrium signal exceeds the noise floor of the VDF distribution more significantly in the Mach 8 case. By comparing the energy spectra across sampling durations in Fig. 1, it can be seen that the significant modes are quite stable across sampling durations while the noise floor in the higher modes progressively decreases with additional sampling time. This is particularly apparent as the third mode emerges from the noise floor only after 33 μs of sampling for the Mach 2 case. A significant fraction of the truncated energy in both cases is the result of truncating noise that approximates an infinite dimensional white spectra. While the fraction of energy retained is significantly higher in the Mach 8 case, the noise floor falls at a similar absolute magnitude of for both cases because a similar number of particles are sampled onto the same VDF mesh.

Sonine polynomials from Chapman-Enskog theory (Appendix A.3) were used to compare the estimated normalized heat-flux from both the analytical perturbative expansion and POD to the full-VDF based heat flux in Fig. 1. While the POD and full-VDF heat fluxes agree well, the Sonine expansion in 25 modes severely amplifies the VDF noise due to mode amplitudes being biased to amplify noise in the high velocity region of the VDF. Interestingly, if the full mesh VDF is first projected into the first

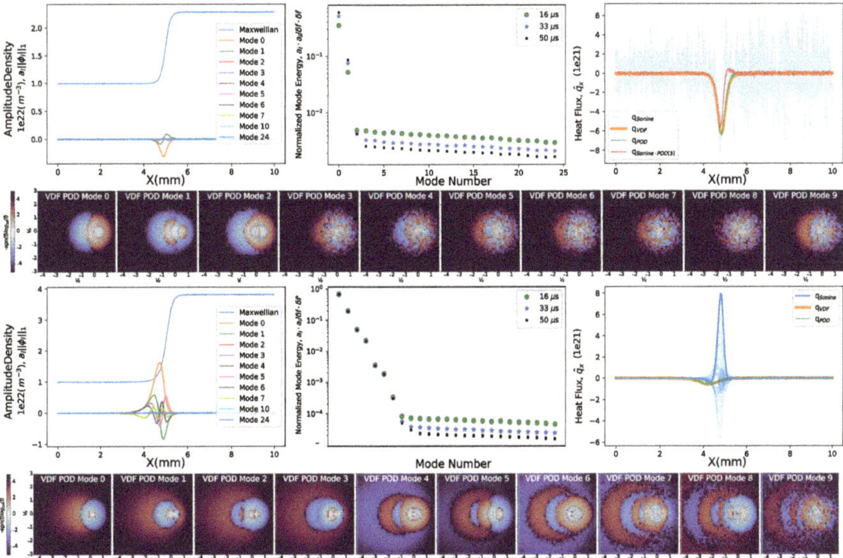

Fig. 1 Mach 2 (top) and Mach 8 (bottom) results: relative mode amplitudes (upper left), mode energy spectrum (upper middle), non-dimensionalized heat flux, $\hat{q}$ (upper right), and 2D summation of top-10 mode basis functions by energy (lower)

3 POD modes and then expanded in Sonine polynomials, this noise is significantly suppressed due to the global averaging that contributes to the POD mode smoothness deep into the tail of the distribution. A noticeable discrepancy remains between this projected Sonine expansion and the POD solution that cannot be attributed to the POD projection as the 3-Mode POD heat flux is indistinguishable from the 25-Mode POD on this scale. This discrepancy also corresponds to the region of highest discrepancy in Navier-Stokes solutions to the shock structure and may be the result of a breakdown of the assumption of small perturbation in this VDF tail region. For Mach 8, the heat-flux comparisons again matched well between the full VDF and POD projection. However, while the noise is significantly reduced for this case, the Sonine basis resulted in divergent heat flux. As additional Sonine basis functions are progressively added, the downstream heat flux resulted in oscillations with progressively higher mode counts. This is likely due to the alternating polynomial sign of the Sonine weighting amplified by the very heavy tails of the Mach 8 distributions. However, further investigations into this phenomena are beyond the scope of this work.

The rapid decay of the energy spectra in both cases is the hallmark of a problem amenable to construction of efficient low-Rank ROMs. This is noteworthy as "advection" dominated ROMs are challenging due to the *Kolmogorov n-width* problem [12]. Though not traditional advection, the equilibrium VDF does translate and scale relative to a fixed VDF mesh through the shock-wave. By first applying the affine transformation to the particle distribution that removes these drift and temperature variations, the anticipated low-Rank structure in the perturbative expansions underpinning Chapman-Enskog theory [13] is recovered. This suggests an alternative route for ROM construction for problems previously tackled with similar perturbative expansions that falls between linear projection and fully nonlinear neural network methods. Such transformations can inherit the benefits of improved interpretability and closed form construction while expanding the applicability of an otherwise linear low-rank basis.

4 Conclusion and Future Work

The rapid mode decay suggests a few-mode perturbative expansion based ROM may provide reasonable shock structures calculations. While expected for the Mach 2 case where fluid solutions retain reasonable accuracy, low rank structure of Mach 8 results is more compelling evidence that a linear POD-Galerkin ROM may be feasible for the family of 1D shock structures. The first 3 significant modes in Mach 2 and 7–8 in Mach 8 are likely sufficient for single Mach input. However, initial attempts at projecting the Mach 2 modes into the Mach 8 basis results in significant residuals due to the heavy tails throughout the Mach 8 basis. Each basis is therefore insufficient for flows in the other condition. While 10–11 concatenated modes would still be quite low-rank, the overall rank needed to represent a family of intermediate Mach shock structure solutions within this finite Mach number range requires further exploration. The selection of sufficient intermediate conditions to accurately build such a basis also remains unexplored.

For some moderate Mach range, full 1D ROMs should next be constructed. This will require coupling the 0D Boltzmann ROM for collisional relaxation with some collisionless transport method. Due to the imposed translational invariance in the VDF ROM basis, it would be particularly interesting to apply this ROM to non-stationary solutions where long DSMC sampling used to construct modes would be challenging to converge. Though the transport portion could use particles methods like DSMC, development of a fluid based moment-method in the data-driven basis may be considerably more efficient. However, if coarsely resolved particle transport is sufficient to project and denoise VDF solutions back into the ROM basis, a hybrid algorithm may have additional advantages such as simpler integration of boundary conditions or domain decomposed coupling with traditional DSMC or deviational particle flow regions.

The low rank structure of the solutions also has interesting implications for the analysis of other learned closure models [6, 11]. Relatively few linearly independent non-equilibrium observables are sufficient to uniquely specify an entire VDF approximation for these problems. Once reconstructed, any other observable function of the VDF such as moments and transport fluxes could then be approximated with bounded approximation error. This low rank structure likely underpins the observed successes in more general learned closures.

Beyond 1D, using VDF transformation that align to the dominant eigenvectors of the particle covariance, as in presentation [9], rather than isotropic temperature would be a relatively straightforward extension to multiple dimensions. Whether this results in emergence of a similar low-Rank structure is much less clear. The Kolmogorov n-Width problem may re-emerge for multi-parametric solution families like oblique or curved shocks. Similar issues should be expected for solid interfaces if the surface VDF discontinuity is significant. The closer the flow conforms to assumptions underpinning continuum Navier-Stokes solutions, the more likely that the perturbation expansion logic will translate into a similar low-Rank structure. This makes these ROMs likely to be most useful in extending the domain of validity of near equilibrium fluid solutions into non-equilibrium flows. Whether fully nonlinear mode compression becomes practically necessary as the degree of non-equilibrium increases is an interesting question, but in all these cases, these low rank structures are likely to remain useful in analyzing the accuracy and model capacity requirements of more complex methods.

Appendix A.1   Test Case Setup Details

See Table 2.

Standard VHS parameters, T_{ref}=273 K, α=0.35, and d_{ref}=5.74E-10 were used with a 1E7 macroparticle weight. The solution was initialized in a 1000 cell (1 cm) long 1D domain with $1\,\text{mm}^2$ cross section as a discontinuity at the center of the domain given the pre-shock (0) and post-shock (f) states.

Appendix A.2   Streaming POD

Consider the data block $\mathbf{X} \in \mathcal{R}^{n \times m}$ representing the full time-averaged meshed VDF solutions where n is the state dimension of the VDF mesh time averaged solutions (41^3) and m is the number of VDFs (1000). The matrix $\mathbf{X}$ can be replaced with a low-rank approximation truncating to the first $r \ll m$ largest singular values as $\tilde{\mathbf{X}} = \tilde{\mathbf{U}}\tilde{\Sigma}\tilde{\mathbf{V}}^*$ where $\tilde{\mathbf{U}}$ is $n \times r$, $\tilde{\Sigma}$ is $r \times r$ leading singular values, and $\tilde{\mathbf{V}}$ is the $r \times m$

Table 2 Xenon shock initial conditions

Case	Mach 2	Mach 8
Density, $n_0(\text{m}^{-3})$	1.0E22	1.0E22
Density, $n_f(\text{m}^{-3})$	2.2857E22	3.8209E22
Temp., T_0 (K)	293.0	293.0
Temp., T_f (K)	608.891	6115.517
Velocity, v_0(m/s)	351.71	1406.84
Velocity, v_f(m/s)	153.873	368.196

amplitudes for the r modes in each $m = 1000$ cells. While $\mathbf{X}$ is not unmanageably large at 551mb for this simple 1D problem in double precision, following Ref. [7], a streaming version of the POD portion of the dynamic mode decomposition algorithm can be adapted that avoids both storing the large matrix and solving large eigenvalue problems. In this case, $\tilde{\mathbf{U}}$ can be built progressively by appending appending orthogonal projections of solution vectors, $e = x_i - (\mathbf{U}\mathbf{U}^T)x_i$, equivalent to the Gram-Schmidt basis that would be constructed if the full $\mathbf{X}$ were available. This basis is allowed to grow and the outer product of the low-rank approximated vectors $\tilde{x}_i = \mathbf{U}^T x_i$ are accumulated in an $r \times r$ matrix $\mathbf{G}$ by padding $\mathbf{G}$ with 0s every time r is increased. Because the particle noise results in large residuals, the rank grows with each snapshot, x_i, added. When the rank exceeds some $r > r_{max}$, the $\mathbf{G}$ matrix can be used to re-compress the solutions by solving an $r \times r$ eigenvalue problem for the r_{min} leading eigenvectors $\tilde{\mathbf{V}}$ of $\mathbf{G}$ as $\tilde{\tilde{\mathbf{G}}} = \tilde{\mathbf{V}}^T \tilde{\mathbf{G}} \tilde{\mathbf{V}}$ and $\tilde{\tilde{\mathbf{U}}} = \tilde{\mathbf{V}}^T \tilde{\mathbf{U}}$. This periodic re-compression preserves a computational cost proportional to the square of r_{max} which remains relatively tractable for low-rank problems.

Appendix A.3   Sonine Polynomial Expansion

This appendix reviews construction of a Sonine polynomial basis derived from the descriptions provided in Ref. [13]. Starting from the perturbative expansion of the VDF, $f = f_0(1 + \Phi_1 + \Phi_2 + \dots)$, the perturbation, Φ, follows the functional form in Eq. 1. The forms of A_i and B_{ij} that respect the symmetries and invariances of the collision operator is are then shown in Eq. 2. Due to the symmetric nature of the 1D flow, only A_x for q_x and B_{xx} for τ_{xx} are therefore nonzero.

$$\Phi_1 = \quad -\frac{1}{n}\left[\sqrt{2kT/m}\,A_j\frac{\partial}{\partial x_i}(\ln T) + B_{jk}\frac{\partial \bar{c}_j}{\partial x_k}\right] \tag{1}$$

$$A_i = A(\mathcal{C}, T)\mathcal{C}_i, \qquad B_{ij} = B(\mathcal{C}, T)\left(\mathcal{C}_i\mathcal{C}_j - \tfrac{1}{3}\mathcal{C}^2\right) \tag{2}$$

$$S_m^{(n)} = \qquad \sum_{p=0}^{n}\frac{(m+n)!}{(m+p)!p!(n-p)!}(-x)^p = L_n^m \tag{3}$$

$$f = \qquad f_0(1 + A_x + B_{xx})$$

$$= \quad f_0\left(\begin{array}{l}1 + \sum_{r=0}^{\infty} a_r S_{3/2}^{(r)}[\mathcal{C}^2]\mathcal{C}_x + \\ \sum_{r=1}^{\infty} b_r S_{5/2}^{(r-1)}[\mathcal{C}^2]\left(\mathcal{C}_x\mathcal{C}_x - \tfrac{1}{3}\mathcal{C}^2\right)\end{array}\right) \tag{4}$$

The use of Sonine polynomials, $S_m^{(n)}$ for expanding A_i and B_{ij} was developed to leverage orthogonality that enables rapid convergence of the evaluation of transport properties. Sonine terminology has since been supplanted by 'associated Laguerre polynomials', L_n^m, where the subscript and superscript have been exchanged as shown in Eq. 3.

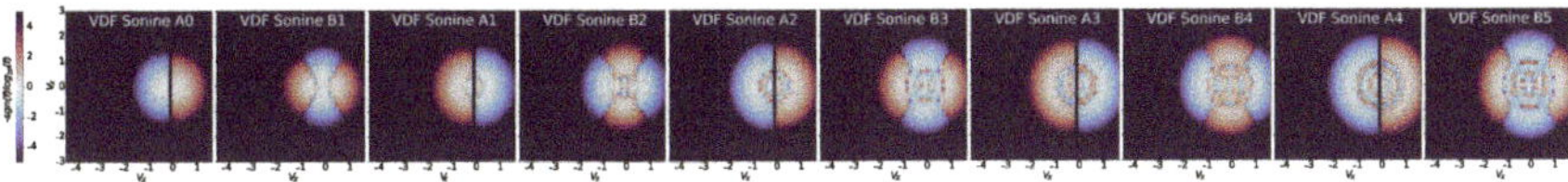

Fig. 2 2D summation of first 5 Sonine mode basis functions for A_x and B_{xx} modes

This gives perturbed distribution function, Eq. 1, in terms of the Sonine series expanded around the equilibrium distribution, f_0, shown in Eq. 4. For the 1D flow, the distribution function can then be expanded as a power series in terms of coefficients a_i and b_i and orthogonal mode shapes $f_0^{1/2} S_{3/2}^{(r)} \mathcal{C}_x$ and $f_0^{1/2} S_{5/2}^{(r-1)} \left(\mathcal{C}_x \mathcal{C}_x - \frac{1}{3}\mathcal{C}^2 \right)$ as shown in Fig. 2 where the mode normalization can be absorbed into the series coefficients, a_r and b_r.

References

1. Alekseenko A, Martin RS, Wood A (2022) Fast evaluation of the Boltzmann collision operator using data driven reduced order models. J Comput Phys 470:111526
2. Araki S, Martin RS, Bilyeu DL, Brieda L, Hill C, Choi M (2022) Current capabilities of AFRL's spacecraft simulation tool. In: 37th international electric propulsion conference, Boston, MA
3. Bird GA (1970) Aspects of the structure of strong shock waves. Phys Fluids 13(5):1172–1177
4. Bird G (1994) Molecular gas dynamics and the direct simulation of gas flows. Oxford University Press
5. Brunton SL, Kutz JN (2022) Data-driven science and engineering, 2nd edn. Cambridge University Press
6. Garg G, Mankodi TK, Esmaeilifar E, Myong RS (2024) Neural network-based finite volume method and direct simulation Monte Carlo solutions of non-equilibrium shock flow guided by nonlinear coupled constitutive relations. Phys Fluids 36(10):106113. https://doi.org/10.1063/5.0223654
7. Hemati MS, Williams MO, Rowley CW (2014) Dynamic mode decomposition for large and streaming datasets. Phys Fluids 26(11):111701
8. Martin R, Koo J (2015) Thermophysics universal research framework—infrastructure release. Air Force Research Laboratory (AFRL/RQRS), Version 1.0.0 edn. aFRL-RQ-ED-OT-2015-108
9. Martin RS (2013) Multidimensional effects on conservative particle merging. In: DSMC, Santa Fe, NM, aFRL-RQ-ED-VG-2013-241
10. Martin RS, Eckhardt D (2019) Denoising quasi-steady state particle simulations. In: 31st rarefied gas dynamics, AIP conference proceedings, vol 2132, issue 1, p 090010
11. Nair AS, Sirignano J, Panesi M, MacArt JF (2023) Deep learning closure of the Navier–Stokes equations for transition-continuum flows. AIAA J 61(12):5484–5497. https://doi.org/10.2514/1.J062935
12. Peherstorfer B (2022) Breaking the Kolmogorov barrier with nonlinear model reduction. Notices of the American Mathematical Society
13. Vincenti W, Kruger Jr C (2022) Introduction to physical gas dynamics. Krieger Publishing Company

Numerical Study on the One-Way Flow Between Plates of Different Temperatures with Periodic Distribution of Accommodation Coefficient

Ryuichiro Imazu, Hiroshi Sugimoto, and Maika Itou

Abstract This research numerically explores the feasibility of a one-way flow between two finite-sized planar plates, each with a different uniform temperature. In the case of infinite plates, a periodic distribution of the accommodation coefficient enables a one-way flow. An actual plate inevitably has its edges, and in the rarefied gas, the edge in a non-uniform temperature field induces localized fast flow—thermal edge flow. The present results show that the one-way flow by the distribution of the accommodation coefficient is also possible for finite-sized plates as well as infinite plates. The one-way flow survives under strong thermal edge flow vortices. We also report the optimal aspect ratio between the plate spacing and the period length of the distribution that maximizes the one-way flow as a function of the Knudsen number.

Keywords Thermal transpiration flow · Thermal edge flow · Knudsen pump

1 Introduction

The temperature field plays a pivotal role in various phenomena in rarefied gas. We have long observed various gas flows around stationary bodies solely due to the interaction of gaseous temperature and the surface. With recent advances in microdevices, numerous researchers have worked vigorously on developing devices such as micropumps and gas mixture separation that utilize these phenomena. The concept underlying these devices is thermal transpiration flow, where the cold gas adjacent to the solid wall moves toward the hot gas part along the wall. Some devices induce this by varying the temperature of the wall, and some use the geometrical shapes and arrangement of the hot/cold parts, sometimes accompanied by the distribution of the accommodation coefficient. One of the authors recently found a possibility of a similar flow in an elementary geometry—a channel between two flat plates—by the

R. Imazu · H. Sugimoto (✉) · M. Itou
Department of Aeronautics and Astronautics, Kyoto University, Kyoto, Japan
e-mail: sugimoto.hiroshi.7s@kyoto-u.ac.jp

© The Author(s) 2026

M. Grabe et al. (eds.), *Rarefied Gas Dynamics*, Springer Aerospace Technology,
https://doi.org/10.1007/978-3-032-00094-1_54

numerical simulation [5]. The one-way flow through the channel occurs if one of the plates is heated and both plates have a periodic distribution of the accommodation coefficient. The simple structure and the separation of hot and cold parts enhance easy fabrication of the device.

At the same time, the flow speed is slower than other standard thermal transpiration flows. Actual plates must have their edges due to their finite size, and the edges most likely induce strong localized gas flow—the thermal edge flow [4]—which may disrupt the structure of the weak one-way flow. Therefore, we numerically analyze the one-way flow induced by the periodic distribution of the accommodation coefficient using the finite-sized plates to evaluate its feasibility in real devices.

2 Problem Setups

The basic 2D geometry that induces the one-way flow in [5] is shown in Fig. 1a. A rarefied gas fills the gap of height D between a heated plate (uniform temperature T_1) and an unheated plate (uniformly T_0). Both plates have a periodic distribution of the accommodation coefficient, which varies alternately between 1 (i.e., diffuse reflection) and some $\alpha(< 1)$ at a distance of L. In this setup, the gas flows along the plate if the distributions of the accommodation coefficient on two plates have a phase difference Λ. To account for the geometry's periodicity, we perform our calculation in the green region in the figure.

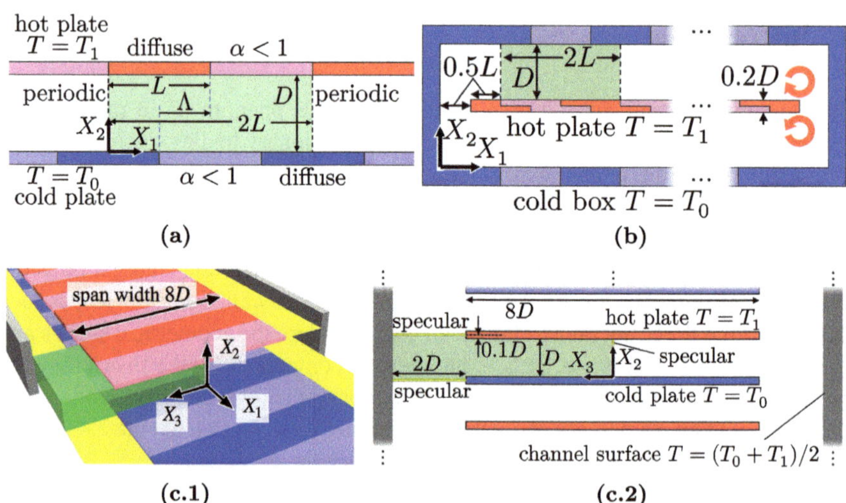

Fig. 1 Setups of the problems. **a** The basic setup in [5]. **b** Streamwise edge setup. **c.1** Spanwise edge setup. **c.2** A cross-sectional view of (**c.1**) on the $X_2 X_3$ plane

We introduce the ends of the plate to confirm the feasibility of the one-way flow by finite-sized plates. In the 2D model shown in Fig. 1b, a heated plate of thickness $0.2D$ is confined in an unheated box of $7.5L \times 2.2D$ to make a closed loop. The upper and lower channels correspond to the units, one shown in green and lined up to induce counterclockwise flow, but we have two ends of the heated plate. We assume the accommodation coefficient around the edge is unity. Therefore, the two edges induce the thermal edge vortices [4] (indicated by the red circle arrows in the figure). We simulate the rarefied gas flow to see if a one-way flow can go through thermal edge vortices.

We also have the edges in the spanwise direction for actual finite-sized plates. In response, we consider a spanwise edge setup Fig. 1c.1. The setup requires 3D simulation. Considering the periodicity in X_1 and the symmetry in X_3 of the geometry, we calculate the gas flows in the green region in the figure. A cross-sectional view of this on the $X_1 X_2$ plane is the same as in Fig. 1a, and that on the $X_2 X_3$ plane is shown in Fig. 1c.2. A vertical array of plates (thickness $0.2D$) of finite span width $8D$ is put in a diffusely reflecting channel $-6D < X_3 < 6D$, whose surface temperature is $(T_0 + T_1)/2$. The temperatures of the array plates are alternately set to T_0 and T_1. For the sake of simplicity, we assume the flow is symmetric at $X_2 = -0.1D$ and $1.1D$. We observe if the one-way flow in the central part of the plate survives the disturbance of thermal edge vortices on the edges in the spanwise direction.

3 Governing Equation

The governing equation is the BGK model:

$$
\left.
\begin{aligned}
&\frac{\partial F}{\partial t} + \xi \cdot \frac{\partial F}{\partial X} = J\,[F], \qquad J\,[F] = A_C n\,(F_E - F), \\
&F_E\,(n, v, T; \xi) = \frac{n}{(2\pi \kappa T/m)^{3/2}} \exp\left(\frac{-\,|\,\xi - v\,|^2}{2\kappa T/m}\right).
\end{aligned}
\right\}
\tag{1}
$$

Here, t is the time, X is the Cartesian coordinate, ξ is the molecular velocity, $F(t, X, \xi)$ is the velocity distribution function, m is the molecular mass, κ is the Boltzmann constant, J is the collision term, and A_C is a constant whose products with the number density n equals the collision frequency. F_E is the equilibrium distribution determined by n, the flow velocity v, and the temperature T of the gas. The distribution function F defines them through

$$
\left.
\begin{aligned}
&n = \int_{|\xi|<\infty} F \, d\xi, \qquad nv = \int_{|\xi|<\infty} \xi F \, d\xi, \\
&\frac{3}{2}\kappa n T = \int_{|\xi|<\infty} \frac{m}{2}\,|\,\xi - v\,|^2 F \, d\xi, \qquad p = \kappa n T,
\end{aligned}
\right\}
\tag{2}
$$

where p is the pressure. The boundary condition is the Maxwell condition

$$F\left(t, \boldsymbol{X}_w, \boldsymbol{\xi} \cdot \boldsymbol{n}_w > 0\right) = \alpha F_E\left(\sigma_w, \boldsymbol{0}, T_w; \boldsymbol{\xi}\right) + (1 - \alpha)F\left(t, \boldsymbol{X}_w, \boldsymbol{\xi} - 2\left(\boldsymbol{\xi} \cdot \boldsymbol{n}_w\right)\boldsymbol{n}_w\right),$$

$$\sigma_w = -\sqrt{\frac{2\pi}{\kappa T_w/m}}\, \int_{\boldsymbol{\zeta} \cdot \boldsymbol{n}_w < 0} \boldsymbol{\zeta} \cdot \boldsymbol{n}_w F\left(t, \boldsymbol{X}_w, \boldsymbol{\zeta}\right)\, \mathrm{d}\boldsymbol{\zeta},$$

where T_w is the wall temperature, $\boldsymbol{n}_w$ is the unit normal directed to the gas, and α is the accommodation coefficient. The case $\alpha = 1$ is the diffuse reflection, and $\alpha = 0$ is the symmetry condition. The mean free path l_0, the most probable speed of molecule c_0, and the corresponding Knudsen number Kn at the reference stationary state (number density n_0 and temperature T_0) are expressed by

$$c_0 = \sqrt{\frac{2\kappa T_0}{m}}, \qquad l_0 = \frac{2c_0}{\sqrt{\pi}A_C n_0}, \qquad \mathrm{Kn} = \frac{l_0}{D}.$$

4　Method of Analysis

We use a deterministic numerical scheme that uses the *spherical designed grid* and the *traveling finite volume (TFV) method* [2]. The *spherical design* [6] is a set of points $\boldsymbol{a}_k$ $(k = 1, 2, \ldots, N_\Omega)$ arranged nearly uniformly on a unit sphere $\mathbb{S}^2$. The points $\boldsymbol{a}_k$ are selected so that the average value of arbitrary t-th order polynomial $P_t(\boldsymbol{x})$ on these points is the average of $P_t(\boldsymbol{x})$ on $\mathbb{S}^2$. An example is shown in Fig. 2a. Then, we approximate the integral of $f(\boldsymbol{x})$ on $\mathbb{S}^2$, thus the integrals in (2) by, e.g.,

$$n = \int_{\boldsymbol{a} \in \mathbb{S}^2}\left[\int_0^\infty F\left(\xi \boldsymbol{a}\right)\xi^2 \mathrm{d}\xi\right]\mathrm{d}\Omega\left(\boldsymbol{a}\right) \sim \frac{4\pi}{N_\Omega}\sum_{k=1}^{N_\Omega}\int_0^\infty F\left(\xi \boldsymbol{a}_k\right)\xi^2 \mathrm{d}\xi. \tag{3}$$

Therefore, we define discretized molecular velocity

$$\xi_j \boldsymbol{a}_k \qquad \left(j = 1, 2, \ldots, N_\xi,\ k = 1, 2, \ldots, N_\Omega\right) \tag{4}$$

in the simulation. This grid gives exceptionally accurate results compared to the standard rectangular grid, shown in Fig. 2b. For a fixed error size, the spherical designed grid (indicated in red) requires a smaller number of the grid points $N_\sharp = N_\xi N_\Omega$ by few orders compared with the rectangular grid (indicated in black). Since the computational resources—the memory size and computational time—are closely related to $N_\sharp$, the present method is much faster than the standard rectangular grid. The spherical designed grid enables us to carry out 3D flow simulation by personal computers.

The *traveling finite volume method* is a numerical method for the Boltzmann equation (1) with nearly complete conservation of mass, momentum, and energy. To

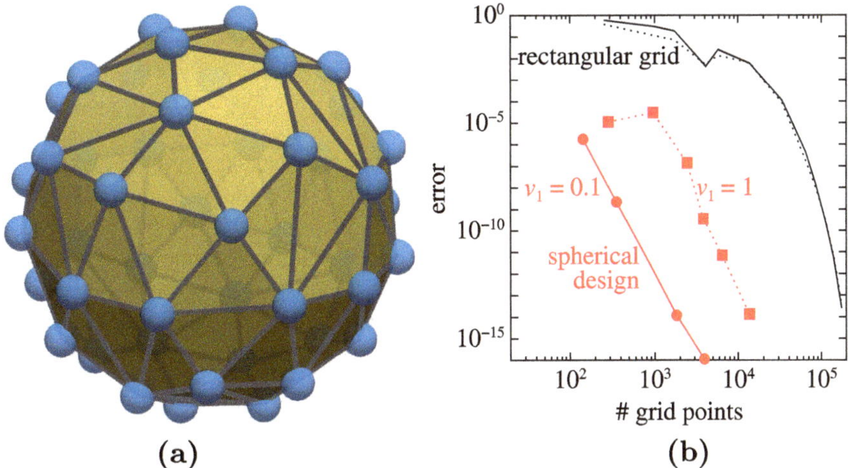

Fig. 2 **a** An example of the spherical design [6]. The case for $N_\Omega = 48$ and t = 9. **b** The errors in the numerical integration of $\int \Phi(\xi)\, d\xi = 1 \left[\Phi(\xi) = a\left(1 + \frac{7}{5}|\xi - v^2|\right)\exp\left(-\frac{7}{5}|\xi - v|^2\right), v = (v_1, 0, 0)\right]$. Solid line: $v_1 = 0.1$ and dotted line: $v_1 = 1$. Black: The Simpson's formula and uniform rectangular grid for ξ. Red: The spherical designed grid

this end, we replace (t, X) with the characteristic coordinate $(t, \chi + \xi t)$. Then we have

$$\left(\frac{\partial G}{\partial t}\right)_{\chi, \xi:\, \text{fix}} = J[G], \qquad G(t, \chi, \xi) = F(t, \chi + \xi t, \xi). \tag{5}$$

We apply the finite volume method [1] for χ-space. During the time step Δ_t, the finite volume cells travels in X-space with each molecular velocity (4). Therefore, we also prepare fixed finite volume cells in X-space, and reconstruct the X-cell data from the χ-cell data at every time step. By repeating the reconstruction and time advance by the finite volume scheme, we obtain the time evolution of X-cell data by (1). The advantage of the present method is (i) we do not need numerical flux by (5) because it is essentially a bundle of ordinary differential equations (i.e., we do not have the CFL condition), and (ii) we do not need the numerical derivative across the discontinuity [3] of the velocity distribution function because the solution of the ordinary differential equation (5) is continuous in t.

5 Results and Discussion

5.1 Infinite Plates Problem

First, we show the present result in Fig. 3a. The parameters are mostly common to [5]: $\Lambda/L = 0.5$, $\alpha = 0.5$, Kn $= 0.5$, and $L/D = 1$. We choose vanishingly small temperature difference, $T_1/T_0 - 1 = \varepsilon = 10^{-4}$, because (i) the gas flow obtained in [5] is roughly proportional to ε, (ii) small ε is easy for future experimental verification, and (iii) the present TFV method becomes accurate for small variation of the flow field. The reference number density n_0 is the average number density in the region. The white lines are the streamlines obtained as the contour of the stream function Ψ defined by

$$\Psi(X_1, X_2) = \frac{1}{n_0 c_0 D} \int_0^{X_2} [nv_1](X_1, y)\, \mathrm{d}y.$$

Despite the differences in the molecular models and temperature setup from Ref. [5] (which employs the hard-sphere model and $T_1/T_0 = 3$), temperature and flow velocity distribution show no qualitative difference: the temperature of the gas close to the diffuse section approaches that of the wall, resulting in a vortex flow near the discontinuity of the accommodation coefficient. Some streamlines connect the vortices and go through between the plates. Thus, Fig. 3a clearly shows that the one-way flow also exists in the BGK model, just as in the hard-sphere molecules.

The nondimensional mass flux through the channel $Q = \Psi(X_1, D)$ does not depend on X_1 in the steady state. Figure 4 shows Q/ε versus Kn for several aspect ratios L/D. For a fixed value of L/D, Q takes its maximum value at some Kn. For $L/D \lesssim 1$, the Knudsen number is close to 0.5. Reference [5] also reports the same Knudsen number Kn ~ 0.5 for hard-sphere molecules for $L/D = 1$. For larger L/D, Q takes a maximum at higher Knudsen numbers. The maximum value of Q varies with L/D: it takes a maximum at $L/D \sim 2$ or 4. The flow field at $L/D = 2$ and Kn $= 1$ is shown in Fig. 3b. The meandering of the streamlines becomes gentler compared with $L/D = 1$ in Fig. 3a. Therefore, unless otherwise stated, we consider the case with Kn ~ 1 and $L/D = 2$ in the subsequent calculations.

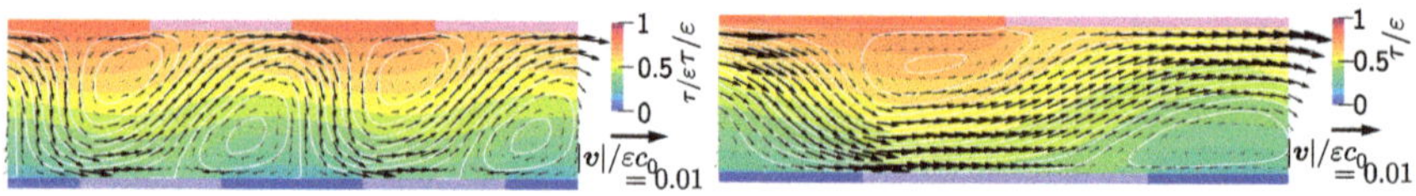

Fig. 3 Temperature and flow velocity distributions for the basic setup. $\Lambda/L = 0.5$, $\alpha = 0.5$ and $T_1/T_0 - 1 = \varepsilon = 10^{-4}$. Colors show the temperature (T/T_0 or $\tau = T/T_0 - 1$), arrows show the flow velocity, and white lines show the streamlines. **a** The case for Kn $= 0.5$ and $L/D = 1$: the same as Fig. 15 of [5] except the value of T_1/T_0. **b** The case for Kn $= 1$ and $L/D = 2$

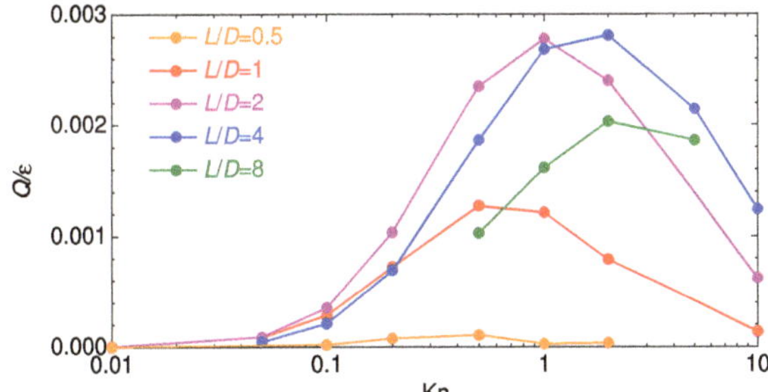

Fig. 4 Nondimensional mass flux versus Kn for several aspect ratios for the basic setup with $\Lambda/L = 0.5$, $\alpha = 0.5$ and $\varepsilon = 10^{-4}$

5.2 Effect of Streamwise Edge

Figure 5 shows the flow fields in the streamwise edge setup (Fig. 1b). We find four strong vortices around $X_1/L \sim 0.5$ and 7. The direction of the flow coincides with those by the thermal edge flow in Fig. 1b. Despite these strong vortices, we observe one-way flows similar to that in Fig. 3b in the central part of the upper and lower channels; the flow image in the upper channel is reversed in the X_1 direction, which accords with the distribution of the accommodation coefficient. The non-dimensional mass flux Q in the upper channel in $1 < X_1/L < 6.5$ is approximately $-2.1\varepsilon \times 10^{-3}$, which is of the same order of magnitude as that observed in Fig. 3b ($2.8\varepsilon \times 10^{-3}$). To summarize, we have confirmed that the plate edge in the streamwise direction disturbs only slightly the one-way flow induced by the distribution of the accommodation coefficient.

5.3 Effect of Spanwise Edge

Figure 6 shows the flow velocity and temperature field projections for the spanwise edge setup (Fig. 1c.1) in several cross-sections. Figure 7 presents some of the cross-section images separately. Panel (a) is the image at $X_1 = 0$. The thermal edge flow induced by the edges forms a strong vortex; however, the vortex flow does not pen-

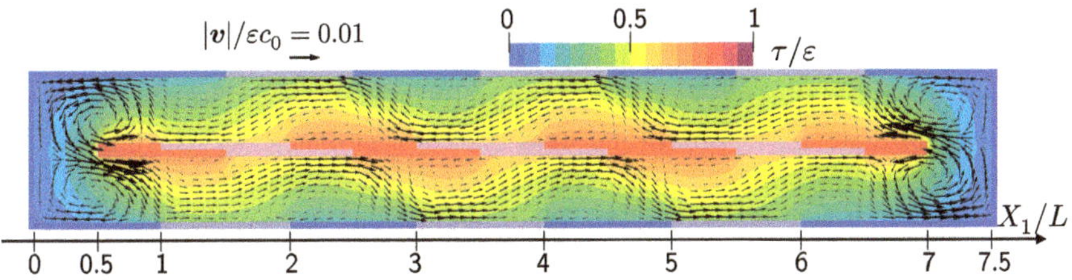

Fig. 5 Temperature and flow velocity distribution for the spanwise edge setup (Fig. 1b). Each unit corresponds to $\Lambda/L = 0.5$, $\alpha = 0.5$, $\varepsilon = 10^{-4}$, Kn $= 1$ and $L/D = 2$ shown in Fig. 3b

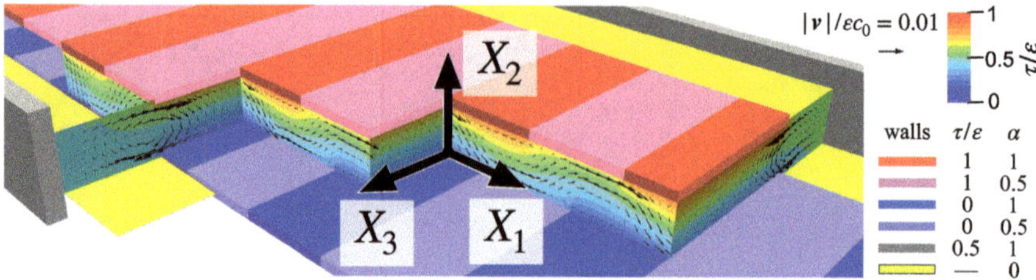

Fig. 6 Projections of the flow velocity and temperature distribution for the spanwise edge setup (Fig. 1c.1) on several cross-sections. The plates correspond to the basic setup Fig. 1a for $\Lambda/L = 0.5$, $\alpha = 0.5$, $\varepsilon = 10^{-4}$, Kn = 0.5 and $L/D = 2$

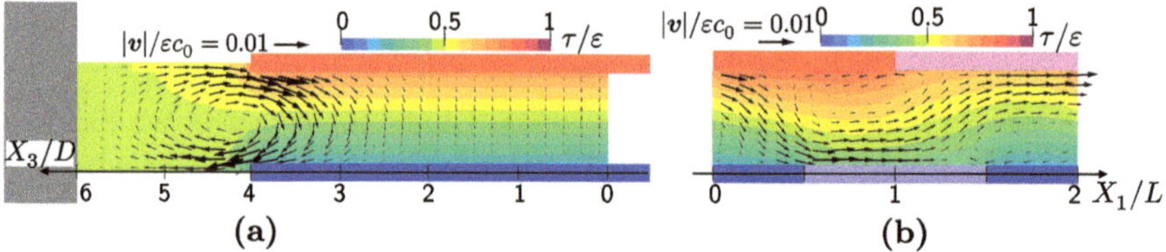

Fig. 7 Projections of the flow velocity and temperature distributions for the spanwise edge setup (Fig. 1c.1) on two cross-sections. The parameters are the same as in Fig. 6. **a** The spanwise cross-section $X_1 = 0$. **b** The streamwise cross-section at the center of the plates $X_3 = 0$

etrate the gap between hot and cold plates. The cross-section image at $X_3 = 0$ is panel (b). The one-way flow similar to that in Fig. 3b occurs here. To summarize, the spanwise edge induces a strong thermal edge vortex but does not stop the one-way flow at the central part. Therefore, we can expect the occurrence of the one-way flow if the plate has a sufficiently large width in the spanwise direction.

6 Summary

We report numerical research results on the feasibility of the one-way flow [5] between flat plates of different temperatures induced by the periodic distribution of the accommodation coefficient. The BGK model gives almost the same qualitative results as the hard-sphere molecules. The nondimensional mass flux (i.e., the flow speed) through the gap reaches its maximum value when the period of the accommodation coefficient distribution is 4–8 times the distance between the plates. The mean free path of the molecule should ideally be the same as the plate distance. We also investigate the effects of the finite size of real plates. Although the edges of the plates induce strong thermal edge flow and vortices, their effects are confined to some narrow areas around the edge. A sufficient width of the one-way flow device can weaken the influence of thermal edge flow. To summarize, the one-way flow between the flat plates with the periodic distribution of the accommodation coefficient in real devices is highly feasible, judging from the analysis based on the Maxwellian boundary condition.

Acknowledgements　This work is supported by the JSPS grant-in-aid 24K07303.

References

1. LeVeque RJ (2002) Finite volume methods for hyperbolic problems. Cambridge texts in applied mathematics. Cambridge University Press, Cambridge
2. Morikawa Y, Sugimoto H (2021) A new fast numerical method for rarefied gas simulation by spherically designed grid and traveling finite volume (in Japanese). In: Nagare, vol 40, pp 394–397
3. Sone Y (2007) Molecular gas dynamics. Birkhäuser
4. Sone Y, Yoshimoto M (1997) Demonstration of a rarefied gas flow induced near the edge of a uniformly heated plate. Phys Fluids 9(11):3530–3534. https://doi.org/10.1063/1.869461
5. Sugimoto S, Sugimoto H (2022) Thermal transpiration flows induced by differences in accommodation coefficients. Phys Fluids 34(4):042005. https://doi.org/10.1063/5.0084455
6. Womersley RS (2018) Efficient spherical designs with good geometric properties. Springer International Publishing, Cham, pp 1243–1285. https://doi.org/10.1007/978-3-319-72456-0_57

Simulation of the Neutral Gas Flow Within an Electric Thruster by the Kinetic Fokker-Planck Method and Comparison with Force Balance Measurements

Jens Schmidt and Leo Basov

Abstract This study investigates the neutral gas flow in the DEEVA electric thruster, developed within DLR. The thruster uses electron-cyclotron-resonance (ECR) discharge and a magnetic nozzle to accelerate plasma, supporting noble and reactive gases as propellants. Simulations using the Fokker-Planck method in SPARTA were conducted for argon and xenon at different flow rates to understand asymmetries in the neutral gas plume. Results showed that the original injector design caused plume asymmetry due to pressure gradients, while a refined injector produced a symmetric plume. Experimental cold gas thrust measurements partially matched simulation results and analytical models, though deviations occurred for xenon at higher flow rates. Findings support injector optimization and provide initial conditions for plasma simulations, improving discharge modeling and thruster performance. Future work will focus on density measurements and studies with diatomic propellants.

Keywords Rarefied gas dynamics · Fokker-Planck-method · DSMC · Electric propulsion

1 Introduction

Within the German Aerospace Center's (DLR) *Decentralized Energy supplied Electric Propulsion* (DEEP) project, a microwave-heated electric thruster is developed [4]. The plasma is created using an electron-cyclotron-resonance (ECR) discharge and accelerated using a magnetic nozzle. This allows the usage of different propellants such as noble gases but will also allow the use of reactive gases like air. As a first step to calculate the plasma properties and understand the neutral gas distri-

J. Schmidt (✉) · L. Basov
German Aerospace Center, Institute of Aerodynamics and Flow Technology, Göttingen, Germany
e-mail: jens.schmidt@dlr.de

L. Basov
Applied and Computational Mathematics, RWTH Aachen, Aachen, Germany

© The Author(s) 2026

M. Grabe et al. (eds.), *Rarefied Gas Dynamics*, Springer Aerospace Technology,
https://doi.org/10.1007/978-3-032-00094-1_55

bution within the discharge tube, a Fokker-Planck simulation of the the neutral gas flow is performed. This is motivated by an observed asymmetry of the plasma plume during operation which can not be explained by any effects on the charged particles. Since the power coupling efficiency of the ECR discharge and the confinement of the plasma depend highly on the local number density, the channel flow and the plume of the thruster should initially be simulated for neutral gas. Therefore a simulation was performed to understand the neutral gas distribution and how the injector design will influence it. In addition, simulations are performed with a different injector design. The simulations are performed for argon and xenon to understand the influence of the molecular weight on the neutral gas distribution. To validate the results of the simulation, the thrust is calculated from the flow and compared with experimental measurements of the cold gas thrust as well as estimates from an analytical model. The simulation is performed for argon and xenon at three different volume flow rates of $\dot{V} = 1, 10, 45$ sccm which are representative for the thruster operation and ignition.

2   Description of the DEEVA Thruster

The DEEVA (DLR Electrodeless ECR Via magnetic nozzle Acceleration) thruster, as shown schematically in Fig. 1 consists of 4 major components: the gas injector, the discharge tube, the slotted antenna and the magnet system.   The neutral gas is injected into the discharge tube by the gas injector and then ionized due to the microwaves emitted by the two slits of the slotted antenna. Due do the magnetic field, the electrons perform circular motions around the magnetic field line. The electrons can be heated in resonance if they are excited at the same frequency at which they are gyrating around the magnetic field lines, which is dependent on the local magnetic field strength. The thruster is usually operated at a frequency of $f = 2.4 - 2.5$ GHz leading to a necessary magnetic field strength of $B = 0.875$ T.

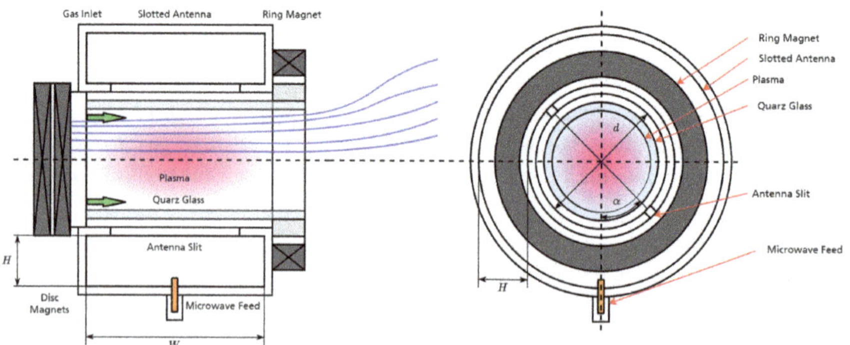

Fig. 1   Scheme of the DEEVA thruster as seen from the side (left) and front (right)

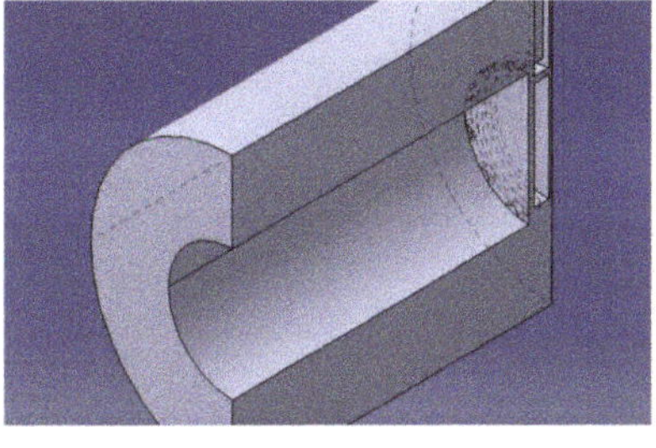

(a) Old gas injector geometry

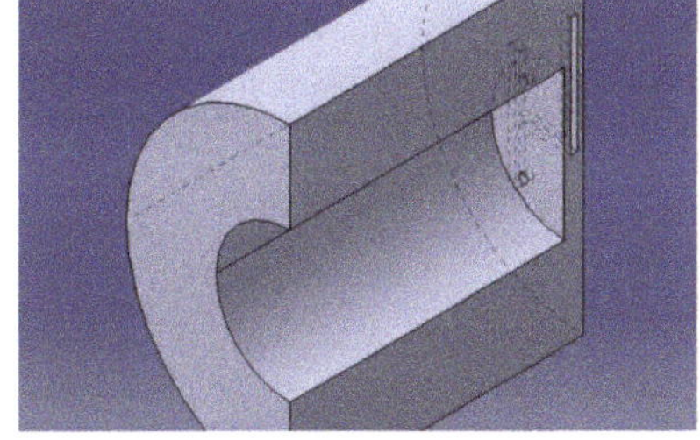

(b) New gas injector geometry

Fig. 2 CAD model of the simulated geometries for the old (left) and new (right) gas injector design and discharge channel

Due to the magnetic field configuration, the plasma is then confined to reduce wall losses and expanded along the divergent magnetic field lines utilizing the field of the magnetic nozzle. Since the electric field strength of the microwave is stronger at the outer edge of the discharge tube closer to the slits, the number density of the neutral gas at the outer edge of the tube should be higher. In addition, the neutral gas distribution should be symmetric to not introduce any asymmetries into the plasma plume. This can be ensured by a proper design of the neutral gas injection which is studied numerically and experimentally. For this study, two injector geometries were compared numerically, both shown in Fig. 2, furthermore referred to as 'old' and 'new'. In the 'old' injector geometry, the gas is fed from the top and flows along the outer radius to the injector holes as seen in Fig. 2a. In the 'new' geometry, the gas first flows from the top to the center and then is spread radially in the four injector openings, resulting in a more equal pressure distribution.

3 Numerical Method

The simulations were performed using the kinetic cubic Fokker-Planck (FP) model [2]. The method is being continuously developed at DLR Göttigen as an extension to the open-source DSMC code SPARTA by Sandia National Labs [1, 3]. It is important to note that this is an extension of the code, not the DSMC method. Unlike Direct-Simulation Monte Carlo (DMSC), the FP method does not resolve the particle-particle interactions of the Boltzmann collision operator directly. Rather it uses a continuous stochastic process which leads to the computational cost being independent of the Knudsen number Kn. This allows the method to be computationally more efficient than DSMC in areas of low Kn numbers like in the inflow structure used in the thruster under investigation. The signal-to-noise ratio of both methods should be similar, as the noise scales with the inverse of the square root of the particle number. The FP method is necessary, as there is a strong pressure (and density) gradient within the discharge channel due to the gas injection at almost

Table 1 Boundary conditions and species properties from bird [1]

Boundary	Condition		
1–4	Outflow		
5–7	Inflow		
8	Wall (Diffuse Reflection with $T_w = 273.15$ K)		
Gas	m/kg	d_{ref}/m	ω
Argon	$6.63 \cdot 10^{-26}$	$3.657 \cdot 10^{-10}$	0.81
Xenon	$218.0 \cdot 10^{-27}$	$5.74 \cdot 10^{-10}$	0.85

atmospheric pressure and the expansion into high vacuum. The Knudsen number is $Kn \approx 10^{-5}$ in the discharge tube, and $Kn \approx 1$ in the plume, therefore there is a transition from a collisional into the collisionless regime, which makes a method necessary which is able to work in both regimes. The simulation was performed for Argon and Xenon as a neutral gas at different given volume flow rates $\dot{V}$ representing different operational points. The volume flow rates used were 1 sccm, 10 sccm and 45 sccm. The simulation was performed using only one half of the geometry to save computational time, since the geometry is symmetric along the middle axis. The downstream boundaries as well as the outside of the simulation domain were defined as outflow. The inflow in the injector was calculated using the given volume flow and calculating the number density in the inlet surface for a assumed temperature of $T = 300$ K. Variable hard sphere (VHS) parameters of the used species required for the simulation as well as the boundary conditions are given in Table 1. The resulting thrust force can be found by integrating over the local density $\rho(\vec{x})$ and velocity $\vec{v}(\vec{x})$ at the position $\vec{x}$ of the inlet and outlet of the control volume shown in Fig. 3 by

$$F = \oiint \rho(\vec{x})\vec{v}(\vec{x})\vec{v}(\vec{x})dS = \dot{m} \oiint \vec{v}(\vec{x})dS \tag{1}$$

to consider non-uniformities of the flow that might occur due to an asymmetry of the gas injection.

4 Experimental Setup

To confirm the numerical results, cold gas thrust measurements of the DEEVA thruster with argon and xenon were performed using DLRs DEPB thrust balance within the STG-MT vacuum chamber, which has previously been tested with other thrusters[5]. The thrust balance features quartz glass rods as load-bearing springs to minimizhe thermal drift, a Sartorius® WZA224-N load cell for precision, and an eddy current brake to dampen oscillations. The calibration is performed using fine

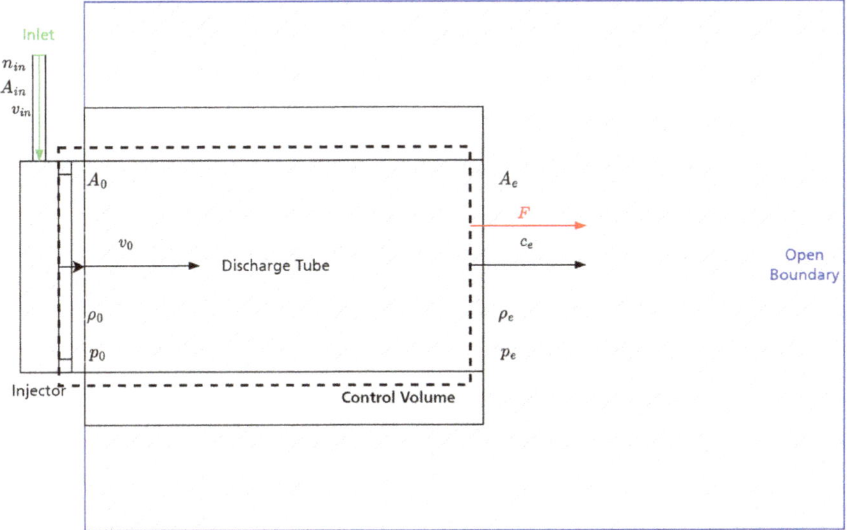

Fig. 3 Control volume used for the simulation and calculation of the thrust and boundaries. Boundary 4 is not shown. The area filled with hatches indicates the simulation domain

weights to ensure accurate and repeatable thrust measurements. This will lead to further understand the influence of the injector geometry and assist in optimization of the neutral flow injection into the discharge channel.

5 Analytical Model

The thrust is calculated with the analytical model under the assumption that the flow in the tube is subsonic. The exit velocity c_e can then be calculated from the isentropic relation

$$c_e = \sqrt{2c_p(T_e - T_0)} = \sqrt{\frac{\kappa}{\kappa - 1}(1 - (p_a/p_0)^{\frac{\kappa-1}{\kappa}}}$$ (2)

In which the isentropic coefficient was assumed to be a constant at $\kappa = 1.67$ for noble gases, the reservoir pressure was assumed to be $p_0 = 150000$ Pa and the ambient pressure was assumed to be $p_a = 0.001$ Pa, while the gas in the reservoir is assumed to be at room temperature of $T_0 = 300$ K. The thrust can then be calculated from

$$F = \dot{m}c_e$$ (3)

While this model is very simple, it can give a first-order estimate of the expected neutral gas thrust. The mass flow $\dot{m} = \rho\dot{V}$ was calculated from the volume flow $\dot{V}$ under the assumption of a ideal gas at a temperature of $T_0 = 300$ K to calculate the density.

6 Results

The simulation revealed details about the neutral gas flow within the discharge tube. As it can be seen for the results with the old geometry, in Fig. 4b, there is a pressure gradient in the gas injector as the gas flow is injected from the top and distributing within the radial channel. Since the pressure is higher at the top inlet ports, the flow velocity and density in the ports is also significantly higher. Due to the collisions downstream in the discharge channel, there is also a gradient in the exiting neutral gas plume, resulting in the observed asymmetry of the plume. With the new injector geometry and a central feeding of the gas into the discharge channel, the plume is symmetric (Fig. 5). In addition, it was observed for both the simulation results and the experimental measurements, that the cold gas thrust increases with the volume flow rate. The comparison of these results for argon and xenon with the analytical model are shown in Fig. 6, where the cold gas thrust F as calculated from the analytical model (circles), the measured experimental results (diamonds) and the calculated numerical results (squares) are shown. As it can be seen for argon in Fig. 6a, the agreement between the analytical model and the experimental and simulated results is quite good for argon and within the measurement uncertainty of the thrust measurements. For xenon, there are more differences observed. The analytical model increases linear with the volume flow rate, while for the experiment and simulation a logarithmic behaviour is observed. The results from the simulation and the experiment are within the measurement uncertainty for the lower flow rates but disagree for a very high flow rate. For the analytical model by the linear behaviour and the one-dimensional calculation not modelling all physics correctly. For the lower flow rates, this difference is significant enough. The disagreement between experiment and simulation for the higher flow rate can not be explained at the moment, except that Xenon at low temperatures due to the expansion might not be modelled correctly (Fig. 7).

7 Conclusion

Within this work, the neutral gas flow within an electric thruster was simulated using the Fokker-Planck-method extension in SPARTA. Comparative measurements of the cold gas thrust were performed for argon and xenon. The results will allow to improve the design of the gas injector and the modelling of the plasma discharge itself, since the structure of the magnetic field and microwave radiation are known from independent measurements and simulations. Therefore, the knowledge of the local number density can directly assist in the improvement of the simple discharge models. Additionally, the results for the inflow from the injector can be used as an initial condition for particle-in-cell simulations. In addition, it was observed that the improved gas injector will produce a homogeneous distribution of the neutral gas in the plume. The comparison with the experimental results showed good agreement within the mea-

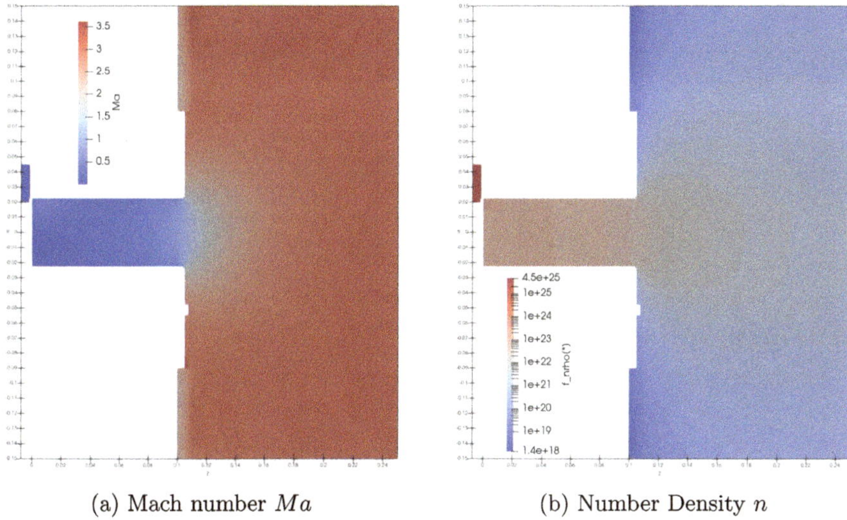

(a) Mach number Ma (b) Number Density n

Fig. 4 Mach number and number density in the old geometry for argon at a flowrate of $\dot{V} = 1$ sccm

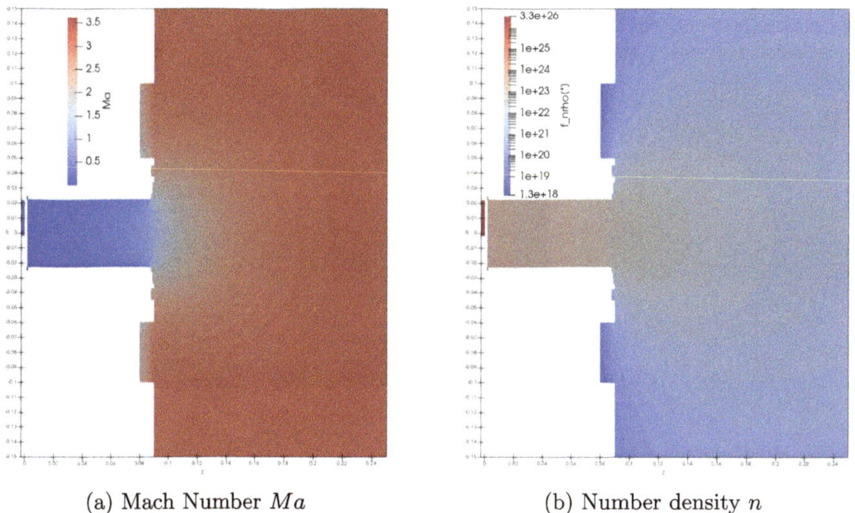

(a) Mach Number Ma (b) Number density n

Fig. 5 Mach number and number density in the new geometry for argon at a flowrate of $\dot{V} = 1$ sccm

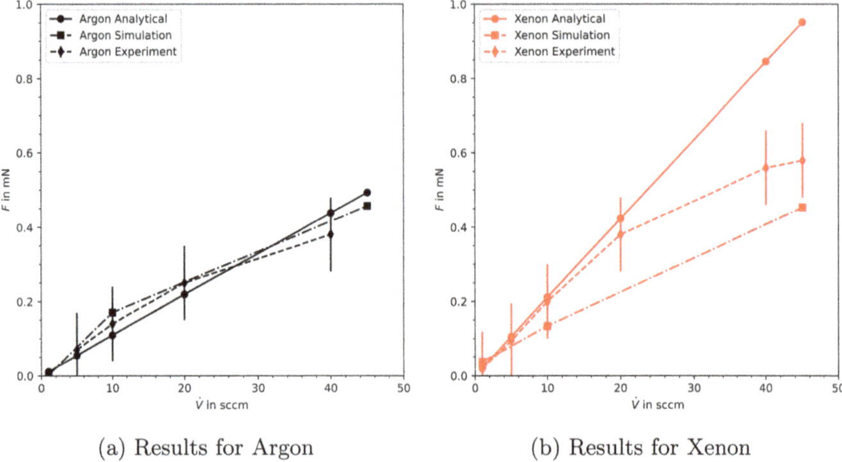

(a) Results for Argon   (b) Results for Xenon

Fig. 6 Comparison of the cold gas thrust measurements (diamonds) with the analytical model (circles) and the numerical results (squares)

Fig. 7 Thruster plume during operation with argon at a volume flow rate of 2 sccm and a microwave power of $P_{MW} = 60$ W

surement uncertainty of the thrust balance for for argon, but disagreement for xenon. In future, the number density in the plume should be measured using a Patterson probe to understand the neutral gas distribution and compare it with the simulation results. In addition, more operation points shall be studied at a lower measurement uncertainty and for diatomic or multispecies propellants.

References

1. Bird GA (1994) Molecular gas dynamics and the direct simulation of gas flows. Oxford University Press
2. Gorji MH, Torrilhon M, Jenny P (2011) Fokker-Planck model for computational studies of monatomic rarefied gas flows. J Fluid Mech 680:574–601
3. Plimpton SJ, Moore SG, Borner A, Stagg AK, Koehler TP, Torczynski JR, Gallis MA (2019) Direct simulation Monte Carlo on petaflop supercomputers and beyond. Phys Fluids 31(8):086101
4. Schmidt J, Simon J, Grabe M (2022) The DEEP thruster concept. In: IEPC (ed) 37th international electric propulsion conference, Boston, pp 1–9. https://elib.dlr.de/148314/
5. Schmidt J, Simon J, Neumann A (2023) Thrust measurements of the RIT-10 thruster with a new type of thrust balance. Int J Energetic Mater Chem Propulsion 22(4) (Begel House Inc)

DSMC: A Statistical Mechanics Perspective

Alejandro L. Garcia

Abstract This paper presents a perspective in which Direct Simulation Monte Carlo (DSMC) is viewed not in its traditional role as an algorithm for solving the Boltzmann equation but as a numerical method for statistical mechanics. First, analytical techniques such as the collision virial and Green-Kubo relations, commonly used in molecular dynamics, are used to study the numerical properties of the DSMC algorithm. The stochastic aspect of DSMC, which is often viewed as unwanted numerical noise, is shown to be a useful feature for problems in statistical physics, such as Brownian motion and thermodynamic fluctuations. Finally, it is argued that fundamental results from statistical mechanics can provide guardrails when applying machine learning to DSMC.

Keywords Direct simulation Monte Carlo · Statistical mechanics · Molecular dynamics · Kinetic theory · Fluctuations

1 MD and DSMC—Brothers Separated at Birth

Three scientific eras dawned in the middle of the last century: the atomic age, the space age, and the age of the computer. Numerical algorithms were developed for the study of fluids under extreme conditions, such as nuclear explosions and orbital reentry, that were beyond the realm of conventional laboratory experiments. An early example is the Metropolis Monte Carlo method, which in 1953 showed that a particle representation could be used to map phase space and using statistical mechanics extract a fluid's thermodynamic properties, such as the equation of state [39]. It was closely followed by molecular dynamics (MD), introduced in 1959 by Alder and Wainwright, which uses a similar particle representation but includes dynamics, allowing it to measure transport properties such as viscosity [1].

A. L. Garcia (✉)
San Jose State University, San Jose, CA, USA
e-mail: Alejandro.Garcia@sjsu.edu
URL: http://www.algarcia.org

© The Author(s) 2026

M. Grabe et al. (eds.), *Rarefied Gas Dynamics*, Springer Aerospace Technology,
https://doi.org/10.1007/978-3-032-00094-1_56

In 1963, Graeme Bird submitted to *Physics of Fluids* a paper entitled "Approach to Translational Equilibrium in a Rigid Sphere Gas." In this article he introduced a stochastic collision algorithm that would become the cornerstone of Direct Simulation Monte Carlo (DSMC). The paper was rejected (G.A. Bird, personal communication). The referee had failed to recognize that DSMC differed significantly from MD; after revising it to explain these differences, Bird's paper was accepted [12].

DSMC went on to become the dominant numerical method in aerospace engineering for rarefied gas dynamics while MD established popularity in the field of statistical mechanics for the simulation of microscopic systems. For many years, these communities remained separate, yet recently they have found common interests in topics such as nanofluidics and granular flows. This paper presents several short stories in which DSMC is used by researchers, such as myself, whose background is in statistical mechanics and molecular dynamics. As we shall see, by approaching DSMC from an unconventional perspective, a number of new avenues are opened. This narrative is intentionally personal, since Graeme Bird, to whom this paper is dedicated, was a colleague and friend for nearly 30 years. In the interest of space, a summary of the DSMC algorithm is omitted; see [4] for an introduction and [14, 15, 20] for full treatments.

2 But Why Ideal Gas Law?

In 1989 I joined the faculty at San Jose State University, located in Silicon Valley. About that time I met Berni Alder and started working with him at the nearby Livermore Lab. Alder took an immediate interest in DSMC and we eventually published 16 papers together, most of them on ways to extend the DSMC algorithm. Yet when I first explained DSMC to Berni, he was puzzled and asked "But why does it give the ideal gas law?" Graeme Bird did not address this question in his books on DSMC [13–15] since he always viewed the algorithm as numerically solving the Boltzmann equation, a perspective that was reinforced by a proof by Wagner [48].

In their seminal MD simulations Alder and Wainwright [2] measured the equation of state for hard spheres (mass m, diameter d) as

$$p = nkT + \tfrac{1}{3}m\Gamma\Theta \tag{1}$$

where p, n, T are pressure, number density, and temperature; k is the Boltzmann constant. The collision rate is Γ and

$$\Theta = \langle \Delta\mathbf{v}_\alpha \cdot \mathbf{r}_{\alpha\beta} \rangle \geq 0 \tag{2}$$

is the virial. For a collision between particles α and β, $\Delta\mathbf{v}_\alpha$ is the change in velocity for particle α and $\mathbf{r}_{\alpha\beta} = \mathbf{r}_\alpha - \mathbf{r}_\beta$ is their separation. The virial is an average over collisions and depends on the particle diameter since $|\mathbf{r}_{\alpha\beta}| = d$. In (1) the pressure is the sum of translational and collisional contributions to momentum transfer.

In DSMC collisions occur within cells and the probability of a collision is *independent* of the positions of the particles, so $\langle \mathbf{r}_{\alpha\beta} \rangle = 0$ and $\langle \Delta \mathbf{v}_\alpha \cdot \mathbf{r}_{\alpha\beta} \rangle = \langle \Delta \mathbf{v}_\alpha \rangle \cdot \langle \mathbf{r}_{\alpha\beta} \rangle = 0$ so $\Theta = 0$. Since the virial contribution is zero, the equation of state for DSMC is the ideal gas law. Note that Wagner's proof is for infinitesimal collision cells yet we have the ideal gas law even for finite cell size due to symmetry in the selection of collision partners in DSMC.

Alder found this unsatisfactory, so, together with Frank Alexander, we formulated the Consistent Boltzmann Algorithm (CBA) [5]. It is identical to DSMC except that collisions change both the velocities *and* the positions of colliding particles. Specifically, after computing the post-collision velocities the positions are shifted as

$$\mathbf{r}_\alpha \Rightarrow \mathbf{r}_\alpha + d\,\hat{\mathbf{a}}_\alpha \quad ; \quad \mathbf{r}_\beta \Rightarrow \mathbf{r}_\beta - d\,\hat{\mathbf{a}}_\alpha \tag{3}$$

where $\hat{\mathbf{a}}_\alpha = \Delta \mathbf{v}_\alpha / |\Delta \mathbf{v}_\alpha|$ is a unit vector in the direction of the apse line (line between centers) for a hard sphere collision with this change in velocity. Simulations verified that CBA gives the correct pressure and reasonably accurate transport coefficients up to moderately high densities.

The Consistent Universal Boltzmann Algorithm (CUBA) makes the post-collision displacement a function of density and temperature, which allows one to choose the desired equation of state [6, 34]. For example, by choosing the van der Waals equation, we were able to simulate the condensation of vapor into liquid droplets. However, for simulating liquids, we found that CUBA was not competitive with MD in terms of computational cost.

Today, the most popular variants of DSMC for dense gases are based on the Enskog equation [27, 40]. Instead of using a CBA displacement, the correct virial Θ can be obtained by using a collision probability that depends on both the relative velocity and the relative position for a collision pair [25]. In other words, the Enskog perspective is that particle positions affect collisions, while in CBA, collisions affect particle positions.

3 A Lesson on Transport from Molecular Dynamics

In his first book, Graeme Bird said of DSMC, "the results are remarkably insensitive …and no deleterious effects are generally present …if the cell size approaches the mean free path or if the time step approaches the mean collision time" [13]. Nevertheless, the demonstration program in this book for the Rayleigh problem sets the collision cell size to half a mean free path and the time step to a fifth of a mean free collision time.

A common way of quantifying the error due to cell size and time step is to measure viscosity or thermal conductivity and, at low Knudsen number, compare with Chapman-Enskog theory. These transport coefficients can be obtained by measuring the momentum (or heat) flux in a DSMC simulation with a linear velocity (or temperature) gradient [19]. From the perspective of non-equilibrium thermodynamics, the transport coefficient is an Onsager coefficient that relates the thermodynamic flux to the thermodynamic force due to the gradient [24, 41].

Yet in the early days of molecular dynamics, such non-equilibrium measurements were not feasible since simulations were limited to a few hundred particles. Instead, transport properties were measured in simulations of equilibrium systems using results from statistical mechanics.

The basic idea is to apply the fluctuation-dissipation theorem, which connects statistical fluctuations to the rate of dissipation. The best known example is Brownian motion in which the mobility, μ, of a particle is obtained from measuring the variance of its position in time as

$$\mu = \frac{1}{6kT} \lim_{t \to \infty} \frac{\langle |\mathbf{r}(t) - \mathbf{r}(0)|^2 \rangle}{t} \tag{4}$$

By contrast, a non-equilibrium measurement obtains the mobility $\mu = v_T/F_{\text{ext}}$ from the terminal velocity, v_T, for the particle under an external force of magnitude F_{ext}.

Using Einstein-Helfand theory one can measure the viscosity of a fluid from molecular trajectories as [3]

$$\eta = \frac{1}{2VkT} \lim_{t \to \infty} \frac{\langle [G(t) - G(0)]^2 \rangle}{t} \quad \text{with} \quad G(t) = m \sum_{i=1}^{N} \dot{x}_i(t) y_i(t) \tag{5}$$

where N is the number of molecules, V is the system volume, x_i, y_i are components of $\mathbf{r}_i$, and $\dot{x} \equiv dx/dt$. This allows us to measure the viscosity (and other transport coefficients) from *equilibrium* simulations. A related approach uses the off-diagonal stress tensor component,

$$J(t) = \dot{G}(t) = m \sum_{i=1}^{N} \dot{x}_i(t) \dot{y}_i(t) + \frac{1}{2} \sum_{i=1}^{N} \sum_{i=j}^{N} F_{ij}^x (y_i - y_j) \tag{6}$$

where $\mathbf{F}_{ij}$ is the force between molecules. Our previous result now takes the more familiar form from Green-Kubo theory,

$$\eta = \frac{1}{VkT} \int_0^\infty \langle J(t) J(0) \rangle \, dt \tag{7}$$

Notice from (6) that J is the sum of a kinetic (ballistic) contribution and a contribution due to the transfer of momentum by intermolecular forces.

For hard sphere collisions Wainwright wrote [49]

$$J(t) = m \sum_{i=1}^{N} \dot{x}_i(t) \dot{y}_i(t) + \sum_{c}^{\text{collisions}} \frac{m}{d^2} (\Delta \mathbf{v}_\alpha \cdot \mathbf{r}_{\alpha\beta})(x_\alpha - x_\beta)(y_\alpha - y_\beta) \delta(t - t_c) \tag{8}$$

where t_c is the time of a collision. He was able to evaluate (7) and express it as the sum of three contributions, $\eta = \eta^K + \eta^{K\times C} + \eta^C$. The kinetic contribution η^K is precisely the Chapman-Enskog viscosity. The cross term $\eta^{K\times C}$ and the collision term η^C are dense gas corrections.

In DSMC, the collision term η^C is *not* zero due to the finite distance between the collision partners (i.e., the finite collision cell size). Unlike the equation of state contribution due to the virial (see (2)), this contribution to the viscosity does not average to zero over DSMC collisions within a cell. In fact, following a similar analysis as Wainwright one finds

$$\eta^C = \frac{m^2 \Gamma}{2kT} \left\langle (y_\alpha - y_\beta)^2 \right\rangle \left\langle (\Delta \dot{x}_\alpha - \Delta \dot{x}_\beta)^2 \right\rangle \tag{9}$$

For cubic collision cells of size ℓ the average $\left\langle (y_\alpha - y_\beta)^2 \right\rangle = \ell^2/6$; at equilibrium $\left\langle (\Delta \dot{x}_\alpha - \Delta \dot{x}_\beta)^2 \right\rangle = 4kT/3m$. In DSMC the cross term $\eta^{K\times C}$ is zero and so the hard sphere viscosity is [7]

$$\eta = \eta^K + \eta^C = \frac{5}{16d^2} \sqrt{\frac{mkT}{\pi}} \left(1 + \frac{16}{45\pi} \frac{\ell^2}{\lambda^2} \right) \tag{10}$$

where λ is the mean free path. Note that the error goes as ℓ^2; for a cell size of one mean free path the viscosity deviates from its Chapmann-Enskog value by 11%. A similar analysis (with a similar result) may be applied to obtain the error in thermal conductivity due to finite cell size. Finally, this approach also gives the error in the transport coefficients due to a finite time step τ; in DSMC this error goes as τ^2 [31, 33].

In his last book Bird wrote "(In) early DSMC programs …the mean spacing between collision pairs was too large in comparison with the mean free path and this led to the introduction of sub-cells …(and) an option to select the collision pairs from the nearest-neighbour pairs within the cell" [15]. He also mentions that "The time step is set to a specified small fraction of the sampled mean collision time." A challenge for modern DSMC codes is to minimize the distance between colliding particles and to minimize the time between ballistic movement steps while avoiding unphysical bias errors (see Sect. 5).

4 Fluctuations—One Man's Garbage …

My introduction to DSMC came in 1983 as I was finishing my doctoral studies at The University of Texas at Austin in what is now The Ilya Prigogine Center for Studies in Statistical Mechanics and Complex Systems. At that time there were theoretical results, confirmed by light scattering experiments, indicating that hydrodynamic fluctuations had weak, long-ranged correlations in non-equilibrium systems (e.g., under a temperature gradient).

In his first book, Bird describes fluctuations in DSMC as an unavoidable annoyance but that with a sufficiently large sample size the desired mean value can be measured to the desired accuracy [13]. One commonly held misconception was that these fluctuations came from the Monte Carlo nature of the collision algorithm in DSMC. Yet the very same fluctuations are also found in molecular dynamics, which has deterministic dynamics.

Due to computer memory constraints, the number of particles (or "simulators") in a DSMC calculation is typically a small fraction of the number of physical molecules. For example, at standard temperature and pressure, the mean free path in air is roughly 60 nanometers and a cubic mean free path contains about 6000 molecules. In contrast, a typical DSMC simulation uses 20 particles per cubic mean free path, that is, each DSMC particle represents $F_N = 300$ physical molecules. This number scales as the Knudsen number so it is much higher for rarefied flows.

Statistical mechanics gives us the standard deviation of hydrodynamic variables, allowing us to estimate the expected statistical standard error in particle simulations [35]. For example, in a cell with N_c particles, the variance in each component of fluid velocity is $\langle \delta u_x^2 \rangle = kT/mN_c$ [36]. For both MD and DSMC the fractional error in estimating u_x from S independent statistical samples is

$$E = \frac{\sqrt{\langle \delta u_x^2 \rangle}/\sqrt{S}}{|u_x|} \approx \frac{1}{\sqrt{SN_c}} \frac{1}{\mathrm{Ma}} \tag{11}$$

where Ma is the Mach number of the flow. This result illustrates why nanofluidic flows, in which Ma $\ll 1$, are much harder to measure in DSMC compared to hypersonic flows. In DSMC N_c is smaller than the number of physical molecules by a factor of F_N, the variance in fluid velocity increases by the same factor.

After accounting for this amplification factor, we find that hydrodynamic fluctuations in DSMC are physically correct [30] and yield hydrodynamic information. For example, from the time correlation of density fluctuations one may measure the sound speed and viscosity from the location and width of the Brillouin peak, mimicking the technique used by light scattering [10, 21]. This dynamic structure factor is a useful way to compare DSMC models (e.g. internal energy relaxation) with laboratory measurements and theoretical predictions [22].

The first numerical observations of long-ranged correlations of hydrodynamic fluctuations were made using DSMC for systems with a temperature gradient [37] and a velocity gradient [29]. The "giant fluctuation" phenomenon, first reported in diffusive mixing experiments [46], is due to such non-equilibrium correlations for fluctuations of concentration and fluid velocity, as confirmed by DSMC [26]. Most recently DSMC simulations have been used to study the impact of thermal fluctuations on turbulence [28, 38]. They confirm theoretical predictions [9] and numerical fluctuating hydrodynamic observations [11] that molecular fluctuations *dominate* the turbulent energy cascade at the Kolmogorov length scale.

Bird wrote in his last book, "While the fluctuations are unphysical when F_N is large, they are physically realistic ...(with) a one-to-one correspondence between real and simulated molecules. This is another instance of DSMC going beyond the Boltzmann equation because fluctuations are neglected in the Boltzmann model." His last three papers were on Brownian motion [16–18].

5 Exorcising the Demons

In 1887 Maxwell presented a thought experiment, "...conceive of a being whose faculties are so sharpened that he can follow every molecule in its course ...so as to allow only the swifter molecules to pass from (chamber) A to B, and only the slower molecules to pass from B to A. He will thus, without expenditure of work, raise the temperature of B and lower that of A, in contradiction to the second law of thermodynamics." The DSMC algorithm has full information and control of the particle's dynamics and modern implementations have many complex stages. How can we be sure that we do not have a Maxwell demon hiding in our simulations? In other words, how do we test that our simulations satisfy the Second Law of Thermodynamics? Boltzmann's H-theorem is not helpful, since, due to fluctuations, it only applies in the limit $N \to \infty$.

Consider an isolated system with microstate dynamics [44]

$$\dot{p}_i = \sum_{j}^{\text{states}} (p_j W_{ji} - p_i W_{ij}) \tag{12}$$

where p_i is the probability of state i and W_{ij} is the transition rate from i to j. At equilibrium $\dot{p}_i = 0$. If $W_{ij} = W_{ji}$, then we have *microscopic reversibility*. At equilibrium we have *detailed balance* ($p_j W_{ji} = p_i W_{ij}$) and *ergodicity* ($p_i = p_j$) if and only if we have microscopic reversibility. Finally, the dynamics satisfies the Second Law ($\dot{S} = -k \sum_j \dot{p}_j \ln p_j \geq 0$) if and only if we have ergodicity [44].

Detailed balance is a necessary condition for microscopic reversibility, which is a sufficient condition for the dynamics to obey the Second Law. Yet, it is not uncommon to find DSMC implementations that do *not* satisfy detailed balance. For example, some of the early surface scattering models were ad hoc empirical fits to experimental data, which led Cercignani and Lampis to formulate a model that imposed detailed balance [23]. Many implementations of inflow/outflow boundaries also fail to satisfy detailed balance [45]. If machine learning methods are used to handle some elements in a DSMC calculation, such as chemical reactions and internal degrees of freedom, we may not know if detailed balance is satisfied [8]. And even when the algorithm in a DSMC code is theoretically sound, there is always the possibility that the implementation has a 'bug' (coding error).

In testing for the presence of Maxwell demons in a DSMC code, equilibrium hydrodynamic fluctuations can serve as a coal mine canary. For example, at equilibrium, the number of particles in each cell is independently Poisson distributed, and this result is independent of the collision model, boundary conditions, etc. Statistical mechanics provides further relations for the variances and correlations of fluid velocity, temperature, internal energy, and other variables [32, 36]. While this is a necessary but not sufficient statistical test, it is a sensitive one that can reveal malevolent problems. Ensuring the correctness of simulation data is increasingly important as we enter the age of artificial intelligence. Given the AI systems' voracious appetite for such data, we must avoid having machines learn the wrong lessons.

6 Concluding Remarks

Graeme Bird and Berni Alder met for the first time at the 22nd Rarefied Gas Dynamics Symposium in Sydney, Australia, nearly 40 years after their original publications on DSMC and MD. In the 25 years since that meeting, we find increasing cooperation between the two algorithms' communities. One example is the study of rotational relaxation using all-atom molecular dynamics simulations combined with DSMC [47]. Yet perhaps the best example is the shared heritage of LAMPPS [43] and SPARTA [42], two of the most widely used codes for molecular dynamics and direct simulation Monte Carlo. Although separated at birth, MD and DSMC are now the patriarchs overseeing a large family of particle-based fluid algorithms.

Acknowledgements The author acknowledges support from the US Department of Energy, Office of Science, Office of Advanced Scientific Computing Research, Applied Mathematics Program under contract no. DE-AC02-05CH11231.

References

1. Alder BJ, Wainwright TE (1959) Studies in molecular dynamics. i. General method. J Chem Phys 31(2):459–466
2. Alder BJ, Wainwright TE (1960) Studies in molecular dynamics. ii. Behavior of a small number of elastic spheres. J Chem Phys 33(5):1439–1451
3. Alder B, Gass D, Wainwright T (1970) Studies in molecular dynamics. viii. The transport coefficients for a hard-sphere fluid. J Chem Phys 53(10):3813–3826
4. Alexander FJ, Garcia AL (1997) The direct simulation Monte Carlo method. Comput Phys 11(6):588–593
5. Alexander FJ, Garcia AL, Alder BJ (1995) A consistent Boltzmann algorithm. Phys Rev Lett 74(26):5212
6. Alexander FJ, Garcia AL, Alder BJ (1997) The consistent Boltzmann algorithm for the van der Waals equation of state. Phys A Statistical Mech Appl 240(1–2):196–201
7. Alexander FJ, Garcia AL, Alder BJ (1998) Cell size dependence of transport coefficients in stochastic particle algorithms. Phys Fluids 10(6):1540–1542
8. Ball ND, MacArt JF, Sirignano J (2024) Online optimisation of machine learning collision models to accelerate direct molecular simulation of rarefied gas flows. arXiv preprint arXiv:2411.13423

9. Bandak D, Goldenfeld N, Mailybaev AA, Eyink G (2022) Dissipation-range fluid turbulence and thermal noise. Phys Rev E 105(6):065113

10. Baras F, Mansour MM, Garcia A, Mareschal M (1995) Particle simulation of complex flows in dilute systems. J Comput Phys 119(1):94–104

11. Bell JB, Nonaka A, Garcia AL, Eyink G (2022) Thermal fluctuations in the dissipation range of homogeneous isotropic turbulence. J Fluid Mech 939:A12

12. Bird G (1963) Approach to translational equilibrium in a rigid sphere gas. Phys Fluids 6:1518–1519

13. Bird G (1976) Molecular gas dynamics. Oxford engineering science series, Clarendon Press. https://books.google.com/books?id=OAgpAQAAMAAJ

14. Bird G (1994) Molecular gas dynamics and the direct simulation of gas flows. No. v. 1 in molecular gas dynamics and the direct simulation of gas flows. Clarendon Press. https://books.google.com/books?id=Bya5QgAACAAJ

15. Bird G (2013) The DSMC method. CreateSpace Independent Publishing Platform. https://books.google.com/books?id=0YdnngEACAAJ

16. Bird G (2015) The rate of energy transfer from air as an initially stationary particle acquires Brownian motion. Nano Energy 11:463–470

17. Bird G (2016) Brownian relaxation of an inelastic sphere in air. Phys Fluids 28(6)

18. Bird G (2017) The possibility of harvesting useful energy from the thermal motion of air molecules. NanoWorld J 3(1):18–22

19. Bird G, Gallis M, Torczynski J, Rader D (2009) Accuracy and efficiency of the sophisticated direct simulation Monte Carlo algorithm for simulating noncontinuum gas flows. Phys Fluids 21(1)

20. Boyd I, Schwartzentruber T (2017) Nonequilibrium gas dynamics and molecular simulation. Cambridge Aerospace Series. Cambridge University Press. https://books.google.com/books?id=8PdFDgAAQBAJ

21. Bruno D, Frezzotti A, Ghiroldi GP (2017) Rayleigh-Brillouin scattering in molecular oxygen by CT-DSMC simulations. Eur J Mech-B/Fluids 64:8–16

22. Bruno D, Giovangigli V (2022) Internal energy relaxation processes and bulk viscosities in fluids. Fluids 7(11):356

23. Cercignani C, Lampis M (1971) Kinetic models for gas-surface interactions. Transport Theory Statistical Phys 1(2):101–114

24. De Groot S, Mazur P (2013) Non-equilibrium thermodynamics. Dover Books on Physics, Dover Publications. https://books.google.com/books?id=mfFyG9jfaMYC

25. Donev A, Alder BJ, Garcia AL (2008) Stochastic hard-sphere dynamics for hydrodynamics of nonideal fluids. Phys Rev Lett 101(7):075902

26. Donev A, Bell JB, de La Fuente A, Garcia AL (2011) Diffusive transport by thermal velocity fluctuations. Phys Rev Lett 106(20):204501

27. Frezzotti A (1997) A particle scheme for the numerical solution of the enskog equation. Phys Fluids 9(5):1329–1335

28. Gallis M, Bitter N, Koehler T, Torczynski J, Plimpton S, Papadakis G (2017) Molecular-level simulations of turbulence and its decay. Phys Rev Lett 118(6):064501

29. Garcia A, Mansour MM, Lie G, Clementi E (1987) Hydrodynamic fluctuations in a dilute gas under shear. Phys Rev A 36(9):4348

30. Garcia A, Penland C (1991) Fluctuating hydrodynamics and principal oscillation pattern analysis. J Statistical Phys 64:1121–1132

31. Garcia A, Wagner W (2000) Time step truncation error in direct simulation Monte Carlo. Phys Fluids 12(10):2621–2633

32. Garcia A (2007) Estimating hydrodynamic quantities in the presence of microscopic fluctuations. Commun Appl Math Comput Sci 1(1):53–78

33. Hadjiconstantinou NG (2000) Analysis of discretization in the direct simulation Monte Carlo. Phys Fluids 12(10):2634–2638

34. Hadjiconstantinou NG, Garcia AL, Alder BJ (2000) The surface properties of a van der Waals fluid. Phys A Statistical Mech Appl 281(1–4):337–347

35. Hadjiconstantinou NG, Garcia AL, Bazant MZ, He G (2003) Statistical error in particle simulations of hydrodynamic phenomena. J Comput Phys 187(1):274–297
36. Landau LD, Lifshitz EM (2013) Statistical physics, vol 5. Elsevier
37. Mansour MM, Garcia AL, Lie GC, Clementi E (1987) Fluctuating hydrodynamics in a dilute gas. Phys Rev Lett 58(9):874
38. McMullen RM, Krygier MC, Torczynski JR, Gallis MA (2022) Navier-stokes equations do not describe the smallest scales of turbulence in gases. Phys Rev Lett 128(11):114501
39. Metropolis N, Rosenbluth AW, Rosenbluth MN, Teller AH, Teller E (1953) Equation of state calculations by fast computing machines. J Chem Phys 21(6):1087–1092
40. Montanero JM, Santos A (1997) Simulation of the enskog equation a la bird. Phys Fluids 9(7):2057–2060
41. Öttinger H (2005) Beyond equilibrium thermodynamics. Wiley. https://books.google.com/books?id=Prh9moT1WzMC
42. Plimpton SJ, Moore S, Borner A, Stagg A, Koehler T, Torczynski J, Gallis M (2019) Direct simulation Monte Carlo on petaflop supercomputers and beyond. Phys Fluids 31(8)
43. Thompson AP, Aktulga HM, Berger R, Bolintineanu DS, Brown WM, Crozier PS, In't Veld PJ, Kohlmeyer A, Moore SG, Nguyen TD et al (2022) Lammps—a flexible simulation tool for particle-based materials modeling at the atomic, meso, and continuum scales. Comput Phys Commun 271:108171
44. Thomsen JS (1953) Logical relations among the principles of statistical mechanics and thermodynamics. Phys Rev 91(5):1263
45. Tysanner MW, Garcia AL (2005) Non-equilibrium behaviour of equilibrium reservoirs in molecular simulations. Int J Numerical Methods Fluids 48(12):1337–1349
46. Vailati A, Giglio M (1997) Giant fluctuations in a free diffusion process. Nature 390(6657):262–265
47. Valentini P, Zhang C, Schwartzentruber TE (2012) Molecular dynamics simulation of rotational relaxation in nitrogen: implications for rotational collision number models. Phys Fluids 24(10)
48. Wagner W (1992) A convergence proof for bird's direct simulation Monte Carlo method for the Boltzmann equation. J Statistical Phys 66:1011–1044
49. Wainwright T (1964) Calculation of hard-sphere viscosity by means of correlation functions. J Chem Phys 40(10):2932–2937

Ab Initio-Based Collision Model for High-Temperature Molecular Gas Flows

Ashirbad Mallick and Tapan K. Mankodi

Abstract The present work focuses on a new ab initio-based collision model for calculating collision cross-sections from the data generated using the quasi-classical trajectory (QCT) method. A large number of trajectories were run on state-of-the-art potential energy surfaces to get the collision cross-section data. The model is expected to be better than the existing phenomenological collision models and produce more realistic results for studying re-entry flows. The new model is employed in the in-house direct simulation Monte Carlo (DSMC) code and used to study the effect of the collision models on Couette flow simulations and steady shock structure flow. The results are compared with those employing the widely used variable hard sphere (VHS) model.

Keywords Ab initio model · Re-entry flows · Direct simulation Monte Carlo

1 Introduction

The flow physics of re-entry of a space vehicle travelling at hypersonic speeds, especially at rarefied conditions, requires the study of the inelastic energy exchange, chemical and ionisation reactions that occur near the vehicle's surface. Analysis of such non-equilibrium flow involves studying the basic process of collision. Such rarefied flows are often studied using a widely used method known as the direct simulation Monte Carlo (DSMC) [2] method. Traditionally, the DSMC algorithm employs phenomenological collision models to estimate cross-sections and post-collision velocities. Such models are developed to obtain accurate relations of various transport properties as a function of temperature. However, the range over which such models can accurately predict the transport properties is often limited. In addition, phenomenological models usually rely on experimental values of transport properties that are often difficult to estimate at extremely high temperatures.

A. Mallick (✉) · T. K. Mankodi
Department of Mechanical Engineering, Indian Institute of Technology Guwahati, Guwahati, Assam, India
e-mail: m.ashirbad@iitg.ac.in

© The Author(s) 2026
M. Grabe et al. (eds.), *Rarefied Gas Dynamics*, Springer Aerospace Technology,
https://doi.org/10.1007/978-3-032-00094-1_57

2 Ab Initio-Based Collision Cross-Section Model

A new ab initio-based collision model has been proposed as a part of the present work to estimate the cross-section more appropriately. These collision cross-section values are found using quasi-classical trajectory (QCT) [1, 11] simulations, which use highly accurate CASSCF-CASPT2-based potential energy surfaces (PES) [10]. The QCT cross-section data were already computed and taken from the supplementary material of [6]. The QCT code developed for calculating the cross-sections has been validated extensively in previous reports [7, 8] on dissociation cross-sections for various colliding species.

It was observed that the total collision cross-sections from the QCT method [6] were invariant with the rotational quantum levels since they did not affect the scattering angle distributions at a given vibrational level. However, the effect of vibrational quantum levels on the collision cross-section was significant. Such effects have generally been ignored in the phenomenological models that are employed in DSMC.

The collision cross-section in the new ab initio-based collision model is a function of the relative speed as well as the vibrational levels of the molecule(s), which can be described for an atom-molecule collision using a simple fit given as (for atom-molecule interaction):

$$\sigma_{ab-initio} = (pv_1 + q)V_R^{(lv_1+m)} \tag{1}$$

and molecule-molecule interaction as

$$\sigma_{ab-initio} = (a(v_2 + v_3) + bv_2v_3 + c)V_R^{d(v_2+v_3)+ev_2v_3+f} \tag{2}$$

where v_1, v_2 and v_3 are the vibrational levels of the molecules, V_R is the relative velocity, and, $p, q, l, m, a, b, c, d, e$ and f are model parameters.

Figure 1 shows the comparison of the fitting function of the new model with the data obtained from the QCT simulations for $O_2 - O$, $N_2 - O$, $O_2 - O_2$ and $N_2 - N_2$ at various combinations of vibrational quantum levels. It is clear from the figure that the above expression of the fit is a close match with the collision cross-section data for different atom-molecule and molecule-molecule interactions from the QCT data [6].

The ab initio-based collision model was implemented in the in-house DSMC code. This was used to study the flow physics in the planar Couette flow problem and the effect of non-equilibrium on the shock structure profiles for nitrogen gas at various Mach numbers. The results obtained from the simulation employing the ab initio-based collision model were compared with those employing the extensively used VHS model [2]. In all these simulations, N_2 was used as the gas medium.

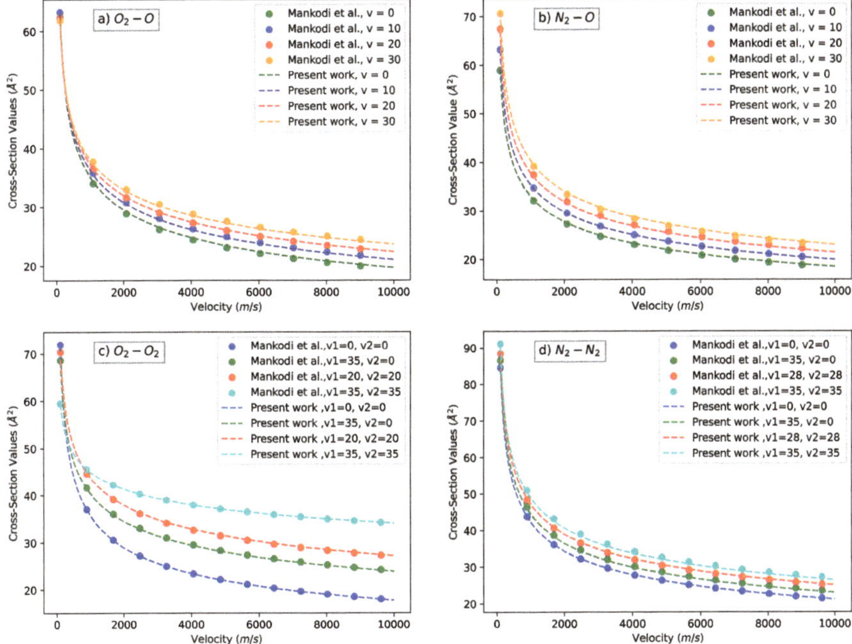

Fig. 1 Collision cross-section area comparison by using the ab initio-based model with the QCT computed data for the cases of **a**) $O_2 - O$, **b**) $N_2 - O$, **c**) $O_2 - O_2$ and **d**) $N_2 - N_2$

3 Couette Flow

The ab initio-based model was used to study Couette flow with nitrogen gas (N_2) flowing in a planar channel. The DSMC simulations employing the traditional VHS and ab initio-based collision models were carried out with the constant parameter Larsen-Borgnakke model ($Z_R = 5$ and $Z_V = 50$) [3] for handling rotational and vibrational relaxation. The walls in the horizontal channel were separated by 1 m ($L = 1$ m). The lower wall was assumed to be stationary, and the upper wall was made to move at a velocity of 300 m/s. The translational temperature of the lower and upper walls was kept at 2000 K and 3000 K, respectively [5]. The values of the translational (T_{trans}) and the vibrational (T_{vib}) temperatures differed for the upper and lower walls, resulting in a non-equilibrium boundary condition. The vibrational temperatures at the lower and upper walls were 3000 K and 2000 K, respectively. It was assumed that the rotational temperatures were in equilibrium with the translational temperatures at the wall.

Figure 2 shows the variation of normalised translational temperature, vibrational temperature and the horizontal velocity component along the channel height obtained from the DSMC simulations employing the two collision models. It can be inferred from the figures that a good agreement at most regions (central core) of the plots

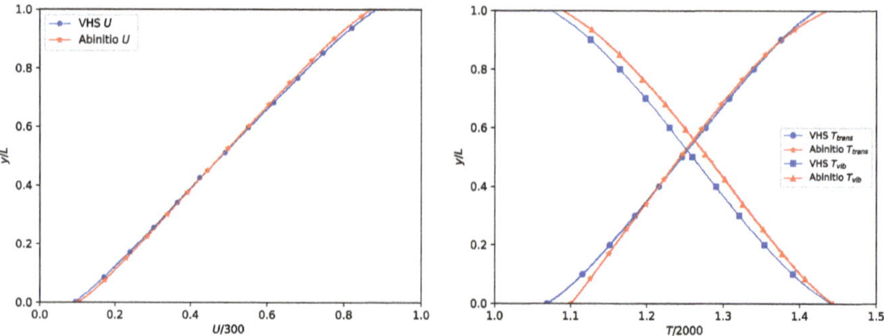

Fig. 2 Comparison of the horizontal component of velocity (left) and translation temperature and vibrational temperature (right) in Couette flow problem employing the VHS and ab initio-based collision model

can be seen for T_{trans}, T_{vib} and U predicted by the ab initio and the VHS models. However, we see significant differences in the temperature profiles in the regions closer to the lower and upper walls. The effect of the ab initio-based cross-section is more prominent in the temperature jump values compared to the velocity slip. At the lower wall, the temperature jump in the translational temperature is higher in the simulation employing the ab initio-based collision model than that obtained from the VHS model. The vibrational temperature at the lower wall is higher than the translational temperature, leading to a high degree of non-equilibrium. This results in a viscosity lower than that estimated by the VHS model [9]. In contrast, at the upper wall, the vibrational temperature is less than the translational temperature, leading to higher viscous effects as predicted by the cross-section data, leading to more collisions and a lower translational temperature jump.

4　Shock Structures in Nitrogen Gas

In addition to the Couette flow problem, simulations were conducted to study the interplay between non-equilibrium and collision models on the Mach 15 nitrogen shock structure. The driven and driver conditions for the Mach 15 shock structure were calculated using the Rankine-Hugoniot relations for vibrationally active diatomic gases, as shown in Table 1.

DSMC simulations were run for two models of the Larsen-Borgnakke inelastic collision model: (a) constant parameters ($Z_R = 5$ and $Z_V = 50$), and (b) temperature-dependent model [4]. The latter allows us to understand the combined effect of the elastic collision and inelastic energy models on the shock structure and the vibrational relaxation [12]. Figure 3 shows various temperature (translational, rotational and vibrational) profiles across the shock structure employing the two collision mod-

Table 1 Driver and driven conditions calculated using the Rankine-Hugoniot relations

Upstream		Downstream	
p_1	1.0 Pa	p_2	274.11 Pa
T_1	273.15 K	T_2	9907.03 K
U_1	5054.40 m/s	U_2	672.17 m/s

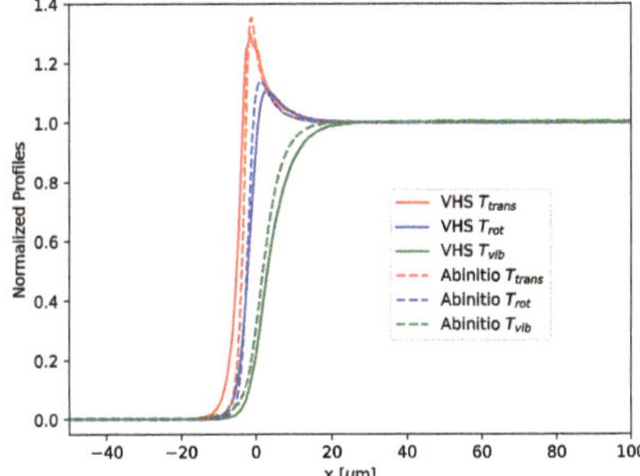
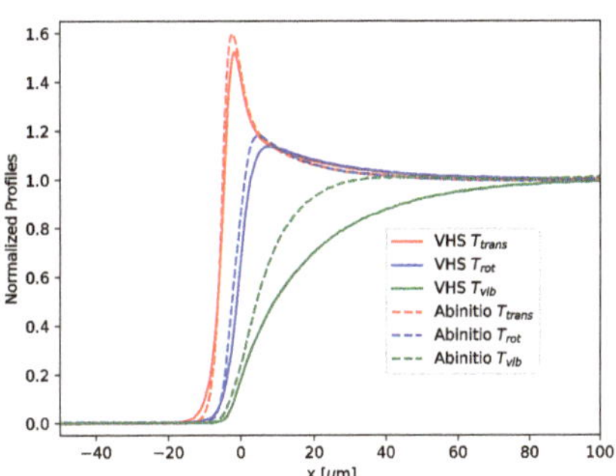

Fig. 3 Comparison of temperature profiles obtained from DSMC simulations with the VHS and ab initio-based collision models and employing constant parameters LB model (left) and temperature-dependent LB model (right)

els, the constant parameter LB model and the temperature-dependent LB model, respectively. It is clear from the temperature profiles that the peak translational temperature was higher for the shock structure obtained from the DSMC simulation employing the ab initio-based model compared to those obtained from the simulations employing the VHS model. In addition, the shock structures obtained using the ab initio-based collision model were much steeper and narrower than those obtained using the VHS model. This effect is even more pronounced in the case of simulation employing the temperature-dependent LB model. This is attributed to the vibrational favouring when calculating the collision cross-section using the ab initio-based collision model.

Higher vibrational levels in the ab initio-based collision model lead to higher collision cross-section values. This leads to more collisions and higher translational temperatures in the shock structure. The higher amount of collisions also results in more inelastic collisions, leading to higher values of rotational and vibrational temperatures. This also leads to faster relaxation, resulting in a narrower shock structure. This is even more pronounced in the case of simulations employing the temperature-dependent LB model. Higher translational temperatures in the simulation using the ab initio-based collision model lead to lower rotational (Z_{rot}) and vibrational (Z_{vib}) relaxation numbers compared to those calculated in the VHS model. This further leads to faster relaxation in the shock structure in the simulation employing the ab initio-based collision model compared to the VHS model. The difference in the shock structure profile can also be attributed to the phenomenological nature of the

VHS model, rendering an inaccurate viscosity-temperature relation, especially at high temperatures and higher degrees of non-equilibrium.

5 Conclusion

The present work introduces a new ab initio-based collision model which employs collision cross-section data computed using the QCT method. According to this model, the collision cross-section is a function of vibrational levels in addition to the relative velocity of the colliding particles. The in-house DSMC code employed the new model for simulating planar Couette flow with non-equilibrium boundary conditions and nitrogen shock structures at high Mach numbers. The results were compared with those obtained using the phenomenological VHS model. Contrasting temperature jump conditions at the lower and upper walls were observed from the two models, pointing to the contrasting effect of different kinds of non-equilibrium conditions on collision cross-section and viscous effects. In the case of the shock structure study, it was found that the profiles obtained from the simulation employing the ab initio-based collision model were narrower with a higher peak temperature and faster translational-rotational and translational-vibrational relaxation compared to those obtained using the VHS model. The interplay between the chemistry modelling and collision model will be investigated for the re-entry flows as part of future work. In addition, the work will be extended to study hypersonic flow past re-entry vehicles with realistic ambient conditions at various altitudes.

References

1. Bender JD, Nompelis I, Valentini P, Doraiswamy S, Schwartzentruber TE, Candler GV, Paukk Y, Yang KR, Varga Z, Truhlar DG (2014) Quasiclassical trajectory analysis of the N_2+N_2 reaction using a new ab initio potential energy surface. In: 11th AIAA/ASME joint thermophysics and heat transfer conference, p 2964
2. Bird GA (1994) Molecular gas dynamics and the direct simulation of gas flows (Clarendon Press; Oxford University Press)
3. Borgnakke C, Larsen PS (1975) Statistical collision model for Monte Carlo simulation of polyatomic gas mixture. J Comput Phys 18(4):405–420
4. Boyd ID, Schwartzentruber TE (2017) Nonequilibrium gas dynamics and molecular simulation, vol 42. Cambridge University Press
5. Casseau V, Palharini RC, Scanlon TJ, Brown RE (2016) A two-temperature open-source CFD model for hypersonic reacting flows, part one: zero-dimensional analysis. Aerospace 3(4):34
6. Mankodi TK, Bhandarkar UV, Myong R (2020) Collision cross sections and nonequilibrium viscosity coefficients of N_2 and O_2 based on molecular dynamics. Phys Fluids 32(3)
7. Mankodi TK, Bhandarkar UV, Puranik BP (2017) An ab initio chemical reaction model for the direct simulation Monte Carlo study of non-equilibrium nitrogen flows. J Chem Phys 147(8):084305
8. Mankodi TK, Bhandarkar UV, Puranik BP (2018) Dissociation cross-sections for high energy O_2–O_2 collisions. J Chem Phys 148(14)

9. Nidharia M, Mane S, Mankodi TK (2024) Ab initio-based two-temperature transport property model for hypersonic non-equilibrium flows. Phys Fluids 36(12)
10. Paukku Y, Yang KR, Varga Z, Truhlar DG (2013) Global ab initio ground-state potential energy surface of N_4. J Chem Phys 139(4)
11. Truhlar DG, Muckerman JT (1979) Reactive scattering cross sections iii: quasiclassical and semiclassical methods. In: Atom-molecule collision theory. Springer, Heidelberg, pp 505–566
12. Valentini P, Zhang C, Schwartzentruber TE (2012) Molecular dynamics simulation of rotational relaxation in nitrogen: implications for rotational collision number models. Phys Fluids 24(10)

Impact of Electronic State-Specific Modeling on High-Enthalpy Nitrogen Flows

Yuzhe Zhang, Qizhen Hong, Xiaoyong Wang, Chao Yang, and Quanhua Sun

Abstract An electronic state-to-state (StS) approach is developed to numerically simulate the nonequilibrium radiative metric measured in experiments conducted at the Electric Arc Shock Tube (EAST) facility. The present StS model solves for the electronic states of strong radiators by integrating a collisional-radiative model into the master equations, while considering both rotational nonequilibrium and vibrational nonequilibrium under a multi-temperature assumption. The StS results for radiation behind the shock waves are compared with predictions from the four-temperature quasi-steady-state (4T-QSS) model. It is found that the StS model accurately predicts the spectrally-integrated radiance in the Vis/NIR range for experimental shot 40, but it overestimates both the peak and post-peak values in the UV range for shot 37. Moreover, both the StS and 4T-QSS models exhibit discrepancies when compared to the experimental nonequilibrium radiative metrics for both shots. The analysis of electronic state populations of N_2 reveals a post-shock region where the QSS assumption breaks down, and the populations deviate notably from the Boltzmann distribution until reaching equilibrium.

Keywords High-enthalpy flows · State-to-state modeling · Quasi-steady-state · Nonequilibrium radiation

Y. Zhang · Q. Hong (✉) · X. Wang · Q. Sun
State Key Laboratory of High Temperature Gas Dynamics, Institute of Mechanics, Chinese Academy of Science, Beijing, China
e-mail: hongqizhen@imech.ac.cn

Q. Sun
e-mail: qsun@imech.ac.cn

Y. Zhang · Q. Sun
School of Engineering Science, University of Chinese Academy of Sciences, Beijing, China

C. Yang
Wide Range Flight Engineering Science and Application Center, Institute of Mechanics, Chinese Academy of Science, Beijing, China

M. Grabe et al. (eds.), *Rarefied Gas Dynamics*, Springer Aerospace Technology,
https://doi.org/10.1007/978-3-032-00094-1_58

1 Introduction

In hypersonic flows, intense shock waves convert kinetic energy into thermal energy, resulting in very high gas temperatures within the shock layer. This gas environment is characterized by significant thermochemical nonequilibrium phenomena occurring simultaneously [13], including internal energy excitation promoting the population of elevated internal energy levels and chemical reactions (e.g., dissociation and ionization) leading to the emergence of atoms and free electrons. To numerically simulate the physicochemical processes described above, the multi-temperature model, particularly the two-temperature (2T: translational-rotational and vibrational-electron-electronic temperatures) model, has been widely employed to calculate high-enthalpy flowfields. Additionally, a quasi-steady-state (QSS) approximation is commonly used to estimate the populations of molecular excited states, which are crucial for accurately predicting spectral features and radiative intensity [11]. In our previous work [18], we conducted a comprehensive assessment of the performance of both the 2T and four-temperature (4T: translational, rotational, vibrational, and electron-electronic temperatures) models for high-enthalpy shock-heated nitrogen flows, using various modeling parameters. It was found that the 4T model provides better accuracy in predicting experimental temperatures compared to the 2T model. Despite this advantage, discrepancies persisted between the calculations and experimental nonequilibrium radiative metrics when using the 4T-QSS model, even after incorporating various modeling options. This situation motivates the present study to explore the impact of more detailed modeling on the prediction of high-temperature radiation.

As pointed out by several studies [6, 8–10], the QSS assumption lacks sufficient accuracy due to the decoupling of molecular excited state population calculations from the flowfield solution process. The QSS assumption approximates the net population rate of molecular states as zero, which becomes invalid under conditions of significant thermochemical nonequilibrium. To address this limitation, the theoretically more accurate state-to-state (StS) model has been used instead to directly track the population of internal energy states by solving the master equations [1, 7, 9, 10]. Notably, the vibrational StS model, incorporating first-principle-calculated vibrational StS kinetics, has been shown to accurately predict post-shock vibrational temperatures in undiluted oxygen [17], and the vibrational-electronic StS model has proven effective in accurately predicting radiative intensities across both infrared and ultraviolet spectral ranges in shock-heated CO flows [2]. Therefore, this study aims to evaluate the impact of electronic state-specific modeling on predicting high-temperature nitrogen radiation and to quantify the discrepancies introduced by the QSS assumption. The organization of this paper is as follows: Sect. 2 describes the thermochemistry models. In Sect. 3, post-shock simulations are performed using the present electronic StS model, with the calculated radiative properties compared to experimental data [3] and the 4T-QSS results. To further evaluate the impact of electronic state-specific modeling, the electronic state populations of N_2 calculated from both the StS model and the QSS assumption are compared. Conclusions are drawn in the last section.

Table 1 Rate parameters for the electron impact excitation of atomic N

i	Reaction	A (cm^3/mol/s)	n	E (K)	T_c	Ref.
1	$N(^4S^o) + e^- \leftrightarrow N(^2D^o) + e^-$	1.05E+15	0.20	27669	T_{ee}	[13]
2	$N(^4S^o) + e^- \leftrightarrow N(^2P^o) + e^-$	3.05E+14	0.21	41500	T_{ee}	[13]
3	$N(^2D^o) + e^- \leftrightarrow N(^2P^o) + e^-$	2.70E+14	0.27	13831	T_{ee}	[13]

The rate coefficient is expressed in the generalized Arrhenius form, $k_f(T_c) = AT_c^n \exp(-E/T_c)$, where T_c represents the controlling temperature

2 Thermochemistry Modeling

To simulate the high-enthalpy nitrogen shock-heated flows, we solve the one-dimensional Euler equations using the electronic StS (collisional-radiative, CR) model in the Poshax3 code [16]. In the electronic StS model, five species of N_2, N_2^+, N, N^+, and e^- are considered, and the relevant electronic states of strong radiators are treated as distinct pseudo-species in master equations, i.e., $N_2(X^1\Sigma_g^+, A^3\Sigma_u^+, B^3\Pi_g, C^3\Pi_u)$, $N_2^+(X^2\Sigma_g^+, A^2\Pi_u, B^2\Sigma_u^+, C^2\Sigma_u^+)$, and $N(^4S^o, ^2D^o, ^2P^o)$. The other electronic states of N_2, N_2^+, and N are assumed to follow the Maxwell-Boltzmann distribution referencing the nearest lower pseudo-species under the electronâŁ"electronic temperature T_{ee}, e.g., the number density of $N_2(b'^1\Sigma_u^+)$ (the upper state of the Birge-Hopfield 2 band) is derived from the population of $N_2(C^3\Pi_u)$. In contrast, for N^+ and e^-, only the ground electronic state is considered due to their negligible contribution to radiation. Overall, the StS model incorporates 13 electronic states from different species and accounts for 57 state-specific reactions in the master equations. The rate coefficients for these reactions are inherited from the CR model presented in our previous work [18], except that we now incorporate three additional electron impact excitation reactions for atomic N (Table 1) to predict the non-Boltzmann populations of excited N species. Heavy-particle impact excitations of N are neglected, as their rate coefficients are much smaller than those of electron-atom collisions [13], and predissociation of $N_2(C^3\Pi_u)$ is also excluded due to significant uncertainties in the rate coefficients at high temperatures (above 10,000 K). Moreover, rotational and vibrational nonequilibrium are accounted for by employing the relaxation times proposed by Park [14] and Kim and Jo [12] in the present work, respectively. The modified Marrone-Treanor (MMT) model [4], which accounts for non-Boltzmann distribution effects on dissociation rates and vibrational energy removal, is employed to address the vibration-dissociation coupling. As demonstrated in our previous work [18], the MMT model provides more accurate predictions than the Park and Marrone-Treanor models for both experimental temperatures and nonequilibrium radiative metrics.

Additionally, we conduct 4T-QSS simulations for comparison. Details of the 4T-QSS model can be found in Ref. [18], except that the CR model used in the QSS analysis is now the same as that of the present StS model.

Table 2 EAST shot conditions (shock-tube diameter 10.16 cm) and investigated spectral ranges

Shot No.	Velocity (km/s)	Pressure (Torr)	Range (nm)	Major features
37	7.15	0.2	UV 200–360	$N_2(2^+)$, $N_2^+(1^-)$
40	6.88	0.2	Vis/NIR 480–900	$N_2(1^+)$

3 Results

The aforementioned electronic state-specific modeling is applied to nitrogen shock-heated flows corresponding to two different shots (see Table 2) in EAST experiments (campaign 62). The nonequilibrium radiative metric, defined as the average radiance within ± 2 cm of the peak [3], is calculated based on the StS and 4T-QSS results, respectively. Radiative spectrum calculations are conducted with the open-source code Photaura [16], and further details, along with modifications to the code, can be found in Ref. [18]. Figure 1 shows the nonequilibrium radiative metric in the UV and Vis/NIR ranges for shots 37 and 40, respectively. In both shots, deviations between the predictions and measurements are observed for the detailed spectrum. Specifically, in the UV spectrum for shot 37, the present StS model shows a slightly better agreement with the experimental peaks of the 2nd Positive band of N_2 compared to the 4T-QSS model. However, significant discrepancies exist concerning the 1st Negative band of N_2^+ (peak at about 330.8 and 358.2 nm), which suggests that the population of $N_2^+(B^2\Sigma_u^+)$ (upper state of this band) is significantly underestimated. Additionally, the large discrepancies observed around 350 nm might also be due to CN contamination present in the experiment [5]. In the Vis/NIR spectrum for shot 40, both models underestimate the radiative intensity of the 1st Positive band of N_2, primarily due to an underestimation of the $N_2(B^3\Pi)$ (upper state of this band), while overestimating the intensity contributed by atomic N (with several Ritz wavelengths marked in the plot). A similar underestimation of the 1st Positive band of N_2 was recently observed using the vibronic state-specific model proposed by Pereira et al. [15], whereas their predictions were an order of magnitude further lower than the present StS results.

Although varying degrees of discrepancies persist in the detailed spectra, the present StS model performs better in predicting the spectrally-integrated radiance versus position. Spatial radiance is the integral of spectral intensity over considered wavelengths, representing the total radiation from both discrete and continuous emissions by atomic or molecular radiators. Note that the computed values have been spatially adjusted to ensure that the point where the radiance begins to rise abruptly aligns with the measured values [15], since the exact position of the shock wave cannot be directly determined from the spectra obtained in the EAST experiments. As shown in Fig. 2, the StS results agree well with experimental data in the Vis/NIR range for shot 40, although it still overestimates both the peak and post-peak values in the UV range for shot 37. In contrast, the 4T-QSS model performs slightly better in shot 37, while it underestimates the peak values in shot 40. The spatial radiance

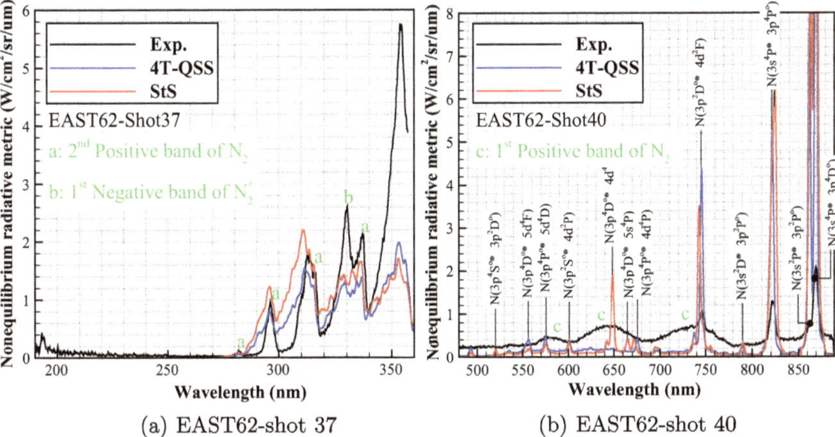

Fig. 1 Comparison of the nonequilibrium radiative metrics between the StS and 4T-QSS result for shots 37 and 40, respectively

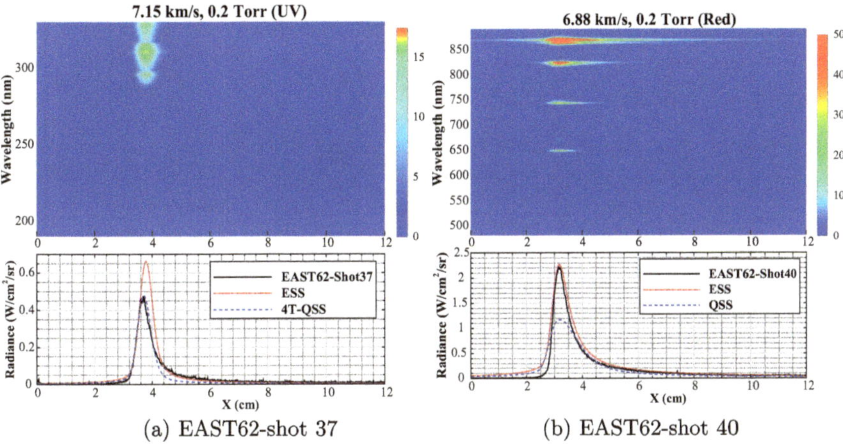

Fig. 2 The StS 3D spectral intensity maps (radiance versus position and wavelength) and spectrally-integrated radiance versus position for the UV range (over 199.05–329.99 nm) for shot 37 and the Vis/NIR range (over 480.04–889.86 nm) for shot 40

clearly illustrates the nonequilibrium overshoot behind the shock wave and the subsequent relaxation toward equilibrium, as also seen in the calculated 3D spectral intensity maps. The persistent discrepancies between the StS calculations and experimental data in both the nonequilibrium radiative metric and spatial radiance indicate that the currently adopted StS kinetics may lack sufficient detail and accuracy for the investigated temperature ranges (additionally, as noted by Pereira et al. [15], the shock tube-related phenomena, such as the absorption of radiation emitted by the driver gas, may also contribute to these discrepancies). While state-of-the-art values

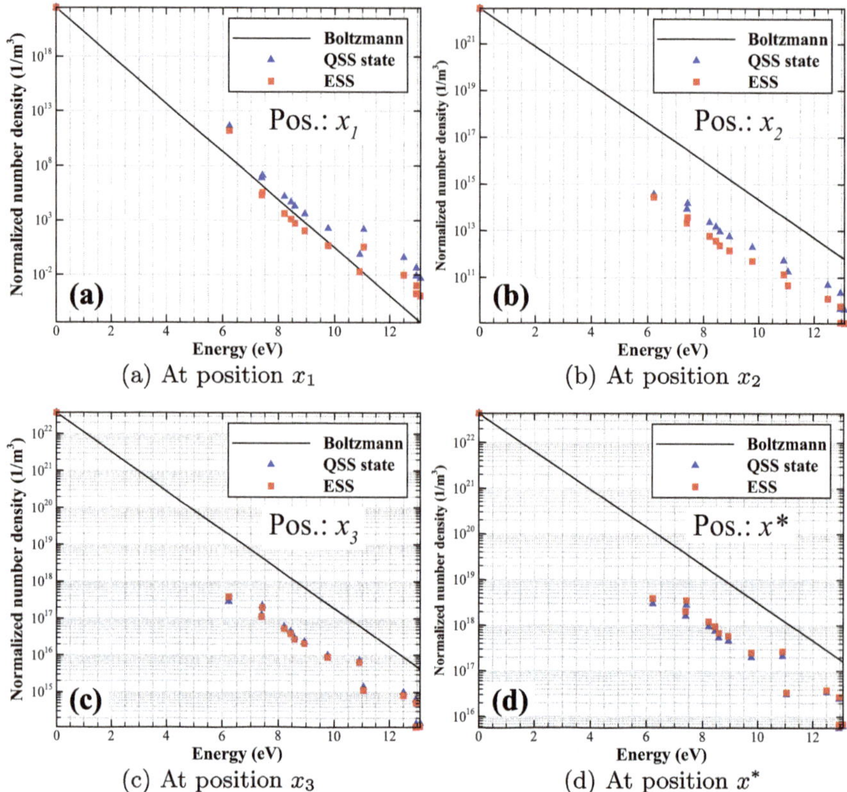

(a) At position x_1

(b) At position x_2

(c) At position x_3

(d) At position x^*

Fig. 3 Comparison of the normalized population distribution (the ratio of number density to degeneracy) of the electronic states of N_2 at different positions (see text) for shot 37, where x^* represents the position of the peak T_{ee}

are used herein for the StS rate coefficients available in the literature, many relevant rate coefficients remain unavailable and exhibit (significant) uncertainties. Therefore, uncertainty quantification of the StS rate coefficients is ongoing, with the aim of improving specifically the accuracy of the most relevant rate coefficients through *ab initio* calculations.

To further investigate the impact of electronic state-specific modeling, we quantify the discrepancies introduced by the QSS assumption. While widely used to estimate internal state populations quickly, the QSS assumption approximates the net population rates of excited states as being much smaller than both the sum of incoming rates and the sum of outgoing rates [13], which may become invalid under significant thermochemical nonequilibrium conditions. For a fair comparison, the QSS populations (unrelated to the aforementioned 4T-QSS results) are derived from macroscopic quantities obtained from the StS results and compared to the StS populations. Figure 3 shows the populations of N_2 at various post-shock locations, including x_i ($i = 1, 2, 3,$

representing three spatial positions at equal intervals between the shock front and x^*) and x^* (the position of the peak T_{ee}). For reference, the Boltzmann distributions at the local T_{ee} are also presented. At position x_1, near the shock front where internal energy states remain largely unexcited, the QSS populations of all excited electronic states are systematically higher than the StS populations. This discrepancy arises because electronic energy excitation processes dominate in this region, causing the sum of incoming rates to become the dominant term in the net population rate, thereby leading to the breakdown of the QSS assumption. The dominance of electronic energy excitation also reflects on the overpopulation effect for both QSS and StS populations when compared to the Boltzmann distribution, especially for states above $N_2(C^3\Pi_u)$ (around 11.05 eV). As the flow moves downstream to position x_2, the difference between QSS and StS populations persists but with a slight reduction, and both models predict excited state populations lower than the Boltzmann distribution. This underpopulation effect is attributed to the significant dissociation and ionization of electronic excited states at position x_2, where elevated controlling temperatures enhance these processes, thereby accelerating reaction rates and significantly depleting the excited states (more energetically favorable for reaction) of N_2. As the vibrational and electron-electronic temperatures (thus the controlling temperatures for reactions) increase while the flow approaches position x^*, the incoming rates of excited states by energy excitation and the outgoing rates by dissociation and ionization become more balanced. This dynamic interplay facilitates the establishment of QSS conditions, as illustrated by the population distributions observed at positions x_3 (panel c) and x^* (panel d), while both QSS and StS populations remain lower than the Boltzmann distribution.

4 Conclusion

This paper presented an electronic StS approach, with incorporating both rotational and vibrational nonequilibrium, for simulating nitrogen shock-heated flows and radiation. It was found that the StS model accurately predicts the spatial radiance in the Vis/NIR range for EAST shot 40 while it overestimates both the peak and post-peak values in the UV range for shot 37. Additionally, the StS model showed slightly better agreement with the experimental radiative peaks (in nonequilibrium radiative metric) of the 2nd Positive band of N_2 compared to the 4T-QSS model for shot 37. In the Vis/NIR spectrum for shot 40, both models underestimated the radiative intensity of the 1st Positive band of N_2 while overestimating the intensity contributed by atomic N. The persisted discrepancies between the electronic state-specific modeling and experimental data indicate the need for a comprehensive uncertainty analysis of the StS kinetic mechanisms in future work. Furthermore, based on the StS results, the populations of N_2 were obtained using the QSS assumption and compared with those derived directly from the StS results. The comparison revealed that differences between the StS and QSS models mainly occur in the region behind the shock front, where electronic energy excitation processes dominate, causing the sum of incoming rates to become the dominant term in the net population rate and leading to the breakdown of the QSS assumption. However, as the flow approaches and moves beyond the peak T_{ee} position, the QSS assumption generally holds as the incoming and outgoing rates of electronic excited states become more balanced.

Acknowledgements This study was supported by the Strategic Priority Research Program of the Chinese Academy of Sciences (grant number XDB0620201).

References

1. Aiken TT, Boyd ID (2024) Collisional-radiative modeling of shock-heated nitrogen mixtures. J Appl Phys 135(9)
2. Aliat A, Chikhaoui A, Kustova E (2003) Nonequilibrium kinetics of a radiative co flow behind a shock wave. Phys Rev E 68(5):056306
3. Brandis AM, Cruden BA (2018) Shock tube radiation measurements in nitrogen. In: 2018 joint thermophysics and heat transfer conference, p 3437
4. Chaudhry RS, Boyd ID, Candler GV (2020) Vehicle-scale simulations of hypersonic flows using the MMT chemical kinetics model. In: AIAA aviation 2020 forum, p 3272
5. Glenn AB, Varley O, Collen P, McGilvray M (2024) Radiation measurements of shockwaves in synthetic air and pure nitrogen. In: AIAA SCITECH 2024 forum, p 0226
6. Hao J, Wang J, Lee C (2017) State-specific simulation of oxygen vibrational excitation and dissociation behind a normal shock. Chem Phys Lett 681:69–74
7. Hong Q, Wang X, Hu Y, Lin X, Sun Q (2020) Rebuilding experimental nonequilibrium radiation in shock-heated Martian-like mixture flows using electronic state-to-state approach. Int J Modern Phys B 34(14n16):2040084
8. Istomin V, Kustova E, Prutko K (2024) Different approaches for simulating heat and radiative fluxes in planetary entry problems. In: AIP conference proceedings, vol 2996. AIP Publishing
9. Jo SM, Munafò A, Panesi M (2024) Self-consistent flow-radiation coupling for hypersonic atmospheric entry. In: AIAA SCITECH 2024 forum, p 1686
10. Johnston CO, Panesi M (2018) Impact of state-specific flowfield modeling on atomic nitrogen radiation. Phys Rev Fluids 3(1):013402
11. Johnston CO (2006) Nonequilibrium shock-layer radiative heating for Earth and Titan entry. Ph.D. thesis, Virginia Polytechnic Institute and State University
12. Kim JG, Jo SM (2021) Modification of chemical-kinetic parameters for 11-air species in re-entry flows. Int J Heat Mass Transfer 169:120950
13. Park C (1989) Nonequilibrium hypersonic aerothermodynamics. Wiley, New York
14. Park C (2004) Rotational relaxation of N_2 behind a strong shock wave. J Thermophysics Heat Transfer 18(4):527–533
15. Pereira É, Loureiro J, da Silva ML (2023) Vibronic state-specific modelling of high-speed nitrogen shocked flows. Part ii: shock tube simulations. arXiv preprint arXiv:2308.05164
16. Potter D (2011) Modelling of radiating shock layers for atmospheric entry at Earth and Mars. Ph.D. thesis, School of Mechanical and Mining Engineering, The University of Queensland
17. Wang X, Hong Q, Yang C, Sun Q (2023) Uncertainty quantification in state-specific modeling of thermal relaxation and dissociation of oxygen. AIAA J 61(6):2734–2738
18. Zhang Y, Hong Q, Wang X, Yang C, Sun Q (2024) Assessment of thermochemical models on nonequilibrium flowfield and radiation in shock-heated nitrogen. Comput Fluids 289:106533

Exploitation of Continuum and Kinetic Theory Approaches for the Simulation of Particle Beam Experiments

Surya Kiran Peravali, Amit K. Samanta, Muhamed Amin, Philipp Neumann, Jochen Küpper, and Michael Breuer

Abstract X-ray free-electron lasers (XFELs) promise to allow for atomically resolved imaging of isolated nanoparticles through single-particle diffractive imaging (SPI) [1]. Achieving this requires nanoparticle beams with controlled dimensions, typically generated using aerosol injectors operating under varying gas-flow conditions. Numerical simulations play a vital role in understanding and optimizing these injection systems. We present a multi-scale simulation framework that models gas dynamics across continuum-, transition-, and free-molecular-flow regimes as well as particle translation. Leveraging computational fluid dynamics (CFD), direct simulation Monte Carlo (DSMC), and corresponding hybrid methods, the framework provides a foundation for improving aerosol-injection techniques. It was validated against experiments over wide temperature (4–300 K) and size (10–300 nm) ranges.

Keywords DSMC · Nanoparticle injection · Continuum assumption · Transition regime · Rarefied flow · single-particle imaging · Particle injection

S. K. Peravali · M. Breuer
Professur für Strömungsmechanik, Helmut-Schmidt-Universität/Universität der Bundeswehr Hamburg, Hamburg, Germany

S. K. Peravali (✉) · A. K. Samanta · J. Küpper
Center for Free-Electron Laser Science CFEL, Deutsches Elektronen-Synchrotron DESY, Hamburg, Germany
e-mail: peravals@hsu-hh.de

A. K. Samanta · J. Küpper
Center of Ultrafast Imaging, Universität Hamburg, Hamburg, Germany

J. Küpper
Department of Physics, Universität Hamburg, Hamburg, Germany

M. Amin
Laboratory of Computational Biology, National Heart, Lung and Blood Institute, National Institutes of Health, Bethesda, MD, USA

P. Neumann
Department of Informatics, High Performance Computing & Data Science, Universität Hamburg, Hamburg, Germany

Deutsches Elektronen-Synchrotron DESY, Hamburg, Germany

© The Author(s) 2026

M. Grabe et al. (eds.), *Rarefied Gas Dynamics*, Springer Aerospace Technology,
https://doi.org/10.1007/978-3-032-00094-1_59

1 Introduction

X-ray free-electron lasers (XFELs) have revolutionized structural biology and material sciences by enabling high-resolution imaging of nanoscale objects and their ultrafast dynamics. The technique of single-particle diffractive imaging (SPI) leverages intense, femtosecond XFEL pulses to capture atomically resolved structures of isolated nanoparticles without requiring large, highly ordered crystals [1, 2]. This opens new avenues for studying the structure and dynamics of complex systems, including biomolecules, viruses, and nanomaterials.

A critical requirement for successful SPI experiments is the ability to deliver well-focused nanoparticle beams with controlled dimensions and minimal divergence. To achieve this objective, aerodynamic lens stacks (ALS) [3–5] and cryogenically cooled buffer-gas cells (BGC) [6] are employed as particle injection systems. These injectors utilize gas dynamics to guide particles from an ambient-pressure environment into the high-vacuum chamber. However, optimizing these processes to produce particle beams with the desired properties remains a significant challenge.

Numerical simulations play an indispensable role in addressing this bottleneck, providing insights into the complex interplay of gas dynamics and particle behavior across various flow regimes. From continuum to free-molecular flows, simulations allow researchers to predict particle trajectories and evaluate injector performance under a wide range of operating conditions. Recent advancements in computational modeling techniques, such as computational fluid dynamics (CFD), the direct simulation Monte Carlo (DSMC) method, and hybrid CFD-DSMC approaches, have significantly enhanced our ability to simulate gas-particle interactions with high accuracy.

In this study, we present a multi-scale numerical framework designed to model the transport of nanoparticles in gas flows under diverse experimental conditions. This framework integrates CFD, DSMC, and hybrid methodologies to capture the carrier gas dynamics and particle migration over a broad spectrum of temperatures (4–300 K) and particle sizes (10–300 nm). By incorporating various force models influencing particle trajectories, the framework provides a comprehensive tool for simulating and optimizing nanoparticle injection systems.

The proposed methodologies were validated against experimental data, demonstrating their applicability and reliability for SPI experiments. Additionally, we discuss benchmark results and highlight potential areas for future improvements, including refinements in drag force models and coupling techniques. This work aims to bridge the gap between simulation and experiment, offering a pathway toward more efficient and precise nanoparticle injection systems for XFEL applications.

2 Methodology

This study investigates the gas-particle dynamics in different aerosol injectors using a two-step approach. First, the steady-state flow field is computed, followed by particle trajectory calculations based on interpolated flow quantities and the corresponding forces on the particles. A decoupled approach is adopted as the particle volume fraction is low, ensuring a negligible effect of the particles on the gas flow.

For continuum and slip regimes ($Kn < 0.1$), the flow field is predicted using `rhoCentralFoam`, a compressible CFD-solver in OpenFOAM [7]. For rarefied regimes ($Kn > 0.1$), the DSMC method implemented in SPARTA [8] is employed to capture molecular-scale effects. To address flows in the transition regime between these two cases, a one-way coupled hybrid CFD-DSMC approach is utilized. The coupling is based on a continuum-breakdown criterion, defined as the maximum of the global Knudsen number and Boyd's gradient-based local Knudsen numbers— calculated from density, temperature, and velocity fields – enabling efficient domain decomposition and accurate multiscale resolution of the flow field [9, 10]. In the DSMC domain, molecular collisions are modeled using the Variable Hard Sphere (VHS) model, while internal energy exchange is handled via the Larsen–Borgnakke (LB) model with constant relaxation. For nitrogen, a rotational relaxation number of 0.2 is used, whereas for helium it is set to zero, consistent with its monoatomic nature. Vibrational and chemical reaction modes are assumed to be frozen in the present study. The numerical parameters for the VHS and LB models are taken from standard literature sources [11].

Particle transport is modeled using the Langevin equation, incorporating drag ($\mathbf{F}_{drag}$) and Brownian motion forces ($\mathbf{F}_b$) [12]. The governing equation reads

$$\frac{d}{dt}\left(m_p\,\mathbf{u}_p\right) = \mathbf{F}_{drag} + \mathbf{F}_b \, , \tag{1}$$

with the mass of the particle m_p, the particle velocity vector $\mathbf{u}_p$, and the time t. Drag models incorporate Stokes-Cunningham corrections for slip-flow regimes [13] and molecular drag models for rarefied conditions [14, 15], with parameters optimized for accuracy across varying flow conditions. The Brownian motion is modeled using a Gaussian white-noise approach [12, 16]. A detailed assessment of the applicability and accuracy of these drag models across different Knudsen and Mach number regimes, including corrections for high rarefaction, has been provided in an earlier work [10], where their influence on particle dynamics in hybrid simulations was systematically evaluated.

This integrated multi-scale simulation methodology, combining hybrid CFD-DSMC gas flow predictions—was previously validated for gas-only nozzle flows [9] and for full gas-particle simulations in ALS-relevant configurations [10], demonstrating both accuracy and computational efficiency. The complete hybrid framework, including the coupling strategy and selection criteria for numerical parameters such as CFD discretization schemes, DSMC cell size, number of particles per cell, and sampling strategies, is detailed in these studies [9, 10].

3 Results and Discussion

The behavior of nanoparticle beams was analyzed for three distinct cases: (1) an aerodynamic lens system (ALS), (2) a buffer-gas cell (BGC), and (3) the combined BGC-ALS system. Simulation results were compared to experimental data to evaluate beam widths and focal positions for each case. This comparison highlights the influence of aerodynamic focusing and gas-particle collisions on beam shaping and focusing efficiency and it validates the accuracy of the numerical framework developed in this work.

3.1 Case I: Aerodynamic-Lens-Stack Injector at Room Temperature

The aerodynamic lens stack (ALS) is composed of a series of orifices that focus particle beams by accelerating and concentrating them into a narrow stream. Particles are transferred into the gas phase, collimated, and subsequently extracted into a vacuum. The ALS design used in this study follows the configuration previously described in Worbs [5].

Figure 1 illustrates the flow field for nitrogen at room temperature, and corresponding particle trajectories for 69 nm-diameter polystyrene particles at an inlet pressure of $p_{in} = 50$ Pa. The hybrid-CFD-DSMC method predicted the flow field, represented by the axial velocity (v_z) along the main flow direction, while particle trajectories were computed using the Stokes-Cunningham drag model combined with Brownian forces. At the ALS exit $(z = 0)$, the gas expands at supersonic speeds into the vacuum chamber $(z > 0)$. The simulated particle trajectories converge to a minimum beam width within the vacuum chamber.

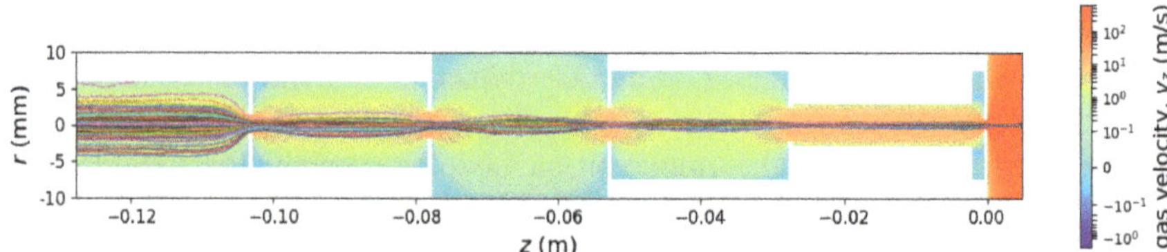

Fig. 1 Simulated trajectories of 69 nm-diameter polystyrene particles (colored lines) passing through the ALS. The axial velocity of the gas-flow field at $p_{in} = 50$ Pa is illustrated on a logarithmic colorscale

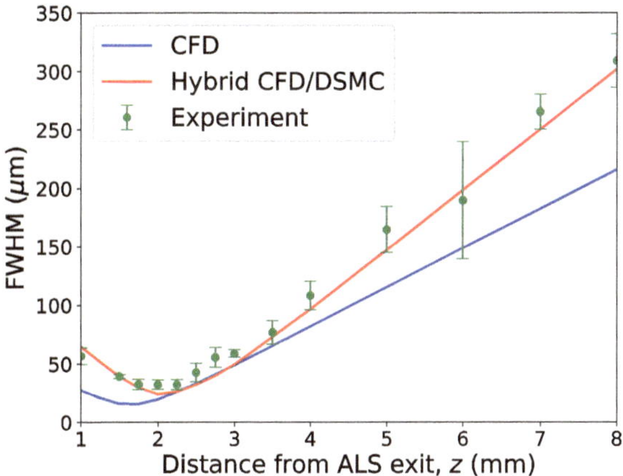

Fig. 2 Simulated evolution of the particle beam size (FWHM) for 69 nm-diameter polystyrene spheres at an inlet pressure of $p_{in} = 50$ Pa, compared to experimental data [17]

Beam widths were measured at positions starting from $z = 1$ mm, where the particle beam exhibits a Gaussian-like distribution [17]. The beam width, defined by the full-width at half-maximum (FWHM[1]), was compared with experimental data in Fig. 2. The hybrid CFD/DSMC simulations accurately capture the focusing-defocusing behavior and focus position, showing excellent agreement with experimental results, and they perform better than pure CFD simulations.

3.2 Case II: Cryogenic Buffer-Gas Cell

The cryogenic buffer-gas cell (BGC) was designed to rapidly cool nanoparticles, such as biomolecules and proteins, below their glass-temperature at ~ 133 K within nano- or microseconds [6]. This shockfreezing process is sufficiently fast to preserve the biomolecules structural integrity. The cooled particles are extracted into a collimated particle stream under high-vacuum conditions. In this injector, cold helium gas, down to 4 K, serves as the carrier gas.

Figure 3a illustrates the helium flow field within the cryogenic buffer-gas cell (BGC)—simulated using the hybrid CFD-DSMC methodology for a helium inlet flow rate of 25 ml$_n$/min. The flow is visualized by streamlines, colored according to the Mach-number magnitude. Helium entered the BGC through the inlet located at the top of the thin inlet disk on the left side of the main cell. Particles of 220 nm diameter were injected through a capillary transport tube [6], passed through the BGC starting from outlet 1, and were eventually extracted through outlet 2. Molecular-drag-force models [14, 15] were applied to predict the particle behavior under these conditions.

---

[1] Difference between two points on the curve where the value is half of the max. value.

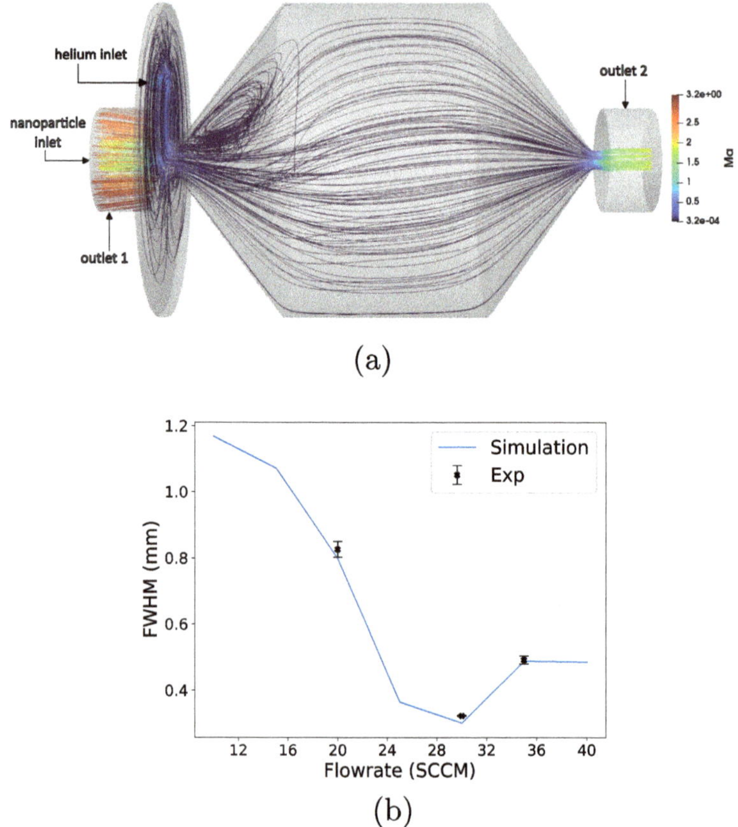

Fig. 3 **a** Streamlines of the flow field inside the cryogenic buffer-gas cell, colored by the Mach-number magnitude, as simulated using the hybrid DSMC/CFD method. **b** Predicted particle beam widths (blue spheres) for 220 nm-diameter polystyrene spheres as a function of the helium flow rate, compared to experimental data (black x-cross and error bars) [6]

Figure 3b shows the particle beam widths (FWHM) of 220 nm polystyrene particles measured 10 mm downstream of outlet 2 as a function of the helium flow rate, capturing the effects of varying pressures and velocities. The simulation results align well with experimental data [6], accurately reproducing the observed trends.

3.3 *Case III: Aerodynamic Lens with Cryogenic Cooling*

Delivering smaller particles with diameters less than 20 nm, e.g., proteins, still poses a challenge due to their low inertia and high diffusivity. These particles tend to

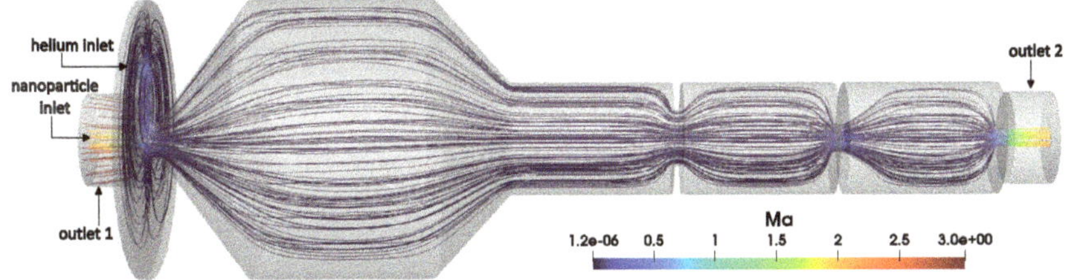

Fig. 4 Streamlines of the flow field through the combined setup of BGC and ALS, colored by the Mach-number magnitude, as simulated using the hybrid DSMC/CFD method

closely follow the gas streamlines, resulting in only limited focusing. Furthermore, their motion is strongly influenced by Brownian dynamics, leading to substantial diffusivity. This behavior not only widens the particle beam, but also causes significant particle losses, making it even more difficult to produce well-collimated particle streams for SPI experiments. One effective strategy to address these challenges involves reducing the carrier-gas temperature, which enhances the injection of small particles [17]. Further experimental progress was demonstrated for the focusing of particles smaller than 100 nm by combining an ALS with the BGC.

Figure 4 depicts the helium flow field within the BGC-ALS system, simulated using the hybrid CFD-DSMC methodology at a helium inlet flow rate of 15 ml_n/min. The flow field is represented by streamlines, with colors indicating the Mach-number magnitude. Helium enters the system through the inlet on the upper side of the thin inlet disk at the left of the main cell. Particles of 88 nm diameter were introduced through outlet 1 via a capillary transport tube [17], traversed the BGC-ALS system, and were subsequently extracted through outlet 2. Particle trajectories were computed using molecular-drag-force models [14, 15], ensuring accurate predictions of particle dynamics under these conditions.

Figure 5 presents the evolution of particle beam widths (FWHM) at various distances downstream of the BGC-ALS exit (final lens) for helium flow rates of 10 ml_n/min and 15 ml_n/min. The beam widths are compared with experimental measurements taken 6, 8, and 10 mm downstream the exit [17]. The simulation results show very good agreement with the experimental data, accurately capturing the particle beam focusing behavior.

4 Conclusions

The behavior of nanoparticle beams in the multi-scale regime was investigated for three configurations: An aerodynamic-lens-stack system (ALS), a buffer-gas cell (BGC), and the combined buffer-gas cell and the aerodynamic-lens-stack system (BGC-ALS). The simulation results for beam widths and focus positions were com-

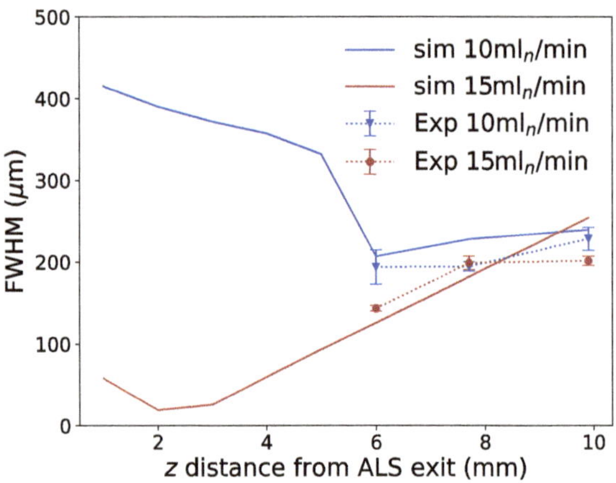

Fig. 5 Simulated evolution of the particle beam size (FWHM) for 88 nm-diameter polystyrene spheres at different flow rates of helium, compared to experimental data [17]

pared with experimental data, showing good agreement and thus validating the accuracy of the developed numerical framework.

The results highlight the strengths and limitations of each configuration. The ALS effectively focuses nanoparticle beams at low pressures, while the BGC enables controlled cooling and collimation of the particles. The BGC-ALS combination demonstrates superior performance by leveraging the strengths of both systems, achieving enhanced beam quality and improved focus.

The numerical framework developed in this study provides a robust tool for modeling and optimizing nanoparticle injection systems, offering insights into their design and performance under various operation conditions. Future work will extend the framework to include more complex particle interactions and it will incorporate machine-learning techniques for further optimization.

Acknowledgements This work was supported by the HELMHOLTZ Data Science Graduate School for the Structure of Matter (DASHH, HIDSS-0002), by the Helmut-Schmidt University Hamburg, and by Deutsches Elektronen-Synchrotron (DESY@HGF).
We also acknowledge dtec.bw – Digitalization and Technology Research Center of the Bundeswehr (project hpc.bw) for provision of computational resources on HSUper. dtec.bw is funded by the European Union—NextGenerationEU.

References

1. Neutze R, Wouts R, Van der Spoel D, Weckert E, Hajdu J (2000) Potential for biomolecular imaging with femtosecond X-ray pulses. Nature 406:752–757. https://doi.org/10.1038/35021099
2. Spence J, Chapman N (2014) The birth of a new field. Philos Trans R Soc Lond B Biol Sci 369:1647. https://doi.org/10.1098/rstb.2013.0309
3. Roth N, Horke DA, Lübke J, Samanta AK, Estillore AD, Worbs L, Pohlman N, Ayyer K, Morgan A, Fleckenstein H, Domaracky M, Erk B, Passow C, Correa J, Yefanov O, Barty A,

Bajt S, Kirian RA, Chapman HN, Küpper J (2024) New aerodynamic lens injector for single particle diffractive imaging. Nucl Instrum Methods Phys Res A 1058:168820. https://doi.org/10.1016/j.nima.2023.168820

4. Roth N, Awel S, Horke DA, Küpper J (2018) Optimizing aerodynamic lenses for single-particle imaging. J Aerosol Sci 124:17–29. https://doi.org/10.1016/j.jaerosci.2018.06.010

5. Worbs L, Roth N, Lübke J, Estillore AD, Xavier PL, Samanta AK, Küpper J (2021) Optimizing the geometry of aerodynamic-lens stack injectors for single-particle diffractive imaging. J Appl Cryst 54(6):1730–1737

6. Samanta AK, Amin M, Estillore AD, Worbs L, Horke DA, Küpper J (2020) Controlled beams of shock-frozen, isolated, biological and artificial nanoparticles. Structural Dyn 7(2):024304. https://doi.org/10.1063/4.0000004

7. OpenFOAM—The open source CFD toolbox. https://www.openfoam.com

8. SPARTA—Stochastic parallel rarefied-gas time-accurate analyzer. https://www.sparta.github.io

9. Peravali SK, Jafari V, Samanta AK, Amin M, Neumann P, Küpper J, Breuer M (2024) Accuracy and performance evaluation of low density internal and external flow predictions using CFD and DSMC. Comput Fluids 279:106346. https://doi.org/10.1016/j.compfluid.2024.106346

10. Peravali SK, Samanta AK, Amin M, Neumann P, Küpper J, Breuer M (2025) An improved simulation methodology for nanoparticle injection through aerodynamic lens systems. Phys Fluids 37:033380. https://doi.org/10.1063/5.0260295

11. Bird GA (1994) Molecular gas dynamics and the direct simulation of gas flows. Clarendon Press, Oxford

12. Welker S, Amin M, Küpper J (2022) CMInject: Python framework for numerical simulation of nanoparticle injection pipelines. Comput Phys Commun 270:108138. https://doi.org/10.1016/j.cpc.2021.108138

13. Cunningham E, Larmor J (1910) On the velocity of steady fall of spherical particles through a fluid medium. Proc R Soc Lond A 83:357–365

14. Epstein PS (1924) On the resistance experienced by spheres in their motion through gases. Phys Rev 23:710–723

15. Baines MJ, Williams IP, Asebiomo AS, Agacy RL (1965) Resistance to the motion of a small sphere moving through a gas. Mon Not R Astron Soc 130:63–74. https://doi.org/10.1093/mnras/130.1.63

16. Li A, Ahmadi G (1992) Dispersion and deposition of spherical particles from point sources in a turbulent channel flow. Aerosol Sci Techn 16:209

17. Worbs L (2022) Toward cryogenic beams of nanoparticles and proteins. Ph.D. thesis, Universität Hamburg, Hamburg, Germany

Calibration of the Larsen-Borgnakke Model for NO-O Vibrational Relaxation

Shubham Thirani, Deborah A. Levin, and Ingrid J. Wysong

Abstract Modeling of Nitric Oxide (NO) vibrational states in hypersonic reentry flows is necessary to interpret emission and absorption spectra. To that end, the vibrational relaxation number, Z_v^C, is an important parameter in the Larsen-Borgnakke (LB) model which is used to express vibrational relaxation rates for collisions. In this work, we use rates obtained from Quasi-Classical Trajectory (QCT) calculations of Andrienko et al. [2] for the NO-O vibrational relaxation process to arrive at a new fit for the Z_v^C parameter. This new fit for Z_v^C is implemented in Direct Simulation Monte Carlo (DSMC) calculations for the case of hypersonic flow over a cylinder where the effect of the new fit on NO macroparameters and vibrational state populations is studied.

Keywords Larsen-Borgnakke · Vibrational relaxation · Nitric oxide infrared emission

1 Introduction

NO emission measurements in the infrared (IR) region from hypersonic ground testing facilities [4, 5] may provide important information about non-equilibrium flow chemistry. A fundamental aspect related to the modeling of NO vibrational states in the ground electronic state is the modeling of relaxation mechanisms for the vibrational energy mode of NO. In this work, we focus our efforts on the vibrational relaxation number, Z_v^C, used to model the NO-O vibrational relaxation process. Based on the continuum approach, Z_v^C is defined as the ratio of the vibrational relaxation time constant, τ_v, and the mean collision time, τ_c, that is,

S. Thirani (✉) · D. A. Levin
Department of Aerospace Engineering, University of Illinois Urbana-Champaign, Urbana, IL, USA
e-mail: thirani2@illinois.edu

I. J. Wysong
AFRL Retired, Seattle, WA, USA

© The Author(s) 2026

M. Grabe et al. (eds.), *Rarefied Gas Dynamics*, Springer Aerospace Technology,
https://doi.org/10.1007/978-3-032-00094-1_60

$$Z_v^C := \frac{\tau_v}{\tau_c}, \tag{1}$$

where the superscript C emphasizes the connection with continuum approaches. In this work, we obtain a new fit for the Z_v^C parameter while using vibrational state-to-state rates obtained from QCT calculations of Andrienko et al. [2] for the NO-O collisional mechanism and assess its impact on flow macroparameters and IR emission spectra in the expansion regions of the flowfield.

2 Isothermal Bath Calculations

The LB model [11] defines the probabilities for energy transfer from the translational mode to internal energy modes during inelastic binary collisions. In their DSMC work Gimelshein and Wysong [8] discuss why even though it may be important to use a state-to-state model for vibrational relaxation in O_2 or N_2 collisions, the LB model works very well for the vibrational relaxation of NO. Since NO relaxation is a rapid process mediated by attractive forces, the final vibrational level tends to be independent of the initial vibrational level. Gimelshein and Wysong [8] modeled the NO-O vibrational relaxation process using the calibrated fit for the Z_v^C parameter developed by Kulakhmetov et al. [10] for the O_2-O vibrational relaxation process as,

$$Z_v^C = \begin{cases} 52.46 + 0.04911 T_T - 3.128 \times 10^{-5} T_T^2 & \text{if } T_T \leq 1000 \text{ K} \\ \exp\left[5.228 - 0.1405 \ln(T_T)\right] & \text{if } T_T > 1000 \text{ K,} \end{cases} \tag{2}$$

where T_T is the local translational temperature. Following the framework described by Kulakhmetov et al. [10], 0-D thermal relaxation studies for constant ro-translational temperatures, T_{RT}, and variable vibrational temperatures, T_V, were performed for the NO-O non-reacting vibrational relaxation process to obtain τ_v assuming that the vibrational relaxation process proceeds only through the following mechanism,

$$\text{NO}(x) + \text{O} \leftrightarrow \text{NO}(y) + \text{O}, \quad x, y \in [0, 48] \text{ and } x \neq y, \tag{3}$$

where x and y represent different NO vibrational levels. The forward and backward mechanisms described in Eq. 3 generate a set of coupled first order differential equations, termed the master equations, which can be solved to track the evolution of the NO vibrational state populations with time. The forward rates were obtained from Ref. [2] and detailed mass balance was employed to obtain the backward rates. The 0-D simulations were initialized with a mixture of NO and O at a total number density of 10^{21} m^{-3} and equal mole fractions.

Assuming a Landau-Teller form for vibrational relaxation, the evolution of the relaxation parameter, Λ, was obtained as a function of time from the solution of the master equation, where

$$\Lambda(t, T_{RT}) = \exp\left(-\frac{1}{\tau_v}\right) = \frac{\overline{E_v(t)} - E_{eq,v}(T_{RT})}{E_v^o - E_{eq,v}(T_{RT})} \tag{4}$$

and t is the instantaneous time, $\overline{E_v(t)}$ is the average vibrational energy at time t, $E_{eq,v}(T_{RT})$ is the vibrational energy at thermal equilibrium with temperature T_{RT}, and E_v^o is the initial vibrational energy. Furthermore, the instantaneous vibrational temperature, T_V, was evaluated using the partition functions as

$$\overline{E_v(t)} = \frac{\Sigma_v E_v(v) \exp\left(-\frac{E_v(v)}{k_B T_V}\right)}{\Sigma_v \exp\left(-\frac{E_v(v)}{k_B T_V}\right)} \tag{5}$$

where k_B is Boltzmann's constant. We use Eq. 5 to calculate E_v^o using the T_V value at $t = 0$ and $E_{eq,v}(T_{RT})$ by assigning $T_V = T_{RT}$. The vibrational energy of the vth vibrational state, $E_v(v)$, is expressed as

$$E_v(v) = \omega_e(v + 0.5) - \omega_e x_e(v + 0.5)^2 + \omega_e y_e(v + 0.5)^3 \tag{6}$$

where ω_e is the vibrational frequency, and x_e and y_e are the first and second anharmonicity constants [9].

Figure 1 presents the results of the 0-D simulations initialized with the different T_{RT} values given in the figure where all cases had an initial T_V of 500 K. Given the definition of τ_v in Eq. 4, the projections of the intersection points of the $\Lambda = 1/e$ line with the Λ curves on the t axis gives the τ_v value for each 0-D thermal relaxation case. Furthermore, τ_c may be written as [3, 8],

$$\tau_c = \frac{1}{2\sqrt{\pi}d^2\sqrt{2k_B T_{RT}/m_R}(T_{\text{ref}}/T_{RT})^{\omega-0.5}} , \tag{7}$$

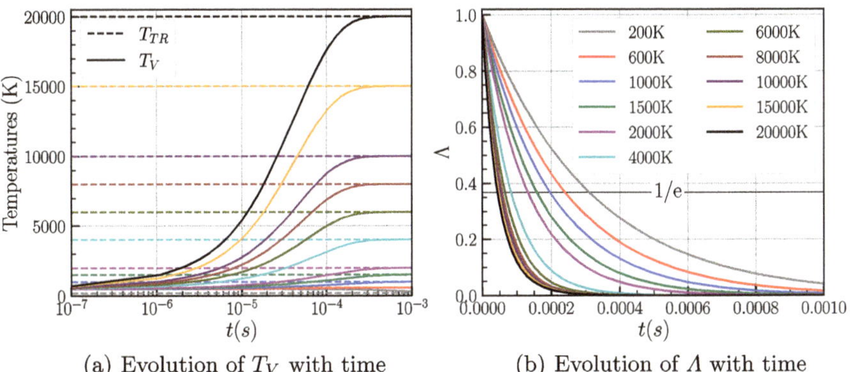

(a) Evolution of T_V with time

(b) Evolution of Λ with time

Fig. 1 Evolution of **a** T_V and **b** Λ with time for different initial T_{RT}

where d, ω and T_{ref} are the variable hard sphere (VHS) diameter, exponent and reference temperature, respectively, and m_R is the reduced mass for the NO-O collision pair. The values of τ_v derived from Fig. 1b and Eqs. 1 and 7 are then used to calculate Z_v^C.

The variation of the product of the partial pressure of the target molecule, p_{xx}, and τ_v with respect to different initial T_{RT} is shown in Fig. 2a where xx refers to NO (green and blue curves) and O_2 (red curve) using the results of Kulakhmetov et al. [10] for the O_2-O mixture. The relaxation times derived here follow trends noted by Torres and Schwartzentruber [13] for comparisons of their NO+O vibrational relaxation times with rates previously available in literature. Figure 2b illustrates the variation in Z_v^C with respect to different initial T_{RT} values with blue and red squares indicating our current calculations and the calculations performed by Kulakhmetov et al. [10], respectively. The red curve is generated using the fit for Z_v^C defined in Eq. 2 and the dashed purple curve represents the new fit based on our current solutions of the master equation, that is,

$$Z_v^C = 21.084 + \frac{3476.852}{122.958 + (T_{RT}/408.665)^3} .$$

(8)

The 0-D solutions indicate that NO-O collisions experience faster vibrational relaxation compared to those of O_2-O, consistent with QCT simulations [2, 10].

To test our implementation of Eq. 8 in the DSMC solver, we performed 0-D DSMC simulations (see Fig. 2a) that can be compared with the master equation results. We also implemented the correction proposed by Gimelshein et al. [7] to resolve the inconsistency between the continuum and discrete [3] definitions of the vibrational relaxation number. The results from 0-D DSMC simulations show a close match with the master equation results demonstrating the consistent implementation of the new fit in the DSMC solver.

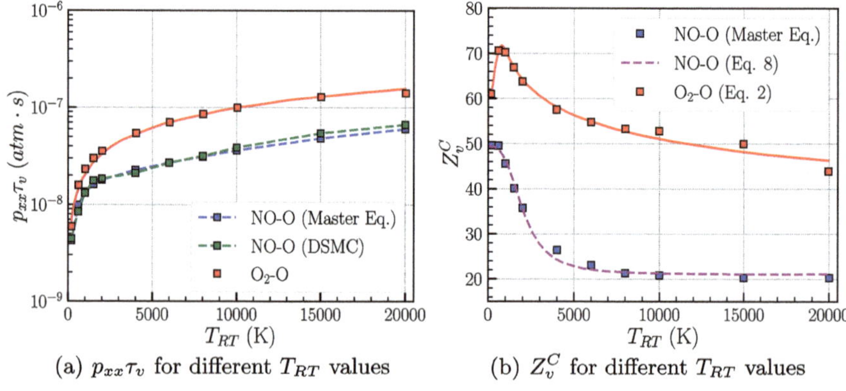

(a) $p_{xx}\tau_v$ for different T_{RT} values

(b) Z_v^C for different T_{RT} values

Fig. 2 Variation of $p_{xx}\tau_v$ and Z_v^C as a function of T_{RT} for NO-O and O_2-O collisions. Note that pressure is varied with T_{RT} for the same mole fraction of NO or O_2 for all cases considered

3 Impact on Emission Spectra

To assess the impact of faster NO-O vibrational relaxation times predicted by the fit for Z_v^C described in Eq. 8, we simulate a hypersonic flow over a cylinder as shown in Fig. 3. Table 1 presents the freestream conditions, which are based on the work of Yanes et al. [15], and DSMC computational parameters. Knudsen number is defined as the ratio of the freestream mean free path and the radius of the cylinder and FNum is the number of physical molecules represented by a simulated molecule. The cylinder is modeled as an isothermal wall at 300 K and has a diameter of 31.75 mm. The domain has a total of 1.84×10^8 simulation particles, with at least 200 particles in each collision cell. A five species air mixture with 19 chemical reactions [8] were modeled in the DSMC simulations. A high fidelity non-equilibrium model that uses the Bias model [8] to account for vibrational favoring in dissociation reactions was used. Furthermore, the high fidelity model used discrete vibrational energy levels so as to model internal modes more accurately. The forced harmonic oscillator [1]

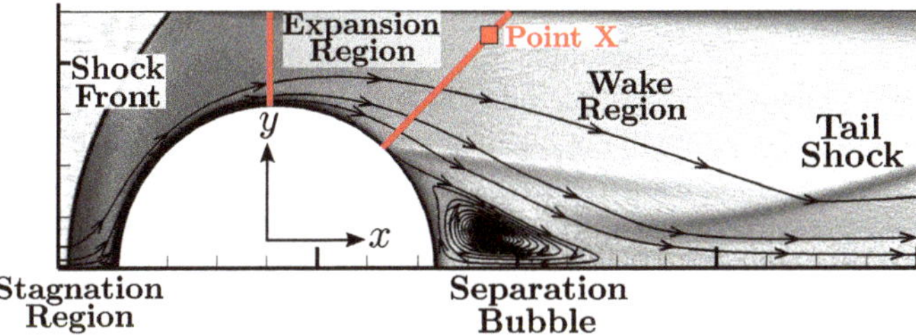

Fig. 3 Schematic showing important flow features for hypersonic flow over a cylinder

Table 1 Freestream conditions and DSMC computational parameters

Parameter	Value
Velocity (m/s)	3950
Temperature (K)	752
Mach Number	7.2
Number density (m^{-3})	7.233×10^{22}
Mean free path (m)	2.092×10^{-5}
Knudsen number	1.318×10^{-3}
Reynolds number** (m^{-1})	4.152×10^5
Sampling cell size (m)	1.0×10^{-4}
Number of time steps sampled	50, 000
Time step (s)	1.0×10^{-9}
FNum	3.0×10^{11}
Domain size ($x \times y$)	10×2.5 cm

*Reynolds number based on freestream viscosity computed using the VHS model [3] parameters given by Gimelshein and Wysong [8]

model was used to calculate vibrational relaxation rates for O_2 and N_2 for collisional processes involving all species present in the flow, and energy redistribution after a reaction was performed using the LB [11] model. We consider the two sets of rates described in Eqs. 2 and 8 to model the NO-O vibrational relaxation mechanism.

Flowfield macroparameters are extracted along the vertical line at the shoulder and the 45° line in the expansion region (red lines in Fig. 3). We are not showing comparisons of solutions in the stagnation region and along the wake line since both sets of Z_v^C give similar predictions in these regions. Figure 4 shows the temperatures and NO mole fractions along the vertical and 45° lines with solutions from the Z_v^C fits given by Eqs. 2 and 8 denoted as 'Old Z_v^C' and 'New Z_v^C', respectively. We note that there are negligible differences in the bulk temperatures with the use of the new Z_v^C fit which is expected as the bulk temperatures are dominated by contributions from N_2 molecules which have mole fractions in the range of $0.67 - 0.78$ throughout the domain. Furthermore, even though there are negligible differences in NO mole fraction values, the calculation performed with the new Z_v^C fit predicts NO vibrational temperatures $\sim$300 K and $\sim$400 K lower than those computed with the old Z_v^C fit along the vertical and 45° lines respectively. This difference in solutions is small near the cylinder wall and peaks at the domain boundary.

Figure 5a presents NO vibrational level populations at a point 17 mm from the cylinder wall, Point X in Fig. 3, which is characterized by a $\sim$400 K difference in NO vibrational temperatures predicted by the two sets of Z_v^C. The distributions marked as 'DSMC' are obtained by sampling and binning NO simulation particles over the discrete vibrational energy levels directly from the DSMC simulations once the flowfield is fully developed. It is to be noted that though the sampling process involved gathering statistics for higher vibrational states we are reporting distributions only up to a vibrational level of six due to the higher statistical noise associated with the higher states. The population distributions marked as 'Boltz, NO T_v' are the Boltzmann population distributions evaluated at the NO vibrational temperature at Point X. As can be seen, the population distributions obtained with the DSMC sampling procedure are similar to those obtained using the Boltzmann population distribution

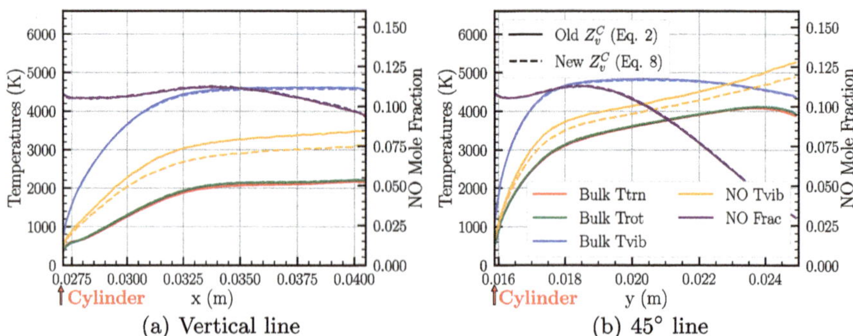

(a) Vertical line　　　　(b) 45° line

Fig. 4 Comparison of bulk temperatures, NO vibrational temperatures and NO mole fractions obtained by the two sets of Z_v^C approximations

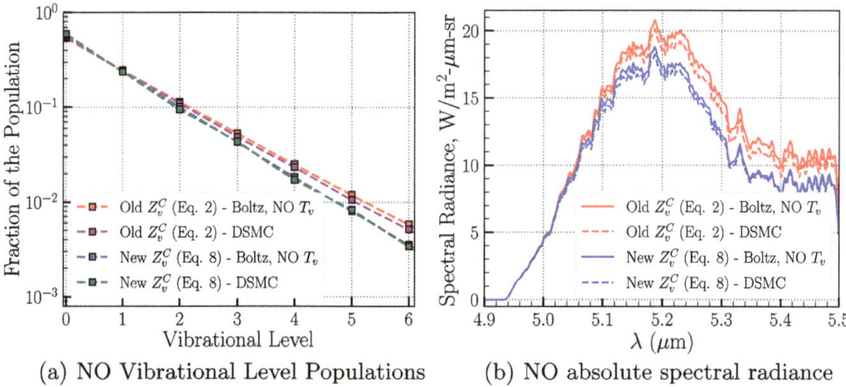

(a) NO Vibrational Level Populations

(b) NO absolute spectral radiance

Fig. 5 NO **a** vibrational level populations and **b** absolute spectral radiance computed at Point X along the 45° line

function for both sets of Z_v^C considered in this work. We also note that the population distributions are noticeably different between the two sets of Z_v^C considered with the larger slope for the distribution predicted using the 'New Z_v^C' fit signifying lower NO vibrational temperatures as compared to that predicted using the 'Old Z_v^C' fit.

We used a modified version of the Nonequilibrium Radiative Transport and Spectra Program (NEQAIR) [6, 14], which computes emission and absorption intensities for relevant IR transitions based on user-specified vibrational state populations, to compute emission spectra. Figure 5b presents absolute spectral radiance calculated along a line-of-sight with constant flow properties passing through Point X and along the spanwise direction with a path length of 95.25 mm. The translational, rotational and vibrational temperatures of NO at Point X are provided as input to NEQAIR along with NO vibrational state populations shown in Fig. 5a. A Smeared Gaussian lineshape [6] was used to model the instrument lineshape in NEQAIR which was also used by Thirani et al. [12] to model NO IR emission spectra in the NASA Electric Arc Shock Tube facility.

We observe that the set of spectra generated using flowfield solutions predicted by the 'New Z_v^C' fit have lower radiance values as compared to those generated using flowfield solutions predicted by the 'Old Z_v^C' fit. This is a direct result of lower NO vibrational temperature values predicted by the 'New Z_v^C' fit. Furthermore, since the spectra generated using the 'Boltz, NO T_v' populations are similar to those generated with 'DSMC' populations, reasonable estimates of emission spectra can be obtained while using the Boltzmann distribution evaluated at NO vibrational temperature for flowfields similar to that considered in this work. This does, however, necessitate an accurate estimate of the vibrational temperature.

4 Conclusion

A new fit for the vibrational relaxation number, Z_v^C, for NO-O vibrational state-to-state collisions was obtained in this work and its impact was evaluated for a hypersonic flow over a cylinder. We observed that the faster relaxation rates led to only negligible differences in the bulk temperatures and NO mole fractions while leading to lower NO vibrational temperatures in regions where the flow is expanding. Furthermore, the maximum difference in NO vibrational temperatures between the two set rates was $\sim$400 K. This difference in NO vibrational temperatures is reflected in absolute radiance values for the IR emission spectra which can be generated using the Boltzmann population distribution computed using an accurate estimate of NO vibrational temperatures.

Acknowledgements The research is supported by the Air Force Office of Scientific Research through AFOSR Grant No. FA9550-19-1-0342 and computational resources are provided by the National Science Foundation's ACCESS program with the project number TG-PHY210065. The authors are thankful to Dr. Daniil Andrienko and Alexander Fangman for providing the QCT rates for NO-O vibrational relaxation mechanisms. The authors are also thankful to Dr. Sergey F. Gimelshein for help with running the 0-D thermal relaxation bath cases in the DSMC solver.

References

1. Adamovich IV, Macheret SO, Rich JW, Treanor CE (1996) Vibrational energy transfer rates using a forced harmonic oscillator model. J Thermophysics Heat Transfer 12(1):57–65
2. Andrienko DA, Boyd ID (2018) State-resolved characterization of nitric oxide formation in shock flows. AIAA SciTech 2018 forum
3. Bird GA (1994) Molecular gas dynamics and the direct simulation of gas flows. Clarendon, Oxford, England, UK
4. Chang E, Streicher JW, Krish A, Hanson RK (2022) High-temperature infrared-based diagnostic for nitric oxide using tunable diode laser absorption spectroscopy. AIAA SciTech 2022 Forum
5. Cruden BA, Tibere-Inglesse AC (2024) Radiative emission in incident air shocks from 3–7 km/s. AIAA Aviation Forum and Ascend 2024
6. Cruden BA, Brandis AM (2014) Updates to the neqair radiation solver. In: 6th international workshop on radiation of high temperature gases in atmospheric entry. St. Andrews, Scotland
7. Gimelshein NE, Gimelshein SF, Levin DA (2002) Vibrational relaxation rates in the direct simulation Monte Carlo method. Phys Fluids 14(12):4452–4455
8. Gimelshein SF, Wysong IJ (2019) Nonequilibrium air flow predictions with a high-fidelity direct simulation Monte Carlo approach. Phys Rev Fluids 4(3):033405
9. Herzberg G (1950) Molecular spectra and molecular structure 1. Spectra of diatomic molecules. D. Van Nostrand Company, New York
10. Kulakhmetov MF, Sebastiao IB, Alexeenko A (2016) Adapting vibrational relaxation models in DSMC and CFD to ab-initio calculations. In: 46th AIAA thermophysics conference
11. Larsen PS, Borgnakke C (1975) Statistical collision model for Monte Carlo simulation of polyatomic gas mixture. J Comput Phys 18(4):405–420
12. Thirani S, Karpuzcu IT, Levin DA (2024) Experimental validation of modeling efforts to estimate nitric oxide emission in high Mach number flows. AIAA Scitech 2024 Forum

13. Torres E, Schwartzentruber TE (2025) Characteristic vibrational and rotational relaxation times for air species from first-principles calculations. J Thermophysics Heat Transfer 0(0):1–27
14. Whiting EE, Park C, Liu Y, Arnold JO, Paterson JA (1996) Neqair 96 user manual. Tech rep, NASA Ames Research Center, Moffett Field, CA
15. Yanes NJ, Austin JM (2019) Nitric oxide spectroscopic measurements in a hypervelocity stagnation flow. In: AIAA Paper 2019–0794

Adaptive Particle Discretization Methods for Multiscale Non-equilibrium Flows

Simone Lauterbach, Stefanos Fasoulas, and Marcel Pfeiffer

Abstract Highly accurate simulation results for complex flow fields with strong nonequilibrium and noncontinuum effects, such as present during atmospheric entry processes, can be obtained by the Direct Simulation Monte Carlo (DSMC) method. However its use is limited by the increasing computational cost in denser regions. In these cases, efficiency-increasing methods are required. Alternative particle methods, such as Fokker-Planck (FP) or Bhatnagar-Gross-Krook (BGK) can achieve similar accuracy to DSMC in regions with lower Knudsen numbers, but with a lower computational demand. Additionally, these methods allow for a straightforward coupling to DSMC. Several optimization techniques have been implemented in the simulation code PICLas, including adaptive particle weights and a mesh refinement. In addition, various equilibrium breakdown criteria have been implemented for an efficient coupling between DSMC and other particle-based methods. All implemented adaption techniques are evaluated in terms of their accuracy and efficiency, in comparison to a non-adapted DSMC simulation.

Keywords DSMC · BGK · FP · Adaptive methods · Efficiency increase

1 Introduction

The efficient treatment of multiscale flows poses a significant challenge for many applications in the field of aerospace science. During the atmospheric entry process, the spacecraft enters the atmosphere in the rarefied regime, with a fast transition into denser regions at lower altitudes. Additionally, while the free-stream of the flow on the vehicle can be well in the continuum regime, rarefaction effects can become important in the wake of the flow. The Direct Simulation Monte Carlo (DSMC) method can accurately treat non-equilibrium effects, but has a too high computational demand for the transitional and continuum regime [1]. Alternative particle-based continuum methods, such as the Bhatnagar-Gross-Krook (BGK) method [2] and the Fokker-

S. Lauterbach (✉) · S. Fasoulas · M. Pfeiffer
University of Stuttgart, Stuttgart, Germany
e-mail: lauterbachs@irs.uni-stuttgart.de

M. Grabe et al. (eds.), *Rarefied Gas Dynamics*, Springer Aerospace Technology,
https://doi.org/10.1007/978-3-032-00094-1_61

Planck (FP) [3, 4] method have gained importance in the recent years. These methods have less stringent constraints on the temporal and spatial discretization, as the mean collision time and the mean free path do not need to be resolved. At the same time, the methods have a similar accuracy compared to DSMC in the transition regime and a straightforward coupling to DSMC is possible, as the numerical algorithm for the methods differs only in the collision treatment. Therefore, interface regions are not required in the coupling procedure.

Apart from the use of alternative numerical methods, the efficiency of the particle methods can be further improved by a meaningful choice of the underlying simulation set-up. Parameters, such as the particle weight or the cell size, can be optimized in such a way that a higher discretization and thus a higher computational effort is used only in the regions of interest, while still ensuring physical validity throughout the simulation domain.

Several adaptive procedures for all particle-based methods have been implemented in the open-source gas and plasma simulation tool PICLas [5], including an automatic adaption of the particle weight per cell and an anisotropic refinement of the mesh by the use of virtual subcells. Furthermore, various continuum breakdown criteria have been implemented to allow for an efficient coupling between DSMC and other available particle methods.

The implemented methods are presented in the following in more detail.

2 Numerical Methods

The Boltzmann equation is the basis for the description of the microscopic gas state and the change in the particle distribution function $f(\mathbf{x}, \mathbf{v}, t)$ in dependence of the particle positions $\mathbf{x}$, velocities $\mathbf{v}$ and the time t, together with the collision term $\frac{\partial f}{\partial t}|_{\text{Coll}}$. External forces are neglected in this equation:

$$\frac{\partial f}{\partial t} + \mathbf{v} \cdot \nabla_{\mathbf{x}} f = \frac{\partial f}{\partial t}|_{\text{Coll}} \tag{1}$$

An analytical solution of the Boltzmann equation is not possible, therefore particle methods are used to approximate the Boltzmann collision integral. The physical gas flow is represented by the movement and collision of weighted simulation particles. Particle motion and interactions are treated independent of each other, with the motion step following Newton's laws. The particle methods in PICLas are all of stochastic nature, but differ in the collision treatment. In a DSMC simulation, binary particle pairs are determined, for which a collision probability is calculated and compared with a random number in order to decide whether an interaction takes place. In the BGK method, the Boltzmann collision integral is approximated by the relaxation towards a target distribution function f^t, chosen in a way to achieve the correct Prandtl number:

$$\frac{\partial f}{\partial t}\Big|_{\text{Coll}} = \nu(f^t - f) \tag{2}$$

The FP approach approximates the Boltzmann equation by a Fokker-Planck equation, in dependence of a drift A_i and a diffusion term D_{ij}:

$$\frac{\partial f}{\partial t}\Big|_{\text{Coll}} = -\sum_{i=1}^{3} \frac{\partial}{\partial v_i}(A_i f) + \frac{1}{2}\sum_{i=1}^{3}\sum_{j=1}^{3} \frac{\partial^2}{\partial v_i \partial v_j}(D_{ij} f) \tag{3}$$

2.1 Coupled Particle-Methods

As the algorithms of the particle methods in PICLas differ only in the approximation of the Boltzmann collision integral, a coupling of the methods can be performed directly by a cell-local switch of the collision routine. The particles carry all information through the simulation domain, therefore no data exchange or interface region is necessary between the methods. The coupling strongly depends on a meaningful continuum breakdown criterion for an optimal compromise between accuracy and efficiency. In PICLas, multiple coupling parameter can be employed individually or in combination with each other. The simulations are in all cases initialized with the more efficient BGK or FP method. Afterwards, the domain division is updated dynamically throughout the simulation to capture emerging non-equilibrium regions by the more accurate DSMC method.

The most simple criterion in PICLas is a density threshold, above which the more efficient particle method is employed. In a slightly more sophisticated setup, the global Knudsen number Kn_G of each cell can be calculated from the mean free path λ_c and a characteristic length L:

$$Kn_G = \frac{\lambda_c}{L} \tag{4}$$

If Kn_G exceeds a given value in a simulation cell, the DSMC method is used, while the remaining part of the simulation domain is treated by the BGK or FP method. Both the density threshold and the global Knudsen number typically lead to similar predictions of non-equilibrium without a significant increase in computational demand. However, both criteria are not representative of local effects of non-equilibrium and the definition of the macroscopic length scale in the global Knudsen number remains somewhat arbitrary.

A more suitable estimator for local continuum breakdown is the gradient-length-local Knudsen number Kn_L as defined by Burt and Boyd [6]. Here, continuum breakdown is defined to occur if the gradients of the flow variables Q are large:

$$Kn_L = \frac{\lambda_c}{Q}\left|\frac{dQ}{dx}\right| \tag{5}$$

In PICLas, the maximum value for the density, velocity and temperature gradients are used. However, a disadvantage of this criterion is the increase in the computational demand due to the gradient calculations.

An alternative parameter, that is also representative of local effects of non-equilibrium, is the simplified Chapman-Enskog parameter. Continuum breakdown is defined to occur if the maximum value of the normalized heat flux vector and shear stress tensor exceed a given threshold [7]:

$$B = \max(\left| q_i^* \right|, \left| \tau_{ij}^* \right|) \tag{6}$$

In addition to the above defined breakdown parameter, an estimator for thermal non-equilibrium is used in the simulation of di- and poly-atomic species:

$$\Delta T = \sqrt{\frac{(T_{x,y,z} - T_{\text{tot}})^2 + \xi_\text{R}(T_\text{R} - T_{\text{tot}})^2 + \xi_\text{V}(T_\text{V} - T_{\text{tot}})^2}{(3 + \xi_\text{R} + \xi_\text{V})T_{\text{tot}}^2}} \tag{7}$$

2.2 Adaptive Particle Weights

Preliminary simulation results can be used to automatically determine the optimal particle weight per simulation cell in PICLas. The adaption aligns the particle weight with the number density in the cell, while also ensuring that the necessary particle discretization is met throughout the simulation domain. To avoid jumps in the particle weights between neighboring cells, an averaging filter is used on the initial weight distribution.

When a particle moves to a new cell with a lower weight, it is cloned with a probability, based on the weighting factor before and after the particle movement step, w_{old} and w_{new}:

$$P_{\text{clone}} = \frac{w_{\text{old}}}{w_{\text{new}}} - \text{INT}\left(\frac{w_{\text{old}}}{w_{\text{new}}}\right) \tag{8}$$

Clones are introduced over the course of multiple time steps and are inserted in random order to promote collisions between nearly created clones and other particles. To avoid unphysical effects, such as the Avalanche phenomenon [8], more than one clone per particle is not allowed.

If a particle moves to a cell with a higher weight, the deletion probability is calculated from the respective cloning probability for the opposite moving direction:

$$P_{\text{delete}} = 1 - P_{\text{clone}} \tag{9}$$

The deletion rate per time step is restricted to half of the particle flux to a new cell, in order to limit the loss of information.

In the calculation of the collision probability, an average weighting factor is used for particles with different weights. Energy and momentum are not conserved in each collision but on average over many collisions.

2.3 Anisotropic Mesh Adaption

Coarse simulation meshes in PICLas can be handled dynamically by an automatized cell refinement. Virtual subcells are introduced to reduce the separation distance between collision partners for DSMC and to improve the spatial resolution of the moment estimators for BGK and FP. In contrast to adaptive remeshing, the virtual subcells are considered only for the collision and relaxation step, which reduces the overall memory requirement.

Mesh cells undergo refinement if the local mean free path in the cell for DSMC or the gradients of the density, velocity and temperature for BGK and FP exceed a predefined threshold. In contrast to an Octree refinement [9], the implemented mesh adaption is not recursive and allows to re-use the virtual subcell structure for a given number of iterations. Thus, the overall computational demand of the method is reduced. The number of virtual subcells per mesh cell is calculated from the number of simulation particles in the cell and a predefined minimum particle number:

$$N_{\text{cell}} = \text{INT}\left[\left(\frac{N_{\text{part}}}{N_{\text{min}}} \right)^{1/3} \right] \tag{10}$$

For BGK and FP, a minimum number of 12 particles per cell has proven to result in meaningful macroscopic coefficients.

As the particle methods in PICLas make use of unstructured meshes, the virtual subcells are defined around equally spaced interpolation points in the the $[-1, 1]^3$ reference space. Particles are first mapped to the reference space by a Jacobian mapping, in which all subcell and position operations are performed. This mapping allows for a more efficient handling of the mesh adaption, as the calculation of particle distances would otherwise be costly, especially for larger numbers of interpolation points per cell.

To ensure that the minimum requirement of simulation particles per cell is met, the subcells can be merged along a space-filling Peano curve. This cell merge is of particular importance in BGK or FP simulations to achieve qualitatively good results. The Peano curve initially connects the cells in planar slices and only after the slices are merged completely, the third dimension is taken into account. A favored direction, which is only dissolved in the last step, emerges naturally from this procedure. In the current implementation, the merge order is based on the relative size of the directional temperature gradient, such that the cells along the direction of the highest gradient are always merged last. In addition, the merge procedure along the Peano curve can be used to generate an anisotropic refinement in contrast to the isotropic Octree scheme.

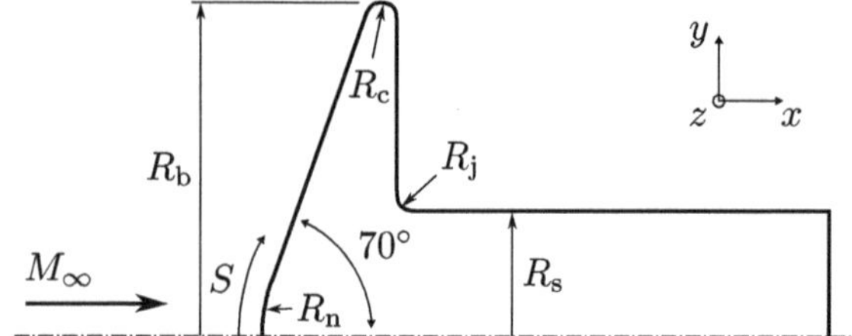

Fig. 1 Schematic representation of the simulated 70° blunted cone that is taken as input for PICLas. Simulated is the inflow of molecular nitrogen, with an angle $\alpha = 0°$

3 Hypersonic Flow Around a 70° Blunted Cone

As a test case for the adaptive methods, the hypersonic flow of nitrogen around a 70° blunted cone, as illustrated in Fig. 1, is chosen. To allow for a comparison to experimental heat flux values, the free-stream conditions are chosen according to [10], with $v_\infty = 1633.8$ m/s, $T_\infty = 15.29$ K and $n_\infty = 1.002 \times 10^{-22}$ m^3/s. The flow case has a Mach number of $M_\infty = 20$ and a Knudsen number $K_n = 0.005$. For the surface, full thermal accommodation is assumed, with a constant wall temperature of $T_w = 300$ K. To reduce the overall computational effort, the symmetric flow on the cone is made use of in a 2D-axisymmetric simulation setup. First, a highly resolved and non-adapted reference DSMC simulation is performed, with a time step $\Delta t = 2 \times 10^{-9}$ s and weighting factor $\omega = 1 \times 10^9$. The test case is further evaluated by ESFP, ESBGK and the respective coupled simulations with DSMC. Here, the same time step as for the DSMC case is used. However, the simulation cases are initialized with a higher weighting factor $\omega = 1 \times 10^{11}$. The coupling of DSMC with ESBGK or ESFP is performed based on the local Knudsen number. As a threshold, $Kn_L = 0.1$ was chosen, above which the DSMC method is utilized.

This value has proven to still produce good results, but no longer led to a significant reduction in the computational effort. In addition, the newly implemented automatic determination of the particle weights and the automatic gradient-based mesh refinement is employed for these simulation cases.

The accuracy of the adapted simulation cases is evaluated by comparing the results to the non-adapted simulations, as well as to the reference DSMC simulation and experimental values. In Fig. 2 the heat flux values for the various ESBGK and ESFP simulation cases are compared. The non-adapted ESBGK result overestimates the heat flux at the front of the cone, while good agreement can be observed for both the adapted ESBGK and ESBGK-DSMC simulation with the DSMC results and the experimental values. As for the ESBGK cases, the non-adapted ESFP simulation overestimates the heat flux at the front significantly. The use of the adaptive methods in both the ESFP and the coupled ESFP-DSMC simulations led to an increase in the accuracy, however the heat flux values still are still not in perfect agreement with the DSMC results.

To evaluate the efficiency of the various methods, the computational time and the required number of simulation particles for a resolved simulation are compared in Table 1. The simulation time for the non-adapted DSMC method is utilized as

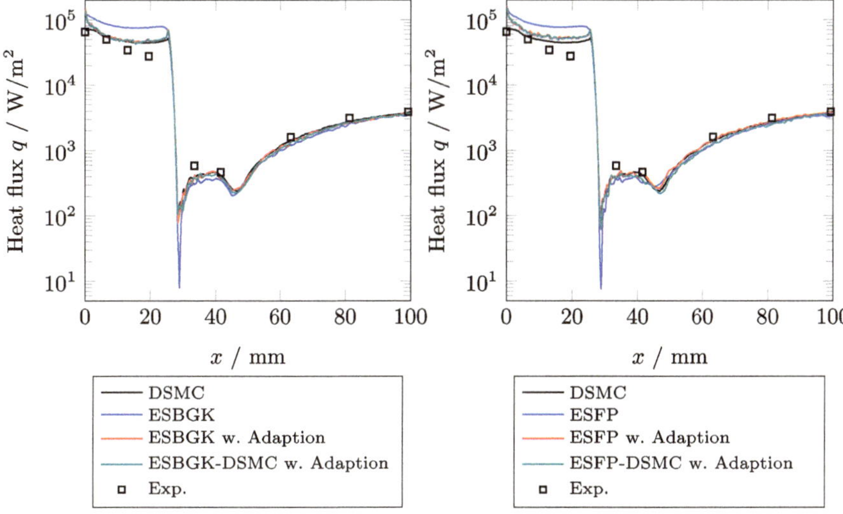

Fig. 2 Comparison of the measured and calculated heat flux over the length of the surface for the flow around the 70° blunted cone

Table 1 Comparison of the total simulation particle number and the required computational time for the different methods

	N_{part}	CPU h$_{\text{DSMC}}$/CPU h
DSMC	$1.24 \times 10^{+8}$	1
ESBGK	$1.81 \times 10^{+6}$	223.5
ESBGK w. adaption	$1.01 \times 10^{+6}$	242.9
ESBGK-DSMC w. adaption	$1.00 \times 10^{+6}$	252.3
ESFP	$1.81 \times 10^{+6}$	258.7
ESFP w. adaption	$9.96 \times 10^{+5}$	232.5
ESFP-DSMC w. adaption	$9.98 \times 10^{+5}$	264.4

a reference point. While a large particle number was required to achieve properly resolved DSMC results, approximately two order of magnitudes fewer particles could be employed in the ESBGK and ESFP simulations. These methods are not based on binary particle collisions and therefore do not require the resolution of the mean free path. With this lower particle number, the computational effort could be significantly reduced in comparison to the DSMC reference. With adaptive particle weights, the simulations showed a further increase in efficiency when compared to the non-adapted ESBGK and ESFP simulation.

Overall, the adapted ESBGK and ESBGK-DSMC showed the best performance in the test case, with a good overall agreement to the reference DSMC results, but a significantly reduced computational effort.

4 Conclusion

Several adaptive procedures for the particle methods are implemented into PICLas, including an anisotropic refinement of the simulation mesh and an automatized determination of the optimal particle weight for each cell. In addition, the coupling between different particle methods is extended from a density-based criterion by multiple, more sophisticated continuum-breakdown criteria.

The implementation and functionality of the adaptive methods are verified by the simulation of the hypersonic flow of nitrogen around a 70° blunted cone. The application of the adaptive methods allowed for a significant reduction of the computational time, while ensuring good overall agreement to experimental heat flux values.

In future work, the adaptive weighting scheme will be extended for the handling of trace species, especially for chemical reactions. In addition, the efficiency of the mesh adaption process will be further increased, in such a way, that the optimal number of subcells is directly calculated at the start of the simulation, without the need for a cell merging process.

Acknowledgements The authors gratefully acknowledge funding provided by Airbus Defence and Space and by ArianeGroup. This project has received funding from the European Research Council (ERC) under the European Union's Horizon 2020 research and innovation programme (grant agreement No. 899981 MEDUSA).

References

1. Bird G (1994) Molecular gas dynamics and the direct simulation of gas flows. Oxford University Press
2. Bhatnagar P, Gross E, Krook M (1954) A model for collision processes in gases. I. Small amplitude processes in charged and neutral one-component systems. Phys Rev 94:511
3. Gorji M, Jenny P (2014) An efficient particle Fokker-Planck algorithm for rarefied gas flows. J Comput Phys 262:325–343
4. Pfeiffer M, Gorji M (2017) Adaptive particle–cell algorithm for Fokker-Planck based rarefied gas flow simulations. Comput Phys Commun 213:1–8
5. Fasoulas S, Munz C, Pfeiffer M, Beyer J, Binder T, Copplestone S, Mirza A, Nizenkov P, Ortwein P, Reschke W (2019) Combining particle-in-cell and direct simulation Monte Carlo for the simulation of reactive plasma flows. Phys Fluids 31
6. Burt J, Boyd I (2009) A hybrid particle approach for continuum and rarefied flow simulation. J Comput Phys 228:460–475
7. Garcia A, Alder B (1998) Generation of the Chapman-Enskog distribution. J Comput Phys 140:66–70
8. Galitzine C, Boyd I (2015) An adaptive procedure for the numerical parameters of a particle simulation. J Comput Phys 281:449–472
9. Pfeiffer M, Mirza A, Fasoulas S (2013) A grid-independent particle pairing strategy for DSMC. J Comput Phys 246:28–36
10. Allegre J, Bisch D, Lengrand J (1997) Experimental rarefied heat transfer at hypersonic conditions over 70-degree blunted cone. J Spacecraft Rockets 34:724–728

An Effective Temperature Formulation for Predicting State-Specific Reaction Rates and Simulating One-Dimensional Hydrogen-Air Detonations in DSMC

Shrey Trivedi, Ahren W. Jasper, John K. Harvey, and Jacqueline H. Chen

Abstract In this paper, simulations of hydrogen-air detonations are performed in the continuum regime using Direct Simulation Monte Carlo (DSMC). Continuum codes assuming thermal equilibrium or even with a multi-temperature approach may be inadequate for simulating conditions with strong thermal and chemical non-equilibrium, and hence DSMC is utilized here. There are two aspects of this study: (1) One-dimensional (1-D) detonation waves are simulated in the continuum regime, under near-equilibrium conditions using the established TCE model and (2) a novel Effective Temperature (T_{eff}) formulation accounting for vibrational non-equilibrium, based on the equipartition principle, is tested in DSMC. For detonation analysis, a preheated case at Mach number M = 3.0, with reactants at an initial temperature of 900 K and initial pressure of 0.3 atm is simulated. The results are compared with the Zel'dovich-von Neumann-Döring (ZND) solution obtained from the Shock and Detonation (SDT) toolbox. It is found that DSMC results in a robust and steady detonation structure and shows excellent agreement with the ZND solution. For conditions with non-equilibrium effects, reaction rates specific to individual ro-vibrational states need to be incorporated. To this effect, the use of T_{eff} within the TCE model results in good agreement of state-specific reaction rates with those obtained from the Quasi-Classical Trajectory (QCT) calculations.

Keywords DSMC · 1D detonation · Hydrogen-air reacting flows

S. Trivedi (✉) · J. H. Chen
Sandia National Laboratories, Livermore, CA, USA
e-mail: strived@sandia.gov

A. W. Jasper
Argonne National Laboratories, Lemont, IL, USA

J. K. Harvey
University of Cambridge, Cambridge, UK

© The Author(s) 2026

M. Grabe et al. (eds.), *Rarefied Gas Dynamics*, Springer Aerospace Technology,
https://doi.org/10.1007/978-3-032-00094-1_62

639

1 Introduction

The Direct Simulation Monte Carlo (DSMC) method proposed by Bird [2] has been used extensively for solving rarefied flow problems in the upper atmosphere. However, DSMC is also well-suited for combustion problems with significant non-equilibrium effects or when molecular transport is not adequately treated, for example, in Scramjets or Rotating Detonation Engines. In DSMC, the motion of molecules is solved in a time-accurate manner in two phases, namely a move phase and a collision phase [1]. These are carried out in alternate time steps. In the move phase, the particles move in a straight line unless hitting a surface or leaving the domain. In the collision phase, the sample colliding partners are appropriately selected and their interaction is treated through collision models. These two phases effectively solve the Boltzmann equation. However, the DSMC time step and cell size must be comparable to the mean collision time and the mean free path respectively. This makes DSMC computationally expensive.

The objective of this paper is to present hydrogen-air detonation using DSMC in the continuum regime. To the best of our knowledge, no such analysis has been performed within the framework of DSMC. Combustion studies using DSMC are rare [16–18]. Furthermore, DSMC analyses of flows in the continuum or near-continuum regime are also limited [20, 21]. Some preliminary hydrogen-oxygen detonation studies exist [3, 7, 25] utilizing a less detailed reaction mechanism. The hydrogen-oxygen system has a lower activation energy, making it easier to ignite, and it has a more steady detonation structure. Detonations show rapid discontinuous changes, and the effect of non-equilibrium can be prominent. It has been shown, for example, that the non-equilibrium effects can change the ignition delay time in Scramjets [8]. DSMC can account for molecular phenomena such as transport processes, internal energy relaxation, and chemical reactions, directly through molecular collisions. With accurate molecular data based on first principles, DSMC is a robust tool to gain insights into the fundamental problem of reacting flows under non-equilibrium conditions.

In cases of non-equilibrium, collisional cross sections can be evaluated from ab-initio data or using models based on molecular dynamics [16]. Models based on Quasi-Classical Trajectory (QCT) calculations are being developed for internal energy relaxation and dissociation reactions [9, 11, 15, 24]. However, these calculations require potential surfaces for each of the collision pairs which are yet ready for all of the reactions describing hydrogen-air chemistry. To this effect, a simpler effective temperature model is presented in this paper, based on the equipartition principle [10], to account for the vibrational non-equilibrium. For treating collisions between particles, the Variable Soft Sphere (VSS) model [5] is used in this analysis. For post-collisional redistribution of total energy, the Larsen-Borgnakke [4] model is commonly utilized and is used here. The open access code SPARTA [14] is employed in this analysis.

2 Effective Temperature Formulation

Effective temperature models approximate rate constants for non-thermal reactions using a single effective temperature. Here, the non-thermal rate constant k^* for a bimolecular reaction with a distinct reactant vibrational energy E_{vib}, reactant rotational temperature T_{rot}, and relative translational temperature T_{trans} is evaluated from the thermal rate constant k as follows

$$k^*(E_{vib}, T_{rot}, T_{trans}) \sim k(T_{eff}). \tag{1}$$

Here, $k(T)$ is the known thermal rate constant fit to an Arrhenius expression. It can be obtained either from experimental evaluations or from QCT calculations. The effective temperature formulation used in this study was validated for similar reactions in recent work [10] and uses the equipartition principle to evaluate T_{eff} as

$$T_{eff} = (T_{vib} n_{vib} + T_{rot} n_{rot}/2 + T_{trans} n_{trans}/2)/(n_{vib} + n_{rot}/2 + n_{trans}/2) \tag{2}$$

The variable n_x is the number of modes of type x. For example, for H + O_2, $n_{vib} = 1$, $n_{rot} = 2$, and $n_{trans} = 3$.

In DSMC since the simulation proceeds through the particle energies in different modes, i.e., E_{trans}, E_{vib}, and E_{rot}. To convert the collision energy, for example, E_{vib} to T_{vib}, the following expressions is used.

$$T_{vib} = E_{vib}/k_B n_{vib} \tag{3}$$

Note that the temperature in DSMC is evaluated from the energy of a collision pair, and not the cell-averaged values. Finally, the total collision energy (TCE) model is used to evaluate the reaction probabilities. The default TCE models only makes use of the translational temperature in evaluating the reaction probabilities. However, the implementation of effective temperature in DSMC is achieved by replacing the translational temperature in the default TCE model with the effective temperature T_{eff} (Eq. 2).

3 Results

3.1 Reaction Rates Using Effective Temperature Formulation

Reaction 1: H + O_2 → OH + O The reaction H + O_2 → O + OH is well known to be an important chain branching reaction in the hydrogen-air reaction mechanism. There are many Potential Energy Surfaces (PES) for this reaction in the literature [23]. For this analysis, the PES from the reference [10] is chosen. Validation of this PES is provided through the comparison of the predicted thermal rate from QCT calculations with an experimental recommendation from Varga et al. [22]. Results from an earlier set of QCT calculations from Miller et al. [13] is also provided. The cross-sections obtained from the QCT calculations are integrated over all vibrational and rotational levels to calculate the equilibrium reaction rates. These integrated rates and the experimental results are shown in Fig. 1.

A good agreement can be observed for all cases, with current QCT giving a slightly better agreement with the experiments. Note that experimental results are limited to $\sim$3000 K whereas the temperature of interest for non-equilibrium conditions could exceed that value. Under such conditions, QCT can be utilized to evaluate reaction rates, especially when the effect of other internal modes (rotational and vibrational) needs to be taken into account. The integrated reaction rates from the current QCT calculations can be fitted to a thermal rate constant 5.078e–13 (T / 300 K) 1.635exp (– 2458 K / T) cm^3/s.

Next, vibrational states were included in the analysis to account for the non-thermal state with vibrational non-equilibrium. In the most detailed scheme, the vibrational and rotational quantum numbers of $O_2(v, j)$ as well as the translational temperatures of O_2 and H should be used (with T_{trans} for the reaction to be the arithmetic average of the two translational temperatures of the reactants). In this analysis, the rovibrational coupling has been ignored, and the rotational temperature is assumed to be equilibrated with translational temperature. The trans-rotational temperature is simply referred to as T in the rest of the analysis. In summary, we use v, and T or equivalently E_{vib}, and T.

The results for T = 4000 K and E_{vib} corresponding to $v = 0 - 9$ are shown in Fig. 2a. Note that O_2 has bound states up to $v = 41$. The dot symbols in this figure show the reaction rates directly from QCT. The dashed line shows the effective temperature model based on the mean vibrational temperature (Eq. 3) plugged simply into the Arrhenius formulation. The orange line shows the reaction rates from DSMC, obtained by using T_{eff} within the TCE model. The effective temperature model stays in good agreement with the QCT results. DSMC performs well at lower vibrational energy, but degrades slightly as Evib exceeds 10 kcal/mol. However, the values remain within a factor of 2–3 from the QCT results which can be considered a good

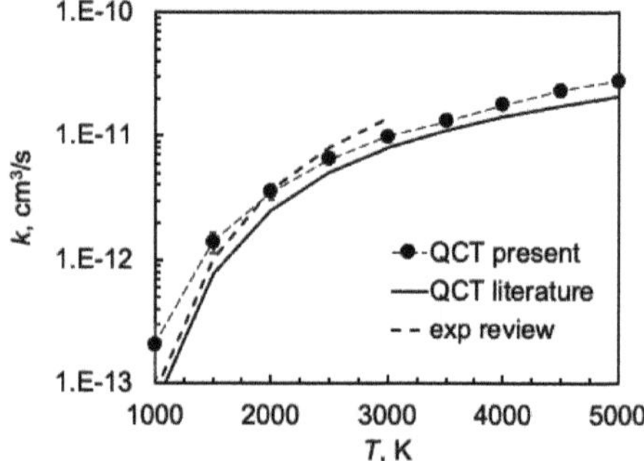

Fig. 1 Reaction rates obtained from the QCT calculations for reaction H + $O_2 \rightarrow$ OH + O from this work (solid dots), and Miller et al. [13] (solid line). The dashed line represents the measurements from Varga et al. [22]

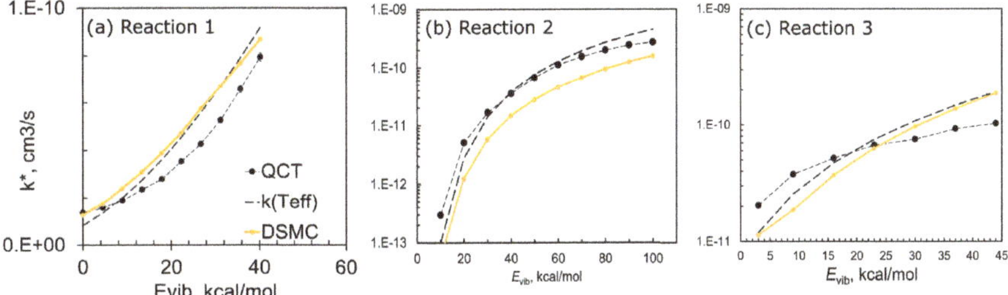

Fig. 2 Reaction rates for the reactions **a** $H + O_2 \rightarrow OH + O$, **b** $H_2O + H \rightarrow H_2 + OH$, and **c** $O_2 + H \rightarrow H_2 + O_2$, with vibrational non-equilibrium. Black dots show results from QCT calculations, black dashed line shows the Arrhenius formulation with the effective temperature model, orange dashed lines show the DSMC results using T_{eff} in the TCE model

agreement. This is also consistent with the findings of a previous analysis of CH_4 based reactions in Ref. [10].

Reaction 2 and 3 Results for the reaction $H_2O + H \rightarrow H_2 + OH$ are shown in Fig. 2b. For this reaction, T = 1000 K. The $n_{vib} = 3$ for H_2O. This can be accounted for in SPARTA through a separate file containing information about the different vibrational modes. For this reaction, the T_{eff} plugged into the Arrhenius form matches closely with the QCT results, whereas the DSMC results are slightly underpredicted for the entire range of the E_{vib} considered. Nevertheless, the results from the effective temperature model remain well within an order of magnitudefrom QCT.

The reaction rates for another reaction, $HO_2 + H \rightarrow H_2 + O_2$, are shown in Fig. 2c. In this case, the results from the T_{eff} model are underpredicted compared to QCT for lower vibrational energies, and overpredicted for higher vibrational energy, matching only at $E_{vib} \sim 20$ kcal/mol. Once again, the rates differ by a factor of two to three.

These results demonstrate that the effective temperature model adequately captures the reaction rates of the hydrogen-air reactions. Although the rates from the T_{eff} model differ slightly, this is acceptable since the uncertainly in the experimental results can be of an order of magnitude or higher in the range that is relevant to combustion/detonation. In the future, reaction rates at more values of temperature, and at a higher range of Evib will be tested.

3.2 1D Detonation Using T_{eff}

A quasi one-dimensional (1D) hydrogen-air detonation wave was simulated by setting a domain with $L_x >> L_y, L_z$, with L_x set equal to 50 mm. The domain had an inflow of an unburnt hydrogen-air stoichiometric mixture at Mach number M = 3.0. The reactants are at T = 900 K and P = 0.3 atm. A temperature and pressure discontinuity was initialized close to the middle of the domain. At $24.6 < x < 25.0$ mm,

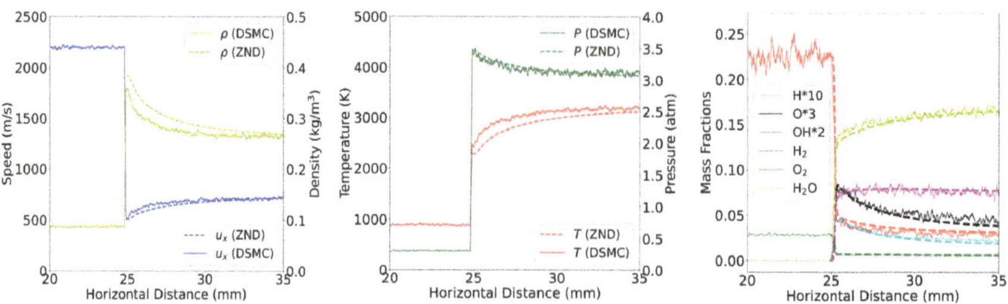

Fig. 3 Temperature, pressure, density, velocity, and species mass fractions from the 1D hydrogen-air detonation using the T_{eff} model. Solid lines show the results for DSMC whereas the dashed lines correspond to the equilibrium ZND solution

the pressure and temperature values were set corresponding to the post-shock values of the hydrogen-air mixture at M = 3.0. This region mimics the induction zone in a detonation wave, in which the pressure and temperature remain constant for a short period. At x > 25.0 mm, the temperature and pressure were set corresponding to the post-detonation values obtained by solving the so-called equilibrium ZND model using the Shock and Detonation Toolbox (SDT) [12]. The gas in the region was the combustion exhaust, i.e., H_2O and N_2. More details on this setup can be found in the reference [19], along with the results from the default TCE model. Note that the given conditions correspond to a slightly overdriven detonation wave. A 9-species 21-reaction mechanism of Burke et al. [6] was used to describe the chemistry.

The profiles for temperature, pressure, density, velocity, and species mass fractions are shown in Fig. 3. The dashed lines in this figure show the profiles from the equilibrium ZND solution. The figure shows that the profiles from the effective temperature model closely match ZND solution. This is not unexpected since this case is close to equilibrium (see Sect. 2). Nevertheless, it confirms that the effective temperature model is implemented correctly and is able to reproduce the expected results for this case. In future, cases with higher degrees of non-equilibrium will be tested with this model. The induction zone length (abbreviated as L_{ind}) from the default TCE model and the one using T_{eff} is shown in Table 1. The value of L_{ind} is evaluated by calculating the distance between the peak pressure and the peak thermicity. However, computing thermicity requires calculation of gradients, and hence can be noisy for individual snapshots. Thus, time-averaging was performed

Table 1 Comparison of induction zone length between DSMC and the ZND solution from the default TCE model and one using effective temperature T_{eff}

Case	L_{ind} DSMC (mm)	L_{ind} ZND (mm)	Percentage diff. (%)
Default TCE	0.275	0.295	7.0
T_{rff}	0.285	0.295	3.6

between t = 15 μs and 30 μs and L_{ind} is based on time-averaged thermicity. The value of L_{ind} in both cases is very close to its ZND counterpart for both cases.

4 Conclusions

This paper demonstrates simulating hydrogen-air detonations in the continuum region using DSMC. The results from DSMC are compared with the equilibrium ZND solution. It is shown that the default TCE model captures the temperature, pressure, density, velocity, and species mass fraction profiles accurately. Even though rapid changes occur across the detonation front that result in non-equilibrium effects inherent to these problems, the effects of non-equilibrium do not appear to be significant in this case, and the DSMC results match closely with the equilibrium ZND solutions.

For cases with higher degree of non-equilibrium, reaction models accounting for vibrational energy becomes important (assuming the rotational temperature quickly equilibrates with the translational temperature). To this effect, an effective temperature (T_{eff}) model based on the equipartition of energy is proposed [10]. The DSMC implementation of it involves using the TCE model but using T_{eff} instead of translational temperature. The reaction rates from quasi-classical trajectory (QCT) calculations accounting for individual vibrational levels is compared with those obtained from the T_{eff} model, and a good agreement is obtained for select reactions with differences of the order of two to three. This is acceptable since combustion reactions often have an order of magnitude or more uncertainty. Finally, a 1D detonation wave is simulated with the T_{eff} model, with results matching closely with those from the ZND model.

This paper sets the framework for simulating hydrogen-air detonations with higher degree of non-equilibrium in the continuum regime. Since DSMC can account for transport properties, internal energy exchange, and chemical reactions at a molecular level, it is well-suited for detonation reacting flow problems where rapid changes occur and non-equilibrium can play an important role. Currently, DSMC is limited by the availability and accuracy of relevant molecular data. However, it should be acknowledged this method is still in early stages of development for combustion studies.

Acknowledgements The work is supported by the US Department of Energy, Office of Basic Energy Sciences, Division of Chemical Sciences, Geosciences, and Biosciences. Sandia National Laboratories is a multimission laboratory managed and operated by National Technology and Engineering Solutions of Sandia, LLC., a wholly owned subsidiary of Honeywell International, Inc., for the U.S. Department of Energy's National Nuclear Security Administration under contract DE-NA-0 0 03525. The computational resources were provided by National Energy Research Scientific Computing Center (NERSC) and Oak Ridge Leadership Computing Facility (OLCF).

References

1. Babinsky H, Harvey JK (2011) Shock wave-boundary-layer interactions, vol 32. Cambridge University Press
2. Bird GA (1994) Molecular gas dynamics and the direct simulation of gas flows. Oxford University Press
3. Bondar YA, Maruta K, Ivanov MS (2011) Hydrogen-oxygen detonation study by the DSMC method. In: AIP conference proceedings, vol 1333. American Institute of Physics, pp 1209–1214
4. Borgnakke C, Larsen PS (1975) Statistical collision model for Monte Carlo simulation of polyatomic gas mixture. J Comput Phys 18(4):405–420
5. Boyd ID, Schwartzentruber TE (2017) Nonequilibrium gas dynamics and molecular simulation, vol 42. Cambridge University Press
6. Burke MP, Chaos M, Ju Y, Dryer FL, Klippenstein SJ (2012) Comprehensive H_2/O_2 kinetic model for high-pressure combustion. Int J Chem Kinet 44(7):444–474
7. Chou CH, Pan KL (2024) Computational investigation of chemical and non-equilibrium effects on the Richtmyer-Meshkov instability. Proc Combust Inst 40(1):105335
8. Fiévet R, Voelkel S, Koo H, Raman V, Varghese PL (2017) Effect of thermal nonequilibrium on ignition in scramjet combustors. Proc Combust Inst 36(2):2901–2910
9. Gimelshein S, Wysong I (2017) DSMC modeling of flows with recombination reactions. Phys Fluids 29(6):067106
10. Jasper AW, Sivaramakrishnan R, Klippenstein SJ (2019) Nonthermal rate constants for ch4* + x → ch3 + hx, x = h, o, oh, and o2. J Chem Phys 150(11):114112
11. Kulakhmetov M, Gallis M, Alexeenko A (2016) Ab initio-informed maximum entropy modeling of rovibrational relaxation and state-specific dissociation with application to the $O_2 + O$ system. J Chem Phys 144(17):174302
12. Lawson J, Shepherd J (2021) https://shepherd.caltech.edu/EDL/PublicResources/sdt/
13. Miller JA, Garrett BC (1997) Quantifying the non-RRKM effect in the H + O2 $\rightleftarrows$ OH + O reaction. Int J Chem Kinet 29(4):275–287
14. Plimpton SJ, Gallis MA. Sparta. http://sparta.sandia.gov
15. Sebastião IB, Kulakhmetov M, Alexeenko A (2017) DSMC study of oxygen shockwaves based on high-fidelity vibrational relaxation and dissociation models. Phys Fluids 29(1):017102
16. Sebastião IB, Qiao L, Alexeenko A (2018) Direct simulation Monte Carlo modeling of H_2-O_2 deflagration waves. Combust Flame 198:40–53
17. Trivedi S, Cant RS, Harvey JK (2021) Molecular level simulations of combustion processes using the DSMC method. Combust Theory Modell 25(2):351–363
18. Trivedi S, Harvey JK, Stewart Cant R (2022) Molecular level simulations of hydrogen-air flame at high pressures. Proc Combust Inst
19. Trivedi S, Salinas JS, Harvey JK, Poludnenko AY, Chen JH (2025) Simulations of hydrogen-air detonations using Direct Simulation Monte Carlo: *under review*. Combust Flame
20. Trivedi S, Salinas JS, Lee M, Desai S, Gallis MA, Harvey JK, Chen J. Comparison of DSMC and multi-temperature simulations of thermal nonequilibrium in a supersonic Kelvin-Helmholtz shear layer. AIAA SCITECH 2024 Forum
21. Valentini P, Grover MS, Verhoff AM, Bisek NJ (2023) Near-continuum, hypersonic oxygen flow over a double cone simulated by direct simulation monte carlo informed from quantum chemistry. J Fluid Mech 966:A32
22. Varga T, Olm C, Nagy T, Zsély IG, Valkó E, Pálvölgyi R, Curran HJ, Turányi T (2016) Development of a joint hydrogen and syngas combustion mechanism based on an optimization approach. Int J Chem Kinet 48(8):407–422
23. Voelkel S, Raman V, Varghese P (2016) Effect of thermal nonequilibrium on reactions in hydrogen combustion. Shock Waves 26(5):539–549

24. Wysong I, Gimelshein S, Bondar Y, Ivanov M (2014) Comparison of direct simulation Monte Carlo chemistry and vibrational models applied to oxygen shock measurements. Phys Fluids 26(4):043101
25. Yang C, Huang H, Sun Q (2019) A microscopic study of one-dimensional H_2-O_2 detonation using DSMC method. In: AIP conference proceedings, vol 2132. AIP Publishing LLC, p 070013

Plasma Flows and Processes

Effect of Microwave Coupling and Magnetic Field Topology on Electron Temperature and Thrust on the Example of Two Electric Propulsion Concepts

C. E. Schäfer, J. Schmidt, F. Plettenberg, M. Grabe, Y. A. Chan, J. Martinez-Schramm, K. Bräumer, K. Holste, and P. J. Klar

Abstract Electric propulsion concepts involving electron cyclotron resonance excitation for plasma generation eliminate the need for electrodes inside the plasma. Concepts, employing magnetic nozzles to accelerate the entire plasma for thrust production, remove the necessity for a neutralizer. Two thruster concepts realizing such plasma production and acceleration are investigated: the well-known MINOTOR thruster prototype of ONERA and a new prototype under development at DLR called DEEVA. It is assumed that converging parts in the magnetic field lines along the plume direction lead to the weaker performance of the DEEVA prototype compared to the MINOTOR prototype. This study investigates the effects of magnetic field configurations on the DEEVA prototype, examining both a converging-diverging character and a strictly diverging configuration of the DEEVA magnetic field topology. The effects of these magnetic field variations are compared with the MINOTOR prototype using Langmuir probe measurements and thrust balance measurements. By observing trends resulting from operational variations of the prototypes - such as changes in input power, frequency settings, and volume flow - we can infer the influence of the magnetic field topology on plasma parameters and thrust. Potential improvements for thruster development are discussed based on the measured magnetic field topologies.

Keywords Electric propulsion · ECRT · Magnetic nozzle · Langmuir probe · Thrust balance

C. E. Schäfer (✉)
Propulsion Department, OHB System AG, Bremen, Germany
e-mail: clara.schaefer@ohb.de

J. Schmidt · F. Plettenberg · M. Grabe · Y. A. Chan · J. Martinez-Schramm
Spacecraft Department, German Aerospace Center, Göttingen, Germany

K. Bräumer · K. Holste · P. J. Klar
Institute of Experimental Physics I, Justus Liebig University, Giessen, Germany

© The Author(s) 2026

M. Grabe et al. (eds.), *Rarefied Gas Dynamics*, Springer Aerospace Technology,
https://doi.org/10.1007/978-3-032-00094-1_63

1 Introduction

Typical issues with ion thruster technologies include electrode and grid erosion, along with the requirement for additional neutralizer devices [3]. Using electron cyclotron resonance (ECR) excitation for plasma generation, which eliminates the need for electrodes, offers a promising solution. Additionally, employing a magnetic nozzle (MN) to accelerate the entire plasma for thrust production removes the need for a neutralizer. The effects of electrodeless microwave coupling on plasma properties, combined with the magnetic field topology of a new thruster that achieves electrodeless plasma production and acceleration, are currently unknown. This study aims to fill this knowledge gap by comparing two thruster concepts. We examine the DEEVA (*DLR Electrodeless ECR Via microwave plasma Accelerator*) thruster, developed by the German Aerospace Center, which uses ECR for plasma production and a magnetic nozzle for acceleration. This is compared with a prototype of the well-known MINOTOR (*Magnetic Nozzle Electron Cyclotron Resonance Thruster*) concept, developed by the Office national d'études et de recherches aérospatiales (ONERA) [8]. Both prototypes employ a magnetic nozzle, are of similar size, and are designed to operate within a similar frequency and power range. However, they differ in terms of the concept employed for microwave coupling into the plasma as well as in the arrangement of the permanent magnets delivering the magnetic field topology. Comparing both thrusters under the same operating conditions (same power, volume flow and frequency setting) allows us to carry out detailed studies of the impact of the microwave coupling method and the magnetic field topology on plasma parameters.

As former studies have shown, the magnetic field topology of the DEEVA prototype includes a converging part in the magnetic field lines between the ECR zone and the thruster exit plane [11]. This is assumed to be the reason for the DEEVA prototype's performance not yet being comparable to the MINOTOR prototype [10, 11]. To investigate this assumption, we vary the magnetic field topology of DEEVA by realigning the permanent magnets. In one configuration, the disc magnets and the ring magnet are aligned in an attracting manner. This setup, referred to as *DEEVA-attractive*, was investigated in earlier studies [10, 11]. The second magnetic field setup aligns the magnets in a repulsive manner, and is therefore denoted as *DEEVA-repulsive*. In the repulsive configuration, a strictly diverging magnetic field character is expected. Previous studies have shown a correlation between electron temperature and ion energy for both the MINOTOR and DEEVA prototypes [5, 10]. This correlation is based on the assumption of an ambipolar electric field in the magnetic nozzle due to electron dynamics. Analyzing data from different diagnostic methods and directly comparing the plasma parameters of both thrusters enables us to assess the impact of different magnetic field configurations on performance.

2　Methods and Experimental Set Up

The experiments are conducted in the vacuum facility *Simulationsanlage für Treib-strahlen Göttingen - Miniatur Triebwerke* (STG-MT) at the DLR in Göttingen. The chamber has a length of 1 m and a diameter of 1.1 m. It is equipped with two backing pumps, a rotary vane pump, and a roots pump yielding a base pressure of 10^{-3} mbar. For lower pressure ranges a turbomolecular pump is added allowing pressure ranges of 10^{-6} mbar and background pressures during thruster operation in the range of 10^{-5} mbar. For the operation of the thrusters, we use the microwave signal generator KU-SG 2.45-250A of the company Kuhne Elecronics GmbH as well as the Bronkhorst mass control unit (MCU) for maximum 50 sccm air. The gases in use are xenon, in case of MINOTOR, and argon, in case of DEEVA. The mapping of each thruster is performed as follows: For all operation points the plasma is measured with Langmuir probe (LP) in $d = 100 \pm 0.5$ cm distance to the thruster exit. The thrust is investigated by means of thrust balance (TB) measurements. Multiple measurements performed at the same thruster setting using both the LP and the TB yield the experimental uncertainty in the results.

The operating principle of both the ECR thrusters with MN is the ionization of the propellant via ECR and the acceleration by a divergent magnetic field [2, 8, 14]. The prototypes are described in various publications [9, 10]. In presence of a magnetic field, charged particles (electrons and ions) are trapped along the magnetic field lines in such a way that they are circulating (gyrating) around the field lines, due to Lorentz forces [14]. If an electromagnetic wave with that cyclotron frequency is applied, the electron is resonantly accelerated by the electric field of the wave. It absorbs the energy of the electromagnetic wave, gains kinetic energy and increases the impact ionization process rate [2, 14]. Given the magnetic field of 87.5 mT, one comes to a microwave frequency of 2.45 GHz for resonant excitation. The acceleration of the produced ions is due to the gradient in the magnetic field strength. The magnetic moment μ of the charged particles (forming due to the gyration of the particles about the magnetic field lines) and the mass of the particle m, can be used to formulate an acceleration ($\dot{v}_{\parallel}$) opposite to the gradient direction $\nabla_{\parallel} B$ [6, 14]:

$$m\dot{v}_{\parallel} = -\mu\nabla_{\parallel}B. \tag{1}$$

The electrons therefore experience a force in opposite direction of the gradient in magnetic field strength, leading to a charge separation and the build up of an ambipolar electric field. It is a well-established assumption of MN's, that this electrostatic field converts the thermal energy motion of the electrons into ion kinetic energy. Therefore the electron dynamics plays a crucial role in configuring the electrostatic field in the plume, responsible for ion acceleration and ultimately thrust generation [4].

The magnetic field plays a crucial role in the functioning of the thruster, more specifically in the generation and acceleration of the quasi-neutral plasma. For this reason, the thruster's magnetic field is measured in three spatial directions using a

3D Hall probe from the company Projekt Elektronik GmbH. The range of the probe in use is $\pm200\,$mT and the linearity error is given as $\pm0.1\,$mT. In addition to the downstream measurements, the magnetic field in the vicinity of the ECR zone is also measured. This is possible by moving the probe in all three spatial directions (x, y, z) with linear stages. We will present results in the x, y plane after confirming rotational symmetry.

One of the most technically simple, yet difficult to interpret, diagnostics tools is the Langmuir probe (LP) [14]. A single electrode probe is introduced into the plasma and the voltage U is varied with respect to a reference potential. The current-voltage characteristic obtained, delivers a set of plasma parameters. For this study the crucial electron energy distribution function EEDF ($f(E)$)—based on the second derivative of the measured current voltage characteristic $\frac{d^2 I_e}{dU^2}$ is determined [1]. By taking the moment of the distribution function we obtain the electron temperature T_e of the plasma:

$$T_e = \frac{2}{3 n_e} \int_0^\infty f(E)\, E \, dE. \tag{2}$$

The thrust balance DEPB (*DLR Electric Propulsion Thrust Balance*) used in this investigation is described in Refs. [7] and [13]. It is an inverted double pendulum, consisting of two pendulums connected by a plate with a given stiffness and elasticity. The TB consists of an aluminium table, casing and eight quartz-glass rods used as flexible bearings for the thruster table. To dampen possible high frequency oscillations an eddy current brake is built in. A micrometer-screw is used to establish the connection between the moving thrust balance table and the Sartorius® WZA224 load cell [13]. An accuracy of $0.2\,$mN and a repeatability of $0.1\,$mN for the DEPB are given [12]. In our study, multiple measurements are conducted for each operation point of the thruster to determine the mean thrust based on delta measurements. A "delta measurement" involves running the thruster for several minutes and then switching it off. The resulting step change in the force measurement curve is identified as the produced thrust. This process is repeated multiple times, and the thrust values are averaged. The standard deviation of these values is represented as error bars in the following.

3 Magnetic Field and Plasma Characteristics

The three magnetic field configurations can be seen in Fig. 1. The ECR zones, where a magnetic field strength of $|\vec{B}| = 87.5\,$mT is achieved, is marked by a white line for all three configurations. For all three set ups, $x = 0$ marks the thruster exit plane. As one can see in case of MINOTOR the ECR condition is achieved close to the back plate. The magnetic field then decreases fast, leading to a high gradient in magnetic field strength. In case of DEEVA-attractive, a diverging-converging-diverging character

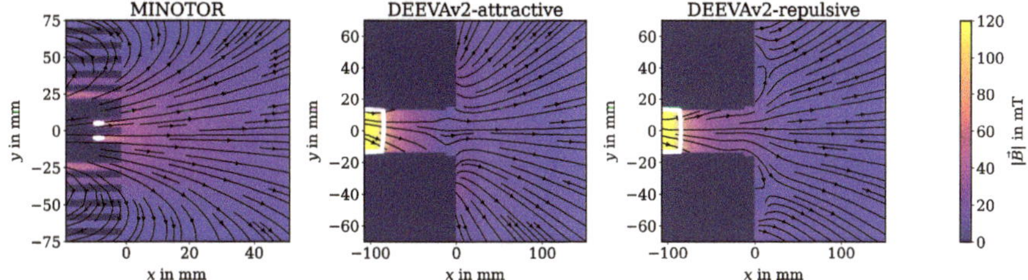

Fig. 1 Magnetic field configurations of the ECR thrusters. **a** MINOTOR is pictured on the left, dark area at $x < 0$ marks the mask where the probe could not access the thruster. $x = 0$ marks the thruster exit plane. The white dots mark the magnetic field strength of $|\vec{B}| = 87.5$ mT, the ECR zone. DEEVA in the center is shown in attractive configuration (**b**). On the right DEEVA is shown in repulsive configuration (**c**). Again, the dark area at $x < 0$ marks the mask where the probe could not access the slotted antenna. $x = 0$ marks the thruster exit plane and the white line $|\vec{B}| = 87.5$ mT, therefore the ECR zone

of the magnetic field can be observed. Since the converging part could lead to a reflection of particles, it is decided to test a different magnetic field configuration; the DEEVA-repulsive configuration is depicted in Fig. 1 on the right. Here we see again, the ECR condition is fulfilled in the back of the thruster vessel, near to the gas inlet. In contrast to DEEVA-attractive, we observe in this configuration a strictly diverging character of the magnetic field lines.

The effect of magnetic field topologies on the electron temperature with changes in operational points is shown in Fig. 2. On the left, we see the input power variation. The center plot shows the frequency variation, and the right plot depicts the variation in volume flow. The electron temperature for the MINOTOR prototype is higher across all operational parameter variations. For the two DEEVA configurations, DEEVA-attractive and DEEVA-repulsive, the repulsive configuration shows higher values in both power and frequency variations. For power variation, there is not a significant change in electron temperature for all prototypes. However, for frequency variation, higher values are observed around the design frequency of 2450 MHz for MINOTOR (a) and DEEVA-repulsive (c). In contrast, DEEVA-attractive (b) shows little change with frequency variation. For all three prototypes, electron temperature decreases with increasing volume flow. The values for the DEEVA-attractive configuration are the lowest, while those for the MINOTOR are the highest.

As shown in Fig. 3, the overall thrust for the MINOTOR prototype is higher, reaching up to $T_T = 1.7$ mN. In contrast, the thrust values for both magnetic field topologies of the DEEVA thruster do not exceed 0.3 mN. There is no significant difference in thrust between the two DEEVA configurations. We observe an increase in thrust for all three prototypes with increasing input power. For MINOTOR, there is a clear optimum at 2450 MHz. DEEVA-attractive also shows an optimum at this frequency. DEEVA-repulsive operates at a constant thrust value over a broader frequency range. For volume flow variation, thrust decreases with increasing volume flow for all prototypes. However, DEEVA-attractive shows a slight increase in thrust around 2 sccm before decreasing as well.

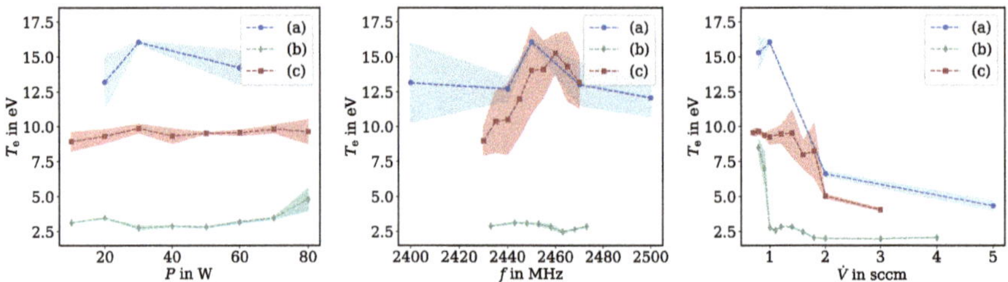

Fig. 2 Comparison between the determined electron temperatures T_e for the MINOTOR (**a**), DEEVA-attractive (**b**) and DEEVA-repulsive (**c**) configurations. The plot on the left shows power variations at a frequency set of 2450 MHz and a volume flow of 1 sccm of argon (in case of DEEVA) and xenon (in case of MINOTOR). The center plot depicts frequency variation at power input of 30 W and a volume flow of 1 sccm. The right plot shows the volume flow variation (of again argon, in case of DEEVA; and xenon, in case of MINOTOR) at the constant frequency 2450 MHz and input power of 30 W. The filled space shows the standard deviation of the resulting values from multiple measurements

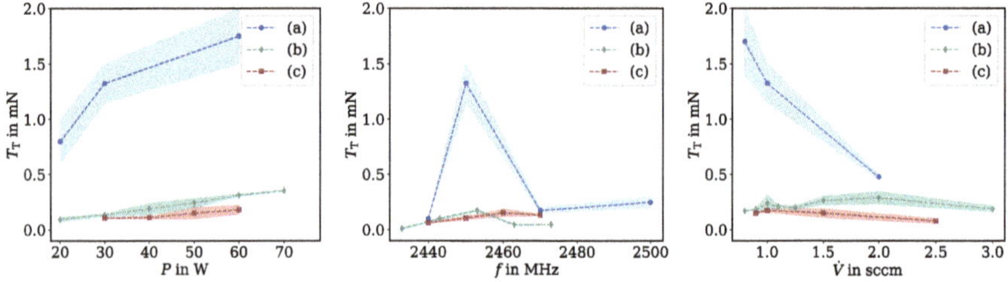

Fig. 3 Comparison between the determined thrust T_T for the MINOTOR (**a**), DEEVA-attractive (**b**) and DEEVA-repulsive (**c**) configurations. The plot on the left shows power variations at a frequency set of 2450 MHz and a volume flow of 1 sccm of argon (in case of DEEVA) and xenon (in case of MINOTOR). The center plot depicts frequency variation at power input of 30 W and a volume flow of 1 sccm. The right plot shows the volume flow variation (of again argon, in case of DEEVA; and xenon, in case of MINOTOR) at the constant frequency 2450 MHz and input power of 30 W. The filled space shows the standard deviation of the resulting values from multiple measurements

4　Discussion

First and foremost, it is important to note that MINOTOR is operated with xenon, while both DEEVA configurations are operated with argon. This choice is based on preliminary measurements showing significantly higher ion energies when DEEVA attractive is operated with argon [10]. This comparative study must be considered in light of this difference. Xenon is heavier than argon, and thus, the same ion energy results in higher thrust for xenon. Measurements of the magnetic field topologies revealed that the DEEVA-repulsive configuration indeed achieves a strictly diverging magnetic field. The higher gradient observed after the thruster exit in MINOTOR should result in greater electron acceleration (see Eq. 1). With the changes in magnetic field topology in the DEEVA-repulsive case, aiming for high gradients while avoiding excessive convergence in magnetic field lines, we achieve higher electron temperatures compared to the attractive configuration, and they are almost comparable to those of the MINOTOR prototype.

We can consider possible explanations for our findings that the determined thrust is not significantly higher in the DEEVA-repulsive configuration compared to the DEEVA-attractive configuration and not as high as in the MINOTOR case, despite the higher electron temperatures suggesting otherwise. Firstly, the ion energies must be investigated and correlated with the electron temperatures presented in this study. Secondly, the ion current, which determines the total number of ions ejected from the thruster need to be determined. The third explanation involves thrust generated by neutral gas. Alongside investigating ion energy and ion current to determine thrust production, it's crucial to consider the effects of neutral gas on thrust. As reported in Ref. [15], thrust calculated using probe measurements underestimates total thrust. By comparing calculated thrust with thrust balance results, we can estimate the efficiency of the magnetic nozzle designs employed in this study.

Due to the early development stage of the new DEEVA thruster, it is uncertain whether it can achieve performance levels comparable to the optimized MINOTOR thruster in the future. However, the DEEVA concept introduces electrodeless microwave coupling into the plasma and electrodeless acceleration with a magnetic nozzle, potentially paving the way for long-term space missions employing ECR thrusters.

References

1. Demidov V, Ratynskaia S, Rypdal K (2002) Electric probes for plasmas: the link between theory and instrument. Rev Sci Instrum 73(10):3409–3439
2. Goebel D, Katz I (2008) Fundamentals of electric propulsion: ion and Hall thrusters, vol 1. Wiley
3. Holste K, Dietz P, Scharmann S, Keil K, Henning T, Zschätzsch D, Reitemeyer M, Nauschütt B, Kiefer F, Kunze F et al (2020) Ion thrusters for electric propulsion: Scientific issues developing a niche technology into a game changer. Rev Sci Instrum 91(6):061101
4. Kim J, Chung KJ, Takahashi K, Merino M, Ahedo E (2022) Kinetic electron cooling in magnetic nozzle: experiments and modeling. Plasma Sources Sci Technol
5. Lafleur T, Cannat F, Jarrige J, Elias P, Packan D (2015) Electron dynamics and ion acceleration in expanding-plasma thrusters. Plasma Sources Sci Technol 24(6):065013
6. Longmier B, Bering E, Carter M, Cassady L, Chancery W, Díaz FC, Glover T, Hershkowitz N, Ilin A, McCaskill G et al (2011) Ambipolar ion acceleration in an expanding magnetic nozzle. Plasma Sources Sci Technol 20(1):015007
7. Neumann A, Simon J, Schmidt J (2021) Thrust measurement and thrust balance development at dlr's electric propulsion test facility. EPJ Techn Instrum 8(1):17
8. Packan D, Elias P, Jarrige J, Vialis T, Correyero S, Peterschmitt S, Porto J, Merino M, Sánchez-Villar A, Ahedo E et al (2019) H2020 minotor: magnetic nozzle electron cyclotron resonance thruster. In: 36th international electric propulsion conference
9. Peterschmitt S, Packan D (2019) Comparison of waveguide coupled and coaxial coupled ecra magnetic nozzle thruster using a thrust balance. In: IEPC 2019

10. Schaefer C, Schmidt J, Plettenberg F, Chan Y, Grabe M, Marinez-Schramm J, Holste K, Klar PJ (2024) Comparative study of electron temperature and ion energy in two different magnetic nozzle thruster designs. In: 39th international electric propulsion conference
11. Schaefer C, Schmidt J, Plettenberg F, Chan Y, Grabe M, Marinez-Schramm Heinz TJ, Holste K, Klar PJ (2024) Investigation of the correlation between microwave coupling and ion energy on the example of two thruster concepts. In: Space propulsion conference
12. Schmidt J (2023) DLR electric propulsion balance documentation
13. Schmidt J, Simon J, Neumann A (2022) Thrust measurements of the rit-10 thruster with dlr's new depb thrust balance. In: 8th international conference on space propulsion, pp 1–7
14. Stroth U (2011) Plasmaphysik. Springer
15. Vialis T, Jarrige J, Aanesland A, Packan D (2018) Direct thrust measurement of an electron cyclotron resonance plasma thruster. J Propuls Power 34(5):1323–1333

Spatially Resolved Collisional-Radiative Model for a Rarefied Plasma Plume

Bruno Fontaine, Federico Bariselli, Pietro Parodi, Pedro Jorge,
Damien Le Quang, and Thierry Magin

Abstract The VKI DRAG-ON facility was designed to replicate Very Low Earth Orbit flow conditions, to study Air-Breathing Electric Propulsion intakes and gas-surface interaction physics. Its Particle Flow Generator produces a low-density, low-temperature argon plasma plume. This study investigates the flow's thermochemical non-equilibrium by comparing synthetic spectra from Collisional-Radiative modeling to Optical Emission Spectroscopy measurements. A Lagrangian reactor approach is used to account for the reacting species advection. The results also show how the assumption on the Electron Energy Distribution Function shape can affect the diagnostics. Model-experiment comparisons reveal potential for plasma diagnostics and suggest targeted chemical model refinements. The results also show how the assumption on the Electron Energy Distribution Function shape can affect the diagnostics.

Keywords Optical emission spectroscopy · Lagrangian reactor

B. Fontaine (✉) · F. Bariselli · P. Parodi · P. Jorge · D. Le Quang · T. Magin
Aeronautics and Aerospace Department, von Karman Institute for Fluid Dynamics, Sint-Genesius-Rode, Belgium
e-mail: bruno.fontaine@vki.ac.be

B. Fontaine · T. Magin
Aéro-Thermo-Mécanique, Université Libre de Bruxelles, Brussels, Belgium

B. Fontaine
Aérospatiale et Mécanique, Université de Liège, Liège, Belgium

P. Parodi
Department of Mathematics, KU Leuven, Leuven, Belgium

P. Jorge
Materials and Chemistry, Vrije Universiteit Brussel, Brussels, Belgium

© The Author(s) 2026

M. Grabe et al. (eds.), *Rarefied Gas Dynamics*, Springer Aerospace Technology,
https://doi.org/10.1007/978-3-032-00094-1_64

659

1 Introduction

The VKI DRAG-ON facility, commissioned in 2023, was designed to replicate the Very Low Earth Orbit (VLEO) flow conditions to study Air-Breathing Electric Propulsion (ABEP) [1] intakes and gas-surface interaction physics [7]. It operates in free molecular flow conditions, at pressures of the order of 10^{-5} mbar, using a LTA-100 Particle Flow Generator (PFG) from ThrustMe [12] to generate a partially ionized argon or oxygen plume with ion speeds of 6–10 km/s.

Since ion beams interact with walls through sheath potentials, as opposed to the mostly neutral flow at VLEO, corrections for wall effects require knowledge of the electron temperature and density. Characterizing these parameters is therefore essential for extrapolation to flight conditions. Optical Emission Spectroscopy (OES) is of particular interest for its simple setup and non-invasive nature. It provides line-of-sight (LoS) integrated emission intensities, which can be used to infer plasma parameters such as electron temperature and density [2, 9]. Their relation to the emission line intensities is established through a Collisional-Radiative (CR) model, describing the chemical reactions and radiative processes. The CR model depends on the Electron Energy Distribution Function (EEDF) shape, which is often assumed to be Maxwellian. Other EEDF shapes can however be appropriate in some cases, including low-density plasmas [5, 6].

CR models are usually applied in a 0-D, Quasi-Steady-State (QSS) approach. However, for flows where chemical and flow timescales are comparable, the advection of the reacting species should be considered. A Lagrangian reactor approach provides an efficient alternative to a full flow-chemistry coupling, by transporting a thermochemical reactor along pre-computed streamlines without detailed chemistry, then recomputing the gas composition at each position [3].

In this study, we aim to advance the understanding of the non-equilibrium plasma dynamics in the VKI DRAG-ON facility by comparing spatially resolved OES measurements to predictions from a spatially resolved CR model using a Lagrangian reactor approach, for the PFG running with argon gas. We aim to identify the key factors influencing the model results and discrepancies with the experiment, such as electron temperature and density at the outlet of the PFG and the EEDF shape. In Sect. 2 we detail the experimental setup and OES data processing, in Sect. 3 we describe the modeling approach, and finally in Sect. 4 we compare results to assess model sensitivity and fidelity.

2 Optical Emission Spectroscopy

The DRAG-ON facility consists of two vacuum chambers linked by a duct that serves as test section for ABEP intake systems. Each chamber is equipped with a rotary vane mechanical pump and a turbomolecular pump, achieving an ultimate pressure of

10^{-8} mbar (when no gas is fed to the PFG). Each chamber is instrumented with hot-cathode and cold-cathode pressure gauges, and the main chamber has side windows for optical access. The PFG is aligned with the center axis of the main chamber (0.9 m length, 0.61 m diameter), and operated at 3 sccm argon flow, 60 W RF-power, 1 A focusing coil current, and floating electrode bias. During operation, the chamber pressure was 2.7×10^{-5} mbar.

Figure 1 shows the OES setup. The light emitted by the plasma is collected with a 12 mm reflective collimator (RC12SMA-P01, ThorLabs) mounted on a 2-D motorized motion stage with 0.03 mm positioning accuracy. A 1 m optical fiber (MS53L01, ThorLabs) conducts the light to the high-sensitivity spectrometer (Maya2000Pro, OceanOptics) with a 600–1000 nm range, a circular aperture 50 µm in diameter, a blazed grating with 600 lines/mm, and a Hamamatsu S11510 CCD detector. The spectrometer was calibrated in wavelengths by the manufacturer. Relative intensity calibration was performed by replicating the setup in front of a black body source (R1500T, AMATEK Land) at 1723 K.

Spatial emission measurements were conducted by moving the optics relative to the plume's axis at the PFG outlet. Scans were performed at axial positions $x = 14$, 21, 28, and 35 mm, with a transverse step size of 5 mm up to $z = \pm20$ mm, and a step of 10 mm beyond. The OceanView software was used for data acquisition. At each measurement point, 25 scans were performed for averaging, with 400 ms integration time each, with dark current correction.

Figure 2 shows a typical spectrum, with background subtracted and setup spectral response correction. The emission lines are identified to argon transitions (red labels, with levels indexed by energy starting from the ground state Ar(0)). They are integrated spectrally for comparison to the CR model results. To reduce noise, they are fitted to a Gaussian, typical for instrumental broadening [2].

Figure 3 shows the intensity spatial profiles for the 811 nm line. Others have similar profiles. The results show good symmetry and repeatability across campaigns. The uncertainty level, including fitting error and calibration noise, is small (average standard deviation of 3% and maximum of 10% for all lines, at all positions). The size of the window is a limitation of the current setup, as it prevents measurements far from the centerline, preventing Abel inversion.

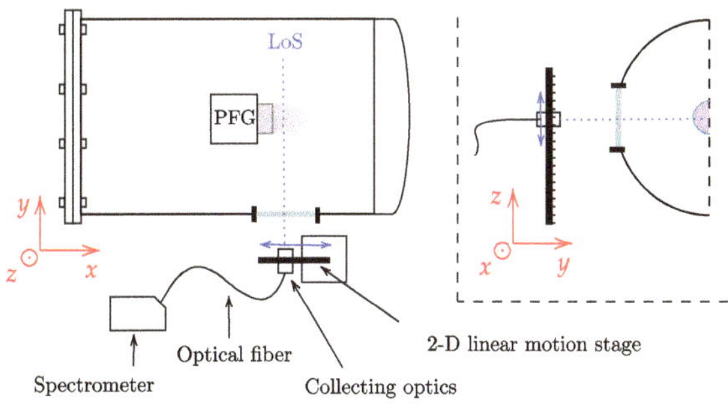

Fig. 1 Schematic experimental setup top view (*left*) and front view (*right*)

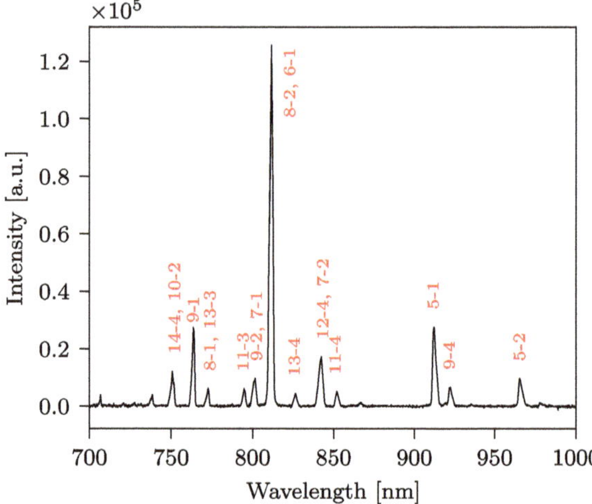

Fig. 2 Typical argon spectrum, at $x = 14$ mm and $z = 0$ mm

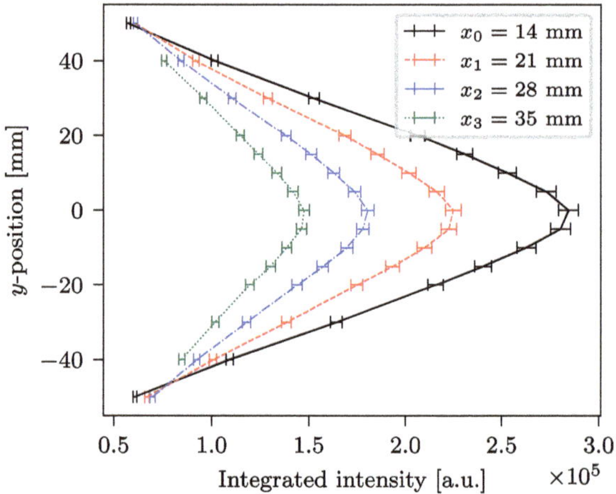

Fig. 3 Vertical intensity profiles (LoS integrated) for the line at 811 nm

3 Plume Model

The model predicts the intensity of emission lines by coupling a flow model with a Collisional-Radiative (CR) model using a Lagrangian reactor approach.

The governing equation for species conservation along streamlines is [3]:

$$\frac{\mathrm{d}Y_i}{\mathrm{d}s} = \frac{\dot{\omega}_i}{\rho U}, \quad \forall i \in \mathcal{S}, \tag{1}$$

where $\mathcal{S}$ is the set of species (here pseudo-species representing the excited states of argon), s (m) the curvilinear coordinate along the streamline, $Y_i = \rho_i/\rho$ (–) the mass fraction of species i, ρ (kg/m^3) the total mass density, U (m/s) the flow velocity

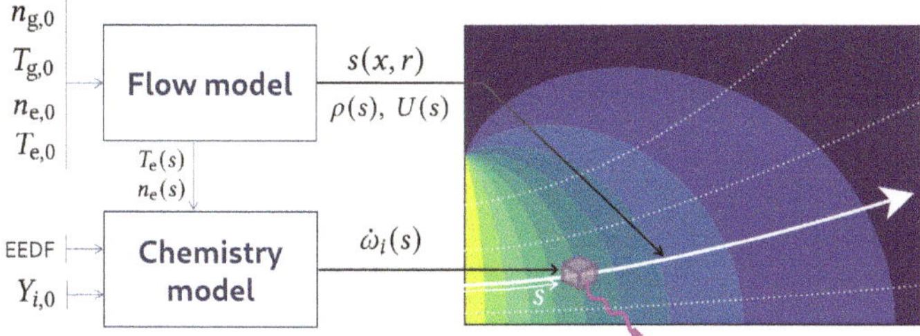

Fig. 4 Modeling approach and input parameters.

norm, and $\dot{\omega}_i$ (kg/(m^{3}s)) the net chemical production rate of species i. The approach (Fig. 4) requires a *flow model*, which takes as inputs the conditions inside the PFG $n_{g,0}$, $T_{g,0}$, $n_{e,0}$ and $T_{e,0}$ respectively the neutral density, neutral temperature, electron density and electron temperature. Through the flow model, these inputs fix the streamlines and flow conditions along them. A *chemistry model* then recomputes the species concentrations according to Eq. (1) with a Lagrangian reactor advected along the streamlines starting from an initial distribution $Y_{i,0}$, with local reaction rates depending on the local electron temperature, electron density and shape of the EEDF.

Flow model: The plume is modeled as a collisionless thermal expansion of neutral argon, with analytical velocity and density fields [4], interacting with a field of hot electrons. The electron temperature and density fields were computed with a Particle-in-Cell (PIC) simulation of DRAG-ON [10], with uniform outlet conditions $T_{e,0} = 5$ eV and $n_{e,0} = 2 \times 10^{15}$ m^{-3}. For the purpose of this study, the results were extrapolated to different outlet conditions with a linear scaling.

CR model: It includes the 31 first energy levels of argon, extracted from the NIST Atomic Spectral Database (ASD) [8]. Neglecting heavy interactions and assuming the plasma to be optically thin, the reactions included in the model are electron impact excitation/de-excitation and spontaneous radiative decay:

$$\text{Ar}(l) + \text{e}^- \underset{K_{ul}}{\overset{K_{lu}}{\rightleftharpoons}} \text{Ar}(u) + \text{e}^-, \quad \text{Ar}(u) \xrightarrow{A_{ul}} \text{Ar}(l) + h\nu,$$

where $u > l$ are respectively the upper and lower states. Rate coefficients for radiative decay A_{ul} (s^{-1}) are extracted from NIST-ASD, selecting all allowed transitions involving levels in S with an experimentally observed wavelength and a known A_{ul} (77 reactions). Electron-impact rates K_{lu} (m^3/s) are computed as a weighted integral of cross sections with the EEDF, using cross sections from [13] (465 reactions). The integrals are pre-fitted to a modified Arrhenius form for efficiency. The de-excitation coefficients K_{ul} are computed with the principle of detailed balance. Finally, the net production rate for each level i is:

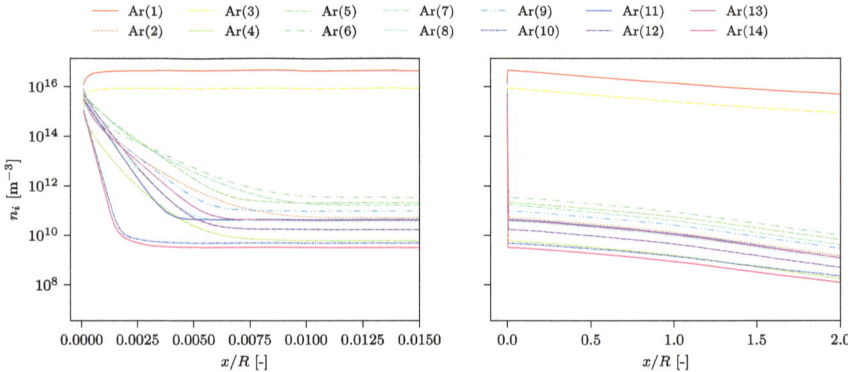

Fig. 5 Evolution of the densities of the 14 first excited states (centerline), with $n_{g,0} = 10^{19}$ m^{-3}, $T_{g,0} = 500$ K, $n_{e,0} = 5 \times 10^{14}$ m^{-3}, $T_{e,0} = 10$ eV, Maxwellian EEDF

$$\dot{\omega}_i / M = \sum_{i<j} A_{ji} n_j - \sum_{i>j} A_{ij} n_i + \sum_{j \neq i} K_{ji} n_e n_j - \sum_{j \neq i} K_{ij} n_e n_i. \qquad (2)$$

with $M = 6.63 \times 10^{-26}$ kg the atomic mass of argon.

Solving Eq. (1) along a streamline gives the evolution of excited-state populations. Figure 5 shows results for the centerline, for typical input parameters. The initial conditions are chosen as a Boltzmann distribution at $0.2 \times T_{e,0}$ to start from a high state of excitation, as expected in the PFG ionization chamber.

We identify two main regions in the flow. First is a radiative length scale ($s/R <$ 0.015), where the highly excited configuration from the PFG quickly decays, except for Ar(1) and Ar(3), the known argon metastables. After that, the flow appears to be chemically frozen, with a constant distribution of excited states along the streamline. For other plasma conditions than shown in Fig. 5, we also note a reactive length scale in-between these two regions, of varying length depending on the T_e and n_e distributions.

4 Comparison of Model and Experiment

In this section, we reconstruct the experimental results from the model results. A comparison criterion is defined to perform a sensitivity analysis on the inputs.

The line intensities for each transition considered are predicted in the model from the rate coefficient A_{ul} and the number density of the upper excited state n_u, with $I_{ul}(x, r) \propto A_{ul} n_u(x, r)/\lambda_{ul}$. For lines involving multiple transitions, contributions are summed. LoS-integrated model intensities are generated for the positions $x_i \in \mathcal{X}$ (axial positions with experimental data) by applying the forward Abel transform, assuming axisymmetry and the medium to be optically thin. OES intensities being relative, we normalize them with

$$\hat{I}_l^m(x, z) = \frac{I_l^m(x, z)}{\frac{1}{N_c} \sum_{x_i \in \mathcal{X}} \sum_{k \in \mathcal{L}} I_k^m(x_i, 0)},$$ (3)

where $m \in \{\text{OES},\text{CR}\}$, l is the line index in set $\mathcal{L}$ of 12 identified lines, and $N_c = \sum_{\mathcal{X}} \sum_{\mathcal{L}} 1$ is the total number of intensity values measured on the centerline.

A comparison criterion ξ (–) is defined to quantify the discrepancy between the OES and model results:

$$\xi = \sqrt{\frac{1}{N} \sum_{l \in \mathcal{L}} \sum_{x_i \in \mathcal{X}} \sum_{z_j \in \mathcal{Z}_i} \frac{\left(\hat{I}_l^{\text{OES}}(x_i, z_j) - \hat{I}_l^{\text{CR}}(x_i, z_j)\right)^2}{\left(\hat{I}_l^{\text{OES}}(x_0, 0)\right)^2}},$$ (4)

with $\mathcal{Z}_i$ the set of z positions in the scan at x_i, and $N = \sum_{\mathcal{L}} \sum_{\mathcal{X}} \sum_{\mathcal{Z}_i} 1$ is the total number compared values. The resulting comparison criterion ξ is a function of the model parameters only, and equals zero for a perfect match.

A sensitivity analysis was performed on ξ with small perturbations to identify key parameters. The initial conditions $Y_{i,0}$ were parametrized using a Boltzmann distribution at $2 \times T_{\text{el},0}$. The neutral density in the PFG $n_{\text{g},0}$ is fixed as it only affects absolute intensities. The model showed low sensitivity to T_g ($\leq 1\%$ variation between 300 and 600 K) or to $T_{\text{el},0}$ ($\leq 7\%$ between 1 and 15 eV, likely damped by the rapid radiative decay). However, higher sensitivity was found for the parameters governing electron impact excitation rates (EEDF shape, $T_{\text{e},0}$, $n_{\text{e},0}$). We therefore performed a mapping with respect to these parameters. The tested EEDF shapes were chosen to have very different tails (power-law for Kappa [11], exponential for Maxwellian, super-exponential for Druyvesteyn [5]).

Figure 6 shows ξ maps over ($T_{\text{e},0}$, $n_{\text{e},0}$) for each EEDF. These maps could in principle reveal whether one EEDF shape leads to better agreement with the experiment.

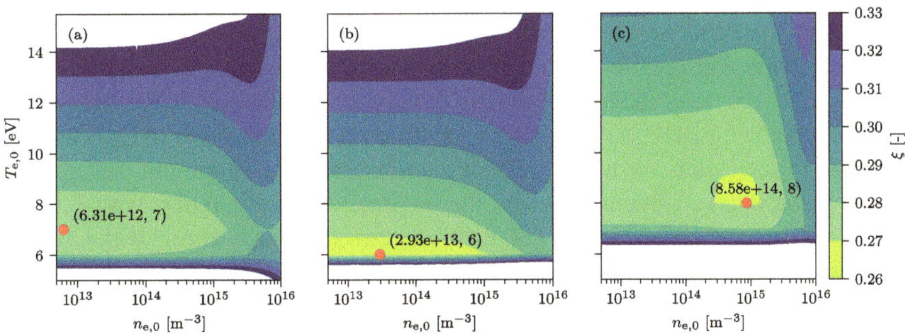

Fig. 6 Mapping of $\xi(n_{\text{e},0}, T_{\text{e},0})$ for each EEDF, with **a** Maxwellian, **b** Druyvesteyn and **c** Kappa ($\kappa = 2$), and location of minimum discrepancy ξ^* at ($T_{\text{e},0}^*$, $n_{\text{e},0}^*$)

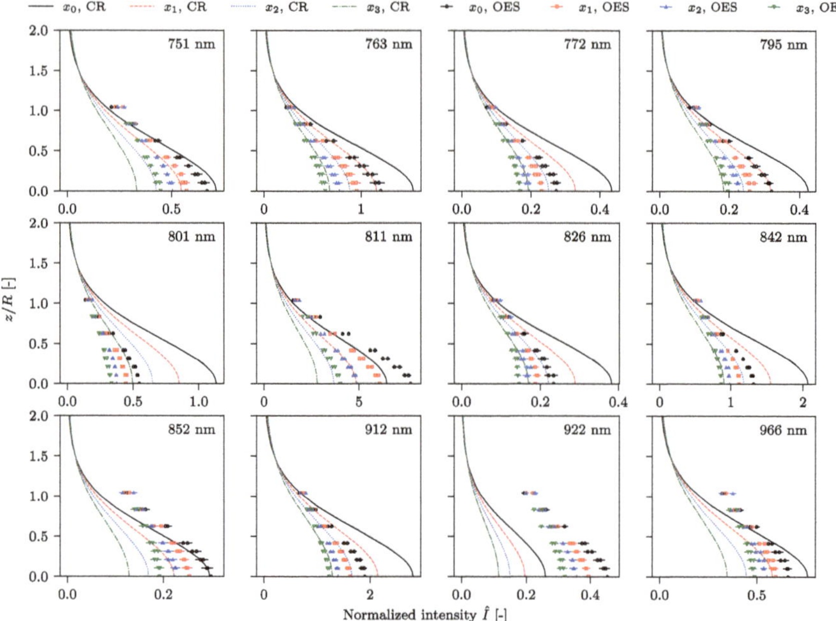

Fig. 7 Comparison of OES and model normalized intensity profiles for all lines, for a Maxwellian EEDF, $T_{e,0}^* = 10$ eV, $n_{e,0}^* = 3 \times 10^{15}$ m^{-3}, $T_{el,0}^* = 2$ eV, and $T_{g,0}^* = 500$ K

Here, however, all three yield similar minimum ξ^*. Nonetheless, the ξ contours differ depending on the choice of EEDF, and so does the location of the minimum discrepancy. In particular, we note that a less populated tail leads to a higher sensitivity on T_e. This can also be said for the sensitivity on n_e, although the sensitivity on this parameter remains relatively low for all EEDF shapes.

The obtained estimates of the electron temperature $T_{e,0}^*$, identified in Fig. 6, align well with invasive measurements [7]. This supports the potential for non-invasive diagnostics in DRAG-ON for the electron temperature. For the electron density, the optimal n_e^* are scattered across several orders of magnitude depending on the choice of EEDF shape. The diagnostics of this plasma parameter is therefore not conclusive at this stage because of the low sensitivity of xi on n_e and the correlation between n_e^* and the unknown EEDF shape. Figure 7 shows a visual comparison of OES and CR normalized intensities for a Maxwellian distribution at the minimum discrepancy. It shows reasonable agreement, given the large model uncertainties, with Einstein coefficients uncertainties reaching 50% [8] and cross sections varying widely across databases, but shows further model refinenements could be incorporated to reduce discrepancies.

5 Conclusions and Future Work

This study advances the understanding of non-equilibrium plasma dynamics in the VKI DRAG-ON facility by comparing the first spatially resolved OES measurements in the facility to a CR model using a Lagrangian reactor approach. A promising first problem inversion identified an electron temperature in the expected range, demonstrating the potential of our approach for non-invasive diagnostics in DRAG-ON, which could ultimately be applied during ABEP intake tests or gas-surface interaction studies. Remaining discrepancies between model and experiment suggest future model improvements such as extended PIC databases for different PFG conditions and inclusion of radiation trapping by metastable states. Particular attention should also be given to assumptions on the EEDF. While the EEDF shape had little impact on the extracted electron temperature under the present test conditions, it did affect the sensitivity of the model to the electron temperature and density, suggesting that it remains a critical parameter for plasma diagnostics. Langmuir probe measurements or Boltzmann equation simulations could provide alternatives to the analytical shapes used here. Future work will also focus on plasma diagnostics by problem inversion with uncertainty quantification tools.

Acknowledgements Bruno Fontaine is a F.R.S.-FNRS Research Fellow.

References

1. Andreussi T et al (2022) The AETHER project: development of air-breathing electric propulsion for VLEO missions. CEAS Space J 14(4):717–740
2. Bariselli F et al (2022) Characterization of an Air-Xenon operated electric thruster plume through optical emission spectroscopy. In: 9th international workshop on radiation of high temperature gases for space missions. Azores, Portugal
3. Boccelli S et al (2019) Lagrangian diffusive reactor for detailed thermochemical computations of plasma flows. Plasma Sources Sci Technol 28(6):065002
4. Cai C, Boyd ID (2007) Collisionless gas expanding into vacuum. J Spacecr Rockets 44(6):1326–1330
5. Druyvesteyn MJ, Penning FM (1940) The mechanism of electrical discharges in gases of low pressure. Rev Modern Phys 12(2):87–174
6. Gudmundsson JT (2001) On the effect of the electron energy distribution on the plasma parameters of an argon discharge: a global (volume-averaged) model study. Plasma Sources Sci Technol 10(1):76–81
7. Jorge PDC et al (2025) Relevance of an RF plasma source to satellite material testing in VLEO conditions. Under review at CEAS Space Journal (preprint available on Research Square)
8. Kramida A et al (2022) NIST atomic spectra database
9. Lee D et al (2022) Two-dimensional electron temperature and density profiles of Hall thruster plume plasmas using tomographically reconstructed optical emission spectroscopy. Plasma Sources Sci Technol 31(12):125004
10. Parodi P et al (2024) Particle-in-cell simulation of the VKI DRAG-ON facility. In: AIAA SciTech forum

11. Pierrard V, Lazar M (2010) Kappa distributions: theory and applications in space plasmas. Solar Phys 267(1):153–174
12. Rafalkyi D, Dudin S (2024) RF plasma flow generator for LEO environment simulation: use cases. In: 38th international electric propulsion conference
13. Zatsarinny O, Wang Y, Bartschat K (2014) Electron-impact excitation of argon at intermediate energies. Phys Rev A 89(2)

Evaluation of the Ambipolar Diffusion Approximation Using an Eulerian Boltzmann-Poisson-BGK Solver

Marisa Petrusky and Iain D. Boyd

Abstract Because first principles modeling of a weakly ionized gas is computationally expensive, the ambipolar diffusion approximation is often invoked. This approximation assumes the ionized gas undergoes ideal ambipolar diffusion, and is conventionally considered valid if the electron Debye length is orders of magnitude smaller than the neutral mean free path. However, this criterion is not intrinsic to ambipolar diffusion theory. An improved set of criteria for evaluating the validity of the ambipolar diffusion approximation is proposed, then verified with theory and simulation. An Eulerian Boltzmann-Poisson-BGK solver, which circumvents consideration of statistical noise present in particle methods, is used to simulate an unsteady expanding weakly ionized gas undergoing ideal ambipolar diffusion. Potential consequences of incorrectly applying the ambipolar diffusion approximation are demonstrated and discussed.

Keywords Weakly ionized gas · Plasma · Ambipolar diffusion approximation · Boltzmann solver

1 Introduction

The presence of charged particles in rarefied gas flows warrants careful consideration of modeling methods. All charged particles exert electrostatic forces on each other, and if a sufficient degree of charge separation occurs bulk electric fields form. First principles modeling of electrostatic effects requires solving Maxwell's equations in addition to treating electrons as an independent species. The latter is computationally expensive as electrons' lightweight mass necessitates a significantly smaller simulation time step for propagation. To simplify calculations the ambipolar diffusion approximation is often invoked. The approximation assumes positive ions and electrons travel at the same bulk velocity. In particle-based kinetic methods such as

M. Petrusky (✉) · I. D. Boyd
Ann and H.J. Smead Department of Aerospace Engineering Sciences, University of Colorado Boulder, Boulder, CO, USA
e-mail: marisa.petrusky@colorado.edu

© The Author(s) 2026

M. Grabe et al. (eds.), *Rarefied Gas Dynamics*, Springer Aerospace Technology,
https://doi.org/10.1007/978-3-032-00094-1_65

direct simulation Monte Carlo (DSMC) [2], the approximation is typically enforced by transporting electrons with the corresponding ion they were created with [1], or by propagating electrons at the mean ion velocity [4]. In computational fluid dynamics (CFD) codes, the approximation is enforced by setting the electron flux equal to the ion flux in each fluid element [10]. The validity of the ambipolar diffusion approximation is typically assessed by checking whether the electron Debye length $\lambda_{D,e}$ is much less than the neutral mean free path λ_n, where the full definition of the Debye length λ_D is given by

$$\lambda_D = \sqrt{\frac{\epsilon_0 k_B/e^2}{n_e/T_e + \sum_i z_i^2 n_i/T_i}} \tag{1}$$

where subscripts e and i denote electron and ionic species properties. ϵ_0 is the permittivity of free space, k_B is the Boltzmann constant, e is the elementary charge, n is number density, T is translational temperature, and z is the charge number. If the motion of ions is negligible compared to electrons, the ion term can be neglected, and Eq. 1 defines the electron Debye length, $\lambda_{D,e}$.

Despite its near ubiquitous use in both kinetic and continuum simulation of rarefied flows, the criterion $\lambda_{D,e} \ll \lambda_n$ is not intrinsic to ambipolar diffusion theory and thus may be insufficient justification for the approximation. It may also falsely predict when electrostatic interactions are negligible. In this study, a new set of criteria is proposed for assessing the validity of the ambipolar diffusion approximation in continuum flows. These criteria are verified through a series of unsteady weakly ionized gas simulations using a Boltzmann-Poisson-BGK solver, which circumvents consideration of statistical noise present in particle simulations. A case is tested where $\lambda_{D,e} \ll \lambda_n$ is violated. Additionally, simulation results with the ambipolar diffusion approximation and with direct electric field modeling are compared to assess how well the ambipolar diffusion approximation captures ideal ambipolar diffusion.

2 Review of the Derivation of Ambipolar Diffusion

The ambipolar diffusion approximation is based on, but not strictly interchangeable with, the phenomena of ambipolar diffusion. In a weakly ionized plasma, fast electrons will separate from ions, leading to a negative charge deficit. To remain quasineutral the plasma develops an electric field within itself to slow electrons and accelerate ions such that they diffuse at the same rate and have the same bulk velocity. From the fluid equations of motion for a single species plasma one can derive the ambipolar electric field $\mathbf{E}_a$

$$\mathbf{E}_a \equiv \frac{D_i - D_e}{\mu_i + \mu_e} \frac{\nabla n}{n} \tag{2}$$

$$\mu_j \equiv \frac{|q_j|}{m_j \nu_{j,n}} \tag{3}$$

$$D_j \equiv \frac{k_B T_j}{m_j \nu_{j,n}} \tag{4}$$

where the subscript j denotes either species quantities, n is the quasineutral plasma density, m_j is the mass, q_j is electrical charge, and $\nu_{j,n}$ is the collision frequency with neutrals. The mobility and Einstein relation charged species diffusion coefficients are μ_j and D_j, respectively [6].

Several assumptions are made throughout this derivation. By starting from the fluid equations the flow is implied to be under continuum conditions. Second, the gas must be sufficiently weakly ionized such that charged particles primarily collide with neutrals rather than each other. This ensures particles diffuse in a random-walk manner consistent with Fick's Law, rather than the collective and often oscillatory motion of strongly and fully ionized plasmas. Finally, the weakly ionized gas must be a "plasma" by the following definition:

1. $\lambda_D \ll L$, where L is the characteristic size of the system
2. $N_D \gg 1$, where N_D is the number of plasma particles within a radius of λ_D
3. $\omega\tau > 1$, where ω is the frequency of typical plasma oscillations, and τ is the mean time between collisions between plasma particles and neutrals

Most critical to this study is the first item, known as quasineutrality. Plasmas are able to shield out charge disturbances, such that far from the disturbance, the plasma is charge neutral. While "charge neutral" refers to the net charge of an ionized gas being exactly zero at a point, "quasineutral" refers to the plasma as a whole being sufficiently close to charge neutral, such that $n_i \simeq n_e$. Therefore, it is possible that within a quasineutral plasma at length scales $\sim \lambda_D$, local charge neutrality may not be maintained.

3 Methodology

An Eulerian grid-based Boltzmann-Poisson-BGK solver (known as DK) is used for this study [8]. A detailed explanation of the solver, as well as citations to previous work, can be found in [5]. To investigate charged species diffusion with as few external variables as possible a simplified weakly ionized gas expansion setup is designed in one physical dimension and one velocity dimension (1D1V). A uniform background neutral gas is initialized inside a periodic domain of length $L = 1.0$ m, and an ionization fraction is imposed at the center of the domain from $0.25L \leq x \leq 0.75L$. Both the neutral and the ionized gas are at thermodynamic equilibrium at 500 K. At time $t = 0$, the ionized gas expands into the background neutral gas. Each simulation is allowed to run until a sufficient number of collisions have occurred such that any ambipolar fields have time to form. The number of normalized collision times is given by $n_{coll} = t/\tau_{j,n}$, where $\tau_{j,n}$ refers to the average time between collisions between a particle of type j and a neutral. Argon is used as the test gas. No ionization or recombination is allowed.

Both hard sphere and variable hard sphere collision models [2] are tested, and simulations are run with and without Coulomb and charge exchange (CEX) collisions. The final conclusions made in this study are invariant of collision model and the addition of Coulomb and CEX collisions. The plots shown in the following section are from simulations with hard sphere collisions with only momentum exchange (MEX) collisions between charged particles and neutrals. This represents the closest possible simulation setup to ideal ambipolar diffusion. The configuration space grids ensure at least 4 points per λ_i and at least 32 points per $\lambda_{D,e}$. Velocity space is discretized with 256 points. These grids ensure sufficient resolution of derived electrostatic properties [5].

4 Results

Results presented here are assessed in terms of the following. First, given a set of flow conditions, one must know whether ambipolar diffusion is expected. The Phelps diffusion length ratios [11] were derived to determine charged species diffusion type in gas discharge vessels where the diffusion length Λ is calculated from the length of the domain. Its ratio with the ion mean free path (Λ/λ_i) determines degree of ionized gas collisionality, and its ratio with the electron Debye length $(\Lambda/\lambda_{D,e})$ determines how well the ionized gas behaves like a plasma (i.e. a restatement of quasineutrality). The ionized gas expansion setup relaxes several assumptions of Phelps' derivation. Thus, a comprehensive set of simulations are run testing multiple diffusion length ratios, which ultimately found the Phelps ratios can successfully predict charged species diffusion type for the setup described above. In this setup, ratios of $\Lambda/\lambda_{D,e} \geq 100$ and $\Lambda/\lambda_i \geq 100$ are predicative of ambipolar diffusion, and ratios of $\Lambda/\lambda_{D,e} \leq 1$ and $\Lambda/\lambda_i \geq 10$ are predicative of free diffusion. Results are also assessed in terms of normalized charge separation $(= |n_i - n_e|/n_i)$, which measures local charge neutrality. For the purposes of this study, local charge neutrality is considered achieved when the normalized charge separation value is below 0.01 in a given cell.

4.1 Time Evolution of Ambipolar Diffusion

Ambipolar diffusion theory suggests time must elapse between the start of diffusion and the point when local charge neutrality is established everywhere. During this time the electric field described by Eq. 2 establishes itself the moment charge neutrality is violated, decaying as the plasma neutralizes itself, becoming zero when local charge neutrality is restored everywhere.

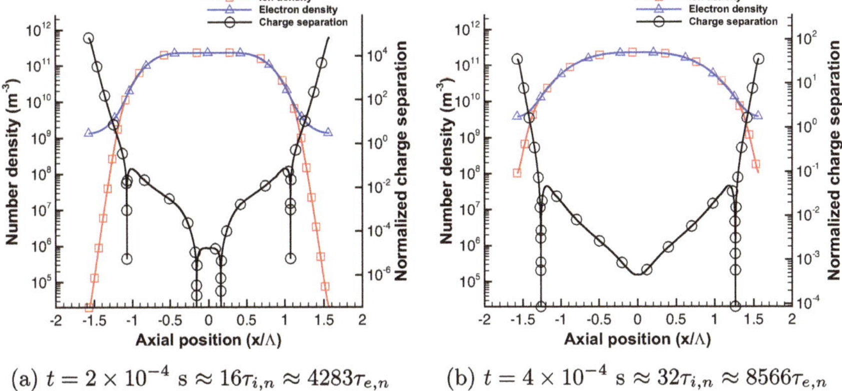

(a) $t = 2 \times 10^{-4}$ s $\approx 16\tau_{i,n} \approx 4283\tau_{e,n}$ (b) $t = 4 \times 10^{-4}$ s $\approx 32\tau_{i,n} \approx 8566\tau_{e,n}$

Fig. 1 Number density and normalized charge separation profiles during ambipolar diffusion at midpoint and end of simulation ($\Lambda/\lambda_i = 100$, $\Lambda/\lambda_{D,e} = 100$)

Figure 1 shows the time evolution of ion and electron number density and normalized charge separation profiles for a plasma undergoing ambipolar diffusion. The initial number densities of the background and ionized gas were $n_n = 6.12 \times 10^{20}$ m^{-3} and $n_i = n_e = 2.35 \times 10^{11}$ m^{-3}, corresponding to $\Lambda/\lambda_i = 100$ and $\Lambda/\lambda_{D,e} = 100$. It is evident that the rate at which local charge neutrality is achieved everywhere occurs on the order of tens of ion collision times and thousands of electron collision times. At the end of the simulation (Fig. 1b) local charge neutrality is approximately achieved everywhere except at the far boundaries of the simulation ($|x/\Lambda| > 1.25$). Thus, the ambipolar diffusion approximation would only be valid in this simulation at the given point in time for $|x/\Lambda| < 1.25$. Ultimately, unsteady simulations with processes that occur on time scales on the order of collision times should be mindful of the time dependence of ambipolar diffusion, especially when using the ambipolar diffusion approximation.

4.2 Electron Debye Length Versus the Mean Free Path

As previously discussed, the criterion of the electron Debye length being orders of magnitude smaller than the neutral mean free path is not inherent to the ambipolar diffusion derivation, nor is it consistent with the Phelps diffusion length criteria for ambipolar diffusion. One could alternatively take the limit as $\lambda_n \to \infty$, in which case the plasma is not collisional with neutrals, and the charged species are in the ion inertia regime.

This is confirmed with simulation results in Fig. 2a and b. A simulation case is set up with a Λ/λ_i ratio of 10,000 and a $\Lambda/\lambda_{D,e}$ ratio of 100, corresponding to $n_n = 6.12 \times 10^{22}$ m^{-3} and $n_i = n_e = 2.35 \times 10^{11}$ m^{-3}. The hard sphere collision

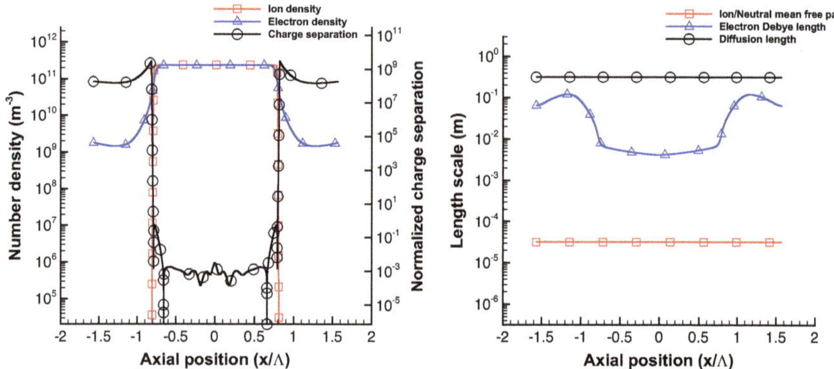

(a) Number density and charge separation profiles.

(b) Characteristic length scales for evaluating ambipolar diffusion.

Fig. 2 Number density (left) and characteristic length scale profiles (right) for ambipolar diffusion case where $\Lambda/\lambda_{D,e} = 100$, $\Lambda/\lambda_i = 10,000$, but $\lambda_{D,e} \gg \lambda_n$, at $t = 4 \times 10^{-5}$ s $\approx 323\tau_{i,n} \approx 37,418\tau_{e,n}$

model ensures the neutral mean free path is identical to the ion mean free path. This case violates the conventional criterion for the validity of the ambipolar diffusion approximation by two orders of magnitude. Figure 2a shows the number density and charge separation profiles, while Fig. 2b compares relevant length scales. At all points in the flow, $\lambda_{D,e} > \lambda_i$ by at least two orders of magnitude. However, at the points where $\Lambda/\lambda_{D,e} = 100$ ($|x/\Lambda| < 0.8$), local charge neutrality is achieved, indicating ambipolar diffusion is occurring. Thus, the Phelps criteria for charged species diffusion is a more accurate predictor of ambipolar diffusion than $\lambda_{D,e} \ll \lambda_n$.

4.3 Comparison with Kinetic Simulations Invoking the Ambipolar Diffusion Approximation

To assess the validity of the ambipolar diffusion approximation as it pertains to capturing actual ambipolar diffusion, an ambipolar diffusion case ($\Lambda/\lambda_i = 100$, $\Lambda/\lambda_{D,e} = 10,000, n_n = 6.12 \times 10^{20}$ m^{-3}, $n_i = n_e = 2.35 \times 10^{15}$) is run using both the DK solver and the Stochastic PArallel Rarefied-gas Time-accurate Analyzer (SPARTA) DSMC code [12], the latter of which implements the ambipolar diffusion approximation such that local charge neutrality is always guaranteed. Each SPARTA run follows the same simulation initialization conditions described in Sect. 3, with the exception of the grid points per Debye length and velocity space requirements. The SPARTA results shown are time averaged over 1000 runs, and use a particle weight of 10^6. By simulating only one physical grid cell in the perpendicular directions SPARTA can be run in 1D with three velocity dimensions. It is permissible to compare 1D1V DK to 1D3V SPARTA if the flow does not have significant energy transfer between translational energy modes. This is confirmed by benchmarking a 1D1V and 1D3V DK simulation.

Figure 3a demonstrates that even for a plasma well into the ambipolar diffusion regime the approximation does not capture the same number density profiles as DK's direct electric field modeling. This is best understood by looking at the axial velocity profiles for both codes (Fig. 3b). When the ambipolar diffusion approximation is implemented the shared axial velocity profile for ions and electrons averages to zero across the domain, in part because no ambipolar particles have traveled past $|x/\Lambda| > 0.9$. Meanwhile, the DK ion and electron velocity profiles begin to linearly increase from $0.6 < |x/\Lambda| < 1.5$ before dropping off towards the boundaries of the domain. At the beginning of the simulation, the fastest electrons would have escaped from the center of the domain. Once the ambipolar field establishes itself, any electrons leaving the center of the domain will do so at the ambipolar diffusion rate with a proportional number of ions. However, the formation of an electric field will also allow the fastest ions to accelerate to the boundaries. Only a small portion of ions will do this, as the peak ion axial velocity corresponds with a number density of around 10^{10} m^{-3}. In spite of this, the ion acceleration mechanism allows just enough ions to catch up with electrons to allow the domain to achieve local charge neutrality and equalize charged species bulk velocities sooner.

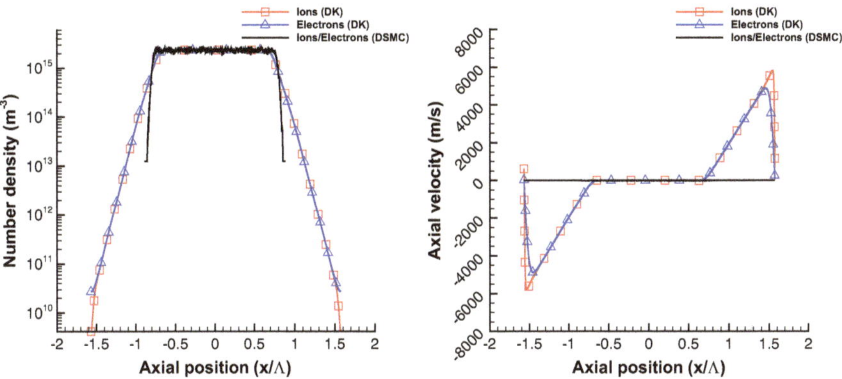

(a) Number density profiles for DK charged species and SPARTA ambipolar diffusion approximation plasma particles.

(b) Axial velocity profiles for DK charged species and DSMC ambipolar diffusion approximation plasma particles.

Fig. 3 Number density (left) and axial velocity profiles (right) for ambipolar diffusion approximation (DSMC) versus direct electric field modeling (DK) for ambipolar diffusion case with $(\Lambda/\lambda_i = 100, \Lambda/\lambda_{D,e} = 10{,}000)$ at $t = 5 \times 10^{-5}$ s $\approx 4\tau_{i,n} \approx 755\tau_{e,n}$.

5 Conclusions

Altogether, for a weakly ionized gas flow in the continuum regime, the ambipolar diffusion approximation can be justified if the following conditions hold:

1. The gas meets the three fundamental criteria for a plasma,
2. Charged particles primarily collide with neutrals rather than each other,
3. The ratio between the electron Debye length and the diffusion length scale (proportional to the characteristic length scale of the problem) has a ratio of at least $\Lambda/\lambda_{D,e} = 100$, and
4. The flow has had sufficient time for the ambipolar field to restore local charge neutrality (generally, on the order of many ion collision times).

These criteria do not guarantee results obtained with the ambipolar diffusion approximation will be identical to those from direct electric field modeling, particularly in unsteady flows. They also require confidence in plasma properties a priori, especially in the electron number density distributions used to calculate the Debye length, which are heavily influenced by electrostatic effects. Overall, these criteria represent a significant improvement from the conventional criterion used to justify the ambipolar diffusion approximation: $\lambda_{D,e} \ll \lambda_n$. During free diffusion, ions and electrons should be simulated as their own separate species. If the flow is within the transition regime between free and ambipolar diffusion direct electric field modeling is recommended as charged species diffusion dynamics cannot readily be predicted.

Whether the ambipolar diffusion approximation is valid under non-continuum conditions requires further investigation. Because the derivation of ambipolar diffusion begins with the fluid equations of motion, a breakdown of continuum behavior may result in irregularities in the plasma diffusion which might make the ambipolar diffusion approximation invalid. This has been investigated recently in hypersonic flows with particle-based methods [7, 9] and fluid codes [3] and will be investigated with the DK solver in upcoming studies.

References

1. Bird GA (1986) Direct simulation of typical AOTV entry flows. In: AIAA/ASME 4th joint thermophysics and heat transfer conference. AIAA, Boston. https://doi.org/10.2514/6.1986-1310
2. Bird GA (1994) Molecular gas dynamics and the direct simulation of gas flows. Clarendon Press, Oxford
3. Blanco A, Josyula E (2020) Numerical modeling of hypersonic weakly ionized external flowfields with Poisson's equation. AIAA J 58(8). https://doi.org/10.2514/1.J059307
4. Boyd ID (1997) Monte Carlo simulation of nonequilibrium flow in a low-power hydrogen arcjet. Phys Fluids 9:3086–3095. https://doi.org/10.1063/1.869474
5. Chan WR, Boyd ID (2022) Enabling direct kinetic simulation of dense plasma plume expansion for laser ablation plasma thrusters. J Electr Propuls 1(26). https://doi.org/10.1007/s44205-022-00030-x

6. Chen FF (2016) Introduction to plasma physics and controlled fusion, 3rd edn. Springer International Publishing Switzerland, Cham
7. Farbar ED, Boyd ID (2010) Modeling of the plasma generated in a rarefied hypersonic shock layer. Phys Fluids 22(10):106101. https://doi.org/10.1063/1.3500680
8. Hara K (2015) Development of grid-based direct kinetic method and hybrid kinetic-continuum modeling of hall thruster discharge plasmas. Dissertation, University of Michigan
9. Kaplan CR, Oran ES (2022) Analysis of the ambipolar diffusion approximation in hypersonic ionizing reentry flows. J Thermophys Heat Trans 36(3). https://doi.org/10.2514/1.T6416
10. Lee JH (1984) Basic governing equations for the flight regimes of aeroassisted orbital transfer vehicles. In: AIAA 19th thermophysics conference. AIAA, Snowmass. https://doi.org/10.2514/6.1984-1729
11. Phelps AV (1990) The diffusion of charged particles in collisional plasmas: free and ambipolar diffusion at low and moderate pressures. J Res Natl Inst Stand Technol 95(407):407–431. https://doi.org/10.6028/jres.095.035
12. Plimpton S, Moore S, Borner A, Stagg A, Koehler T, Torcyznski J, Gallis M (2019) Direct simulation Monte Carlo on petaflop supercomputers and beyond. Phys Fluids 31:086101. https://doi.org/10.1063/1.5108534

A Structure-Preserving Solver for Particle-Wave Interaction in Non-uniform Magnetized Plasmas

Kun Huang, Irene M. Gamba, and Chi-Wang Shu

Abstract We present a struture-preserving solver for particle-wave interaction in magnetized plasmas. The solver combines a conservative local discontinuous Galerkin (LDG) scheme for the interaction part with a trajectory averaging method for the Hamiltonian flow part. The proposed LDG scheme is an extension of the conservative scheme we developed in 2023. The trajectory averaging method significantly reduces computational cost by taking advantage of the multiscale feature of this system. By introducing a novel concept "trajectory bundle", we transform a continuous topological problem into a discrete graph theory problem. Numerical examples for a non-uniform magnetized plasma in an infinitely long symmetric cylinder is presented. It is verified that the connection-proportion algorithm allows to distinguish different trajectory bundles, and the proposed DG scheme rigorously preserves all the conservation laws.

Keywords Quasilinear theory · Discontinuous Galerkin method · Hamiltonian dynamics · Multiscale method

1 Introduction

The interaction between charged particles and waves is a critical aspect of plasma behavior, which occurs across multiple time scales. In order to simulate the waves within reasonable computational cost, waves are modeled as quasi-particles (plasmons) under WKB approximation [4], which have Hamiltonian equations of motion in the phase space. At the same time, their interaction with particles are modeled by the quasilinear theory [6, 10], which renders an additional collision-like operator: the quasilinear interaction operator.

K. Huang (✉) · I. M. Gamba
The University of Texas at Austin, Austin, TX, USA
e-mail: k_huang@utexas.edu

C.-W. Shu
Brown University, Providence, RI, USA

© The Author(s) 2026

M. Grabe et al. (eds.), *Rarefied Gas Dynamics*, Springer Aerospace Technology,
https://doi.org/10.1007/978-3-032-00094-1_66

As an approximation of the Vlasov-Maxwell system, our kinetic system of electrons and plasmons inherits the conservation properties, which relies on the delicate structure of the "collision" operators and the Poisson bracket. The goal of this paper is to develop numerical solvers that preserve the structure of the underlying physical systems.

The only precedent work on structure-preserving schemes for quasilinear theory is our paper [8] in 2023, where a continuous Galerkin method was adapted to discretize the diffusion operator for electrons. We generalize the idea and propose a novel discontinuous Galerkin version for quasilinear theory.

The trajectory-averaging technique [3] has been successful in tackling problems with simple advection field lines. For more complicated problems, there is the ray-tracing method [1], where sample trajectories are calculated with the Hamilton's equation, and numerical averaging is performed on these trajectories. However, such approach is expensive, not conservative, and not compatible with finite element or finite difference solvers. To tackle these problems, instead of discretizing the averaging operator directly, we propose a structure-preserving method based on the discretization of untraditional test spaces.

This paper is organized as follows. In Sect. 2 we describe the set up of the problem. After that we derive the weak form of trajectory-averaged equation and the connection-proportion algorithm in Sect. 3. The conservative LDG method for quasilinear theory will be introduced in Sect. 5. Section 6 contains the numerical results.

2 The Kinetic System for Particles and Plasmons

Consider an infinitely long cylinder $\Omega_x = (0, R) \times (0, 2\pi) \times \mathbb{R}_z$. At any given location $(r, \phi, z) \in \Omega_x$, there is a set of local orthonormal basis $(\mathbf{e}_z, \mathbf{e}_r, \mathbf{e}_\phi)$. Suppose a plasma is confined in the cylinder, embedded in a magnetic field along the z-axis, $\mathbf{B}(\mathbf{x}) = B(r)\mathbf{e}_z$.

We focus on the (z, ϕ)-symmetric case for simplicity, which means any function $g(\mathbf{x})$ depend only on r.

The momentum $\mathbf{p}$ of a particle at (r, ϕ, z) can be decomposed as $\mathbf{p} = p_z\mathbf{e}_z + p_r\mathbf{e}_r + p_\phi\mathbf{e}_\phi \in \mathbb{R}_p^3$. Define $p_\parallel = \frac{\mathbf{B}}{|\mathbf{B}|} \cdot \mathbf{p}$ and $p_\perp = \left|\left(I - \frac{\mathbf{B}}{|\mathbf{B}|} \otimes \frac{\mathbf{B}}{|\mathbf{B}|}\right) \cdot \mathbf{p}\right|$. Then for highly magnetized plasma, averaging over the fast gyro-motion renders a particle probability density function independent of azimuthal angle φ, i.e. $f(\mathbf{p}, \mathbf{x}, t) = f(p_\parallel, p_\perp, r, t)$. The momentum $\mathbf{k}$ of a plasmon (the wave vector of a wave packet) can also be decomposed as $\mathbf{k} = k_z\mathbf{e}_z + k_r\mathbf{e}_r + k_\phi\mathbf{e}_\phi \in \mathbb{R}_k^3$. Let us define $q_\phi = k_\phi \cdot r$, then (k_r, q_ϕ, k_z) and (r, ϕ, z) are a set of canonical coordinates. Given (z, ϕ)-symmetry, the plasmon probability density function can be written as $N(\mathbf{k}, \mathbf{x}, t) = N(k_r, q_\phi, k_z, r, t)$.

The following system governs the evolution of the particle and plasmon probability density functions:

$$\begin{cases} \partial_t f + v_z \partial_z f &= \nabla_p \cdot (D[N] \cdot \nabla_p f), \\ \partial_t N + \{\omega, N\} &= \Gamma[f]N. \end{cases} \tag{1}$$

The advection term $v_z \partial_z f$ comes from the gyro-averaged Vlasov equation, it vanishes since f does not depend on z. The Poisson bracket $\{\omega, N\} = \frac{\partial N}{\partial r}\frac{\partial \omega}{\partial k_r} - \frac{\partial \omega}{\partial r}\frac{\partial N}{\partial k_r}$ is a result of WKB approximation. Let us denote it as $\mathcal{T}N$, then obviously the operator $\mathcal{T}$ is anti-symmetric. The quasilinear diffusion term and the reaction term accounts for particle-plasmon interaction [4].

Testing the equations with $\varphi(\mathbf{p}, \mathbf{x})$ and $\eta(\mathbf{k}, \mathbf{x})$, we obtain the following weak form:

$$\iint_{px} \varphi\partial_t f + \iint_{kx} \eta\partial_t N = \iint_{kx} N\mathcal{T}\eta + \iiint_{pkx} (\eta\mathcal{L}E - \omega\mathcal{L}\varphi)\mathcal{L}fN\mathcal{B}, \tag{2}$$

where $\mathcal{E}(\mathbf{p})$ is the kinetic energy of a single particle,

$$\mathcal{E}(\mathbf{p}) = \gamma(\mathbf{p})mc^2 = \sqrt{m^2c^4 + p^2c^2},$$

the directional differential operator $\mathcal{L}$ is defined as,

$$\mathcal{L}g := k_\parallel \frac{\partial E}{\partial p_\perp}\frac{p}{p_\perp}\frac{\partial g}{\partial p_\parallel} + \left(\omega - k_\parallel \frac{\partial E}{\partial p_\parallel}\right)\frac{p}{p_\perp}\frac{\partial g}{\partial p_\perp}, \tag{3}$$

and the emission/absorption kernel $\mathcal{B}$ reads,

$$\mathcal{B}(\mathbf{p}, \mathbf{k}, \mathbf{x}) := \sum_l \frac{1}{\omega^2(\mathbf{k}, \mathbf{x})} U_l(\mathbf{p}, \mathbf{k}, \mathbf{x})\delta(\omega(\mathbf{k}, \mathbf{x}) - k_\parallel v_\parallel - l\omega_c(\mathbf{x})/\gamma(\mathbf{p})). \tag{4}$$

The dispersion relation $\omega(\mathbf{k}, \mathbf{x})$ is determined by the electron density $n_e(\mathbf{x})$. For details we refer the readers to the appendix of [8].

Theorem 1 *(unconditional conservation [8])*
   If $f(\mathbf{p}, \mathbf{x}, t)$ and $N(\mathbf{k}, \mathbf{x}, t)$ solve Eq. (1) with emission/absorption kernel replaced by $\mathcal{B}_\varepsilon$, then for any $\mathcal{B}_\varepsilon$ we have the following conservation laws:

$$\partial_t \mathcal{M}_{tot} = \partial_t \left((f, 1)_{px} + (N, 0)_{kx}\right) = 0,$$
$$\partial_t \mathcal{P}_{z,tot} = \partial_t \left((f, p_z)_{px} + (N, k_z)_{kx}\right) = 0,$$
$$\partial_t \mathcal{E}_{tot} = \partial_t \left((f, \mathcal{E})_{px} + (N, \omega)_{kx}\right) = 0.$$

3 Trajectory Average

As can be observed in Eq. (1), there are three different time scales associated with the model, τ_{diff}, τ_{adv} and τ_{reac}. According to Kiramov and Breizman [9], the advection process is much faster than the other two, $\tau_{\text{adv}} \ll \tau_{\text{diff}} \sim \tau_{\text{reac}}$. In practice, particle-wave interaction is of interest, rather than plasmon advection. Hence it is reasonable to eliminate the Poisson bracket term through trajectory averaging.

Consider the Liouville-reaction equation

$$\frac{\partial N(k, x, t)}{\partial t} + \frac{1}{\varepsilon} \{N(k, x, t), H(k, x)\} = \Gamma(k, x, t) N(k, x, t). \tag{5}$$

We define the advection operator T as follows,

$$Tg := \nabla_k H \cdot \nabla_x g - \nabla_x H \cdot \nabla_k g. \tag{6}$$

Denoting the orthogonal projection from $L^2(\Omega_b)$ onto $L^2(\Omega_b) \cap \ker T$ as $\mathcal{P}$, we have $\mathcal{P}u \in L^2(\Omega_b) \cap \ker T$, and $(\mathcal{P}u, \eta)_{\Omega_b} = (u, \eta)_{\Omega_b}$ for any $\eta \in L^2(\Omega_b) \cap \ker T$. According to [3], the lowest order approximation N_0 satisfies

$$\partial_t N_0 = \mathcal{P}(\Gamma N_0). \tag{7}$$

We call $\mathcal{P}$ the trajectorial averaging operator, and Eq. (7) is the averaged equation. The weak form of the averaged problem can be stated as follows:

Find an $N_0 \in L^2(\Omega_b) \cap \ker T$ such that

$$(\partial_t N_0, \eta)_{\Omega_b} = (\Gamma N_0, \eta)_{\Omega_b}, \tag{8}$$

for any $\eta \in L^2(\Omega_b) \cap \ker T$.

Since $\tau_{\text{adv}} \ll \tau_{\text{diff}}$, the diffusion equation for particle probability density function in (1) can be approximated with

$$\partial_t f = \nabla_p \cdot (D[N_0] \cdot \nabla_p f)$$

Testing the averaged system with $\varphi(\mathbf{p}, \mathbf{x}) \in \mathcal{G} = H^1(\mathbb{R}_p^3) \otimes L^2(\Omega_x)$ and $\eta(\mathbf{k}, \mathbf{x}) \in \mathcal{N} = \ker T \cap L^2(\mathbb{R}_k^3 \times \Omega_x)$ yields

$$\iint_{px} \varphi \partial_t f + \iint_{kx} \eta \partial_t N = \iiint_{pkx} (\eta \mathcal{L} E - \omega \mathcal{L} \varphi) \mathcal{L} f N B, \tag{9}$$

where we have used N instead of N_0. It can be verified that the unconditional conservation property is preserved even after trajectorial average.

4 Trajectory Bundle: Definition and Construction

To discretize Eq. (9), it is necessary to construct a series of finite dimensional spaces approximating the test space $\ker\mathcal{T} \cap L^2(\mathbb{R}_k^3 \times \Omega_x)$. Hence in this section we introduce the concept of trajectory bundle, and propose the connection-proportion algorithm to construct basis functions.

For details, we refer the readers to [7].

5 The Conservative Scheme Based on LDG Method

As has been proved in Theorem 1, the averaged system admits three conservation laws. In this section, we seek for a discretization that preserves all of them rigorously. The idea here is analogous to the one introduced in our previous work [8]: to replace some quantities with their projection in the test spaces. In what follows, we generalize the idea from continuous Galerkin method to local discontinuous Galerkin method.

As has been discussed in [2], the LDG weak form can also be written in bilinear form using discrete gradient operator ∇_h:

$$(\partial_t f, \varphi)_\Omega + (\nabla_h f, D \cdot \nabla_h \varphi)_\Omega = 0$$

The test space for particle probability density function, $\mathcal{G}_h$, consists of discontinuous piecewise polynomials. The test space for plasmon probability density function, $\mathcal{N}_h$, consists of indicator functions of trajectory bundles.

5.1 The Conservative Scheme

Recall the definition of directional differential operator

$$\mathcal{L}g := k_\| \frac{\partial E}{\partial p_\perp} \frac{p}{p_\perp} \frac{\partial g}{\partial p_\|} + \left(\omega - k_\| \frac{\partial E}{\partial p_\|}\right) \frac{p}{p_\perp} \frac{\partial g}{\partial p_\perp}.$$

We propose a discretized operator as follows,

$$\mathcal{L}_h g_h := k_{\|,h} \left(\partial_{\perp,h} E_h\right) \cdot \frac{p}{p_\perp} \left(\partial_{\|,h} g_h\right) + \left(\omega_h - k_{\|,h} \left(\partial_{\|,h} E_h\right)\right) \frac{p}{p_\perp} \left(\partial_{\perp,h} g_h\right), \quad (10)$$

where $k_{\|,h} := \Pi_{kx,h} k_\|$, $\omega_h := \Pi_{kx,h}\omega$, $E_h := \Pi_{px,h} E$, $\partial_{\|,h} g_h := \frac{\mathbf{B}}{|\mathbf{B}|} \cdot \nabla_{p,h} g_h$, $\partial_{\perp,h} g_h := \left|\left(I - \frac{\mathbf{B}}{|\mathbf{B}|} \otimes \frac{\mathbf{B}}{|\mathbf{B}|}\right) \cdot \nabla_{p,h} g_h\right|$, and the discrete gradient operator $\nabla_{p,h}$ is defined in [5].

Based on the above discrete operator $\mathcal{L}_h$, we propose an unconditionally conservative semi-discrete weak form [7].

Theorem 2 *If $f_h(t) \in \mathcal{G}_h$ and $N_h(t) \in \mathcal{N}_h$ satisfy that*

$$\iint_{px} \varphi_h \partial_t f_h + \iint_{kx} \eta_h \partial_t N_h = \iiint_{pkx} N_h \mathcal{B}_\varepsilon \left(\eta_h \mathcal{L}_h E_h - \omega_h \mathcal{L}_h \varphi_h \right) \mathcal{L}_h f_h, \qquad (11)$$

for any $\varphi_h(t) \in \mathcal{G}_h$ and $\eta_h(t) \in \mathcal{N}_h$, then we have discrete conservation laws:

$$\partial_t \mathcal{M}^h_{tot} := \partial_t \left((f_h, 1)_{px} + (N_h, 0)_{kx} \right) = 0.$$
$$\partial_t \mathcal{P}^h_{z,tot} := \partial_t \left((f_h, \Pi_{px,h} p_z)_{px} + (N_h, \Pi_{kx,h} k_z)_{kx} \right) = O(e^{-L^2}),$$
$$\partial_t \mathcal{E}^h_{tot} := \partial_t \left((f_h, \Pi_{px,h} E)_{px} + (N_h, \Pi_{kx,h} \omega)_{kx} \right) = 0.$$

6 Numerical Results

Analogous to [8], only anomalous Doppler resonance with $l = 1$ is considered. For dispersion relation $\omega(\mathbf{k}, \mathbf{x})$ we take the lowest branch.

Consider a magnetized plasma with non-uniform electron density $n_e(r) = n_0 \left[1 - \left(\frac{r}{R_{max}} \right)^2 \right]$, embedded in a uniform external magnetic field $\mathbf{B}(r) = B_0 \mathbf{e}_z$.

Analogous to [8], we only compute the "bump" part of electron probability density function $f(\mathbf{p}, \mathbf{x}, t)$, which takes the following initial configuration:

$$f(p_\parallel, p_\perp, r)\big|_{t=0} = \left[10^{-5} \frac{1}{\sqrt{\pi}} \exp \left(-\left(\frac{p_\parallel}{m_e c} - 20 \right)^2 - \left(\frac{p_\perp}{m_e c} \right)^2 \right) \right] \frac{n_0}{m_e^3 c^3} \left(\frac{1}{R_{max}} \right)^3.$$

Meanwhile, the plasmon probability density function $N(\mathbf{k}, \mathbf{x}, t)$ is initialized as follows:

$$N\left(r, k_r, q_\phi, k_z \right)\big|_{t=0} = 10^{-5} \frac{n_0 m c^2}{(\omega_0/c)^3} \frac{1}{\hbar \omega_0} \left(\frac{1}{R_{max}} \right)^3.$$

Trajectory bundles As shown in Fig. 1, the triangular partition of the (r, k_r)-domain is done by dividing every rectangular mesh into two. The connection-proportion algorithm successfully distinguishes different trajectory bundles in the inverse image of the same box(in this case, interval).

Electron pdf In the first row of Fig. 2 we show the evolution of electron probability density function $f(p_\parallel, p_\perp, r, t)$ at $r = \frac{31 R_{max}}{40}$. In the second row of Fig. 2 the electron probability density function at the same time point $t = 3 \times 10^6 \frac{1}{2\pi\omega_0}$ in different positions is presented.

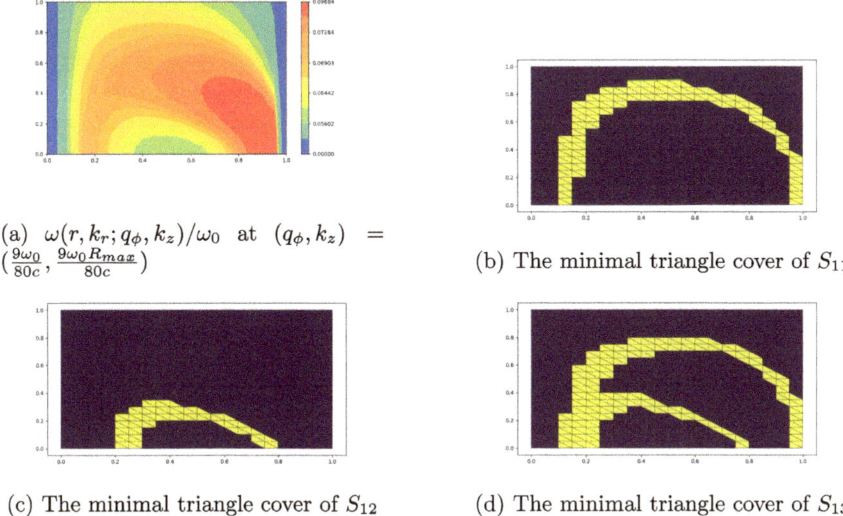

(a) $\omega(r, k_r; q_\phi, k_z)/\omega_0$ at $(q_\phi, k_z) = \left(\frac{9\omega_0}{80c}, \frac{9\omega_0 R_{max}}{80c}\right)$

(b) The minimal triangle cover of S_{11}

(c) The minimal triangle cover of S_{12}

(d) The minimal triangle cover of S_{13}

Fig. 1 Trajectory bundles and their minimal triangle covers. The x-axis represents r/R_{max}, and the y-axis represents $k_r c/\omega_0$. Since the Hamiltonian is symmetric for $\pm k_r$, we only plot half of the domain. Note that S_{11} and S_{12} are two trajectory bundles generated by the same Hamiltonian range interval. And S_{13} is not a single strip because it contains a saddle point

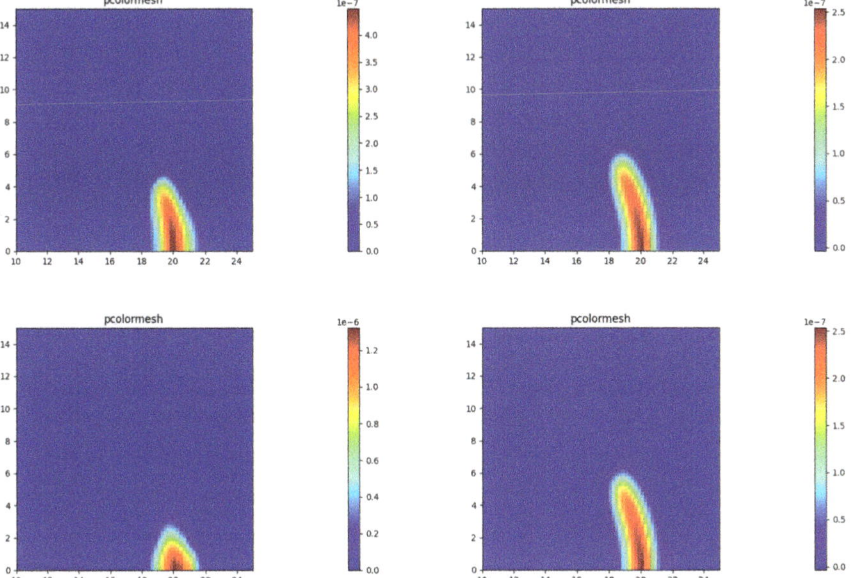

Fig. 2 Electron distribution

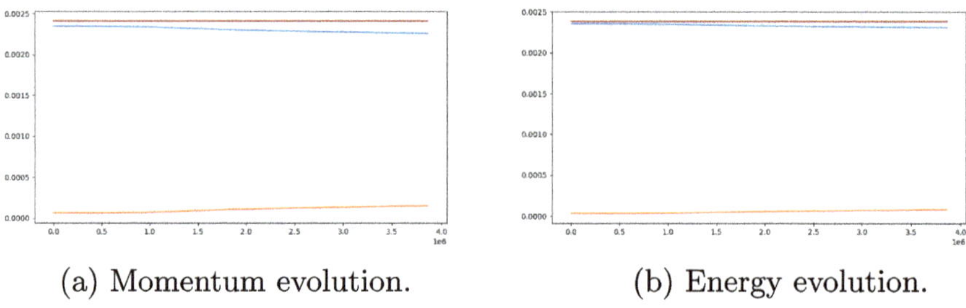

(a) Momentum evolution.   (b) Energy evolution.

Fig. 3 Conservation laws

Conservation verification For the evolution of the electron-plasmon system momentum and energy, see Fig. 3. With $T_{max} = 3.86 \times 10^6 \frac{1}{2\pi\omega_0}$, we have the following relative errors: $e_{rel}(\mathcal{M}_{tot,h}) = 4.6 \times 10^{-16}$, $e_{rel}(\mathcal{P}_{\parallel,tot,h}) = 1.8 \times 10^{-14}$ and $e_{rel}(\mathcal{E}_{tot,h}) = 5.5 \times 10^{-16}$.

7 Conclusion

In this paper, we introduced a structure-preserving solver for particle-wave interaction in a cylinder with radially non-uniform plasma density. We preserve all the conservation laws during "collision" by adopting the unconditionally conservative weak form. For the fast periodic advection of plasmons in the phase space, we proposed a novel averaging method, which significantly reduces the computational cost without violating the Hamiltonian structure. All these properties have been verified by numerical experiments, where we observe different diffusion rate in different positions.

In the future, we plan to extend the solver for more sophisticated models, enabling the full-scale simulation of runaway electrons.

Acknowledgements The authors thank and gratefully acknowledge the support from the Oden Institute of Computational Engineering and Sciences and the University of Texas Austin. This project was supported by funding from NSF DMS: 2009736, NSF grant DMS-2309249 and DOE DE-SC0016283 project Simulation Center for Runaway Electron Avoidance and Mitigation.

References

1. Aleynikov P, Breizman B (2015) Stability analysis of runaway-driven waves in a tokamak. Nuclear Fusion 55(4):043014
2. Arnold DN, Brezzi F, Cockburn B, Marini LD (2002) Unified analysis of discontinuous galerkin methods for elliptic problems. SIAM J Numer Anal 39(5):1749–1779

3. Bostan M (2010) Transport equations with disparate advection fields. Application to the gyrokinetic models in plasma physics. J Differ Eq 249(7):1620–1663
4. Breizman BN, Aleynikov P, Hollmann EM, Lehnen M (2019) Physics of runaway electrons in tokamaks. Nuclear Fusion 59(8):083001
5. Di Pietro DA, Ern A (2011) Mathematical aspects of discontinuous Galerkin methods, vol 69. Springer Science & Business Media
6. Drummond W, Pines D (1962) Non-linear stability of plasma oscillations
7. Huang K (2023) A numerical and analytical study of kinetic models for particle-wave interaction in plasmas. Ph.D. thesis
8. Huang K, Abdelmalik M, Breizman B, Gamba IM (2023) A conservative galerkin solver for the quasilinear diffusion model in magnetized plasmas. J Comput Phys 488:112220
9. Kiramov D, Breizman B (2021) Reduced quasilinear treatment of energetic electron instabilities in nonuniform plasmas. In: APS division of plasma physics meeting abstracts, vol 2021, pp JP11–068
10. Vedenov A, Velikhov E, Sagdeev R (1961) Nonlinear oscillations of rarified plasma. Nuclear Fusion 1(2):82

Potential Fluctuations of Emissive Sheaths in Collisional Boundary Layer Plasmas Using a Kinetic Approach

Davut Vatansever and Deborah A. Levin

Abstract Emissive plasma sheaths are subject to various instabilities, including two-stream (TS) and self-spike (SS) oscillations, both of which play critical roles in the sheath dynamics. This study extends our previous work by examining the effects of these instabilities on potential field fluctuations and the electron velocity distribution functions (EVDFs) in boundary layer plasmas. Using a fully kinetic PIC-DSMC approach, we demonstrate that the TS instability induces strong oscillations in the sheath potential, exceeding the sheath potential by up to tenfold in large domains. In contrast, the SS instability produces sawtooth oscillations with instantaneous, non-amplifying disturbances. Additionally, the TS instability significantly alters the EVDFs, creating highly non-Maxwellian distributions, while the SS instability causes instantaneous shifts in the EVDF peaks during SS ejections.

1 Introduction

Emissive plasma sheaths are subject to various instabilities, including two-stream (TS) electron instability, self-spike (SS) oscillations, electrostatic instabilities, and gradient-driven instabilities. The TS instability arises in the presence of counter-streaming electron populations [3, 6]. It can be triggered by variations in the domain length, plasma density, electron temperature, or potential bias [7]. The SS instability occurs in the presence of significant charge-exchange (CEX) collisions within the sheath. These collisions lead to oscillations in the virtual cathode (VC), which manifest as spikes in the plasma sheath [1]. Our simulations demonstrate the occurrence of TS and SS instabilities, both independently and simultaneously. When a sufficiently long domain size ($30\,\lambda_D$) with collisions is considered, interactions between the SS and TS instabilities are observed, where the latter dominates due to stronger

D. Vatansever (✉) · D. A. Levin
Department of Aerospace Engineering, University of Illinois Urbana-Champaign, Urbana, IL, USA
e-mail: davutv2@illinois.edu

D. A. Levin
e-mail: deblevin@illinois.edu

© The Author(s) 2026

M. Grabe et al. (eds.), *Rarefied Gas Dynamics*, Springer Aerospace Technology,
https://doi.org/10.1007/978-3-032-00094-1_67

and more frequent fluctuations. These interactions result in the weakening of the SS spikes due to strong fluctuations induced by the TS instability.

In our previous work [8], we explored the mechanisms behind the development of TS and SS instabilities, characterized the wave-number frequency spectrum of their oscillations, and analyzed the interaction between these two instabilities. In the present study, we focus on the effects of these instabilities on potential field fluctuations and electron velocity distribution functions (EVDFs), probed both inside and outside the VC region in a boundary layer plasma. This work is an extension to our results where we revealed that the TS instability induces significantly larger potential fluctuations compared to the SS instability (sawtooth oscillations), for which the fluctuations are instantaneous at spike times and not amplifying.

2   Numerical Method and Simulation Cases

1D3V simulations were conducted with our fully kinetic GPU-based solver, CHAOS [5]. CHAOS incorporates fully kinetic particle approaches, namely the Particle-In-Cell (PIC) and Direct Simulation Monte Carlo (DSMC) methods, which together allow for the self-consistent calculation of the electric field at each time step while accurately capturing collisions between heavy charged particles and neutrals. The PIC module determines the spatial distribution of the electric field by solving Poisson's equation,

$$\nabla^2 \phi = \frac{e(n_i - n_e - n_{ee})}{\epsilon_0} \tag{1}$$

$$E = -\nabla \phi \tag{2}$$

where ϕ is electric potential, ϵ is permittivity of free space in vacuum, e is elementary charge, E is electric field which accelerates charged species, n_i, n_e, and n_{ee} are number density of ions, plasma electrons and wall emitted electrons, respectively. The DSMC approach is used to model ion-neutral and neutral-neutral collisions in the presence of neutrals in addition to plasma. By coupling the DSMC and PIC modules, the solver simulates CEX collisions and analyzes their impact on the electric field. The presence of neutrals leads to significant alterations in the sheath structure, as CEX ions are generated and become trapped within the VC formed by strong electron emission from the wall.

Figure 1a and b illustrate the emissive sheath structure in a hypersonic boundary layer plasma and the corresponding computational setup. The near-wall region, spanning only tens of Debye lengths, is simulated by placing an electron-emitting wall at z = 0 and injecting plasma from z = L. The wall is biased at 0 V, with Dirichlet boundary conditions of 2V is applied at z = L.

A Maxwellian argon plasma with a temperature of 0.5 eV flows into the domain from z = L, into a vacuum for the collisionless cases and into a Maxwellian neutral

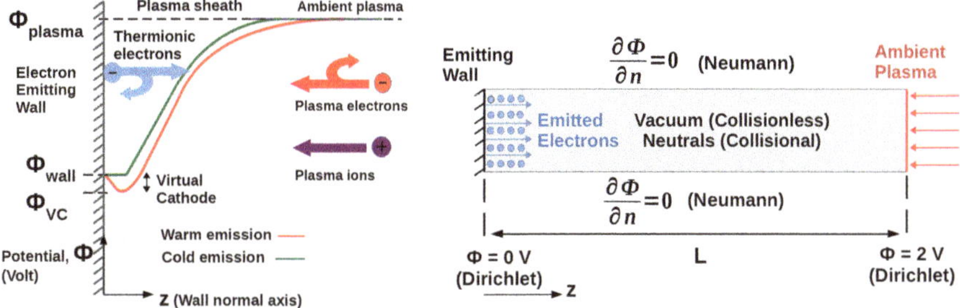

(a) A SCL emissive sheath structure. (b) Computational domain.

Fig. 1 Emissive sheath structure and computational setup for simulations

Table 1 Conditions for different sheath simulation cases

Case	Collisions	Domain length (mm)	Neutral density	Neutral temperature
1A	Off	$0.4 = 15\,\lambda_D$	–	–
2A	Off	$0.8 = 30\,\lambda_D$	–	–
3A	Off	$1.6 = 60\,\lambda_D$	–	–
1B	On	$0.4 = 15\,\lambda_D$	1×10^{19} m^{-3}	300 K
2B	On	$0.4 = 15\,\lambda_D$	1×10^{20} m^{-3}	300 K
3B	On	$0.4 = 15\,\lambda_D$	1×10^{20} m^{-3}	2000 K
2C	On	$0.8 = 30\,\lambda_D$	1×10^{20} m^{-3}	300 K

background for the collisional cases. Warm electrons, emitted from the wall at $z = 0$, are continuously injected with a half-Maxwellian distribution corresponding to a temperature of 0.17 eV. The plasma density is 8×10^{16} m^{-3}, with equal ion and electron densities. Plasma ions have a bulk speed of 8 km/s, equivalent to the ram speed of a hypersonic vehicle [2]. Wall-emitted electron density is set to ten times that of the plasma electrons to form a space-charge limited sheath [3], which includes the VC region as shown in Fig. 1a. The full descriptions of the PIC and DSMC methods implemented in CHAOS can be found in [4, 5], while the specific computational parameters used in this study are detailed in [8].

Seven plasma sheath cases, summarized in Table 1, were analyzed to study the effects of domain length, collisionality, neutral density, and temperature. The collisionless cases (1A, 2A, 3A) used domain lengths of 0.4, 0.8, and 1.6 mm to explore the impact of domain size on TS instabilities. The collisional cases (1B, 2B, 3B, 2C) included varying neutral densities of 10^{19} and 10^{20} m^{-3}, with neutral temperatures of 300 K (0.026 eV) and 2000 K (0.17 eV) for the high-density scenario. Cases 1B, 2B, and 3B focus on the shortest domain length (0.4 mm) to isolate SS instability from TS effects, while case 2C demonstrates their combined impact. The present study considers only ion-neutral collisions, as their mean free paths are significantly shorter than those of other relevant collision processes, namely, Coulomb collisions and electron-neutral collisions, under the conditions investigated [8].

3 Potential Fluctuations Created by Two-Stream Instability

Large-amplitude plasma oscillations develop early in simulations when the domain size is sufficiently large to support long-wavelength disturbances, as in cases 2A and 3A. Figure 2 shows time snapshots of the velocity phase space of plasma and emitted electrons in case 3A, represented by red and blue dots, respectively. At 0.125 μs, plasma and emitted electron streams begin to occupy the domain and interact. A distinct shear layer forms between the emitted and plasma electrons, leading to disturbances in the plasma electron phase space. At initial times up to 0.175 μs, as seen from the top row, the mixing of the two populations is minimal, and their phase spaces remain largely separable.

By 0.5 μs, the two populations have fully mixed, as shown in the bottom-left frame of the figure. Beyond this time, large-amplitude plasma oscillations becomes apparent. The bottom row illustrates the continued evolution and mixing, with increasingly complex phase space structures emerging over time. These fluctuations in the phase space also generate large potential fluctuations, further amplifying the dynamic behavior of the sheath.

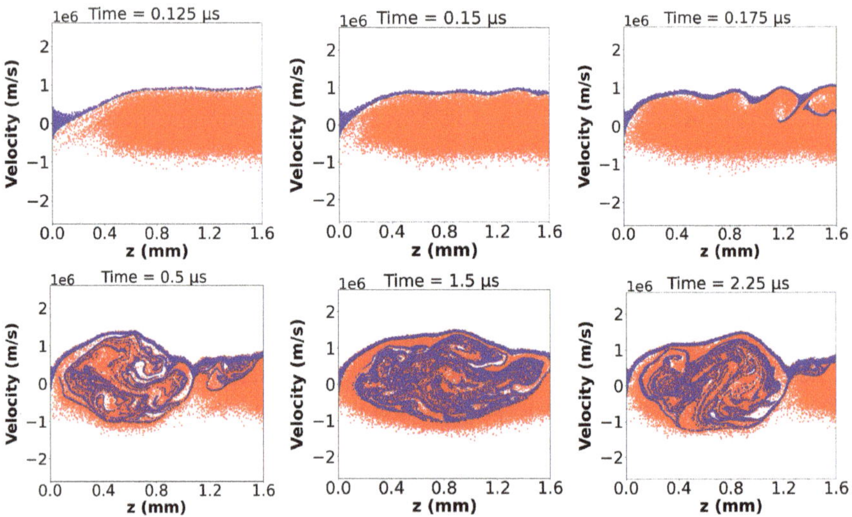

Fig. 2 Time snapshots of velocity phase spaces of plasma (red) and emitted (blue) electrons from 0.1 μs to 2.25 μs for Case 3A, where $L = 60\,\lambda_D$

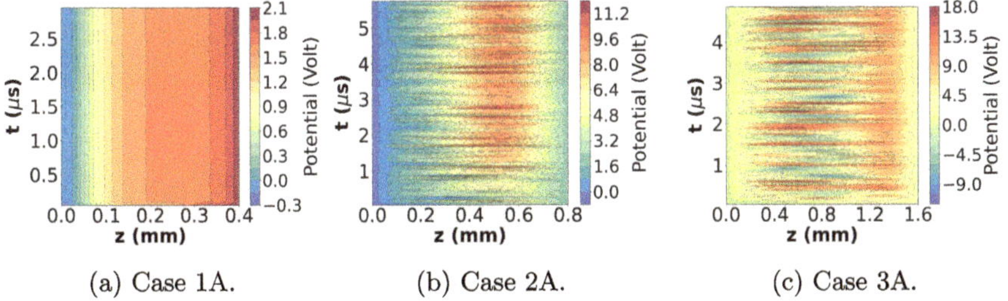

(a) Case 1A.						(b) Case 2A.						(c) Case 3A.

Fig. 3 Spatio-temporal fluctuations in electric potential field for the cases 1A ($15\,\lambda_D$), 2A ($30\,\lambda_D$), and 3A ($60\,\lambda_D$). The case with the shortest domain length reaches a quasi-steady state. In contrast, strong periodic oscillations are observed for cases 2A and 3A

Figure 3 shows the time evolution of the spatial distribution of the electric potential field for the three collisionless cases simulated. Development of the TS instability with increased domain length causes spatio-temporal fluctuations in the electric potential. Note that for all the three cases, the wall is negatively biased at 0 V with respect to the plasma potential of 2 V. Case 1A with a $15\,\lambda_D$ length domain is at quasi-steady state and does not exhibit temporal and spatial oscillations. As the length of the domain is increased from $15\,\lambda_D$ to $30\,\lambda_D$, strong periodic oscillations can be observed in the downstream region, especially after 1 μs. In case 1A, the potential monotonically increases from the dip of the VC to the plasma boundary on the right hand side (from -0.2 to 2 V). However, the development of the TS instability creates electron turbulences and disrupts the monotonically increasing profile observed in case 1A. For case 2A, the figure shows that the time-varying potential always remains above the plasma potential in the downstream between 0.4 mm $< z <$ 0.8 mm. As the domain length is further increased, we start to see even stronger spatio-temporal fluctuations in the plasma field. Due to increased electron mixing in case 3A, even negative potential values can be observed in the downstream region. The strength of the oscillations is higher in this case with the plasma potential oscillating between $-$10 and 18 V in comparison to the fluctuations between -0.2 and 11 V in the case 2A. The presence of periodic structures can be observed in these two unstable cases. For example in case 3A, starting from 0.5 μs, nine different peaks can be observed in the z-t domain of the plasma potential. It is also noticeable that the potential fluctuations are weaker in the VC, z $<$ 0.03 mm, for both cases.

4 Potential Fluctuations Created by Self-spike Instability

In this section, we examine the impact of ion-neutral collisions on the plasma sheath potential fluctuations caused by the trapping of slow CEX ions within the VC. In our published work [8], we explained the detailed transient CEX entrapment and SS ejection processes. The periodic ejections of CEX ions from the VC induce sawtooth

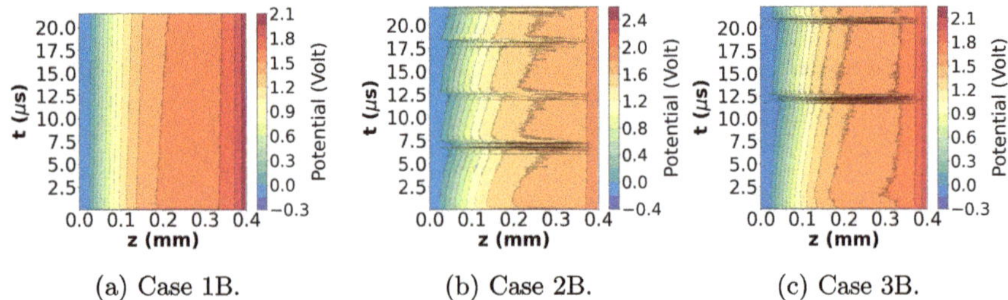

Fig. 4 Sawtooth fluctuations in electric potential field for the collisional cases 1B, 2B, and 3B

oscillations in the potential field, significantly influencing the sheath dynamics. The ejection of trapped ions from the VC occurs with different periodicities depending on the neutral temperature and density, leading to varying fluctuation patterns in the potential field.

Figure 4a shows that the potential field for case 1B exhibits no spikes due to its low collisionality, as the neutral density is an order of magnitude lower than in the other collisional cases. However, a subtle spatio-temporal gradient in the potential field can still be observed in case 1B compared to the collisionless case 1A (Fig. 3a). In contrast, sawtooth oscillations are clearly visible in the potential fields for cases 2B and 3B, as shown in Fig. 4b and c, respectively. These oscillations arise from the higher collisionality and increased CEX ion generation. In case 2B, the colder neutral temperature (300 K) results in slower CEX ion velocities, causing more ions to become trapped in the VC compared to case 3B, where the neutral temperature is 2,000 K. This difference in ion behavior produces distinct periodic spike patterns in the potential field, with spike periods of approximately 5 μs for case 2B and 8 μs for case 3B.

5 The Effect of Instabilities on EVDFs

First, we analyze the effect of the TS instability on the EVDFs. In general, fluctuations caused by the TS instability in the VC region are less significant than those observed in the region downstream of the VC. VC region is dominated by the more populous space charged thermionic electron causing the EVDF to largely reflect thermionic electron behavior. Hence, the EVDFs probed in the VC shows similar behavior for all cases, except during a SS time. However, in the region downstream of the VC where there is a large population of plasma electrons, the EVDFs can be seen to have large TS fluctuations even at earlier times.

The top row of Fig. 5 shows that at 0.1 μs there are two peaks in the EVDF for each case corresponding to thermionic and plasma electrons. As the time progresses to 1.5 μs, the EVDF of the short domain case 2B remains almost constant whereas additional peaks occur for the 30 λ_D cases of 2A and 2C. This is due to electron

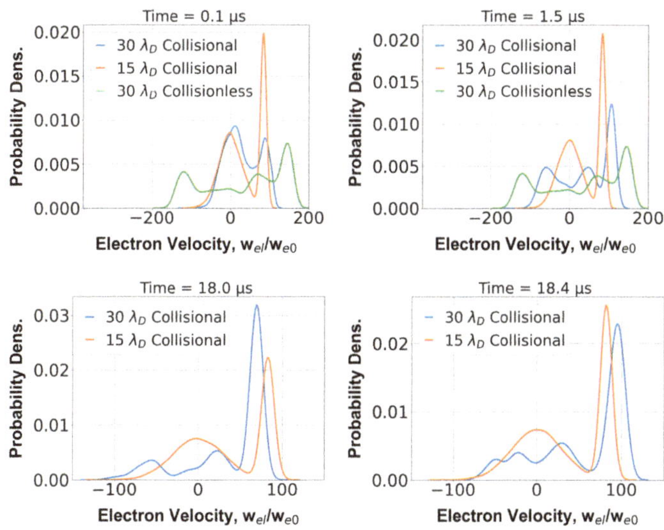

Fig. 5 Comparison of EVDFs in the downstream region of $0.3 < z < 0.4$ mm for long ($30\,\lambda_D$) and short ($15\,\lambda_D$) collisional (2C and 2B) and the long ($30\,\lambda_D$) collisionless (2A) cases

mixing caused by the TS instability in the $30\,\lambda_D$ cases. The top row of the figure shows that the EVDFs of the cases 2A and 2C become highly non-Maxwellian and their distributions differ from each other. However, it is difficult to attribute this difference to the CEX collisions and SS instability because the TS electron instability is more dominant in this region as well as the fact that in this region the ions are not trapped. At longer times, as seen from the bottom row of Fig. 5, the EVDF of case 2C shows highly non-Maxwellian distribution due to electron mixing process while that of case 2B has still a bimodal distribution.

Next, we present the effect of the SS instability by comparing the short domain cases 1A and 2B. The EVDFs from cases 1A and 2B for long times in both the VC (top row of Fig. 6) and downstream regions (bottom row of Fig. 6) are presented. In the VC region, the EVDFs of both cases look similar, with the difference that the EVDF of case 2B at 4μs starts to deviate from earlier times due to accumulation of trapped CEX ions in the VC. For times up to 4μs, EVDFs of cases 1A and 2B show similar bimodal distributions. In the VC, the peak of the EVDF differs due to the oscillations triggered by the SS for case 2B as shown in the top row of Fig. 6. Within the downstream region, despite maintaining a bimodal distribution, the EVDF of case 2B undergoes alterations in both peak widths and heights as shown in the bottom row of Fig. 6. These changes are observed specifically during the occurrence of a SS, with the most significant shift observed in the second peak, corresponding to emitted electrons.

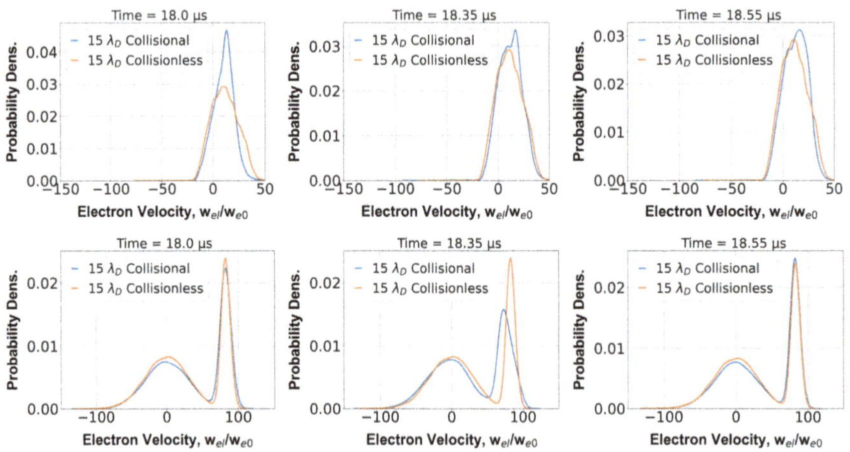

Fig. 6 Comparison of EVDFs during spike times. Top and bottom rows are in the VC and the downstream region of $0.3 < z < 0.4$ mm, respectively for the 15 λ_D length collisional (2B) versus collisionless (1A) cases

6 Conclusion

The plasmadynamics of an emitting electron wall for conditions similar to the near-wall region of a hypersonic boundary layer have been predicted using a fully kinetic PIC-DSMC approach. The TS instability in the sheath leads to stronger oscillations in the potential field, with fluctuations far exceeding the sheath potential. In sufficiently large plasma domains (60 λ_D), the potential in the sheath can oscillate up to 10 times the sheath potential, demonstrating the considerable influence of TS instability on sheath dynamics. For the SS instability, the potential field exhibits sawtooth oscillations, with disturbances occurring instantaneously and not resulting in amplified fluctuations following the SS ejection. We also show that both instabilities significantly impact the EVDFs: the TS instability results in highly non-Maxwellian distributions, while the SS instability causes instantaneous shifts in the EVDF peaks within the VC and bulk plasma regions during SS ejection.

Acknowledgements This project is supported by the Lockheed Martin Corporation, LMC Reference : S20-002-001/UIUC Grant C0764. We thank NCSA for providing us with the computational resources on Delta petascale computing facility.

References

1. Campanell MD, Umansky MV (2017) Improved understanding of the hot cathode current modes and mode transitions. Plasma Sources Sci Technol 26(12):124002. https://doi.org/10.1088/1361-6595/aa97a9

2. Hanquist KM, Hara K, Boyd ID (2017) Detailed modeling of electron emission for transpiration cooling of hypersonic vehicles. J Appl Phys 121(5):053302
3. Hara K, Hanquist K (2018) Test cases for grid-based direct kinetic modeling of plasma flows. Plasma Sources Sci Technol 27(6):065004. https://doi.org/10.1088/1361-6595/aac6b9
4. Jambunathan R, Levin DA (2017) Advanced parallelization strategies using hybrid MPI-CUDA octree DSMC method for modeling flow through porous media. Comput Fluids 149:70–87
5. Jambunathan R, Levin DA (2018) Chaos: an octree-based PIC-DSMC code for modeling of electron kinetic properties in a plasma plume using MPI-CUDA parallelization. J Comput Phys 373:571–604
6. Sydorenko D, Smolyakov A, Kaganovich I, Raitses Y (2007) Effects of non-maxwellian electron velocity distribution function on two-stream instability in low-pressure discharges. Phys Plasmas 14(1):013508. https://doi.org/10.1063/1.2435315
7. Thorne KS, Blandford RD (2017) Modern classical physics: optics, fluids, plasmas, elasticity, relativity, and statistical physics. Princeton University Press, New Jersey
8. Vatansever D, Nuwal N, Levin DA (2024) Numerical investigations of spatiotemporal dynamics of space-charge limited collisional sheaths. Phys Plasmas 31(9):093508. https://doi.org/10.1063/5.0216487

Challenges Using Rarefied Atmospheric Gases as Propellant for an RAM-EP System

Jana Zorn, Ole Forchheim, Kristof Holste, and Peter J. Klar

Abstract Many challenges have to be faced in order to develop an air-breathing electric propulsion system using a radio frequency ion thruster (RIT). The underlying assumptions of gas-surface-interactions for DSMC simulations of the intake give a wide spread of results in operating points available for the thruster. Thus, it is necessary to replicate the VLEO conditions experimentally in the future. Furthermore, precise experimental characterization of the propulsion system using atmospheric propellants is essential for mission success, although this also presents significant challenges. The background pressure of the test facility can cause an overestimation of the measured atomic-to-molecular ratio by a factor of 10. The sputtering of the beam dump can cause a back-streaming of carbon particles to the thruster of approx. 3.4×10^{13} particles/(s mA). Moreover, the reactive propellant causes significant material degradation of all materials in direct contact with the plasma.

Keywords ABEP · RAM-EP · Challenges · RIT · RF-neutralizer · Air · Air-breathing

1 Introduction

Air-breathing electric propulsion systems (ABEP, or RAM-EP) are receiving considerable attention recently. They shall provide a cheap and sustainable tool for long-term missions in the very low earth orbit (VLEO). An intake system collects the residual atmospheric gas present in VLEO. Since density and composition of the atmosphere serving as propellant strongly fluctuate with orbit and sun activity it is very challenging to design an intake that always guarantees a sufficient mass flow to the thruster and thus a constant drag compensation [1]. As a consequence of the significant fluctuations in atmospheric conditions, the thruster must be capable of maintaining optimal performance across a wide range of mass flows while operating within a constrained power envelope, ensuring consistent thrust delivery [1].

J. Zorn (✉) · O. Forchheim · K. Holste · P. J. Klar
Justus Liebig University, Giessen, Germany
e-mail: jana.zorn@physik.uni-giessen.de

© The Author(s) 2026

M. Grabe et al. (eds.), *Rarefied Gas Dynamics*, Springer Aerospace Technology, https://doi.org/10.1007/978-3-032-00094-1_68

Thus, the thruster system must be characterized with high precision for all possible conditions. However, achieving this required precision in terrestrial experimental environments is a significant challenge. For example, the various chemical interactions and sputtering processes between materials that are in direct contact with the atmospheric plasma can change the performance significantly. Furthermore, charge exchange processes and other facility effects impact the experimental data and can cause unreliable results.

2   Intake Design

Designing a functional intake system to collect enough residual gas for the propulsion system is quite challenging due to the strong fluctuations of the atmospheric conditions. An important unknown is the gas-surface interaction (GSI) between the intake material and the residual gas at a velocity of approximately 8 km/s. Especially the interaction between atomic oxygen and surface material of the intake is not well investigated yet. Atomic oxygen can get physically adsorbed or react chemically and thus change the surface significantly, resulting in varying GSI (diffuse, specular, or Cercignani-Lampis-Lord (CLL) model). The scattering process is significantly influenced by the material, surface smoothness, projectile type, velocity and angle of incidence [9]. It primarily determines the performance for a chosen configuration of the intake system, as can be seen in the results of a DSMC simulation in Fig. 1. As experimentally shown, the scattering of an atomic oxygen beam with 8 km/s hitting various materials is most likely described with the CLL model [8]. Most importantly, scattering processes of atomic oxygen and molecular nitrogen with intake materials and their surfaces must be studied under VLEO conditions, in order to develop a reliable intake system. None of the attempts reported so far have been able to meet all VLEO requirements [1]. The beam typically consists of ions instead of neutrals in most terrestrial approaches [5, 11]. The replication of atmospheric conditions in a laboratory setting to accurately investigate the GSI remains a major challenge.

3   Propulsion System

A further crucial, yet demanding aspect of an ABEP system is the propulsion system itself, which is primarily composed of a thruster and a neutralizer. Both devices need to be well characterized in order to deliver thrusts larger than the drag acting on the satellite. However, testing light molecular propellants leads to several challenges that need to be considered for obtaining reliable results.

Experimental setup: Several diagnostics are installed in the test facility "Obelix" (Fig. 2). The vacuum chamber has a volume of 4800 L and is equipped with one turbo-molecular pump and one cryogenic pump. A base pressure in the low 10^{-7} mbar range

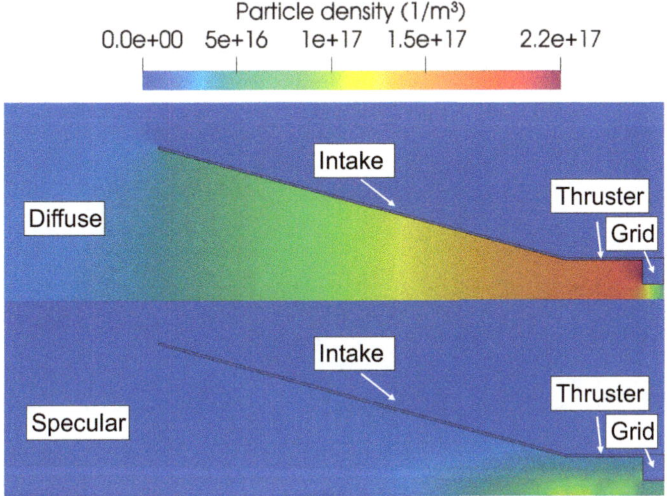

Fig. 1 Intake simulation assuming two different gas-surface interactions (GSI)

can be reached, enabling a pressure during the experiments in the mid 10^{-6} mbar range. At the end of the chamber a mass spectrometry setup is installed consisting of a double-focusing electromagnet. A Faraday cup at the end of the magnet is used for ion detection. Two additional turbo-molecular pumps with a total pumping speed of 66 L/s are attached to the mass spectrometry setup. A RIT-10 with a 10 cm diameter discharge vessel and a three grid system was employed as thruster. A radio frequency neutralizer (RFN) is employed in addition to the RIT-10. A graphite plate with an orifice of 6 mm and a thickness of 1 mm is used as ion catcher. A second ion catcher is positioned in close proximity to the gas inlet, consisting of nine thin tungsten wires (0.3 mm in diameter) of varying lengths. To measure beam dump sputtering a witness plate is installed in the facility. It is placed 31.5 cm behind the thrusters exit plane and 13.5 cm away from the center line of the beam. It consists of a 94 mm^2 small piece of silicon wafer, from which 26 mm^2 are covered with tape. This yields a total collection area of 68 mm^2, where sputtered material may be deposited.

Influence of background pressure: When performing any kind of beam diagnostics it is of great importance to keep the pressure inside the test facility as low as possible to avoid misleading results [3, 13]. To study the effect of the background pressure on the beam composition, the thruster was kept stable in one operating point at 8 sccm nitrogen and 100 mA of beam current. The pressure in the main facility was kept constant in the medium 10^{-6} mbar range during the experiment, while the background pressure in the beam line towards the detector (see Fig. 2) was changed between 5×10^{-6} mbar and 7×10^{-5} mbar. The measured atomic-to-molecular ratio for different background pressures at the detector behind the analyzing magnet is shown in Fig. 3. The atomic-to-molecular ratio increases by almost a decade

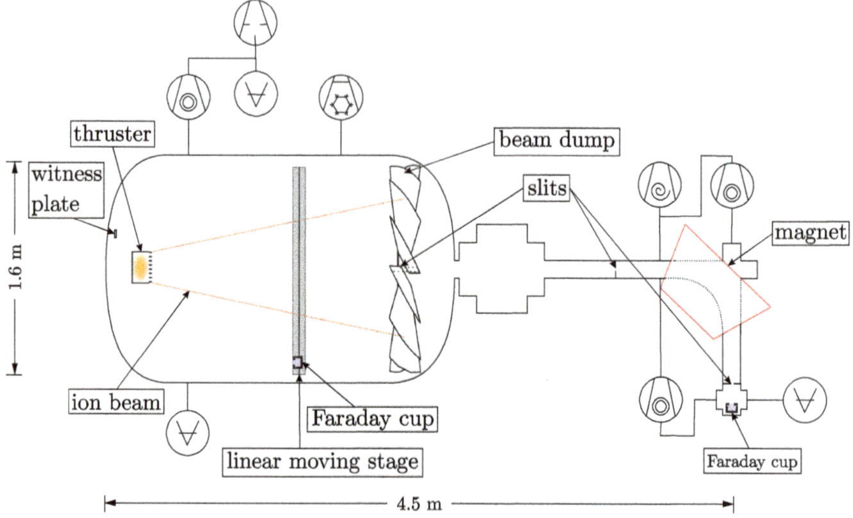

Fig. 2 Schematic drawing of the test facility "Obelix" with diagnostics.

from 0.3 % to 2.5 % going to higher pressures. The cause of this phenomenon can be attributed to scatter and charge exchange processes along the beam path. An elevated background pressure increases the likelihood of these processes occurring. Thus, a correction of the measured data is needed when limited in pumping speed. The correction for scattering and CEX losses can be calculated using

$$\left(\frac{N^+}{N_2^+}\right)_{corr} = \frac{1}{1 - n \cdot \sigma \cdot z} \cdot \left(\frac{N^+}{N_2^+}\right)_{measure}. \tag{1}$$

n is the neutral particle density in the detector line or facility, σ is the cross-section for CEX ($2.48 \times 10^{-19}\,\mathrm{m^2}$ for N_2^+-N_2 and $5.44 \times 10^{-20}\,\mathrm{m^2}$ for N^+-N_2 [10]) or scattering between the different species and z is the length of the detector line or

Fig. 3 Measured ratio of N^+/N_2^+ in the plume for different pressures (black squares) and corrected values (red circles)

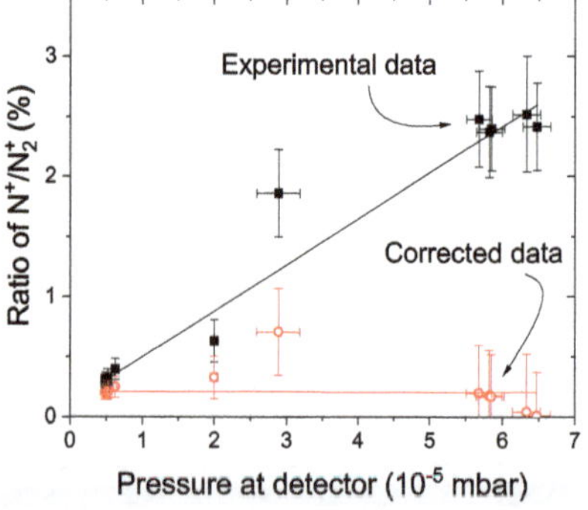

the facility. For simplicity, the cross sections for scattering are for neutral atoms and molecules. Furthermore, it was assumed that the background gas only consists of molecular nitrogen. The cross section for scattering was calculated with the geometrical cross section using the molecular and atomic radii ($4.16 \times 10^{-19}\,\mathrm{m}^2$ for N_2-N_2 and $2.01 \times 10^{-19}\,\mathrm{m}^2$ for N-N_2) [2, 4].

The corrected data tends towards a constant ratio, in this case 0.22 % for all background pressures. The deviation of the corrected and the measured data increases strongly when increasing the pressure in the region of the detector. A linear fit to the deviation, followed by calculation of the pressure for a deviation of 0 %, yields a result of $2.7 \times 10^{-6}\,\mathrm{mbar}$. Thus, at pressures below $3 \times 10^{-6}\,\mathrm{mbar}$, the beam composition measured at the end of the beam line agrees with that right after the grid, i.e., scattering and CEX along the beam path are negligible. Nevertheless, a correction of the measured data, independent of the diagnostics, will be necessary at higher pressures. Both CEX and collisions must be taken into account for the correction, as both have a non-negligible impact on the data reliability.

Beam dump sputtering: Another challenge for precisely characterizing an EP thruster using atmospheric propellants are increased sputtering effects inside the facility that can disturb or falsify measurements. Mostly graphite beam dumps are used at the end of test chambers to reduce sputtering effects by ion beams. However, since the energy threshold for sputtering is very low for all materials when sputtered by oxygen or nitrogen (maximum threshold is 52 eV for C [7]) there is a significant amount of back-streaming of sputtered material towards the thruster, the diagnostics, and the pumps, as was already demonstrated for xenon as propellant [6]. Back-streaming conductive material towards the thruster can cause a conductive layer inside the discharge vessel. Thus, a fraction of the power used for driving the rf coil of the RIT will be lost in this thin layer due to eddy current losses and will not be available to generate and excite the plasma. The efficiency will be significantly reduced, contingent on layer thickness.

To measure the back-streaming flux of graphite particles towards the thruster a small piece of silicon wafer is used, as shown in Fig. 2. The RIT-10 was used for a total of 55 h working at different operating points with both gases and an extraction voltage on the first grid of 1400 V, extracting beam currents between 10 and 150 mA. After the test, the layer thickness on the witness plate was measured and investigated with SEM and EDX. The layer thickness was determined to 77 nm and a very smooth surface layer has formed, with only some minor defects. The EDX measurement yields several elements in the deposited layer. Most of the found elements was silicon from the wafer (68 %), but also a lot of carbon from the beam dump (19 %), some oxygen (5 %) and also some iron from the facility (2 %) and alumina from the diagnostics (1 %). To calculate the back-streaming flux to the thruster some simplifications are made. To derive the particle number on the witness plate, a face-centered unit cell of the sputtered material (closed pacing of spheres) was assumed. With the radius of a carbon atom (70 pm) [2], the volume of the unit cell was calculated to $V_{\mathrm{unitcell}} = 7.8 \times 10^{-21}\,\mathrm{mm}^3$. The probe volume, consisting of the open area times the layer thickness, is $V_{\mathrm{probe}} = 5.2 \times 10^{-3}\,\mathrm{mm}^3$. Assuming 4 atoms per unit cell, the number of particles on the probe was calculated to $N_{probe} = 2.7 \times 10^{18}$. The sputter yield for oxygen atoms at 1400 eV on a carbon target is $Y_{oxygen} = 0.69$, for nitrogen it is $Y_{nitrogen} = 0.63$ [7]. A mean sputter yield of $Y_{mean} = 0.66$ is assumed. Taking into account the total duration of the experiment and all the beam currents extracted, it can be calculated that $N_{ext} = 1.3 \times 10^{23}$ particles hit the beam dump. Thus, a total of $N_{sputter} = 8.7 \times 10^{22}$ sputtered carbon particles leaving the beam dump. To calculate the theoretical number on the sample, the angle Θ between the surface normal and the sample and the solid angle $d\Omega$ are required. In a very simplified calculation, it is also assumed that all incident particles hit a very small surface element of the beam dump at normal incident. Thus, the divergence of the beam and CEX processes, as well as collisions with the background gas, are neglected. Θ can be determined via the tangent of the distances of the sample to the beam dump to $\Theta = 17°$. With the Lambert's cosine law,

$$N_{theo} = N_{sputter} \cos(\Theta) \frac{A_{witness}}{l^2}, \tag{2}$$

where l is the distance between the probe and the center of the beam dump, the theoretical particle density on the witness plate can be calculated to $N_{theo} = 5.7 \times 10^{18}$ particles on the witness plate.

The difference of a factor two between the calculated particle number from the measurement and the calculated particle number from the theoretical estimation can be explained by all the simplifications made. Nevertheless, it shows that this simple theoretical estimate at least can be used to determine the correct order of magnitude of the back-streaming of carbon atoms from the beam dump forwards the thruster. Subsequently the anticipated particle flux onto the thruster can be calculated using $\Theta = 0°$. Thus, a particle flux of 3.4×10^{15} particles/s from the beam dump back to the thruster is obtained. If we now consider that the average extracted particle flux from the thruster in the experiment was 6.4×10^{17} particles/s, it indicates that 0.5 % of the extracted particle flux returns to the thruster as sputtered carbon atoms. The dependence between extracted current and back-sputtering carbon to the thruster is shown in Fig. 4 for oxygen as well as nitrogen. Due to the higher sputter yield the back-streaming for oxygen is slightly larger than for nitrogen. Up to 3.4×10^{13} carbon particles/(s mA) are deposited onto the thruster.

Material durability: The materials of the thruster and neutralizer, which are in direct contact with the plasma, are also exposed to corrosive and erosive effects. The ion catcher material inside an RF neutralizer can experience significantly erosion effects in the presence of an elevated electron temperature [12]. The ion catcher is negatively biased to actively attract ions so that an equal amount of negative charges can be extracted through the orifice. The choice of ion catcher material is crucial for the lifetime of the ABEP system. Tungsten and carbon have the highest sputter threshold for nitrogen and oxygen. Consequently, both materials were subjected to testing within the RF neutralizer under these very challenging conditions imposed by oxygen and nitrogen. After testing the neutralizer at different operating points for 13 h using oxygen and 8.5 h using nitrogen, the catcher materials were subjected to a

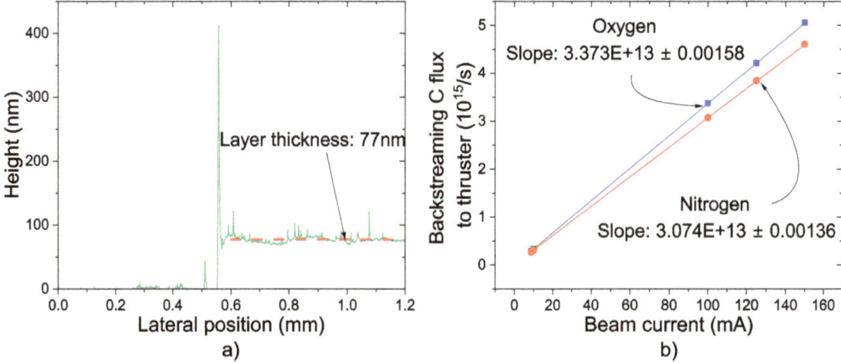

Fig. 4 Layer thickness of the back sputtered graphite. Calculated back-streaming of the sputtered beam dump to the thruster for oxygen and nitrogen

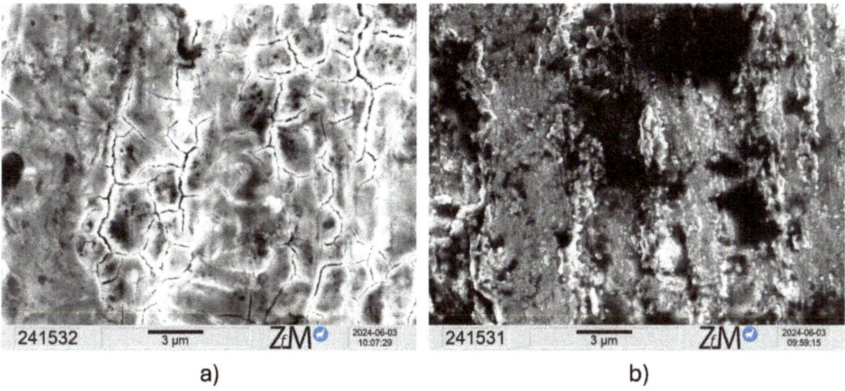

Fig. 5 SEM images of tungsten wires **a** pristine state, **b** after use in O_2 plasma

series of evaluations utilizing SEM and EDX. It is evident that both materials experienced significant damage as a result of exposure to oxygen and nitrogen plasma. The surface of both materials has undergone a significant degree of roughening following the exposure period, with tungsten exhibiting a particularly pronounced alteration (Fig. 5). A large amount of tungsten was found on the graphite plate, on both the SEM and EDX measurements. In addition, carbon was found on the tungsten wires. This indicates a high sputter rate of both materials during the experiments, which leads to the assumption of a high electron temperature of the plasma inside the RF neutralizer in combination with the varying negative catcher potential. Furthermore, non-conductive layers were found on the surface of the tungsten wires, indicating a tungsten-oxide layer that formed during the experiment. This non-conductive oxide layer leads to a smaller effective catcher area, leading to smaller extraction currents for the same operating point. These material issues need to be solved in order to guarantee quasi-neutrality on the satellite.

4   Conclusion

The design of the intake is of critical importance, as its configuration is contingent upon the poorly understood GSI that are present in VLEO. On-ground experiments are inevitable to accurately validate intake simulations. To accurately measure the performance of the thruster system, regardless of whether it is the neutralizer or the thruster, the background pressure of the test facility must be below 3×10^{-6} mbar to get reliable and accurate results. Due to the low sputter thresholds for atmospheric gases, the back-streaming of carbon onto the thruster with approximately 0.5 % of the extracted beam current is a huge challenge. Conductive layers form in the discharge vessel and disturb the inductive power coupling between rf-coil and plasma. Furthermore, sputtering and degradation of materials in direct contact with the plasma (e.g. grids or ion catcher) can shorten the mission lifetime.

References

1. Andreussi T, Ferrato E, Giannetti V (2022) A review of air-breathing electric propulsion: from mission studies to technology verification. J Electr Propuls 1(1)
2. Cordero B, Gómez V, Platero-Prats AE, Revés M, Echeverría J, Cremades E, Barragán F, Alvarez S (2008) Covalent radii revisited. Dalton Trans 21:2832
3. Désangles V, Packan D, Jarrige J, Peterschmitt S, Dietz P, Scharmann S, Holste K, Klar PJ (2023) ECRA thruster advances: 30 W and 200 W prototypes latest performances. J Electr Propuls 2(1)
4. Ismail AF (2015) Gas separation membranes. Springer International Publishing AG, Cham (description based on publisher supplied metadata and other sources)
5. Jorge P, Magin T, Hubin A, Le Quang D (2022) Modeling and simulation of VKI's new low density facility
6. Lim G, Tran H, Levin DA, Chew HB, Nishii K (2024) Kinetic simulation of carbon sputtering by a gridded ion thruster with a molecular dynamics based model. In: Proceedings of the 38th international electric propulsion conference
7. Matsunami N, Yamamura Y, Itikawa Y, Itoh N, Kazumata Y, Miyagawa S, Morita K, Shimizu R, Tawara H (1984) Energy dependence of the ion-induced sputtering yields of monatomic solids. Atomic Data Nuclear Data Tables 31(1):1–80
8. Minton TK, Schwartzentruber TE, Xu C (2021) On the utility of coated POSS-polyimides for vehicles in very low earth orbit. ACS Appl Mater Interf 13(43):51673–51684
9. Murray VJ, Pilinski MD, Smoll EJ, Qian M, Minton TK, Madzunkov SM, Darrach MR (2017) Gas-surface scattering dynamics applied to concentration of gases for mass spectrometry in tenuous atmospheres. J Phys Chem C 121(14):7903–7922
10. Phelps AV (1991) Cross sections and swarm coefficients for nitrogen ions and neutrals in N2 and argon ions and neutrals in Ar for energies from 0.1 eV to 10 keV. J Phys Chem Refer Data 20(3):557–573
11. Rafalskyi D, Dudin S (2024) RF plasma flow generator for LEO environment simulation: use cases. In: IEPC-2024-337, international electric propulsion conference (IEPC), Toulouse, France

12. Tisaev M, Karadag B, Ferrato E, Andreussi T, Lucca Fabris A (2024) Development and standalone testing of the air-breathing Microwave Plasma CAThode (AMPCAT). Acta Astronaut 214:722–736
13. Walker MLR, Victor AL, Hofer RR, Gallimore AD (2005) Effect of backpressure on ion current density measurements in hall thruster plumes. J Propuls Power 21(3):408–415

Astronomical Line Polarization Through Gas Transport

Boy Lankhaar

Abstract Magnetic fields are critical to the dynamics of the interstellar medium (ISM), influencing star formation and interstellar gas dynamics. This paper explores a novel way to probe magnetic fields through molecular line polarization mechanism, that is based on the experimentally well-established Senftleben-Beenakker (SB) effect. In molecular line polarization through SB effects, gaseous transport processes such as viscous strain and thermal gradients, cause molecules to align through directional collisions, subsequently polarizing their emission lines. This polarization mechanism differs from the Goldreich-Kylafis (GK) effect, as it is not dependent on optical depth nor non-thermal excitation. We derive a theoretical framework to model molecular alignment due to the SB effect. We discuss the applicability of SB polarization as a tool to trace magnetic fields in turbulent media and accretion disks.

1 Introduction

Magnetic fields permeate the interstellar medium, shaping astrophysical dynamics. This work focuses on molecular interstellar media, that exist in gas condensations with densities of 100 to 10^9 cm^{-3}, and temperatures ranging from 10 to 1000 Kelvin. Although ionization fractions are low, collisional coupling transfers magnetic influence to neutral gas, significantly affecting dynamics, as magnetic energy can rival or exceed turbulent and thermal energy densities. Thus, measuring magnetic fields in molecular media is crucial for a comprehensive understanding of interstellar processes.

Magnetic fields in molecular media can be indicated through the observation of molecular line polarization. Molecular line emission is polarized to degrees of some per cents and is observable in the strongest emission lines. The origin of molecular line polarization is commonly attributed to the Goldreich-Kylafis (GK) effect [7]. The GK effect explains this polarization as arising from molecular excitation in an

B. Lankhaar (✉)
Department of Space, Earth and Environment, Onsala Space Observatory, Chalmers University of Technology, Onsala, Sweden
e-mail: lankhaar@chalmers.se

M. Grabe et al. (eds.), *Rarefied Gas Dynamics*, Springer Aerospace Technology,
https://doi.org/10.1007/978-3-032-00094-1_69

anisotropic medium, where the anisotropic radiation field aligns molecules. Molecular alignment is initially oriented in the direction of the radiation field anisotropy, but due to the precession of the molecule around the magnetic field, the alignment is re-oriented to the magnetic field direction. When magnetic precession is faster than other molecular interactions, emission polarizes parallel or perpendicular to the magnetic field. The GK effect is effective at near-unity optical depth and low gas densities, when radiative excitation dominates over collisions. The GK effect has been leveraged to trace magnetic fields towards (high-mass) star-forming regions, protoplanetary disks, the circumstellar envelopes of evolved stars, and molecular clouds in external galaxies.

Molecular alignment and subsequent line polarization can be induced not only by radiative processes but also by collisions [16, 25]. This study explores additional conditions under which directional collisions induce molecular alignment in an interstellar gas. To this end, we consider the experimentally well-established and characterized Senftleben-Beenakker (SB) effect [3, 23]. In the SB effect, fluids, that are under the influence of a magnetic field, acquire anisotropic transport properties. The anisotropies in the transport processes are the consequence of the molecular constituents of the fluid aligning along the transport direction, through directional collisions between particles that have anisotropic potential. The magnetic field impacts this process, as particles also precess around it. If this precession is fast compared to the collision time, then the alignment will be re-oriented to the magnetic field. In this way, molecular alignment along the magnetic field direction is retained, while it is effectively quenched in other cases. The resulting anisotropic transport coefficients, that are a function of the magnetic field strength divided by the pressure (i.e., the magnetic precession frequency divided by the collision frequency), have been well-characterized through experimental means for hydrodynamical transport tensors such as the viscosity tensor [13], and the thermal conductivity tensor [12], for a wide variety of molecules (for a full review, see [4, 21]). While tensor-anisotropies under magnetic fields is interesting from a fluid-dynamical perspective, their measurements have also served as experimental benchmarks for the testing of collisional modeling, specifically their dependence on the (anisotropic) intermolecular potential energy surfaces [10]. The large body of experimental and theoretical literature on the SB effect has been condensed in the monographs of Ref. [21].

Given the low densities of interstellar gas, where magnetic precession rates typically exceed collision rates [17], SB effects should be prevalent. Effectively, this means that under the influence of gaseous viscous strains, or thermal gradients, the molecular constituents of the gas should acquire alignment, leading to the polarization of their spectral lines. This paper investigates the conditions under which SB effects significantly and detectably polarize molecular emission. Molecular emission polarized through SB effects provides for a robust polarization mechanism that, in contrast to the GK effect, is not dependent on line optical depth, and does not require non-thermal excitation of the molecules. The molecular line polarization due to SB effects can be used as a robust tracer of the ambient interstellar magnetic field. In addition, it may also be possible to relate the line polarization due to SB effects to fluid dynamical properties of interstellar gases that cannot be measured otherwise.

This paper is built up as follows. In Sect. 2 we review the theory of alignment and polarization of molecules and their spectral lines. In Sect. 3, we discuss astrophysical processes conducive to SB effects and identify regions in the interstellar medium where anisotropic transport phenomena lead to significantly polarized line emission. Section 4 highlights potential limitations and model deficiencies. We conclude in Sect. 5.

2 Polarization and Alignment Due to the SB Effect

Due to the Senftleben-Beenakker (SB) effect, gas transport coefficients, of e.g. thermal transport or viscous strain, become anisotropic under the influence a magnetic field. The anisotropic transport properties are the result of alignment manifesting in the molecular constituents. Therefore, it is implicit in the theory of the SB effect, that a fluid also becomes birefringent under the effect of a directional flow. Leveraging this insight, Hess [8, 9] discussed the possibility to characterize the SB effect through experimental polarization observations, and in-so-doing derived the proper kinetic theory, starting from the Waldmann-Snider equations, of the alignment of molecules under directional flows.

We briefly revise concepts from the description of quantum molecular alignment, that are commonly applied to transfer of polarized line radiation in astrophysical gases. We assume that a molecule, with eigenstates of definite (total) angular momentum, j, precesses fast around a magnetic field, $\boldsymbol{B}$, which determines its symmetry axis: its eigenstates can be accurately described by the total angular momentum and its projection onto the magnetic field direction, $|jm\rangle$. We describe the molecular population in terms of its irreducible density matrix operator elements, $\hat{\rho}_{jk} = \sum_m \left[C_{m\,-m\,0}^{j\,j\,k} \right]^2 |jm\rangle \langle jm|$, where the symbol $C_{m\,-m\,0}^{j\,j\,k}$ is a Clebsch-Gordan coefficient. The rank of the irreducible tensor element, k, runs from 0 to $2j$. The $k = 0$ element is related to the total population of the quantum state j, while for $k > 0$, the elements of k which are even are commonly referred to as alignment elements, and those for which k is odd are referred to as the orientation elements. We note the expectation value of irreducible density matrix operator as $\rho_{jk} = \langle \hat{\rho}_{jk} \rangle$. The relative alignment and orientation of a specific quantum level we denote by $\sigma_{jk} = \rho_{jk}/\rho_{j0}$. Alignment is present in a population of molecules if their magnetic sublevels are populated unequally, but symmetrically, for levels with equal absolute projection, $|m|$. Orientation is present in a population of molecules if there is an imbalance between the population of magnetic sublevels of equal absolute projection $|m|$. Both alignment and orientation can be produced through collisions, yet in gaseous transport, alignment elements are more potently excited and accordingly have more effect on the transport properties [21]. Thus in the following, we set population elements, ρ_{jk}, of uneven k to zero: our molecules are aligned but not oriented.

We now discuss the transfer of radiation through an aligned molecular medium. We describe the propagation of radiation, with spectral energy flux density, $I_\nu(\hat{k})$ at frequency ν, in the direction $\hat{k}$. Since alignment of the molecules is with respect to the magnetic field direction, it is most advantageous to denote the transfer of radiation in the two polarization modes parallel and perpendicular to the magnetic field rejection onto the propagation direction, $\hat{e}_\parallel$ and $\hat{e}_\perp = \hat{k} \times \hat{e}_\parallel$. For brevity, we denote these $I_\parallel$ and $I_\perp$, dropping in the notation the dependence on frequency and direction. We are interested in the production of polarization, $p_l = [I_\parallel - I_\perp]/[I_\parallel + I_\perp]$, by the molecular medium. We consider the transfer of radiation at radio-to-millimeter frequencies, where radiation scattering is negligible, and astronomical observations of molecules are commonly made. In the optically thin limit, the polarization fraction of a transition between quantum states $j \to j - 1$, that characterizes astrophysically common diatomic molecules, can be related to the degree of molecular alignment, and tends to [16, 19], $p_l \simeq -\frac{3}{4\sqrt{5}}\sqrt{\frac{(j+1)(2j+3)}{j(2j-1)}}\sigma_{j2}\ \sin^2\theta$, where $\cos\theta = \hat{k} \cdot \hat{b}$ is the projection of the radiation direction onto the magnetic field direction.

Reference [9] modeled the population of magnetic sublevels within a level j, which he denotes $N_{jm} = \langle|jm\rangle\langle jm|\rangle$, under the influence of a viscous strain. We restate equation (22) of Ref. [9] in terms of the relative alignment of the molecular quantum level, under the influence of a viscous strain $\sigma_{j2} = \sqrt{\frac{6j(j+1)}{(2j-1)(2j+3)}}\left[\frac{\eta}{p_0}\frac{\omega_{T\eta}}{\omega_\eta}\left(\hat{b}\cdot\overline{\nabla v^T}\hat{b}\right)\right]$, where η and p_0 are the dynamical viscosity and pressure, and $\overline{\nabla v^T}$ is the traceless symmetric part of the strain-rate tensor. The ratio $\omega_{T\eta}/\omega_\eta$ is a measure for the anisotropy of the interaction potential of the molecular collision complex. We can now derive the expected linear polarization fraction of an optically thin $j \to j - 1$ transition, that is the result of the viscous strain in the gas that it traces, $p_l^\eta = -\frac{3}{4}\sqrt{\frac{6}{5}\frac{j+1}{2j-1}}\left[\frac{\eta}{p_0}\frac{\omega_{T\eta}}{\omega_\eta}\left(\hat{b}\cdot\overline{\nabla v^T}\hat{b}\right)\right]\sin^2\theta$. In this way, we have related the polarization degree of line radiation that emerges from astrophysical gas, to the viscous strain in the direction of the magnetic field that this gas is under.

3   Astrophysical Conditions Leading to Viscous Polarization

Viscous dissipation of MHD turbulence. Turbulence is ubiquitous in the interstellar medium, and spans eight orders of magnitude from its driving scale to the viscous scale, where dissipation occurs. Given the complexity of turbulence, we take an order-of-magnitude approach to estimate its effects on molecular alignment and polarization. To estimate the polarization fraction due to turbulent strains, we focus on dissipation at the viscous scale, as at these scales, the internal states of the gaseous particles become excited. The dynamical viscosity scales as $\eta \sim \rho c_s \ell_{\mathrm{mfp}}$, with the gas density, ρ, the sound velocity c_s and the mean-free-path of the molecular gas constituents, ℓ_{mfp}. The pressure scales as $p_0 \sim \rho c_s^2$. Thus, the fraction $\frac{\eta}{p_0} \sim \frac{\ell_{\mathrm{mfp}}}{c_s}$. Now, we specialize the viscous strain term, $\hat{b}\cdot\overline{\nabla v^T}\hat{b}$, to a turbulent medium. At first glance, one would expect this term to average out to 0 in the case of isotropic turbulence; after all, the directional strain-rate of circular eddies averages to zero. However, it is well established that turbulence in the interstellar medium is magnetized, and therefore is anisotropic. Eddies are elongated in the direction along the magnetic field, $\ell_\parallel > \ell_\perp$, and their relative elongation increases going to lower scales. While the precise scaling relations are still under debate, a common result is the scaling of $\ell_\parallel \sim \ell_\perp^{2/3} L_{\rm inj}$ [2]. Thus, with such elongated eddies, the highest rate-of-strain, and thus the strongest alignment, is expected perpendicular to the magnetic field direction: $\hat{b} \cdot \overline{\nabla v^T} \hat{b} \simeq -\frac{1}{2}\frac{v_\eta}{\ell_\eta}$, where v_η and ℓ_η are the expected velocity and dimension at the viscous scale. We estimate this fraction by order-of-magnitude. We assume that turbulence in the interstellar medium follows roughly two scaling regimes: above the sonic scale, turbulence follows a Burgers scaling, while below the sonic scale, we assume that turbulence follows a Kolmogorov scaling [22]. Adopting these scalings and extrapolating to the sonic scale, we find that the strain at the viscous scale can be expressed as $\frac{v_\eta}{\ell_\eta} \sim \frac{[\ell_\eta/\ell_s]^{1/3}c_s}{\ell_{\rm mfp}^{3/4}\ell_s^{1/4}} \sim c_s[\ell_{\rm mfp}\ell_s]^{-1/2}$. Now we can proceed to make an estimate for the polarization fraction that is due to anisotropic turbulent viscous dissipation. First for a general $j \to j-1$, transition, we find $p_l^{\eta,\rm turb} \sim \frac{3}{8}\sqrt{\frac{6}{5}}\frac{j+1}{2j-1}\frac{\omega_{T\eta}}{\omega_\eta}\sqrt{\frac{\ell_{\rm mfp}}{\ell_s}}$. Now, specializing in $j = 1$, adopting $\omega_{\eta T}/\omega_\eta \sim 0.1$ [9], and using typical values from Ref. [22]: $\ell_{\rm mfp} \sim 10^{13}$ cm and $\ell_s \sim 10^{17}$ cm, we obtain $p_l^{\eta,\rm turb} \sim 0.1\%$. Since $p_l^{\eta,\rm turb} \propto \ell_{\rm mfp}^{1/2}$, denser regions with shorter mean free paths should exhibit greater angular momentum polarization. Notably, such regions are also more influenced by gravitational forces, which can further enhance directional alignment, as we will explore in the next section.

In principle, all constituents of the interstellar medium should be impacted by turbulent angular momentum polarization, and exhibit polarization degrees of the order 0.1%. However, modern telescopes are sensitive to these low degrees of polarization only for those species that are associated with brightest emission lines, such as CO. We expect this polarization mechanism to be dominant when molecular lines are thermally excited, at densities above the critical density, when collisional transitions occur frequently with respect to the Einstein A-coefficient of the transition under investigation. For subthermally excited molecules, alternative polarization effects such as through the GK effect [7] likely dominate.

Viscous dissipation in the α-disk. The α-disk model was introduced by Ref. [24] as a theoretical model of an accretion disk around a compact object. In the α-disk, the transport of angular momentum, thus facilitating accretion, is due to so-called turbulent viscosity. Whereas for regular microscopic kinematic viscosity, $\nu \sim c_s\ell_{\rm mfp}$, turbulence can enhance this viscosity, when describing global processes that do not resolve the eddy-like motions that cause strains on smaller scales. In an accretion disk, the effective turbulent kinematic viscosity is thus parameterized according to its typical velocities and dimensions, as $\nu = \alpha c_s H$, where H is the disk scale height and $\alpha \leq 1$.

We restate the estimate of the polarization fraction due to viscous strain, putting in the proper relations according to α-disk theory of a thin accretion disk. For the ratio between the viscosity and the pressure, we put $\eta/p_0 \sim \nu/c_s \sim \alpha H/c_s$. Furthermore, in a thin disk, we have $H \sim c_s/\Omega_K$, where $\Omega_K = \sqrt{GM/R^3}$ is the Keplerian rate of rotation at radial distance R, around a compact object of mass M (G is the gravitational constant). We let the velocity field be fully Keplerian: $\left(\hat{b} \cdot \overline{\nabla v^T} \hat{b}\right) \sim \Omega_K$. Then, the line polarization fraction of a $j \to j-1$ transition due to viscous strain in an α-disk is, $p_l^\alpha \sim -\frac{3}{4}\sqrt{\frac{6}{5}}\frac{j+1}{2j-1}\frac{\omega_{T\eta}}{\omega_\eta}\alpha$. This leaves us with a rather simple relation between the α-parameter, that is a proxy for the disk activity, and the polarization fraction for molecular lines excited in them. For a typical molecule possessing a dipole, we have $\omega_{\eta T}/\omega_\eta \sim 0.1$ for collisions with its main collision partner H_2. Typical estimates place α in the range $\sim 10^{-2} - 0.6$ [1]. Thus, for a $1 \to 0$ transition, we have $p_l^\alpha = 1.6\alpha$. For the most active accretion disks, $p_l^\alpha = 10\%$, while typical disks of $\alpha \sim 0.1$ yield polarization fractions $p_l^\alpha \sim 0.2\%$.

Protoplanetary disks. A protoplanetary disk forms around a low-mass protostar at the late stages of its formation. The physical conditions, dynamics, evolution and chemistry of the disk determine the environment where planets form. While α-disk prescriptions were popular in early modeling of protoplanetary disks, it is now believed that accretion processes in protoplanetary disks are mainly due to the launching of magnetically driven winds [20], even though the magnetic fields that would drive such winds have not been directly measured [15, 27]. The low levels of turbulence that have been observed towards a variety of protoplanetary disks [5] put stringent upper limits on the $\alpha \lesssim 10^{-2}$, making viscosity subdominant as an agent of angular momentum transport. Concurrently, it also puts an upper limit on the predicted angular momentum polarization due to viscous strain in an alpha disk, and constrains $p_l^\alpha \lesssim 0.1\%$. There have been deep observations to probe the polarized emission from protoplanetary disk molecules, which have yielded marginal detections of the polarized emission at the level of some percents for ^{12}CO and its main isotopologue ^{13}CO [26]. If real, such high polarization fractions cannot be the product of the angular momentum polarization due to SB effects, but instead are likely due to GK effects [17, 18].

Active Galactic Nuclei (AGN). Accretion disks around AGN are crucial for feeding supermassive black holes and influencing galaxy evolution. As an example of an AGN disk, we turn to the closeby active galaxy NGC 1068. While it has been attempted to fit the accretion disk of NGC 1068 to the mold of an α-disk, the α-disk model proved inconsistent as it implied strong gravitational instabilities. In an attempt to supersede the simple α-disk prescription, models that include gas clumpiness are favored [14]. At any rate, high-resolution ALMA observations show intricate dynamics towards the nuclear region of NGC 1068, with a flip in rotation direction in the inner disk [11]. Such dynamics are associated with very high degrees of shearing, that should be reflected in the angular momentum polarization of the molecular constituents. The site of the rotation flip should be associated with high polarization degrees of several per cents due to SB polarization in the thermal lines. SB polarization through gaseous transport should be dominant in dense gas where GK polarization is quenched through collisions. In this way, resolved polarization observations of thermal lines would offer a unique view of the magnetic fields of in the highly dynamic nuclear gas.

NGC 1068 belongs to a class of active galaxies that hosts strong water megamaser emissions. NGC 1068 is the first galaxy where water megamasers have been found to have polarized emission [6], which has lead to interesting constraints on the accretion dynamics in the innermost regions and its relation to magnetic fields. Water masers should be polarized to degrees $p_l^{H_2O} \sim 3\% \left[\frac{\alpha}{0.1}\right] \left[\frac{\omega_{\eta T}/\omega_\eta}{0.1}\right] \left[\frac{\tau_m}{10}\right]$, adopting representative values for α and $\omega_{\eta T}/\omega_\eta$ and the maser optical depth, τ_m. This mechanism of polarization is about as effective as through anisotropic pumping [19].

4 Model Assumptions and Parameters

We examined molecular alignment in astrophysical transport processes and its observability through polarized molecular line emission. Molecular alignment is a prerequisite for the Senftleben-Beenakker (SB) effect, where anisotropic transport coefficients arise under a magnetic field. This effect is well understood both experimentally and theoretically, emerging naturally from the Boltzmann transport equations when applied to rotating molecules. The specialization of these equations to rotational states, known as the Waldmann-Snider equations, requires assumptions for solvability. Typically, a near-equilibrium molecular distribution is assumed, with first-order corrections accounting for gas strains and thermal gradients. Anisotropic transport effects arise when molecular rotation interacts with a magnetic field, but the equilibrium assumption weakens as the Knudsen number increases. Experiments have validated SB effects, particularly by measuring anisotropic transport coefficients at moderately low Knudsen numbers [12, 13]. Reference [9] introduced polarization spectroscopy as a tool for characterizing SB effects, laying the theoretical foundation that this work builds upon. The degree of polarization can be linked to the Knudsen number: $\frac{\eta}{p_0}\nabla \bar{v}^T \sim \mathcal{M}\frac{\ell_{mfp}}{L} \sim \mathcal{M}$ Kn. The moderate polarization fractions that we predict puts Kn $\lesssim 1$, supporting the validity of our approach. However, since SB effects emerge at viscous scales, direct observations are beyond current telescope capabilities, and our estimates rely on extrapolation. If an observational experiment can be set up that pristinely traces SB polarization in a spectral line, this would yield a direct probing of gas motions at viscous scales.

Another key quantity in polarization estimates is the ratio $\omega_{\eta T}/\omega_\eta$, which strongly depends on the molecular species. Although the interstellar medium (ISM) is composed almost entirely of H2, its symmetric structure and high rotational constant make it difficult to observe. Radio astronomers therefore rely on tracer species like CO (with typical abundances of $\sim 10^{-4}$) to characterize the molecular ISM. This means that when considering the ratio $\omega_{\eta T}/\omega_\eta$, the relevant collision-complex is the one between the tracer species and H_2. While collisions between ISM molecules and H_2 are under active and ongoing investigation, the properties that are commonly extracted from these collision complexes are the state-to-state collisional transition rate coefficients. While the collision-complex anisotropy is an extremely relevant variable to these calculations, it's representation in the $\omega_{\eta T}/\omega_\eta$ factor is not commonly extracted and reported from the calculations. As such, no modern theoretical constraints on $\omega_{\eta T}/\omega_\eta$ for different ISM species exist. Similarly, experimental efforts have not been directed to constrain the transport properties of different molecules embedded in H_2 gases, as experimentalists have directed most of their attention to pure gases [21]. This leaves the ratio $\omega_{\eta T}/\omega_\eta$ largely unconstrained.

5 Conclusion

We have introduced a novel mechanism for molecular line polarization in astrophysical environments, through the theoretically and experimentally well established Senftleben-Beenakker (SB) effect. SB effects show that molecules that constitute a gas that is under viscous strains or thermal gradients, align themselves along the direction of gaseous transport. In environments such as interstellar media, due to the presence of magnetic fields, alignment is re-oriented to the magnetic field, and thus spectral emission from aligned molecules carries information on the magnetic field. In contrast to other mechanisms of interstellar molecular line polarization, such as through the Goldreich-Kylafis (GK) effect, line polarization due to SB effects does not require molecules to be excited subthermally, or excited under the influence of an anisotropic radiation field. Rather, line polarization due to SB effects require some gas strain to be present to be operative. Exploratory calculations suggest that gas strains due to turbulent dissipation can lead to polarization fractions of a fraction of a per cent, while the gas strain that is present in viscously evolving accretion disks can lead to molecular line polarization of some percents.

References

1. Balbus SA, Hawley JF (1998) Instability, turbulence, and enhanced transport in accretion disks. Rev Mod Phys 70(1):1
2. Beattie JR et al (2024) Magnetized compressible turbulence with a fluctuation dynamo and reynolds numbers over a million. arXiv preprint arXiv:2405.16626
3. Beenakker JJ, Scoles G, Knaap HF, Jonkman R (1962) The influence of a magnetic field on the transport properties of diatomic molecules in the gaseous state. Phys Lett 2
4. Beenakker J, McCourt F (1970) Magnetic and electric effects on transport properties. Annu Rev Astron Astrophys 21(1):47–72
5. Flaherty KM et al (2018) Turbulence in the TW Hya disk. Astrophys J 856(2):117
6. Gallimore JF, Impellizzeri CV, Aghelpasand S, Gao F, Hostetter V, Lankhaar B (2024) The discovery of polarized water vapor megamaser emission in a molecular accretion disk. Astrophys J Lett 975(1):L9
7. Goldreich P, Kylafis ND (1981) On mapping the magnetic field direction in molecular clouds by polarization measurements. Astrophys J 243:L75–L78
8. Hess S (1969) Flow birefringence of polyatomic gases. Phys Lett A 30(4):239–240

9. Hess S (1973) Flow-birefringence in gases an example of the kinetic theory based on the boltzmann-equation for rotating particles. In: The Boltzmann equation: theories and applications. Springer
10. Hutson JM, McCourt FR (1984) Close-coupling calculations of transport and relaxation cross sections for H_2 in Ar. J Chem Phys 80(3):1135–1149
11. Impellizzeri CV et al (2019) Counter-rotation and high-velocity outflow in the parsec-scale molecular torus of NGC 1068. Astrophys J Lett 884(2):L28
12. Knaap H, Beenakker J (1967) Heat conductivity and viscosity of a gas of non-spherical molecules in a magnetic field. Physica 33(3):643–670
13. Korving J et al (1967) The influence of a magnetic field on the transport properties of gases of polyatomic molecules: part I, viscosity. Physica 36(2):177–197
14. Kumar P (1999) The structure of the central disk of NGC 1068: a clumpy disk model. Astrophys J 519(2):599
15. Lankhaar B, Teague R (2023) Three-dimensional magnetic field imaging of protoplanetary disks using Zeeman broadening and linear polarization observations. Astron Astrophys 678:A17
16. Lankhaar B, Vlemmings W (2020) Collisional polarization of molecular ions: a signpost of ambipolar diffusion. Astron Astrophys 638:L7
17. Lankhaar B, Vlemmings W (2020) Portal: three-dimensional polarized (sub)millimeter line radiative transfer. Astron Astrophys 636:A14
18. Lankhaar B, Vlemmings W, Bjerkeli P (2022) Tracing the large-scale magnetic field morphology in protoplanetary disks using molecular line polarization. Astron Astrophys 657:A106
19. Lankhaar B et al (2024) Maser polarization through anisotropic pumping. Astron Astrophys 683:A117
20. Lesur GR (2021) Systematic description of wind-driven protoplanetary discs. Astron Astrophys 650:A35
21. McCourt FR, Beenakker JJ, Kohler WE, Kuscer I (1991) Nonequilibrium phenomena in polyatomic gases: dilute gases. Oxford University Press
22. McKee CF, Ostriker EC (2007) Theory of star formation. Annu Rev Astron Astrophys 45(1):565–687
23. Senftleben H (1930) Einfluss eines magnetfieldes auf das wärmeleitungsvermögen fon paramagnetischen gasen. Phys Z 31:822
24. Shakura N, Sunyaev R (1976) A theory of the instability of disk accretion on to black holes and the variability of binary X-ray sources, galactic nuclei and quasars. Mon Not R Astron Soc 175(3):613–632
25. Sinha M, Caldwell C, Zare R (1974) Alignment of molecules in gaseous transport: alkali dimers in supersonic nozzle beams. J Chem Phys 61(2):491–503
26. Teague R et al (2021) Discovery of molecular-line polarization in the disk of TW Hya. Astrophys J 922(2):139
27. Vlemmings W, Lankhaar B et al (2019) Stringent limits on the magnetic field strength in the disc of TW Hya-ALMA observations of CN polarisation. Astron Astrophys 624:L7

GPSR Compliance
The European Union's (EU) General Product Safety Regulation (GPSR) is a set
of rules that requires consumer products to be safe and our obligations to
ensure this.

If you have any concerns about our products, you can contact us on

ProductSafety@springernature.com

In case Publisher is established outside the EU, the EU authorized
representative is:

Springer Nature Customer Service Center GmbH
Europaplatz 3
69115 Heidelberg, Germany